AF323750

STOCHASTIC SYSTEMS

Theory and Applications

STOCHASTIC SYSTEMS

Theory and Applications

V. S. Pugachev

I. N. Sinitsyn

Russian Academy of Sciencies

Translated by

I. V. Sinitsyna

World Scientific
New Jersey • London • Singapore • Hong Kong

Published by

World Scientific Publishing Co. Pte. Ltd.

P O Box 128, Farrer Road, Singapore 912805

USA office: Suite 1B, 1060 Main Street, River Edge, NJ 07661

UK office: 57 Shelton Street, Covent Garden, London WC2H 9HE

British Library Cataloguing-in-Publication Data
A catalogue record for this book is available from the British Library.

STOCHASTIC SYSTEMS
Theory and Applications

ISBN 981-02-4742-7

Printed in Singapore by World Scientific Printers (S) Pte Ltd

PREFACE

The monograph is dedicated to the systematic presentation of the applied advanced theory and analytical methods for the stochastic systems, i.e. the dynamical systems described by the finite- and infinite-dimensional stochastic differential, difference, integral, integrodifferential etc. equations. It is based on the results of the fundamental research performed in the Institute for Informatic Problems of the Russian Academy of Sciences in the context of the scientific program "Stochastic Systems" and the lecture courses delivered by the authors in the domestic and the foreign Technical Universities.

The book may be used as the textbook for the applied mathematics faculties of the universities. The unified procedure, thorough selection of the examples and the problems (over 500) and the large applications make this book useful for the graduates, the post-graduates and the lecturers. The book is of considerable interest also for the mathematicians who deal with the stochastic equations and their applications.

The construction of general stochastic systems theory is based on the equations for the multi-dimensional characteristic functions and functionals. The stochastic differential equations with the arbitrary processes with the independent increments are studied. The equations with the Wiener and the Poisson processes are considered as a special case. The methods of the parametrization of the multi-dimensional distributions in the nonlinear stochastic systems based on the moments, the semiinvariants, the quasimoments and on the consistent orthogonal expansions are stated systematically. Special attention is paid to linear and reducible to linear stochastic systems theory based on the canonical representations (canonical expansions and integral canonical representations). Most attention has been concentrated on the structural theory of the composed stochastic systems on the grounds of the conditional probability measures.

Chapter 1 is devoted to the mathematical models of the dynamic systems under conditions of the random disturbances and their characteristics. The linear and the nonlinear continuous, discrete and continuous-discrete systems described by the stochastic equations in the finite- and the infinite-dimensional spaces are considered. Special attention is paid to the composed stochastic systems.

The main notions of probability distributions theory of the random variables, the random processes and the random functions are stated in Chapter 2. After determining the probability spaces, the conditional probabilities, probabilities in the finite and the infinite products of the spaces are considered. The conditions of the existence of the regular probabilities are established. Further the different probability measures of the random functions and the probabilities of events connected with the random functions are studied in detail. The last sections are devoted to the distributions of the separable random functions, to the derivation of the criteria of the continuity and the differentiability of the random functions.

Chapter 3 contains the account of the basic notions connected with the expectation of the random variables in Banach and topological linear spaces, the properties of the operators of the moments of the second and higher orders, the properties of the characteristic functionals and the correspondence between the probability measures and the characteristic functionals are established. The conditional operators of the moments and their properties are studied in detail. Special section is devoted to the elements of the probabilities distribution in the infinite-dimensional spaces. The properties of the normal distribution on the product of two spaces, the conditional normal distributions and the normal distributions in Hilbert spaces are considered. Approximate representation of the one- and the multi-dimensional distributions based on the consistent orthogonal expansions is treated. Special attention is paid to the semiinvariant and the quasimoments. A systemical presentation of general theory of the canonical expansions is given. Practical ways of the canonical expansions constructing are described. The joint canonical expansions are considered.

Chapter 4 is devoted to the selected topics of stochastic calculus. The theory of the convergence, the continuity and the differentiability is considered. The integrability of the random functions over nonrandom and random measures is discussed. The theory of the stochastic Itô integrals, the symmetrized (Stratonovich) integral and the θ-integral is outlined systematically. The stochastic integrals over the Wiener and the Poisson measures are studied. The connection between the processes with the independent increments and the Itô integral is established. The elements of the theory of the generalized random functions are stated. A systematic account to the integral canonical representations is given. The connection of the integral canonical representations with the spectral one is established. Practical ways of constructing the integral canonical representations and, in particular, the methods of the shaping filters are considered. The joint integral canonical representations are studied.

The general theory of the finite- and the infinite-dimensional stochastic systems is given in Chapter 5. At the beginning the differentiation of the Itô process is set out in detail. The Itô formulae for the differentiation of the composite functions in the cases of the Wiener, the Poisson and the general random process with the independent increments have a significant place. The theory of the stochastic integral, differential and integrodifferential Itô equations including the methods of reducing the state equations to the stochastic differential equations, change of variables, sufficient conditions of the existence and the uniqueness of the mean square solutions are considered. The central problem of stochastic system theory – the problem of calculation of the multidimensional distributions is stated. The elements of numerical integration of the stochastic differential equations are given. The theory of the one- and the multi-dimensional distributions in the continuous, discrete and continuous-discrete systems in the finite-dimensional spaces are stated in detail. For such systems the known equations (Fokker–Plank–Kolmogorov, Feller–Kolmogorov, Pugachev and their generalizations) are derived. The problems of one-dimensional and multi-dimensional distributions theory on the basis of the equations for the correspondent characteristic functionals in composite stochastic systems including the stochastic systems in the Hilbert spaces are considered. The main statements of the structural theory of the composed stochastic systems on the basis of conditionally probability measures are outlined. The last section contains the applications to optimal online filtering problems. Simple derivation of the Kushner–Stratonovich and the Zakai equations for the normal white noises is presented. Corresponding equations for the composite stochastic systems are given in Problems.

Chapter 6 contains the account of the main methods of linear stochastic systems theory in connection with continuous, discrete and continuous-discrete systems. The spectral and correlation methods, the methods of general theory of linear stochastic systems are outlined. The correlation methods based on the canonical representations are considered with great attention. The general theory of the canonical representations of the linear transformations is given. The separate subsections are devoted to the solution of the linear operator equations by the canonical representations methods. The applications to the offline spectral and correlation analysis and the shaping linear filters design are given. Software for the analytical modeling problems is described. The Kalman–Bucy and the Liptser–Shiryaev online filters are considered.

Chapter 7 is devoted to the approximate methods of nonlinear finite- and infinite-dimensional stochastic systems theory based on the parametrization of the distributions. The normal approximation, statistical and the equivalent linearization methods are considered. The normalization of the composite stochastic systems is treated. General problem of parametrization of the multi-dimensional distributions is stated. The methods of the moments, the semiinvariants, the quasimoments and the orthogonal expansions are given. The practical problems of reducing the number of equations for the automatic derivation of the equations and the corresponding software for distribution parameters are considered. The solution of the structural parametrization problem of the multi-dimensional distributions is outlined on the basis of the ellipsoidal approximation method. The canonical representations methods for the nonlinear stochastic systems are considered. The basic methods are given for stochastic systems with the normal white noises. Corresponding methods for the composite stochastic systems are given in Problems. The applications of the given methods to the nonlinear stochastic systems offline and online analysis, modeling and filtering problems in mechanics, physics, biology, chemistry, finance and control are stated. Specialized software is described.

Chapters 1, 5–7 give efficient toolkit for the analytical modeling and the design of the stochastic systems. To facilitate the mastering of the stated methods about 500 thoroughly selected examples and problems are given in the correspondent subsections, sections and chapters of this book.

Appendices 1–7 contain auxiliary materials which include the information about tables of stochastic Itô differentials for the typical composite functions, the Hermite polynomials of a vector argument, the polynomials orthogonal to gamma and χ^2-distributions, the tables of the coefficients of statistical linearization for the typical composite functions, integrands evaluations in the equations of the basic methods of distribution parametrization, the tables of a basic characteristics and transformation the structural rules of the linear systems and the tables of some definite integrals and special functions.

For information about the notions and theorems from vatious parts of mathematics used in the book we advise (Korn and Korn 1968).

Bibliographical notes and References are given at the end of the book. The authors do not pretend in any way to provide a complete bibliography. Only those sources are given in the list which are cited in the text. Basic notations are presentented before Index.

The book is divided into chapters, sections and subsections. The formulae (examples) are numerated with indication of the number of the chapter and the section, for instance, (1.2.3) means that formula (3) belongs to section 2 of chapter 1. The problems are numerated only with indication of the number of chapter. At each Appendix the separate numeration is accepted.

For the convenience of the readers the formulations of all basic results and statements are given in the separate theorems and the corollaries or in the text italics. The beginnings and the ends of the evaluations, the proofs and the discussions which lead to certain results are indicated by triangular indices ▷ and ◁.

The book may be used for studying stochastic systems theory by the different categories of the readers. When reading this book it is necessary to know the foundations of probability theory, for instance, the book (Pugachev 1984) and the basis of functional analysis, for instance, (Pugachev and Sinitsyn 1999). A reader who is the master of the foundations of the random functions theory and the stochastic differential equations may study directly Chapters 5–7 after Chapter 1. Finally, for studying the finite-dimensional stochastic systems theory in the simplest statement it is sufficient to study Chapter 1 (Sections 1.1–1.5), Chapter 2 (Sections 2.1.–2.10), Chapter 3 (Sections 3.1, 3.3–3.5, 3.9), Chapters 4-7 (excluding the subsections concerning the infinite-dimensional systems).

The book may be used for the lecturers on probability theory and stochastic calculus, on statistical foundations of system analysis, information, radioengineering, control and communication theory, reliability theory of structures, dynamics of machines and mechanisms, and some fragments in specialized courses, for example, analytical mechanics of stochastic systems, statistical theory of oscillations and stability and etc.

Systematical statement of the methods of the canonical expansions theory and the integral canonical representations in the finite- and the infinite-dimensional spaces and their applications are given in the book for the first time (Sections 3.9, 4.8, 6.3, 6.4, 7.7). This material may be used for the special courses on applied mathematics. Many examples and problems are original, have complex character and therefore may be used for the students and the post-graduates works.

The suggestion that the authors could write a series of book for World Scientific was put forward by Editor Dr. Sen Hu in 1998. Dr. Jitan Lu performed the final work as desk editor during 2001. The authors are particularly indebted to them.

The authors are very much obliged to Mrs. I.V.Sinitsyna for her translation of the manuscript and the list of References preparation, Mrs. E.N.Fedotova for the camera ready copy of the manuscript preparation, Mrs. I.V.Makarenkova and Mr. V.V.Belousov for the figures preparation.

The authors also owe thanks to M.E.Shaikin for his valuable remarks concerning the book as a whole, to A.V.Bosov, A.V.Borisov, B.G.Dostupov, P.S.Igumnov, I.E.Kazakov, E.R.Korepanov, V.N.Pugachev, V.I.Shin, V.I.Sinitsyn, M.I.Tleubergenov whose help considerably improved the manuscript.

The authors acknowledges the help of our colleagues in Russia and abroad. We are also grateful to the Institute for Informatics Problems, the Russian Academy of Sciences, the Russian Foundation of Fundamental Research, the International Institute for Applied Systems Analysis (Austria, Laxenburg), the Delft Technical University (The Netherlands), the Institute of Thermodynamics (Czech Republic) for their support.

Moscow   V.S.Pugachev
1997–2001   I.N.Sinitsyn

                                      Institute for Informatics Problems
                                         Russian Academy of Sciences

CONTENTS

CHAPTER 1
DYNAMICAL SYSTEMS AND THEIR CHARACTERISTICS

Chapter 1 provides an accessible introduction to stochastic systems theory. In Section 1.1 basic deterministic mathematical models of continuous linear and nonlinear dynamical systems and their characteristics are studied. Linear differential systems are outlined in Section 1.2. Section 1.3 is dedicated to the theory of structural transforms for deterministic dynamical systems. Basic linear and nonlinear stochastic differential systems are introduced in Section 1.4. Deterministic and stochastic models of discrete system are considered in Section 1.5. In Section 1.6. general notions of stochastic composed systems are presented. Basic problems of stochastic systems theory are stated.

1.1. Continuous Systems

1.1.1. Mathematical Models of Continuous Systems

A set of interacting subjects of any nature is called *a system*. The whole external world or any part of it, human society, any field of national economy, a plant, a flying vehicle, a ship , a computer network, an organism of a man or of an animal, etc. may serve as examples of systems.

In order to use mathematical methods for the study of any functioning system, it is necessary to design a mathematical model of this system. First, it is necessary for this purpose to determine a set of variables which may serve as quantitative characteristics of system functioning, and then establish relations among these variables which approximately describe the functioning of the system.

Any system, besides the whole external world, interacts with surroundings, receives something from without and after some operations returns something into the surroundings, in particular, to other systems. This is the functioning of the system. A system may receive from without and give into the surroundings various substances, subjects, information, control actions. It is usually said that a system receives some input data and produces output data. So, for instance, a plant receives input streams of raw materials and various means of production, work quota, managing instructions of higher organizations, and gives output production, waste products, materials and means which are transmitted to other systems and documentation. A controlled air or space vehicle receives (from a pilot or from an automatic control system) control actions, i.e. the state of its final control elements (of rudders and of throttles of propulsion engines) - as a function of time. As a result some forces and their moments acting on a vehicle are created which change the orientation of its axes and the direction of its flight. Owing to this fact the trajectory of its flight and law of motion of its axes are obtained. In addition to the control actions of a pilot or of a control system an aircraft flying in atmosphere is subject also to random variations of the wind velocity vector along the trajectory owing to atmospheric turbulence. These variations produce the random oscillations of aerodynamic forces and their moments acting on the aircraft. As a result random oscillations of aircraft arise, the effect of the so-called *rough air*. Thus, receiving random variations of aerodynamic forces and their moments due to the atmospheric turbulence the aircraft responds to them by random oscillations.

The first step in designing a mathematical model of a system is the mathematical description of what a system receives and what it produces. This description consists in fixing two sets of variables, one for describing external actions on the system, the other for describing the response of the system.

The variables which describe external actions on a system are called its *inputs* and those which describe the action of the system on environment and, in particular, on other systems are called the *outputs* of the system.

Besides inputs and outputs we often have to introduce some auxiliary variables for designing a mathematical model of a system that may include the variables characterizing the actions of various parts of the system on each other (interactions of various parts of the system). All these variables which characterize the state of the system at a given time instant are usually called *state variables* of the system.

In sequel we shall call the *input* of a system the set of all its inputs, the *output* of the system the set of all its outputs, and the *state vector*, the set of all state variables of the system.

The set of all possible inputs of a system is called its *space of inputs*, the set of all possible outputs of the system is called its *space of outputs*, and the set of all possible state vectors of the system is called its *state space*.

Inputs, outputs and the state vector of a system as functions of time characterize the *functioning* of the system, or as is often said its *behaviour.*

After the determination of the inputs, the outputs and the state vector of a system it remains only to establish the relations among these variables for obtaining its mathematical model. These relations may be either deterministic or contain some elements of uncertainty. In the latter case the statistical (probability) approach is usually applied ascribing a random character and corresponding distributions to all uncertain variables.

Thus we come to the following definition of a mathematical model of a system. The set of four elements is called the *mathematical model of a system*: (i) the state space; (ii) the space of the inputs; (iii) the space of the outputs; (iv) the relations which connect the inputs, the outputs and the state vector of the system.

Strictly speaking the notions of the input, the output and the state vector are not related to the system itself but to its mathematical model. In fact, it is impossible to characterize the real state of any system, all external influences on it and all its actions on environment and, in particular, on other systems by any set of mathematical notions and *a fortiori* by a finite set of the variables. Therefore speaking about the inputs, the outputs and the state of a system we always mean the inputs, the outputs and the state of the mathematical model of the system.

E x a m p l e 1.1.1. The mathematical model of the electric circuit, consisting of the resistor R and the capacitor C (Fig.1.1.1), is the set of equations which follows from Ohm and Kirchhoff laws

$$u_1 = Ri, \quad Cdu_2/dt = i, \quad u_1 + u_2 = x,$$

where i is the electric current, u_1 is the voltage drop across the resistor, u_2 is the capacitor voltage, and x is the input voltage.

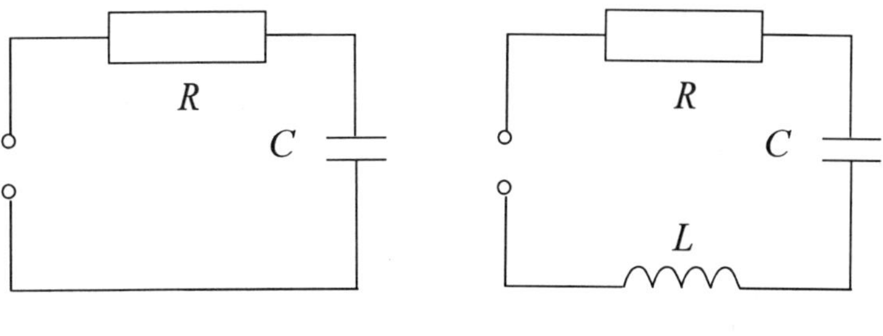

Fig. 1.1.1   Fig. 1.1.2

Since the variables u_1 and i may be expressed in terms of u_2 and x from the first and the third equations: $u_1 = x - u_2$, $i = u_1/R = (x - u_2)/R$, the state of the circuit may be characterized by one variable $z = u_2$. Then we obtain the following differential equation:

$$T\dot{z} + z = x, \quad T = RC, \tag{I}$$

which describes the circuit state as a function of time. Assuming the capacitor voltage $u_2 = z$ as an output of the circuit y, we shall have $y = z$.

The set of scalar time functions serves as the spaces of the inputs and of the outputs, and the real axis serves as a state space.

E x a m p l e 1.1.2. Assuming the voltage drop across the resistor u_1 as an output of the circuit of the previous example we shall have the equation of the previous example for the state variable $z = u_2$ and the formula for the output signal $y = x - z$. Eliminating from this formula and the state equation the state variable z, we obtain the equation which connects the input and the output of the circuit:

$$T\dot{y} + y = T\dot{x}.$$

The spaces of the inputs, of the outputs and the state space are the same as in Example 1.1.1.

E x a m p l e 1.1.3. The set of equations

$$u_1 = Ri, \quad C du_2/dt = i, \quad L di/dt = u_3, \quad u_1 + u_2 + u_3 = x,$$

serves as a mathematical model of an oscillatory circuit (Fig.1.1.2), where in addition to the notations of Examples 1.1.1 and 1.1.2 the variable u_3 is the voltage drop across the inductance L.

As the variables u_1 and u_3 may be expressed in terms of i, u_2 and x from the first and the fourth equations: $u_1 = Ri$, $u_3 = x - Ri - u_2$, the state of the circuit may be characterized by two variables $z_1 = i$, $z_2 = u_2$. Then we obtain the differential equations of the system state

$$L\dot{z}_1 = x - Rz_1 - z_2, \quad C\dot{z}_2 = z_1.$$

Assuming the capacitor voltage $u_2 = z_2$ as an output of the circuit y we shall have $y = z_2$. Eliminating the state variables z_1 and z_2, we obtain the equation connecting the input and the output of the circuit:

$$T^2\ddot{y} + 2\xi T\dot{y} = x, \quad T^2 = LC, \quad 2\xi = R\sqrt{C/L}.$$

The set of scalar time functions serve as the spaces of the inputs and the outputs and the plane serves as a state space.

E x a m p l e 1.1.4. The second Newton law serves as a mathematical model of the motion of a free particle having the mass m

$$m\ddot{y} = x.$$

The force x which acts on the particle is the input, and the position vector of the particle y is the output. The state of the particle at every instant is determined by its coordinates and its velocity vector. Thus the six-dimensional vector $z = \{y, \dot{y}\}$ is the state vector of the particle. The set of all three-dimensional vector functions of time is a space of inputs. The space of outputs represents the set of twice differentiable three-dimensional functions of time. The six-dimensional Euclidean space serves as a state space.

E x a m p l e 1.1.5. The set of three dynamic Euler equations

$$I_x \dot{\omega}_x + (I_z - I_y)\omega_y \omega_z = M_x,$$
$$I_y \dot{\omega}_y + (I_x - I_z)\omega_z \omega_x = M_x,$$
$$I_z \dot{\omega}_z + (I_y - I_x)\omega_x \omega_y = M_z,$$

and three kinematic Euler equations

$$\omega_x = \dot{\psi} \sin\theta \sin\varphi + \dot{\theta} \cos\varphi,$$
$$\omega_y = \dot{\psi} \sin\theta \cos\varphi - \dot{\theta} \sin\varphi,$$
$$\omega_z = \dot{\psi} \cos\theta + \dot{\varphi},$$

serves as a mathematical model of motion of a solid with one fixed point relative to fixed axes, where I_x, I_y, I_z are inertia moments of the solid relative to its principal inertia axes xyz, ω_x, ω_y, ω_z the projections of the vector of angular velocity of the solid on the principal inertia axes, M_x, M_y, M_z the moments of forces acting on the solid relative to the principal inertia axes, and ψ, θ, φ the angles determining the orientation of the principal inertia axes relative to the fixed axes $\xi\eta\zeta$. The moments $x_1 = M_x$, $x_2 = M_y$, $x_3 = M_z$, acting on the solid are the inputs, and three Euler angles $y_1 = \psi$, $y_2 = \theta$, $y_3 = \varphi$ are the outputs. The state of the solid at every time instant is determined by three Euler angles and three projections of the angular velocity. Thus the six-dimensional vector with components

$$z_1 = \psi, \;\; z_2 = \theta, \;\; z_3 = \varphi, \;\; z_4 = \omega_x, \;\; z_5 = \omega_y, \;\; z_6 = \omega_z.$$

is a state vector. The space of inputs, the space of outputs and the state space are the same as in Example 1.1.4.

A model of a system is called *deterministic* if one definite realization of the output corresponds with every realization of its input, i.e. if its output is the result of some mapping of the input space into the output space. All the models considered in Examples 1.1.1-1.1.5 are deterministic.

A model of a system is called *stochastic* if a definite distribution of its output corresponds with every realization of its input.

One may design many different models for the same system. According to the degree of detailing of the characteristics of system behaviour and the quantity of factors which we take into account some models will be simpler, the others more complicated. The more factors are taken into account in a model the more complete and more accurate, in principle, is the description of the system behaviour. But the accuracy of a complicated model may be illusory. Because of the limitations of accessible information, in particular, because of inaccuracy of initial data which are used for the design and modeling, an excessively complex model may be less accurate than a simple one. Therefore the degree of complexity of a model must be coordinated with accessible information which may be used for designing the model and for its application.

On the contrary the same model may describe different systems, for example, a linear differential equation of the second order with positive constant coefficients serves as the model of an oscillatory circuit (Example 1.1.3) and as the model of small oscillations of a pendulum in viscous medium (see Example 1.4.2).

A characteristic feature of complex systems is that as a rule none of the models can reproduce all system functions with sufficient accuracy. Some models may be better for one application and other models may be better for other applications. But none of the models can be the best for all applications. Therefore we usually design several models for complex systems and use one model for some purposes (for studying some functions of a system) and the other models for other purposes. Some of these models may be deterministic and the others stochastic. So, for instance, the model of a plant which takes into account only the mean number of workers participating daily in production and the mean quantity of materials received daily is deterministic. A model of the same plant which takes into account random daily variations of the number of workers due to truancies for various reasons and random daily variations of materials received by the plant is stochastic. Besides, it is characteristic for complex systems such as a plant, a branch of industry, the economy of a region or of a country that they consist of great number of simpler systems (subsystems). Consequently it is impossible to control them without corresponding organization inside the system itself, without organization of control of every separate subsystem and without determining definite interactions between all the subsystems. As a result a control system for such a system is first, hierarchic and secondly, distributed among the elements of the system. It is necessary to organize the systems first and to introduce the elements of control system into every element of the system with fixing corresponding hierarchy of control, these we shall call *large scale systems*.

The inputs and the outputs of any system represent functions of time. If they are determined for all instants beginning from some initial instant then the system is called *continuous*. Accordingly, in this case the system is called *continuous*. If the realizations of the inputs and the outputs are determined only for discrete sets of instants then the system is called a *discrete* or a *sampled-data system*. In this case the system itself is usually called a *discrete* or *sampled-data system*. However, discrete models are often used for the description of the behaviour of continuous systems. In particular discrete models are always used for calculations on digital computers whether the corresponding systems are continuous or discrete.

The values of the input and the output at every instant may be scalar, finite-dimensional vectors or elements of some functional spaces (i.e. functions of some variables, for example, of the coordinates of points in the space). In the latter case the system (its model) is called *distributed* (or a system with *distributed parameters*). If the values of the inputs and the outputs at every instant are scalar variables then the system is called *one-dimensional*. If the values of the input or of the output or of both of them at every instant are finite-dimensional vectors, then the system is called *finite-dimensional*.

As regards a state vector, its value at a given instant may also be scalar, a finite-dimensional vector or a function of some variables, even for the systems whose values of the input and the output at every instant are scalar or finite-dimensional vectors.

Later on we shall always assume that the mathematical model of a system under study is designed and speak about the system meaning the model of this system. In particular, when speaking about the characteristics of the system we shall mean the characteristics of its mathematical model.

1.1.2. Characteristics of Continuous Systems

The main characteristic of a system is its *operator* which determines the mechanism of forming the output corresponding to a given input.

The operator of a deterministic system relates a definite output to any possible input. Thus *the operator of a deterministic system maps the space of the inputs into the space of the outputs.*

The operator of a stochastic system relates a distribution of the output to any possible input. Thus *the operator of a stochastic system maps the space of the inputs into the space of all possible distributions in the space of the outputs.*

Let $x(t)$ be the input of a deterministic system which represents a scalar (Fig.1.1.3a) or n-dimensional vector function of time t, $y(t)$ being the output which represents an m-dimensional vector function of t (Fig.1.1.3b). We denote by the letter A the system operator. Then the relation between the input and the output of a deterministic system may be written in the form

$$y(t) = Ax(t). \tag{1.1.1}$$

This short formula includes the whole totality of mathematical operations which we have to fulfill in order to determine the function $y(t)$ for a given function $x(t)$.

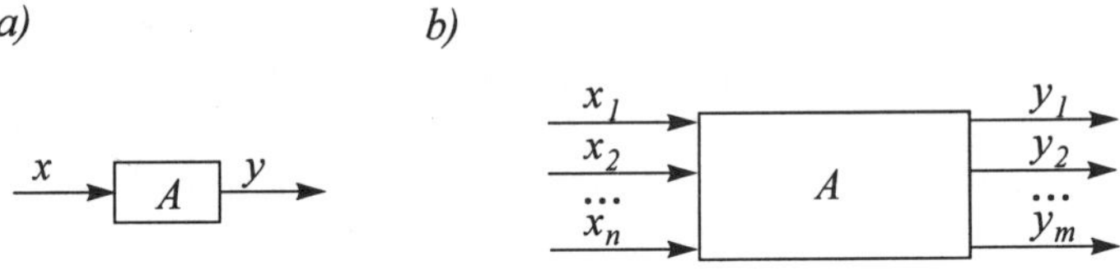

Fig. 1.1.3

A deterministic system is called *nonanticipative* (or physically realizable) if the value of its output $y(t)$ at every instant t does not depend on the values of its input $x(\tau)$ at $\tau > t$. Thus the value of the output of a nonanticipative system $y(t)$ at every instant t is a functional of the input $x(\tau)$ in the time interval $t_0 \leq \tau \leq t$. Here t_0 is the instant at which the system starts functioning.

A stochastic system is called *nonanticipative* (or physically realizable) if the distribution of the value of its output $Y(t)$ at any instant t does not depend on the values of its input $x(\tau)$ at $\tau > t$.

Let $x(t)$ be some fixed input of a deterministic system which we refer as a nonperturbed one. Let $y(t)$ be the corresponding output process. Another input $x'(t)$ we shall call a perturbed input and its corresponding output $y'(t)$ we shall call a perturbed output.

The deviations of the input and the output from nonperturbed ones we shall define as $\Delta x(t) = x'(t) - x(t)$ and $\Delta y(t) = y'(t) - y(t)$. Deterministic system (1.1.1) is called *stable relative to the given nonperturbed process* if the deviation $\Delta y(t)$ remains as small as possible at given sufficiently small deviation $\Delta x(t)$ of the input, i.e. for any $\varepsilon > 0$ there exists such $\delta = \delta(\varepsilon) > 0$ at $t > t_0$ that $|\Delta y(t)| < \varepsilon$ at $t > t_0$ when $\sup |\Delta x(t)| < \delta$ at all $t \geq t_0$.

A stochastic system is called *stable relative to the given nonperturbed process almost surely* (with probability 1) if the deviation of its output $\Delta Y(t)$ is small as possible with the probability 1 at any sufficiently small deviation of the input $\Delta x(t)$. A stochastic system is called *stable relative to the given nonperturbed process in probability* if at any $\varepsilon > 0$ there exists such $\delta = \delta(\varepsilon) > 0$ that $\lim\limits_{t \to \infty} P\left(\sup\limits_{t > t_0} |\Delta Y(t)| > \varepsilon \right) = 0$ at all $\Delta x(t)$ satisfying the condition $\sup\limits_{t > t_0} |\Delta x(t)| < \delta$. A stochastic system is called *stable relative to the given nonperturbed process in p-mean, $p > 0$*, if the expectation $\mathrm{E}|\Delta Y(t)|^p$ remains as small as possible at all sufficiently small deviations of the input $\Delta x(t)$.

From the known Chebyshev inequality follows that a *stochastic system is stable in probability if it is p-stable*. Similarly, *from the stability almost surely follows the stability in probability. From p-stability at a given p follows p-stability at all smaller p.* The inverse is not true in general.

R e m a r k. Speaking about the stability of a system we always mean in practical problems the stability of all realizations of the processes in the system. From this point of view the notion of the almost surely stability is of great importance for applications. But in practical problems we often have to restrict ourselves to the stability in *mean* $(p = 1)$ or in *mean square* $(p = 2)$. The class of systems stable in mean square, and the class of systems almost surely stable are the subclasses of systems stable in probability and both of these subclasses have a nonempty intersection. Special practical interest presents the notion of the *asymptotic stability* in *p-mean* when $E |\Delta Y(t)|^p \to 0$ at $t \to \infty$.

A deterministic system is called *linear* if for any numbers N, c_1, ..., c_N and any functions $x_1(t)$, ..., $x_N(t)$

$$A\left\{ \sum_{\nu=1}^{N} c_\nu x_\nu(t) \right\} = \sum_{\nu=1}^{N} c_\nu A x_\nu(t). \qquad (1.1.2)$$

This defining property of linear systems is usually called *the superposition principle*. Therefore the linear systems may be defined as the systems for which the superposition principle is valid.

The operator A possessing property (1.1.2) is called *linear*. Thus a *deterministic system is linear if and only if its operator is linear.*

A system is called *nonlinear* if the superposition principle does not hold. The operator of a deterministic nonlinear system is always nonlinear. Models of Examples 1.1.1-1.1.4 are linear whereas in Example 1.1.5 the model is nonlinear.

In practical problems we usually manage to represent the model of any technical system with the finite-dimensional input, the output and the state vector in the form of a connection of a finite number of typical linear systems (Table A.5.1) and nonlinear elements without memory. We speak about the nonlinear elements without memory since the output of a nonlinearity in a technical system depends practically always only on the instantaneous value of its input at the same instant and is independent of the past values of the input. Namely the value of the output of a nonlinear element at any instant t represents a known function of the value of its input at the same instant t and is independent of the values of the input before the moment t:

$$y(t) = \varphi(x(t), t), \qquad (1.1.3)$$

The input and the output of a nonlinear element are denoted here by $x(t)$ and $y(t)$. The function φ completely characterizes the element without memory and therefore is assumed as its characteristic. In the general case the input of a nonlinear element in any system is a vector whose components represent some of the state variables and perhaps some of the inputs of the system (Section 1.6).

The most part of linear systems at practice represent systems whose behaviour is described by ordinary differential equations which follow from the laws of mechanics, physics, chemistry, biology etc. By introducing additional state variables if necessary these equations may be represented in the standard Cauchy form, i.e. reduced, to the set of equations of the first order solved relative to the derivatives. Adding then equations, similar to Eqs. (1.1.3) for nonlinear elements of the system we get a set of nonlinear differential equations in the Cauchy form which describes the nonlinear system.

Let us denote by $z(t) = [z_1(t) \dots z_p(t)]^T$ (T being the operator of matrix transposition) the state vector of a system, by $x(t) = [x_1(t) \dots x_n(t)]^T$ its input vector and by $y(t) = [y_1(t) \dots y_m(t)]^T$ its output vector. Then the differential equations describing a system behaviour may be written in the form

$$\dot{z} = f(z, x, t), \quad y = g(z, t), \qquad (1.1.4)$$

where f is a p-dimensional vector function of the vectors z, x and of time t, and g is an m-dimensional vector function of the vector z and time t.

A system with finite-dimensional state vector, input and output described by an ordinary differential equation and by a formula for the output of form (1.1.4) we shall call for brevity *a differential system*. Eqs. (1.1.4) with the given initial state of the system $z_0 = z(t_0)$ determine the operator of the system completely and uniquely.

R e m a r k 1. Using δ-function we may always take the initial conditions equal to zero by including them within the input vector. Therefore, Eqs. (1.1.4) with the initial condition $z(t_0) = z_0$ may be written in the form

$$\dot{z} = f(z, x, t) + z_0 \delta(t - t_0), \quad y = g(z, t) \qquad (1.1.5)$$

and be integrated with zero initial conditions $z = 0$ at $t = t_0 + \varepsilon$ with any arbitrarily small $\varepsilon > 0$. Assuming z_0 as an additional set of components of the input vector we may affirm that Eqs. (1.1.5) with zero initial conditions completely determine the operator of the system.

R e m a r k 2. The function $g(z, t)$ in formula (1.1.4) for the output does not depend explicitly on the input x. But in some cases we have to replace the function $g(z, t)$ by the function $g(z, x, t)$, explicitly depending on the input x. In special cases the output of a system may coincide with some of the state variables. In such a case we suppose without loss of generality that the first m components of the vector z represent the outputs and write the second Eq. (1.1.4) or (1.1.5) in the form $y = [I_m \ 0]z$ where I_m is the unit matrix of order m, and 0 represents the $m \times (p - m)$ matrix with zero elements.

1.1.3. Weighting Function of One-Dimensional Linear System

Let us consider an one-dimensional linear system. Its input as a continuous function may be represented by the expansion in terms of infinitesimal instantaneous impulses

$$x(t) = \int_{-\infty}^{\infty} x(\tau)\delta(t - \tau)d\tau. \qquad (1.1.6)$$

Hence on the basis of the superposition principle (Subsection 1.1.2) we obtain the following expression of the output:

$$y(t) = Ax(t) = \int_{-\infty}^{\infty} x(\tau)A_t\delta(t - \tau)d\tau, \qquad (1.1.7)$$

where the index t at the operator A under the sign of integral shows that this operator acts on the function $\delta(t - \tau)$ considered as a function of t at fixed value of τ. Formula (1.1.7) shows that it is sufficient for finding the response of a linear system to an arbitrary input $x(t)$ to know its response to the unit instantaneous impulse $\delta(t - \tau)$ acting on it at an arbitrary moment τ. This response depends on the variables t and τ, i.e. on the moment of action τ of the impulse and the current moment t: $g(t, \tau) = A_t \delta(t - \tau)$. The function $g(t, \tau)$, determined by this formula is a complete characteristic of a linear system and is called its *weighting function* or *impulse response function*. Thus *the weighting function of a linear system represents its response at the instant t to the unit impulse acting at the instant τ.*

Using the notion of a weighting function we may write relation (1.1.7) between the input and the output of a linear system in the form

$$y(t) = Ax(t) = \int_{-\infty}^{\infty} g(t, \tau)x(\tau)d\tau . \qquad (1.1.8)$$

Thus *the operator of any linear system may be represented in the form of a linear integral operator.*

E x a m p l e 1.1.6. In order to find the weighting function of the circuit of Example 1.1.1 which sometimes is called as an *aperiodic link* (Table A.5.1) we integrate the last equation of this example supposing that the input voltage is applied at the instant t_0. Then we obtain at $t > t_C$

$$y(t) = \frac{1}{T}e^{-t/T} \int_{t_0}^{t} e^{\tau/T} x(\tau)d\tau . \qquad (I)$$

Comparing this formula with (1.1.8) and taking into account that $x(\tau) = 0$ at $\tau < t_0$, we find

$$g(t, \tau) = \frac{1}{T}e^{-(t-\tau)/T} \mathbf{1}(t - \tau) , \qquad (II)$$

where $\mathbf{1}(s)$ is the unit step function.

E x a m p l e 1.1.7. For the circuit of Example 1.1.2 supposing that $x(\tau) = 0$ at $\tau \leq t_0$, we obtain

$$y(t) = e^{-t/T} \int_{t_0}^{t} e^{\tau/T} \dot{x}(\tau)d\tau$$

$$= x(t) - \frac{1}{T}e^{-t/T}\int_{t_0}^{t} e^{\tau/T}x(\tau)d\tau = \int_{t_0}^{t} \left[\delta(t-\tau) - \frac{1}{T}e^{-(t-\tau)/T} \right] x(\tau)d\tau.$$

$$\text{(I)}$$

Comparing this formula with (1.1.8) we find

$$g(t,\tau) = \delta(t-\tau) - \frac{1}{T}e^{-(t-\tau)/T}\mathbf{1}(t-\tau). \qquad \text{(II)}$$

E x a m p l e 1.1.8. For the oscillatory circuit of Example 1.1.3

$$y(t) = \frac{1}{\omega_0 T^2}\int_{t_0}^{t} e^{-\xi(t-\tau)/T}\sin\omega_0(t-\tau)x(\tau)d\tau, \qquad \text{(I)}$$

where $\omega_0 = \sqrt{1-\xi^2}/T$. Hence we find

$$g(t,\tau) = \frac{1}{\omega_0 T^2}e^{-\xi(t-\tau)/T}\sin\omega_0(t-\tau)\mathbf{1}(t-\tau)d\tau. \qquad \text{(II)}$$

Let us consider a linear finite-dimensional system with n inputs and m outputs. On the basis of the superposition principle the action of every input on a finite-dimensional linear system may be considered separately and then the responses of the system to all the inputs summarized for each output.

In order to find the response at the kth output of a linear system to the action of the hth input only, we may consider this system as a system with one input and one output. Then for calculating the kth output of a system under the action of the hth input only (in the absence of signals at other inputs) it is sufficient to know the corresponding weighting function $g_{kh}(t,\tau)$. This weighting function represents a response at the kth output of a system at the instant t to the instantaneous unit impulse acting at the hth input at the instant τ, in the absence of the signals at other inputs. The aggregate of the weighting functions $g_{kh}(t,\tau)$ ($k = 1, \ldots, m$; $h = 1, \ldots, n$), corresponding to all the inputs and all the outputs is a complete characteristic of a finite-dimensional linear system.

Knowing the weighting functions of a finite-dimensional linear system corresponding to all the inputs and all the outputs we may calculate its response at every output to a signal acting at any one of the inputs by formula (1.1.8). Performing the summation of the obtained results

for all the inputs for every separate output we find all the outputs of the
considered linear system corresponding to simultaneous action of all the
inputs. Thus the outputs $y_1(t)$, ... , $y_m(t)$ of a linear system with n inputs and m outputs are expressed in terms of its inputs $x_1(t)$, ... , $x_n(t)$
by the formula

$$y_k(t) = \sum_{h=1}^{n} \int_{-\infty}^{\infty} g_{kh}(t,\tau)x_h(\tau)d\tau \quad (k = 1, \ldots, m). \tag{1.1.9}$$

Introducing the matrix

$$g(t,\tau) = \begin{bmatrix} g_{11}(t,\tau) & g_{12}(t,\tau) & \cdots & g_{1n}(t,\tau) \\ g_{21}(t,\tau) & g_{22}(t,\tau) & \cdots & g_{2n}(t,\tau) \\ \cdots & \cdots & \cdots & \cdots \\ g_{m1}(t,\tau) & g_{m2}(t,\tau) & \cdots & g_{mn}(t,\tau) \end{bmatrix}, \tag{1.1.10}$$

we may write this formula in the form

$$y(t) = Ax(t) = \int_{-\infty}^{\infty} g(t,\tau)x(\tau)d\tau, \tag{1.1.11}$$

where $x(t) = [\,x_1(t)\ldots x_n(t)\,]^T$, $y(t) = [\,y_1(t)\ldots y_m(t)\,]^T$ are the column
vectors of the input and the output signals. The matrix function $g(t,\tau)$
determined by formula (1.1.10) is called *the weighting function* of a linear
system with n inputs and m outputs.

According to the definition in Subsection 1.1.2

$$g(t,\tau) = 0 \ \text{ at } \ \tau > t \tag{1.1.12}$$

for a nonanticipative linear system, and formula (1.1.11) takes the form

$$y(t) = Ax(t) = \int_{t_0}^{t} g(t,\tau)x(\tau)d\tau, \tag{1.1.13}$$

where t_0 is the instant at which the system starts functioning.

It is clear that the linear systems of Examples 1.1.6-1.1.8 are nonanticipative as they represent real circuits. Therefore they satisfy condition (1.1.12).

E x a m p l e 1.1.9. If the capacitor voltage y_1 and the current in the circuit y_2 are considered in Example 1.1.3 as the outputs of the oscillatory circuit we may add to the equations of Example 1.1.3 the relation $y_2 = z_1$ or $y_2 = C\dot{y}_1$. Hence it is clear that the weighting function found in Example 1.1.8 serves as one of the elements of the matrix weighting function of the oscillatory circuit considered as a system with one input and two outputs. To find the other element we differentiate the expression of the first output obtained in Example 1.1.8:

$$y_1(t) = \frac{1}{\omega_0 T^2} \int_{t_0}^{t} e^{-\xi(t-\tau)/T} \sin \omega_0(t-\tau) x(\tau) d\tau \tag{I}$$

and multiply the result by C. Then taking into account $T^2 = LC$ we get

$$y_2(t) = \frac{1}{L} \int_{t_0}^{t} e^{-\xi(t-\tau)/T} \left[\cos \omega_0(t-\tau) - \frac{\xi}{\omega_0 T} \sin \omega_0(t-\tau) \right] x(\tau) d\tau. \tag{II}$$

Thus the weighting function of the considered system represents a column matrix

$$g(t,\tau) = \left[\begin{array}{c} g_{11}(t,\tau) \\ g_{21}(t,\tau) \end{array} \right], \tag{III}$$

where

$$g_{11}(t,\tau) = \frac{1}{\omega_0 T^2} e^{-\xi(t-\tau)/T} \sin \omega_0(t-\tau) \mathbf{1}(t-\tau),$$

$$g_{21}(t,\tau) = \frac{1}{L} e^{-\xi(t-\tau)/T} \left[\cos \omega_0(t-\tau) - \frac{\xi}{\omega_0 T} \sin \omega_0(t-\tau) \right] \mathbf{1}(t-\tau).$$

1.1.4. Stability of Linear Systems

For the stability of linear system (1.1.11) it is necessary and sufficient that its weighting function satisfy the condition:

$$\int_{-\infty}^{\infty} |g(t,\tau)|\, d\tau < \infty. \tag{1.1.14}$$

$\triangleright$ For a stable linear system and continuous inputs formula (1.1.8) follows immediately from the known in functional analysis the Riesz theorem about the general form of a continuous linear functional on the space of continuous functions. The value of the output $y(t)$ at fixed t represents a linear functional of the input $x(t)$. From the definition of stability follows that for a stable linear system this functional is continuous. Therefore, on the basis on the Riesz theorem

$$y(t) = \int_{-\infty}^{\infty} x(\tau)d_\tau h(t,\tau), \qquad (1.1.15)$$

where $h(t,\tau)$ is a function of the variable τ of bounded variation at any t, and the index τ at the sign of the differential indicates that the integration is performed with respect to the function $h(t,\tau)$ considered as a function of τ at fixed t. If $h(t,\tau)$ has at any t the derivative with respect to τ, probably containing a linear combination of δ-functions, then putting $g(t,\tau) = \partial h(t,\tau)/\partial\tau$, we get (1.1.8). The condition of the boundedness of the variation of $h(t,\tau)$ relative to τ gives the necessary and sufficient condition of stability (1.1.9). If the variation of $h(t,\tau)$ relative to τ is unbounded then the system is unstable. We see that the Riesz theorem justifies formula (1.1.8) only for stable systems. But it is evidently valid for classes of nonstable systems such, for instance, as described by linear differential equations having no stable solutions. Formula (1.1.8) is easily obtained for such systems by the usual method of variation of constants for solving nonhomogeneous linear differential equations. ◁

Similarly from the Riesz theorem, the general form of a continuous linear functional on the space of continuous functions with continuous derivatives up to the Nth order follows formula (1.1.8) for the case of N-times continuously differentiable inputs, and in this case $g(t,\tau)$ may also contain a linear combination of the derivatives of δ-functions up to the Nth order. The necessary and sufficient condition (1.1.14) of the stability of a linear system is in this case replaced by the condition

$$\int_{-\infty}^{\infty} \left| \int_{-\infty}^{\tau} (\tau - \sigma)^{N-1} g(t,\sigma)d\sigma \right| d\tau < \infty. \qquad (1.1.16)$$

For the stability of a finite-dimensional linear system it is necessary and sufficient that all the elements of the matrix $g(t,\tau)$ satisfy conditions similar to (1.1.14):

$$\int_{-\infty}^{\infty} |g_{kh}(t,\tau)|\, d\tau < \infty \quad (k=1,\ldots,m; h=1,\ldots,n). \qquad (1.1.17)$$

1.1.5. Stationary Systems

A systems is called *stationary* if at any shift in time of its input without changing its shape (i.e. at the replacement of $x(t)$ by $x(t-T)$ at any T) the output is similarly shifted in time without changing its shape (i.e. $y(t)$ is replaced by $y(t-T)$).

It is easy to see that the system described by Eqs. (1.1.4) or (1.1.5) is stationary if and only if the right-hand sides of these equations, i.e. the functions $f(z,x,t)$ and $g(z,t)$ do not depend explicitly on time: $f(z,x,t) = f(z,x)$, $g(z,t) = g(z)$.

1.1.6. Transfer Function of Stationary Linear System

From the definition of a stationary system (Subsection 1.1.5) it follows that the weighting function of a stationary linear system depends only on the difference of its arguments. Therefore, according to the definition the response of a stationary linear system at the instant t to the unit impulse acting at the instant τ coincides with its response at the instant $t - \tau$ to the unit impulse which acts at zero instant, i.e. $g(t,\tau) = g(t-\tau,0)$ at all t, τ. Putting $g(t-\tau,0) = w(t-\tau)$ we shall have $g(t,\tau) = w(t-\tau)$.

The main peculiarity of nonanticipative stationary and stable linear systems is the fact that *any exponential function e^{st} acting at the input for an indefinitely long time it multiplies by some coefficient without changing its shape.* Therefore, putting in Eq. (1.1.13) in the case of an one-dimensional system $x(\tau) = e^{s\tau}$, $g(t,\tau) = w(t-\tau)$, $t_0 = -\infty$ we get

$$y(t) = \int_{-\infty}^{t} w(t-\tau)e^{s\tau}\,d\tau = e^{st}\int_{-\infty}^{t} w(t-\tau)e^{-s(t-\tau)}\,d\tau = e^{st}\int_{0}^{\infty} w(\sigma)e^{-s\sigma}\,d\sigma\,.$$

Denoting the coefficient at e^{st} by $\Phi(s)$,

$$\Phi(s) = \int_{0}^{\infty} w(\sigma)e^{-s\sigma}\,d\sigma\,, \tag{1.1.18}$$

we obtain $y(t) = \Phi(s)e^{st}$. This formula proves our statement and shows that the coefficient $\Phi(s)$ at the exponential function in the expression of the output depends on the parameter s. Accordingly this coefficient is called *the transfer function* of the stationary linear system. Formula

(1.1.18) shows that the transfer function exists only at those real s for which the above integral converges.

R e m a r k. The statement proved above is also valid for complex values of the parameter s whose real part is greater than some negative number. In order to be convinced of that it is sufficient to recall that on the basis of the superposition principle the response of a linear system to a complex input signal represents a complex time function whose real and imaginary parts are equal to the responses of the system to the real and imaginary parts of the input respectively. Certainly, the function $\Phi(s)$ has in general a complex value if the imaginary part of the parameter s is different from zero. It means that *a stationary linear system preserves the shape of harmonic oscillations with the amplitude varying according to the exponential law, amplifing its amplitude and shifting the phase. The gain in the amplitude is equal to* $|\Phi(s)|$ *and the shift of the phase is equal to* $\arg\Phi(s)$.

Formula (1.1.18) shows that a transfer function of a stationary linear system $\Phi(s)$ is the Laplace transformation of its weighting function. Using the terms of operational calculus one may say that *a transfer function of a stationary linear system represents a transform of its weighting function.* This circumstance makes it possible to use the tables of the formulae of the operational calculus for determining the transfer functions of the stationary linear systems by their weighting functions (and vice versa) and thus to avoid the calculations of the integrals (1.1.18).

Notice that formula (1.1.18) derived for the one-dimensional systems also determines the transfer function of a stationary linear system with n inputs and m outputs. In order to be convinced of that it is sufficient to use formula (1.1.18) separately for any input-output pair and write the obtained nm relations in a matrix form. Hence the transfer function of a finite-dimensional system is determined as the $m \times n$ matrix whose elements are the transfer functions from all the inputs to all the outputs considered separately.

E x a m p l e 1.1.10. The formulae of Examples 1.1.1-1.1.3 show that the electric circuits considered in these examples represent stationary linear systems. Accordingly the weighting functions of the circuits found in Examples 1.1.6-1.1.9 depend only on $t - \tau$.

E x a m p l e 1.1.11. In order to find the transfer function of the circuit of Example 1.1.1 let us substitute the expression of its weighting function obtained in Example 1.1.6 into Eq. (1.1.18). After integration we get

$$\Phi(s) = \frac{1}{T}\int_0^\infty e^{-\sigma/T - s\sigma}\,d\sigma = \frac{1}{Ts+1}. \tag{I}$$

Strictly speaking this formula determines $\Phi(s)$ only at $\mathrm{Re}\,\{s\} > -1/T$. But the last part of this formula represents a function of the complex variable s determined at all s except the pole at the point $s = -1/T$. Thus the last part of the obtained formula gives the analytical extension of the transfer function determined by formula (1.1.18) only at those s for which the integral converges to the whole complex plane.

E x a m p l e 1.1.12. Substituting the expression of the weighting function of the circuit (I) in Example 1.1.2 obtained in Example 1.1.7 into Eq. (1.1.18) we find the transfer function of this circuit

$$\Phi(s) = \int_0^\infty \left[\delta(\sigma) - \frac{1}{T}e^{-\sigma/T} \right] e^{-s\sigma}\,d\sigma = 1 - \frac{1}{Ts+1} = \frac{Ts}{Ts+1}\,. \qquad \text{(I)}$$

E x a m p l e 1.1.13. Substituting the expression of the weighting function of the oscillatory circuit in Example 1.1.3 obtained in Example 1.1.8 into Eq. (1.1.18) we find the transfer function of the circuit

$$\Phi(s) = \frac{1}{\omega_0 T^2} \int_0^\infty e^{-\xi\sigma/T - s\sigma} \sin\omega_0\sigma\,d\sigma = \frac{1}{T^2 s^2 + 2\xi Ts + 1}\,. \qquad \text{(I)}$$

E x a m p l e 1.1.14. Substituting into Eq. (1.1.18) the expression of the weighting function of the oscillatory circuit considered as a system with two outputs obtained in Example 1.1.9 we find that the transfer function of this circuit is the 2×1-matrix with entries

$$\Phi_{11}(s) = \frac{1}{T^2 s^2 + 2\xi Ts + 1}\,, \quad \Phi_{21}(s) = \frac{Cs}{T^2 s^2 + 2\xi Ts + 1}\,. \qquad \text{(I)}$$

1.1.7. Frequency Response of Stationary Linear System

Restricting ourselves to purely imaginary values of the parameter s, $s = i\omega$ we get the transfer function of an one-dimensional stationary linear system $\Phi(i\omega)$ as a function of the angular frequency of harmonic oscillations $e^{i\omega t}$ acting on the input of the system. In this case $\Phi(i\omega)$ determines the gain in the amplitude $|\Phi(i\omega)|$ and the phase shift $\arg\Phi(i\omega)$ of the output harmonic oscillations as the function of the frequency ω. In exactly the same way in the case of a finite-dimensional system the elements of the matrix $\Phi(i\omega)$ determine the gains in amplitudes and the phases shifts of harmonic oscillations passing from every input to every output of the system.

The transfer function of a system considered as a function of purely imaginary parameter $s = i\omega$ (i.e. restricted to the imaginary axis of the complex plane) is called *a frequency response* of a stationary linear system.

The property of a stationary linear system to transfer harmonic oscillations without changing their shape, only multiplying the amplitude by $|\Phi(i\omega)|$ and shifting the phase by $\arg \Phi(i\omega)$ which gives the opportunity to study the systems by purely algebraic methods underlies *the frequency response method* which serves as the convenient and widely used method for studying any stationary linear systems (Section 1.7). Due to widespread computerization the method of frequency response as the design method for the synthesis of stationary linear systems is now in the background giving way to modern computational methods. Knowing the frequency response of a stationary linear system it is easy to calculate its steady state response to any input which may be expanded into elementary harmonic oscillations (i.e. be represented by the Fourier series or the Fourier integral).

Suppose that the input x of a stable stationary linear system may be represented by the Fourier integral

$$x(t) = \int\limits_{-\infty}^{\infty} c(i\omega)e^{i\omega t}\,d\omega . \tag{1.1.19}$$

Then on the basis of the superposition principle the steady state output y (corresponding to the infinitely long action of the input x) is determined by the formula

$$y(t) = \int\limits_{-\infty}^{\infty} \Phi(i\omega)c(i\omega)e^{i\omega t}\,d\omega . \tag{1.1.20}$$

In particular, representing the δ-function $\delta(t-\tau)$ by Fourier integral

$$\delta(t - \tau) = \frac{1}{2\pi} \int\limits_{-\infty}^{\infty} e^{i\omega(t-\tau)}\,d\omega , \tag{1.1.21}$$

we express the system response to the input $\delta(t - \tau)$, i.e. its weighting function in terms of the frequency response

$$g(t,\tau) = w(t - \tau) = \frac{1}{2\pi} \int\limits_{-\infty}^{\infty} \Phi(i\omega)e^{i\omega(t-\tau)}\,d\omega$$

or

$$w(u) = \frac{1}{2\pi} \int\limits_{-\infty}^{\infty} \Phi(i\omega) e^{i\omega u}\, d\omega \,. \qquad (1.1.22)$$

Since any bounded continuous input acting during the finite time interval may be represented by the Fourier integral we may calculate by means of frequency response the steady state outputs of stable stationary linear systems practically for any inputs.

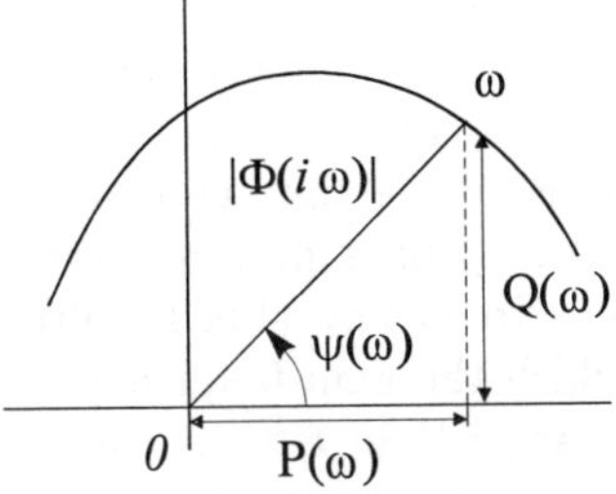

Fig. 1.1.4

The complex number $\Phi(i\omega)$ may be expressed by means of a vector (Fig.1.1.4) on the complex plane. The trajectory of this vector is called the *hodograph* of frequency response or *the gain-phase* frequency response characteristic. Let us introduce in complex plane and denote by $|\Phi(i\omega)|$ and $\psi(\omega) = \arg \Phi(i\omega)$ the *gain-frequency* and *phase-frequency* characteristics correspondingly. The gain-phase frequency characteristic presents graphic mapping of the relation between $|\Phi(i\omega)|$ and $\psi(\omega)$.

The complex number $\Phi(i\omega)$ may be represented also in an ordinary algebraic form

$$\Phi(i\omega) = P(\omega) + iQ(\omega) \,, \qquad (1.1.23)$$

$$P(\omega) = |\Phi(i\omega)|\cos\psi(\omega)\,, \quad Q(\omega) = |\Phi(i\omega)|\sin\psi(\omega)\,, \qquad (1.1.24)$$

$$|\Phi(i\omega)|^2 = P^2(\omega) + Q^2(\omega)\,, \quad \psi(\omega) = \operatorname{arctg}\frac{Q(\omega)}{P(\omega)}\,, \qquad (1.1.25)$$

where $P(\omega)$ and $Q(\omega)$ are real functions of the frequency ω. The function $P(\omega)$ is called *a real frequency characteristic*, and $Q(\omega)$ is *an imaginary frequency characteristic* of a system. From (1.1.18) and (1.1.23) follow the formulae

$$w(\sigma) = \frac{1}{\pi} \int\limits_{-\infty}^{\infty} P(\omega)\cos\omega\sigma\, d\omega = -\frac{1}{\pi} \int\limits_{-\infty}^{\infty} Q(\omega)\sin\omega\sigma\, d\omega \ \ (\sigma > 0)\,.$$

$$(1.1.26)$$

Formulae (1.1.26) show that nonanticipative linear stationary system is fully characterized only by its real frequency characteristic or only by its imaginary frequency characteristic. Knowing a real and an imaginary frequency characteristics of such a system we may find by means of these formulae its weighting function, and consequently any other characteristics.

In Table A.5.2 the gain-frequency and the phasefrequency characteristics are given for the typical linear stationary systems (Table A.5.1).

The frequency ω_p for which the gain-frequency characteristic has the maximum is called *the resonance frequency* as at this frequency the harmonic oscillations passing through the system receive the largest amplification. Depending on the character of the frequency response a stationary linear system may have no resonance frequency (Fig.1.1.5a), may have one resonance frequency (Fig.1.1.5b) or several frequencies. The frequency ω_c at which the gain-frequency characteristic decreasing passes from the values larger than 1 to the values smaller than 1 and at further decreasing of the frequency it remains smaller than 1 is called *a cut-off frequency* (Fig.1.1.5b).

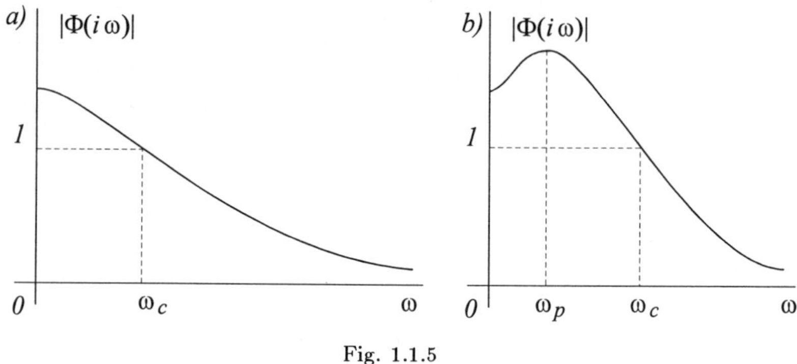

Fig. 1.1.5

The frequency response characteristics are mapped in the logarithmic scale, putting $\lg\omega$ as x-coordinate and $\lg|\Phi(i\omega)|$ or on y-coordinate for the gain-characteristics and or $\arg\Phi(i\omega)$ the phase characteristics. The frequency response characteristics of a system constructed in such a way are usually called *the logarithmic frequency characteristics*.

The special units for measuring amplification are used. As an amplification unit is assumed such an amplification at which the power of a signal increases into 10 times. This unit is called Bell. Thus the amplification of a system is equal to 2 Bell if it intensifies the power of a signal into 100 times. While intensifying the amplitude of the oscillations equal to 1 Bell the value $\lg |\Phi(i\omega)|^2 = 2\lg |\Phi(i\omega)|$ is equal to 1. Consequently, the amplification of the amplitude oscillations expressed in terms of Bells numerically equals to $2\lg |\Phi(i\omega)|$.

The value

$$L(\omega) = 20\lg |\Phi(i\omega)| \,, \tag{1.1.27}$$

considered as the function of $\lg\omega$ is called *the logarithmic gain-frequency* (decibel-log-frequency) characteristics of a system. The value

$$\psi(\omega) = \arg \Phi(i\omega) \,, \tag{1.1.28}$$

considered as the function $\lg\omega$ is called *the logarithmic phase-frequency* (log phase-frequency) characteristics of a system. While using logarithmic frequency characteristics it is more convenient to construct the gain-phase characteristic in Cartesian coordinates putting on coordinate axes the values $\psi(\omega)$ in the degrees and $L(\omega)$ in decibels (Fig.1.1.6).

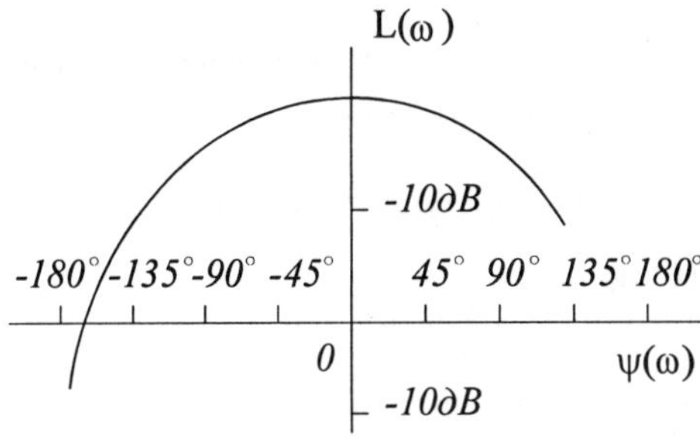

Fig. 1.1.6

1.2. Linear Differential Systems

1.2.1. Equations of Linear System and Its Weighting Function

In the case of linear system Eqs. (1.1.4) are naturally linear and consequently have the form

$$\dot{z} = az + a_1 x + a_0 \,, \quad y = bz + b_0 \,, \tag{1.2.1}$$

where a is a square matrix of order p, a_1 is a $p \times n$-matrix, a_0 is a p-dimensional vector, b is a $m \times p$-matrix, b_0 is an m-dimensional vector.

In the general case a, a_1, a_0, b and b_0 may depend on time t. In the special case of a stationary linear system a, a_1, a_0, b and b_0 are constant.

In some cases the state vector may be eliminated from the equations of the system (1.2.1). In this case we obtain a linear differential equation generally of higher than the first order connecting the output y with the input x. Then we may study the behaviour of the system without showing interest in its state. But for studying systems using computers it is always convenient to represent their differential equations in the standard Cauchy form. For this purpose we have to transform these equations into a set of the first order equations by introducing some additional variables. These additional variables are usually assumed as the system state variables. In order to find the weighting function of a nonanticipative linear system described by differential equations we integrate the first Eq. of (1.2.1).

$\triangleright$ Let $u(t, \tau)$ be the solution of the homogeneous equation

$$\dot{u} = au \qquad (1.2.2)$$

satisfying the initial condition $u(\tau, \tau) = I$, i.e. the fundamental matrix of solutions of Eq. (1.2.2). By the change of variables $z = u(t, t_0)v$ the first Eq. of (1.2.1) is reduced to the form

$$\dot{u}v + u\dot{v} = auv + a_1 x + a_0$$

or by virtue of Eq. (1.2.2)

$$u\dot{v} = a_1 x + a_0 \, .$$

R e m a r k. It is known from differential equations theory that the determinant of the matrix $u(t, t_0)$ is expressed by

$$\Delta = |u(t, t_0)| = \exp\left\{\int_{t_0}^{t} \operatorname{tr} a(\tau)d\tau\right\}, \qquad (1.2.3)$$

where $\operatorname{tr} a(\tau)$ is the trace of the matrix $a(\tau)$. To derive this formula it is sufficient to notice that according to the differentiation rule of a determinant and the properties of the determinants $\dot{\Delta} = \Delta \operatorname{tr} a(\tau)$. The solution of this equation with the initial condition $\Delta(t_0) = |u(t_0, t_0)| = 1$ is determined by formula (1.2.3). It follows from this formula that $\Delta \neq 0$ (we take for granted that the integral in (1.2.3) is finite at all t, t_0 owing to which the exponential function does not vanish).

Hence, taking into account that the matrix u is always invertible we get

$$\dot{v} = u^{-1}(a_1 x + a_0) \, .$$

Integrating this equation with the initial condition $v(t_0) = u(t_0, t_0)^{-1} z(t_0) = z_0$ gives

$$v(t) = z_0 + \int_{t_0}^{t} u(\tau, t_0)^{-1} \left[a_1(\tau) x(\tau) + a_0(\tau) \right] d\tau, \tag{1.2.4}$$

and

$$z(t) = u(t, t_0) v(t) = u(t, t_0) z_0$$
$$+ u(t, t_0) \int_{t_0}^{t} u(\tau, t_0)^{-1} \left[a_1(\tau) x(\tau) + a_0(\tau) \right] d\tau . \ \triangleleft \tag{1.2.5}$$

Let us prove that at any t_0, τ, t, $t_0 \leq \tau \leq t$,

$$u(t, t_0) = u(t, \tau) u(\tau, t_0) . \tag{1.2.6}$$

$\triangleright$ According to the definition $u(t, t_0)$ represents the solution of Eq. (1.2.2) with the initial condition $u(t_0, t_0) = I$. The value of this solution at $t = \tau$ is equal to $u(\tau, t_0)$. Consequently, $u(t, t_0)$ may be considered as the solution of the same Eq. (1.2.2) with the initial condition $u = u_0 = u(\tau, t_0)$ at $t = \tau$. On the other hand, the solution of Eq. (1.2.2) equal to u_0 at $t = \tau$ is determined by $u = u(t, \tau) u_0$. Thus by virtue of the uniqueness of the solution of Eq. (1.2.2) with a given initial condition $u(t, t_0) = u(t, \tau) u_0 = u(t, \tau) u(\tau, t_0)$ which proves (1.2.6). $\triangleleft$

One the basis of Eq. (1.2.6) $u(t, t_0) u(\tau, t_0)^{-1} = u(t, \tau)$ and formula (1.2.5) for $z(t)$ takes the form

$$z(t) = u(t, t_0) z_0 + \int_{t_0}^{t} u(t, \tau) \left[a_1(\tau) x(\tau) + a_0(\tau) \right] d\tau . \tag{1.2.7}$$

Finally, substituting this expression into the second Eq. (1.2.1) we find

$$y(t) = b(t) u(t, t_0) z_0 + \int_{t_0}^{t} b(t) u(t, \tau) a_1(\tau) x(\tau) d\tau$$
$$+ b(t) \int_{t_0}^{t} u(t, \tau) a_0(\tau) d\tau + b_0(t) . \tag{1.2.8}$$

Hence it is clear that the system considered represents a linear system with the additive additional signal at the output equal to

$$y_0(t) = b(t)u(t,t_0)z_0 + b(t)\int_{t_0}^{t} u(t,\tau)a_0(\tau)d\tau + b_0(t)\,, \qquad (1.2.9)$$

and with the weighting function

$$g(t,\tau) = b(t)u(t,\tau)a_1(\tau)\mathbf{1}(t-\tau)\,, \qquad (1.2.10)$$

where $\mathbf{1}(s)$ is the unit step function equal to 1 at $s > 0$ and to 0 at $s < 0$. Formula (1.2.10) shows that the weighting function of a linear differential system may be easily found if the fundamental matrix of solutions of the corresponding homogeneous Eq. (1.2.2) is known.

E x a m p l e 1.2.1. In the case of a stationary linear system all the coefficients of Eqs. (1.2.1) and (1.2.2) are constant and the matrix of fundamental solutions $u(t,\tau)$ is determined by the known formula

$$u(t,\tau) = \begin{bmatrix} \alpha_{11}e^{s_1(t-\tau)} & \alpha_{12}e^{s_2(t-\tau)} & \dots & \alpha_{1p}e^{s_p(t-\tau)} \\ \alpha_{21}e^{s_1(t-\tau)} & \alpha_{22}e^{s_2(t-\tau)} & \dots & \alpha_{2p}e^{s_p(t-\tau)} \\ \dots & \dots & \dots & \dots \\ \alpha_{p1}e^{s_1(t-\tau)} & \alpha_{p2}e^{s_2(t-\tau)} & \dots & \alpha_{pp}e^{s_p(t-\tau)} \end{bmatrix}\alpha^{-1}\,.$$

Here $s_1, \dots, s_p$ are the roots of the characteristic equation $|a - sI| = 0$ (by $|A|$ we denote the determinant of the matrix A), and the elements α_{ij} of the matrix α are determined by the sets of linear algebraic equations

$$(a - s_k I)\left[\alpha_{1k}\dots\alpha_{pk}\right]^{T} = 0 \ (k = 1, \dots, p)$$

in the case of different roots of the characteristic equation $s_1, \dots, s_p$ at $t > \tau$. So formula (1.2.10) gives:

$$g(t,\tau) = w(t-\tau) = b\begin{bmatrix} \alpha_{11}e^{s_1(t-\tau)} & \dots & \alpha_{1p}e^{s_p(t-\tau)} \\ \dots & \dots & \dots \\ \alpha_{p1}e^{s_1(t-\tau)} & \dots & \alpha_{pp}e^{s_p(t-\tau)} \end{bmatrix}\alpha^{-1}\mathbf{1}(t-\tau)\,.$$

1.2.2. Finding Weighting Function by Adjoint System Method

Eq. (1.2.2) with the initial condition $u = I$ at $t = \tau$ determines the fundamental matrix of solutions $u(t,\tau)$ as a function of t at fixed τ. But formula (1.2.8) shows that for calculating the output of a system at the moment t it is necessary to find $u(t, \tau)$ as a function of τ at fixed t. For solving this problem *the adjoint system method* is usually applied.

▷ Putting $u = u(t, t_0)$,

$$v = v(t, t_0) = u(t, t_0)^{-1T}, \qquad (1.2.11)$$

we have at any t_0 and $t > t_0$ $uv^T = I$. Differentiating this formula we get: $\dot{u}_t v^T + u\dot{v}_t^T = 0$. So yielding by Eqs. (1.2.2), (1.2.11) we have

$$u\dot{v}_t^T = -auv^T = -a \text{ or } \dot{v}_t^T = -u^{-1}a. \qquad (1.2.12)$$

Hence by virtue of Eq (1.2.11) we obtain the differential equation for v: $\dot{v}_t^T = -v^T a$. Now we notice that according to (1.2.6) and (1.2.11) $u(t, \tau) = u(t, t_0)v(\tau, t_0)^T$. Differentiating this formula with respect to τ and taking into account that according to (1.2.12) $\dot{v}_\tau(\tau, t_0)^T = -v(\tau, t_0)^T a(\tau)$, we find

$$\dot{u}_\tau(t, \tau) = u(t, t_0)\dot{v}_\tau(\tau, t_0)^T = -u(t, t_0)v(\tau, t_0)^T a(\tau) = -u(t, \tau)a(\tau).$$

Thus $u(t, \tau)$ as a function of τ at fixed t is determined at $\tau < t$ by the equation

$$\dot{u}_\tau(t, \tau) = -u(t, \tau)a(\tau) \qquad (1.2.13)$$

with the initial condition $u(t, t) = I$ at $\tau = t$. ◁

Transposing Eq. (1.2.13) we obtain

$$\dot{u}_\tau(t, \tau)^T = -a(\tau)^T u(t, \tau)^T. \qquad (1.2.14)$$

This equation is the adjoint equation to Eq. (1.2.2). Therefore $u(t, \tau)^T$ as a function of τ at fixed t is determined at $\tau < t$ by Eq. (1.2.14) adjoint to Eq. (1.2.2) with the condition $u(t, t)^T = I$ at the end. So $u(t, \tau)$ may *be found by the integration of the adjoint* Eq. (1.2.14) *in reversed time with the initial condition* $u(t, t) = I$.

1.2.3. Transformation of Linear System Equations into Cauchy Form

If the input-output relation of a system is represented in the form of a differential equation of higher than the first order then it usually contains not only the derivatives of the output but the derivatives of the input as well. Therefore, the problem arises to transform the equation of a linear system into a set of equations of the first order which do not contain the derivatives of the input. Now we shall show how to do this.

Let us consider a nonanticipative system whose behaviour is described by the linear differential equation of the form

$$\sum_{k=0}^{n} a_k y^{(k)} = \sum_{k=0}^{m} b_k x^{(k)}, \tag{1.2.15}$$

where the coefficients a_0, a_1, ..., a_n, b_0, b_1, ..., b_m depend in the general case on time t, and $m \leq n$. It is evidently sufficient to consider the case of $m = n$. The case of $m < n$ is the special case where $b_n = \cdots = b_{m+1} = 0$.

▷ Let us introduce new variables

$$z_1 = y - q_0 x, \quad z_{k+1} = \dot{z}_k - q_k x \ (k = 1, \ldots, n-1), \tag{1.2.16}$$

where q_0, q_1, ..., q_{n-1} are some functions which we shall determine from the condition that the resulting equations of the first order contain no derivatives of the input x. From Eqs. (1.2.16) using the known formula for the derivatives of the product of two functions

$$(uv)^{(r)} = \sum_{p=0}^{r} C_r^p u^{(p)} v^{(r-p)}, \quad C_r^p = \frac{r!}{p!(r-p)!}, \tag{1.2.17}$$

we find

$$y = z_1 + q_0 x,$$

$$y^{(s)} = z_{s+1} + \sum_{r=0}^{s} (q_{s-r} x)^{(r)} = z_{s+1} + \sum_{r=0}^{s} \sum_{p=0}^{r} C_r^p q_{s-r}^{(r-p)} x^{(p)} \tag{1.2.19}$$

$$(s = 1, \ldots, n-1).$$

After changing the order of summation we get

$$y^{(s)} = z_{s+1} + \sum_{p=0}^{s} x^{(p)} \sum_{r=p}^{s} C_r^p q_{s-r}^{(r-p)} \ (s = 1, \ldots, n-1). \tag{1.2.20}$$

Differentiating formula (1.2.19) corresponding to $s = n - 1$ gives

$$y^{(n)} = z_{n+1} + \sum_{r=0}^{n-1} (q_{n-r-1} x)^{(r+1)} = \dot{z}_n + \sum_{r=1}^{n} \left(q_{n-r}^{(r)} x \right)^{(r)}$$

$$= \dot{z}_n + x \sum_{i=1}^{n} q_{n-i}^{(i)} + \sum_{p=1}^{r} x^{(p)} \sum_{r=p}^{n} C_r^p q_{n-r}^{(r-p)}. \tag{1.2.21}$$

Substituting the expressions (1.2.18), (1.2.20) and (1.2.21) into Eq. (1.2.15) we obtain

$$a_n \dot{z}_n + \sum_{l=1}^{n} a_{l-1} z_l + x \left\{ a_n \sum_{l=1}^{n} q_{n-l}^{(l)} + \sum_{s=0}^{n-1} a_s \sum_{l=0}^{s} q_{s-l}^{(l)} \right\}$$

$$+ \sum_{p=l}^{n} x^{(p)} \sum_{s=p}^{n} \sum_{r=p}^{s} C_r^p a_s q_{s-r}^{(r-p)} = \sum_{p=0}^{n} b_p x^{(p)} . \qquad (1.2.22)$$

Equating the coefficients at the corresponding derivatives of the input x in the left- and right-hand sides of Eq. (1.2.22) we obtain the equations for the functions q_k:

$$\sum_{s=p}^{n} \sum_{r=p}^{s} C_r^p a_s q_{s-r}^{(r-p)} = b_p \quad (p = 1, \ldots, n) . \qquad (1.2.23)$$

The last of these equations corresponding to $p = n$ gives

$$q_0 = a_n^{-1} b_n . \qquad (1.2.24)$$

For solving the other Eqs. (1.2.23) notice that

$$\sum_{s=p}^{n} \sum_{r=p}^{s} C_r^p a_s q_{s-r}^{(r-p)} = \sum_{s=p}^{n} \sum_{l=0}^{s-p} C_{p+l}^p a_s q_{s-p-l}^{(l)}$$

$$= \sum_{l=0}^{n-p} \sum_{s=p+l}^{n} C_{p+l}^p a_s q_{s-p-l}^{(l)} = \sum_{l=0}^{n-p} \sum_{h=0}^{n-p-l} C_{p+l}^p a_{p+l+h} q_h^{(l)}$$

$$= \sum_{h=0}^{n-p} \sum_{l=0}^{n-p-h} C_{p+l}^p a_{p+l+h} q_h^{(l)} = a_n q_{n-p} + \sum_{h=0}^{n-p-l} \sum_{l=0}^{n-p-h} C_{p+l}^p a_{p+l+h} q_h^{(l)} .$$

$$\qquad (1.2.25)$$

Substituting this expression into (1.2.23) and putting $n - p = k$ we get

$$a_n q_k + \sum_{h=0}^{k-1} \sum_{l=0}^{k-h} C_{n-k+l}^{n-k} a_{n-k+h+l} q_h^{(l)} = b_{n-k} \quad (k = 1, \ldots, n-1) .$$

$$\qquad (1.2.26)$$

From (1.2.26) we obtain the recursive formula for determining the functions q_1, $\ldots, q_{n-1}$:

$$q_k = a_n^{-1} \left[b_{n-k} - \sum_{h=0}^{k-1} \sum_{l=0}^{k-h} C_{n-k+l}^{n-k} a_{n-k+h+l} q_h^{(l)} \right] \quad (k = 1, \ldots, n-1) .$$

$$\qquad (1.2.27)$$

After determining in such a way the functions $q_0, q_1, \ldots, q_{n-1}$ we transform Eq. (1.2.22) into the form

$$a_n \dot{z}_n + \sum_{l=1}^{n} a_{l-1} z_l = \left\{ b_0 - a_n \sum_{l=1}^{n} q_{n-l}^{(l)} + \sum_{s=0}^{n-1} a_s \sum_{l=0}^{s} q_{s-l}^{(l)} \right\} x \, . \qquad (1.2.28)$$

Solving this equation for $\dot{z}_n$ gives together with Eqs. (1.2.16) corresponding to $k = 1, \ldots, n-1$ the equation forming the set of the first order equations

$$\dot{z}_k = z_{k+1} + q_k x \ (k = 1, \ldots, n-1) \, ,$$
$$\dot{z}_n = - \sum_{l=1}^{n} a_n^{-1} a_{l-1} z_l + q_n x \, , \qquad (1.2.29)$$

where

$$q_n = a_n^{-1} \left[b_0 - a_n \sum_{l=1}^{n} q_{n-l}^{(l)} - \sum_{s=0}^{n-1} a_s \sum_{l=0}^{s} q_{s-l}^{(l)} \right]$$
$$= a_n^{-1} \left[b_0 - \sum_{h=0}^{n-1} a_h q_h - \sum_{s=1}^{n} a_s \sum_{l=1}^{s} q_{s-l}^{(l)} \right] \, . \qquad (1.2.30)$$

Noticing that

$$\sum_{s=1}^{n} a_s \sum_{l=1}^{s} q_{s-l}^{(l)} = \sum_{l=1}^{n} \sum_{s=l}^{n} a_s q_{s-l}^{(l)} = \sum_{l=1}^{n} \sum_{h=0}^{n-l} a_{h+l} q_h^{(l)} = \sum_{h=0}^{n-1} \sum_{l=1}^{n-h} a_{h+l} q_h^{(l)} \, ,$$
$$(1.2.31)$$

we transform the formula (1.2.30) for q_n into the form

$$q_n = a_n^{-1} \left[b_0 - \sum_{h=0}^{n-1} \sum_{l=0}^{n-h} a_{h+l} q_h^{(l)} \right] \, . \qquad (1.2.32)$$

It is evident that this formula coincides with (1.2.27) at $k = n$. Thus all the functions $q_1, \ldots, q_n$ are determined by (1.2.27).

The case of $m < n$ may be considered as the special case where $b_n = \cdots = b_{m+1} = 0$. In this case formulae (1.2.24) and (1.2.27) give

$$q_0 = q_1 = \cdots = q_{n-m-1} = 0 \, ,$$

$$q_{n-m} = a_n^{-1} b_m \, , \quad q_k = a_n^{-1} \left[b_{n-k} \right.$$

$$- \sum_{h=n-m}^{k-1} \sum_{l=0}^{k-h} C_{n-k+l}^{n-k} a_{n-k+h+l} q_h^{(l)} \right] \quad (k = n-m+1, \ldots, n). \quad \triangleleft \quad (1.2.33)$$

Thus we transformed Eq. (1.2.15) into the set of the first-order Eqs. (1.2.29) which do not contain the derivatives of the input x. The components of the vectors $z_1, \ldots, z_n$ in Eq. (1.2.29) represent the state variables of the system. The output of the system y is determined from the first Eq. (1.2.16)

$$y = z_1 + q_0 x. \qquad (1.2.34)$$

It is evident that Eqs. (1.2.29) and (1.2.34) represent the special case of Eqs. (1.2.1) where $a_0 = 0$, $b_0 = 0$,

$$a = \begin{bmatrix} 0 & 1 & 0 & \ldots & 0 \\ 0 & 0 & 1 & \ldots & 0 \\ \ldots & \ldots & \ldots & \ldots & \ldots \\ 0 & 0 & 0 & \ldots & 1 \\ -a_n^{-1} a_0 & -a_n^{-1} a_1 & -a_n^{-1} a_2 & \ldots & a_n^{-1} a_{n-1} \end{bmatrix}, \quad a_1 = \begin{bmatrix} q_1 \\ q_2 \\ \ldots \\ q_{n-1} \\ q_n \end{bmatrix},$$

$$(1.2.35)$$

$$b = [1\, 0\, \ldots\, 0], \quad b_1 = q_0. \qquad (1.2.36)$$

In the case of $m = n$ the output y depends explicitly on the input x. But this is a rare case since $m < n$ practically always yields $q_0 = 0$. The expression (1.2.34) of the output y does not contain x in this case.

In the special case of a stationary system all the coefficients a_1, $\ldots, a_n$, b_0 $b_1, \ldots, b_n$ are constant and thus $q_0, q_1, \ldots, q_n$ determined successively by formulae (1.2.24) and (1.2.27) are constant. Therefore $q_n^{(l)} = 0$ $(h = 0, 1, \ldots, n; l = 1, 2, \ldots)$ and formulae (1.2.27) become considerably simpler in this case:

$$q_0 = a_n^{-1} b_n, \quad q_k = a_n^{-1} \left(b_{n-k} - \sum_{h=0}^{k-1} a_{n-k+h} q_h \right) \quad (k = 1, \ldots, n).$$

$$(1.2.37)$$

Formulae (1.2.33) in this case take the form

$$q_0 = q_1 = \cdots = q_{n-m-1} = 0, \quad q_{n-m} = a_n^{-1} b_m,$$

$$q_k = a_n^{-1} \left(b_{n-k} - \sum_{h=n-m}^{k-1} a_{n-k+h} q_h \right) \quad (k = n-m+1, \ldots, n). \quad (1.2.38)$$

Notice that all previous calculations and consequently formulae (1.2.27), (1.2.33), (1.2.37) and (1.2.38) for q_0, q_1, ..., q_n are valid both for the scalar and the vector input x and the output y. In the latter case a_0, a_1, ..., a_n, b_0, b_1, ..., b_n, q_0, q_1, ..., q_n represent matrices of the corresponding dimensions.

E x a m p l e 1.2.2. In order to transform the equation

$$a_3 \dddot{y} + a_2 \ddot{y} + a_1 \dot{y} + a_0 y = b_2 \ddot{x} + b_1 \dot{x} + b_0 \tag{I}$$

into a set of the first order equations we put according to (1.2.16) and (1.2.33) $y = z_1$, $z_2 = \dot{z}_1 - q_1 x$, $z_3 = \dot{z}_2 - q_2 x$. Then by (1.2.33) we find

$$q_0 = 0, \quad q_1 = a_3^{-1} b_2, \quad q_2 = a_3^{-1}(b_1 - a_0 q_1 - 2a_1 \dot{q}_1),$$

$$q_3 = a_3^{-1}(b_0 - a_1 q_1 - a_2 \dot{q}_1 - a_3 \ddot{q}_1 - a_2 q_2 - a_3 \dot{q}_2).$$

As a result we obtain Eqs. (1.2.29) which has in this case the form

$$\dot{z}_1 = z_2 + q_1 x, \quad \dot{z}_2 = z_3 + q_2 x, \quad \dot{z}_3 = -a_3^{-1}(a_0 z_1 + a_1 z_2 + a_2 z_3) + q_3 x, \quad y = z_1. \tag{II}$$

1.2.4. Inverse Systems

The system *inverse* to a system A is such a system whose output is equal to the input of the system A when its input is equal to the output of A (under the corresponding initial conditions). Thus the inverse system performs the transformation of signals inverse to that performed by a given system A. It is evident that if the system B is inverse to the system A then the system A is inverse to the system B. In other words, A and B represent two *reciprocal* or *mutually inverse* systems. The amplifier with the gain k and the amplifier with the gain $1/k$, the differentiator and the integrator may serve as examples of pairs of mutually inverse systems.

If a system with the input x and the output y is described by Eq. (1.2.15) then the inverse system is described by the same differential equation in which y serves as its input and x serves as its output.

If $m = n$, then the reciprocal systems described by differential equation (1.2.15) are of the same type (i.e. both differential systems). After transforming the equation into a set of the first order equations, as we have already seen in Subsection 1.2.3, the output of each of these two systems will contain its input with some coefficient and the remaining part of the output will be determined by the corresponding set of the first order differential equations.

If $m = 0$ then the output of the inverse system is given by

$$x = b_0^{-1} \sum_{k=0}^{n} a_k y^{(k)} . \tag{1.2.39}$$

It goes without saying that for the existence of the inverse system in this case it is necessary that the inequality $b_0 \neq 0$ be valid at all t in the case of an one-dimensional system and that the matrix b_0 be invertible at all t in the case of a finite-dimensional system. Under this condition the inverse system performs the linear differential operation

$$L = b_0^{-1} \sum_{k=0}^{n} a_k D^k , \quad D = d/dt . \tag{1.2.40}$$

▷ If $0 < m < n$ then the operations which are performed by the inverse system include the $(n - m)$-tiple differentiation of the input. As a result the output of the inverse system contains a linear combination of the input and its derivatives up to the $(n - m)^{\text{th}}$ order. In order to determine the inverse system it is necessary in this case to find the remaining part of its output. For this purpose we put

$$x = z + z_1 = z + \sum_{l=0}^{n-m} c_l y^{(l)} \tag{1.2.41}$$

and determine the coefficients $c_0, c_1 , \ldots , c_{n-m}$ in such a way that the variable z contains no linear combination of the input y and its derivatives.

Using the known formula (1.2.17) of the derivatives of the product of two functions we find from (1.2.41)

$$x^{(s)} = z^{(s)} + \sum_{l=0}^{n-m} \sum_{r=0}^{s} C_s^r c_l^{(s-r)} y^{(l+r)}$$

$$= z^{(s)} + \sum_{r=0}^{s} \sum_{l=0}^{n-m} C_s^r c_l^{(s-r)} y^{(l+r)} = z^{(s)} + \sum_{r=0}^{s} \sum_{k=r}^{n-m+r} C_s^r c_{k-r}^{(s-r)} y^{(k)} .$$

$$\tag{1.2.42}$$

After this evident transformation of the double sum it is expedient to change again the order of summation in order to obtain the coefficient at $y^{(k)}$ in the explicit form.

For determining the limits of the inner sum we notice that $0 \le r \le k \le n-m+r$ whence it follows that $r \le k$ and $r \ge k-n+m$. On the other hand, $r \le s$ and $r \ge 0$. Since the inequalities $k > s$ and $k-n+m < 0$ are possible we come to the conclusion that $\max(0, k-n+m) \le r \le \min(k, s)$. As a result we obtain

$$x^{(s)} = z^{(s)} + \sum_{k=0}^{n-m+s} \sum_{r=\max(0,k-n+m)}^{\min(k,s)} C_s^r c_{k-r}^{(s-r)} y^{(k)},$$

$$\sum_{s=0}^{m} b_s x^{(s)} = \sum_{s=0}^{m} b_s z^{(s)} + \sum_{s=0}^{m} b_s \sum_{k=0}^{n-m+s} \sum_{r=\max(0,k-n+m)}^{\min(k,s)} C_s^r c_{k-r}^{(s-r)} y^{(k)}.$$

$$(1.2.43)$$

After changing the order of summation and taking into account that $k \le n - m + s$ implies $s \ge k - n + m$ and that $s \ge 0$ we get

$$\sum_{s=0}^{m} b_s x^{(s)} = \sum_{k=0}^{m} b_k z^{(k)}$$

$$+ \sum_{k=0}^{n} \sum_{s=\max(0,k-n+m)}^{m} \sum_{r=\max(0,k-n+m)}^{\min(k,s)} C_s^r b_s c_{k-r}^{(s-r)} y^{(k)}. \qquad (1.2.44)$$

Let us substitute this expression into Eq. (1.2.15) and equate the coefficients at $y^{(m)}, \ldots, y^{(n)}$ in the left- and the right-hand sides of the equality obtained. Then taking into account that $\min(k, s) = s$ at $s \le m$, $k \ge m$ we get

$$\sum_{s=\max(0,k-n+m)}^{m} \sum_{r=\max(0,k-n+m)}^{s} C_s^r b_s c_{k-r}^{(s-r)} = a_k \ (k = m, \ldots, n).$$

$$(1.2.45)$$

In order to solve these equations for $c_0, c_1, \ldots, c_{n-m}$ we change once more the order of summation. Remembering that $r \le s$ we get

$$\sum_{r=\max(0,k-n+m)}^{m} \sum_{s=r}^{m} C_s^r b_s c_{k-r}^{(s-r)} = a_k \ (k = m, \ldots, n) \qquad (1.2.46)$$

or putting $s = r + h$

$$\sum_{r=\max(0,k-n+m)}^{m} \sum_{h=0}^{m-r} C_{r+h}^r b_{r+h} c_{k-r}^{(h)} = a_k \ (k = m, \ldots, n). \qquad (1.2.47)$$

At $k = n$ the sum contains only one item corresponding to $r = m$, $h = 0$. Therefore, we have at $k = n$ the relation $b_m c_{n-m} = a_n$. Whence we find

$$c_{n-m} = b_m^{-1} a_n \, . \tag{1.2.48}$$

Separating in the item corresponding to $r = m$ and taking into account that h has only one value $h = 0$ when $r = m$ we transform Eqs. (1.2.45), $k < n$ into the form

$$b_m c_{k-m} + \sum_{r=\max(0,k-n+m)}^{m-1} \sum_{h=0}^{m-r} C_{r+h}^r b_{r+h} c_{k-r}^{(h)} = a_k \; (k = m, \, \dots , \, n-1) \, . \tag{1.2.49}$$

As the sum here contains only c_l with l greater than $k - m$ ($k - r > k - m$ for all items since $r < m$), these equations can be solved for c_{k-m}. As a result we obtain

$$c_{k-m} = b_m^{-1} \left[a_k - \sum_{r=\max(0,k-n+m)}^{m-1} \sum_{h=0}^{m-r} C_{r+h}^r b_{r+h} c_{k-r}^{(h)} \right] \; k = m, \dots , n-1 \, . \tag{1.2.50}$$

After determining c_{n-m} by (1.2.48) we may find successively c_{n-m-1} , $\cdots$, c_1, c_0 using formula (1.2.50). It goes without saying that the condition $b_m \neq 0$ at all t is necessary for this purpose in the case of an one-dimensional system and the invertibility of the matrix b_m at all t in the case of a finite-dimensional system.

After determining the coefficients c_0, c_1 , $\dots$, c_{n-m} by (1.2.41) and substituting the expression (1.2.41) into Eq. (1.2.15) the items which contain $y^{(m)}$, $\dots$, $y^{(n)}$ in the left- and right-hand sides of Eq. (1.2.15) will be cancelled and we shall obtain the differential equation for z:

$$\sum_{k=0}^{m} b_k z^{(k)} = \sum_{k=0}^{m-1} \tilde{a}_k y^{(k)} \, , \tag{1.2.51}$$

where

$$\tilde{a}_k = a_k - \sum_{s=\max(0,k-n+m)}^{m-1} \sum_{r=\max(0,k-n+m)}^{\min(k,s)} C_s^r b_s c_{k-r}^{(s-r)} \tag{1.2.52}$$

$$(k = 0, 1, , \dots , m - 1) \, .$$

Changing the order of summation here and taking into account that $\max_{s} \min(k, s) = k$ at $k \leq m - 1$, $s \leq m$ and that $r \leq s$ we get

$$\tilde{a}_k = a_k - \sum_{r=\max(0,k-n+m)}^{k} \sum_{s=r}^{m} C_s^r b_s c_{k-r}^{(s-r)} \, , \tag{1.2.53}$$

or

$$\tilde{a}_k = a_k - \sum_{r=\max(0,k-n+m)}^{k} \sum_{h=0}^{m-r} C_{r+h}^{r} b_{r+h} c_{k-r}^{(h)} \quad (k = 0, 1, , \ldots, n - m),$$

$$(1.2.54)$$

Formula (1.2.41) shows that *the system inverse to the system described by* Eq. (1.2.15) *at* $0 < m < n$ *represents the parallel connection of the system which performing the differential operation*

$$L = \sum_{k=0}^{n-m} c_k D^k , \quad D = d/dt , \qquad (1.2.55)$$

and the system described by Eq. (1.2.51) (the system described by Eq. (1.2.51) is indicated in Fig.1.2.1 by the letter L_1). ◁

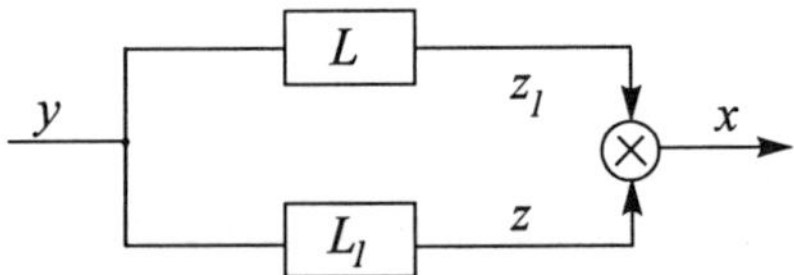

Fig. 1.2.1

Formulae (1.2.48) and (1.2.50) show that the inverse system can exist only for such a system whose input and output are of the same dimension.

In the special case of a stationary system formulae (1.2.50) and (1.2.54) take the form

$$c_{k-m} = b_m^{-1} \left[a_k - \sum_{r=\max(0,k-n+m)}^{m-1} b_r c_{k-r} \right] \quad (k = m, \ldots, n - 1),$$

$$\tilde{a}_k = a_k - \sum_{r=\max(0,k-n+m)}^{k} b_r c_{k-r} \quad (k = 0, 1, \ldots, m - 1).$$

$$(1.2.56)$$

E x a m p l e 1.2.3. For the system described by Eq (I) of Example 1.2.2 the inverse system represents a parallel connection of the system performing the differential operation $L = c_0 + c_1 D$ and the system described by the differential equation

$$b_2 \ddot{z} + b_1 \dot{z} + b_0 z = \tilde{a}_1 \dot{y} + \tilde{a}_0 y ,$$

where

$$c_1 = b_2^{-1} a_3 , \quad c_0 = b_0^{-1}(a_2 - b_1 c_1 - 2b_2 \dot{c}_1) ,$$

$$\tilde{a}_0 = a_0 - b_0 c_0 - b_1 \dot{c}_0 - b_2 \ddot{c}_0 , \quad \tilde{a}_1 = a_1 - b_0 c_1 - b_1 \dot{c}_0 - b_2 \ddot{c}_0 - b_1 c_1 - 2b_2 \dot{c}_1 .$$

1.2.5. Transfer Function of Stationary Linear System

In order to find the transfer function of a stationary linear system described by Eq. (1.2.1) it is necessary to put $a_0 = 0$, $b_0 = 0$, and to substitute into Eq. (1.2.1) the exponential function e^{st} instead of every component of the input x_h and the function $\Phi_h(s)e^{st}$ instead of the output y where $\Phi_h(s)$ is the hth column of the matrix transfer function of the system $\Phi(s)$ $(h = 1 , \ldots , n)$. At the same time it is necessary to put $z = \Psi_h(s)e^{st}$. After cancelling the equation by e^{st} we find $\Psi_h(s)$ and $\Phi_h(s)$. Instead of this we may find directly $\Psi_h(s)$ and $\Phi_h(s)$ putting in Eq. (1.2.1) $x = Ie^{st}$, $y = \Phi e^{st}$, $z = \Psi(s)e^{st}$. As a result we obtain after cancelling by e^{st}

$$s\Psi(s) = a\Psi(s) + a_1 , \quad \Phi(s) = b\Psi(s) .$$

Solving these equations for $\Psi(s)$ and $\Phi(s)$ we get

$$\Psi(s) = -(a - sI)^{-1} a_1 , \quad \Phi(s) = -b(a - sI)^{-1} a_1 . \tag{1.2.57}$$

Hence it is clear that *the transfer function of a stationary linear system described by linear differential equations with constant coefficients represents a rational function of the complex variable s.*

In the special case of an one-dimensional system described by Eq. (1.2.15) with constant coefficients at $m \leq n$ using formulae (1.2.35) and (1.2.36) for the matrices a, a_1, b we find after simple calculations

$$\Phi(s) = \frac{b_m s^m + b_{m-1} s^{m-1} + \cdots + b_1 s + b_0}{a_n s^n + a_{n-1} s^{n-1} + \cdots + a_1 s + a_0} . \tag{1.2.58}$$

Later on it will be convenient to write the differential equations with constant coefficients compactly using differential operators. For this purpose we introduce the polynomials

$$F(D) = a_n D^n + a_{n-1} D^{n-1} + \cdots + a_1 D + a_0 = \sum_{k=0}^{n} a_k D^k ,$$

$$H(D) = b_m D^m + b_{m-1} D^{m-1} + \cdots + b_1 D + b_0 = \sum_{k=0}^{m} b_k D^k . \tag{1.2.59}$$

Then the differential Eq. (1.2.15) may be written compactly as

$$F(D)y = H(D)x \,. \tag{1.2.60}$$

Formula (1.2.58) for the transfer function of an one-dimensional system will take the form

$$\Phi(s) = H(s)/F(s) \,. \tag{1.2.61}$$

The second formula of (1.2.57) determines the transfer function of a stationary linear system described by Eqs. (1.2.1) at all values of s, besides those which coincide with the roots of the characteristic equation

$$|a - sI| = 0 \,. \tag{1.2.62}$$

But in a physical sense this transfer function does exist not at all values of s. Really not at all values of s, the formula

$$y(t) = \Phi_h(s)e^{st} \tag{1.2.63}$$

determines the steady response of the system to an exponential disturbance acting only at the hth input. The response of the system to an exponential disturbance acting only at the hth input represents the general solution of Eqs. (1.2.1) while (1.2.63) determines only a particular solution. In order to find the general solution of Eqs. (1.2.1) the general solution of the respective homogeneous equations

$$\dot{z} = az \,, \quad y = bz \tag{1.2.64}$$

should be added to the particular solution (1.2.63).

It is known from differential equations theory that if the roots of the characteristic Eq. (1.2.62) $s_1 \,, \ldots \,, s_p$ are all different then the general solution of the homogeneous equations corresponding to (1.2.1) represents a linear combination of the exponential functions $\gamma_1 e^{s_1 t}$, $\ldots, \gamma_p e^{s_p t}$ with arbitrary coefficients where $\gamma_1 \,, \ldots \,, \gamma_p$ are the m-dimensional vectors determined by the matrices a, a_1, b in the expression of the weighting function obtained in Example 1.2.1. Thus in the case where the characteristic Eq. (1.2.62) has no multiple roots the general solution of Eqs. (1.2.1) at $x_h = s^{st}$ and zero values of all other x_h is determined by the formula

$$y = \Phi_h(s)e^{st} + c_1\gamma_1 e^{s_1 t} + \cdots + c_p\gamma_p e^{s_p t} \,, \tag{1.2.65}$$

where $c_1, \ldots, c_p$ are arbitrary constants. Hence it follows that the steady response independent of the initial conditions of the system considered to the exponential signal e^{st} at the hth input exists only in such a case where all the roots $s_1, \ldots, s_p$ of the characteristic equation (1.2.62) have negative real parts, and the real part of the parameter s is negative or equal to zero. If these conditions are not fulfilled then the response of the system to the signal e^{st} increases indefinitely with time. But in this case we may also speak about the steady response of the system to the signal e^{st} if the first item in the right-hand side of (1.2.65) increases when $t \to \infty$ quicker than all the other items. In this case for sufficiently long time t of system functioning its response to the signal e^{st} will be determined practically only by the first item in the right-hand side of (1.2.65).

In order to determine the transfer functions from the hth input to all the output of the system we divide formula (1.2.65) by e^{st}:

$$y/e^{st} = \Phi_h(s) + c_1\gamma_1 e^{(s_1-s)t} + \cdots + c_p\gamma_p e^{(s_p-s)t}. \qquad (1.2.66)$$

If the real parts of all the differences $s_1 - s, \ldots, s_p - s$ are negative then all the exponential functions in (1.2.66) tend to vanish when $t \to \infty$. In this case the transfer function of the system representing the ratio of its response to the infinitely long acting signal Ie^{st} at the input to e^{st} is determined only by the first item in formula (1.2.66), i.e. is equal to $\Phi_h(s)$. If at least one of the numbers $s_1 - s, \ldots, s_p - s$ has positive real part then the corresponding exponential functions in (1.2.66) increases indefinitely when $t \to \infty$ and consequently, the transfer function of the system does not exist.

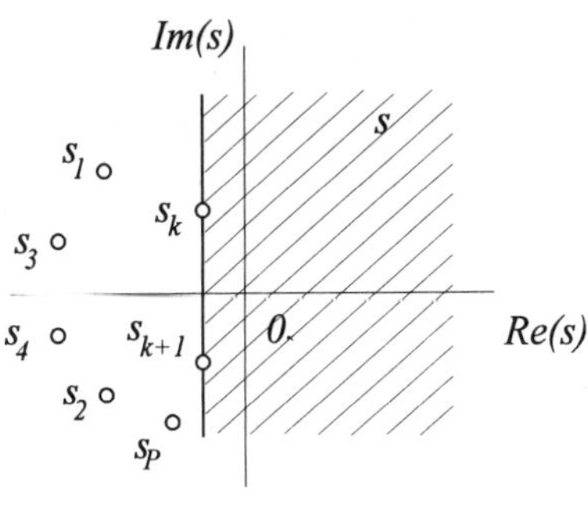

Fig. 1.2.2

Thus *the transfer function of a stationary linear system whose behaviour is described by differential equations exists only in the region of* *values of s whose real parts are larger than the real parts of all the roots of the characteristic equation $s_1, \ldots, s_p$.* In other words *the transfer function (1.2.61) of system (1.2.60) exists only in the half-plane of the complex parameter s to the right of the vertical line passing through the root of the characteristic equation with the largest real part. On the left of this line and on the line itself the transfer function (1.2.61) does not exist in spite of the second formula (1.2.57) formally determining it at all values of s, besides the points (Fig.1.2.2) $s_1, \ldots, s_p$.* This conclusion is also valid in the case of multiple roots of the characteristic equation, since at any $r > 0$ the product $t^r e^{(s_k - s)t}$ tends to zero when $t \to \infty$, if the real part of the parameter s is greater than the real part of the root s_k of the characteristic equation, and increases indefinitely if the real part of the parameter s is smaller than the real part of the root s_k.

If follows from the condition of stability of a linear system (Subsection 1.1.4) that *for the stability of a stationary linear differential system it is necessary and sufficient that the real parts of all the roots of the characteristic* Eq. (1.2.62) *be negative. The transfer function $\Phi(s)$ of a stable system exists at purely imaginary values of s. Therefore the frequency response of a stable stationary linear system always exists.* For unstable systems the frequency response does not exist and may be used only in formal evaluations as a rational function of the complex variable $s = i\omega$ existing at all s which do not coincide with its poles $s_1, \ldots, s_p$.

E x a m p l e 1.2.4. From the differential equations of Examples 1.1.1-1.1.3 the formulae for the transfer functions derived in Examples 1.1.11-1.1.13 may be directly obtained using (1.2.58).

It follows from the previous discussion that *the transfer function of a stationary linear system described by differential equations is a rational function of the complex variable s* (in the case of a finite-dimensional system all the elements of the matrix transfer function are the rational functions of s). The inverse statement is also true: *to any stationary linear system with rational transfer function corresponds a differential equation (a linear equation with constant coefficients) which connects its input and output.*

If the transfer function $\Phi(s)$ of an one-dimensional system is rational then it may be represented in the form of the ratio of two polynomials $\Phi(s) = H(s)/F(s)$. From (1.2.60) and (1.2.61) it follows that in this case the input x and the output y of the system are connected by Eq. (1.2.60). Thus *in order to find the differential equation of an one-dimensional stationary linear system with rational transfer function the variable s in the numerator and denominator of the expression of the transfer function* *should be replaced by the operator of differentiation with respect to time* $D = d/dt$. The differential operators obtained represent respectively the right-hand (with the input) and the left-hand (with the output) sides of the differential equation.

It is easy to see that in order that all the coefficients of Eq. (1.2.60) (or what is the same of Eq. (1.2.15)) be real it is necessary and sufficient that all purely imaginary and complex roots of the polynomials $F(s)$ and $H(s)$ be pair-wise conjugate.

R e m a r k. The differential equation of a stationary linear system may also be found from a given frequency response if it represents a rational function of frequency ω. It is sufficient for this purpose to remember that the frequency response represents the value of the transfer function at $s = i\omega$. It is clear therefore that for finding the differential equation of the system it is necessary to represent its frequency response in the form of the ratio of two polynomials relative to $i\omega$, $\Phi(i\omega) = F(i\omega)^{-1} H(i\omega)$. Then the differential equation of the system is obtained in the same way as in the case of a given transfer function. To obtain the model described by a linear differential equation with constant coefficients for a system whose frequency response is found from experiments this frequency response should first be approximated by a rational function. After this the differential equation of a model may be found by the above stated method.

If the transfer function $\Phi(s)$ of a finite-dimensional system is rational then all the elements $\Phi_{pq}(s)$ of the pth row of the matrix $\Phi(s)$ may be represented in the form $H_{pq}(s)/F_p(s)$ $(p = 1, \ldots, m; q = 1, \ldots, n)$ by reducing to a common denominator where $H_{pq}(s)$ and $F_p(s)$ are polynomials. Introducing the matrix $H(s)$ with the elements $H_{pq}(s)$ and the diagonal matrix $F(s)$ with the elements $F_1(s), \ldots, F_m(s)$ we represent the transfer function of the system in the form $\Phi(s) = F(s)^{-1} H(s)$. Then the differential equation which connects the input x and the output y will take the form of Eq. (1.2.60). Here the coefficients $a_0, a_1, \ldots, a_n$ in expression (1.2.59) of the polynomial $F(s)$ will be the diagonal matrices of order m.

E x a m p l e 1.2.5. Consider the case where the transfer function of a system with two inputs and two outputs is determined by

$$\Phi(s) = \begin{bmatrix} \dfrac{k_{11}}{s + \alpha_1} & \dfrac{k_{12}}{s + \alpha_1} \\[2mm] \dfrac{k_{21}}{s + \alpha_2} & \dfrac{k_{22}}{s + \alpha_2} \end{bmatrix}. \tag{I}$$

Here all the elements of every row of the matrix have a common denominator and consequently, according to the above method

$$F(s) = \begin{bmatrix} s + \alpha_1 & 0 \\ 0 & s + \alpha_2 \end{bmatrix}, \ H(s) = \begin{bmatrix} k_{11} & k_{12} \\ k_{21} & k_{22} \end{bmatrix} \qquad \text{(II)}$$

and the differential equations of the system have the form

$$\dot{y}_1 + \alpha_1 y_1 = k_{11} x_1 + k_{12} x_2 , \ \dot{y}_2 + \alpha_2 y_2 = k_{21} x_1 + k_{22} x_2 . \qquad \text{(III)}$$

It is clear that in the case of a finite-dimensional system all the coefficients of its differential equations are real if and only if all the purely imaginary and complex roots of the polynomials $F_k(s)$, $H_{kh}(s)$ ($k = 1, \ldots, m; h = 1, \ldots, n$) are pair-wise conjugate numbers.

The outlined method of obtaining the differential equation of a finite-dimensional system gives one of the simplest forms of this equation as only one component of the output enters into every equation of the obtained set of equations. The problem of finding the differential equation of a finite-dimensional system given by its rational transfer function has in general no unique solution. An infinite set of various differential equations connecting the input and the output corresponds to the same finite-dimensional system. To understand this it is sufficient to notice that in the case of a finite-dimensional system with m outputs any equation obtained by multiplying Eq. (1.2.15) from the left by an arbitrary nonsingular $m \times m$ matrix describes the behaviour of the same system.

The problem of transforming the differential equation of a multi-dimensional system into one of the simplest forms is of great practical importance for structural theory of stochastic systems.

1.3. Connections of Systems

1.3.1. Basic Systems Connections

Let us consider the basic connections of elementary (simple) systems and their characteristics. The basic types of the connections in composite systems are the connections *in series, in parallel* and *by feedback loop.*

A series connection of the systems is called such a connection when the output of every system is connected with the input of the following system, i.e. when the output of each system serves as the input for the following system (Fig.1.3.1a). Here it is supposed that the connected systems possess the *directed action*. It means that while connecting the output of one system with the input of another system the characteristics of the first system do not change, i.e. that at a given input $x(t)$ of a system its output variable represents the same function $y(t)$ no matter whether the output of this system is connected with the input of another system or not.

a)   b)

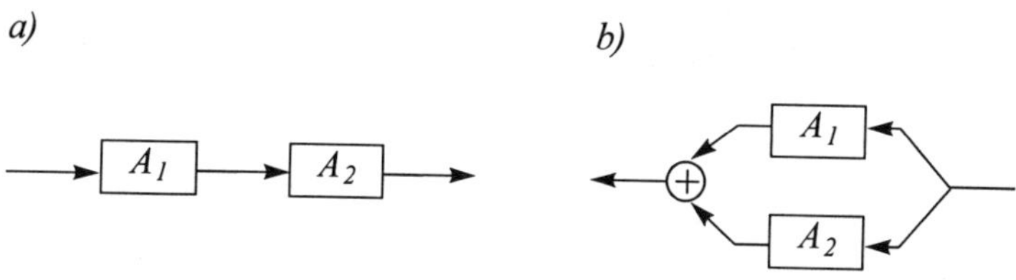

Fig. 1.3.1

A parallel connection of the systems is called such a connection at which the input is supplied simultaneously on several systems and their outputs are summarized (Fig.1.3.1b).

A feedback connection is called a connection of the system output with its input (Fig.1.3.2a). If the output of the system directly set on its output without any transformation then its feedback is called a *rigid*. A rigid feedback may be *positive* or *negative* in accordance with the fact whether the output of the system is summarized with its input or is subtracted from it. If some system is included into the loop of a feedback transforming the output main system then such feedback is called *flexible* (Fig.1.3.2b).

a)   b)

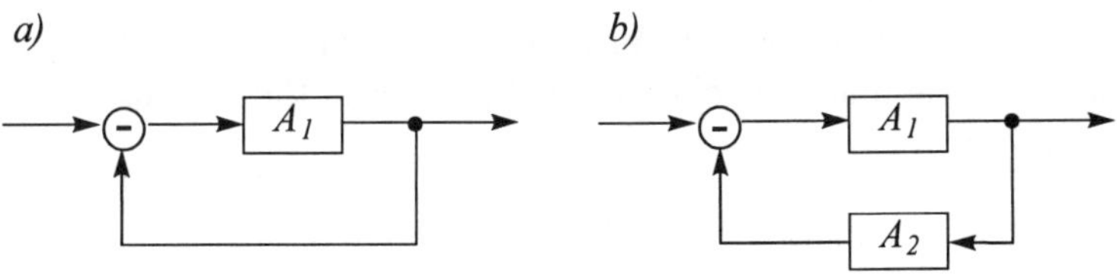

Fig. 1.3.2

For the convenience we shall always consider a negative flexible feedback what evidently does not loss the generality as the change of the sign may be always included into the transformation performed by the system which is inserted into the loop of a feedback. It is evident that a rigid feedback may be considered as a special case of a flexible feedback when the system inserted into the loop of a feedback represents an ideal servomechanism (a negative rigid feedback) or as an amplifying circuit with a gain coefficient equal to -1 (a positive rigid feedback).

In practice the description methods of the systems in the form of the operators, weighting, transfer, frequency and other describing functions give the opportunity to construct convenient stream diagram of the mathematical models of the linear systems which in turn may be considered as equivalent visual (graphical) models. Usually two types of graphical models are distinguished: the block diagrams and the oriented graphs.

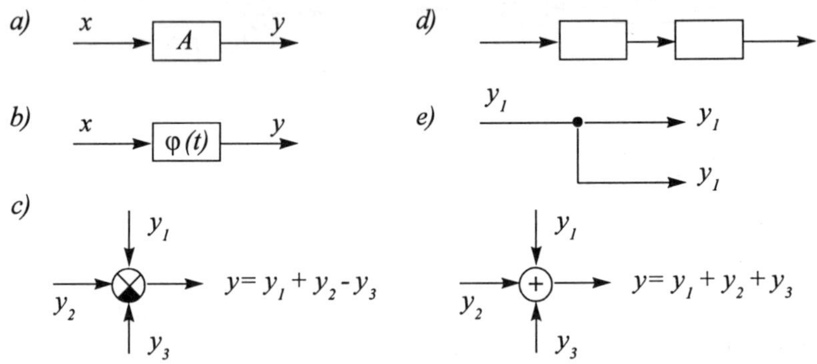

Fig. 1.3.3

A block diagram is called a graphical image of the system structure or its blocks. Here *the structure* is regarded the aggregate of the blocks and the connections representing the loops. The elements of a block diagram are (Fig.1.3.3):

• Linear dynamic *links* presented by the blocks with the operators describing the functions filled in the blocks or in the description applied (Fig.1.3.3a).

• Linear and nonlinear functional generators (Fig.1.3.3b).

• The summators presented by the circles divided into the sectors. Generally the summators realize the algebraic summation. Some times we use $\pm$ signs in circles (Fig.1.3.3c).

• Couplings which transmit the directed action without changing (Fig.1.3.3d).

• Branching points on the coupling lines (Fig.1.3.3e).

Equally with the block diagrams *the oriented graphs* are often used (Fig.1.3.4a,b,c). The main elements of an oriented graph are edge, node and arrow. The oriented graphs do not need special notations for summators and couplings as in the block diagrams.

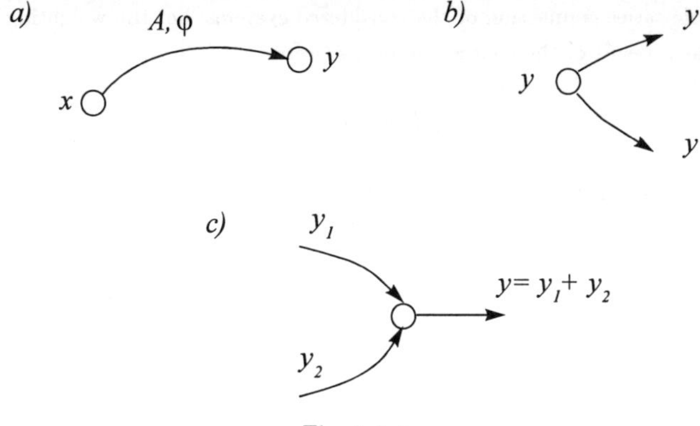

Fig. 1.3.4

Transformation rules for block diagrams and graphs are given in Subsection 1.3.4 and Appendix 6.

1.3.2. Weighting Functions of Linear Systems Typical Connections

Consider at first a series connection of two linear systems whose weighting functions $g_1(t,\tau)$ and $g_2(t,\tau)$ are given (Fig.1.3.5). Find the weighting function $g(t,\tau)$ of a composite system.

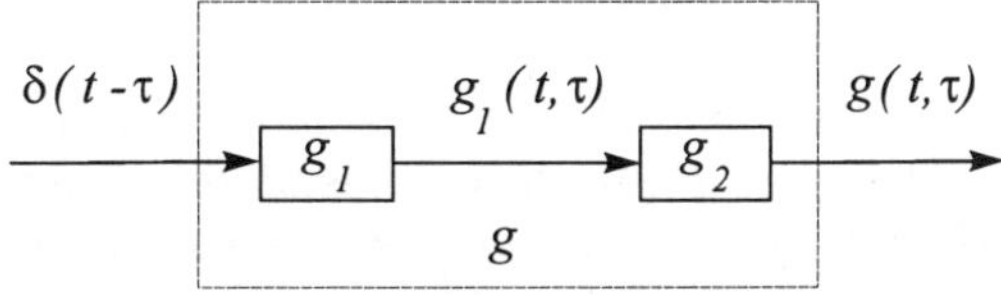

Fig. 1.3.5

▷ For this purpose we find its response on the unit impulse. As a result of the action of the unit impulse on the input of the first system its output will represent its weighting function $g_1(t,\tau)$. Consequently, the input of the second system is equal

$$x_2(t) = g_1(t,\tau). \qquad (1.3.1)$$

Here the input of the second system according to general formula (1.1.8) is expressed by formula

$$y_2(t) = \int_{-\infty}^{\infty} g_2(t,\sigma)x_2(\sigma)\, d\sigma = \int_{-\infty}^{\infty} g_2(t,\sigma)g_1(\sigma,\tau)\, d\sigma. \qquad (1.3.2)$$

But the output of the second system represents also in a given case the output of the whole series connection of the considered systems, i.e. the weighting function obtained as a result of the connection of a system

$$y_2(t) = g(t, \tau). \tag{1.3.3}$$

Comparing this formula with (1.3.2) we obtain the following formula for the weighting function of a series connection of two linear systems

$$g(t, \tau) = \int_{-\infty}^{\infty} g_2(t, \sigma) g_1(\sigma, \tau)\, d\sigma. \lhd \tag{1.3.4}$$

Thus *the weighting function of two connected in series linear systems is determined by formula (1.3.4)*.

In the special case of nonanticipative systems we have $g_1(\sigma, \tau) = 0$ at $\sigma < \tau$ and $g_2(t, \sigma) = 0$ at $\sigma > t$. Consequently, for nonanticipative systems an integrand in (1.3.4) differs from zero only in the limits $\tau \leq \sigma \leq t$ and formula (1.3.4) takes the form

$$g(t, \tau) = \int_{\tau}^{t} g_2(t, \sigma) g_1(\sigma, \tau)\, d\sigma \tag{1.3.5}$$

This formula is valid at $t \geq \tau$. At $t < \tau$ the integrand in (1.3.4) is equal to zero at all σ. Thus we come to the conclusion that *a series connection of the nonanticipative systems always gives the nonanticipative system*.

Using formulae (1.3.4) and (1.3.5) sequentially we may find the weighting function of the system obtained as a result of a series connection of any number of linear systems.

The result of a series connection of the linear systems depends on the order of their connection. Really, changing the places of the systems subject to connect in Fig.1.3.5 we must also change the places of the weighting functions g_1 and g_2 in formula (1.3.4). As a result the weighting function of a series connection of two systems will be determined by the following formula:

$$g'(t, \tau) = \int_{-\infty}^{\infty} g_1(t, \sigma) g_2(\sigma, \tau)\, d\sigma. \tag{1.3.6}$$

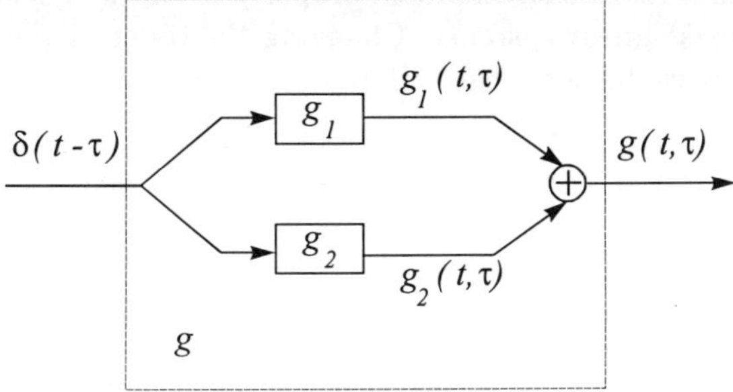

Fig. 1.3.6

In general case this expression does not coincide with (1.3.4). It is easy to make sure on the example of a series connection of the amplifier with variable gain coefficient and the integrator.

Let us consider now a parallel connection of the linear systems which have the known weighting functions g_1 and g_2 (Fig.1.3.6).

$\triangleright$ For determining the weighting function $g(t,\tau)$ of a composite system we shall find its response on the unit impulse $\delta(t-\tau)$. Accounting that while delivering the unit impulse $\delta(t-\tau)$ on the inputs of connected systems the outputs are equal to the corresponding weighting functions $g_1(t,\tau)$ and $g_2(t,\tau)$ we obtain

$$g(t,\tau) = g_1(t,\tau) + g_2(t,\tau). \quad \triangleleft \qquad (1.3.7)$$

Thus *at the parallel connection of the linear systems their weighting functions are added.*

For studying the systems with a feedback we use the notion of the reciprocal (mutually inverse) systems (Subsection 1.2.4). Consider a series connection of two reciprocal linear systems. The weighting function of the linear system inverse relatively to the system with the weighting function $g(t,\tau)$ we shall denote by the symbol $g^-(t,\tau)$. The weighting function of a series connection of two reciprocal systems is equal to $\delta(t-\tau)$. On the other hand the weighting function of a series connection may be calculated using formula (1.3.4). As a result we obtain

$$\int_{-\infty}^{\infty} g^-(t,\sigma)g(\sigma,\tau)\,d\sigma = \delta(t-\tau). \qquad (1.3.8)$$

This relation represents the main property of the weighting functions of the reciprocal linear systems. Changing the order of the connection of the systems we have another relation

$$\int_{-\infty}^{\infty} g(t,\sigma)g^-(\sigma,\tau)\,d\sigma = \delta(t-\tau). \tag{1.3.9}$$

Thus *the weighting functions of two reciprocal linear systems satisfy two integral relations (1.3.8) and (1.3.9) which in general do not coincide.*

The weighting function $g(t,\tau)$ of the system with a feedback mapped in Fig.1.3.7 may be defined as in the previous cases after observing the passage the unit impulse. But this way is rather difficult and leads to an integral equation relative to the sought weighting function $g(t,\tau)$. It is rather easy to find the weighting function of an inverse system (Subsection 1.2.4).

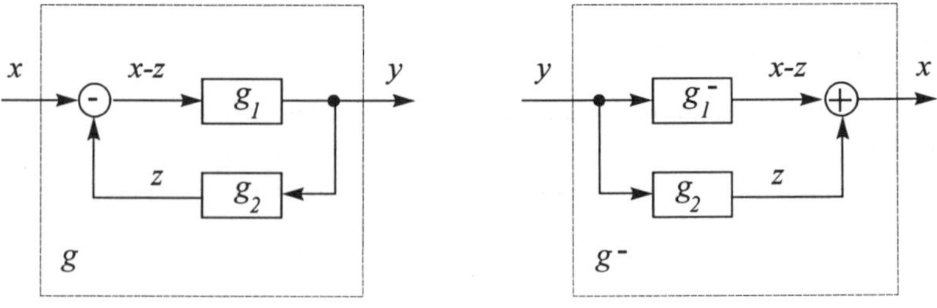

Fig. 1.3.7   Fig. 1.3.8

▷ It is evident that the system which is inverse relative to the system with a feedback is the parallel connection of the system relative to the system in the direct loop and the system in the feedback loop (Fig.1.3.8). Really, denoting the input of the considered system by x, the output by y and the output of the system in a feedback loop by z we see that $x - z$ will be the input of the system in the direct loop and y will be its output. While supplying the output y on the input of the parallel connection of the systems with the weighting functions g_1 and g_2 we shall obtain on the output of the first system the signal $x - z$, and on the output of the second system the signal z. Adding these signals will give on the output of a parallel connection the signal x (Fig.1.3.8). Thus a series connection of the systems shown in Fig.1.3.7 and 1.3.8 give an ideal servomechanism. Consequently, these systems are reciprocal. Applying formula (1.3.7) for the weighting function of the parallel connection of the linear systems we find

$$g^-(t,\tau) = g_1^-(t,\tau) + g_2(t,\tau). \triangleleft \tag{1.3.10}$$

This formula determines the weighting function of the linear system which is inverse relative to the system with a feedback shown on Fig.1.3.7.

1.3.3. Connections of Stationary Linear Systems

At first let us consider a series connection of two stationary linear systems with transfer functions $\Phi_1(s)$ and $\Phi_2(s)$ (Fig.1.3.9). Suppose that on the input of this series connection a signal representing an exponential function of time e^{st} acts indefinitely for a long time. This signal while passing through the first system is multiplied by the transfer function $\Phi_1(s)$. Thus on the output of the first system and on the input of the second system will be the function $\Phi_1(s)e^{st}$. By virtue of the linearity of the second system its output will be equal to the product of its response on the exponential signal e^{st} by the constant multiplier $\Phi_2(s)$, i.e. $\Phi_1(s)\Phi_2(s)e^{st}$. Thus *a series connection of the stationary linear systems gives the stationary linear system whose transfer function $\Phi(s)$ is equal to the product of the transfer functions of connected systems*

$$\Phi(s) = \Phi_1(s)\Phi_2(s). \tag{1.3.11}$$

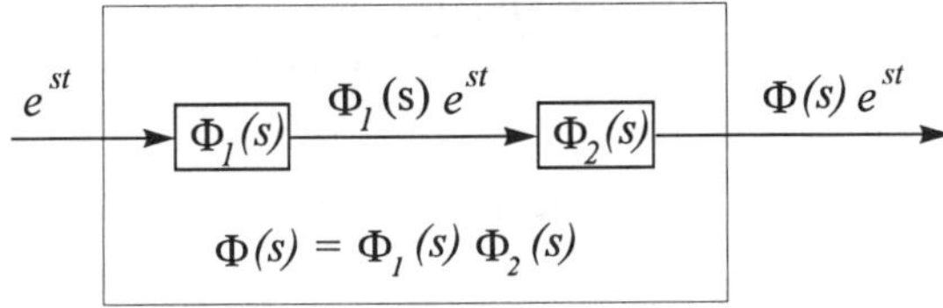

Fig. 1.3.9

This formula is easily spread to a series connection of any number of the stationary linear systems. Namely, if the transfer functions of n connected in a series systems are equal to $\Phi_1(s),\ldots,\Phi_n(s)$ then the transfer function of the connection $\Phi(s)$ is determined by formula

$$\Phi(s) = \Phi_1(s)\Phi_2(s)\cdots\Phi_n(s) = \prod_{k=1}^{n} \Phi_k(s). \tag{1.3.12}$$

Thus *the result of a series connection of the stationary linear systems does not depend on the order of their connection.*

As the modulus of the product of the complex numbers is equal to the product of the moduluses of the factors, and the argument is equal to the sum of the arguments of the factors then from (1.3.12) follow the formulae for the gain- and phase-frequency characteristics of a series connection

$$|\Phi(i\omega)| = \prod_{k=1}^{n} |\Phi_k(i\omega)|, \qquad (1.3.13)$$

$$\lg|\Phi(i\omega)| = \sum_{k=1}^{n} \lg|\Phi_k(i\omega)|, \quad \arg\Phi(i\omega) = \sum_{k=1}^{n} \arg\Phi_k(i\omega). \qquad (1.3.14)$$

Formulae (1.3.14) show that *at a series connection of the stationary linear systems their decibel-log and log phase-frequency characteristics are summarized.*

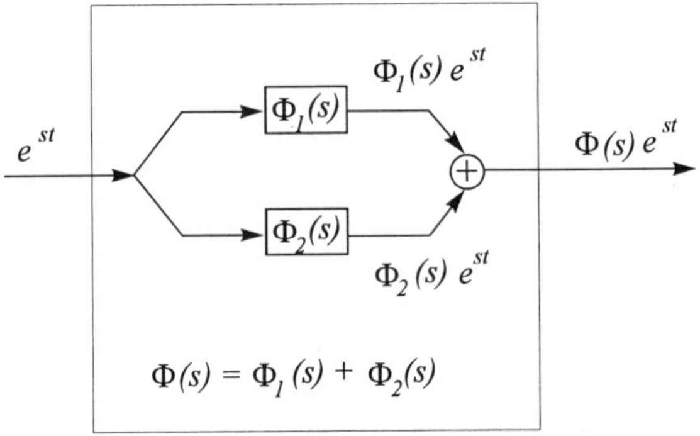

Fig. 1.3.10

For determining the transfer function of the parallel connection of the stationary linear systems with the weighting functions $\Phi_1(s)$ and $\Phi_2(s)$ (Fig.1.3.10) we suppose that on the input of this connection the signal representing the exponential time function e^{st} acts indefinitely for a long time. As a result the outputs of the connected systems will be equal to $\Phi_1(s)e^{st}$ and $\Phi_2(s)e^{st}$ correspondingly. The output of the connection will be equal to the sum of these expressions. Consequently, *the transfer function of a parallel connection of the stationary linear system is equal to the sum of the transfer functions of connected systems:*

$$\Phi(s) = \Phi_1(s) + \Phi_2(s). \qquad (1.3.15)$$

Evidently, this formula is also spread to any number of parallelly connected systems:

$$\Phi(s) = \sum_{k=1}^{n} \Phi_k(s). \tag{1.3.16}$$

Let us pass to the determination of the transfer functions of the stationary linear systems closed by a feedback. At first we consider the system consisting of the stationary linear system with the transfer function $\Phi_1(s)$ closed by the nonnegative feedback containing a stationary linear system with the transfer function $\Phi_2(s)$ (Fig.1.3.11). For determining the transfer function $\Phi(s)$ of this system let us consider an inverse system which has the transfer function $1/\Phi(s)$ (Subsection 1.3.2). This inverse system represents a parallel connection of the system with the transfer function $1/\Phi_1(s)$ and the system with the transfer function $\Phi_2(s)$. Therefore using formula (1.3.16) we get

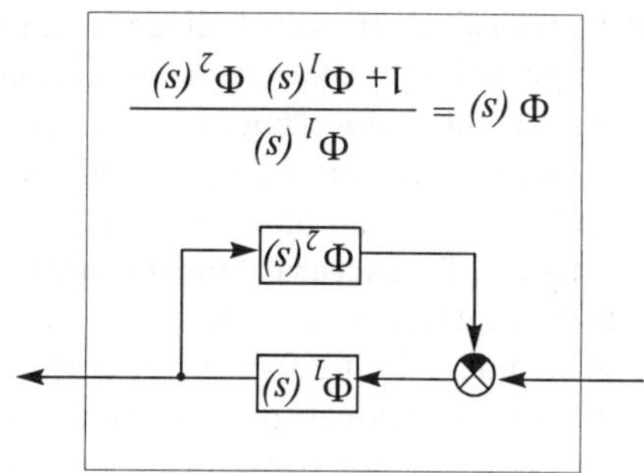

Fig. 1.3.11

$$\frac{1}{\Phi(s)} = \frac{1}{\Phi_1(s)} + \Phi_2(s) = \frac{1 + \Phi_1(s)\Phi_2(s)}{\Phi_1(s)}.$$

From here we find the transfer function of the system with a feedback:

$$\Phi(s) = \frac{\Phi_1(s)}{1 + \Phi_1(s)\Phi_2(s)}. \tag{1.3.17}$$

At a rigid negative feedback it is necessary to put $\Phi_2(s) = 1$ in formula (1.3.17).

Thus *at any connections of the stationary linear systems we always obtain the stationary linear systems, whose transfer functions and frequency response characteristics are determined by means of elementary algebraic operations on given transfer functions (on frequency response characteristics respectively) of connected systems.*

At all possible connections of the stationary linear systems described by ordinary differential equations as a result we always obtain stationary linear systems whose behaviour is also described by ordinary differential equations. Really, in this case the transfer functions of all connecting systems represent the rational functions of s. Therefore all derived formulae for the transfer functions of the connections determine the transfer function of the connection $\Phi(s)$ in the form of the rational function of s. After presenting this function in the form of the ratio of two polynomials and replacing s by the operator of the differentiation with respect to time D we shall find the operators $F(D)$ and $H(D)$ and Eq. (1.2.60) for the input – output relation.

1.3.4. Structural Transformation Rules

It is expedient to transform the block diagram into more convenient form by means of the partition of the composite systems into more simple ones (decomposition), the connection of the simple systems into one (aggregation) and another transformations for the sake of simplicity of the structure on retention of the number and the order of included systems. The main principle of structural transformations consists in the fact that all the transformations of the block diagram of the linear system (Subsection 1.3.1) should be performed in such a way that all the inputs and the outputs of each transforming section of the block diagram will remain invariant. With regard to Subsections 1.3.2 and 1.3.3 from this principle follows the rules of the structural transformations of linear systems given in Appendix 6.

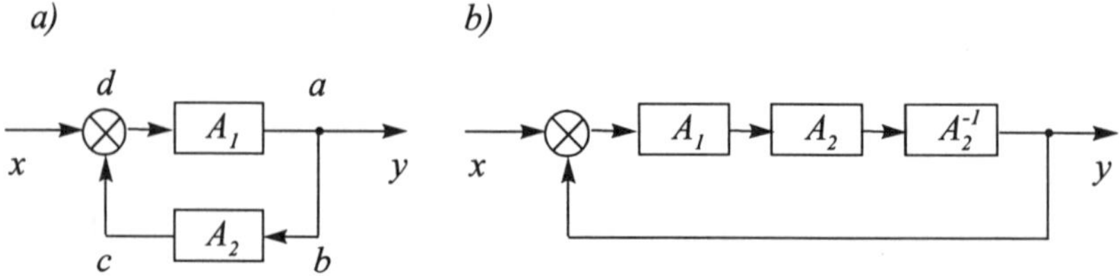

Fig. 1.3.12

E x a m p l e 1.3.1. For transforming the flexible feedback by the rigid one (Fig.1.3.12a) we interchange the branching point and A_2 system in a feedback (Fig.1.3.12b).

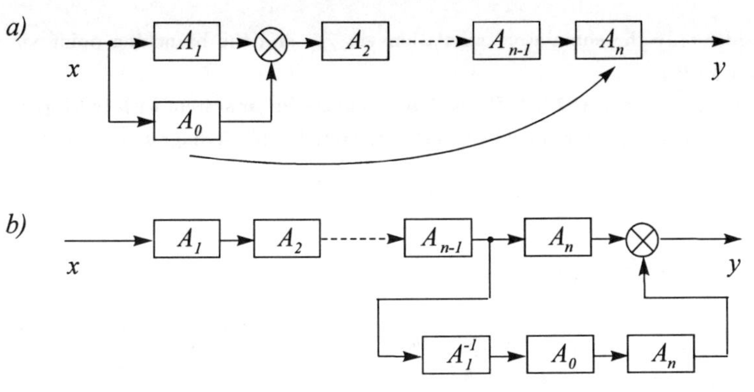

Fig. 1.3.13

E x a m p l e 1.3.2. Consider the translation of the system A_0 along the connected in a series stationary linear systems $A_1 , \ldots , A_n$ (Fig.1.3.13a). As the stationary linear systems connected in a series may be arbitrary interchanged then the mutually inverse (reciprocal) systems introduced by the translation of the branching point and the summator via the one and the same system are mutually compensated. So for the translation of the parallel loop from A_n to A_1 or vica versa it is necessary to consider the translation of the branching point via A_1 and summator through A_n (Fig.1.3.13b).

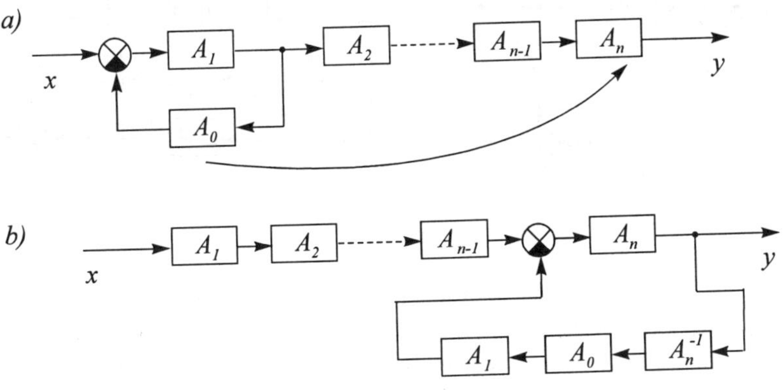

Fig. 1.3.14

E x a m p l e 1.3.3. For the system (Fig.1.3.14a) analogously we conclude that for the translation of A_0 in a feedback from A_1 to A_n and vica versa along the connected in a series stationary linear systems $A_2 , \ldots , A_n$ it is necessary to consider only the summator translation via A_1 and the branching point via A_n (Fig.1.3.14b).

E x a m p l e 1.3.4. Block diagram of the linear system with two inputs and one output (Fig.1.3.15a) may be transformed into the block diagram on (Fig.1.3.15b).

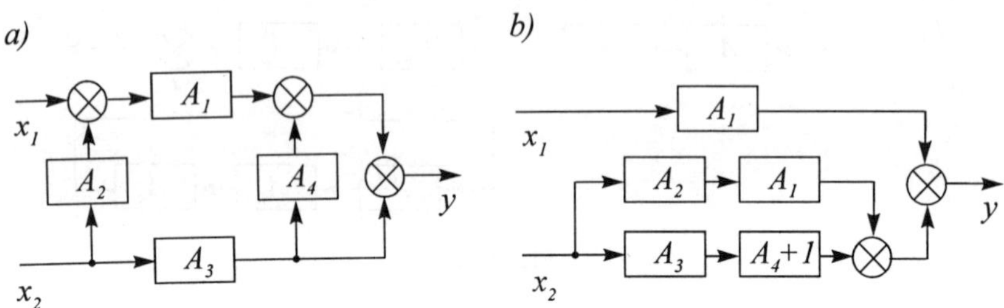

Fig. 1.3.15

E x a m p l e 1.3.5. For the block diagram of the linear system with one input and two outputs (Fig.1.3.16a) the transformed block diagram is given on Fig.1.3.16b.

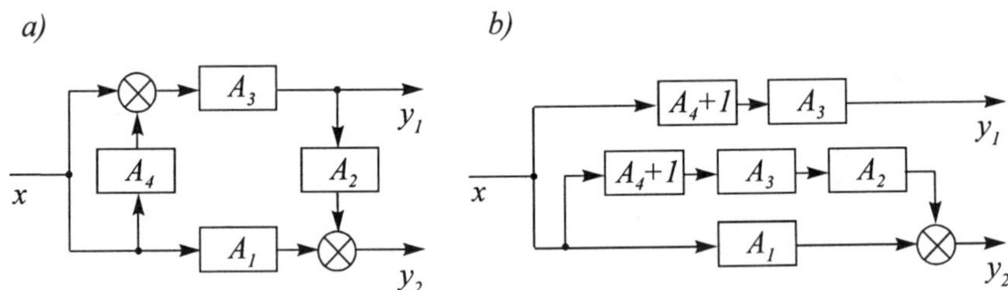

Fig. 1.3.16

E x a m p l e 1.3.6. Fig.1.3.17a presents the block diagram of the linear system with a parallel loop and the additional branching point in one of the loops. Fig.1.3.17b and Fig.1.3.17c show the sequential stages of the summator translation.

E x a m p l e 1.3.7. Block diagram of the composed linear system with a feedback loop having the internal summator is shown on Fig.1.3.18a. The sequential stages of the summator translation are given on Fig.1.3.18b and 1.3.18c.

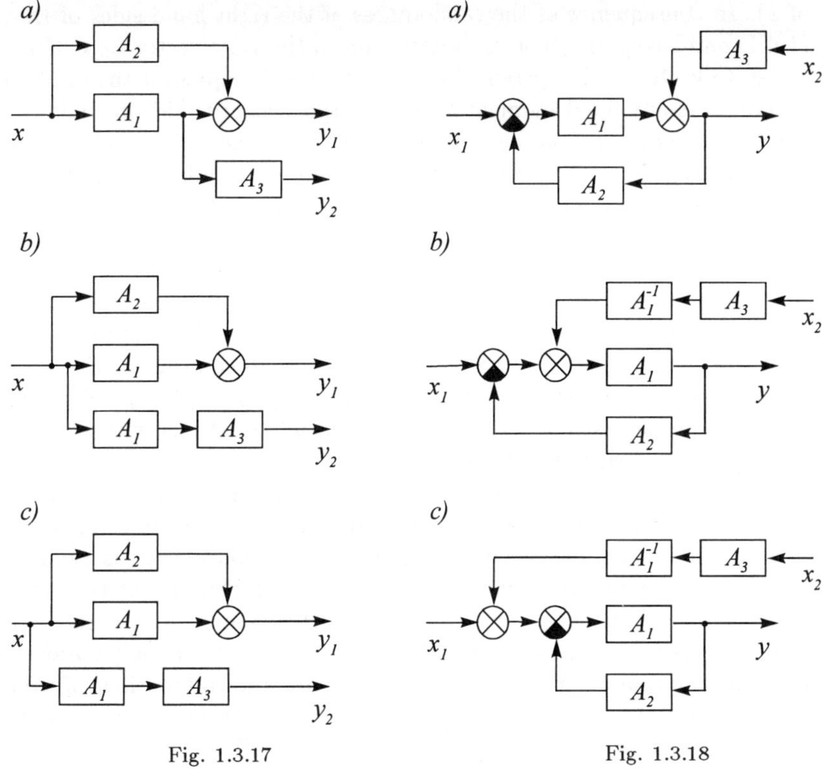

Fig. 1.3.17   Fig. 1.3.18

1.4. Stochastic Differential Systems

1.4.1. General Form of Equations of Stochastic Differential Systems

Stochastic models of systems involve the action of various random factors. While using models described by differential equations the inclusion of random factors lead to the equations which contain random functions, i.e. such functions whose values at given values of the arguments are random variables (Chapter 2).

Eqs. (1.1.4) for a stochastic system must be replaced in the general case by the equations

$$\dot{Z} = F(Z, x, t), \quad Y = G(Z, t), \tag{1.4.1}$$

where $F(z, x, t)$ and $G(z, t)$ are random functions of the p-dimensional vector z, n-dimensional vector x and time t (as a rule G is independent of x). In consequence of the randomness of the right-hand sides of Eqs. (1.4.1) and also perhaps of the initial value of the state vector $Z_0 = Z(t_0)$ the state vector of the system Z and the output Y represent the random variables at any fixed moment t. This is the reason to denote them by capital letters as well as the random functions in the right-hand sides of Eqs. (1.4.1). The state vector of the system $Z(t)$ and its output $Y(t)$ considered as functions of time t represent random functions of time t (in the general case vector random functions). In every specific trial the random functions $F(z, x, t)$ and $G(z, t)$ are realized in the form of some deterministic functions $f(z, x, t)$ and $g(z, t)$ and these realizations determine the corresponding realizations $z(t)$, $y(t)$ of the state vector $Z(t)$ and the output $Y(t)$ satisfying the differential equations (which are the realizations of Eqs. (1.4.1.): $\dot{z} = f(z, x, t)$, $y = g(z, t)$. Thus we come to the necessity to study the differential equations with random functions in the right-hand sides.

In practical problems the randomness of the right-hand sides of the differential equations arises usually from the fact that they represent known functions some of whose arguments are considered as random variables or as random functions of time t and perhaps of the state and the output of the system. But in the latter case these functions are usually replaced by the random functions of time which are only obtained by assuming that their arguments Z and Y are known functions of time corresponding to the nominal regime of system functioning. In practical problems such an assumption usually provides sufficient accuracy.

Thus we may restrict ourselves to the case where all uncertain variables in the right-hand sides of differential equations may be considered as random functions of time. Then Eqs. (1.4.1) may be written in the form

$$\dot{Z} = f(Z, x, N_1(t), t), \quad Y = g(Z, N_2(t), t), \qquad (1.4.2)$$

where f and g are known functions whose arguments include random functions of time $N_1(t)$ and $N_2(t)$. The initial state vector of the system Z_0 in practical problems is always a random variable independent of the random functions $N_1(t)$ and $N_2(t)$ (independent of random disturbances acting on the system).

Every realization $\left[n_1(t)^T n_2(t)^T \right]^T$ of the random function $[N_1(t)^T N_2(t)^T]^T$ determines the corresponding realizations $f(z, x, n_1(t), t)$, $g(z, n_2(t), t)$ of the functions $f(z, x, N_1(t), t)$, $g(z, N_2(t), t)$, and in accordance with this Eqs. (1.4.2) determine respective realizations $z(t)$ and $y(t)$ of the state vector of the system $Z(t)$ and its output $Y(t)$.

E x a m p l e 1.4.1. The equations of motion in Examples 1.1.4 and 1.1.5 describe the stochastic models of motion of various mechanical systems if the forces and the moments in the right-hand sides of these equations represent random functions of time.

1.4.2. Stochastic Linear Differential Systems

The differential equations of a linear stochastic system differ from Eqs. (1.2.1) of a linear deterministic system by additional random items:

$$\dot{Z} = aZ + a_1 x + a_0 + a_2 N_1(t), \quad Y = bZ + b_0 + b_1 N_2(t). \qquad (1.4.3)$$

In these equations $N_1(t)$ and $N_2(t)$ are random functions of time (in general they are vector random functions).

Introducing the composite vector random function $N(t) = \left[N_1(t)^T N_2(t)^T \right]^T$ and block-matrices $a_2' = [a_2 \, 0]$, $b_1' = [0 \, b_1]$ where 0 is the matrix whose elements are all equal to zero we represent the random items of Eqs.(1.4.3) in the form $a_2 N_1(t) = a_2' N(t)$, $b_1 N_2(t) = b_2' N(t)$. Therefore without loss of generality we may omit the indices at random functions and write Eqs. (1.4.3) in the form

$$\dot{Z} = aZ + a_1 x + a_0 + a_2 N(t), \quad Y = bZ + b_0 + b_1 N(t). \qquad (1.4.4)$$

In practical problems the deviations of a nonlinear system from the required regime may sometimes be assumed as sufficiently small. In such a case the equations of the system may often be linearized relative to random deviations from the required regime and relative to random disturbances acting on the system. In such cases the nonlinear equations describing the behaviour of the system are replaced by approximate linear equations in the deviations. It gives the opportunity to study the nominal regime of system functioning using the deterministic nonlinear model (1.1.4) and then to study random deviations from the nominal regime using the much simpler linear stochastic models (1.4.4).

1.4.3. Linear Systems with Parametric Noises

In practical problems we have to deal with linear systems in which the noises depend linearly on the state vector of the system. In such cases we have to use for the description of the behaviour of a system linear differential equations with fluctuating coefficients. The fluctuations of the coefficients of the differential equations of a linear system are usually called *parametric noises*. For such systems the matrices a_2 and b_1 in Eqs. (1.4.4) depend not only on time, but are also linear functions of the state vector of a system. Thus Eqs. (1.4.4) in the case of parametric noises are replaced by the equations

$$\dot{Z} = aZ + a_1 x + a_0 + \left(a_{20} + \sum_{k=1}^{p} a_{2k} Z_k \right) N(t),$$

$$Y = bZ + b_0 + \left(b_{10} + \sum_{k=1}^{p} b_{1k} Z_k \right) N(t). \tag{1.4.5}$$

E x a m p l e 1.4.2. The motion of a pendulum with vibrating suspension point in viscous medium is described by the equation

$$A\ddot{\varphi} + B\dot{\varphi} + mgl(1 + N_2)\sin\varphi + mgl N_1 \cos\varphi = 0, \tag{I}$$

where φ is the deviation angle of a pendulum from the vertical, A is the centroidal moment of inertia, B is the coefficient of moment of forces of viscous friction, mgl is the static moment, $gN_1 = gN_1(t)$ and $gN_2 = gN_2(t)$ are the horizontal and vertical components of the acceleration vector of the suspension point representing random functions of time.

At small angles φ putting in (I) $\sin\varphi \approx \varphi$, $\cos\varphi \approx 1$ and assuming as state variables $\varphi = Z_1$, $\dot{\varphi} = Z_2$ we obtain a linear stochastic differential system with additive and parametric noises whose equation has the form

$$\begin{bmatrix} \dot{Z}_1 \\ \dot{Z}_2 \end{bmatrix} = \begin{bmatrix} 0 & 1 \\ -\omega_0^2 & -2\varepsilon \end{bmatrix} \begin{bmatrix} Z_1 \\ Z_2 \end{bmatrix} + \omega_0^2 \left(\begin{bmatrix} 0 & 0 \\ -1 & 0 \end{bmatrix} + \begin{bmatrix} 0 & 0 \\ 0 & -1 \end{bmatrix} Z_1 \right) \begin{bmatrix} N_1 \\ N_2 \end{bmatrix},$$
$$\tag{II}$$

where $\omega_0^2 = mgl/A$, $2\varepsilon = B/A$.

1.4.4. Stochastic Bilinear Differential Systems

The systems in which the vectors Z, Y, N_1, N_2 in Eqs. (1.4.2) allowing the decomposition of the form

$$Z = \begin{bmatrix} Z'^T & Z''^T \end{bmatrix}^T, \quad Y = \begin{bmatrix} Y'^T & Y''^T \end{bmatrix}^T, \quad N_{1,2} = \begin{bmatrix} N'_{1,2}{}^T & 0 \end{bmatrix},$$

are called so, with the components Z', Y' and Z'', Y'' satisfying the linear equation of the form Eq. (1.4.3) and the bilinear equations, respectively:

$$\dot{Z}' = aZ' + a_1 x + a_0 + a_2 N_1', \quad Y' = bZ' + b_0 + b_1 N_2',$$

$$\dot{Z}'' = \left(A_0 + \sum_{k=1}^{p'} A_k Z_k'\right) Z'', \quad Y'' = \left(B_0 + \sum_{k=1}^{p'} B_k Z_k'\right) Z''. \quad (1.4.6)$$

Here a, a_1, a_2, A_0, A_k, b_0, b_1, B_0, B_k are the matrices, a_0, b_0 are the vectors of the suitable dimension.

The bilinear stochastic systems described by Eqs. (1.4.6) are an important special case of the stochastic systems with random parameters,

$$\dot{Z} = f(Z, x, N_1(t), U, t), \quad Y = g(Z, N_2(t), U, t), \quad (1.4.7)$$

U being a vector of random parameters.

E x a m p l e 1.4.3. The equations of a body motion around a fixed axis in viscous media have the form

$$\frac{d^2\varphi}{dt^2} = -\frac{\Lambda}{I}\frac{d\varphi}{dt}. \quad (I)$$

Here I is an axial moment of inertia, $\dot{\varphi}$ is an angular velocity, Λ/I is specific coefficient of viscous friction which is a random time function satisfying to linear differential equation

$$\dot{\lambda} = -a\lambda + hN, \quad \lambda = \Lambda/I, \quad (II)$$

where a and h are some positive coefficients, $N = N(t)$ is a random time function. Denoting $Z_1 = \dot{\varphi}$, $Z_2 = \lambda$ we reduce the Eqs. (I) and (II) to the following two-dimensional bilinear stochastic differential equations:

$$\dot{Z}_1 = -Z_1 Z_2, \quad \dot{Z}_2 = -aZ_2 + hN. \quad (III)$$

1.5. Stochastic Discrete Systems

1.5.1. Discrete Systems

It is known that for a discrete insertion of the inputs into a system it is necessary to modulate some parameters of the input impulses in accordance with their values. A device forming the sequence of the impulses which depends on an input is called *an impulse element*. Usually the impulses form remains invariable at the modulation. We may distinguish three main types of the impulses modulation:

- pulse-amplitude modulation (PAM) at which the amplitude of the impulses a_k depends linearly (with the gain ξ_{PAM}) on the input value at the moment of the impulse action t_k, $a_k = a_k(t_k) = \xi_{\text{PAM}} x(t_k)$;

- pulse-width modulation (PWM) at which the duration of the impulse T_i linearly depends (with the gain ξ_{PWM}) on the input value at the moment of the impulse action, $T_i(t_k) = \xi_{\text{PWM}} x(t_k)$;

- time-impulse modulation (TIM) at which the time shift T_s of an impulse linearly depends (with the gain ξ_{TIM}) on the input value at a definite time moment, $T_s = \xi_{\text{TIM}} x(t_k)$.

At PAM and PWM the modulating signal changes the area (i.e. the intensity) of the impulses, and at TIM the area of the impulse remains constant. The dependence of modulating parameter of the impulses produced by an impulse element on the correspondent discrete input values is called *a characteristic* of an impulse element. The latter may be linear or nonlinear. The impulse element with linear characteristic is linear, and with a nonlinear characteristic is nonlinear. The impulse elements vary also in the form and the character of the impulses modulation, in impulses frequency and their relative duration. Usually the impulse elements are periodically generate one impulse at each period. The period of the impulses succession T_r is called *a period of repetition* (*a tact*) of a discrete system. The variable $\omega_r = 2\pi/T_r$ represents *a repetition rate*. The parameters γ, $\gamma = T_i/T_r$ and $1 - \gamma$ are called *a relative duration* and *impulses ratio*.

1.5.2. Characteristics of Discrete Linear Systems

Let us denote $g_k(t)$ the response of a linear discrete system on a short-time input signal equal to 1 and acting only during the kth impulse action. Then its response on a short-time signal equal to $x(t_k)$ and which acts only during the time of the kth impulse action will be equal to $g(t_k)x(t_k)$ on the basis of the superposition principle. The response of a linear discrete system on the whole sequence of the impulses modulated by the input disturbance $x(t_k)$ on the basis of the superposition principle (Subsection 1.1.2) will be determined by the formula

$$y(t) = \sum_{k=-\infty}^{\infty} g_k(t)x(t_k). \qquad (1.5.1)$$

The functions $g_k(t)$ characterize completely a linear discrete system as knowing these functions we may calculate the response of a discrete linear system on any input disturbance $x(t)$. The functions $g_k(t)$ determine the part of the input values acting at different time moments t_k in the formation of the output at any time moment t. That is why the functions $g_k(t)$ are called *the weighting coefficients* of a linear discrete system.

Any nonanticipative (physically realizable) discrete system (1.5.1) may response at a given moment t only on the signal acting on it at the previous instant. Therefore for any nonanticipative discrete system

$$g_k(t) = 0 \quad \text{at} \quad t < t_k \quad (k = 0, \pm 1, \pm 2, \ldots). \tag{1.5.2}$$

In particular, for the nonanticipative linear discrete system for which $x(t_k) = 0$ till the instant t_0 formula (1.5.1) may be rewritten in the form

$$y(t) = \sum_{t_0 \leq t_k \leq t} g_k(t) x(t_k). \tag{1.5.3}$$

Here the inequality under the sign of the sum shows that the summation is spread only if the instant t_k of the impulses action are included in the interval $[t_0, t]$.

For determining the weighting function $g(t, \tau)$ of a discrete linear system it is sufficient to find its response on an unit impulse acting at the moment τ. Assuming $x(t) = \delta(t - \tau)$ in formula (1.5.1) we shall get the following formula for $g(t, \tau)$ of a discrete linear system:

$$g(t, \tau) = \sum_{k=-\infty}^{\infty} g_k(t) \delta(t_k - \tau). \tag{1.5.4}$$

Thus *the weighting function of any discrete linear system represents a series of δ-functions.* And vice versa, *any linear system whose weighting function is given by formula (1.5.4) is discrete.* Really, substituting expression (1.5.4) into formula (1.1.11) and taking into consideration the rule of δ-functions integration we get formula (1.5.1). It follows from this formula that the input disturbance $x(t)$ acts on a system only at the time moments t_k.

After determining the weighting function of a discrete linear system by formula (1.5.4) we may find by formulae of Subsection 1.1.7 its response on the exponential signal and the frequency response. Setting in formulae of Subsection 1.2.6 the weighing functions equal to a series of δ-functions we obtain the formulae for the weighting coefficients of different connections of the composed linear discrete systems. Finally

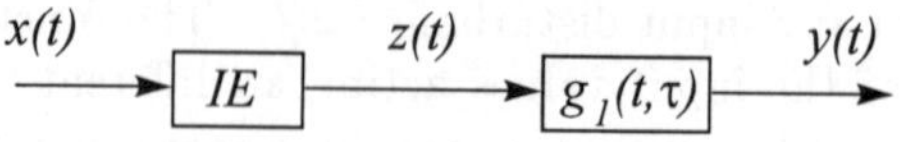

$$x(t) \longrightarrow \boxed{IE} \xrightarrow{\ z(t)\ } \boxed{g_1(t,\tau)} \xrightarrow{\ y(t)\ }$$

Fig. 1.5.1

the formulae of Subsection 1.2.2 determine the systems conjugated with the discrete linear systems.

Let us consider a discrete linear system representing a series connection of an impulse element (IE) and a continuous linear system with the weighting function $g_1(t, \tau)$ (Fig.1.5.1). Let us denote a function describing the form of the impulses generated by an impulse element in terms of $\eta(t)$. In the case of PAM the output of the linear impulse element $z(t)$ will be

$$z(t) = \sum_{k=-\infty}^{\infty} x(t_k)\eta(t - t_k). \qquad (1.5.5)$$

This function is the input of the continuous linear system with the weighting function $g_1(t, \tau)$. Consequently, on the basis of general formula (1.1.8) the output of considered discrete linear system is determined by formula

$$y(t) = \int_{-\infty}^{\infty} g_1(t,\tau)z(\tau)d\tau = \sum_{k=-\infty}^{\infty} x(t_k) \int_{-\infty}^{\infty} g_1(t,\tau)\eta(\tau - t_k)d\tau.$$

The function $\eta(\tau - t_k)$ differs from zero only at the interval $t_k \leq \tau \leq t_k + T_i$ where T_i is the duration of an impulse, therefore the latter formula may be rewritten in the form

$$y(t) = \sum_{k=-\infty}^{\infty} x(t_k) \int_{t_k}^{t_k+T_i} g_1(t,\tau)\eta(\tau - t_k)d\tau.$$

Comparing the given formula with (1.5.1) we get:

$$g_k(t) = \int_{t_k}^{t_k+T_i} g_1(t,\tau)\eta(\tau - t_k)d\tau = \int_{0}^{T_i} g_1(t,t_k + \sigma)\eta(\sigma)d\sigma. \qquad (1.5.6)$$

This formula shows that at PAM the weighting coefficients of a series connection of an impulse element and a continuous linear system depend on the form of the impulses generated by an impulse element and the weighting function of the continuous part of a system.

In the case of PWM we shall assume for the simplicity that an impulse element generates the rectangular impulses of the constant amplitude a whose duration is proportional to the input values at the corresponding instants. Then the input $z(t)$ of a continuous part will be equal to a at the intervals of time $[t_k, t_k + \xi x(t_k)]$, where $\xi = \xi_{\text{PWM}}$ and to zero out of these intervals. Consequently, we get the following expression for the output of the considered linear discrete system:

$$y(t) = a \sum_{k=-\infty}^{\infty} \int_{t_k}^{t_k + \xi x(t_k)} g_1(t, \tau) d\tau \, . \qquad (1.5.7)$$

At small duration of the impulses relation (1.5.7) may be approximately considered as a linear one and the weighting function $g_1(t, \tau)$ may be assumed at each integral in (1.5.7) practically constant and equal to its value $g_1(t, t_k)$ at $t = t_k$. After factoring it out from the sign of the integral formula (1.5.7) takes the form

$$y(t) \approx a\xi \sum_{k=-\infty}^{\infty} g_1(t, t_k) x(t_k) \, , \quad \xi = \xi_{\text{PWM}} \, . \qquad (1.5.8)$$

Here the value $a\xi$ represents the product of the impulse amplitude by its duration at the unit input and it determines the intensity of the impulses. Thus *at PWM a series connection of an impulse element and a continuous system with the weighting function $g_1(t, \tau)$ may be considered as a discrete linear system whose weighting coefficients are determined by formula*

$$g_k(t) = a\xi g_1(t, t_k), \quad \xi = \xi_{\text{PWM}} \, . \qquad (1.5.9)$$

Analogously we get that *for TIM a series connection of an impulse element and a continuous system may be considered approximately as a discrete linear system if the maximal possible time shift of an impulse is sufficiently small so that the weighting function of a continuous part of the system $g_1(t, \tau)$ is approximately linear function τ in the range of possible values of time shift.*

Now let us consider the peculiarities of the linear discrete systems which contain digital computers. We are interested in an output only at the definite time moments t_l' $(l = 0, \pm 1, \pm 2, \ldots)$ Putting in formula (1.5.1) $t = t_l'$ we shall find the output values of a system at the time moments t_l' $(l = 0, \pm 1, \pm 2, \ldots)$ which we are interested in:

$$y(t_l') = \sum_{k=-\infty}^{\infty} g_k(t_l') x(t_k) \, . \qquad (1.5.10)$$

For brevity setting $x_k = x(t_k)$, $y_l = y(t'_l)$, $g_k(t'_l) = g_{lk}$ we can rewrite formula (1.5.10) finally in the form

$$y_l = \sum_{k=-\infty}^{\infty} g_{lk} x_k \quad (l = 0, \pm 1, \pm 2, \ldots). \qquad (1.5.11)$$

In the case when the instants of an output fixing coincide with the instants of the impulses action, $t'_l = t_l$ ($l = 0, \pm 1, \pm 2, \ldots$), formula (1.5.11) at $t = t_l$ will give the following expression for the values of an output of a nonanticipative linear discrete system:

$$y_l = \sum_{k=0}^{l} g_{lk} x_k \quad (l = 0, 1, 2, \ldots). \qquad (1.5.12)$$

Formula (1.5.12) and the analogous formulae for the discrete systems with many inputs and outputs represent, in particular, a digital computer as a discrete linear ystem because the results of the calculations depend linearly on the initial data. In this case the initial data introduced at each step of the calculations are the input signals and the results of the calculations are the outputs signals. The weighting coefficients g_{lk} determine the program of the calculations.

At the sufficiently large impulses repetition rate any discrete system may be considered approximately as a continuous one. Really, the response of any system on a short-term unit signal acting during one period of the impulses repetition T_r at infinitesimal T_r is an infinitesimal of the order T_r. Therefore the function $g_k(t)$ at small $T_r = \Delta t$ may be expressed by formula

$$g_k(t) \approx g(t, t_k) T_r = g(t, t_k) \Delta t. \qquad (1.5.13)$$

Here $g(t, t_k)$ is some function independent of T_r. So, for instance, in the case of a series connection of an impulse element and a continuous linear system with the weighting function $g_1(t, \tau)$ we can express the dependence of the impulses of a given form on the period of the repetition of the impulses assuming the function $\eta(t)$ in the form $\eta(t) = h(t/T_r)$, where $h(\xi)$ is a function different from zero only at the interval $0 \leq \xi \leq \gamma = T_i/T_r$. Then formula (1.5.6) may be written in such a way:

$$g_k(t) = \int_0^{T_i} g_1(t, t_k + \sigma) h\left(\frac{\sigma}{T_r}\right) d\sigma = T_r \int_0^{\gamma} g_1(t, t_k + \xi T_r) h(\xi) d\xi.$$

At sufficiently small period of the impulses repetition T_r the weighting function g_1 in the integral will be practically constant at the integration interval close to its value $g_1(t, t_k + 0)$. Factoring it out the sign of the integral and putting

$$g(t, t_k) = g_1(t, t_k + 0) \int\limits_0^{\gamma} h(\xi) d\xi \,, \qquad (1.5.14)$$

we get formula (1.5.13). Substituting expression (1.5.13) into (1.5.1) we shall have

$$y(t) \approx \sum_{k=-\infty}^{\infty} g(t, t_k) x(t_k) \Delta t \,. \qquad (1.5.15)$$

Here a relative error of this equality tends to zero at $\Delta t = T_r \to 0$. But at $\Delta t \to 0$ the sum in (1.5.15) tends to integral (1.1.18). Therefore *at sufficiently small T_r a discrete linear system may be considered approximately as a continuous linear system with the weighting function* $g(t, t_k) = g_k(t)/T_r$.

In the special case of a series connection of an impulse element and a linear stationary system described by a differential equation the condition of approximate constancy of the weighting function of a continuous part of a system over the period of the impulses repetition T_r is equivalent to the condition of the proximity to the unit of all the variables $e^{\nu_1 T_r}, \ldots, e^{\nu_n T_r}$ where $\nu_1, \ldots, \nu_n$ are the roots of a characteristic equation of continuous part of the system. The continuous part of the system may be presented at a given case in the form of a series connection of typical links. Here the roots of a characteristic equations $\nu_1, \ldots, \nu_n$ will be in inverse proportion to the time constants of the typical links which enter into a continuous part of a system. Consequently, the condition of the smallness of the variables $\nu_1 T_r, \ldots, \nu_n T_r$ from a physical point of view is the condition of the smallness of the period of the T_r impulses repetition in comparison with the time constants of a continuous part of a system. Thus *a series connection of an impulse element and a continuous linear stationary system may be considered as a continuous linear stationary system if the period of the impulses repetition is sufficiently small in comparison with the time constants of a continuous part of a system.*

E x a m p l e 1.5.1. Find the weighting coefficients of a linear discrete system representing a series connection of an impulse element which generates the unit rectangular impulses of T_i duration with the period T_r of the repetition and an aperiodic link (Example 1.1.6). Substituting into formula (1.5.6) expression (I) of the weighting function of an aperiodic link and taking into account that $\eta(\sigma) = 1$ at $0 \leq \sigma \leq T_i$ we obtain

$$g_k(t) = \frac{k}{T} \int\limits_0^{t-t_k} e^{-(t-t_k-\sigma)/T}\, d\sigma = k \left(1 - e^{-(t-t_k)/T} \right) \quad \text{at } t_k < t < t_k + T_i,$$

$$\text{(I)}$$

$$g_k(t) = \frac{k}{T} \int\limits_0^{T_i} e^{-(t-t_k-\sigma)/T}\, d\sigma = k e^{-(t-t_k)/T} \left(e^{T_i/T} - 1 \right) \quad \text{at } t > t_k + T_i.$$

$$\text{(II)}$$

1.5.3. Discrete Linear Stationary Systems

A discrete system is called *a stationary* one if at time shift of an input time function without changing its form on time interval multiple the period of the impulses repetition, the output shifts in time on the same interval without changing its form. From this definition follows that *a discrete system may be stationary only if the values of the input follow each other in equal time intervals. If the values of the input are inserted into a system in unequal time intervals then a discrete system is nonstationary.*

The weighting coefficient $g_h(t)$ of a discrete linear system according to the definition (Subsection 1.5.2) is its response on the short-time input (i.e. the input of the duration smaller than the period of the impulses T_r repetition) equal to 1 at the instant $t_h = hT_r$. If a linear discrete system is stationary then at the shift of the short-time unit input in time at interval kT_r its response will shift in time at the same interval without changing the form, i.e. will be equal $g_h(t - kT_r)$. At the same time the response of a system on the unit input acting at the instant $t_{h+k} = (h + k)T_r$ will be equal to its weighting coefficient $g_{h+k}(t)$. Consequently, at any integer k and h for the linear stationary discrete system $g_{h+k}(t) = g_h(t - kT_r)$. Putting here $h = 0$ we obtain at any k

$$g_k(t) = g_0(t - kT_r) \quad (k = 0, \pm 1, \pm 2, \ldots). \qquad (1.5.16)$$

Thus *the weighting coefficients of a linear stationary discrete system represent the same function shifted in time on the intervals multiple to the period of the impulses repetition.*

In particular for a linear discrete system (Fig.1.5.1) representing a series connection of an impulse element which generates the sequence of equidistant impulses and a continuous linear stationary system with the weighting function $w(t - \tau)$ formula (1.5.6) has the form

$$g_k(t) = g_0(t - kT_r) = \int_0^{T_i} w(t - kT_r - \sigma)\eta(\sigma)d\sigma . \qquad (1.5.17)$$

Putting in (1.5.16) $t = t_l = lT_r$ we obtain

$$g_{lk} = g_k(t_l) = g_0((l - k)T_r) \ (k, l = 0, \pm 1, \pm 2, \ldots) . \qquad (1.5.18)$$

Consequently, *the weighting coefficients g_{lk} for a discrete linear stationary system depend only on the difference of indexes.* Latter on we shall denote them as w_m $(m = 0, \pm 1, \pm 2, \ldots)$. For the nonanticipative stationary linear discrete systems $g_0(t) = 0$ at $t < 0$, and consequently,

$$w_m = 0 \text{ at } m < 0 . \qquad (1.5.19)$$

On the basis of formulae (1.5.17) and (1.5.18) we conclude that *the weighting coefficients w_m of a series connection of an impulse element and a continuous linear stationary system are determined by formula*

$$w_m = \int_0^{T_i} w(mT_r - \sigma)\eta(\sigma)d\sigma \ (m = 0, \pm 1, \pm 2, \ldots) . \qquad (1.5.20)$$

Formula (1.2.17) which expresses the sequence of the output values in terms of the sequence of the input values for a stationary linear discrete system takes the form

$$y_l = \sum_{k=-\infty}^{\infty} w_{l-k} x_k = \sum_{m=-\infty}^{\infty} w_m x_{l-m} . \qquad (1.5.21)$$

Let us consider now the response of the system on the exponential input $x(t) = s^{st}$. In this case $x_{l-m} = e^{s(l-m)T_r}$ and formula (1.5.21) gives

$$y_l = \sum_{m=-\infty}^{\infty} w_m e^{s(l-m)T_r} = e^{slT_r} \sum_{m=-\infty}^{\infty} w_m e^{-smT_r} . \qquad (1.5.22)$$

Thus the *response of a discrete linear stationary system on the exponential input e^{st} at the instant $t = lT_r$ is equal to the value of this input at the same instant multiplied by the function*

$$\Phi(s) = \sum_{m=-\infty}^{\infty} w_m e^{-smT_r}, \qquad (1.5.23)$$

dependent only on the parameter s. The function $\Phi(s)$ is a *transfer function of a discrete linear stationary system.*

Setting in formula (1.5.23) $s = i\omega$ we find a *frequency characteristic of a system*

$$\Phi(i\omega) = \sum_{m=-\infty}^{\infty} w_m e^{-imT_r\omega}. \qquad (1.5.24)$$

Formula (1.5.24) shows that a frequency characteristic of a linear stationary discrete system is a periodic function of the frequency ω with the period $\Omega = 2\pi/T_r$. Formula (1.5.24) determines a frequency characteristic of a linear stationary discrete system in the form of the Fourier series. The coefficients of the series are the weighting coefficients of a system. Consequently, using the known formula for the coefficients of the Fourier series we shall express the weighting coefficients of the considered system in terms of its frequency characteristic:

$$w_m = \frac{1}{\Omega} \int_0^{\Omega} \Phi(i\omega)e^{imT_r\omega}\,d\omega = \frac{T_r}{2\pi} \int_0^{2\pi/T_r} \Phi(i\omega)e^{imT_r\omega}\,d\omega. \qquad (1.5.25)$$

For a nonanticipative discrete linear stationary system the integral in (1.5.25) is equal to zero at $m < 0$ and formula (1.5.23) takes the form

$$\Phi(s) = \sum_{m=0}^{\infty} w_m e^{-smT_r}. \qquad (1.5.26)$$

Formulae (1.5.23) and (1.5.21) show that the transfer functions of linear stationary discrete systems are the functions of the variable $z = e^{sT_r}$. Denoting the transfer function of a nonanticipative stationary discrete system considered as a function of the parameter z in terms of $\Psi(z)$ we can rewrite formula (1.5.26)

$$\Psi(z) = \sum_{m=0}^{\infty} w_m z^{-m}. \qquad (1.5.27)$$

Here according to the definition

$$\Phi(s) = \Psi\left(e^{sT_r}\right) , \quad \Psi(z) = \Phi\left(\frac{1}{T_r}\ln z\right) . \tag{1.5.28}$$

It is easy to see that all previous formulae are also valid for the transfer functions of typical connections of discrete linear stationary systems (Section 1.3.1).

In some cases it is convenient to consider the transfer function of a discrete linear stationary system as the function of the following parameters v and z:

$$v = \frac{z-1}{z+1} = \frac{e^{sT_r}-1}{e^{sT_r}+1} , \quad z = \frac{1+v}{1-v} . \tag{1.5.29}$$

Let us denote a transfer function considered as the function of v in terms of $\Omega(v)$. Then from (1.5.28) we shall obtain the following relations between the transfer functions Ω, Ψ and Φ:

$$\Omega(v) = \Psi\left(\frac{1+v}{1-v}\right) = \Phi\left(\frac{1}{T_r}\ln\frac{1+v}{1-v}\right) , \tag{1.5.30}$$

$$\Psi(z) = \Omega\left(\frac{z-1}{z+1}\right) , \quad \Phi(s) = \Omega\left(\frac{e^{sT_r}-1}{e^{sT_r}+1}\right) . \tag{1.5.31}$$

E x a m p l e 1.5.2. Find the transfer function of the discrete system (Fig.1.5.2) whose impulse element generates the rectangular impulses of the duration T_i with the period of the repetition T_r. At first we shall find the transfer function $\Psi_1(z)$ of the open-loop system. Substituting into formula (1.5.22) the expression from Example 1.5.1 for the weighting coefficients of the open-loop impulse system and accounting that $l - k = m$ we obtain the formula for the transfer function $\Psi_1(z)$:

$$\Psi_1(z) = k\left(e^{\gamma T_r/T} - 1\right)\sum_{m=1}^{\infty} e^{-mT_r/T} z^{-m}$$

$$= k z_1 \frac{z_1^{-\gamma} - 1}{z - z_1} , \quad z_1 = e^{-T_r/T} , \quad \gamma = T_i/T_r . \tag{I}$$

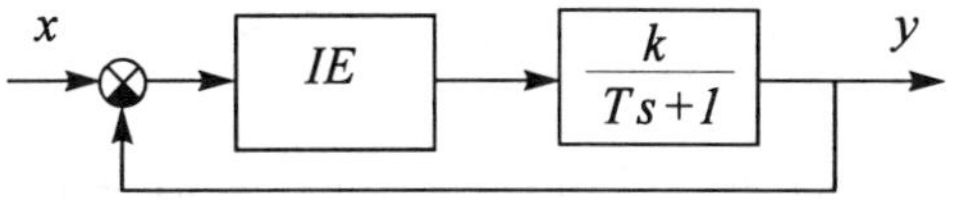

Fig. 1.5.2

Substituting expression (I) into formula (1.3.17) we shall find the transfer function of the closed-loop system:

$$\Psi(z) = \frac{\Psi_1}{1 + \Psi_1(z)} = \frac{kz_1(z_1^{-\gamma} - 1)}{z - z_1(k + 1 - kz_1^{-\gamma})}. \tag{II}$$

The replacement of the variable z by v according to formula (1.5.29) in expressions (I) and (II) gives for the open-loop system

$$\Omega_1(v) = kz_1(z_1^{-\gamma} - 1)\frac{1 - v}{1 - z_1 + (1 + z_1)v}. \tag{III}$$

Hence the transfer function $\Omega(v)$ of the closed-loop system will be

$$\Omega(v) = \frac{\Omega_1(v)}{1 + \Omega_1(v)} = \frac{kz_1(z_1^{-\gamma} - 1)(1 - v)}{1 - z_1 + kz_1(z_1^{-\gamma} - 1) + \left[1 + z_1 - kz_1(z_1^{-\gamma} - 1)\right]v}. \tag{IV}$$

For determining the output of linear stationary systems with time varying output in the interval between the impulses we may use general formula (1.5.1) replacing in the weighting coefficients $g_k(t)$ by expression (1.5.16). It is practically convenient here to determine the instant t between two adjacent impulses by relative time increment after each impulse ε. In other words, any instant in the interval between the impulses $lT_r < t < (l + 1)T_r$ may be determined by formula $t = lT_r + \varepsilon T_r$ $(0 < \varepsilon < 1)$. Substituting this expression into formula (1.5.1) and assuming

$$\left.\begin{aligned} y(lT_r + \varepsilon T_r) &= y_l(\varepsilon), \\[2mm] g_k(lT_r + \varepsilon T_r) &= g_0((l - k)T_r + \varepsilon T_r) = w_{l-k}(\varepsilon) \end{aligned}\right\} \tag{1.5.32}$$

for a nonanticipative discrete linear stationary system we obtain

$$y_l(\varepsilon) = \sum_{k=-\infty}^{\infty} w_{l-k}(\varepsilon)x_k .$$

After changing the index $m = l - k$ we reduce this formula to the form

$$y_l(\varepsilon) = \sum_{m=0}^{\infty} w_m(\varepsilon)x_{l-m} \quad (0 \leq \varepsilon < 1; l = 0, \pm 1, \pm 2, \ldots). \tag{1.5.33}$$

In the case of the exponential input we have $x(t) = e^{st}$, $x_{l-m} = e^{s(l-m)T_r}$ and formula (1.5.33) gives

$$y_l(\varepsilon) = e^{-slT_r} \sum_{m=0}^{\infty} w_m(\varepsilon)e^{-smT_r} . \qquad (1.5.34)$$

The function which enters into this formula,

$$\Phi(s,\varepsilon) = \sum_{m=0}^{\infty} w_m(\varepsilon)e^{-smT_r} \quad (0 \le \varepsilon < 1), \qquad (1.5.35)$$

is called *the transfer function* of a discrete linear stationary system with a continuous output. At $s = i\omega$ formula (1.5.35) determines a frequency characteristic of a linear stationary discrete system with a continuous output. The dependence of the transfer function and the frequency characteristic on ε reflects the fact that a discrete system may be stationary only relative to the shifts in time multiple to the period of the impulses repetition, and cannot be stationary in full sense, i.e. relatively to any shifts in time. As on the basis of the second formula (1.5.32) $w_m(0) = g_0(mT_r) = w_m$, the transfer function of a discrete linear stationary system coincides with the transfer function $\Phi(s)$ introduced earlier by formula (1.5.26): $\Phi(s,0) \equiv \Phi(s)$.

Considering the transfer function of a discrete linear stationary system with a continuous output as a function of the parameter $z = e^{sT_r}$ we may rewrite formula (1.5.35) with account (1.5.28) in the form

$$\Psi(z,\varepsilon) = \sum_{m=0}^{\infty} w_m(\varepsilon)z^{-m} \quad (0 \le \varepsilon < 1). \qquad (1.5.36)$$

Formula (1.5.26) can be obtained from general formula (1.1.18) after substituting in it the weighting function $w(\sigma)$ by a series of δ-functions:

$$w(\sigma) = \sum_{m=0}^{\infty} w_m\delta(\sigma - mT_r). \qquad (1.5.37)$$

In other words *for a discrete linear stationary system the transfer function is the Laplace transform of its weighting function.* The Laplace transformation in (1.5.37) is called *the Laplace discrete transformation* of the numerical sequence w_m. Analogously, the function $\Psi(z)$ determined by formula (1.5.27) is usually called the *z-transformation* of the sequence of the weighting coefficients w_m of a linear stationary discrete system.

E x a m p l e 1.5.3. Find the weighting coefficients $w_m(\varepsilon)$ and the transfer function $\Psi(z,\varepsilon)$ of a discrete linear system from Examples 1.5.1 and 1.5.2. Putting $t = lT_r + \varepsilon T_r$, $t_k = kT_r$, $l - k = m$ we get for the weighting coefficients $w_m(\varepsilon)$:

$$
w_0(\varepsilon) = g_k(kT_r + \varepsilon T_r) = \begin{cases} k(1 - z_1^\varepsilon) & \text{at} \ \ 0 \le \varepsilon \le \gamma, \\ kz_1^\varepsilon(z_1^{-\gamma} - 1) & \text{at} \ \ \gamma < \varepsilon < 1, \end{cases} \tag{I}
$$

$$
w_m(\varepsilon) = g_k(lT_r + \varepsilon T_r) = kz_1^{m+\varepsilon}(z_1^{-\gamma} - 1) \ (m = 1, 2, \ldots). \tag{II}
$$

Substituting these expressions in formula (1.5.36) we find

$$
\Psi(z,\varepsilon) = \begin{cases} k(1 - z_1^\varepsilon) + kz_1^\varepsilon(z_1^{-\gamma} - 1) \displaystyle\sum_{m=1}^\infty z_1^{m_z - m} \\[2mm] \quad = k(1 - z_1^\varepsilon) + kz_1^{1+\varepsilon}\dfrac{z_1^{-\gamma} - 1}{z - z_1} \ \text{at} \ 0 \le \varepsilon \le \gamma, \\[4mm] kz_1^\varepsilon(z_1^{-\gamma} - 1) \displaystyle\sum_{m=0}^\infty z_1^m z^{-m} = kz_1^\varepsilon\dfrac{(z_1^{-\gamma} - 1)z}{z - z_1} \\[2mm] \qquad\qquad \text{at} \ \gamma < \varepsilon < 1. \end{cases} \tag{III}
$$

From the first part of formulae (III) at $\varepsilon = 0$ we obtain (I) from Example 1.5.2.

1.5.4. Discrete Stationary Systems Described by Difference Equations

While studying the behaviour of the discrete systems only at the definite equidistant instants it is expedient to introduce instead of the derivatives the finite differences of the functions. *A difference of the first order or first difference* of the function $f(t)$ at the change of an argument on a given variable T_r is called its increment $\Delta f(t) = f(t + T_r) - f(t)$. Assuming $t = kT_r$ and introducing for brevity the denotion $f_k = f(kT_r)$ $(k = 0, \pm 1, \pm 2, \ldots)$ we obtain the following expression:

$$
\Delta f_k = f_{k+1} - f_k \ (k = 0, \pm 1, \pm 2, \ldots).
$$

Using the notion of a function difference we may determine its differences of any orders by a recurrent relation:

$$
\Delta^l f_k = \Delta^{l-1} f_{k+1} - \Delta^{l-1} f_k \ (l = 1, 2, 3, \ldots). \tag{1.5.38}
$$

It is evident that all the differences of the function may be expressed in the terms of its values f_k, f_{k+1}, $f_{k+2}, \ldots$ according to the following general formula:

$$
\Delta^l f_k = \sum_{h=0}^l (-1)^{l-h} C_l^h f_{k+h}, \tag{1.5.39}
$$

where C_l^h are the binomial coefficients.

The relation connecting the values of the unknown function f_k and its differences of different orders Δf_k, $\Delta^2 f_k$, ..., is called *the difference equation*. If this relation is linear then it represents *the linear* difference equation. A difference equation may contain either the values of an unknown function and its differences of the different orders at one and the same value of the argument $t = kT_r$ or the values of an unknown function at the different equidistant values of the argument. A difference equation which contains the differences of an unknown function till the order n or the values of an unknown function from f_k till f_{k+n} is called *a difference equation of the nth order*.

In the case when the behaviour of a discrete linear system is described by the difference equation this equation may be always written in the form

$$a_n y_{k+n} + a_{n-1} y_{k+n} + \cdots + a_1 y_{k+1} + a_0 y_k$$

$$= b_m x_{k+m} + b_{m-1} x_{k+m-1} + \cdots + b_1 x_{k+1} + b_0 x_k . \tag{1.5.40}$$

In order a discrete linear system described by Eq. (1.5.40) be stationary, it is necessary and sufficient that all the coefficients of the equation be constant.

Eq. (1.5.40) may be written in an operator form. For this purpose we introduce the operator of the shift ∇ which determines a shift of a function on the repetition period T_r: $\nabla f(t) = f(t + T_r)$. Putting $t = kT_r$ we get $\nabla^l f_k = f_{k+l}$ $(l = 1, \ldots, n)$. It means that Eq. (1.5.40) may be presented in the polynomial form relative to ∇:

$$F(\nabla) y_k = H(\nabla) x_k , \tag{1.5.41}$$

where

$$F(\nabla) = a_n \nabla^n + a_{n-1} \nabla^{n-1} + \cdots + a_1 \nabla + a_0 ,$$

$$H(\nabla) = b_m \nabla^m + b_{m-1} \nabla^{m-1} + \cdots + b_1 \nabla + b_0 .$$

Let us find the transfer function of a discrete linear stationary system described by Eq. (1.5.41). According to the definition of the transfer function $\Psi(z)$ of a discrete linear stationary system (Subsection 1.5.3) its response on the exponential input $x(t) = e^{st}$, $x_k = e^{skT_r}$ is determined by formula $y_k = \Psi(z) e^{skT_r}$. Substituting this expression and the corresponding expression of the input variable $x_k = e^{skT_r}$ into Eq. (1.5.41) we reduce it to the form

$$\Psi(z) F(\nabla) e^{skT_r} = H(\nabla) e^{skT_r} . \tag{1.5.42}$$

But

$$\nabla e^{skT_r} = \nabla e^{st_k} = e^{s(t_k + T_r)} = e^{sT_r} e^{skT_r} = z e^{skT_r} \, ,$$

$$\nabla^l e^{skT_r} = z^l e^{skT_r} \quad (l = 1, 2, \ldots) \, .$$

Consequently, in Eq. (1.5.42) the shift operator ∇ may be replaced by the variable z. We obtain the following linear algebraic equation for the sought transfer function $\Psi(z)$:

$$\Psi(z) F(z) e^{skT_r} = H(z) e^{skT_r} \, .$$

After reducing this equation by e^{skT_r} and solving it we shall obtain the following formula for the transfer function of the system described by difference Eq. (1.5.41):

$$\Psi(z) = F(z) / H(z) \, . \tag{1.5.43}$$

Thus *the transfer function of a discrete linear stationary system whose behaviour is described by difference equation is always a rational function of the variable $z = e^{sT_r}$.*

R e m a r k. The transfer functions of a discrete linear stationary system described by the difference equations are determined in just the same way as the transfer functions of a continuous linear stationary systems described by the differential equations. In the case of a continuous system the differention operator in a differential equation is replaced by a parameter of the exponential function s, and in the case of a discrete system a shift operator in a difference equation is replaced by the variable $z = e^{sT_r}$. In both cases as a result we get an algebraic equation for a transfer function of a system. The operators of the differention D and the shift operator ∇ are connected by the same relation as the variables s and z: $\nabla = e^{T_r D}$ and $z = e^{sT_r}$.

Now let us determine the transfer function $\Psi(z)$ of a discrete linear stationary system (Fig.1.5.1) which represents a series connection of an impulse element and a continuous stationary system described by Eq. (1.3.33) at $n > m$. In the case when the polynomial $F(D)$ has only simple roots $\nu_1, \ldots, \nu_r$ the weighting function of the system is determined by formula (1.3.45). Accounting formulae (1.5.27) and (1.3.47) we shall have

$$\Psi(z) = \sum_{m=0}^{\infty} w_m z^{-m} = \sum_{m=1}^{\infty} \sum_{i=1}^{n} \rho_i \frac{H(\nu_i)}{F'(\nu_i)} z_i^m z^{-m}$$

$$= \sum_{i=1}^{n} \rho_i \frac{H(\nu_i)}{F'(\nu_i)} \cdot \sum_{m=1}^{\infty} z_i^m z^{-m} = \sum_{i=1}^{n} \rho_i \frac{H(\nu_i)}{F'(\nu_i)} \frac{z_i z^{-1}}{1 - z_i z^{-1}}$$

$$= \sum_{i=1}^{n} \frac{H(\nu_i)}{F'(\nu_i)} \frac{\rho_i z_i}{z - z_i} \, , \tag{1.5.44}$$

where the following notations are introduced:

$$z_i = e^{\nu_i T_r} \, , \quad \rho_i = \int_0^{T_r} e^{-\nu_i \sigma} \eta(\sigma) d\sigma \quad (i = 1, \ldots, n) \, .$$

E x a m p l e 1.5.4. Under the conditions of Examples 1.5.2 and 1.5.3 at the rectangular impulses $\eta(\sigma) = 1$ find the transfer function $\Psi(z)$ and the corresponding difference equation. We have

$$F(s) = Ts + 1 \, , \quad F'(s) = T \, , \quad \nu_1 = \nu = -1/T \, , \quad H(s) = kz,$$

$$\rho_1 = \int_0^{T_i} e^{\sigma/T} d\sigma = T\left(e^{T_i/T} - 1\right) \, , \quad z_1 = e^{-T_r/T} \, . \tag{I}$$

Consequently,

$$\Psi(z) = \frac{k}{T} \frac{\rho_1 z_1}{z - z_1} = kz_1 \frac{z_1^{-\gamma} - 1}{z - z_1} \, , \quad \gamma = \frac{T_i}{T_r} \tag{II}$$

Here the difference equation responding to (II) may be written in the form

$$(\nabla - z_1)y_l = kz_1 \left(z_1^{-\gamma} - 1\right) x_l \tag{III}$$

or

$$y_{l+1} - z_1 y_l = kz_1 \left(z_1^{-\gamma} - 1\right) x_l \quad (l = 0, 1, 2, \ldots) \, . \tag{IV}$$

1.5.5. Discrete Linear Systems Described by Difference Equations

In applications for a mathematical description of the models of discrete linear systems we usually use the following difference equations:

$$z_{k+l} = \sum_{r=0}^{l-1} a_{rk} z_{k+r} + \sum_{s=0}^{l} a_{1sk} x_{k+s} + a_{0k} \, ,$$

$$y_k = b_k z_k + b_{0k} \, , \quad l \geq 1, \tag{1.5.45}$$

or

$$z_{k+1} = c_k z_k + q_k x_k + c_{0k}, \, y_k = b_k z_k + b_{0k} \tag{1.5.46}$$

respectively for the state vector z and the output y. It is assumed in Eqs. (1.5.45) and (1.5.46) that the vectors x, z, y have the dimensions n, p, m respectively; a_{rk} being $(p \times p)$-matrix; a_{1sk} being $(p \times n)$-matrix; a_{0k} and b_{0k} being the vectors of p and m dimension respectively.

For the stationary systems the coefficients which enter into Eqs. (1.5.45) and (1.5.46) do not depend on k: $a_{rk} = a_r$, $a_{1sk} = a_{1s}$, $a_{0k} = a_0$, $b_k = b$, $b_{0k} = b$. Model (1.5.45) is called *a discrete linear model of the order l* ($l = 1, 2, \ldots$); model (1.5.46) is called *a discrete linear model of the first order*.

Analogously as in Subsection 1.2.3 one can show that linear discrete system (1.5.45) of any order may be reduced to a linear model of the first order (1.5.46). Here we restrict ourselves to the consideration of an equation only for the state vector z.

▷ Let us introduce new variables $\bar{z}_{1k}$, $\ldots$, $\bar{z}_{lk}$ after determining them by the following relations:

$$z_k = \bar{z}_{1k} + q_{1k}x_k \,, \quad \bar{z}_{r,n+1} = \bar{z}_{r+1,k} + q_{r+1,k}x_k \quad (r = 1, \ldots, l-1),$$
$$(1.5.47)$$

where q_{1k}, q_{lk} are yet unknown (matrix in general case). Then we shall have

$$z_{k+1} = \bar{z}_{1,k+1} + q_{1,k+1}x_{k+1} = \bar{z}_{2k} + q_{2k}x_k + q_{1,k+1}x_{k+1} \qquad (1.5.48)$$

and generally

$$z_{k+r} = \bar{z}_{r+1,k} + q_{r+1,k}x_k + q_{r,k+1}x_{k+1} + \cdots + q_{1,k+r} \quad (r = 1, \ldots, l-1).$$
$$(1.5.49)$$

After replacing k by $k+1$ in formulae (1.5.48) and (1.5.49) at $r = l-1$, we obtain

$$z_{k+l} = \bar{z}_{l,k+1} + q_{l,k+1}x_{k+1} + \cdots + q_{1,k+l}x_{k+l} = z_{l,k+1} + \sum_{s=1}^{l} q_{l-s+1,k+s}x_{k+s} \,.$$
$$(1.5.50)$$

Substituting expressions (1.5.49) and (1.5.50) into (1.5.45) we shall have

$$\bar{z}_{l,k+1} + \sum_{s=1}^{l} q_{l-s+1,k+s}x_{k+s} = \sum_{r=0}^{l-1} a_{rk}\left(\bar{z}_{r+1,k} + q_{r+1,k}x_k + \right.$$

$$\left. \cdots + q_{1,k+r}x_{k+r}\right) + \sum_{s=0}^{l} b_{sk}x_{k+s} = \sum_{r=0}^{l-1} a_{rk}\bar{z}_{r+1,k}$$

$$+ \sum_{s=0}^{l-1} x_{k+s} \sum_{r=s}^{l-1} a_{rk} q_{r-s+1,k+s} + \sum_{s=0}^{l} b_{sk} x_{k+s} \cdot \tag{1.5.51}$$

Let us determine now q_{1k} , $\ldots$, q_{lk} in such a way that Eq. (1.5.51) will not contain x_{k+1} , $\ldots$, x_{k+l}. Comparing the coefficients at x_{k+l} in the left- and the right-hand sides of Eq. (1.5.51) we obtain $q_{1,k+l} = b_{lk}$. Hence after replacing k by $k-l$ we find $q_{1k} = b_{l,k-l}$. Comparing the coefficients at x_{k+s} $(s = 1 , \ldots , l-1)$ in the left- and right-hand sides of Eq. (1.5.51) we get

$$q_{l-s+1,k+s} = \sum_{r=s}^{l-1} a_{rk} q_{r-s+1,k+s} + b_{sk} \cdot \tag{1.5.52}$$

Replacing k by $k-s$ we find from equality (1.5.52)

$$q_{l-s+1,k} = b_{s,k-s} + \sum_{r=s}^{l-1} a_{r,k-s} q_{r-s+1,k} \cdot \tag{1.5.53}$$

Let us introduce now instead of the indexes s and r new indexes i and j putting $l - s + 1 = i, r - s + 1 = j$. Then we shall have $s = l - i + 1, r = l - i + j$, and formula (1.5.53) will take the form

$$q_{ik} = b_{l-i+1,k-l+i-1} + \sum_{h=1}^{i-1} a_{l-i+h,k-l+i-1} q_{hk} \cdot \tag{1.5.54}$$

After determining in such a way q_{1k} , $\ldots$, q_{lk} we shall find the coefficient $q_{l+1,k}$ at x_k in equality (1.5.52):

$$q_{l+1,k} = b_{0k} + \sum_{r=0}^{l-1} a_{rk} q_{r+1,k} \cdot \tag{1.5.55}$$

As a result Eq. (1.5.52) with account of equality (1.5.55) will take the form

$$\bar{z}_{l,k+1} = \sum_{r=0}^{l-1} a_{rk} \bar{z}_{r+1,k} + q_{l+1,k} x_n \cdot \tag{1.5.56}$$

Equalities (1.5.47) and (1.5.56) represent a system of the difference equations of the first order determining $\bar{z}_{1k}$, $\ldots$, $\bar{z}_{lk}$. Put now

$$\bar{z}_k = \begin{bmatrix} \bar{z}_{1k} \\ \bar{z}_{2k} \\ \cdots \\ \bar{z}_{lk} \end{bmatrix}, \quad c_k = \begin{bmatrix} 0 & I & 0 & \cdots & 0 \\ 0 & 0 & I & \cdots & 0 \\ \cdots & \cdots & \cdots & \cdots & \cdots \\ 0 & 0 & 0 & \cdots & I \\ a_{0k} & a_{1k} & a_{2k} & \cdots & a_{k-1,k} \end{bmatrix}, \quad q_k = \begin{bmatrix} q_{2k} \\ q_{3k} \\ \cdots \\ q_{k+1,k} \end{bmatrix},$$

$$\tag{1.5.57}$$

where 0 is a quadratic matrix of the same dimension as matrices a_{0k}, a_{1k}, $\ldots$, $a_{k-1,k}$ whose all elements are equal to zero, and I being an unit matrix of the same dimension. Then Eqs. (1.5.47) and (1.5.56) with account of (1.5.57) will be written in the form of Eq. (1.5.46). ◁

Analogously to formulae (1.2.7) and (1.2.8) we find for linear discrete systems

$$z_k = u_{k0}z_0 + \sum_{r=0}^{k} u_{kr} a_{1r} x_r + a_{0r} , \tag{1.5.58}$$

$$y_k = b_k u_{k0} z_0 + \sum_{r=0}^{k} b_k u_{kr} a_{1r} x_r + b_k \sum_{r=0}^{k} u_{kr} a_{0r} + b_{0k}. \tag{1.5.59}$$

The matrix of the weighting coefficients g_{kr} is connected with the matrix of the fundamental solutions u_{kr} by the formula similar (1.2.10)

$$g_{kr} = b_k u_{kr} a_{1r} \mathbf{1}(k - r) . \tag{1.5.60}$$

1.5.6. Discrete Nonlinear Systems Described by Difference Equations

For the description of nonlinear discrete systems the following two main types of nonlinear difference equations are used:

$$z_{k+l} = f_k(z_k , \ldots , z_{k+l-1}, x_k) , \quad y_k = g_k(z_k, x_k) , \quad l = 1, 2, \ldots . \tag{1.5.61}$$

$$z_k = f_k(z_k, x_k) , \quad y_k = g_k(z_k, x_k) , \tag{1.5.62}$$

where z_k and y_k are the vectors of the dimension p and m respectively; f_k is the p-dimension vector function; g_k is the m-dimension function. For nonlinear stationary discrete models we have $f_k = f$, $g_k = g$ independently on k in Eqs. (1.5.61), (1.5.62).

Nonlinear model (1.5.61) is reduced to model (1.5.62). It is sufficient for this purpose to introduce the sequence of the composite vectors $\{\bar{z}_k\}$, $\bar{z}_k = \left[y_k^T \ldots y_{k+l-1}^T \right]^T$ and to add the identical equations $y_{k+1} = y_{k+1}$, $y_{k+l-1} = y_{k+l-1}$ which may be written in the form $\bar{z}_{k+1,1} = \bar{z}_{k2}$, $\bar{z}_{k+1,l-1} = \bar{z}_{kl}$. These equations together with Eqs. (1.5.61) may be written in the form Eqs. (1.5.62) for the sequence of the composite vectors $\{\bar{z}_k\}$.

1.5.7. Stochastic Discrete Linear and Nonlinear Systems

Taking into consideration the results of Subsection 1.5.6 we get the following nonlinear stochastic difference equations respond to Eqs. (1.4.1), (1.4.2)

$$Z_{k+1} = F_k(Z_k, x_k), \quad Y_k = G_k(Z_k), \tag{1.5.63}$$

$$Z_{k+1} = f_k(Z_k, x_k, N_{1k}), \quad Y_k = g_k(Z_k, N_{2k}). \tag{1.5.64}$$

For a stochastic discrete linear system Eqs. (1.5.64) may be presented as

$$\begin{aligned} Z_{k+1} &= a_k Z_k + a_{1k} x_k + a_{0k} + a_{2k} N_{1k}, \\ Y_k &= b_k Z_k + b_{0k} + b_{1k} N_{2k}. \end{aligned} \tag{1.5.65}$$

Hence after introducing composite vector $N_k = \left[N_{1k}^T \, N_{2k}^T \right]^T$, block matrices $a_{2k} = [a_{2k}\, 0]$, $b_{1k} = [0\, b_{1k}]$ and putting $a_{2k} N_{1k} = a_{2k} N_k$, $b_{1k} N_{2k} = b_{2k} N_k$ we get after omitting indexes from Eqs. (1.5.65)

$$\begin{aligned} Z_{k+1} &= a_k Z_k + a_{1k} x_k + a_{0k} + a_{2k} N_k, \\ Y_k &= b_k Z_k + b_{0k} + b_{1k} N_k. \end{aligned} \tag{1.5.66}$$

For the discrete systems with the parametric noises N_{1k} and N_{2k} (Subsection 1.4.3) Eqs. (1.5.63), (1.5.64) may be written in the form:

$$Z_{k+1} = a_k Z_k + a_{1k} x_k + a_{0k} + \left(a_{20k} + \sum_{i=1}^{p} a_{2ik} Z_{ik} \right) N_{1k},$$

$$Y_k = b_k Z_k + b_{0k} + \left(b_{10k} + \sum_{i=1}^{p} b_{1ik} Z_{ik} \right) N_{2k}, \tag{1.5.67}$$

where $Z_k = \left[z_{1k} \ldots Z_{pk} \right]^T$.

For stochastic discrete bilinear systems (Subsection 1.4.4) we have analoguously the following set of equations:

$$\begin{aligned} Z'_{k+1} &= a_k Z'_k + a_{1k} x_k + a_{0k} + a_{2k} N'_{2k}, \\ Y'_k &= b_k Z'_k + b_{0k} + b_{1k} N'_{2k}, \end{aligned} \tag{1.5.68}$$

where

$$Z''_{k+1} = \left(A_{0k} + \sum_{i=1}^{p'} A_{ik} Z'_{ik} \right) Z''_k, \quad Y''_k = \left(B_{0k} + \sum_{i=1}^{p'} B_{ik} Z'_{ik} \right) Z''_k$$

and $Z_k = \left[Z'^T_k \, Z''^T_k \right]^T$, $Y_k = \left[Y'^T_k \, Y''^T_k \right]^T$, $N_{1,2k} = \left[N'^T_{1,2} \, 0 \right]$.

Typical discrete linear stochastic systems for the case $Y_k = Z_k$, $N_{1k} = N_{2k} = N_k$ are given in Table A.5.3.

1.6. Stochastic Composed Systems. Basic Problems of Stochastic System Theory

1.6.1. Definitions

Let us generalize the structural approach given in Section 1.3 for the stochastic composed systems. Following Subsection 1.1.1 we call *a system* any ensemble of interacting subjects of any nature. The set of all external influences on a system, including the influences of surrounding medium, is called *the input* of the system. The set of all essential features of the behaviour of a system is called its *output*. In general case various features of phenomena in a system and the surrounding medium can serve as the inputs and the outputs. For instance, the inputs and the outputs of control systems represent scalar or vector functions, those of finite automata are logical variables, those of service (queueing) systems are streams of events, those of recognition systems are images, sounds of speech, situations etc. It is thus necessary to consider inputs and outputs of systems as elements of any abstract spaces.

A relation between elements of two spaces X and Y called respectively *the space* of *the inputs* and *the space of the outputs* can serve as mathematical model of a system. Given an output $x \in X$, the system elaborates such an output $y \in Y$ that the pair (x, y) belongs to the relation characterizing the system.

To apply the methods of probability theory (Chapters 2–4) it is necessary to consider only systems having input-output relation of a probabilistic nature.

We call *stochastic* a system whose input-output relation is a family of probability distributions in the space of outputs Y depending upon the input $x \in X$. Such a system elaborates its output $y \in Y$ in accordance with probability distribution on Y corresponding to a given $x \in X$.

To develop the stochastic system theory it is necessary to suppose that the spaces of inputs X and outputs Y are measurable. We shall thus suppose that a σ- algebra $\mathcal{A}$ of sets is defined in the space of inputs X, and a σ-algebra $\mathcal{B}$ is defined in the space of outputs Y.

The probability distribution in the space of outputs Y of a stochastic system for a given input $x \in X$ is determined by a conditional probability measure $\mu_y(\mathcal{E}_y \mid x)$ representing the probability of an output belonging to a set $\mathcal{E}_y \in \mathcal{B}$ for a given $x \in X$ (Section 2.3). The measure $\mu_y(\mathcal{E}_y \mid x)$ is a nonnegative σ-additive function of a set $\mathcal{E}_y$ defined on the σ-algebra $\mathcal{B}$ for each $x \in X$ and a function of point x measurable with respect to the σ-algebra $\mathcal{A}$ for each set $\mathcal{E}_y \in \mathcal{B}$.

It is natural to take the conditional probability measure (CPM) $\mu_y(E_y \mid x)$ as the main characteristic of a stochastic system. Considering several stochastic systems we add corresponding subscripts to μ. For instance CPM of the system A we denote by μ_A.

R e m a r k. At practice less detailed probability characteristics are used e.g. conditional moments, quasi moments, one- and multi-dimensional densities and characteristic functions (Chapters 2 and 3).

1.6.2. Main Types of Stochastic Systems Connections

The parallel connection of systems A and B is such a connection in which A and B have the same input x, and the pair (y, x) of their outputs serves as output of the composed system (Fig.1.6.1). The parallel connection of the systems A and B we denote $A \oplus B$. It is evident that the parallel connection is commutative, $B \oplus A = A \oplus B$.

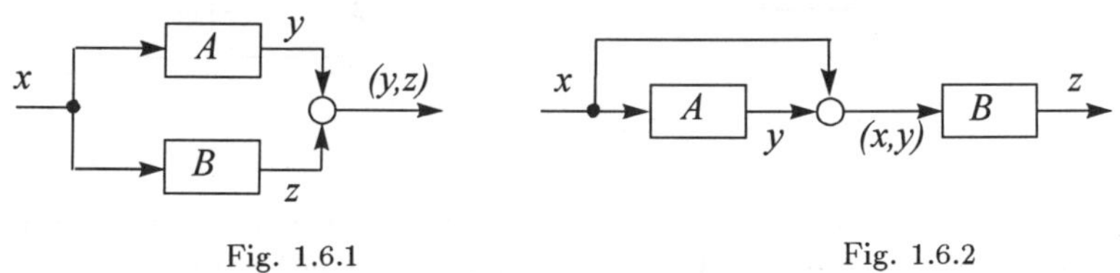

Fig. 1.6.1   Fig. 1.6.2

The case in which only some of the components of the input of the system A coincide with some components of the input of the system B is evidently covered by above definition the common input x of A and B representing the set of all components of the inputs of A and B. In this case the measures μ_A and μ_B are independent on some of the components of x.

The connection of systems A and B in series $A \otimes B$ is such a connection in which the input of B contains the output of A (Fig.1.6.2). It is clear that the connection in series is not commutative in general: $B \otimes A \neq A \otimes B$. The usual connection in series considered in Section 1.3 in which the input of the second system is the output of the first one represents, evidently, the special case of the general connection in series with input-output relation of the second system independent of the input x of the first system. On the other hand the general case can be reduced to this special case the pair (x, y) being considered as the output of the parallel connection of the system A and the identity system I (Fig.1.6.3), $A \otimes B = (A \oplus I) \otimes B$.

The elementary feedback of a system A is such a connection of A with itself in which the input of A contains its output. Fig.1.6.4a shows

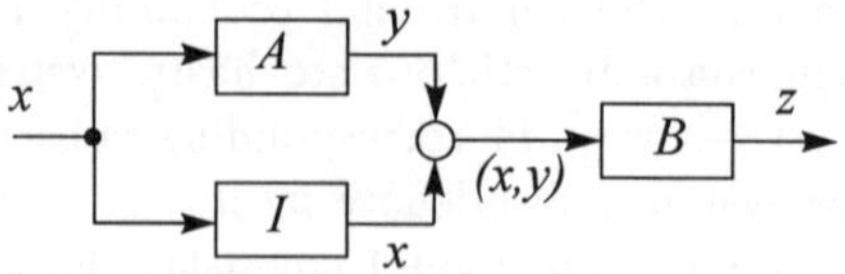

Fig. 1.6.3

a system A whose input represents an element (x, y) of the Cartesian product of the spaces $X \times Y$ and the output is the element z of the space Y. Fig1.6.4b shows the elementary feedback of the system A. The elementary feedback of the system A we denote by $[A]$.

a)   b)

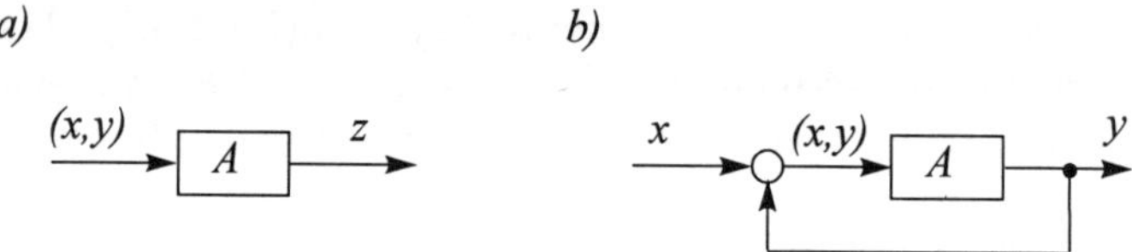

Fig. 1.6.4

The connection of a system with itself in which its input contains only some of the components of its output is, evidently, the special case of the general elementary feedback in which the input-output relation of A is independent of some of the components of y.

The feedback of a system A via a system B is such a connection in which the input of the system A contains its output transformed by the system B (Fig.1.6.5). This connection we denote by $[A]_B$.

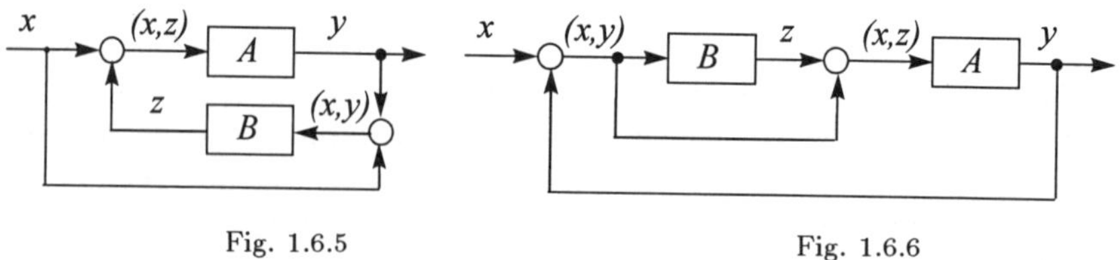

Fig. 1.6.5   Fig. 1.6.6

It is easily understood that all possible connections of systems represent combinations of parallel connections, series connections and feedbacks.

E x a m p l e 1.6.1. It is clear that the feedback of the system A via the system B is equivalent to the elementary feedback of the connection of the systems B and A in series $[A]_B = [B \otimes A]$ (Fig.1.6.6).

E x a m p l e 1.6.2. The system N shown in Fig.1.6.7 is

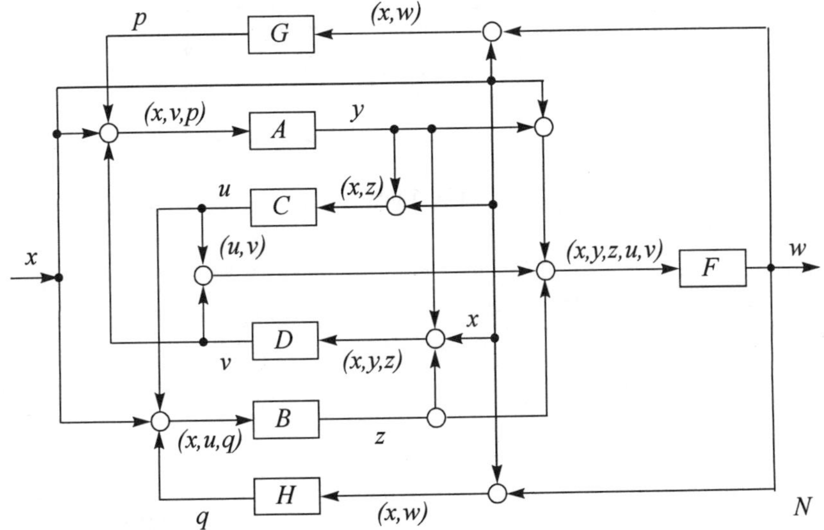

Fig. 1.6.7

$$N = [(G \oplus H) \otimes ([D \otimes ([C \otimes (A \oplus B \oplus I_u)] \oplus) I_v] \otimes F],$$

where I_u and I_v representing the systems with the input (x, u, v, p, q) and the outputs u and v respectively (i.e. identity systems for the signals u and v non-passing other components of the inputs). Figs.1.6.8–1.6.10 show consequent steps of equivalent transformations.

$$L = [L \otimes (A \oplus B \oplus I_u)]$$

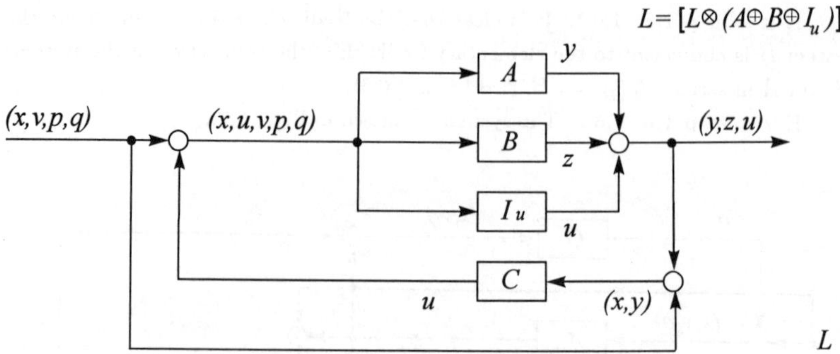

Fig. 1.6.8

$$M = [D \otimes (L \oplus I_v)]$$

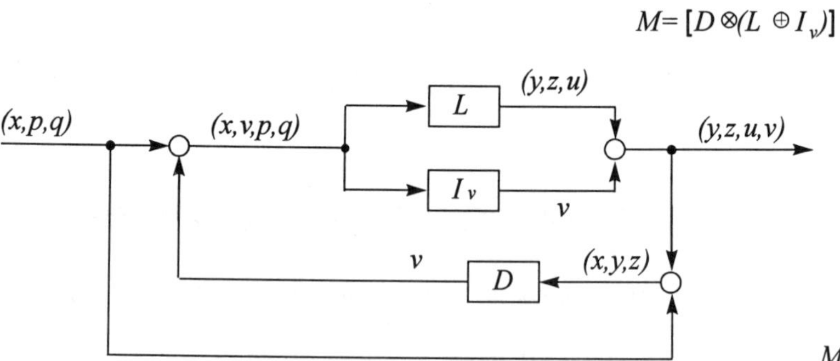

Fig. 1.6.9

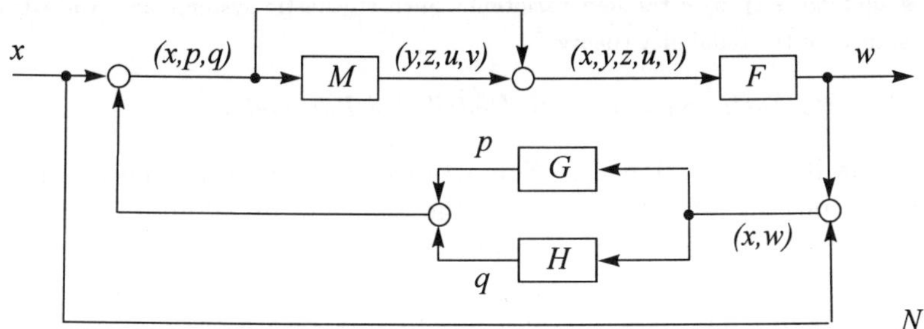

$$N = [(G \oplus H) \otimes (M \otimes F)]$$

Fig. 1.6.10

The main feature of connections of stochastic systems is the possibility of random breaking and arising of some connections. The connections in a stochastic system are thus *stochastic* themselves. The random variations of the connections in the stochastic system cause the corresponding random variations of its CPM. The random variations of the stochastic system CPM may be also caused by any random variations of its inner state.

It is clear that the stream of events causing random variations of the state of the stochastic system A, in particular, of connections among its components, can be considered as the output of some other stochastic system B and at the same time as the additional component of the input of A. Thus the study of the system A with random variations of its CPM can be reduced to the study of the connection in series of two stochastic systems with definite (i.e. nonrandom) CPM. One is the system B generating the random stream of events causing the variations of CPM of the system A, and other is the system A.

CPM for typical connections of stochastic systems with the known CPM of components will be considered in Section 5.8.

E x a m p l e 1.6.3. Let us consider the stochastic system with random structure described by Eq. (1.4.1), $N_1(t)$ being step-wise random function with the same finite set values $\{s_1, \ldots, s_n\}$ for all t. To each fixed value s_k of the random function $N_1(t)$ corresponds the defined structure of the system described by the following equations:

$$\dot{Z} = f(Z, s_k, N_2(t), t), \quad Y = g(Z, s_k, N_2(t), t), \tag{I}$$

where $N_2(t)$ being the vector random function composed by all components of random function $N(t)$ in Eqs. (1.4.1) except $N_1(t)$. While $N_1(t)$ passing from one value to another at random time the system structure changes and the state vector undergo step-wise random variation. Such stochastic systems are studied, for instance, in the reliability theory.

1.6.3. Stochastic Systems with Distributed Parameters

As it was mentioned in Subsections 1.1.1 and 1.6.1 the input and the output signal values may be the elements of infinite-dimensional (abstract) spaces. Let us consider the case when X, Y and system operator A depends on Cartesian point coordinates: $X = X(\xi, \eta, \zeta, t)$, $Y = Y(\xi, \eta, \zeta, t)$, $A = A(\xi, \eta, \zeta, t)$. In this case a system is called *a system with distributed parameters*. Processes in such systems are characterized by the state variables which include the space variables also. Using the vector time-space variable we can generalize the notions and characteristics introduced for usual systems on stochastic systems with the lumped parameters (Sections 1.1–1.5).

Characteristics of some typical 1-dimensional (1D) linear systems with distributed parameters are given in Table A.5.4.

1.6.4. Stochastic Composed Systems

As it is already mentioned in Subsection 1.1.2 the notion of the quality of a real composed system is very important. It permits to describe quantitatively the main properties of a real system both on the stages of its development and testing and under the conditions of its exploitation. Usually the qualitative indices are introduced for the quantitative description of the quality of a real system. The quantitative index depends both on the system characteristics and on an input. The calculation of the quantitative indices gives the opportunity to determine the suitability of a system functioning, to compare different systems, to determine the dependence of these indices on the characteristics of a system and on an input.

In general stochastic systems theory the following general principle of the system analysis quality is usually used. Any deviation of the output Y of a composed system from the required signal Y_r causes the losses which are characterized in each concrete case by some loss function $l = l(Y, Y_r)$. The quality of a composed system in mean for a given realization of the required output Y_r at all possible realizations of real output Y which are correspondent to Y_r is estimated by a conditional expectation of the loss function at a given realization of the required output

$$\rho(A \mid Y_r) = E\left[l(Y, Y_r) \mid Y_r\right] . \qquad (1.6.1)$$

This variable is usually called *conditional risk*. It depends on the operator A determining Y and on Y_r also. Finally, the mean quality of a composed system at all possible realizations Y, Y_r is characterized by an expectation of a conditional risk which is equal to the unconditional expectation of the loss function:

$$R(A) = E\left[\rho(A \mid Y_r)\right] = E\left[l(Y, Y_r)\right] . \tag{1.6.2}$$

The variable $R(A)$ characterizes *the mean losses* or *the mean risk* and is a generalized index of the quality of a composed system. Evidently, that $R(A)$ includes different partial indices of the quality of the system $R_1(A)$, $R_2(A)$, ... The establishment of the relations between $R(A)$ and $R_1(A)$, $R_2(A)$, ... for the most part is an informal problem and is determined by a specific character of solving problem.

Let us state some typical characteristics of the composed stochastic systems.

1^0 The mean square error $Z = Y - Y_r$ of a finite-dimensional system may be obtained from (1.6.2) at the choice of the loss function in the form $l(Y, Y_r) = (Y - Y_r)^T(Y - Y_r)$. In this case we have

$$R(A) = E\left[l(Y, Y_r)\right]. \tag{1.6.3}$$

Here $\Gamma_Z = EZ^T Z_2$, $Z = Y - Y_r$. Hence in one-dimensional case $R(A) = EZ^T Z = D_Z + m_Z$.

2^0 The general mean square error of a finite-dimensional system may be found at the loss function of the form $l(Y, Y_r) = (Y - Y_r)^T Q(Y - Y_r)$ where Q is a nonnegative-definite matrix. In this case Γ_Z should be substituted by $Q\Gamma_Z$ in (1.6.3).

3^0 The index of the accumulation of the loss at the exponential loss function $l(Y, Y_r) = 1 - \exp\left[(Y - Y_r)^T Q_1(Y - Y_r)\right]$ is determined by formula $R(A) = E\left[l(Y, Y_r)\right] = 1 - E\{\exp[-(Y - Y_r)^T Q_1(Y - Y_r)]\}$.

4^0 At the quality index of an one-dimensional system in the form of the probability of the error nonexcursion $Z = Y - Y_r$ from the given interval $[-a, a]$ we have the mean losses $R(A) = E\left[l(Y, Y_r)\right] = P\left\{|Y - Y_r| > a\right\}$.

1.6.5. Problems of Stochastic Systems Theory

The stochastic systems considered in Subsections 1.4.1, 1.5.7, 1.6.3 and their connections may be described by the following operator equation:

$$Y = AX , \tag{1.6.4}$$

$A = A(N, t)$ being differential, difference, integral etc operator, N is the perturbation of operator A. Such a representation is useful in particular for the stochastic systems with the random parameters depending upon the time-space variable and input.

The stochastic systems description in the terms of a conditional probability measure (Subsection 1.6.1) at $X = x$, $N = n$

$$\mu_y = \mu_y(\mathcal{E}_y|n, x), \tag{1.6.5}$$

is more general than desciption (1.6.1). This description is usually used for the stochastic systems with the distributed parameters.

Main problems of the stochastic systems theory based on *a priori* data about probability characteristics are as follows:

• find the output probability characteristics at known probability characteristics of A, N and X (direct problem);

• find the input probability characteristics at known probability characteristics of A, N and Y (inverse problem);

• find the probability characteristics of some components of X, Y, A, N at known probability characteristics of other components of X, Y, A, N (combined problem).

The stability (Subsections 1.1.2 and 1.1.4) is one of the dynamic systems. The notion of the quality allows to pose and solve the problem of the quantitave measuring of the properties of a real dynamic system correctly. The reliability, the accuracy, the cost etc may serve as these properties. The calculations of the quality indexes give the opporunity to determine the ability of a real dynamic system to functioning, to compare different systems, to establish the dependence of the quality indexes on the characteristics of a system. A quantative estimate of the quality is one of the main problems of stochastic systems theory.

Using the notion of the quality for a composed stochastic system in whole and its subsystems it is not difficult to formulate two main problems the stochastic composed systems theory:

• find the probabilistic characteristics of the quality index of an output at the known characteristics of a composed system and of an input (direct problem);

• find the probability characteristics of the quality index of an input at the known characteristicsof the system and of an output (inverse problem).

Along with two main problems there are also combined ones when it is required to determine the characteristics of other components by means of the known probability characteristics of the quality indices of the part of the components of the inputs and the output and the subsystems. The problems connected with the decomposition, the aggregation, the transformation of the structure of a composed system and the theory of the equivalence of different signals and systems are of great practical importance.

These problems will be stated in details and solved with the help of the exact and approximate probability methods given in Chapters 2–7.

Main problems of stochastic systems theory based on *a posterior* data about probability characteristics are:
- the offline estimation problems,
- the online (real time) estimation problems,
- the combined estimation problems.

The offline estimation problems are based on the post-data processing and are usually solved by standard methods of mathematical statistics of random variables, processes and functions. The online estimation problems are based on the online data processing. These problems need specific methods. Only some of the latter problems will be considered in Chapters 5–7.

Problems

1.1. Derive the transfer functions given in Table A.5.1 and vice versa derive the weighting functions for the corresponding transfer functions.

1.2. Derive the transfer functions given in Table A.5.4 and vice versa derive the weighting functions for the corresponding transfer functions.

1.3. For a stable stationary linear system

$$\dot{y} = ay + bx, \quad y = [y_1 y_2]^T, \quad x = [x_1 x_2]^T, \tag{I}$$

where a is a constant square matrix, b is a scalar constant, y is the output, show that the matrix of the transfer functions has the form

$$\Phi(s) = -(a - sI)^{-1}b = \frac{1}{\Delta(s)} \begin{bmatrix} -(a_{22} - s) & a_{12} \\ a_{21} & -(a_{11} - s) \end{bmatrix} b, \tag{II}$$

$$\Delta(s) = s^2 - (a_{11} + a_{22})s + a_{11}a_{22} - a_{12}a_{21}.$$

1.4. Show that for a stable stationary system of Problem 1.3 at $a_{11} = a_{22} = -\varepsilon$, $a_{12} = -a_{21} = \omega$ $(\varepsilon, \omega > 0)$ the matrix of fundamental solutions $u(t, \tau)$

and the matrix of the transfer functions $\Phi(s)$ are determined by the formulae

$$u(t,\tau) = e^{-\varepsilon(t-\tau)}\begin{bmatrix} \cos\omega(t-\tau) & \sin\omega(t-\tau) \\ -\sin\omega(t-\tau) & \cos\omega(t-\tau) \end{bmatrix}, \tag{I}$$

$$\Phi(s) = \frac{b}{(s+\varepsilon)^2+\omega^2}\begin{bmatrix} s+\varepsilon & \omega \\ -\omega & s+\varepsilon \end{bmatrix}. \tag{II}$$

1.5. Let us consider a stationary linear mechanical system with one degree of freedom

$$\begin{bmatrix} \dot{q} \\ \dot{p} \end{bmatrix} = \begin{bmatrix} 0 & 1/A \\ -C & -B/A \end{bmatrix}\begin{bmatrix} q \\ p \end{bmatrix} + \begin{bmatrix} 0 \\ Q \end{bmatrix}, \tag{I}$$

where the generalized coordinate $q = y_1$ and the generalized impulse $p = A\dot{q} = y_2$ are the components of the state vector $y = [y_1 y_2]^T$, $x = [0,Q]^T$ is the input, y is the output of the system, $A(A > 0)$ B, C are constant coefficients determining inertial, dissipative and position forces respectively. Show that the matrix of fundamental solutions $u(t,\tau)$ and the matrix of transfer functions $\Phi(s)$ have the form

$$u(t,\tau) = e^{-\varepsilon(t-\tau)}$$

$$\times \begin{bmatrix} \cos\omega_c(t-\tau) + \frac{\varepsilon}{\omega_c}\sin\omega_c(t-\tau) & \frac{1}{A\omega_c}\sin\omega_c(t-\tau) \\ -\frac{A\omega_0^2}{\omega_c}\sin\omega_c(t-\tau) & \cos\omega_c(t-\tau) - \frac{\varepsilon}{\omega_c}\sin\omega_c(t-\tau) \end{bmatrix}, \tag{II}$$

$$\Phi(s) = \begin{bmatrix} 0 & \dfrac{1}{As^2+Bs+C} \\ 0 & \dfrac{As}{As^2+Bs+C} \end{bmatrix}, \quad (2\varepsilon = B/A, \omega_c^2 = \omega_0^2 - \varepsilon^2, \omega_0^2 = C/A). \tag{III}$$

1.6. Show that for a linear nonstationary system

$$\dot{y}_1 = y_2, \quad \dot{y}_2 = -ct^{-2}y_1 + x, \quad t > 0, \tag{I}$$

the elements of the matrix of fundamental solutions $u(t,\tau)$ at $2\gamma = \sqrt{|1-4c|}$, $c < 1/4$ are determined by the formulae:

$$u_{11}(t,\tau) = \frac{1}{4\gamma}\left(\frac{t}{\tau}\right)^{1/2}\left[(1+2\gamma)\left(\frac{\tau}{t}\right)^{\gamma} - (1-2\gamma)\left(\frac{t}{\tau}\right)^{\gamma}\right],$$

$$u_{12}(t,\tau) = \frac{(\tau t)^{1/2}}{2\gamma}\left[\left(\frac{t}{\tau}\right)^{\gamma} - \left(\frac{\tau}{t}\right)^{\gamma}\right],$$

$$u_{21}(t,\tau) = \frac{1-4\gamma^2}{8\gamma(t\tau)^{1/2}}\left[\left(\frac{\tau}{t}\right)^{\gamma} - \left(\frac{t}{\tau}\right)^{\gamma}\right], \tag{II}$$

$$u_{22}(t,\tau) = \frac{1}{4\gamma}\left(\frac{\tau}{t}\right)^{1/2}\left[(1+2\gamma)\left(\frac{t}{\tau}\right)^{\gamma} - (1-2\gamma)\left(\frac{\tau}{t}\right)^{\gamma}\right].$$

Find $u(t, \tau)$ at $c > 1/4$.

    1.7. Prove that for a nonstationary linear system

$$\dot{y}_1 = y_2, \quad \dot{y}_2 = -t^{-2}y_1 - t^{-1}y_2 + x, \quad t > 0, \tag{I}$$

the matrix of fundamental solutions $u(t, \tau)$ is equal to

$$u(t, \tau) = \begin{bmatrix} \cos \ln \frac{t}{\tau} & \tau \sin \ln \frac{t}{\tau} \\ -\frac{1}{t} \sin \ln \frac{t}{\tau} & \cos \ln \frac{t}{\tau} \end{bmatrix}. \tag{II}$$

    1.8. Prove that the weighting function $g(t, \tau)$ of a nonstationary linear system

$$\ddot{y} + t^{-1}\dot{y} + (1 - n^2 t^{-2})y = x, \quad y = z, \tag{I}$$

is determined by:

$$g(t, \tau) = \frac{\pi}{2} \left[J_n(\tau)N_n(t) - N_n(t)J_n(t) \right] \tau, \tag{II}$$

where $J_n(\tau)$ and $N_n(t)$ are the Bessel functions of the first and the second kind respectively.

    1.9. For a stationary linear system

$$\dot{z} = az + bx, \quad y = z, \quad z = \left[z_1 z_2 z_3 \right]^T, \quad x = \left[x_1 x_2 x_3 \right]^T, \tag{I}$$

where a is a constant square matrix, b a scalar constant, y the output, show that the elements of the matrix of transfer fucntions $\Phi(s) = -(a - sI))^{-1}b$ at $\Delta(s) = |a - sI|$ are determined by the formulae

$$\begin{aligned}
\Phi_{11}(s) &= -b\left[s^2 - s(a_{22} + a_{33}) + a_{22}a_{33} - a_{23}a_{32} \right]/\Delta(s), \\
\Phi_{12}(s) &= b(-sa_{12} + a_{12}a_{33} - a_{32}a_{13})/\Delta(s), \\
\Phi_{13}(s) &= -b(sa_{13} + a_{12}a_{23} - a_{13}a_{22})/\Delta(s), \\
\Phi_{21}(s) &= b(-sa_{21} + a_{21}a_{33} - a_{31}a_{23})/\Delta(s), \\
\Phi_{22}(s) &= -b\left[s^2 - s(a_{11} + a_{33}) + a_{11}a_{33} - a_{13}a_{31} \right]/\Delta(s), \\
\Phi_{23}(s) &= b(-sa_{23} + a_{11}a_{23} - a_{21}a_{13})/\Delta(s), \\
\Phi_{31}(s) &= -b(sa_{31} + a_{21}a_{32} - a_{31}a_{22})/\Delta(s), \\
\Phi_{32}(s) &= b(-sa_{32} + a_{11}a_{32} - a_{12}a_{31})/\Delta(s), \\
\Phi_{33}(s) &= -b\left[s^2 - s(a_{11} + a_{22}) + a_{11}a_{22} - a_{12}a_{21} \right]/\Delta(s).
\end{aligned} \tag{II}$$

1.10. Show that for the stationary system of Problem 1.9 at $a_{11} = a_{31} = a_{13} = a_{32} = 0$, $a_{12} = a_{23} = 1$, $a_{21} = -\omega_0^2$, $a_{22} = -2\varepsilon$, $a_{33} = -\alpha$ the elements of the matrix of transfer functions are determined by

$$\Phi_{11}(s) = -b\left[s^2 + (2\varepsilon + \alpha)s + 2\alpha\varepsilon\right]/\Delta(s),$$
$$\Phi_{12}(s) = -b(s + \alpha)/\Delta(s), \quad \Phi_{13}(s) = -b/\Delta(s),$$
$$\Phi_{21}(s) = b\omega_0^2(s + \alpha)/\Delta(s), \quad \Phi_{22}(s) = -bs(s + \alpha)/\Delta(s),$$
$$\Phi_{23}(s) = -bs/\Delta(s), \quad \Phi_{31} = \Phi_{32} = 0,$$
$$\Phi_{33}(s) = -b(s^2 + 2\varepsilon s + \omega_0^2)/\Delta(s), \quad \Delta(s) = -(s + \alpha)\left[\omega_0^2 + (2\varepsilon + s)s\right].$$
$$\tag{I}$$

Check that the elements of the matrix of fundamental solutions $u(t, \tau)$ are given by

$$u_{13}(t, \tau) = (-1/\omega_c^2)\left[u_{11}(t, \tau) + \alpha/\omega_0 u_{21}(t, \tau) - u_{33}(t, \tau)\right],$$
$$u_{31} = u_{32} = 0, \quad u_{33}(t, \tau) = \exp\{-\alpha(t - \tau)\},$$
$$u_{23}(t, \tau) = (-1/\omega_c^2)\left\{u_{21}(t, \tau)(1 - 2\varepsilon\alpha/\omega_0^2) - \alpha\left[u_{11}(t, \tau) - u_{33}(t, \tau)\right]\right\},$$
$$\tag{II}$$

the functions u_{11}, u_{12}, u_{21}, u_{22} being determined in Problem 1.5.

1.11. Show that for the stationary linear mechanical system with n degrees of freedom the differential equations of motion have the following form:

(a) In Lagrangian variables

$$A\ddot{q} + (B + B')\dot{q} + (C + C')q = Q, \tag{I}$$

where $q = [q_1 \ldots q_n]^T$ is the vector of generalized coordinates, and $Q = [Q_1 \ldots Q_n]^T$ is the vector of generalized forces. In this case the vector $y = q$ represents the output, and the vector $x = Q$ represents the input, A is the symmetric matrix of the inertia coefficients, B is the symmetric matrix of the dissipative or acceleration forces, B' is the antisymmetric matrix of gyro forces, C is the symmetric matrix of the conservative position forces, and C' is the antisymmetric matrix of nonconservative position forces.

(b) In canonical variables

$$\begin{bmatrix} \dot{q} \\ \dot{p} \end{bmatrix} = \begin{bmatrix} 0 & A^{-1} \\ -(C + C') & -(B + B')A^{-1} \end{bmatrix} \begin{bmatrix} q \\ p \end{bmatrix} + \begin{bmatrix} 0 \\ Q \end{bmatrix}, \tag{II}$$

where q and $p = A\dot{q}$ are the vector of generalized coordinates and the vector of generalized impulses respectively. In this case $[q^T p^T]^T$ represents the state vector of the system, the input and output being the same as in the case of the Lagrangian variables.

1.12. Show that for the stable stationary system of Problem 1.11 at $n = 2$

$$\begin{bmatrix} A_{11} & A_{12} \\ A_{12} & A_{22} \end{bmatrix}\begin{bmatrix} \ddot{q}_1 \\ \ddot{q}_2 \end{bmatrix} + \left(\begin{bmatrix} B_{11} & B_{12} \\ B_{12} & B_{22} \end{bmatrix} + \begin{bmatrix} 0 & B'_{12} \\ -B'_{12} & 0 \end{bmatrix}\right)\begin{bmatrix} \dot{q}_1 \\ \dot{q}_2 \end{bmatrix}$$

$$+ \left(\begin{bmatrix} C_{11} & C_{12} \\ C_{12} & C_{22} \end{bmatrix} + \begin{bmatrix} 0 & C'_{12} \\ -C'_{12} & 0 \end{bmatrix}\right)\begin{bmatrix} q_1 \\ q_2 \end{bmatrix} = \begin{bmatrix} Q_1 \\ Q_2 \end{bmatrix} \qquad \text{(I)}$$

the elements of the matrix of transfer functions $\Phi(s)$ are determined by the formulae

$$\Phi_{11}(s) = \left\{ A^-_{11}(|C| + C'^2_{12}) - s[C_{11}(C_{22} + s) \right.$$
$$\left. - (B_{12} + B'_{12})(C_{12} - C'_{12})] \right\}/\Delta(s),$$

$$\Phi_{12}(s) = -\left\{ A^-_{12}(|C| + C'^2_{12}) - s[C_{11}(B_{12} - B'_{12}) \right.$$
$$\left. - (B_{11} + s)(C_{12} - C'_{12})] \right\}/\Delta(s),$$

$$\Phi_{21}(s) = -\left\{ -A^-_{12}(|C| + C'^2_{12}) + s[(C_{12} + C'_{12})(B_{22} + s) \right.$$
$$\left. - C_{22}(b_{12} + B'_{12})] \right\}/\Delta(s),$$

$$\Phi_{22}(s) = \left\{ A^-_{22}(|C| + C'^2_{12}) + s[(C_{12} - C'_{12})(B_{12} - B'_{12}) \right.$$
$$\left. - C_{22}(B_{11} + s)] \right\}/\Delta(s),$$

$$\qquad \text{(II)}$$

where $\Delta(s) = s^4 + r_1 s^3 + r_2 s^2 + r_3 s + r_4$,

$$r_1 = B_{11}A^-_{22} + B_{22}A^-_{11} - 2B_{12}A^-_{12},$$

$$r_2 = \left[B_{11}A^-_{22} - (B_{12} + B'_{12})A^-_{12} \right]\left[B_{22}A^-_{11} - (B_{12} - B'_{12})A^-_{12} \right]$$
$$+ (C_{11}A^-_{22} + C_{22}A^-_{11} - 2C_{12}A^-_{12}),$$

$$r_3 = \left[B_{11}A^-_{22} - (B_{12} + B'_{12})A^-_{12} \right]\left[(B_{12} - B'_{12})A^-_{22} - B_{22}A^-_{12} \right]$$
$$+ \left[B_{22}A^-_{11} - (B_{12} - B'_{12})A^-_{12} \right]\left[C_{11}A^-_{22} - (C_{12} + C'_{12})A^-_{12} \right]$$
$$+ \left[(B_{12} - B'_{12})A^-_{22} - B_{22}A^-_{12} \right]\left[C_{11}A^-_{12} - (C_{12} + C'_{12})A^-_{11} \right]$$
$$+ \left\{ \left[B_{11}A^-_{22} + (B_{12} + B'_{12})A^-_{12} \right] + \left[(B_{12} + B'_{12})A^-_{11} - B_{11}A^-_{22} \right] \right\}$$
$$\times \left[C_{22}A^-_{22} - (C_{12} - C'_{12})A^-_{22} \right], \quad r_4 = (|C| + C'^2_{12})/|A|,$$

$$\qquad \text{(III)}$$

$$A^-_{ij} = A_{ij}/|A| \ (i,j = 1,2), \quad |A| = A_{11}A_{22} - A^2_{12}, \quad |C| = C_{11}C_{22} - C^2_{12}.$$

1.13. Show that if the matrices $A^{-1}B$ and $A^{-1}C$ in Problem 1.11 are commute and $B' = 0$, $C' = 0$, then there exists always a linear transformation of the output $y = L\zeta$ satisfying the conditions

$$L^{-1}AL = I, \quad L^T BL = \text{diag}\,(2\varepsilon_1, \ldots, 2\varepsilon_n),$$
$$L^T CL = \text{diag}\,(\omega^2_1, \ldots, \omega^2_n), \qquad \text{(I)}$$

reducing the set of equations of motion to n independent linear differential second-order equations

$$\ddot{\zeta}_h + 2\varepsilon_h\dot{\zeta}_h + \omega_h^2\zeta_h = u_h\,, \quad (h = 1,\ldots,n)\,, \tag{II}$$

where $u_h = u_h(t)$ is the hth element of the column-matrix $L^T Q$. Write the corresponding equations at $B = 2\varepsilon A$, $B = \eta C$.

1.14. Consider a mechanical system with n degrees of freedom described by Lagrange equations

$$\frac{d}{dt}\left(\frac{\partial L}{\partial \dot{q}}\right) - \frac{\partial L}{\partial q} = Q\,, \tag{I}$$

where $q = [\,q_1 \ldots q_n\,]^T$ is the vector of generalized coordinates, $L = T - \Pi$ is the Lagrange function, $T = T(q,\dot{q},t)$ is the kinetic energy, $\Pi = \Pi(q)$ is the potential energy; $Q = [\,Q_1 \ldots Q_n\,]^T$, $Q = Q(q,\dot{q},t)$ is the vector of generalized nonconservative forces. Show that Eq. (I) may be transformed into the set of Hamilton equations in canonical variables q and $p = \partial T/\partial \dot{q}$

$$\dot{q} = \frac{\partial H}{\partial p}\,, \quad \dot{p} = -\frac{\partial H}{\partial q} + Q\,. \tag{II}$$

Here $H = H(q,p,t)$ being the Hamilton function, $H = p^T\dot{q} - T$.

1.15. The vertical oscillations of the body of a car while moving on the rough road are described by the equations

$$M\ddot{y} + 2\psi(\dot{y} - \dot{q}) + 2\varphi(y - q) = 0\,,$$

where M is the car mass, $\psi(\dot{y} - \dot{q})$ and $\varphi(y - q)$ are the nonlinear functions determining damping and restoring forces respectively, $q = q(\sigma)$ is the function characterizing the microprofile of the road in the vertical plane, $\sigma = vt$, v being the car velocity. We assume that both pairs of wheels of the car undergo the same vertical displacement at every instant due to the roughness of the road (i.e. the road profile in the vertical plane has a very long correlation interval). Find the transfer function of the car at equalities $\psi(\dot{y} - \dot{q}) = b(\dot{y} - \dot{q})$ and $\varphi(y - q) = c(y - q)$, b and c being constant coefficients, assuming q as an input and y as an output.

1.16. The equations of motion of the body with one degree of freedom equipped with the nonlinear dynamic damper have the form

$$\mu\ddot{x} + b\dot{x} + cx + \psi_d(\dot{x} - \dot{x}_d) + \varphi_d(x - x_d) = F(t)\,,$$

$$\mu_d\ddot{x}_d - \psi_d(\dot{x} - \dot{x}_d) - \varphi_d(x - x_d) = 0\,,$$

where x and x_d are the displacements of the body and of the damper, μ and μ_d their masses, b and c the coefficients of viscous friction and of restoring forces respectively, $\psi_d(\dot{x}-\dot{x}_d)$ and $\varphi_d(x-x_d)$ the functions characterizing the forces of viscous friction and the restoring forces, and $F(t)$ the disturbing force. In the special case where $\psi_d(\dot{x}-\dot{x}_d=b_d(\dot{x}-\dot{x}_d)$ and $\varphi_d(x-x_d)=c_d(x-x_d)$, with b_d, c_d being constant coefficients find the transfer function of the system considering the force $F(t)$ as an input, and the displacements x and x_d of the body and damper as the components of the output.

1.17. Show that the nonlinear equation

$$y^{(n)} + \sum_{k=n-m}^{n-1} a_k(y, y', \ldots, y^{(n-m-1)}, t)y^{(k)} + \varphi(y, y', \ldots, y^{(n-m-1)}, t)$$

$$= \sum_{n=0}^{m} b_h(y, y', \ldots, y^{(n-m-1)}, t)x^{(k)}, \quad (m < n) \tag{I}$$

by the change of variables

$$z_1 = y, \quad z_{k+1} = z_k', \quad (k = 1, \ldots, n-m-1),$$

$$z_{k+1} = z_k' - q_k(z_1, \ldots, z_{n-m}, t)x, \quad (k = n-m, \ldots, n-1) \tag{II}$$

is reduced to the following set of equations:

$$z_k' = z_{k+1}, \quad (k = 1, \ldots, n-m-1),$$
$$z_k' = z_{k+1} + q_k(z_1, \ldots, z_{n-m}, t)x, \quad (k = n-m, \ldots, n-1),$$
$$z_n' = -\sum_{l=n-m-1}^{n} a_{l-1}(z_1, \ldots, z_{n-m}, t)z_l - \varphi(z_1, \ldots, z_{n-m}, t) \tag{III}$$
$$+ q_n(z_1, \ldots, z_{n-m}, t)x.$$

Here functions $q_{n-m}, \ldots, q_n$ at $a_n = I$ are defined by (1.2.37).

1.18. Prove the following properties for the adjoint linear systems:

(i) the weighting functions of two mutually adjoint systems satisfy the equality $g^*(t, \tau) = g(\tau, t)$;

(ii) a system adjoint with a nonanticipative one is anticipative;

(iii) a system adjoint with two connected in series linear systems is a reverse series connection of two corresponding adjoint systems;

(iv) a system adjoint with two connected in parallel linear systems is a parallel connection of corresponding adjoint systems;

(v) systems adjoint with two mutually inverse (reciprocal) systems are mutually inverse (reciprocal), i.e. $\left(g^-\right)^* = \left(g^*\right)^-$;

(vi) a system adjoint with a linear feed back system may received from original one by the change of all its elements by corresponding adjoint systems. Notice that for mapping adjoint systems with backward direction of signals we need to change the branching point by summator and vice versa.

1.19. Prove that the adojoint system with the linear second order system

$$a_2\ddot{y} + a_1\dot{y} + a_0 y = x$$

is described by equations

$$\frac{d^2}{dt^2}(a_2\eta) - \frac{d}{dt}(a_1\eta) + a_0\eta = \xi.$$

CHAPTER 2
RANDOM VARIABLES, PROCESSES AND FUNCTIONS

This Chapter is devoted to the advenced theory of probability distributions of random variables, processes and functions. In Section 2.1 probability spaces, measures and their properties are determined. Definitions of vector random continuous, discrete and mixed variables, probability spaces and measures are given. Special attention is paid to the equivalence of random variables. Various types of the convergence, properties of distribution functions and densities are considered. In Section 2.3 theory of conditional distributions of vector random variables is given. In Sections 2.4 and 2.5 probabilities in finite and infinite products of spaces are studied. Definitions and main properties of regular conditional probabilities are considered in Section 2.6. Section 2.7 is dedicated to various definitions and main properties of random processes and functions. Theory of multi-dimensional distributions is studied in detail. Markov normal and conditionally normal processes are considered. Problems of events streams and "crossing" problems are outlined. Theory of the separable random functions is presented in Section 2.8. Analytical properties of random processes are studied in Sections 2.9 and 2.10. Some sufficient conditions of realizations continuity and differentiability are established.

2.1. Probability Spaces

2.1.1. Space of Elementary Events

As it is known from elementary probability theory some trial whose results may be different in consequence of the action of the randomnesses at a given identical complex of conditions is always considered. Any qualitative result of a trial about which we may say that it appeared or not is called *an event*. Some set of possible events is connected with each trial. It is supposed that each of these events has quite definite probability. The probabilities of all possible events at a given trial form *a probability distribution* connected with a given trial.

There exists the analogy with the notions of an event and a set. For understanding this analogy it is sufficient to notice that all main relations between the sets and the actions on them are identical to the following relations between the events and the actions on them.

- If the event A obligatory appears at the occurrence of the event B then we say that the event B composes a part or *a subevent* of the event A and we write $A \supset B$ or $B \subset A$.

- If $A \subset B$ and $B \subset A$, i.e. the event A appears if and only if the event B appears then we say that the events A and B are *equivalent* and write $A = B$.

- *The union* or *the sum* $\cup A_\alpha$ of a countable set of the events $\{A_\alpha\}$ is called the event consisting in the occurrence of at least one of these events.

- *The intersection* $\cap A_\alpha$ of a countable set of the events $\{A_\alpha\}$ is the joint occurrence of all these events.

- The event $\overline{A}$, *complementary* to the event A is called the non-occurrence of the event A.

After complementing a set of the events by the impossible event $\emptyset$ and by the certain event Ω we obtain the opportunity to transfer all the actions upon the sets and the relations between them on the events.

In the case when the set of all possible events at a given trial is countable each event practically represents a set of some simplest events. Among the sets of all possible events at a given trial we may always single out the so called "indivisible" event which does not contain any subevents besides the impossible event and itself and any event may be obtained by union operation. These simplest "indivisible" events are called *elementary* events. Introducing the notion of an elementary event we receive the opportunity to consider any event as some set of elementary events and to establish in such a way the complete identity of the events and the sets. Here the set of all elementary events represents the certain event Ω and an empty set $\emptyset$ – the impossible event. The set Ω of all elementary events is called *the space of elementary events*. Every elementary event ω represents a point of the space of elementary events Ω.

E x a m p l e 2.1.1. If only one event A is connected with a given trial then the space of elementary events Ω represents a two-point set: $\{\omega_1, \omega_2\}$, $\omega_1 = A$, $\omega_2 = \overline{A}$.

E x a m p l e 2.1.2. If the considered trial consists of n trials and we are interested only in such ones where the occurrence or nonoccurrence of some event A (Bernoulli scheme) takes place then all possible sequences $A_1, \dots, A_n$ where $A_1 = \cdots = A_n = A$ or $\overline{A}$ are elementary events. The space of elementary events Ω contains 2^n points.

E x a m p l e 2.1.3. In artillery the striking of a shell at a given area or a point x is an elementary event ω. Here the space of elementary events Ω represents an Euclidean space. In the problem of studying a scalar random variable or a finite-dimensional random vector the corresponding finite-dimensional space serves as the space of elementary events Ω.

E x a m p l e 2.1.4. If the considered trial represents unbounded sequence of the trials in Bernoulli scheme then the set of all infinite sequences $\{A_k\}$, $A_k = A$ or $\overline{A}$ serves as the space of elementary events Ω.

E x a m p l e 2.1.5. In problems where the results of a trial are described by an infinite sequence of the numerical random variables $\{X_n\}$ all possible numerical sequences $\{x_n\}$, $\omega = \{x_n\}$, serve as elementary events, and the space of the elementary Ω represents an infinite-dimensional space of all possible numerical sequences.

E x a m p l e 2.1.6. In problems of control theory, in particular, in queuing we have to deal with a sequence of the events. The sequence of the homogeneous events is completely characterized by the random moments of the events occurrence. Therefore we may consider that a stream of the events represents an unbounded nondecreasing sequence of real random variables $\{T_n\}$. All possible unbounded non-decreasing sequences of the real numbers $\{t_n\}$, $\omega = \{t_n\}$ are elementary events, and the space of elementary events Ω represents an infinite- dimensional space of all these sequences.

E x a m p l e 2.1.7. While studying the random function $X(t)$ we may assume as elementary events its possible realizations $\omega = x(t)$. The corresponding functional space is the space of elementary events Ω.

In the case of uncountable set Ω all its singletons $\{\omega\}$, $\omega \in \Omega$ may serve as elementary events but it is improper to state that any event may be obtained by the operation of uniting elementary events. For instance, an uncountable subset in Ω cannot be obtained by countable uniting its singleton subsets. Therefore while constructing probability models in the case of uncountable Ω not all possible subsets Ω are given but some systems of the subsets which are closed relative to countable unions and countable intersections of a system elements. In such a way we come to the notion of σ-algebras of the events.

2.1.2. Probability Spaces

As it was mentioned in Subsection 2.1.1 the set of the events $\mathcal{S}$ for which it is necessary to determine the probabilities must represent an algebra in the case of finite number of elementary events and a σ-algebra in the case of an infinite set of elementary events. The probability of the event $P(A)$, $A \in \mathcal{S}$ here represents a nonnegative additive (in a finite case) or σ-additive function of a set in the space of elementary events.

Thus in order to determine the probability distribution it is necessary to assign the space of elementary events Ω, σ-algebra of the space $\mathcal{S}$ (an algebra in a finite case) in this space and to determine on this σ-algebra (an algebra) the probability $P(A)$.

From the general properties of σ-additive functions and numerical measures given on σ-algebra known from functional analysis follow the properties of the probability:

(i) $0 \leq P(A) \leq 1$ for all $A \in \mathcal{S}$;

(ii) the probability of the impossible event $\varnothing$ is equal to zero, $P(\varnothing) = 0$;

(iii) if A, $B \in \mathcal{S}$ and $A \subset B$ then $P(A) \leq P(B)$;

(iv) the probability is continuous, i.e. for any countable monotonous sequence of the sets $\{A_n\} \subset \mathcal{S}$, $A = \lim A_n$,

$$P(A) = \lim P(A_n);$$

(v) the probability is semiadditive, i.e. for any countable sequence of the sets $\{A_n\} \subset \mathcal{S}$

$$P(\cup A_n) \leq \sum P(A_n);$$

(vi) the probability of the certain event Ω is equal to 1, $P(\Omega) = 1$.

It is evident that a set of elementary events which have the probability different from zero is finite or countable.

The space of elementary events Ω with the distinguished σ-algebra of the sets $\mathcal{S}$ in it and the given probability $P(A)$ on this σ-algebra is called *the probability space* and is denoted by $(\Omega, \mathcal{S}, P)$. Here it is postulated that the probability $P(A)$ represents a normed nonnegative measure, i.e. a nonnegative σ-additive function of a set satisfying the condition $P(\Omega) = 1$.

E x a m p l e 2.1.8. In Example 2.1.1 $\Omega = \{\omega_1, \omega_2\}$, $\mathcal{S} = \{\varnothing, A, \overline{A}, \Omega\}$ and for setting P it is sufficient to put $P(A) = p$, $0 \leq p \leq 1$, $q = 1 - p$.

E x a m p l e 2.1.9. In Example 2.1.2 the aggregate of all subsets 2^n of the point space of elementary events Ω is the algebra of the events $\mathcal{S}$, the probability $P(S)$ is equal $\sum_S p^m q^{n-m}$, where p is the probability of the occurrence of the event A in one trial, $q = 1 - p$, m is the number of the events A_k equal to A in the sequence $\omega = (A_1, \ldots, A_n)$ corresponding to a given point ω of the space Ω. The summation is spread over all elementary events ω which enter into S.

E x a m p l e 2.1.10. In Example 2.1.3 the algebra $\mathcal{S}$ represents the σ-algebra of Borel sets of the correspondent finite-dimensional space, and the probability $P(A)$ is expressed by formula

$$P(A) = \int\limits_A f(x)dx \tag{I}$$

for such numerical function $f(x) \geq 0$ that $P(\Omega) = 1$.

If a nonnegative finite or σ-finite measure ν is determined in the measurable space $(\Omega, \mathcal{S})$ then the probability $P = P(A)$ may be presented in the form of the sum of the ν-continuous and ν-singular measures, and the ν-continuous part represents an integral over the measure ν with respect to some function $f(\omega)$

$$P(A) = \int\limits_A f(\omega)\nu(d\omega) + P_s(A). \tag{2.1.1}$$

When the probability $P(A)$ is ν-continuous, $P_s(A) \equiv 0$ and

$$P(A) = \int\limits_A f(\omega)\nu(d\omega). \tag{2.1.2}$$

As it is known from the functional analysis in this case the function $f(\omega)$ represents Radon-Nikodym derivative of the probability P with respect to the measure ν and is called *a probability density with respect to measure ν* in the space $(\Omega, \mathcal{S}, P)$.

Any subset N of the set $N_0 \in \mathcal{S}$ for which $P(N_0) = 0$ is called *a zero set*. If zero set N belongs to the σ-algebra $\mathcal{S}$ of the probability space $(\Omega, \mathcal{S}, P)$ then, evidently $P(N) = 0$. But zero set does not need belonging to $\mathcal{S}$ (if $N_0 \subset \mathcal{S}$ then the subsets of the set N_0 do not necessarily belong to $\mathcal{S}$). On such zero sets the probability $P(N)$ is not determined. The probability space $(\Omega, \mathcal{S}, P)$ is called *a complete* one if the σ-algebra $\mathcal{S}$ contains all zero sets of this space. If the probability space $(\Omega, \mathcal{S}, P)$ is not complete then it may be always supplemented after adding in $\mathcal{S}$ all zero sets and all the unions of zero sets with the sets from $\mathcal{S}$ and determining the probability P on the supplemented sets.

2.2. Random Variables

2.2.1. Definitions

In elementary probability theory *a random variable* is a variable which assumes as a result of a trial one of possible values and it is impossible (before a trial) to know what value it will assume namely. Here it is supposed that in the space of possible values of each random variable there exists a probability distribution. In order to come to an exact definition of a random variable notion we shall refer to the examples from elementary probability theory and random functions theory.

E x a m p l e 2.2.1. The number of the occurrences of the event A (in one trial) in Examples 2.1.1 and 2.1.8 represents a random variable which assumes the value 1 when A appears and the value 0 when A does not appear. It is easy to notice that this random variable represents a function of a point in the probability space $(\Omega, \mathcal{S}, P)$ of Example 2.1.8 equal to 1 at $\omega = \omega_1$ and 0 at $\omega = \omega_2$.

E x a m p l e 2.2.2. The number of the occurrences of the event A at n triales in Bernoulli scheme represents a random variable with possible values $0, 1, \ldots, n$. This random variable is a point function in the probability space $(\Omega, \mathcal{S}, P)$ of Example 2.1.9 equal to m at all points ω to which correspond the sequences $A_1, \ldots, A_n$ containing m letters A and $n - m$ letters $\overline{A}$ $(m = 0, 1, \ldots, n)$.

E x a m p l e 2.2.3. The finite-dimensional random vector X may be considered as the point function $x(\omega) = \omega$ in the probability space $(\Omega, \mathcal{S}, P)$ of Example 2.1.10. Each component X_k of the random vector X represents also the point function $x_k(\omega)$ in the probability space $(\Omega, \mathcal{S}, P)$.

E x a m p l e 2.2.4. Infinite-dimensional random vector X (the sequence $\{X_k\}$ of the scalar random variables) may be also considered as the point function $x(\omega) = \omega$ in the space Ω of all numerical sequences $\{x_k\}$. It is evident that any component of an infinite-dimensional random vector is the function of a point in the space Ω (x_m at any fixed m is the function of the sequence $\{x_k\}$). The same is true, in particular, for the sequence of the events in Example 2.1.6.

The considered examples turn us to the following definition of a random variable. *A random variable* with values in a measurable space $(X, \mathcal{A})$ is called $(\mathcal{S}, \mathcal{A})$-measurable function of a point in the probability space $(\Omega, \mathcal{S}, P)$ mapping Ω into X and determined everywhere in Ω besides may be the set of zero probability P. The value of a random variable at any given point ω of the space Ω (i.e. the value which it assumes when as a result of a trial the elementary event ω appears) is called *a realization* of this random variable.

It is evident that any random variable in $(X, \mathcal{A})$ is at the same time a random variable in $(X, \mathcal{A}')$ where $\mathcal{A}'$ is any σ-algebra in X containing $\mathcal{A}$, $\mathcal{A}' \subset \mathcal{A}$.

We shall often denote a random variable by the same letter as the space of its values. While considering the set of the random variables with the values in one and the same space the denotations of these variables will be supplemented with the indexes. So, for instance, X, X_α represent the random variables with the values in the space X. If X, X_α are the random variables in $(X, \mathcal{A})$ then according to the definition $X = x(\omega)$, $X_\alpha = x_\alpha(\omega)$ where $x(\omega)$, $x_\alpha(\omega)$ are $(\mathcal{S}, \mathcal{A})$-measurable functions mapping Ω into X.

2.2.2. Probability Measure

By virtue of $(\mathcal{A}, \mathcal{S})$-measurability of a random variable $X = x(\omega)$ in $(X, \mathcal{A})$ the probability of the fact that it will take the value from the set A is determined by the following formula:

$$P(X \in A) = P(x^{-1}(A)), \quad A \in \mathcal{A},$$

where as usual $x^{-1}(A)$ is an inverse set A in the space Ω.

Formula

$$\mu_x(A) = P(X \in A) = P(x^{-1}(A)), \quad A \in \mathcal{A}, \qquad (2.2.1)$$

determines the probability μ_x induced by the random variable $X = x(\omega)$ in the space of its values. The triple $(X, \mathcal{A}, \mu_x)$ represent the probability space induced into $(X, \mathcal{A})$ by the random variable $X = x(\omega)$.

The space X of the values of the random variable $X = x(\omega)$ is called its *phase space* or *the space of the realizations*. The probability μ_x in the phase space is called *the probability measure* or the distrubution of the random variable X. The probability space $(X, \mathcal{A}, \mu_x)$ is called *the phase probability space* of the random variable $X = x(\omega)$. It is evident that any probability space may be considered as a phase space of some random variable. If the random variable $X = x(\omega)$ is considered in its phase probability set then the argument ω in its denotion naturally is not indicated. In such cases we shall denote it simply by X and its realizations by x.

If the nonnegative finite or σ-finite measure ν is determined in the space $(X, \mathcal{A})$ then the probability measure μ_x of the random variable X may be presented by formula

$$\mu_x(A) = \int_A f(x)\nu(dx) + \mu_x^s(A), \quad A \in \mathcal{A}, \qquad (2.2.2)$$

where $\mu_x^s(A)$ is the ν-singular measure. In the special case when the measure μ_x is ν-continuous, $\mu_x^s(A) = 0$, we have

$$\mu_x(A) = \int_A f(x)\nu(dx), \quad A \in \mathcal{A}. \tag{2.2.3}$$

Here $f(x)$ is Radon–Nikodym derivative of the probability measure μ_x with respect to the measure ν which is called *a probability density* of the random variable X with respect to the measure ν.

R e m a r k. In elementary probability theory the probability density of a random variable is always considered in a finite-dimensional phase space (Example 2.1.10). In problems which are connected with one random variable it is always convenient to assume its phase probability space as the probability space. But in the problems connected with the set of the random variables (possibly with different phase spaces) it is expedient to study all of them in one probability space.

E x a m p l e 2.2.5. The two-points set $\{0,1\}$ serves as the phase space of the random variable in Example 2.2.1, and the probability measure is completely determined by its value in one of these two points: $\mu_x(\{1\}) = p$, $\mu_x(\{0\}) = q$, $p + q = 1$.

E x a m p l e 2.2.6. The set $\{0, 1, \ldots, n\}$ serves as the phase space of the number of the occurrences of an event in n trials in Example 2.2.2, and the probability measure μ_x is completely determined by its values in these points: $\mu_x(\{m\}) = C_n^m p^m q^{n-m}$ $(C_n^m = n!/m!(n-m)!, m = 0, 1, \ldots, n)$.

E x a m p l e 2.2.7. In Examples 2.2.3 and 2.2.4 the space of elementary events Ω itself serves as the phase space of the complex (correspondingly real) random vector X, and the complex plane (correspondingly real axis) serves as the common phase space of its components.

2.2.3. Equivalence of Random Variables

Let $X = x(\omega)$ be a random variable with its phase space X. The inverse mapping $x^{-1}(\mathcal{A})$ of the σ-algebra $\mathcal{A}$ represents the σ-algebra which is a part of the σ-algebra $\mathcal{S}$ in the space of elementary events Ω, $x^{-1}(\mathcal{A}) \subset \mathcal{S}$. In the general case $x^{-1}(\mathcal{A})$ does not coincide with $\mathcal{S}$. Thus the variable $X = x(\omega)$ induces in the space Ω some σ-algebra $\mathcal{S}_x = x^{-1}(\mathcal{A})$, as a rule more poor than the σ-algebra $\mathcal{S}$. Therefore the random variable $Y = y(\omega)$ in the measurable space $(Y, \mathcal{B})$ determined on the same probability space $(\Omega, \mathcal{S}, P)$will be not obligatory $(\mathcal{S}_x, \mathcal{B})$-measurable (from the fact that $y^{-1}(B) \in \mathcal{S}$ does not follow that $y^{-1}(B) \in \mathcal{S}_x$). So if the random variable Y in $(Y, \mathcal{B})$ whose phase space

Y being a separable B-space is $(\mathcal{S}_x, \mathcal{B})$-measurable then it is $(\mathcal{A}, \mathcal{B})$-measurable function of the random variable X.

The random variables which are equal almost surely, i.e. with the probability 1, are called *equivalent*. Thus two statements are valid.

Theorem 2.2.1. *If the random variable $Y = y(\omega)$ in $(Y, \mathcal{B})$ whose phase space Y being a separable B-space is $(\mathcal{S}_x, \mathcal{B})$-measurable then it is equivalent to some $(\mathcal{A}, \mathcal{B})$-measurable function of the random variable $X = x(\omega)$.*

Theorem 2.2.2. *If the random variable $Y = y(\omega)$ in $(Y, \mathcal{B})$ is $(\mathcal{A}, \mathcal{B})$-measurable function of the random variable $X = x(\omega)$ (or is equivalent to such a function) then it is $(\mathcal{S}_x, \mathcal{B})$-measurable. In this case there are none restrictions on the phase space of the random variable Y.*

It is evident that all equivalent random variables have one and the same probability measure.

If the random variable $Y = y(\omega)$ in $(Y, \mathcal{B})$ is $(\mathcal{A}, \mathcal{B})$-measurable function of the random variable $X = x(\omega)$, $Y = \varphi(X)$ then it is a random variable in the probability space $(X, \mathcal{A}, \mu_x)$. Therefore the probability measure μ_y of this random variable may be expressed in terms of the probability measure μ_x of the random variable X by formula (2.2.1)

$$\mu_y(B) = \mu_x(\varphi^{-1}(B)), \quad B \in \mathcal{B}. \tag{2.2.4}$$

The same formula determines a probability measure of any random variable Y equivalent to $(\mathcal{A}, \mathcal{B})$-measurable function $\varphi(X)$ of the random variable X.

2.2.4. Two Types of Convergence of Random Sequences

The sequence of the random variables $\{X_k\}$, $X_k = x_k(\omega)$ with the values in the topological space is called *a.s. convergent* (almost surely, with probability 1) to the random variable $X = x(\omega)$ if the sequence of the functions $\{x_k(\omega)\}$ converges to $x(\omega)$ almost everywhere on Ω relatively to the measure P. The sequence of the random variables $\{X_k\}$, $X_k = x_k(\omega)$ with the values in the separable B-space X is called *convergent in probability* to the random variable $X = x(\omega)$ if at any $\varepsilon > 0$

$$\lim_{k \to \infty} P(\| X_k - X \| > \varepsilon) = \lim_{k \to \infty} P(\{\omega : \| x_k(\omega) - x(\omega) \| > \varepsilon\}) = 0.$$

The following statements are valid.

Theorem 2.2.3. *For the a.s. (respectively in probability) convergence of the sequence of the random variables $\{X_k\}$ with the values in the separable B-space X it is necessary and sufficient that this sequence will be fundamental a.s. (respectively in probability).*

Theorem 2.2.4. *Any a.s. convergent sequence of the random variables $\{X_k\}$ with the values in the separable B-space is also convergent in probability.*

The inverse is in general case not true. But from any sequence of random variables convergent in probability we may separate such sequence which a.s. converges.

E x a m p l e   2.2.8. Let X_n be the random variable with two possible values 0 and n whose probabilities are equal to $1 - n^{-2}$ and n^{-2} respectively. The sequence of the random variables $\{X_n\}$ corresponding to $n = 1, 2, \ldots$, converges in probability to 0 as at any $\varepsilon > 0$

$$P(|X_n| \geq \varepsilon) = P(X_n = n) = \frac{1}{n^2} \to 0 \quad \text{at} \quad n \to \infty .$$

This sequence also a.s. converges as at any $\varepsilon, 0 < \varepsilon \leq 1$,

$$P(|X_n| \geq \varepsilon \quad \text{if even at one} \quad n \geq p) \leq$$

$$\leq \sum_{n=p}^{\infty} P(X_n = n) = \sum_{n=p}^{\infty} \frac{1}{n^2} \to 0 \quad \text{at} \quad p \to \infty .$$

E x a m p l e   2.2.9. Let X_n be the random variable with two possible values 0 and 1 whose probabilities are equal to $1 - n^{-1}$ and n^{-1} respectively. Let us consider the sequence of such random variables $\{X_n\}$ and suppose that any finite subset of the variables of this sequence represents an aggregate of independent random variables. This sequence converges to 0 in probability as at any $\varepsilon, 0 < \varepsilon \leq 1$, $P(|X_n| \geq \varepsilon) = P(X_n = 1) = \frac{1}{n} \to 0$ at $n \to \infty$. But the sequence $\{X_n\}$ does not a.s. converge to 0 as at any $\varepsilon, 0 < \varepsilon \leq 1$ and at any natural p

$$P(|X_n| < \varepsilon \quad \text{at all} \quad n \geq p)$$

$$= \lim_{m \to \infty} P(|X_n| < \varepsilon \quad \text{at all} \quad n, p \leq n \leq p + m)$$

$$= \lim_{m \to \infty} \prod_{n=p}^{p+m} P(X_n = 0) = \lim_{m \to \infty} \frac{p-1}{p} \frac{p}{p+1} \ldots \frac{p+m-1}{p+m} = 0 .$$

After choosing from $\{X_n\}$ the subsequence $\{X_{2^k}\}$ we obtain the sequence a.s. convergent to 0. Really at any $\varepsilon, 0 < \varepsilon \leq 1$ and at $p \to \infty$ we have

$$P(|X_{2^k}| \geq \varepsilon \quad \text{if even at one} \quad k \geq p)$$

$$\le \sum_{k=p}^{\infty} P(\,|X_{2^k}| \ge \varepsilon) = \sum_{k=p}^{\infty} P(X_{2^k} = 1) = \sum_{k=p}^{\infty} \frac{1}{2^k} = \frac{1}{2^{p-1}} \to 0\,.$$

2.2.5. Finite-Dimensional Random Variables

Let us consider now a special case when the phase space X of the random variable $X = x(\omega)$ represents a finite-dimensional vector space. The variable X is in this case either a scalar random variable (if X is one-dimensional) or a random vector. In the latter case after choosing in X an arbitrary basis $\{e_1, \ldots e_n\}$ we may present a random variable in the form $X = \sum X_k e_k$ and treat it as the finite array $X_1, \ldots, X_n$ of the scalar random variables which are the components of the vector X. It is evident that the components of a random vector may be always considered as rectangular Cartesian coordinates of a random point in the correspondent Euclidean space.

It is settled to consider the inequalities between the vectors of one and the same finite-dimensional space similar as the inequalities between all their components. So, for instance, $a < b$, $a = \{a_k\}$, $b = \{b_k\}$ means $a_k < b_k$ for all k; $a < x < b$, or $x \in (a, b)$ means that the point with the coordinates $\{x_k\}$ is inside of the rectangle (a, b) whose projections on the coordinates axis are the intervals (a_k, b_k). The σ-algebra $\mathcal{A}$ of the sets in the finite-dimensional space X is usually assumed as the σ-algebra of Borel sets, i.e. the σ-algebra induced by a class of open intervals (rectangles) of the space X.

Let μ_x be the probability measure of the random variable X. The function of a point

$$F(x) = P(X \in (-\infty, x)) = \mu_x((-\infty, x)) \tag{2.2.5}$$

is called *a distribution function* of the random variable X.

From general properties of the probability follow immediately the properties of a distribution function:

(i) $0 \le F(x) \le 1$;

(ii) $F(x) = 0$ if even one of the components of the vector x is equal to $-\infty$;

(iii) $F(x) = 1$ if all the components of the vector x are equal to ∞;

(iv) $F(x)$ represents a nondecreasing function of each component of the vector x;

(v) $F(x)$ is continuous from the left on each component of the vector x and there exist the limits $F(x + 0) = \lim F(x + \varepsilon)$, $\varepsilon > 0$ when the components of the vector ε different from zero tend to zero (some components of the vector ε may be equal to zero).

Let us denote by $\Delta_I^{(k)}$ the operator of the finite difference which corresponds any function of the vector x with its increment under changes of the kth component x_k within the interval I. By virtue of the additivity of the probability measure μ_x the expressions $\Delta_{[\alpha,\beta)}^{(1)} F$, $\Delta_{[\alpha,\beta]}^{(1)} F$, $\Delta_{(\alpha,\beta]}^{(1)} F$, $\Delta_{(\alpha,\beta)}^{(1)} F$ determine the probabilities of the occurences of the random vector X respectively into the rectangles with the sides $[\alpha,\beta)$, $[\alpha,\beta]$, $(\alpha,\beta]$, (α,β) on the axis x_1 and with the sides $(-\infty, x_k)$ $(k = 2, \ldots, n)$ on other axes. Generally the expression of the form $\Delta_{I_1}^{(k_1)} \ldots \Delta_{I_p}^{(k_p)} F$ for any different $k_1, \ldots, k_p = 1, \ldots, n$ and all $p = 1, \ldots, n$ represents the probability of the occurrence of the random vector X into the rectangle $R_{k_1, \ldots, k_p}^{(p)}$ with the sides $I_1, \ldots, I_p$ on the axes $x_{k_1}, \ldots, x_{k_p}$ and with the sides $(-\infty, x_m)$ $(m \neq k_1, \ldots, k_p)$ on the other axes:

$$P\left(X \in R_{k_1, \ldots, k_p}^{(p)}\right) = \mu_x\left(R_{k_1, \ldots, k_p}^{(p)}\right) = \Delta_{I_1}^{(k_1)} \ldots \Delta_{I_p}^{(k_p)} F. \qquad (2.2.6)$$

Putting $p = n$, $k_1 = 1, \ldots, k_n = n$ and contracting the intervals $I_1, \ldots, I_n$ respectively into the points $x_1, \ldots, x_n$ we receive the formula for the probability of the occurrence of the random vector X into a given point x with the coordinates $x_1, \ldots, x_n$:

$$P(X = x) = \mu_x(\{x\}) = \lim_{I_k \to x_k} \Delta_{I_1}^{(1)} \ldots \Delta_{I_n}^{(n)} F. \qquad (2.2.7)$$

At $n > 1$ from formula (2.2.6) follows one more important property of a distribution function of the random vector:

(vi) $\Delta_{I_1}^{(k_1)} \ldots, \Delta_{I_p}^{(k_p)} F \geq 0$ at all $p, k_1, \ldots, k_p$ $(p, k_1, \ldots, k_p = 1, \ldots, n; k_i \neq k_j$ at $i \neq j)$ for any intervals $I_1, \ldots, I_p$.

From formula (2.2.5) and (i)–(vi) properties we get the following statements.

Theorem 2.2.5. *The probability measure of the finite-dimensional random variable determines completely its distribution function.*

Theorem 2.2.6. *The distribution function of the finite-dimensional random variable completely determines its probability measure.*

So any function $F(x)$ possessing properties (i)–(v) represents the distribution function of some random variable X in $(X, \mathcal{A})$. After applying formula (2.2.2) which determines Lebesgue expansion of the probability measure to the case when A is the rectangle $(-\infty, x)$ and ν is Lebesgue measure in the n-dimensional space X we obtain the corresponding expansion of the distribution function of the random variable X:

$$F(x) = \int\limits_{-\infty}^{x} f(u)du + F_s(x) \,. \tag{2.2.8}$$

Here the integral should be considered as the nth multiple Lebesgue integral over all components of the vector x. Let us denote by $\{x_m\}$ the set of the points X which have the probability different from zero and put $p_m = P(X = x_m) = \mu_x(x_m)$. After isolating from the singular function $F_s(x)$ the jumps function

$$F_d(x) = \sum_{\{m:x_m \in (-\infty, x)\}} p_m \,,$$

we shall present F_s in the form $F_s(x) = F_d(x) + F_{sc}(x)$, where $F_{sc}(x)$ is a continuous singular component. Using δ-function the jumps function $F_d(x)$ may be expressed in the form

$$F_d(x) = \int\limits_{-\infty}^{x-0} \sum p_m \delta(u - x_m)du \,,$$

where the sum is spread over all the points x_m with the probabilities different from zero. After this formula (2.2.8) will take the form

$$F(x) = \int\limits_{-\infty}^{x-0} f^*(u)du + F_{sc}(x) \,,$$

where is denoted

$$f^*(x) = f(x) + \sum p_m \delta(x - x_m) \,. \tag{2.2.9}$$

If $F_s(x) = 0$ in (2.2.8) then the random function X is called *a continuous* one and $f(x)$ represents its probability density. If $f(x) = F_{sc}(x) = 0$ then the random variable X is called *a discrete*. Formula (2.2.9) at $f(x) = 0$ determines in this case its probability density which represents a linear combination of the δ-functions. If $F_{sc}(x) = 0$ and $f(x) \neq 0$ and if only one of the probability p_m differs from zero, then X

represents a random variable of *a mixed type*. Formula (2.2.9) determines the probability density of a random variable of a mixed type.

The probability density of a random variable is determined only almost everywhere. Therefore the values of the probability density (or its regular part) of a finite-dimensional random function in the points of its discontinuity are not essential and usually are not indicated (as, for instance, in the case of an uniform or exponential distribution).

2.3. Conditional Probabilities

2.3.1. Definitions

As it is known from elementary probability theory if $P(B) \neq 0$, $B \in \mathcal{S}$ then *a conditional probability of any event $A \in \mathcal{S}$ relative to the event B (at a given event B)* is determined by the following formula:

$$P(A|B) = \frac{P(AB)}{P(B)} . \qquad (2.3.1)$$

The event A is called *independent* of B if $P(A|B) = P(A)$. If $P(A|B) \neq P(A)$ then A *depends* on B. It is evident that if A does not depend on B then B does not depend on A also. Thus two events are always mutually dependent or independent.

The events $A_1 , \ldots , A_n$ are called *independent* if each of them does not depend on any of others and any other intersections. The equality $P(A_1 , \ldots , A_n) = P(A_1) \ldots P(A_n)$ serves as a necessary and sufficient condition of the independence of the events $A_1 , \ldots , A_n$.

Let $\{B_k\} \subset \mathcal{S}$ be an arbitrary sequence of mutually exclusive events, $\mathcal{B}$ be the σ-algebra $\mathcal{B} \subset \mathcal{S}$ generated by this sequence. It is evident that any event $B \in \mathcal{B}$ represents the union of some set of the events B_k, $B = \bigcup_i B_{k_i}$. Therefore for any events $A \in \mathcal{S}$ and $B \in \mathcal{B}$

$$P(AB) = P(\bigcup_i AB_{k_i}) = \sum_i P(AB_{k_i}) = \sum P(B_{k_i})P(A|B_{k_i}) . \quad (2.3.2)$$

In the special case when $B = \Omega$ from this formula follows a formula of total probability which is a well known from elementary probability theory.

Let us introduce now a function of a point in the space of elementary events Ω at fixed $A \in \mathcal{S}$ after determining it by the equality

$$P_{\mathcal{B}}(A|\omega) = P(A|B_k) \quad \text{at} \quad \omega \in B_k$$

on those sets B_k for which $P(B_k) \neq 0$ and remaining it indefinite on those sets B_k for which $P(B_k) = 0$. It is evident that $P_{\mathcal{B}}(A|\omega)$ represents P-integrable elementary function ω at any fixed event A which is determined on Ω almost everywhere to relative P. The last part of formula (2.3.2) represents an integral over the set B with respect to the elementary function $P_{\mathcal{B}}(A|\omega)$ over the measure P. Therefore formula (2.3.2) may be presented in the form

$$P(AB) = \int\limits_{B} P_{\mathcal{B}}(A|\omega)P(d\omega), \quad B \in \mathcal{B}. \qquad (2.3.3)$$

Now notice that for calculating the integral in this formula there is no need to know the probability P on the whole σ-algebra $\mathcal{S}$. It is sufficient to assign P on the smaller σ-algebra $\mathcal{B}$. Therefore denoting by $P_{\mathcal{B}}$ the contraction of the probability P on the σ-algebra $\mathcal{B}$, $P_{\mathcal{B}}(B)$ $= P(B)$, $B \in \mathcal{B}$ we can rewrite formula (2.3.3) in the form

$$P(AB) = \int\limits_{B} P_{\mathcal{B}}(A|\omega)P_{\mathcal{B}}(d\omega), \quad B \in \mathcal{B}. \qquad (2.3.4)$$

Formula (2.3.1) shows that at any fixed event B, $P(B) \neq 0$ the conditional probability $P(A|B)$ possesses all the properties of the probability, i.e. it represents itself the probability on $\mathcal{S}$. Hence it follows that $P_{\mathcal{B}}(A|\omega)$ considered as a function of the set A is the σ-additive function.

2.3.2. Properties of Conditional Probability

Let $\mathcal{B}$ be an arbitrary σ-algebra which is contained in $\mathcal{S}$, $\mathcal{B} \subset \mathcal{S}$. As $P(AB) \leq P(B) = P_{\mathcal{B}}(B)$ at any $A \in \mathcal{S}$, $B \in \mathcal{B}$ then $P(AB) = 0$ if $P_{\mathcal{B}}(B) = 0$. It means that the probability $P(AB)$ considered as the function of the set B at fixed A is $P_{\mathcal{B}}$-continuous. Consequently, according to Radon–Nikodym theorem which is known from functional analysis at any fixed A there exists $P_{\mathcal{B}}$-integrable function $P_{\mathcal{B}}(A|\omega)$ in terms of which $P(AB)$ is expressed by integral (2.3.4). This function is determined on Ω almost everywhere relative to $P_{\mathcal{B}}$ (shortly: $P_{\mathcal{B}}$ (a.e.)), and consequently, relative to P and possesses the following properties:

(i) $0 \leq P_{\mathcal{B}}(A|\omega) \leq 1$, $P_{\mathcal{B}}$ (a.e.);

(ii) if $A_1 \subset A_2$, A_1, $A_2 \in \mathcal{S}$, then $P_{\mathcal{B}}(A_1|\omega) \leq P_{\mathcal{B}}(A_2|\omega)$ $P_{\mathcal{B}}$ (a.e.) and $P_{\mathcal{B}}(\varnothing|\omega) = 0$, $P_{\mathcal{B}}(\Omega|\omega) = 1$ $P_{\mathcal{B}}$ (a.e.);

(iii) if $\{A_k\}$ is an arbitrary sequence of mutually exclusive $A_k \in \mathcal{S}$, then $P_{\mathcal{B}}$ (a.e.)

$$P_{\mathcal{B}}(\cup A_k|\omega) = \sum P_{\mathcal{B}}(A_k|\omega). \qquad (2.3.5)$$

Thus $P_{\mathcal{B}}(A|\omega)$ represents σ-additive function of the set on $\mathcal{S}$ with the values in the space of $P_{\mathcal{B}}$-integrable functions $L_1(\Omega, \mathcal{B}, P_{\mathcal{B}})$.

The function $P_{\mathcal{B}}(A|\omega)$ is called *a conditional probability relative to the σ-algebra $\mathcal{B}$*. It is determined for all the events $A \in \mathcal{S}$ and each A represents the $P_{\mathcal{B}}$-integrable function of the point in Ω determined almost everywhere relative to P.

It follows from the σ-additivity of the conditional probability (2.3.5) that it is a continuous function of a set with the values in the space $L_1(\Omega, \mathcal{S}, P_{\mathcal{B}})$. It means that for any monotonous sequence of the sets $\{A_n\}$, $A = \lim A_n$ the following property is valid:

(iv) $P_{\mathcal{B}}(A|\omega) = \lim P_{\mathcal{B}}(A_n|\omega)$ almost at all ω. $\qquad (2.3.6)$

R e m a r k. It does not follow in any way from the established relations that $P_{\mathcal{B}}(A|\omega)$ at fixed value of ω represents the probability on $(\Omega, \mathcal{S})$. The fact is that these exceptional sets of zero measure P for which the derived relations are not valid depend on the sets A, A_1, A_2 and the sequence of the sets $\{A_k\}$. Therefore there may exist one of the value ω at which all these relations will be valid simultaneously for all the sets A, A_1, A_2 and all the sequences of the sets $\{A_k\}$. Namely it is necessary for the fact that the function $P_{\mathcal{B}}(A|\omega)$ will be the probability on $(\Omega, \mathcal{S})$ at fixed ω.

2.3.3. Regular Conditional Probability

If the conditional probability $P_{\mathcal{B}}(A|\omega)$ represents the probability on $\mathcal{S}$ almost at all ω then it is called *regular*. In general case when the conditional probability $P_{\mathcal{B}}(A|\omega)$ is not regular then the *regular conditional probability* in the space $(\Omega, \mathcal{S})$ relative to the σ-algebra $\mathcal{B}$ is called a function of the variable ω and the set $A \in \mathcal{S}$ coinciding with $P_{\mathcal{B}}(A|\omega)$ almost everywhere at any fixed event A and represents the probability on $\mathcal{S}$ almost at any fixed ω. If a regular conditional probability exists then we shall always consider namely regular conditional probability. Therefore we shall not introduce a special denotion for it.

It is evident that a conditional probability at any A represents a scalar random variable as it is a numerical measurable function of a point in Ω (the same concerns to a regular conditional probability). It assumes a definite numerical value only after a trial when a definite elementary event ω appears (or in the assumption that as a result of a trial a given elementary event ω appears). If $A \in \mathcal{B}$ then

$$P(AB) = P_{\mathcal{B}}(AB) = \int_{AB} P_{\mathcal{B}}(d\omega) = \int_{B} \mathbf{1}_A(\omega) P_{\mathcal{B}}(d\omega) \,.$$

After comparing this expression with (2.3.4) and bearing in mind that both expressions $P(AB)$ are valid at all $B \in \mathcal{B}$ we a.e. get $P_{\mathcal{B}}(A|\omega) = \mathbf{1}_A(\omega)$ if $A \in \mathcal{B}$ and by virtue of the σ-additivity of the function of the set $\mathbf{1}_A(\omega)$ at any fixed ω in a given case a conditional probability is regular. Thus *the contraction of the conditional probability $P_{\mathcal{B}}(A|\omega)$ on σ-algebra $\mathcal{B}$ represents always a regular conditional probability*.

If $\mathcal{B} = \mathcal{S}$ then $A \in \mathcal{B}$ if $A \in \mathcal{S}$. Therefore $P_{\mathcal{S}}(A|\omega) = \mathbf{1}_A(\omega)$ a.e. for any set $A \in \mathcal{S}$. Thus *a conditional probability relative to the complete σ-algebra $\mathcal{S}$ of the probability space is regular and represents an indicator of a set*.

In another case when $\mathcal{B}$ consists only of the empty set $\varnothing$ and the whole Ω formula (2.3.2) at $B = \Omega$ gives $P(A) = P_{\mathcal{B}}(A|\omega)P(\Omega) = P_{\mathcal{B}}(A|\omega)$. Thus *a conditional probability relatively to $\mathcal{B} = \{\varnothing, \Omega\}$ coincides with the probability P, and consequently, is also regular*.

2.3.4. Conditional Distribution of Random Variable

Let us consider now the random variable $Z = z(\omega)$ in some measurable space $(Z, \mathcal{C})$ determined on the probability space $(\Omega, \mathcal{S}, P)$. The function

$$\mu_{\mathcal{B}}(C|\omega) = P_{\mathcal{B}}(z^{-1}(C)|\omega)\,, \quad C \in \mathcal{C}\,, \tag{2.3.7}$$

is called *a conditional distribution of the random variable $Z = z(\omega)$ relative to the σ-algebra $\mathcal{B}$*. If $\mu_{\mathcal{B}}(C|\omega)$ represents a regular conditional probability (i.e. the probability on $\mathcal{C}$ almost at all ω) then it is called *a conditional probability measure or a regular conditional distribution of the random variable $Z = z(\omega)$ relative to the σ-algebra $\mathcal{B}$*.

Let $X = x(\omega)$ be a random variable in $(X, \mathcal{A})$, $\mathcal{S}_x$ be the σ-algebra induced by it, $\mathcal{S}_x = x^{-1}(\mathcal{A})$ and $P_{\mathcal{S}_x}(A|\omega)$ be *the conditional probability relative to this σ-algebra*. As at fixed $A \in \mathcal{S}$ it is $P_{\mathcal{S}_x}$-integrable, and consequently, is also measurable relatively to $\mathcal{S}_x$ then it represents a measurable numerical function of the random variable X relatively to the σ-algebra $\mathcal{A}$:

$$P_{\mathcal{S}_x}(A|\omega) = Q_x(A|x(\omega))\,. \tag{2.3.8}$$

Thus a conditional probability of any event is determined in this case by the value x which assumes the random variable X as a result of a trial and does not depend on the fact what namely of the elementary events ω realizing this values x appears in this case.

A conditional probability relative to the random variable $X = x(\omega)$ in the space $(\Omega, \mathcal{S})$ is called a conditional probability relative to the σ-algebra $\mathcal{S}_x$ induced by this random variable. Let $Y = y(\omega)$ be now a random variable in $(Y, \mathcal{B})$. The function

$$\mu_y(B|x) = Q_x(y^{-1}(B)|x)\,, \quad B \in \mathcal{B}\,, \tag{2.3.9}$$

is called *a conditional distribution of the random variable Y relative to the random variable X.* The conditional distribution $\mu_y(B|x)$ of the random variable Y relative to the variable $X = x(\omega)$ and the conditional distribution $\mu_{\mathcal{S}_x}(B|\omega)$ of the variable Y relative to the σ-algebra $\mathcal{S}_x$ induced by the random variable $X = x(\omega)$ are connected by the following relation:

$$\mu_{\mathcal{S}_x}(B|\omega) = \mu_y(B|x(\omega))\,, \quad B \in \mathcal{B}\,. \tag{2.3.10}$$

If $\mu_y(B|x)$ represents a regular conditional probability (i.e. the probability on $\mathcal{B}$ almost at all x) then it is called *a conditional probability measure or a regular conditional distribution of the random variable Y relative X (at a given value x of the variable X).* It is evident that if there exists the regular conditional probability $Q_x(A|x)$ then the regular conditional distribution $\mu_y(B|x)$ of the random variable Y relative to X exists also.

R e m a r k. The function of the set $Q_x(y^{-1}(B)|x)$, $B \in \mathcal{B}$ represents the contraction of the conditional probability $Q_x(A|x)$ on the σ-algebra $\mathcal{S}_y = y^{-1}(\mathcal{B})$. Therefore for the existence of a regular conditional distribution of the random variable Y there is no need in the existence of the regular conditional probability $Q_x(A|x)$. It is sufficient that a regular contraction of the conditional probability $Q_x(A|x)$ on the σ-algebra $\mathcal{S}_y = y^{-1}(\mathcal{B})$ induced in the space Ω of the random variable $Y = y(\omega)$ exists also.

2.4. Probabilities in Finite Product of Spaces

2.4.1. Definitions

Let us consider the space representing the product $(\Omega_1 \times \Omega_2, \mathcal{S}_1 \times \mathcal{S}_2)$ of the measurable spaces $(\Omega_1, \mathcal{S}_1)$, $(\Omega_2, \mathcal{S}_2)$. Suppose that in this product of the spaces the probability $P(A)$, $A \in \mathcal{S}_1 \times \mathcal{S}_2$ is determined. The contraction $P_1(A_1)$ of the probability $P(A)$ on the σ-algebra

$S_1 \times \Omega_2$ evidently represents the probability in the space (Ω_1, S_1). Similarly $P_2(A_2)$ of the probability on the σ-algebra $\Omega_1 \times S_2$ represents the probability in the space (Ω_2, S_2). Thus the probability given on the product of the spaces induces definite probability in each of these spaces.

The probabilities $P_1(A_1)$, $A_1 \in S_1$ and $P_2(A_2)$, $A_2 \in S_2$ induced in the spaces (Ω_1, S_1) and (Ω_2, S_2) by the probability P given on the product of the spaces $(\Omega_1 \times \Omega_2, S_1 \times S_2)$ are called *the projections* of the probability P on the subspaces (Ω_1, S_1) and (Ω_2, S_2).

Let us study the conditional probabilities in $(\Omega_1 \times \Omega_2, S_1 \times S_2)$ relative to the σ-algebras $\Omega_1 \times S_2$ and $S_1 \times \Omega_2$. As $P_{\Omega_1 \times S_2}(\Omega_1 \times A_2)$ $= P_2(A_2)$ depends only on the set A_2 of the space Ω_2 then the integration in formula (2.3.4) is performed by the variable ω_2 and the set $B \in \Omega_1 \times S_2$ represents the rectangle $B = \Omega_1 \times A_2$, $A_2 \in S_2$. Therefore formula (2.3.4) for this case takes the form

$$P(A \cap (\Omega_1 \times A_2)) = \int_{A_2} P_{\Omega_1 \times S_2}(A|\omega_2) P_2(d\omega_2)\,, \quad A_2 \in S_2\,. \qquad (2.4.1)$$

Thus the conditional probability in the product of the spaces relatively to the σ-algebra $\Omega_1 \times S_2$ depends only on the coordinate ω_2 of the point ω of this space. Analogously, the conditional probability relatively to the σ-algebra $S_1 \times \Omega_2$ depends only on the coordinate ω_1.

The contraction of the conditional probability $P_{\Omega_1 \times S_2}(A|\omega_2)$ considered as the function of the set A on the σ-algebra $S_1 \times \Omega_2$,

$$P_1(A_1|\omega_2) = P_{\Omega_1 \times S_2}(A_1 \times \Omega_2|\omega_2)$$

we shall call *a conditional probability in the space* Ω_1. The same formula

$$P_2(A_2|\omega_1) = P_{S_1 \times \Omega_2}(\Omega_1 \times A_2|\omega_1)$$

determines *a conditional probability in the space* Ω_2.

2.4.2. Probability in Two Spaces Product

$\triangleright$ Putting in (2.4.1) $A = A_1 \times \Omega_2$ and accounting that $(A_1 \times \Omega_2) \cap (\Omega_1 \times A_2) = A_1 \times A_2$ we obtain

$$P(A_1 \times A_2) = \int_{A_2} P_1(A_1|\omega_2) P_2(d\omega_2) \qquad (2.4.2)$$

and by the symmetry

$$P(A_1 \times A_2) = \int_{A_1} P_2(A_2|\omega_1)P_1(d\omega_1). \qquad (2.4.3)$$

We see that the probability P in the product of the spaces $(\Omega_1 \times \Omega_2, \, \mathcal{S}_1 \times \mathcal{S}_2)$ determines the probabilities and the conditional probabilities in the spaces $(\Omega_1, \mathcal{S}_1)$ and $(\Omega_2, \mathcal{S}_2)$. And vice versa, if the probability in $(\Omega_1, \mathcal{S}_1)$ and the conditional probability in $(\Omega_2, \mathcal{S}_2)$ (or the probability in $(\Omega_2, \mathcal{S}_2)$ and the conditional probability in $(\Omega_1, \mathcal{S}_1)$) are known then by formula (2.4.3) (correspondingly (2.4.2)) we may determine the probability P in the product of the spaces $(\Omega_1 \times \Omega_2, \mathcal{S}_1 \times \mathcal{S}_2)$ on the set of all measurable rectangles. By virtue of the σ-additivity of an integral and a conditional probability the probability P may be continued on the σ-algebra generated by the class of measurable rectangles, i.e. on $\mathcal{S}_1 \times \mathcal{S}_2$. ◁

Theorem 2.4.1. *The probability given on the product of the spaces determines uniquely the probabilities and the conditional probabilities in these spaces, and vice versa a given probability in one space and a conditional probability in the other space determine uniquely the probability in the product of these spaces.*

If the conditional probability $P_1(A_1|\omega_2)$ is regular then the extension of the probability P in the product of the spaces $\Omega_1 \times \Omega_2$ given by formula (2.4.2) are determined by

$$P(C) = \int P_2(d\omega_2) \int \mathbf{1}_C(\omega_1, \omega_2)P_1(d\omega_1|\omega_2), \quad C \in \mathcal{S}_1 \times \mathcal{S}_2. \quad (2.4.4)$$

Similarly if the conditional probability $P_2(A_2|\omega_1)$ is regular then the extension of P determined by formula (2.4.3) has the form

$$P(C) = \int P_1(d\omega_1) \int \mathbf{1}_C(\omega_1, \omega_2)P_2(d\omega_2|\omega_1), \quad C \in \mathcal{S}_1 \times \mathcal{S}_2. \quad (2.4.5)$$

Let us consider now two random variables $X = x(\omega)$ and $Y = y(\omega)$ given on the same probability space $(\Omega, \mathcal{S}, P)$. Suppose that X is a random variable in $(X, \mathcal{A})$ and Y in $(Y, \mathcal{B})$. The pair of the random variables X, Y represents the random variable $Z = z(\omega)$ in the product of the spaces $(X \times Y, \mathcal{A} \times \mathcal{B})$. The random variables $X = x(\omega)$, $Y = y(\omega)$ and $Z = z(\omega)$ induce in the spaces $(X, \mathcal{A})$, $(Y, \mathcal{B})$ and $(Z, \mathcal{C}) = (X \times Y, \mathcal{A} \times \mathcal{B})$ the conditional measures determined according to (2.2.1) by the following formulae:

$$\mu_x(A) = P(x^{-1}(A)), \quad \mu_y(B) = P(y^{-1}(B)), \quad \mu(C) = P(z^{-1}(C)),$$

and the conditional probabilities

$$\mu_x(A|y) = Q_y(x^{-1}(A)|y), \quad \mu_y(B|x) = Q_x(y^{-1}(B)|x).$$

Here Q_y is the conditional probability in $(\Omega, \mathcal{S})$ relative to the random variable Y, and Q_x is the conditional probability in $(\Omega, \mathcal{S})$ relative to random variable X. The probabilities $\mu_x(A)$ and $\mu_x(A|y)$ are determined on the σ-algebra $\mathcal{A}$, the probabilities $\mu_y(B)$ and $\mu_y(B|x)$ – on $\mathcal{B}$, and μ – on $\mathcal{C} = \mathcal{A} \times \mathcal{B}$.

Let us find the projections of the probability μ in $(X \times Y, \mathcal{A} \times \mathcal{B})$ on the subspaces $(X, \mathcal{A})$ and $(Y, \mathcal{B})$. According to the definition the projection μ_1 of the probability μ on the subspace $(X, \mathcal{A})$ represents the contraction of the probability μ on the σ-algebra $\mathcal{A} \times Y$:

$$\mu_1(A) = \mu(A \times Y) = P(x^{-1}(A) \cap y^{-1}(Y))$$
$$= P(x^{-1}(A)) - P(x^{-1}(A) \cap (\Omega \backslash y^{-1}(Y))).$$

But $P(x^{-1}(A) \cap (\Omega \backslash y^{-1}(Y))) \le P(\Omega \backslash y^{-1}(Y))$, and $P(\Omega \backslash y^{-1}(Y)) = 0$ as $y^{-1}(Y)$ represents the domain of the random variable $Y = y(\omega)$ which may differ from Ω only on the set of zero probability relatively to P. Consequently, $\mu_1(A) = P(x^{-1}(A)) = \mu_x(A)$.

Thus the projections of the probability μ on the subspaces $(X, \mathcal{A})$ and $(Y, \mathcal{B})$ of the space $(X \times Y, \mathcal{A} \times \mathcal{B})$ coincide with the conditional measures of the random variables X and Y correspondingly:

$$\left. \begin{array}{ll} \mu_x(A) = \mu(A \times Y), & A \in \mathcal{A}, \\ \mu_y(B) = \mu(X \times B), & B \in \mathcal{B}. \end{array} \right\} \qquad (2.4.6)$$

The following statements are valid.

Theorem 2.4.2. *The conditional probabilities generated in $(X, \mathcal{A})$ and $(Y, \mathcal{B})$ by the probability μ coincide with the conditional distributions of the random variables X and Y.*

Theorem 2.4.3. *The values of the conditional measure μ on the measurable rectangles of the product of the spaces $(X \times Y, \mathcal{A} \times \mathcal{B})$ are expressed in terms of the distributions and the conditional distributions of the random variables X and Y by formula*

$$\mu(A \times B) = \int_B \mu_x(A|y)\mu_y(dy) = \int_A \mu_y(B|x)\mu_x(dx). \qquad (2.4.7)$$

Theorem 2.4.4. *If there exist the regular conditional distributions $\mu_x(A|y)$ and $\mu_y(B|x)$ then the extension of the measure μ on the σ-algebra $\mathcal{C} = \mathcal{A} \times \mathcal{B}$ are expressed by formula*

$$\mu(C) = \int \mu_x(dx) \int \mathbf{1}_C(x, y)\mu_y(dy|x)$$
$$= \int \mu_y(dy) \int \mathbf{1}_C(x, y)\mu_x(dx|y), \quad C \in \mathcal{C}. \tag{2.4.8}$$

Theorem 2.4.5. *The value of the conditional probability $\mu_x(A|y)$ at any $A \in \mathcal{A}$ represents Radon–Nikodym derivative of the measure $\mu(A \times B)$ considered as the function of the set B over the measure $\mu_y(B)$, and the value of the conditional measure $\mu_y(B|x)$ at any $B \in \mathcal{B}$ represents Radon–Nikodym derivative of the measure $\mu(A \times B)$ considered as the function of the set A over the measure $\mu_x(A)$:*

$$\mu_x(A|y) = \frac{d\mu(A \times y)}{d\mu_y(y)}, \quad \mu_y(B|x) = \frac{d\mu(x \times B)}{d\mu_x(x)}. \tag{2.4.9}$$

$\triangleright$ Let us suppose now that the nonnegative σ-finite measures ν_x and ν_y are determined in the spaces $(X, \mathcal{A})$ and $(Y, \mathcal{B})$ and that the probability measure μ in the product of the spaces $(X \times Y, \mathcal{A} \times \mathcal{B})$ is absolutely continuous relative to the product of the measures ν_x and ν_y. So we have

$$\nu(C) = \int \nu_x(dx) \int \mathbf{1}_C(x, y)\nu_y(dy) = \int \nu_y(dy) \int \mathbf{1}_C(x, y)\nu_x(dx).$$

According to the Radon–Nikodym theorem for any set $C \in \mathcal{C} = \mathcal{A} \times \mathcal{B}$ we get

$$\mu(C) = \int_C f(z)\nu(dz),$$

where $f(z) = f(x, y)$ is the probability density of the random variable Z, i.e. the joint probability density of the random variables X, Y over the measure ν. The integral over the product of the spaces in the previous formula is equal to either of two iterated integrals over the measures ν_x and ν_y:

$$\mu(C) = \int \nu_x(dx) \int \mathbf{1}_C(x, y)f(x, y)\nu_y(dy)$$
$$= \int \nu_y(dy) \int \mathbf{1}_C(x, y)f(x, y)\nu_x(dx), \quad C \in \mathcal{C}.$$

Applying this formula to the case when C represents the measurable rectangle $C = A \times B$, $A \in \mathcal{A}$, $B \in \mathcal{B}$ we shall have

$$\mu(A \times B) = \int_A \nu_x(dx) \int_B f(x,y)\nu_y(dy) = \int_B \nu_y(dy) \int_A f(x,y)\nu_x(dx)\,.$$

$$(2.4.10)$$

Putting here $B = Y$ we obtain

$$\mu_x(A) = \mu(A \times Y) = \int_A \nu_x(dx) \int f(x,y)\nu_y(dy)\,.$$

Hence it is clear that the probability measure of the random variable X is ν_x-continuous and the probability density $f_x(x)$ of the random variable X over the measure ν_x is determined by formula

$$f_x(x) = \int f(x,y)\nu_y(dy)\,. \qquad (2.4.11)$$

By the symmetry we conclude that the probability measure of the random variable Y is ν_y-continuous and the probability density $f_y(y)$ of the random variable Y over the measure ν_y is determined by formula

$$f_y(y) = \int f(x,y)\nu_x(dx)\,. \qquad (2.4.12)$$

We rewrite the first part of formula (2.4.7) by means of (2.4.10) in the form

$$\int_B \nu_y(dy) \int_A f(x,y)\nu_x(dx) = \int_B \mu_x(A|y)f_y(y)\nu_y(dy)\,.$$

As this equality is valid for any set $B \in \mathcal{B}$ then a.e.

$$\int_A f(x,y)\nu_x(dx) = \mu_x(A|y)f_y(y)\,.$$

Hence we find

$$\mu_x(A|y) = \frac{1}{f_y(y)} \int_A f(x,y)\nu_x(dx) = \int_A \frac{f(x,y)}{f_y(y)}\nu_x(dx)\,, \quad f_y(y) \neq 0\,.$$

This formula shows that the conditional distribution of the random variable X is ν_x-continuous and there exists the conditional probability density of the random variable X relative to Y over the measure ν_x

$$f_x(x|y) = \frac{f(x,y)}{f_y(y)}\,. \tag{2.4.13}$$

This conditional probability density is a.e. determined relative to the products of the measures ν_x and ν_y. The formula

$$\mu_x(A|y) = \int_A f_r(x|y)\nu_x(dx)$$

shoes that in a given case the conditional distribution $\mu_x(A|y)$ is determined almost everywhere for all the sets $A \in \mathcal{A}$, i.e. is regular.

By the symmetry we conclude that in the considered case the conditional distribution of the random variable Y is regular, ν_y-continuous and the conditional probability density of the variable Y relative to X over the measure ν_y is given by formula

$$f_y(y|x) = \frac{f(x,y)}{f_x(x)}\,. \quad \triangleleft \tag{2.4.14}$$

Thus we have the following statement.

Theorem 2.4.6. *If there exists a joint probability density of the random variables X, Y over some product of the measures then their conditional distributions are regular and there are exist their probability densities determined by formulae (2.4.11) and (2.4.12) and the conditional probability densities determined by formulae (2.4.13) and (2.4.14).*

Combining formulae (2.4.13) and (2.4.14) we get *the multiplication theorem of probability densities:*

$$f(x,y) = f_x(x)f_y(y|x) = f_y(y)f_x(x|y)\,. \tag{2.4.15}$$

2.4.3. Calculation of Probability in Finite Product of Spaces

Let us consider now the product of n measurable spaces $(\Omega_k, \mathcal{S}_k)$ $(k = 1, \ldots, n)$. Suppose that in this product of the spaces $(\prod \Omega_k, \prod \mathcal{S}_k)$ the probability P is determined. Putting

$$(\overline{\Omega}_1, \overline{\mathcal{S}}_1) = \left(\prod_{k=1}^{m} \Omega_{r_k}, \prod_{k=1}^{m} \mathcal{S}_{r_k} \right), \quad (\overline{\Omega}_2, \overline{\mathcal{S}}_2) = \left(\prod_{k=m+1}^{n} \Omega_{r_k}, \prod_{k=m+1}^{n} \mathcal{S}_{r_k} \right),$$

where $r_1, \ldots, r_n$ – an arbitrary permutation of the numbers $1, \ldots, n$; $m = 1, \ldots, n-1$. Let us present $(\prod \Omega_k, \prod \mathcal{S}_k)$ in the form of the product of two spaces $(\overline{\Omega}_1 \times \overline{\Omega}_2, \overline{\mathcal{S}}_1 \times \overline{\mathcal{S}}_2)$. Then after applying the previous definitions and formulae we find the projections of the probability P on $(\overline{\Omega}_1, \overline{\mathcal{S}}_1)$ and $(\overline{\Omega}_2, \overline{\mathcal{S}}_2)$ and the conditional probabilities in $(\overline{\Omega}_1, \overline{\mathcal{S}}_1)$ and $(\overline{\Omega}_2, \overline{\mathcal{S}}_2)$ at all possible presentations $(\prod \Omega_k, \prod \mathcal{S}_k)$ in the form of the product of two spaces $(\overline{\Omega}_1 \times \overline{\Omega}_2, \overline{\mathcal{S}}_1 \times \overline{\mathcal{S}}_2)$. The probabilities and the conditional probabilities in all $(\overline{\Omega}_1, \overline{\mathcal{S}}_1)$ and $(\overline{\Omega}_2, \overline{\mathcal{S}}_2)$ obtained in such a way are *consistent* in such a sense that the probability (a conditional probability) in $(\Omega_{r_1} \times \cdots \times \Omega_{r_m}, \mathcal{S}_{r_1} \times \cdots \times \mathcal{S}_{r_m})$ is a projection of the probability (correspondingly of a conditional probability) in $(\Omega_{r_1} \times \cdots \times \Omega_{r_k}, \mathcal{S}_{r_1} \times \cdots \times \mathcal{S}_{r_k})$ at any $k > m$.

$\triangleright$ Let at $k > m$

$$P_{r_1, \ldots, r_m}(A) = P(A \times \Omega_{r_{m+1}} \times \cdots \times \Omega_{r_n}),$$

$$P_{r_1, \ldots, r_k}(B) = P(B \times \Omega_{r_{k+1}} \times \cdots \times \Omega_{r_n}), \tag{2.4.16}$$

$$P_{r_1, \ldots, r_m}(A) = P_{r_1, \ldots, r_k}(A \times \Omega_{r_{m+1}} \times \cdots \times \Omega_{r_k}). \tag{2.4.17}$$

Further applying formula (2.4.2) to the cases $\overline{\Omega}_1 = \Omega_{r_1} \times \cdots \times \Omega_{r_m}$, $\overline{\Omega}_2 = \Omega_{r_{k+1}} \times \cdots \times \Omega_{r_h}$ and $\overline{\Omega}_1 = \Omega_{r_1} \times \cdots \times \Omega_{r_k}$, $\overline{\Omega}_2 = \Omega_{r_{k+1}} \times \cdots \times \Omega_{r_h}$ at $k < h \leq n$ we get

$$P_{r_1, \ldots, r_m, r_{k+1}, \ldots, r_h}(A \times C)$$

$$= \int_C P_{r_1, \ldots, r_m}(A | \omega_{r_{k+1}}, \ldots, \omega_{r_h}) dP_{r_{k+1}, \ldots, r_h},$$

$$P_{r_1, \ldots, r_h}(B \times C) = \int_C P_{r_1, \ldots, r_k}(B | \omega_{r_{k+1}}, \ldots, \omega_{r_h}) dP_{r_{k+1}, \ldots, r_h}$$

$$\tag{2.4.18}$$

at any sets $A \in \mathcal{S}_{r_1} \times \cdots \times \mathcal{S}_{r_m}$, $B \in \mathcal{S}_{r_1} \times \cdots \times \mathcal{S}_{r_k}$ and $C \in \mathcal{S}_{r_{k+1}} \times \cdots \times \mathcal{S}_{r_h}$. And as according to just proved

$$P_{r_1,\ldots,r_m,r_{k+1},\ldots,r_h}(A \times C) = P_{r_1,\ldots,r_h}(A \times \Omega_{r_{m+1}} \times \cdots \times \Omega_{r_k} \times C),$$

then at almost all $\omega_{r_{k+1}}, \ldots, \omega_{r_h}$

$$P_{r_1,\ldots,r_m}(A|\omega_{r_{k+1}}, \ldots, \omega_{r_h})$$

$$= P_{r_1,\ldots,r_k}(A \times \Omega_{r_{m+1}} \times \cdots \times \Omega_{r_k}|\omega_{r_{k+1}}, \ldots, \omega_{r_h}). \qquad (2.4.19)$$

If all the conditional probabilities are regular then at any k and h, $k < h$ the conditional probability $P_{r_1,\ldots,r_k}(C|\omega_{r_{k+1}}, \ldots, \omega_{r_h})$ may be considered as the probability in $(\Omega_{r_1} \times \cdots \times \Omega_{r_k}, \mathcal{S}_{r_1} \times \cdots \times \mathcal{S}_{r_k})$. So we determine the conditional probabilities for it in its turn in $(\Omega_{r_1} \cdots \times \Omega_{r_m}, \mathcal{S}_{r_1} \times \cdots \times \mathcal{S}_{r_m})$ $(m = 1, \ldots, k-1)$ which we shall denote yet by $P'_{r_1,\ldots,r_m}(A|\omega_{r_{m+1}}, \ldots, \omega_{r_h})$. They are determined by formula (2.4.2):

$$P_{r_1,\ldots,r_k}(A \times B|\omega_{r_{k+1}}, \ldots, \omega_{r_h}) = \int_B P'_{r_1,\ldots,r_m}(A|\omega_{r_{m+1}}, \ldots, \omega_{r_h})$$

$$\times P_{r_{m+1},\ldots,r_k}(d\omega_{r_{m+1}} \ldots d\omega_{r_k}|\omega_{r_{k+1}}, \ldots, \omega_{r_h}).$$

We notice that the conditional probabilities $P'_{r_1,\ldots,r_m}(A|\omega_{r_{m+1}}, \ldots, \omega_{r_h})$ coincide with the conditional probabilities $P_{r_1,\ldots,r_m}(A|\omega_{r_{m+1}}, \ldots, \omega_{r_h})$.

Thus parallel with the relations of form (2.4.18) the regular conditional probabilities satisfy the relations of the form

$$P_{r_1,\ldots,r_k}(A \times B|\omega_{r_{k+1}}, \ldots, \omega_{r_h}) = \int_B P_{r_1,\ldots,r_m}(A|\omega_{r_{m+1}}, \ldots, \omega_{r_h})$$

$$\times P_{r_{m+1},\ldots,r_k}(d\omega_{r_{m+1}} \times \cdots \times \omega_{r_k}|\omega_{r_{k+1}}, \ldots, \omega_{r_h})$$

$$(2.4.20)$$

for all m, $k > m$ and $h > k$ and all permutations $r_1, \ldots, r_n$ of numbers $1, \ldots, n$. Besides consistent conditions (2.4.17)–(2.4.20) the probabilities and the conditional probabilities satisfy also the condition of the symmetry. At any permutation of low indexes they do not change as at this permutation we get the same probabilities in the same space. ◁

Thus we proved the following statements.

Theorem 2.4.7. *The probability P in the product of the spaces $(\prod \Omega_k, \prod \mathcal{S}_k)$ determines the family of the probabilities and the conditional probabilities in the spaces $(\Omega_k, \mathcal{S}_k)$ $(k = 1, \ldots, n)$ and in all*

*the products of the spaces composed of them which satisfies conditions
(2.4.17)–(2.4.19), and additionally conditions (2.4.20) in the case when
all conditional probabilities are regular.*

Theorem 2.4.8. *Knowing the probability in one of the spaces
$(\Omega_k, \mathcal{S}_k)$ or in any product of $m < n$ spaces $(\Omega_k, \mathcal{S}_k)$ and the correspondent conditional probabilities we may find the probability P on
the set of all measurable rectangles of the product of the spaces by sequential application of formula (2.4.18), and further extend it on the
σ-algebra $\prod \mathcal{S}_k$ generated by the set of all measurable rectangles.*

R e m a r k. In particular, if the probability $P_1(A_1)$ is given in $(\Omega_1, \mathcal{S}_1)$
and the conditional probabilities $P_2(A_2|\omega_1)$, $P_3(A_3|\omega_1, \omega_2), \ldots, P_n(A_n|\omega_1,$
$\ldots, w_{n-1})$ are given in $(\Omega_2, \mathcal{S}_2), (\Omega_3, \mathcal{S}_3), \ldots, (\Omega_n, \mathcal{S}_n)$ correspondingly, then
using formula $(2.4.18)$ sequentially we may determine the probability P at first on
the class of the rectangles and further on to extend it on the σ-algebra $\prod \mathcal{S}_k$.

If n random variables $X_k = x_k(\omega)$ are given on one and the same
probability space in the spaces $(X_k, \mathcal{A}_k)$ $(k = 1, \ldots, n)$ correspondingly
then the aggregate of these variables represents the random variable Z
$= z(\omega)$ in the product of the spaces $(\prod X_k, \prod \mathcal{A}_k)$. After partitioning the set of the random variables $\{X_1, \ldots, X_n\}$ by all possible ways
into two groups $X = (X_{r_1}, \ldots, X_{r_m})$ and $Y = (X_{r_{m+1}}, \ldots, X_{r_n})$,
where $r_1, \ldots, r_n$ is any permutation of the indexes $1, \ldots, n$ and m
$= 1, \ldots, n-1$ we ensure that the distributions and the conditional distributions of the random variables $X_1, \ldots, X_n$ and all their combinations
satisfy the relations of type (2.4.16)–(2.4.19), and in the case when all
conditional distributions are regular they satisfy the relations of type
(2.4.20).

Theorem 2.4.9. *If there exists the probability density of the random variable $Z = (X_1, \ldots, X_n)$ over some product of the measures in
$(\prod X_k, \prod \mathcal{A}_k)$ then the conditional distributions of all random variables
$X_1, \ldots, X_n$ and all their possible combinations are regular and there are
exist the probability densities and the conditional probability densities
over the correspondent measures of the random variables $X_1, \ldots, X_n$
and all their possible combinations. Here at any partition of the set of
the variables $(X_{r_1}, \ldots, X_{r_n})$ into two groups $X = (X_{r_1}, \ldots, X_{r_m})$ and
$Y = (X_{r_{m+1}}, \ldots, X_{r_h})$ relations (2.4.11)–(2.4.15) are valid.*

2.4.4. Independence of Random Variables

The random variable Y is called *independent* of the random variable
X if $\mu_y(B|x) = \mu_y(B)$ almost at all x relative to μ_x. If the measure μ_x of

the set z for which $\mu_y(B|z) \neq \mu_y(B)$ differs from zero then the variable Y depends on X. If follows from formula (2.4.7) that if the random vaiable Y does not depend on X then

$$\mu(A \times B) = \mu_x(A)\mu_y(B) = \int_B \mu_x(A)\mu_y(dy)\,. \qquad (2.4.21)$$

On the other hand the same formula (2.4.7) gives

$$\mu(A \times B) = \int_B \mu_x(A|y)\mu_y(dy)\,. \qquad (2.4.22)$$

As these both formulae are valid for any set $B \in \mathcal{B}$ then $\mu_x(A|y) = \mu_x(A)$ almost at all y relative to μ_y.

Thus if the random variable Y does not depend on X then the random variable X also does not depend on Y. Two random variables are always mutually dependent or mutually independent. Equality

$$\mu(A \times B) = \mu_x(A)\mu_y(B) \ \text{ for all } \ A \in \mathcal{A},\ B \in \mathcal{B} \qquad (2.4.23)$$

or $\mu = \mu_x \times \mu_y$ serves as a necessary and sufficient condition of the independence of two random variables X and Y. If there exists a joint probability density of random variables X and Y then equality

$$f(x,y) = f_x(x)f_y(y)\,, \qquad (2.4.24)$$

may also serve as necessary and sufficient condition of the independence.

The random variables $X_1,\ldots,X_n$ are called *independent* if each of them does not depend on any one of them. The equality

$$\mu(A_1 \times \ldots \times A_n) = \mu_1(A_1)\ldots\mu_n(A_n),\ \ A_k \in \mathcal{A}_k \ \ (k = 1,\ldots,n),$$
$$(2.4.25)$$

where $\mu_k(A_k)$ is the probability measure of the random variable X_k ($k = 1,\ldots,n$), and μ is the joint probability measure of the random variables $X_1,\ldots,X_n$, serves as a necessary and sufficient condition of the independence of the random variables $X_1,\ldots,X_n$.

In the case when there exists probability densities equality

$$f(x_1,\ldots,x_n) = f_1(x_1)\ldots f_n(x_n)\,, \qquad (2.4.26)$$

where $f_k(x)$ is the probability density of the random variable X_k ($k = 1,\ldots,n$), and $f(x_1,\ldots,x_n)$ is a joint probability density of the random variables $X_1,\ldots,X_n$, serves also as a necessary and sufficient condition of the independence of the random variables $X_1,\ldots,X_n$.

Mutual independence is not sufficient for the independence of the random variables $X_1,\ldots,X_n$.

2.5. Probabilities in Infinite Product of Spaces

2.5.1. Introductory Remarks

Let $\{\Omega_t\}$, $t \in T$ be an arbitrary set of the spaces. If we choose
the point ω_t from each space Ω_t then the aggregate of the chosen points
correspondent to all $t \in T$ is assumed as the point ω of the product of
the spaces $\Omega^T = \prod_{t \in T} \Omega_t$, $\omega = \{\omega_t, t \in T\}$. If $\Omega_t = R$ for any $t \in T$ then
the numbers ω_t at fixed t represent the coordinates of the point ω.

If in each of the spaces $\Omega_{t_1}, \ldots, \Omega_{t_n}$ we choose some sets A_{t_k}
$\subset \Omega_{t_k}$ then the set of all the points ω of the product of the spaces
Ω^T whose coordinates $\omega_{t_1}, \ldots, \omega_{t_n}$ satisfy the conditions $\omega_{t_k} \in A_{t_k}$ (k
$= 1, \ldots, n$) represents a rectangular *cylinder* in the space Ω^T with the
base $A_{t_1} \times \cdots \times A_{t_n}$ in the product of spaces $\Omega_{t_1} \times \cdots \times \Omega_{t_n}$. It is evident
that any cylinder with the base $A_{t_1} \times \cdots \times A_{t_n}$ in $\Omega_{t_1} \times \cdots \times \Omega_{t_n}$ may be
considered as a rectangle with the base $A_{t_1} \times \cdots \times A_{t_n} \times \Omega_{t_{n+1}} \times \cdots \times \Omega_{t_m}$
in $\Omega_{t_1} \times \cdots \times \Omega_{t_m}$ at any $m > n$.

Analogously the points set of the product of the spaces Ω^T satisfying
the condition $\{\omega_{t_1}, \ldots, \omega_{t_n}\} \in B$, where B is an arbitrary set in Ω_{t_1}
$\times \cdots \times \Omega_{t_n}$ represents a cylinder with the base B in $\Omega_{t_1} \times \cdots \times \Omega_{t_n}$.

If the σ-algebra of the sets $\mathcal{S}_t$ is given in each space Ω_t then the
product of the σ-algebras $\mathcal{S}^T = \prod_{t \in T} \mathcal{S}_t$ is determined as the minimal
σ-algebra which contains all measurable rectangles of the product of
the spaces Ω^T, i.e. the rectangles with the measurable sides $A_{t_k} \in \mathcal{S}_{t_k}$
($k = 1, \ldots, n$)correspondent to all n, $t_1, \ldots, t_n \in T$ and to all the
sets $A_{t_k} \in \mathcal{S}_{t_k}$ ($k = 1, \ldots, n$). Thus the product of the σ-algebras $\mathcal{S}^T$
represents the σ-algebra generated by the class of the rectangles with
the measurable sides in Ω^T.

Let us study the structure of the σ-algebra $\mathcal{S}^T$. It is evident that
the cylinders in Ω^T with "one-dimensional" bases $A_{t_1} \in \mathcal{S}_{t_1}$ in Ω_{t_1} at
fixed t_1 form the σ-algebra as their bases form the σ-algebra. For brevity
we shall denote this σ-algebra in the product of the spaces Ω^T similarly
as the σ-algebra of the basis of these cylinders, i.e. $\mathcal{S}_{t_1}$. The cylinders
with all possible bases $A_{t_1} \times \cdots \times A_{t_n}$, $A_{t_k} \in \mathcal{S}_{t_k}$ ($k = 1, \ldots, n$) in
$\Omega_{t_1} \times \cdots \times \Omega_{t_n}$ at any fixed n, $t_1, \ldots, t_n$ generate in infinite product of
the spaces Ω^T the σ-algebra of the cylinders with the measurable basis
in $(\Omega_{t_1} \times \cdots \times \Omega_{t_n}, \mathcal{S}_{t_1} \times \cdots \times \mathcal{S}_{t_n})$. Shortly we shall denote this σ-algebra
in the same way as the σ-algebra of their basis, i.e. $\mathcal{S}_{t_1} \times \cdots \times \mathcal{S}_{t_n}$. It is
evident that $\mathcal{S}_{t_1} \times \cdots \times \mathcal{S}_{t_n} \subset \mathcal{S}^T = \prod_{t \in T} \mathcal{S}_t$ at any n, $t_1, \ldots, t_n$.

Just in the same way for any subset L of the set T, $L \subset T$ we shall denote by $\mathcal{S}^L = \prod_{l \in L} \mathcal{S}_l$ the σ-algebra of the space Ω^T generated by the cylinders with the bases in $\Omega_{l_1} \times \cdots \times \Omega_{l_n}$, $l_1, \ldots, l_n \in L$. Evidently $\mathcal{S}^L \subset \mathcal{S}^T$ for any subset $L \subset T$. We may prove the following statement.

Theorem 2.5.1. *The union of the σ-algebras $\mathcal{S}^L$ correspondent to all countable subsets L of the set T represents the σ-algebra which coincides with $\mathcal{S}^T$.*

Hence it follows immediately from Theorem 2.5.1 that any measurable set A, $A \in \mathcal{S}^T$ of the infinite product of the spaces $(\Omega^T, \mathcal{S}^T)$ is measurable relatively to the σ-algebra $\mathcal{S}^L$ which corresponds to some countable subset L of the set T. It means that any set $A \in \mathcal{S}^T$ consists of the points which have the restrictions imposed on no more than the countable set of the coordinates ω_l, $l \in L$, and the other coordinates ω_t, $t \overline{\in} L$ may have any values from the correspondent spaces $\omega_t \in \Omega_t$. Such sets may be called *cylinders with the countable-dimensional bases*.

2.5.2. Distributions in Infinite Product of Spaces

Let us consider the existence of the distributions in infinite products of the spaces Ω^T. The following theorem will be proved.

Theorem 2.5.2. *The probabilities and all possible regular conditional probabilities in each of the spaces $(\Omega_t, \mathcal{S}_t)$, $t \in T$ determine uniquely the probability P in infinite product of these spaces $(\Omega^T, \mathcal{S}^T)$*
$$= \left(\prod_{t \in T} \Omega_t, \prod_{t \in T} \mathcal{S}_t \right).$$

$\triangleright$ Let $(X, \mathcal{A})$ be an arbitrary measurable space, $x(\omega)$ be a $(\mathcal{S}^T, \mathcal{A})$ measurable function in infinite product of the spaces Ω^T with the values in X. Suppose that there exists a countable class of the sets $\{F_k\}$ generating the σ-algebra $\mathcal{A}$. Then $\{x^{-1}(F_k)\}$ will be a countable class of the sets in Ω^T generating the σ-algebra $x^{-1}(\mathcal{A}) \subset \mathcal{S}^T$ induced in Ω^T by the function $x(\omega)$. There exists such sequence of the countable subsets $\{L_k\}$ of the set T, that $x^{-1}(F_k) \in \mathcal{S}^{L_k}$ $(k = 1, 2, \ldots)$. We put $L = \cup L_k$. Then we shall have $x^{-1}(F_k) \in \mathcal{S}^L$ at all k. Thus there exists such countable set $L \subset T$ that the σ-algebra $\mathcal{S}^L$ contains all the sets $x^{-1}(F_k)$ $(k = 1, 2 \ldots)$. As $x^{-1}(\mathcal{A})$ is the minimal σ-algebra containing all $x^{-1}(F_k)$ then $x^{-1}(\mathcal{A}) \subset \mathcal{S}^L$. Consequently, the function $x(\omega)$ is $(\mathcal{S}^L, \mathcal{A})$-measurable. Thus any measurable function $x(\omega)$ in Ω^T with the values in the measurable space $(X, \mathcal{A})$ in which the σ-algebra $\mathcal{A}$ is generated by the countable class of the sets is $(\mathcal{S}^L, \mathcal{A})$-measurable at some countable $L \subset T$, i.e. depends on no more than a countable set of the coordinates of the point ω in Ω^T. In particular, any measurable functional in Ω^T depends on no more than a countable set of the coordinates of the point ω in Ω^T.

Now we pass to the construction of the probability in Ω^T. Suppose that at each space $(\Omega_t, \mathcal{S}_t)$ the probability $P_t(A_t)$ and the regular conditional probabilities $P_{t,l_1,\ldots,l_n}(A_t|\omega_{l_1},\ldots,\omega_{l_n})$ correspondent to all $l_1,\ldots,l_n \in T$ ($n = 1, 2, \ldots$) are given. Then applying formula (2.4.18) sequentially we find the probabilities in all possible finite products of the spaces Ω_t:

$$
\left.
\begin{aligned}
&P_{t_1,t_2}(A_{t_1} \times A_{t_2}) = \int\limits_{A_{t_1}} P_{t_2,t_1}(A_{t_2}|\omega_1)P_{t_1}(d\omega_1)\,, \\[2mm]
&P_{t_1,\ldots,t_n}(A_{t_1} \times \cdots \times A_{t_n}) \\[1mm]
&= \int\limits_{A_{t_1}\times\cdots\times A_{t_{n-1}}} P_{t_n,t_1,\ldots,t_{n-1}}(A_{t_n}|\omega_1,\ldots,\omega_{n-1}) \\[2mm]
&\quad\quad \times P_{t_1,\ldots,t_{n-1}}(d\omega_1 \times \cdots \times d\omega_{n-1})
\end{aligned}
\right\}
\qquad (2.5.1)
$$

$$
(t_1,\,\ldots,\,t_n \in T;\ \ n = 3, 4, \ldots)\,.
$$

After determining the probability P_{t_1,t_2} on the measurable rectangles $\Omega_{t_1} \times \Omega_{t_2}$ by the first formula (2.5.1) at first we shall extend it on the σ-algebra $\mathcal{S}_{t_1} \times \mathcal{S}_{t_2}$. Then using the second formula (2.5.1) at $n = 3$ we detemine the probability P_{t_1,t_2,t_3} on the measurable rectangles $\Omega_{t_1} \times \Omega_{t_2} \times \Omega_{t_3}$ and extend it on the σ-algebra $\mathcal{S}_{t_1} \times \mathcal{S}_{t_2} \times \mathcal{S}_{t_3}$. Continuing this process we determine consequently the probabilities $P_{t_1,\ldots,t_n}$ at $n = 4, 5, \ldots$.. It is easy to understand that the conditional probabilities $P_{t,l_1,\ldots,l_n}(A_t|\omega_{l_1,\ldots,l_n})$ $(t, l_1,\,\ldots,\,l_n \in T; n = 1, 2, \ldots)$ should be consistent in such a way that they will give one and the same probability $P_{t_1,\ldots,t_n}$ in the product of the spaces $\Omega_{t_1} \times \cdots \times \Omega_{t_n}$ $(n = 2, 3, \ldots)$ in any ways. Then the obtained set of the probabilities in all finite products of the spaces $(\Omega_t, \mathcal{S}_t)$ will satisfy the conditions of the consistence and the symmetry.

Let us assign now the function of the set P on the set of all measurable rectangles of the infinite product of the spaces Ω^T by formula

$$
P(R) = P_{t_1,\ldots,t_n}(A_{t_1} \times \cdots \times A_{t_n})\,,
\qquad (2.5.2)
$$

where R is the rectangle with the sides $A_{t_1},\,\ldots,\,A_{t_n}$ which lie in the spaces $\Omega_{t_1},\,\ldots,\,\Omega_{t_n}$ correspondingly. This function is additive as the bases of any finite set of the rectangles may be considered as belonging to one and the same finite product of the spaces, for instance $\Omega_{t_1} \times \cdots \times \Omega_{t_n}$, and the probability $P_{t_1,\ldots,t_n}$ is additive in $\left(\Omega_{t_1} \times \cdots \times \Omega_{t_n}, \mathcal{S}_{t_1} \times \cdots \times \mathcal{S}_{t_n}\right)$. Furthermore if we restrict ourselves to measurable rectangles whose all bases belong to one and the same finite product of the spaces $\Omega_{t_1} \times \cdots \times \Omega_{t_n}$ then the function $P(R)$ is σ-additive by virtue of the σ-additivity of the probability $P_{t_1, \ldots, t_n}$. Consequently, the function $P(R)$ is the function on the set of all measurable rectangles of the space $(\Omega^T, \mathcal{S}^T)$ which have the basis in one and the same finite product of the spaces $\Omega_{t_1} \times \cdots \times \Omega_{t_n}$. According to the theorem from functional analysis about the measure extension it may be uniquely extended over the σ-algebra $\mathcal{S}_{t_1} \times \cdots \times \mathcal{S}_{t_n}$ of all measurable cylinders of the space Ω^T with the bases in $\Omega_{t_1} \times \cdots \times \Omega_{t_n}$. It is evident that the extension is determined by formula

$$P(C_{t_1, \ldots, t_n}) = P_{t_1, \ldots, t_n}(B_{t_1, \ldots, t_n}), \qquad (2.5.3)$$

where $C_{t_1, \ldots, t_n}$ is the cylinder in Ω^T with the measurable base $B_{t_1, \ldots, t_n}$ in $(\Omega_{t_1} \times \cdots \times \Omega_{t_n}, \mathcal{S}_{t_1} \times \cdots \times \mathcal{S}_{t_n})$.

Let us consider now the problem of the possibility of the function $P(R)$ extension determined on the class of the measurable rectangles of the space $(\Omega^T, \mathcal{S}^T)$ by formula (2.5.2) on the whole σ-algebra $\mathcal{S}^T$.

The class $\mathcal{C}$ of all measurable rectangles is a semi-algebra and satisfies the conditions of the known from functional analysis theorem about the measure extension.

In order that it may be possible to use the theorem about the measure extension for determining the probability on the whole σ-algebra $\mathcal{S}^T$ it remains only to prove the σ-additivity of the function of the set P on the class of all measurable rectangles which is determined by formula (2.5.2). For this purpose it is sufficient to show that P is continuous in zero, i.e. that for any decreasing sequence of the rectangles $\{R_n\}$, $R_p \supset R_q$ at $p < q$ with the empty intersection $\lim P(R_n) = 0$. Hence it is sufficient to consider the case when the set of the spaces $\{\Omega_{t_k}\}$ in which lie the sides of the rectangles R_n is infinite (the case of a finite set of the spaces $\{\Omega_{t_k}\}$ we have already considered). We denote by $A_n^{(k)}$ the side of the rectangle R_n in the space Ω_{t_k} and put $B_n^{(k)} = A_n^{(k+1)} \times \cdots \times A_n^{(r_n)}$ where r_n is the number of the sides of the rectangle R_n. Applying formula (2.4.18) we may present formula (2.5.2) at $R = R_n$ in the form

$$P(R_n) = \int\limits_{A_n^{(1)}} P(B_n^{(1)}|\omega_1) P_{t_1}(d\omega_1) = \int 1_{A_n^{(1)}}(\omega_1) P(B_n^{(1)}|\omega_1) P_{t_1}(d\omega_1),$$

where $P(B_n^{(1)}|\omega_1)$ is a conditional probability in the space $(\Omega_{t_2} \times \cdots \times \Omega_{t_{r_n}}, \mathcal{S}_{t_2} \times \cdots \times \mathcal{S}_{t_{r_n}})$. The sequence of the values of the functions $\left\{1_{A_n^{(1)}}(\omega_1) P(B_n^{(1)}|\omega_1)\right\}$ almost at all ω_1 relatively to the measure P_{t_1} represents a nonincreasing sequence of the numbers bounded below by zero as $B_p^{(1)} \supset B_q^{(1)}$ at

$p < q$. Consequently, it a.e. converges to some function $f(\omega_1)$. As all the functions $P(B_n^{(1)}|\omega_1)$ are a.e. bounded above by the P_{t_1}-integrable function $\varphi(\omega_1) \equiv 1$ then according to the Lebesgue theorem from functional analysis the function $f(\omega_1)$ is P_{t_1}-integrable and

$$\int f(\omega_1) P_{t_1}(d\omega_1) = \lim \int \mathbf{1}_{A_n^{(1)}}(\omega_1) P(B_n^{(1)}|\omega_1) P_{t_1}(d\omega_1) = \lim P(R_n).$$

If $\lim P(R_n) \neq 0$ then there exists such $\varepsilon > 0$ that $P(R_n) > \varepsilon$ at all n. As $P_{t_1}(\Omega_{t_1}) = 1$ then it follows that there exists such point $\overline{\omega}_1$ in Ω_{t_1} that $f(\overline{\omega}_1) \geq \varepsilon$. It is evident that $\overline{\omega}_1 \in A_n^{(1)}$ at all n as $f(\omega_1) = 0$ if ω_1 does not belong to $A_n^{(1)}$ beginning with some n. But the sequence of the conditional probabilities $\left\{ P(B_n^{(1)}|\overline{\omega}_1) \right\}$ is nonincreasing and $\lim P(B_n^{(1)}|\overline{\omega}_1 = f(\overline{\omega}_1)$. Therefore from $f(\overline{\omega}_1) \geq \varepsilon$ follows $P(B_n^{(1)}|\overline{\omega}_1) \geq \varepsilon$ at all n. Further on the basis of (2.4.20) we have

$$P(B_n^{(1)}|\overline{\omega}_1) = \int\limits_{A_n^{(2)}} P(B_n^{(2)}|\overline{\omega}_1, \omega_2) P_{t_1,t_2}(d\omega_2|\overline{\omega}_1), \qquad (2.5.4)$$

where $P(B_n^{(2)}|\overline{\omega}_1, \omega_2)$ is a conditional probability in $(\Omega_{t_3} \times \cdots \times \Omega_{t_{r_n}}, \mathcal{S}_{t_3}, \ldots, \mathcal{S}_{t_{r_n}})$. From $P(B_n^{(1)}|\overline{\omega}_1) \geq \varepsilon$ follows the existence of such point $\overline{\omega}_2$ in Ω_{t_2} that $P(B_n^{(2)}|\overline{\omega}_1, \overline{\omega}_2) \geq \varepsilon$ at all n and $\overline{\omega}_2 \in A_n^{(2)}$ at all n. Continuing this process we shall get the sequence of the points $\{\overline{\omega}_k\}$ in the space Ω_{t_k} $(k = 1, 2, \ldots)$. Here $\overline{\omega}_k \in A_n^{(k)}$. Consequently, all the points ω of the infinite product of the space Ω^T which have the coordinates $\overline{\omega}_1, \ldots, \overline{\omega}_p$ in the spaces $\Omega_{t_1}, \ldots, \Omega_{t_p}$ belong to all rectangles R_n whose sides lie in the spaces $\Omega_{t_1}, \ldots, \Omega_{t_p}$. Hence it follows that the points of the space Ω^T which have the coordinates $\overline{\omega}_k$ in all the spaces Ω_{t_k} $(k = 1, 2, \ldots)$ belong to all the rectangles R_n, i.e. the intersection of all the rectangles R_n is not empty.

So at $\lim P(R_n) \neq 0$ the intersection of all the rectangles R_n cannot be emply. Consequently, $\lim P(R_n) = 0$ if $\cap R_n = \varnothing$ what proves the continuity of the function P in zero and at the same time its σ-additivity on the set of all measurable rectangles of the space Ω^T. Thus formula (2.5.2) determines the probability in the space Ω^T on the set of all measurable rectangles $\mathcal{C}$. Hence there exists an unique extension of the probability P determined by formula (2.5.2) on the σ-algebra $\mathcal{S}^T$ generated by the set $\mathcal{C}$ of all measurable rectangles.

It follows from formula (2.5.2) that the probability $P_{t_1, \ldots, t_n}$ represents the projection of the probability P on the subspace $(\Omega_{t_1} \times \cdots \times \Omega_{t_n}, \mathcal{S}_{t_1}, \ldots, \mathcal{S}_{t_n})$ of the space $(\Omega^T, \mathcal{S}^T)$. Hence it follows that the probability P_L in the subspace $(\Omega^L, \mathcal{S}^L)$ of the space $(\Omega^T, \mathcal{S}^T)$, $L \subset T$ determined in such a way is the projection of the probability P on $(\Omega^L, \mathcal{S}^L)$.

The requirement of the regularity of all conditional probabilities in each space is essential as formula (2.5.4) and all analogous formulae which are necessary for unbounded extension of the process in given proof of the continuity of P in zero have the sense only if all conditional probabilities $(\Omega_t, \mathcal{S}_t)$ are regular. ◁

R e m a r k. Instead of assigning the probability and consistent regular conditional probabilities in each space $(\Omega_t, \mathcal{S}_t)$ and then finding by formulae (2.5.1) a consistent set of the probabilities in all finite products of the spaces $(\Omega_t, \mathcal{S}_t)$ we may assign currently consistent set of the probabilities $P_{t_1, \ldots, t_n}$ in the finite products of the spaces. If there exist all regular conditional probabilities $P_{t,l_1, \ldots, l_n}(A_t | \omega_{l_1}, \ldots, \omega_{l_n})$ $(l_1, \ldots, l_n \in T, n = 1, 2, \ldots)$ in each of the spaces $(\Omega_t, \mathcal{S}_t)$ then a consistent set of the probabilities $P_{t_1, \ldots, t_n}$ in all finite products of the spaces $(\Omega_t, \mathcal{S}_t)$ determines uniquely the probability in infinite product of the spaces $(\Omega^T, \mathcal{S}^T)$.

2.6. Existence of Regular Conditional Probabilities

2.6.1. Definitions

Let us consider the n-dimensional random vector $X = x(\omega)$. Suppose that the conditional distribution $\mu_{\mathcal{F}}(A | \omega)$ of the random variable X relatively to some σ-algebra $\mathcal{F}$ of the sets of the probability space $(\Omega, \mathcal{S}, P)$, $\mathcal{F} \subset \mathcal{S}$ is determined on the σ-algebra $\mathcal{A} = \mathcal{B}^n$ of Borel sets of the n-dimensional space X. Let us try to regularize it, i.e. to find the function $\mu'_{\mathcal{F}}(A | \omega)$ which at any $A \in \mathcal{A}$ coincides with $\mu_{\mathcal{F}}(A | \omega)$ almost at all ω and represents the probability on $\mathcal{A}$ almost at each ω. We shall denote by the letter r the points of the space X with the rational coordinates, and by the letter R – the rectangles with the apexes in the points r. Let us determine the function on the set of the points r

$$F_{\mathcal{F}}(r | \omega) = \mu_{\mathcal{F}}((-\infty, r) | \omega). \qquad (2.6.1)$$

In consequence of the properties of the conditional probabilities the function $F_{\mathcal{F}}(r | \omega)$ possesses the following properties:

(i) $0 \le F_{\mathcal{F}}(r | \omega) \le 1$ at all ω besides may be some P-zero set N_r;

(ii) $F_{\mathcal{F}}(r | \omega) = 0$ if even one of the components of the vector r is equal to $-\infty$;

(iii) $F_{\mathcal{F}}(r | \omega) = 1$ if all the components of the vector r are equal to ∞ at all ω besides may be some P-zero set N_0;

(iv) at all ω besides may be some P-zero set N_r there exist the limits

$$\lim F_{\mathcal{F}}(r - r' | \omega) = F_{\mathcal{F}}(r | y), \quad \lim F_{\mathcal{F}}(r + r' | \omega) = F_{\mathcal{F}}(r + 0 | \omega),$$

when the components of the positive vector r' (with rational components) different from zero tend to zero;

(v) for any rectangle R with the sides $I_1, \ldots, I_n$ $\Delta_{I_1}^{(1)}$ $\ldots \Delta_{I_n}^{(n)} F_{\mathcal{F}}(r|\omega) \geq 0$ at all ω besides may be some P-zero set N_R.

The following statement is valid.

Theorem 2.6.1. *The finite-dimensional random variables have always regular conditional distributions.*

Corollary 2.6.1. *All finite-dimensional random variables have regular conditional distributions relative to any random variables.*

$\triangleright$ As the set of the points r with the rational coordinates is countable then the set of the rectangles R with the rational coordinates of the apexes is countable also. Consequently, the set

$$N = N_0 \cup \left(\bigcup_r N_r\right) \cup \left(\bigcup_r N_r'\right) \cup \left(\bigcup_R N_R\right)$$

represents the P-zero set of the values ω as it is a countable union of zero sets. The function $F_{\mathcal{F}}(r|\omega)$ at all ω besides $\omega \in N$ possesses properties (i)–(v) simultaneously for all r and all the rectangles R. While extending it over the whole set X on the continuity in the left we determine the set of the distribution functions in the space X

$$F_{\mathcal{F}}(x|\omega) = \lim_{r \uparrow x} F_{\mathcal{F}}(r|\omega) \qquad (2.6.2)$$

for all ω, besides $\omega \in N$. Each such distribution function assigns uniquely the probability measure $\mu_{\mathcal{F}}'(A|\omega)$ on the σ-algebra $\mathcal{A}$ of Borel sets of the space X. Thus at each ω besides $\omega \in N$ we determined the probability measure $\mu_{\mathcal{F}}(A|\omega)$. In order to make sure that this measure is a required regular conditional distribution of the variable X relatively to $\mathcal{F}$ it is sufficient to make sure of the fact that for any $A \in \mathcal{A}$

$$\mu_{\mathcal{F}}'(A|\omega) = \mu_{\mathcal{F}}(A|\omega) \quad \text{almost at all} \quad \omega \,. \; \triangleleft \qquad (2.6.3)$$

2.6.2. Kolmogorov Theorem and Its Generalizations

Now let $(\Omega_1, \ldots, \Omega_m)$ be the finite-dimensional spaces, $(\mathcal{A}_1, \ldots, \mathcal{A}_m)$ be the σ-algebras of Borel sets in them. Suppose that the probability P is determined in the product of these spaces. Let us determine in probability space $(\prod \Omega_k, \prod \mathcal{A}_k, P)$ the random vectors $X = x(\omega)$ with the components $x_1(\omega) = \omega_{s_1}, \ldots, x_k(\omega) = \omega_{s_k}$ and Y

$= y(\omega)$ with the components $y_1(\omega) = \omega_{s_{k+1}}, \ldots, y_{m-k}(\omega) = \omega_{s_m}$ where $(s_1, \ldots, s_m)$ is any permutation of the numbers $1, \ldots, m$. Later on while speaking about the finite-dimensional and countable-dimensional random variables we shall always bear in mind the random variables in the spaces $(R^n, \mathcal{B}^n)$ $(n \leq \infty)$. The following Kolmogorov theorem is true.

Theorem 2.6.2. *If each of the spaces $(\Omega_t, \mathcal{S}_t)$, $t \in T$ is a finite-dimensional with the σ-algebra of Borel sets, then any consistent set of the probabilities $P_{t_1, \ldots, t_n}$ in all finite products of the spaces $(\Omega_t, \mathcal{S}_t)$ determines uniquely the probability P in the infinite product of the spaces $(\Omega^T, \mathcal{S}^T)$.*

▷ Now let us consider the random vector $X = x(\omega)$ with the countable set of the components $X_k = x_k(\omega)$ $(k = 1, 2, \ldots)$. Suppose that on the σ-algebra of Borel sets $\mathcal{A}$ of the phase space of this random variable X its conditional distribution $\mu_{\mathcal{F}}(A|\omega)$ relative to some σ-algebra $\mathcal{F} \subset \mathcal{S}$ of the probability space $(\Omega, \mathcal{S}, P)$ is assigned. The space $(X, \mathcal{A})$ may be considered as the countable product of one-dimensional spaces $(R^\infty, \mathcal{B}^\infty)$:

$$X = R^\infty = \prod_{k=1}^{\infty} R_k, \quad \mathcal{A} = \mathcal{B}^\infty = \prod_{k=1}^{\infty} \mathcal{B}_k,$$

where $R_k = R$ is the real axis; $\mathcal{B}_k = \mathcal{B}$ is the σ-algebra of Borel sets of this axis $(k = 1, 2, \ldots)$. In this case we may ensure that any countable-dimensional random variables have conditional distributions. In other words there exist the regular conditional probabilities determined on the correspondent σ-algebras of Borel sets in any countable-dimensional probability spaces. ◁

Thus we obtain the first generalization of the Kolmogorov theorem.

Theorem 2.6.3. *If each of the spaces Ω_t, $t \in T$ is countable-dimensional, i.e. is the space of the sequences of real numbers, and the σ-algebra $\mathcal{A}_t$ represents in it the σ-algebra of Borel sets then the consistent set of the probabilities $P_{t_1, \ldots, t_n}$ in all finite products of the spaces $(\Omega_t, \mathcal{A}_t)$ determines uniquely the probability in the infinite product of the spaces $(\Omega^T, \mathcal{A}^T)$.*

▷ Finally let us consider the random variable $X = x(\omega)$ in arbitrary measurable space $(X, \mathcal{A})$ and its conditional distribution $\mu_{\mathcal{F}}(A|\omega)$ relative to some σ-algebra $\mathcal{F} \subset \mathcal{S}$ of the sets of the probability space $(\Omega, \mathcal{S}, P)$. Suppose that there exists such finite- or countable-dimensional random variable $(\Omega, \mathcal{A}, \mu_x)$ in the probability space and that the σ-algebra $\mathcal{A}_u = u^{-1}(\mathcal{B}^n)$ induced by it coincides with $\mathcal{A}$. According to (2.2.4) formula

$$\nu_{\mathcal{F}}(B|\omega) = \mu_{\mathcal{F}}(u^{-1}(B)|\omega), \quad B \in \mathcal{B}^n, \tag{2.6.4}$$

determines the conditional distribution of the variable $U = u(\omega)$. In conformity with the proved theorems there exists the regular conditional distribution $\nu'_{\mathcal{F}}(B|\omega)$ of the variable U. Naturally the question arises: whether is it possible to determine the regular conditional distribution $\mu'_{\mathcal{F}}(A|\omega)$ of the variable X by formula (2.6.4). As by supposition any set $A \in \mathcal{A}$ is the inverse of some set $B \in \mathcal{B}^n$, $A = u^{-1}(B)$ then formula

$$\mu'_{\mathcal{F}}(A|\omega) = \nu'_{\mathcal{F}}(B|\omega), \quad A = u^{-1}(B), \tag{2.6.5}$$

determines $\mu'_{\mathcal{F}}(A|\omega)$ for all sets $A \in \mathcal{S}$. But in the general case this formula determines $\mu'_{\mathcal{F}}(A|\omega)$ uniquely not for all $A \in \mathcal{S}$. For the existence of the regular conditional distributions of the random variable X it is sufficient that such a finite- or countable-dimensional function $u(x)$ in $(X, \mathcal{A}, \mu_x)$ be measurable relative to $\mathcal{A}$ and the corresponding σ-algebra of Borel sets $\mathcal{B}^n$ will exist such that its range Δ_u will be Borel set and the σ-algebra $\mathcal{A}_u = u^{-1}(\mathcal{B}^n)$ induced by it will coincide with $\mathcal{A}$.

As any probability space may be considered as the phase probability space of some random variable then for the existence of the regular conditional probabilities in the probability space $(\Omega, \mathcal{S}, P)$ it is sufficient the existence of finite- or countable-dimensional random variable having Borel range and inducing the σ-algebra $\mathcal{S}$. $\triangleleft$

Hence we get the second generalization of the Kolmogorov theorem.

Theorem 2.6.4. *If there exists a finite- or countable-dimensional random variable with Borel range in each of the probability spaces (Ω_t, S_t, P_t), $t \in T$ which induces the correspondent σ-algebra S_t then the consistent set of the probabilities $P_{t_1, \ldots, t_n}$ in all finite products of the spaces (Ω_t, S_t) determines uniquely the probability in the infinite product of the spaces (Ω^T, S^T).*

2.7. Probability Measure of Random Function

2.7.1. Definitions

In elementary probability theory a random function is called such a function whose value at each given value of an argument is a random variable. It follows from this definition that a random function represents a set of the random variables correspondent to all values of the argument from its range (the domain of a random function). As a result of a trial a random function at each value of an argument assumes some concrete possible value. The aggregate of these values correspondent to all values of an argument represents some concrete function. Thus while observing a random function some function is obtained in each trial. In different trials we receive different functions. Each function which may be obtained as a result of observation of a random function is called *a realization* or *a sample function*. Each realization of a random function represents a concrete, deterministic function of the same argument.

Here the scalar and finite-dimensional functions of a scalar independent variable will be essentially studied. Such random functions are usually called the *stochastic processes* and an argument is time t. In the case when the point in more than one dimensional space serves as an argument the random functions are called the *stochastic fields*. The stochastic fields dependent on time are called the *dynamic stochastic fields*. Therefore in general theory of stochastic systems the random functions should be assumed as the elements of arbitrary functional spaces.

The general definition of a random variable given in Section 2.2 includes also the random functions. According to this definition a random function represents a random variable with the values in some functional space. But for the constructive definition of a probability measure of a random function it is convenient to consider it as the set of random variables with the values in sufficiently simple spaces (on the real axis, on the complex plane or in the finite-dimensional space). Therefore in applications the following definition of a random function is assumed.

A *random function* in the measurable space $(X, \mathcal{A})$ is called an infinite set of the random variables $\{X_t, t \in T\}$ in $(X, \mathcal{A})$, i.e. the set of $(\mathcal{S}, \mathcal{A})$-measurable functions of the point $\{x_t(\omega), t \in T\}$ in some probability space $(\Omega, \mathcal{S}, P)$. In other words, a random function $x(t, \omega)$ in $(X, \mathcal{A})$ is function which maps the product of the spaces $T \times \Omega$ in X and whose section in any point $t \in T$ is $(\mathcal{S}, \mathcal{A})$-measurable.

Depending on the character of the research it is expedient to use different notations of the random functions. If the most essential fact is that the values of a random function at fixed values of the argument $t \in T$ represent the random variables then we denote the random functions by $X_t = x_t(\omega)$, $Y_t = y_t(\omega)$ and so on (t is considered as a parameter). If the most essential fact is that the realizations of a random function are the functions of the argument $t \in T$ then we shall denote the random functions by $X(t) = x(t, \omega)$, $Y(t) = y(t, \omega)$ and so on.

2.7.2. Measurable Random Functions

The definition of a random function requires only the measurability of the sections at any t of the function $x(t, \omega)$ and does not require its measurability as the function of two variables. The random function $X(t) = x(t, \omega)$ is called *measurable* if the σ-algebra $\mathcal{F}$ is determined in the space T and the function $x(t, \omega)$ represents $(\mathcal{F} \times \mathcal{S}, \mathcal{A})$-measurable function mapping $T \times \Omega$ into X. This definition is noncontradictory as all the sections of a measurable function are measurable and consequently, a measurable random function is always a random function in the sense of the first definition (Subsection 2.7.1).

2.7.3. Multi-Dimensional Distributions of Random Function

If the probability measure $\mu_t(A)$ of the value of the random function X_t for each value $t \in T$ at this t and a consistent set of the regular conditional probability measures $\mu_{t,l_1,\ldots,l_n}(A|x_{l_1,\ldots,l_n})$ relative to its values $X_{l_1}, \ldots, X_{l_n}$ at all others values of the argument t, $l_1, \ldots, l_n \in T$, $n = 1, 2, \ldots$ are given then as it is known from Section 2.6 the probability measure μ of the random function X_t on the σ-algebra $\mathcal{A}^T$ of the space X^T is determined uniquely. It is necessary at first to determine by formulae (2.5.1) the multi-dimensional distributions $\mu_{t_1,\ldots,t_n}$ of the random function X_t, i.e. the joint probability measures of its values $X_{t_1}, \ldots, X_{t_n}$ at all n, $t_1, \ldots, t_n$:

$$
\left.
\begin{aligned}
&\mu_{t_1,t_2}(A_{t_1} \times A_{t_2}) = \int_{A_{t_1}} \mu_{t_2,t_1}(A_{t_2}|x_1)\mu_{t_1}(dx_1), \\
&\mu_{t_1,\ldots,t_n}(A_{t_1} \times \cdots \times A_{t_n}) \\
&\quad = \int_{A_{t_1} \times \cdots \times A_{t_{n-1}}} \mu_{l_n,l_1,\ldots,l_{n-1}}(A_{t_n}|x_1,\ldots,x_{n-1}) \\
&\qquad \times \mu_{t_1,\ldots,t_{n-1}}(dx_1 \times \cdots \times dx_{n-1}) \quad n = 3, 4, \ldots.
\end{aligned}
\right\} \tag{2.7.1}
$$

Then we need to determine the probability measure μ of the random function X_t on the measurable rectangles of the space $(X^T, \mathcal{A}^T)$ by formula (2.5.2):

$$
\mu(R) = \mu_{t_1,\ldots,t_n}(A_{t_1} \times \cdots \times A_{t_n}). \tag{2.7.2}
$$

And finally we need to extend this probability measure on the σ-algebra $\mathcal{A}^T$ (in formula (2.7.2) R is a rectangle with the sides $A_{t_1}, \ldots, A_{t_n}$, $R = \{x : x_{t_1} \in A_{t_1}, \ldots, x_{t_n} \in A_{t_n}\}$. Formula (2.7.3) determines the extension of the probability measure μ of the random function X_t on the σ-algebra $\mathcal{A}_{t_1} \times \cdots \times \mathcal{A}_{t_n}$ of all the cylinders of the space X^T with the measurable basis in one and the same finite product of the spaces $(X_{t_1} \times \cdots \times X_{t_n}, \mathcal{A}_{t_1} \times \cdots \times \mathcal{A}_{t_n})$:

$$\mu(C_{t_1,\ldots,t_n}) = \mu_{t_1,\ldots,t_n}(B_{t_1,\ldots,t_n}). \tag{2.7.3}$$

Here $B_{t_1,\ldots,t_n} \in \mathcal{A}_{t_1} \times \cdots \times \mathcal{A}_{t_n}$ is the base of the cylinder $C_{t_1,\ldots,t_n}$ situated in $X_{t_1} \times \cdots \times X_{t_n}$.

R e m a r k. In a given case all the spaces $(X_t, \mathcal{A}_t)$ coincide with the space of the values $(X, \mathcal{A})$ of the random function X_t. At any $t_1, \ldots, t_n$ the space $(X_{t_1} \times \cdots \times X_{t_n}, \mathcal{A}_{t_1} \times \cdots \times \mathcal{A}_{t_n})$ represents the product of n identical spaces $(X, \mathcal{A})$. This product is usually denoted by $(X^n, \mathcal{A}^n)$. The indexes $t_1, \ldots, t_n$ in the detailed denotion play the same role as the different letters in the denotion of different coordinate axes and each of them represents the real axis R.

Instead of assigning the distributions and the regular conditional distributions of the values a random function we may directly assign the consistent set of the multi-dimensional distributions $\mu_{t_1,\ldots,t_n}$, $t_1, \ldots, t_n$ $\in T$ ($n = 1, 2, \ldots$). According to the Kolmogorov theorem (Section 2.6) we conclude.

Theorem 2.7.1. *The consistent set of the multi-dimensional distributions of a scalar or finite-dimensional vector random function $X(t)$ determines uniquely the probability measure of this random function in its functional phase space X^T on the σ-algebra $\mathcal{A}^T$ generated by all Borel rectangles (i.e. by the rectangles whose sides are Borel sets in the values space of the random function X.*

Theorem 2.7.1 is also generalized according to Theorem 2.6.3, on the vector random function with the countable set of the components.

Finally, Theorem 2.6.4 gives more general proposal.

Theorem 2.7.2. *In order that the joint set of the multi-dimensional distributions of the random function $X(t)$ in the measurable space $(X, \mathcal{A})$ will determine uniquely its probability measure on $(X^T, \mathcal{A}^T)$ it is sufficient that in $(X, \mathcal{A})$ there exists the finite- or countable-dimensional function $u(x)$ measurable relative to $\mathcal{A}$ and the correspondent σ-algebra of Borel sets with Borel range Δ_u which induces the σ-algebra $\mathcal{A}$ in X.*

Naturally the question arises: whether any consistent set of the multi-dimensional distributions $\mu_{t_1,\dots,t_n}$ determines some random function. The Kolmogorov theorem gives a positive answer. Really, according to the Kolmogorov theorem any consistent set of multi-dimensional distributions determines uniquely the probability μ in the space $(X^T, \mathcal{A}^T)$. The triple $(X^T, \mathcal{A}^T, \mu)$ represents the probability space, any random variable in this space is a measurable function of a point of a given space (for example a functional or an operator).

In elementary probability theory the one-dimensional and multi-dimensional distributions are usually assigned by the correspondent distribution functions, the densities and the characteristic functions.

E x a m p l e 2.7.1. The multi-dimensional distribution functions of the random function $X(t) = \varphi(t, U)$ where $\varphi(t, U)$ is a given function of the variable t and the finite-dimensional random variable U are determined by formula

$$F_n(x_1, \dots, x_n; t_1, \dots, t_n) = \int_{A_n} f(u)\,du \quad (n = 1, 2, \dots), \qquad (I)$$

where $f(u)$ is the probability density of the random variable U and $A_n = \{u : \varphi(t_1, u) < x_1, \dots, \varphi(t_n, u) < x_n\}$.

Any multi-dimensional distribution of a random function determines uniquely its multi-dimensional distributions of lower dimension. In general case none of the multi-dimensional distributions determine distributions of highest dimensions. But there exist the random functions for which any of the multi-dimensional distributions determines the whole sequence of its multi-dimensional distributions.

For the random function $X(t)$ with the independent values the random variables $X_{t_1}, \dots, X_{t_n}$ are independent at any $t_1, \dots, t_n \in T$ and at any natural n. Therefore the following relation takes place for the random function $X(t)$ with the independent values for $(n = 1, 2, \dots)$:

$$f_n(x_1, \dots, x_n; t_1, \dots, t_n) = f_1(x_1; t_1) f_1(x_2; t_2) \dots f_1(x_n; t_n). \quad (2.7.4)$$

Thus *all multi-dimensional distributions of the random function with the independent values are determined uniquely by its one-dimensional distributions.*

Another type of the random functions whose all multi-dimensional distributions are determined by one of them are Markov random processes (Section 2.7.4).

2.7.4. Markov Random Processes

At first we cite the information from the elementary Markov processes theory. Before giving the definition of a Markov random process we shall give the definition of a Markov sequence of the random variables. A sequence of the random variables $\{X_p\}$ is called *a Markov random sequence* if at any finite sequence $p_1 < \cdots < p_n$ the conditional distribution of the variable X_{p_n} relative to the variables $X_{p_1}, \ldots, X_{p_{n-1}}$ depends only on $X_{p_{n-1}}$, i.e. coincides with the conditional distribution of the variable X_{p_n} relative to $X_{p_{n-1}}$.

The random variable $X_t = X(t)$ of continuous scalar variable t is called *a Markov random process* if at any choice of the values sequence of the argument $\{t_p\}$, $t_{p-1} < t_p$ the sequence of the random variables $\{X_{t_p}\}$ is Markov one. A Markov random process $X(t)$ possesses such property that at its given value x at some instant τ, $X(\tau) = x$ the distribution of its value X_t at any consequent instant $t > \tau$ is determined uniquely by its value x at the instant τ and does not depend absolutely on its realization until the instant τ. In other words, to all the realizations of a Markov process $X(t)$ which take one and the same value x at τ, corresponds *a conditional distribution* $X(t)$ at any instant $t > \tau$ independent of the realization until the instant τ.

As it is known (Subsection 2.4.2) the n-dimensional density for the random process $X(t)$ according to the multiplication densities theorem is determined by the following formula:

$$f_n(x_1, \ldots, x_n; t_1, \ldots, t_n) = f_1(x_1; t_1) f_2(x_2; t_2 | x_1; t_1)$$

$$\times f_3(x_3; t_3 | x_1, x_2; t_1, t_2) \ldots f_n(x_n; t_n | x_1, \ldots, x_{n-1}; t_1, \ldots, t_{n-1}).$$

$$(2.7.5)$$

Here $f_k(x_k; t_k | x_1, \ldots, x_{k-1}; t_1, \ldots, t_{k-1})$ is a conditional density of the value X_{t_k} of the process $X(t)$ at its values given $x_1, \ldots, x_{k-1}$ for $t = t_1, \ldots, t_{k-1}$ ($k = 2, 3, \ldots$). But for a Markov process $X(t)$ according to the definition we have

$$f_k(x_k; t_k | x_1, \ldots, x_{k-1}; t_1, \ldots, t_{k-1}) = f_k(x_k; t_k | x_{k-1}; t_{k-1}). \quad (2.7.6)$$

Consequently, the n-dimensional density of a Markov random process $X(t)$ by virtue of (2.7.5) and (2.7.6) is determined by formula

$$f_n(x_1, \ldots, x_n; t_1, \ldots, t_n)$$

$$= f_1(x_1; t_1) f_2(x_2; t_2 | x_1; t_1) \ldots f_n(x_n; t_n | x_{n-1}; t_{n-1}). \quad (2.7.7)$$

Expressing the conditional densityes $f_k(x_k; t_k | x_{k-1}; t_{k-1})$ in terms of the one-dimensional and the two-dimensional densities we get for ($n = 3, 4, \ldots$)

$$f_n(x_1, \ldots, x_n; t_1, \ldots, t_n)$$

$$= \frac{f_2(x_1, x_2; t_1, t_2) f_2(x_2, x_3; t_2, t_3) \ldots f_2(x_{n-1}, x_n; t_{n-1}, t_n)}{f_1(x_2; t_2) f_1(x_3; t_3) \ldots f_1(x_{n-1}; t_{n-1})} \qquad (2.7.8)$$

Thus *all multi-dimensional distributions of a Markov process are determined uniquely by its two-dimensional distributions.*

In applications the conditional density $f_2(x_k; t_k | x_{k-1}; t_{k-1})$ ($k = 2, 3, \ldots$) is called *a transfer density* and is denoted simply as $f(x_k; t_k | x_{k-1}; t_{k-1})$.

The multi-dimensional distributions of the random function X_t with given first probability density $f_1(x; t)$ and the transfer density $f(x; t | y; s)$ are determined according to (2.7.1) by formula

$$\mu_{t_1, \ldots, t_n}(A_{t_1} \times \cdots \times A_{t_n}) = \int\limits_{A_{t_1}} dx_1 \int\limits_{A_{t_2}} dx_2 \ldots \int\limits_{A_{t_n}} f_1(x_1; t_1) f(x_2; t_2 | x_1; t_1)$$

$$\times f(x_3; t_3 | x_2; t_2) \cdots f(x_n; t_n | x_{n-1}; t_{n-1}) dx_n. \qquad (2.7.9)$$

The extension $\mu_{t_1, \ldots, t_n}$ over all measurable sets of the n-dimensional space $(X_{t_1} \times \cdots \times X_{t_n}, A_{t_1} \times \cdots \times A_{t_n})$ is expressed by formula

$$\mu_{t_1, \ldots, t_n}(B)$$

$$= \int\limits_B \cdots \int f_1(x_1; t_1) f(x_2; t_2 | x_1; t_1) \cdots f(x_n; t_n | x_{n-1}; t_{n-1}) dx_1 \ldots dx_n.$$

$$(2.7.10)$$

R e m a r k. After finding the multi-dimensional distributions the probability measure of the random function $X(t)$ is determined on the set of measurable rectangles of the space $(X^T, \mathcal{A}^T)$ by formula (2.7.2). Let us denote by $C_{t_1, \ldots, t_n}$ the cylinder in X^T with the base $B_{t_1, \ldots, t_n}$ in the n-dimensional space $(X_{t_1} \times \cdots \times X_{t_n})$. The formula $\mu(C_{t_1, \ldots, t_n}) = \mu_{t_1, \ldots, t_n}(B_{t_1, \ldots, t_n})$ determines the extension of the probability measure of the random function X_t over the set of all measurable cylinders of the functional space X^T. On the basis of the theorem about the extension of a measure for any set $A \in \mathcal{A}^T$ at any $\varepsilon > 0$ we may find such a cylinder $C_A \supset A$ (with a finite-dimensional base) that $\mu(C_A) - \varepsilon \leq \mu(A) \leq \mu(C_A)$. Thus the known theorem about the measure extension to approximate with any degree of accuracy the value of the probability measure of the random function on any measurable set of the functional space $(X^T, \mathcal{A}^T)$ by its value on some cylinder.

2.7.5. Normal Random Process

At first we cite for further study the known from elementary theory of random functions the formulae for the multi-dimensional normal densities of the distributions.

Firstly we consider the scalar random process $X_t = X(t)$ with zero expectation whose multi-dimensional distributions are normal, and consequently, they are determined by the following probability densities f_n and by the characteristic functions g_n:

$$f_n = f_n(x_1, \ldots, x_n; t_1, \ldots, t_n) = [(2\pi)^n |K_n|]^{-1/2} \exp\left\{-\frac{1}{2} x^T K_n^{-1} x\right\},$$

$$g_n = g_n(\lambda_1, \ldots, \lambda_n; t_1, \ldots, t_n) = \exp\left\{-\frac{1}{2} \lambda^T K_n \lambda\right\} \quad (n = 1, 2, \ldots),$$

$$(2.7.11)$$

where

$$K_n = \begin{bmatrix} K(t_1, t_1) & K(t_1, t_2) & \ldots & K(t_1, t_n) \\ K(t_2, t_1) & K(t_2, t_2) & \ldots & K(t_2, t_n) \\ \vdots & \vdots & \ldots & \vdots \\ K(t_n, t_1) & K(t_n, t_2) & \ldots & K(t_n, t_n) \end{bmatrix},$$

$K(t, s)$ is the covariance function of the random process $X_t = X(t)$, and $|K_n|$ is the determinant of the matrix K_n. Such random functions is called *normally distributed*. It is easy to verify that the set of normal multi-dimensional distributions is consistent. Formula (2.7.3) determines the probability measure of the scalar random process X_t on the set of all measurable cylinders of the functional space $(X^T, \mathcal{A}^T)$:

$$\mu(C_{t_1,\ldots,t_n}) = [(2\pi)^n |K_n|]^{-1/2} \int\limits_{B_{t_1,\ldots,t_n}} \exp\left\{-\frac{1}{2} x^T K_n^{-1} x\right\} dx,$$

$$(2.7.12)$$

where $B_{t_1,\ldots,t_n}$ is the base of the cylinder $C_{t_1,\ldots,t_n}$. By expression of such a type we may approximate the value of the probability measure of the considered scalar random process on any measurable set $A \in \mathcal{A}^T$ with any degree of the accuracy.

It is convenient to determine the normal distribution for the finite-dimensional vector random process by the characteristic function:

$$g_{t_1,\ldots,t_n}(\lambda_1, \ldots, \lambda_n) = \exp\left\{i\lambda^T m_n - \frac{1}{2} \lambda^T K_n \lambda\right\} \quad (n = 1, 2, \ldots).$$

$$(2.7.13)$$

Here

$$\lambda = \left[\, \lambda_1^T \lambda_2^T \ldots \lambda_n^T \,\right]^T \,, \quad m_n = \left[\, m_x(t_1)^T m_x(t_2)^T \ldots m_x(t_n)^T \,\right]^T \,,$$

$$K_n = \left[\begin{array}{cccc} K_x(t_1,t_1) & K_x(t_1,t_2) & \ldots & K_x(t_1,t_n) \\ K_x(t_2,t_1) & K_x(t_2,t_2) & \ldots & K_x(t_2,t_n) \\ \ldots & \ldots & \ldots & \ldots \\ K_x(t_n,t_1) & K_x(t_n,t_2) & \ldots & K_x(t_n,t_n) \end{array}\right] ,$$

where the matrices-columns λ and m_n and the matrix K_n should be considered as the block matrices. If all the matrices K_n are nonsingular then we may also determine a normal distribution of a random process by means of the multi-dimensional densities of the form

$$f_n(x_1, \ldots, x_n; t_1, \ldots, t_n)$$

$$= \left[(2\pi)^n |K_n|\right]^{-1/2} \exp\left\{-\frac{1}{2}(u_n^T - m_n^T)K_n^{-1}(u_n - m_n)\right\} \quad (n = 1, 2, \ldots).$$

$$(2.7.14)$$

Here in addition to the previous denotions $u_n = \left[\, x_1^T x_2^T \ldots x_n^T \,\right]^T$.

Analogously the consistency of the multi-dimensional distribution functions of a finite-dimensional random processes is verified.

2.7.6. Conditionally Normal Random Processes

Let us consider an important class of random processes for which the first probability density $f_1(x;t)$ of the random function $X_t = X(t)$ and the conditional probability densities $f_n = f_n(x;t|y_1, \ldots, y_n; s_1, \ldots, s_n)$ of its value at a given t relative to its values at all others values t are normal and at $n = 1, 2, \ldots$ are given by formulae

$$f_1(x;t) = (2\pi c_0)^{-1/2} \exp\left\{-\frac{1}{2}c_0 x^2\right\},$$

$$f_n(x;t|y_1, \ldots, y_n; s_1, \ldots, s_n) = (2\pi c_n)^{-1/2} \exp\left\{-\frac{1}{2}c_n(x - a_n)^2\right\},$$

$$(2.7.15)$$

Here $c_0 = \left[K(t,t)\right]^{-1}$; $c_n = |K_n|/|\bar{K}_n|$, where K_n is the matrix in which the variables $t_1, \ldots, t_n$ are substituted by the variables $s_1, \ldots, s_n$; $\bar{K}_n$ is the bordered matrix K_n equal to

$$\bar{K}_n = \left[\begin{array}{cccc} K(t,t) & K(t,s_1) & \ldots & K(t,s_n) \\ K(s_1,t) & K(s_1,s_1) & \ldots & K(s_1,s_n) \\ \vdots & \vdots & \ldots & \vdots \\ K(s_n,t) & K(s_n,s_1) & \ldots & K(s_n,s_n) \end{array}\right] ,$$

$a_n = \sum_{k=1}^{n} d_k y_k$, where the coefficients $d_1, \ldots, d_n$ satisfy the linear equations

$$\sum_{k=1}^{n} K(s_m, s_k) d_k = K(s_m, t) \quad (m = 1, \ldots, n).$$

It is easy to check that formulae (2.7.1) give the same normal multidimensional distributions of the random process X_t. Consequently, the random process X_t is normally distributed and its probability measure is determined by formula (2.7.12).

R e m a r k. These results are easily generalized on the random processes with any consistent set of the multi-dimensional probability densities $f_n(x_1, \ldots, x_n; t_1, \ldots, t_n)$ (or with any consistent set of the conditional distributions $f_n(x; t | y_1, \ldots, y_n; s_1, \ldots, s_n)$, and in this case $f_n(x_1, \ldots, x_n; t_1, \ldots, t_n) = f_1(x_1; t_1) \, f_1(x_2; t_2 | x_1; t_1) \ldots f_{n-1}(x_n; t_n | x_1, \ldots, x_{n-1}; t_1, \ldots, t_{n-1}))$. The probability measure of a random process of the set of all measurable cylinders in the functional space $(X^T, \mathcal{A}^T)$ is determined in the general case by formula

$$\mu(C_{t_1, \ldots, t_n}) = \int \ldots \int_{B_{t_1, \ldots, t_n}} f_n(x_1, \ldots, x_n; t_1, \ldots, t_n) dx_1 \ldots dx_n.$$

$$(2.7.16)$$

By the formula of such a type we may also approximate with any degree of the accuracy the value of the probability measure of the random process on any measurable set of the space $(X^T, \mathcal{A}^T)$.

2.7.7. Streams of Events

Let us consider some notions of the theory of the stochastic streams of the events which are important for the stochastic systems of queueing.

A *stream* of the homogeneous events may be characterized by the distributions of the random instants of the events occurrences. Let us assume T_0 as the initial instant and suppose that the probability measure $\nu_1(A)$ of the moment T_1 of the occurrence of the first event and the conditional probability measures $\nu_n(A | t_1, \ldots, t_{n-1})$ of the instants of the occurrences T_n ($n = 2, 3, \ldots$) of all following events are known. All these conditional distributions are regular as they are determined on the σ-algebra $\mathcal{B}$ of Borel sets of the real axis R (Section 2.4). Considering T_n as a random function of an integer argument n we may determine its multi-dimensional distributions by formula

$$\nu_{1 \ldots n}(A_1 \times \cdots \times A_n) = \int_{A_1} \nu_1(d\tau_1) \int_{A_2} \nu_2(d\tau_2 | d\tau_1)$$

$$\times \ldots \times \int_{A_{n-2}} \nu_{n-2}(d\tau_{n-2}|\tau_1, \ldots, \tau_{n-3})$$

$$\times \int_{A_{n-1}} \nu_n(A_n|\tau_1, \ldots, \tau_{n-1})\nu_{n-1}(d\tau_{n-1}|\tau_1, \ldots, \tau_{n-2}) \qquad (2.7.17)$$

and for any system of numbers $k_1 < k_2 < \ldots < k_n$

$$\nu_{k_1 \ldots k_n}(A_1 \times \cdots \times A_n) = \nu_{1 \ldots k_n}(R'_{k_1, \ldots, k_n}). \qquad (2.7.18)$$

Here $R'_{k_1, \ldots, k_n}$ is the rectangle in the space $(R_{k_1} \times \cdots \times R_{k_n}, \mathcal{B}_{k_1} \times \cdots \times \mathcal{B}_{k_n})$. After determining according to (2.7.2) the probability measure ν on all measurable rectangles of the infinite-dimensional space $(R^\infty, \mathcal{B}^\infty)$ by formula

$$\nu(R_{k_1, \ldots, k_n}) = \nu_{k_1 \ldots k_n}(A_1 \times \cdots \times A_n), \qquad (2.7.19)$$

where $R_{k_1, \ldots, k_n}$ is a measurable rectangle in $(R^\infty, \mathcal{B}^\infty)$ with the base $A_1 \times \cdots \times A_n$ in $(R_{k_1} \times \cdots \times R_{k_n}, \mathcal{B}_{k_1} \times \cdots \times \mathcal{B}_{k_n})$ we may extend uniquely this measure on the σ-algebra $\mathcal{B}^\infty$ of the space R^∞.

Let us consider a special case of a homogeneous stream in conformity to the recurrent stream of the events. Recall that a stream of the events is called *recurrent* if all the intervals between adjacent events are the independent random variables. Below $(A - \tau)$ means that as usual a set of the numbers obtained by the subtraction of τ from all the numbers contained in A. Then for such a stream of the events with one and the same distribution ν of all the intervals between adjacent events we have $\nu_1(A) = \nu(A)$, $\nu_n(A|t_1, \ldots, t_{n-1}) = \nu_n(A - t_{n-1})$ $(n = 2, 3, \ldots)$ and formula $(2.7.17)$ for $\nu_{1 \ldots n}$ will take the form

$$\nu_{1 \ldots n}(A_1 \times \cdots \times A_n)$$

$$= \int_{A_1} \nu(d\tau_1) \int_{A_2 - \tau_1} \nu(d\tau_2) \ldots \int_{A_{n-1} - \tau_{n-2}} \nu(A_n - \tau_{n-1})\nu(d\tau_{n-1}).$$

$$(2.7.20)$$

If there exists the probability density $f(t)$ of the time interval between the adjacent events

$$\nu(A) = \int_A f(t)dt,$$

then formula (2.7.20) may be rewritten in the form

$$\nu_{1 \ldots n}(A_1 \times \cdots \times A_n) = \int_{A_1} f(\tau_1)d\tau_1 \int_{A_2} f(\tau_2 - \tau_1)d\tau_2 \ldots \int_{A_n} f(\tau_n - \tau_{n-1})d\tau_n.$$

$$(2.7.21)$$

It is convenient to connect the stream of the events with the random function $X(t)$ which represents the number of the events occuring at the time interval $[t_0, t]$. Evidently, all the realizations of this random function are the step functions which increase by unit jump at each instant of the occurrence of the next event (by n if the moments of the occurrence of n events coincide). Let us find the multi-dimensional distributions $\mu_{t_1,\ldots,t_n}$ of the random function $X(t)$. As at any moment t the random function $X(t)$ may have only integer non-negative values then for determining $\mu_{t_1,\ldots,t_n}$ it is sufficient to find the probabilities

$$p_{t_1,\ldots,t_n}(m_1,\ldots,m_n) = P(\bigcap_{k=1}^{n} \{X(t_k) = m_k\}) \qquad (2.7.22)$$

for all $m_1 = 0, 1, \ldots$, $m_2 = m_1, m_1 + 1, \ldots$, $m_n = m_{n-1}, m_{n-1} + 1, \ldots$. But these probabilities are easily determined by means of the known distributions $\nu_{k_1,\ldots,k_n}$ of the instants of the events occurrences found before. It is easy to see that

$$p_{t_1,\ldots,t_n}(m_1,\ldots,m_n) = \nu_{m_1,m_1+1,m_2,m_2+1,\ldots,m_n,m_n+1}([t_0, t_1]$$

$$\times (t_1, t_2](t_2, t_3] \times \cdots \times (t_{n-1}, t_n] \times (t_n, \infty)). \qquad (2.7.23)$$

For the recurrent stream of the events formula (2.7.23) takes the form

$$p_{t_1,\ldots,t_n}(m_1,\ldots,m_n) = \int_{t_0}^{t_1} \nu(d\tau_1) \int_{0}^{t_1-\tau_1} \nu(d\tau_2) \ldots \int_{0}^{t_1-\tau_{m_1-1}} \nu(d\tau_{m_1})$$

$$\times \int_{t_1-\tau_{m_1}}^{t_2-\tau_{m_1}} \nu(d\tau_{m_1+1}) \int_{0}^{t_2-\tau_{m_1+1}} \nu(d\tau_{m_1+2})$$

$$\times \ldots \times \int_{t_{n-1}-\tau_{m_{n-1}}}^{t_n-\tau_{m_{n-1}}} \nu(d\tau_{m_{n-1}+1}) \ldots \int_{0}^{t_n-\tau_{m_{n-1}}} \nu(d\tau_{m_n}) \int_{t_n-\tau_{m_n}}^{\infty} \nu(d\tau_{m_n+1}).$$

$$(2.7.24)$$

In the special case of the stationary *Poisson stream of events* $f(t) = \lambda e^{-\lambda t}$ where λ is the intensity of a stream (the mean number of the events in an unit of time) formula (2.7.24) gives

$$p_{t_1,\ldots,t_n}(m_1,\ldots,m_n)$$

$$= \frac{\lambda^{m_n}(t_1 - t_0)^{m_1}(t_2 - t_1)^{m_2-m_1}\ldots(t_n - t_{n-1})^{m_n-m_{n-1}}}{m_1!(m_2 - m_1)!\ldots(m_n - m_{n-1})!} e^{-\lambda(t_n-t_0)},$$

$$(2.7.25)$$

$$(0 \leq m_1 \leq m_2 \leq \cdots \leq m_n \leq \cdots).$$

Formula (2.7.25) may be easily derived directly taking into account that in the case of Poisson stream of the events variables $X(t_1) = m_1$, $X(t_k) - X(t_{k-1}) = m_k - m_{k-1}$ $(k = 2, \ldots, n)$ are independent.

After finding the probabilities $p_{t_1,\ldots,t_n}(m_1, \ldots, m_n)$ the multi-dimensional distributions of the random function $X(t)$ will be determined on the sets of the measurable rectangles in the correspondent finite-dimensional spaces $(X_{t_1} \times \cdots \times X_{t_n}, \mathcal{A}_{t_1} \times \cdots \times \mathcal{A}_{t_n})$ by formula

$$\mu_{t_1,\ldots,t_n}(A_{t_1} \times \cdots \times A_{t_n}) = \sum_{m_k \in A_{t_k}} p_{t_1,\ldots,t_n}(m_1, \ldots, m_n). \qquad (2.7.26)$$

It is easy to verify that these multi-dimensional distributions are consistent. By the Kolmogorov theorem the multi-dimensional distributions $\mu_{t_1,\ldots,t_n}$ determine uniquely the probability measure of the random function $X(t)$ on the σ-algebra $\mathcal{A}^T$ of its phase space X^T.

2.7.8. Excursions of Random Functions

The so-called excursion problems of the random function are of great importance for the systems reliability theory. The final aim of this problem is the determination of the probability of the fact that the random function X_t will not cross the given limits (a_t, b_t) in some region T, $t \in T$. Formulae (2.7.9) and (2.7.12) give the opportunity to determine the probability of the fact that the random function X_t will not cross the limits (a_t, b_t) in any finite set of the points $t_1, \ldots, t_n$:

$$P(a_{t_k} < X_{t_k} < b_{t_k}, k = 1, \ldots, n)$$

$$= \int_{a_{t_1}}^{b_{t_1}} \cdots \int_{a_{t_n}}^{b_{t_n}} f_n(x_1, \ldots, x_n; t_1, \ldots, t_n) dx_1 \ldots dx_n. \qquad (2.7.27)$$

If the realizations of the random function are "sufficiently smooth" (Section 2.8) then we may hope that this formula will permit to determine the probability of the absence of the excursions of the random function X_t in the region T with sufficient accuracy at sufficiently dense disposition of the points $t_1, \ldots, t_n$.

2.7.9. Probabilities of Events Connected with Random Functions

The multi-dimensional distributions of the random function determine its probability density only on the σ-algebra $\mathcal{A}^T$. If the random process is a random variable not only in $(X^T, \mathcal{A}^T)$ but in the measurable space $(X^T, \mathcal{F})$ with more wide σ-algebra $\mathcal{F} \supset \mathcal{A}^T$ then the multi-dimensional distributions in general case do not determine the probability measure of the random process on the sets from $\mathcal{F}$ which do not belong to $\mathcal{A}^T$.

The results of the study of the structure of the σ-algebra in infinite product of the spaces (Section 2.5) show that any set from the σ-algebra $\mathcal{A}^T$ in the phase space of the random function represents a set of the functions $x(t)$ on which the restrictions are imposed in not more than countable set of the points t. Therefore the σ-algebra $\mathcal{A}^T$ is sufficiently rich and contains all interesting events only in the case of the countable set T when the random process may be presented in the form of the sequence of the random variables. If the set T is uncountable then many important events do not enter into σ-algebra $\mathcal{A}^T$ and their probabilities in the general case cannot be calculated by means of the multi-dimensional distributions of the random process. The σ-algebra $\mathcal{A}^T$ is rather poor for solving such problems.

In order to approximate the probabilities of the events restricting the values of the random process in uncountable set of the points t by means of the multi-dimensional distributions it is necessary to narrow the class of studied random processes after imposing some definite restrictions on the behaviour of their realizations. For example, the excursions problem considered in Subsection 2.7.6 refers to the class of the problems which require such restrictions. The generalization of this problem is the problem of the probabilities calculation for the events of the type

$$\{\omega : x_t(\omega) \in A_t \quad \text{at all} \quad t\} = \bigcap_{t \in T} x_t^{-1}(A_t). \qquad (2.7.28)$$

In the general case the sets of such a type may not belong to the σ-algebra $\mathcal{S}$ in the main probability space in view of the uncountability of T. Even if a random process is a random variable in more wide measurable space $(X^T, \mathcal{F})$, $\mathcal{F} \subset \mathcal{A}^T$ and the sets of the considered type belong to the σ-algebra $x^{-1}(\mathcal{F}) \subset \mathcal{S}$, and consequently, are measurable relative to $\mathcal{S}$ and have the definite probabilities then in this case the multi-dimensional distributions give none information about these probabilities (the values of the probability measure of the random function on these sets). Meanwhile in applications the multi-dimensional distributions serve yet as the unique instrument for the constructive determination of the probabilities of the events connected with the random processes. Let us illustrate this statement by the examples.

E x a m p l e 2.7.2. The probability space $(\Omega, \mathcal{S}, P)$ represents the real axis with the σ-algebra of Borel sets and a normal distributions of the probabilities with mean in the point $\omega = 0$ and the variance equal to one. Let us consider the random function of the scalar argument $X_t = x_t(\omega) = \omega t$. It is evident that all multi-dimensional distributions of this scalar function are normal distributions on the real axis and on the lines $x_1/t_1 = x_2/t_2 = \cdots = x_n/t_n$ $(n = 2, 3, \ldots)$, i.e. the normal distributions with the covariance matrices which are equal to $\{t_p t_q\}_{p,q=1,\ldots,n}$. These distributions determine uniquely the probability measure of the considered random function in the functional space $(X^T, \mathcal{A}^T)$. This measure is determined in such a way only on the sets which restrict a given random function almost in countable set of the points. Meanwhile the probability of the absence of the excursions of this random function at the interval $(-T, T)$ beyond the interval $(-a, a)$ is calculated in this case quite elementary:

$$P(|X_t| < a \quad \text{at all} \quad t, |t| < T)$$

$$= P(\{\omega : |\omega t| < a, |t| < T\}) = \sqrt{\frac{2}{\pi}} \int_0^{a/T} e^{-\frac{1}{2}z^2} dz = 2\Phi\left(\frac{a}{T}\right). \qquad (I)$$

Similarly the probabilities of such events as $A_t = \{X_t$ is continuous at the point $t\}$, $B = \{X_t$ is continuous at all $t\}$, are easily calculated $P(A_t) = P(B) = 1$. It means that the considered random process X_t is a random variable not only in the space $(X^T, \mathcal{A}^T)$ but in the space $(X^T, \mathcal{F})$ also with more rich σ-algebra $\mathcal{F}$ and that the considered events enter into the σ-algebra $\mathcal{F}$ (i.e. the probability measure of the function X_t is determined for them).

E x a m p l e 2.7.3. Let us consider now another random process in the probability space of the previous example

$$Y_t = y_t(\omega) = \begin{cases} \omega t & \text{at} \quad t \neq \frac{1}{n\omega}, n = 1, 2, \ldots \\ b & \text{at} \quad t = \frac{1}{n\omega}, n = 1, 2, \ldots \end{cases} \qquad (I)$$

As at any fixed t the probability of the occurrence of the realization with $\omega = 1/nt$ at any is n equal to zero then all the multi-dimensional distributions of the random process Y_t coincide with the correspondent multi-dimensional distributions of the random function of the previous example. But the probability of the absence of the excursions beyond the interval $(-a, a)$ at $|t| < T$ of this random process is different:

$$
P(|X_t| < a \text{ at all } t, |t| < T) = \begin{cases} 0 & \text{at} \quad a < b \\ \\ 2\Phi(\frac{a}{T}) & \text{at} \quad a \geq b. \end{cases} \tag{II}
$$

Analogously the probabilities of the events A_t and B, $P(A_t) = 1$ at $t \neq 0$, $P(A_0) = P(B) = 0$ differ from those in Example 2.7.2. Thus the random process Y_t is also a random variable in the space $(X^T, \mathcal{S})$ with more wide σ-algebra $\mathcal{S} \supset \mathcal{A}^T$ to which all the considered events belong. But the values of the probability measure of the random function Y_t of these events differ from the values of the probability measure of the random process X_t. Besides notice that $P(X_t = Y_t$ at a given $t) = 1$ for any t while $P(X_t = Y_t$ at all $t, |t| < T) = 0$ whatever T is small (there is none pair of the realizations $x(t)$, $y(t)$ of the random functions X_t, Y_t which should coincide in arbitrarily small vicinity of the origin).

E x a m p l e 2.7.4. The probability space $(\Omega, \mathcal{S}, P)$ represents the plane with the σ-algebra of Borel sets and with circular normal distribution with the center in the origin and with the unit variance. Let us consider the random process $X_t = x_t(\omega) = \omega_1 \sin t + \omega_2 \cos t$. All the multi-dimensional distributions of this random process are normal with the centers at the origin and with the covariance matrices $\{\cos(t_p - t_q)\}_{p,q=1,\ldots,n}$. In a given case the probability of the absence of the excursion of the interval $(-a, a)$ on any interval of t is determined by formula

$$
P(|X_t| < a \text{ at all } t, t_0 < t < t_1) = P(\{\omega : \omega_1^2 + \omega_2^2 < a^2\}) = 1 - e^{-\frac{1}{2}a^2}. \tag{I}
$$

The probabilities of the events A_t and B of the previous examples in a given case are equal to $P(A_t) = P(B) = 1$ as all the realizations of the random process X_t are continuous on the whole real axis. Consequently, the random process in this example is a random variable not only in $(X^T, \mathcal{A}^T)$ but in the space $(X^T, \mathcal{F})$ with more rich σ-algebra $\mathcal{F} \supset \mathcal{A}^T$ also.

E x a m p l e 2.7.5. Let us consider the random process Y_t on the probability space of the previous example,

$$
Y_t = y_t(\omega) = \begin{cases} \omega_1 \sin t + \omega_2 \cos t & \text{at } t \neq \dfrac{1}{n(\omega_1^2 + \omega_2^2)} \\ \\ b \cos n(\omega_1^2 + \omega_2^2) & \text{at } t = \dfrac{1}{n(\omega_1^2 + \omega_2^2)} \end{cases}, \quad (n = \pm 1, \pm 2, \ldots). \tag{I}
$$

This random process has the same multi-dimensional distributions that the random process X_t of the previous example. But the probabilities of all the events considered in the previous example have other values for it. The probability of the absence of the excursion of the interval $(-a, a)$ in any time interval (t_0, t_1) which contains the origin is determined by formula

$$P(|X_t| < a \text{ at all } t, t_0 < t < t_1) = \begin{cases} 0 & \text{at} \quad a < b, \\ 1 - e^{-\frac{1}{2}a^2} & \text{at} \quad a \geq b. \end{cases} \tag{II}$$

The probabilities of others events considered in the previous example in a given case are equal $P(A_t) = 1$ at $t \neq 0$, $P(A_0) = P(B) = 0$. Notice that in latter two examples as in Examples 2.7.2 and 2.7.3 we have for random processes X_t and Y_t correspondingly

$$\begin{aligned} P(X_t = Y_t \text{ at given } t) &= 1 \text{ at any } t, \\ P(X_t = Y_t \text{ at any } t, |t| < T) &= 0 \text{ at any } T. \end{aligned} \tag{III}$$

In order to obtain the possibility of calculating the probabilities of the events which restrict the values of a random process in uncountable set of the points by means of its multi-dimensional distribution it is natural to require that almost all the realizations of a random function (with the exception of the set of zero probability) will be completely determined by their values in the countable set of the points. This condition may be mathematically formulated by different ways. One of them leads to the notion of the separability of the random process (Section 2.8). The other one which is more natural is based on different possible definitions of the continuity of the random processes (Section 2.9).

2.8. Separable Random Functions

2.8.1. Definitions

Let us introduce the notions of the separability and the separant which are new for elementary theory of random functions. We shall suppose that T and X represent the topological spaces with the countable vicinities of all the points, the σ-algebra $\mathcal{A}$ in X is generated by the topology of the space X, and the probability space $(\Omega, \mathcal{S}, P)$ is complete (Section 2.1). The random function $X(t) = x(t, \omega)$ in $(X, \mathcal{A})$ is called *separable* if there exists such countable set $L \subset T$ and such zero set $N \in \mathcal{S}$, $P(N) = 0$ that at any $\omega \bar{\in} N$ the value of the realization $x(t, \omega)$ of the random function $X(t)$ at any $t \in T$ represents the limit of the sequence of its values $\{x(l_k, \omega)\}$ at the points of some sequence $\{l_k\} \subset L$ convergent to t. The countable set L is called *a separant* of the random function $X(t)$.

For expressing the definition of the separability of the random function $X(t) = x(t,\omega)$ by a formula we notice that from the existence for any $t \in T$ and $\omega \bar\in N$ of such a sequence $\{l_k\} \subset L$ that $l_k \to t$, $x(l_k,\omega) \to x(t,\omega)$ follows that whatever is the vicinity V_t of the point t the value $x(t,\omega)$ of the random function $X(t)$ at $\omega \bar\in N$ is the limit point of the set $x(LV_t,\omega)$. In other words $x(t,\omega)$ belongs to the closure of this set $[x(LV_t,\omega)]$, i.e. to the closure of the image of the set LV_t, $x(t,\omega) \in [x(LV_t,\omega)]$ at $\omega \bar\in N$. As this condition is fulfilled for any vicinity V_t of the point t then

$$x(t,\omega) \in \bigcap_{V_t}[x(LV_t,\omega)] \quad \text{at all} \quad t \in T, \ \omega \bar\in N, \qquad (2.8.1)$$

where the intersection is taken in all vicinities V_t of the point t. Thus *if the random function $X(t) = x(t,\omega)$ is separable then it satisfies condition (2.8.1)*. It is evident that vice versa, *any random function satisfying condition (2.8.1) is separable*. It follows from the fact that the point $x(t,\omega)$ is the limit point of any set $[x(LV_t,\omega)]$ and has a countable defining system of the vicinities. Consequently, (2.8.1) may serve as the definition of the separability of the random function.

R e m a r k. From the existence of the sequence $\{l_k\} \subset L$ convergent to t for any $t \in T$ it follows that the countable set $L \subset T$ is dense in T. Therefore the random function may be separable only in that case when its domain T is separable. The separant L does not cease to be separant if we add to it some countable set of the points from T. Thus the separant is not unique.

2.8.2. Multi-Dimensional Distributions for Separable Random Function

Let us establish the connection between the property of the function separability and the probabilities of the events which restrict its values in the uncountable set of the points.

▷ Let G be an arbitrary open set of the space T, $G \subset T$. If the random function $X(t) = x(t,\omega)$ is separable then on the basis of (2.8.1) for all $t \in G$, $\omega \bar\in N$ $x(t,\omega) \in [x(LG,\omega)]$, and consequently, $[x(G,\omega)] \subset [x(LG,\omega)]$. On the other hand $[x(G,\omega)] \supset [x(LG,\omega)]$ as $G \supset LG$. So $[x(G,\omega)] \subset [x(LG,\omega)]$ for any open set $G \supset LG$ and $\omega \bar\in N$ and

$$\{\omega : [x(G,\omega)] \neq [x(LG,\omega)]\} \subset N. \qquad (2.8.2)$$

Let now F be an arbitrary closed set of the space X, $F \subset X$. Evidently if $x(t,\omega) \in F$ at a given ω and all $t \in U$, $U \subset T$ then $[x(U,\omega)] \subset F$ and vice versa from $[x(U,\omega)] \subset F$ follows $x(t,\omega) \in F$ at all $t \in U$. Therefore for any subset U of the set T we have

$$\{\omega : x(t,\omega) \in F, t \in U\} = \{\omega : [x(U,\omega)] \subset F\}. \qquad (2.8.3)$$

Notice that any point ω of the set $\{\omega : x(t,\omega) \in F\}$, $t \in G$ belongs also to the set $\{\omega : x(t,\omega) \in F, t \in LG\}$ as if $x(t,\omega) \in F$ for all $t \in G$ then a fortiori $x(t,\omega) \in F$ for $t \in LG$. Consequently, $\{\omega : x(t,\omega) \in F, t \in G\} \subset \{\omega : x(t,\omega) \in F, t \in LG\}$. Further on from (2.8.2) and (2.8.3) follow also the inclusions:

$$\{\omega : x(t,\omega) \in F, t \in LG\} \backslash \{\omega : x(t,\omega) \in F, t \in G\}$$

$$= \{\omega : [x(LG,\omega)] \subset F, [x(G,\omega)]\bar{\subset}F\}$$

$$\subset \{\omega : [x(G,\omega)] \neq [x(LG,\omega)]\} \subset N.$$

The difference of the sets $\{\omega : x(t,\omega) \in F, t \in LG\}$ and $\{\omega : x(t,\omega) \in F, t \in G\}$ represents zero set for any closed set $F \subset X$ and for any open set $G \subset T$. By virtue of the completeness of the probability space $(\Omega, \mathcal{S}, P)$ it follows from here that

$$P(\{\omega : x(t,\omega) \in F, t \in LG\} \backslash \{\omega : x(t,\omega) \in F, t \in G\}) = 0.$$

Consequently, the set $\{\omega : x(t,\omega) \in F, t \in G\}$ is measurable, $\{\omega : x(t,\omega) \in F, t \in G\} \in \mathcal{S}$ and

$$P(\{\omega : x(t,\omega) \in F, t \in G\}) = P(\{\omega : x(t,\omega) \in F, t \in LG\}).$$

The right-hand side of this equality may be calculated by means of the multi-dimensional distributions of the random function $X(t)$. $\lhd$

Thus we proved the following statement.

Theorem 2.8.1. *If the random function $X(t) = x(t,\omega)$ is separable then the probabilities of the events $\{X(t) \in F, t \in G\}$ which restrict it in the uncountable set of the points are determined by its multi-dimensional distributions for any closed sets $F \subset X$ and for any open sets $G \subset T$ in particular for $G = T$.*

R e m a r k. From the inclusion

$$\{\omega : x(t,\omega) \in F, t \in LG\} \backslash \{\omega : x(t,\omega) \in F, t \in G\} \subset N,$$

it does not follow in any way that the left-hand side set is the inverse of some zero set of the phase probability space $(X^T, \mathcal{A}^T, \mu)$ of the random function $X(t)$. The inclusion only proves that any separable random function is a random variable not only in $(X^T, \mathcal{A}^T)$ but in some measurable space $(X^T, \mathcal{F})$ with more rich σ-algebra $\mathcal{F} \supset \mathcal{A}^T$ and that the sets of the functions $\{x(t) : x(t) \in F,\ t \in LG\} \setminus \{x(t) : x(t) \in F,\ t \in G\}$ and $\{x(t) : x(t) \in F,\ t \in G\}$ belong to this more wide σ-algebra $\mathcal{F}$.

If follows from (2.8.2) that for real random functions when X is the extended real axis $\bar{R}$ we have at all $\omega \bar{\in} N$:

$$\sup_{t \in G} x(t, \omega) = \sup_{l \in LG} x(l, \omega), \quad \inf_{t \in G} x(t, \omega) = \inf_{l \in LG} x(l, \omega), \qquad (2.8.4)$$

$$\overline{\lim_{t' \to t}} x(t', \omega) = \overline{\lim_{l \in L}} x(l, \omega), \quad \varliminf_{t' \to t} x(t', \omega) = \varliminf_{l \in L, l \to t} x(l, \omega). \qquad (2.8.5)$$

These formulae are the direct corollary of the fact that $\sup_{t \in U} x(t, \omega)$ and $\inf_{t \in U} x(t, \omega)$ are the largest and the smallest of the numbers correspondingly which enter into the set $[x(U, \omega)]$. Thus we get

Corollary 2.8.1. *If the real random function $X(t) = x(t, \omega)$ is separable then the left-hand sides of all formulae (2.8.4) and (2.8.5) are the measurable functions of variable ω, i.e. are the real scalar random variables.*

2.8.3. Separable Random Functions and Their Realizations

Theorem 2.8.2. *If each of the topological spaces T and X has a countable base then for any random function $X(t) = x(t, \omega)$ in $(X, \mathcal{A})$ there exist such countable set $L \subset T$ and such family of zero sets $\{N_t, t \in T\}$, $P(N_t) = 0$ that at any $t \in T$*

$$x(t, \omega) \in \bigcap_{V_t} [x(LV_t, \omega)] \quad for\ all \quad \omega \bar{\in} N_t. \qquad (2.8.6)$$

R e m a r k. Let us explain the difference between the condition of separability (2.8.1) and condition (2.8.6). Condition (2.8.1) requires that *each* realization of a random function may be with the exception of the set of the realizations which has zero summary probability *at all* $t \in T$ will be determined by its values in the countable set of the points $L \subset T$ (zero set N does not depend on t). Condition (2.8.6) means that *at any fixed* t almost all realizations of the random function are determined by their values at the points of the countable set $L \subset T$. Zero set N depends on t in consequence of the fact that the union of all sets N_t may be not only zero but may coincide with the whole space Ω. In the latter case there is none realization of the random function which at all t will be determined by its values in the countable set of the points.

$\triangleright$ Primarily we shall show that for any random function $X(t) = x(t, \omega)$, $t \in T$ for any open set $G \subset T$ and for any closed set $F \subset X$ there exists such countable set $L_{G,F} \subset G$ that at any $t \in G$ the set $N_{G,F,t} = \{\omega : [x(L_{G,F}, \omega)] \subset F,\ x(t, \omega)\bar{\in}F\}$ is zero, $P(N_{G,F,t}) = 0$. Let us take an arbitrary point $l_1 \in G$ and put

$$\sup_{t \in G} P(\{\omega : x(l_1, \omega) \in F,\ x(t, \omega)\bar{\in}F\}) = p_1.$$

We assign an arbitrary $\varepsilon > 0$ and choose the second point $l_2 \in G$ in such a way that $P(\{\omega : x(l_1, \omega) \in F,\ x(l_2, \omega)\bar{\in}F\}) > p_1 - \varepsilon/2$. Continuing this process we suppose that the points $l_1, \ldots, l_n$ are determined. We put

$$\sup_{t \in G} P(\{\omega : [x(l_k, \omega)] \in F,\ k = 1, \ldots, n,\ x(t, \omega)\bar{\in}F\}) = p_n$$

and choose the point $l_{n+1} \in G$ so that

$$P(\{\omega : x(l_k, \omega) \in F,\ k = 1, \ldots, n,\ x(l_{n+1}, \omega)\bar{\in}F\}) > p_n - \varepsilon/2^n.$$

In such a way we shall determine the sequence of the points $\{l_k\} = L_{G,F} \in G$. As the sets $A_n = \{\omega : x(l_k, \omega) \in F, k = 1, \ldots, n, x(l_{n+1}, \omega)\bar{\in}F\}$ $(n = 1, 2, \ldots)$ are mutually exclusive then

$$\sum_{n=1}^{n} p_n - \varepsilon < \sum_{n=1}^{\infty} P(A_n) \le 1.$$

Hence it follows that $p_n \to 0$ and $\sup_{t \in G} P(\{\omega : x(L_{G,F}, \omega) \subset F, x(t, \omega)\bar{\in}F\}) = 0$. Let us notice that by virtue of the closure of the set F from $x(L_{G,F}, \omega) \subset F$ follows $[x(L_{G,F}, \omega)] \subset F$. Therefore from the latter equality we establish that for any fixed $t \in G$ the set $N_{G,F,t} = \{\omega : [x(L_{G,F}, \omega)] \subset F, x(t, \omega)\bar{\in}F\}$ is zero what proves our statement.

Now let the sequence of open sets $\{G_k\}$ form the base of the topological space T, and the sequence of open sets $\{D_k\}$ forms the base of the topological space X. It means that any open set $D \subset X$ represents an arbitrary union of some sets D_k and any closed set $F \subset X$ is an arbitrary intersection of some (closed) sets $F_k = \bar{D}_k$. It is evident that the union of the countable sets L_{G_k, F_h} which correspond to all sets G_k and F_h,

$$L = \bigcup_{k,h=1}^{\infty} L_{G_k, F_h}$$

is countable as a countable union of the countable sets. The union of zero sets correspondent to all sets $N_{G_k,F_h,t}$ correspondent to all sets G_k containing a given t and to all sets F_h,

$$N_t = \bigcup_{\substack{k,h=1 \\ G_k \ni t}}^{\infty} N_{G_k,F_h,t}$$

at any $t \in T$ is zero one as a countable union of zero sets. As open set $G \subset T$ is an union of some sets G_k, $G = \bigcup G_k$ then $LG \supset LG_k \supset L_{G_k F_h}$ for any $G_k \subset G$ and any F_h. Therefore for any F_h, any $t \in G$ and any $t \in G_k$ take place of the inclusions:

$$\{\omega : [x(LG,\omega)] \subset F_h, \ x(t,\omega)\bar{\in}F_h\}$$

$$\subset \{\omega : [x(L_{G_k F_h},\omega)] \subset F_h, \ x(t,\omega)\bar{\in}F_h\} = N_{G_k,F_h,t} \subset N_t.$$

As any closed set $F \subset X$ is the intersection of some closed sets F_h, $F = \bigcap F_h$, and consequently, $F \subset F_h$, $\bar{F} = \bigcup \bar{F}_h$ then for any open set $G \subset T$ and the closed set $F \subset X$ at any $t \in G$ we obtain

$$\{\omega : [x(LG,\omega)] \subset F, \ x(t,\omega)\bar{\in}F\}$$

$$= \{\omega : [x(LG,\omega)] \subset F, \ x(t,\omega) \in \bar{F}\}$$

$$= \bigcup_{F_h \supset F} \{\omega : [x(LG,\omega)] \subset F, \ x(t,\omega) \in \bar{F}_h\}$$

$$\subset \bigcup_{F_h \supset F} \{\omega : [x(LG,\omega)] \subset F_h, \ x(t,\omega) \in \bar{F}_h\} \subset N_t.$$

Thus the sets $\{\omega : [x(LG,\omega)] \subset F, \ x(t,\omega)\bar{\in}F\}$ correspondent to all open sets $G \subset T, t \in G$ and to all closed sets $F \subset X$ at any fixed $t \in T$ belong to one and the same zero set N_t. Consequently, any unions of such sets belong also to N_t. In particular, for any $t \in G$

$$\bigcup_F \{\omega : [x(LG,\omega)] \subset F, \ x(t,\omega)\bar{\in}F\} \subset N_t,$$

where the union is taken on all closed sets $F \subset X$.

Now it remains only to notice that

$$\{\omega : x(t,\omega) \bar{\in} [x(LG,\omega)]\} = \bigcup_F \{\omega : [x(LG,\omega)] \subset F, \ x(t,\omega)\bar{\in}F\}.$$

Really, if $\omega_0 \in \{\omega : x(t,\omega) \bar{\in} [x(LG,\omega)]\}$ then $x(t,\omega_0)\bar{\in} [x(LG,\omega_0)]$. Putting $F_0 = [x(LG,\omega_0)]$ we get

$$\omega_0 \in \{\omega : [x(LG,\omega)] \subset F_0, \ x(t,\omega)\bar{\in}F_0\}$$

$$\subset \bigcup_{F} \{\omega : [x(LG, \omega)] \subset F_0, \ x(t, \omega) \bar{\in} F\}.$$

And vice versa if $[x(LG, \omega_0)] \subset F, \ x(t, \omega_0) \bar{\in} F\}$ for some closed set F then $\omega_0 \in \{\omega : x(t, \omega) \bar{\in} [x(LG, \omega')]\}$. $\triangleleft$

R e m a r k. Theorem 2.8.2 assumes the following equivalent formulation: at any fixed $t \in T$, for any open set $G, t \in G$ the inclusion $\{\omega : x(t, \omega) \bar{\in} [x(LG, \omega)]\} \subset N_t$ takes place, i.e. $x(t, \omega) \bar{\in} [x(LG, \omega)]$ if $\omega \bar{\in} \bar{N}_t$. As it is valid for any open set $G, t \in G$ then it is valid for any vicinity $G = V_t$ of the point t. Consequently, if $\omega \bar{\in} N_t$ then $x(t, \omega)$ belongs to all the sets $[x(LV_t, \omega)]$ and namely belongs to their intersection what proves (2.8.6).

2.8.4. Equivalent Random Functions and Their Realizations

Naturally the statements in Subsection 2.8.3 suggests that after correcting only zero set of the realizations at each $t \in T$, $t \bar{\in} L$ in such a way that each realization at all t will satisfy the condition $x(t, \omega) \in \bigcap [x(LV_x, \omega)]$, we may make a random function a separable one without changing its multi-dimensional distributions. But before proving this fact it is necessary to introduce one new notion. Two random functions $X(t) = x(t, \omega)$ and $Y(t) = y(t, \omega)$ are called equivalent if their values at any fixed t are equivalent random variables, i.e. if $P(X(t) = Y(t)) = 1$ at any fixed $t \in T$. It is evident that the equivalent random functions have the same multi-dimensional distributions, and consequently, one and the same probability measure μ in their common phase space $(X^T, \mathcal{A}^T)$.

It should be noticed that the equivalent random functions in $(X, \mathcal{A})$ may be not equivalent random variables in $(X^T, \mathcal{A}^T)$ as the uncountable union of zero sets in Ω for which $X(t) \neq Y(t)$ correspondent to all $t \in T$ may be not only zero but may coincide with the whole space Ω. Thus two equivalent random functions may have no realizations at all which coincide.

E x a m p l e 2.8.1. The random processes $X(t)$ and $Y(t)$ of Examples 2.7.2 and 2.7.3 are equivalent as $P(X(t) = Y(t) = 1)$ at any t. But they have no common realizations as $P(X(t) = Y(t)$ at all $t, |t| < T) = 0$ for the infinitesimal T. The probability measures of these random processes coincide on the σ-algebra $\mathcal{A}^T$ of the phase space X^T. But as we have already seen they do not coincide beyond the bounds of the σ-algebra $\mathcal{A}^T$. The random process $X(t)$ is separable as all its realizations are continuous, and consequently, are completely determined by their values on the countable sets of the rational points. The random process $Y(t)$ is not separable. But correcting at each t only that set of the realizations of zero probability for which $\omega = 1/nt$ $(n = 1, 2, \ldots)$, namely substituting the value

$y(t, 1/nt) = b$ by $\tilde{y}(t, 1/nt) = (1/nt)t = 1/n$ we shall obtain the separable random process $\tilde{Y}(t)$ which coincides identically with $X(t)$.

E x a m p l e   2.8.2. The same refers to the random processes $X(t)$ and $Y(t)$ of Examples 2.7.3 and 2.7.4. Correcting at each t only set of the realizations of zero probability of the random process $Y(t)$ for which $\omega_1^2 + \omega_2^2 = 1/nt$ ($n = 1, 2, \ldots$), namely, substituting the value $y(t, \omega) = b\cos(1/t)$ by the value $\tilde{y}(t, \omega) = (1/nt)\sin[t + \text{arctg}\,(\sqrt{(1/nt)^2 - \omega_1^2}/\omega_1)]$, we shall obtain the separable random process $\tilde{Y}(t)$ which coincides identically with $X(t)$.

E x a m p l e   2.8.3. The step random process $X(t)$ which describes the stream of the events (Subsection 2.7.5) is separable as any its realization evidently is completely determined by its values on the set of the rational t. But it is easy to construct the nonseparable random processes with the same multi-dimensional distributions which are equivalent to $X(t)$.

The examples given in Subsection 2.8.4 show that the extension of the probability measure of the random function beyond the bounds of the σ-algebra $\mathcal{A}^T$ or its supplement $\overline{\mathcal{A}^T}$ may be not unique.

As it is known from functional analysis the set A of the topological space is called *bicompact* if we may single out a finite covering from its any covering by open sets. From this definition passing to the complements we conclude that if a system of the closed subsets of a bicompact set has an empty intersection then an empty finite-intersection will be obligatory appear in this system. Therefore any *centered* system of the subsets of the bicompact set has the intersection which according to the definition has no finite empty intersections. Now we suppose that in addition to properties of Subsection 2.8.1 the random function $X(t) = x(t, \omega)$ possesses the property that the set $[x(L, \omega)]$ where L is a countable set, $L \subset T$ is a bicompact set at any $\omega \in \Omega$.

The following statement is true.

Theorem 2.8.3. *If each of the topological spaces T and X has the countable base and the random function $X(t) = x(t, \omega), t \in T$ possesses such a property that the set $[x(L, \omega)]$ is bicompact for the correspondent countable set $L \subset T$ at any $\omega \in \Omega$ then there exists the separable random function $Y(t) = y(t, \omega)$ which is equivalent to $X(t) = x(t, \omega)$ (The Doob theorem).*

▷ Let us construct the new random function $Y(t) = y(t, \omega)$ assuming $y(t, \omega) = x(t, \omega)$ at all $t \in L$ and at all $t \in T, t\bar{\in}L$ at which $\omega\bar{\in}N_t$. For those $t \in T$, $t\bar{\in}L$ for which $\omega \in N_t$ we substitute the value of the realization $x(t, \omega)$ by an arbitrary value $y(t, \omega) \in \bigcap_{V_t}[x(LV_t, \omega)]$. It is possible as the set $\bigcap_{V_t}[x(LV_t, \omega)]$ at any t and ω is not empty as the intersection of the centered system of the closed subsets of the bicompact set $[x(L, \omega)]$. It is evident that the random function $Y(t) = y(t, \omega)$ is separable as all its realizations at all $t \in T$ satisfy the condition $y(t, \omega) \in \bigcap[x(LV_t, \omega)] = \bigcap[y(LV_t, \omega)]$. On the other hand the random function $Y(t)$ is equivalent to the random function $X(t)$ as $\{\omega : y(t, \omega) \neq x(t, \omega)\} \subset N_t$ at any $t \in T$ and as a result $P(Y(t) = X(t)) = 1$ at any $t \in T$. ◁

2.9. Continuous Random Functions

2.9.1. Definitions

Let us generalize the notion of the continuity in the mean square which is known from elementary probability theory and consider the random functions with the values in the separable B-space X with the topology generated by the norm. Here as before we shall suppose that T represents a separable topological space, and the probability space $(\Omega, \mathcal{S}, P)$ is complete.

The random function $X(t) = x(t, \omega)$, $t \in T$ is called *continuous in probability at the point* $t \in T$ if at any $\varepsilon > 0$

$$\lim_{t' \to t} P(\{\omega : \| x(t', \omega) - x(t, \omega) \| \geq \varepsilon\}) = 0. \qquad (2.9.1)$$

The random function $X(t) = x(t, \omega)$ is called *continuous in probability on* T if it is continuous in probability at all $t \in T$.

The random function $X(t) = x(t, \omega)$ is called *almost sure (a.s.) continuous (continuous with probability 1) at the point* $t \in T$ if

$$P(\{\omega : x(t', \omega) \to x(t, \omega) \quad \text{at} \quad t' \to t\}) = 1. \qquad (2.9.2)$$

The random function $X(t) = x(t, \omega)$ is called *a.s. continuous (continuous with probability 1) on* T if it is a.s. continuous at all $t \in T$.

The random function $X(t) = x(t, \omega)$ is called *continuous* if almost all its realizations are continuous

$$P(\{\omega : x(t, \omega) \text{ continuous on } T\}) = 1. \qquad (2.9.3)$$

R e m a r k. It follows from the general properties of the convergence in probability and almost sure (Section 2.2) that the random function continuous almost sure is also continuous in probability. In the general case the inverse is not valid. It is also evident that the continuous random function is a.s. continuous, and consequently, and in probability. But from the a.s. continuity of the random function does not follow in any way that it is continuous (uncountable union of zero sets $N_t = \{\omega :$ $x(t',\omega) \not\to x(t,\omega)$ at $t \to t'\}$, $t \in T$ may be not zero set).

E x a m p l e   2.9.1. The random process $X(t)$ of Example 2.7.2 is continuous as all its realizations are continuous. Consequently, it is a.s. continuous and is continuous in probability. The same refers to the random process $X(t)$ of Example 2.7.3.

E x a m p l e   2.9.2. The random process $Y(t)$ of Example 2.7.2 is a.s. continuous, and consequently, continuos in probability at any $t \neq 0$. Therefore it is a.s. continuous and also conutinuous in probability at any interval T which does not contain the origin. But it is not continuous as at any interval (t_0, t_1), $t_0 > 0$, $t_1 > 2t_0$ all its realizations correspondent to $|\omega| < 1/t_0$ are discontinuous what implies

$$P(\{\omega : y(t,\omega) \text{ discontinuous at } (t_0,t_1)\}) = 2\Phi(1/t_0) > 0.$$

In a given case the uncountable union of zero sets $N_t = \{\omega : y(t',\omega) \not\to y(t,\omega)\}$ correspondent to all $t \in (t_0, t_1)$ is not a zero set. The same refers to the random process $Y(t)$ of Example 2.7.5.

2.9.2. Separant for Continuous Random Function

Theorem 2.9.1. *If the random function $X(t) = x(t,\omega)$ is continuous in probability on T and separable then any countable set L dense in T is a separant.*

$\triangleright$ Suppose at first that the random function $X(t) = x(t,\omega)$ is continuous in probability on T and let L be an arbitrary everywhere dense countable subset of the set T. From the continuity $x(t,\omega)$ in probability it follows that at any t for any subsequence $\{l_k\} \subset L$, $l_k \to t$ the sequence of the random variables $\{x(l_k,\omega)\}$ converges in probability to the random variable $x(t,\omega)$. We may single out from it the subsequence $\{x(l_{k_n},\omega)\}$ convergent to $x(t,\omega)$ almost sure (Section 2.2). After denoting by N_t a zero set of points ω on which $x(l_{k_n},\omega) \not\to x(t,\omega)$ we shall have

$$x(t,\omega) \in \bigcap_{V_t}[x(LV_t,\omega)], \quad \text{if} \quad \omega \bar{\in} N_t.$$

Thus any random function continuous in probability satisfies condition $(2.8.6)$ for an arbitrary countable everywhere dense set L, $L \subset T$. In the general case as we have already seen in Section 2.8 the countable set $L \subset T$ cannot be chosen arbitrarily.

Now we suppose that $X(t) = x(t,\omega)$ is continuous in probability and is separable, L_0 is any its separant, N_0 is the correspondent zero set. It is evident that the countable union of zero sets N_l corresponding to all $l \in L_0$ is zero set

$$N = \bigcup_{l \in L_0} N_l, \quad P(N) = 0.$$

According to the just proved above $x(l, \omega) \in [x(LG, \omega)]$ for all $l \in L_0$ and for all open sets $G \subset T$, $l \in G$, and consequently, $[x(L_0 G, \omega)] \subset [x(LG, \omega)]$ at all $\omega \bar{\in} N$. Thus

$$\bigcap_{V_t} [x(L_0 V_t, \omega)] \subset \bigcap_{V_t} [x(LG, \omega)] \quad \text{at all} \quad t \in T \quad \text{and all} \quad \omega \bar{\in} N.$$

On the other hand by virtue of the separability of $x(t, \omega)$

$$x(t, \omega) \in \bigcap_{V_t} [x(L_0 V_t, \omega)] \quad \text{at all} \quad t \in T \quad \text{and all} \quad \omega \bar{\in} N_0.$$

Consequently we get

$$x(t, \omega) \in \bigcap_{V_t} [x(L V_t, \omega)] \quad \text{at all} \quad t \in T \quad \text{and all} \quad \omega \bar{\in} N_0 \bigcup N.$$

This proves that an arbitrary everywhere dense countable set $L \subset T$ serves as a separant of the random $X(t) = x(t, \omega)$. ◁

2.9.3. Measurability and Separability of Continuous Random Function

It is very important for many problems of probability theory that the random function $X(t) = x(t, \omega)$ will be measurable, i.e. that the function of two variables $x(t, \omega)$ will be a measurable mapping of the product of the spaces $(T \times \Omega, \mathcal{F} \times \mathcal{S})$ in $(X, \mathcal{A})$. It turns out that if the random function $X(t)$ is continuous in probability then we may make it measurable and separable after changing at each t only zero set of its realizations.

So we shall prove the following statement.

Theorem 2.9.2. *If the random function $X(t) = x(t, \omega)$ with the values in the separable B-space X defined on the separable space T presented as the limit of the increasing sequence of the compact sets $\{T_n\}$ is continuous in probability then there exists the measurable separable random function $Y(t) = y(t, \omega)$ which is equivalent to it.*

▷ Suppose that the random function $X(t) = x(t, \omega)$ is continuous in probability on T and that T may be presented as the limit of increasing sequence of the compact sets $\{T_n\}$, $T = \lim T_n = \bigcup T_n$. It follows from the continuity $x(t, \omega)$ in probability that at any $\varepsilon, \delta > 0$ any point $t \in T$ has such vicinity $V_t(\varepsilon, \delta)$ that for all $t' \in V_t(\varepsilon, \delta)$

$$P(\{\omega : \| x(t', \omega) - x(t, \omega) \| \geq \varepsilon\}) < \delta. \qquad (2.9.4)$$

Let $L = \{l_k\}$ be an arbitrary everywhere dense countable set $L \subset T$. We assign arbitrary sequences of the positive numbers $\{\varepsilon_n\}$, $\{\delta_n\}$ convergent to zero and determine the vicinities of all the points $l_k \in L$ by the formula $V_k^n = V_{l_k}(\varepsilon_n, \delta_n 2^{-n})$ $(k, n = 1, 2, \ldots)$. Then we shall have by virtue of (2.9.4)

$$P(\{\omega : \| x(t, \omega) - x(l_k, \omega) \| \geq \varepsilon_n\}) < \delta_n 2^{-n} \qquad (2.9.5)$$

at any $t \in V_k^n$. The vicinities V_k^n of the points l_k at any n we may choose in such a way that each of the first n vicinities $V_1^n, \ldots, V_n^n$ will contain only one of the points $l_1, \ldots, l_n$. By virtue of the compactness of the set T_n it is covered by the finite number of the vicinities V_k^n at a given n. We denote by q_n the largest of the indexes k for the vicinities V_k^n which enter into this finite mapping and denote at each n mutually disjoint sets:

$$U_1^n = V_1^n, \quad U_k^n = V_k^n \bigcap_{p=1}^{k-1} \overline{U_p^n} \quad (k = 2, 3, \ldots; \ n = 1, 2, \ldots).$$

All the sets V_k^n and U_k^n are measurable, $V_k^n, U_k^n \in \mathcal{F}$, as the σ-algebra $\mathcal{F}$ according to our general agreement is generated by the topology of the space T, and consequently, contains the vicinities of all the points $t \in T$. Therefore the functions

$$y_n(t, \omega) = \sum_{k=1}^{q_n} x(l_k, \omega) \mathbf{1}_{U_k^n}(t) \quad (n = 1, 2, \ldots) \qquad (2.9.6)$$

are $(\mathcal{F} \times \mathcal{S}, \mathcal{A})$-measurable. Hence in accordance with the fact that each of the sets $U_1^n, \ldots, U_n^n$ by the construction contains only one of the points $l_1, \ldots, l_n$, $l_k \in U_k^n$ $(k = 1, \ldots, n)$ at all $\omega \in \Omega$ the equalities are valid

$$y_n(l_k, \omega) = x(l_k, \omega) \quad (k = 1, \ldots, n). \qquad (2.9.7)$$

Let us take now an arbitrary fixed $t \in T$. By virtue of (2.9.5) and (2.9.6) at any n the inequality $\| y_n(t, \omega) - x(t, \omega) \| < \varepsilon_n$ is fulfilled for all ω except some set A_t^n which has the probability smaller than $\delta^n 2^{-n}$, $P(A_t^n) < \delta_n 2^{-n}$. Putting

$$B_t^n = \bigcup_{p=n}^{\infty} A_t^p \quad (n = 1, 2, \ldots),$$

we shall have by virtue of the property of semi-additivity of the probability that

$$P(B_t^n) \leq \sum_{p=n}^{\infty} P(A_t^p) < \sum_{p=n}^{\infty} \delta_p 2^{-p} < \delta_n 2^{1-n} < \delta_n.$$

We assume arbitrary $\varepsilon, \delta > 0$ and choose n in such a way that $\varepsilon_n < \varepsilon$, $\delta_n < \delta$. Then at all $p \geq n$ and $\omega \bar{\in} B_t^n$ the inequality $\| x(t,\omega) - y_p(t,\omega) \| < \varepsilon$ will be valid. It means that at any fixed $t \in T$ the sequence of the functions $\{y_n(t,\omega)\}$ converges to $x(t,\omega)$ uniformly on Ω except the set B_t^n of an infinite measure $P(B_t^n) < \delta$. In other words, the sequence $\{y_n(t,\omega)\}$ at any $t \in T$ converges to $x(t,\omega)$ almost uniformly on Ω. Consequently, it a.e. converges to $x(t,\omega)$ on Ω relatively to the measure P. We denote by N_t the zero set of those ω for which $y_n(t,\omega) \not\to x(t,\omega)$ at a given t. Then we shall get $\lim y_n(t,\omega) = x(t,\omega)$ in all points (t,ω) of the product of the spaces $T \times \Omega$ which do not belong to the set $Q = \{(t,\omega) : \omega \in N_t, t \in T\}$. But $\lim y_n(t,\omega)$ may exist also in some points of the set Q. Therefore we determine the function

$$y(t,\omega) = \lim y_n(t,\omega). \qquad (2.9.8)$$

This function is $(\mathcal{F} \times \mathcal{S}, \mathcal{A})$-measurable as the limit of the sequence of $(\mathcal{F} \times \mathcal{S}, \mathcal{A})$-measurable functions $\{y_n(t,\omega)\}$. Consequently, its domain D_y is measurable, $D_y \in \mathcal{F} \times \mathcal{S}$. We extend the function $y(t,\omega)$ over all Ω by putting $y(t,\omega) = y_m(t,\omega)$ at $(t,\omega) \bar{\in} D_y$ and arbitrary m. The function $y(t,\omega)$ determined in such a way on the whole space $T \times \Omega$ is $(\mathcal{F} \times \mathcal{S}, \mathcal{A})$-measurable as it represents the sum of two $(\mathcal{F} \times \mathcal{S}, \mathcal{A})$-measurable functions

$$y^{(1)}(t,\omega) = \begin{cases} y(t,\omega) & \text{at } (t,\omega) \in D_y, \\ 0 & \text{at } (t,\omega) \bar{\in} D_y, \end{cases} \qquad (2.9.9)$$

$$y^{(2)}(t,\omega) = \begin{cases} 0 & \text{at } (t,\omega) \in D_y, \\ y_m(t,\omega) & \text{at } (t,\omega) \bar{\in} D_y. \end{cases} \qquad (2.9.10)$$

According to the proved $y(t,\omega) = x(t,\omega)$ at any $t \in T$ if $\omega \bar{\in} N_t$. Consequently, the random functions $X(t) = x(t,\omega)$ and $Y(t) = y(t,\omega)$ at any $t \in T$ a.s. coincide, i.e. they are equivalent. It remains to show that $Y(t) = y(t,\omega)$ is separable. It is sufficient for this purpose to notice that by virtue of (2.9.7) and (2.9.8) $y(t,\omega)$ coincides with $x(t,\omega)$ at all points $l_k \in L$, $y(l_k,\omega) = x(l_k,\omega)$ at all $\omega \in \Omega$ in consequence of which the value of the function $y(t,\omega)$ at $(t,\omega) \in D_y$ represents the limit of the sequence of its values $\{y(l_{k_n},\omega)\}$ at the points of some sequence $\{l_{k_n}\} \subset L$, $l_{k_n} \to t$. And at $(l,\omega) \bar{\in} D_y$ we have $y(t,\omega) = y_m(t,\omega) = x(l_k,\omega) = y(l_k,\omega)$ if $t \in U_k^m$. Consequently, any realization of the random function $Y(t) = y(t,\omega)$ at all t satisfies the definition of the separability. Zero set N in a given case is empty. $\triangleleft$

2.9.4. Sufficient Criteria of Continuity

The continuity of a random function in probability imposes the restrictions only on its two-dimensional distribution as the set $\{\omega :$

$\| x(t',\omega) - x(t,\omega) \| \geq \varepsilon \}$ depends only on the values of a random function at two values of an argument. Knowing the two-dimensional distributions μ_{t_1,t_2} of the random function $X(t) = x(t,\omega)$ we may represent definition (2.9.1) of the continuity in probability in the form

$$\lim_{t' \to t} \mu_{t,t'}(\{(x,x') : \| x' - x \| \geq \varepsilon\}) = 0. \qquad (2.9.11)$$

As a result of this the continuity of a random function in probability is easily established. As regards to the continuity almost sure and the continuity these properties of a random function are not directly determined by its multi-dimensional distributions. Meanwhile it is practically very important to be able to determine whether a given random function is continuous or not. Practically a convenient criterion of the continuity of a random function may be obtained if a random function is determined in the finite region of the finite-dimensional space.

Theorem 2.9.3. *Let $X(t) = x(t,\omega)$ be a separable random function with the values in the separable B-space determined in the finite region T of the n-dimensional space. In order the random function $x(t,\omega)$ be continuous it is sufficient that there will exist such nondecreasing positive continuous in zero functions $\varepsilon(h)$, $\delta(h)$ determined on some interval $(0,\eta)$, $\eta > 1/2$, and satisfying the conditions*

$$\sum_{p=1}^{\infty} \varepsilon(2^{-p}) < \infty, \quad \sum_{p=1}^{\infty} 2^{np}\varepsilon(2^{-p}) < \infty, \qquad (2.9.12)$$

that at any $h \in (0,\eta)$

$$P(\{\omega : \| x(t+\tau,\omega) - x(t,\omega) \| \geq \varepsilon(h)\}) < \delta(h) \qquad (2.9.13)$$

at all $t, t+\tau \in T, \| \tau \| = h$ (The Kolmogorov–Loève criterion).

$\triangleright$ For proving that at the fulfillment of conditions (2.9.12) and (2.9.13) almost all realizations of the random function $x(t,\omega)$ are continuous it is sufficient to show that almost at all ω

$$\sup_{\substack{t,t' \in T \\ \| t'-t \| < h}} \| x(t',\omega) - x(t,\omega) \| \to 0 \quad \text{at} \quad h \to 0.$$

But according to Theorem 2.8.2 we have for the separable $x(t,\omega)$ almost at all ω

$$\sup_{\substack{t,t' \in T \\ \| t'-t \| < h}} \| x(t',\omega) - x(t,\omega) \| = \sup_{\substack{l,l' \in L \\ \| l'-l \| < h}} \| x(l',\omega) - x(l,\omega) \|,$$

where L is a separant. Therefore we may restrict our consideration to the study of the realizations $x(t,\omega)$ only on the separant L, $t \in L$.

Now we notice that from condition (2.9.13) it follows the continuity $x(t,\omega)$ in probability as at any $\varepsilon > 0$ for all h such that $\varepsilon(h) < \varepsilon$,

$$\{\omega : \| \, x(t+\tau,\omega) - x(t,\omega) \, \| \geq \varepsilon\}$$

$$\subset \{\omega : \| \, x(t+\tau,\omega) - x(t,\omega) \, \| \geq \varepsilon(h)\},$$

in consequence of which $P(\{\omega : \| \, x(t+\tau,\omega) - x(t,\omega) \, \| \geq \varepsilon\}) \to 0$ at $\tau \to 0$. Therefore any countable set L dense in T is a separant for $x(t,\omega)$. In particular, the set L of all the points of the region T with the dyadic-rational coordinates may serve as a separant, i.e. with the coordinates of the type $k2^{-p}$, where k is any integer, and p is any natural number. Thus we may restrict oneselves by the values $x(t,\omega)$ on the set L on the points T with dyatic-rational coordinates.

Let us introduce the variables

$$Z_p = z_p(\omega) = \sup \| \, x(k_1 2^{-p}, \ldots, (k_q+1)2^{-p}, \ldots, s, k_n 2^{-p}, \omega)$$

$$-x(k_1 2^{-p}, \ldots, k_q 2^{-p}, \ldots, k_n 2^{-p}, \omega) \, \| \quad (p = 1, 2, \ldots),$$

where the upper bound is taken over all the points $l = (k_1 2^{-p}, \ldots, k_n 2^{-p}) \in T$ and over all q from 1 till n. It is evident that $Z_p \geq \varepsilon(2^{-p})$ if and only if when even one of the variables

$$Y^{p,q}_{k_1,\ldots,k_n} = \| \, x(k_1 2^{-p}, \ldots, (k_q+1)2^{-p}, \ldots, k_n 2^{-p}, \omega)$$

$$-x(k_1 2^{-p}, \ldots, k_q 2^{-p}, \ldots, k_n 2^{-p}, \omega) \, \|$$

is larger or equal $\varepsilon(2^{-p})$. For calculating the number of the variables $Y^{p,q}_{k_1,\ldots,k_n}$, we denote by T_m Lebesgue measure of the projection of the set T on the axis m $(m = 1, \ldots, n)$. Then the number of different values of the mth coordinate of the type $k_m 2^{-p}$ in the region T does not exceed $2^p T_m$. So the number of points $(k_1 2^{-p}, \ldots, k_n 2^{-p})$ in the region T does not exceed $2^{np} T_1 \ldots T_n$. As to each such point corresponds n variables $Y^{p,q}_{k_1,\ldots,k_n}$ $(q = 1, \ldots, n)$ then the number of the random variables $Y^{p,q}_{k_1,\ldots,k_n}$ does not exceed $n2^{np} T_1 \ldots T_n$. So on the basis of (2.9.13) we find

$$P(Z_p \geq \varepsilon(2^{-p})) \leq \sum P(Y^{p,q}_{k_1,\ldots,k_n} \geq \varepsilon(2^{-p})) < n2^{np} T_1 \ldots T_n \delta(2^{-p}).$$

Let us introduce the events

$$A_p = \{Z_p \geq \varepsilon(2^{-p})\} = \{\omega : z_p(\omega) \geq \varepsilon(2^{-p})\},$$

$$B_q = \bigcup_{p=q}^{\infty} A_p, \quad N = \lim B_q = \bigcap_{q=1}^{\infty} B_q.$$

According to the proved above $P(A_p) < n2^{np} T_1 \ldots T_n \delta(2^{-p})$. Consequently,

$$P(B_q) \le \sum_{p=q}^{\infty} P(A_p) < n T_1 \ldots T_n \sum_{p=q}^{\infty} 2^{np} \delta(2^{-p}),$$

$$P(N) \le P(B_q) < n T_1 \ldots T_n \sum_{p=q}^{\infty} 2^{np} \delta(2^{-p}).$$

As it is valid at any $q \ge 1$ then by virtue of (2.9.12) $P(N) = 0$. Thus the set N is zero. If $\omega \in \bar{N}$ then as $\bar{N} = \bigcup \bar{B}_q$ then $\omega \in \bar{B}_q$ at any q. But $\bar{B}_q = \bigcap_{p=q}^{\infty} \bar{A}_p$. Therefore ω belongs to all $\bar{A}_p$ at $p \ge q$. It means that $Z_p = z_p(\omega) < \varepsilon(2^{-p})$ at any point $\omega \bar{\in} N$ at all sufficiently large p.

Let us take now two arbitrary points $l,\ l'$ with dyadic-rational coordinates, $l, l' \in L$. If $\| l' - l \| < 2^{-q}$ then there exists such point l_0 with the coordinates $k_1 2^{-q}, \ldots, k_n 2^{-q}$ that all the components of the vectors of the differences $l - l_0$ and $l' - l_0$ is smaller than 2^{-q}. Here the difference of the mth coordinates of the points l and l_0 will be expressed by the dyadic number of the type

$$\pm \sum_{p=q+1}^{q_m} \tau_{mp} 2^{-p} \quad (m = 1, \ldots, n),$$

where each of the numbers τ_{mp} is equal to 0 and 1. If we perform the transition from the point l_0 into l by the sequential steps of the length $\tau_{mp} 2^{-p}$ $(p = q+1, q+2, \ldots)$ at first on one axis further on the other and so on, on the nth axis and use repeatedly the inequality of the triangular then we get

$$\| x(l, \omega) - x(l_0, \omega) \| \le \sum_{m=1}^{n} \sum_{p=q+1}^{q_m} z_p(\omega) < n \sum_{p=q+1}^{\infty} z_p(\omega).$$

The similar inequality we obtain for $\| x(l', \omega) - x(l_0, \omega) \|$. Consequently,

$$\| x(l', \omega) - x(l, \omega) \| \le \| x(l', \omega) - x(l_0, \omega) \|$$

$$+ \| x(l, \omega) - x(l_0, \omega) \| < 2n \sum_{p=q+1}^{\infty} z_p(\omega).$$

As this inequality is valid for all $l, l' \in L$, $\| l' - l \| < 2^{-q}$ then

$$\sup_{\substack{l,l' \in L \\ \| l' - l \| < 2^{-q}}} \| x(l', \omega) - x(l, \omega) \| < 2n \sum_{p=q+1}^{\infty} z_p(\omega).$$

But according to the proved above $z_p(\omega) < \varepsilon(2^{-p})$ at all sufficiently large p if $\omega \bar{\in} N$. Consequently,

$$\sup_{\substack{l,l' \in L \\ \| l' - l \| < 2^{-q}}} \| x(l', \omega) - x(l, \omega) \| < 2n \sum_{p=q+1}^{\infty} \varepsilon(2^{-p}) \tag{2.9.14}$$

at any $\omega \bar{\in} N$ if q is sufficiently large. It remains to notice that the right-hand side of this inequality by virtue of (2.9.12) tends to zero at $q \to \infty$ and that $P(N) = 0$. This fact proves the continuity of almost all realizations of the random function $X(t) = x(t, \omega)$. ◁

Condition (2.9.13) imposes the restriction on the two-dimensional distribution of a random function. Knowing the two-dimensional distributions μ_{t_1, t_2} we may rewrite this condition in the form

$$\mu_{t,t+\tau}(\{(x, x') : \| x' - x \| \geq \varepsilon(h)\}) < \delta(h), \quad \| \tau \| = h. \tag{2.9.15}$$

Theorem 2.9.4. *For establishing the continuity of a random function it is sufficient to find such functions $\varepsilon(h)$ and $\delta(h)$ satisfying condition (2.9.12) that the two-dimensional distribution of a random function will satisfy inequality (2.9.15).*

There are many pairs of the functions satisfying conditions (2.9.12). In particular, the functions

$$\varepsilon(h) = ah^\lambda, \quad \delta(h) = bh^{n+\mu} \tag{2.9.16}$$

at any $a, b, \lambda, \mu > 0$ satisfy conditions (2.9.12). Therefore in many cases we may find the functions $\varepsilon(h)$ and $\delta(h)$ in the form of (2.9.16).

E x a m p l e 2.9.3. Let us find the conditions at which the real separable centered random function $X(t)$ with the normal two-dimensional distribution is continuous in the finite region T. Taking into account that the random variable $Y_\tau = X(t+\tau) - X(t)$ in this case is normally distributed and has zero expectation and the variance $\sigma^2(t, \tau) = K(t+\tau, t+\tau) - 2K(t, t+\tau) + K(t, t)$ we obtain for the probability in (2.9.13) the following formula:

$$P(|X(t+\tau) - X(t)| \geq \varepsilon(h)) = 1 - 2\Phi\left(\frac{\varepsilon(h)}{\sigma(t,\tau)}\right). \tag{I}$$

Using the inequality which is easily derived for Laplace function $\Phi(u)$ at $u > 0$ and at any natural k, namely

$$1 - 2\Phi(u) < \frac{k!2^{2k+1}}{\sqrt{2\pi}}\frac{1}{u^{2k+1}},$$

we come to the inequality

$$P(|X(t+\tau) - X(t)| \geq \varepsilon(h)) < \frac{k!2^{2k+1}}{\sqrt{2\pi}}\left[\frac{\sigma(t,\tau)}{\varepsilon(h)}\right]^{2k+1}. \qquad \text{(II)}.$$

Hence it is clear that if

$$\sigma(t,\tau) \leq ch^{\nu}, \quad \|\ \tau\ \| = h, \qquad \text{(III)}$$

at some $c, \nu > 0$ then the random function $X(t)$ is continuous. Really, putting in (2.9.16) $\varepsilon(h) = ah^{\lambda}, \lambda < \nu$ and choosing k in such a way that $(2k+1)(\nu-\lambda) > n$ we may assume

$$\delta(h) = \frac{k!2^{2k+1}}{\sqrt{2\pi}}\left(\frac{c}{a}\right)^{2k+1}h^{(2k+1)(\nu-\lambda)}.$$

The functions $\varepsilon(h)$ and $\delta(h)$ chosen in such a way have the form of (2.9.16) at

$$b = \frac{k!2^{2k+1}}{\sqrt{2\pi}}\left(\frac{c}{a}\right)^{2k+1}, \quad \mu = (2k+1)(\nu-\lambda) - n$$

and therefore satisfy conditions (2.9.12). The obtained inequality (II) shows that condition (2.9.13) is satisfied in this case. Thus for the continuity of the real separable function with a normal distribution it is sufficient that its covariance function will satisfy condition (III) at some $c, \nu > 0$. It is valid both for a scalar and for a finite-dimensional vector argument t. If the argument t is scalar then in the previous formulae $n\ =\ 1.$ In particular the covariance function $k(\tau)$ $= De^{-\alpha|\tau|}(\cos\omega\tau + \gamma\sin\omega|\tau|)$ satisfies condition (III) at $\nu\ =\ 1/2$. Consequently, a stationary separable random function with such covariance function and with the normal two-dimensional distribution is continuous, i.e. almost all its realizations are continuous. In the special cases $\gamma = 0$ and $\gamma = \omega = 0$ we come to the conclusion that all separable random functions with the exponential and exponential-cosine covariance functions are continuous if they have the normal two-dimensional distribution.

2.10. Differentiable Random Functions

2.10.1. Definitions

Generalizing the notion of the differentiability known from elementary probability theory we come to the following definitions.

Let $X(t) = x(t, \omega), t \in T$ be a random function of the n-dimensional vector t whose phase space represents a separable B-space. We denote by $e_1, \ldots, e_n$ the unit vectors of the coordinates axes in T and put

$$x_m^h(t, \omega) = \frac{1}{h}[\, x(t + e_m h, \omega) - x(t, \omega)\,] \quad (m = 1, \ldots, n). \qquad (2.10.1)$$

The random function $X(t) = x(t, \omega)$ is called *differentiable in probability at the point $t \in T$* if there exist such random functions $X_m'(t) = x_m'(t, \omega)$ $(m = 1, \ldots, n)$ that at any $\varepsilon > 0$

$$\lim_{h \to 0} P(\{\omega : \|\, x_m^h(t, \omega) - x_m'(t, \omega)\, \| \geq \varepsilon\}) = 0. \qquad (2.10.2)$$

The random function $X(t) = x(t, \omega)$ is called *differentiable in probability on T* if it is differentiable in probability at all $t \in T$.

The random function $X(t) = x(t, \omega)$ is called *a.s. differentiable (differentiable with probability 1) at the point $t \in T$* if there exists such random function $X_m'(t) = x_m'(t, \omega)$ that

$$P(\{\omega : x_m^h(t, \omega) \to x_m'(t, \omega) \quad \text{at} \quad h \to 0\}) = 1. \qquad (2.10.3)$$

The random function $X(t) = x(t, \omega)$ is called *a.s. differentiable (differentiable with probability 1) on T* if it is a.s. differentiable at all $t \in T$.

The random function $X(t) = x(t, \omega)$ is called *differentiable* if almost all its realizations are differentiable:

$$P(\{\omega : x(t, \omega) \quad \text{differentiable on} \quad T\}) = 1. \qquad (2.10.4)$$

R e m a r k. It follows from the general properties of the convergence in probability and a.s. (Section 2.2) that the random function a.s. differentiable is also differentiable in probability. The inverse in the general case is not valid. It is also evident that the differentiable random function is differentiable a.s., and consequently, in probability. But it does not follow from the a.s. differentiability of a random function that it is differentiable as the uncountable union of zero sets $N_t = \{\omega : x_m^h(t, \omega) \not\to x_m'(t, \omega)\}$ at $h \to 0$ even at one $m\}$, $t \in T$ may be not zero set.

E x a m p l e 2.10.1. The random process $X(t)$ of Example 2.7.2 is differentiable. Consequently, it is differentiable a.s. and in probability. The same refers to the random process $X(t)$ of Example 2.7.4.

E x a m p l e 2.10.2. The random process $Y(t)$ of Example 2.7.3 is a.s. differentiable, and consequently, in probability at any $t \neq 0$. Therefore it is a.s. differentiable and in probability at any interval T which does not contain the origin. But it is not differentiable on any interval (t_0, t_1), $t_0 > 0$, $t_1 > 2t_0$ as none of its realizations corresponding to $|\omega| < 1/t_0$ is differentiable on this interval and $P(\{\omega : |\omega| < 1/t_0\}) = 2\Phi(1/t_0) > 0$. The same refers to the random process $Y(t)$ of Example 2.7.5.

As for the convergence in probability of the sequence of the random functions it is necessary and sufficient that the sequence should be fundamental in probability then condition (2.10.2) may be substituted by the condition

$$\lim_{h,l \to 0} P(\{\omega : \| x_m^h(t,\omega) - x_m^l(t,\omega) \| \geq \varepsilon\}) = 0. \qquad (2.10.5)$$

As it is easily seen this condition imposes the restriction only on the three-dimensional distribution of the random function $X(t) = x(t,\omega)$. Knowing its three-dimensional distribution μ_{t_1,t_2,t_3} we may present condition (2.10.5) in the form

$$\lim_{h,l \to 0} \mu_{t,t+e_m h,t+e_m l}\left(\left\{(x_1, x_2, x_3) : \left\| \frac{x_2 - x_1}{h} - \frac{x_3 - x_1}{l} \right\| \geq \varepsilon\right\}\right) = 0.$$
$$(2.10.6)$$

As a result it is always easy to determine whether a given random function is differentiable in probability or not. As regards to a.s. differentiability and the differentiability of a random function then these properties cannot be established directly by means of its multi- dimensional distributions. But there exist sufficient conditions of the differentiability of a separable random function which are analogous to conditions (2.9.12) and (2.9.13).

2.10.2. Sufficient Conditions of Differentiability

Suppose the separable random function $X(t) = x(t,\omega)$ with the values in the separable B-space determined in the finite region T of the n-dimensional space satisfies conditions (2.9.12), (2.9.13) and besides that at any h, $0 < h < \eta$ satisfies the conditions

$$P(\{\omega : \| x(t + e_m h + e_r h, \omega) - x(t + e_m h, \omega) - x(t + e_r h, \omega)$$

$$+x(t,\omega)\ \| \geq \varepsilon_1(h)\}) < \delta_1(h) \quad (m,r = 1,\ldots,n). \tag{2.10.7}$$

Here $\varepsilon_1(h)$ and $\delta_1(h)$ are the nondecreasing positive continuous functions such that

$$\sum_{p=1}^{\infty} 2^p \varepsilon_1(2^{-p}) < \infty, \quad \sum_{p=1}^{\infty} 2^{np}\delta_1(2^{-p}) < \infty. \tag{2.10.8}$$

We shall prove that almost all realizations of the random function $X(t) = x(t,\omega)$ have continuous partial derivatives over all the components of the vector t. For this purpose we construct for each $m = 1,\ldots,n$ the sequence of the finite-valued functions ($p = 1,2,\ldots$):

$$y_m^p(t,\omega) = 2^p\big[\,x(k2^{-p} + e_m 2^{-p},\omega) - x(k2^{-p},\omega)\,\big]$$

$$\text{at}\quad k2^{-p} \leq t < (k + e_1 + \cdots + e_n)2^{-p} \quad (k2^{-p} \in T), \tag{2.10.9}$$

where k is the n-dimensional vector with integer components. The supposition will be proved if we show that the sequence $\{y_m^p(t,\omega)\}$ converges to the continuous function $x'_m(t,\omega)$ whose integral with respect to the variable t_m in any bounds coincides with the increment of the function $x(t,\omega)$ on the correspondent interval.

$\triangleright$ It follows from (2.10.9) that

$$\sup_{t\in T} \|\ y_m^{p-1}(t,\omega) - y_m^p(t,\omega)\ \|$$

$$< n2^{-p}\sup \|\ x((k + e_m + e_r)2^{-p},\omega)$$

$$-x((k + e_m)2^{-p},\omega) - x((k + e_r)2^{-p},\omega) + x(k2^{-p},\omega)\ \|,$$

where the upper bound is taken on all $r = 1,\ldots,n$ and on all the points $k2^{-p} \in T$ for which all written values of the argument of the function $x(t,\omega)$ belongs to T. In order to be sure in this fact it is sufficient to partition the cube $k2^{-p+1} \leq t < (k + e_1 + \cdots + e_n)2^{-p+1}$ into 2^n cubes with the side 2^{-p} and express the difference $y_m^{p-1}(t,\omega) - y_m^p(t,\omega)$ in each of the obtained cubes in terms of the second differences of the function $x(t,\omega)$ in the appexes of these cubes. Introducing the random variables

$$V_m^p = v_m^p(\omega) = \sup_{t\in T} \|\ y_m^{p-1}(t,\omega) - y_m^p(t,\omega)\ \|, \tag{2.10.10}$$

$$U_{mrk}^p = u_{mrk}^p(\omega) = \|\ x((k + e_m + e_r)2^{-p},\omega)$$

$$-x((k+e_m)2^{-p},\omega) - x((k+e_m)2^{-p},\omega) + x(k2^{-p},\omega) \ \|, \quad (2.10.11)$$

we get

$$P(V_m^p \geq n2^p\varepsilon_1(2^{-p})) \leq P(\sup U_{mrk}^p \geq \varepsilon_1(2^{-p}))$$

$$\leq \sum P(U_{mrk}^p \geq \varepsilon_1(2^{-p})).$$

Here the summation is extended over all variables U_{mrk}^p. But according to (2.10.2) we have $P(U_{mrk}^p \geq \varepsilon_1(2^{-p})) < \delta_1(2^{-p})$. We calculate the number of the random variables U_{mrk}^p at given m and p. Let T_m be Lebesgue measure of the projection of the region T on the axis e_m. Then it is evident that the number of the points $k2^{-p}$ in the region T does not exceed $2^{np}T_1 \ldots T_n$ and to each point $k2^{-p} \in T$ it corresponds n random variables U_{mrk}^p at $r = 1, \ldots, n$. Consequently, the number of the random variables U_{mrk}^p does not exceed $nT_1 \ldots T_n 2^{np}$ and

$$P(V_m^p \geq n2^p\varepsilon_1(2^{-p})) < nT_1 \ldots T_n 2^{np}\delta_1(2^{-p}).$$

Now we introduce the events

$$C_m^p = \{\sup U_{mrk}^p \geq \varepsilon_1(2^{-p})\} \supset \{\omega : v_m^p(\omega) \geq n2^p\varepsilon_1(2^{-p})\},$$

$$D_m^q = \bigcup_{p=q}^{\infty} C_m^p, \quad N_m = \bigcap_{p=1}^{\infty} D_m^p.$$

Analogously as in Subsection 2.9.4 we ensure in the fact that $P(N_m) = 0$ and that for any $\omega \bar{\in} N_m$ we have $v_m^p(\omega) \leq n2^p \sup u_{mrk}^p(\omega) < n2^p\varepsilon_1(2^{-p})$ at all sufficiently large p. Consequently, at any $\omega \bar{\in} N_m$ and all sufficiently large q we have

$$\sup_{t \in T} \| \ y_m^q(t,\omega) - y_m^{q+s}(t,\omega) \ \| \leq \sum_{p=q+1}^{q+s} \sup_{t \in T} \| \ y_m^{p-1}(t,\omega) - y_m^p(t,\omega) \ \|$$

$$= \sum_{p=q+1}^{q+s} v_m^p(\omega) < n \sum_{p=q+1}^{\infty} 2^p \ eps_1(2^{-p}). \quad (2.10.12)$$

As the latter expression tends to zero at $q \rightarrow \infty$ by virtue of (2.10.8) then the sequence $\{y_m^q(t,\omega)\}$ at $\omega \bar{\in} N_m$ is fundamental and consequently, converges uniformly to some limit function $x_m'(t,\omega)$.

For proving the continuity of $x_m'(t,\omega)$ it is sufficient to notice that by virtue of uniform convergence of the sequence $\{y_m^q(t,\omega)\}$, $\omega \bar{\in} N_m$ at any $\varepsilon > 0$ there exists such natural q_ε that

$$\| \ y_m^q(t,\omega) - x_m'(t,\omega) \ \| < \varepsilon/3 \quad \text{at all} \ \ t \in T \ \ \text{and} \ \ q \geq q_\varepsilon.$$

On the other hand, by virtue of (2.10.9) at all $t, t' \in T$ for which all the components of the vector $t' - t$ are smaller than 2^{-q},

$$y_m^q(t',\omega) - y_m^q(t,\omega) = 2^q[\, x((k+e_m)2^{-q},\omega)$$

$$-x(k2^{-q},\omega) - x((k_1+e_m)2^{-q},\omega) + x(k_1,\omega)\,].$$

Here k_1 is the vector with integer components which differ from the correspondent components of the vector k no more than 1. Consequently, for all such t, t' we find

$$\| \, y_m^q(t',\omega) - y_m^q(t,\omega) \, \| \le n2^q \sup u_{mrk}^q(\omega) < n2^q\varepsilon(2^{-q})$$

at $\omega \bar{\in} N_m$. After choosing q so large that $n2^q\varepsilon_1(2^{-q}) < \varepsilon/3$ we get

$$\| \, x_m'(t',\omega) - x_m'(t,\omega) \, \| \le \| \, x_m'(t',\omega) - y_m^q(t',\omega) \, \|$$

$$+ \| \, y_m^q(t',\omega) - y_m^q(t,\omega) \, \| + \| \, y_m^q(t,\omega) - x_m'(t,\omega) \, \| < \varepsilon$$

at all $t, t' \in T$ for which all the components of the vector $t' - t$ is smaller than 2^{-q}. It proves the continuity of the function $x_m'(t,\omega)$.

Finally, it follows from (2.10.9) that at any $q \ge p$ and at any $\omega \bar{\in} N_m$ the following equalities take place

$$\int_{k2^{-p}}^{(k+le_m)2^{-p}} y_m^p(t,\omega)dt = x((k+le_m)2^{-p},\omega) - x(k2^{-p},\omega),$$

$$\int_{k2^{-p}}^{(k+le_m)2^{-p}} x_m'(t,\omega)dt = x((k+le_m)2^{-p},\omega) - x(k2^{-p},\omega).$$

Thus at $\omega \bar{\in} N_m$ the integral over the function $x_m'(t,\omega)$ with respect to any segment parallel to the vector e_m which connects the points with dyadic-rational coordinates is equal to the increment of the function $x(t,\omega)$ on this segment. By virtue of the continuity of the functions $x(t,\omega)$ and $x_m'(t,\omega)$ at $\omega \bar{\in} N \bigcup N_m$ this property is extended on any segment of the line parallel to the vector e_m which is contained in the region T. Consequently, the function $x_m'(t,\omega)$ represents a partial derivative of the function $x(t,\omega)$ with respect to the mth component of the vector t. ◁

So we proved the following statment.

Theorem 2.10.1. *While fulfilling conditions* (2.9.12), (2.9.13), (2.10.7) *and* (2.10.8) *all the realizations of the separable random function* $X(t) = x(t,\omega)$ *besides, may be, the set* $N \bigcup N_1 \bigcup \ldots \bigcup N_n$ *of zero* *probability are continuous and have continuous partial derivatives with respect to all components of the vector t.*

Conditions (2.10.7) impose the restriction only on the four-dimensional distribution of a random function, and in the case of the scalar argument t – only on the three-dimensional distribution. Therefore the fulfillment of this condition is easily verified. Knowing the four-dimensional distribution μ_{t_1,t_1,t_3,t_4} of the random function $X(t)$ we may present condition (2.10.7) in the form

$$\mu_{t,t+e_mh,t+e_rh,t+e_mh+e_rh}(\{(x_1, x_2, x_3, x_4) : \| \, x_1 + x_4$$

$$-x_2 - x_3 \, \| \geq \varepsilon_1(h)\}) < \delta_1(h) \quad (m, r = 1, \ldots, n). \tag{2.10.13}$$

In the special case of the scalar t we have $m = r = 1$ and condition (2.10.13) takes the form

$$\mu_{t,t+h,t+2h}(\{(x_1, x_2, x_3) : \| \, x_1 + x_3 - 2x_2 \, \| \geq \varepsilon_1(h)\}) < \delta_1(h). \tag{2.10.14}$$

Among the functions $\varepsilon_1(h)$ and $\delta_1(h)$ satisfying condition (2.10.8) it is expedient to choose the following functions:

$$\varepsilon_1(h) = ah^{1+\lambda}, \quad \delta_1(h) = bh^{n+\mu}, \quad a, b, \lambda, \mu > 0. \tag{2.10.15}$$

E x a m p l e 2.10.3. Let us consider the centered real random function $X(t) = x(t, \omega)$ with the normal four-dimensional distribution. In this case all random variables

$$Y_{mr} = X(t + e_mh + e_rh) - X(t + e_mh) - X(t + e_rh) + X(t) \tag{I}$$

are normally distributed, their expectations are equal to zero, and the covariances are determined by formula

$$\sigma_{mr}^2(t, h) = K(t + e_mh + e_rh, t + e_mh + e_rh)$$

$$+K(t + e_mh, t + e_mh) + K(t + e_rh, t + e_rh) + K(t, t)$$

$$+2K(t + e_mh + e_rh, t) + 2K(t + e_mh, t + e_rh)$$

$$-2K(t + e_mh + e_rh, t + e_mh) - 2K(t + e_mh + e_ry, t + e_rh)$$

$$-2K(t + e_mh, t) - 2K(t + e_rh, t). \tag{II}$$

After determining the variances of the variables Y_{mr} we get for the probabilities in (2.10.7) the expressions

$$P(|\,Y_{mr}\,| \ge \varepsilon_1(h)) = 1 - 2\Phi\Big(\frac{\varepsilon_1(h)}{\sigma_{mr}(t,h)}\Big). \qquad \text{(III)}$$

Hence similarly as in Example 2.9.3 we obtain that the condition

$$\sigma_{mr}(t,h) < ch^{1+\nu} \quad (m,r = 1,\ldots,n) \qquad \text{(IV)}$$

at some $c, \nu > 0$ is sufficient for the fact that almost all realizations of the random function $X(t)$ will have continuous first derivatives with respect to all components of the vector t. In particular, the covariance function $k(\tau) = De^{-\alpha|\tau|}(\cos\omega\tau + \gamma\sin\omega|\tau|)$ satisfies condition (IV) at $\nu = 1/2$ only at $\gamma = \alpha/\omega$. At other values of γ it does not satisfy (IV)). Thus almost all realizations of the normal random function $X(t)$ with such covariance function at $\gamma = \alpha/\omega$ have continuous derivatives. It goes without saying that the obtained result does not give grounds for the conclusion that at $\gamma \ne \alpha/\omega$ the random function $X(t)$ cannot have almost all differentiable realizations.

2.10.3. Trace Probabilities

The space X^T on which the probability measure of the random function $X(t)$ is determined by its multi-dimensional distributions represents a space of all functions of the variable t, $t \in T$, with the values in the space X. Hence any measurable set of the functions includes not only the realizations of the random function $X(t)$ but a set of other functions with very irregular behaviour as any set from the σ-algebra $\mathcal{A}^T$ restricts the function only in countable set of the points t. Therefore the space X^T is too large. It is always convenient to consider the random variables in the minimal phase space. In particular, as the phase space of a random function it is expedient to assume the set of all its possible realizations or the minimal class of the functions which contains all its realizations. Namely on this phase space it is necessary to determine the probability measure of a random function. So, for instance, for a separable random function it is natural to determine the probability measure in the space of the functions satisfying the condition of separability (2.8.1).

Let $(\Omega, \mathcal{S}, P)$ be the probability space, $\Omega' \subset \Omega$ be a set on which the probability P should be transferred. In the general case the set Ω' does not belong to the σ-algebra $\mathcal{S}$ in consequence of which the probability P is not determined on it. We denote by $\mathcal{S}'$ the class of the subsets of the set Ω' which represents the intersection of the sets of the σ-algebra $\mathcal{S}$ with Ω', $\mathcal{S}' = \{B : B = \Omega'A, A \in \mathcal{S}\}$. It is easy to see that $\mathcal{S}'$ represents the σ-algebra of the sets of the space Ω'. The class $\mathcal{S}'$ contains the empty set $\emptyset$ and space Ω'. In parallel with any set $B = \Omega'A$, $A \in \mathcal{S}$, $\mathcal{S}'$ contains its complement $\Omega\backslash B$ in Ω' as $\Omega'\backslash B = \Omega'\backslash\Omega'A = \Omega'\bar{A}$ and $\bar{A} \in \mathcal{S}$. Finally, for any sequence of the sets $\{B_k\}$, $B_k \in \mathcal{S}'$ we have $B_k = \Omega'A_k$, $A_k \in \mathcal{S}$ and $\bigcup B_k = \Omega'\bigcup A_k \in \mathcal{S}'$ as $\bigcup A_k \in \mathcal{S}$. The σ-algebra $\mathcal{S}'$ of the sets of the space Ω' is called *the trace* of the σ-algebra $\mathcal{S}$ on Ω'. It is easy to understand that the transfer of the probability P on the σ-algebra $\mathcal{S}'$ is possible in principle only in the case when $P(C) = 1$ for any set $C \in \mathcal{S}$ which contains Ω', $C \supset \Omega'$. In this case evidently $P(D) = 0$ for any set $D \in \mathcal{S}$ which does not intersect with Ω'. If this condition is fulfilled then the probability P' on the σ-algebra $\mathcal{S}'$ is determined uniquely by formula

$$P'(B) = P'(\Omega'A) = P(A), \quad A \in \mathcal{S}. \tag{2.10.16}$$

$\triangleright$ Really, for any sets $A_1, A_2 \in \mathcal{S}$ such that $B = \Omega'A_1 = \Omega'A_2$ the sets $A_1\backslash A_2$ and $A_2\backslash A_1$ do not intersect with Ω' in consequence of which $P(A_1\backslash A_2) = P(A_2\backslash A_1) = 0$ and $P(A_1) = P(A_2)$. It proves the uniqueness of the definition of P'. Further on it is evident that $P'(\emptyset) = 0$ and $P'(\Omega') = P(C) = 1$, $C \supset \Omega'$. Finally for any mutually exclusive sets $B_k = \Omega'A_k$, $A_k \in \mathcal{S}$ $(k = 1, 2, \ldots)$ the sets A_k may be also assumed as mutually exclusive as otherwise their intersections do not intersect with Ω' and they may be rejected. Then we shall have

$$P'(\bigcup B_k) = P(\bigcup A_k) = \sum P(A_k) = \sum P'(B_k). \tag{2.10.17}$$

Equalities (2.10.17) prove the σ-additivity of P'. Thus formula (2.10.16) determines uniquely the probability P' on the σ-algebra $\mathcal{S}'$. $\triangleleft$

The probability P' on the measurable space $(\Omega', \mathcal{S}')$ determined in such a way is called *the trace* of the probability P on this space or *the trace probability*. Usually the trace probability P' is determined by means of the notion of *the outer measure* which is known from functional analysis. *The outer probability* P_0 is determined for any set $B \subset \Omega$ by formula

$$P_0(B) = \inf_{B \subset C \in \mathcal{S}} P(C). \tag{2.10.18}$$

The trace probability P' represents the contraction of the outer probability P_0 defined by (2.10.18) on the class of the sets $\mathcal{S}'$.

$\triangleright$ Really, for any set $B \in \mathcal{S}'$ we have $B = \Omega'A \subset A$, $A \in \mathcal{S}$ in consequence of which $P_0(B) \leq P'(A)$. On the other hand, for any set $C \in \mathcal{S}$, $C \supset B = \Omega'A$

we have $P(A) \leq P(C) + P(A\backslash C) = P(C)$ as $A\backslash C$ do not intersect with Ω'. Consequently, $P'(A) \leq P_0(B)$. From the obtained opposite inequalities follows the equality $P_0(B) = P_0(\Omega' A) = P'(A)$ for any $B \in \mathcal{S}'$. The condition of the existence of the trace probability may be written by means of the outer probability in the form $P_0(\Omega') = 1$. $\triangleleft$

Now let μ_x be a probability measure of the random function $X(t)$ determined in the space $(X^T, \mathcal{A}^T)$; $Y(t)$ be a separable random function equivalent to $X(t)$. We denote by H the set of all realizations of the separable random function $Y(t)$. Suppose that the set $C \in \mathcal{A}^T$ contains H. The set H restricts involved functions only in the countable set of the points for instance $L_C \subset T$. At each point $l \in L_C$ only zero set of the realizations of the random function $X(t)$ does not coincide with the correspondent realizations of the random function $Y(t)$. The countable union of these zero sets correspondent to all $l \in L_C$ is also a zero set. Thus almost all realizations of the random function $X(t)$ coincide with the correspondent realizations of the random function $Y(t)$ on L_C. Consequently, any set $C \in \mathcal{A}^T$ which contains all the realizations of the random function $Y(t)$ contains also almost all the realizations of the random function $X(t)$. Therefore $\mu(C) = 1$ for any set of the functions C which contains the space H of all the realizations of the random function $Y(t)$. It means that the transfer of the probability measure μ_x on the space H is possible. We denote by $\mathcal{B}$ the σ-algebra of the sets of the space H which represents the trace of the σ-algebra $\mathcal{A}^T$ on H, $\mathcal{B} = \{B : B = HA, A \in \mathcal{A}^T\}$. Then the probability measure μ_y of the separable random function $Y(t)$ on the basis of (2.10.16) is determined by formula

$$\mu_y(B) = \mu_x(A), \quad B = HA, \quad A \in \mathcal{A}^T. \tag{2.10.19}$$

As to any random function corresponds a separable random function with the same probability measure we have proved the following statement.

Theorem 2.10.2. *Any probability measure* $(X^T, \mathcal{A}^T)$ *obtained by means of the multi-dimensional distributions of a random function may be transferred to the space* H *of the separable functions. Excluding all nonseparable random functions with given multi-dimensional distributions we may remain for study only equivalent separable random functions.*

R e m a r k. Analogously, if the random function $X(t)$ is determined in the finite region T of the finite-dimensional space and its two-dimensional distribution μ_{t_1,t_2} satisfies conditions (2.9.15) and (2.9.12), the problem of the definition of the probability measure of the equivalent continuous random function $X(t)$ in the space C of the continuous functions is solved. Finally, if the random function $X(t)$ determined in the finite region T of the finite-dimensional space satisfies conditions (2.9.15), (2.9.12), (2.10.7) and (2.10.8) then the probability measure of the equivalent continuous differentiable random function $Y(t)$ in the space C^1 of the differentiable functions is defined analogously.

P r o b l e m s

2.1.   Under the conditions of Subsection 2.2.1 prove that if (Ω, S, P) is a probability space, then $(\Omega, \bar{S}, P)$ is complete if the sets from $\bar{S}$ may be represented in the form $A \cup N$, $A \in S$, $N \in \mathcal{N}$, where $\mathcal{N}$ is a class of all zero sets. Hence it is necessary to extend the probability P from S to $\bar{S}$ putting $P(A \cup N) = P(A)$ for any set $A \cup N \in \bar{S}$.

2.2. Under the conditions of Subsection 2.8.1 prove that if the random variable Y in $(Y, \mathcal{B})$ whose phase space Y is the separable B-space be $(\mathcal{S}_y, \mathcal{B})$-measurable then it is $(\mathcal{A}, \mathcal{B})$-measurable random function.

2.3. Under the conditions of Subsection 2.8.1 prove that if the random variable $Y = y(\omega)$ in $(Y, \mathcal{B})$ is $(\mathcal{A}, \mathcal{B})$-measurable functions of the random variable $X = x(\omega)$ then it is $(\mathcal{S}_x, \mathcal{B})$-measurable.

2.4. Under the conditions of Subsection 2.2.5 prove property (v) of the distribution function F of the finite-dimensional random variable.

2.5. Under the conditions of Subsection 2.2.5 prove that any function $F(x)$ possessing properties (i)–(v) represents a distribution function of some random variable X in $(X, \mathcal{A})$.

2.6. Prove that if two finite-dimensional random vectors X and Y are functionally connected, $Y = \varphi(X)$, and the function $\varphi(x)$ is measurable and that the density $f_1(x)$ of the random vector X is known then the distribution function of the random vector Y is determined as

$$F_2(y) = \int\limits_{\varphi(x)<y} f_1(x)dx . \tag{I}$$

2.7. Prove that if the random vectors X and $Y = \varphi(X)$ have one and the same dimension, and the equation at any y has the unique solution x in the range X (in the region where the density $f_1(x)$ different from 0) then

$$f_2(y) = f_1(\varphi^{-1}(y)) \, |J(y)| , \tag{I}$$

$$J(y) = \frac{\partial \varphi^{-1}}{\partial y}(y) = \frac{\partial(\varphi_1^{-1}, \ldots, \varphi_n^{-1})}{\partial(y_1, \ldots, y_n)} = \begin{bmatrix} \dfrac{\partial \varphi_1^{-1}}{\partial y_1} & \cdots & \dfrac{\partial \varphi_1^{-1}}{\partial y_n} \\ \cdots & \cdots & \cdots \\ \dfrac{\partial \varphi_n^{-1}}{\partial y_1} & \cdots & \dfrac{\partial \varphi_n^{-1}}{\partial y_n} \end{bmatrix}.$$

2.8. We assign that the dimension m of the random vector $Y = \varphi(X)$ is smaller than the dimension n of the random vector X, $m < n$, the equation $y = \varphi(x)$ at all y has the unique solution relatively to these m coordinates of the vector x in the region $f_1(x) > 0$. Let x' be a vector formed by these m coordinates; u be a vector formed by the remained $n - m$ coordinates x. We denote in terms of $x' = \varphi^{-1}(y, u)$ the solution of the equation $y = \varphi(x', u)$. Then we have the following expression for the density Y:

$$f_2(y) = \int\limits_{-\infty}^{\infty} f_1(\varphi^{-1}(y, u), u) \, |J(y, u)| \, du, \tag{I}$$

$$J(y, u) = \frac{\partial(\varphi_1^{-1}, \ldots, \varphi_m^{-1})}{\partial(y_1, \ldots, y_m)}.$$

2.9. Let X and Y be random finite-dimensional vectors of the same dimension with the known joint density $f_2 = f_2(x, y)$. Prove that the density of the sum $Z = X + Y$ is expressed by the formulae

$$f_1(z) = \int\limits_{-\infty}^{\infty} f_2(x, z - x)dx = \int\limits_{-\infty}^{\infty} f_2(z - y, y)dy. \tag{I}$$

Derive for the case of the independent items $f_2(x, y) = g_1(x)g_2(y)$ the formula of the densities convolution $f_1 = g_1 * g_2$:

$$f_1(z) = \int\limits_{-\infty}^{\infty} g_1(x)g_2(z - x)dx = \int\limits_{-\infty}^{\infty} g_1(z - y)g_2(y)dy. \tag{II}$$

2.10. The current value of the input random variable X of nonlinear system is determined by the distribution function $F(x)$ and the density $f(x)$. The output Y is connected with the input X by the nonlinear dependence of the type

$$Y = \begin{cases} -a & \text{at} & X < -a, \\ X & \text{at} & |X| \le a, \\ a & \text{at} & X > a. \end{cases} \tag{I}$$

Show that the distribution function $F_y(y)$ and the probability density of the variable Y are determined by formulae

$$F_y(y) = \begin{cases} 0 & \text{at} \quad y \leq -a, \\ F(y) & \text{at} \quad y \in (-a, a], \\ 1 & \text{at} \quad y \geq a; \end{cases} \tag{II}$$

$$f_y(y) = f(y) + F(-a+0)\delta(y+a) + [1 - F(a)]\delta(y-a),$$

$$y \in [-a, a] \ (\text{and } f_y(y) = 0, \ y \notin [-a, a]). \tag{III}$$

2.11.   Prove that if $X = [X_1 \ldots X_n]^T$ is the n-dimensional random vector whose distribution function $F_1(x_1)$ of the coordinate x_1 and the conditional distribution functions $F_2(x_2|x_1), \ldots, F_n(x_n|x_1, \ldots, x_{n-1})$ are continuous then the random variables $Y_1 = F_1(X_1), Y_2 = F_2(X_2|X_1), \ldots, Y_n = F_n(X_n|X_1, \ldots, X_{n-1})$ are independent.

2.12. Prove that if the random variables $X_1, \ldots, X_n$ are independent then the random variables $Y_1 = \varphi_1(X_1), \ldots, Y_n = \varphi_n(X_n)$ at any measurable functions $\varphi_1, \ldots, \varphi_n$ are also independent.

2.13. Let the random function $X(t), t \geq 0$ has the separant $\{t_j\}$. Prove that the inequalities take place for almost all ω:

(i) $\lim\limits_{n\to\infty} \inf\limits_{|t_j-t|<1/n} X(t_j, \omega) \leq X(t, \omega) \leq \lim\limits_{n\to\infty} \sup\limits_{|t_j-t|<1/n} X(t_j, \omega);$

(ii) $\sup\limits_{a<t<b} X(t, \omega) = \sup\limits_{a<t_j<b} X(t_j, \omega), \ \inf\limits_{a<t<b} X(t, \omega) = \inf\limits_{a<t_j<b} X(t_j, \omega);$

(iii) for any $\varepsilon, \delta > 0 \quad \sup\limits_{|t_i-t_j|<\delta} |X(t_i, \omega) - X(s_j, \omega)| \leq \varepsilon$ if and only if $\sup\limits_{|t-s|<\delta} |X(t, \omega) - X(s, \omega)| \leq \varepsilon.$

2.14. Show that the n-dimensional distribution of the random process $X(t) = U$, where U is a random variable with a given density $f_u(u)$ is determined by the density $f_n(x_1, \ldots, x_n; t_1, \ldots, t_n) = f_u(x_1)\delta(x_2 - x_1)\delta(x_3 - x_2) \ldots \delta(x_n - x_{n-1})$ and by the distribution $F(x_1, \ldots, x_n; t_1, \ldots, t_n) = F_u(\min\{x_1, \ldots, x_n\}).$

2.15. Show that the n-dimensional distribution function of the random process $X(t) = \varphi(t)U$, where $\varphi(t)$ is a given positive function, U is a random variable which is distributed according to the normal law $N(0, 1)$ is equal to

$$F_n(x_1, \ldots, x_n; t_1, \ldots, t_n) = \Phi\left(\min\left(\frac{x_1}{\varphi(t_1)}, \ldots, \frac{x_n}{\varphi(t_n)}\right)\right).$$

CHAPTER 3
MOMENTS, CHARACTERISTIC FUNCTIONS, FUNCTIONALS AND CANONICAL EXPANSIONS

The main object of this Chapter to give the advanced basics of the correspondent methods of the stochastic systems theory in Chapters 5–7. Section 3.1 contains the main definitions and detailed presentation of the properties of the expectations of the random variables, random processes and random functions in the functional spaces. On theoretical grounds of Section 3.1 the theory of the second order and higher orders moments is considered in detail in Section 3.2. Section 3.4 is devoted to the theory of the conditional moments. The main notions and the properties of the characteristic functions and the functionals are derived in Section 3.5. The correspondence between the probability measures and the characteristic functionals is established. In Section 3.6 the sequences of the probability measures and the characteristic functionals are studied. The relationship of the consistent sets of the multi-dimensional distributions and the characteristic functionals is established. Subsection 3.6 is dedicated to the distributions theory in Banach spaces. The main properties of a normal distribution in the linear spaces are considered in Section 3.7. The normal distributions on the product of two spaces, the conditional normal distributions and the normal distributions in Hilbert space are studied. Section 3.8 contains detailed treatment of the orthoronal expansions of probability densities. Last Section 3.9 contains systematic treatment of the canonical expansions.

3.1. Expectation of Random Function

3.1.1. Definitions

At first let us consider the random variable X with the finite set of possible realizations $x_1, \ldots, x_N$ in an arbitrary linear space. We denote by $p_1, \ldots, p_N$ the probabilities of the realizations $x_1, \ldots, x_N$ correspondingly. The sum of the products of all realizations by their probabilities is called *an expectation* of the random variable. This definition is a natural generalization of an elementary definition of an expectation of a numerical random variable. We denote an expectation of a random variable X by a symbol EX or m_x. Then we obtain the fomula for an expectation of the random variable X

$$m_x = EX = \sum_{k=1}^{N} x_k p_k \,. \tag{3.1.1}$$

Now we consider the random function $X(t)$, $t \in T_1$. Suppose that at any value of the argument $t \in T_1$ the value of the random function $X_t = X(t)$ has the finite expectation EX_t. The set of the expectations of the variables X_t correspondent to all $t \in T_1$ forms the function $m_x(t)$. This function in elementary theory of the random functions is called *an expectation* of the random function $X(t)$. Thus an expectation of the random function $X(t)$ is a function $m_x(t)$ whose value at each given $t \in T_1$ represents an expectation of a value of the random function $X(t)$ and hence $EX(t) = m_x(t)$. An expectation of the real random function $X(t)$ is expressed in terms of its one-dimensional density by formula

$$m_x(t) = EX(t) = \int_{-\infty}^{\infty} x f_1(x;t) dx \,. \tag{3.1.2}$$

The advanced theory of the probability distributions stated in Chapter 2 is applicable to the random variables with the values in any measurable spaces. For further development of the probability theory it is necessary to impose some restrictions on the phase spaces of the random variables. Therefore we restrict ourselves to the random variables in the linear spaces.

It follows from formula (3.1.1) that for any nonrandom vector c in an arbitrary linear space $Ec = c$. Really, the vector c may be considered as a random variable with unique possible realization c whose probability is equal to 1.

If the random variable $X = x(\omega)$ with the finite set of the realizations is defined on the probability space $(\Omega, \mathcal{S}, P)$ then $x(\omega)$ represents a simple function of an elementary event ω

$$x(\omega) = \sum_{k=1}^{N} x_k \mathbf{1}_{A_k}(\omega) \,, \tag{3.1.3}$$

where $A_1, \ldots, A_N$ are mutually exclusive measurable sets $A_k \in \mathcal{S}$, $P(\bigcup A_k) = 1$. In this case $p_k = P(A_k)$ $(k = 1, \ldots, N)$. Recalling from the functional analysis the definition of an integral with respect to a simple function we may rewrite formula (3.1.1) in the form

$$m_x = EX = \int x(\omega)P(d\omega)\,. \qquad (3.1.4)$$

Thus *an expectation of the random variable $X = x(\omega)$ with a finite set of possible realizations determined on the probability space $(\Omega, \mathcal{S}, P)$ represents an integral with respect to this random variable over the measure P extended on the whole* space *of the elementary events Ω.*

The cited definition of an expectation is extended over the random variables with a finite set of the realizations $\{x_k\}$ in an arbitrary topological linear space. Formula (3.1.4) shows that for the extension of a notion of an expectation over the random variables with the uncountable set of the realizations it is necessary to restrict oneselves to the random variables in such spaces in which the notion of an integral has the sense. First of all we shall consider the random variables in Banach spaces (B-spaces).

So *the expectation of the random variable $X = x(\omega)$ in the B-space X determined on the probability space $(\Omega, \mathcal{S}, P)$ is called the vector $m_x = EX \in X$ which is determined by formula (3.1.4).*

For the existence of an expectation of the random variable $X = x(\omega)$ it is necessary and sufficient that the function $x(\omega)$ should be P-integrable, i.e. should belong to the space $L_1(\Omega, \mathcal{S}, P)$ of the functions with the values in X. To each function $x(\omega) \in L_1(\Omega, \mathcal{S}, P)$ formula (3.1.4) corresponds the vector $m_x \in X$. Thus we may speak about an *operator of an expectation* E mapping $L_1(\Omega, \mathcal{S}, P)$ into the phase space of the random variable X. In the special case when X is a complex plane an operator of an expectation represents a functional in $L_1(\Omega, \mathcal{S}, P)$.

Formula (3.1.4) is reduced to the form (3.1.5) by the change of variables $x = x(\omega)$ with the account to formula (2.2.1) for the probability measure μ_x of the random variable X

$$m_x = EX = \int x\mu(dx)\,. \qquad (3.1.5)$$

If there exists the probability density $f(x)$ of the random variable X over some measure ν formula (3.1.5) takes the form

$$m_x = EX = \int x f(x)\nu(dx)\,. \qquad (3.1.6)$$

E x a m p l e 3.1.1. Let X be a finite-dimensional real random vector. Suppose that a part μ^s of the probability measure μ of the random vector X singular relative to Lebesgue measure ν is concentrated on a finite or countable set of the points $\{x_k\}$. Then using expansion (2.2.2) of the measure μ we may present formula (3.1.5) in the form

$$m_x = EX = \int xf(x)dx + \sum x_k p_k \, . \tag{I}$$

Here p_k is the value of the measure μ on the singleton x_k, $p_k = \mu_x(x_k)$ ($k = 1, 2, \ldots$), i.e. the probability of the value x_k of the vector X. In this case the vector X represents a random variable of the mixed type and its distribution function $F(x)$ does not contain a continuous singular component, $F_{sc}(x) \equiv 0$. It is easy to see that the expectation of the random vector X may be expressed in the considered case by the formula

$$m_x = EX = \int xf^*(x)dx \, , \tag{II}$$

where $f^*(x)$ is the probability density of the vector X determined by formula (2.2.9).

E x a m p l e 3.1.2. Suppose the phase space of the scalar random variable X is a functional B-space whose elements are the functions of the variable $t \in T_1$ with the values in the real one-dimensional space. In this case the variable X is the random function $X(t) = x(t, \omega)$ and its expectation is determined on the basis of (3.1.4) by formula

$$m_x = EX(t) = \int x(t, \omega)P(d\omega) \, . \tag{I}$$

Hence it is clear that an expectation of the random function $X(t)$ represents the function $m_x = m_x(t)$ whose value at each t is equal to an expectation of the value of the random function $X(t)$ at a given t.

3.1.2. Properties of Expectation

As an expectation of a random variable represents an integral all the properties of the expectations are direct consequence of (i)–(xiv) known properties of the integrals which are studied in functional analysis (Pugachev and Sinitsyn 1999).

(i) It follows from property (i) of an integral that *if the random variable Y represents the product of the random variable X by the number a, $Y = aX$ then*

$$EY = E(aX) = aEX \, . \tag{3.1.7}$$

Thus the nonrandom multiplier may be factored out from the sign of an expectation.

(ii) It follows from property (ii) of an integral that *an expectation of the sum of the random variables is equal to the sum of their expectations*:

$$E\sum_{\nu=1}^{n} X_\nu = \sum_{\nu=1}^{n} EX_\nu . \qquad (3.1.8)$$

And generally it follows from property (iii) of an integral that *if Y represents a linear combination of the random variables* $X_1 , \ldots , X_n$,

$$Y = \sum_{\nu=1}^{n} a_\nu X_\nu , \qquad (3.1.9)$$

then

$$m_y = EY = \sum_{\nu=1}^{n} a_\nu EX_\nu . \qquad (3.1.10)$$

This formula shows that an operator of an expectation is linear.

(iii) *If X is a random variable with a finite set of the realizations* $x_1 , \ldots , x_N$, *and T is an arbitrary linear operator mapping its phase space X into another linear space Y, and* $x_1 , \ldots , x_N \in D_T$ *then the random variable $Y = TX$ has a finite set of the realizations* $y_k = Tx_k$ $(k = 1 , \ldots , N)$ and we have

$$m_y = E(TX) = TEX = Tm_x . \qquad (3.1.11)$$

Now we consider an arbitrary random variable X in B-space X. Let T be a bounded linear operator mapping X into another B-space Y. If there exists EX then on the basis of the theorem about the permutability of an integral with bounded linear operator (property (v) of an integral) there exists an expectation of the random variable $Y = TX$ and here relation (3.1.11) is valid.

Now suppose that a phase space of the random variable X is separable. Let T be a closed linear operator mapping X into another separable B-space Y. On the basis of the theorem about the permutability of an integral with a closed linear operator if the domain Δ_x of the random variable $X = x(\omega)$ belongs to the domain D_T of the operator T, $\Delta_x \subset D_T$ and there exist the expectations of the random variables X and $Y = TX$ then $m_x = EX \in D_T$ and formula (3.1.11) is valid. Formula (3.1.11) expresses the property of the permutability of an operator of an expectation with a linear operator.

(iv) It follows from property (vii) of an integral that *the expectations of the equivalent random variables coincide: $EY = EX$ if Y is equivalent to X.*

(v) Property (viii) of an integral shows that *an expectation of the random variable X in B-space exists if and only if there exists an expectation of its norm $\| X \|$ and here*

$$\| EX \| \leq E \| X \| . \tag{3.1.12}$$

(vi) For the real numerical random variables $X = x(\omega)$ and $Y = y(\omega)$ it follows from properties (ix) and (x) of an integral that *$EX \leq EY$ if the variable X with the probability 1 does not exceed Y and that an expectation a.s. of a nonnegative random variable cannot be negative.*

(vii) Property (xi) of an integral shows that *$E \| X \| = 0$ if and only if the random variable X is equivalent to zero (i.e. is equal to 0 with the probability 1).*

(viii) It follows from property (xii) of an integral that *if $\| X \| < c$ a.s. then $\| EX \| < c$.*

(ix) It follows from the theorem about a monotonous convergence that *if a nondecreasing sequence of the nonnegative random variables $\{X_n\}$, $X_n = x_n(\omega)$, a.s. converges to the random variable $X = x(\omega)$ then the sequence of the expectations $\{EX_n\}$ converges to EX,*

$$\lim EX_n = EX , \tag{3.1.13}$$

independently of the fact whether EX exist or not (in the latter case $EX = \infty$).

(x) On the basis of the theorem about the term-wise integration of the series *if the series $\sum\limits_{p=1}^{\infty} Y_p$ whose terms are the real random variables $Y_p = y_p(\omega)$ $(p = 1, 2, \ldots)$ absolutely a.s. converges to the random variable $X = x(\omega)$ then*

$$EX = \sum_{p=1}^{\infty} EY_p . \tag{3.1.14}$$

(xi) It follows from the Fatou lemma that *for any sequence of the real random variables $\{X_n\}$ lower bounded by the random variable U which has the finite expectation EU,*

$$E(\underline{\lim} X_n) \leq \underline{\lim} EX_n . \tag{3.1.15}$$

Similarly, if the sequence of the real random variables $\{X_n\}$ is upper bounded by the random variable U which has the finite expectation then

$$\overline{\lim}\, EX_n \leq E(\overline{\lim}\, X_n)\,. \tag{3.1.16}$$

(xii) It follows from the theorem about the majorizable sequence that *if the sequence of the real random variables $\{X_n\}$ a.s. converges to the random variable X and is lower and upper bounded by the random variables U and V which have the finite expectations, $U \leq X_n \leq V$, EU, $EV < \infty$ then all random variables X_n and X have the finite expectations and the sequence $\{EX_n\}$ converges to $\{EX\}$, i.e. formula* (3.1.13) *is valid.*

(xiii) It follows from the general Lebesgue theorem about the transition to the limit under the sign of an integral that *if the sequence of the random variables $\{X_n\}$ with one and the same phase B-space X a.s. converges or in probability to the random variable X and all its variables X_n are bounded by a norm of the random variable U which has a finite expectation, $\| X_n \| \leq U$, $EU < \infty$ then there exists EX and relation* (3.1.13) *is valid.*

(xiv) *If the series $\sum\limits_{p=1}^{\infty} Y_p$ a.s. converges or converges in probability to the random variable X and all its finite segments are bounded by a norm of the random variable U which has a finite expectation then there exists EX and formula* (3.1.14) *is valid.*

R e m a r k. The expectations of the random variables in any B-space possess properties (i)–(v), (vii), (viii), (xiii) and (xiv). Properties (vi), (ix)–(xii) are inherent only to the expectation of the real random variables.

E x a m p l e 3.1.3. If all the realizations of the random function $X(t) = x(t,\omega)$ belong to the space of the numerical continuous functions $C([a, b])$, i.e. $X(t)$ represents a random variable with the phase space $C([a, b])$ then according to property (v) of the expectations its expectation $m_x(t)$ exists if and only if there exists an expectation of the random variable $Y = \sup\limits_{t\in[a,b]} |X(t)|$,

$$m_y = EY = \int \sup\limits_{t\in[a,b]} |x(t,\omega)|\, P(d\omega) < \infty\,. \tag{I}$$

The differentiation operator $D = d/dt$ is a closed linear operator in $C([a, b])$ and its domain is a set of differentiable functions which is dense everywhere in $C([a, b])$. If all the realizations of the random function $X(t)$ are differentiable and there exist the expectations of the random function $X(t)$ and its derivative $X'(t)$

then by the theorem about the permutability of an operator of an expectation with a closed linear operator the expectation $m_x(t)$ of the random function $X(t)$ represents a differentiable function of the variable t and $m_{x'}(t) = EX'(t) = m_x'(t)$.

E x a m p l e 3.1.4. If all the realizations of the random function $X(t) = x(t,\omega)$ belong to the space of the differentiable functions $C^1([a,b])$, i.e. $X(t)$ represents a random variable with the phase space $C^1([a,b])$ then its expectation $m_x(t)$ exists if and only if the random variable $Z = \sup\limits_{t\in[a,b]} |X(t)| + \sup\limits_{t\in[a,b]} |X'(t)|$ has the expectation

$$m_z = \int \sup_{t\in[a,b]} |x(t,\omega)|\ P(d\omega) + \int \sup_{t\in[a,b]} |x_t'(t,\omega)|\ P(d\omega) < \infty. \qquad (I)$$

The differentiation operator $D = d/dt$ is a bounded operator in $C^1([a,b])$. Therefore using the theorem about the permutability of an operator of an expectation with bounded linear operator we obtain again the result of the previous example. The difference is in the fact that the requirement of the existence of an expectation of the derivative $X'(t)$ in a given case is the result of the condition of the existence of an expectation of the random function $X(t)$.

E x a m p l e 3.1.5. If all the realizations of the random function $X(t) = x(t,\omega)$ belong to the space $L_p(T,\mu)$, i.e. $X(t)$ represents a random variable with the phase space $L_p(T,\mu)$ then an expectation of the random function $X(t)$ exists if and only if the random variable

$$U = \left[\int \|\,x(t,\omega)\,\|^p\ \mu(dt)\right]^{1/p} \qquad (I)$$

has the expectation equal to

$$m_u = \int \left[\int \|\,x(t,\omega)\,\|^p\ \mu(dt)\right]^{1/p} P(d\omega) < \infty. \qquad (II)$$

R e m a r k. The mentioned examples show that the definition of an expectation of a random variable in B-space impose rather rigid restrictions on the random functions considered as the random variables with the functional phase spaces. Therefore it is necessary for the theory of random functions to extend the notion of an expectation over the random variables in more general linear spaces.

3.2. Expectation of Random Variable in Topological Linear Space

3.2.1. Definitions

We shall consider now the random variables in topological linear spaces with a weak topology determined by the linear functionals. Let $X = x(\omega)$ be the random variable whose phase space X represents a topological linear space with a weak topology which is determined by the set F_c of the linear functionals. We shall suppose that the space X possesses the weak completeness and the set of the functionals F_c is sufficiently rich for the fact that from $fx = 0$ for all $f \in F_c$ follows that $x = 0$. These conditions are sufficient for the fact that the weak integrals with respect to the functions with the values in the space X would be worth while. The vector $m_x \in X$ determined as a weak integral with respect to the function $X = x(\omega)$ over the measure P by formula (3.1.4) is called *an expectation of the random variable $X = x(\omega)$ in a weakly complete topological linear space X*. As it is possible to introduce a weak topology in any B-space by means of the continuous linear functions the given definition of an expectation refers also to the random variables in a weakly complete B-spaces.

R e m a r k. As it is known from functional analysis any integrable function with the values in B-space is weakly integrable and its weak integral coincides with a strong one. Therefore any random variable in B-space which has an expectation in the sense of the definition of Subsection 3.1.1 has the same expectation in the sense of the definition of the given subsection. But not every random variable in a weakly complete B-space which has an expectation in the sense of the latter definition has an expectation in the sense of the definition of Subsection 3.1.1. Thus our new definition of an expectation is more general than the definition of Subsection 3.1.1.

E x a m p l e 3.2.1. Let $X(t) = x(t, \omega)$, $t \in T_1$ be a random function determined in the probability space $(\Omega, \mathcal{S}, P)$ whose all realizations belong to some linear space of the numerical functions X. We denote by f_l a linear functional which associated to any function $x(t) \in X$ its value at the point $t = l$, $f_l x(t) = x(l)$. We assume as a set of the linear functionals F_c a set of all finite linear combinations of the functionals f_l correspondent to all $l \in T_1$. Then any functional $f \in F_c$ will be expressed by formula $fx(t) = \sum_{k=1}^{l} f_k x(t_k)$ at some $l, f_1, \ldots, f_l, t_1, \ldots, t_l$. It is evident that the set of the functionals F_c constructed in such a way is sufficiently complete as from $\sum_{k=1}^{l} f_k x(t_k) = 0$ at all $l, f_1, \ldots, f_l, t_1, \ldots, t_l$ follows that

$x(t) \equiv 0$. We determine in X a weak topology by means of the set of the functionals F_c. In other words, we determine in X the vicinities of zero

$$\left\{ x(t) \ : \ \left| \sum_{k=1}^{l_1} f_k^1 x(t_k^1) \right| < \varepsilon_1 \, , \, \ldots \, , \, \left| \sum_{k=1}^{l_n} f_k^n x(t_k^n) \right| < \varepsilon_n \right\} \, , \qquad \text{(I)}$$

correspondent to all possible n, l_1, $\ldots$, l_n, f_k^p, $t_k^p \in T_1$ ($k = 1$, $\ldots$, l_p; $p = 1$, $\ldots$, n), ε_1, $\ldots$, $\varepsilon_n > 0$. It is easy to see that the convergence of the sequence of the functions $\{x_n(t)\} \subset X$ to $x(t)$ in this topology represents the convergence at each fixed $t \in T_1$. Really, from $f x_n(t) \to f x(t)$ at all $f \in F_c$ taking $f = f_s$ we get $x_n(s) \to x(s)$ at any $s \in T_1$. Hence it follows that for any sequence $\{x_n(t)\}$ there exists the limit function $x(t)$. Therefore the space X may be complemented by including in it the limits of all fundamental in this topology sequences of the functions $\{x_n(t)\}$, $x_n(t) \in X$. As a result X will become a weak complete topological linear space.

Now we assume that the function $x(t, \omega)$ of the variable ω with the values in X is weakly P-integrable. It means that there exists the sequence of the simple functions $\{x^n(t, \omega)\}$ of the variable ω of the form of (3.1.3):

$$x^n(t, \omega) = \sum_{k=1}^{N_n} x_k^n(t) 1_{A_k^n}(\omega) \, , \quad A_k^n \in \mathcal{S} \, , \qquad \text{(II)}$$

which weakly determines $x(t, \omega)$, i.e. weakly convergent to $x(t, \omega)$ almost at all ω relatively to the measure P and satisfying the condition

$$\int |f x^n(t, \omega) - f x^m(t, \omega)| \, P(d\omega) \to 0 \quad \text{at} \quad n, m \to \infty \qquad \text{(III)}$$

for all $f \in F_c$. Hence taking into account that the convergence in our topology represents the convergence at all $t \in T_1$ we conclude that the function $x(t, \omega)$ of the variable ω is wealky P-integrable if and only if the numerical function $x(t, \omega)$ at any fixed t is P-integrable. It means that the expectation of the random function $X(t) = x(t, \omega)$ exists if and only if the value of the random function at any $t \in T_1$ has an expectation.

E x a m p l e   3.2.2.   Let us consider the function $X(t) = x(t, \omega) = \sin t/\omega$, $0 \leq t \leq T_1$ where ω is a random function uniformly distributed at the interval $[0, 1]$. In a given case the space $(\Omega, \mathcal{S}, P)$ represents the interval $[0, 1]$ with the σ-algebra of Borel sets and Lebesgue measure P. The random function $X(t)$ represents a random variable in B-space of the differentiable functions $C^1([0, T_1])$. The variable has an expectation in the sense of the definition of a given section as

$$|m_x(t)| = \left| \int_0^1 \sin \frac{t}{\omega} d\omega \right| \leq 1 \quad \text{at any} \quad t \, . \qquad \text{(I)}$$

But this random variable has none expectation in the sense of the definition of Subsection 3.1.1 as the random variable

$$Z = \sup_{t\in[0,T_1]} \left|\sin\frac{t}{\omega}\right| + \sup_{t\in[0,T_1]} \left|\frac{1}{\omega}\cos\frac{t}{\omega}\right| > \frac{1}{\omega} \tag{II}$$

has no expectation.

E x a m p l e 3.2.3. If the numerical function $x(t,\omega)$ of two variables $t \in T_1$ and $\omega \in \Omega$ is integrable relative to the product of some nonnegative measure σ determined on the σ-algebra $\mathcal{J}$ of the sets of the space T_1 and the probability P then on the basis of the Fubini theorem known from functional analysis all the realizations of the random function $x(t) = x(t,\omega)$ will be σ-integrable. Therefore $x(t,\omega)$ has an expectation almost at all t and

$$\int P(d\omega) \int x(t,\omega)\sigma(dt) = \int \sigma(dt) \int x(t,\omega)P(d\omega). \tag{I}$$

Hence it follows that the expectation $m_x(t)$ of a random function represents the σ-integrable function and

$$E \int X(t)\sigma(dt) = \int m_x(t)\sigma(dt). \tag{II}$$

Thus in a given case an operator of an expectation is permutable with an integral.

3.2.2. Properties of Expectation

In the correspondence with the properties of the weak integrals, the properties (i), (ii) of the expectation (the linearity of an operator of an expectation) remain valid at our new definition. In exactly the same way property (iv) remains valid (the identity of the expectations of equivalent random variables). Finally, property (iii) remains valid with inessential changes (the permutability of an operator of an expectation with a linear operator). We cite the formulation of the correspondent properties which follow directly from the theorems of functional analysis about the permutability of a weak integral with a linear operator.

If T is a continuous linear operator mapping a phase space of the random variable X into another weakly complete topological linear space Y and there exists an expectation of the random variable X then there exists also an expectation of the random variable $Y = TX$ and in this case relation (3.1.11) is valid.

If T is a closed linear operator mapping a phase space of the random variable X into another weakly complete topological linear space Y and there exists an expectation of the random variable $V = \{X, TX\}$ (with the phase space $X \times Y$) then there exist the expectations of the random variables X and $Y = TX$ and here relation (3.1.11) is valid.

At the general definition of an expectation the corresponding properties about the transition to the limit under the sign of an expectation and termwise application of an operator of an expectation to the infinite series (properties (xiii) and (xiv) of the expectations) remain also valid.

If the sequence of the random variables $\{X_n\}$ weakly converges a.s. or in probability to the random variable X, there exist the expectations of all the variables X_n and X and for any linear functional $f \in F_c$ there exists such positive random variable U_f with the finite expectation that $|fX_n| \leq U_f$ a.s. then the sequence of the expectations $\{EX_n\}$ weakly converges to the expectation of the random variable X, $EX = \lim(\text{weakly})\, EX_n$.

If the series $\sum\limits_{p=1}^{\infty} Y_p$ whose terms are the random variables with finite expectations weakly converges a.s. or in probability to the random variable X which has the expectation and for any linear functional $f \in F_c$ there exists the positive random variable U_f with the finite expectation majorizing in modulus all finite segments of the series $\sum\limits_{p=1}^{\infty} fY_p$ then the series of the expectations $\sum\limits_{p=1}^{\infty} EY_p$ weakly converges to the expectation of the random variable X.

3.3. Moments

3.3.1. Definitions

In elementary probability theory the moment of the second order Γ_x and the covariance matrix K_x of the random vector X are determined by following formulae:

$$\Gamma_x = \left[\gamma_{pq}^x\right] = EXX^*, \tag{3.3.1}$$

$$K_x = \left[K_{pq}^x\right] = EX^0 X^{0*}, \quad X^0 = X - m_x. \tag{3.3.2}$$

Here the asterisk means the operation of the transposition of a matrix with the change of all its complex elements by the correspondent conjugated numbers. There exists the dependence between m_x, Γ_x and K_x:

$$\Gamma_x = K_x + m_x m_x^*. \tag{3.3.3}$$

We introduce the notions of a cross second order moment (matrix) Γ_{xy} and a covariance (cross-covariance matrix) K_{xy} for two random vectors X and Y:

$$\Gamma_{xy} = \left[\gamma_{pq}^{xy}\right] = EXY^*, \tag{3.3.4}$$

$$K_{xy} = \left[K_{pq}^{xy}\right] = EX^0Y^{0*}, \tag{3.3.5}$$

and hence

$$\Gamma_{xy} = K_{xy} + m_x m_y^*. \tag{3.3.6}$$

In elementary theory of random functions the second order moment of the random function $X(t)$, $t \in T_1$ (matrix $\Gamma_x(t_1, t_2)$) and the covariance function (matrix $K_x(t_1, t_2)$) are determined by following formulae:

$$\Gamma_x(t_1, t_2) = \left[\gamma_{pq}^x(t_1, t_2)\right] = EX(t_1)X(t_2)^*, \tag{3.3.7}$$

$$K_x(t_1, t_2) = \left[K_{pq}^x(t_1, t_2)\right] = EX^0(t_1)X^0(t_2)^*, \tag{3.3.8}$$

and hence

$$\Gamma_x(t_1, t_2) = K_x(t_1, t_2) + m_x(t_1)m_x(t_2)^*. \tag{3.3.9}$$

The cross-moment of the second order (matrix $\Gamma_{xy}(t_1, t_2)$) and the cross-covariance function (matrix $K_{xy}(t_1, t_2)$) for two random functions $X(t)$ and $Y(t)$, $t \in T_1$ are determined by formulae:

$$\Gamma_{xy}(t_1, t_2) = \left[\gamma_{pq}^{xy}(t_1, t_2)\right] = EX(t_1)Y(t_2)^*, \tag{3.3.10}$$

$$K_{xy}(t_1, t_2) = \left[K_{pq}^{xy}(t_1, t_2)\right] = EX^0(t_1)Y^0(t_2)^*, \tag{3.3.11}$$

and hence

$$\Gamma_{xy}(t_1, t_2) = K_{xy}(t_1, t_2) + m_x(t_1)m_y(t_2)^*. \tag{3.3.12}$$

Now we suppose that the phase space X of the random variable $X = x(\omega)$ determined on the probability space $(\Omega, \mathcal{S}, P)$ is B-space or a weakly complete topological linear space. Let us take an arbitrary linear functional f in X. It belongs to the space $F = X^*$ adjoint with X. We consider the random variable $X\overline{fX}$. Evidently as its phase space serves the space X. If there exists an expectation of this random variable then it represents the vector from X dependent on $f \in F$. Thus formula

$$\Gamma_x f = EX\overline{fX} = \int x(\omega)\overline{fx(\omega)}P(d\omega) \tag{3.3.13}$$

determines the operator Γ_x mapping F into X. The domain D_{Γ_x} of this operator evidently will be some subspace of the space F.

The operator Γ_x is called *an operator of the second order moment* of the random variable X. The operator of the second order moment of the centered random variable $X^0 = X - m_x$ is called *a covariance operator* of the random variable X.

On the basis of (3.3.13) the covariance operator K_x of the random variable X is determined by formula

$$K_x f = EX^0 \overline{fX^0} = \int [x(\omega) - m_x] \overline{f[x(\omega) - m_x]} P(d\omega). \quad (3.3.14)$$

By the change of variables $x = x(\omega)$ with the account of formula (2.2.1) which determines the probability measure μ_x of the random variable X formulae (3.3.13) and (3.3.14) are reduced correspondingly to the following form

$$\Gamma_x f = EX\overline{fX} = \int x\overline{fx}\mu_x(dx), \quad (3.3.15)$$

$$K_x f = EX^0 \overline{fX^0} = \int (x - m_x)\overline{f(x - m_x)}\mu_x(dx). \quad (3.3.16)$$

It is easy to see that a covariance operator of a random variable is expressed in terms of its expectation and an operator of the second order. Really, on the basis of the properties of the expectations

$$EX^0\overline{fX^0} = EX\overline{fX} - m_x\overline{fEX} - \overline{fm_x}EX + m_x\overline{fm_x} = EX\overline{fX} - m_x\overline{fm_x},$$

whence

$$K_x f = \Gamma_x f - m_x\overline{fm_x}. \quad (3.3.17)$$

This formula shows the following.

Theorem 3.3.1. *For the existence of a covariance operator of a random function it is necessary and sufficient the existence of the operators of an expectation and the second order moment. The domain D_{K_x} of a covariance operator coincides with the domain D_{Γ_x} of the operator Γ_x.*

In the special case when the variable X has a finite or a countable set of the realizations $\{x_k\}$ formulae (3.3.13) and (3.3.14) which determine the operators Γ_x and K_x take the form

$$\Gamma_x f = \sum_k x_k\overline{fx_k}p_k, \quad K_x f = \sum_k (x_k - m_x)\overline{f(x_k - m_x)}p_k, \quad (3.3.18)$$

where p_k is the probability of the case that X will take the value x_k $(k = 1, 2, \ldots)$. As the domain of the operators Γ_x and K_x in this case serves the whole space F.

E x a m p l e 3.3.1. Let us consider the n-dimensional random vector X. With the understanding that the vectors are presented in the form of matrices-columns, the operation of the transposition is marked by the sign T, and the operation of the transposition with simultaneous change of all complex elements by the correspondent conjugated numbers is marked by the asterisk. Then the result of the action of any linear functional f on the vector X may be written in the form of the product of the transposed the n-dimensional vector f on the vector X, $fX = f^T X$ where in the left-hand side f means the functional, X means the vector, and in the right-hand side f and X are the correspondent matrices-columns. Formula (3.3.14) gives here $K_x f = EX^0(\overline{f^T X^0})$. Hence it follows that $K_x f$ represents a matrix-column with the components $(K_x f)_p = \sum_{q=1}^{n} \overline{f_q} E X_p^0 \overline{X_q^0}$. Here $K_x f$ may be considered as the product of the covariance matrix K_x of the vector X, $K_x = [k_{pq}]$, $k_{pq} = E X_p^0 \overline{X_q^0}$ $(p, q = 1, \ldots, n)$ by the matrix-column $\overline{f}$, $K_x f = K_x \overline{f}$. Thus *the covariance operator of a random vector is determined by its covariance matrix.*

In exactly the same way we come to the conclusion that an operator of the second order moment of the random vector X is determined by the matrix of the second order moments of its components $\Gamma_x = [\gamma_{pq}]$ where $\gamma_{pq} = E X_p \overline{X_q}$ $(p, q = 1, \ldots, n)$. $\Gamma_x f$ represents the product of the matrix Γ_x by the matrix-column $\overline{f}$. In this case formula (3.3.17) gives the known relation between the centered and initial second orders moments and an expectation of a random vector. This relation has a matrix form $K_x = \Gamma_x - m_x m_x^*$. It is easy to see that all these results are also valid for a random vector with the countable set of components.

Now we notice that if $X_p = x_p(\omega)$ $(p = 1, \ldots, n)$ are scalar random variables given in the probability space $(\Omega, \mathcal{S}, P)$ then their second order moments exist if and only if the functions $x_p(\omega)$ belong to H-space $L_2(\Omega, \mathcal{S}, P)$ as

$$\gamma_{pq} = E X_p \overline{X_q} = \int x_p(\omega)\overline{x_q(\omega)} P(d\omega) = (x_p, x_q).$$

Thus a set of all scalar random variables with the finite second order moments determined in one and the same probability space represents a H-space with the scalar product $(X, Y) = EX\overline{Y}$. In particular, a set of all centered scalar random variables with the finite variance is a H-space in which as a scalar product serves a covariance moment of the random variable.

From the property of the scalar products follows that a determinant of a matrix of the second order moments of a finite-dimensional random vector is a real nonnegative number which is equal to zero if and only if some linear combination of the vector components is equivalent to zero. The covariance matrix possesses the same property.

E x a m p l e 3.3.2. Consider the random function $X = X(t)$ whose all realizations belong to some linear space of the scalar function X. We determine in this space a weak topology by means of the set F_c of the linear functionals of Example 3.3.1. Then for any $f \in F_c$ we shall have $K_x f = \sum_{k=1}^{l} \overline{f_k} E X^0(t) \overline{X^0(t_k)}$.

Hence it is clear that the covariance operator of the random function $X(t)$ is completely determined by its covariance function $K_x(t,t') = E X^0(t) \overline{X^0(t')}$. Analogously we come to the conclusion that an operator of the second order moment of the random function $X(t)$ is completely determined by its initial second order moment $\Gamma_x(t,t') = E X(t) \overline{X(t')}$. Formula (3.3.17) gives in this case the known relation between the covariance matrix, the initial second order moment and the expectation of the random function. It is easily to see that if the functions $K_x(t,t')$ and $\Gamma_x(t,t')$ exist the domain of the operators K_x and Γ_x contains F_c.

E x a m p l e 3.3.3. If $X = X(t)$ is a random function whose all the realizations belong to some linear space X of the n-dimensional vector functions then as the set of linear functionals F_c determining a weak topology in X we may assume the set of all finite linear combinations of the functionals f_{qs} $(q = 1, \ldots, n; s \in T_1)$, $f_{qs} x(t) = x_q(s)$, i.e. the functionals which to any vector function $x(t) \in X$ corresponds the value of any of its components at some value s of the argument t. Then we shall have $K_x f = \sum_{k=1}^{l} \overline{f_k} E X^0(t) \overline{X^0_{q_k}(t_k)}$. Thus $K_x f$ represents a vector function from the space X with the components $(K_x f)_p = \sum_{k=1}^{l} \overline{f_k} E X^0_p(t) \overline{X^0_{q_k}(t_k)}$.

Hence it is clear that a covariance operator of the vector random function $X(t)$ is determined by a matrix of the covariance and cross-covariance functions of its components

$$K_x(t,t') = \begin{bmatrix} K_{11}(t,t') & K_{12}(t,t') & \ldots & K_{1n}(t,t') \\ K_{21}(t,t') & K_{22}(t,t') & \ldots & K_{2n}(t,t') \\ \ldots & \ldots & \ldots & \ldots \\ K_{n1}(t,t') & K_{n2}(t,t') & \ldots & K_{nn}(t,t') \end{bmatrix},$$

where

$$K_{pq}(t,t') = E X^0_p(t) \overline{X^0_q(t')} \ (p,q = 1, \ldots, n).$$

Similarly we convince in the fact that an operator of the second order moment of the vector random function $X(t)$ is determined by the matrix $\Gamma_x(t,t')$ of the initial second order moments of its components. Formula (3.3.17) gives in this case the known relation between the covariance function, the initial second order moments and the expectations. This relation has the matrix form $K_x(t,t') = \Gamma_x(t,t')$ $- m_x(t)m_x(t')^*$, where $m_x(t)^* = \left[\overline{m_1(t)} \ldots \overline{m_n(t)}\right]$, $m_p(t) = EX_p(t)$ $(p = 1, \ldots, n)$. Analogously as in Example 3.3.2 if the matrix functions $K_x(t,t')$ and $\Gamma_x(t,t')$ exist then the domain of the operator K_x and Γ_x contains F_c.

Let $X = x(\omega)$ and $Y = y(\omega)$ be the random variables in the spaces X and Y correspondingly, determined in the probability space $(\Omega, \mathcal{S}, P)$. If each of the spaces X and Y represent a B-space or a weakly topological linear space, and F and G are the correspondent adjoint spaces of the linear functionals then formula

$$\Gamma_{xy} g = EX\overline{gY} = \int x(\omega)\overline{gy(\omega)}P(d\omega) \qquad (3.3.19)$$

determines the operator Γ_{xy} mapping G into X. The operator Γ_{xy} determined by formula (3.3.19) is called *a cross-operator of the second order moment* of the random variables X and Y.

A cross-operator of the second order moment of the centered random variables $X^0 = X - m_x$ and $Y^0 = Y - m_y$ is called *a cross-covariance operator* of the random variables X and Y. It follows from (3.3.19) that the cross-covariance operator K_{xy} of the random variables X and Y is determined by formula

$$K_{xy} g = EX^0\overline{gY^0} = \int [x(\omega) - m_x]\overline{g[y(\omega) - m_y]}P(d\omega). \qquad (3.3.20)$$

It is easily derived the relation between K_{xy}, Γ_{xy}, m_x and m_y from (3.3.19) and (3.3.20) analogous to (3.3.17)

$$K_{xy} g = \Gamma_{xy} g - m_x\overline{gm_y}. \qquad (3.3.21)$$

Hence it follows the following statement.

Theorem 3.3.2. *For the existence of the operator K_{xy} it is necessary and sufficient the existence of m_x, m_y and Γ_{xy}. The domains of the operators Γ_{xy} and K_{xy} coincide.*

If X and Y are the random variables with the finite or countable set of the realizations $\{x_k, y_k\}$ then

$$\Gamma_{xy} g = \sum_k x_k\overline{gy_k}p_k, \qquad K_{xy} g = \sum_k (x_k - m_k)\overline{g(y_k - m_y)}p_k, \qquad (3.3.22)$$

where p_k is the probability of the occurrence of the kth pair of the realizations $(x_k y_k)$ $(k = 1, 2, \ldots)$ of the variables X, Y. In particular, it follows from (3.3.18) and (3.3.22) that for the nonrandom vectors $X = c$, $Y = d$ we have $\Gamma_x = c\overline{f}c$, $\Gamma_{xy} = c\overline{g}d$, $K_x = K_{xy} = 0$. It follows also from (3.3.20) that $K_{xy} = 0$ if even one of the vectors X and Y be not random.

The random variables X and Y are called *uncorrelated* if $K_{xy}f = 0$ for all $f \in F$ (i.e. $K_{xy} = 0$). If $K_{xy}f \neq 0$ even for one $f \in F$ then the variables X and Y are called *correlated*.

$\triangleright$ Suppose that the random variables X and Y are independent. In this case the probability maesure μ_z of the composite random variable $Z = (X, Y)$ is equal to the product of the probability measures μ_x, μ_y of the variables X, Y. Let $\varphi(x)$ be an arbitrary scalar μ_x-integrable function in the space X, and ψ be an arbitrary μ_y-integrable function in the space Y with the values in some B-space or in weakly complete topological linear space. In this case according to the definition of an integral in functional analysis the product $\varphi(x)\psi(y)$ is integrable over the product μ_z of the measures μ_x and μ_y, and on the basis of the Fubini theorem we have

$$E\varphi(X)\psi(Y) = \int \varphi(x)\psi(y)\mu_z(dx \times dy)$$

$$= \int \varphi(x)\left\{ \int \psi(y)\mu_y(dy) \right\} \mu_x(dx) = \int \varphi(x)\mu_x(dx) \cdot \int \psi(y)\mu_y(dy)$$

$$= E\varphi(X)E\psi(Y). \triangleleft \qquad\qquad (3.3.23)$$

Thus we proved the following theorem.

Theorem 3.3.3. *If the random variable U represents a function of the random variable X, and the random variable V represents a function of the random variable Y and one of the variables U, V is scalar then in the case of independent variables X and Y an expectation of the product of the variables U and V is equal to the product of their expectations.*

Putting, in particular, $\varphi(X) = X^0 = X - m_x$, $\psi(Y) = \overline{gY^0} = \overline{gY} - \overline{gm_y}$ we get $K_{xy}g = EX^0\overline{gY^0} = EX^0 E\overline{gY^0} = 0$. Thus *the independent random variables are always uncorrelated*. The inverse as it is known from elementary probability theory is not always true.

E x a m p l e 3.3.4. If $X = X(t)$, $t \in T_1$ is a scalar random function whose all the realizations belong to the linear space X, and $Y = Y(s)$, $s \in S$ is a scalar random function whose all the realizations belong to the linear space Y then putting in X and Y the weak topologies of Example 3.3.2 by means of the linear functionals F_c and G_c similarly as in Example 3.3.2 we come to the conclusion that the operators Γ_{xy} and K_{xy} are determined by the mixed initial moment of the second order $\Gamma_{xy}(t,s) = EX(t)\overline{Y(s)}$ and by the cross- covariance function $K_{xy}(t,s) = EX^0(t)\overline{Y^0(s)}$ of the functions X and Y.

E x a m p l e 3.3.5. If $X = X(t)$, $t \in T_1$ is the n-dimensional vector random function, and $Y = Y(s)$, $s \in S$ is the m-dimensional random function then putting in the functional phase spaces X and Y the topologies of Example 3.3.3 we come to the conclusion that the operators Γ_{xy} and K_{xy} are determined by the matrices $\Gamma_{xy}(t,s)$ and $K_{xy}(t,s)$ of the mixed initial moments of the second order and cross-covariance functions of the components of the vector functions $X(t)$ and $Y(s)$.

3.3.2. Properties of Second Order Moments

In elementary probability theory the following properties of the second order moments for the random variables X and Y are known:
- $\Gamma_x^* = \Gamma_x$ (hermitivity);
- the matrices Γ_x and K_x are nonnegatively determined (i.e. for any complex vector u) $u^T \Gamma_x \bar{u} \geq 0$ and $u^T K_x \bar{u} \geq 0$;
- $\Gamma_{yx} = \Gamma_{xy}^*$ and $K_{yx} = K_{xy}^*$.

In elementary theory of random functions for the random functions $X(t)$ and $Y(t)$, $t \in T_1$ the analogous properties take place:
- $\Gamma_x(t_1,t_2) = \Gamma_x(t_1,t_2)^*$;
- for any $t_1, \ldots, t_N \in T_1$ and any complex vectors $u_1, \ldots, u_N$

$$\sum_{p,q=1}^{N} u_p^T \Gamma_x(t_p,t_q)\bar{u}_q \geq 0 \quad \text{and} \quad \sum_{p,q=1}^{N} u_p^T K_x(t_p,t_q)\bar{u}_q \geq 0;$$

- $\Gamma_{yx}(t_2,t_1) = \Gamma_{xy}(t_1,t_2)^*$ and $\Gamma_{yx}(t_2,t_1) = K_{xy}(t_1,t_2)^*$.

Let us study the main properties of the operators of the second order moment and the covariance operators.

(i) From property (ii) of the expectations follows that *operators Γ_x, K_x, Γ_{xy} and K_{xy} are the adjoint linear operators.*

(ii) From property (iii) of the expectations *for any continuous linear functional f follows*

$$f\Gamma_x f = E(fX)(\overline{fX}) = E\,|fX|^2 \geq 0\,. \tag{3.3.24}$$

Thus the operator of the second order moment and the covariance operator are positive $\Gamma_x > 0$, $K_x > 0$.

(iii) From property (iii) of the expectations follows also that *for any $f_1, f_2 \in F_c$*

$$f_1 \Gamma_x f_2 = E(f_1 X)(\overline{f_2 X}) = \overline{f_2 \Gamma_x f_1}\,. \tag{3.3.25}$$

Thus the operator of the second order moment and covariance operator possess the symmetry property (3.3.25).

Analogously we have *for any functionals $f \in F_c$, $g \in G_c$*

$$f\Gamma_{xy}g = E(fX)(\overline{gY}) = \overline{g\Gamma_{xy}f} \,. \tag{3.3.26}$$

(iv) According to property (i) of the expectations *if $Y = aX$, $U = cZ$ then $\Gamma_y = |a|^2 \Gamma_x$, $K_y = |a|^2 K_x$, $\Gamma_{yu} = a\overline{c}\Gamma_{xz}$, $K_{yu} = a\overline{c}K_{xz}$.*

(v) From property (ii) of the expectations follows that *if*

$$Y = \sum_{\nu=1}^{n} a_\nu X_\nu \,, \quad Z = \sum_{\nu=1}^{n} b_\nu X_\nu \,, \tag{3.3.27}$$

then

$$\left.\begin{aligned}
\Gamma_y &= \sum_{\nu,\mu=1}^{n} a_\nu \overline{a_\mu} \Gamma_{\nu\mu} \,, \quad &K_y &= \sum_{\nu,\mu=1}^{n} a_\nu \overline{a_\mu} K_{\nu\mu} \,, \\
\Gamma_{yz} &= \sum_{\nu,\mu=1}^{n} a_\nu \overline{b_\mu} \Gamma_{\nu\mu} \,, \quad &K_{yz} &= \sum_{\nu,\mu=1}^{n} a_\nu \overline{b_\mu} K_{\nu\mu} \,.
\end{aligned}\right\} \tag{3.3.28}$$

Here $\Gamma_{\nu\mu}$ and $K_{\nu\mu}$ are a cross-operator of the second order moment and a cross-covariance operator of the random variables X_ν and X_μ $(\nu,\mu = 1, \ldots, n)$ correspondingly.

(vi) From property (iii) of the permutability of an operator of an expectation with a linear operator follows the relations between the operators of the second order moment and the covariance operators of the random variables connected by the linear transformations. Let T be a linear operator mapping a phase space of the random variable X into the linear space Y, and S be a linear operator mapping a phase space of the random variable Z into the linear space U. Suppose that there exist the adjoint operators T^* and S^*. We introduce the random variables $Y = TX$ and $U = SZ$. From property (iii) it follows that *if X is a random variable with a finite set of the realizations $x_1, \ldots, x_N \in D_T$ then $Y = TX$ is a random variable with the realizations $y_1 = Tx_1, \ldots, y_N = Tx_N$, and the operators Γ_y, Γ_{yx}, Γ_{xy}, K_y, K_{yx} and K_{xy} are expresed by the formulae:*

$$\left.\begin{aligned}
\Gamma_y &= T\Gamma_x T^* \,, \quad &\Gamma_{yx} &= T\Gamma_x \,, \quad &\Gamma_{xy} &= \Gamma_x T^* \,, \\
K_y &= TK_x T^* \,, \quad &K_{yx} &= TK_x \,, \quad &K_{xy} &= K_x T^* \,.
\end{aligned}\right\} \tag{3.3.29}$$

The domain of the operators Γ_y, Γ_{xy}, K_y and K_{xy} serves the domain D_{T^*} of the operator T^*, and the operators Γ_{yx} and K_{yx}, similarly as Γ_x

and K_x, are determined on the whole space F. Really, denoting by G the space of the linear functionals in Y adjoint with Y we obtain for any $g \in D_{T^*} \subset G$

$$\Gamma_y g = EY\overline{gY} = ETX\overline{(gT)X} = TEX\overline{(T^*g)X} = T\Gamma_x T^* g \,. \qquad (3.3.30)$$

Analogously other formulae (3.3.29) are proved.

In exactly the same way if X and Z are *random variables with a finite set of the pairs of the realizations* $(x_1, z_1), \ldots, (x_N, z_N)$, $x_1,$ $\ldots, x_N \in D_T$, $z_1, \ldots, z_N \in D_S$ *then a cross-operator of the random variables* $Y = TX$ *and* $U = SZ$ *are expressed by the formulae*

$$\Gamma_{yu} = T\Gamma_{xz}S^* \,, \quad K_{yu} = TK_{xz}S^* \,, \qquad (3.3.31)$$

and its domain serves D_{S^*}.

If X and Y are B-spaces or weakly complete topological linear spaces and the operator T is continuous then follows that if there exist Γ_x and K_x then the operators Γ_y, Γ_{yx}, Γ_{xy}, K_y, K_{yx} and K_{xy} exist also and as the domain of the operators Γ_{yx} and K_{yx} serves the subspace D_{Γ_x}, and as the domain of the operators Γ_y, Γ_{xy}, K_y and K_{xy} serves the subspace $\{g : g \in D_{T^*}, T^*g \in D_{\Gamma_x}\}$.

Analogously, if X, Y, Z and U are B-spaces or weakly complete topological linear spaces, and the operators T and S are continuous then from the existence of the operators Γ_{xz} and K_{xz} follows the existence of the operators Γ_{yu}, K_{yu} whose domain serves the subspace $\{k : k \in D_{S^*}, S^*k \in D_{\Gamma_{xy}}\}$ of the space K adjoint with U.

Let X and Y be the separable B-spaces, T be a closed linear operator and $\Delta_x \subset D_T$ then from the existence of the operators Γ_x and Γ_{yx}, K_x and K_{yx} with the common domain D and nonempty intersection of the domain Δ_{T^*} of the operator T^* with D follows the existence of the operators Γ_y, Γ_{xy}, K_y and K_{xy} with the domain $\{g : g \in D_{T^*}, T^*g \in D\}$ and formulae (3.3.29). Really, in this case for any $f \in D$ there exist the expectations $\Gamma_x f = EX\overline{fX}$, $\Gamma_{yx}f = EY\overline{fX} = ETX\overline{fX}$, and consequently, $\Gamma_x f \in D_T$ and $\Gamma_{yx}f = T\Gamma_x f$. Similarly, for any $g \in D_{T^*}$, $T^*g \in D$ there exist the expectations

$$\begin{aligned}
\Gamma_{xy}g &= EX\overline{gY} = EX\overline{(T^*g)X} = \Gamma_x T^* g \,, \\
\Gamma_y g &= EY\overline{gY} = ETX\overline{(T^*g)X} \,,
\end{aligned} \qquad (3.3.32)$$

and here $\Gamma_x T^* g \in D_T$ and $\Gamma_y g = T\Gamma_x T^* g$.

Analogously, if X, Y, Z and U are separable B-spaces, T and S are closed linear operators, $\Delta_x \subset D_T$, $\Delta_z \subset D_S$ then from the existence *of the operators* Γ_{xz}, Γ_{yz}, K_{xz}, K_{yz} *and nonempty intersection of the domain* Δ_{S^*} *of the operator* S^* *with* $D = D_{\Gamma_{xz}} = D_{K_{xz}}$ *follows the existence of the operators* Γ_{yu}, K_{yu} *with the domain* $\{k : k \in D_{S^*}, S^*k \in D\}$ *and formulae (3.3.31) also.*

Suppose that X *and* Y *are weakly complete topological linear spaces and the operator* T *is closed in the topologies of these spaces. We denote by* $V = \{X, TX\}$ *a random variable whose phase space serves the graph of the operator* T *then from the existence of the operators* Γ_{vx}, K_{vx} *and nonempty intersection of the domain* Δ_{T^*} *of the adjoint operator* T^* *with the domain* D *of the operator* Γ_{vx} *follows the existence of the operators* Γ_x, Γ_{yx}, K_x, K_{yx} *with the domain* D *and the operators* Γ_y, Γ_{xy}, K_y, K_{xy} *with the domain* $\{g : g \in D_{T^*}, T^*g \in D\}$, *and formulae (3.3.29) also.* Really, in this case for any $f \in D$ there exist the expectations of the form (3.3.32) and $\Gamma_x f \in D_T$ and $\Gamma_{yx} f = T\Gamma_x f$. And in exactly the same way for any $g \in D_{T^*}$, $T^*g \in D$ there exist the expectations

$$\Gamma_{xy}g = EX\overline{gY} = EX\overline{(T^*g)X} = \Gamma_x T^*g, \quad \Gamma_y g = EY\overline{gY} = ETX\overline{(T^*g)X},$$

and $\Gamma_x T^*g \in D_T$ and $\Gamma_y g = T\Gamma_x T^*g$.

Analogously, *if* X, Y, Z *and* U *are weakly complete linear spaces, the operator* T *is closed in weak topological spaces* X, Y *and the operator* S *is closed in weak topologies of the spaces* Z, U, *and* $\Delta_x \subset D_T$, $\Delta_z \subset D_S$ *then from the existence of the operators* Γ_{vz}, K_{vz} *and nonempty intersection* Δ_{S^*} *with* $D = D_{\Gamma_{vz}}$ *follows the existence of the operators* Γ_{xz}, Γ_{yz}, K_{xz}, K_{yz} *with the domain* D *and the operators* Γ_{xu}, Γ_{yu}, K_{xu}, K_{yu} *with the domain* $\{k : k \in D_{S^*}, S^*k \in D\}$, *and relations (3.3.31) also.*

(vii) From property (iv) of the expectations follows that *the operators* Γ *and* K *of the equivalent random variables coincide,* $\Gamma_x = \Gamma_y$, $K_x = K_y$, $\Gamma_{xz} = \Gamma_{yu}$, $K_{xz} = K_{yu}$ *if the variable* X *is equivalent to* Y *and* Z *is equivalent to* U.

(viii) From property (v) of the expectations follows that *if* $E \parallel X \parallel^2 < \infty$ *then the operators* Γ_x *and* K_x *exist and bounded (and consequently, continuous).* Really, as $\parallel X\overline{fX} \parallel \leq \parallel f \parallel \cdot \parallel X \parallel^2$ for any bounded functional $f \in F_c$ there exists an expectation of the variable $X\overline{fX}$ and

$$\parallel \Gamma_x f \parallel = \parallel EX\overline{fX} \parallel \leq E \parallel f \parallel E \parallel X \parallel^2 . \tag{3.3.33}$$

From P-integrability of the function $\parallel x(\omega) \parallel^2$ follows P-integrability of $x(\omega) \times \overline{fx(\omega)}$. Hence it follows that the operator Γ_x is determined everywhere on F_c, is bounded and its norm does not exceed $E \parallel X \parallel^2$.

Similarly, *from the fact that $E \parallel X \parallel^2 < \infty$, $E \parallel Y \parallel^2 < \infty$ follows the existence and the boundedness of the operators Γ_{xy}, K_{xy} determined everywhere on F_c.* For proving it is sufficient to remember that variable $E \parallel X \parallel \cdot \parallel Y \parallel$ represents a scalar product of the elements $\parallel X \parallel = \parallel x(\omega) \parallel$ and $\parallel Y \parallel = \parallel y(\omega) \parallel$ in H-space of the scalar random variables (Example 3.3.1), and to use Buniakowski–Schwarz inequality.

R e m a r k. In the special case when the phase spaces of the random variables X and Y represent H-space this result may be essentially strong. Namely, from $E \parallel X \parallel^2 < \infty$, $E \parallel Y \parallel^2 < \infty$ follows the complete continuity of the operators Γ_x, Γ_{xy}, K_x and K_{xy}. Really, let $X = x(\omega) \in L_2(\Omega, \mathcal{S}, P)$ and $\{x^n(\omega)\}$ be the sequence of simple functions convergent to $x(\omega)$ almost everywhere relatively to the measure P. Such sequence exists always, $\{x^n(\omega)\}$ converges to $x(\omega)$ in $L_2(\Omega, \mathcal{S}, P)$ in consequence of which

$$\int \parallel x^n - x \parallel^2 dP \to 0 \quad \text{and} \quad \int \parallel x^n \parallel^2 dP \to \int \parallel x \parallel^2 dP,$$

i.e. $E \parallel X^n - X \parallel^2 \to 0$ and $E \parallel X^n \parallel^2 \to E \parallel X \parallel^2$, $X^n = x^n(\omega)$. Denoting in terms of Γ_n the operator of the second order moment of a random variable

$$X^n = x^n(\omega) = \sum_{k=1}^{N_n} x_k^n 1_{A_k^n}(\omega),$$

and accounting that in H-space $\overline{fx} = (f, x)$ where f is some vector of H-space we shall have

$$\Gamma_n f = EX^n(f, X^n) = \sum_{k=1}^{N_n} x_k^n(f, x_k^n)P(A_k^n).$$

As a result of the fact that in the case of H-space X the spaces F_c and $\overline{F_c}$ coincide with X, the operators Γ and K are linear in H-space. In the case of more general linear spaces they are linear in the space $\overline{F_c}$ as $\overline{fx} = \overline{f}x$ where $\overline{f} \in \overline{F_c}$. The latter formula shows that at any n the operator Γ_n is finite-dimensional, and consequently, is quite continuous. Further, at any $f \in X$

$$\parallel \Gamma_x f - \Gamma_n f \parallel = \parallel E\left[X(f, X) - X^n(f, X^n) \right] \parallel$$

$$\leq E \parallel (X - X^n)(f, X) \parallel + E \parallel X^n(f, X - X^n) \parallel$$

$$\leq \parallel f \parallel E(\parallel X \parallel + \parallel X^n \parallel) \parallel X - X^n \parallel .$$

On the basis on Buniakowski–Schwarz inequality we get

$$\parallel \Gamma_x f - \Gamma_n f \parallel \leq \parallel f \parallel \left(\sqrt{E \parallel X \parallel^2} + \sqrt{E \parallel X^n \parallel^2} \right) \sqrt{E \parallel X^n - X \parallel^2} .$$

We assign now an arbitrary $\varepsilon > 0$ and choose n in such a way that

$$\sqrt{E\,\|X^n\|^2} < \sqrt{E\,\|X\|^2} + \varepsilon\,, \quad \sqrt{E\,\|X^n - X\|^2}$$

$$< \frac{\varepsilon}{2\sqrt{E\,\|X\|^2} + \varepsilon}\,.$$

Then we shall have $\|\,\Gamma_x f - \Gamma_n f\,\| < \varepsilon\,\|\,f\,\|$, hence it follows that $\|\,\Gamma_x - \Gamma_n\,\| < \varepsilon$. Thus in a given case there exist such set of completely continuous operators $\{\Gamma_n\}$ that at any $\varepsilon > 0$ we have $\|\,\Gamma_x - \Gamma_n\,\| < \varepsilon$ for some n. Hence it follows that the operator Γ_x is completely continuous. Analogously the complete continuity of the operator Γ_{xy} is proved.

(ix) From property (vii) of the expectations follows that *if $\Gamma_x = 0$ then the random variable fX is equivalent to zero at any $f \in F$*. Really, in this case $E\,|fX|^2 = f\Gamma_x f = 0$. But from here in no way follows that the random variable X is equivalent to zero.

E x a m p l e   3.3.7. Consider the random function

$$X = x(t, \omega) = \begin{cases} 0 & \text{at} \quad t \neq n\omega\,, \\ 1 & \text{at} \quad t = n\omega\,, \end{cases}$$

where ω is a normally distributed random variable. This random function represents a random variable in B-space $B((-\infty, \infty))$ of the bounded functions in which the norm of the function $x(t)$ is equal to $\sup\limits_{t} |x(t)|$. Evidently, that for any functional from the set F_c of Example 3.3.2 we have $fX = \sum\limits_{k=1}^{l} f_k x(s_k, \omega) = 0$ if ω does not coincide with any of the numbers $s_1/n, \ldots, s_l/n$ $(n = 1, 2, \ldots)$. Therefore $P(fX = 0) = 1$ at any $f \in F_c$. But none of the realizations of the considered random function is equal to zero in $B((-\infty, \infty))$ as all the realizations have the norm equal to 1 in consequence of which $P(X = 0) = 0$.

In the special case when *the random function $Z(f) = fX$ of the variable $f \in F$ is separable, $\Gamma_x = 0$ if and only if X is equivalent to 0.* Really, let L be a separant of the random function $Z(f)$. Then if $\Gamma_x = 0$ then the equality takes place

$$P(fX = 0 \ \text{ at all } \ f \in F) = P(fX = 0 \ \text{ at all } \ f \in L) = 1\,.$$

Hence it follows that $fX = fx(\omega) = 0$ for all $f \in F$ and almost at all ω, i.e. $x(\omega) = 0$ almost at all ω. It means that the random variable X equal to $x(\omega)$ is equivalent to zero.

(x) From property (xiii) of the expectations follows that *if the sequence of the random variables $\{X_n\}$ in B-space converges a.s. or in probability to the random variable at all n and $EU^2 < \infty$ then there exist the operators*

$$\Gamma_n f = EX_n\overline{fX_n}\,, \quad K_n f = EX_n^0\overline{fX_n^0} \quad (n = 1, 2, \ldots)\,.$$

Here Γ_x and K_x are determined on the whole space F_c of the continuous linear functionals, and $\Gamma_x f = \lim \Gamma_n f$, $K_x f = \lim K_n f$ for all $f \in F_c$. As $\| X^n\overline{fX_n} \| = |fX_n| \, \| X_n \| \leq \| f \| \cdot \| X_n \|^2 \leq \| f \| \, U^2$ for any continuous linear functional f then from the existence of EU^2 follows the existence of all the operators Γ_n determined everywhere on F_c. As the sequence of the random variables $\left\{ X_n\overline{fX_n} \right\}$ for any $f \in F_c$ converges a.s. or in probability to the random variable $X\overline{fX}$ then for any $f \in F_c$ there exists the expectation $EX\overline{fX} = \Gamma_x f$ and the sequence $\{\Gamma_n f\}$ converges to $\Gamma_x f$.

Analogously, from property (xiii) of the expectations follows that *if the sequence of the random variables $\{X_n\}$ in a weakly complete topological linear space weakly converges a.s. or in probability to the random variable X there exist the operators Γ and K for all random variable X_n and X with the domain F_c and for any $f \in F_c$ there exists such nonnegative random variable U_f, $EU_f^2 < \infty$ that $|fX_n| \leq U_f$ then the sequences $\{\Gamma_n f\}$ and $\{K_n f\}$ weakly converge to $\Gamma_x f$ and $K_x f$ correspondingly*. Really, the sequence of the random variables $\left\{ X_n\overline{fX_n} \right\}$ in a given case for any $f \in F_c$ weakly converges a.s. or in probability to $X\overline{fX}$ and for any $f, g \in F_c$ we have $\left| gX_n\overline{fX_n} \right| = |fX_n| \, |gX_n| \leq U_f U_g$ and $EU_f U_g \leq \sqrt{EU_f^2 EU_q^2} < \infty$. Consequently, by property (xiii) of the expectations the sequence $\{\Gamma_n f\}$ weakly converge to $\Gamma_x f$ for any $f \in F_c$.

(xi) From property (xiv) of the expectations follows that *if the series $\sum Y_p$ whose elements are the random variables Y_p in B-space converges a.s. and in probability to the random variable X and all its finite segments are majored on the norm by the random variable U, $EU^2 < \infty$ then there exist the operators Γ_{pq}, K_{pq} $(p, q = 1, 2, \ldots)$, Γ_x and K_x, $\Gamma_{pq} f = EY_p\overline{fY_p}$, $K_{pq} f = EY_p^0\overline{fY_p^0}$, determined everywhere on F_c and*

$$\Gamma_x f = \sum_{p,q=1}^{\infty} \Gamma_{pq} f\,, \quad K_x f = \sum_{p,q=1}^{\infty} K_{pq} f\,, \tag{3.3.34}$$

Here the series converge at any $f \in F_c$.

In exactly the same way *if the series $\sum Y_p$ whose terms are the random variables Y_p in weakly complete topological linear space weakly converges a.s. or in probability to the random variable X, if there exist all the operators Γ_{pq}, K_{pq}, Γ_x and K_x and for any continuous linear functional $f \in F_c$ there exists such random variable U_f, $EU_f^2 < \infty$ that*

$$\left| \sum_{p=1}^{n} fY_p \right| < U_f \text{ at all } n \text{ then formulae (3.3.34) in which the series weakly}$$

converge to $\Gamma_x f$ and $K_x f$ are valid.

E x a m p l e 3.3.8. If X is a finite-dimensional random vector then from the properties (ii) and (iii) of the operators Γ_x and K_x follows the positiveness of quadratic form correspondent to the matrices Γ_x and K_x, and the symmetry of these matrices $\gamma_{qp} = \overline{\gamma_{pq}}$, $k_{qp} = \overline{k_{pq}}$. From formulae (3.3.29) we may derive the correspondent formulae of the transformation of the matrices of the second order moments and the covariance matrices at linear transformation of the random vector X. In particular, we obtain the following formulae for the covariance matrix of the random vector $Y = AX$ and a cross-covariance matrix of the vectors Y and X: $K_y = AK_x A^*$, $K_{yx} = AK_x$. Similarly relations (3.3.31) give the following formula for the cross- covariance matrix of the random vectors $Y = AX$ and $U = BZ$: $K_{yu} = AK_{xz}B^*$.

E x a m p l e 3.3.9. In the case where $X = X(t)$, $t \in T_1$ is the n-dimensional vector random function from properties (ii) and (iii) of the operators Γ_x and K_x follow the positive definiteness and the symmetry of the matrix functions

$$\Gamma_x(t,t') \text{ and } K_x(t,t'): \sum_{p,q=1}^{l} f_p^T \Gamma_x(t_p,t_q)\overline{f_q} \geq 0, \ \sum_{p,q=1}^{l} f_p^T K_x(t_p,t_q)\overline{f_q} \geq 0$$

for any l, any vectors $f_1, \ldots, f_l$ and $t_1, \ldots, t_l \in T$, $\Gamma_x(t,t') = \Gamma_x(t,t')^*$, $K_x(t,t') = K_x(t,t')^*$. Analogously in the case of two vector random functions $X = X(t)$ and $Y = Y(s)$ formula (3.3.26) gives the following relation between the matrices $K_{xy}(t,s)$ and $K_{yx}(s,t)$: $K_{yx}(s,t) = K_{xy}(t,s)^*$.

Let the random function $X(t) = x(t,\omega)$ be measurable and the conditions of the integrability are fulfilled. We introduce the m-dimensional vector random function

$$Y(s) = \int_{T_1} g(s,t)X(t)\sigma(dt). \tag{I}$$

In this case relations (3.3.29) give the following formulae for the covariance function of the random function $Y(s)$ and the cross- covariance function of the random functions $Y(s)$ and $X(t)$:

$$K_y(s,s') = \int_{T_1}\int_{T_1} g(s,t)K_x(t,t')g(s',t')^*\sigma(dt)\sigma(dt'), \tag{II}$$

$$K_{yx}(s,t) = \int\limits_{T_1} g(s,t')K_x(t',t)\sigma(dt') . \tag{III}$$

Notice that $\Gamma_y(s,s')$ and $\Gamma_{yx}(s,t)$ are determined by the similar formulae. In the special case when $n = m = 1$ from these formulae follow the correspondent formulae for the covariance function of the scalar random functions. Analogously from (3.3.31) follows the formula for the cross-covariance function of the random functions $Y(s)$ and $U(l) = \int\limits_V h(l,v)Z(v)\tau(dv)$:

$$K_{yu}(s,l) = \int\limits_{T_1} \int\limits_V g(s,t)K_{xz}(t,v)h(l,v)^*\sigma(dt)\tau(dv) . \tag{IV}$$

3.3.3. White Noise

As it is known from elementary theory of random processes the random process $X(t)$ with zero expectation and the covariance function which contains as a multiplier the δ-function, $.m_x(t) = 0$, $K_x(t_1,t_2) = \nu(t_1)\delta(t_1 - t_2)$ is called *a white noise in the broad sense*. Taking into the consideration that $\delta(t_1 - t_2) = 0$ at $t_1 \neq t_2$ the multiplier $\nu(t_1)$ may be replaced by the multiplier $\nu(t_2)$ or by the symmetric multiplier $\sqrt{\nu(t_1)\nu(t_2)}$. The multiplier $\nu(t)$ at the δ-function is called *an intensity* of the white noise $X(t)$. The intensity of a scalar white noise is essentially positive. The intensity of a vector white noise represents nonnegatively determined symmetrical matrix. It is evident that the variance of a white noise is infinite, and its values in two quite close points are noncorrelated.

For solving practical problems it is expedient to substitute a random process by a white noise. We may do it only in that case when the least interval between the values of an argument at which the values of an random process are practically uncorrelated is sufficiently small. This interval which is called *a correlation interval* . If the variable $|K_x(t_1,t_2)| / K_x(t_1,t_1)$ for the scalar random process $X(t)$ may be assumed practically equal to zero at $|t_1 - t_2| > \tau_k$ and the variable τ_k is sufficiently small then the random process $X(t)$ may be assumed as a nonstationary white noise with the intensity equal to

$$\nu(t) = \int\limits_{-\infty}^{\infty} K_x(t,t + \tau)d\tau . \tag{3.3.35}$$

The correlation interval τ_k for the scalar random process $X(t)$ is determined by formula

$$\tau_k = \frac{1}{2} \max_t \int\limits_{-\infty}^{\infty} \frac{K_x(t, t+\tau)}{K_x(t, t)} d\tau . \tag{3.3.36}$$

The vector random process may be assumed as a white noise if all its components may be assumed as white noises.

3.3.4. Moments of Higher Orders

In elementary probability theory for more detailed characteristic of real scalar and vector random variables the following formulae for initial and central moments of higher orders are used:

$$\mu_r = E(X^0)^r , \quad \alpha_r = EX^r , \quad (r = 1, 2, \ldots) \tag{3.3.37}$$

$$\alpha_r = \alpha_{r_1, \ldots, r_n} = EX_1^{r_1} \ldots X_n^{r_n} ,$$

$$\mu_r = \mu_{r_1, \ldots, r_n} = E(X_1^0)^{r_1} \ldots (X_n^0)^{r_n} , \tag{3.3.38}$$

$$(|r| = r_1 + \cdots + r_n , \quad |r| = 1, 2, \ldots) .$$

The vector variable $r = (r_1, \ldots, r_n)$ in formula (3.3.38) is called a *multi-index*. The following formulae of the connection have place for the scalar random variables between initial and the centred moments:

$$m_x = \alpha_1, \quad \alpha_r = \sum_{p=0}^{r} C_r^p \mu_p m_x^{r-p}, \quad C_r^p = \frac{r!}{p!(r-p)!} , \tag{3.3.39}$$

$$\mu_r = \sum_{p=0}^{r} (-1)^{r-p} C_r^p \alpha_p m_x^{r-p} . \tag{3.3.40}$$

Analogously the higher moments for the real scalar random functions $X(t)$, $t \in T_1$ are determined in elementary theory of random functions

$$\alpha_r(t_1, \ldots, t_r) = EX(t_1) \ldots X(t_r)$$

$$= \int\limits_{-\infty}^{\infty} \cdots \int\limits_{-\infty}^{\infty} x_1 \ldots x_r f_r(x_1, \ldots, x_r; t_1, \ldots, t_r) dx_1 \ldots dx_r , \tag{3.3.41}$$

$$\mu_r(t_1, \ldots, t_r) = EX^0(t_1)\ldots X^0(t_r)$$

$$= \int_{-\infty}^{\infty} \cdots \int_{-\infty}^{\infty} [x_1 - m_x(t_1)]\ldots[x_r - m_r(t_r)]$$

$$\times f_r(x_1, \ldots, x_r; t_1, \ldots, t_r)dx_1\ldots dx_r. \tag{3.3.42}$$

The mixed initial and central moments of the order r, $(|r| = r_1 + \cdots + r_n)$ for the real scalar random functions $X_1(t), \ldots, X_n(t)$ are calculated by formulae

$$\alpha_{r_1,\ldots,r_n}(t_1^{(1)}, \ldots, t_{r_1}^{(1)}, \ldots, t_1^{(n)}, \ldots, t_{r_n}^{(n)})$$

$$= E\left\{X_1(t_1^{(1)})\ldots X_1(t_{r_1}^{(1)})\ldots X_n(t_1^{(n)})\ldots X_n(t_{r_n}^{(n)})\right\}, \tag{3.3.43}$$

$$\mu_{r_1,\ldots,r_n}(t_1^{(1)}, \ldots, t_{r_1}^{(1)}, \ldots, t_1^{(n)}, \ldots, t_{r_n}^{(n)})$$

$$= E\left\{\left[X_1(t_1^{(1)}) - m_1(t_1^{(1)})\right]\ldots\left[X_1(t_{r_1}^{(1)}) - m_1(t_{r_1}^{(1)})\right]\right.$$

$$\left.\ldots\left[X_n(t_1^{(n)}) - m_n(t_1^{(n)})\right]\ldots\left[X_n(t_{r_n}^{(n)}) - m_n(t_{r_n}^{(n)})\right]\right\}. \tag{3.3.44}$$

The operators of higher orders moments are determined in the same way as an operator of the second order moment and a covariance operator. Here for the simplicity we restrict ourselves to the random variables in the real linear spaces. Let $X = x(\omega)$ be a random variable in the real linear space X. Formula

$$m_r(f_1, \ldots, f_{r-1}) = EX(f_1 X)\ldots(f_{r-1} X)$$

$$= \int x(\omega)f_1 x(\omega)\ldots f_{r-1}x(\omega)P(d\omega)$$

$$= \int x f_1 x \ldots f_{r-1}x\mu_x(dx) \tag{3.3.45}$$

determines the operator m_r mapping the product of the spaces F^{r-1} into X. Here $f_1, \ldots, f_{r-1} \in F$ are arbitrary linear functionals on X. An operator defined by formula (3.3.45) is called *an operator of the rth order moment* of the random variable X. The operator m_r^0 of the rth order moment of the centred random variable $X^0 = X - m_x$ is called *an operator of the centered moment of the rth order* of the random variable

X. The functional $(f_1, \ldots, f_r) \to R$ on the product of the spaces F^r determined by formula

$$f_r m_r(f_1, \ldots, f_{r-1}) = E(f_1 X) \ldots (f_r X), \qquad (3.3.46)$$

is called *a moment form of the rth order* of the random vriable X, and the functional

$$f_r m_r^0(f_1, \ldots, f_r) \to f_r m_r^0(f_1, \ldots, f_{r-1}) = E(f_1 X^0) \ldots (f_r X^0)$$
$$(3.3.47)$$

is called *a central moment form of the rth order* of the random variable X. At $r = 1$ the moment form of the first order coincides with the variable $f m_x$, $f \in F_c$. It is evident that an operator of the rth order moment is the $(r-1)$-linear operator, and the moment form of the rth order is the r-linear functional. We recall that the function m of the variables $f_1, \ldots, f_m$ is called *m-linear* if it is linear on each variable at fixed valued of the others.

It is easy to derive the formulae which express the moment operators and the moment forms in terms of the operators of the central moments and the central moment forms and vice versa. From the general properties of the expectations similarly as in Subsections 3.3.2 and 3.3.4 the main properties of the moments operators and the moment forms are derived.

E x a m p l e 3.3.14. If X is a real n-dimensional random vector and

$$f_i X = \sum_{k=1}^{n} f_{ik} X_k, \quad i = 1, 2, \ldots, \qquad (\text{I})$$

then

$$m_r(f_1, \ldots, f_{r-1}) = \sum_{k_1, \ldots, k_{r-1}=1}^{n} f_{1 k_1} \ldots f_{r-1, k_{r-1}} E X X_{k_1} \ldots X_{k_{r-1}}.$$
$$(\text{II})$$

Thus an operator of the rth order moment of a finite- dimensional random vector is determined by the polylinear moment form of the rth order moments of all its components.

E x a m p l e 3.3.15. If $X = X(t)$ is the n-dimensional real vector random function and

$$f_i X = \sum_{k=1}^{l} f_{ik} X_{q_k}(t_k), \qquad (\text{I})$$

then

$$m_r(f_1, \ldots, f_{r-1})$$

$$= \sum_{k_1=1}^{l_1} \cdots \sum_{k_{r-1}=1}^{l_{r-1}} f_{1k_1} \ldots f_{r-1,k_{r-1}} EX(t)X_{q_{k_1}}(t_{k_1}) \ldots X_{q_{k_{r-1}}}(t_{k_{r-1}}).$$

$$(\mathrm{II})$$

Thus the operator of the rth order moment of the n- dimensional vector random function is determined by the polylinear form of the moments and cross moments of the rth order of all its components.

3.4. Conditional Moments

3.4.1. Definitions

In elementary probability theory the different conditional characteristics of the random variables and vectors are used. So *a conditional expectation of a given function $\varphi(X)$ of a random scalar or vector variable X at a given value of a random scalar or vector variable Y* is determined by formula

$$E\left[\varphi(X)|y\right] = \int\limits_{-\infty}^{\infty} \varphi(x)f_2(x|y)dx, \qquad (3.4.1)$$

where $f_2(x|y)$ is a conditional density of the variable X at a given value y of the random variable Y.

As a special case from (3.4.1) follows a formula for a conditional expectation of a random variable at a given value of the variable X:

$$E\left[X|y\right] = \int\limits_{-\infty}^{\infty} xf_2(x|y)dx. \qquad (3.4.2)$$

The denotion $f_2(x|y)$ is used for the density of the random variable X dependent on the parameter y also in the case when y is not the value of some random variable Y. Here formulae (3.4.1) and (3.4.2) determine the expectations $\varphi(X)$ and X as the functions of the parameter y. The expectation of the random variable X as the function of the parameter y on which the distribution X depends is called *a regression of X on y*. In the special case when the parameter y is a possible value of some random variable Y the regression X on y represents a conditional expectation X at $Y = y$.

Knowing a conditional expectation we may determine all conditional moments of the random variables.

A conditional expectation of the random variable $Z = \varphi(X)$ considered as the function of the random variable Y, $E[\varphi(X)|Y]$ is called *a conditional expectation of the random variable $Z = \varphi(X)$ relatively to Y*. As the conditional expectations and the conditional moments of different orders of the random variables relatively to random variable Y are the random variables themselves then in its turn we may determine the expectations and the moments of different orders. Finally, we present the formula for total expectation which is often used in elementary probability theory

$$E\varphi(X, Y) = E[E[(X, Y)|Y]]. \qquad (3.4.3)$$

Hence at $\varphi(X, Y) = X$ we find

$$EX = E[E[X|Y]]. \qquad (3.4.4)$$

Now we suppose that on the probability space $(\Omega, \mathcal{S}, P)$ the random variable $X = x(\omega)$ whose phase space serves a measurable linear space $(X, \mathcal{A})$ is given. Let us introduce similarly as in Subsection 3.2.1 a weak topology in X by means of sufficiently complete set of the linear functionals F_c and admit that X is complete in this topology (i.e. is weakly complete topological linear space). Suppose that the conditional probability $P_{\mathcal{F}}(A|\omega)$, $A \in \mathcal{S}$ relatively to some σ-algebra $\mathcal{F} \subset \mathcal{S}$ is regular. *A conditional expectation of a random variables $X = x(\omega)$ relatively to σ-algebra $\mathcal{F}$ is called a weak integral*

$$E_{\mathcal{F}}(X|\omega) = \int x(\omega')P_{\mathcal{F}}(d\omega'|\omega). \qquad (3.4.5)$$

In the special case when X is a separable B-space, and F_c is a space of continuous linear functionals on X integral (3.4.5) may be considered as a strong integral (if $x(\omega')$ is $P_{\mathcal{F}}(A|\omega)$-integrable at a given value $\omega \in \Omega$).

The measure $P_{\mathcal{F}}(A|\omega)$ at all $A \in \mathcal{S}$ is deteremined almost at all ω relatively to the measure P and is a measurable function ω relatively $\mathcal{F}$. On the basis of the theorems about the measurable functions $f E_{\mathcal{F}}(X|\omega)$ for any functional $f \in F_c$ represents the function ω measurable relatively to $\mathcal{F}$ determined almost at all ω relatively to the measure P. Consequently, the inverse images of the sets $\{x : |f_1 x - f_1 x_0|$

$< \varepsilon_1, \ldots, |f_n x - f_n x_0| < \varepsilon_n \}$ of the space X correspondent to all n, $x_0 \in X$, $f_1, \ldots, f_n \in F_c$, $\varepsilon_1, \ldots, \varepsilon_n > 0$ determined by the function $E_{\mathcal{F}}(X|\omega)$ and together with them the inverse images of all the sets of the σ-algebra $\mathcal{A}'$ generated by this class of the sets belong to the σ-algebra $\mathcal{F}$. It means that the function $E_{\mathcal{F}}(X|\omega)$ is $(\mathcal{F}, \mathcal{A}')$-measurable, i.e. is a random variable in $(X, \mathcal{A}')$ determined on the probability space $(\Omega, \mathcal{F}, P_{\mathcal{F}})$ where $P_{\mathcal{F}}$ is the contraction of the probability P on $\mathcal{F}$.

If the σ-algebra $\mathcal{F}$ is induced by some random variable $Y = y(\omega)$ then $\mathcal{F} = \mathcal{S}_y$, i.e. represents the inverse image of the correspondent σ-algebra $\mathcal{B}$ of the sets of the phase space of the variable Y. Then formula (3.4.5) determines *a conditional expectation of the random variable X relatively to the random variable Y*. In this case as it was shown in Section 2.3 $P_{\mathcal{F}}(A|\omega)$ is a measurable function of the random variable $Y = y(\omega)$ relatively to $\mathcal{B}$. Consequently, a conditional expectation of the random variable X relatively to Y represents the $(\mathcal{B}, \mathcal{A}')$-measurable function of the random variable Y. Therefore we shall denote a conditional expectation of the variable X relatively to Y by a symbol $E(X|Y)$ or shortly, $m_{X|Y}$. By the change of variables $x = x(\omega)$ with account of formula (2.3.7) for a conditional probability measure formula (3.4.5) in this case is reduced to the form

$$m_{X|Y} = E(X|Y) = \int x \mu_x(dx|Y) \,. \qquad (3.4.6)$$

This formula determines a conditional expectation of the random variable X relatively to Y in the case of the existence of a regular conditionally distribution $\mu_x(A|y)$. Substituting in (3.4.6) the random variable Y by its realization y we obtain conditional expectation $m_{X|y} = E(X|y)$ of the variable X at a given value y of the variable Y.

3.4.2. Properties of Conditional Moments

The conditional expectations possess all the properties of the expectations which are established in Section 3.1 with the only difference that each of the presented statements is valid only almost for all ω (or y) relatively to the measure P (correspondingly μ_y) and may not be valid for some zero set of the values ω (correspondingly y). Alongside with these general properties the conditional expectations possess a number of specific properties.

At first we prove the following statement.

Theorem 3.4.1. *If a conditional expectation $E_{\mathcal{F}}(X|\omega)$ represents a weakly $P_{\mathcal{F}}$-integrable function and the random variable X has an expectation then for any set $\mathrm{B} \in \mathcal{F}$*

$$\int_{\mathrm{B}} E_{\mathcal{F}}(X|\omega)P_{\mathcal{F}}(d\omega) = \int_{\mathrm{B}} x(\omega)P(d\omega). \qquad (3.4.7)$$

▷ For the proof we rewrite this formula in the form

$$\int_{\mathrm{B}} P_{\mathcal{F}}(d\omega)\int x(\omega')P_{\mathcal{F}}(d\omega'|\omega) = \int_{\mathrm{B}} x(\omega)P(d\omega). \qquad (3.4.8)$$

It is easy to see that formula (3.4.8) is valid in the special case when $x(\omega)$ represents an indicator of any set $A \in \mathcal{S}$, $x(\omega) = \mathbf{1}_A(\omega)$. Really, in a given case

$$\int_{\mathrm{B}} P_{\mathcal{F}}(d\omega)\int \mathbf{1}_A(\omega')P_{\mathcal{F}}(d\omega'|\omega) = \int_{\mathrm{B}} P_{\mathcal{F}}(A|\omega)P_{\mathcal{F}}(d\omega).$$

But the latter integral is equal to $P(AB)$, and consequently,

$$\int_{\mathrm{B}} P_{\mathcal{F}}(d\omega)\int \mathbf{1}_A(\omega')P_{\mathcal{F}}(d\omega'|\omega) = P(AB)\int_{\mathrm{B}} \mathbf{1}_A(\omega)P(d\omega).$$

From the validity of formula (3.4.8) for the indicators of all measurable sets follows its validity for any simple function $x(\omega)$. By the theorem from functional analysis about a monotonous convergence follows the validity of formula (3.4.8) for any nonnegative function $x(\omega)$ (independently of the fact whether the integrals in it are finite or infinite). As any complex function $x(\omega)$ may be presented in the form $x(\omega) = x_1(\omega) - x_2(\omega) + ix_3(\omega) - ix_4(\omega)$ where x_1, x_2, x_3 and x_4 are nonnegative real-valued functions then formula (3.4.8) is also valid for any complex function $x(\omega)$ and from the existence of one integral in it follows the existence of another. Now let $x(\omega)$ be the function with the values in any weakly complete topological linear space X for which there exist weak integrals in formulae (3.4.5) and (3.4.8).

For any continuous linear functional $f \in F_c$ formula (3.4.8) is valid also for the complex function $fx(\omega)$:

$$\int_{\mathrm{B}} P_{\mathcal{F}}(d\omega)\int fx(\omega')P_{\mathcal{F}}(d\omega'|\omega) = \int_{\mathrm{B}} fx(\omega)P(d\omega).$$

From the weak $P_{\mathcal{F}}(A|\omega)$-integrability of the function $x(\omega')$, the $P_{\mathcal{F}}$-integrability of the function $E_{\mathcal{F}}(X|\omega)$ and the weak P-integrability of the function $x(\omega)$ follow three equalities:

$$\int\limits_{B} P_{\mathcal{F}}(d\omega) \int fx(\omega')P_{\mathcal{F}}(d\omega'|\omega) = f \int\limits_{B} P_{\mathcal{F}}(d\omega) \int x(\omega')P_{\mathcal{F}}(d\omega'|\omega),$$

$$\int\limits_{B} fx(\omega)P(d\omega) = f \int\limits_{B} x(\omega)P(d\omega),$$

$$f \int\limits_{B} P_{\mathcal{F}}(d\omega) \int x(\omega')P_{\mathcal{F}}(d\omega'|\omega) = f \int\limits_{B} x(\omega)P(d\omega). \lhd$$

In consequence of the fact that the measure $P_{\mathcal{F}}$ represents the contraction of the measure P on the σ-algebra $\mathcal{F}$, and coincides with P on all the sets in $\mathcal{F}$, from the $P_{\mathcal{F}}$-integrability of the conditional expectation $E_{\mathcal{F}}(X|\omega)$ follows its P-integrability and the coincidence of both integrals. Therefore formula (3.4.7) may be rewritten in the form

$$\int\limits_{B} E_{\mathcal{F}}(X|\omega)P(d\omega) = \int\limits_{B} x(\omega)P(d\omega), \quad B \in \mathcal{F}. \tag{3.4.9}$$

In particular, putting $B = \Omega$ and taking into the consideration that the integrals at $B = \Omega$ represent the expectations of the correspondent random variables we get

$$E\left[E_{\mathcal{F}}(X|\omega)\right] = EX. \tag{3.4.10}$$

This formula is called *a formula of a total expectation*. If σ-algebra $\mathcal{F}$ is induced by the random variable Y then $E_{\mathcal{F}}(X|\omega) = E(X|Y)$ and formula (3.4.10) takes the form

$$E\left[E(X|Y)\right] = EX. \tag{3.4.11}$$

Now we consider a special case when the function $x(\omega)$ is measurable relatively to the σ-algebra $\mathcal{F}$, i.e. $(\mathcal{F}, \mathcal{A})$ is measurable. In this case integral (3.4.5) coincides with the integral over the function $x(\omega')$ with respect to contracted on the σ-algebra $\mathcal{F}$ measure $P_{\mathcal{F}}(A|\omega)$. As for any set $B \in \mathcal{F}$ we have $P_{\mathcal{F}}(B|\omega) = \mathbf{1}_B(\omega)$ almost at all ω then

$$E_{\mathcal{F}}(X|\omega) = \int x(\omega')P_{\mathcal{F}}(d\omega'|\omega) = x(\omega)$$

almost at all ω. Thus if the function $x(\omega)$ is $(\mathcal{F}, \mathcal{A})$-measurable then almost at all ω there exists an expectation of the random variable $X = x(\omega)$ relatively to $\mathcal{F}$ which coincides with this random variable:

$$E_{\mathcal{F}}(X|\omega) = x(\omega) \quad \text{almost at all} \quad \omega. \tag{3.4.12}$$

We shall prove also the following theorem.

Theorem 3.4.2. *If a scalar random variable $Z = z(\omega)$ is measurable relatively to $\mathcal{F}$ then for any random variable $X = x(\omega)$ for which there exists $E_{\mathcal{F}}(ZX|\omega)$, almost at all ω*

$$E_{\mathcal{F}}(ZX|\omega) = z(\omega)E_{\mathcal{F}}(X|\omega) = ZE_{\mathcal{F}}(X|\omega). \tag{3.4.13}$$

$\triangleright$ Firstly we consider a case when $z(\omega)$ represents an indicator of the set $\mathrm{B} \in \mathcal{F}$, $z(\omega) = \mathbf{1}_{\mathrm{B}}(\omega)$. Then

$$E_{\mathcal{F}}(ZX|\omega) = \int \mathbf{1}_{\mathrm{B}}(\omega')x(\omega')P_{\mathcal{F}}(d\omega'|\omega) = \int_{\mathrm{B}} x(\omega')P_{\mathcal{F}}(d\omega'|\omega).$$

As $P_{\mathcal{F}}(\mathrm{B}|\omega) = \mathbf{1}_{\mathrm{B}}(\omega)$ almost at all ω then the latter integral coincides with

$$\int x(\omega')P_{\mathcal{F}}(d\omega'|\omega) = E_{\mathcal{F}}(X|\omega)$$

almost at all $\omega \in \mathrm{B}$ and is equal to 0 almost at all $\omega \overline{\in} \mathrm{B}$, i.e.

$$\int_{\mathrm{B}} x(\omega')P_{\mathcal{F}}(d\omega'|\omega) = \mathbf{1}_{\mathrm{B}}(\omega)E_{\mathcal{F}}(X|\omega)$$

almost at all ω ,what proves formula (3.4.13). Consequently, formula (3.4.13) is also valid for all simple functions $z(\omega)$ relatively to $\mathcal{F}$.

Now let $z(\omega)$ be an arbitrary nonnegative measurable function relatively to $\mathcal{F}$. After presenting it as the limit of the increasing sequence of the simple functions $\{z^n(\omega)\}$ we get the random variable ZX as the weak limit of the consequence of the random variables $\{Z^n X\}$ which a.s. converges and for any functional $f \in F_c$ the sequence $\{Z^n f X\}$ is modulus bounded of the random variable $Z\,|fX|$ which has a finite conditional expectation. By property (xiii) of the expectations (Subsection 3.1.3) follows that $E_{\mathcal{F}}(ZX|\omega) = \lim(\text{weak})E_{\mathcal{F}}(Z^n X|\omega)$. And as formula (3.4.13) is valid for the simple functions $Z^n = z^n(\omega)$ then almost at all ω we get

$$\lim(\text{weak})\, E_{\mathcal{F}}(Z^n X|\omega) = \lim(\text{weak})\, Z^n E_{\mathcal{F}}(X|\omega) = ZE_{\mathcal{F}}(X|\omega).$$

This proves formula (3.4.13) for the nonnegative random variables Z. Accounting that any complex random variables may be expressed in terms of four nonnegative real random variables we ensure in the validity of formula (3.4.9) in the general case. ◁

In particular, when the σ-algebra $\mathcal{F}$ is induced by the random variable $Y = y(\omega)$ in $(Y, \mathcal{B})$ and is measurable relatively to $\mathcal{F}$ the scalar random variable Z according to the proved at the end of Section 2.2 represents a measurable relatively to σ-algebra $\mathcal{B}$ function of the random variable Y, $Z = \varphi(Y)$. As $E_{\mathcal{F}}(X|\omega) = E(X|Y)$ then formula (3.4.13) takes the form

$$E(\varphi(Y)X|Y) = \varphi(Y)E(X|Y), \qquad (3.4.14)$$

and this equality is valid with probability 1. By the way formula (3.4.14) is a direct consequence of property (i) of the expectations as according to this property for any fixed realization y of the random variable Y for which the conditional probability measure $\mu_x(A|y)$ is determined $E(\varphi(y)X|y) = \varphi(y)E(X|y)$. As $\mu_x(A|y)$ is determined almost for all $y \in Y$ then equality (3.4.14) is valid with probability 1.

It is easy to understand that if a total expectation represents a result of a total average of a random variable then a conditional expectation represents the result of its partial average. The more poor is the σ-algebra $\mathcal{F}$ the more rough is the average of a random variable which is performed by a conditional expectation. In the limiting case when $\mathcal{F} = \mathcal{S}$ a conditional expectation coincides with a random variable itself as $P_y(A|\omega) = 1_A(\omega)$ almost at all ω (Section 2.3). In the other limiting case when $\mathcal{F}$ contains only the whole space and an empty set, $\mathcal{F} = \{\Omega, \varnothing\}$, $P_{\mathcal{F}}(A|\omega) = P(A)$ and a conditional expectation coincides with an unconditional one, $E_{\mathcal{F}}(X|\omega) = EX$.

On the basis of these arguments it is easy to come to the following conclusion.

Theorem 3.4.3. *If we apply to a random variable an operation of a conditional expectation twice: once relatively to the σ-algebra $\mathcal{F}$ and in the other case relatively to more poor σ-algebra $\mathcal{F}' \subset \mathcal{F}$ then the result will be equal to a conditional expectation relatively to $\mathcal{F}'$ independently of the order of these two averages:*

$$E_{\mathcal{F}}X = E_{\mathcal{F}}(E_{\mathcal{F}'}(X|\omega)|\omega) = E_{\mathcal{F}'}(E_{\mathcal{F}}(X|\omega)|\omega) = E_{\mathcal{F}'}(X|\omega). \quad (3.4.15)$$

▷ For the rigorous proof of these equalities we notice that the function $E_{\mathcal{F}'}(X|\omega)$ is $(\mathcal{F}, \mathcal{A}')$-measurable in consequence of which on the basis of (3.4.12) we have $E_{\mathcal{F}}(E_{\mathcal{F}'}(X|\omega)|\omega) = E_{\mathcal{F}'}(X|\omega)$ almost at all ω. Suppose now temporarily that $X = x(\omega)$ is a scalar random function. After applying formula (3.4.7) to the random variable $Y = y(\omega) = E_{\mathcal{F}}(X|\omega)$ determined on the probability space $(\Omega, \mathcal{F}, P_{\mathcal{F}})$ we obtain

$$\int_B E_{\mathcal{F}'}(E_{\mathcal{F}}(X|\omega)|\omega)P_{\mathcal{F}'}(d\omega) = \int_B E_{\mathcal{F}}(X|\omega)P_{\mathcal{F}}(d\omega)$$

for any set $B \in \mathcal{F}'$. And as $\mathcal{F}' \subset \mathcal{F}$ then $B \in \mathcal{F}$ and on the basis of the same formula (3.4.7)

$$\int_B E_{\mathcal{F}}(X|\omega)P_{\mathcal{F}}(d\omega) = \int_B x(\omega)P(d\omega) = \int_B E_{\mathcal{F}'}(X|\omega)P_{\mathcal{F}'}(d\omega) \, .$$

Thus

$$\int_B E_{\mathcal{F}'}(E_{\mathcal{F}}(X|\omega)|\omega)P_{\mathcal{F}'}(d\omega) = \int_B E_{\mathcal{F}'}(X|\omega)P_{\mathcal{F}'}(d\omega) \, .$$

Hence in consequence of the arbitrariness of $B \in \mathcal{F}'$ and the uniqueness of Radon–Nikodym derivative of $P_{\mathcal{F}'}$-continuous function (with the accuracy till $P_{\mathcal{F}'}$-equivalence) we conclude that almost at all ω the functions $E_{\mathcal{F}'}(E_{\mathcal{F}}(X|\omega)|\omega)$ and $E_{\mathcal{F}'}(X|\omega)$ coincide. This proves the second equality (3.4.15) for the scalar random variable $X = x(\omega)$. In order to prove this equality in the general case we write it in the form

$$\int P_{\mathcal{F}'}(d\omega|\omega) \int x(\omega'')P_{\mathcal{F}}(d\omega''|\omega') = \int x(\omega')P_{\mathcal{F}'}(d\omega'|\omega) \, . \qquad (3.4.16)$$

Putting here $x(\omega) = \mathbf{1}_A(\omega)$, $A \in \mathcal{S}$ we get

$$\int P_{\mathcal{F}}(A|\omega')P_{\mathcal{F}'}(d\omega'|\omega) = P_{\mathcal{F}'}(A|\omega) \qquad (3.4.17)$$

for all $A \in \mathcal{S}$ and almost at all ω. Now we determine on the product of the measurable spaces $(\Omega^2, \mathcal{S}^2)$, $\Omega^2 = \Omega \times \Omega$, $\mathcal{S}^2 = \mathcal{S} \times \mathcal{S}$ the following set of the measures:

$$\nu(C|\omega) = \int P_{\mathcal{F}'}(d\omega'|\omega) \int \mathbf{1}_C(\omega', \omega'')P_{\mathcal{F}}(d\omega''|\omega') \quad (C \in \mathcal{S}^2) \, .$$

It is easy to see that the contraction of the measure $\nu(C|\omega)$ with the σ-algebra $\mathcal{S}^2$ on the σ-algebra $\Omega \times \mathcal{S} \subset \mathcal{S}^2$ at any fixed ω coincides with the integral in (3.4.17).

In order to make sure of this fact it is sufficient to take the rectangular set C of the type $\Omega \times A$, $A \in \mathcal{S}$. Consequently, the contraction of the measure $\nu(C|\omega)$ on $\Omega \times \mathcal{S}$ coincides with $P_{\mathcal{F}'}(A|\omega)$ almost at all ω.

Now let $X = x(\omega)$ be a random variable in an arbitrary weakly complete topological linear space for which almost at all ω there exist $E_{\mathcal{F}}(X|\omega)$, $E_{\mathcal{F}'}(X|\omega)$ and $E_{\mathcal{F}'}(E_{\mathcal{F}}(X|\omega)|\omega)$. By virtue of the inequality

$$\nu(A \times B|\omega) \le P_{\mathcal{F}'}(A|\omega), \quad A, B \in \mathcal{S},$$

from weak $P_{\mathcal{F}'}(A|\omega)$-integrability of the function $x(\omega'')$ follows its weak $\nu(C|\omega)$-integrability. Here we have

$$\int x(\omega'')\nu(d\omega' \times d\omega''|\omega) = \int P_{\mathcal{F}'}(d\omega'|\omega) \int x(\omega'')P_{\mathcal{F}}(d\omega''|\omega').$$

And as the function $x(\omega'')$ is measurable relatively to the σ-algebra $\Omega \times \mathcal{S}$ then the measure $\nu(C|\omega)$ may be substituted by its contraction on the σ-algebra $\Omega \times \mathcal{S}$ which according to the proved above coincides with $P_{\mathcal{F}'}(A|\omega)$, $A \in \mathcal{S}$ almost at all ω. Then we get

$$\int x(\omega'')\nu(d\omega' \times d\omega''|\omega) = \int x(\omega'')P_{\mathcal{F}'}(d\omega''|\omega).$$

The obtained equalities prove the validity of formula (3.4.16) and together with it the validity of the second equality (3.4.5) almost at all ω for any random variables and weakly complete topological linear spaces. $\triangleleft$

If a regular conditional probability relatively to the σ-algebra $\mathcal{F}$ does not exist then the conditional probability $P_{\mathcal{F}}(A|\omega)$ represents the measure on the σ-algebra $\mathcal{S}$ with the values in B-space $L_1(\Omega, \mathcal{S}, P)$ of the scalar P-integrable functions (Section 2.3). In this case the integral in (3.4.5) has the sense for the scalar function $x(\omega')$ as a strong integral over $x(\omega')$ with respect to the vector measure $P_{\mathcal{F}}(A|\omega)$ and represents an element of the B-space $L_1(\Omega, \mathcal{S}, P)$. Thus formula (3.4.5) determines also a conditional expectation for the scalar random variables in the case when the conditional probability $P_{\mathcal{F}}(A|\omega)$ is not regular.

It is easy to see that all the properties of the conditional expectations established above remain also for this generalization of the notion of a conditional expectation . Formula (3.4.7) which is valid both for a regular and an irregular conditional probability $P_{\mathcal{F}}(A|\omega)$ for any set $B \in \mathcal{F}$ shows that if a regular conditional probability $P_{\mathcal{F}}(A|\omega)$ exists then both definitions give the values of the conditional expectation $E_{\mathcal{F}}(X|\omega)$ which coincide almost at all ω.

Finally, if a regular conditional probability $P_{\mathcal{F}}(A|\omega)$ does not exist then for calculating a conditional expectation of a scalar random variable we may directly use formula (3.4.7). It is sufficient to notice for this purpose that in the consequence of the coincidences of the measure P and its contraction $P_{\mathcal{F}}$ on the σ-algebra $\mathcal{F}$ on the sets B from $\mathcal{F}$ the integral in the right-hand side of (3.4.7) represents the $P_{\mathcal{F}}$-continuous function of the set on $\mathcal{F}$ and the conditional expectation $E_{\mathcal{F}}(X|\omega)$ may be determined as Radon–Nikodym derivative of this integral with $P_{\mathcal{F}}$. As a result of the uniqueness of Radon–Nikodim derivative this new definition gives almost at all ω the same value of the conditional expectation as integral (3.4.5).

3.4.3. Properties of Conditional Moments Operators

Basing on the notion of a conditional expectation we may determine the conditional moments operators of different orders, the conditional moment forms and the conditional characteristic functionals of the random variables. From the properties of the conditional expectations follows that all the properties of the operators of the second order moments are also valid with probability 1 (almost at all ω) for the conditional operators of the second orders moments. From formulae (3.4.7), (3.4.12)–(3.4.16) follow the specific properties of the conditional moments of the second and the higher orders.

3.5. Characteristic Functions and Functionals

3.5.1. Definitions

In elementary probability theory the expectation of the random variable $e^{i\lambda^T X}$ considered as a function of a real variable λ is called *the characteristic function* of the real random variable X. The characteristic function of a random variable X is expressed in terms of its density by formula

$$g(\lambda) = E e^{i\lambda^T X} = \int_{-\infty}^{\infty} e^{i\lambda^T x} f(x)dx . \qquad (3.5.1)$$

The dimension of the variable λ coincides with the dimension of the variable X. We recall the reader some properties of characteristic functions.

  • A characteristic function is continuous and $|g(\lambda)| \leq 1$, $g(0) = 1$, $g(-\lambda) = \overline{g(\lambda)}$.

• A characteristic function is positive-definite: for any λ_1, ..., λ_N and any complex ξ_1, ..., ξ_N

$$\sum_{p,q=1}^{N} g(\lambda_p - \lambda_q)\xi_p \bar{\xi}_q \geq 0.$$

• A characteristic function $g_2(\mu)$ of the random vector $Y = AX + a$ is expressed in terms of the characteristic function $g_1(\lambda)$ of the variable X by the formula $g_2(\mu) = e^{i\mu^T a} g_1(A^T \mu)$.

• A characteristic function of a projection of a random vector on any subspace is equal to the contraction of its characteristic function on this subspace.

• A characteristic function $g(\lambda)$ of the sum of the independent random variables X_1, ..., X_n is equal to the product of the characteristic functions $g_k(\lambda)$ $(k = 1$, ..., $n)$: $g(\lambda) = \prod_{k=1}^{n} g_k(\lambda)$.

• If X_1, ..., X_n are independent random variables then the characteristic function $g(\lambda)$, $\lambda = [\lambda_1 \ldots \lambda_n]^T$ of the composite random vector $X = [X_1 \ldots X_n]^T$ is equal to the product of the characteristic functions $g_k(\lambda_k)$ $(k = 1$, ..., $n)$: $g(\lambda) = \prod_{k=1}^{n} g_k(\lambda_k)$.

In elementary theory of random functions for the random function $X(t)$, $t \in T_1$ the sets of one-dimensional and multi-dimensional characteristic functions $g_1 = g_1(\lambda; t)$ and $g_n = g_n(\lambda_1$, ..., $\lambda_n; t_1$, ..., $t_n)$ with the parameters t and t_1, ..., t_n are determined correspondingly by

$$g_1(\lambda; t) = E e^{i\lambda^T X(t)}, \tag{3.5.2}$$

$$g_n(\lambda_1, \ldots, \lambda_n; t_1, \ldots, t_n)$$

$$= E \exp\left\{ i\lambda_1^T X(t_1) + \cdots + i\lambda_n^T X(t_n) \right\} \quad (n = 2, 3, \ldots). \tag{3.5.3}$$

Let $X = x(\omega)$ be random variable in the real linear space X and F be the space of all continuous linear functionals on X dual with X. The functional

$$g_x(f) = E e^{ifX} = \int e^{ifx(\omega)} P(d\omega) = \int e^{ifx} \mu(dx) \tag{3.5.4}$$

on the space F $(f \in F)$ is called *a characteristic functional* of the random variable X. Any random variable X in any real linear space has a characteristic functional as for such X variable fX is a real random variable and $\left|e^{ifx}\right| = 1$.

E x a m p l e 3.5.1. If X is the n-dimensional random vector then

$$fX = \sum_{k=1}^{n} f_k X_k, \quad \text{and} \quad g_x(f) = E \exp\left\{i \sum_{k=1}^{n} f_k X_k\right\} \tag{I}$$

represents a function of the n-dimensional vector f with the components f_1, $\ldots$, f_n. By virtue of (3.5.1) this function is a characteristic function of the random vector X.

E x a m p l e 3.5.2. If $X = X(t)$ is a real random function we get

$$fX = \sum_{k=1}^{l} f_k X(t_k) \quad \text{and} \quad g_x(f) = E \exp\left\{i \sum_{k=1}^{l} f_k X(t_k)\right\}. \tag{I}$$

The right-hand side of the latter formula represents a characteristic function of the l-dimensional random vector with the components $X_k = X(t_k)$ $(k = 1, \ldots, l)$. Thus a characteristic functional of a scalar random function determined on the subspace F_c of the space F is the aggregate of all the multi-dimensional characteristic functions of this random function.

E x a m p l e 3.5.3. In the case of the n-dimensional real vector random function $X = X(t)$ while introducing in the space of its realizations the topology of Example 3.2.3 we obtain $g_x(f) = E \exp\left\{i \sum_{k=1}^{l} X_{q_k}(t_k)\right\}$. Thus a characteristic functional of a vector random function determined on F_c represents the aggregate of the characteristic functions of all finite-dimensional vectors composed of the values of its components.

E x a m p l e 3.5.4. Suppose that a real random function $X = X(t) = x(t, \omega)$ is measurable and integrable on $T_1 \times \Omega$ relatively to the product of the measure P on Ω and some measure σ on T, and all its realizations belong to the space $X = L_p$, $p < \infty$. In this case any continuous linear functional f on L_p is determined by formula:

$$fX = \int f(t)X(t)\sigma(dt), \tag{I}$$

where $f(t) \in L_q$, $p^{-1} + q^{-1} = 1$. Consequently, a characteristic functional of a random variable $X(t) = x(t, \omega)$ whose all the realizations belong to the space L_p of the real functions of the variable t on the subspace F_c is determined by the formula

$$g_x(f) = E \exp \left\{ i \int f(t) X(t) \sigma(dt) \right\}$$
$$= \int \exp \left\{ i \int f(t) x(t, \omega) \sigma(dt) \right\} P(d\omega). \tag{II}$$

E x a m p l e 3.5.5. In exactly the same way as in the previous Example we come to the conclusion that a characteristic functional of the measurable $\sigma \times P$-integrable n-dimensional real vector random function $X(t) = x(t, \omega)$ whose all the realizations belong to the space L_p of the n-dimensional vector functions $p < \infty$ is determined on the space of the continuous linear functionals F_c by the formula

$$g_x(f) = E \exp \left\{ i \sum_{k=1}^{n} \int f_k(t) X_k(t) \sigma(dt) \right\}$$
$$= \int \exp \left\{ i \sum_{k=1}^{n} \int f_k(t) x_k(t, \omega) \sigma(dt) \right\} P(d\omega), \tag{I}$$

where $f = f(t)$ is the vector function with the components $f_1(t), \ldots, f_n(t)$ which belongs to the space L_q, $p^{-1} + q^{-1} = 1$.

3.5.2. Properties of Characteristic Functional ˙

Let us study the main properties of characteristic functionals.

(i) It is clear directly from definition (3.5.4) that *as fX is a real random variable at any f then*

$$|g_x(f)| \leq g_x(0) = 1, \quad g_x(-f) = \overline{g_x(f)}. \tag{3.5.5}$$

(ii) *At any $f_1, \ldots, f_n \in F$ for any complex numbers $\xi_1, \ldots, \xi_n$ we* have

$$\sum_{p,q=1}^{n} g_x(f_p - f_q) \xi_p \bar{\xi}_q$$

$$= E \sum_{p,q=1}^{n} \xi_p \bar{\xi}_q e^{i(f_p X - f_q X)} = E \left| \sum_{p=1}^{n} \xi_p e^{i f_p X} \right|^2 \geq 0. \tag{3.5.6}$$

Thus a characteristic functional possesses a property of positive definiteness. By the way it is easy to see that inequality and the second formula (3.5.5) are the results of (3.5.6). In fact, from (3.5.6) at $n = 1$ follows $g_x(0) \geq 0$, and at $n = 2$, $f_1 = 0$, $f_2 = f$ holds

$$g_x(0)(|\xi_1|^2 + |\xi_2|^2) + g_x(f) \overline{\xi_1} \xi_2 + g_x(-f) \xi_1 \overline{\xi_2} \geq 0.$$

But the left-hand side may be real only at $g_x(-f) = \overline{g_x(f)}$. Taking this fact into account and putting $\xi_1 = 1$, $\xi_2 = -\exp\{-i \arg g_x(f)\}$ we get $2g_x(0) - 2|g_x(f)| \geq 0$. Hence inequality (3.5.6) follows.

(iii) *We introduce in the space F a weak topology by means of the set Φ_c of the linear functional of the type fx, $x \in X$, $\Phi_c = \{\varphi : \varphi f = fx, x \in X\}$. Then a characteristic functional is continuous in this topology in the sense that $g_x(f_n) \to g_x(f)$ for any convergent sequence $\{f_n\} \subset F$, $f = \lim f_n$.* It is evidently so because $f_n \to f$ in our topology if and only if $f_n x \to fx$ at any $x \in X$, and from $e^{if_n x(\omega)} \to e^{ifx(\omega)}$, $\left|e^{if_n x(\omega)}\right| = 1$ at all ω according to the Lebesgue theorem follows $g_x(f_n) \to g_x(f)$.

(iv) *Let T be a bounded (continuous) linear operator mapping a phase space of the random variable X into another real linear space Y and $\Delta_x \subset D_T$. Suppose that there exists an adjoint operator T^*. Then a characteristic functional of the random variable $Y = TX$ is determined by the formula $g_y(g) = Ee^{igY} = Ee^{igTX}$, where g is any linear functional on Y, $g \in G$.* But for any $g \in T^*$ by the definition of the adjoint operator $gT = T^*g$. Consequently, the contraction of a characteristic functional of the random variable Y on the subspace D_{T^*} of the space G dual with Y is expressed by formula

$$g_y(g) = Ee^{i(T^*g)X} = g_x(T^*g). \tag{3.5.7}$$

This formula determines the transformation of a characteristic functional of the random variable at its linear transformation.

(v) It follows directly from (3.5.4) that *the characteristic functional of the random variable $Y = X + a$ is expressed by formula*

$$g_y(f) = Ee^{if(X+a)} = e^{ifa}g_x(f). \tag{3.5.8}$$

(vi) *If $f = \sum_{k=1}^{n} \alpha_k f_k$, where $f_1, \ldots, f_n \in F$ are arbitrarily chosen linear functionals and $\alpha_1, \ldots, \alpha_n$ are real variables then*

$$g_x\left(\sum_{k=1}^{n} \alpha_k f_k\right) = Ee^{i\sum_{k=1}^{n} \alpha_k f_k X}. \tag{3.5.9}$$

Introducing the random variables

$$Z_k = f_k X \quad (k = 1, \ldots, n) \tag{3.5.10}$$

we have

$$g_x\left(\sum_{k=1}^{n}\alpha_k f_k\right) = Ee^{\,i\sum_{k=1}^{n}\alpha_k Z_k}. \tag{3.5.11}$$

This formula shows that the value of a characteristic functional of the random variable X on any linear combination of the fixed functionals $f_1,\ldots,f_n \in F$ represents a characteristic function of the real random vector with the components $Z_k = f_k X$ $(k=1,\ldots,n)$.

E x a m p l e 3.5.6. If m-dimensional real random vector Y represents the result of the affine transformation of the random vector X, $Y = AX + b$ then on the basis of (3.5.7) and (3.5.8) the characteristic function of the vector Y is expressed in terms of the characteristic function of the vector X by formula

$$g_y(f) = e^{if a} g_x(A^T f). \tag{I}$$

E x a m p l e 3.5.7. If $X(t) = x(t,\omega), t \in T_1$ is a measurable real random function which is integrable in $T_1 \times \Omega$ relatively to product of the measure σ on T_1 and the probability P on Ω whose all the realizations belong to the space L_p, $p < \infty$ and

$$Tx = \int \varphi(u,t)x(t)\sigma(dt), \quad u \in U, \tag{I}$$

is the operator mapping L_p in the space L_h, $h < \infty$ then the adjoint operator T^* is determined by formula

$$T^*g = \int \varphi(u,t)g(u)\chi(du). \tag{II}$$

It maps the space L_k, $h^{-1} + k^{-1} = 1$ dual with the space L_h, in the space L_q, $p^{-1} + q^{-1} = 1$ dual with the space L_p. In this case a characteristic functional of the random function

$$Y(u) = TX = \int \varphi(u,t)X(t)\sigma(dt) \tag{III}$$

according to (3.5.7) is determined by the following formula:

$$g_y(g) = g_x\left(\int \varphi(u,t)g(u)\chi(du)\right). \tag{IV}$$

3.5.3. Relation between Characteristic Function and Moments

In elementary probability theory the following formulae are derived from (3.5.1) for the scalar random variable X on the basis of Maclaurin formula in supposition that there exist the moments of the r-order:

$$\alpha_r = i^{-r} g^{(r)}(0) \quad r = 1, 2, \ldots, \tag{3.5.12}$$

$$g(\lambda) = 1 + \sum_{r=1}^{\nu} \frac{i^r \alpha_r}{r!} \lambda^r + R_\nu, \tag{3.5.13}$$

$$\mu_r = i^{-r} \left[\frac{d^r}{d\lambda^r} e^{-i\lambda m_x} g(\lambda) \right]_{\lambda=0}, \tag{3.5.14}$$

$$e^{-i\lambda m_x} g(\lambda) = 1 + \sum_{r=2}^{\nu} \frac{i^r \mu_r}{r!} \lambda^r + R'_\nu. \tag{3.5.15}$$

Here R_ν and R'_ν are the residuals in Maclaurin formula. Analogously we get the relation formulae for the n-dimensional random vector:

$$\alpha_{h_1, \ldots, h_n} = i^{-r} \left[\frac{\partial^r g(\lambda)}{\partial \lambda_1^{h_1} \ldots \partial \lambda_n^{h_n}} \right]_{\lambda=0}, \tag{3.5.16}$$

$$\mu_{h_1, \ldots, h_n} = i^{-r} \left[\frac{\partial^r}{\partial \lambda_1^{h_1} \ldots \partial \lambda_n^{h_n}} e^{-i\lambda^T m_x} g(\lambda) \right]_{\lambda=0}, \tag{3.5.17}$$

$$g(\lambda) = 1 + \sum_{r=1}^{\nu} i^r \sum_{h_1 + \cdots + h_n = r} \frac{\alpha_{h_1, \ldots, h_n}}{h_1! \ldots h_n!} \lambda_1^{h_1} \ldots \lambda_n^{h_n} + R_\nu, \tag{3.5.18}$$

$$e^{-i\lambda^T m_x} g(\lambda) = 1 + \sum_{r=2}^{\nu} i^r \sum_{h_1 + \cdots + h_n = r} \frac{\mu_{h_1, \ldots, h_n}}{h_1! \ldots h_n!} \lambda_1^{h_1} \ldots \lambda_n^{h_n} + R'_\nu, \tag{3.5.19}$$

3.5.4. Expansion of Characteristic Functional

Now we suppose that the random variable X has the operators of the moments till $(n + 1)$-order inclusively. In this case we have

$$g_x(f) = \int e^{ifx} \mu_x(dx) = \int \left(1 + \sum_{r=1}^{n} \frac{i^r}{r!} (fx)^r \right.$$

$$\left. + \frac{i^{n+1}}{(n+1)!} (fx)^{n+1} e^{i\theta fx} \right) \mu_x(dx), \tag{3.5.20}$$

where θ is the constant in Lagrange formula. Hence taking into account (3.2.46) we obtain

$$g_x(f) = 1 + \sum_{r=1}^{n} \frac{i^r}{r!} fm_r(f, \ldots, f) + \rho_n(f). \qquad (3.5.21)$$

Here $fm_r(f, \ldots, f)$ is so called *a moment form* of *the rth order* of the random variable X in which all the arguments $f_1, \ldots, f_r$ are equal to f, and $\rho_n(f)$ is the residual for which the following estimate is valid

$$|\rho_n(f)| \leq \frac{1}{(n+1)!} \int |fx|^{n+1} \mu_x(dx). \qquad (3.5.22)$$

If the random variable X has the operators of the moments of all orders and the integral in (3.5.22) at all $f \in F$ increases slower than $(n+1)!$ at $n \to \infty$ then $\rho_n(f) \to 0$ and formula (3.5.21) takes the form

$$g_x(f) = 1 + \sum_{r=1}^{\infty} \frac{i^r}{r!} fm_r(f, \ldots, f). \qquad (3.5.23)$$

3.5.5. Correspondence between Probability Measures and Characteristic Functionals

As we saw in Section 3.5.1 each random variable in a real topological linear space has a characteristic functional. To a given probability measure in a real linear space it corresponds the unique characteristic functional. Naturally the question arises: *in what degree a characteristic functional determines a distribution of a random variable, i.e. whether the unique probability measure corresponds to a given characteristic functional.* At first we shall give a positive answer to this question for the finite-dimensional random variables and then we extend this statement over a general case.

In order to answer to this question we show how we may construct a probability measure by a given characteristic functional and prove its uniqueness.

Lemma 3.5.1. *Let X be the n-dimensional random vector, $g_x(f)$ be its characteristic functional (function), then a probability measure μ_x of the vector X on any finite rectangular R is determined by formula*

$$\mu_x(R) = \lim_{F \to \infty} \int_{-F}^{F} \cdots \int_{-F}^{F} \prod_{k=1}^{n} \frac{e^{-if_k a_k} - e^{-if_k b_k}}{2\pi i f_k} g_x(f_1, \ldots, f_n) df_1 \ldots df_n,$$

$$(3.5.24)$$

where $(a_k b_k)$ $(k = 1, \ldots, n)$ are the sides of the rectangular R.

▷ For proving formulae (3.5.24) we substitute under the sign of the integral expression (3.5.4) for $g_x(f)$. Noticing that in a given case $fx = \sum_{k=1}^{n} f_k x_k$ and denoting

$$I_F = \int_{-F}^{F} \cdots \int_{-F}^{F} \prod_{k=1}^{n} \frac{e^{-if_k a_k} - e^{-if_k b_k}}{2\pi i f_k} g_x(f_1, \ldots, f_n) df_1 \ldots df_n,$$

we get

$$I_F = \int_{-F}^{F} \cdots \int_{-F}^{F} \prod_{k=1}^{n} \frac{e^{-if_k a_k} - e^{-if_k b_k}}{2\pi i f_k} df_1 \ldots df_n \int e^{i \sum_{k=1}^{n} f_k x_k} \mu_x(dx).$$

$$(3.5.25)$$

But the function

$$\left\{ \prod_{k=1}^{n} \frac{e^{-if_k a_k} - e^{-if_k b_k}}{2\pi i f_k} \right\} e^{i \sum_{k=1}^{n} f_k x_k}$$

$$= \prod_{k=1}^{n} \frac{e^{if_k(x_k - a_k)} - e^{if_k(x_k - b_k)}}{2\pi i f_k} \tag{3.5.26}$$

is measurable, bounded, and consequently is integrable with respect to the variables $f_1, \ldots, f_n, x_1, \ldots, x_n$ relatively to the product of Lebesgue measure in the n-dimensional cube $(-F, F)^n$ and the measure μ_x. Therefore on the basis of the Foubini theorem from functional analysis the order of the integration in the obtained formula may be changed and we may integrate at first over the variables $f_1, \ldots, f_n$ and after that over x. Then from (3.5.25) and (3.5.26) we get

$$I_F = \int \mu_x(dx) \int_{-F}^{F} \cdots \int_{-F}^{F} \prod_{k=1}^{n} \frac{e^{if_k(x_k - a_k)} - e^{if_k(x_k - b_k)}}{2\pi i f_k} df_1 \ldots df_n$$

$$= \int \left\{ \prod_{k=1}^{n} \int_{-F}^{F} \frac{e^{if_k(x_k - a_k)} - e^{if_k(x_k - b_k)}}{2\pi i f} df \right\} \mu_x(dx)$$

$$= \int \left\{ \prod_{k=1}^{n} \int_{0}^{F} \frac{\sin f(x_k - a_k) - \sin f(x_k - b_k)}{\pi f} df \right\} \mu_x(dx). \tag{3.5.27}$$

It remains to pass to the limit at $F \to \infty$. We have

$$\lim_{F \to \infty} \int_0^F \frac{\sin fz}{\pi f} df = \frac{1}{2} \operatorname{sgn} z \quad (\operatorname{sgn} 0 = 0),$$

$$\lim_{F \to \infty} \int_0^F \frac{\sin f(x-a) - \sin f(x-b)}{\pi f} df = \begin{cases} 0 & \text{at} & x \overline{\in} (a,b), \\ \frac{1}{2} & \text{at} & x = a, x = b, \\ 1 & \text{at} & x \in (a,b). \end{cases}$$

As at any $F > 0$

$$0 \le \int_0^F \frac{\sin fz}{\pi f} df \le \int_0^\pi \frac{\sin u}{\pi u} du < 1, \tag{3.5.28}$$

then by virtue of (3.5.28) the integrand in (3.5.27) enclosed in parentheses at all F is absolute-values bounded by the μ_x-integrable function $\varphi(x) = 2$. Therefore the limit transition in (3.5.27) may be fulfilled under the sign of the integral. Then accounting that the measure μ_x of the perimeter of the rectangle R is equal to zero we obtain

$$\lim_{F \to \infty} I_F = \int \mathbf{1}_R(x)\mu_x(dx) = \mu_x(R), \tag{3.5.29}$$

what proves formula (3.5.24). ◁

Using Lemma 3.5.1 we prove the following statement.

Theorem 3.5.1. *A characteristic functional of a finite-dimensional random variable determines completely and uniquely its probability measure.*

▷ Any rectangle may be represented as the limit of a monotonous sequence of the rectangles of the considered type and the measure μ_x is continuous. Therefore a characteristic functional of a random vector X determines uniformly its probability measure μ_x on the set of all the rectangles of the n-dimensional space X. In the consequence of the σ-additivity of the measure μ_x it may be uniquely extend of over the σ-algebra which is generated by the set of the rectangles, i.e. on the σ-algebra of Borel sets of the phase space of the random vector X. When the characteristic function $g_x(f)$ of the random vector X is integrable (over Lebesgue) because of the boundedness of the second multiplier under the integral (3.5.24) the whole integrand in (3.5.24) is integrable. Therefore formula (3.5.24) may be rewritten in the form

$$\mu_x(R) = \int_{-\infty}^\infty \cdots \int_{-\infty}^\infty \prod_{k=1}^n \frac{e^{-if_k a_k} - e^{-if_k b_k}}{2\pi i f_k} g_x(f_1, \ldots, f_n) df_1 \ldots df_n$$

$$= \frac{1}{(2\pi)^n} \int g_x(f) df \int_R e^{-ifx} dx . \tag{3.5.30}$$

In consequence of the boundedness of the function $g_x(f)e^{-ifx}$ and its integrability over f it is an integrable function of two vectors f and x. Therefore we may change the order of the integration in formula (3.5.30). Then we obtain

$$\mu_x(R) = \frac{1}{(2\pi)^n} \int_R dx \int e^{-ifx} g_x(f) df .$$

Hence it is clear that in a given case there exists the probability density $f_x(x)$ of the random vector X which is determined by formula

$$f_x(x) = \frac{1}{(2\pi)^n} \int e^{-ifx} g_x(f) df . \tag{3.5.31}$$

Then formula (3.5.4) takes the form

$$g_x(f) = \int e^{ifx} f_x(x) dx . \tag{3.5.32}$$

And vice versa if there exists the probability density $f_x(x)$ of the random vector X over the Lebesgue measure then its characteristic functional $g_x(f)$ is determined by formula (3.5.32). On the basis of the known Fourier theorem $f_x(x)$ is expressed in terms of $g_x(f)$ by formula (3.5.31) and the function $g_x(f)$ is integrable. $\triangleleft$

Corollary 3.5.1. *A characteristic function of a finite-dimensional random vector is integrable if and only if there exists a probability density of this vector over the Lebesgue measure.*

$\triangleright$ Let now X be a random variable in an arbitrary real topological linear space, and $g_x(f)$, $f \in F$ be its characteristic functional. According to property (vi) of a characteristic functional (Section 3.5.2) the expression

$$g_z(\alpha_1 , \ldots , \alpha_n) = g_x \left(\sum_{k=1}^n \alpha_k f_k \right) \tag{3.5.33}$$

at any fixed $f_1 , \ldots , f_n \in F$ represents a characteristic function of a random vector with the components $Z_1 = f_1 X, \ldots Z_n = f_n X$. In accordance with Theorem 3.5.1 it determines completely and uniquely the probability measure μ of this random vector on the σ-algebra of Borel sets of the correspondent n-dimensional space, and in particular, on the rectangles $\{z : z_1 \in A_1 , \ldots , z_n \in A_n\}$. But the random vector Z represents a function of the random variable X and as the prototype of the rectangle $\{z : z_1 \in A_1 , \ldots , z_n \in A_n\}$ in a phase space of the random variable X serves the set $\{x : f_1 x \in A_1 , \ldots , f_n x \in A_n\}$. Consequently, on the basis of formula (2.2.4) we have

$$\mu_z(\{z : z_1 \in A_1 , \ldots , z_n \in A_n\}) = \mu_x(\{x : f_1 x \in A_1 , \ldots , f_n x \in A_n\}).$$
$$(3.5.34)$$

Hence it is clear that the characteristic functional $g_x(f)$ together with the measure μ_z of the vector Z determines also uniquely the probability measure μ_x of the random variable X on the sets of the type $\{x : f_1 x \in A_1 , \ldots , f_n x \in A_n\}$ at given n, $f_1 , \ldots , f_n \in F$. As $n, f_1 , \ldots , f_n$ are arbitrary then a characteristic functional determines a probability measure of the random variable X on such sets which are correspondent to all $n, f_1 , \ldots , f_n \in F$ and to all Borel sets $A_1 , \ldots , A_n$. After determining the measure μ_x on the class of such sets we may extend it uniquely on the σ-algebra generated by this class of the sets $\triangleleft$

This fact proves the following statement.

Theorem 3.5.2. *A characteristic functional of any random variable X in any real linear space determines uniquely its probability measure on the minimal σ-algebra which contains all the sets of the type $\{x : f_1 x \in A_1 , \ldots , f_n x \in A_n\}$. In particular, if in a phase space of the random variable X a weak topology is determined by means of the set of the linear functionals F then it will be the σ-algebra generated by the base of the topological space X.*

R e m a r k 1. Thus there exists one-to-one correspondence between the probability measures in a real linear space and the characteristic functionals. The probability distribution in a real topological linear space may be assumed both as a probability measure and a characteristic functional (naturally, if as the σ-algebra in this case is assumed σ-algebra generated by the considered class of the sets).

R e m a r k 2. Formula (3.5.23) which expresses a characteristic functional in terms of the moments shows that in the case when there exist the operators of the moments of all orders of the random variable X, and the moments of the random variables $|fX|$ at all $f \in F$ increase with the extending of the order r slower than the factorials $r!$ the sequence of the operators of the moments determines uniquely a characteristic functions, and consequently, a probability measure of the random variable X. Thus the probability density in a real linear space may be assigned also by means of an unbounded sequence of the operators of the moments.

3.6. Sequences of Probability Measures and Characteristic Functions. Distributions in Banach Spaces

3.6.1. Helly-Bray Theorem

It was shown in Section 3.5 that a characteristic functional $g(f)$ of any random variable in any real linear space X possesses such properties as:

- a positive definiteness;
- a continuity with respect to any topology of the space F in such sense that for any sequence $\{f_n\}$ weakly convergent to f ($f_n x \to f x$ for all $x \in X$), $g(f_n) \to g(f)$;
- $g(0) = 1$.

The question arises: whether any functional possessing these properties is a characteristic functional of some random variable? In order to answer this question we have to study some properties of the sequences of the probability measures and the characteristic functionals which are correspondent to them. At first we prove the following statement.

Theorem 3.6.1 (The Helly-Bray theorem). *If the sequence of the probability measures $\{\mu_p\}$ in the space X converges to the probability measure μ then the sequence of the integrals with respect to any bounded continuous measurable function $y(x)$ with the values in the separable B-space over the measures μ_p converges to the integral with respect to $y(x)$ over the measure μ.*

$\triangleright$ As it is known from the theory of the measurable functions with the values in the B-space there exists a sequence of the elementary functions $y^n(x) = \sum_{k=1}^{\infty} y_k^n 1_{D_k^n}(x)$ $(n = 1, 2, \ldots)$ uniformly convergent to $y(x)$. On the basis of property (xii) of an integral with respect to the bounded function over a finite measure for all p the inequalities are valid

$$\left\| \int y^n \, d\mu_p - \int y \, d\mu_p \right\| \leq \sup_x \| y^n(x) - y(x) \|,$$

$$\left\| \int y^n \, d\mu - \int y \, d\mu \right\| \leq \sup_x \| y^n(x) - y(x) \|.$$

Let us take an arbitrary $\varepsilon > 0$ and choose n in such a way that the inequality $\sup_x \| y^n(x) - y(x) \| < \varepsilon/3$ will be fulfilled. After that we choose p_0 in such a way that at all $p \geq p_0$ the inequalities will be fulfilled

$$|\mu_p(D_k^n) - \mu(D_k^n)| < \frac{\varepsilon}{3 N_n \max_k \| y_k^n \|} \quad (k = 1, \ldots, N_n).$$

Then we shall have at all $p \geq p_0$

$$\left\| \int y \, d\mu_p - \int y \, d\mu \right\| \leq \left\| \int y \, d\mu_p - \int y^n \, d\mu_p \right\|$$

$$+ \left\| \int y^n \, d\mu_p - \int y^n \, d\mu \right\| + \left\| \int y^n \, d\mu - \int y \, d\mu \right\| < \varepsilon . \ \triangleleft$$

3.6.2. Direct and Inverse Levi Theorems

As the result of the Helly-Bray theorem we get the direct and inverse Levi theorems.

Theorem 3.6.2 (The direct Levi theorem). *If the sequence of the probability measures $\{\mu_p\}$ in a real linear space X converges to the probability measure μ then the sequence of the correspondent characteristic functionals $\{g_p\}$ converges to the characteristic functional g which is determined by the measure μ.*

Theorem 3.6.3 (The inverse Levi theorem). *If a sequence of the characteristic functions $\{g_p\}$ of the n-dimensional random variables converges to the function g continuous in zero then the sequence of the correspondent probability measures $\{\mu_p\}$ converges to some probability measure μ and g is a characteristic function which is determined by it.*

$\triangleright$ Let $F_p(x)$ be a distribution function correspondent to the measure μ_p (Section 2.2). Show that we may single out the subsequence $\{F_{p_k}(x)\}$ convergent to some distribution function $F(x)$ at all the points of its continuity from the sequence $\{F_p(x)\}$. For this purpose we take any countable set $D_c = \{x_m\}$ dense in X, for instance, a set of all the points with rational coordinates. As $0 \leq F_p(x_1) \leq 1$ at all p then we may single out from the sequence $\{F_p(x)\}$ such a subsequence $\{F_{p1}(x)\}$ that the numerical sequence $\{F_{p1}(x_1)\}$ will be convergent. Analogously we may single out from the sequence $\{F_{p1}(x)\}$ such a subsequence $\{F_{p2}(x)\}$ that the numerical sequence $\{F_{p2}(x_2)\}$ will be convergent. Continuing this process at any r we single out from the sequence $\{F_{p,r-1}(x)\}$ such a subsequence $\{F_{pr}(x)\}$ that the numerical sequence $\{F_{pr}(x_r)\}$ will be convergent. The diagonal sequence $\{F_{pp}(x)\}$ will converge at all the points of the set D_c as the sequence $\{F_{pp}(x_r)\}$ at any r is a subsequence of the sequence $\{F_{pr}(x_r)\}$ at $p > r$. Suppose that the function $F_{pp}(x)$ represents the function $F_{mp}(x)$ from the initial sequence $\{F_m(x)\}$ $(p = 1, 2, \ldots)$. We denote the limit function $F_{D_c}(x) = \lim_{p \to \infty} F_{m_p}(x)$ at $x \in D_c$ and define the function $F(x) = \lim_{x_r \to x} F_{D_c}(x_r)$, where $\{x_r\}$ is the sequence of the points of the set D_c convergent below to the point x, $x_r < x$. We show that the sequence $\{F_{m_p}(x)\}$ converges to $F(x)$ at all the points of the continuity $F(x)$. From the inequality $F_{m_p}(x') \leq F_{m_p}(x) \leq F_{m_p}(x'')$ at $x' < x < x''$, $x', x'' \in D_c$ follows that

$$F_{D_c}(x') \leq \underline{\lim} F_{m_p}(x) \leq \overline{\lim} F_{m_p}(x) \leq F_{D_c}(x'').$$

After taking two sequences $\{x'_k\}$ and $\{x''_k\}$ of the points of the set D_c, $x'_k < x < x''_k$ which are convergent to x we get

$$F(x) \leq \underline{\lim} F_{m_p}(x) \leq \overline{\lim} F_{m_p}(x) \leq F(x+0).$$

Hence it is clear that at any point x of the continuity $F(x)$ there exists the limit $\lim F_{m_p}(x) = F(x)$.

We shall show that $F(x)$ is a distribution function of some random variable. It is sufficient for this purpose to prove that the measure μ (Section 2.2) determined by the function $F(x)$ is a probability measure, i.e. is normalized $\mu(X) = 1$. First of all we notice that $\mu(X)$ cannot be more than 1 as for any finite rectangle R whose perimeter has zero measure μ, $\mu(R) = \lim \mu_{m_p}(R) \leq 1$ and by virtue of the continuity of the measure $\mu(X)$ represents the limit of the sequence $\{\mu(R_k)\}$ for any increasing sequence of the rectangles $\{R_k\}$ for which $\lim R_k = \bigcup R_k = X$. Suppose that $\mu(X) = c < 1$ and prove that it leads to the contradiction. Let us take an arbitrary ε in the interval $0 < \varepsilon < (1-c)/2$. By virtue of the continuity of the limit function $g(f)$ in zero there exists such n-dimensional cube $A = (-a, a)^n$ that

$$\frac{1}{(2a)^n} \left| \int_A g(f)\, df \right| > 1 - \varepsilon > c + \varepsilon. \tag{3.6.1}$$

On the other hand,

$$\int_A g(f)\, df = \lim \int_A g_{m_p}(f)\, df$$

$$= \lim \int_A df \int e^{ifx} \mu_{m_p}(dx) = \lim \int \left(\int_A e^{ifx}\, df \right) \mu_{m_p}(dx).$$

As $\left| e^{ifx} \right| = 1$ then at any x

$$\left| \int_A e^{ifx}\, df \right| \leq (2a)^n. \tag{3.6.2}$$

At any x outside the cube $B = (-b, b)^n$ of the space X

$$\int_A e^{ifx}\,df = \int_{-a}^{a} e^{if_h x_h}\,df_h$$

$$\times \int_{-a}^{a} \ldots \int_{-a}^{a} \exp\left\{ i \sum_{k=1}^{n}{}^{(h)} f_k x_k \right\} df_1 \ldots df_{h-1} df_{h+1} \ldots df_n$$

$$= \frac{2}{x_h} \sin a x_h \int_{-a}^{a} \ldots \int_{-a}^{a} \exp\left\{ i \sum_{k=1}^{n}{}^{(h)} f_k x_k \right\} df_1 \ldots df_{h-1} df_{h+1} \ldots df_n,$$

where x_h is any of the coordinate of the point x superior to the modulus b; the index h upward the sum means that the item correspondent to $k = h$ does not inter into the sum. Hence we obtain the estimate

$$\left| \int_A e^{ifx}\,dx \right| \leq \frac{2}{b}(2a)^{n-1} \quad \text{at any} \quad x \bar{\in} B. \tag{3.6.3}$$

Using estimate (3.6.2) at $x \in B$ and estimate (3.6.3) at $x \bar{\in} B$ we come to the inequality

$$\left| \int_A g_{m_p}(f)\,df \right| \leq \int_B \left| \int_A e^{ifx}\,df \right| \mu_{m_p}(dx)$$

$$+ \int_{\bar{B}} \left| \int_A e^{ifx}\,df \right| \mu_{m_p}(dx) \leq (2a)^n \mu_{m_p}(B) + \frac{2}{b}(2a)^{n-1}.$$

Now we choose $b > 2/(a\varepsilon)$. After this accounting that $\mu(B) = \lim \mu_{m_p}(B) \leq c$ we choose p_b in such a way that at all $p \geq p_b$ would be $\mu_{m_p}(B) < c + \varepsilon/2$. Then we get

$$\frac{1}{(2a)^n} \left| \int_A g_{m_p}(f)\,df \right| < c + \varepsilon \quad \text{at all} \quad p \geq p_b,$$

$$\frac{1}{(2a)^n} \left| \int_A g(f)\,df \right| = \frac{1}{(2a)^n} \left| \lim \int g_{m_p}(f)\,df \right| \leq c + \varepsilon.$$

The last inequality contradicts (3.6.1). Consequently, $\mu(X) = 1$, i.e. μ is a probability measure. On the basis of the direct Levi theorem the sequence of the characteristic functions $\{g_{m_p}(f)\}$ converges to the characteristic function determined by the measure μ. But $\{g_{m_p}(f)\}$ converges to $g(f)$. Consequently, $g(f)$ is a characteristic function correspondent to the probability measure μ.

It remains to prove that the whole sequence of the measures $\{\mu_p\}$ converges also to μ. If it was not so then after rejecting form $\{\mu_p\}$ the subsequence $\{\mu_{m_p}\}$ and arguing analogously we should single out from the obtained sequence the subsequence which is convergent to some probability measure μ' different from μ, and the function g would be a characteristic function corespondent to the measure μ'. But by virtue of the uniqueness of the probability measure μ' cannot differ from μ. This fact proves the convergence of the measures $\{\mu_p\}$ to μ in the finite-dimensional case. ◁

3.6.3. Finite-Dimensional Case of Bochner Theorem

Now we are able to give an answer to the question raised at the beginning of Subsection 3.6.1. At first we shall consider a finite- dimensional case and probe the following statement.

Theorem 3.6.4. *Any continuous positively determined function of the n-dimensional vector represents a characteristic function of some n-dimensional random vector.*

▷ Let $g(f)$ be a positively determined continuous function of the n-dimensional vector f and $g(0) = 1$. We determine in the region F the following function:

$$g_F(f) = \begin{cases} g(f) \prod_{k=1}^{n} \left(1 - \frac{|f_k|}{F}\right) & \text{at} & |f_k| < F, \\ 0, & \text{if even one} & |f_k| > F. \end{cases}$$

By virtue of property (3.5.5) of positively determined functions $g_F(f)$ is bounded and therefore integrable. Consequently, it may be expressed by the Fourier formula

$$g_F(f) = \int e^{ifx} p_F(x)\, dx, \quad fx = \sum_{k=1}^{n} f_k x_k, \tag{3.6.4}$$

where the integral should be assumed as its limit on the n-dimensional rectangle which is symmetric relatively to the origin, and

$$p_F(x) = \frac{1}{(2\pi)^n} \int e^{-ifx} g_F(f)\, df.$$

Now we notice that the function $p_F(x)$ is nonnegative as

$$\int e^{-ifx} g_F(f)\, df = \int\limits_{-F}^{F} \cdots \int\limits_{-F}^{F} e^{-i(f_1 x_1 + \cdots + f_n x_n)}$$

$$\times g(f_1, \ldots, f_n) \prod_{k=1}^{n} \left(1 - \frac{|f_k|}{F}\right) df_1 \ldots df_n$$

$$= \frac{1}{F^n} \int\limits_{-F/2}^{F/2} \cdots \int\limits_{-F/2}^{F/2} e^{-i[(f_1' - f_1'')x_1 + \cdots + (f_n' - f_n'')x_n]}$$

$$\times g(f_1' - f_1'', \ldots, f_n' - f_n'')\, df_1' df_1'' \ldots df_n' df_n'' \geq 0$$

(by virtue of the positive definiteness of $g(f)$ all the Riemann sums approximating this $2n$-multiple integral are nonnegative). From here and from (3.6.4) at $f = 0$ it follows that the function $p_F(x)$ is integrable and the conditions of the normalization are fulfilled

$$\int p_F(x)\, dx = g_F(0) = g(0) = 1.$$

Thus the function $p_F(x)$ possesses all the properties of the probability density, and $g_F(x)$ by virtue of (3.5.32) represents a characteristic function determined by this probability density.

Now let us take an arbitrary unboundedly increasing sequence $\{F_n\}$ of the values F, $F_n > 0$, $F_n \to \infty$ at $n \to \infty$. The correspondent sequence of the characteristic functions $\{g_{F_n}\}$ converges to the continuous function g. Consequently, by virtue of the inverse Levi theorem the sequence of the correspondent probability measures

$$\mu_{F_n}(A) = \int\limits_{A} p_{F_n}(x)\, dx \quad (n = 1, 2, \ldots) \tag{3.6.5}$$

converges to some probability measure μ and the function g represents a characteristic function which is determined by the measure μ,

$$g(f) = \int e^{ifx} \mu(dx). \quad \lhd \tag{3.6.6}$$

Theorem 3.6.5 (The Bochner theorem). *The function $\varphi(z)$ of the finite-dimensional vector z is continuous and positively determined if and only if it is presented in the form*

$$\varphi(z) = \int e^{izx}\nu(dx), \quad zx = \sum_{k=1}^{n} z_k x_k, \tag{3.6.7}$$

where ν is some nonnegative finite measure.

▷ According to Theorem 3.6.4 the function $g(z) = \varphi(z)/\varphi(0)$ represents a characteristic function of some n-dimensional random vector X and therefore is expressed by formula (3.6.6). Introducing the finite nonnegative measure $\nu = \varphi(0)\mu$ we get (3.6.7) what proves the necessity of the condition. The sufficiency is the direct result of property (3.5.6) and the continuity of the characteristic functions. By virtue of the uniqueness of the probability measure correspondent to a given characteristic functional the measure ν in the presentation of (3.6.7) is unique. ◁

3.6.4. General Case of Bochner Theorem

Now let X be an arbitrary real linear space, F be some set (space) of the linear functionals on X (in the general case F is the subspace of the space of all linear functionals dual with X). Suppose that the space F contains sufficiently many functionals for the fact that from $fx = 0$ at all $f \in F$ follows $x = 0$ and introduce into F a weak topology of Section 3.5 by means o the vicinities of zero $\{f : |fx_1| < \varepsilon_1, \ldots, |fx_n| < \varepsilon_n\}$, correspondent to all n, $x_1, \ldots, x_n \in X$, $\varepsilon_1, \ldots, \varepsilon_n > 0$. Let $g(f)$ be a continuous in this topology positively determined functional and $g(0) = 1$. The function $g_f(\alpha) = g(\alpha f)$ of the real variable α at any fixed $f \in F$ is continuous and positively determined. Therefore on the basis of Theorem 3.6.5 it is a characteristic function of some real random variable dependent on f, i.e. is an one-dimensional characteristic function of some random function $Z(f)$. Similarly, the function

$$g_{f_1,\ldots,f_n}(\alpha_1, \ldots, \alpha_n) = g\left(\sum_{k=1}^{n} \alpha_k f_k\right) \tag{3.6.8}$$

of the real variables $\alpha_1, \ldots, \alpha_n$ at any fixed $f_1, \ldots, f_n \in F$ is continuous and positively determined in consequence of which it represents a characteristic function of some n-dimensional random vector. The components of this vector are the values of the random vector. The components of this vector are the values of the random function $Z(f)$ at $f = f_1, \ldots, f_n$

as in the case when all α_k besides one of them are equal to zero, the functions $g_{f_1,\ldots,f_n}$ coincide with g_f at the correspondent values f.

Thus formula (3.6.8) determines the consistent set of the multi-dimensional characteristic functions of some random function $Z(f)$ and the correspondent set of its multi-dimensional distributions $\mu_{f_1,\ldots,f_n}$. On the basis of the Kolmogorov theorem (Section 2.6) this set of the multi-dimensional distributions determines uniquely the probability measure μ_z of the random function $Z(f)$ in the infinite product of the measurable spaces $(R^F, \mathcal{B}^F)$ where R is the real axis R, and $\mathcal{B}$ is the σ-algebra of Borel sets on it.

Now we show that *almost all the realizations of the random function $Z(f)$ are linear functionals in consequence of which the measure μ_z may be transferred from R^F to the subspace of the linear functionals on F.*

$\triangleright$ It is sufficient for this purpose to show that the measure μ_z of any set $\mathcal{B}^F$ which contains all linear functions is equal to 1. Let C be an arbitrary set which contains all linear functions of the variable f, $C \in \mathcal{B}^F$. This set restricts the functions which enter into it only in the countable set of the points, for instance, $S_C = \{f_r\}$. If between the vectors $f_r \in S_C$ there are no linear dependences then the set C cannot impose any restrictions upon the functions which enter into it at the points f_r as there exist the linear functions with arbitrary values in linearly independent points. Consequently, in this case $C = R^F$ and $\mu_z(C) = 1$. If there exist the linear dependences between the vectors of the set S_C

$$\sum_{k=1}^{n} c_{nk} f_k = 0 \quad (n = 1, 2, \ldots) \tag{3.6.9}$$

(if the vector f_n is not the linear combination of $f_1, \ldots, f_{n-1}$ then $c_{n1} = \cdots = c_{nn} = 0$) and introducing the sets of the functions

$$B_n = \left\{ z(f) : \sum_{k=1}^{n} c_{nk} z(f_k) = 0 \right\} \quad (n = 1, 2, \ldots), \tag{3.6.10}$$

we ensure in the fact that along with the set Φ of all linear functions the set C contains also the intersection of all the sets B_n, $C \supset \bigcap B_n$. Really, as C restricts the functions which enter into it only in the countable set of the points S_C then along with all linear functions C contains also all the functions satisfying the conditions of the linearity only on S_C. But $\bigcap B_n$ is in fact the set of all the functions satisfying the conditions of the linearity on S_C.

It remains to notice that $\mu_z\left(\bigcap_{n=1}^{m} B_n \right) = 1$ at any finite m. Really, the characteristic function of the random vector U with the components U_n

$= \sum_{k=1}^{n} c_{nk} Z(f_k)$ $(n = 1, \ldots, m)$ is determined by formula

$$g_u(\alpha_1, \ldots, \alpha_m) = g_{f_1,\ldots,f_m}\left(\sum_{n=1}^{m} \alpha_n c_{n1}, \ldots, \alpha_m c_{mm}\right).$$

But by virtue of (3.6.8) and (3.6.9)

$$g_{f_1,\ldots,f_m}\left(\sum_{n=1}^{m} \alpha_n c_{n1}, \ldots, \alpha_m c_{mm}\right) = g\left(\sum_{k=1}^{m} f_k \sum_{n=k}^{m} \alpha_n c_{nk}\right)$$

$$= g\left(\sum_{n=1}^{m} \alpha_n \sum_{k=1}^{n} c_{nk} f_k\right) = g(0) = 1.$$

Thus the characteristic function of the random vector U is identically equal to 1 and the measure completely concentrated in zero corresponds to it

$$P(U = 0) = P\left(\sum_{k=1}^{n} c_{nk} Z(f_k) = 0, \ n = 1, \ldots, m\right) = 1, \qquad (3.6.11)$$

i.e. $\mu_z(\bigcap_{n=1}^{m} B_n) = 1$. As it is valid at any m then $\mu_z(\bigcap_{n=1}^{\infty} B_n) = \lim \mu_z\left(\bigcap_{n=1}^{m} B_n\right)$ $= 1$. And as $\bigcap B_n \subset C$ and $\mu_z(\bigcap B_n) \le \mu_z(C)$, then $\mu_z(C) = 1$ what proves the possibility of the transfer of the measure μ_z on the subspace Φ of the linear functions of the variable $f \in F$. According to (2.10.16) the obtained measure μ on Φ is determined by formula $\mu(A) = \mu(C\Phi) = \mu_z(C)$, $C \in \mathcal{B}^F$. As the σ-algebra $\mathcal{B}^F$ is generated by the class of the sets $\{z(f) : z(f_1) \in A, \ldots, z(f_n) \in A_n\}$ correspondent to all natural n, to all $f_1, \ldots, f_n \in F$ and to all Borel sets $A_1, \ldots, A_n$ of the real axis R^1, and $\Phi = \{z(f) : z(f) = \varphi f\}$ then the measure μ in the space Φ is determined on the minimal σ-algebra $\mathcal{F}$ which contains the sets $\{\varphi : \varphi f_1 \in A_1, \ldots, \varphi f_n \in A_n\}$ correspondent to all finite sets of the functionals $f_1, \ldots, f_n$ from F and to all Borel sets $A_1, \ldots, A_n$ of the real exis. ◁

Thus in the general case of an arbitrary real linear space X the answer to the question raised at the beginning of Subsection 3.6.1 is obtained negative. *Any continuous in a weak topology positively determined functional $g(f)$, $f \in F$ in the general case determines the probability measure μ in the space Φ of all linear functionals which act on F but not in the initial space X. As the space X may be considered as the subspace of the space Φ then the functional $g(f)$ in the general case determines the probability measure μ in the space Φ which is wider than X.*

The additional conditions are necessary in order that the measure μ determined on Φ by the functional $g(f)$ will be concentrated on the subspace X of the space Φ, and more exactly, in order that for any set $C \in \mathcal{F}$ which contains X is $\mu(C) = 1$. In this case the measure μ may be transferred into X as the correspondent trace probability (Section 2.5). So the measure μ is determined on the σ-algebra $\mathcal{A}$ of the space X generated by the sets of the type $\{x : f_1 x \in A_1, \ldots, f_n x \in A_n\}$ correspondent to all natural n, to all $f_1, \ldots, f_n \in F$ and to all Borel sets $A_1, \ldots, A_n$ of the real axis R^1. Here almost all the realizations of the random function $Z(f)$ are determined by the formula $z(f) = \varphi f = fx$, $x \in X$. Consequently, almost sure $Z(f) = fX$ is valid where X is a random variable with the probability measure μ.

A characteristic functional of the random variable X by virtue of property (vi) of the characteristic functionals and formula (3.6.8) coincides with $g(f)$ on all finite-dimensional subspaces of the space F, and consequently, coincides with $g(f)$ on the whole space F. Really, if it does not coincide with $g(f)$ in some point f_1 then it will not coincide with $g(f)$ on any finite-dimensional subspace which contains f_1, in particular on the one-dimensional subspace $\alpha_1 f_1$. Thus we get the following result.

Theorem 3.6.6. *In the case when the probability measure in Φ determined by a given continuous positively determined functional $g(f)$ turns out to be concentrated on the subspace X of the space Φ, $g(f)$ is a characteristic functional of some random variable X in the space X and uniquely determines the probability measure of the random variable.*

R e m a r k. The probability measure μ determined by the functional $g(f)$ is concentrated on X if the space Φ coincides with X, i.e. when any linear functional φ on F is representable in the form $\varphi f = fx$. In special case when the space X itself represents a space of linear functionals on some linear space Y we may provide the equality $\Phi = X$ assuming $F = Y$ (i.e. assuming as F the set of all linear functionals on X of the type $fx = xy$ at some $y \in Y$).

E x a m p l e 3.6.1. Let us consider the space C of continuous real functions of the finite-dimensional vector t, $t \in T_1$. We put $X = C$. The set F we determine as the set of all linear combinations of the values of the function from C in the finite set of the points:

$$fx = \sum_{\nu=1}^{n} f_\nu x(t_\nu). \tag{I}$$

It is evident that the value φf of any linear functional on F at the point f is a linear function of the correspondent variables $f_1, \ldots, f_n$. Consequently, the general form of a linear functional on F is determined by the following formula:

$$\varphi f = \sum_{\nu=1}^{n} \varphi_\nu^n(t_1, \ldots, t_n) f_\nu. \tag{II}$$

We put $f^1 x = \sum_{\nu=1}^{n-1} f_\nu x(t_\nu)$, $f^2 x = f_n x(t_n)$. According to (3.6.11) the values of the functional φ at the points f^1 and f^2 are equal

$$\varphi f^1 = \sum_{\nu=1}^{n-1} \varphi_\nu^{n-1}(t_1, \ldots, t_{n-1}) f_\nu, \quad \varphi f^2 = \varphi_1^1(t_n) f_n. \tag{III}$$

As the sum $f^1 + f^2$ represents the functional f determined by formula (I) then in accordance with (II) $\varphi(f^1 + f^2) = \sum_{\nu=1}^{n} \varphi_\nu^n(t_1, \ldots, t_n) f_\nu$. On the other hand $\varphi(f^1 + f^2) = \varphi f^1 + \varphi f^2$ by virtue of the linearity of the functional φ. Thus we obtain the following recurrent relations:

$$\varphi_n^n(t_1, \ldots, t_n) = \varphi_1^1(t_n),$$

$$\varphi_\nu^n(t_1, \ldots, t_n) = \varphi_\nu^{n-1}(t_1, \ldots, t_{n-1}) \quad (\nu = 1, \ldots, n-1; \ n = 2, 3, \ldots). \tag{IV}$$

Setting in the second relation sequentially $\nu = n - 1, n - 2, \ldots, 1$ and accounting the first one we find $\varphi_\nu^n(t_1, \ldots, t_n) = \varphi(t_\nu)$ $(\nu = 1, \ldots, n, \ n = 1, 2, \ldots)$, where $\varphi(t) = \varphi_1^1(t)$. Thus any linear functional on F is determined by the formula $\varphi f = \sum_{\nu=1}^{n} f_\nu \varphi(t_\nu)$, where $\varphi(t)$ is any function. The space Φ in a given case may be identified with the space R^{T_1} of all the functions of the variable t. The space C of the continuous functions represents a subspace of the space $\Phi = R^{T_1}$. Therefore not every continuous in the weak topology of the space F positively determined functional $g(f)$ on F is a characteristic functional of some random function with the continuous realizations.

E x a m p l e 3.6.2. The results of Example 3.6.1 show that if we take as the initial space X the space of all functions of the variable t while leaving the space F invariable then any linear functional on F will take the form $\varphi f = \sum_{\nu=1}^{n} f_\nu x(t_\nu)$, where $x(t_\nu)$ is some function which belongs to the space X. Consequently, in this case $\Phi = X$ a positively determined functional $g(f)$ on f serves as a characteristic functional of some random function $X(t)$.

E x a m p l e 3.6.3. Consider a case when X represents the B-space L_p of the scalar functions, $1 < p < \infty$. We assume as F the space L_q, $q = p/(p-1)$ of continuous linear functionals on L_p dual with L_p. The space L_p is reflexive in such a sense that the set of all continuous on L_q linear functionals coincides with L_p. But the space of all (not only continuous) linear functionals L_q does not coincide with L_p. Therefore not every continuous positively determined functional $g(f)$ on L_q determines the probability measure on L_p.

3.6.5. Probability Distributions in Banach Spaces. Mourier Theorem

Now let us consider the case when X represents the B-space. It is natural to assume as the set F of the linear functionals on X topologically dual with X. In the general case the space Φ of all linear functionals on F_c does not coincide with X. According to Theorem 3.6.6 not every continuous positively determined functional $g(f)$ on F_c determines the probability measure on X. In order the functional would determine the probability measure in X it must satisfy some additional conditions.

Notice that at any fixed $f_1, \ldots, f_n \in F$ the function

$$g_{f_1,\ldots,f_n}(\alpha_1, \ldots, \alpha_n) = g\left(\sum_{k=1}^n \alpha_k f_k\right) \qquad (3.6.12)$$

of the real variables $\alpha_1, \ldots, \alpha_n$ determines not only the n-dimensional distribution $\mu_{f_1,\ldots,f_n}$ of the random function $Z(f)$ (Subsection 3.6.4) but also the distribution

$$\mu^x_{f_1,\ldots,f_n}(\{x : (f_1 x, \ldots, f_n x) \in B\})$$

$$= \mu_{f_1,\ldots,f_n}(\{z(f) : (z(f_1), \ldots, z(f_n)) \in B\}) \qquad (3.6.13)$$

on the σ-algebra $\mathcal{A}_{f_1,\ldots,f_n}$ of the sets$\{x : (f_1 x, \ldots, f_n x) \in B\}$ of the space X correspondent to all the Borel sets B of the n-dimensional space R^n. The aggregate of all such sets correspondent to all finite sets of the vectors $f_1, \ldots, f_n \in F$ as it is easy to see forms the algebra $\mathcal{A}_0$ of the sets of the space X. Formula (3.6.13) at all n, $f_1, \ldots, f_n \in F$ determines on this algebra $\mathcal{A}_0$ the nonnegative additive function of the set μ_x. If the function μ_x is σ-additive on $\mathcal{A}_0$ then according to the theorem about the extension of the measure we may extend it till the measure on the σ-algebra $\mathcal{A}$ generated by the class of all sets $\{x : (f_1 x, \ldots, f_n x) \in B\}$. But from the theorem of Subsection 3.6.4 we may conclude that the function μ_x in the general case will be not σ-additive on $\mathcal{A}_0$. It will be σ-additive only in that case when the distribution μ in the space $(\Phi, \mathcal{F})$ is completely concentrated on the subspace $(X, \mathcal{A})$, $\mathcal{A} = X\mathcal{F}$.

The arguments given above show that for finding the conditions at which a continuous positively determined functional $g(f)$, $f \in F$

determines the probability measure on X it is sufficient to know at what conditons the function of the set μ_x is the σ-additive on the algebra $\mathcal{A}_0$. For short notation we determine the sets $A_N^p = \{x : |f_p x| \leq N\}$ and put

$$B_N^n = \bigcap_{p=1}^{n} A_N^p = \{x : \sup_{p \leq n} |f_p x| \leq N\} \quad (N, n = 1, 2, \ldots). \quad (3.6.14)$$

All these sets belong to the algebra $\mathcal{A}_0$ and the function μ_x is determined on it.

Let us prove the following statement.

Theorem 3.6.7 (The Mourier theorem). *In the case when X is the B-space and dual with X the B-space F_c is separable (in this case X is also separable) for the σ-additivity of the function μ_x on $\mathcal{A}_0$ it is necessary and sufficient the existence of such a countable set $\{f_n\}$ which is dense on the unit sphere of the space F_c that at any $\varepsilon > 0$, at all natural n and all sufficiently large natural N the condition will be fulfilled*

$$\mu_x(\{x : \sup_{p \leq n} |f_p x| > N\}) < \varepsilon. \quad (3.6.15)$$

$\triangleright$ For proving the necessity of condition (3.6.15) we suppose that μ_x is the measure on $\mathcal{A}_0$. Then we may extend it uniquely till the measure on the σ-algebra $\mathcal{A}$ generated by a weak topology of the space X. But in the separable B-space σ-algebra generated by a weak topology coincides with the σ-algebra of Borel sets generated by a strong (i.e. natural metric) topology. Consequently, the sets $B_N = \{x : \|x\| \leq N\}$ belong to the σ-algebra $\mathcal{A}$ and the function μ_x is determined for them. As $\{\bar{B}_N\}$ is a decreasing sequence of the sets and $\lim \bar{B}_N = \bigcap \bar{B}_N = \emptyset$ then by virtue of the continuity of the function μ_x we have $\lim \mu_x(\bar{B}_N) = 0$. Hence at any $\varepsilon > 0$ there exists such N_ε that $\mu_x(\bar{B}_N) < \varepsilon$ at all $N \geq N_\varepsilon$. But by virtue of the fact that the set $\{f_n\}$ is dense on the unit sphere of the space F_c, $\sup_p |f_p x| = \|x\|$. Consequently, $B_N = \{x : \|x\| \leq N\} = \lim_{n \to \infty} \{x : \sup_{p \leq n} |f_p x| \leq N\} = \lim_{n \to \infty} B_N^n$. Thus at any N the sets B_N^n form a decreasing sequence convergent to B_N, and the additional sets $\bar{B}_N^n$ form an increasing sequence convergent to $\bar{B}_N$. Therefore from $\mu_x(\bar{B}_N) < \varepsilon$ follows $\mu_x(\bar{B}_N^n) < \varepsilon$ at all n and at all $N \geq N_\varepsilon$.

For proving the sufficiency of condition (3.6.15) we show that at its fulfillment the function μ_x determined on the algebra $\mathcal{A}_0$ is continuous in zero, i.e. $\lim \mu_x(A_n) = 0$ for any decreasing sequence of the sets $\{A_n\} \subset \mathcal{A}_0$ with empty intersection $\bigcap A_n = \emptyset$. From here will follows the σ-additivity of μ_x on $\mathcal{A}_0$.

In this case it is sufficient to prove our statement for the sets $A_n \in \mathcal{A}_0$ closed in the weak topology of the space F_c (weakly closed). Really, any Borel set B in the finite-dimensional space may be represented as the limit of a monotonous increasing sequence of closed sets $\{F_n\}$, $F_n \subset B$. Here the correspondent set $A = \{x : (g_1 x, \ldots, g_m x) \in B\} \in \mathcal{A}_0$ will be the limit of monotonously increasing sequence of weakly closed sets $C_n = \{x : (g_1 x, \ldots, g_m x) \in F_n\} \in \mathcal{A}_0$. As the function μ_x is σ-additive, consequently, is also continuous on the σ-algebra $\mathcal{A}_{g_1,\ldots,g_m}$ of all the sets $\{x : (g_1 x, \ldots, g_m x) \in B\}$ at fixed $g_1, \ldots, g_m$ then $\mu_x(A) = \lim \mu_x(C_n)$. Hence it follows that for any decreasing sequence of the sets $\{A_n\} \subset \mathcal{A}_0$ and any sequence $\{\varepsilon_n\}$, $\varepsilon_n > 0$, $\varepsilon_n \to 0$ at $n \to \infty$ we may find such decreasing sequence of weakly closed sets $\{C_n\} \subset \mathcal{A}_0$, $C_n \subset A_n$ that $0 \leq \mu_x(A_n) - \mu_x(C_n) < \varepsilon_n$. Now we suppose that $\bigcap A_n = \emptyset$. Then it will be $\bigcap C_n = \emptyset$ and if $\lim \mu_x(C_n) = 0$ then at any $\varepsilon > 0$ there exists such n_ε that $\mu_x(C_n) < \varepsilon/2$ at all $n \geq n_\varepsilon$. After choosing n'_ε that $\varepsilon_n < \varepsilon/2$ at all $n \geq n'_\varepsilon$ we get $\mu_x(A_n) < \mu_x(C_n) + \varepsilon_n < \varepsilon$ at all $n \geq n_\varepsilon, n'_\varepsilon$. Hence it will follow that $\lim \mu_x(A_n) = 0$. Thus it is sufficient to prove the continuity of μ_x in zero for the sequences of weakly closed sets $\{C_n\}$.

Let $\{C_n\}$ be an arbitrary decreasing sequence of weakly closed sets, $C_n \in \mathcal{A}_0$, with empty intersection $C_{n+1} \subset C_n$, $\bigcap C_n = \emptyset$. Prove that $\mu_x(C_n) \to 0$ at $n \to \infty$. For this purpose we suppose that $\mu_x(C_n) > \varepsilon$ at all n for some $\varepsilon > 0$ and show that in this case the intersection of the sets C_n cannot be empty. At first we show that at any sufficiently large N the intersection of any set C_n with the set $B_N = \{x : \|x\| \leq N\}$ cannot be empty.

As $C_n \in \mathcal{A}_0$ then $C_n = \{x : (g_1 x, \ldots, g_m x) \in F_n\}$ at some m, $g_1, \ldots, g_m \in F_C$ and the closed set F_n of the m-dimensional space R^m. If $C_n B_N = \emptyset$ then

$$r = \inf_{x \in C_n} \|x\| > N. \tag{3.6.16}$$

In fact, in consequence of the fact that the linear functionals $g_1, \ldots, g_m \in F_C$ are continuous in the strong topology of the space X generated by a norm the relations $z_1 = g_1 x, \ldots, z_m = g_m x$ establish a continuous mapping of X on R^m. Therefore the set C_n as the inverse set of the closed set F_n at this mapping is closed in a strong topology of the space X. Consequently, in C_n there exists the point x_0 in which the continuous nonnegative function $\varphi(x) = \|x\|$ achieves its accurate lower bound r and if $r \leq N$ then the intersection $C_n B_N$ cannot be empty.

Now we consider the function $\varphi(z_1, \ldots, z_m)$ which is $\inf \|x\|$ on the set $\{x : g_1 x = z_1, \ldots, g_m x = z_m\}$:

$$\varphi(z_1, \ldots, z_m) = \inf_{g_1 x = z_1, \ldots, g_m x = z_m} \|x\|. \tag{3.6.17}$$

As this function is continuous $\varphi(\alpha z_1, \ldots, \alpha z_m) = \alpha \varphi(z_1, \ldots, z_m)$ at any $\alpha > 0$ and $\varphi(0, \ldots, 0) = 0$ then the set $\{z_1, \ldots, z_m : \varphi(z_1, \ldots, z_m) < c\}$ represents a simply connected region of the space R^m which contains the origin. Therefore it follows from (3.6.16) that if $C_n B_N = \varnothing$ then the set F_n does not intersect with the set $\{z_1, \ldots, z_m : \varphi(z_1, \ldots, z_m) < r\}$ (i.e. is disposed outside of this set). The set $D_C = \{z_1, \ldots, z_m : \varphi(z_1, \ldots, z_m) < c\}$ at any $c > 0$ is convex. Really, let $\{z_1', \ldots, z_m'\} \in D_c$, $\{z_1'', \ldots, z_m''\} \in D_c$. From definition (3.6.17) of the function $\varphi(z_1, \ldots, z_m)$ follows that at any $\varepsilon > 0$ there exist such $x', x'' \in X$ that $g_k x' = z_k'$, $g_k x'' = z_k''$ $(k = 1, \ldots, m)$ and $\|x'\| < \varphi(z_1', \ldots, z_m') + \varepsilon$, $\|x''\| < \varphi(z_1'', \ldots, z_m'') + \varepsilon$. Taking $\varepsilon < c - \max\{\varphi(z_1', \ldots, z_m'), \varphi(z_1'', \ldots, z_m'')\}$ we have $\|x'\|, \|x''\| < c$. Putting $x = \lambda x' + (1 - \lambda)x''$, $0 < \lambda < 1$ and $z_k = \lambda z_k' + (1 - \lambda)z_k''$ $(k = 1, \ldots, m)$, we get $g_k x = z_k$ $(k = 1, \ldots, m)$ and $\|x\| \leq \lambda\|x'\| + (1 - \lambda)\|x''\| < c$. Consequently, $\varphi(z_1, \ldots, z_m) < c$, i.e. $\{z_1, \ldots, z_m\} \in D_c$ what proves the convexity of the set D_c.

It follows from the proved above that the set $\{z_1, \ldots, z_m : N < \varphi(z_1, \ldots, z_m) < r\}$ represents the difference of two convex simply connected regions which contain the origin. Therefore in R^m there exists the convex polyhedron G with finite number of s bounds which contains the set $D_N^+ = \{z_1, \ldots, z_m : \varphi(z_1, \ldots, z_m) \leq N\}$ and completely disposed inside the set $D_r = \{z_1, \ldots, z_m : \varphi(z_1, \ldots, z_m) < r\}$. Here none of the bounds of this polyhedron by virtue of its convexity intersects with the convex set D_N^+. It is evident that the set F_n is completely disposed out of the polyhedron G. Let the equations of its bounds have the form $\sum_{k=1}^{m} a_{1k} z_k = b_l$, $b_l > 0$, $(l = 1, \ldots, s)$. Then for all the points disposed inside the polyhedron G, $\sum_{k=1}^{m} a_{lk} z_k < b_l$ $(l = 1, \ldots, s)$, and for all the points of the set F_n we have $\sum_{k=1}^{m} a_{lk} > b_l$ at least for someone l, $1 \leq l \leq s$. Passing to the inverse sets in the space X under mapping $z_1 = g_1 x, \ldots, z_m = g_m x$ we conclude that the set $C_n = \{x : (g_1 x_1, \ldots, g_m x) \in F_n\}$ belongs to the union of the sets $\{x : \sum a_{lk} g_k x > b_l\}$ $(l = 1, \ldots, s)$: $C_n \subset \bigcup_{l=1}^{s} \{x : \sum_{k=1}^{m} a_{lk} g_k x > b_l\}$, and none of them intersects with the sphere $B_N = \{x : \|x\| \leq N\}$. We put for brevity $h_l = \sum_{k=1}^{m} a_{lk} g_k$ $(l = 1, \ldots, s)$ then we shall have

$$C_n \subset \bigcup_{l=1}^{s} \{x : h_l x > b_l\}. \qquad (3.6.18)$$

Now we notice that if the set $\{x : h_l x > b_l\}$ does not intersect with the sphere B_N then it follows $b_l > N\|h_l\|$. Really, by virtue of the completeness of the B-space X there exists the vector x_0 in X for which $\|x_0\| = 1$, $h_l x_0 = \|h_l\|$. Putting at an arbitrary $\delta > 0$ $x_\delta = (b_l + \delta)x_0/\|h_l\|$, we shall have $\|x_\delta\| = (b_l + \delta)/\|h_l\|$ and $h_l x_\delta = b_l + \delta$. Consequently, the point x_δ belongs to the set $\{x : h_l x > b_l\}$ and if $b_l \leq N\|h_l\|$ then by virtue of the arbitrariness $\delta > 0$ $\|x_\delta\| \leq N$. But if $b_l > N\|h_l\|$ then $N\|h_l\| < b_l < h_l x \leq \|h_l\| \|x\|$, for any $x \in \{x : h_l x > b_l\}$

and consequently, $\|x\| > N$. Introducing the linear functionals $h'_l = h_l / \|h_l\|$, belonging to the unit sphere of the space F_c we obtain from (3.6.18) and from the fact that $b_l > N\|h_l\|$ the following relations:

$$C_n \subset \bigcup_{l=1}^{s} \{x : h'_l x > b_l/\|h_l\|\} \subset \bigcup_{l=1}^{s} \{x : |h'_l x| > N\}, \qquad (3.6.19)$$

$$\mu_x(C_n) \le \mu_x \left(\bigcup_{l=1}^{s} \{x : |h'_l x| > N\} \right). \qquad (3.6.20)$$

As according to the condition the set $\{f_p\}$ is dense on the unit sphere of the space F_c then for each h'_l there exists convergent to it the sequence $\{f_{lr}\} \subset \{f_p\}$, $f_{lr} = f_{p_{lr}} \to h'_l$ at $r \to \infty$ $(l = 1, \ldots, s)$. Here by virtue of the continuity of the functional $g(f)$ at $r \to \infty$ we have

$$g_{f_{1r},\ldots,f_{sr}}(\alpha_1, \ldots, \alpha_s) = g\left(\sum_{l=1}^{s} \alpha_l f_{lr} \right)$$

$$\to g\left(\sum_{l=1}^{s} \alpha_l h'_l \right) = g_{h'_1,\ldots,h'_s}(\alpha_1, \ldots, \alpha_s). \qquad (3.6.21)$$

But $g_{f_1,\ldots,f_s}(\alpha_1, \ldots, \alpha_s)$ at any $f_1, \ldots, f_s \in F_c$ represents a characteristic function of some s-dimensional random vector. Therefore by virtue of the inverse Levi theorem it follows from (3.6.21) that the sequence of the probability measures correspondent to the characteristic functions $g_{f_{1r},\ldots,f_{sr}}(\alpha_1, \ldots, \alpha_s)$ converges to the probability measure correspondent to the characteristic function $g_{h'_1,\ldots,h'_s}(\alpha_1, \ldots, \alpha_s)$. Consequently, at any $N > 0$

$$\mu_x \left(\bigcup_{l=1}^{s} \{x : |h'_l x| > N\} \right) = \lim_{r \to \infty} \mu_x \left(\bigcup_{l=1}^{s} \{x : |f_{lr} x| > N\} \right). \qquad (3.6.22)$$

At any r we have $\bigcup_{l=1}^{s} \{x : |f_{lr} x| > N\} \subset \overline{B_N^{n_r}}$, where n_r is the largest of the numbers of the functionals $f_{1r} = f_{p_{1r}}, \ldots, f_{sr} = f_{p_{sr}}$, $n_r = \max\{p_{1r}, \ldots, p_{sr}\}$. And as $\mu_x(\overline{B_N^n}) < \varepsilon$ at all n thenwe get from (3.6.22)

$$\mu_x \left(\bigcup_{l=1}^{s} \{x : |f_{lr} x| > N\} \right) \le \mu(\overline{B_N^{n_r}}) < \varepsilon. \qquad (3.6.23)$$

Accounting (3.6.23) from (3.6.20) and (3.6.22) we get $\mu_x(C_n) < \varepsilon$ what contradicts the supposition that $\mu_x(C_n) > \varepsilon$. Thus the intersection $B_N C_n$ cannot be empty if there exists such $\varepsilon > 0$ that $\mu_x(C_n) > \varepsilon$ at all n. But $\{B_N C_n\}$ is a centered system of closed subsets of the set B_N in a weak topology of the space X. And the set B_N is bounded and consequently is compact in a weak topology. So the intersection $\bigcap\limits_n B_N C_n$ cannot be empty, and it means that the intersection $\bigcap\limits_n C_n$ also cannot be empty. Thus we proved that if there exists such $\varepsilon > 0$ that $\mu_x(C_n) > \varepsilon$ at all n then $\bigcap\limits_n C_n \neq \varnothing$. Hence it follows that $\lim \mu_x(C_n) = 0$ if $\bigcap\limits_n C_n = \varnothing$ what proves the sufficiency of condition (3.6.15) for the continuity of μ_x in zero on $\mathcal{A}_0$, and consequently, for the σ-additivity of μ_x on $\mathcal{A}_0$. Thus the sufficiency (3.6.15) is proved. $\triangleleft$

3.7. Normal Distributions in Linear Spaces

3.7.1. Properties of Normal Distribution

Let X be a real linear space with a weak topology of Section 3.6 generated by a set of linear functionals F and with the σ- algebra of the sets $\mathcal{A}$ generated by a class of the sets $\{x : (f_1 x, \ldots, f_n x) \in B\}$ at all finite sets $f_1, \ldots, f_n \in F$ and all Borel sets B of the correspondent finite-dimensional spaces. The distribution of the random variable X in $(X, \mathcal{A})$ is called *a normal* if a characteristic functional of this random variable has the form

$$g_x(f) = \exp\left\{ ifm_x - \frac{1}{2} f K_x f \right\}, \quad f \in F, \qquad (3.7.1)$$

where m_x and K_x are the expectation and the covariance operator of the random variable X. We prove the following statements.

Theorem 3.7.1. *If the random variable X is normally distributed then the random vector Z with the components $Z_1 = f_1 X, \ldots, Z_n = f_n X$ at any finite set of the functionals $f_1, \ldots, f_n \in F$ is normally distributed.*

Theorem 3.7.2. *If at any $f \in F$ the random variable $Z = fX$ is normally distributed then the random variable X is normally distributed.*

$\triangleright$ Really, on the basis of formula (3.5.9) a characteristic function of the random vector Z is determined by the formula

$$g_z(\alpha_1, \ldots, \alpha_n) = g_x \left(\sum_{p=1}^{n} \alpha_p f_p \right)$$

$$= \exp \left\{ i \sum_{p=1}^{n} \alpha_p f_p m_x - \frac{1}{2} \sum_{p,q=1}^{n} \alpha_p \alpha_q f_p K_x f_q \right\}.$$

But $f_p m_x$ represents an expectation of the random variable $Z_p = f_p X$, consequently, $g_z(\alpha_1, \ldots, \alpha_n)$ is a characteristic function of a normally distributed random vector, what proves Theorem 3.7.1.

Further, a characteristic function of the variable Z in Theorem 3.7.2 is determined by the formula

$$g_z(\alpha) = \exp \left\{ i\alpha m_z - \frac{1}{2}\alpha^2 D_z \right\} = \exp \left\{ i\alpha m_z - \frac{1}{2}\alpha^2 f K_x f \right\}.$$

On the other hand, $g_z(\alpha) = E e^{i\alpha Z} = E e^{i\alpha f X}$. Hence it is clear that a characteristic functional of the random variable X in a given point $f \in F$ is equal

$$g_x(f) = E e^{i f X} = g_z(1) = \exp \left\{ i f m_x - \frac{1}{2} f K_x f \right\}.$$

As it is valid at any $f \in F$ then $g_x(f)$ at any $f \in F$ is expressed by formula (3.7.1). ◁

Let us introduce in the space F a weak topology (Section 3.5) determining it by the vicinities of zero $\{f : |f x_1| < \varepsilon_1, \ldots, |f x_n| < \varepsilon_n\}$ at all n, $x_1, \ldots, x_n \in X$, $\varepsilon, \ldots, \varepsilon_n > 0$. As the characteristic functional is continuous in this topology in the sense that for any convergent sequence $\{f_n\}$, $f_n \to f$ (i.e. $f_n x \to f x$ at any $x \in X$), $g_x(f_n) \to g_x(f)$ then form (3.7.1) follows that a covariance operator of a normally distributed random variable X satisfies the condition

$$f_n K_x f_n \to f K_x f \tag{3.7.2}$$

for any sequence $\{f_n\} \subset F$ convergent to f. But on the basis of Theorem 3.6.6 it is not sufficient. In order the functional (3.7.1) will be a characteristic functional of some random variable in $(X, \mathcal{A})$ in the general case the covariance operator K_x must satisfy some additional conditions. These conditions require that the probability measure on the space Φ of all linear functional on F determined by the characteristic functional $g_x(f)$ will be equal to 1 on any measurable set which contains the space X. Let us prove the following statement.

Theorem 3.7.3. *As the result of any measurable continuous (in weak topologies) linear transformations of normally distributed random variables we always obtain normally distributed random variables.*

▷ Let Y be another real linear space with a weak topology determined by the linear set of nonlinear functionals G and with the σ-algebra of the sets $\mathcal{B}$ generated by the class of the sets of the type $\{y : (g_1 y, \ldots, g_n y) \in B\}$ where B are Borel sets of the correspondent space. As usual we suppose that the set of nonlinear functionals is sufficiently complete for the case when from $gy = 0$ at all $g \in G$ follows $y = 0$. We introduce a weak topology determined by the vicinities of zero $\{g : |gy_1| < \varepsilon_1, \ldots, |gy_n| < \varepsilon_n\}$ at all n, $y_1, \ldots, y_n \in Y$, $\varepsilon_1, \ldots, \varepsilon_n > 0$ in the space G similarly as in F. Let T be the continuous $(\mathcal{A}, \mathcal{B})$-measurable linear operator mapping X into Y, b be an arbitrary vector of the space Y. The variable $Y = TX + b$ represents a random variable in $(Y, \mathcal{B})$. On the basis of formulae (3.5.7) and (3.5.8) which determine the transformation of a characteristic functional at linear transformation of a random variable and a shift, a characteristic functional of the variable Y is determined by the formula

$$g_y(g) = e^{igb} \exp\left\{ i(T^*g)m_x - \frac{1}{2}(T^*g)K_x(T^*g) \right\},$$

where T^* is a linear operator adjoint with T mapping G into F. But $(T^*g)x = g(Tx)$ for any $x \in X$ and $g \in G$ by the definition of the adjoint operator. By virtue of the continuity of the operator T the domain of the adjoint operator T^* coincides with G. Therefore with the account of formula (3.1.11) and (3.3.29) for the transformations of an expectation and a covariance operator at linear transformation we have the following relations:

$$(T^*g)m_x = g(Tm_x) = g(m_y - b), \;\; (T^*g)K_x(T^*g) = g(TK_xT^*)g = gK_yg,$$

$$g_y(g) = \exp\left\{ igm_y - \frac{1}{2}gK_yg \right\}.$$

Hence it is clear that the random variable $Y = TX + b$ is normally distributed. ◁

3.7.2. Normal Distribution on Product of Two Spaces

Theorem 3.7.4. *If the random variable $Z = (Y, X)$ in the product of the real linear spaces $X \times Y$ is normally distributed then each of the random variables X and Y is normally distributed. Here the random variables X and Y are independent if and only if they are uncorrelated.*

▷ Let us study a normal distribution on the product of two real linear spaces X and Y with weak topologies determined by the sets of the linear functionals with the σ-algebras $\mathcal{A}$ and $\mathcal{B}$ generated by the classes of the sets $\{x : (f_1 x, \ldots, f_n x) \in A\}$ and $\{y : (g_1 y, \ldots, g_n y) \in B\}$ at all n, $f_1, \ldots, f_n \in F$, $g_1, \ldots, g_n \in G$ and at all Borel sets A, B of the correspondent finite-dimensional spaces.

We determine a set of the linear functionals H on the direct product of the spaces $Z = X \times Y$ assuming as H the set of the linear functionals of the form $hz = fx + gy$ at all $f \in F$, $g \in G$. Here each linear functional $h \in H$ may be considered as the pair (f, g), $f \in F$, $g \in G$. Consequently, the space H represents a direct product of the spaces F and G, $H = F \times G$. So determined set H of linear functionals on Z is complete: from $hz = 0$ at all $h \in H$ follows $z = 0$. By means of the set H we determine similarly as for the spaces X and Y the weak topologies in Z and H and the σ-algebra $\mathcal{C}$ in Z. We may prove that the σ-algebra $\mathcal{C}$ contains the product of the σ-algebras $\mathcal{A}$ and $\mathcal{B}$, $\mathcal{A} \times \mathcal{B} \subset \mathcal{C}$.

Let us consider a normally distributed random variable Z in the space $(Z, \mathcal{C})$. According to definition (3.7.1) its characteristic functional

$$g_z(h) = \exp\left\{ ihm_z - \frac{1}{2}hK_zh \right\}, \quad h \in H, \qquad (3.7.3)$$

uniquely determines the probability measure μ_z of the variable Z on the σ-algebra $\mathcal{C}$ (Section 3.5), and consequently, according to the proved earlier also on $\mathcal{A} \times \mathcal{B}$. By Theorem 2.4.1 μ_z determines the distributions μ_x and μ_y on the σ-algebras $\mathcal{A}$ and $\mathcal{B}$ in the spaces X and Y (the projections of the distribution μ_z on the subspaces X and Y of the space Z). Hence the expectation m_z and the covariance operator K_z of the variable Z evidently exist if and only if there exist the expectations m_x, m_y and the covariance operators K_x, K_y of the variables X, Y. As along with them there exist also the cross covariance operators K_{xy}, K_{yx} of the variables X, Y (Section 3.2) and for any $h, h' \in H$, $h = (f, g)$, $h' = (f', g')$ the equalities are fulfilled

$$hm_z = fm_x + gm_y,$$

$$hK_zh' = E(hZ^0)(h'Z^0) = E(fX^0 + gY^0)(f'X^0 + g'Y^0)$$

$$= fK_xf' + fK_{xy}g' + gK_{yx}f' + gK_yg', \quad gK_{yx}f' = f'K_{xy}g,$$

then formula (3.7.3) may be presented in the form

$$g_z(h) = g_z(\{f, g\})$$

$$= \exp\left\{ ifm_x + igm_y - \frac{1}{2}(fK_xf + 2fK_{xy}g + gK_yg) \right\}. \qquad (3.7.4)$$

Hence it is clear that the distributions μ_x and μ_y of the random variables X and Y are normal. $\triangleleft$

3.7.3. Conditional Distributions

Let us pass to the definition of the conditional distributions of the random variables X and Y. Beforehand we prove the following proposal.

Theorem 3.7.5. *For any random variable X in the linear space $(X, \mathcal{A})$ which has the expectation m_x and the covariance operator K_x the distribution μ_x is completely concentrated on the set $m_x + \Delta_{K_x}$, where $\Delta_{K_x} \subset X$ is the domain of the operator K_x.*

$\triangleright$ It is sufficient to show that if $\Delta_{K_x} \neq X$ then $\mu_x(A) = 1$ for any set $A \in \mathcal{A}$, $A \supset m_x + \Delta_{K_x}$. According to Theorem 2.5.1 condition $A \in \mathcal{A}$ imposes the restrictions on no more than the countable set of the variables of the type fx. In other words, for any $A \in \mathcal{A}$ there exists such countable set $S_A = \{f_n\} \subset F$ that only the variables $f_n x$ $(n = 1, 2, \ldots)$ are subject to the restriction. For any $f \bar{\in} S_A$ the variable fx, $x \in A$ may have any values. It follows from $m_x + \Delta_{K_x} \subset A$ that $f_n(x - m_x) = 0$ at all $f_n \in S_A$ for any vector $x \in m_x + \Delta_{K_x}$. Really, suppose that $f_n(x - m_x) = c \neq 0$ for $f_n \in S_A$. As Δ_{K_x} contains along with any vector ξ the vector $\eta_\alpha = \alpha\xi$ at any α then from $f_n \xi = c$ it follows $f_n \eta_\alpha = \alpha c$. Putting $x_\alpha = m_x + \eta_\alpha$ we shall have $x_\alpha \in m_x + \Delta_{K_x}$ and $f_n(x_\alpha - m_x) = \alpha c$ at any α. While changing α the variable αc runs the whole complex plane. And as $m_x + \Delta_{K_x} \subset A$ then $x_\alpha \in A$ at all α and the variables $f_n x_\alpha$ are not subject to some restrictions what contradicts to the supposition that $f_n \in S_A$. Noticing that from $x \in m_x + \Delta_{K_x}$ follows the existence of such $f \in F$ that $x - m_x = K_x f$ we come to the conclusion that $f_n K_x f = 0$ for all $f_n \in S_A$. As $K_x f \in \Delta_{K_x}$ for any f then $f_n K_x f = 0$ for all $f \in F$, $f_n \in S_A$. In particular, $E|f_n X^0|^2 = f_n K_x f_n = 0$. Hence it follows that $f_n X^0 = 0$ with probability 1, i.e. $\mu_x(\{x : f_n x = f_n m_x\}) = 1$ $(n = 1, 2, \ldots)$ or

$$\mu_x(\{x : f_n x \neq f_n m_x\}) = 0 \quad (n = 1, 2, \ldots),$$

$$\mu_x\left(\bigcup_n \{x : f_n x \neq f_n m_x\}\right) \leq \sum_n \mu_x(\{x : f_n x \neq f_n m_x\}) = 0.$$

Passing to the opposite events we find

$$\mu_x\left(\bigcap_n \{x : f_n x = f_n m_x\}\right) = 1.$$

Finally noticing that $\{x : f_n x = f_n m_x\} \subset A$ at any $f_n \in S_A$ by virtue of the fact that the point $x = m_x$ belongs to the set $m_x + \Delta_{K_x} \subset A$, and consequently, the value $f_n m_x$ of the variable $f_n x$ is admissible for the set A we come to the conclusion that

$$\mu_x(A) \geq \mu_x \left(\bigcap_n \{x : f_n x = f_n m_x\} \right) = 1. \triangleleft$$

R e m a r k. If $\Delta_{K_x} \neq X$ (and consequently, $m_x + \Delta_{K_x} \neq X$) then the distribution μ_x of the variable X may be transferred on the space $(X', \mathcal{A}')$, $X' = m_x + \Delta_{K_x}$, $\mathcal{A}' = \mathcal{A}(m_x + \Delta_{K_x})$ as the trace probability (Section 2.10). But the space X' in the general case will not be linear (it is linear only in the case when $m_x \in \Delta_{K_x}$).

Theorem 3.7.6. *For any random variables X and Y in the linear spaces $(X, \mathcal{A})$ and $(Y, \mathcal{B})$ correspondingly equations*

$$x = K_x f, \quad y = K_{yx} f, \quad f \in F, \tag{3.7.5}$$

uniquely determine in the parametric form a continuous measurable linear operator L:

$$L K_x = K_{xy}, \tag{3.7.6}$$

which maps the space Δ_{K_x} into Y, $y = Lx$.

$\triangleright$ Really, for any $f_1, f_2 \in F$, $f_1 \neq f_2$, $K_x f_1 = K_x f_2$ we have $K_x(f_1 - f_2) = EX^0 \overline{(f_1 - f_2)X^0} = 0$ and $(f_1 - f_2)K_x(f_1 - f_2) = E|(f_1 - f_2)X^0|^2 = 0$. Thus $(f_1 - f_2)X^0 = 0$ with probability 1 and $K_{yx}(f_1 - f_2) = EY^0\overline{(f_1 - f_2)X^0} = 0$. Hence we find $K_{yx} f_1 = K_{yx} f_2$ what proves the uniqueness of the mapping $X \to Y$ determined by Eqs. (3.7.5). The linearity of this mapping follows from the adjoint linearity of the operators K_x and K_{yx} (Section 3.2).

Now we show that the operator L determined by Eqs. (3.7.5) is continuous. For this purpose we notice that at any $g \in G$ and any $\varepsilon > 0$ $|gLx - gLx_0| < \varepsilon$ for all $x_0 = K_x f_0$ and $x = K_x f$ for which f belongs to the vicinities $\{f : |f x_1 - f_0 x_1| < \varepsilon_1\}$, $x_1 = K_{xy} g$ of the point f_0. It follows from the fact that by virtue of Eqs. (3.7.5) the following equalities are valid:

$$|gLx - gLx_0| = |gK_{yx} f - gK_{yx} f_0| = |fK_{xy} g - f_0 K_{xy} g|.$$

Futher we notice that the range Δ_{K_x} of the operator K_x may be always assumed as a complete linear space (in the weak topology of the phase space X of the random variable X). In fact by virtue of the supposition about the weak completeness of the space X any fundamental sequence $\{x_n\} \subset \Delta_{K_x}$ has the limit

$x_0 \in X$, $f x_n \to f x_0$ at all $f \in F$. But $x_n = K_x f_n$, and as $f K_x f_n = \overline{f_n K_x f}$ then $\overline{f_n} x \to f x_0$ at any $x \in \Delta_{K_x}$ and $\lim f_n x$, evidently, represents the result of the action of some linear functional f_0 on x. Consequently, the sequence of the functionals $\{f_n\}$ converges to f_0 in such a sense that $f_n x \to f_0 x$ at all $x \in \Delta_{K_x}$. If $f_0 \in F$ then $K_x f_0 \in \Delta_{K_x}$ and as $K_x f_0 = x_0$ by virtue of the continuity of the operator K_x then $x_0 \in \Delta_{K_x}$.

The continuity K_x in the weak topologies of the spaces F and X follows from the fact that by virtue of the equality $f_1 K_x f = \overline{f K_x f_1}$ the inverse set of vicinity $\{x' : |f_1 x' - f_1 x| < \varepsilon\}$ of the point $x = K_x f$ at any $\varepsilon > 0$ and $f, f_1 \in F$ contains the vicinity $\{f' : |f' x_1 - f x_1| < \varepsilon\}$, $x_1 = K_x f_1$ of the point f.

If $f_0 \bar{\in} F$ for some fundamental sequences $\{x_n\} \subset \Delta_{K_x}$ then extending the space F on account of the inclusion into it all such f_0 and extending the operator K_x on the continuity we get $x_0 \in \Delta_{K_x}$ for all fundamental sequences $\{x_n\} \subset \Delta_{K_x}$, $x_0 = \lim x_n$. Thus Δ_{K_x} may be always considered as a closed set. But then $K_{xy} g \in \Delta_{K_x}$ at any $g \in G$. Otherwise for the functional $g \in G$ for which $K_{xy} g$ does not belong to Δ_{K_x} we may find such linear functional $f \in F$ that $f x = 0$ at any $x \in \Delta_{K_x}$ and $f K_{xy} g \neq 0$. But it is impossible as

$$|f K_{xy} g|^2 = |E(f X^0)(\overline{g Y^0})|^2 \le E|f X^0|^2 E|g Y^0|^2 = f K_x f \cdot g K_y g$$

and from $f x = 0$ at $x \in \Delta_{K_x}$ will follow $f K f = 0$ and $K_{xy} g = 0$. The obtained contradiction proves that $x_1 = K_{xy} g \in \Delta_{K_x}$ at any $g \in G$. But then there exists such linear functional $f_1 \in F$ that $x_1 = K_x f_1$ and

$$|f x_1 - f_0 x_1| = |f K_x f_1 - f_0 K_x f_1| = |f_1 K_x f - f_1 K_x f_0| = |f_1 x - f_1 x_0|.$$

Consequently, $|g L x - g L x_0| < \varepsilon$ at any $\varepsilon > 0$, $g \in G$, $x_0 \in \Delta_{K_x}$ and all x in the vicinity $\{x : x \in \Delta_{K_x}, |f_1 x - f_1 x_0| < \varepsilon\}$ of the point x_0 where $f_1 \in F$ is determined by the equation $K_x f_1 = K_{xy} g$. This fact proves the continuity of the operator L.

At last we prove the measurability of the operator L determined by Eqs. (3.7.5). For this purpose it is sufficient to notice that as the inverse set $\{y : (g_1 y, \ldots, g_n y) \in B\} \in \mathcal{B}$ for any $g_1, \ldots, g_n \in G$ and for any Borel set B of the space R^n serves the set $\{x : (g_1 L x, \ldots, g_n L x) \in B\}$. This set belongs to the σ-algebra $\mathcal{A}$ by virtue of the fact that the operator L is continuous. Consequently, $g_p L x$ represents the result of the action of a continuous linear functional $f_p = L^* g_p \in F$ on x ($p = 1, \ldots, n$). Remind that in a weak topology only those linear functionals are continuous by means of which this weak topology is determined. Therefore from the continuity $f_p = L^* g_p$ in the weak topology of the space X it follows that $f_p \in F$. So it follows directly from Eqs. (3.7.5) that the operator L satisfies Eq. (3.7.6). ◁

Now we may pass directly to calculating conditional distributions of the random variables X and Y in the case of a normal distribution of the variable $Z = (X, Y)$.

Let us prove the following statement.

Theorem 3.7.7. *At a normal distribution of the random variable $Z = (X, Y)$ there exist the regular conditional distributions of the random variables X, Y and these distributions are normal.*

$\triangleright$ Let us consider the linear mapping S

$$u = (x, y) \to v = Su = (x, y'), \ \ y' = y - Lx$$

of the space $Z = X \times Y$ into itself. This mapping evidently is continuous and measurable as the operator L according to the proved is continuous and for any $h = (f, g) \in H$

$$hv = fx + gy' = fx + gy - gLx = (f - L^*g)x + gy = h'u,$$

where $h' = (f - L^*g, g) \in H$. As a result of this the set $\{v : (h_1 v, \ldots, h_n v) \in B\} \in C$ has the inverse set $\{u : (h'_1 u, \ldots, h'_n u) \in B\} \in C$. Therefore the random variable

$$Z' = (m_x, 0) + SZ^0 = (X, Y'), \ \ \ Y' = Y^0 - LX^0$$

in the space $Z = X \times Y$ is normally distributed. We find a cross covariance operator $K_{y'x}$ of the variables Y' and X. Putting $W = LX^0$ we find

$$K_{y'x}f = EY'(fX^0) = EY^0(fX^0) - E(LX^0)(fX^0) = K_{yx} - K_{wx}.$$

But according to (3.3.29) $K_{wx} = LK_x$ therefore $K_{y'x} = K_{yx} - LK_x$. Hence it follows from Eq. (3.7.6) that $K_{y'x} = 0$. Thus the random variables Y' and X are uncorrelated, and consequently, by virtue of the normality of the distribution of the variable Z' they are also independent. Therefore the probability measure of the random variable $Z' = (X, Y')$ is determined by the formula (Section 2.4)

$$\mu_{z'}(C) = \int \mu_x(dx) \int \mathbf{1}_C(x, y)\mu_{y'}(dy), \ \ \ C \in C, \tag{3.7.7}$$

where μ_x and $\mu_{y'}$ are the probability measures of the random variables X and Y'. Now we notice that the mapping S is convertible as to a given value $v = \{x, y'\}$ corresponds the unique value $u = S^{-1}v = \{x, y\}, \ y = y' + Lx$. Therefore on the basis of formula (2.2.4) from (3.7.7) it follows the expression for the values of the probability measure of the random variable Z on the measurable rectangles $A \times B$, $A \in \mathcal{A}$, $B \in \mathcal{B}$:

$$\mu_z(A \times B) = \mu_z(\{(x, y) : x \in A, y \in B\})$$

$$= \mu_{z'}(\{(x, y') : x \in A, y' \in B - m_y - L(x - m_x)\})$$

$$= \int_A \mu_{y'}(B - m_y - L(x - m_x))\mu_x(dx).$$

Hence we conclude that there exists a conditional distribution of the random variables Y relatively to X connected with probability measure of the random variable Y' by formula

$$\mu_y(B|x) = \mu_{y'}(B - m_y - L(x - m_x)), \quad B \in \mathcal{B}. \tag{3.7.8}$$

This formula shows that $\mu_y(B|x)$ at any fixed value x of the random variable X represents a measure determined on the σ-algebra $\mathcal{B}$. Consequently, a conditional distribution of the random variable Y relatively to X is regular (Section 2.3). Finally, it follows from (3.7.8) that a conditional distribution of the random variable Y relatively to X is normal as the random variable Y' is normally distributed. By the symmetry we conclude that in the considered case there exists a regular conditional distribution of the random variable Y relatively to X and this distribution is normal. ◁

Finally, let us determine the conditional expectations $m_{x|y}$, $m_{y|x}$ and the conditional covariance operators $K_{x|y}$, $K_{y|x}$ of the random variables X and Y. As at any fixed value x of the variable X we have $Y = m_y + Y' + L(x - m_x)$, $m_y' = 0$, hence

$$m_{y|x} = m_y + L(x - m_x),$$

and the conditional covariance operator $K_{y|x}$ of the variable Y coincides with the covariance operator $K_{y'}$ of the variable Y'. For calculating $K_{y'}$ we notice that formula $Y = m_y + LX^0 + Y'$ expresses Y as the sum of two uncorrelated random variables $U = LX^0$ and $m_y + Y'$. Therefore on the basis of the properties of the covariance operators (Section 3.3) we have

$$K_y = LK_{xy} + K_{y'} = LK_x L^* + K_{y'} = K_{yx}L^* + K_{y'},$$

$$K_{y'} = K_y - LK_{xy} = K_y - LK_x L^* = K_y - K_{yx}L^*.$$

Thus the conditional expectations and the conditional covariance operators of the random variables X and Y are determined by the following formulae

$$m_{y|x} = m_y + L(x - m_x), \tag{3.7.9}$$

$$K_{y|x} = K_y - L K_{xy} = K_y - L K_x L^* = K_y - K_{yx} L^*, \tag{3.7.10}$$

where L is a linear operator determined by Eq. (3.7.6); L_1 is a linear operator determined by analogous equation

$$L_1 K_y = K_{xy}. \tag{3.7.11}$$

By the symmetry we have

$$m_{x|y} = m_x + L_1(y - m_y),$$

$$K_{x|y} = K_x - L_1 K_{yx} = K_x - L_1 K_y L_1^* = K_x - K_{xy} L_1^*.$$

Knowing the conditional expectations and the correlation operators of the variables X and Y it is not difficult to write the expressions of their conditional characteristic functionals:

$$\left. \begin{aligned} g_{y|x}(g) &= e^{i(L^*g)(x-m_x)} \exp\left\{igm_y - \tfrac{1}{2}gK_{y|x}g\right\}, \\ g_{x|y}(f) &= e^{i(L_1^*f)(y-m_y)} \exp\left\{ifm_x - \tfrac{1}{2}fK_{x|y}f\right\}. \end{aligned} \right\} \tag{3.7.12}$$

We prove now the inverse theorem.

Theorem 3.7.8. *If a distribution of the variable X is normal and there exists a regular normal conditional distribution of the variable Y relatively to X, such that its conditional expectation is determined by the first formula of* (3.7.9) *and Eq.* (3.7.6), *and a conditional covariance operator does not depend on x then a distribution of the random variable $Z = (X, Y)$ in the product of the spaces $(X \times Y, \mathcal{A} \times \mathcal{B})$ (i.e. a joint distribution of the random variables X and Y) is normal.*

▷ For proving we use the formula of total expectation (3.3.11). Then accounting that the second exponential multiplier in the first formula (3.7.12) does not depend on x we obtain

$$g_z(f, g) = E e^{i(fX+gY)} = E\left[e^{ifX} E(e^{igY}|X)\right] = \int e^{ifx} g_{y|x}(g)\mu_x(dx)$$

$$= e^{-i(L^*g)m_x} \exp\left\{igm_y - \frac{1}{2}gK_{y|x}g\right\} \int e^{i(f+L^*g)x}\mu(dx).$$

The latter integral is equal to $g_x(f + L^*g)$. After expressing this integral by formula (2.7.1) (on the basis of the normality of the measure μ_x) we get by virtue of (3.7.10) formula (3.7.4) for $g_z(\{f, g\})$ what proves the normality of the distribution of the variable Z. ◁

Theorems 3.7.7 and 3.7.8 may be summarized in the following form.

Theorem 3.7.9. *If joint distribution of the random variables X and Y is normal then the distributions of the random variables X and Y are normal and there exist their regular normal conditional distributions relatively each other and the regressions of the correspondent centered random variables one on another are linear, and their conditional covariance operators $K_{x|y}$ and $K_{y|x}$ do not depend on the values of y and x correspondingly.*

E x a m p l e 3.7.1. For a finite dimensional normally distributed random vector X formula (3.7.1) takes the form

$$g_x(f) = \exp\left\{ if^T m_x - \frac{1}{2} f^T K_x f \right\}, \qquad (I)$$

where f is a vector of the same dimension as X represented in the form of a matrix-column, and K_x is a covariance matrix of the random vector X. So the probability measure of the n-dimensional normally distributed random variable X is

$$\mu_x(A) = \frac{1}{\sqrt{(2\pi)^n |K_x|}} \int\limits_A \exp\left\{ -\frac{1}{2}(x^T - m_x^T) K_x^{-1}(x - m_x) \right\} dx. \quad (II)$$

This formula is valid when the rank r of the matrix K_x is equal to n. If $r < n$ then X represents a linear function of the r-dimensional random vector U whose probability measure μ_u is expressed by an analogous formula with K_u of the rank r.

E x a m p l e 3.7.2. For the finite-dimensional real normally distributed vector random function $X(t)$, $t \in T_1$ assuming F as an ordinary set of the linear functional of the type

$$fx = \sum_{p=1}^{n} f_p^T x(t_p), \qquad (I)$$

we obtain from (3.7.1) the following expression of a characteristic functional

$$g_x(f) = \exp\left\{ i \sum_{p=1}^{n} f_p^T m_x(t_p) - \frac{1}{2} \sum_{p,q=1}^{n} f_p^T K_x(t_p, t_q) f_q \right\}. \qquad (II)$$

From the property of the normality of the distribution for all linear functionals over a normally distributed random variable follows the normality of the distributions of any linear combinations of the values of a random function. Hence it follows that all multi-dimensional distributions of a normally distributed random functions are normal.

Let $X(t)$, $t \in T_1$ and $Y(s)$, $s \in S$ be two finite-dimensional vector random functions. Eqs. (3.7.5) which determine the linear operator L take the form

$$x(t) = \sum_{p=1}^{n} K_x(t, t_p) \bar{f}_p,$$

$$y(s) = Lx(t) = \sum_{p=1}^{n} K_{yx}(s, t_p) \bar{f}_p = \sum_{p=1}^{n} L_t K_x(t, t_p) \bar{f}_p.$$

Here the index t at the operator L shows that it acts upon $K_x(t, t_p)$ as the function of the variable t at the fixed t_p. As the latter equality is valid for all functionals $f \in F$, i.e. for all n, for all sets $t_1, \ldots, t_n \in T_1$ and for all vectors $f_1, \ldots, f_n$ then the operator L satisfies equation

$$L_t K_x(t, \tau) = K_{yx}(s, \tau), \tag{3.7.13}$$

at all $s \in S$ and $\tau \in T_1$. In the case of the finite-dimensional vector t and an integrable random function $X(t)$ the operator L usually represents a linear integral operator. In the correspondence with this fact Eq. (3.7.13) represents the integral equation

$$\int_{T_1} l(s, t)^T K_x(t, \tau) dt = K_{yx}(s, \tau) \quad (\tau \in T_1, \ s \in S), \tag{3.7.14}$$

where $l(s, t)$ is a kernel of the integral operator L. Formulae (3.7.9) and (3.7.10) for a conditional expectation and a conditional covariance operator of the random function $Y(s)$ relatively to the random function $X(t)$ take in this case the form

$$m_{y|x}(s) = m_y(s) + \int_{T_1} l(s, t)^T m_x(t) dt, \tag{3.7.15}$$

$$K_{y|x}(s_1, s_2) = K_y(s_1, s_2) - \int_{T_1} l(s_1, t)^T K_{xy}(t, s_2) dt$$

$$= K_y(s_1, s_2) - \iint_{T_1 T_1} l(s_1, t_1)^T K_x(t_1, t_2)\overline{l(s_2, t_2)} dt_1 dt_2$$

$$= K_y(s_1, s_2) - \int_{T_1} K_{yx}(s_1, t)\overline{l(s_2, t)} dt. \qquad (3.7.16)$$

3.7.4. Normal Distribution in H-space

Now let us study the normally distributed random variables in a separable H-space. As the set of linear functionals F is natural to assume the set of all continuous linear functionals on X. As any continuous linear functional f on H-space X may be represented in the form of the scalar product $fx = (x, f)$, $f \in X$ then we may consider that $F = X$. Then the covariance operator K_x of the random variable X will be the self-adjoint positive linear operator which maps X into X. Its linearity follows from the fact that $K_x f = EX^0(\overline{fX^0}) = EX^0(f, X^0)$. The positiveness follows from the general property (ii) of the operators of the second order moment (Section 3.3):

$$(K_x f, f) = f K_x f = E(X^0, f)(f, X^0) = E|(X^0, f)|^2 \geq 0.$$

Finally, the self-adjointness follows from the equality

$$(K_x f, g) = E(X^0, g)(f, X^0) = E\overline{(X^0, f)(g, X^0)}$$

$$= \overline{(K_x g, f)} = (f, K_x g).$$

According to definition (3.7.1) the random variable X with the values in the real H-space is normally distributed if its characteristic functional is determined by formula

$$g_x(f) = \exp\left\{ i(m_x, f) - \frac{1}{2}(K_x f, f) \right\}. \qquad (3.7.17)$$

Theorem 3.7.10. *In order formula* (3.7.17) *would determine a characteristic functional of a random variable it is necessary that the operator K_x be completely continuous.*

▷ To prove this we notice that as it is known from functional analysis each self-adjoint positive operator K_x may be represented in terms of the spectral expansion:

$$K_x = \int_0^\infty t E(dt),$$

where $E(dt)$ is the unit expansion generated by the operator K_x which is determined on the σ-algebra of Borel sets of the real axis (equal to zero on all Borel sets of the negative semiaxis by virtue of the positiveness of the operator K_x). The positive square root of the operator K_x accordingly to the definition of a function of the self-adjoint operator is determined by formula

$$K_x^{1/2} = \int\limits_0^\infty \sqrt{t}\, E(dt).$$

This formula shows that $K_x^{1/2}$ is also a positive self-adjoint operator. Therefore at any $f \in X$

$$(K_x f, f) = (K_x^{1/2} f, K_x^{1/2} f) = \| K_x^{1/2} f \|^2. \tag{3.7.18}$$

Now let $\{f_n\}$ be an arbitrary sequence weakly convergent to $f \in X$. It means that $(f_n, x) \to (f, x)$ or what is the same $(f_n - f, x) \to 0$ at any $x \in X$. From the continuity of $g_x(f)$ and property (3.7.2) of a covariance operator of a normally distributed random variable follows that

$$(f_n - f) K_x (f_n - f) = (K_x (f_n - f), f_n - f) \to 0.$$

So taking into account (3.7.18) we conclude that $\| K_x^{1/2} f_n - K_x^{1/2} f \| \to 0$. Thus operator $K_x^{1/2}$ sets up the correspondence between weakly convergent sequence $\{f_n\}$ and the convergent sequence $\{K_x^{1/2} f_n\}$. Hence it follows that the operator $K_x^{1/2}$ is competely continuous. Finally, from the complete continuity of the operator $K_x^{1/2}$ it follows the complete continuity of the operator K_x. $\triangleleft$

$\triangleright$ In order to find the additional conditions to which the covariance operator K_x should satisfy we recall that the spectrum of completely continuous operator K_x consists only of a finite or a countable set of the eigenvalues $\{\lambda_n\}$ whose unique limit point may be only 0. Let φ_n, $\|\varphi_n\| = 1$ be an eigenvector of the operator K_x correspondent to the eigenvalues λ_n (as usual we suppose that any multiple eigenvalues is repeated in the sequence $\{\lambda_n\}$ as many times as linearly independent eigenvectors correspond to it). It is known that in this case the eigenvectors of the operator K_x form the orthonormal basis $\{\varphi_n\}$ in the space X as X by the supposition is separable. Consider the random variables $Z_n = (X^0, \varphi_n)$. They are uncorrelated and their variances are equal to the eigenvalues of the operator K_x as $E Z_n = 0$ and

$$EZ_p Z_q = E(X^0, \varphi_p)(X^0, \varphi_q) = (K_x \varphi_q, \varphi_p)$$

$$= \lambda_q (\varphi_q, \varphi_p) = \lambda_q \delta_{pq}.$$

The variables Z_n have a normal joint distribution. Consequently, from their noncorrelatedness it follows their independence.

Suppose that $g_x(f)$ serves as a characteristic functional of some random variable X, and consequently, determines the probability measure μ_x on the σ-algebra of Borel sets of H-space X. As

$$\{x : \|x^0\|^2 < u\} = \left\{x : \sum_{p=1}^{\infty}(x^0, \varphi_p)^2 < u\right\} \subset \left\{x : \sum_{p=1}^{n}(x^0, \varphi_p)^2 < u\right\},$$

where $x^0 = x - m_x$ then the sets

$$A_n = \left\{x : \sum_{p=1}^{n}(x^0, \varphi_p)^2 < u\right\} \quad (n = 1, 2, \ldots)$$

at any u form a decreasing sequence convergent to $A = \{x : \|x^0\|^2 < u\}$. Therefore by virtue of the continuity of the measure

$$\mu_x(A) = \lim_{n \to \infty} \mu_x(A_n). \tag{3.7.19}$$

But $\mu_x(A_n)$ represents the distribution function $F_n(u)$ of the scalar random variable $U_n = \sum_{p=1}^{n}(X^0, \varphi_p)^2 = \sum_{p=1}^{n} Z_p^2$, and $\mu_x(A)$ is the distribution function $F(u)$ of the variable $U = \sum_{p=1}^{\infty}(X^0, \varphi_p)^2 = \|X^0\|^2$. It follows from (3.7.19) that $F_n(u) \to F(u)$ at any u. According to Theorem 3.1.1 it follows that the sequence of the characteristic functions $\{g_n(t)\}$ of the random variables U_n at all t converges to the characteristic function $g(t)$ of the random variable U. We calculate the characteristic function $g_n(t)$ of the variable U_n. By virtue of the independence of the random variables Z_n we have $g_n(t) = Ee^{itU_n} = Ee^{itZ_1^2} \ldots Ee^{itZ_n^2}$. As the random variable Z_p is normally distributed its expectation is equal to zero, and the variance is equal to λ_p then

$$Ee^{itZ_p^2} = \frac{1}{\sqrt{2\pi\lambda_p}} \int_{-\infty}^{\infty} e^{itz^2 - (z^2/2\lambda_p)}dz = \frac{1}{\sqrt{1 - 2it\lambda_p}}.$$

To derive this formula it is sufficient to differentiate the integral over the parameter t and after that to perform the integration by parts. As a result we shall obtain a linear differential equation of the first order for the function $\varphi_p(t) = E e^{it Z_p^2}$. After integrating this equation at the initial condition $\varphi_p(0) = 1$ we get the required result. Consequently,

$$g_n(t) = \left[\prod_{p=1}^{n} (1 - 2it\lambda_p) \right]^{-1/2}. \tag{3.7.20}$$

From the convergence of the sequence $\{g_n(t)\}$ at all t it follows the convergence of the product $\prod_{p=1}^{\infty} (1 - 2it\lambda_p)$ at all t. But the product converges at any t if and only if the series $\sum \lambda_p$ converges. Consequently, the series composed of the eigenvalues of the operator K_x must be convergent. The positive self-adjoint operators possessing such property are called operators of trace class. $\triangleleft$

Thus we proved the following theorem.

Theorem 3.7.11. *In order formula (3.7.17) would determine a characteristic functional of a normally distributed random variable in a separable H-space X it is necessary that the operator K_x be of trace class.*

Theorem 3.7.12. *For any normally distributed random variable in a H-space there exists a finite expectation of the square of its norm $E\|X\|^2 < \infty$.*

$\triangleright$ Really, at any $x \in X$ we have $\sum_{p=1}^{n} (x, \varphi_p)^2 \to \|x\|^2$. Hence on the basis of property (ix) of the expectations follows

$$E \sum_{p=1}^{n} Z_p^2 = E \sum_{p=1}^{n} (X^0, \varphi_p)^2 \to E\|X^0\|^2.$$

As $E(X^0, \varphi_p)^2 = (K_x \varphi_p, \varphi_p) = \lambda_p$ and the series $\sum \lambda_p$ converges then

$$E\|X\|^2 = \lim \sum_{p=1}^{\infty} E(X^0, \varphi_p)^2 = \sum_{p=1}^{\infty} \lambda_p.$$

Hence it follows that $E\|X^0\|^2 = \|m_x\|^2 + E\|X^0\|^2 < \infty$. $\triangleleft$

Theorem 3.7.13. *The trace property of operator K_x is sufficient for the case that formula (3.7.17) determines a characteridtics functional of a normally distributed random variable in the separable H-space X.*

▷ It was shown in Subsection 3.6.5 that the functional $g_x(f)$ determines in the space X the function of the set μ_x on the algebra of the sets $\{x : ((x, f_1), \ldots, (x, f_n)) \in B\}$, correspondent to all finite sets of the vectors $f_1, \ldots, f_n \in X$ and to all Borel sets B of the finite-dimensional spaces, and the contraction of this function μ_x on any σ-algebra $\mathcal{A}_{f_1,\ldots,f_n}$ of the sets of the type $\{x : ((x, f_1), \ldots, (x, f_n)) \in B\}$ at fixed set of the vectors $f_1, \ldots, f_n \in X$ represents the probability measure on $\mathcal{A}_{f_1,\ldots,f_n}$. We take as the fixed set $f_1, \ldots, f_n \in X$ the set of the eigenvectors $\{\varphi_1, \ldots, \varphi_n\}$ of the operator K_x, correspondent to the first n eigenvalues $\lambda_1, \ldots, \lambda_n$ and use Chebyshev inequality. It is known that for an arbitrary nonnegative measurable function $u(S)$ of the random variable S and arbitrary $\alpha, \varepsilon > 0$ we may write

$$\mu_s(\{s : u(s) > \varepsilon\}) \le \frac{1}{\varepsilon_\alpha} \int u^\alpha(s)\mu_s(ds) \le \frac{1}{\varepsilon_\alpha} Eu^\alpha(S).$$

This is Chebyshev inequality. As a result we get

$$\mu_x\left(\left\{x : \sum_{p=1}^{n}(x, \varphi_p)^2 > N^2\right\}\right) \le \frac{1}{N^2}\sum_{p=1}^{n} E(X, \varphi_p)^2, \qquad (3.7.21)$$

where X is a random variable for which the contraction of the function μ_x on the σ-algebra $\mathcal{A}_{\varphi_1,\ldots,\varphi_n}$ serves as the probability distribution. For determining the expectations in this formula we shall find a characteristic function of a joint probability distribution of the random variables $Z_1 = (X, \varphi_1), \ldots, Z_n = (X, \varphi_n)$. According to formula (3.5.9) this characteristic function is obtained from (3.7.17) at $f = \sum_{p=1}^{n} \alpha_p\varphi_p$ (Section 2.4):

$$g_{z_1,\ldots,z_n}(\alpha_1, \ldots, \alpha_n) = \exp\left\{i\sum_{p=1}^{n} \alpha_p(m_x, \varphi_p)\right.$$

$$\left. -\frac{1}{2}\sum_{p,q=1}^{n} \alpha_p\alpha_q(K_x\varphi_p, \varphi_q)\right\} = \exp\left\{\sum_{p=1}^{n}\left[i\alpha_p(m_x, \varphi_p) - \frac{1}{2}\alpha_p^2\lambda_p\right]\right\}.$$

Hence it is clear that the random variables $Z_1, \ldots, Z_n$ are independent, normally distributed with expectations $(m_x, \varphi_1), \ldots, (m_x, \varphi_n)$, and the variances $\lambda_1, \ldots, \lambda_n$. Therefore $E(X, \varphi_p)^2 = EZ_p^2 = (m_x, \varphi_p)^2 + \lambda_p$. Substituting this expression into (3.7.21) and taking into account that by virtue of the Bessel inequality $\sum_{p=1}^{n}(m_x, \varphi_p)^2 \le \|m_x\|^2$, we get

$$\mu_x\left(\left\{x : \sum_{p=1}^{n}(x, \varphi_p)^2 > N^2\right\}\right) < \frac{1}{N^2}\left(\|m_x\|^2 + \sum_{p=1}^{n}\lambda_p\right).$$

In consequence of the convergence of the series $\sum \lambda_p$ we conclude that at any $\varepsilon > 0$ there exists such N_ε that at all n and $N \geq N_\varepsilon$

$$\mu_x\left(\left\{x : \sum_{p=1}^{n}(x, \varphi_p)^2 > N^2\right\}\right) < \varepsilon. \tag{3.7.22}$$

Now we notice that the set of all finite linear combinations of the eigenvectors φ_p of the operator K_x with the rational coefficients is countable and dense in X and therefore the set of the vectors

$$f_n = \sum_{p=1}^{N_n} \alpha_{np}\varphi_p, \quad \alpha_{np} = \frac{r_{np}}{\sqrt{r_{n1}^2 + \cdots + r_{nN_n}^2}},$$

correspondent to all possible finite sets of the rational numbers $r_{n1}, \ldots, r_{nN_n}$ is countable and dense on the unit sphere of the space X. Here in consequence of the fact that $\sum_{p=1}^{N_n} \alpha_{np}^2 = 1$ for any $x \in X$ the following inequality holds:

$$(x, f_n)^2 = \left\{\sum_{p=1}^{N_n} \alpha_{np}(x, \varphi_p)\right\}^2 \leq \sum_{p=1}^{N_n}(x, \varphi_p)^2.$$

Hence it follows that

$$\sup_{q \leq n}(x, f_q)^2 \leq \sum_{p=1}^{\sup_{q \leq n} N_q} (x, \varphi_p)^2,$$

and at any narutal N we have

$$\{x : \sup_{q \leq n} |(x, f_q)| > N\} \subset \{x : \sum_{p=1}^{\sup_{q \leq n} N_q} (x, \varphi_p)^2 > N^2\}.$$

Using inequality (3.7.22) we obtain

$$\mu_x(\{x : \sup_{q \leq n} |(x, f_q)| > N\}) < \varepsilon$$

at all n and at all sufficiently large N. According to the Mourier theorem (Subsection 3.7.5) $g_x(f)$ is a characteristic functional of some random variable in H-space X and deteremines uniquely the probability measure μ_x of this random variable on the σ-algebra of Borel sets of the space X. ◁

E x a m p l e 3.7.3. Let $K_x = \sum\limits_{p=1}^{n} \lambda_p(\cdot, \varphi_p)\varphi_p, \quad \lambda_p > 0$, where $\{\varphi_p\}$ is an arbitrary sequence of the vectors of H-space X then $(K_x f, f) = \sum\limits_{p=1}^{n} \lambda_p(f, \varphi_p)^2$ and formula (3.7.17) gives

$$g_x(f) = \exp\left\{ i(m_x, f) - \frac{1}{2} \sum_{p=1}^{n} \lambda_p(f, \varphi_p)^2 \right\}. \tag{I}$$

In this case for any vector f orthogonal to the subspace G generated by the vectors $\varphi_1, \ldots, \varphi_n$ we have $(f, \varphi_1) = \cdots = (f, \varphi_n) = 0$. Consequently, the random variable $Z = (X, f)$ has the expectation (m_x, f) and zero variance. It means that the distribution of the random variable X is completely concentrated on the hyperplane $m_x + G$.

E x a m p l e 3.7.4. The functional $g(f) = e^{-1/2\|f\|^2}$ determines a normal distribution on any σ-algebra $\mathcal{A}_{f_1, \ldots, f_n}$ of the sets of the type $\{x : ((x, f_1), \ldots, (x, f_n)) \in B\}$ at the fixed linearly independent $f_1, \ldots, f_n$. Really, putting $f = \sum\limits_{p=1}^{n} \alpha_p f_p$ we obtain

$$g(f) = \exp\left\{ -\frac{1}{2} \sum_{p,q=1}^{n} \alpha_p \alpha_q (f_p, f_q) \right\}. \tag{I}$$

The right-hand side of this formula represents a characteristic function of some normally distributed n-dimensional random vector $Z = \{Z_1, \ldots, Z_n\}$. The expectation of such vector is equal to 0, and the elements of its covariance matrix K_z are the scalar products (f_p, f_q). As $f_1, \ldots, f_n$ are linearly independent then the determinant of the matrix K_z is strictly positive in consequence of which the distribution of the vector Z is not singular. Let us consider the random variable

$$X = \sum_{p=1}^{n} \left(\sum_{q=1}^{n} k_{pq}^{-} Z_q \right) f_p, \tag{II}$$

where k_{pq}^{-} are the elements of the matrix K_z^{-1}. This is the normally distributed random variable in H-space X whose all realizations belong to the n-dimensional subspace G formed by the vectors $f_1, \ldots, f_n$. A characteristic functional of the variable X is determined by the evident formula

$$g_x(f) = E e^{i(X,f)} = E \exp\left\{ i \sum_{q=1}^{n} \left[\sum_{p=1}^{n} k_{pq}^{-}(f, f_p) \right] Z_q \right\}$$

$$= g_z \left(\sum_{p=1}^{n} k_{p1}^{-}(f, f_p), \ldots, \sum_{p=1}^{n} k_{pn}^{-}(f, f_p) \right)$$

$$= \exp \left\{ -\frac{1}{2} \sum_{p,q=1}^{n} (f_p, f_q) \sum_{r,s=1}^{n} k_{rp}^{-} k_{sq}^{-}(f, f_r)(f, f_s) \right\}. \tag{III}$$

But

$$\sum_{p,q=1}^{n} (f_p, f_q) \sum_{r,s=1}^{n} k_{rp}^{-} k_{sq}^{-}(f, f_r)(f, f_s)$$

$$= \sum_{r,s=1}^{n} (f, f_r)(f, f_s) \sum_{p=1}^{n} k_{rp}^{-} \sum_{q=1}^{n} k_{sq}^{-}(f_q, f_p). \tag{IV}$$

Accounting that (f_q, f_p) are the elements of the matrix K_z, and k_{sq}^{-} are the elements of the inverse matrix K_z^{-1} in consequence of which $\sum_{q=1}^{n} k_{sq}^{-}(f_q, f_p) = \delta_{sp}$ we get

$$\sum_{p,q=1}^{n} (f_p, f_q) \sum_{r,s=1}^{n} k_{rp}^{-} k_{sq}^{-}(f, f_r)(f, f_s) = \sum_{r,s=1}^{n} k_{rs}^{-}(f, f_r)(f, f_s). \tag{V}$$

The latter expression represents a square of the norm of the projection of the vector f on the subspace G. Therefore the obtained expression for the characteristic functional of the random variable X may be rewritten in the form $g_x(f) = e^{-1/2\|Pf\|^2}$, where P is an operator of the projection on the subspace G. In particular, for any vector $f \in G$

$$g_x(f) = e^{-1/2\|f\|^2} = g(f). \tag{VI}$$

The contraction $g(f)$ on any finite-dimensional subspace of the space X similarly as the contraction of a characteristic functional of a normally distributed random variable determines a normal distribution on this finite-dimensional subspace.

If we take the orthonormal vectors $f_1, \ldots, f_n$ then the random variables $Z_1, \ldots, Z_n$ will be independent and their variances will be equal to 1. Therefore

$$\mu_x \left(\left\{ x : \sum_{p=1}^{n} (x, f_p)^2 \leq N^2 \right\} \right) < \mu_x \left(\bigcap_{p=1}^{n} \{ x : |(x, f_p)| \leq N \} \right)$$

$$= \prod_{p=1}^{n} \mu_x(\{ x : |(x, f_p)| \leq N \}) = [2\Phi(N)]^n. \tag{VII}$$

For Laplace function $\Phi(N)$ at any N we have $0 < 2\Phi(N) < 1$. So at $n \to \infty$ we get from (VII)

$$\mu_x\left(\left\{x : \sum_{p=1}^{n}(x, f_p)^2 \leq N^2\right\}\right) \to 0.$$

Thus as $\sum_{p=1}^{\infty}(x, f_p)^2 = \|x\|^2$ we conclude that μ_x cannot be a probability distribution on X.

3.8. Approximate Representation of Distributions

3.8.1. Pearson Curves System

In applications it is often required to represent a density or a distribution function of a random variable approximately by a suitable analytical expression. At practice any simplification of an analytical representation of a density leads inevitably to the parametrization. The system of the distribution curves of Pearson (Pearson 1894, 1895, 1902) is widely used for an approximate analytical representation of the densities of scalar random variables. Pearson noticed that majority of the densities $y = f(x)$ satisfies the following ordinary differential equation:

$$\frac{dy}{dx} = \frac{x + a}{b_0 + b_1 x + b_2 x^2}$$

where a, b_0, b_1, b_2 are some constants. In particular, normal, exponential, Student, β-, γ-, χ- and χ^2- distributions satisfy this equation. Varying the parameters a, b_0, b_1, b_2 one may obtain a great variety of distribution curves.

The most general way of approximate representation of distributions is their representation in the form of linear combinations of some given functions, in particular, in the form of various truncated series.

3.8.2. Orthogonal Expansion of Density

Let $w(x)$ be some density in the r-dimensional space R^r for which all the moments exist. A system of pairs of polynomials $p_\nu(x)$, $q_\nu(x)$ $(\nu = 0, 1, 2, \ldots)$ is called *a biorthonormal system with the weight* $w(x)$ if

$$\int_{-\infty}^{\infty} w(x)p_\nu(x)q_\mu(x)dx = \delta_{\nu\mu} = \begin{cases} 0 & \text{at} \quad \mu \neq \nu, \\ 1 & \text{at} \quad \mu = \nu. \end{cases} \tag{3.8.1}$$

A system of pairs of polynomials $p_\nu(x)$, $q_\nu(x)$ $(\nu = 0, 1, 2, \ldots)$ is called *a biorthogonal system with the weight* w if condition (3.8.1) is fulfilled only at $\mu \ne \nu$. Any biorthogonal system of pairs of polynomials $\{p_\nu(x), q_\nu(x)\}$ may be reduced to a biorthonormal system by dividing the polynomials $p_\nu(x)$, $q_\nu(x)$ by numbers α_ν, β_ν respectively whose product is equal to the integral in (3.8.1) at the corresponding ν and $\mu = \nu$. It is evident that at every ν one of the numbers α_ν, β_ν may be chosen arbitrarily.

In the special case where $q_\nu(x) \equiv p_\nu(x)$ $(\nu = 0, 1, 2, \ldots)$ condition (3.8.1) takes the form

$$\int_{-\infty}^{\infty} w(x)p_\nu(x)p_\mu(x)dx = \delta_{\nu\mu} \, . \tag{3.8.2}$$

In this case the system of polynomials $\{p_\nu(x)\}$ is *orthonormal* if it satisfies condition (3.8.2) at all ν, μ and *orthogonal* if it satisfies the condition (3.8.2) only at $\mu \ne \nu$. Any orthogonal system of polynomials $\{p_\nu(x)\}$ may be normalized by dividing $p_\nu(x)$ by the square root of the integral in (3.8.2) at the corresponding ν and $\mu = \nu$.

It is evident that the existence of all the moments for the density $w(x)$ is necessary and sufficient for the existence of all the integrals in (3.8.1) and (3.8.2).

It is expedient to use multi-index numeration of the polynomials $p_\nu(x)$, $q_\nu(x)$ in such a way that the sum of the coordinates $|\nu| = \nu_1 + \cdots + \nu_r$ of a multi-index subscript $\nu = [\nu_1 \ldots \nu_r]$ be equal to the degree of the polynomials $p_\nu(x)$ and $q_\nu(x)$. Then the number of linearly independent polynomials of a given degree $m = |\nu|$ will be equal to the number of independent monomials of degree m, i.e. to $C_{r+m-1}^m = (r+m-1)!/[m!(r-1)!]$. It is evident that a multi-index numeration may always be replaced by the usual one.

▷ Let $f(x)$ be the density of some r-dimensional random vector X for which the moments of all orders exist. Let us try to represent the density $f(x)$ by an expansion of the form

$$f(x) = w(x) \sum_{\nu_1, \ldots, \nu_r = 0}^{\infty} c_\nu p_\nu(x) \, . \tag{3.8.3}$$

To determine the coefficients c_ν of expansion (3.8.3) we multiply (3.8.3) termwise by $q_\mu(x)$ and integrate over the whole r-dimensional space R^r. As a result taking into account (3.8.1) we shall have

$$\int_{-\infty}^{\infty} f(x)q_\mu(x)dx = \sum_{\nu_1,\ldots,\nu_r=0}^{\infty} c_\nu \int_{-\infty}^{\infty} w(x)p_\nu(x)q_\mu(x)dx = c_\mu\,. \qquad (3.8.4)$$

Consequently,

$$c_\nu = \int_{-\infty}^{\infty} f(x)q_\nu(x)dx = Eq_\nu(X) = q_\nu(\alpha)\,, \qquad (3.8.5)$$

where $q_\nu(\alpha)$ represents a linear combination of the moments of the random variable X which is obtained by replacing all the monomials $x_1^{k_1}\ldots x_r^{k_r}$ in $q_\nu(x)$ by the corresponding moments $\alpha_{k_1}\ldots\alpha_{k_r}$. Thus all the coefficients c_ν of expansion (3.8.3) are expressed in terms of the moments of the random vector X. Here $c_0 p_0(x) = 1$ always since by (3.8.1), $p_0(x)$ and $q_0(x)$ are mutually inverse constants (the polynomials of zero degree).

To construct expansion (3.8.3) it is convenient to choose the density $w(x)$ in such a way that all its moments of the first and the second order coincide with the corresponding moments of the density $f(x)$ of the random vector X. Then for all the polynomials $q_\nu(x)$ of the first and the second degree $\bigl(|\nu| = \nu_1 + \cdots + \nu_r = 1, 2\bigr)$ we shall have by virtue of (3.8.1)

$$c_\nu = \int_{-\infty}^{\infty} f(x)q_\nu(x)dx = \int_{-\infty}^{\infty} w(x)q_\nu(x)dx$$

$$= \frac{1}{p_0(x)} \int_{-\infty}^{\infty} w(x)p_0(x)q_\nu(x)dx = 0\,.$$

In consequence of this expansion (3.8.3) takes the form

$$f(x) = w(x)\left[1 + \sum_{k=3}^{\infty}\sum_{|\nu|=k} c_\nu p_\nu(x)\right]\,. \qquad (3.8.6)$$

In some cases it is convenient to express the coefficients c_ν in (3.8.6) not in terms of the moments of the random vector X as in (3.8.5) but in terms of its characteristic function. Minding that any moment of the vector X is obtained by the differentiation of its characteristic function $g(\lambda)$ with respect to the corresponding coordinates of the vector $i\lambda$ followed by equalizing λ to zero (Subsection 3.5.3) we may represent formula (3.8.5) in the form

$$c_\nu = [q_\nu(\partial/i\partial\lambda)g(\lambda)]_{\lambda=0} . \triangleleft \qquad (3.8.7)$$

R e m a r k. Formula (3.8.6) as well as formula (3.8.3) gives the expansion of the function $f(x)/\sqrt{w(x)}$ in terms of the unit vectors $\left\{ \sqrt{w(x)}p_\nu(x) \right\}$ of the function space $L_2(R^r)$, which form a biorthonormal system of the vectors together with the vectors $\left\{ \sqrt{w(x)}q_\nu(x) \right\}$. On the basis of the general theorems of functional analysis expansion (3.8.3) of the function $f(x)/\sqrt{w(x)}$ in case

$$\int\limits_{-\infty}^{\infty} \frac{f^2(x)}{w(x)}dx < \infty$$

m.s. converges to $f(x)/\sqrt{w(x)}$. The sequence of functions $\{\sqrt{w(x)}p_\nu(x)$ is complete in $L_2(R^r)$. The sequences of Hermite functions (Appendix 2) may serve as examples of complete systems of functions in the corresponding spaces $L_2(R^r)$.

Formula (3.8.6) determines the orthogonal expansion of the density $f(x)$. Partial sums of the series (3.8.6) may be used to approximate the distribution defined by $f(x)$ with any degree of accuracy also in the case where the random variables X has no moments. It is sufficient in this case to replace the distribution $f(x)$ by the corresponding truncated distribution

$$f_D(x) = f(x)\mathbf{1}_D(x) / \int\limits_{D} f(x)dx , \qquad (3.8.8)$$

approximating $f(x)$ with sufficient accuracy and then to approximate $f_D(x)$ by a partial sum of series (3.8.6).

A finite segment of expansion (3.8.6) may be a function which assumes small negative values at some x and nevertheless may give a good approximation to $f(x)$.

Retaining in (3.8.6) only the polynomials not higher than N^{th} degree we obtain the approximate formula for the density $f(x)$:

$$f(x) \approx f^*(x) = w(x) \left[1 + \sum_{k=3}^{N} \sum_{|\nu|=k} c_\nu p_\nu(x) \right] . \qquad (3.8.9)$$

The function $f^*(x)$ approximating the density $f(x)$ by virtue of (3.8.5) is completely determined by the moments of a random variable of up to the Nth order. In this case the moments of the function $f^*(x)$ up to the Nth order coincide with the corresponding moments of the random variable X.

$\triangleright$ Really, multiplying the last equality in (3.8.9) by $q_\mu(x)$ and using (3.8.1) and (3.8.5) we get

$$\int_{-\infty}^{\infty} f^*(x)q_\mu(x)dx = c_\mu = q_\mu(\alpha) \tag{3.8.10}$$

at $|\mu| \leq N$. Thus the expectations of all the polynomials $q_\nu(X)$ of not higher than N^{th} degree calculated by means of the approximating function $f^*(x)$ coincide with the expectations of the corresponding polynomials $q_\nu(X)$ calculated by means of the true density $f(x)$. But the polynomials $q_\nu(x)$ of any given degree k are linearly independent and their number coincides with the number of the moments of the kth order. Therefore from the coincidence of the expectations of the polynomials $q_\nu(X)$ it follows also the coincidence of the moments of the function $f^*(x)$ and of the density $f(x)$ up to the Nth order. Denoting the moments of the approximating function $f^*(x)$ by $\alpha^*_{k_1,\ldots,k_r}$ and $\mu^*_{k_1,\ldots,k_r}$ we may write the obtained result in the form

$$\alpha^*_{k_1,\ldots,k_r} = \alpha_{k_1,\ldots,k_r}, \quad \mu^*_{k_1,\ldots,k_r} = \mu_{k_1,\ldots,k_r} \text{ at } k_1 + \cdots + k_r \leq N. \triangleleft \tag{3.8.11}$$

As regards the moments of higher orders of the approximating function $f^*(x)$ they are expressed in terms of the moments up to the Nth order from the equations

$$q_\mu(\alpha^*) = 0 \text{ at } |\mu| > N. \tag{3.8.12}$$

In order to make (3.8.12) sure, it is sufficient to notice that by virtue of (3.8.9), (3.8.5) and (3.8.1)

$$q_\mu(\alpha^*) = \int_{-\infty}^{\infty} f^*(x)q_\mu(x)dx = 0, \quad |\mu| > N. \tag{3.8.13}$$

Later on we may also need the formulae expressing the moments of the random variable X of up to the Nth order in terms of the coefficients of the approximate expression (3.8.9) of its density.

$\triangleright$ Multiplying (3.8.9) by $x_1^{k_1} \ldots x_r^{k_r}$, integrating with respect to $x_1, \ldots, x_r$ and taking into account that the moments of the approximating function $f^*(x)$ up to the Nth order coincide with the corresponding moments of the random variable X we get

$$\alpha_{k_1, \ldots, k_r} = \alpha_{k_1, \ldots, k_r}^w + \sum_{k=1}^{N} \sum_{|\nu|=k} c_\nu p_{\nu, k_1, \ldots, k_r}(\alpha^w)$$

$$(k_1, \ldots, k_r = 0, 1, \ldots, N; \; k_1 + \cdots + k_r = 3, \ldots, N), \qquad (3.8.14)$$

where by the superscript w the moments of the density $w(x)$ are indicated,

$$p_{\nu, k_1, \ldots, k_r}(x) = x_1^{k_1} \ldots x_r^{k_r} p_\nu(x), \qquad (3.8.15)$$

and $p_{\nu, k_1, \ldots, k_r}(\alpha^w)$ is obtained from $p_{\nu, k_1, \ldots, k_r}(x)$ in the same way as $q_\nu(\alpha)$ was obtained from $q_\nu(x)$ in (3.8.5) (i.e. by means of the replacement of all the monomials $x_1^{h_1} \ldots x_r^{h_r}$ by the corresponding moments $\alpha_{h_1, \ldots, h_r}^w$ of the density $w(x)$). $\triangleleft$

Formula (3.8.14) gives the expressions of the moments of the random variable X of up to the Nth order in terms of the moments of the density $w(x)$ and the coefficients of the approximate representation $f^*(x)$ of the density $f(x)$ random variable X.

R e m a r k. The functions $p_\nu(x)$ and $q_\nu(x)$ are not obligatory the polynomials. They may be any functions satisfying the conditions of the biorhtonormality (3.8.1) and the condition of the existence of all integrals (3.8.5). Expansion (3.8.6) is also valid in this, more general case. But if the functions $q_\nu(x)$ are not polynomials then despite the coincidence of the first and the second orders moments of the distributions $w(x)$ and $f(x)$ the coefficients c_ν will not be equal to zero at $|\nu| = 1$ and 2 as a consequence of which the summation over k in expansion (3.8.6) will begin from $k = 1$.

Sometimes the expansion in powers of the derivatives of some density $w(x)$ which has the derivatives and the moments of all orders is used

$$f(x) = \sum_{\nu=0}^{\infty} c_\nu w^{(\nu)}(x). \qquad (3.8.16)$$

Here in the case of the n-dimensional vector $x = [\, x_1 \ldots x_n \,]^T$ the index ν represents a vector of the same dimension $\nu = [\nu_1 \ldots \nu_n]$, the derivative $w^{(\nu)}(x)$ is assumed as a partial derivative $\partial^{|\nu|} w(x)/\partial x_1^{\nu_1} \ldots \partial x_n^{\nu_n}$, and the summation is performed on all the coordinates of the vector ν from 0 till ∞. Here $p_\nu(x) = w^{(\nu)}(x)/w(x)$ and the functions $q_\nu(x)$ are the polynomials. These polynomials in the case of the scalar x are determined by the following recurrent formula:

$$q_\nu(x) = (-1)^\nu \frac{x^\nu}{\nu!} - \sum_{\mu=0}^{\nu-1} (-1)^{\nu-\mu} \frac{\alpha_{\nu-\mu}^w q_\mu(x)}{(\nu-\mu)!}, \qquad (3.8.17)$$

where α_λ^w are the moments of the density $w(x)$ equal to

$$\alpha_\lambda^w = \int_{-\infty}^{\infty} x^\lambda w(x) dx \quad (\lambda = 1, 2, \ldots). \qquad (3.8.18)$$

$\triangleright$ For proving we notice that $p_0(x) = 1$, and consequently, we may assume $q_0(x) = 1$. Further the following theorems are derived by the integration by parts:

$$\int_{-\infty}^{\infty} x^\lambda w^{(\nu)}(x) dx = 0 \text{ at } \lambda < \nu, \quad \int_{-\infty}^{\infty} x^\nu w^{(\nu)}(x) dx = (-1)^\nu \nu!, \quad (3.8.19)$$

$$\int_{-\infty}^{\infty} x^\lambda w^{(\nu)}(x) dx = (-1)^\nu \frac{\lambda!}{(\nu-\lambda)} \alpha_{\lambda-\nu}^w \text{ at } \lambda > \nu. \qquad (3.8.20)$$

Formulae (3.8.19) and (3.8.20) are valid at the condition that all the derivatives $w^{(\nu)}(x)$ decrease at $|x| \to \infty$ too fast than at any $\lambda > 0$ we have $\lim_{|x| \to \infty} x^\lambda w^{(\nu)}(x) = 0$. On the basis of (3.8.20) any polynomial of μ degree is orthogonal (with the weight $w(x)$) to all the functions $p_\nu(x)$ at any $\nu > \mu$. Therefore at any ν it is sufficient to construct the polynomial $q_\nu(x)$ which is orthogonal to the functions $p_0(x) = 1, p_1(x), \ldots, p_{\nu-1}(x)$. Putting $q_1(x) = c_1 x + b_{10}$ on the basis of cited formulae (3.8.19) and (3.8.20) and from the condition of the biorthonormality we get from (3.8.1) at $\mu, \nu = 0, 1$

$$\int_{-\infty}^{\infty} w(x) p_0(x) q_1(x) dx = c_1 \alpha_1^w + b_{10} = 0, \qquad (3.8.21)$$

$$\int_{-\infty}^{\infty} w(x) p_1(x) q_1(x) dx = -c_1 = 1. \qquad (3.8.22)$$

From (3.8.21) and (3.8.22) we find $c_1 = -1$, $b_{10} = \alpha_1^w$. After this we put $q_2(x) = c_2 x^2 + b_{20} + b_{21} q_1(x)$ and find by means of the cited formulae c_2, b_{20} and b_{21} from the condition of the biorthonormality (3.8.1) at ν, $\mu = 0, 1, 2$: $c_2 = 1/2!$, $b_{20} = -\alpha_2^w/2!$, $b_{21} = \alpha_1^w$. Suppose that continuing in such a way we found the polynomials $q_1(x), \ldots, q_{\nu-1}(x)$ satisfying together with the functions $p_\mu(x)$ the condition of the biorthonomality (3.8.1). After putting $q_\nu(x)$

$$= c_\nu x^\nu + \sum_{\mu=0}^{\nu-1} b_{\nu\mu} q_\mu(x)$$ from the condition of the orthogonality $q_\nu(x)$ to $p_0(x)$

$= 1, \ldots, p_{\nu-1}(x)$ and from condition (3.8.1) at $\mu = \nu$ we find c_ν
$= (-1)^\nu/\nu!$, $b_{\nu\mu} = (-1)^{\nu-\mu-1}\alpha_{\nu-\mu}^w/(\nu-\mu)!$ $(\mu = 0, 1, \ldots, \nu-1)$. $\triangleleft$

Analogously from the monomials $x_1^k \ldots x_n^k$ $(k_1, \ldots, k_n = 0, 1, \ldots)$ we may obtain the polynomials $q_\nu(x)$ satisfying together with the functions $p_\nu(x) = w^{(\nu)}(x)/w(x)$ to condition of the biorthonormality (3.8.1) in the case of the n-dimensional vector variable x.

3.8.3. Hermite Polynomial Expansion of Density

The system of Hermite polynomials $\{H_\nu(x), G_\nu(x)\}$ (see Appendix 2) represents a biorthogonal system of polynomials with the weight representing a normal density with zero expectation. In order to obtain a biorthonormal system polynomials corresponding to the normal density with the expectation m and the covariance matrix K we may assume on the basis of formula (6) Appendix 2

$$p_\nu(x) = \frac{1}{\nu_1! \ldots \nu_r!} H_\nu(x - m), \quad q_\nu(x) = G_\nu(x - m). \tag{3.8.23}$$

Then formula (3.8.9) approximately expressing the density of a random variable X with the same expectation m and covariance matrix K will take the form

$$f(x) \approx f^*(x) = w_N(x) \left[1 + \sum_{k=3}^{N} \sum_{|\nu|=k} \frac{c_\nu H_\nu(x - m)}{\nu_1! \ldots \nu_r!} \right]. \tag{3.8.24}$$

Formula (3.8.5) determining the coefficients of this expansion takes the form

$$c_\nu = \int_{-\infty}^{\infty} f(x) G_\nu(x - m) dx = E G_\nu(X - m) = G_\nu(\mu), \tag{3.8.25}$$

where $G_\nu(\mu)$ is the linear combination of central moments of X obtained
as a result of replacement in $G_\nu(x - m)$ of every monomial of the form
$(X_1 - m_1)^{h_1} \ldots (X_r - m_r)^{h_r}$ by the corresponding moment $\mu_{h_1, \ldots, h_r}$.
According to the result of Subsection 3.8.1 all the moments (both ini-
tial and central) of the function $f^*(x)$ approximating the density $f(x)$
up to the Nth order coincide with the corresponding moments of the
density $f(x)$, and the moments of higher orders of the function $f^*(x)$
are expressed in terms of the moments of up to the Nth order from the
relations

$$G_\nu(\mu^*) = 0 \ \text{ at } \ |\nu| > N, \tag{3.8.26}$$

where by the asterisk the moments of the function $f^*(x)$ are indicated.

The coefficients c_ν of $H_\nu(x - m)/(\nu_1! \ldots \nu_r!)$ in the Hermite polynomial expansion of the density $f(x)$ are called *the quasimoments* of the
random variable X. The number $|\nu| = \nu_1 + \cdots + \nu_r$ is called *the order*
of the quasimoment c_ν. According to (3.8.25) the quasimoment of the
k^{th} order represents a linear combination of central moments up to the
k^{th} order.

3.8.4. Relations between Quasimoments and Semiinvariants

For expressing the quasimoments of a random variable in terms of
its semiinvariants we may certainly substitute in formula (3.8.25) the
expressions of the centered moments in terms of the semiinvariants. In
this case following formulae are used:

$$\ln g(\lambda) = \sum_{r=1}^{\nu} i^r \sum_{h_1 + \cdots + h_n = r} \frac{\kappa_{h_1, \ldots, h_n}}{h_1! \ldots h_n!} \lambda_1^{h_1} \ldots \lambda_n^{h_n} + R_\nu'', \tag{3.8.27}$$

$$\kappa_{h_1, \ldots, h_n} = i^{-r} \left[\frac{\partial^r \ln g(\lambda)}{\partial \lambda_1^{h_1} \ldots \partial \lambda_n^{h_n}} \right]_{\lambda=0} \ (h_1 + \cdots + h_n = r), \tag{3.8.28}$$

$$\kappa_{h_1, \ldots, h_n} = \mu_{h_1, \ldots, h_n} - \sum_{p=2}^{[(h_1 + \cdots + h_n)/2]} \frac{(-1)^p}{p} h_1! \ldots h_n!$$

$$\times \sum_{\nu_{11} + \cdots + \nu_{np}} \prod_{k=1}^{p} \frac{\mu_{\nu_{1k}, \ldots, \nu_{nk}}}{\nu_{1k}! \ldots \nu_{nk}!} \tag{3.8.29}$$

$$\mu_{h_1, \ldots, h_n} = \kappa_{h_1, \ldots, h_n} + \sum_{p=r}^{[(h_1 + \cdots + h_p)/2]} \frac{h_1! \ldots h_n!}{p!}$$

$$\times \sum_{\nu_{11}+\ldots+\nu_{1p}=h_1} \sum_{\nu_{n1}+\ldots+\nu_{np}=h_n} \prod_{k=1}^{p} \frac{\kappa_{\nu_{1k},\ldots,\nu_{nk}}}{\nu_{1k}!\ldots\nu_{nk}!}$$

$$(h_1,\ldots,h_n = 0,1,2,\ldots, h_1+\cdots h_n \geq 4). \qquad (3.8.30)$$

The integral part of the correspondent number is marked here by square brackets. But in the general case the result is achieved by means of very complicated and cumbersome calculations. Therefore we shall try to express directly the quasimoments c_ν in terms of the semiinvariants κ_r.

$\triangleright$ Let X be the r-dimensional random vector with an expectation m, a covariance matrix K and a density $f(x)$. We put $u = i\lambda$ in formula (2) of Appendix 2 for the generating function of Hermite polynomials $G_\nu(x)$ corresponding to this random vector and replace the variable x by $x - m$. Then taking into account that $(u^T - m^T)\lambda = \lambda^T(x - m)$, we shall have

$$\exp\left\{i\lambda^T(x-m) + \frac{1}{2}\lambda^T K\lambda\right\} = \sum \frac{(i\lambda_1)^{\nu_1}\ldots(i\lambda_r)^{\nu_r}}{\nu_1!\ldots\nu_r!} G_\nu(x-m),$$
$$(3.8.31)$$

where the summation extends to all nonnegative values of $\nu_1,\ldots,\nu_r$, and $\nu = [\nu_1\ldots\nu_r]$. After multiplying this equality by the density $f(x)$ of the random vector X and integrating with respect to x we shall get by (3.8.25)

$$e^{-i\lambda^T m+\lambda^T K\lambda/2}g(\lambda) = 1 + \sum_{k=3}^{\infty}\sum_{|\nu|=k} \frac{(i\lambda_1)^{\nu_1}\ldots(i\lambda_r)^{\nu_r}}{\nu_1!\ldots\nu_r!}c_\nu, \qquad (3.8.32)$$

where $g(\lambda)$ is the characteristic function of X. Thus $e^{-i\lambda^T m+\lambda^T K\lambda/2}g(\lambda)$ as a function of $i\lambda$ serves as a generating function for the quasimoments as well as $g(\lambda)$ serves as a generating function for the initial moments, $e^{-i\lambda^T m}g(\lambda)$ serves as a generating function for the central moments, and $\ln g(\lambda)$ serves as a generating function for the semiinvariants. Substituting into (3.8.32) the expression of $g(\lambda)$ in terms of the semiinvariants,

$$g(\lambda) = e^{\ln g(\lambda)} = \exp\left\{\sum_{h=1}^{\infty}\sum_{|s|=h} \frac{(i\lambda_1)^{s_1}\ldots(i\lambda_r)^{s_r}}{s_1!\ldots s_r!}\kappa_s\right\}, \qquad (3.8.33)$$

and remembering that the semiinvariants of the first and the second order coincide with the expectations and the elements of the covariance matrix of X respectively we may write

$$\exp\left\{\sum_{h=3}^{\infty}\sum_{|s|=h} \frac{(i\lambda_1)^{s_1}\ldots(i\lambda_r)^{s_r}}{s_1!\ldots s_r!}\kappa_s\right\} = 1+\sum_{k=3}^{\infty}\sum_{|\nu|=k} \frac{(i\lambda_1)^{\nu_1}\ldots(i\lambda_r)^{\nu_r}}{\nu_1!\ldots\nu_r!}c_\nu$$

or

$$\sum_{p=1}^{\infty} \frac{1}{p!} \left[\sum_{h=3}^{\infty} \sum_{|s|=h} \frac{(i\lambda_1)^{s_1} \ldots (i\lambda_r)^{s_r}}{s_1! \ldots s_r!} \kappa_s \right]^p = \sum_{k=3}^{\infty} \sum_{|\nu|=k} \frac{(i\lambda_1)^{\nu_1} \ldots (i\lambda_r)^{\nu_r}}{\nu_1! \ldots \nu_r!} c_\nu .$$

Hence it is directly clear that the quasimoments c_ν of the third, the fourth and the fifth order coincide with the corresponding semiinvariants $c_\nu = \kappa_\nu$ ($|\nu| = 3, 4, 5$). If follows from the fact that the expansion in brackets in the left-hand side begins with the terms of the third degree in consequence of which only the item of the first degree $p = 1$ in the left-hand side may give the terms of the third, the fourth and the fifth degrees. As regards the terms of the sixth and higher degrees in the left-hand side for which $|\nu| \geq 6$ they are present in the items corresponding to $p \leq [\,|\nu|\,/3\,]$. Therefore, after raising the expression in the brackets to the corresponding powers and collecting the terms of the same degrees we obtain

$$\frac{c_\nu}{\nu_1! \ldots \nu_r!} = \sum_{p=1}^{[\,|\nu|\,/3\,]} \frac{1}{p!} \sum_{q_1 + \cdots + q_p = \nu} \frac{\kappa_{q_1} \ldots \kappa_{q_p}}{q_{11}! \ldots q_{pr}!} \quad (\,|\nu| = 6, 7, \ldots), \quad (3.8.34)$$

where $q_1 = \left[\, q_{11} \ldots q_{1r} \,\right]^T, \ldots, q_p = \left[\, q_{p1} \ldots q_{pr} \,\right]^T$ as well as $\nu = \left[\, \nu_1 \ldots \nu_r \,\right]$ are r-dimensional multi-index subscripts and the inner sum is extended over all the values of $q_1, \ldots, q_p,\ |q_1|, \ldots, |q_p| \geq 3$, whose sum is equal to the vector $\nu = \left[\, \nu_1 \ldots \nu_r \,\right]$. The first term in the right-hand side of formula (3.8.34) corresponding to $p = 1$ is equal to κ_ν. ◁

In exactly the same way, after taking logs of both parts of the equality (3.8.32) and expanding the logarithm in the right-hand side into the Maclaurin series, we obtain the expression of the semiinvariants in terms of the quasimoments.

3.8.5. Edgeworth Series

The items in formula (3.8.34) corresponding to various p may be different in value, the largest value having the items containing the largest number of multipliers, i.e. corresponding to $p = [\,|\nu|\,/3\,]$, and the least value having the first item κ_ν. To understand this let us consider the case where a random variable X represents the sum of a large number n of independent random variables. In this case all the semiinvariants are of order n (the semiinvariants of the sum of independent random variables are equal to the sums of the corresponding semiinvariants of the items). Therefore the first item κ_ν in expression (3.8.34) of the quasimoments c_ν is of order n while the pth item is of order n^p. This enables us to calculate the quasimoments c_ν approximately neglecting the semi-invariants of higher orders in contradistinction to formula (3.8.25) which requires the knowledge of the moments μ_ν (and consequently also the semiinvariant κ_ν) for the calculation of c_ν. In connection with this fact the idea arises to rearrange the items in the expansion of $f(x)$ in terms of the Hermite polynomials in such a way that all the terms of the same order relative to n be collected together.

$\triangleright$ Substituting expression (3.8.34) of the quasimoments c_ν into the Hermite polynomial expansion of $f(x)$ we obtain

$$f(x) = w_N(x) \left[\sum_{h=3}^{\infty} \sum_{|\nu|=h} \sum_{p=1}^{[h/3]} \frac{1}{p!} \sum_{q_1+\cdots+q_o=\nu} \frac{\kappa_{q_1} \dots \kappa_{q_p}}{q_{11}! \dots q_{pr}!} H_\nu(x-m) \right].$$

$$(3.8.35)$$

Let us estimate the order of the polynomial $H_\nu(x-m)$. For this purpose we notice that every differentiation in formula (4) of Appendix 2 gives the corresponding component of the vector K_x^{-1} as a factor. But all the elements of the matrix K are of the same order n and the mean square deviations of the components of a random X are of order $n^{1/2}$. Therefore $K^{-1}(x-m)$ is of order $n^{-1/2}$, in consequence of which $H_\nu(x-m)$ is of order $n^{-(\nu_1+\cdots+\nu_r)/2} = n^{-|\nu|/2}$. Thus the term $\kappa_{q_1} \dots \kappa_{q_p} H_\nu(x-m)$ is of order $n^{p-k/2}$. In consequence of this all the items with the same value of $p-k/2$ are of the same order. To collect all the items of the same order let us consider even and odd values of k separately and denote by S_1 the sum of all the items with odd k and by S_2 the sum of all the items with even k. Then we shall have

$$S_1 = \sum_{l=1}^{\infty} \sum_{p=1}^{[(2l+1)/3]} \frac{1}{p!} L_{2l+1,p} \, , \quad S_2 = \sum_{l=2}^{\infty} \sum_{p=1}^{[2l/3]} \frac{1}{p!} L_{2l,p} \, , \qquad (3.8.36)$$

where we put for brevity

$$L_{k,p} = \sum_{|\nu|=k} \sum_{q_1+\cdots+q_p=\nu} \frac{\kappa_{q_1} \dots \kappa_{q_p}}{q_{11}! \dots q_{pr}!} H_\nu(x-m) \, . \qquad (3.8.37)$$

The quantity $L_{2l+1,p}$ is of order $n^{-l-1/2+p}$, and the quantity $L_{2l,p}$ is of order n^{-l+p}. Therefore al the items with the same value of $s = l - p$ are of the same order. In order to collect all those items let us pass from the summation index p to the new index $s = l - p$. As a result we shall have

$$S_1 = \sum_{l=1}^{\infty} \sum_{s=l-[(2l+1)/3]}^{l-1} \frac{1}{(l-s)!} L_{2l+1,l-s} \,,$$

$$S_2 = \sum_{l=2}^{\infty} \sum_{s=l-[2l/3]}^{l-1} \frac{1}{(l-s)!} L_{2l,l-s} \,. \tag{3.8.38}$$

Let us change the order of summation in these double sums. Taking into account that by virtue of the inequalities $l - [(2l+1)/3] \le s \le l - 1$ (or $l - [(2l)/3]$ $\le s \le l - 1$) at fixed integer s the inequalities $s - l \le l \le 3s + 1$ are valid for l and that the smallest value of s in S_1 is equal to $1 - [(2 \cdot 1 + 1)/3] = 0$, and in S_2 it is equal to $1 - [2 \cdot 1/3] = 1$, we obtain

$$S_1 = \sum_{s=0}^{\infty} \sum_{l=s+1}^{3s+1} \frac{1}{(l-s)!} L_{2l+1,l-s} \,, \quad S_2 = \sum_{s=1}^{\infty} \sum_{l=s+1}^{3s} \frac{1}{(l-s)!} L_{2l,l-s} \,.$$

$$\tag{3.8.39}$$

Finally replacing the summation index l by the index $p = l - s$ we rewrite the obtained equalities in the form

$$S_1 = \sum_{s=0}^{\infty} \sum_{p=1}^{2s+1} \frac{1}{p!} L_{2s+2p+1,p} \,, \quad S_2 = \sum_{s=1}^{\infty} \sum_{p=1}^{2s} \frac{1}{p!} L_{2s+2p,p} \,. \tag{3.8.40}$$

Here all the items of the inner sum at a given s are of the same order $n^{-s-1/2}$ in S_1 and of the same order n^{-s} in S_2. Substituting the obtained expressions into the formula for $f(x)$ we get

$$f(x) = w_N(x)$$

$$\times \left[1 + \sum_{s=0}^{\infty} \sum_{p=1}^{2s+1} \frac{1}{p!} \sum_{|\nu|=2s+2p+1} \sum_{q_1+\cdots+q_p=\nu} \frac{\kappa_{q_1}\ldots\kappa_{q_p}}{q_{11}!\ldots q_{pr}!} H_\nu(x-m) \right.$$

$$\left. + \sum_{s=1}^{\infty} \sum_{p=1}^{2s} \frac{1}{p!} \sum_{|\nu|=2s+2p} \sum_{q_1+\cdots+q_p=\nu} \frac{\kappa_{q_1}\ldots\kappa_{q_p}}{q_{11}!\ldots q_{pr}!} H_\nu(x-m) \right]. \tag{3.8.41}$$

In this formula every term of the first sum over s represents the sum of all the items of order $n^{-s-1/2}$, and every term of the second sum over s represents the sum of all the items of order n^{-s}. $\lhd$

Expansion (3.8.41) is called *the Edgeworth series* by the name of the scientist who first obtained this expansion for a scalar random variable (Edgeworth 1905).

Edgeworth series (3.8.41) represents an asymptotic expansion of the density $f(x)$ relative to n^{-1}. Cramèr showed that in the case of a scalar random variable X an error of a finite segment of series (3.8.41) has the order of the first rejected term (i.e. of the first rejected item of the sum over s_0 (Cramèr 1946).

The use of the Edgeworth series allows us to include into a segment of the expansion considerably more Hermite polynomials with the account of the moments up to a given order than the ordinary orthogonal expansion. Namely, retaining the moments (or the semiinvariants) of up to the Nth order we have in the segment of the Edgeworth series Hermite polynomials of up to the $(3N - 6)^{\text{th}}$ degree instead of the Nth degree in the ordinary orthogonal expansion. As a result the accuracy of approximation to the true distribution is increased. Therefore the Edgeworth series is often used in practice. There are good reasons for this because the bulk of random variables encountered in application belong to the class of random variables which may be considered as the sums of a great number of independent items.

Similarly as in the case of an orthogonal expansion while approximating a density $f(x)$ by a segment $f^*(x)$ of the Edgeworth series with the account of the moments of up to the Nth order, the moments of approximating function $f^*(x)$ of up to the Nth order coincide with the corresponding moments of the density $f(x)$, and the moments of orders higher than $3N - 6$ are determined from the relations $G_\nu(\mu^*) = 0$. The moments of orders from $N + 1$ up to $3N - 6$ are obtained from the equation $G_\nu(\mu^*) = \gamma_\nu$, where γ_ν is the sum of all the coefficients at $H_\nu(x - m)/(\nu_1! \ldots \nu_r!)$ in the segment of Edgeworth series (3.8.41). In order to show this we multiply the segment of series (3.8.41) by $G_\tau(x - m)$ and integrate it with respect to x. Then we shall obtain in the left-hand side $G_\tau(\mu^*)$, and in the right-hand side by virtue of (3.8.1) we get γ_τ. It is evident that $\gamma_\tau = c_\tau$ at $|\nu| \leq N$, $\gamma_\nu = 0$ at $|\nu| > 3N - 6$; at $N < |\nu| \leq 3N - 6$ the value γ_ν contains only the part of the items of expression (3.8.34) of the quasimoments.

3.8.6. Consistent Biorthogonal Systems of Polynomials

It is expedient to use in some sense consistent biorthogonal systems of polynomials for simultaneous approximate representation of all the multi-dimensional distributions of a random function.

Let $\{w_n(x_1, \ldots, x_n; t_1, \ldots, t_n)\}$ be a consistent family of multi-dimensional densities where $x_1, \ldots, x_n$ are r-dimensional vector variables. We shall connect with every density w_n a biorthonormal system of the polynomials $\{p_{\nu_1,\dots,\nu_n}(x_1,\dots,x_n;t_1,\dots,t_n)$ and $q_{\nu_1,\dots,\nu_n}(x_1,\dots,x_n;t_1,\dots,t_n)\}$. Naturally, these polynomials depend on the variables $t_1, t_2,\dots$ as parameters. This is indicated in our notations.

We shall call biorthonormal systems of polynomials $\{p_{\nu_1,\dots,\nu_n}(x_1,\dots,x_n;t_1,\dots,t_n), q_{\nu_1,\dots,\nu_n}(x_1,\dots,x_n;t_1,\dots,t_n)\}$ $(n=1,2\dots)$ *consistent* if they satisfy the conditions

$$\text{(i)} \quad \int_{-\infty}^{\infty} w_n(x_1,\dots,x_n;t_1,\dots,t_n)p_{\nu_1,\dots,\nu_n}(x_1,\dots,x_n;t_1,\dots,t_n)$$

$$\times\, dx_n = w_{n-1}(x_1,\dots,x_{n-1};t_1,\dots,t_{n-1})p_{\nu_1,\dots,\nu_{n-1}}(x_1,\dots,x_{n-1};$$

$$t_1,\dots,t_{n-1})\delta_{0,\nu_n}, \text{ where } \delta_{0,0}=1,\ \delta_{0,\nu_n}=0 \text{ at } \nu_n\neq 0;$$

$$\text{(ii)} \quad q_{\nu_1,\dots,\nu-1,0}(x_1,\dots,x_n;t_1,\dots,t_n) = q_{\nu_1,\dots,\nu_{n-1}}(x_1,$$

$$\dots,x_{n-1};t_1,\dots,t_{n-1});$$

$$\text{(iii)} \quad \sum_{k=0}^{\nu_{n-1}} p_{\nu_1,\dots,\nu_{n-1}-k,k}(x_1,\dots,x_{n-1},x_{n-1};t_1,\dots,t_{n-1},t_{n-1})$$

$$= p_{\nu_1,\dots,\nu_{n-1}}(x_1,\dots,x_{n-1};t_1,\dots,t_{n-1});$$

$$\text{(iv)} \quad q_{\nu_1,\dots,\nu_{n-1},\nu_n}(x_1,\dots,x_{n-1},x_{n-1};t_1,\dots,t_{n-1},t_{n-1})$$

$$= q_{\nu_1,\dots,\nu_{n-1}+\nu_n}(x_1,\dots,x_{n-1};t_1,\dots,t_{n-1}).$$

R e m a r k. As the polynomials p_ν and q_ν may contain factors as arbitrary mutually inverse constants we may replace conditions (iii) and (iv) by the conditions

$$\text{(v)} \quad \sum_{k=0}^{\nu_{n-1}} p_{\nu_1,\dots,\nu_{n-1}-k,k}(x_1,\dots,x_{n-1},x_{n-1};t_1,\dots,t_{n-1},t_{n-1})$$

$$\times\, \gamma_{\nu_1,\dots,\nu_{n-1}-k,k}^{-1} = p_{\nu_1,\dots,\nu_{n-1}}(x_1,\dots,x_{n-1};t_1,\dots,t_{n-1})\gamma_{\nu_1,\dots,\nu_{n-1}};$$

$$\text{(vi)} \quad \gamma_{\nu_1,\dots,\nu_{n-1},\nu_n}q_{\nu_1,\dots,\nu_{n-1},\nu_n}(x_1,\dots,x_{n-1},x_{n-1};t_1,\dots,t_{n-1},$$

$$t_{n-1}) = \gamma_{\nu_1,\dots,\nu_{n-1}+\nu_n}q_{\nu_1,\dots,\nu_{n-1}+\nu_n}(x_1,\dots,x_{n-1};t_1,\dots,t_{n-1}).$$

Consistent in this sense the biorthonormal systems of the polynomials exist. As an example of such the systems of polynomials

$$p_{\nu_1,\dots,\nu_n}(x_1,\dots,x_n;t_1,\dots,t_n)$$

$$= H_{\nu_1,\dots,\nu_n}(x_1-m(t_1),\dots,x_n-m(t_n))/(\nu_{11}!\dots\nu_{nr}!),\quad (3.8.42)$$

$$q_{\nu_1,\dots,\nu_n}(x_1,\dots,x_n;t_1,\dots,t_n)$$

$$= G_{\nu_1,\dots,\nu_n}(x_1-m(t_1),\dots,x_n-m(t_n)),$$

may serve where $H_{\nu_1,\dots,\nu_n}(x_1,\dots,x_n)$, $G_{\nu_1,\dots,\nu_n}(x_1,\dots,x_n)$ are Hermite polynomials $(n=1,2,\dots)$ (Appendix 2). A consistent family of the multi-dimensional normal densities serves in this case as a family of the densities w_n:

$$w_n(x_1, \ldots, x_n; t_1, \ldots, t_n)$$

$$= \left[(2\pi)^{rn} |K_n|\right]^{1/2} \exp\left\{-\frac{1}{2}(x^{(n)T} - m_n^T)K_n^{-1}(x^{(n)} - m_n)\right\},$$

$$x^{(n)} = \left[x_1^T \ldots x_n^T\right]^T, \quad m_n = \left[m(t_1)^T \ldots m(t_n)^T\right]^T, \qquad (3.8.43)$$

where

$$K_n = \begin{bmatrix} K(t_1, t_1) & K(t_1, t_2) & \ldots & K(t_1, t_n) \\ K(t_1, t_2)^T & K(t_2, t_2) & \ldots & K(t_2, t_n) \\ \ldots & \ldots & \ldots & \ldots \\ K(t_1, t_n)^T & K(t_2, t_n)^T & \ldots & K(t_n, t_n) \end{bmatrix},$$

and $m(t)$ and $K(t, t')$ are the expectation and the covariance function of some random function. The conditions of consistency (i), (ii) and (iv) are satisfied in this case at $r = 1$ by virtue of theorems 1, 2, and 4 of Appendix 2. In order to make sure that condition (3) is also satisfied at $r = 1$, we rewrite equality (16) of Appendix 2 in the form

$$\sum_{k=0}^{\nu_{n-1}} \frac{1}{\nu_1! \ldots (\nu_{n-1-k})! k!} H_{\nu_1, \ldots, \nu_{n-1}-k, k}(x_1, \ldots, x_{n-1}, x_{n-1})$$

$$= \frac{1}{\nu_1! \ldots \nu_{n-1}!} h_{\nu_1, \ldots, \nu_{n-1}}(x_1, \ldots, x_{n-1}). \qquad (3.8.44)$$

But this is condition (iii) at $r = 1$. Using the relations of conditions (i)– (iv) successively r times we make sure that conditions (v)–(iv) are also satisfied in the case of r-dimensional vector variables x_1, $x_2 \ldots$ and subscripts $\nu_1, \nu_2 \ldots$.

The polynomial vector variables $H_{\nu_1, \ldots, \nu_n}(x_1, \ldots, x_n)$, $G_{\nu_1, \ldots, \nu_n}(x_1, \ldots, x_n)$ depend in this case on $t_1, \ldots, t_n$ as the matrix K_n generating them depends on $t_1, \ldots, t_n$.

3.8.7. Consistent Orthogonal Expansions of Multi-Dimensional Densities

Let $\{f_n(x_1, \ldots, x_n; t_1, \ldots, t_n)\}$ be the consistent family of the multi-dimensional densities of an r-dimensional vector random function $X(t)$ which has the moments of all orders, $\{p_{\nu_1, \ldots, \nu_n}(x_1, \ldots, x_n; t_1,$

$\ldots, t_n)\}$, $\{q_{\nu_1, \ldots, \nu_n}(x_1, \ldots, x_n; t_1, \ldots, t_n)\}$ $(n = 1, 2, \ldots)$ be the consistent biorthonormal systems of polynomials corresponding to consistent multi-dimensional densities $w_n(x_1, \ldots, x_n; t_1, \ldots, t_n)$ which have the same moments of the first and the second order as the random function $X(t)$. Representing every multi-dimensional density $f_n(x_1, \ldots, x_n; t_1, \ldots, t_n)$ of the random function $X(t)$ by the orthogonal expansion (3.8.6) in terms of the polynomials $p_{\nu_1, \ldots, \nu_n}(x_1, \ldots, x_n; t_1, \ldots, t_n)$ we obtain *the consistent orthogonal expansions* of all its multi-dimensional densities:

$$f_n(x_1, \ldots, x_n; t_1, \ldots, t_n) = w_n(x_1, \ldots, x_n; t_1, \ldots, t_n)$$

$$\times \left[1 + \sum_{k=3}^{\infty} \sum_{|\nu_1| + \cdots + |\nu_n| = k} c_{\nu_1, \ldots, \nu_n}(t_1, \ldots, t_n) \right.$$

$$\left. \times p_{\nu_1, \ldots, \nu_n}(x_1, \ldots, x_n; t_1, \ldots, t_n) \right], \qquad (3.8.45)$$

where the coefficients $c_{\nu_1, \ldots, \nu_n}(t_1, \ldots, t_n)$ are determined on the basis of (3.8.7) and (3.8.5) by the formula

$$c_{\nu_1, \ldots, \nu_n}(t_1, \ldots, t_n) = \int_{-\infty}^{\infty} \cdots \int_{-\infty}^{\infty} f_n(x_1, \ldots, x_n; t_1, \ldots, t_n)$$

$$\times q_{\nu_1, \ldots, \nu_n}(x_1, \ldots, x_n; t_1, \ldots, t_n) dx_1 \ldots dx_n$$

$$= [q_{\nu_1, \ldots, \nu_n}(\partial/i\partial\lambda_1, \ldots, \partial/i\partial\lambda_n; t_1, \ldots, t_n)$$

$$\times g_n(\lambda_1, \ldots, \lambda_n; t_1, \ldots, t_n)]_{\lambda_1 = \cdots = \lambda_n = 0}. \qquad (3.8.46)$$

Here $g_n(\lambda_1, \ldots, \lambda_n; t_1, \ldots, t_n)$ is the n-dimensional characteristic function of the random function $X(t)$.

From the condition (ii) of consistency of the polynomial systems

$$\{p_{\nu_1, \ldots, \nu_n}(x_1, \ldots, x_n; t_1, \ldots, t_n), \ q_{\nu_1, \ldots, \nu_n}(x_1, \ldots, x_n; t_1, \ldots, t_n)\}$$

at $n = 1, 2, \ldots$ and the condition (i) of consistency of the densities f_n it follows that

$$c_{\nu_1, \ldots, \nu_{n-1}, 0}(t_1, \ldots, t_n) = c_{\nu_1, \ldots, \nu_{n-1}}(t_1, \ldots, t_{n-1}) \qquad (3.8.47)$$

at all $\nu_1, \ldots, \nu_{n-1}$. After this from condition (i) of consistency of polynomial system it follows that expansion (3.8.45) for the density f_{n-1} is obtained by integration with respect to x_n of expansion (3.8.45) for the density f_n. Hence, by induction it follows that expansions (3.8.45) are consistent in such a sense that expansion (3.8.45) for f_m is obtained by integration with respect to $x_{m+1}, \ldots, x_n$ of expansion (3.8.45) for f_n at any m.

From the condition (iv) of consistency of polynomial systems and from the equality

$$f_n(x_1, \ldots, x_n; t_1, \ldots, t_{n-1}, t_{n-1})$$

$$= f_{n-1}(x_1, \ldots, x_{n-1}; t_1, \ldots, t_{n-1})\delta(x_n - x_{n-1}). \qquad (3.8.48)$$

it follows that

$$c_{\nu_1, \ldots, \nu_{n-1}, \nu_n}(t_1, \ldots, t_{n-1}, t_{n-1}) = c_{\nu_1, \ldots, \nu_{n-1}+\nu_n}(t_1, \ldots, t_{n-1}). \qquad (3.8.49)$$

After this from the condition (iii) and from the validity of formula (3.8.48) for the densities w_n and w_{n-1} it follows that at $t_n = t_{n-1}$ formula (3.8.45) gives expression (3.8.48) for the density f_n where f_{n-1} is represented by the corresponding expansion (3.8.45). Thus expansions (3.8.45) of multi-dimensional densities of a random function are also consistent in such a sense that formula (3.8.48) is valid for them. This formula may also be considered as one of the conditions of consistency of multi-dimensional distributions.

Retaining in (3.8.45) polynomials of no higher than the Nth degree we obtain a consistent approximate representation of all the multi-dimensional distributions of the random function $X(t)$. As was pointed out in Subsection 3.8.2 we may use this approximate representation also in such cases where the random function $X(t)$ has no moments at all, the moments involved in (3.8.46) being in such cases the moments of the functions f_n^* approximating the densities f_n.

3.8.8. Consistent Hermite Polynomial Expansions of Multi-Dimensional Densities

Taking as the densities w_n the normal multi-dimensional densities determined by the expectation $m(t)$ and the covariance function $K(t, t')$ of a random function $X(t)$ we obtain the consistent Hermite polynomial expansions of the multi-dimensional densities of the random function $X(t)$:

$$ f_n(x_1, \ldots, x_n; t_1, \ldots, t_n) $$

$$ = [(2\pi)^{rn} |K_n|]^{-1/2} \exp\left\{ -\frac{1}{2}(x^{(n)T} - m_n^T)K_n^{-1}(x^{(n)} - m_n) \right\} $$

$$ \times \left\{ 1 + \sum_{k=3}^{\infty} \sum_{|\nu_1 + \cdots + \nu_n| = k} \frac{c_{\nu_1, \ldots, \nu_n}(t_1, \ldots, t_n)}{\nu_{11}! \ldots \nu_{nr}!} H_{\nu_1, \ldots, \nu_n}(x_1 - m(t_1), \right. $$

$$ \left. \ldots, x_n - m(t_n)) \right\} \quad (n = 1, 2, \ldots). \tag{3.8.50} $$

Formula (3.8.46) for the coefficients $c_{\nu_1, \ldots, \nu_n}(t_1, \ldots, t_n)$ takes in this case the form

$$ c_{\nu_1, \ldots, \nu_n}(t_1, \ldots, t_n) $$

$$ = \int_{-\infty}^{\infty} \cdots \int_{-\infty}^{\infty} f_n(x_1, \ldots, x_n; t_1, \ldots, t_n) G_{\nu_1, \ldots, \nu_n}(x_1 - m(t_1), $$

$$ \ldots, x_n - m(t_n)) dx_1 \ldots dx_n = [G_{\nu_1, \ldots, \nu_n}(\partial/i\partial\lambda_1 - m(t_1), \ldots, \partial/i\partial\lambda_n $$

$$ -m(t_n)) g_n(\lambda_1, \ldots, \lambda_n; t_1, \ldots, t_n)]_{\lambda_1 = \cdots = \lambda_n = 0} $$

$$ = [G_{\nu_1, \ldots, \nu_n}(\partial/i\partial\lambda_1, \ldots, \partial/i\partial\lambda_n) g_n(\lambda_1, \ldots, \lambda_n; t_1, \ldots, t_n) $$

$$ \times \exp\left\{ -i\lambda_1^T m(t_1) - \cdots - i\lambda_n^T m(t_n) \right\}]_{\lambda_1 = \cdots = \lambda_n = 0}. \tag{3.5.51} $$

3.9. Canonical Expansions

3.9.1. Introductory Remarks

In elementary probability theory it is convenient to operate with the random vectors which possess uncorrelated conponents. In connection with this fact the problem of the decorrelation of the conponents of a random vector arises. This problem is equivalent to the problem of transforming the covariance matrix K_x of the random vector $X = [X_1 \ldots X_n]^T$ to the diagonal form. Really, it is known from linear algebra that any Hermitian (symmetric) matrix K_x may be transformed to the diagonal form by an unitary (orthogonal) transformation. Let $\lambda_1, \ldots, \lambda_n$ be the eigenvalues of the matrix K_x (nonnegative by virtue of the nonnegative definiteness of K_x) and $\varphi_1, \ldots, \varphi_n$ be the correspondent orthonormal eigenvectors. Then the following relations hold:

$$ K_x \varphi_p = \lambda_p \varphi_p, \quad (\varphi_p, \varphi_q) = \varphi_q^* \varphi_p = \delta_{pq} \quad (p, q = 1, \ldots, n). \tag{3.9.1} $$

So if $\Psi = [\,\varphi_1 \ldots \varphi_n\,]$ be a square matrix whose columns serve the eigenvectors $\varphi_1, \ldots, \varphi_n$ then evidently the random vector $U = \Psi^* X^0$ where $X^0 = X - m_x$ has uncorrelated components. Hence follows the formula for the components of the vector U:

$$U_p = \varphi^* X^0 = (X^0, \varphi_p) \ (p = 1, \ldots, n).$$

Consequently, we come to the following eigenvector expansions of X and K_x with respect to the basis $\{\varphi_1, \ldots, \varphi_n\}$:

$$X = m_x + \sum_{p=1}^{n} U_p \varphi_p, \tag{3.9.2}$$

$$K_x = \sum_{p=1}^{n} \lambda_p \varphi_p \varphi_p^*. \tag{3.9.3}$$

R e m a r k. The random variables $U_p \varphi_p$ in the eigenvector espansion (3.9.2) are called *the main components* of the random vector. This name is explained by the fact that among the projections of the vector X on all possible directions the projection on the eigenvector φ_1 has the maximal variance equal to the largest eivenvalue, and for any s among the projections of the vector $X^0 - U_1 \varphi_1 - \cdots - U_{s-1} \varphi_{s-1}$ on all possible directions the projection on the eigenvector φ_s has the maximal variance equal to s^{th} eigenvalue λ_s:

$$\lambda_1 = \sup_{|\varphi|=1} \varphi^* K_x \varphi, \quad \lambda_s = \sup_{|\varphi|=1} \varphi^* \left(K_x - \sum_{p=1}^{s-1} \lambda_p \varphi_p \varphi^* \right) \varphi.$$

The computations of the eigenvalues and eigenvectors usually are cumbersome. Meanwhile the covariance matrix K_x may be easily reduced to the diagonal form by a variety ways if we refuse of an unitarity transformation. In this case the representation of X and its covariance matrix K_x correspondingly in the form

$$X = m_x + \sum_{p=1}^{n} V_p x_p, \tag{3.9.4}$$

$$K_x = \sum_{p=1}^{n} D_p x_p x_p^* \tag{3.9.5}$$

is called a *canonical expansion of the random vector X and its covariance matrix K_x*. Here the random coefficients $V_1 , \ldots , V_n$ are the uncorrelated random variables with the variances $D_1 , \ldots , D_n$, and the nonrandom vectors $x_1 , \ldots , x_n$ are the coordinate vectors.

From the nonuniqueness of the transformation of a covariance matrix to a diagonal form follows that for any random vector with a finite second order moment there exists nonnumerable variety of canonical expansions.

R e m a r k. The equation $u^T K_x u = c$ at the nonnegative determined matrix K_x with the real elements represents an equation of ellipsoid (ellipses at $n = 2$). Therefore the problem of finding canonical expansion (3.9.2) and (3.9.4) is equipotent to the problem of finding the directions conjugated relatively to the ellipsoid and reducing its equation to a canonical form. It is clear from this geometric interpretation why such an expansion is called *a canonical one*.

In elementary theory of random functions which possess the finite moments of the second order *a canonical expansion of the random* function $X = X(t)$ is called an expansion of the type

$$X(t) = m_x(t) + \sum_{\nu=1}^{\infty} V_\nu x_\nu(t), \qquad (3.9.6)$$

$$K_x(t,t') = \sum_{\nu=1}^{\infty} D_\nu x_\nu(t) x_\nu(t')^* . \qquad (3.9.7)$$

Here V_1, $V_2, \ldots$ are the uncorrelated random *coefficients of the canonical expansion* possessing zero expectation and the variances D_1, $D_2, \ldots$; $x_1(t)$, $x_2(t), \ldots$ are known *coordinate functions* of the canonical expansion.

Before formulating the main statements of the general theory of canonical expansions of the random variable X let us consider in auxiliary H-space associated with X.

3.9.2. Auxiliary H-space

Let X be an arbitrary linear space. On the space of all linear functionals on X we consider the linear subspace F and assign the topology τ_F in X in which all the functionals $f \in F$ are continuous. By the special role of the subspace F which provides X by the structure of the topological linear space further on we shall denote everywhere F in terms of F_c , recalling the reader that all the elements in F_c are continuous functionals. In terms of $\mathcal{A}$ we shall denote the σ-algebra in X generated by the sets of the type $\{x \in X : (f_1 x , \ldots , f_n x) \in B\}$ responding to all entire n, to all $f_1 , \ldots , f_n \in F_c$ and to all Borel sets $B \subset R^n$.

Let X be a random variable in $(X, \mathcal{A})$ with the expectation $m_x \in X$ and the covariance operator $K_x \in L(F_c, X)$; as usual the set of all linear mapping from F_c into X is denoted as $L(F_c, X)$. Let us denote as N_0 the null-space of the operator K_x, i.e. the set $N_0 = \{f \in F_c : K_x f = 0\}$. Accounting the functionals f_1, $f_2 \in F_c$ equivalent if $f_1 - f_2 \in N_0$ we get the partition of F_c into the classes of the equivalent elements $N_f = f + N_0$ where $f \in F_c$. As from $f_1 - f_2 \in N_0$ follows $K_x(f_1 - f_2) = 0$, then consequently, $K_x f_1 = K_x f_2$ for the equivalent elements f_1, f_2. It means that on the factor-space F_c/N_0 the operator K_x may be found by formula $K_x \bar{f} = K_x f$ where $f \in \bar{f} \equiv N_f$. Moreover, if $g \in F_c$ and $g \in \bar{g} = N_g$ then $g K_x f = \overline{f K_x g} = \overline{f K_x \bar{g}} = \bar{g} K_x f$ and we may determine the action of $\bar{g} \in F_c/N_0$ on the vector $K_x f \in X$ according to the rule $\bar{g} K_x f = g K_x f$. Thus the variable $\bar{g} K_x \bar{f} = g K_x f$ is determined correctly both on the elements f, $g \in F_c$ and on the elements $\bar{f}, \bar{g} \in F_c/N_0$.

E x a m p l e 3.9.1. Suppose that the spaces X and F_c are both n-dimensional. It is clear that F_c is the space X dual with X^*. The operator K_x in the fixed basis spaces X and X^* is presented by $n \times n$-matrix (Example 3.2.1). If K_x is of the n-rank then $K_x f = 0$ if and only if $f = 0$ and $N_0 = \{0\}$, and the spaces F_c and F_c/N_0 are naturally isomorphic. If K_x is of the $m < n$-rank then N_0 is the $(n - m)$-dimensional subspace in F_c, the classes $N_f = f + N_0$ are obtained by the shift of N_0 on the vector $f \in F_c$ and the dimension F_c/N_0 is equal to m.

E x a m p l e 3.9.2. Let $X = L_2([0, T_1])$ be a space of the real functions which are square-integrable on the segment $[0, T_1]$ over Lebesgue measure. Here the space X^* dual with X is isomorphic to X and we put $F_c = L_2([0, T_1])$. The random variable with the values in $X = L_2([0, T_1])$ is the set $\{X(t), t \in [0, T_1]\}$ of the scalar random variables with the parametric set $[0, T_1]$. Let $K_x(t, t') = D \cos \lambda(t - t')$, $D > 0$, $\lambda = 2\pi/T_1$ be the covariance matrix of the process $X(t)$. The covariance operator K_x is an integral operator of the form

$$K_x f = \int_0^{T_1} K_x(t, t') f(t') dt' \tag{I}$$

with square-integrable kernel $K_x(t, t')$. From the condition $K_x f = 0$ we obtain that the null-space N_0 of the operator K_x consists of all f which are orthogonal to the functions $\sin \lambda t$, $\cos \lambda t$ in the metric $(f, g) = \int_0^{T_1} f(t) g(t) dt$. Correspondingly,

F_c/N_0 consists of the linear combinations $a_1 \cos \lambda t + a_2 \sin \lambda t$, a_1, $a_2 \in R$ what is evident from the presentation

$$f(t) = a_0 + \sum_{k=1}^{\infty} (a_k \cos k\lambda t + b_k \sin k\lambda t) \qquad \text{(II)}$$

for any $f \in F_c$. The space F_c/N_0 is two-dimensional here.

E x a m p l e 3.9.3. Let in Example 3.9.2 be $K_x(t, t') = D e^{-\alpha |t-t'|}$, $D > 0$, $\alpha > 0$. From the condition $K_x f = 0$ follows that $f = 0$. Thereby $N_0 = \{0\}$ and F_c/N_0 is isomorphic to F_c.

Now we determine in the product $(F_c/N_0) \times (F_c/N_0)$ the functional (f, g) assuming

$$(f, g) = f K_x g . \qquad (3.9.8)$$

This functional possesses all the properties of a scalar product and we shall suppose that F_c/N_0 is endowed with this scalar product. After supplying F_c/N_0 by the norm $\| f \|^2 = (f, f)$ we get H-space H_x correspondent to the random variable X. The space H_x plays an important role in the general theory of the canonical expansions.

As the functional f is continuous in the topology τ_{F_c} the formula $V = f X^0$ determines an isometric mapping of H-space H_x of the random variables $X^0 = X - m_x$ on H-space H_v for the random variables $f X^0$, $f \in H_x$ and $E f X^0 = 0$, it the scalar product in H_v is given by the formula

$$(U, V) = E U \bar{V} . \qquad (3.9.9)$$

$\triangleright$ Really, to any f, $g \in H_x$ correspond the random variables $V = f X^0$, $W = g X^0 \in H_v$. And vice versa any random variables V, $W \in H_v$ may be presented in the form $V = f X^0$, $W = g X^0$, f, $g \in H_x$ and on the basis of (3.9.8) we have $(V, W) = E V \overline{W} = E(f X^0)(\overline{g X^0}) = f K_x g = (f, g)$. In particular, $\| V \| = \| f \|$. Thus the formula $V = f X^0$ establishes the mapping of the whole space H_x on the whole space H_v with the retention of the norm, i.e. an isometric mapping of H_x on H_v. $\triangleleft$

The covariance operator K_x is continuous in the weak topology of the space X determined by a set of the linear functionals F_c, i.e. in the topology determined by the vicinities of zero $\{x : |f_1 x| < \varepsilon_1,$ $\ldots, |f_n x| < \varepsilon_n\}$, correspondent to all n, $f_1, \ldots, f_n \in F_c$, $\varepsilon_1,$ $\ldots, \varepsilon_n > 0$.

$\triangleright$ Really, if the sequence $\{f_n\} \subset H_x$ converges to f, $\| f_n - f \| \to 0$ then $(f', f_n) \to (f', f)$ or $f' K_x f_n \to f' K_x f$ for any $f' \in H_x$. As any functional

$g \in F_c$ enters into some element f' of the space H_x (it is the representative of some class f' of the equivalent elements) then $g K_x f_n \in g K_x f$ for any $g \in F_c$. It means that the operator K_x is weakly continuous on H_x. ◁

E x a m p l e 3.9.4. Let X be a linear space of the vector functions $X(t)$ determined on the region $D \subset R^p$ with the values in R^n which are square-integrable with respect to some measure μ, i.e. $X = L_2(D, R^n, \mu)$. As F_c we assume the space X^* dual with X, by virtue of the isomorphism X and X^* we have $F_c = L_2(D, R^n, \mu)$. Let $K_x(t, t')$ be a kernel of the covariance operator K_x of the random variable X in the space X. In the space H_x associated with the random variable X we assign the scalar product

$$(f, g) = \iint\limits_{D \times D} f(t)^T K_x(t, t') \overline{g(t')} \mu(dt) \mu dt') . \tag{I}$$

The isometric mapping $V = f X^0$ from H_x into H_v is determined here by the formula

$$V = \int\limits_D f(t)^T X_0(t) \mu(dt) . \tag{II}$$

E x a m p l e 3.9.5. Let $X(t)$, $t \in D \subset R^n$ be a scalar m.s. continuous stationary random function with the covariance function $K_x(\tau)$, $\tau \in D$. Let X be a linear space generated by all the realizations of the random function $X(t)$, and F_c be a linear space generated by the functionals of the form $f x = \sum\limits_{p=1}^{N} f_p x(t_p)$, $x \in X$ responding to all integer N, to all scalars $f_1, \ldots, f_N$ and to all $t_1, \ldots, t_n$ from D. It is evident that F_c may be identified with the space of the functions $f(t)$, $t \in D$ if we put formally

$$f x = \int\limits_D f(t)^T x(t) dt , \quad f(t) = \sum\limits_{p=1}^{N} f_p \delta(t - t_p) . \tag{I}$$

The covariance function $k_x(\tau)$ is a continuous positively determined function therefore as it is known from functional analysis it is uniquely representable by Bochner theorem in the form

$$k_x(\tau) = \int\limits_{-\infty}^{\infty} e^{i\lambda^T \tau} \sigma(d\lambda) , \tag{II}$$

where σ is a nonnegative finite measure. Therefore for any function $f(t) \in F_c$ expression (I) of the previous example has in a given case the form

$$\int\limits_{-\infty}^{\infty} \int\limits_{-\infty}^{\infty} k_x(t - t') f(t) \overline{f(t')} dt dt' = \sum\limits_{p,q=1}^{N} k_x(t_p - t_q) f_p \overline{f_q}$$

$$= \int_{-\infty}^{\infty} \sum_{p,q=1}^{N} f_p \overline{f_q} e^{i\lambda^T(t_p-t_q)}\sigma(d\lambda) = \int_{-\infty}^{\infty} \left| \sum_{p=1}^{N} f_p e^{i\lambda^T t_p} \right|^2 \sigma(d\lambda). \qquad \text{(III)}$$

After constructing H-space H_x the formula

$$\varphi(\lambda) = \int_{-\infty}^{\infty} f(t) e^{i\lambda^T t} dt \qquad \text{(IV)}$$

determines the isometric mapping H_x on the subspace of H-space $L_2(R^n, \mathcal{B}^n, \sigma)$ of the functions of the variable t formed by the supplement of the space of all the functions $\varphi(\lambda) = \sum_{p=1}^{N} f_p e^{i\lambda^T t_p}$. For any elements $f(t)$ and $g(t)$ of the space H_x the scalar product will be expressed by the formula

$$(f,g) = \int_{-\infty}^{\infty} \int_{-\infty}^{\infty} k_x(t-t') f(t)\overline{g(t')} dt dt' = \int_{-\infty}^{\infty} \varphi(\lambda)\overline{\psi(\lambda)}\sigma(d\lambda) \qquad \text{(V)}$$

where

$$\varphi(\lambda) = \int_{-\infty}^{\infty} f(t) e^{i\lambda^T t} dt, \quad \psi(\lambda) = \int_{-\infty}^{\infty} g(t) e^{i\lambda^T t} dt. \qquad \text{(VI)}$$

The space H_v in a given case represents H-space which contains all linear combinations of the values of the random functions $X^0(t)$

$$V = \int_{-\infty}^{\infty} f(t) X^0(t) dt = \sum_{p=1}^{N} f_p X^0(t_p) \qquad \text{(VII)}$$

and their m.s. limits.

E x a m p l e 3.9.6. Let $X(t), t \in D$ be a white noise of the intensity $\nu(t)$ with the covariance function $K_x(t,t') = \nu(t)\delta(t-t')$. In the space $F_c = C(D)$ of the continuous function on D we shall determine according to general theory the scalar product

$$(f,g) = \int_D \nu(t) f(t)\overline{g(t)} dt. \qquad \text{(I)}$$

As from $K_x f = \nu(t)\overline{f(t)} = 0$ it follows $f = 0$ then $N_0 = \{0\}$ and F_c/N_0 is isomorphic to F_c. The space H_x associated with the white noise $X(t), t \in D$ is the space of all continuous functions and the limits of the fundamental sequences of continuous functions. The isometric mapping H_x on H_v is given by the formula

$$V = \int f(t)X^0(t)dt\,, \quad f \in C(D)\,, \tag{II}$$

and any element from H_v is the mean square limit of the fundamental sequences of such variables.

3.9.3. Direct Theorem about Canonical Expansions

Let us consider a general theory of the canonical expansions in the linear space. Suppose that the space H_x (and consequently, H_v also) is separable. Let $\{f_\nu\}$ be any orthonormal basis in H_x, $(f_\nu, f_\mu) = \delta_{\nu\mu}$. As it is known from functional analysis an arbitrary vector $f \in H_x$ is expressed in terms of the expansion

$$f = \sum_{\nu=1}^{\infty} (f, f_\nu) f_\nu\,. \tag{3.9.10}$$

As at $n \to \infty$

$$\| f - \sum_{\nu=1}^{n} (f, f_\nu) f_\nu \|^2 = E \left| fX^0 - \sum_{\nu=1}^{n} (f, f_\nu) f_\nu X^0 \right|^2 \to 0\,,$$

then at any $f \in H_x$ the random variable fX^0 is expressed in terms of m.s. convergent expansion $fX^0 = \sum_{\nu=1}^{n} (f, f_\nu) f_\nu X^0$. The random variables $V_\nu = f_\nu X^0 \in H_v$ $(\nu = 1, 2, \ldots)$ are uncorrelated and have zero expectations and unit variances as

$$EV_\nu \overline{V_\mu} = E(f_\nu X^0)(\overline{f_\mu X^0}) = f_\nu K_x f_\mu = (f_\nu, f_\mu) = \delta_{\nu\mu}\,.$$

Further $(f, f_\nu) = f K_x f_\nu$. Assuming $x_\nu = K_x f_\nu$ we shall have $(f, f_\nu) = f x_\nu$ and the obtained expansion will take the form $fX = f m_x + \sum_{\nu=1}^{\infty} V_\nu f x_\nu$. Hence we get the following expansion of the random variable X:

$$X = m_x + \sum_{\nu=1}^{\infty} V_\nu x_\nu\,. \tag{3.9.11}$$

This expansion m.s. converges to the random variable X in the weak topology of the space X.

Thus we obtain the expansion of the random variable X into the uncorrelated items. Such expansions are called *canonical*. The vectors $x_\nu = K_x f_\nu$ are called *the coordinate vectors* of the canonical expansion. From the fact that $(f_\nu, f_\mu) = f_\nu K_x f_\mu = f_\nu x_\mu$ follows that the coordinate vectors x_ν and the basis vectors f_ν satisfy the condition of *the biorthogonality* $f_\nu x_\mu = \delta_{\nu\mu}$. Besides that from the symmetry of K_x follows that $f_\nu x_\mu = \overline{f_\mu x_\nu}$.

▷ The orthogonal basis $\{f_n\}$ may be constructed starting from any complete sequence of the vectors $\{\varphi_n\} \subset H_x$ after applying to it an ordinary way of the orthonormalization. But in practice there is no need that the vectors f_ν should be normalized. It is sufficient that they should be orthogonal. In this case instead of (3.9.10) we have

$$f = \sum_{\nu=1}^{\infty} D_\nu^{-1}(f, f_\nu)f_\nu , \qquad (3.9.12)$$

where $D_\nu = \| f_\nu \|^2$. In order to obtain the orthogonal basis $\{f_\nu\}$ from a given complete system of the vectors $\{\varphi_\nu\}$ and at the same time to determine the sequence of the coordinate vectors $\{x_\nu\}$ we put

$$f_1 = \varphi_1 , \quad f_\nu = \alpha_{\nu 1}f_1 + \cdots + \alpha_{\nu,\nu-1}f_{\nu-1} + \varphi_\nu \quad (\nu = 2, 3, \ldots), \quad (3.9.13)$$

$$x_\nu' = K_x f_\nu \quad (\nu = 1, 2, \ldots), \qquad (3.9.14)$$

and find sequentially α_{21} from the condition $f_2 x_1' = 0$, then α_{31} and α_{32} from the conditions $f_3 x_1' = f_3 x_2' = 0$ and generally $\alpha_{\nu 1}, \ldots, \alpha_{\nu,\nu-1}$ from the conditions $f_\nu x_1' = \cdots = f_\nu x_{\nu-1}' = 0$. Here automatically (by virtue of the symmetry of the operator K_x) will be satisfied the conditions $f_1 x_\nu' = \cdots = f_{\nu-1} x_\nu' = 0$. Then we find the numbers

$$D_\nu = f_\nu x_\nu' = f_\nu K_x f_\nu = \| f_\nu \|^2 \quad (\nu = 1, 2, \ldots) \qquad (3.9.15)$$

and finally the coordinate vectors

$$x_\nu = D_\nu^{-1} x_\nu' = D_\nu^{-1} K_x f_\nu \quad (\nu = 1, 2, \ldots). \qquad (3.9.16)$$

Formulae (3.9.13)–(3.9.16) give the opportunity to find sequentially the pairs of the vectors $\{f_n, x_n\}$, $f_n \in H_x$, $x_n \in X$. The random variables V_ν in the canonical expansion (3.9.11) will have in this case the variances equal to the correspondent numbers D_ν.

It follows from (3.9.12) that for any $f, g \in H_x$ we have (f, g) $= \sum_{\nu=1}^{\infty} D_\nu^{-1}(f, f_\nu)(f_\nu, g)$ or accounting that on the basis of (3.9.14) (f, f_ν) $= \overline{f K_x f_\nu} = D_\nu \overline{f x_\nu}, (f_\nu, g) = \overline{(g, f_\nu)} = D_\nu \overline{g x_\nu},$

$$f K_x g = \sum_{\nu=1}^{\infty} D_\nu (f x_\nu)(\overline{g x_\nu}) \quad \text{at all} \quad f, g \in H_x . \tag{3.9.17}$$

Consequently, the covariance operator K_x is expressed by the expansion

$$K_x = \sum_{\nu=1}^{\infty} D_\nu x_\nu (\cdot \, \overline{x_\nu}) , \tag{3.9.18}$$

weakly convergent point-wise on the whole H_x (as usual, the place of an argument of the covariance operator is indicated by a point). This expansion represents *a canonical expansion of a covariance operator* of the random variable X. In the special case of the orthonormal basis $\{f_\nu\}$ all the numbers D_ν are equal to 1. ◁

Thus we get the following statement.

Theorem 3.9.1. (The direct Pugachev theorem). *To each canonical expansion (3.9.11) of the random variable X corresponds canonical expansion (3.9.18) of its covariance operator.*

E x a m p l e 3.9.7. Let X be the n-dimensional vector with the expectation m_x and the covariance matrix K_x. If the rank of the matrix K_x is equal to m then $H_x = F_c/N_0$ represents the m-dimensional space with the scalar product $(f, g) = f K_x \overline{g}$. After taking in H_x any linearly independent vectors $\varphi_1, \ldots, \varphi_m$ by formulae (3.9.13)–(3.9.16) we find the orthogonal vectors $f_1, \ldots, f_m$ of the variance $D_\nu = f_\nu K_x f_\nu$ of the uncorrelated random variables $V_1, \ldots, V_m$ and the coordinate vectors $x_1, \ldots, x_m$, $x_\nu = D_\nu^{-1} K_x \overline{f_\nu}$. The random vector X and its covariance matrix K_x will be expressed by the following canonical expansions:

$$X = m_x + \sum_{\nu=1}^{m} V_\nu x_\nu , \quad K_x = \sum_{\nu=1}^{m} D_\nu x_\nu x_\nu^* . \tag{I}$$

As $\varphi_1, \ldots, \varphi_m$ we may take any vectors from the correspondent classes of the equivalent vectors then the unit vectors of any axes of the coordinates in the space X whose projections on F_c/N_0 are linear independent may serve as such vectors.

E x a m p l e 3.9.8. Find a canonical expansion of the finite-dimensional random function $X(t)$ of the finite-dimensional vector argument t, $t \in T_1$. Let

$m_x(t)$ be its expectation, $K_x(t, t')$ be its covariance function, X be a space of all functions t, $t \in T_1$; F_c be a set of the linear functionals

$$fx = \sum_{p=1}^{N} f_p^T x(t_p) = \int \left[\sum_{p=1}^{N} f_p^T \delta(t - t_p) \right] x(t)dt\,, \tag{I}$$

correspondent to all N, t_1, ..., $t_N \in T_1$ and to all vectors (matrix-columns) f_1, ..., f_N. We saw in Example 3.2.1 that a weak convergence in the space X represents a convergence of the functions at each $t \in T_1$. After uniting into one class N_0 all the functionals $f \in F_c$ for which

$$K_x f = \sum_{p=1}^{N} K_x(t, t_p)f_p = \int K_x(t, \tau) \sum_{p=1}^{N} f_p^T \delta(\tau - t_p)d\tau = 0\,, \tag{II}$$

we find the factor-space F_c/N_0 and supplying it on the norm generated by a scalar product

$$(f, g) = fK_x g = \iint f(t)^T K_x(t, t')\overline{g(t')}dtdt'\,, \tag{III}$$

we construct the functional H-space H_x. Suppose that the covariance matrix $K_x(t, t')$ is continuous. In this case H_x is separable as F_c/N_0 is dense in H_x, and any functional from F_c may be approximated with any degree of the accuracy in the metric of the space H_x by the functional of the same type with rational f_p, t_p. After applying formulae (3.9.13)–(3.9.16) to an arbitrary complete consequence of the functions $\{\varphi_\nu(t)\} \subset H_x$, we obtain the sequence of the orthogonal functions $\{f_\nu(t)\}$,

$$(f_\nu, f_\mu) = \iint f_\nu(t')K_x(t, t')\overline{f_\mu(t')}dtdt' = D_\nu \delta_{\nu\mu}\,, \tag{IV}$$

and the sequence of the coordinate functions

$$x_\nu(t) = \frac{1}{D_\nu} \int K_x(t, t')\overline{f_\nu(t')}dt' \quad (\nu = 1, 2, \ldots)\,. \tag{V}$$

The random variables V_ν are determined by formula

$$V_\nu = \int f_\nu(t)^T X^0(t)dt = \int X^0(t)^T f_\nu(t)dt \quad (\nu = 1, 2, \ldots)\,. \tag{VI}$$

As it was proved the random function $X(t)$ is expressed by the canonical expansion

$$X(t) = m_x(t) + \sum_{\nu=1}^{\infty} V_\nu x_\nu(t), \tag{VII}$$

m.s. convergent at each $t \in T_1$. Then the covariance matrix $K_x(t, t')$ will be expressed by the canonical expansion

$$K_x(t, t') = \sum_{\nu=1}^{\infty} D_\nu x_\nu(t) x_\nu(t')^*, \qquad \text{(VIII)}$$

convergent to all $t, t' \in T_1$.

It is known from the general theory that the functionals $f_\nu(t)$ represent the classes of the equivalent functions but not functions itself. It goes without saying that in applications instead of the classes of the functions we take quite definite functions $f_\nu(t)$. Hence it should be noticed that these functions are determined with the accuracy only till arbitrary items $h_\nu(t)$ for which

$$K_x h_\nu = \int K_x(t, t') \overline{h_\nu(t')} dt' = 0 . \qquad \text{(IX)}$$

E x a m p l e 3.9.9. Find a joint canonical expansion of a random function and its derivatives. If under the conditions of the previous example the random function $X(t)$ has m.s. derivatives $\dot{X}(t), \ldots, X^{(n)}(t)$ then the linear combinations of the derivatives of the δ-function till the order n inclusively will enter into the space H_x. In this case the functions $f_\nu(t)$ may also contain the linear combinations of the derivatives of the δ-function till the order n inclusively. Hence not only the random function $X(t)$ but its derivatives are expressed in terms of the canonical expansions

$$X^{(p)}(t) = m_x^{(p)}(t) + \sum_{\nu=1}^{\infty} V_\nu x_\nu^{(p)}(t), \ t \in T_1 , \ (p = 0, 1, \ldots, n), \qquad \text{(I)}$$

which m.s. converge at each $t \in T_1$. Similarly all the derivatives of the covariance function $\partial^{p+q} K_x(t, t')/\partial t^p \partial t'^q$ $(p, q = 0, 1, \ldots, n)$ will be expressed in terms of the canonical expansions

$$\frac{\partial^{p+q} K_x(t, t')}{\partial t^p \partial t'^q} = \sum_{\nu=1}^{\infty} D_\nu x_\nu^{(p)}(t) x_\nu^{(q)}(t)^* , \quad (p, q = 0, 1, \ldots, n), \qquad \text{(II)}$$

convergent at all $t, t' \in T_1$. In the case of a vector argument the operation $\partial/\partial t$ may be substituted by any differential operation in a space of the argument t, in particular, by the operation of the differentiation over any component of the vector t.

E x a m p l e 3.9.10. Find a canonical expansion of a scalar white noise (Subsection 3.3.3). Let $X(t)$ be a white noise of the intensity $\nu(t)$. According to the results of Example 3.9.6 as an orthogonal basis in H_x we may take any complete sequence of the functions $\{f_n(t)\}$ which are orthogonal on a given interval T_1:

$$(f_n, f_m) = \int f_n(t)^T \nu(t) f_m(t) dt = D_n \delta_{nm} . \tag{I}$$

For constructing such sequence by means of formulae (3.9.13) and (3.9.14) we may take any complete sequence of the continuous functions $\{\varphi_n(t)\}$. Formula (3.9.16) for the coordinate functions in this case gives $x_n(t) = D_n^{-1} \nu(t) f_n(t)$. The canonical expansion of a white noise will be m.s. convergent in such sense that any stochastic integral which contains this white noise will be expressed in terms of the series convergent in mean square:

$$\int f(t)^T X(t) dt = \sum_{n=1}^{\infty} \frac{V_n}{D_n} \int f(t)^T \nu(t) \overline{f_n(t)} dt , \quad f(t) \in H_x . \tag{II}$$

Hence in particular it follows that the Wiener process $W(t)$ at any finite interval $[0, T_1]$ is expressed in terms of a canonical expansion:

$$W(t) = \sum_{n=1}^{\infty} V_n g_n(t) , \quad g_n(t) = \frac{1}{D_n} \int_0^t f_n(\tau) d\tau , \tag{III}$$

where $\{f_n(\tau)\}$ is a complete sequence of the orthogonal functions at the interval $[0, T_1]$. The random variables V_n are determined as the stochastic integrals of the form

$$V_n = \int_0^{T_1} \overline{f_n(t)} dW(t) . \tag{IV}$$

So the canonical expansions of the stationary white noise $V(t) = \dot{W}(t)$ and its covariance function $K_v(t, t') = \nu \delta(t - t')$ at the time interval $[0, T_1]$ are determined by the following m.s. convergent series:

$$V(t) = \frac{1}{T_1} \sum_{n=-\infty}^{\infty} V_n e^{i\lambda_n t} , \quad K_v(t, t') = \frac{\nu}{T_1} \sum_{n=-\infty}^{\infty} e^{i\lambda_n(t-t')} \tag{V}$$

where $\lambda_n = 2\pi n / T_1$; $D_n = D[V_n] = \nu T_1$.

The canonical expansions of the nonstationary white noise $V(t)$ and its covariance function $K_v(t, t') = \nu_1(t) \nu_1(t') \delta(t - t')$ have the following form:

$$V(t) = \frac{\nu_1(t)}{T_1} \sum_{n=-\infty}^{\infty} V_n e^{i\lambda_n t} , \quad K_v(t, t') = \frac{\nu_1(t) \nu_1(t')}{T_1} \sum_{n=-\infty}^{\infty} e^{i\lambda_n(t-t')} ,$$

$$\tag{VI}$$

where $\lambda_n = 2\pi n / T_1$; $D_n = D[V_n] = T_1$.

E x a m p l e 3.9.11. Under the conditions of Example 3.9.3 at $D = 1$ and the time interval $[0, T_1]$ the canonical expansion takes place:

$$X(t) = m_x(t) + \sum_{n=1}^{\infty} \frac{2V_n}{T_1 + \lambda_n} \sin\left[\omega_n\left(t - \frac{T_1}{2}\right) + \frac{\pi n}{2}\right],\qquad \text{(I)}$$

where $D_n = D[V_n] = \lambda_n / (T_1 + \lambda_n)$; $\omega_1, \omega_2, \ldots$ are the positive roots of the transcendental equation: $\mathrm{tg}\,\omega T_1 = -2\alpha\omega/(\alpha^2 - \omega^2)$; λ_n are the eigenvalues of the integral equation

$$\int\limits_0^{T_1} e^{-\alpha|t-t'|}\varphi(t')dt' = \lambda\varphi(t).\qquad \text{(II)}$$

E x a m p l e 3.9.12. It is not difficult to make sure that for the covariance function $K_x(t, t') = \nu(t)\nu(t')e^{-\alpha|t-t'|}$ at the time interval $[0, T_1]$ the canonical expansion has the following form:

$$X(t) = m_x(t) + \sum_{n=1}^{\infty} \frac{2V_n\nu(t)}{T_1 + \lambda_n} \sin\left[\omega_n\left(t - \frac{T_1}{2}\right) + \frac{\pi n}{2}\right],\qquad \text{(I)}$$

where λ_n, ω_n, D_n are determined in the previous example.

E x a m p l e 3.9.13. Consider the random function X with the expectation $m_x = m_x(t)$ and the covariance function

$$K_x(t, t') = \begin{cases} q_1(t')q_2(t) & \text{at} \ \ t < t', \\ q_1(t)q_2(t') & \text{at} \ \ t > t', \end{cases}\qquad \text{(I)}$$

where $q_1 = q_1(t)$ and $q_2 = q_2(t)$ are such that the relation q_2/q_1 is a monotonous increasing function. Find its canonical expansion at the interval $T_2 - T_1 \leq t \leq T_2$. We put for some $\alpha > 0$ that

$$\xi = \frac{1}{2\alpha}\ln\frac{q_1(T_2 - T_1)q_2(t)}{q_2(T_2 - T_1)q_1(t)},\ \eta = \frac{1}{2\alpha}\ln\frac{q_1(T_2 - T_1)q_2(T_2)}{q_2(T_2 - T_1)q_1(T_2)}.\qquad \text{(II)}$$

The variable ξ is monotonously increasing function of t and to the interval $T_2 - T_1 \leq t \leq T_2$ corresponds the interval $0 \leq \xi \leq \eta$. We determine the function $Q(\xi)$ by the equality $Q^2(\xi) = q_1(t)q_2(t)$. Then the change of the variable t by ξ gives the

opportunity to present $K_x(t, t')$ in the form $K_x(t, t') = Q(\xi)Q(\xi')e^{-\alpha|\xi-\xi'|}$.
Using the results of the previous example we obtain

$$X(t) = m_x(t) + \sum_{n=1}^{\infty} \frac{2V_n\sqrt{q_1(t)q_2(t)}}{\eta + \lambda_n} \sin\left[\omega_n\sigma(t) + \frac{\pi n}{2}\right], \qquad \text{(III)}$$

where $\lambda_n = 2\alpha/(\alpha^2 + \omega_n^2)$, $D_n = D[V_n] = \lambda_n(\eta + \lambda_n)$,

$$\sigma(t) = \frac{1}{2\alpha}\ln\left\{\frac{q_2(t)}{q_1(t)}\sqrt{\frac{q_1(T_2 - T_1)q_1(T_2)}{q_2(T_2 - T_1)q_2(T_2)}}\right\}, \qquad \text{(IV)}$$

and ω_n are the positive roots of the equation $\operatorname{tg}\eta\omega = -2\alpha\omega/(\alpha^2 - \omega^2)$.

3.9.4. Inverse Theorem about Canonical Expansions

Theorem 3.9.2 (The inverse Pugachev theorem). *If none of the
vectors x_ν belongs to the closed linear span of the other vectors x_ν, i.e.
to the subspace X_ν closed in the weak topology of the space X formed
by the sequence $\{x_\mu\}_{\mu\neq\nu}$ then to canonical expansion (3.9.18) of the covariance operator K_x corresponds canonical expansion (3.9.11) of the
random variable X.*

▷ Really, if $x_\nu \overline{\in} X_\nu$ at none of ν then at each ν there exists the linear functional $f_\nu \in F_c$ whose zero subspace contains X_ν but does not contain x_ν. This
functional may be always chosen in such a way that $f_\nu x_\nu = 1$ (for this purpose it is
sufficient to assume $f_\nu = g_\nu x_\nu$ where g_ν is an arbitrary linear functional with zero
subspace which contains X_ν but not x_ν). Thus the condition of Theorem is sufficient for the existence of the sequence of the linear functions $\{f_\nu\} \in F_c$ satisfying
the condition $f_\nu x_\mu = \delta_{\nu\mu}$. To this sequence corresponds the orthogonal sequence
of the vectors of the space H_x. In fact if the covariance operator K_x is expressed
by canonical expansion (3.9.18) then from $f_\nu \in F_c$, $f_\nu x_\mu = \delta_{\nu\mu}$ follows that
$K_x f_\nu = \sum_{\lambda=1}^{\infty} D_\lambda x_\lambda(\overline{f_\nu x_\lambda}) = D_\nu x_\nu$. Hence it is clear that the elements f_ν of the
space F_c may be replaced by the correspondent classes of the equivalent elements, i.e.
by the vectors of the space H_x and then $f_\mu K_x f_\nu = D_\nu f_\mu x_\nu = D_\nu \delta_{\nu\mu}$. On the
basis of Theorem 3.9.1 the orthogonal sequence of the vectors $\{f_\nu\} \subset H_x$ generates
canonical expansion (3.9.11) of the random variable X with the coordinate vectors
x_ν. ◁

3.9.5. Way to Construct Canonical Expansions

To find the systems of the linear functionals $\{f_n\} \subset F_c$, satisfying the condition of the biorthogonality $f_\nu x_\mu = \delta_{\nu\mu}$, together with the coordinate vectors x_ν of canonical expansion (3.9.18) we may use the following method which is known from functional analysis. At first we take an arbitrary sequence $\{\varphi_n\} \subset F_c$ for which at any n the following condition is fulfilled,

$$\det \begin{bmatrix} \varphi_1 x_1 & \varphi_2 x_1 & \cdots & \varphi_n x_1 \\ \varphi_1 x_2 & \varphi_2 x_2 & \cdots & \varphi_n x_2 \\ \cdots & \cdots & \cdots & \cdots \\ \varphi_1 x_n & \varphi_2 x_n & \cdots & \varphi_n x_n \end{bmatrix} = 1 . \qquad (3.9.19)$$

The mentioned functionals exist always as for any nonzero vector $x \in X$ in F_c there exists such a functional φ that $\varphi x \neq 0$, and consequently, it may be always chosen in such a way that $\varphi x = 1$. After determining the sequence $\{\varphi_n\} \subset F_c$ we assume that

$$\psi_1 = \varphi_1 , \quad \psi_\nu = \alpha_{\nu 1}\psi_1 + \cdots + \alpha_{\nu,\nu-1}\psi_{\nu-1} + \varphi_\nu \quad (\nu = 2, 3, \ldots) , \qquad (3.9.20)$$

and find $\alpha_{\nu\mu}$ sequentially from the conditions $\psi_\nu x_\mu = 0$ at $\mu < \nu$. As a result we obtain

$$\alpha_{\nu 1} = -\varphi_\nu x_1 , \quad \alpha_{\nu\mu} = -\varphi_\nu x_\mu - \sum_{\lambda=1}^{\mu-1} \alpha_{\nu\lambda}\psi_\lambda x_\mu \begin{pmatrix} \mu = 2, \ldots, \nu - 1 \\ \nu = 2, 3, \ldots \end{pmatrix}$$

$$(3.9.21)$$

and $\psi_\nu x_\nu = 1$ $(\nu = 1, 2, \ldots)$. Really, from (3.9.20) with the account that $\psi_\lambda x_\mu = 0$ at $\mu < \lambda$ we find

$$\psi_\nu x_1 = \alpha_{\nu 1}\psi_1 x_1 + \varphi_\nu x_1 , \quad \psi_\nu x_\nu = \sum_{\lambda=1}^{\nu-1} \alpha_{\nu\lambda}\psi_\lambda x_\nu + \varphi_\nu x_\nu ,$$

$$\psi_\nu x_\mu = \sum_{\lambda=1}^{\mu-1} \alpha_{\nu\lambda}\psi_\lambda x_\mu + \alpha_{\nu\mu}\psi_\mu x_\mu + \varphi_\nu x_\mu \quad (\mu = 2, \ldots, \nu - 1) .$$

Putting $\psi_\nu x_\mu = 0$ at $\mu < \nu$ and excluding the unknown $\alpha_{\nu 1}, \ldots, \alpha_{\nu,\nu-1}$ from the obtained ν equations we shall have

$$(\psi_1 x_1) \ldots (\psi_n x_\nu) = \begin{vmatrix} \psi_1 x_1 & 0 & \cdots & 0 & \varphi_\nu x_1 \\ \psi_1 x_2 & \psi_2 x_2 & \cdots & 0 & \varphi_\nu x_2 \\ \cdots & \cdots & \cdots & \cdots & \cdots \\ \psi_1 x_{\nu-1} & \psi_2 x_{\nu-1} & \cdots & \psi_{\nu-1} x_{\nu-1} & \varphi_\nu x_{\nu-1} \\ \psi_1 x_\nu & \psi_2 x_\nu & \cdots & \psi_{\nu-1} x_\nu & \varphi_\nu x_\nu \end{vmatrix} .$$

We replace here zeros by the correspondent expressions $\psi_\lambda x_\mu$ and subtract from the $(\nu-1)^{\text{th}}$ column all the previous columns multiplied by $\alpha_{\nu-1,1}, \ldots, \alpha_{\nu-1,\nu-2}$ correspondingly and then subtract from the $(\nu-2)^{\text{th}}$ column all the previous ones multiplied by $\alpha_{\nu-2,1}, \ldots, \alpha_{\nu-2,\nu-3}$ correspondingly and so on, from the second column the first one multiplied by α_{21}. By virtue of (3.9.20) and (3.9.19) we get

$$
(\psi_1 x_1)\ldots(\psi_\nu x_\nu) =
\begin{vmatrix}
\varphi_1 x_1 & \varphi_2 x_1 & \cdots & \varphi_\nu x_1 \\
\varphi_1 x_2 & \varphi_2 x_2 & \cdots & \varphi_\nu x_2 \\
\cdots & \cdots & \cdots & \cdots \\
\varphi_1 x_\nu & \varphi_2 x_\nu & \cdots & \varphi_\nu x_\nu
\end{vmatrix} = 1 \, .
$$

Putting here sequentially $\nu = 2, 3, \ldots$ and accounting that $\psi_1 x_1 = \varphi_1 x_1 = 1$ we find $\psi_\nu x_\nu$ at all ν. After this we derive formulae (3.9.21) from the obtained equations. Finally, after determining the sequence $\{\psi_n\}$ we put that

$$
f_\nu = \sum_{\lambda=\nu}^{\infty} \beta_{\nu\lambda} \psi_\lambda \quad (\nu = 1, 2, \ldots), \tag{3.9.22}
$$

and calculate $\beta_{\nu\lambda}$ sequentially from the condition $f_\nu x_\mu = \delta_{\nu\mu}$. As a result we get

$$
\beta_{\nu\nu} = 1, \quad \beta_{\nu\mu} = -\sum_{\lambda=\nu}^{\mu-1} \beta_{\nu\lambda} \psi_\lambda x_\mu \quad \begin{pmatrix} \mu = \nu+1, \nu+2, \ldots \\ \nu = 1, 2, \ldots \end{pmatrix} . \tag{3.9.23}
$$

3.9.6. Canonical Expansion of Random Variable in H-space

Theorem 3.9.3. *If the random variable X in H-space possesses such a property that $E \parallel X^0 \parallel^2 < \infty$ (or what is the same $E \parallel X \parallel^2 < \infty$) then it may be presented by a canonical expansion in powers of the eigenvectors of its covariance operator a.s and m.s. convergent.*

▷ Let X be random variable whose phase space is H-space X. The space of the continuous linear functionals dual with X may be identified with X as any continuous linear functional on X is representable in the form $fx = (x, f)$ where $f \in X$. Assuming $F_c = X$ we express a covariance operator of the random variable X by formula

$$
K_x f = EX^0(f, X^0) \, . \tag{3.9.24}
$$

Suppose that $E \parallel X^0 \parallel^2 < \infty$. In this case as it was proved (Section 3.3) the covariance operator K_x is the positively self-adjoint completely continuous operator. Therefore it has a finite or a countable set of the eigenvalues $\{\lambda_n\}$ and the correspondent orthonormal sequence of the eigenvectors $\{\varphi_n\}$:

$$
K_x \varphi_n = \lambda_n \varphi_n \, , \quad (\varphi_\nu, \varphi_\mu) = \delta_{\nu\mu} \, . \tag{3.9.25}
$$

Here for any vector $f \in X$ take place the expansions

$$f = f_0 + \sum_{\nu=1}^{\infty} (f, \varphi_\nu) \varphi_\nu , \quad K_x f_0 = 0 , \tag{3.9.26}$$

$$K_x f = f_0 + \sum_{\nu=1}^{\infty} \lambda_\nu (f, \varphi_\nu) \varphi_\nu . \tag{3.9.27}$$

Hence it is clear that the range Δ_{K_x} is the separable H-space with the orthonormal basis of the operator K_x. Introducing into Δ_{K_x} another scalar product

$$(f, g)_1 = g K_x f = (K_x f, g) = E(X^0, g)(f, X^0)$$

and supplying Δ_{K_x} by the limits of all fundamental sequences in the metric generated by this scalar product we obtain an auxiliary H-space H_x. In essence the latter space does not differ from the space H_x constructed in Subsection 3.9.2. Really, the characteristic property is only in the fact that each element in a new space H_x represents not a class of all the elements X equivalent relatively to K_x but only one element of this which belongs to Δ_{K_x}.

The orthonormal basis $\{\varphi_n\}$ in Δ_{K_x} formed by the eigenvectors of the operator K_x appears to be an orthogonal basis in H_x as $(\varphi_\nu, \varphi_\mu)_1 = (K_x \varphi_\nu, \varphi_\mu) = \lambda_\nu (\varphi_\nu, \varphi_\mu) = \lambda_\nu \delta_{\nu\mu}$. Therefore according to the direct Pugachev theorem the sequence $\{\varphi_n\}$ determines a canonical expansion of the random variable X

$$X = m_x + \sum_{\nu=1}^{\infty} X_\nu \varphi_\nu , \tag{3.9.28}$$

where $X_\nu = (X^0, \varphi_\nu)$ are the uncorrelated random variables with zero expectations and the variances which are equal to the correspondent eigenvalues λ_ν. Formula (3.9.27) gives here the correspondent canonical expansion of the covariance operator of the random variable X.

Expansion (3.9.28) according to the proved theorem weakly m.s. converges to X and expansion (3.9.27) weakly converges to $K_x f$. But according to the theorem about the spectral expansion of completely continuous self-adjoint operator which is known from functional analysis series (3.9.27) converges strongly in the metric of the space X. It turns out that in this case canonical expansion (3.9.28) of the random variable X also strongly a.s. and m.s. converges in the metric of the space X. In fact almost all the realizations of the centered random variable X^0 belong to the space Δ_{K_x} and this means that they belong also to the space H_x. Consequently, almost all the realizations x^0 of the random variable X^0 may be presented by the expansions in powers of the basis $\{\varphi_n\}$:

$$x^0 = \sum_{\nu=1}^{\infty} (x^0, \varphi_\nu) \varphi_\nu . \tag{3.9.29}$$

Hence it is clear that the sequence of the random variables

$$X^n = \sum_{\nu=1}^{n} X_\nu \varphi_\nu = \sum_{\nu=1}^{n} (X^0, \varphi_\nu)\varphi_\nu \quad (\nu = 1, 2, \ldots) \tag{3.9.30}$$

a.s. converges to the random variable X^0. This proves the strong a.s. convergence of canonical expansion (3.9.28). As all random variables X^n are a.s. bounded on the norm of the random variable X^0, $\| X^n \|^2 = \sum_{\nu=1}^{n} |X_\nu|^2 \leq \sum_{\nu=1}^{\infty} |X_\nu|^2 = \| X^0 \|^2$, and $E \| X^0 \|^2 < \infty$ then according to property (xiii) of the expectation (Section 3.1) $E\|X^0\|^2 = \lim E \| X^n \|^2$. But

$$E\|X^n\|^2 = E\left(\sum_{\nu=1}^{n} X_\nu\varphi_\nu, \sum_{\mu=1}^{n} X_\mu\varphi_\mu \right) = \sum_{\nu,\mu=1}^{n} EX_\nu X_\mu(\varphi_\nu, \varphi_\mu) = \sum_{\nu=1}^{n} \lambda_\nu.$$

Consequently, the series $\sum \lambda_\nu$ converges and

$$E\|X^0\|^2 = \sum_{\nu=1}^{\infty} \lambda_\nu . \tag{3.9.31}$$

Finally, from the conditions

$$\left(X^0 - \sum_{\nu=1}^{n} X_\nu\varphi_\nu, \varphi_\mu \right) = (X^0, \varphi_\mu) - X_\mu = 0 \quad (\mu = 1, \ldots, n) \tag{3.9.32}$$

it follows that

$$E\left\| X^0 - \sum_{\nu=1}^{n} X_\nu\varphi_\nu \right\|^2 = E\left(X^0 - \sum_{\nu=1}^{n} X_\nu\varphi_\nu, X^0 \right) = E \| X^0 \|^2$$

$$- \sum_{\nu=1}^{n} EX_\nu(\varphi_\nu, X^0) = E \| X^0 \|^2 - \sum_{\nu=1}^{n} E |X_\nu|^2 = E \| X^0 \|^2 - \sum_{\nu=1}^{n} \lambda_\nu . \tag{3.9.33}$$

On the basis of (3.9.31) this variable tends to 0 at $n \to \infty$ what proves the strong m.s. convergence of expansion (3.9.28). ◁

E x a m p l e 3.9.14. Let X be a finite-dimensional random vector, K_x be its covariance matrix. In a given case the condition $E \| X \|^2 = \operatorname{tr} K_x < \infty$ is always fulfilled if there exists K_x. In this case there exists a finite set of the eigenvalues $\lambda_1, \ldots, \lambda_N$ (not obligatory different) and the correspondent set of the orthonormal vectors $\varphi_1, \ldots, \varphi_n$. The random vector X may be presented by the expansion in powers of the eigenvectors (3.9.28) where $X_\nu = (X^0, \varphi_\nu)$ $= \sum_p X_p^0 \overline{\varphi}_{\nu p}$ are uncorrelated random variables with zero expectations and variances equal to the correspondent eigenvalues λ_ν.

E x a m p l e 3.9.15. Let $X(t)$, $t \in T_1$ be a finite-dimensional vector random function with the realizations in H-space $L_2(T_1, \mu)$. If its covariance function satisfies the condition

$$E \parallel X^0 \parallel^2 = \int \mathrm{tr}\, K_x(t, t) \mu(dt) < \infty, \tag{I}$$

then its covariance operator

$$K_x f = \int K_x(t, t') f(t') \mu(dt') \tag{II}$$

is completely continuous and the random function $X(t)$ may be presented by canonical expansion

$$X(t) = m_x(t) + \sum_{\nu=1}^{\infty} X_\nu \varphi_\nu(t), \quad t \in T_1, \tag{III}$$

in powers of the eigenfunctions $\varphi_\nu(t)$ which are determined by the integral equation

$$\int K_x(t, t') \varphi_\nu(t') \mu(dt') = \lambda_\nu \varphi_\nu(t), \quad t \in T_1. \tag{IV}$$

Here $\{\lambda_\nu\}$ is a nonincreasing sequence of the eigenvalues. Expansion (III) is one of the possible canonical expansions of the random function $X(t)$. The uncorrelated random coefficients of this expansion are determined by the formula

$$X_\nu = (X^0, \varphi_\nu) = \int X^0(t)^T \overline{\varphi_\nu(t)} \mu(dt). \tag{V}$$

Expansion (III) a.s. and m.s. converges in the sense that

$$\int \left\| X^0(t) - \sum_{\nu=1}^{n} X_\nu \varphi_\nu(t) \right\|^2 \mu(dt) \to 0 \quad \text{at} \quad n \to \infty. \tag{VI}$$

3.9.7. Joint Canonical Expansion of Two Random Variables

We saw in Subsection 3.9.6 that in the case of the separable H_x there exists an uncountable set of the canonical expansions of the random variable X. Any orthonormal basis in H_x gives the correspondent canonical expansion of the random variable X. Naturally the idea to find such an orthogonal basis in H_x which would give a canonical expansion of two random variables with the same coordinate vectors appears. This problem is quite analogous to the known problem about simultaneous reducing of two matrices to a diagonal form.

▷ Let X and Y be two random variables with one and the same space X whose covariance operators K_x and K_y are determined on one and the same linear space F_c. Suppose that zero subspace N_0 of the operator K_x is completely contained in zero subspace of the operator K_y, $N_0 \subset \{f : f \in F_c, K_y f = 0\}$. In this case the operator K_y is also determined on the H-space H_x constructed in Subsection 3.9.2 as each class of the functionals $f \in F_c$ equivalent relatively K_y contains the class N_f of functionals equivalent relatively to K_x and consequently, K_y has one and the same value for all $f \in N_f$. The operator K_x mapping H_x into X has evidently the inverse operator K_x^{-1} ($K_x f = 0$ at $f \in H_x$ only in the case when $f = 0$). Let us consider the operator $A = K_x^{-1} K_y$. It is easy to see that this is a self-adjoint operator in H_x as for any $f, g \in H_x$

$$(Af, g) = \overline{(g, Af)} = \overline{g K_x A f} = \overline{g K_x K_x^{-1} K_y f} = \overline{g K_y f}$$

$$= f K_y g = f K_x K_x^{-1} K_y g = f K_x A g = (f, Ag) \tag{3.9.34}$$

and $(Af, f) = f K_y f \geq 0$.

Suppose that a spectrum of the operator A consists only of the eigenvalues different from zero (and consequently, positive) which form the countable set $\{\gamma_\nu\}$. Then there exists an orthonormal sequence of the correspondent eigenvectors $\{\varphi_\nu\}$, $(\varphi_\nu, \varphi_\mu) = \delta_{\nu\mu}$ and for any $f \in H_x$ the following formula is valid

$$Af = \sum_{\nu=1}^{\infty} \gamma_\nu (f, \varphi_\nu) \varphi_\nu. \tag{3.9.35}$$

In a given case the expansion of the unit of the operator A is concentrated on the countable set of the points γ_ν: $\mathcal{E}(B) = \sum_{\gamma_\nu \in B} P_\nu$, where $P_\nu = (\cdot, \varphi_\nu) \varphi_\nu$ is a projector on the vector φ_ν. Hence it follows (3.9.35). As the operator A on the supposition has no zero values then its eigenvectors form a basis in H_x, and consequently, any vector $f \in H_x$ may be presented by the expansion

$$f = \sum_{\nu=1}^{\infty} (f, \varphi_\nu) \varphi_\nu. \tag{3.9.36}$$

If $\gamma = 0$ is an eigenvalue of the operator A then formulae (3.9.35) and (3.9.36) will be also valid if into the sequence $\{\varphi_\nu\}$ we include (orthonormal) eigenvectors which correspond to zero eigenvalue.

According to Subsection 3.9.3 a weakly m.s. convergent canonical expansion of the random variable X corresponds to the orthonormal sequence $\{\varphi_\nu\} \subset H_x$:

$$X = m_x + \sum_{\nu=1}^{\infty} V_\nu u_\nu \,, \tag{3.9.37}$$

where $V_\nu = \varphi_\nu X^0$ are uncorrelated random variables with zero expectations and unit variances, and

$$u_\nu = K_x \varphi_\nu \quad (\nu = 1, 2, \ldots)\,. \tag{3.9.38}$$

Here the covariance operator of the random variable X is expressed by the canonical expansion

$$K_x = \sum_{\nu=1}^{\infty} u_\nu(\overline{\cdot u_\nu}) \,, \tag{3.9.39}$$

which weakly converges in any point of the space H_x.

We prove that the random variable Y is expressed by a canonical expansion with the same coordinate vectors u_ν:

$$Y = m_y + \sum_{\nu=1}^{\infty} W_\nu u_\nu \,, \tag{3.9.40}$$

where $W_\nu = \varphi_\nu Y^0$ are the uncorrelated random variables with zero expectations and the variances equal to the correspondent eigenvalues γ_ν of the operator A. Really,

$$EW_\nu \overline{W}_\mu = E(\varphi_\nu Y^0)(\overline{\varphi_\mu Y^0}) = \varphi_\nu K_y \varphi_\mu = (A\varphi_\nu, \varphi_\mu) \,.$$

As φ_ν represents an eigenvector of the operator A which corresponds to the eigenvalue γ_ν then

$$A\varphi_\nu = \gamma_\nu \varphi_\nu \quad (\nu = 1, 2, \ldots)\,, \tag{3.9.41}$$

and we get $EW_\nu \overline{W}_\mu = \gamma_\nu(\varphi_\nu, \varphi_\mu) = \gamma_\nu \delta_{\nu\mu}$. Consequently, expansion (3.9.40) is a canonical one. For proving its weak m.s. convergence to Y we notice that by virtue of (3.9.38) $fu_\nu = fK_x\varphi_\nu = (f, \varphi_\nu)$ at all $f \in H_x$. Therefore for any $f \in H_x$

$$E\left| fY^0 - \sum_{\nu=1}^{n} W_\nu f u_\nu \right|^2 = E\left| \left\{ f - \sum_{\nu=1}^{n}(f, \varphi_\nu)\varphi_\nu \right\} Y^0 \right|^2$$

$$= \left\{ f - \sum_{\nu=1}^{n}(f, \varphi_\nu)\varphi_\nu \right\} K_y \left\{ f - \sum_{\nu=1}^{n}(f, \varphi_\nu)\varphi_\nu \right\}$$

or by virtue of (3.9.34)

$$E \left| fY^0 - \sum_{\nu=1}^{n} W_\nu f u_\nu \right|^2 = \left(f - \sum_{\nu=1}^{n} (f, \varphi_\nu)\varphi_\nu, \; Af - \sum_{\nu=1}^{n} (f, \varphi_\nu) A\varphi_\nu \right).$$

Hence taking into the consideration (3.9.41) we get

$$E \left| fY^0 - \sum_{\nu=1}^{n} W_\nu f u_\nu \right|^2 = \left(f - \sum_{\nu=1}^{n} (f, \varphi_\nu)\varphi_\nu, \; Af - \sum_{\nu=1}^{n} \gamma_\nu (f, \varphi_\nu)\varphi_\nu \right).$$

$$(3.9.42)$$

From formulae (3.9.42) and the convergence of expansions (3.9.35), (3.9.36) in the metric of the space H_x follows weak m.s. convergence of canonical expansion (3.9.40) of the random variable Y.

Finally, from canonical expansion (3.9.40) of the random variable Y we may determine the correspondent canonical expansion of its covariance operator:

$$K_y = \sum_{\nu=1}^{\infty} \gamma_\nu u_\nu (\overline{\cdot u_\nu}). \; \triangleleft \qquad (3.9.43)$$

Theorem 3.9.4. *The random variables X and Y in a separable H_x space may be represented by canonical expansions (3.9.36) and (3.9.40) with the same coordinate vectors if the spectrum of the operator $A = K_x^{-1} K_y$ consists only of the countable set of the eigenvalues.*

▷ The condition of the proved theorem is fulfilled for instance, when there exists such number $\gamma_0 > 0$ that the operator $B = A - \gamma_0 P$ where P is the operator of the projection on the subspace Δ_A of the values of the operator A is completely continuous. Really, in this case the spectrum of the operator B consists only of the eigenvalues from a finite or a countable set $\{\lambda_n\}$ to which the orthonormal sequence of the eigenvectors $\{\psi_n\}$ corresponds:

$$B\psi_\nu = \lambda_\nu \psi_\nu \quad (\nu = 1, 2, \ldots). \qquad (3.9.44)$$

Here for any vector $f \in H_x$ we have

$$f = \sum_{\nu=1}^{\infty} [(f, \psi_\nu)\psi_\nu + (f, \psi_\nu')\psi_\nu'], \qquad (3.9.45)$$

$$Bf = \sum_{\nu=1}^{\infty} \lambda_\nu (f, \psi_\nu)\psi_\nu, \qquad (3.9.46)$$

where $\{\psi_\nu'\}$ is an orthonormal sequence of the eigenvectors correspondent to zero eigenvalue of the operator B (the space H_x supposed to be separable). After disposing the eigenvectros ψ_ν and ψ_ν' in one sequence $\{\varphi_n\}$ we notice that all of them serve as eigenvectors of the operator A. In this case the eigenvalues of the operator A equal to the number $\gamma_0 + \lambda_\nu$ respond to the eigenvectors ψ_ν belonging to the subspace Δ_B of the values of the operator B; the eigenvalue of the operator A equal to γ_0 corresponds to the eigenvectors ψ_ν' belonging to zero subspace N_B of the operator B but not belonging to zero subspace N_A of the operator A; zero eigenvalue of the operator A responds to the eigenvectors ψ_ν' belonging to N_A. The equation $A\varphi = \gamma\varphi$ may be rewritten in the form:

$$K_y\varphi = \gamma K_x\varphi. \qquad (3.9.47)$$

Consequently, the problem of finding the eigenvalues and the eigenvectors of the operator A is reduced to the calculation of the values γ at which Eq. (3.9.47) has the solutions different from zero and to the definition of the correspondent solutions. Assuming in (3.9.47) $\gamma = \gamma_\nu \neq 0$, $\varphi = \varphi_\nu$ we may rewrite formula (3.9.38) for the coordinate vectors u_ν in the form

$$u_\nu = K_x\varphi_\nu = \gamma_\nu^{-1} K_y\varphi_\nu \quad (\nu = 1, 2, \ldots). \vartriangleleft \qquad (3.9.48)$$

R e m a r k. The conditions of Theorem 3.9.4 are fulfilled if there exists such number $\gamma_0 > 0$ that the operator $B = A - \gamma_0 P$ where P is the operator of the projection on the subspace Δ_A of the values of the operator A is completely continuous.

E x a m p l e 3.9.16. Let us consider the random vectors X and Y in the n-dimensional Euclidean space. Let K_x and K_y be their covariance matrices and $\{f : K_x f = 0\} \subset \{f : K_y f = 0\}$. If K_x has the rank $m \leq n$ then H_x represents the m-dimensional factor-space of Example 3.9.1. Eq. (3.9.47) in this case has m eigenvalues $\gamma_1, \ldots, \gamma_m$ different from zero and the correspondent orthogonal eigenvectors $u_1, \ldots, u_m$ in H_x. Consequently, the vectors X and Y are expressed by the joint canonical expansions:

$$X = m_x + \sum_{\nu=1}^{m} V_\nu u_\nu, \quad Y = m_y + \sum_{\nu=1}^{m} W_\nu u_\nu. \qquad (\mathrm{I})$$

The variances of the random variables $V_1, \ldots, V_m$ are equal to 1, and the variances of the variables $W_1, \ldots, W_m$ are equal to $\gamma_1, \ldots, \gamma_m$ correspondingly. The covariance matrices of the random vectors X and Y are expressed by the canonical expansions

$$K_x = \sum_{\nu=1}^{m} u_\nu u_\nu^*, \quad K_y = \sum_{\nu=1}^{m} \gamma_\nu u_\nu u_\nu^*. \tag{II}$$

E x a m p l e 3.9.17. Consider two m.s. continuous finite-dimensional vector random functions $X(t)$ and $Y(t)$, $t \in T_1$. As F_c we assume the same set of the linear functionals as in Subsection 3.9.3. Then we obtain the same H-space H_x. If zero subspace of the operator K_x is contained in zero subspace of the operator K_y and the operator $A = K_x^{-1} K_y$ in H_x has the spectrum which consists only of the countable set of the eigenvalues then these eigenvalues γ_ν and the correspondent eigenfunctions $\varphi_\nu(t)$ are determined by the integral equation

$$\int K_y(t,t')\overline{\varphi(t')}dt' = \gamma \int K_x(t,t')\overline{\varphi(t')}dt'. \tag{I}$$

According to the proved theorem the random functions $X(t)$ and $Y(t)$ are expressed in a given case by the joint canonical expansions:

$$X(t) = m_x(t) + \sum_{\nu=1}^{\infty} V_\nu u_\nu(t), \quad Y(t) = m_y(t) + \sum_{\nu=1}^{\infty} W_\nu u_\nu(t), \tag{II}$$

where

$$u_\nu(t) = \int K_x(t,t')\overline{\varphi_\nu(t')}dt' = \frac{1}{\gamma_\nu} \int K_y(t,t')\overline{\varphi_\nu(t')}dt', \tag{III}$$

and the second expression is valid only at $\gamma_\nu \neq 0$. The variances of the random variables V_ν are equal to 1, and the variances of the variables W_ν are equal to the correspondent eigenvalues γ_ν. Here the covariance functions of the random functions $X(t)$ and $Y(t)$ are expressed by the canonical expansions

$$K_x(t,t') = \sum_{\nu=1}^{\infty} u_\nu(t)u_\nu(t')^*, \, K_y(t,t') = \sum_{\nu=1}^{\infty} \gamma_\nu u_\nu(t)u_\nu(t')^*. \tag{IV}$$

Canonical expansions (II) m.s. converge at each $t \in T_1$, and expansions (IV) at all $t, t' \in T_1$.

E x a m p l e 3.9.18. If $X(t)$ and $Y(t)$ are two finite-dimensional vector random functions which have m.s. derivatives till the order n inclusively, zero subspace of the operator K_x is contained in zero subspace of the operator K_y (at the same F_c and H_x that in Example 3.9.8) and the spectrum of the operator $A = K_x^{-1} K_y$ in H_x consists only of the countable set of the eigenvalues γ_ν then after determining γ_ν and the correspondent eigenfunctions $\varphi_\nu(t)$ from Eq. (I) of Example 3.9.17 we get expansions (II) and (III) which may be termwise differentiated n times with respect to t and n times with respect to t'.

E x a m p l e 3.9.19. Let $X(t)$ and $Y(t)$ under the conditions of Example 3.9.17 represent the scalar random functions of the scalar variable t, $t \in [0, T_1]$ with the exponential covariance functions: $K_x(t, t') = De^{-\alpha|\tau|}$ and $K_y(t, t') = \Delta e^{-\beta|\tau|}$, $\tau = t - t'$. Eq. (I) has the form

$$\Delta \int_0^{T_1} e^{-\beta|t-\tau|} \varphi(\tau)d\tau = \gamma D \int_0^{T_1} e^{-\alpha|t-\tau|} \varphi(\tau)d\tau. \tag{I}$$

It is easy to verify by direct calculations that the operator $A = K_x^{-1} K_y$ in the space H_x is determined by formula

$$Af = K_x^{-1} K_y f = \frac{\beta\Delta}{\alpha D} f(t)$$

$$+ \frac{(\alpha - \beta)\Delta}{2\alpha D} \int_0^{T_1} \left[(\alpha + \beta)e^{-\beta|t-\tau|} + e^{-\beta\tau} + e^{-\beta(T_1-\tau)} \right] f(\tau)d\tau. \tag{II}$$

In fact for any function $f(t) \in H_x$ we have

$$K_y f = \Delta \int_0^{T_1} e^{-\beta|t-\tau|} f(\tau)d\tau = K_x g = D \int_0^{T_1} e^{-\alpha|t-\tau|} g(\tau)d\tau,$$

where $g(t)$ is the right-hand side of formula (II). It is clear from (II) that the operator $B = A - \gamma_0 I$ at $\gamma_0 = \beta\Delta/\alpha D$ is completely continuous and as a result of this the spectrum of the operator A consists only of the countable set of the eigenvalues $\{\gamma_\nu\}$. It is easy to check out by the substitution that the eigenvalues γ_ν at which Eq. (I) has the solutions different from zero are determined by formula

$$\gamma_\nu = \frac{\beta\Delta}{\alpha D} \frac{\alpha^2 + \omega_\nu^2}{\beta^2 + \omega_\nu^2} \quad (\nu = 1, 2, \ldots), \tag{IV}$$

where ω_ν are the positive roots of the transcendental equation

$$\mathrm{tg}\,\omega T_1 = \frac{2(\alpha + \beta)(\omega^2 - \alpha\beta)\omega}{(\omega^2 - \alpha\beta)^2 - (\alpha + \beta)^2\omega^2}, \tag{V}$$

and the correspondent solutions of Eq. (I) are the orthonormalized eigenfunctions

$$\varphi_\nu(t) = c_\nu\left[(\alpha+\beta)w_\nu(t) + \delta(t) + w_\nu(T_1)\delta(t-T_1)\right], \qquad \text{(VI)}$$

where

$$c_\nu = \frac{\omega_\nu}{\sqrt{D(\beta^2+\omega_\nu^2)}}\left\{T_1 + \frac{(\alpha+\beta)(\omega_\nu^2+\alpha\beta)}{(\alpha^2+\omega_\nu^2)(\beta^2+\omega_\nu^2)}\right\}^{-1/2}, \qquad \text{(VII)}$$

$$w_\nu(t) = \cos\omega_\nu t - \frac{\omega_\nu^2-\alpha\beta}{(\alpha+\beta)\omega_\nu}\sin\omega_\nu t, \quad w_\nu(T_1) = \pm1. \qquad \text{(VIII)}$$

Finally, the coordinate functions of the joint canonical expansion (IV) of the random functions $X(t)$ and $Y(t)$ have the form

$$u_\nu(t) = D\int_0^{T_1} e^{-\alpha|t-\tau|}\varphi_\nu(\tau)d\tau$$

$$= \frac{\Delta}{\gamma_\nu}\int_0^{T_1} e^{-\beta|t-\tau|}\varphi_\nu(\tau)d\tau = Dc_\nu\frac{2\alpha(\alpha+\beta)}{\alpha^2+\omega_\nu^2}w_\nu(t). \qquad \text{(IX)}$$

Similarly as in Subsection 3.9.3 it is sufficient to find the orthogonal eigenvectors φ_ν of the operator $A = K_\nu^{-1}K_y$ which are not obligatory normalized. Then the variances of the random variables V_ν in expansion (3.9.37) will be equal to correspondent numbers $D_\nu = \|\varphi_\nu\|^2 = \varphi_\nu K_x\varphi_\nu$, the variances of the random variables W_ν in expansion (3.9.37) will be equal to the correspondent numbers $\Delta_\nu = \gamma_\nu D_\nu = \varphi_\nu K_y\varphi_\nu$, and the coordinate vectors are determined by formula

$$u_\nu = D_\nu^{-1}K_x\varphi_\nu = \Delta_\nu^{-1}K_y\varphi_\nu. \qquad \text{(X)}$$

Here instead of (3.9.39) and (3.9.43) for the covariance operators K_x and K_y we get formulae

$$K_x = \sum_{\nu=1}^{\infty} D_\nu u_\nu(\cdot\,\overline{u_\nu}), \quad K_y = \sum_{\nu=1}^{\infty} \Delta_\nu u_\nu(\cdot\,\overline{u_\nu}). \qquad \text{(XI)}$$

P r o b l e m s

3.1. Show that a scalar white noise may be obtained by the limit transition from any random function $X(t)$ whose covariance function $K_x(t, t')$ decreases sufficiently quickly with increasing $|t - t'|$. Namely if $|K_x(t, t')| < \sigma(t)\sigma(t')e^{-\alpha|t-t'|}$ then the random function $Y(t)$ whose covariance function is determined by the formula

$$K_y(t, t') = hK_x\left(\frac{t + t'}{2} + h\frac{t - t'}{2}, \frac{t + t'}{2} - h\frac{t - t'}{2}\right) \quad \text{at } h > 0$$

at the limit $h \to \infty$ turns into a white noise. Give the generalization in the case of the random functions in the finite-dimensional spaces.

3.2. Prove that for the reduction of the scalar random function $Y(t) = b_0(t) + b_1(t)X(t)$ to the stationary by linear transformation it is necessary and sufficient that its covariance function will depend on the difference of the arguments

$$K_x(t, t') = K_y(t, t')/\sqrt{D_y(t)D_y(t')} = r_y(t - t').$$

Give the generalization on the case of the functions in the finite-dimensional and infinite-dimensional spaces.

3.3. Prove that for the reduction to the stationary random function $Y(t) = b_0(t) + b_1(t)X(\varphi(t))$ by linear transformation with the simultaneous change of an argument it is necessary and sufficient that the ratio ε of the partial derivatives of its correlation function may be expressed in the form of the ratio of the values of some function ψ $(\varphi = \dot\psi)$ taken with inverse sign at the values of the argument equal to t and t' correspondingly:

$$\varepsilon = \frac{\partial R_y(t, t')/\partial t}{\partial R_y(t, t')/\partial t'} = -\frac{\psi(t)}{\psi(t')}, \quad R_y(t, t') = \frac{K_y(t, t')}{\sqrt{D_y(t)D_y(t')}}.$$

Give the generalization on the case of the random functions in the finite-dimensional spaces.

3.4. Using the result of Problem 3.3 prove that the random function $Y(t)$ with the covariance function

$$K_y(t, t') = c^2 \exp\{-h^2(t^2 + t'^2) - l^2(t - t')^2[a + b(t + t')]^2\}$$

is reduced to the stationary one at $\varphi(t) = at + bt^2$.

3.5. Using the result of Problem 3.3 prove that the random function $Y(t)$ with the covariance function

$$K_y(t, t') = D \exp\{\mu(t + t') - \alpha|t - t'|\}$$

is reduced to the stationary by the transformation $X(s) = e^{-\mu t}Y(t)$, $s = t^2$ where $X(s)$ is the stationary random function with the exponential covariance function $R_x(\sigma) = De^{-\alpha|\sigma|}$, $\sigma = s - s'$.

3.6. Let the function $K(t, t')$, t, $t' \in T_1$ be the covariance function of a random process and $R(u)$ be a polynomial with the positive coefficients. Prove that the function $K_1(t, t') = R(K(t, t'))$ is also a covariance function of a random process.

3.7. Show that an expectation and the second initial moment of a random function

$$Y(t) = \int\limits_0^t \sum_{h,l=1}^n a_{hl} X_h(\tau) X_l(\tau) d\tau$$

where $X(t) = [X_1(t) \ldots X_n(t)]^T$ is a normally distributed random process whose components have the expectations $m_l(t)$ and the covariance and cross covariance functions $K_{hl}(t_1, t_2)$, are determined by the formulae

$$m_y(t) = \int\limits_0^t \sum_{h,l=1}^n a_{hl}[m_h(\tau)m_l(\tau) + K_{hl}(\tau, \tau)]d\tau, \tag{I}$$

$$\Gamma_y(t_1, t_2) = \int\limits_0^{t_1}\int\limits_0^{t_2} \sum_{h,l,p,q=1}^n a_{hl}a_{pq}[K_{hl}(\tau_1, \tau_1)K_{pq}(\tau_1, \tau_2)$$

$$+K_{hp}(\tau_1, \tau_2)K_{lq}(\tau_1, \tau_2) + K_{hq}(\tau_1, \tau_2)K_{lp}(\tau_1, \tau_2)$$

$$+K_{hl}(\tau_1, \tau_1)m_p(\tau_1)m_q(\tau_2) + K_{hp}(\tau_1, \tau_2)m_l(\tau_1)m_q(\tau_2)$$

$$+K_{hq}(\tau_1, \tau_2)m_l(\tau_1)m_p(\tau_2) + K_{lp}(\tau_1, \tau_2)m_h(\tau_1)m_q(\tau_2)$$

$$+K_{lq}(\tau_1, \tau_2)m_h(\tau_1)m_p(\tau_2) + K_{pq}(\tau_2, \tau_2)m_h(\tau_1)m_l(\tau_1)]d\tau_1 d\tau_2. \tag{II}$$

3.8. Show that an expectation and a variance of the random function $Y(t, a)$ which expresses the relative time during which the random process $X(t)$ exceeds the given function $a(t)$ at the time interval (t_0, t),

$$Y(t, a) = \frac{1}{t - t_0} \int\limits_{t_0}^t 1[X(\tau) - a(\tau)]d\tau, \tag{I}$$

are determined by the formulae

$$m_y(t) = \frac{1}{t - t_0} \int\limits_{t_0}^t d\tau \int\limits_{t_0}^t f_1(x; \tau)dx, \tag{II}$$

$$D_y(t) = \frac{1}{(t-t_0)^2} \int\limits_{t_0}^{t} \int\limits_{t_0}^{t} d\tau_1 d\tau_2 \int\limits_{a(\tau_1)}^{\infty} \int\limits_{a(\tau_2)}^{\infty} [\, f_2(x_1, x_2; \tau_1, \tau_2)$$

$$-f_1(x_1; \tau_1) f_1(x_1; \tau_2)\,] dx_1 dx_2 \qquad \text{(III)}$$

where $f_1(x; \tau)$, $f_2(x_1, x_2; \tau_1, \tau_2)$ are the one-dimensional and two-dimensional densities of the random process $X(t)$.

3.9. It is known that the number of the intersections of the scalar process $X(t)$ by the given curve $a(t)$ at the time interval (t_0, t) is determined by formula

$$Z(t, a) = \int\limits_{t_0}^{t} |\dot{X}(\tau) - \dot{a}(\tau)| \delta(X(\tau) - a(\tau)) d\tau. \qquad \text{(I)}$$

Prove that the moment α_r $(r = 0, 1, 2, \ldots)$ of the number of the intersections $Z(t)$ is determined by the formula

$$\alpha_r = \int\limits_{t_0}^{t} d\tau_1 \ldots \int\limits_{t_0}^{t} d\tau_r$$

$$\times \int\limits_{0}^{\infty} \ldots \int\limits_{0}^{\infty} y_1 \ldots y_r f_r(a, y_1, \ldots, a, y_r; \tau_1, \ldots, \tau_r) dy_1 \ldots dy_r \qquad \text{(II)}$$

where $f_r(x_1, y_1, \ldots, x_r, y_r; \tau_1, \ldots, \tau_r)$ is the r-dimensional density of the vector random function $[\, X(\tau)\, Y(\tau)\,]^T$, $Y(\tau) = \dot{X}(\tau)$.

3.10. It is known that the number of the stationary points of the process $X(t)$ at the interval (t_0, t), i.e. the points s where $\dot{X}(s) = 0$ is determined by the formula

$$Y_1(t) = \int\limits_{t_0}^{t} \delta(\dot{X}(\tau)) |\ddot{X}(\tau)| d\tau. \qquad \text{(I)}$$

Derive the formula for the moment α_r $(r = 0, 1, 2, \ldots)$ of the number of the stationary points. Give the generalization on the case of the vector process.

3.11. It is known that the number of the exceeding by the scalar process $X(t)$ of the curve $a(t)$ at the time interval (t_0, t) is determined by the formula

$$Y(t, a) = \int\limits_{t_0}^{t} |\dot{X}(\tau) - \dot{a}(\tau)| 1(\dot{X}(\tau) - \dot{a}(\tau)) \delta(X(\tau) - a(\tau)) d\tau. \qquad \text{(I)}$$

Prove that an expectation and a variance of the random function $Y(t)$ are calculated by the formulae

$$m_y(t) = \int_{t_0}^{t} d\tau \int_{0}^{\infty} \eta f_1(a(\tau), \dot{a}(\tau) + \eta; \tau)d\eta, \qquad (\text{II})$$

$$D_y(t) = \int_{t_0}^{t}\int_{t_0}^{t} d\tau_1 d\tau_2 \int_{0}^{\infty}\int_{0}^{\infty} \eta_1\eta_2[\, f_2(a(\tau_1), \dot{a}(\tau_1) + \eta_1, a(\tau_2), \dot{a}(\tau_2)$$

$$+\eta_2; \tau_1, \tau_2) - f_1(a(\tau_1), \dot{a}(\tau_1)+\eta_1; \tau_1)f_1(a(\tau_2), \dot{a}(\tau_2)+\eta_2; \tau_2)\,]d\eta_1 d\eta_2 \quad (\text{III})$$

where $f_1(x, \dot{x}; \tau_1, \tau_2)$ and $f_2(x_1, \dot{x}_1, x_2, \dot{x}_2; \tau_1, \tau_2)$ are the one-dimensional and two-dimensional densities of the vector process $[\, X(t)\, \dot{X}(t)\,]^T$.

3.12. Show that under the conditions of Problems 3.9 for the stationary process $X(t)$ and constant level $a(t) = a$ an approximate expression for the density of the variable $Z(t, a)$ has the form

$$f(z) = f_0(z)\left[1 + \sum_{r=2}^{n} c_r p_r(z)\right], \qquad (\text{I})$$

$$f_0(z) = \sum_{m=0}^{\infty} \frac{\alpha_1^m}{m!}e^{-\alpha_1}\delta(z - m). \qquad (\text{II})$$

Here $p_r(z)$ are the orthonormal polynomials with the weight $f_0(z)$, $p_0(z) = 1$,
$p_r(z) = \sum_{k=0}^{r} a_{rk}z^k$,

$$c_r = \int_{-\infty}^{\infty} f(z)p_r(z)dz = a_{r0} + \sum_{k=1}^{r} a_{rk}\alpha_k \quad (r = 2, 3, \ldots, n), \qquad (\text{III})$$

where $\alpha_1, \alpha_2, \ldots, \alpha_n$ are the moments of $Z(t, a)$.

3.13. Under the conditions of Problem 3.12 restricting ourselves by the moments no higher than the second order show that

$$p_m = P(Z = m) = \frac{\alpha_1^m e^{-\alpha_1}}{m!}\left[1 + \frac{\alpha_2 - \alpha_1 - \alpha_1^2}{2}\right.$$

$$\left. -\frac{(\alpha_2 - \alpha_1 - \alpha_1^2)(1 + 2\alpha_1)}{2\alpha_1^2}m + \frac{\alpha_2 - \alpha_1 - \alpha_1^2}{2\alpha_1^2}m^2\right], \quad m = 0, 1, 2, \ldots.$$

3.14. Generalize the results of Problem 3.12 on nonstationary case and show that the expression for the distribution function of the moment T of the first intersection has the form

$$F(t) = P(T < t) = 1 - P(Z(t, a) = 0)$$

$$= 1 - \left[1 + \sum_{r=2}^{n} c_r(t) p_r(0, t) \right] e^{-\alpha_1(t)} \tag{I}$$

where in the explicit form the dependence of c_r and p_r on time t is indicated. Derive from (I) an approximate formula restricting to the moments of no higher than the second order:

$$F(t) \approx 1 - \left[1 + \frac{\alpha_2(t) - \alpha_1(t) - \alpha_1^2(t)}{2} \right] e^{-\alpha_1(t)} . \tag{II}$$

3.15. Prove that while choosing as the coefficients V_ν of canonical expansion (3.4.11) the linear combinations of the centered random function $X^0(t)$ with the covariance function $K_x(t, t')$ in the discrete series of the points the exact canonical expansion of $X^0(t)$ in this discrete series of the points is obtained. If the number of the points t_h is finite then (3.4.11) will be the finite sum and is expressed by formula

$$X^0(t_\nu) = \sum_{h=1}^{\nu-1} V_h x_\nu(t_\nu) + V_\nu \tag{I}$$

where

$$D_\nu = D[V_\nu] = K_x(t_\nu, t_\nu) - \sum_{h=1}^{\nu-1} D_h \left| x_h(t_\nu) \right|^2 , \tag{II}$$

$$x_\nu(t) = \frac{1}{D_\nu} \left[K_x(t, t_\nu) - \sum_{h=1}^{\nu-1} D_h x_h(t) \overline{x_h(t_\nu)} \right] \quad (\nu = 2, 3, \ldots). \tag{III}$$

3.16. Show that for the scalar centered random function $X(t)$ with the covariance function $K_x(t, t') = D e^{-\alpha |t-t'|}$ formula (I)–(III) of Problem 3.15 at the points $t_h = -(h-1)\tau$ $(h = 1, 2, \ldots)$ take the following form:

$$x_1(t) = e^{-\alpha |t|} , \quad D_1 = D ,$$

$$x_2(t) = \left\{ e^{-\alpha |t_r|} - q e^{-\alpha |t|} \right\} / (1 - q^2), \quad D_2 = D(1 - q^2) ,$$

$$x_l(t) = \left\{ e^{-\alpha |t+(l-1)\tau|} - q e^{-\alpha |t+(l-2)\tau|} \right\} / (1 - q^2) ,$$

$$D_l = D(1 - q^2), \quad l = 3, 4, \ldots, \quad q = e^{-\alpha\tau} ,$$

$$X(t_1) = V_1 , \quad X(t_l) = V_1 q^{l-1} + \cdots + V_{l-1} q + V_l .$$

3.17. Show that the coordinate functions $x_\nu(t)$ and the variances D_ν of the canonical expansion of the scalar random function $X(t)$ and its known covariance function $K_x(t, t')$ on an arbitrary rectangular grid at $t = t_h$, $t' = t_h$ may be represented in the following form

$$D_1 = K_x(t_1, t_1), \quad x_1(t) = \frac{K_x(t, t_1)}{K_x(t_1, t_1)},$$

$$D_\nu = K_x^{(\nu-1)}(t_\nu, t_\nu), \quad x_\nu(t) = \frac{K_x^{(\nu-1)}(t, t_\nu)}{K_x^{(\nu-1)}(t_\nu, t_\nu)}, \quad (\nu = 2, 3, \ldots),$$

where

$$K_x^{(\nu)}(t, t') = K_x^{(\nu-1)}(t, t') - \frac{K_x^{(\nu-1)}(t, t_\nu) K_x^{(\nu-1)}(t_\nu, t')}{K_x^{(\nu-1)}(t_\nu, t_\nu)}.$$

3.18. Construct a canonical expansion of a scalar random function with the covariance function $K_x(t, t')$ ($|t| < T$) by means of the expansion $K_x(t, t')$ in double Fourier series.

3.19. Let the scalar centered random function $X(t)$ be presented in the form of the linear combination of the known function with the random coefficients $X(t) = \sum_{r=1}^{n} A_r \varphi_r(t)$. Show that one of its canonical expansions will be the following

$$X(t) = \sum_{\nu=1}^{n} V_\nu \psi_\nu(t), \tag{I}$$

where the uncorrelated random coefficients V_ν and the coordinate functions $\psi_\nu(t)$ are determined from equations

$$A_1 = V_1, \quad A_r = \sum_{\nu=1}^{r-1} a_{r\nu} V_\nu + V_r \ (r = 2, \ldots, n), \tag{II}$$

$$\begin{cases} \psi_\nu(t) = \varphi_\nu(t) + \sum_{r=\nu+1}^{n} a_{r\nu} \varphi_r(t) \ (\nu = 1, \ldots, n-1), \\ \psi_n(t) = \varphi_n(t). \end{cases} \tag{III}$$

Here $a_{r\nu}$ and the variances D_ν of the random variables V_ν are expressed in terms of the covariance moments $k_{\nu\mu}$ of the random variables A_r:

$$a_{\nu 1} = k_{\nu 1}/D_1,$$

$$a_{\nu\mu} = \left(k_{\nu\mu} - \sum_{\lambda=1}^{\mu-1} a_{\nu\lambda} \bar{a}_{\mu\lambda} D_\lambda \right) / D_\mu \ (\nu, \mu = 2, \ldots, \nu-1), \tag{IV}$$

$$D_1 = k_{11}, \quad D_\nu = k_{\nu\nu} - \sum_{\lambda=1}^{\nu-1} |a_{\nu\lambda}|^2 D_\lambda. \tag{V}$$

3.20. Under the conditions of Problem 3.20 show that if the scalar centered random function $X(t)$ is represented by means of interpolational Lagrange polynomial:

$$X(t) = \sum_{r=1}^{n} X(t_r) \frac{p(t)}{(t-t_r)p'(t_r)}, \quad p(t) = (t-t_1)(t-t_2)\ldots(t-t_n),$$

then in formulae (III)–(V) we shall have

$$k_{\nu\mu} = K_x(t_\nu, t_\mu),$$

$$x_\nu(t) = \frac{p(t)}{(t-t_\nu)p'(t_\nu)} + \sum_{\mu=\nu+1}^{n} \frac{a_{\mu\nu}p(t)}{(t-t_\mu)p'(t_\mu)} \quad (\nu = 1, \ldots, n-1),$$

$$x_n(t) = \frac{p(t)}{(t-t_n)p'(t_n)}.$$

3.21. Let X be a normally distributed random vector, m its expectation, K its covariance matrix. We divide the vector X into two blocks X_1 and X_2, $X = \left[X_1^T X_2^T \right]^T$, and represent in the corresponding block form its expectation $m = \left[m_1^T m_2^T \right]^T$, covariance matrix K and the inverse matrix $C = K^{-1}$,

$$K = \begin{bmatrix} K_{11} & K_{12} \\ K_{21} & K_{22} \end{bmatrix}, \quad C = K^{-1} = \begin{bmatrix} C_{11} & C_{12} \\ C_{21} & C_{22} \end{bmatrix}. \tag{I}$$

Show that conditional density of the vector X_2 at a given value x_1 of the vector X_1 and conditional moments are given by formulae

$$f_2(x_2 \mid x_1) = [(2\pi)^n \mid K \mid / \mid K_{11} \mid]^{-1/2}$$

$$\times \exp\left\{ -\frac{1}{2} \left[u_2^T C_{22} u_2 + u_2^T C_{22}(C_{22}^{-1}C_{21}u_1) + (C_{22}^{-1}C_{21}u_1)^T C_{22}u_2 \right. \right.$$

$$\left. \left. + (C_{22}^{-1}C_{21}u_1)^T C_{22}C_{22}^{-1}C_{21}u_1 \right] \right\} = [(2\pi)^n \mid K \mid / \mid K_{11} \mid]^{-1/2}$$

$$\times \exp\left\{ -\frac{1}{2} \left[u_2^T + (C_{22}^{-1}C_{21}u_1)^T \right] C_{22} \left[u_2 + C_{22}^{-1}C_{21}u_1 \right] \right\}$$

$$= [(2\pi)^n \mid K \mid / \mid K_{11} \mid]^{-1/2}$$

$$\times \exp\left\{ -\frac{1}{2}\left[u_2^T - (K_{21}K_{11}^{-1}u_1)^T \right] C_{22} \left[u_2 - K_{21}K_{11}^{-1}u_1 \right] \right\}, \qquad \text{(II)}$$

$$m_{2|1} = m_2 + K_{21}K_{11}^{-1}(x_1 - m_1), \qquad \text{(III)}$$

$$K_{2|1} = C_{22}^{-1} = K_{22} - K_{21}K_{11}^{-1}K_{12}, \qquad \text{(IV)}$$

$$\left| K_{2|1} \right| = \left| K_{22} - K_{21}K_{11}^{-1}K_{12} \right| = |K| \,/\, |K_{11}| . \qquad \text{(V)}$$

CHAPTER 4
STOCHASTIC INTEGRALS, SPECTRAL AND INTEGRAL CANONICAL REPRESENTATIONS

The main goal of this chapter is to give essentials of stochastic integrals theory and its applications to spectral and integral canonical representations based on stochastic measures theory. Section 4.1 is devoted to theory of m.s. limits. In Section 4.2 theory of m.s. continuity is considered. Special subsection is dedicated to stochastic processes with uncorrelated increments. Theory of m.s. differentiable stochastic functions in finite- and infinite-dimensional spaces is studied in Section 4.3. Sections 4.4 and 4.5 are dedicated to the integration theory over the nonrandom and random measures. Section 4.6 contains the stochastic Itô, Stratonovich and θ-integrals theory. Special attention is paid to the Wiener and Poisson processes and the corresponding measures. Processes with independent increments and the corresponding white noises are outlined. Section 4.7 is devoted to the theory of generalized stochastic functions. Detailed theory of the integral canonical and spectral representations is given in the last Section 4.8.

4.1. Mean Square Limits

4.1.1. Definitions

In accordance with the general notions of the convergence almost everywhere and in measure in Subsection 2.2.4 two types of the convergence of the random variables were defined: almost sure (a.s.) and in probability. Let us present the necessary definitions for the case of the random variables with the values on B-space.

Let $\{X_r\}$, $X_r = x_r(\omega)$ be an arbitrary set of the random variables with the values in B-space dependent on the parameter r which assumes the values from some set R. In the accordance to the definitions of Subsection 2.2.4 the random variable X_r *converges almost sure* (a.s.) to the random variable $X = x(\omega)$ at $r \to r_0$ if

$$P(X_r \to X) = P(\{\omega : x_r(\omega) \to x(\omega)\}) = 1. \qquad (4.1.1)$$

The random variable X_r *converges in probability* to X if at any $\varepsilon > 0$

$$\lim_{r \to r_0} P(\| X_r - X \| \geq \varepsilon) = \lim_{r \to r_0} P(\{\omega : \| x_r(\omega) - x(\omega) \| \geq \varepsilon\}) = 0.$$

$$(4.1.2)$$

For random variables X_r with the values in a linear space with a weak topology given by means of the set of the linear functionals F_c two types of a weak convergence of the random variables are determined: a.s. and in probability. The random variable X_r *weakly a.s. converges* to X if fX_r a.s. converges to fX at any $f \in F_c$. The random variable X_r *weakly converges* to X *in probability* if fX_r converges in probability to fX at any $f \in F_c$.

Similarly in accordance with the notion of the convergence of the sequence of the functions from L_p in p-mean the convergence of the random variables in p-mean is determined. Consider a set of the random variables $\{X_r\}$, $X_r = x_r(\omega)$ with the values in B-space X. It is said that X_r *converges in mean of the order* p or *shortly* in p-*mean* to the random variable $X = x(\omega)$ at $r \to r_0$, if there exist the absolute moments $E\|X\|^p$, $E\|X_r\|^p$ and

$$E\|X_r - X\|^p \to 0 \quad \text{at} \quad r \to r_0. \tag{4.1.3}$$

In particular, X_r *converges in mean square* (*m.s. converges*) to the random variable X, $X_r \xrightarrow{\text{m.s.}} X$ if $E\|X\|^2$, $E\|X_r\|^2 < \infty$ and

$$E\|X_r - X\|^2 \to 0 \quad \text{at} \quad r \to r_0. \tag{4.1.4}$$

Hence the definition of a weak convergence in p-mean for the random variables $\{X_r\}$, $X_r = x_r(\omega)$ with the values in a weak complete topological linear space which a weak topology is determined by means of the set of the linear functionals F_c follows. They say that X_r *weakly converges in mean of the order p* (*in p-mean*) to the random variable X if at any $f \in F_c$ there exist $E|fX|^p$, $E|fX_r|^p$ and

$$E|fX_r - fX|^p \to 0 \quad \text{at} \quad r \to r_0. \tag{4.1.5}$$

In particular, at $p = 2$ we come to the definition of *weak convergence* in mean square: the random variable X_r *weakly converges in mean square* (*weakly m.s. converges*) *to the random variable* X, $X_r \xrightarrow{\text{w.m.s.}} X$ *if at* $f \in F_c$ *there exist* $E|fX|^2$, $E|fX_r|^2$ *and*

$$E|fX_r - fX|^2 \to 0 \quad \text{at} \quad r \to r_0. \tag{4.1.6}$$

From the general theorems of functional analysis follows that any sequence of the random variables convergent in p-mean also converges to the same limit in probability and we may extract from it the subsequence convergent almost sure. Besides that if the sequence of the random variables $\{X_k\}$, $X_k = x_k(\omega)$, $E\|X_k\|^p < \infty$ (or $E|fX_k|^p < \infty$ at all $f \in F_c$) a.s. converges (weakly converges or in probability) to the random variable $X = x(\omega)$ and is majorized on the norm by the random variable $U = u(\omega) > 0$, $EU^p < \infty$, $\|x_k(\omega)\| \le u(\omega)$ almost at all ω (correspondingly by the random variable $U_f = u_f(\omega) > 0$, $EU_f^p < \infty$, $|fx_k(\omega)| \le u_f(\omega)$ almost at all ω) then $E\|X\|^p < \infty$ (correspondingly $E|fX|^p < \infty$ at all $f \in F_c$) and the sequence $\{X_k\}$ converges (weakly converges) in p-mean to the random variable X.

R e m a r k. The sequence of the random variables convergent in p-mean may be not a.s. convergent and vice versa the sequence of the random variables convergent in probability or a.s. may be not convergent in p-mean.

E x a m p l e 4.1.1. Let X_n be a random variable with two possible values 0 and n whose probabilities are equal to $1 - n^{-2}$ and n^{-2} correspondingly. We saw in Example 2.2.10 that the sequece $\{X_n\}$ converges to 0 in probability and almost sure. But it does not m.s. converge to 0 as at all n

$$E|X_n - 0|^2 = 0 \cdot (1 - 1/n^2) + n^2 \cdot 1/n^2 = 1.$$

E x a m p l e 4.1.2. Let X_n be a random variable with two possible values 0 and 1 whose probabilities are equal to $1 - n^{-1}$ and n^{-1} respectively. We saw in Example 2.2.11 that this sequence converges in probability to 0 but does not a.s. converges. But it m.s. converges to 0 as at $n \to \infty$

$$E|X_n - 0|^2 = 0 \cdot (1 - 1/n) + 1 \cdot 1/n = 1/n \to 0.$$

While studying m.s. convergence of the random variables we naturally suppose that there exist the second order moments at all considered random variables. Therefore we restrict ourselves to the random variables which have the operators of the second order moments. Later on it will be necessary to use the following lemma.

Lemma 4.1.1 (Loève). *Let $\{X_r\}$ and $\{Y_s\}$, $r \in R$, $s \in S$ be two sets of the scalar (in the general case complex-valued) random variables. If X_r m.s. converges to X at $r \to r_0$, and Y_s m.s. converges to Y at $s \to s_0$ then there exists $\lim EX_r\bar{Y}_s = EX\bar{Y}$ independent of the fact how the point (r, s) in the product of the spaces $R \times S$ tends to the point (r_0, s_0).*

▷ For proving it is sufficient to notice that

$$EX_r\bar{Y}_s - EX\bar{Y} = EX_r(\bar{Y}_s - \bar{Y}) + E(X_r - X)\bar{Y},$$

$$| EX_r(\bar{Y}_s - \bar{Y}) |^2 \leq E| X_r |^2 E| Y_s - Y |^2,$$

$$E| X_r |^2 \leq 2E| X |^2 + 2E| X_r - X |^2,$$

$$| E(X_r - X)\bar{Y} |^2 \leq E| X_r - X |^2 E| Y |^2,$$

Therefore $EX_r\bar{Y}_s \to EX\bar{Y}$ if $E| X_r - X |^2 \to 0$, $E| Y_s - Y |^2 \to 0$. ◁

Corollary 4.1.1. *If* $X_r \xrightarrow{\text{m.s.}} X$ *then* $EX_r \to EX$ *and* $X_r^0 \xrightarrow{\text{m.s.}} X_0$.

▷ This follows immediately from the formula

$$E| X_r - X |^2 = | EX_r - EX |^2 + E| X_r^0 - X^0 |^2 . \ ◁$$

Corollary 4.1 2. *If* $X_r \xrightarrow{\text{m.s.}} X$, $Y_s \xrightarrow{\text{m.s.}} Y$ *then the cross-covariance moment of the variables* X_r *and* Y_s *converges to the cross-covariance moment of the random variables* X *and* Y *independently of the fact how* (r, s) *tends to* (r_0, s_0).

▷ Really, according to Corollary 4.1.1 $X_r^0 \xrightarrow{\text{m.s.}} X^0$, $Y_s^0 \xrightarrow{\text{m.s.}} Y^0$ therefore $EX_r^0\bar{Y}_s^0 \to EX^0\bar{Y}^0$. ◁

4.1.2. Theorems about m.s. Convergence

Let $\{X_r, r \in R\}$ be a set of the random variables in a weakly complete topological linear space X whose a weak topology is given by means of the space of the linear functionals F_c and Γ_{rs} be a cross-operator of the second order moment of the variables X_r and X_s.

Theorem 4.1.1. *Let the space* Φ *of all linear functionals on* F_c *coincides with* X *then for a weak m.s. convergence* $f X_r$ *to some random variable* X *in the same space it is necessary and sufficient that at any* $f \in F_c$ *would exist the limit* $\lim_{r,s \to r_0} f\Gamma_{rs}f = \varphi(f)$. *This condition implies that at any* $f, g \in F_c$ *there exists the limit* $\lim_{r,s \to r_0} f\Gamma_{rs}g = f\Gamma g$ *where* Γ *is an operator of the second order moment of the variable* X.

▷ In fact according to the Loève lemma it follows from $f X_r \xrightarrow{\text{m.s.}} f X$ that at any $f \in F_c$ and at all $f, g \in F_c$

$$f\Gamma_{rs}g = E(fX_r)(\overline{gX_s}) \to E(fX)(\overline{gX}) = f\Gamma g.$$

In particular at $g = f$ follows the necessity of the condition of the theorem and $\varphi(f) = f\Gamma f$. If $f\Gamma_{rs}f = E(fX_r)(\overline{fX_s}) \to \varphi(f)$ at any $f \in F_c$ then

$$E| fX_r - fX_s |^2 = E| fX_r |^2 + E| fX_s |^2$$

$$-E(fX_r)(\overline{fX_s}) - E(fX_s)(\overline{fX_r}) \to 0. \qquad (4.1.7)$$

But the random variables fX_r represent the functions $fx_r(\omega)$ of the variable ω belonging to H-space $L_2(\Omega, \mathcal{S}, P)$ (Section 3.2), and $\| fx_r \|^2 = E| fX_r |^2$. Therefore from (4.1.7) in consequence of the completeness of H-space it follows the existence of m.s. limit $U_f = \lim_{r \to \infty} fX_r$ at each $f \in F_c$. This limit evidently is a random linear functional on F_c. As according to the condition $\Phi = X$ any linear functional φ on F_c presentable in the form $\varphi f = fx$ at some $x \in X$ then there exists a random function X in the space X such that $U_f = fX$, $f \in F_c$. Thus X_r weakly m.s. converges to some random variable X what proves the sufficiency of the condition. The condition of the coincidence of Φ and X (the reflexibility of the space X relative to F_c) is always fulfilled in important special case when the space X itself represents a set of all linear functionals on some space Y, and $F_c = \{f : fx = xy, y \in Y\}$ (i.e. $F_c = Y$). Really, in this case the set Φ of all linear functions on F_c is the space X. On the basis of Corollaries 4.1.1 and 4.1.2 of the Loève lemma $X_r \xrightarrow{\text{w.m.s.}} X$ if and only if $EX_r \xrightarrow{\text{w}} EX$ and $fK_{rs}g = E(fX_r^0)(\overline{gX_s^0}) \to E(fX^0)(\overline{gX^0}) = fK_xg$ at all $f, g \in F_c$ (i.e. when EX_r converges to EX in weak topology of the space X, and K_{rs} converges to K in weak topology of the space $\bar{L}(F_c, X)$ of adjoint-linear operators mapping F_c into X, with topology given by the vicinities of zero $\{L : |f_1 Lg_1| < \varepsilon_1, \ldots, |f_n Lg_n| < \varepsilon_n\}$, correspondent to all $f_k, g_k \in F_c$, $\varepsilon_k > 0$, $k = 1, \ldots, n$, $n = 1, 2, \ldots$). ◁

The general theorem just proved allows the following formulation.

Theorem 4.1.2. *For the weak m.s. convergence of X_r to X is it necessary and sufficient a weak convergence of the expectations, the covariance and cross-covariance operators of random variables X_r correspondingly to the expectation and the covariance operator of the random variable X.*

We implement Theorems 4.1.1 and 4.1.2 to the random functions. At first let $\{X_r(t)\}$ be a set of the scalar random functions given on the arbitrary set T_1, F_c be a set of the linear functionals of the form

$$fx = \sum_{k=1}^{N} f_k x(t_k), \qquad (4.1.8)$$

correspondent to all natural N, to all sets of the complex numbers $f_1, \ldots, f_N$ and to all $t_1, \ldots, t_N \in T_1$. In this case a weak convergence in the space of the realizations of the random functions $X_r(t)$ represents the convergence at each t. In order to be sure in this fact it is sufficient to take the functional $fx = x(t_1)$ $(N = 1, f_1 = 1, t_1 \in T_1$ in (4.1.8)). Then from $fx_r \to fx$ at all $f \in F_c$ will follow $x_r(t_1) \to x(t_1)$ at all $t_1 \in T_1$. And vice versa if $x_r(t) \to x(t)$ at all $t \in T_1$ then for any functional f of the form (4.1.8) $fx_r \to fx$. It is evident that in a given case the set of all linear functionals on F_c may be identified with the space of the realizations of the random functions $X_r(t)$ as any linear functional on F_c is expressed by formula (4.1.8).

Corollary 4.1.3. *For the m.s. convergence of $X_r(t)$ to the random function $X(t)$ it is necessary and sufficient the existence of the limit*

$$\lim_{r,s \to r_0} \Gamma_{rs}(t,t) = \lim_{r,s \to r_0} EX_r(t)\overline{X_s(t)}$$

at each fixed $t \in T_1$ and here also exists the limit

$$\lim_{r,s \to r_0} \Gamma_{rs}(t,t') = EX(t)\overline{X(t')} = \Gamma(t,t'), \quad t,t' \in T_1.$$

Corollary 4.1.4. *For the m.s. convergence of $X_r(t)$ to $X(t)$ it is necessary and sufficient the convergence of the expectations $m_{x_r}(t)$ to $m_x(t)$ for all $t \in T_1$ and the convergence of the cross-covariance functions $K_{rs}(t,t')$ to $K_x(t,t')$ for all $t, t' \in T_1$.*

Now let $\{X_r(t)\}$ be a set of n-dimensional vector random functions, $t \in T_1$; F_c be a set of linear functionals of the type

$$fx = \sum_{k=1}^{N} f_k^T x(t_k), \tag{4.1.9}$$

correspondent to all natural N, to all n-dimensional vectors $f_1, \ldots, f_N$ with the complex components and to all $t_1, \ldots, t_N \in T_1$. Analogously as in the previous case we ensure that a weak convergence in the space of the realizations of the random functions X_r represents a component-wise convergence of the vector functions at all $t \in T_1$ and the set Φ of all linear functionals on F_c may be identified with the space X of the realizations of the random functions $X_r(t)$.

Corollary 4.1.5. *For the m.s. convergence $X_r(t)$ to some vector random function $X(t)$ it is necessary and sufficient the existence of the limit*

$$\lim_{r,s \to r_0} \operatorname{tr} \Gamma_{rs}(t,t) = \lim_{r,s \to r_0} \operatorname{tr} EX_r(t)X_s(t)^*, \quad t \in T_1.$$

Here there exist the limit

$$\lim_{r,s \to r_0} \Gamma_{rs}(t,t') = EX(t)X(t')^* = \Gamma(t,t'), \quad t,t' \in T_1.$$

Corollary 4.1.6. *For the m.s. convergence $X_r(t)$ to the vector random function $X(t)$ the convergence of the expectations $m_{x_r}(t)$ to $m_x(t)$ and the covariance and cross-covariance functions $K_{rs}(t,t')$ to the covariance function $K_x(t,t')$, t, $t' \in T_1$ is necessary and sufficient.*

E x a m p l e 4.1.3. Consider a set of the random functions $X_r(t)$ with zero expectations and with covariance and cross-covariance functions $K_{rs}(t,t')$ $= D_{rs}e^{-\alpha_r(t-t')}$ at $t > t'$, $D_{rs} = \alpha_r\alpha_s/(\alpha_r + \alpha_s)$, $K_{rs}(t,t') = K_{sr}(t',t)$. But according to Theorem 4.1.5 $X_r(t)$ m.s. converges to some random function $X(t)$ if and only if there exists the limit $\alpha = \lim\alpha_r$. Here $m_x(t) = 0$ and $K_x(t,t') = \alpha/2e^{-\alpha|t-t'|}$.

E x a m p l e 4.1.4. Under the conditions of the previous example if $\lim\alpha_r = \infty$ then $X_r(t)$ m.s. converges to none of the random functions. But as we shall see in Section 4.7, $X_r(t)$ m.s. converges to the generalized random function $X(t)$ which represents a white noise (Subsection 3.2.3).

4.1.3. Theorems about m.s. Convergence of Random Function in Linear Space

Now we pass to the random functions with the values in the arbitrary linear space L with a weak topology given by means of the set of the linear functionals Λ. As usual we suppose that Λ represents a linear space and it is sufficiently rich for the fact that it will follow $l = 0$ from $\lambda l = 0$ at all $\lambda \in \Lambda$. The cross-second moment $\Gamma_{rs}(t,t')$ of the random functions $X_r(t)$ and $X_s(t')$ in this case is a cross-operator of the second order moment of their values at any $t,t' \in T_1$. In the same way their cross-covariance function $K_{rs}(t,t')$ represents a covariance-operator of their values at any $t,t' \in T_1$.

E x a m p l e 4.1.5. A scalar or a finite-dimensional vector random function of two variables $Z(s,t)$, $s \in S$, $t \in T_1$ may be considered as a random function of the variable t with the values in the space L of the scalar or the finite-dimensional vector functions of the variable $s \in S$. As a set of the linear functionals on L we may assume a set of all the functionals of the type $\lambda l = \sum_{k=1}^{N} \lambda_k l(s_k)$. It is expedient to determine the set F_c of the linear functionals on the space of the realizations of such random functions as the set of all functionals of the form (4.1.8) where $f_1, \ldots, f_N$

are the arbitrary functions from Λ; $f_1, \ldots, f_N \in \Lambda$. We are frequently deal with, for instance, the product of the form $g(s,t)X(t)$ or $G(s,t)X(t)$ under the sign of the integral over t where $g(s,t)$ is a determinate function, and $G(s,t)$ is a random function of the variables s, t. In this case it is expedient to consider $Z(s,t) = g(s,t)X(t)$ or $Z(s,t) = G(s,t)X(t)$ as a random function of the variable t with the values in the space of the functions of the variable s.

From Corollaries 4.1.3 and 4.1.4 about the convergence of the scalar random functions we have the following statement.

Corollary 4.1.7. *The sequence $X_r(t)$ weakly m.s. converges to some random function $X(t)$ with the values in L if and only if at any $\lambda \in \Lambda$ there exists the limit*

$$\lim_{r,s \to r_0} \lambda \Gamma_{rs}(t,t) \lambda = \lim_{r,s \to r_0} E[\lambda X_r(t)][\overline{\lambda X_s(t)}]$$

for each fixed $t \in T_1$. Under this condition for all $\lambda, \lambda' \in \Lambda$ there exists the limit

$$\lim_{r,s \to r_0} \lambda \Gamma_{rs}(t,t') \lambda' = E[\lambda X(t)][\overline{\lambda' X(t')}] = \lambda \Gamma(t,t') \lambda', \quad t, t' \in T_1.$$

Corollary 4.1.8. *For a weak m.s. convergence of $X_r(t)$ to $X(t)$ there are necessary and sufficient a weak convergence of the expectations $m_{x_r}(t)$ to $m_x(t)$ and at each $\lambda \in \Lambda$ the existence of the limit*

$$\lim_{r,s \to r_0} \lambda K_{rs}(t,t) \lambda = \lim_{r,s \to r_0} E[\lambda X_r^0(t)][\overline{\lambda X_s^0(t)}].$$

4.2. Mean Square Continuity of Random Function

4.2.1. Definitions

The random function $X(t)$ with the values in B-space defined on the topological space T_1 is called *continuous in mean square* (*m.s. continuous*) *at the point* $t \in T_1$ if at any $\varepsilon > 0$ there exists such vicinity $V_t(\varepsilon)$ of the point t that

$$E\| X(t') - X(t) \|^2 < \varepsilon \quad \text{at all} \quad t' \in V_t(\varepsilon).$$

The random function $X(t)$ is called *continuous in mean square* (*m.s. continuous*) *on* T_1 if it is m.s. continuous at all $t \in T_1$.

The random function $X(t)$ with the values in the linear space L with the weak topology given by the set of the linear functionals Λ is called *weakly continuous in mean square (weakly m.s. continuous) at the point* $t \in T_1$ (on T_1) if at any $\lambda \in \Lambda$ the scalar random function $\lambda X(t)$ is m.s. continuous at the point $t \in T_1$ (on T_1).

4.2.2. Necessary and Sufficient Conditions of m.s. Continuity

Theorem 4.2.1. *For the m.s. continuity of the scalar random function $X(t)$ at the point $t \in T_1$ $\left(\text{on } T_1\right)$ it is necessary and sufficient the continuity of its second order moment $\Gamma_x(t,t')$ at the correspondent point $(t,t) \in T_1 \times T_1$ (in all diagonal points (t,t) of the product $T_1 \times T_1$; and the function $\Gamma_x(t,t')$ is continuous on $T_1 \times T_1$).*

$\triangleright$ The necessity of the condition follows directly from the Loève lemma: from $X(t') \xrightarrow{\text{m.s.}} X(t)$ follows $\Gamma_x(t',t'') = EX(t')\overline{X(t'')} \to E\,|\,X(t)\,|^2 = \Gamma_x(t,t)$ at $t',t'' \to t$. For proving the sufficiency of the condition we notice that from $|\,\Gamma_x(t',t'') - \Gamma_x(t,t)\,| < \varepsilon/3$ at all $t',t'' \in V_t(\varepsilon)$ we have

$$E\,|\,X(t') - X(t)\,|^2 = \Gamma_x(t',t') - \Gamma_x(t,t) -$$

$$-[\,\Gamma_x(t',t) - \Gamma_x(t,t)\,] - [\,\Gamma_x(t,t') - \Gamma_x(t,t)\,] < \varepsilon\,.$$

Hence by virtue of the arbitrariness $\varepsilon > 0$ it follows m.s. continuity $X(t)$ at the point t. If $X(t)$ is m.s. continuous on T_1 then from the Lòve lemma it follows that $\Gamma_x(t',t'') \to \Gamma_x(t_1,t_2)$ at $t' \to t_1$, $t'' \to t_2$, i.e. the continuity $\Gamma_x(t,t')$ at all points of $T_1 \times T_1$. From the relation

$$E\,|\,X(t') - X(t)\,|^2 = |\,m_x(t') - m_x(t)\,|^2 + E\,|\,X^0(t') - X^0(t)\,|^2$$

and just proved statement we conclude that for m.s. continuity of the random functions $X(t)$ at the point t (on T_1) the continuity of its expectation at the point t (on T_1) and the continuity of its covariance function $K_x(t,t')$ at the point (t,t) of the product of the spaces $T_1 \times T_1$, then $K_x(t,t')$ is continuous on $T_1 \times T_1$ are necessary and sufficient. $\triangleleft$

Now let $X(t)$ be a random function with the values in the linear space L in which a weak topology is given by means of the set of the linear functionals Λ. Using Corollaries 4.1.7 and 4.1.8 for the scalar random function $\lambda X(t)$ we come to the following conclusion.

Theorem 4.2.2. *For a weak m.s. continuity of the random function $X(t)$ at the point t (on T_1) it is necessary and sufficient that one of two following conditions would be fulfilled:*

(i) *at any $\lambda \in \Lambda$ the function $\lambda \Gamma_x(t, t')\lambda$ is continuous at the correspondent point (t, t) (at all diagonal points (t, t) of the product of the space $T_1 \times T_1$; then the function $\lambda \Gamma_x(t, t')\mu$ is continuous on $T_1 \times T_1$ at all $\lambda, \mu \in \Lambda$);*

(ii) *the expectation $m_x(t)$ of the random function is weakly continuous at the point t (on T_1) and at any $\lambda \in \Lambda$ the function $\lambda K_x(t, t')\lambda$ is continuous at the correspondent point (t, t) (in all diagonal points of the product of the spaces $T_1 \times T_1$; then the function $\lambda K_x(t, t')\mu$ is continuous on $T_1 \times T_1$ at all $\lambda, \mu \in \Lambda$).*

E x a m p l e 4.2.1. The function of the form $k(\tau) = De^{-\alpha|\tau|}(\cos \omega \tau + \gamma \sin \omega|\tau|)$, $\tau = t - t'$, is continuous at the point $\tau = 0$ (at all $t' = t$). Therefore the random functions with such covariance function are m.s. continuous on the real axis.

E x a m p l e 4.2.2. The function of the form $K(t, t')$ $= \sum_{k=1}^{N} D_k e^{-\alpha_k|t-t'|} \mathbf{1}(t - t_k)\mathbf{1}(t' - t_k)$ has the discontinuities of the first kind at the points $t' = t = t_k$ $(k = 1, \ldots, N)$. Consequently, the random function with such covariance function is not m.s. continuous on the real axis.

4.2.3. Processes with Uncorrelated Increments

As it is known from elementary probability theory a random process $X(t)$ is called *a process with the uncorrelated increments* if for every nonintersecting intervals $[t_1, t_2)$, $[t_3, t_4)$, $t_1 < t_2 \leq t_3 < t_4$ the correspondent increments $X_{t_2} - X_{t_1}$ and $X_{t_4} - X_{t_3}$ of the process $X(t)$ are uncorrelated. It follows from the definition that the increments of such a process on any finite intervals have the finite moments of the second order (and consequently, of the first order). In this case the process $X(t)$ may have no expectation and the second order moment.

But if at some instant t_0 the value of the process with the uncorrelated increments $X(t)$ is equal to zero a.s., $X_{t_0} \overset{\text{a.s.}}{=} 0$ then the process $X(t)$ has the finite expectation and the second order moment as its value X_t at any instant coincides with its increment on the interval $[t_0, t)$ at $t > t_0$ and with its increment on the interval $[t, t_0)$ taken with the inverse sign if $t < t_0$. Later on we shall consider only such processes with the uncorrelated increments. Any process with the uncorrelated increments $X(t)$ is reduced to such a process. Really, the random process $Y(t) = X(t) - X_{t_0}$ represents a process with the uncorrelated increments which possesses the required property $Y_{t_0} = 0$. Thus if $X_{t_0} \overset{\text{a.s.}}{=} 0$ then the value X_t of the process $X(t)$ at any instant t is uncorrelated with its future increments on the intervals which follow the instant t_0 and with its previous increments on the intervals which precede the instant t_0: $EX_t^0(X_{t_2}^{0^*} - X_{t_1}^{0^*}) = 0$ at $t \leq t_1 < t_2$, $t_1 \geq t_0$ or $t_1 < t_2 \leq t$, $t_2 \leq t_0$.

Let us introduce the function

$$k(t) = \begin{cases} EX_t^0 X_t^{0^*} & \text{at} \quad t > t_0 \,, \\ 0 & \text{at} \quad t = t_0 \,, \\ -EX_t^0 X_t^{0^*} & \text{at} \quad t < t_0 \,. \end{cases} \qquad (4.2.1)$$

It is evident that the covariance matrix of the increment $X_{t_2} - X_{t_1}$ of the process $X(t)$ on any interval and covariance function of the process $X(t)$ are determined by the following formulae:

$$E(X_{t_2}^0 - X_{t_1}^0)(X_{t_2}^{0^*} X_{t_1}^{0^*}) = k(t_2) - k(t_1) \,, \qquad (4.2.2)$$

$$K_x(t_1, t_2) = \begin{cases} k(\min(t_1, t_2)) & \text{at} & t_1, t_2 > t_0 \,, \\ 0 & \text{at} & t \leq t_0 \leq t_2, t_2 \leq t_0 \leq t_1 \,, \\ -k(\min(t_1, t_2)) & \text{at} & t_1, t_2 < t_0 \,. \end{cases} \qquad (4.2.3)$$

Let us study the properties of the function $k(t)$. It is clear from the definition of the function $k(t)$ that at $t > t_0$ it represents a covariance matrix (the variance in the case of the scalar process $X(t)$) of the value of the process $X(t)$ at the instant t, at $t < t_0$ it represents a covariance matrix (the variance) of the value of the process $X(t)$ at the instant t taken with the inverse sign. We conclude from (4.2.2) that the increment of the function $k(t)$ on any interval serves as a covariance matrix (variance) of the increment of the process $X(t)$ on this interval. It follows from these facts that $k(t)$ represents a nondecreasing function which is nonnegative at $t > t_0$, nonpositive at $t < t_0$ and equal to zero at $t = t_0$. In the case of the vector process $X(t)$ the nondecrease of $k(t)$ and the nonnegativeness of $k(t)$ at $t > t_0$ and $-k(t)$ at $t < t_0$ are assumed as a nonnegative definiteness of the matrices $k(t_2) - k(t_1)$, $k(t)$ at $t > t_0$ and $-k(t)$ at $t < t_0$. Thus in order the process $X(t)$ will be a process with the uncorrelated increments and its value at $t = t_0$ will be a.s. equal to zero it is necessary and sufficient that its covariance function will be determined by (4.2.3) where $k(t)$ is nondecreasing function, nonnegative at $t > t_0$ nonpositive at $t > t_0$ and equal to zero at $t = t_0$.

It follows from Theorem 4.2.1 that *the random process with the uncorrelated increments is m.s. continuous if and only if the function $k(t)$ is continuous.*

Further on we shall assume that $k(t)$ is not only continuous but is differentiable. In this case formula (4.2.3) may be written at t_1, $t_2 > t_0$ in the form

$$K_x(t_1, t_2) = \int_{t_0}^{\min(t_1,t_2)} \nu(\tau)d\tau, \qquad (4.2.4)$$

where $\nu(t)$ is a nonnegative function which is called the *intensity* of a process with the uncorrelated increments $X(t)$.

4.3. Differentiation of Random Functions

4.3.1. Definitions

The random function $X(t)$ of a scalar variable $t \in T_1$ with the values in B-space is called *differentiable in mean square (m.s. differentiable) at the point* $t \in T_1$ if at a given t there exists m.s. limit

$$\lim_{r \to 0} \frac{X(t+r) - X(t)}{r}.$$

The random function $X(t)$ is called *differentiable in mean square (m.s. differentiable) on* T_1 if it is m.s. differentiable at all $t \in T_1$.

The random function $X'(t)$ determined from the condition

$$\lim_{r \to 0} E \left\| \frac{X(t+r) - X(t)}{r} - X'(t) \right\|^2 = 0, \qquad (4.3.1)$$

is called *a mean square (m.s.) derivative* of the random function $X(t)$.

The random function $X(t)$ of the scalar variable $t \in T_1$ with the values in the linear space L with a weak topology given by a set of the linear functionals Λ is called *a weakly differentiable in mean square (weakly m.s. differentiable) at the point* $t \in T_1$ (on T_1) if at any $\lambda \in \Lambda$ the scalar random function $\lambda X(t)$ is m.s. differentiable at the point $t \in T_1$ (on T_1).

4.3.2. Theorems about m.s. Differentiability

Let the function $X(t)$ be a scalar random function of the scalar variable $t \in T_1$. We determine the functions $X_r(t) = [X(t+r) - X(t)]/r$. According to Corollary 4.1.3 about m.s. convergence of a scalar random function the m.s. derivative equal by the definition to

$X'(t) = \text{m.s.} \lim\limits_{r \to 0} X_r(t) = \text{l.i.m.}\limits_{r \to 0} X_r$ exists on T_1 if and only if there exists the limit $\lim\limits_{r,s \to 0} EX_r(t)\overline{X_s(t)}$. But

$$EX_r(t)\overline{X_s(t')}$$

$$= [\, \Gamma_x(t + r, t' + s) - \Gamma_x(t + r, t') - \Gamma_x(t, t' + s) + \Gamma_x(t, t')\,] \,/\, rs \,.$$

Consequently, $\lim\limits_{r,s \to 0} EX_r(t)\overline{X_s(t)}$ represents the value of the mixed derivative $\partial^2 \Gamma_x(t, t')/\partial t\, \partial t'$ in the diagonal point (t, t). Thus we have following theorem.

Theorem 4.3.1. *The scalar random function $X(t)$ is m.s. differentiable on T_1 if and only if there exists the finite derivative $\partial^2 \Gamma_x(t, t')/\partial t\, \partial t'$ at all diagonal points $t' = t \in T_1$.*

On the basis of the stated in Section 4.1 the m.s. differentiability of the random function $X(t)$ on T_1 implies the existence of the finite limits on $T_1 \times T_1$ everywhere

$$\lim_{r \to 0} EX_r(t)\overline{X(t')} = \lim_{r \to 0} \frac{1}{r}[\, \Gamma_x(t + r, t') - \Gamma_x(t, t')\,] = \frac{\partial \Gamma_x(t, t')}{\partial t},$$

$$\lim_{s \to 0} EX(t)\overline{X_s(t')} = \lim_{s \to 0} \frac{1}{s}[\, \Gamma_x(t, t' + s) - \Gamma_x(t, t')\,] = \frac{\partial \Gamma_x(t, t')}{\partial t'},$$

$$\lim_{r,s \to 0} EX_r(t)\overline{X_s(t')} = \frac{\partial^2 \Gamma_x(t, t')}{\partial t\, \partial t'},$$

equal to $EX'(t)\overline{X(t')}$, $EX(t)\overline{X'(t')}$, $EX'(t)\overline{X'(t')}$ correspondingly. As a result we obtain the following formulae for the cross-second order moments of the random function $X(t)$ and its m.s. derivative and for the second order moment of m.s. derivative $X'(t)$:

$$\Gamma_{x'x}(t, t') = \frac{\partial \Gamma_x(t, t')}{\partial t}, \quad \Gamma_{xx'}(t, t') = \frac{\partial \Gamma_x(t, t')}{\partial t'}, \tag{4.3.2}$$

$$\Gamma_{x'}(t, t') = \frac{\partial^2 \Gamma_x(t, t')}{\partial t\, \partial t'}. \tag{4.3.3}$$

From Theorem 4.1.2 about m.s. convergence it follows the statement about differentiability.

Theorem 4.3.2. *The scalar random function $X(t)$ is m.s. differentiable on T_1 if and only if its expectation $m_x(t)$ is differentiable on T_1,* *and its covariance function $K_x(t, t')$ has the finite second mixed deriva-tive $\partial^2 K_x(t, t')/\partial t\, \partial t'$ in all diagonal points $t' = t \in T_1$. The expectation of m.s. derivative $X'(t)$ is equal to the derivative of the expectation $m_x(t)$ of the random function $X(t)$:*

$$m_{x'}(t) = EX'(t) = m'_x(t), \qquad (4.3.4)$$

and formulae (4.3.2) and (4.3.3) remain valid if we substitute in them the initial moments Γ by the correspondent covariance function K.

Finally, it follows from the m.s. differentiability of the random func-tion $X(t)$ on the basis of the Loève lemma that for any scalar function $Y(t)$, $t \in T_1$, $E\,|\,Y(t)\,|^2 = \Gamma_y(t, t) < \infty$ everywhere on $T_1 \times T_1$ there exists the finite limits $\lim\limits_{r \to 0} EX_r(t)\overline{Y(t')}$ and $\lim\limits_{s \to 0} EY(t)\overline{X_s(t')}$ equal to $EX'(t)\overline{Y(t')}$ and $EY(t)\overline{X'(t')}$ correspondingly. Thus the cross-second order moments of the random functions $X'(t)$ and $Y(t)$ are expressed in terms of the correspondent cross-second order moments of the random functions $X(t)$ and $Y(t)$ by the formulae analogous to (4.3.2):

$$\Gamma_{x'y}(t, t') = \frac{\partial \Gamma_{xy}(t, t')}{\partial t}, \quad \Gamma_{yx'}(t, t') = \frac{\partial \Gamma_{yx}(t, t')}{\partial t'}. \qquad (4.3.5)$$

From Theorem 4.3.2 and formulae (4.3.2)–(4.3.5) by the induction we get the following theorem.

Theorem 4.3.3. *For the existence of m.s. derivative of the scalar random function $X(t)$ till the order N inclusively it is necessary and suf-ficient the existence of the finite mixed derivative $\partial^{2N}\Gamma_x(t, t')/\partial t^N \partial t'^N$ in all diagonal points $t' = t \in T_1$ (or the existence of the finite deriva-tive $m_x^{(N)}(t)$ on T_1 and finite mixed derivative $\partial^{2N}K_x(t, t')/\partial t^N \partial t'^N$ in all diagonal points $t' = t \in T_1$). Moreover, there exist everywhere the finite derivatives $\partial^{p+q}\Gamma_x(t, t')/\partial t^p \partial t'^q$ and $\partial^{p+q}K_x(t, t')/\partial t^p \partial t'^q$ $(p, q = 0, 1, \ldots, N)$ and the following formulae for the expectations and the second order moments of the derivatives $X^{(p)}(t)$ and $X^{(q)}(t)$ take place:*

$$m_{x^{(p)}}(t) = EX^{(p)}(t) = m_x^{(p)}(t), \qquad (4.3.6)$$

$$\Gamma_{x^{(p)}x^{(q)}}(t, t') = \frac{\partial^{p+q}\Gamma_x(t, t')}{\partial t^p \partial t'^q}, \qquad (4.3.7)$$

$$K_{x^{(p)}x^{(q)}}(t, t') = \frac{\partial^{p+q}K_x(t, t')}{\partial t^p \partial t'^q}. \qquad (4.3.8)$$

E x a m p l e 4.3.1. Find the expectation and the covariance function of the derivative of the random function $X(t)$ if

$$m_x(t) = c \sin \omega_0 t \,, \tag{I}$$

$$K_x(t_1, t_2) = De^{-a\,|t_1 - t_2|} \left\{ \cos \omega_0 (t_1 - t_2) + \frac{a}{\omega_0} \sin \omega_0 \, |t_1 - t_2| \right\}. \tag{II}$$

Applying formulae (4.3.2)–(4.3.4) we find

$$m_{y_1}(t) = m_x'(t) = c\omega_0 \cos \omega_0 t \,. \tag{III}$$

$$K_{xx'}(t_1, t_2) = \frac{\partial K_x(t_1, t_2)}{\partial t_2} = D\frac{a^2 + \omega_0^2}{\omega_0} e^{-a|t_1 - t_2|} \sin \omega_0 (t_1 - t_2). \tag{IV}$$

The covariance function of the m.s. derivative $X'(t)$ is equal

$$K_{x'}(t_1, t_2) = \frac{\partial^2 K_x(t_1, t_2)}{\partial t_1 \partial t_2}$$

$$= D(a^2 + \omega_0^2)e^{-a|t_1 - t_2|} \left\{ \cos \omega_0 (t_1 - t_2) - \frac{a}{\omega_0} \sin \omega_0 \, |t_1 - t_2| \right\}. \tag{V}$$

The latter function is continuous at all t_1, t_2. Therefore the considered random function is m.s. differentiable. As to its realizations they may be differentiable or may be not.

E x a m p l e 4.3.2. Find a covariance function of the derivative of the random function $X(t)$ with zero expectation and the covariance function

$$K_x(t_1, t_2) = De^{\mu(t_1 - t_2) - \alpha|t_1 - t_2|}(1 + \alpha \, |t_1 - t_2|). \tag{I}$$

We have

$$K_{xx'}(t_1, t_2) = \frac{\partial K_x(t_1, t_2)}{\partial t_2} = D\alpha^2(t_1 - t_2)e^{\mu(t_1 + t_2) - \alpha|t_1 - t_2|} + \mu K_x(t_1, t_2), \tag{II}$$

$$K_{x'}(t_1, t_2) = \frac{\partial^2 K_x(t_1, t_2)}{\partial t_1 \partial t_2} = D\alpha^2 e^{\mu(t_1 + t_2) - \alpha|t_1 - t_2|}(1 - \alpha \, |t_1 - t_2|)$$

$$+ \mu^2 K_x(t_1, t_2) + 2D\alpha\mu e^{\mu(t_1 + t_2) - \alpha |t_1 - t_2|} \, |t_1 - t_2| \,. \tag{III}$$

This function is continuous at all $t_1 = t_2 = t$ therefore $X(t)$ is m.s. differentiable. The realizations of the function $X(t)$ may be both differentiable and nondifferentiable.

E x a m p l e 4.3.3. The covariance function of Example 4.2.1 has the finite mixed second derivative

$$\frac{\partial^2 K_x(t,t')}{\partial t\,\partial t'} = -k_x''(\tau) = (\alpha^2+\omega^2)e^{-\alpha|\tau|}\left(\cos\omega\tau - \frac{\alpha}{\omega}\sin\omega|\tau|\right),\quad \tau = t-t',$$

at all $t' = t$. So the random function $X(t)$ with such covariance function and differentiable expectation $m_x(t)$ is m.s. differentiable. As we saw in Examples 2.9.3. and 2.10.3, if all three-dimensional distributions of such random function are normal then almost all its realizations are differentiable.

E x a m p l e 4.3.4. Let us consider the stationary random function $X(t) = Ue^{i\Lambda t}$, where U, Λ are independent random variables, $EU = 0$ and Λ has the density $f(\lambda)$. The covariance function k_x is expressed by the formula

$$k_x(\tau) = D\int\limits_{-\infty}^{\infty} e^{i\lambda\tau} f(\lambda)d\lambda,$$

where D is the variance of the random variable U. At $f(\lambda) = \alpha/\pi(\alpha^2 + \lambda^2)$ we obtain $k_x(\tau) = De^{-\alpha|\tau|}$. This function has no second derivative on the diagonal $t' = t$ (at $\tau = 0$). Consequently, $X(t)$ has no m.s. derivative. It is evident that all its realizations are differentiable. Thus we have an example of a random function which is nondifferentiable in mean square whose all the realizations are differentiable.

From Theorem 4.3.1–4.3.3 and Corollary 4.1.5 it follows.

Theorem 4.3.4. *The vector random function $X(t)$ is m.s. differentiable on T_1 if and only if one of the following two equivalent conditions are fulfilled:*

(i) there exists the finite derivative $\partial^2\,\mathrm{tr}\,\Gamma_x(t,t')/\partial t\,\partial t'$ at all diagonal points $t' = t \in T_1$;

(ii) the expectation $m_x(t)$ is differentiable on T_1 and there exists the finite derivative $\partial^2\,\mathrm{tr}\,K_x(t,t')/\partial t\,\partial t'$ on all diagonal points $t' = t \in T_1$.

Under any condition (i) or (ii) there exist the derivatives $\partial\Gamma_x(t,t')/\partial t$, $\partial\Gamma_x(t,t')/\partial t'$, $\partial^2\Gamma_x(t,t')/\partial t\,\partial t'$, $\partial K_x(t,t')/\partial t$, $\partial K_x(t,t')/\partial t'$, $\partial^2 K_x(t,t')/\partial t\,\partial t'$ everywhere on $T_1 \times T_1$ and formulae (4.3.2)–(4.3.5) are valid.

4.3.3. Theorems about m.s. Differentiability of Random Function in Linear Space

Let $X(t)$ be a random function of a scalar variable $t \in T_1$ with the values in the linear space L whose a weak topology is given by a set (a linear space) of the linear functionals Λ (i.e. by the vicinities of zero $\{\, l : |\lambda_1 l| < \varepsilon_1, \ldots, |\lambda_n l| < \varepsilon_n \,\}$ correspondent to all $\lambda_1, \ldots, \lambda_n \in \Lambda$, all $\varepsilon_1, \ldots, \varepsilon_n > 0$ and $n = 1, 2, \ldots$). The value of the second order moment $\Gamma_x(t, t')$ of the random function $X(t)$ at any given $t, t' \in T_1$ represents an element of the space $\bar{L}(\Lambda, L)$ of the adjoint-linear operators mapping Λ into L. Let us introduce in this space a weak topology by means of the vicinities of zero $\{\, B : B \in \bar{L}(\Lambda, L), |\lambda_1 B \mu_1| < \varepsilon_1, \ldots, |\lambda_n B \mu_n| < \varepsilon_n \,\}$ correspondent to all integer n, all $\lambda_1, \mu_1, \ldots, \lambda_n, \mu_n \in \Lambda$ and all $\varepsilon_1, \ldots, \varepsilon_n > 0$. The derivatives of the operator functions $\Gamma_x(t, t')$ we shall assume as the limits of the correspondent relations of the finite differences in this weak topology. Then from the general theorems about m.s. convergence of the random functions with the values in L (Subsection 4.1.3) we obtain the following statements.

Theorem 4.3.5. *For the existence of a weak m.s. derivative of the random function $X(t)$ on T_1 it is necessary and sufficient the existence of the finite derivative $\partial^2 \lambda \Gamma_x(t, t') \lambda / \partial t\, \partial t'$ in all diagonal points $t' = t \in T_1$ at all $\lambda \in \Lambda$ (or the existence of the weak derivative $m'_x(t)$ and the finite derivative $\partial^2 \lambda K_x(t, t') \lambda / \partial t\, \partial t'$ in all diagonal points $t' = t \in T_1$ at all $\lambda \in \Lambda$). Here there exist the derivatives $\partial \Gamma_x(t, t') / \partial t$, $\partial \Gamma_x(t, t') / \partial t'$, $\partial^2 \Gamma_x(t, t') / \partial t\, \partial t'$, $\partial K_x(t, t') / \partial t$, $\partial K_x(t, t') / \partial t'$, $\partial^2 K_x(t, t') / \partial t\, \partial t'$ everywhere on $T_1 \times T_1$ and formulae (4.3.2)–(4.3.5) are valid.*

Theorem 4.3.6. *For the existence of weak m.s. derivatives of the random function $X(t)$ on T_1 till the order N inclusively it is necessary and sufficient the existence of the finite derivative $\partial^{2N} \lambda \Gamma_x(t, t') \lambda / \partial t^N \partial t'^N$ in all diagonal points $t' = t \in T_1$ at all $\lambda \in \Lambda$ (or the existence of the weak derivative $m_x^{(N)}(t)$ and the finite derivative $\partial^{2N} \lambda K_x(t, t') \lambda / \partial t^N \partial t'^N$ in all diagonal points $t' = t \in T_1$ at all $\lambda \in \Lambda$). Here there exist the derivatives $\partial^{p+q} \Gamma_x(t, t') / \partial t^p \partial t'^q$, $\partial^{p+q} K_x(t, t') / \partial t^p \partial t'^q$ $(p, q = 0, 1, \ldots, N)$ everywhere on $T_1 \times T_1$ and formulae (4.3.6)–(4.3.8) are valid.*

R e m a r k. It is easy to see that all proved theorems and formulae are extended over the random functions of the finite-dimensional vector arguments. Here the operator $\partial / \partial t$ may be replaced by the operator of the partial differentiation over any of the components of the vector t or by any other differential operator in the space of the functions of the vector argument.

E x a m p l e 4.3.5. The random field $X(t)$ has m.s. divergence $\operatorname{div} X(t) = \sum_{k=1}^{n} \partial X(t) / \partial t_k$ on $T_1 \subset R^n$ if and only if there exists the finite function

$\text{div}_t\text{div}_{t'}\Gamma_x(t,t')$ at all $t' = t \in T_1$ (or there exist $\text{div}\,m_x(t)$ on T_1 and $\text{div}_t\text{div}_{t'}K_x(t,t')$ at all $t' = t \in T_1$). Here there exist the functions $T_1 \times T_1$ and formulae (4.3.2)–(4.3.5) with the operators $\text{div}_t\Gamma_x(t,t')$, $\text{div}_{t'}\Gamma_x(t, t')$, $\text{div}_t\text{div}_{t'}\Gamma_x(t,t')$, $\text{div}_t K_x(t,t')$, $\text{div}_{t'}K_x(t,t')$, $\text{div}_t\text{div}_{t'}K_x(t,t')$ everywhere on div_t, $\text{div}_{t'}$ instead of $\partial/\partial t$, $\partial/\partial t'$ are valid.

E x a m p l e 4.3.7. The scalar random field $X(t)$ has m.s. gradient $\nabla X(t)$ if and only if there exists the finite matrix $\nabla_t\nabla_{t'}^T\Gamma_x(t,t')$ at all $t' = t \in T_1$ (or there exist $\nabla m_x(t)$ on T_1 and the finite $\nabla_t\nabla_{t'}^T K_x(t,t')$ at all $t' = t \in T_1$). As usual we represent the vector ∇ in the form of the matrix-column, and the symbol T means the operation of the transposition of the matrix. Here there exist the functions $\nabla_{t'}^T\Gamma_x(t,t')$, $\nabla_t\nabla_{t'}^T\Gamma_x(t,t')$, $\nabla_t K_x(t,t')$, $\nabla_{t'}^T K_x(t,t')$, $\nabla_t\nabla_{t'}^T K_x(t,t')$ everywhere on $T_1 \times T_1$ and formulae (4.3.2)–(4.3.5) with the operators ∇_t, $\nabla_{t'}$ instead of $\partial/\partial t$, $\partial/\partial t'$ are valid.

4.4. Integration of Random Function over Nonrandom Measure

4.4.1. Definitions

Now let us consider the random functions given on the space with the measure $(T_1, \mathcal{T}, \mu)$ where $\mathcal{T}$ is some σ-algebra of the sets of the space T_1, and μ is a nonnegative σ-finite measure given on T_1. Let some linear space L be the space of the values of these random functions.

We shall call *a simple random function* a random function with a finite set of the values whose realizations are measurable relatively to the σ-algebra $\mathcal{T}$ (and the correspondent σ-algebra of the sets of the space L). A simple random function $X(t)$ has the form

$$X(t) = \sum_{p=1}^{N} X_p \mathbf{1}_{B_p}(t), \tag{4.4.1}$$

where $X_1,\ldots,X_N$ are the random variables with values in L and $B_1, \ldots, B_N \in \mathcal{T}$ are measurable sets of the space $(T_1, \mathcal{T})$.

A simple random function is called μ-*integrable* if it is different from zero only on the set of the finite measure μ. As usial an integral with respect to a simple function $X(t)$ over the set $A \in \mathcal{T}$ is as

$$Y(A) = \int_A X(t)\mu(dt) = \int_A X\,d\mu = \sum_{p=1}^{N} X_p\mu(AB_p). \tag{4.4.2}$$

This formula determines the integral with respect to a simple random function with the values in any linear space L. In this case the integral $Y(A)$ represents a random function of the set determined on the σ-algebra $\mathcal{T}$ with the values in the same linear space L.

An expectation, the second order moment and the covariance function of the random function $Y(A)$ on the basis of relations (3.1.8), (3.3.15), (3.3.16) for the random functions with values in a linear space are determined by the following formulae:

$$m_y(A) = EY(A) = \sum_{p=1}^{N} EX_p\mu(AB_p),$$

$$\Gamma_y(A, A') = \sum_{p,q=1}^{N} \Gamma_{pq}\mu(AB_p)\mu(A'B_q),$$

$$K_y(A, A') = \sum_{p,q=1}^{N} K_{pq}\mu(AB_p)\mu(A'B_q),$$

where Γ_{pq} and K_{pq} are the cross-operator of the second order moment and the cross covariance operator of the random variables X_p and X_q correspondingly. As EX_p is the value of the expectation $m_x(t)$ of the random function $X(t)$ at any $t \in B_p$, and Γ_{pq} and K_{pq} are the values of the second order moment $\Gamma_x(t, t')$ and the covariance function $K_x(t, t')$ of the random function $X(t)$ at any $t \in B_p$, $t' \in B_q$ correspondingly then the sums in the obtained formulae represent the integrals with respect to simple functions $m_x(t)$, $\Gamma_x(t, t')$ and $K_x(t, t')$. Therefore they may be rewritten in the form

$$m_y(A) = \int_A m_x(t)\mu(dt), \tag{4.4.3}$$

$$\Gamma_y(A, A')=\int_A\int_{A'}\Gamma_x(t,t')\mu(dt)\mu(dt'), \quad K_y(A, A')=\int_A\int_{A'}K_x(t,t')\mu(dt)\mu(dt'). \tag{4.4.4}$$

Suppose now that V is some abstract set. Analogously we obtain the following formula for the cross second order moment of the random function $Y(A)$ and any random function $Z(v)$, $v \in V$:

$$\Gamma_{yz}(A, v) = \int_A \Gamma_{xz}(t, v)\mu(dt). \tag{4.4.5}$$

At $V = T_1$, $v = t' \in T_1$, $Z(t') = X(t')$ formula(4.4.5) gives

$$\Gamma_{yx}(A, t') = \int_A \Gamma_x(t, t')\mu(dt). \tag{4.4.6}$$

Now we suppose that the random function $Z(v)$ is an integral with respect to the simple random function $X_1(u)$, $u \in U$ over some measure ν determined on the σ-algebra $\mathcal{U}$ of the sets of the space U. In this case $V = \mathcal{U}$, $v = B \in \mathcal{U}$, and

$$Z(B) = \int_B X_1(u)\nu(du),$$

$$\Gamma_{yz}(A, B) = \int_A \int_B \Gamma_{xx_1}(t, u)\mu(dt)\,\nu(du). \tag{4.4.7}$$

Formulae (4.4.5)–(4.4.7) certainly remain valid if we replace in them the functions Γ by the correspondent covariance functions K.

Now let us pass to the general definition of an integral with respect to a random function. The random function $X(t)$ is called μ-integrable in mean square (m.s. μ-integrable) if it is presentable almost at all $t \in T$ as m.s. limit of the sequence of μ-integrable simple random functions $\{X^n(t)\}$ and if there exists m.s. limit of the sequence of the integrals $Y^n(A) = \int_A X^n d\mu$ which is independent of the choice of the sequence $\{X^n(t)\}$. This m.s. limit is called *a mean square (m.s.) integral with respect to random function* $X(t)$:

$$Y(A) = \int_A X(t)\mu(dt) = \int_A X\,d\mu = \underset{n \to \infty}{\text{l.i.m.}}\, Y^n(A). \tag{4.4.8}$$

4.4.2. Theorems about m.s. μ-integrability of Random Function

At once we consider a general case of the random function $X(t)$ with the values in the linear space L in which a weak topology is determined by means of the set of the linear functionals Λ. In this case limit (4.4.8) is assumed as a weak m.s. limit. While applying Theorem 4.1.1 to the sequence $\{Y^n(A)\}$ we come to the following conclusion.

Theorem 4.4.1. *For the m.s. μ-integrability of the random function $X(t)$ it is necessary and sufficient the existence of the integral*

$$I_\lambda(A) = \int\!\!\!\int_{A\,A} \lambda\Gamma_x(t,t')\lambda\,\mu(dt)\,\mu(dt')$$

at all $\lambda \in \Lambda$. Here there exist also at all $\lambda, \lambda' \in \Lambda$ the limits

$$\text{l.i.m.}\,E[\lambda Y^n(A)]\,[\overline{\lambda'Y^m(A')}] = \text{l.i.m.}\,\lambda\Gamma_{y^n y^m}(A,A')\lambda' = \lambda\Gamma_y(A,A')\lambda',$$

$$\text{l.i.m.}\,E[\lambda Y^n(A)]\,[\overline{\lambda'Z(v)}] = \text{l.i.m.}\,\lambda\Gamma_{y^n z}(A,v)\lambda' = \lambda\Gamma_{yz}(A,v)\lambda'.$$

From this fact and the validity of formulae (4.4.3)–(4.4.5) for the μ-integrable of simple random functions it follows the validity of these formulae and together with them the validity of formulae (4.4.6) and (4.4.7) for any m.s. μ-integrable random functions.

Now we consider an important case when the integrable function has the form $g(s,t)X(t)$ where $X(t)$ is a finite-dimensional vector random function represented as usual in the form of a matrix-column, and $g(s,t)$ is a nonrandom matrix. The random function of such a type may be assumed as a random function of the variable t with the values in the linear space L of the functions of the variable s. If we take as the set Λ of the linear functionals on L a set of the functionals of the type $\lambda l = \sum_{k=1}^{N} \lambda_k^T l(s_k)$, which are correspondent to all N and $s_1, \ldots, s_N \in S$ and to all $\lambda_1, \ldots, \lambda_N$ then as a weak convergence in L will be the convergence at each $s \in S$ (a point-wise convergence). From Theorem 4.4.1 in a given case it follows that at any $s \in S$ the m.s. integral

$$Y(s) = \int g(s,t)X(t)\mu(dt) \tag{4.4.9}$$

exists if and only if there exists the integral

$$I(s) = \int\!\!\!\int \text{tr}\,[\,g(s,t)\Gamma_x(t,t')g(s,t')^*\,]\,\mu(dt)\,\mu(dt'). \tag{4.4.10}$$

So the following formulae are valid

$$m_y(s) = \int g(s,t)m_x(t)\mu(dt), \tag{4.4.11}$$

$$\Gamma_y(s,s') = \int\!\!\!\int g(s,t)\Gamma_x(t,t')g(s',t')^*\mu(dt)\mu(dt'), \tag{4.4.12}$$

$$\Gamma_{yz}(s, v) = \int g(s, t)\Gamma_{xz}(t, v)\mu(dt), \qquad (4.4.13)$$

$$\Gamma_{yx}(s, t) = \int g(s, t')\Gamma_x(t, t')\mu(dt'). \qquad (4.4.14)$$

If the random function $Z(v)$ represents the m.s. integral

$$Z(v) = \int h(v, u)X_1(u)\mu(du), \qquad (4.4.15)$$

then on the basis of (4.4.13) we obtain

$$\Gamma_{xz}(t, v) = \int \Gamma_{xx_1}(t, u)h(v, u)^*\nu(du),$$

$$\Gamma_{yz}(s, v) = \int\int g(s, t)\Gamma_{xx_1}(t, u)h(v, u)^*\mu(dt)\nu(du). \qquad (4.4.16)$$

The integrals (4.4.10)–(4.4.16) represent the known from functional analysis ordinary abstract Lebesgue integrals. Formulae (4.4.12)–(4.4.16) are also valid for the correspondent covariance operators.

R e m a r k. The condition of the representability of $g(s, t)X(t)$ in the form of m.s. limit of the sequence of simple random functions of the variable t at each $s \in S$ is fulfilled, for instance, in the case when $g(s, t)$ is a measurable function of t at any $s \in S$, the random function $X(t)$ is m.s. continuous and the variable t is finite-dimensional. In fact, any measurable function with the values in the separable B-space (in particular, in the finite-dimensional Euclidean space) may be presented as the limit of the sequence of the simple functions convergent almost everywhere. The m.s. continuous random function of the finite-dimensional variable may be also presented as m.s. limit of the sequence of the simple random functions. And the product of two simple functions is a simple function too.

Concluding we notice that the scalar random function $X(t)$ with the finite second order moment may be considered as the function of the variable t with the values in the B-space $L_2(\Omega, S, P)$ of the functions of the variable ω, and the integral with respect to $X(t)$ may be considered as an ordinary integral with respect to the function with the values in B-space (Bochner integral). The integral with respect to the random function $X(t)$ exists in this sense if and only if its norm $\|X(t)\|$ is integrable,

$$\|X(t)\| = \left[\int |x(t, \omega)|^2 P(d\omega)\right]^{1/2} = [E\,|X(t)|^2]^{1/2} = \sqrt{\Gamma_x(t, t)}.$$

It is evident that if there exists an integral with respect to the random function $X(t)$ in this sense then there also exist the integral with respect to it in m.s. sence as

$$\left| \iint \Gamma_x(t,t')\mu(dt)\mu(dt') \right| \leq \iint |\Gamma_x(t,t')|\mu(dt)\mu(dt')$$

$$\leq \iint \sqrt{\Gamma_x(t,t)\Gamma_x(t',t')}\mu(dt)\mu(dt') = \left[\int \sqrt{\Gamma_x(t,t)}\mu(dt) \right]^2.$$

But the inverse conclusion is not correct as m.s. integral with respect to the random function may also exist in the case when the Bochner integral with respect to it does not exist.

E x a m p l e 4.4.1. Find a covariance function of a random function

$$Y(t) = \int\limits_0^t X(\tau)d\tau, \tag{I}$$

if $X(t)$ has the exponential covariance function $K_x(t_1,t_2) = De^{-\alpha|t_1-t_2|}$. Using formula (4.4.12) we find

$$K_y(t_1,t_2) = D \int\limits_0^{t_1} \int\limits_0^{t_2} e^{-\alpha|\tau_1-\tau_2|} d\tau_1 d\tau_2. \tag{II}$$

At $t_1 < t_2$ formula (II) gives

$$K_y(t_1,t_2) = D \int\limits_0^{t_1} \left\{ \int\limits_0^{\tau_1} e^{-\alpha(\tau_1-\tau_2)} d\tau_2 + \int\limits_0^{\tau_2} e^{-\alpha(\tau_2-\tau_1)} \right\} d\tau_1$$

$$= \frac{2D}{\alpha}t_1 + \frac{D}{\alpha^2} \left[e^{-\alpha t_1} + e^{-\alpha t_2} - e^{-\alpha(t_2-t_1)} - 1 \right]. \tag{III}$$

At $t_1 > t_2$ in consequence of the symmetry of the covariance function the variables from (III) t_1 and t_2 change the places. As a result we get

$$K_y(t_1,t_2) = \frac{2D}{\alpha} \min\{t_1,t_2\} + \frac{D}{\alpha^2} \left[e^{-\alpha t_1} + e^{-\alpha t_2} - e^{-\alpha|t_1-t_2|} - 1 \right]. \tag{IV}$$

E x a m p l e 4.4.2. Let us consider the random variable $X(t)$, $t \in [0, \infty)$ with zero expectation and the covariance function

$$K_x(t, t') = \Gamma_x(t, t') = \exp\left\{ -\left| e^{\alpha t}t - e^{\alpha t'}t' \right| \right\}. \tag{I}$$

In a given case the integral with respect to the random function $X(t)$ on Lebesgue measure does not exist. But m.s. integral with respect to this random function exists as it is m.s. continuous, and consequently, is representable as m.s. limit of the sequence of the simple random functions and

$$\int_0^\infty\!\!\int_0^\infty \Gamma_x(t, t')\, dt\, dt' = 2\int_0^\infty dt \int_0^t \Gamma_x(t, t')dt' = 2\int_0^\infty dt \int_0^t \exp\left\{ -e^{\alpha t}t \right.$$

$$\left. +e^{\alpha t'}t' \right\} dt' < 2\int_0^\infty \exp\left\{ -e^{\alpha t}t \right\} dt \int_0^t \exp\left\{ e^{\alpha t}t' \right\} dt'$$

$$= 2\int_0^\infty [\,1 - \exp\left\{ -e^{\alpha t} \right\}]e^{-\alpha t}dt < 2\int_0^\infty e^{-\alpha t}dt = \frac{2}{\alpha} < \infty. \tag{II}$$

E x a m p l e 4.4.3. In the case of the real determined function $g(t, \tau)$ and the real random function $X(t)$ the formulae for the high orders moments of the random function $Y(t)$ are analogously derived. While restricting for the simplicity to the case of the scalar-valued $g(t, \tau)$ and $X(t)$ we cite the formula for the moment of the rth order of m.s. integral

$$\alpha_r^y(t_1, \ldots, t_r) = \int_T \cdots \int_T g(t_1, \tau_1)\ldots g(t_r, \tau_r)\alpha_r^x(\tau_1, \ldots, \tau_r)d\tau_1 \ldots d\tau_r. \tag{I}$$

This formula is valid not only for the initial moments α but for the centered moments μ. It is necessary and sufficient for the existence of the moment $\alpha_r^y(t_1, \ldots, t_r)$ that the same moment of the random function $X(t)$ will exist and that the integral in the right-hand side of (I) will converge.

4.5. Integration over Random Measure

4.5.1. Definitions

A random function of the set $Z(A)$ determined on the σ-algebra is called *a random measure in the measurable space* $(T_1, \mathcal{T})$ *if it* $\mathcal{T}$ *the property of the* σ-*additivity*

$$Z(\bigcup A_n) = \sum Z(A_n) \quad \text{a.s.} \tag{4.5.1}$$

for any sequence of pair-wise nonintersecting sets $\{A_n\} \subset \mathcal{T}$. The random measures may have the values in any linear spaces.

Let $Z(A)$ be a scalar random measure, $E\,|\,Z(A)\,|^2 < \infty$ for all $A \in \mathcal{T}$. Its expectation $m_Z(A)$ by virtue of (4.5.1) represents a numerical measure (in the general case a complex-valued) determined on the σ-algebra $\mathcal{T}$. The second order moment of the random measure $Z(A)$ determined as $\Gamma_z(A, A') = EZ(A)\overline{Z(A')}$ and considered as the function of the set A at fixed $A' \in \mathcal{T}$ represents also the numerical measure on the σ-algebra $\mathcal{T}$. If we assign $\Gamma_z(A, A')$ as a function of the rectangular set $A \times A'$ in the product of the measurable spaces $(T_1 \times T_1, \mathcal{T} \times \mathcal{T})$ then on the basis of (4.5.1) we get

$$\Gamma_z(\bigcup A_n, \bigcup B_m) = \sum_{n,m=1}^{\infty} \Gamma_z(A_n, B_m)$$

for any sequences of the pair-wise nonintersecting rectangular sets $\{A_n \times B_n\} \subset \mathcal{T} \times \mathcal{T}$, $A_k A_h = \emptyset$, $B_k B_h = \emptyset$ at $h \neq k$.

Thus *the second order moment of the random measure* represents the σ-additive function of the set on the class of all measurable rectangles of the space $(T_1 \times T_1, \mathcal{T} \times \mathcal{T})$. In other words, $\Gamma_z(A, A')$ represents in general case a complex-valued function of the set on the class of measurable rectangles of the space $(T_1 \times T_1, \mathcal{T} \times \mathcal{T})$.

The integral with respect to the random function (4.4.1) with the values in any linear space L over the random measure $Z(A)$ is defined as

$$Y(A) = \int_A X(t)Z(dt) = \int_A X\,dZ = \sum_{p=1}^{N} X_p Z(AB_p).$$

The integral represents a random function of the set with the values in the same space L determined on the σ-algebra $\mathcal{T}$.

Now we suppose that the random functions $X(t)$ and $Z(A)$ are independent, i.e. that their values at any $t \in T_1$ and any $A \in T$ are independent random variables. In this case (Section 3.4) the following equalities are valid

$$EX_p Z(AB_p) = EX_p \cdot EZ(AB_p),$$

$$EX_p \bar{X}_q Z(AB_p)\overline{Z(A'B_q)} = EX_p \bar{X}_q \cdot EZ(AB_p)\overline{Z(A'B_q)}$$

and consequently,

$$m_y(A) = \sum_{p=1}^{N} EX_p m_z(AB_p) = \int_A m_x(t)m_z(dt), \qquad (4.5.2)$$

$$\Gamma_y(A, A') = \sum_{p,q=1}^{N} \Gamma_{pq}\Gamma_z(AB_p, A'B_q) = \int_A \int_{A'} \Gamma_x(t, t')\Gamma_z(dt, dt'). \quad (4.5.3)$$

Let us agree that some fact takes place almost at all $t \in T_1$ relatively to the random measure Z if it does not take place only at t belonging to some set $N \in T$ on which the measure Z is equal to zero almost sure. Evidently, it is identical to say that a given case takes place almost at all $t \in T_1$ relatively to the measure $m_z(A)$ and to the measures $\Gamma_z(A, A')$ correspondent to all fixed $A' \in T$.

The random function $X(t)$ is called *m.s. integrable over the random measure Z* if it is representable almost at all $t \in T_1$ relatively to Z as the m.s. limit of the sequence of the simple functions $Y^n(A) = \int_A X^n dZ$, and if there exists the m.s. limit of the sequence of the integrals independent of the choice of the sequence $\{X^n(t)\}$. This m.s. limit is called an *integral over the random measure Z* with respect to the random function $X(t)$:

$$Y(A) = \int_A X(t)Z(dt) = \int_A X dZ = \underset{n \to \infty}{\text{l.i.m.}} Y^n(A). \qquad (4.5.4)$$

4.5.2. Necessary and Sufficient Conditions of Integrability over Random Measure

Theorem 4.5.1. *For the existence of an integral with respect to the random function $X(t)$ over the random measure Z it is necessary and sufficient the existence of the integral*

$$J_\lambda(A) = \int_A \int_A \lambda \Gamma_x(t, t')\lambda \Gamma_z(dt, dt') \qquad (4.5.5)$$

for all $\lambda \in \Lambda$. Here for all $\lambda, \lambda' \in \Lambda$ there exist the limits l.i.m. $E[\lambda Y^n(A)]$
$= \lambda m_y(A)$, l.i.m. $E[\lambda Y^n(A)][\overline{\lambda' Y^m(A')}] = \lambda \Gamma_y(A, A')\lambda'.$

From this fact and formulae (4.5.2) and (4.5.3) follow the corresponding formulae for the simple random functions

$$m_y(A) = \int\limits_A m_x(t)m_z(dt) = \int\limits_A m_x dm_z, \qquad (4.5.6)$$

$$\Gamma_y(A, A') = \int\limits_{A A'}\!\!\int \Gamma_x(t, t')\Gamma_z(dt, dt'). \qquad (4.5.7)$$

Let us consider a special case of the m.s. continuus scalar or finite-dimensional vector random function $X(t)$. For such random function we have the following statement.

Theorem 4.5.2. *An integral over the random measure Z exists if its second order moment $\Gamma_x(t, t')$ is bounded and the total variation of the measure $\Gamma_z(A, A')$ on the class of measurable rectangles of the product of the spaces $(T_1 \times T_1, T \times T)$ is finite.*

Both Theorems 4.5.1 and 4.5.2 may be proved similarly as in Section 3.4.

4.6. Stochastic Integral

4.6.1. Definitions

The random *measure* $Z(A)$ is called *stochastic* if its expectation is equal to zero and the covariance function depends only on the intersection of the sets-arguments:

$$K_z(A, A') = EZ(A)\overline{Z(A')} = \sigma(AA'). \qquad (4.6.1)$$

It is easy to see that the function $\sigma(A)$ represents a nonnegative measure as at any $A \in T$

$$\sigma(A) = K_z(A, A) = \Gamma_z(A, A) = E|Z(A)|^2 \geq 0$$

and by virtue of the σ-additivity $\Gamma_z(A, A')$ on A

$$\sigma(\bigcup A_k) = \Gamma_z(\bigcup A_k, T_1) = \sum \Gamma_z(A_k, T_1) = \sum \sigma(A_k)$$

for any sequence $\{A_k\} \subset \mathcal{T}$ of mutually nonintersecting sets. From the properties of the additive functions it follows that $\sigma(\varnothing) = 0$, i.e. that the values of the stochastic measure on nonintersecting sets are uncorrelated.

E x a m p l e 4.6.1. Let $W_1(t), \ldots, W_N(t)$ be the independent scalar processes with the uncorrelated increments (Section 4.2.3). The formula

$$Z\left(\prod_{r-1}^{N}[a_r, b_r)\right) = \prod_{r=1}^{N}[W_r(b_r) - W_r(a_r)]$$

determines the stochastic measure on the set of all half open rectangles (parallelepipeds) of the space R^N. Therefore the σ-additivity of the random function $Z(A)$ is an immediate consequence of the σ-additivity of the increments of the processes $W_1, \ldots, W_N$. Besides, by virtue of the independence of the processes $W_1, \ldots, W_N$ for any nonintersecting rectangles $\prod[a_r, b_r)$ and $\prod[c_s, d_s)$ we have

$$EZ^0\left(\prod[a_r, b_r)\right) \overline{Z^0\left(\prod[c_s, d_s)\right)}$$

$$= \prod_{r=1}^{N} E\left[W_r^0(b_r) - W_r^0(a_r)\right]\left[\overline{W_r^0(d_r)} - \overline{W_r^0(c_r)}\right] = 0,$$

as it follows from the fact that the rectangles $\prod[a_r, b_r)$ and $\prod[c_r, d_r)$ are not intersecting if and only if at least for one r the intervals $[a_r, b_r)$ and $[c_r, d_r)$ do not intersect. In means that the increments $W_r^0(b_r) - W_r^0(a_r)$ and $W_r^0(d_r) - W_r^0(c_r)$ are uncorrelated. Thus for this r we have $E[W_r^0(b_r) - W_r^0(a_r)] \times [W_r^0(d_r) - W_r^0(c_r)] = 0$.

The integrals with respect to the random functions over the stochastic measure are called *the stochastic integrals*. By virtue of the noncorrelatedness of the values of the stochastic measure on the nonintersecting sets $\Gamma_z(AB_p, A'B_q) = \sigma(AA'B_pB_q) = 0$ at $q \neq p$, $\Gamma_z(AB_p, A'B_p) = \sigma(AA'B_p)$ and formula (4.5.3) for the stochastic integral with respect to simple random function $X(t)$ takes the form

$$\Gamma_y(A, A') = \sum_{p=1}^{N} \Gamma_{pp}\sigma(AA'B_p). \tag{4.6.2}$$

As Γ_{pp} represents the value of the second order moment $\Gamma_x(t, t')$ at any $t = t' \in B_p$, i.e. the value of the simple function $\Gamma_x(t, t)$ on the set B_p

$(p = 1, \ldots, N)$ then the obtained formula may be written in the form

$$\Gamma_y(A, A') = \int\limits_{AA'} \Gamma_x(t, t')\sigma(dt). \qquad (4.6.3)$$

From Theorem 4.5.1 it follows that the stochastic integral

$$Y(A) = \int\limits_{A} X(t)Z(dt) = \int\limits_{A} X\,dZ \qquad (4.6.4)$$

at independent $X(t)$ and $Z(A)$ exists if and only if at any $\lambda \in \Lambda$ there exists the integral

$$J_\lambda(A) = \int\limits_{A} \lambda\Gamma_x(t, t)\lambda\sigma(dt) = \int\limits_{A} E\,|\,\lambda X(t)\,|^2\sigma(dt). \qquad (4.6.5)$$

In particular, stochastic integral (4.6.4) with respect to the scalar or the finite-dimensional vector random function $X(t)$ exists if and only if there exists the integral

$$J_\lambda(A) = \int\limits_{A} \Gamma_x(t, t)\sigma(dt) = \int\limits_{A} E\,|\,X(t)\,|^2\sigma(dt). \qquad (4.6.6)$$

For stochastic systems theory the stochastic integrals at the definite type of the dependence between the random function $X(t)$ of the scalar argument t and the stochastic measure $Z(A)$ are very important. Namely, those integrals when $X(t)$ at any $t \in [t_0, \infty)$ represents some nonrandom function of the values of the measure $Z(A)$ on Borel sets of the interval $[t_0, t)$. Let $\mathcal{S}_t$ be the σ-algebra of the sets of the space of the elementary events Ω induced by the random variables $Z(A)$ correspondent to all Borel sets $A \subset [t_0, t)$. As it is known (Section 2.2) the function $X(t)$ is at any $t \in [t_0, \infty)$ a measurable function of the values of the measure $Z(A)$ on the measurable sets of the interval $[t_0, t)$ if and only if $X(t)$ at any $t \in [t_0, \infty)$ represents a random variable which is measurable relatively to the σ-algebra $\mathcal{S}_t$. Thus we have to determine an integral over the stochastic measure $Z(A)$ with respect to the random function $X(t) = x(t, \omega)$ whose value at any $t \in [t_0, \infty)$ represents the function ω which is measurable relatively to the σ-algebra $\mathcal{S}_t$.

We impose the additional restrictions on the stochastic measure $Z(A)$ determined on Borel sets of the interval $[t_0, \infty)$, namely, we shall suppose that the conditional expectations of the random variables $Z(A)$ and $Z(A)\overline{Z(A')}$ relatively to the σ-algebra $\mathcal{S}_t$ (Section 3.3) coincide with their unconditional expectations in the case if $A\bigcup[t_0,t) = A'\bigcup[t_0,t) = \emptyset$, i.e. if Borel sets A and A' are entirely disposed to the right of the interval $[t_0,t)$:

$$E_{\mathcal{S}_t}(Z(A)|\omega) = EZ(A) = 0 \quad \text{at} \quad A\bigcap[t_0,t) = \emptyset , \qquad (4.6.7)$$

$$E_{\mathcal{S}_t}(Z(A)\overline{Z(A')}|\omega) = EZ(A)\overline{Z(A')} = \sigma(AA') \quad \text{at}$$

$$A\bigcap[t_0,t) = A'\bigcap[t_0,t) = \emptyset , \qquad (4.6.8)$$

It is easy to see that these conditions take place when the values of the stochastic measure Z on the nonintersecting sets are independent as in this case the conditional distributions of the random variables $Z(A)$ and $Z(A)\overline{Z(A')}$ relatively to the σ-algebra $\mathcal{S}_t$ generated by the values of the measure Z on Borel sets of the interval $[t_0,t)$ not intersecting with A and A', coincide with their unconditional distributions (Sections 2.3 and 2.4).

Let $X(t)$ be a simple random function whose value at any $t \in [t_0,\infty)$ represents a random variable which is measurable relatively to the σ-algebra $\mathcal{S}_t$:

$$X(t) = \sum_{p=1}^{N} X_p \mathbf{1}_{[t_{p-1},t_p)}(t). \qquad (4.6.9)$$

In this case $B_p = [t_{p-1},t_p)$ in (4.4.1) and the value $X_p = X(t_{p-1})$ of $X(t)$ at any $t \in [t_{p-1},t_p)$ is measurable relatively to the σ-algebra $\mathcal{S}_{t_{p-1}}$. Therefore on the basis of properties (3.3.6) and (3.3.9) of the conditional expectations for any p, q, $p > q$ we have

$$E(\lambda X_p)Z(AB_p) = E[E_{\mathcal{S}_{t_{p-1}}}((\lambda X_p)Z(AB_p)|\omega)]$$

$$= E[(\lambda X_p)E_{\mathcal{S}_{t_{p-1}}}(Z(AB_p)|\omega)] = 0,$$

$$E(\lambda X_p)(\overline{\lambda' X_q})Z(AB_p)\overline{Z(A'B_q)}$$

$$= E[E_{\mathcal{S}_{t_{p-1}}}((\lambda X_p)(\overline{\lambda' X_q})Z(AB_p)\overline{Z(A'B_q)}|\omega)]$$

$$= E[(\lambda X_p)(\overline{\lambda' X_q})\overline{Z(A'B_q)}E_{\mathcal{S}_{t_{p-1}}}(Z(AB_p)|\omega)] = 0,$$

as $B_p\bigcap[t_0,t_{p-1}) = \emptyset$ and consequently, $E_{\mathcal{S}_{t_{p-1}}}(Z(AB_p)|\omega) = 0$ by virtue of (4.6.7). At $q = p$

$$E(\lambda X_p)(\overline{\lambda' X_p})Z(AB_p)\overline{Z(A'B_p)}$$

$$= E[(\lambda X_p)(\overline{\lambda' X_p})E_{\mathcal{S}_{t_{p-1}}}(Z(AB_p)\overline{Z(A'B_p)}|\omega)]$$

$$= E[(\lambda X_p)(\overline{\lambda' X_p})\sigma(AA'B_p)] = \lambda\Gamma_{pp'}\lambda'\sigma(AA'B_p),$$

as $E_{\mathcal{S}_{t_{p-1}}}(Z(AB_p)\overline{Z(A'B_p)}|\omega) = \sigma(AA'B_p)$ by virtue of (4.6.8). So we get

$$E\lambda Y(A) = \sum_{p=1}^{N} E(\lambda X_p)Z(AB_p) = 0,$$

$$\lambda\Gamma_y(A, A')\lambda' = E(\lambda Y(A))(\overline{\lambda'Y(A')})$$

$$= \sum_{p,q=1}^{N} E(\lambda X_p)(\overline{\lambda' X_q})Z(AB_p)\overline{Z(A'B_q)} = \sum_{p=1}^{N} \lambda\Gamma_{pp}\lambda'\sigma(AA'B_p).$$

Thus formula (4.6.2) and consequently formula (4.6.3) remains valid in this case too.

4.6.2. Conditions of Existence of Stochastic Integral

Repeating the arguments of Section 4.5 we come to the following conclusions.

Theorem 4.6.1. *For the existence of stochastic integral* (4.6.4) *with respect to the random function $X(t)$ which is representable in the form of m.s. limit of the sequence of the simple random functions of the considered type it is necessary and sufficient the existence of integral* (4.6.5) *at any $\lambda \in \Lambda$.*

Corollary 4.6.1. *In the special case of the scalar or finite-dimensional vector random function $X(t)$ whose value at each $t \in [t_0, \infty)$ is measurable relatively to the σ-algebra $\mathcal{S}_t$, for the existence of stochastic integral* (4.6.4) *it is necessary and sufficient the existence of integral* (4.6.6).

E x a m p l e 4.6.2. In the case of the stochastic measure of Example 4.6.1 and a scalar function of the N-dimensional vector a stochastic integral represents a multiple integral

$$Y = \int_{T_1} \varphi(\tau)Z(d\tau) = \int_{T_1}\cdots\int_{T_1} \varphi(\tau_1, \ldots, \tau_N)dW_1(\tau_1)\ldots dW_N(\tau_N). \quad \text{(I)}$$

The formulae for the expectation and the covariance matrix of stochastic integral have the following form:

$$m_y = \int_{T_1} \ldots \int \varphi(\tau_1, \ldots, \tau_N) m_1'(\tau_1) \ldots m_N'(\tau_N) d\tau_1 \ldots d\tau_N, \qquad \text{(II)}$$

$$K_y = \int_{T_1} \ldots \int |\varphi(\tau_1, \ldots, \tau_N)|^2 \nu_1(\tau_1) \ldots \nu_N(\tau_N) d\tau_1 \ldots d\tau_N, \qquad \text{(III)}$$

where m_r' and ν_r are such that $\int_a^b m_r'(\tau) d\tau$ and $\int_a^b \nu_r(\tau) d\tau$ $(r = 1, \ldots, N)$ represent the expectations and the variances of the increments $W_r(b) - W_r(a)$.

4.6.3. Processes with Independent Increments

In elementary probability theory a random process $X(t)$ is called *a process with the independent increments* if at any N, $t_0 < t_1 < \cdots < t_N$ the random variables $X_{t_0}, X_{t_1} - X_{t_0}, \ldots, X_{t_N} - X_{t_{N-1}}$ are independent.

Let us study the multi-dimensional distributions of a process $X(t)$ with the independent increments. At first we consider $X(t)$ to be a real (scalar or finite-dimensional vector) process.

▷ Since for any $t_1 < t_2$, $X(t_2) = X(t_1) + [X(t_2) - X(t_1)]$ the value of a process with the independent increments at any t_2 represents the sum of two independent random variables – its value at t_1 and its increment on the interval (t_1, t_2). We denote by $g_1(\lambda; t)$ the one-dimensional characteristic function of the process $X(t)$ and by $h(\lambda; t, s)$ the characteristic function of its increment on the interval (t, s). As it is known the characteristic function of the sum of independent random variables is equal to the product of the characteristic functions of the items (Section 3.5). But the characteristic function of the random variable $X(t_2)$ is $g_1(\lambda; t_2)$, the characteristic function of the random variable $X(t_1)$ is $g_1(\lambda; t_1)$, and the characteristic function of the random variable $X(t_2) - X(t_1)$ is $h(\lambda; t_1, t_2)$. Consequently, $g_1(\lambda; t_2) = g_1(\lambda; t_1)h(\lambda; t_1, t_2)$ at any $t_1 < t_2$. Hence we find

$$h(\lambda; t_1, t_2) = \frac{g_1(\lambda; t_2)}{g_1(\lambda; t_1)}, \quad \forall t_1 < t_2. \ \triangleleft \qquad (4.6.10)$$

Theorem 4.6.2. *The characteristic function of the increment of a process with the independent increments is completely determined by its one-dimensional characteristic function, i.e. by its one-dimensional distribution.*

▷ Further at any $t_1 < \cdots < t_n$, $k = 2, \ldots, n$, we have

$$X(t_k) = X(t_1) + [X(t_2) - X(t_1)] + \cdots + [X(t_k) - X(t_{k-1})]$$

and

$$g_n(\lambda_1, \ldots, \lambda_n; t_1, \ldots, t_n) = E\exp\left\{i\lambda_1^T X(t_1) + \cdots + i\lambda_n^T X(t_n)\right\}$$

$$= E\exp\left\{i(\lambda_1^T + \cdots + \lambda_n^T)X(t_1)\right.$$

$$+ i(\lambda_2^T + \cdots + \lambda_n^T)[X(t_2) - X(t_1)] + \cdots + i\lambda_n^T[X(t_n) - X(t_{n-1})]\right\}.$$

By virtue of the multiplication theorem of the expectations this equality may be rewritten in the form

$$g_n(\lambda_1, \ldots, \lambda_n; t_1, \ldots, t_n) = E\exp\left\{i(\lambda_1^T + \cdots + i\lambda_n^T)X(t_1)\right\}$$

$$\times E\exp\left\{i(\lambda_2^T + \cdots + \lambda_n^T)[X(t_2) - X(t_1)]\right\} \times$$

$$\cdots \times E\exp\left\{i\lambda_n^T[X(t_n) - X(t_{n-1})]\right\}.$$

But for $p = 2, \ldots, n$

$$E\exp\left\{i(\lambda_1^T + \cdots + \lambda_n^T)X(t_1)\right\} = g_1(\lambda_1 + \cdots + \lambda_n; t_1),$$

$$E\exp i(\lambda_p^T + \cdots + \lambda_n^T)[X(t_p) - X(t_{p-1})] = g(\lambda_p + \cdots + \lambda_n; t_{p-1}, t_p).$$

Substituting these expressions into the previous equality we find

$$g_n(\lambda_1, \ldots, \lambda_n; t_1, \ldots, t_n) = g_1(\lambda_1 + \cdots + \lambda_n; t_1)$$

$$\times h(\lambda_2 + \cdots + \lambda_n; t_1, t_2) \ldots h(\lambda_n; t_{n-1}, t_n) \ (n = 2, 3, \ldots).$$

Finally, substituting here the expression of $h(\lambda_p + \cdots + i\lambda_n; t_{p-1}, t_p)$ from (4.6.10) we get at $n = 2, 3, \ldots$

$$g_n(\lambda_1, \ldots, \lambda_n; t_1, \ldots, t_n)$$

$$= \frac{g_1(\lambda_1 + \cdots + \lambda_n; t_1)g_1(\lambda_2 + \cdots + \lambda_n; t_2)\ldots g_1(\lambda_n; t_n)}{g_1(\lambda_2 + \cdots + \lambda_n; t_1)g_1(\lambda_3 + \cdots + \lambda_n; t_2)\ldots g_1(\lambda_n; t_{n-1})}. \ \triangleleft \ (4.6.11)$$

Thus, we proved the following statement.

Theorem 4.6.3. *All the multi-dimensional distributions of a process with the independent increments are uniquely determined by its one-dimensional distribution.*

R e m a r k. Thus the processes with the independent increments belong to the class of random functions whose multi-dimensional distributions are all completely determined by one of them. But in contradistinction to random functions with independent values for which any arbitrarily chosen one-dimensional distribution determines uniquely all the multi-dimensional distributions, the one-dimensional distribution of a processes with the independent increments cannot be taken arbitrarily. Formula (4.6.10) shows that *the characteristic function $g_1(\lambda; t)$ may be the one-dimensional characteristic function of a process with the independent increments only if the ratio $g_1(\lambda; s)/g_1(\lambda; t)$ is also a characteristic function at any $t < s$, i.e. a continuous nonnegative definite function equal to 1 at $\lambda = 0$.*

If a process with the independent increments $X(t)$ has the finite second order moment then it is a process with the uncorrelated increments (Subsection 4.2.3). Consequently, the covariance function of a process $X(t)$ with the independent increments is determined at t_1, $t_2 > t_0$ by formula

$$K_x(t_1, t_2) = k(\min(t_1, t_2)),$$

where $k(t)$is a nondecreasing nonnegative function representing the covariance matrix (the variance in the case of a scalar process $X(t)$) of the value of the process $X(t)$ at a given t. Later on we shall assume that the function $k(t)$ is continuous and differentiable in consequence of which the covariance function of the process $X(t)$ is determined at t_1, $t_2 > t_0$ by formula (4.2.3):

$$K_x(t_1, t_2) = k(t_0) + \int_{t_0}^{\min(t_1, t_2)} \nu(\tau)d\tau. \tag{4.6.12}$$

It is not difficult to understand that any process with the independent increments represents a Markov random process (Subsection 2.7.4).

R e m a r k. The processes with the uncorrelated increments are often the processes with independent increments (for instance, the Wiener and the Poisson processes in Subsections 4.6.5 and 4.6.6).

But the processes with the independent increments may be not the processes with the uncorrelated increments (Problem 4.1) and vice versa (Problem 4.2).

In applications the most general form of a process with the independent increments $X(t)$ is the following

$$X(t) = W_0(t) + \int\limits_{R_0^q} c(x)P(t, dx) \,, \tag{4.6.13}$$

where $W_0(t)$ is a Wiener process, $P(t, A)$ is a Poisson process as a function of t and a Poisson stochastic measure as a function of the set A, and $c(t, x)$ is a vector function mapping $R \times R^q$ into the space of values of the process $X(t)$ at each t. The one-dimensional characteristic function $g_1(\lambda; t)$ of the process $X(t)$ determined by formula (4.6.13) given by:

$$\ln g_1(\lambda; t) = -\frac{1}{2}\lambda^T k(t)\lambda + \int\limits_{R_0^q} \left[e^{i\lambda^T c(x)} - 1 \right] \mu(t, x)dx \,, \tag{4.6.14}$$

where $\mu(t, x)$ is determined by the formula

$$\int\limits_{A} \mu(t, x)dx = EP(t, A) \,. \tag{4.6.15}$$

From formulae (4.6.13) and (4.6.15) taking into account that the expectation of a Wiener process is equal to zero we obtain

$$EX(t) = \int\limits_{R_0^q} c(x)\mu(t, x)dx \,. \tag{4.6.16}$$

Then the characteristic function $g_1^0(\lambda; t)$ of the centered process $X^0(t) = X(t) - EX(t)$ will be determined by the formula

$$\ln g_1^0(\lambda; t) = -\frac{1}{2}\lambda^T k(t)\lambda + \int\limits_{R_0^q} \left[e^{i\lambda^T c(x)} - 1 - i\lambda^T c(x) \right] \mu(t, x)dx \,. \tag{4.6.17}$$

R e m a r k. The increment of a process with independent increments on any interval possesses such a property that it may be represented as the sum of any number of independent random variables. For this purpose it is sufficient to partition a given interval into a corresponding number of parts and to represent the increment of the process on this interval as the sum of its increments on the chosen partial intervals. It means that the distribution of the increment of a process with independent increments represents an *infinitely divisible* distribution. It is known from the theory of infinitely divisible distributions that any such distribution may be represented as the composition of a normal and a Poisson distributions, or more exactly, the logarithm of the characteristic function of an infinitely divisible distribution may be represented in the form of the sum of the logarithm of the characteristic function of a normal distribution and some integral. But generally the process $X(t)$ may have no expectation. In the latter case integral (4.6.16) may diverge both because of too rapid increase of the function $\mu(t, x)$ when $x \to 0$, and its too slow decrease when $x \to \infty$. But in all the cases the integral

$$\int_{R_0^q} \frac{|x|^2 \mu(t, x) dx}{1 + |x|^2} ,$$

converges. Therefore, any m.s. continuous process with independent increments $X(t)$ may be represented in the general case by the formula

$$X(t) = W_0(t) + \int_{0 < |x| \le 1} x P^0(t, dx) + \int_{|x| > 1} x P(t, dx)$$

where $W_0(t)$ is a process with continuous realizations with normally distributed increments, $P(t, A)$ is a Poisson process as a function of t and a Poisson measure as a function of A with independent values on nonintersecting sets, and $P^0(t, A) = P(t, A) - E P(t, A)$ is the centered Poisson measure. The one-dimensional characteristic function of a process with independent increments is expressed respectively in the general case by the formula

$$\ln g_1^0(\lambda; t) = \ln g_0(\lambda) + i\lambda^T m(t) - \frac{1}{2}\lambda^T k(t)\lambda$$

$$+ \int_{0 < |x| \le 1} (e^{i\lambda^T x} - 1 - i\lambda^T x)\mu(t, x)dx + \int_{|x| > 1} (e^{i\lambda^T x} - 1)\mu(t, x)dx ,$$

where $g_0(\lambda)$ is an arbitrary characteristic function independent of t.

We presented some results of the theory of processes with independent increments supposing for simplicity that the expectation of the Poisson measure $P(t, A)$ is representable by integral (4.6.15) with the function $\mu(t, x)$, probably containing a linear combination of δ-functions of the variable x.

4.6.4. White Noise in Strict Sense

It was shown in Subsection 4.2.3 that any process with the uncorrelated increments and a differentiable covariance function has the weak m.s. derivative representing a white noise. In particular, any process with the independent increments possessing a differentiable covariance function has a weak m.s. derivative which represents a white noise. The white noise obtained by the differentiation of a process with the independent increments is called *a white noise in the strict sense.*

R e m a r k. As shown in the Examples of Subsection 3.3.3 the white noise may be either an ordinary random function, whose value at every value of the argument represents a random variable, or a generalized random function (Section 4.7) whose realizations have no definite value at any value of the argument. But only white noises obtained by the differentiation of the processes with the uncorrelated but dependent increments may be ordinary random functions. The white noises in the strict sense are always generalized random functions.

4.6.5. Wiener Processes and Measures

A scalar or a vector real random process with independent increments $W(t)$, $t > 0$, is called *a Wiener process* if it satisfies the conditions:

(i) all the realizations $w(t)$ of the process $W(t)$ are continuous and $w(0) = 0$;

(ii) the one-dimensional distribution of the process $W(t)$ is normal;

(iii) the expectation of the process $W(t)$ is equal identically to zero and its covariance function is determined by the formula

$$K_w(t_1, t_2) = \int\limits_0^{\min(t_1,t_2)} \nu(\tau)d\tau \,,$$

where $\nu(t)$ is a nonnegative function representing the intensity of the Wiener process $W(t)$.

Theorem 4.6.4. *Any Wiener process represents a normally distributed random function.*

▷ Really it follows from the definition that the one-dimensional characteristic function of a Wiener process is determined by the formula

$$g_1(\lambda; t) = \exp\left\{-\frac{1}{2}\lambda^T k(t)\lambda\right\} \,, \quad k(t) = \int\limits_0^t \nu(\tau)d\tau\,. \qquad (4.6.18)$$

Substituting this expression into (4.6.15) after simple but rather cumbersome evaluations we find all the finite-dimensional characteristic functions of the Wiener process:

$$g_n(\lambda_1, \ldots, \lambda_n; t_1, \ldots, t_n)$$

$$= \exp\left\{-\frac{1}{2}\,[\,\lambda_1^T\lambda_2^T\ldots\lambda_n^T\,]\begin{bmatrix} k(t_1) & k(t_1) & \ldots & k(t_1) \\ k(t_1) & k(t_2) & \ldots & k(t_2) \\ \ldots & \ldots & \ldots & \ldots \\ k(t_1) & k(t_2) & \ldots & k(t_n) \end{bmatrix}\begin{bmatrix} \lambda_1 \\ \lambda_2 \\ \ldots \\ \lambda_n \end{bmatrix}\right\}.$$

$$(4.6.19)$$

This formula shows that all the finite-dimensional distributions of a Wiener process are normal. ◁

The white noise representing the weak m.s. derivative of a Wiener process is called *a normally distributed white noise*. This definition extends the notion of the normal distribution to white noises representing the generalized random functions.

A Wiener process as a process with the independent increments which possesses zero expectation and the finite second order moment $k(t)$ for each instant t generates a stochastic measure on the real axis with the independent values on nonintersecting intervals. This stochastic measure is determined by the formula $Z((t_1, t_2]) = W(t_2) - W(t_1)$.

Theorem 4.6.5. *Stochastic integral* (4.6.4) *over the Wiener stochastic measure Z and over any interval $[a, b)$, $A = [a, b)$, $a \geq 0$ may be written in the form of the Stieltjes integral:*

$$Y([a, b)) = \int_a^b X(t)Z(dt) = \int_a^b X(t)dW(t), \qquad (4.6.20)$$

it represents the m.s. Lebesgue–Stielties integral.

▷ Really, in the case of the simple random function $X(t)$ and $a \in [t_{m-1}, t_m)$, $b \in [t_{n-1}, t_n)$

$$Y([a, b)) = X_m Z([a, t_m)) + \sum_{p=m+1}^{n-1} X_p Z([t_{p-1}, t_p)) + X_n Z([t_{n-1}, b))$$

$$= X_m[W(t_m) - W(a)] + \sum_{p=m+1}^{n-1} X_p[W(t_p) - W(t_{p-1})]$$

$$+X_n[W(b) - W(t_{n-1})]$$

and, consequently, the stochastic integral $Y([a, b))$ with respect to the arbitrary random function $X(t)$ represents m.s. limit of the sequence of the Lebesgue-Stieltjes sums. ◁

Formulae (4.6.3), (4.6.5) and (4.6.6) in the case of Wiener stochastic measure Z take the form correspondingly

$$\Gamma_y(A, A') = \int\limits_{AA'} \Gamma_x(t, t)\, dt, \qquad (4.6.21)$$

$$J_\lambda(A) = \int\limits_A E\,|\,\lambda X(t)\,|^2 dt, \qquad (4.6.22)$$

$$J(A) = \int\limits_A E\,|\,X(t)\,|^2 dt. \qquad (4.6.23)$$

In the special case of scalar or finite-dimensional random function $X(t)$ measurable for every $t \in [0, \infty)$ with respect to $\mathcal{S}_t$ the sequence of simple random functions $\{X^n(t)\}$ of the same class m.s. convergent to $X(t)$ always exists. In this case the stochastic integral (4.6.20) is called *the Itô stochastic integral* (Subsection 4.6.7).

R e m a r k. Some properties of the Wiener measure $\dot{Z}((t_1, t_2]) = \mu_W$ are given in Problems 4.9–4.22.

4.6.6. Poisson Processes and Measures

In elementary probability theory a random process $P(t)$ is called *a simple Poisson* process if at any t its value is equal to the number of events of a Poisson stream in the time interval $[0, t]$. Any simple Poisson process is a process with the independent increments and its one-dimensional distribution is the Poisson distribution (Problem 4.3). The simple Poisson process is a step random process increasing by unity at any moment of occurance of an event of the Poisson stream. Thus the simple Poisson process may be called *the counter of events of the Poisson stream*.

Let $P(t)$ be a simple Poisson process. The random process

$$W(t) = \sum_{k=1}^{P(t)} X_k, \qquad (4.6.24)$$

where $\{X_k\}$ being a sequence of independent equally distributed random variables is called *the general Poisson process*. It is clear that any general Poisson process represents a (scalar or vector) step random process with random steps at the moments of occurring of events of the corresponding Poisson stream. Evidently it is a process with independent increments (Problems 4.4 and 4.5).

A scalar random function $P(\Delta, A)$ defined on the rectangular sets $\Delta \times A$ of the product of the spaces $[0, \infty) \times R^q$, Δ being an element of the σ-algebra $\mathcal{B}_+$ of Borel sets of $[0, \infty)$ and A an element of the σ-algebra $\mathcal{B}_0$ generated by all the rectangular sets of the space R_q with the origin pricked out is called *a Poisson measure* if its value an any $\Delta \in \mathcal{B}_+$, $A \in \mathcal{B}_0$ represents a random variable with Poisson distribution and the respective centered random function

$$P^0(\Delta, A) = P(\Delta, A) - \mu_P(\Delta, A), \quad \mu_P(\Delta, A) = EP(\Delta, A) \qquad (4.6.25)$$

represents a stochastic measure with independent values at pairwise nonintersecting sets. It is easy to see that $\mu_P(\Delta, A)$ represents a nonnegative measure.

4.6.7. Itô Stochastic Integral

Let $W(t)$ be a scalar process with independent increments with zero expectation and finite second order moment $k(t)$; $X(t)$ be m.s. continuous scalar random process with finite second order moment, whose value at every t is independent of future increments of the process $W(t)$, $W(t_2) - W(t_1)$, $t \le t_1 < t_2$; $\{P_n\}$ being a sequence of partitions of the interval $(a, b]$

$$P_n : (a, b] = \bigcup_{k=1}^{N_n} (t_{k-1}^{(n)}, t_k^{(n)}], \quad t_0^{(n)} = a, \quad t_{N_n}^{(n)} = b,$$

such that $\max_k (t_k^{(n)} - t_{k-1}^{(n)}) \to 0$ at $n \to \infty$. M.s. limit of the sequence of integral sums $\{Y_n\}$,

$$Y_n = \sum_{k=1}^{N_n} X(t_{k-1}^{(n)})[W(t_k^{(n)}) - W(t_{k-1}^{(n)})], \qquad (4.6.26)$$

if this limit exists, is called *an Itô stochastic integral of the random function $X(t)$ with respect to the process with independent increments*

$W(t)$ *extended over the interval* $(a, b]$:

$$Y = \int_a^b X(t)dW(t) = \underset{n \to \infty}{\text{l.i.m.}} \sum_{k=1}^{N_n} X(t_{k-1}^{(n)})[W(t_k^{(n)}) - W(t_{k-1}^{(n)})]. \quad (4.6.27)$$

Theorem 4.6.6. *The Itô stochastic integral* (4.6.27) *exists if and only if the usual Riemann-Stieltjes integral*

$$\int_a^b E\,|X(t)|^2\,dk(t) = E\,|Y|^2 = DY \quad (4.6.28)$$

exists in which case it is equal to the variance DY *of the Itô integral* Y.

▷ By Theorem 4.1.2 of the m.s. convergence the limit $y = \text{l.i.m.}\,Y_n$ exists if and only if there exists the limit $EY_n \bar{Y}_m$ independent of the way in which n, $m \to \infty$. By this theorem the necessity of the condition follows immediately from the equality

$$E\,|Y_n|^2 = \sum_{k=1}^{N_n} E\,\left|X(t_{k-1}^{(n)})\right|^2 \left[k(t_k^{(n)}) - k(t_{k-1}^{(n)})\right].$$

To derive this equality we note that

$$E\,|Y_n|^2 = \sum_{k,l=1}^{N} EX(t_{k-1}^{(n)})\overline{X(t_{l-1}^{(n)})}\left[W(t_k^{(n)}) - W(t_{k-1}^{(n)})\right]$$

$$\times \left[\overline{W(t_l^{(n)})} - \overline{X(t_{l-1}^{(n)})}\right].$$

and by independence of $W(t_k^{(n)}) - W(t_{k-1}^{(n)})$ on $X(t_{l-1}^{(n)})$, $X(t_{k-1}^{(n)})$ and $W(t_l^{(n)}) - W(t_{l-1}^{(n)})$ at any $k > l$,

$$EX(t_{k-1}^{(n)})\overline{X(t_{l-1}^{(n)})}\left[W(t_k^{(n)}) - W(t_{k-1}^{(n)})\right]\left[\overline{W(t_l^{(n)})} - \overline{W(t_{l-1}^{(n)})}\right]$$

$$= EX(t_{k-1}^{(n)})\overline{X(t_{l-1}^{(n)})}\left[\overline{W(t_l^{(n)})} - \overline{W(t_{l-1}^{(n)})}\right]E\left[W(t_l^{(n)}) - W(t_{l-1}^{(n)})\right] = 0,$$

$$E\,\left|X(t_{k-1}^{(n)})\right|^2 \left|W(t_k^{(n)}) - W(t_{k-1}^{(n)})\right|^2$$

$$= E\left|X(t_{k-1}^{(n)})\right|^2 E\left|W(t_k^{(n)}) - W(t_{k-1}^{(n)})\right|^2.$$

Further, by independence of $W(t_{k-1}^{(n)})$ on $W(t_k^{(n)}) - W(t_{k-1}^{(n)})$ yielding

$$E\left|W(t_k^{(n)})\right|^2 = E\left|W(t_{k-1}^{(n)})\right|^2 + E\left|W(t_k^{(n)}) - W(t_{k-1}^{(n)})\right|^2.$$

Consequently,

$$E\left|W(t_k^{(n)}) - W(t_{k-1}^{(n)})\right|^2 = E\left|W(t_k^{(n)})\right|^2 - E\left|W(t_{k-1}^{(n)})\right|^2$$

$$= k(t_k^{(n)}) - k(t_{k-1}^{(n)}). \quad \triangleleft$$

$\triangleright$ To prove the sufficiency of the condition we unite two partitions P_n, P_m of the interval $(a, b]$. Let P be the resulting partition, $P : (a, b] = \sum_{s=1}^{Q}(t_{s-1}, t_s]$, $t_0 = a$, $t_Q = b$ where each point t_s coincides with some $t_s^{(n)}$ or with some $t_l^{(m)}$. Then we obtain

$$EY_n\bar{Y}_m = \sum_{s=1}^{Q} EX(\tau_s)\overline{X(\tau_s')}\left[k(t_s) - k(t_{s-1})\right]$$

at $\tau_s \le t_{k-1}^{(n)}$, $\tau_s' \le t_{l-1}^{(m)}$ if $(t_{s-1}, t_s] \subseteq (t_{k-1}^{(n)}, t_k^{(n)}] \cap (t_{l-1}^{(m)}, t_l^{(m)}]$. By virtue of the m.s. continuity of $X(t)$, $\lim EY_n\bar{Y}_m$ coincides with the limit

$$\lim_{n,m\to\infty} \sum_{s=1}^{Q} E\left|X(t_{s-1})\right|^2 \left[k(t_s) - k(t_{s-1})\right] = \int_a^b E\left|X(t)\right|^2 dk(t). \quad \triangleleft$$

The definition of the Itô stochastic integral is easily extended to the case of a matrix process $X(t)$ and a vector process $W(t)$.

Theorem 4.6.7. *The vector Itô integral* (4.6.27) *exists if and only if all the scalar Itô integrals involved in* (4.6.27) *exist. The necessary and sufficient condition of the existence of the vector integral* (4.6.27) *is the existence of the Riemann–Stieltjes integral*

$$\int_a^b EX(t)dk(t)X(t)^* = EYY^*,$$

*which is equal to the covariance matrix of the vector Itô integral Y. The
existence of the integrals*

$$\int_a^b EX_{sj}(t)\overline{X_{rh}(t)}dk_{jh}(t)$$

*is an immediate consequence of the existence of integral (4.6.28) for all
the elements of the matrix function $X(t)$.*

Let $Y(t)$ be a random process determined by the formula

$$Y(t) = Y(t_0) + \int_{t_0}^t X_1(\tau)d\tau + \int_{t_0}^t X_2(\tau)dW(\tau), \qquad (4.6.29)$$

where the first integral represents a m.s. integral of a random function
$X_1(t)$, and the second integral represents the Itô integral of a random
function $X_2(t)$ with respect to a process with independent increments
$W(t)$. The random functions $X_1(t)$ and $X_2(t)$ are supposed satisfying
the conditions under which both integrals exist at any t in some interval
T. Let $X(t)$ be a random function. The *Itô integral of the random
function $X(t)$ with respect to the process $Y(t)$ in the interval* $(a, b]$ is
defined as the sum of the integrals

$$\int_a^b X(\tau)dY(\tau) = \int_a^b X(\tau)X_1(\tau)d\tau + \int_a^b X(\tau)X_2(\tau)dW(\tau), \quad (4.6.30)$$

where the first integral represents a m.s. integral and the second integral represents the Itô integral. Here, certainly, the random functions
$X(t)X_1(t)$ and $X(t)X_2(t)$ must satisfy the conditions under which both
integrals in the right-hand side of formula (4.6.30) exist.

4.6.8. Stochastic Integrals over Wiener and Poisson Measures

As it is known (Subsection 4.6.5) the stochastic measure of the
Wiener process $W(t)$ is defined by $Z((t_1, t_2]) = W(t_2) - W(t_1)$. So the
Itô stochastic integral may be calculated by formula (4.6.27) because
condition (4.6.28) is fulfilled.

E x a m p l e 4.6.3. Find the variance of the stochastic integral

$$Y = \int_0^T X(\tau)dW(\tau),$$

where $X(t)$ is a random function with the expectation $m_x(t) = ct$ and the covariance function $K_x(t_1, t_2) = De^{\mu(t_1+t_2)-\alpha|t_1-t_2|}$, and $W(t)$ is a standard Wiener process. According to formula (4.6.28) and taking into account that $\nu(t) = 1$

$$E\,|X(t)|^2 = |m_x(t)|^2 + D_x(t) = c^2 t^2 + De^{2\mu t},$$

$$D_y = c^2 \int_0^T \tau^2 d\tau + D \int_0^T e^{2\mu\tau} d\tau = c^2 T^3/3 + D(e^{2\mu T} - 1)/2\mu.$$

Let $P^0(\Delta, A)$ be a centered Poisson measure, $Y(t, u)$ be a real scalar m.s. continuous random function of two variables t and u with finite second order moment independent of $P((t_1, t_2], A)$ at any $t \leq t_1 < t_2$, $u \in R_0^q = R^q \setminus \{0\}$, $A \in \mathcal{B}_0$. Let $\{P_n\}$ be a sequence of partitions of the interval $(a, b]$ and of set $0 < |u| < R$,

$$P_n : (a, b] = \bigcup_{k=1}^{N_n} (t_{k-1}^{(n)}, t_k^{(n)}], \quad t_0^{(n)} = a, \quad t_{N_n}^{(n)} = b,$$

$$\{u : 0 < |u| < R\} = \bigcup_{j=1}^{M_n} A_j^{(n)}, \quad A_j^{(n)} \in \mathcal{B}_0,$$

(4.6.31)

such that $\max_k (t_k^{(n)} - t_{k-1}^{(n)}) \to 0$, $\max_j \sup_{u,u' \in A_j^{(n)}} |u' - u| \to 0$ as $n \to \infty$.

M.s. limit Z^R of the sequence of integral sums

$$Z_n^R = \sum_{k=1}^{N_n} \sum_{j=1}^{M_n} Y(t_{k-1}^{(n)}, u_j^{(n)}) P^0((t_{k-1}^{(n)}, t_k^{(n)}], A_j^{(n)}), \quad u_j^{(n)} \in A_j^{(n)}, \quad (4.6.32)$$

if it exists, is called *a stochastic integral* of the random function $Y(t, u)$ *with respect to the Poisson measure* $P(\Delta, A)$ extended over the interval $(a, b]$ and the set $\{u : 0 < |u| < R\}$,

$$Z^R = \int_a^b \int_{0<|u|<R} Y(t, u)P^0(dt, du)$$

$$= \text{l.i.m.} \sum_{k=1}^{N_n} \sum_{j=1}^{M_n} Y(t_{k-1}^{(n)}, u_j^{(n)} P^0)((t_{k-1}^{(n)}, t_k^{(n)}], A_j^{(n)}) . \qquad (4.6.33)$$

Theorem 4.6.8. *The stochastic integral* (4.6.33) *exists if and only if the Riemann–Stieltjes integral*

$$\int_a^b \int_{0<|u|<R} EY^2(t,u)\mu_P(dt,du) = E(Z^R)^2 = DZ^R , \qquad (4.6.34)$$

exists which is equal then to the variance DZ^R *of the stochastic integral* Z^R.

▷ In the same way as in Subsection 4.6.7 we evaluate $EZ_n^R Z_m^R$:

$$EZ_n^R Z_m^R = \sum_{s,z=1}^{Q} \sum_{p,q=1}^{S} EY(\tau_s,\sigma_p)Y(\tau_r',\sigma_q')$$

$$\times P^0((t_{s-1},t_s],B_p)P^0((t_{r-1},t_r],B_q) . \qquad (4.6.35)$$

Note that $(a,b] = \bigcup_{s=1}^{Q} (t_{s-1},t_s], t_0 = a, t_Q = b, \{u : 0 < |u| < R\} = \bigcup_{p=1}^{S} B_p$
is the partition of $[a,b)$ and $\{u : 0 < |u| < R\}$ obtained by uniting the partitions
P_n and P_m where any t_s coincides either with some point $t_k^{(n)}$ or $t_l^{(m)}$ and where any
B_p is the intersection of some set $A_j^{(n)}$ with some set $A_n^{(m)}$. In (4.6.35) $\tau_s \leq t_{k-1}^{(n)}$
if $(t_{s-1},t_s] \subset (t_{k-1}^{(n)}; t_k^{(n)}], \tau_r' \leq t_{l-1}^{(m)},$ if $(t_{r-1},t_r] \subset (t_{l-1}^{(m)}, t_l^{(m)}]; \sigma_p = u_j^{(n)},$
if $B_p \subset A_j^{(n)}; \sigma_q' = u_h^{(m)},$ if $B_q \subset A_h^{(m)}.$ By the independence of $Y(\tau_s,\sigma_p),$
$Y(\tau_r',\sigma_q'), P^0((t_{r-1},t_r],B_q)$ of $P^0((t_{s-1},t_s],B_p)$ at any $r \leq s$ we have three
equalities:

$$EY(\tau_s,\sigma_p)Y(\tau_r',\sigma_q')P^0((t_{s-1},t_s],B_p)P^0((t_{r-1},t_r],B_q)$$

$$= EY(\tau_s,\sigma_p)Y(\tau_r',\sigma_q')P^0((t_{r-1},t_r],B_q)EP^0((t_{s-1},t_s],B_p) = 0 ,$$

$$EY(\tau_s,\sigma_p)Y(\tau_s',\sigma_q')P^0((t_{s-1},t_s],B_p)P^0((t_{s-1},t_s],B_q) = 0 ,$$

$$EY(\tau_s,\sigma_p)Y(\tau_s',\sigma_p')[P^0((t_{s-1},t_s],B_p)]^2$$

$$= EY(\tau_s,\sigma_p)Y(\tau_s',\sigma_p')E[P^0((t_{s-1},t_s],B_p)]^2 .$$

Taking into account that $P((t_{s-1}, t_s], B_p)$ has the Poisson distribution with the parameter $\mu_P((t_{s-1}, t_s], B_p) = EP((t_{s-1}, t_s], B_p)$ we have $E[P^0((t_{s-1}, t_s], B_p)]^2 = \mu_P((t_{s-1}, t_s], B_p)$. As a result of previous calculations formula (4.6.35) takes the form

$$EZ_n^R Z_m^R = \sum_{\sigma=1}^{Q} \sum_{p=1}^{S} EY(\tau_s, \sigma_p) Y(\tau_r', \sigma_p') \mu_P((t_{s-1}, t_s], B_p) \,.$$

By m.s. continuity of $Y(u, t)$ this equality may be written as

$$EZ_n^R Z_m^R = \sum_{\sigma=1}^{Q} \sum_{p=1}^{S} EY^2(t_{s-1}, u_p) \mu_P((t_{s-1}, t_s], B_p) + o(\tau) \,, \qquad (4.6.36)$$

where τ is the maximum of four numbers: $\max_k(t_k^{(n)} - t_{k-1}^{(n)})$, $\max_l(t_l^{(m)} - t_{l-1}^{(m)})$, $\max_j \sum_{u, u' \in A_j^{(n)}} |u' - u|$, $\max_h \sup_{u, u' \in A_h^{(m)}} |u' - u|$, and u_p being some point of the set B_p. From equality (4.6.36) and Theorem 4.1.2 it follows the necessarity and sufficiency of the existence of integral (4.6.34), hence the existence of the stochastic integral (4.6.33). ◁

Now we shall extend the definition (4.6.33) of the stochastic integral Z^R, $R < \infty$ with respect to a Poisson measure to the case $R = \infty$.

The stochastic integral Z of the random function $Y(t, u)$ with respect to the Poisson measure $P(\Delta, A)$ extended over the interval $(a, b]$ and the space $R^q \backslash \{0\} = R_0^q$, is defined as m.s. limit of the stochastic integral (4.6.32) as $R \to \infty$, if this limit exists:

$$Z = \int_a^b \int_{R_0^q} Y(t, u) P^0(dt, du) = \underset{R \to \infty}{\text{l.i.m.}} \int_a^b \int_{0 < |u| < R} Y(t, u) P^0(dt, du) \,.$$

$$(4.6.37)$$

Theorem 4.6.9. *The stochastic integral (4.6.37) exists if and only if the Riemann–Stieltjes integral*

$$\int_a^b \int_{R_0^q} EY^2(t, u) \mu_P(dt, du) = EZ^2 = DZ \,. \qquad (4.6.38)$$

exists where the integration with respect to u is extended over all space R^q with the origin of R^q pricked out.

▷ In the same way as relation(4.6.3) was established we find for any R and $R' > R$

$$EZ^R Z^{R'} = \int\limits_a^b \int\limits_{0 < |u| < R} EY^2(t, u)\mu_P(dt, du).$$
(4.6.39)

Hence together with Theorem 4.1.2 we get (4.6.37). ◁

The definition of the stochastic integral with respect to a Poisson measure is directly extended to the case of a vector m.s. continuous random function $Y(t, u)$. Formula (4.6.38) in this case is replaced by

$$K_z = EZZ^T = \int\limits_a^b \int\limits_{R_0^q} EY(t, u)Y(t, u)^T \mu_P(dt, du).$$
(4.6.40)

4.6.9. Other Types of Stochastic Integrals

Using other principles of constructing integral sums Y_n, other choices of the points in which the values of the random function $X(t)$ are taken in the intervals $\left[t_{p-1}^{(n)}, t_p^{(n)}\right]$ we obtain other definitions of a stochastic integral. In particular, taking the values of $X(t)$ at the right ends of the intervals $\left[t_{p-1}^{(n)}, t_p^{(n)}\right]$ we define the stochastic integral

$$Y_1 = \int\limits_a^b X(\tau)d_1 W(\tau) = \operatorname*{l.i.m.}_{n \to \infty} \sum_{p=1}^{N_n} X(t_p^{(n)})\left[W(t_p^{(n)}) - W(t_{p-1}^{(n)})\right].$$

After that we may define a stochastic θ-integral for any $\theta \in [0, 1]$ by the following formula:

$$Y_\theta = \int\limits_a^b X(\tau)d_\theta W(\tau) = (1 - \theta)Y + \theta Y_1$$

$$= \operatorname*{l.i.m.}_{n \to \infty} \sum_{p=1}^{N_n} \left[(1 - \theta)X(t_{p-1}^{(n)}) + \theta X(t_p^{(n)})\right]\left[W(t_p^{(n)}) - W(t_{p-1}^{(n)})\right].$$
(4.6.41)

It is evident that the Itô integral represents the θ-integral at $\theta = 0$, and the integral Y_1 represents the θ-integral at $\theta = 1$. At $\theta = 1/2$ stochastic integral (4.6.41) represents *the symmetrized* or *the Stratonovich stochastic integral* .

R e m a r k. The Itô integral possesses a great advantage as compared with other types of stochastic integrals. This advantage consists in the simplicity of evaluating the moments of the integral. As we have already seen the expectation and the variance (the covariance matrix in the vector case) of an Itô integral are evaluated quite simply. To evaluate the expectation of a θ-integral at $\theta \neq 0$ is very difficult in the general case in consequence of the dependence of $X(t_p^{(n)})$ on $W(t_p^{(n)}) - W(t_{p-1}^{(n)})$. (See Problem 4.7).

As in the case of stochastic integrals of nonrandom function the stochastic integrals of random functions may be considered as the integrals whose integrands contain a white noise (in the strict sense) as a factor. Taking into consideration the formal relation $dW(t)/dt = V(t)$ we may write stochastic Itô integral (4.6.27) in the form

$$\int\limits_a^b X(\tau)dW(\tau) = \int\limits_a^b X(\tau)V(\tau)d\tau .$$

Analogously we may write the stochastic θ-integral. Further on, all integrals of the product of some random function by a white noise in the strict sense we shall consider as stochastic Itô integrals.

4.7. Generalized Random Functions

4.7.1. Definitions

Let us consider a linear space Φ of the scalar functions $\varphi(t)$ of a scalar or finite-dimensional vector variable t. Each of these functions differs from zero only at some bounded region and is continuous together with its derivatives of all orders. In functional analysis this space is called *a space of test functions*. The topology in the space of the test functions is usually assigned by the vicinities of zero $\{ \varphi(t) : |\varphi(t)| < \varepsilon, |\varphi'(t)| < \varepsilon, \ldots, |\varphi^{(n)}(t)| < \varepsilon \}$ correspondent to all n and $\varepsilon > 0$ in the case of the scalar real variable t and by analogous vicinities of zero with the change of the derivatives $\varphi', \varphi'', \ldots$ by all partial derivatives with respect to the components of the vector t in the case of the finite-dimensional vector variable t with real components.

Let X be a space of all linear functionals on the space of the test functions Φ. We set up the correspondence between each function $\varphi \in \Phi$ and a linear continuous functional f_φ on X which is determined by formula

$$f_\varphi x = x\varphi \quad \text{at all} \quad x \in X. \tag{4.7.1}$$

We introduce a weak topology into X by means of the set F_c of all linear functionals of such a type. In this case we may identify F_c with Φ as a topological space. In the space X we determine the σ-algebra $\mathcal{A}$ of the sets generated by a class of the sets of the type $\{\, x : (x\varphi_1, \ldots, x\varphi_n) \in B \,\}$ correspondent to all n, to all $\varphi_1, \ldots, \varphi_n \in \Phi$ and to all Borel sets B of the correspondent finite-dimensional spaces R^n.

A generalized random function of the variable t is by definition a random variable in $(X, \mathcal{A})$ or, in other words, *a generalized random function* is a random linear functional on Φ which is measurable relatively to the σ-algebra $\mathcal{A}$.

It is evident that any m.s. integrable with respect to Lebesgue measure ordinary random function $X(t)$ belongs to the class of the generalized random functions as a random functional on Φ of type

$$X\varphi = \int X(t)\varphi(t)dt. \tag{4.7.2}$$

Therefore the generalized random functions are written sometimes similarly as the ordinary random functions of the variable t with the indication of an argument in the brackets, and any random linear function on Φ is written in the form of (4.7.2).

From the general definitions of Sections 3.3 and 3.4 follow the definitions of an expectation and an operator of the second order moment of a generalized random function. *An expectation of a generalized random function* X is called a linear functional m_x on Φ determined by the formula

$$m_x\varphi = E(X\varphi) \quad \text{at all} \quad \varphi \in \Phi. \tag{4.7.3}$$

A second order moment of a generalized random function X is called an operator Γ_x mapping Φ into X which is determined by the formula

$$\Gamma_x\varphi = EX(\overline{X\varphi}), \quad \varphi \in \Phi. \tag{4.7.4}$$

Replacing X in this formula by the centered random variable $X^0 = X - m_x$ we find the covariance operator K_x of a generalized random function.

As $\Gamma_x\varphi$ represents an element of the space X, i.e. linear functional on Φ then the expression $(\Gamma_x\varphi_1)\varphi_2$, $\varphi_1,\varphi_2 \in \Phi$ has the sense. But $(\Gamma_x\varphi_2)\varphi_1 = EX\varphi_1(\overline{X\varphi_2}) = EX_t(\overline{X_{t'}\varphi_{2t'}})\varphi_{1t} = EX_t(\overline{X_{t'}\bar\varphi_{1t}\varphi_{2t'}})$ where by the indexes t, t' the arguments of the functions φ_1, φ_2 are marked and it is shown that the functional X acts in one case on a function of the variable t' at fixed t, and in another case – on the function of the variable t. But the product $\bar\varphi_{1t}\varphi_{2t'} = \overline{\varphi_1(t)}\varphi_2(t')$ belongs to the space Φ' of the test functions of two variables t, t'. After substituting it by an arbitrary function $\bar\varphi_{t,t'} = \overline{\varphi(t,t')} \in \Phi'$ and noticing that all the realizations of the random variable $\overline{X_{t'}\bar\varphi_{t,t'}}$ are the functions of t which belong to the space Φ we come to the conclusion that the following expression has the sense:

$$\Gamma_x\varphi_{t,t'} = EX_t(\overline{X_{t'}\bar\varphi_{t,t'}}). \tag{4.7.5}$$

Thus an operator of the second order moment of a generalized random function determines a linear functional on the space Φ' of the test functions of two variables t, t'. If this functional is continuous in the topology of the space Φ' then it represents a generalized function of the variables t, t'. In the special case for any m.s. integrable random function $X(t)$ according to (4.7.2) and (4.4.14)

$$\Gamma_x\varphi_{t,t'} = \iint \Gamma_x(t,t')\varphi(t,t')\,dt\,dt'. \tag{4.7.6}$$

This fact gives grounds to consider the second order moment and the covariance function of a generalized random function as the generalized functions of two variables and write them similarly as the ordinary functions with the indication of the arguments in the brackets. Here the formulae which determine an operator of the second order moment and a covariance operator of a generalized function $X(t)$ in the same way as for an ordinary random function are whiten in the form

$$\Gamma_x\varphi = \int \Gamma_x(t,t')\overline{\varphi(t')}dt', \quad K_x\varphi = \int K_x(t,t')\overline{\varphi(t')}dt'. \tag{4.7.7}$$

Analogously taking a space of the vector test functions Φ we may determine the vector generalized random functions.

4.7.2. White Noise as Generalized Function

The white noises and their derivatives (Subsection 3.2.3) refer to the generalized random functions. By *a white noise* we call a generalized random function of the scalar variable $X(t)$ with zero expectation and the covariance operator $K_x\varphi = \nu(t)\overline{\varphi(t)}$. It means that the covariance function of a white noise in accordance with (4.7.7) is proportional to the δ-function: $K_x(t,t') = \nu(t)\delta(t-t')$. We show that if $X(t)$ is the n-dimensional vector white noise of the intensity $\nu(t)$ then the random function $Y(t) = f(t)X(t)$ where $f(t)$ is an arbitrary $m \times n$-matrix represents the m-dimensional vector white noise of the intensity $f(t)\nu(t)f(t)^*$.

▷ Really, an expectation of the generalized random function $Y = fX$ evidently is equal to zero. For determining its covariance operator we use the forth formula of (3.3.16). Keeping in mind that as an operator adjoint with the operator of the multiplication by the matrix $f(t)$ the left is the operator of the multiplication by the matrix $f(t)^*$ on the right we obtain for any function $\varphi(t) \in \Phi$

$$K_y\varphi = fK_xf^*\varphi = f(t)\nu(t)f(t)^*\overline{\varphi(t)}.$$

This formula shows that as a covariance operator of the generalized random function $Y = fX$ serves an operator of complex conjugation and the multiplication by the matrix $f(t)\nu(t)f(t)^*$. Hence by the definition of a white noise it follows that the generalized random function $Y(t) = f(t)X(t)$ represents a white noise of the intensity $f(t)\nu(t)f(t)^*$. ◁

The notion of the generalized random function may be also extended over the random function with the values in an arbitrary B-space L. Let L be a space dual with some B-space B. We take a space of the test functions Φ with the values in the space B. Any linear functional X on Φ will be in this case a generalized random function with the values in L. In particular, by a white noise with the values in some B-space L is called a generalized random function X with the covariance operator $K_x\varphi = \nu(t)\varphi(t)$ where $\nu(t)$ is a function with the values in the space $\bar{L}(B, L)$ of the adjoint-linear operators which map B into L.

4.7.3. Convergence of Generalized Random Functions

Formula (4.7.2) determines the integral with respect to the product of any ordinary random function by any test function as in the given case $F_c = \Phi$. Consequently, the space of all linear functionals on F_c coincides with X. So Theorem 4.1.2 is valid for the generalized random functions. Thus we have the following statement.

Theorem 4.7.1. *It is necessary and sufficient for the existence of m.s. limit of the generalized random function X_r at $r \to r_0$ (in the topo-*

logy of the space X) that there will exist the limit $\lim\limits_{r,s\to r_0}(\Gamma_{rs}\varphi)\varphi$ $= \text{l.i.m.}\limits_{r,s\to r_0} E(X_r\varphi)(\overline{X_s\varphi})$ for any $\varphi\in\Phi$. Here there also exists the limit $\varphi_1,\varphi_2\in\Phi$ for any functions $\lim\limits_{r,s\to r_0}(\Gamma_x\varphi_1)\varphi_2 = \text{l.i.m.}\limits_{r,s\to r_0} E(X_r\varphi_2)(\overline{X_s\varphi_1})$ for any functions $\varphi_1,\ \varphi_2\in\Phi$.

E x a m p l e 4.7.1. Consider a sequence of the random functions $\{X_r(t)\}$ of Examples 4.1.3 and 4.1.4. In this case

$$\Gamma_{rs}\varphi = K_{rs}\varphi = \frac{\alpha_r\alpha_s}{\alpha_r^2 + \alpha_s^2}\left[\int_{-\infty}^{t} e^{-\alpha_r(t-t')}\overline{\varphi(t')}dt'\right.$$

$$\left. + \int_{t}^{\infty} e^{-\alpha_s(t'-t)}\overline{\varphi(t')}dt'\right] \to \overline{\varphi(t)} \quad\text{at}\quad \alpha_r,\alpha_s\to\infty.$$

Consequently, $(\Gamma_{rs}\varphi)\varphi \to \int|\varphi(t)|^2dt$. It follows from Theorem 4.7.1 that the sequence of ordinary random functions $\{X_r(t)\}$ which are considered as the generalized random functions m.s. converges to a white noise of unit intensity. Thus a white noise may be considered as m.s. limit (in the topology of the space X) of the sequence of the random functions with the exponential covariance functions.

E x a m p l e 4.7.2. Consider the random function $X(t) = U(\alpha+i\Lambda)e^{i\Lambda t}$ where U and Λ are independent random variables and $EU = 0;\ DU = E|U|^2 = D;\ \Lambda$ has the Cauchy density $f(\lambda) = \alpha/\pi(\alpha^2 + \lambda^2)$. It is clear that $m_x(t)\equiv 0$. Using the formula of the total expectation for the calculation of the covariance function we shall have

$$K_x(t,t') = E[\,E(X(t))\overline{X(t')}|\Lambda\,] = \frac{\alpha}{\pi}\int_{-\infty}^{\infty}\frac{E(X(t)\overline{X(t')}|\lambda)}{\alpha^2 + \lambda^2}\,d\lambda.$$

But

$$E(X(t)\overline{X(t')}|\lambda) = E|U|^2(\alpha^2 + \lambda^2)e^{i\lambda(t-t')} = D(\alpha^2 + \lambda^2)e^{i\lambda(t-t')}.$$

Substituting this expression into the previous formula we get

$$K_x(t,t') = \frac{D\alpha}{\pi}\int_{-\infty}^{\infty} e^{i\lambda(t-t')}d\lambda = 2D\alpha\delta(t - t').$$

Thus the random function $X(t)$ represents a white noise of the intensity $2D\alpha$. In the given case we have an Example of a white noise which is an ordinary random function of a point whose realizations are infinitely differentiable. The considered Example shows that there are the ordinary random functions of a point whose covariance functions are the generalized functions. All the realizations of the random function $X(t)$ in this Example are continuous. But $X(t)$ is not m.s. continuous.

4.7.4. Differentiation of Generalized Random Functions

Now let us consider the question about the differentiation of the generalized random functions. At first assuming the argument t as a scalar one we compare each function $\varphi(t) \in \Phi$ with the function

$$\varphi_h(t) = [\varphi(t) - \varphi(t - h)]/h \qquad (4.7.8)$$

and introduce the ratio operator of the finite differences D_h: $D_h f = [f(t+h) - f(t)]/h$. Then for any ordinary m.s. differentiable random function $X(t)$ we shall have

$$(D_h X)\varphi = \int \frac{X(t+h) - X(t)}{h}\varphi(t)\,dt = -\int X(t)\varphi_h(t)\,dt. \qquad (4.7.9)$$

According to this fact we extend the operator D_h over the generalized random functions putting

$$(D_h X)\varphi = -X\varphi_h. \qquad (4.7.10)$$

Then by virtue of (4.7.9) we get

$$E[(D_h X)\varphi][\overline{(D_l X)\varphi}] = E(X\varphi_h)(\overline{X\varphi_l}) = \Gamma_x(\varphi_{ht}\bar{\varphi}_{lt'}). \qquad (4.7.11)$$

Hence it follows that if the second order moment Γ_x of the generalized random function X is a generalized function of two variables then $\Gamma_x(\varphi_{ht}\bar{\varphi}_{lt'}) \to \Gamma_x(\varphi'_t\bar{\varphi}'_{t'})$ at $h, l \to 0$ for any function $\varphi \in \Phi$. From Theorem 4.1.2 it follows the existence of m.s. limit $\underset{h \to 0}{\text{l.i.m.}}(D_h X)\varphi = -X\varphi'$ for any function $\varphi \in \Phi$. This fact gives grounds to determine m.s. derivative X' of the generalized random function X by the same formula by which the derivative of a nonrandom generalized function is found,

$$X'\varphi = -X\varphi'. \qquad (4.7.12)$$

It follows from Theorem 4.7.1 that

$$m_{x'}\varphi = -m_x\varphi', \qquad (4.7.13)$$

$$\Gamma_{x'}(\varphi_t\bar{\psi}_{t'}) = E(X'\varphi)(\overline{X'\psi}) = \Gamma_x(\varphi'_t\bar{\psi}'_{t'}), \qquad (4.7.14)$$

$$\Gamma_{xx'}(\varphi_t\bar{\psi}_{t'}) = E(X\varphi)(\overline{X'\psi}) = -\Gamma_x(\varphi_t\bar{\psi}'_{t'}). \qquad (4.7.15)$$

It means that if Γ_x is the generalized function t, t' then

$$m_{x'}\varphi = m'_x\varphi, \qquad (4.7.16)$$

$$\Gamma_x(\varphi_t\bar{\psi}_{t'}) = \frac{\partial^2\Gamma_x}{\partial t\partial t'}(\varphi_t\overline{\psi_{t'}}), \qquad (4.7.17)$$

$$\Gamma_{xx'}(\varphi_t\bar{\psi}_{t'}) = \frac{\partial\Gamma_x}{\partial t'}(\varphi_t\overline{\psi_{t'}}) \qquad (4.7.18)$$

for all $\varphi, \psi \in \Phi$. Formulae (4.7.16)–(4.7.18) extend over the generalized random functions in formulae (4.3.4), (4.3.2) and (4.3.3) for expectation, for the second order moment of the derivative of the random function and for the cross-second order moment of this random function and its derivative. As the generalized functions have the derivatives of all orders then it follows from the obtained result that any function whose second order moment is a generalized function has m.s. derivatives of all orders and formulae (4.3.6)–(4.3.8) are valid for them. Finally, we notice that formulae (4.7.14)–(4.7.18) may be expressed in terms of the functions $\varphi(t,t')$ and $\psi(t,t')$ analogously to (4.7.6).

Now we consider the white noise $X(t)$ of unit intensity (Subsection 3.2.3). Its covariance function $\delta(t-t')$ has the generalized derivatives of all orders. According to formulae (4.3.6)–(4.3.8) the expectations of all m.s. derivatives of the white noise are equal to zero, and their covariance functions are determined by the formula

$$K_{x^{(p)}x^{(q)}}(t,t') = (-1)^q\delta^{(p+q)}(t-t') \quad (p,q = 0,1,2,\ldots).$$

Hence it follows that the covariance operator of the derivative $X^{(p)}(t)$ of the white noise and the cross covariance operator of the derivatives $X^{(p)}(t)$ and $X^{(q)}(t)$ may be presented

$$K_{x^{(p)}}\varphi = (-1)^p\overline{\varphi^{(2p)}(t)}, \quad K_{x^{(p)}x^{(q)}}\varphi = (-1)^q\overline{\varphi^{(p+q)}(t)}.$$

E x a m p l e 4.7.3. The derivative of the order p of the random function $X(t)$ in Example 4.7.2 is described by the formula $X^{(p)}(t) = U(i\Lambda)^p(\alpha+i\Lambda)e^{i\Lambda t}$.

The covariance function of this derivative according to (4.3.8) is $(-1)^p 2D\alpha\delta^{(2p)}(t - t')$. Thus we have an example of a white noise whose all realizations of the derivatives are continuous and infinitely differentiable in spite of the fact that neither the white noise $X(t)$ itself nor its derivatives are m.s. continuous and have m.s. derivatives in ordinary sense (they have the generalized m.s. derivatives).

E x a m p l e 4.7.4. The random function $X(t)$ with zero expectation and the covariance function $K_x(t, t') = De^{-\alpha|t-t'|}$ is not m.s. differentiable. But it has the generalized m.s. derivatives of all orders.

4.7.5. Integration of Generalized Functions

Let us pass to the indefinite integrals with respect to the generalized random function. The equation $Y' = X$ determines the primitive Y of the generalized random function X. Really, it follows from (4.7.12) that $Y'\varphi = -Y\varphi'$. Consequently, the equation $Y' = X$ is equivalent to the equation $-Y\varphi' = X\varphi$ which determines Y on the subspace of the space Φ which represents a set of the derivatives of all functions from Φ. This set does not coincide with the whole space Φ. Thus the derivative of any function $\varphi \in \Phi$ belongs to Φ but not any function from Φ is a derivative of some function Φ. Therefore in order to be sure in the existence of the primitive of the random function X it is necessary to show that the founded linear functional Y may be extended over the whole space Φ.

$\triangleright$ Let $\varphi(t)$ be an arbitrary function from Φ, $\psi(t)$ be some fixed function from Φ with an integral different from zero. Without loss of generality we may assume that $\int \psi(\tau)\, d\tau = 1$. We determine the function

$$\varphi_0(t) = \varphi(t) - \psi(t) \int\limits_{-\infty}^{\infty} \varphi(\tau)\, d\tau. \qquad (4.7.19)$$

As by virtue of (4.7.19) $\int\limits_{-\infty}^{\infty} \varphi_0(\tau)\, d\tau = 0$ and $\varphi_*(t) = \int\limits_{-\infty}^{t} \varphi_0(\tau)\, d\tau \in \Phi$ then $\varphi_0(t)$ is a derivative of the functon $\varphi_*(t)$ from Φ and the functional Y: $Y\varphi_0 = Y\varphi_*' = -X\varphi_*$ is determined for it. Putting here the value $Y\psi$ equal to an arbitrary random variable C we may extend Y over the whole space Φ assuming

$$Y\varphi = -X\varphi_* + C \int\limits_{-\infty}^{\infty} \varphi(\tau)\, d\tau \quad \text{for all} \quad \varphi \in \Phi. \qquad (4.7.20)$$

Here for any function $\varphi \in \Phi$ will be $Y'\varphi = -Y\varphi' = X\varphi$ as $\int\limits_{-\infty}^{\infty} \varphi'(\tau)\, d\tau = 0$ in consequence of which φ_* coincides with φ. The random variable C plays the role of the integration constant. $\triangleleft$

Thus we proved the following statement.

Theorem 4.7.2. *The primitive of the generalized random function exists and is determined uniquely with the accuracy till an arbitrary additive random variable.*

It is easy to understand that the expectation of the primitive Y of the generalized random function X is a primitive of its expectation, and the second order moment and the covariance function are determined by the differential equations

$$\partial^2 \Gamma_y / \partial t\, \partial t' = \Gamma_x, \quad \partial^2 K_y / \partial t\, \partial t' = K_x. \tag{4.7.21}$$

$\triangleright$ Let $V(t)$ be a white noise of the intensity $\nu(t)$. According to the proved before its primitive

$$U(t) = \int\limits_{-\infty}^{t} V(t)\, dt \tag{4.7.22}$$

is determined with the accuracy till an arbitrary additive random variable. Let this random variable be equal to zero. Then an expectation of the primitive of a white noise will be equal to zero and its covariance function will be determined by the formula

$$
\begin{aligned}
K_u \varphi_{1t}\varphi_{2t'} &= K_v \int\limits_{-\infty}^{t} \overline{\varphi_{10}} d\tau \int\limits_{-\infty}^{t'} \varphi_{20} d\tau \\
&= \int\limits_{-\infty}^{\infty} \nu(t) dt \int\limits_{-\infty}^{t} \int\limits_{-\infty}^{t'} \overline{\varphi_{10}(\tau)}\varphi_{20}(\tau')d\tau\, d\tau'.
\end{aligned}
\tag{4.7.23}
$$

Without loss of generality we may suppose that the function $\nu(t)$ is different from zero only on the finite interval (otherwise it may be substituted by such a function). Then after changing the order of the integration in (4.7.23) we obtain

$$K_u \varphi_{1t}\varphi_{2t'} = \int\limits_{-\infty}^{\infty} \int\limits_{-\infty}^{\infty} \overline{\varphi_{10}(\tau)}\varphi_{20}(\tau')d\tau\, d\tau' \int\limits_{\max\{\tau,\tau'\}}^{\infty} \nu(\sigma)d\sigma. \tag{4.7.24}$$

Hence it is clear that the primitive of a white noise represents an ordinary random function $U(t)$ with the covariance function

$$K_u(t,t') = \int_{\max\{t,t'\}}^{\infty} \nu(\tau)d\tau. \qquad (4.7.25)$$

The random function

$$W(t) = U(t) - U(t_0) = \int_{t_0}^{t} V(\tau)d\tau \qquad (4.7.26)$$

by virtue of (4.7.22) has zero expectation $m_w = 0$ and its covariance function is equal to

$$K_w(t,t') = K_u(t,t') - K_u(t,t_0) - K_u(t_0,t') + K_u(t_0,t_0)$$

$$= \left[\int_{\max\{t,t'\}}^{\infty} - \int_{t}^{\infty} - \int_{t'}^{\infty} + \int_{t_0}^{\infty}\right] \nu(\tau)d\tau = \int_{t_0}^{\min\{t,t'\}} \nu(\tau)d\tau. \; \triangleleft \qquad (4.7.27)$$

4.7.6. Stochastic Integrals as Integrals Containing White Noise

Now let us consider a random function of a set determined on semi-open in the right intervals by the following formula:

$$Z([t_1,t_2)) = W(t_2) - W(t_1) = U(t_2) - U(t_1) = \int_{t_1}^{t_2} V(\tau)d\tau. \qquad (4.7.28)$$

We shall find its covariance function. If $t_1 < t_2 \le t_3 < t_4$ then on the basis of (4.7.28)

$$K_z([t_1,t_2),[t_3,t_4,)) = K_w(t_2,t_4) - K_w(t_2,t_3) - K_w(t_1,t_4) + K_w(t_1,t_3)$$

$$= \left[\int_{t_0}^{t_2} - \int_{t_0}^{t_2} - \int_{t_0}^{t_1} + \int_{t_0}^{t_1}\right] \nu(\tau)d\tau = 0. \qquad (4.7.29)$$

If $t_1 < t_3 < t_2 < t_4$ or $t_3 \le t_1 < t_2 < t_4$ then we have correspondingly

$$K_z([t_1,t_2),[t_3,t_4,)) = \int_{t_3}^{t_2} \nu(\tau)d\tau, \quad K_z([t_1,t_2),[t_3,t_4,)) = \int_{t_1}^{t_2} \nu(\tau)d\tau.$$

$$(4.7.30)$$

It is clear from (4.7.29) and (4.7.30) that the covariance function of the random function of the set depends only on the intersection of the intervals: for the functon $\sigma(A) = \int_A \nu(\tau)$ we have $K_z([t_1, t_2), [t_3, t_4)) = \sigma([t_1, t_2) \cap [t_3, t_4))$. Besides that the random function Z is σ-additive and $EZ = 0$.

The formula

$$Z(A) = \int_A V(\tau) d\tau \qquad (4.7.31)$$

determines the extension of the random function Z on σ-algebra of Borel sets of the real axis which represents a stochastic measure.

Thus an integral with respect to a white noise represents a stochastic measure. In consequence of this the stochastic integrals with respect to the functions of a real scalar variable t are often written in the form of the ordinary integrals over the variable t with respect to the function which contains a white noise as a multiplier:

$$\int_A X(t) Z(dt) = \int_A X(t) V(t) dt. \qquad (4.7.32)$$

It is possible only in that case when the function $\sigma(A)$ represents an integral with respect to some function $\nu(t)$ (which may contain a linear combination of δ-functions). In particular, the *integrals over the Wiener stochastic measure may be always written as the ordinary integrals with white noise in the integrand. Analogously we may write the stochastic θ-integrals.*

4.8. Integral Canonical and Spectral Representations

4.8.1. Introductory Remarks

In elementary theory of the random functions a representation of a nonstationary scalar random function $X = X(t)$ in the form

$$X(t) = m_x(t) + \int_{\lambda_1}^{\lambda_2} V(\lambda) x(t, \lambda) d\lambda, \qquad (4.8.1)$$

where $V(\lambda)$ is a white noise of the parameter λ which changes at the interval $[\lambda_1, \lambda_2]$ is called its *integral canonical representation*. The functions $x(t, \lambda)$ of the variable t correspondent to the different fixed values of

the parameter λ are called *the coordinate functions* of the integral canonical representation (4.8.1). To the integral canonical representation of
the random function $X = X(t)$ corresponds the following integral representation of its covariance function $K_x(t, t')$:

$$K_x(t, t') = \int_{\lambda_1}^{\lambda_2} \nu(\lambda) x(t, \lambda) x(t', \lambda)^* d\lambda. \qquad (4.8.2)$$

Here $\nu(\lambda)$ is the intensity of the white noise $V(\lambda)$ which possesses zero expectation and the covariance function $K_v(\lambda, \lambda') = \nu(\lambda)\delta(\lambda - \lambda')$.
The canonical expansions of the random functions (Section 3.4) are the
special case of the integral canonical representations.

For the stationary (in a broad sense) random function $X = X(t)$
when the expectation m_x is constant, and the covariance function satisfies the condition $K_x(t, t') = k_x(\tau)$, $\tau = t - t'$ integral canonical representation (4.8.1) has the form

$$X(t) = m_x + \int_0^\infty [U(\omega) \sin \omega t + Z(\omega) \cos \omega t] \, d\omega \qquad (4.8.3)$$

or in a complex form

$$X(t) = m_x + \int_{-\infty}^\infty V(\omega) e^{i\omega t} \, d\omega \qquad (4.8.4)$$

and is called *a spectral representation*.

The foundation of the spectral theory of the stationary random
functions are the known Wiener–Chinchin formulae:

$$s_{1x}(\omega) = \frac{2}{\pi} \int_0^\infty k_x(\tau) \cos \omega \tau d\tau, \qquad (4.8.5)$$

$$s_x(\omega) = \frac{1}{2\pi} \int_{-\infty}^\infty k_x(\tau) e^{i\omega \tau} d\tau, \qquad (4.8.6)$$

$$k_x(\tau) = \int_0^\infty s_{1x}(\omega) \cos \omega \tau d\omega, \qquad (4.8.7)$$

$$k_x(\tau) = \int_{-\infty}^\infty s_x(\omega) e^{i\omega \tau} d\omega, \qquad (4.8.8)$$

which correspond to the spectral representations (4.8.3) and (4.8.4). The spectral density $s_{1x}(\omega)$ or $s_x(\omega)$ is a very important characteristic stationary random function. Knowing the spectral density $s_{1x}(\omega)$ or $s_x(\omega)$ we may find the covariance function $k_x(\tau)$ by formula (4.8.7) or (4.8.8). And vice versa, knowing the covariance function $k_x(\tau)$ we may calculate by formula (4.8.5) or (4.8.6) the spectral density $s_{1x}(\omega)$ or $s_x(\omega)$ correspondingly.

4.8.2. Direct Theorem about Integral Canonical Representations

Now we shall try to construct more general than in Section 3.4 types of the canonical expansions of the random variables in the linear space. Suppose that in the space H_x (Subsection 3.4.2) there exists such a sequence of the expansions of unit $\{\mathcal{E}_n(B)\}$ determined on some σ-algebra $\mathcal{L}$ of the sets of some space Λ and such sequence of the orthogonal vectors $\{\gamma_n\} \subset H_x$ that the set of the linear combinations of the vectors $\mathcal{E}_n(B)\gamma_n$ (correspondent to all $B \in \mathcal{L}$ and to all natural n) is dense in H_x and the subspaces H_n and H_m (formed by a set of the vectors $\{g : g = \mathcal{E}_n(B)\gamma_n, B \in \mathcal{L}\}$ and by a set of the vectors $\{h : h = \mathcal{E}_m(B)\gamma_m, B \in \mathcal{L}\}$) are orthogonal at $m \neq n$. In this case the space H_x represents an orthogonal sum of the subspaces $H_n(n = 1, 2, \ldots)$: $H_x = \sum_{n=1}^{\infty} \oplus H_n$.

Given some expansion of unit $\mathcal{E}_n$ to each vector γ_n corresponds the nonnegative finite measure $\sigma_n(B) = (\mathcal{E}_n(B)\gamma_n, \gamma_n) = \| \mathcal{E}_n(B)\gamma_n \|^2$. Let us take an arbitrary vector $f \in H_x$. It is evident that the function of the set $\mu_n(B) = (f, \mathcal{E}_n(B)\gamma_n)$ is σ_n-continuous. As it is known from functional analysis according to Radon–Nikodym theorem there exists unique (with the accuracy till the σ_n-equivalency) σ_n-integrable function $u_n(\lambda)$ such that

$$\mu_n(B) = (f, \mathcal{E}_n(B)\gamma_n) = \int_B u_n(\lambda)\sigma_n(d\lambda). \qquad (4.8.9)$$

We determine the linear operators

$$T_n = \int u_n(\lambda)\mathcal{E}_n(d\lambda) \quad (n = 1, 2, \ldots) \qquad (4.8.10)$$

and show that the following representation takes place:

$$f = \sum_{\nu=1}^{\infty} T_\nu \gamma_\nu = \sum_{\nu=1}^{\infty} \int u_\nu(\lambda)\mathcal{E}_\nu(d\lambda)\gamma_\nu. \qquad (4.8.11)$$

▷ For this purpose we notice that by virtue of (4.8.9) and (4.8.10)

$$\left(f - \sum_{\nu=1}^{\infty} T_\nu \gamma_\nu \, , \mathcal{E}_n(B)\gamma_n \right) = \int_B u_n(\lambda)\sigma_n(d\lambda) - \sum_{\nu=1}^{\infty} (T_\nu \gamma_\nu \, , \mathcal{E}_n(B)\gamma_n)$$

$$= \int_B u_n(\lambda)\sigma_n(d\lambda) - \sum_{\nu=1}^{\infty} \int u_\nu(\lambda)(\mathcal{E}_\nu(d\lambda)\gamma_\nu \, , \mathcal{E}_n(B)\gamma_\nu) \, .$$

But in the consequence of the orthogonality of the subspaces H_ν and H_n at $\nu \neq n$

$$(\mathcal{E}_\nu(A)\gamma_\nu \, , \mathcal{E}_n(B)\gamma_n) = 0 \quad \text{at all} \quad A, B \in \mathcal{L} \, , \qquad (4.8.12)$$

$$(\mathcal{E}_n(A)\gamma_n \, , \mathcal{E}_n(B)\gamma_n) = (\mathcal{E}_n(AB)\gamma_n, \gamma_n) = \sigma_n(AB) \, . \qquad (4.8.13)$$

Consequently, for all $B \in \mathcal{L}$ and for all natural n we have

$$\int u_n(\lambda)(\mathcal{E}_n(d\lambda)\gamma_n \, , \mathcal{E}_n(B)\gamma_n) = \int_B u_n(\lambda)\sigma_n(d\lambda) \, ,$$

$$\left(f - \sum_{\nu=1}^{\infty} T_\nu \gamma_\nu \, , \mathcal{E}_n(B)\gamma_n \right) = \int_B u_n(\lambda)\sigma_n(d\lambda) - \int_B (u_n(\lambda)\sigma_n(d\lambda)) = 0$$

As the space formed by the vectors correspondent to all $B \in \mathcal{L}$ and to all natural n coincides with H_x then $f - \sum_{\nu=1}^{\infty} T_\nu \gamma_\nu = 0$ what proves the validity of formula (4.8.11). ◁

From formula (4.8.11) and the definition (Subsection 3.4.3) of a scalar product in H_x it follows that

$$0 = \| f - \sum_{\nu=1}^{\infty} T_\nu \gamma_\nu \|^2 = E \left| fX^0 - \sum_{\nu=1}^{\infty} \int u_\nu(\lambda)\mathcal{E}_\nu(d\lambda)\gamma_\nu X^0 \right|^2 \, . \qquad (4.8.14)$$

Consider the random measures

$$Z_\nu(B) = \mathcal{E}_\nu(B)\gamma_\nu X^0 \, , \quad B \in \mathcal{L} \quad (\nu = 1, 2, \ldots) \, . \qquad (4.8.15)$$

By virtue of (4.8.12) they are uncorrelated as $EZ_\nu(B)\overline{Z_\mu(B')} = (\mathcal{E}_\nu(B)\gamma_n u, \, \mathcal{E}_\mu(B')\gamma_\mu) = 0$ at $\mu \neq \nu$. Their covariance functions are determined according to (4.8.13) by the formula

$$K_\nu(B, B') = EZ_\nu(B)\overline{Z_\nu(B')} = (\mathcal{E}_\nu(B)\gamma_n u \, , \mathcal{E}_\nu(B')\gamma_\nu) = \sigma_\nu(BB') \, .$$

Hence it follows that $Z_\nu(B)$ are uncorrelated stochastic measures. Thus we conclude from (4.8.14) that for any $f \in H_x$ the random variable fX^0 is expressed by the sum of uncorrelated m.s. stochastic integrals:

$$fX^0 = \sum_{\nu=1}^{\infty} \int u_\nu(\lambda) Z_\nu(d\lambda) . \tag{4.8.16}$$

Now we notice that by virtue of (4.8.9) $u_\nu(\lambda)$ at any $\lambda \in \Lambda$ represents the value of the linear functional on H_x at the point f : $u_\nu(\lambda) = \varphi_\nu(\lambda)f$. Equality (4.8.16) determines such random linear functional Φ on H_x that $\Phi f = fX^0$ at all $f \in H_x$.

If at all $\lambda \in \Lambda$ to each functional f corresponds such vector $x_\nu(\lambda) \in X$ that $\varphi_\nu(\lambda)f = fx_\nu(\lambda)$ at all $f \in H_x$ then equality (4.8.16) takes the form

$$X = m_x + \sum_{\nu=1}^{\infty} \int x_\nu(\lambda) Z_\nu(d\lambda), \tag{4.8.17}$$

where the integrals represent the weak m.s. stochastic integrals and the series weakly m.s. converges to $X^0 = X - m_x$. Formula (4.8.17) determines an *integral canonical representation of the random variable* X.

It follows from (4.8.11) that in this case for any $f, g \in H_x$

$$(f, g) = \sum_{\nu, \mu=1}^{\infty} \iint f x_\nu(\lambda) \overline{g x_\mu(\lambda')} (\mathcal{E}_\nu(d\lambda)\gamma_\nu , \mathcal{E}_\mu(d\lambda')\gamma_\mu)$$

or by virtue of (4.8.12) and (4.8.13)

$$(f, g) = f K_x g = \sum_{\nu=1}^{\infty} \int f x_\nu(\lambda) \overline{g x_\nu(\lambda)} \sigma_\nu(d\lambda) .$$

This formula gives an *integral canonical representation of the covariance operator*:

$$K_x = \sum_{\nu=1}^{\infty} \int x_\nu(\lambda) \left[\overline{\cdot x_\nu(\lambda)} \right] \sigma_\nu(d\lambda) . \tag{4.8.18}$$

Theorem 4.8.1. *Let (4.8.17) be some integral canonical representation of the random vriable* X. *Then (4.8.18) is the integral canonical representation of its covariance operator* K_x.

It remains to find a formal expression of the set of the coordinate vectors $x_\nu(\lambda)$. As it follows from (4.8.9) the function $u_\nu(\lambda)$ represents the derivative of the Radon–Nikodym measure $\mu_\nu(B) = (f, \mathcal{E}_\nu(B)\gamma_\nu)$ with respect to the measure $\sigma_\nu(B)$. Consequently,

$$
f x_\nu(\lambda) = \frac{d\mu_\nu}{d\sigma_\nu}(\lambda) = \frac{d(f, \mathcal{E}_\nu \gamma_\nu)}{d\sigma_\nu} = \frac{d(f K_x \mathcal{E}_\nu \gamma_\nu)}{d\sigma_\nu}
$$

$$
= \frac{dE(fX^0)(\overline{\mathcal{E}_\nu \gamma_\nu X^0})}{d\sigma_\nu} = \frac{dE(fX^0)\overline{Z}_\nu}{d\sigma_\nu} .
$$

Hence we find

$$
x_\nu(\lambda) = \frac{dE(X^0 \overline{Z}_\nu)}{d\sigma_\nu}(\lambda) = EX^0 \frac{d\overline{Z}_\nu}{d\sigma_\nu}(\lambda) . \tag{4.8.19}
$$

As $Z_\nu(B) = 0$ with the probability 1 for any $B \in \mathcal{L}$ for which $\sigma_\nu(B) = 0$ then the measure Z_ν is a.s. σ_ν-continuous. Consequently, almost for all the realizations of the stochastic measure Z_ν there exists the Radon–Nikodym derivative $dZ_\nu/d\sigma_\nu$.

R e m a r k 1. If in the sequences $\{\mathcal{E}(B)\}$, $\{\gamma_n\}$ there exist the elements $\mathcal{E}(B)$, γ such that only linear combinations of the vectors $\mathcal{E}(B)\gamma$, $B \in \mathcal{L}$ are dense in H_x then in all previous formulae the series contains only one item. In this case the integral canonical representations (4.8.17) and (4.8.18) take the form

$$
X = m_x + \int x(\lambda) Z(d\lambda) , \tag{4.8.20}
$$

$$
K_x = \int x(\lambda) \left[\overline{\cdot x(\lambda)} \right] \sigma(d\lambda) , \tag{4.8.21}
$$

where Z is a stochastic measure determined by the formula $Z(B) = \mathcal{E}(B)\gamma X^0$; σ is its covariance function, $K_z(B, B') = \sigma(BB')$, and the coordinate vectors $x(\lambda)$ are determined by the formula

$$
x(\lambda) = EX^0 \frac{d\overline{Z}}{d\sigma}(\lambda) . \tag{4.8.22}
$$

R e m a r k 2. For the separable space H_x to any orthonormal basis $\{f_n\}$ corresponds the expansion of the unit: $\mathcal{E}(B) = \sum_{k \in B} (\cdot, f_k) f_k$ determined on the σ-algebra of Borel sets of the real semiaxis $(0, \infty)$. Assuming $\mathcal{E}_n(B) = \mathcal{E}(B)$, $\gamma_n = \alpha_n f_n$ where $\{\alpha_n\}$ is an arbitrary sequence of the numbers we get

$$
\mathcal{E}_n(B)\gamma_n = \begin{cases} \alpha_n f_n = \gamma_n , & \text{if } n \in B , \\ 0 , & \text{if } n \bar{\in} B . \end{cases}
$$

Thus the accepted conditions, see Remark 1, in the given case are fulfilled. The stochastic measures $Z_\nu(B)$ and the measures $\sigma_\nu(B)$ are determined by the formulae

$$Z_\nu(B) = \begin{cases} \alpha_\nu f_\nu X^0 = \gamma_\nu X^0 = V_\nu\,, & \text{if } \nu \in B\,, \\ 0\,, & \text{if } \nu \bar\in B\,, \end{cases}$$

$$\sigma_\nu(B) = \|\,\mathcal{E}_\nu(B)\gamma_\nu\,\|^2 = \begin{cases} |\alpha_\nu|^2 = \|\,\gamma_\nu\,\|^2\,, & \text{if } \nu \in B\,, \\ 0\,, & \text{if } \nu \bar\in B\,. \end{cases}$$

Hence it follows that in the given case

$$\frac{dZ_\nu}{d\sigma_\nu}(\lambda) = \frac{V_\nu}{|\alpha_\nu|^2}$$

and (4.8.19) takes the form $x_\nu(\lambda) = |\alpha_\nu|^{-2} EX^0 V_\nu = |\alpha_\nu|^{-2} K_x \gamma_\nu = x_\nu.$
Canonical representations (4.8.17) and (4.8.18) in this case have the form

$$X = m_x + \sum_{\nu=1}^{\infty} V_\nu x_\nu\,, \qquad K_x = \sum_{\nu=1}^{\infty} |\alpha_\nu|^2\, x_\nu(\cdot\overline{x_\nu})\,.$$

Thus *the canonical expansions represent a special case of the integral canonical representations.*

4.8.3. Inverse Theorem about Integral Canonical Representations

Theorem 4.8.2 (The inverse Pugachev theorem). *For obtaining the integral canonical representation of the random variable X it is sufficient to find the H_x-valued measures $\mu_1, \ldots, \mu_N$ determined on some measurable space $(\Lambda, \mathcal{L})$, the nonnegative measures $\sigma_1, \ldots, \sigma_N$ on $(\Lambda, \mathcal{L})$ and the functions $x_1(\lambda), \ldots, x_N(\lambda)$ on Λ (N may be infinite) satisfying the conditions*

$$K_x \mu_p(B) = \int_B x_p(\lambda)\sigma_p(d\lambda)\,, \qquad B \in \mathcal{L}\,, \tag{4.8.23}$$

$$\mu_p(B)x_q(\lambda) = \delta_{pq} \mathbf{1}_B(\lambda)\,, \tag{4.8.24}$$

$$\sum_{p=1}^{N} \int f x_p(\lambda)\mu_p(d\lambda) = f \quad \text{for any} \quad f \in H_x\,, \tag{4.8.25}$$

where the integrals should be assumed as the weak integrals in the topology generated by the set F_c of the linear functionals on X.

$\triangleright$ For proving we notice that the formula

$$\mathcal{E}(B)f = \sum_{p=1}^{N} \int \mathbf{1}_B(\lambda) f x_p(\lambda) \mu_p(d\lambda) \qquad (4.8.26)$$

determines the expansion of the unit on H_x. Really, from (4.8.23) and (4.8.24) it follows that for any sets B, $B' \in \mathcal{L}$

$$(\mu_p(B), \mu_q(B')) = \mu_p(B) K_x \mu_q(B')$$

$$= \int_{B'} \mu_p(B) x_q(\lambda) \sigma_q(d\lambda) = \delta_{pq} \sigma_p(BB') . \qquad (4.8.27)$$

Therefore from (4.8.26) it follows that for any vectors f, $g \in H_x$ and the sets B, B', $B'' \in \mathcal{L}$ the following equalities take place:

$$(\mathcal{E}(B)f, \mathcal{E}(B')g) = \sum_{p=1}^{N} \int \mathbf{1}_{B'}(\lambda) \overline{g x_p(\lambda)} (\mathcal{E}(B)f, \mu_p(d\lambda)) ,$$

$$(\mathcal{E}(B)f, \mu_p(B'')) = \sum_{q=1}^{N} \int \mathbf{1}_B(\lambda) f x_q(\lambda) (\mu_q(d\lambda), \mu_p(B''))$$

$$= \int_{B''} \mathbf{1}_B(\lambda) f x_p(\lambda) \sigma_p(d\lambda) . \qquad (4.8.28)$$

Using the second from these equalities we transform the first one to the form

$$(\mathcal{E}(B)f, \mathcal{E}(B')g) = \sum_{p=1}^{N} \int \mathbf{1}_{BB'}(\lambda) f x_p(\lambda) \overline{g x_p(\lambda)} \sigma_p(d\lambda) .$$

Hence we find that $(\mathcal{E}(B)f, \mathcal{E}(B')g) = (f, \mathcal{E}(BB')g) = (\mathcal{E}(BB')f, g)$. In particular, at $B' = \Lambda$ (as by virtue of (4.8.26) and (4.8.25) $\mathcal{E}(\Lambda) = I$) we get $(E(B)f, g) = (f, E(B)g)$ and therefore $(\mathcal{E}(B)f, \mathcal{E}(B')g) = (f, \mathcal{E}(B)\mathcal{E}(B')g) = (\mathcal{E}(B)\mathcal{E}(B')f, g)$. Thus $\mathcal{E}(BB') = \mathcal{E}(B)\mathcal{E}(B') = \mathcal{E}(B')\mathcal{E}(B)$. Consequently, $\mathcal{E}(B)$ is a projector, and an operator function of the set $\mathcal{E}(B)$ possesses all the properties of the expansion of the unit.

We put $\mathcal{E}_1(B) = \cdots = \mathcal{E}_N(B) = \mathcal{E}(B)$, $\gamma_1 = \mu_1(\Lambda), \ldots, \gamma_N = \mu_N(\Lambda)$. Then we shall have with the account to (4.8.26), (4.8.24) and (4.8.27)

$$\mathcal{E}_p(B)\gamma_p = \sum_{r=1}^{N} \int \mathbf{1}_B(\lambda) \mu_p(\Lambda) x_r(\lambda) \mu_r(d\lambda)$$

$$= \mu_p(B) , \quad (\mathcal{E}_p(B)\gamma_p , \mathcal{E}_q(B')\gamma_q) = \delta_{pq}\sigma_p(BB') . \tag{4.8.29}$$

Finally, by virtue of (4.8.25) and (4.8.27) any vector $f \in H_x$ is expressed by formula (4.8.11) N may be ∞

$$f = \sum_{p=1}^{N} \int f x_p(\lambda)\mu_p(d\lambda) = \sum_{p=1}^{N} \int f x_p(\lambda)\mathcal{E}_p(d\lambda)\gamma_p .$$

Thus the expansions of the unit $\mathcal{E}_p = \mathcal{E}$ and the vectors γ_p satisfy all the conditions of the general theory about the integral canonical representations. Consequently, the random variable X is expressed by the integral canonical representation

$$X = m_x + \sum_{p=1}^{N} \int x_p(\lambda)Z_p(d\lambda) , \tag{4.8.30}$$

where the uncorrelated stochastic measures $Z_1 , \dots , Z_N$ are determined by formula

$$Z_p(B) = \mathcal{E}(B)\gamma_p X^0 = \mu_p(B)X^0 , \tag{4.8.31}$$

with the covariance functions $\sigma_1 , \dots , \sigma_N$ correspondingly. The covariance operator of the random variable X is equal to

$$K_x = \sum_{p=1}^{N} \int x_p(\lambda) \left[\overline{\cdot x_p(\lambda)} \right] \sigma_p(d\lambda) . \triangleleft \tag{4.8.32}$$

Let $X(t)$ be a finite-dimensional vector random function of the finite-dimensional argument t; $K_x(t, t')$ be its covariance function. We find an integral canonical representation of a random function. The inverse Pugachev theorem in given case is formulated in the following way: if at some measurable space $(\Lambda, \mathcal{L})$ there exist N vector measures $\mu_p(t, B)$, N nonnegative measures $\sigma_p(B)$ and N vector functions $x_p(t, \lambda)$ satisfying the conditions

$$\int K_x(t, t')\overline{\mu_p(t', B)}dt' = \int_B x_p(t, \lambda)\sigma_p(d\lambda) , \quad B \in \mathcal{L} , \tag{4.8.33}$$

$$\int \mu_p(t, B)^T x_q(t, \lambda)dt = \delta_{pq}\mathbf{1}_B(\lambda) , \tag{4.8.34}$$

$$\sum_{p=1}^{N} \int \mu_p(t, d\lambda) \int x_p(\tau, \lambda)^T f(\tau)d\tau = f(t) \tag{4.8.35}$$

for every function $f(t) \in H_x$ then the random function $X(t)$ is expressed by the integral canonical expansion of the form

$$X(t) = m_x(t) + \sum_{p=1}^{N} \int x_p(t, \lambda) Z_p(d\lambda) . \qquad (4.8.36)$$

Formula (4.8.31) which determines the stochastic measures Z_p in a given case takes the form

$$Z_p(B) = \int \mu_p(t, B)^T X^0(t) dt \quad (p = 1, \dots, N) . \qquad (4.8.37)$$

Formula (4.8.33) shows that in a given case the measure $\mu_p(t, B)$ almost at all t is the σ_p-continuous and

$$x_p(t, \lambda) = \int K_x(t, t') \overline{\frac{d\mu_p}{d\sigma_p}(t', \lambda)} dt' , \qquad (4.8.38)$$

$$K_x(t, t') = \sum_{p=1}^{N} \int x_p(t, \lambda) x_p(t', \lambda)^* \sigma_p(d\lambda) . \qquad (4.28.39)$$

Besides we notice that condition (4.8.35) may be written in the equivalent form

$$\sum_{p=1}^{N} \int \mu_p(t, d\lambda) x_p(\tau, \lambda)^T = I\delta(t - \tau) , \qquad (4.8.40)$$

where I is the unit matrix.

In particular, when Λ is a finite-dimensional space and the measures $\mu_p(t, B)$ and $\sigma_p(B)$ are absolutely continuous relatively to the Lebesgue measure, formulae (4.8.38), (4.8.40) and (4.8.34) may be presented in the form

$$x_p(t, \lambda) = \frac{1}{\nu_p(\lambda)} \int K_x(t, t') \overline{a_p(t', \lambda)} dt' , \qquad (4.8.41)$$

$$\sum_{p=1}^{N} \int a_p(t, \lambda) x_p(\tau, \lambda)^T d\lambda = I\delta(t - \tau) , \qquad (4.8.42)$$

$$\int a_p(t, \lambda)^T x_q(t, \lambda') dt = \delta_{pq} \delta(\lambda - \lambda') , \qquad (4.8.43)$$

where $\nu_p(\lambda)$ and $a_p(t, \lambda)$ are the Radon–Nikodym derivatives of the measures $\sigma_p(B)$ and $\mu_p(t, B)$ with respect to the Lebesgue measure. The stochastic measures Z_p will be in this case generated by the noncorrelated white noises

$$V_p(\lambda) = \int a_p(t, \lambda)^T X^0(t)dt, \qquad (4.8.44)$$

whose intensities are equal to the correspondent functions $\nu_p(\lambda)$. Integral canonical representation (4.8.36) of the random function $X(t)$ will take the form

$$X(t) = m_x(t) + \sum_{p=1}^{N} \int V_p(\lambda)x_p(t, \lambda)d\lambda, \qquad (4.8.45)$$

and the correspondent representation of the covariance function (4.8.39) will be

$$K_x(t, t') = \sum_{p=1}^{N} \int \nu_p(\lambda)x_p(t, \lambda)x_p(t', \lambda)^* d\lambda. \qquad (4.8.46)$$

Thus for constructing an integral canonical representation of the random function $X(t)$ it is sufficient to find the functions $x_p(t, \lambda)$, $a_p(t, \lambda)$ and $\nu_p(\lambda)$ $(p = 1, \ldots, N)$ satisfying conditions (4.8.41)–(4.8.43). Here the random function $V_p(\lambda)$ determined by formula (4.8.44) will be uncorrelated white noises with the intensities $\nu_p(\lambda)$ and the random function $X(t)$ will be expressed by the integral canonical representation (4.8.45).

E x a m p l e   4.8.1. Let the random function $X = X(t)$ is defined by the linear differential equation $a_1(t)\dot{X} + a_0(t)X = V$ with the white noise V of unit intensity, and $X(t_0) = 0$ and is equal to zero at $t = t_0$. While integrating this equation we find

$$X(t) = q_1(t) \int_{t_0}^{t} \frac{V(\tau)d\tau}{a_1(\tau)q_1(\tau)}, \quad q_1(t) = \exp\left\{-\int_0^t \frac{a_0(\tau)}{a_1(\tau)}d\tau\right\}. \qquad (I)$$

Formula (I) gives an integral canonical representation of the random function X at time interval $t_0 < t < t_0 + T_1$, $T_1 > 0$. The coordinate functions of this integral canonical representation are determined by formula

$$x(t, \lambda) = \frac{q_1(t)}{a_1(\lambda)q_1(\lambda)}1(t - \lambda). \qquad (II)$$

After rewriting formula (I) in the form

$$V(\lambda) = \int\limits_{t_0}^{t_0+T_1} \left[\, a_1(\lambda)\delta'(\lambda - t) + a_0(\lambda)\delta(\lambda - t)\,\right] X(t)dt \qquad \text{(III)}$$

and comparing (III) with formula

$$V(\lambda) = \int\limits_{t_0}^{t_0+T_1} \overline{a(t,\lambda)}X^0(t)dt\,, \qquad \text{(IV)}$$

we see that in a given case $a(t,\lambda) = a_1(\lambda)\delta'(\lambda - t) + a_0(\lambda)\delta(\lambda - t)$.

4.8.4. Integral Canonical Representations and Shaping Filters

In applications the pairs of the matrix functions $x(t,\lambda)$, $a(t,\lambda)$ satisfying conditions (4.8.42) and (4.8.43) are often known. So, for instance, the weighting functions of two any reciprocal systems with N inputs and N outputs $w(t,\tau)$ and $w^-(t,\tau)$ always satisfy the conditions (Subsection 1.2.4)

$$\int w(t,\lambda)w^-(\lambda,t)d\lambda = I\delta(t - t')\,, \qquad (4.8.47)$$

$$\int w^-(\lambda,t)w(t,\lambda')d\lambda = I\delta(\lambda - \lambda')\,. \qquad (4.8.48)$$

Comparing these formulae with (4.8.42) and (4.8.43) we see that if we determine the matrices-columns $x_p(t,\lambda)$ and $a_p(t,\lambda)$ $(p = 1,\ldots,N)$ as the correspondent columns of the matrices $x(t,\lambda) = w(t,\lambda)$ and $a(t,\lambda) = w^-(\lambda,t)^T$ then conditions (4.8.42) and (4.8.43) will be satisfied. Thus *for finding an integral canonical representation of a random function it is sufficient to choose among all possible pairs of the weighting functions of the reciprocal systems such a pair which satisfies condition* (4.8.41) *at some functions* $\nu_p(\lambda)$. In the general case this problem remains still rather difficult and not always has the solution.

We notice that the construction of an integral canonical representation of a random function is equivalent to finding such a linear system which transforms the vector random function $X(t)$ into the vector white noise $V(t)$ of the same dimension with uncorrelated components. The inverse system will form the random function $X(t)$ from the white noise

$V(t)$ in consequence of which it is usually called *a shaping filter* of the random function $X(t)$. Thus *a mathematical problem of finding an integral canonical representation of a random function is equivalent to the problem of finding a shaping filter of a given random function.* For many standard random functions (for instance, stationary) the problem of constructing the integral canonical representations is sufficiently simple (see Sections 6.4 and 6.5).

4.8.5. Joint Integral Canonical Representation of Two Random Variables

We return to the problem of Subsection 3.4.7 and consider two random variables X and Y with one and the same phase space and the covariance operators K_x and K_y. Suppose that zero subspace N_0 of the operator K_x is fully contained in zero subspace of the operator K_y. It was shown in Subsection 3.4.7 that in this case there exists the operator $A = K_x^{-1} K_y$ which represents a positive self-adjoint operator on H_x. As it is known from functional analysis each positive self-adjoint operator may be presented by a spectral representation

$$A = \int_0^\infty \lambda \mathcal{E}(d\lambda), \tag{4.8.49}$$

where $\mathcal{E}(B)$ is a representation of the unit on Borel σ-algebra $\mathcal{B}$ of the real axis $\mathcal{E}(B) = 0$ for any Borel set B of a negative semiaxis by virtue of the positiveness of A generated by the operator A. Let exists such a sequence of the vectors $\{\gamma_\nu\} \subset H_x$ that the set of the linear combinations of the vectors $\mathcal{E}(B)\gamma_\nu$ correspondent to all $B \in \mathcal{B}$ and $\nu = 1, 2, \ldots$, is dense in H_x and the vectors $\mathcal{E}(B)\gamma_\nu$ and $\mathcal{E}(B')\gamma_\mu$ are orthogonal for $\mu \neq \nu$ and any $B, B' \in \mathcal{B}$. We choose such functions $v_\nu(\lambda)$ $(\nu = 1, 2 \ldots)$ that for any vector $f \in H_x$ the equality has the place

$$u_\nu(\lambda) = \frac{d(f, \mathcal{E}(B)\gamma_\nu)}{d(\mathcal{E}(B)\gamma_\nu, \gamma_\nu)} = f v_\nu(\lambda). \tag{4.8.50}$$

Then the random variable X and its covariance operator K_x are expressed in terms of the integral canonical representations

$$X = m_x + \sum_{\nu=1}^\infty \int v_\nu(\lambda) Z_\nu(d\lambda), \tag{4.8.51}$$

$$K_x = \sum_{\nu=1}^\infty \int v_\nu(\lambda) \left[\overline{\cdot v_\nu(\lambda)} \right] \sigma_\nu(d\lambda). \tag{4.8.52}$$

Here $Z_\nu(B) = \mathcal{E}(B)\gamma_\nu X^0$ are the uncorrelated stochastic measures with the covariance functions $\sigma_\nu(B) = (\mathcal{E}(B)\gamma_\nu, \gamma_\nu) = \|\, \mathcal{E}(B)\gamma_\nu\,\|^2$ and

$$v_\nu(\lambda) = EX^0 \frac{d\overline{Z}_\nu}{d\sigma_\nu}(\lambda)\,. \qquad (4.8.53)$$

Theorem 4.8.3. *Let X and Y be two random varaibles with one and the same phase space and covariance operators K_x and K_y. Let the operator $A = K_x^{-1} A_y$ exists and defines the canonical representations (4.8.51), (4.8.52). Then the random variable Y and its covariance operator K_y are expressed in terms of the integral canonical representatios*

$$Y = m_y + \sum_{\nu=1}^{\infty} \int v_\nu(\lambda) U_\nu(d\lambda)\,, \qquad (4.8.54)$$

$$K_y = \sum_{\nu=1}^{\infty} \int v_\nu(\lambda) \left[\,\overline{\cdot v_\nu(\lambda)}\,\right] \lambda\sigma_\nu(d\lambda)\,. \qquad (4.8.55)$$

Here $U_\nu(B) = \mathcal{E}(B)\gamma_\nu Y^0$ are the uncorrelated stochastic measures with the covariance functions

$$\tau_\nu(B) = (A\mathcal{E}(B)\gamma_\nu, \gamma_\nu) = \int\limits_{B} \lambda\sigma_\nu(d\lambda)\,. \qquad (4.8.56)$$

Formulae (4.8.51–(4.8.56) define the joint integral canonical representation of two random variables X and Y.

▷ Really, as $EU_\nu\overline{U_\mu(B')} = \mathcal{E}(B)\gamma_\nu K_y\mathcal{E}(B')\gamma_\mu = \mathcal{E}(B)\gamma_\nu K_x A\mathcal{E}(B')\gamma_\mu = (\mathcal{E}(B)\gamma_\nu, A\mathcal{E}(B')\gamma_\mu)$, and any self-adjoint operator is permutable with its expansion of the unit then

$$EU_\nu(B)\overline{U_\mu(B')} = (\mathcal{E}(BB')\gamma_\nu, A\gamma_\mu) = \int\limits_{0}^{\infty} \lambda(\mathcal{E}(BB')\gamma_\nu, \mathcal{E}(d\lambda)\gamma_\mu)$$

$$= \int\limits_{BB'} \lambda(\mathcal{E}(d\lambda)\gamma_\nu, \gamma_\mu) = \delta_{\nu\mu} \int\limits_{BB'} \lambda\sigma_\nu(d\lambda)\,.$$

Hence it follows that $U_\nu(B) = \mathcal{E}(B)\gamma_\nu Y^0$ are the uncorrelated stochastic measures with the covariance functions $\tau_\nu(B)$. Further as for any $f \in H_x$

$$E\left| fY^0 - \sum_{\nu=1}^{\infty} \int fv_\nu(\lambda) U_\nu(d\lambda) \right|^2$$

$$= \left(Af - A\sum_{\nu=1}^{\infty} \int fV_\nu(\lambda)\mathcal{E}(d\lambda)\gamma_\nu, f - \sum_{\nu=1}^{\infty} \int fv_\nu(\lambda)\mathcal{E}(d\lambda)\gamma_\nu \right) = 0$$

by virtue of the fact that $f = \sum_{\nu=1}^{\infty} \int fv_\nu(\lambda)\mathcal{E}(d\lambda)\gamma_\nu$ then the random variable Y is expressed by formula (4.8.54). Finally from the direct theorem about the integral canonical representations it follows that

$$K_y = \sum_{\nu=1}^{\infty} \int v_\nu(\lambda)\left[\overline{\cdot v_\nu(\lambda)}\right]\tau_\nu(d\lambda).\ \triangleleft$$

R e m a r k 1. In particular if there exists such vector $\gamma \in H_x$ that the set of the linear combinations of the vectors $\mathcal{E}(B)\gamma$ correspondent to all Borel sets B of the real axis is dense in H_x then sums (4.8.51), (4.8.52), (4.8.54) and (4.8.55) contain one item and these formulae take the form

$$X = m_x + \int v(\lambda)Z(d\lambda)\,, \quad K_x = \int v(\lambda)\left[\overline{\cdot v(\lambda)}\right]\sigma(d\lambda)\,, \quad (4.8.57)$$

$$Y = m_y + \int v(\lambda)U(d\lambda)\,, \quad K_y = \int v(\lambda)\left[\overline{\cdot v(\lambda)}\right]\lambda\sigma(d\lambda)\,. (4.8.58)$$

Here $Z(B) = \mathcal{E}(B)\gamma X^0$, $U(B) = \mathcal{E}(B)\gamma Y^0$ are the stochastic measures with the covariance functions $\sigma(B)$, and

$$\tau(B) = \int_B \lambda\sigma(d\lambda)\,. \qquad\qquad (4.8.59)$$

By virtue of (4.8.53) the function $v(\lambda)$ is equal to

$$v(\lambda) = EX^0\frac{d\overline{Z}}{d\sigma}(\lambda) = EY^0\frac{d\overline{U}}{d\tau}(\lambda)\,. \qquad\qquad (4.8.60)$$

In this case when the spectrum of the operator A consists only of the countable set $\{\gamma_\nu\}$ of the eigenvalues γ_ν we have $\mathcal{E}(B) = \sum_{\gamma_\nu \in B}(\cdot,\varphi_\nu)\varphi_\nu$ where φ_ν is the eigenvector correspondent to the eigenvalue γ_ν,

$$\mathcal{E}(B)\varphi_\nu = \begin{cases} \varphi_\nu & \text{at} \quad \gamma_\nu \in B\,, \\ 0 & \text{at} \quad \gamma_\nu \overline{\in} B\,, \end{cases}$$

and the set of the vectors $\mathcal{E}(B)\varphi_\nu$ is dense in H_x. Integral canonical representations (4.8.51), (4.8.52), (4.8.54) and (4.8.55) of the random variables X and Y and their covariance operators turn into the canonical expansions (Subsection 3.4.7) at $V_\nu = \mathcal{E}(B)\varphi_\nu X^0$, $W_\nu = \mathcal{E}(B)\varphi_\nu Y^0$, any Borel B, $\gamma_\nu \in B$ and $u_\nu = v(\gamma_\nu)$.

R e m a r k 2. Let us consider the self-adjoint operator $A = K_x^{-1} K_y$ which is determined as an operator of the equation

$$\int K_x(t,\tau)g(\tau)d\tau = \int K_y(t,\tau)f(\tau)d\tau$$

at a given function $f(t) \in D_A \subset H_x$. We know that A is expressed by a spectral representation (4.8.49). If there exists such a sequence of the functions $\{\gamma_\nu(t)\} \subset H_x$ that the set of the linear combinations of their projections $\mathcal{E}(B)\gamma_\nu$ at all Borel sets B of the real axis is dense in H_x and $\mathcal{E}(B)\gamma_\nu$ and $\mathcal{E}(B')\gamma_\mu$ are orthogonal at $\mu \neq \nu$ and any B, $B' \in \mathcal{B}$ then according to the proved Theorem 4.8.3 the random functions $X(t)$ and $Y(t)$ are jointly expressed in terms of the integral canonical representations

$$\left.\begin{aligned}
X(t) &= m_x(t) + \sum_{\nu=1}^{\infty} \int v_\nu(t,\lambda)Z_\nu(d\lambda)\,, \\
Y(t) &= m_y(t) + \sum_{\nu=1}^{\infty} \int v_\nu(t,\lambda)U_\nu(d\lambda)\,.
\end{aligned}\right\} \qquad (4.8.61)$$

And the following relations take place:

$$\left.\begin{aligned}
K_x(t,t') &= \sum_{\nu=1}^{\infty} \int v_\nu(t,\lambda)v_\nu(t',\lambda)^*\sigma_\nu(d\lambda)\,, \\
K_y(t,t') &= \sum_{\nu=1}^{\infty} \int v_\nu(t,\lambda)v_\nu(t',\lambda)^*\lambda\sigma_\nu(d\lambda)\,,
\end{aligned}\right\} \qquad (4.8.62)$$

where

$$v_\nu(t,\lambda) = EX^0(t)\frac{d\overline{Z}_\nu}{d\sigma_\nu}(\lambda) \quad (\nu = 1,2,\ldots)\,. \qquad (4.8.63)$$

In the special case when there exists the unique function $\gamma(t)$ for which a set of the linear combinations of the functions $\mathcal{E}(B)\gamma$ correspondent to all Borel sets B of the real axis is dense in H_x the sums in (4.8.61) and (4.8.62) contain only one term in consequence of which formulae (4.8.61) and (4.8.62) have the form

$$X(t) = m_x(t) + \int v(t,\lambda)Z(d\lambda)\,, \quad Y(t) = m_y(t) + \int v(t,\lambda)U(d\lambda)\,, \qquad (4.8.64)$$

$$K_x(t,t') = \int v(t,\lambda)v(t',\lambda)^*\sigma(d\lambda)\,, \quad K_y(t,t') = \int v(t,\lambda)v_\nu(t',\lambda)^*\lambda\sigma(d\lambda)\,, \qquad (4.8.65)$$

where

$$v(t, \lambda) = EX^0(t)\frac{d\overline{Z}}{d\sigma}(\lambda)\,. \qquad (4.8.66)$$

4.8.6. Spectral and Integral Canonical Representations of Scalar Random Functions

Theorem 4.8.4. *The spectral representation of a scalar m.s. continuous stationary random function with vector argument is one of its possible integral canonical representations.*

▷ Let $X(t)$ be a m.s. continuous scalar stationary random function of the n-dimensional vector t, $t \in R^n$; $k_x(\tau)$ be its covariance functions; X be a linear space of all the functions of the variable t; F_c be a set of the linear functionals $fx = \sum\limits_{p=1}^{N} f_p x(t_p)$, correspondent to all N, f_1, ..., f_N, t_1, ..., t_N. In this case H_x represents the range of an isometric mapping of a subspace in the H-space $L_2(R^n, \mathcal{B}^n, \sigma)$ formed by the compliteness set of all functions $\varphi(\lambda) = \sum f_p e^{i\lambda^T t_p}$. As the space $L_2(R^n, \mathcal{B}^n, \sigma)$ is separable then H_x is separable too. We determine in H_x the expansion of unit $\mathcal{E}$ by formula

$$\mathcal{E}(B)f = \frac{1}{(2\pi)^n} \int\limits_B e^{-i\lambda^T t}d\lambda \int f(\tau)e^{i\lambda^T \tau}d\tau$$

$$= \frac{1}{(2\pi)^n} \int \mathbf{1}_B(\lambda)\varphi(\lambda)e^{-i\lambda^T t}d\lambda\,. \qquad (4.8.67)$$

We assume $\gamma = \delta(t)$. Then by virtue of (4.8.67) we have

$$\mathcal{E}(B)\gamma(t) = \frac{1}{(2\pi)^n} \int\limits_B e^{-i\lambda^T t}d\lambda = \frac{1}{(2\pi)^n} \int\limits_B \mathbf{1}_B(\lambda)e^{-i\lambda^T t}d\lambda\,.$$

As a set of the simple functions is dense in $L_2(R^n, \mathcal{B}^n, \sigma)$ so the set of linear combinations $\sum\limits_i c_i E(B_i)\gamma$; $c_i \in C$ is also dense in H_x. Consequently, for obtaining an integral canonical representation of a random function we may restrict ourselves to one expansion of unit $\mathcal{E}$ and one vector γ. Formula

$$Z(B) = \mathcal{E}(B)\gamma X^0 = \frac{1}{(2\pi)^n} \int X^0(t)dt \int \mathbf{1}_B(\lambda)e^{-i\lambda^T t}d\lambda \qquad (4.8.68)$$

determines a stochastic measure with the covariance function $\sigma(BB')$, and for any $f \in H_x$ we have

$$(f, \mathcal{E}(B)\gamma) = \int \varphi(\lambda)\mathbf{1}_B(\lambda)\sigma(d\lambda) = \int_B \sigma(d\lambda) \int f(t)e^{i\lambda^T t}dt \,.$$

By virtue of (4.8.9) $u(\lambda) = \int f(t)e^{i\lambda^T t}dt$, and $x(\lambda) = e^{i\lambda^T t}$. Consequently, the random function $X(t)$ is determined by the integral canonical representation of the form (4.8.20):

$$X(t) = m_x + \int e^{i\lambda^T t}Z(d\lambda) \,. \tag{4.8.69}$$

Integral canonical representation (4.8.21) of a covariance operator in this case gives the integral representation of the covariance function $k_x(\tau)$ (the known Bochner theorem)

$$k_x(t_1 - t_2) = \int e^{i\lambda^T(t_1 - t_2)}\sigma(d\lambda) \,. \triangleleft \tag{4.8.70}$$

Formula (4.8.69) gives *a spectral representation* of the stationary function $X(t)$ with vector argument. According to this fact the measure σ in (4.8.70) is called *a spectral measure* of the stationary random function $X(t)$. In particular, when the part of the measure σ which is singular relatively to Lebesgue measure in R^n is concentrated on the countable set of the points we may determine *a spectral density* of the stationary random function $X(t)$:

$$s_x(\lambda) = \frac{\partial^n S_x(\lambda)}{\partial\lambda_1, \ldots, \partial\lambda_n} = \frac{\partial^n \sigma((-\infty, \lambda))}{\partial\lambda_1, \ldots, \partial\lambda_n} \,, \tag{4.8.71}$$

which may contain a linear combination of the δ-functions. The stationary random functions encountered in the problems of practice have always the spectral densities in this sense. Then spectral expansion (4.8.69) may be rewritten in the form

$$X(t) = m_x + \int V(\lambda)e^{i\lambda^T t}d\lambda \,, \tag{4.8.72}$$

where $V(\lambda)$ is a white noise whose intensity is equal to the spectral density $s_x(\lambda)$ of the random function $X(t)$. As the spectral density $s_x(\lambda)$ is integrable (by the finiteness of the spectral measure σ) then it is expressed by the Fourier integral

$$s_x(\lambda) = \frac{1}{(2\pi)^n} \int d\tau \int s_x(\mu)e^{i(\mu^T - \lambda^T)\tau}d\mu \,.$$

But according to (4.8.70) we have

$$\int e^{i\mu^T \tau} s_x(\mu)d\mu = \int e^{i\mu^T \tau} \sigma(d\mu) = k_x(\tau)\,.$$

Consequently, the spectral density of a scalar stationary random function with vector argument is expressed in terms of its covariance function by formula

$$s_x(\lambda) = \frac{1}{(2\pi)^n} \int k_x(\tau)e^{-i\lambda^T \tau}d\tau\,. \tag{4.8.73}$$

E x a m p l e 4.8.2. Find an approximate canonical expansion of a scalar stationary random function of the scalar argument t, $X = X(t)$ possessing the spectral density $s_x(\lambda)$ at the time interval $|t| < T$. Let $(-\Omega, \Omega)$ be an arbitrary interval of the frequencies. We partition this interval into $2N + 1$ equal segments and denote the length of each of them by 2α. We enumerate the segments of the real axis obtained by such a way by the integers from $-N$ to N and denote by λ_ν the mean of the νth segment. Substituting the integral in the infinite limits in integral canonical expansion (4.8.72) by the integral in the limits $(-\Omega, \Omega)$ we present the obtained integral in the form of the sum of $2N + 1$ integrals extended over the segment 2α length. In each integral instead of the multiplier $e^{i\lambda t}$ we substitute its value in the middle of the interval $e^{i\lambda_\nu t}$. Then we obtain the random function

$$X_N(t) = m_x + \sum_{\nu=-N}^{N} V_\nu e^{i\lambda_\nu t}\,, \tag{I}$$

where

$$V_\nu = \int_{\lambda_\nu - \alpha}^{\lambda_\nu + \alpha} V(\lambda)d\lambda \ (\nu = 0, \pm 1, \pm 2, \ldots, \pm N)\,. \tag{II}$$

The variances D_ν of the random variables V_ν by virtue of the spectral representation of the variance X are equal to

$$D_\nu = \int_{\lambda_\nu - \alpha}^{\lambda_\nu + \alpha} s_x(\lambda)d\lambda \ (\nu = 0, \pm 1, \ldots, \pm N)\,.$$

As $V(\lambda)$ is a white noise and the integration intervals in (I) at the correspondent values of the index ν do not overlap then the random variables V_ν are uncorrelated. We estimate the degree of the proximity of $X_N(t)$ to the function $X(t)$. We have

$$X(t) - X_N(t) = \int_{-\infty}^{-\Omega} V(\lambda)e^{i\lambda t}d\lambda + \int_{\Omega}^{\infty} V(\lambda)e^{i\lambda t}d\lambda$$

$$+ \sum_{\nu=-N}^{N} \int_{\lambda_\nu-\alpha}^{\lambda_\nu+\alpha} V(\lambda) \left(e^{i\lambda t} - e^{i\lambda_\nu t} \right) d\lambda \, . \tag{III}$$

As the expectation of the difference $X - X_N$ is equal to zero and the isolated items in the right-hand side of (III) are uncorrelated then the variance of the difference $X - X_N$ equal to the expectation of the modulus square of this difference is equal to the sum of the variances of the integrals entering into (III). Each of these integrals represents a random function expressed by integral canonical representation (III). Therefore we may use formula (II). Consequently,

$$E\left|X(t) - X_N(t)\right|^2 = \int_{-\infty}^{-\Omega} s_x(\lambda)d\lambda + \int_{\Omega}^{\infty} s_x(\lambda)d\lambda$$

$$+ \sum_{\nu=-N}^{N} \int_{\lambda_\nu-\alpha}^{\lambda_\nu+\alpha} s_x(\lambda) \left| e^{i\lambda t} - e^{i\lambda_\nu t} \right|^2 d\lambda \, . \tag{IV}$$

Supposing that the interval $|t| < T$ is fixed we choose the frequency Ω and the number α in such a way that the first two integrals in (IV) would be smaller than the arbitrarily given number ε:

$$\int_{-\infty}^{-\Omega} s_x(\lambda)d\lambda < \varepsilon \, , \quad \int_{\Omega}^{\infty} s_x(\lambda)d\lambda < \varepsilon \, .$$

Taking into account the inequalities $\left| e^{i\lambda t} - e^{i\lambda_\nu t} \right| = \left| e^{i(\lambda-\lambda_\nu)t} - 1 \right| \le |(\lambda - \lambda_\nu)t| < \alpha T$, we find from (IV)

$$E\left|X(t) - X_N(t)\right|^2 < 2\varepsilon + \alpha^2 T^2 \int_{-\Omega}^{\Omega} s_x(\lambda)d\lambda \le 2\varepsilon + \alpha^2 T^2 D_x \, .$$

Hence as ε is arbitrarily small and α may be chosen as infinitesimal it follows that the scalar stationary random function $X(t)$ on any finite interval $|t| < T$ may be m.s. approximated by random function (I) with any accuracy.

4.8.7. Spectral and Integral Canonical Representations of Vector Stationary Random Function

Theorem 4.8.5. *For any vector-valued stationary m.s. continuous random function its integral canonical representation has the same form* (4.8.69), (4.8.70).

▷ Let us consider m.s. continuous vector random function $X(t)$ with the covariance function $k_x(\tau)$. In this case the scalar product in H_x is determined by formula

$$(f, g) = \iint f(t_1)^T k_x(t_1 - t_2)\overline{g(t_2)}dt_1 dt_2 . \tag{4.8.74}$$

As it was shown in Subsection 4.8.6 each component $X_p(t)$ of the vector random function $X(t)$ is expressed by spectral expansion (4.8.69). Consequently, the function $X(t)$ itself may be represented in the form of (4.8.69) where Z is a vector stochastic measure. We find its covariance function. As the diagonal elements of this function serve the spectral measures σ_{pp} of the components $X_p(t)$. For finding its other elements we introduce into H_x the unit expansion $\mathcal{E}(B)$ using the same formula by means of which any vector function $f(t) \in H_x$ was determined in Subsection 4.8.6. As concerns to the vector functions $\gamma_p \in H_x$ correspondent to the different components $X_p(t)$ of the vector random function $X(t)$ then each function γ_p evidently should be determined as the vector function whose all components besides the pth one are equal to zero, and the pth component represents the δ-function. In this case all the components of the vector function $\mathcal{E}(B)\gamma_p$ besides the pth one will be equal to zero, and the pth component will be determined by formula

$$\{\mathcal{E}(B)\gamma_p\}_p = \frac{1}{(2\pi)^n} \int_B e^{-i\lambda^T t}d\lambda .$$

Then accordingly to Subsection 4.3.1 we have

$$Z_p(B) = \mathcal{E}(B)\gamma_p X^0 = \frac{1}{(2\pi)^n} \int X_p^0(t)dt \int \mathbf{1}_B(\lambda)e^{-i\lambda^T t}d\lambda .$$

The cross-covariance function of the components $Z_p(B)$ and $Z_q(B)$ will be determined by formula $EZ_p(B)\overline{Z_q(B')} = [\mathcal{E}(B)\gamma_p] K_x [\mathcal{E}(B')\gamma_q] = (\mathcal{E}(B)\gamma_p, \mathcal{E}(B')\gamma_q)$. Hence taking (4.8.74) into consideration we get

$$EZ_p(B)\overline{Z_q(B')}$$

$$= \iint k_{pq}(t_1 - t_2)\frac{dt_1 dt_2}{(2\pi)^{2n}} \int \mathbf{1}_B(\lambda)e^{-i\lambda^T t_1}d\lambda \int \mathbf{1}_{B'}(\mu)e^{i\mu^T t_2}d\mu ,$$

where $k_{pq}(\tau)$ is a cross-covariance function of the components $X_p(t)$ and $X_q(t)$. The change of the variables $t_2 = t_1 - \tau$ gives

$$EZ_p(B)\overline{Z_q(B')} = \int k_{pq}(\tau)\frac{d\tau}{(2\pi)^n} \int \frac{dt}{(2\pi)^n} \int \mathbf{1}_B(\lambda)e^{-i\lambda^T t}d\lambda$$

$$\times \int \mathbf{1}_{B'}(\mu)e^{-i\mu^T \tau}e^{i\mu^T t}d\mu .$$

But

$$\int \frac{dt}{(2\pi)^n} \int \mathbf{1}_B(\lambda)e^{-i\lambda^T t}d\lambda \int \mathbf{1}_{B'}(\mu)e^{-i\mu^T \tau}e^{i\mu^T t}d\mu$$

$$= \int \left[\int \frac{\mathbf{1}_B(\lambda)}{\sqrt{(2\pi)^n}}e^{-i\lambda^T t}d\lambda \right] \left[\overline{\int \frac{\mathbf{1}_{B'}(\lambda)e^{i\lambda^T \tau}}{\sqrt{(2\pi)^n}}e^{-\lambda^T t}d\lambda} \right] dt$$

$$= \int \mathbf{1}_B(\lambda)\overline{\mathbf{1}_{B'}(\lambda)e^{i\lambda^T \tau}}d\lambda = \int \mathbf{1}_{BB'}(\lambda)e^{-i\lambda^T \tau}d\lambda .$$

Therefore we obtain

$$EZ_p(B)\overline{Z_q(B')} = \int k_{pq}(\tau)\frac{d\tau}{(2\pi)^n} \int \mathbf{1}_{BB'}(\lambda)e^{-i\lambda^T \tau}d\lambda . \qquad (4.8.75)$$

Hence it is clear that the stochastic measures Z_p and Z_q on the nonintersecting sets are uncorrelated and their cross covariance function depends on BB' only. Putting $EZ_p(B)\overline{Z_q(B')} = \sigma_{pq}(BB')$ we determine a covariance function of the vector stochastic measure $Z(B)$ as the matrix function $\sigma(BB')$. This matrix function is called *a spectral measure* of the vector stationary random function $X(t)$. Its element σ_{pq} at $q \neq p$ is called *a cross-spectral measure* of the components $X_p(t)$ and $X_q(t)$. The cross-spectral measure σ_{pq} of the components $X_p(t)$ and $X_q(t)$ is both σ_{pp}-continuous and σ_{qq}-continuous. $\triangleleft$

R e m a r k. Spectral representation of the vector stationary random function (4.8.69) is not its integral canonical representation as the stochastic measure Z in (4.8.69) in this case is a vector one while all stochastic measures in the integral canonical representation of a random variable are scalar.

4.8.8. Way of Constructing Integral Canonical Representation of Vector Stationary Random Function

$\triangleright$ For constructing an integral canonical representation of the stationary random function $X(t)$ we determine in H_x the unit expansion $\mathcal{E}$ using the same formula by means of which the scalar random function $X(t)$ was determined in Subsection 4.8.6:

$$\mathcal{E}(B)f = \frac{1}{(2\pi)^n} \int_B e^{-i\lambda^T t}d\lambda \int f(\tau)e^{i\lambda^T \tau}d\tau$$

$$= \frac{1}{(2\pi)^n} \int \mathbf{1}_B(\lambda)\varphi(\lambda)e^{-i\lambda^T t}d\lambda \,,$$

where $f(t)$ and $\varphi(\lambda)$ should be assumed as the vector functions. We assume $\mathcal{E}_1 = \cdots = \mathcal{E}_m = \mathcal{E}$ and find such orthogonal vectors $\gamma_1 \ldots, \gamma_m \in H_x$ that the vectors $\mathcal{E}_p(B)\gamma_p$ and $\mathcal{E}_q(B')\gamma_q$ at $q \neq p$ will be orthogonal at any Borel sets B and B' of the n-dimensional space R^n. Putting

$$c_p(\lambda) = \int \gamma_p(t)e^{i\lambda^T t}dt \quad (p = 1, \ldots, m), \tag{4.8.76}$$

we get for the vector $\mathcal{E}_p(B)\gamma_p$ the expression

$$\mathcal{E}(B)\gamma_p = \frac{1}{(2\pi)^n} \int \mathbf{1}_B(\lambda)c_p(\lambda)e^{-i\lambda^T t}d\lambda \quad (p = 1, \ldots, m). \tag{4.8.77}$$

It means that by virtue of (4.8.74) we have

$$(\mathcal{E}_p(B)\gamma_p, \mathcal{E}_q(B')\gamma_q) = (\mathcal{E}(B)\gamma_p, \mathcal{E}(B')\gamma_q) = (\mathcal{E}(BB')\gamma_p, \gamma_q)$$

$$= \iint \frac{1}{(2\pi)^n} \left(\int \mathbf{1}_{BB'}(\lambda)c_p(\lambda)^T e^{-i\lambda^T t_1}d\lambda \right) k_x(t_1 - t_2)\overline{\gamma_q(t_2)}dt_1 dt_2$$

$$= \int \frac{d\tau}{(2\pi)^n} \int \mathbf{1}_{BB'}(\lambda)c_p(\lambda)^T k_x(\tau)e^{-i\lambda^T \tau} \left(\int \overline{\gamma_q(t_2)}e^{-i\lambda^T t_2}dt_2 \right) d\lambda \,.$$

Accounting (4.8.75) and (4.8.76) we find

$$\frac{1}{(2\pi)^n} \int k_x(\tau)d\tau \int \mathbf{1}_A(\lambda)e^{-i\lambda^T \tau}d\lambda = \sigma(A), \quad \int \overline{\gamma_q(t_2)}e^{-\lambda^T t_2}dt_2 = \overline{c_q(\lambda)}.$$

Consequently,

$$(\mathcal{E}_p(B)\gamma_p, \mathcal{E}_q(B')\gamma_q) = \int \mathbf{1}_{BB'}(\lambda)c_p(\lambda)^T \sigma(d\lambda)\overline{c_q(\lambda)} \,.$$

The condition of the orthogonality of the vectors $\mathcal{E}_p(B)\gamma_p$ and $\mathcal{E}_q(B)\gamma_q$ at $q \neq p$ gives the following equations for the determination of the vector functions $c_1(\lambda)$, $\ldots, c_m(\lambda)$:

$$\int \mathbf{1}_B(\lambda)c_p(\lambda)^T \sigma(d\lambda)\overline{c_q(\lambda)} = 0 \quad \text{at} \quad q \neq p. \tag{4.8.78}$$

Thus we have $m(m-1)/2$ conditions for determining m^2 of the components $c_{pq}(\lambda)$ of the vector functions $c_1(\lambda), \ldots, c_m(\lambda)$. Later on we shall prove the existence of the solution of Eqs. (4.8.78) (not unique in general) and show how to find it. But now we suppose that some solution $c_1(\lambda), \ldots, c_m(\lambda)$ of Eqs. (4.8.78) is found and that

$$(\mathcal{E}_p(B)\gamma_p, \gamma_p) = \int_B c_p(\lambda)^T \sigma(d\lambda)\overline{c_p(\lambda)}.$$

Then denoting by $\hat{\sigma}_p(B)$ a diagonal matrix with these elements and by $c(\lambda)$ a matrix with the elements $c_{pq}(\lambda)$ we obtain the relation

$$\hat{\sigma}_p(B) = \int_B c(\lambda)\sigma(d\lambda)c(\lambda)^* . \tag{4.8.79}$$

After repeating the previous calculations we shall obtain the following equality for any vector $f \in H_x$:

$$(f, \mathcal{E}_p(B)\gamma_p) = \int_B \varphi(\lambda)^T \sigma(d\lambda)\overline{c_p(\lambda)}.$$

Now we introduce the vector h_f with the components $(f, \mathcal{E}_p(B)\gamma_p)$, $p = 1, \ldots, m$. Then the previous equality will give for the matrix-row

$$h_f^T = \int_B \varphi(\lambda)^T \sigma(d\lambda)c(\lambda)^* = \int_B \varphi(\lambda)^T c(\lambda)^{-1} c(\lambda)\sigma(d\lambda)c(\lambda)^* .$$

It means that by virtue of (4.8.79) and putting $a(\lambda) = c(\lambda)^{-1}$ we get

$$h_f^T = \int_B \varphi(\lambda)^T a(\lambda)\hat{\sigma}(d\lambda) .$$

This equality gives for the pth component $(f, \mathcal{E}_p(B)\gamma_p)$ of the vector h_f the following expression

$$(f, \mathcal{E}_p(B)\gamma_p) = \int_B \varphi(\lambda)^T a_p(\lambda)\hat{\sigma}_p(d\lambda) = \int_B \hat{\sigma}_p(d\lambda) \int f(t)^T a_p(\lambda)e^{i\lambda^T t} dt ,$$

where $a_p(\lambda)$ is a matrix-column with the elements $a_{1p}(\lambda), \ldots, a_{mp}(\lambda)$. Hence it is clear that

$$u_p(\lambda) = \varphi(\lambda)^T a_p(\lambda) = \int f(t)^T a_p(\lambda)e^{i\lambda^T t} dt = fx_p(\lambda), \tag{4.8.80}$$

$$x_p(\lambda) = a_p(\lambda)e^{i\lambda^T t} . \tag{4.8.81}$$

It remains to ensure that a set of linear combinations of the vectors $\mathcal{E}_p(B)\gamma_p$ correspondent to all Borel sets B of the space R^n and $p = 1, \ldots, m$ is dense in H_x. It is sufficient to show for this purpose that for any vector $f \in H_x$ the vector

$$g = \sum_{p=1}^{m} \int u_p(\lambda)\mathcal{E}_p(d\lambda)\gamma_p$$

coincides with f. By virtue of (4.8.77) and (4.8.80) we have

$$g = \sum_{p=1}^{m} \int u_p(\lambda)\mathcal{E}_p(d\lambda)\gamma_p = \frac{1}{(2\pi)^n} \sum_{p=1}^{m} \int \varphi(\lambda)^T a_p(\lambda)c_p(\lambda)e^{-i\lambda^T t}d\lambda .$$

Hence it follows the expression of the rth component of the vector g:

$$g_r = \frac{1}{(2\pi)^n} \sum_{p=1}^{m} \int \sum_{q=1}^{m} \varphi_q(\lambda)a_{qp}(\lambda)c_{pr}(\lambda)e^{-i\lambda^T t}d\lambda$$

$$= \frac{1}{(2\pi)^n} \sum_{q=1}^{m} \int \varphi_q(\lambda) \sum_{p=1}^{m} a_{qp}(\lambda)c_{pr}(\lambda)e^{-i\lambda^T t}d\lambda .$$

But the matrices $a(\lambda)$ and $c(\lambda)$ are reciprocal. Consequently,

$$\sum_{p=1}^{m} a_{qp}(\lambda)c_{pr}(\lambda) = \delta_{qr} , \quad g_r = \frac{1}{(2\pi)^n} \int \varphi_r(\lambda)e^{-i\lambda^T t}d\lambda = f_r(t) .$$

It proves that any vector $f \in H_x$ may be presented by formula (4.8.11) and consequently the set of linear combinations of the vectors $\mathcal{E}_p(B)\gamma_p$, $B \in \mathcal{B}^n$, $p = 1, \ldots, m$ is dense in H_x.

Thus all the conditions of Theorem 4.8.1 are fulfilled. ◁

Consequently, the vector stationary random function $X(t)$ and its covariance function $R_x(\tau)$ may be expressed by the following integral canonical representations:

$$X(t) = m_x(t) + \sum_{p=1}^{m} \int a_p(\lambda)e^{i\lambda^T t}\hat{Z}_p(d\lambda), \qquad (4.8.82)$$

$$k_x(\tau) = \sum_{p=1}^{m} \int a_p(\lambda)a_p(\lambda)^* e^{i\lambda^T \tau}\hat{\sigma}_p(d\lambda), \qquad (4.8.83)$$

where $\hat{Z}_1 , \ldots , \hat{Z}_m$ are the uncorrelated measures

$$\hat{Z}_p(B) = \mathcal{E}_p(B)\gamma_p X^0 = \frac{1}{(2\pi)^n} \int X^0(t)^T dt \int_B c_p(\lambda)e^{-i\lambda^T t} d\lambda , \quad (4.8.84)$$

whose measures covariance functionsare equal $\hat{\sigma}_1 , \ldots , \hat{\sigma}_m$.

Now it remains to show how we may find the vector functions satisfying Eqs. (4.8.78).

▷ It is sufficient for this purpose to show how we may find such uncorrelated stochastic measures $\hat{Z}_1 , \ldots , \hat{Z}_m$ and an invertible matrix $a(\lambda)$ such that the stochastic measures Z_p in spectral expansion (4.8.69) will be expressed by formula

$$Z_p(B) = \sum_{q=1}^{m} \int_B a_{pq}(\lambda)\hat{Z}_q(d\lambda) , \quad (p = 1 , \ldots , m) . \qquad (4.8.85)$$

In order to make sure in that fact it is sufficient to compare (4.8.82) with (4.8.69) and notice that in this case the spectral measure σ of the vector stochastic measure Z will be expressed by formula

$$\sigma(B) = \int_B a(\lambda)\hat{\sigma}(d\lambda)a(\lambda)^* ,$$

where $\hat{\sigma}$ is a diagonal matrix whose covariance functions $\hat{\sigma}_q$ of the stochastic measures $\hat{Z}_q$ $(q = 1 , \ldots , m)$ serve as its elements. It follows from this formula that

$$\int_B c(\lambda)\sigma(d\lambda)c(\lambda)^* = \int_B a(\lambda)^{-1}a(\lambda)\hat{\sigma}(d\lambda)a(\lambda)^* \left[a(\lambda)^{-1} \right]^* = \hat{\sigma}(B) ,$$

and from the diagonality of $\hat{\sigma}$ it follows that the matrix $c(\lambda) = a(\lambda)^{-1}$ satisfies condition (4.8.78).

As the matrix $a(\lambda)$ must satisfy only the condition of the noncorrelatedness of the stochastic measures $\hat{Z}_q$ then $m(m+1)/2$ of its elements may be given arbitrarily. Using this fact we assume $a_{pp}(\lambda) = 1$, $a_{pq}(\lambda) = 0$ at $q > p$ (i.e. we shall find the triangular matrix $a(\lambda)$ with the diagonal elements which are equal to 1. Then (4.8.85) will give $Z_1(B) = \hat{Z}_1(B)$ and

$$Z_p(B) = \sum_{q=1}^{p-1} \int_B a_{pq}(\lambda)\hat{Z}_q(d\lambda) + \hat{Z}_p(B) \quad (p = 2 , \ldots , m) . \qquad (4.8.86)$$

Consequently, $\hat{\sigma}_1(B) = \sigma_{11}(B)$ and by virtue of the noncorrelatedness of $\hat{Z}_1$ with $\hat{Z}_2, \ldots, \hat{Z}_m$ we have

$$\sigma_{p1}(B) = \int_B a_{p1}(\lambda)\hat{\sigma}_1(d\lambda) \quad (p = 2, \ldots, m).$$

Hence by virtue of the $\hat{\sigma}_1$-continuity (more exactly of the σ_{11}-continuity) of all measures σ_{p1} it follows that the functions $a_{p1}(\lambda)$ exist and $a_{p1}(\lambda) = d\sigma_{p1}/d\hat{\sigma}_1(\lambda)$ ($p = 2, \ldots, m$). Now we determine sequentially the stochastic measures $Z_p^{(1)}(B) = Z_p(B)$ and put

$$Z_p^{(r)}(B) = Z_p^{r-1}(B) - \int_B a_{p,r-1}(\lambda)\hat{Z}_{r-1}(d\lambda) \quad \begin{pmatrix} p = r, \ldots, m \\ r = 2, \ldots, m \end{pmatrix}.$$

By virtue of the noncorrelatedness of $Z_p^{(r)}$ and $\hat{Z}_{r-1}$ we may write the following recurrent formula for the spectral and cross spectral measures of the stochastic measures $Z_p^{(r)}$:

$$\sigma_{pq}^{(r)}(B) = \sigma_{pq}^{(r-1)}(B) - \int_B a_{p,r-1}(\lambda)\overline{a_{q,r-1}(\lambda)}\hat{\sigma}_{r-1}(d\lambda).$$

As $Z_r^{(r)}(B) = \hat{Z}_r(B)$ then

$$\hat{\sigma}_r(B) = \sigma_{rr}^{(r-1)}(B) - \int_B |a_{r,r-1}(\lambda)|^2\,\hat{\sigma}_{r-1}(d\lambda) \quad (r = 2, \ldots, m).$$

$$(4.8.87)$$

From (4.8.86) we have

$$Z_p^{(r)}(B) = \sum_{q=r}^{p-1} \int_B a_{pq}(\lambda)\hat{Z}_q(d\lambda) + \hat{Z}_p(B) \quad \begin{pmatrix} p = r+1, \ldots, m \\ r = 1, \ldots, m \end{pmatrix}.$$

Hence because of the noncorrelatedness of $\hat{Z}_r$ with $\hat{Z}_{r+1}, \ldots, \hat{Z}_m$ it follows the formula

$$\sigma_{pr}^{(r)}(B) = \int_B a_{pr}(\lambda)\hat{\sigma}_r(d\lambda).$$

As all the measures $\sigma_{pr}^{(r)}$ are $\hat{\sigma}_r$-continuous (more exactly $\sigma_{rr}^{(r)}$-continuous) then the functions $a_{pr}(\lambda)$ exist and

$$a_{pr}(\lambda) = \frac{d\sigma_{pr}^{(r)}}{d\hat{\sigma}_r}(\lambda) \quad (p = r+1, \ldots, m, \ r = 1, \ldots, m). \qquad (4.8.88)$$

So *formulae (4.8.87) and (4.8.88) give the opportunity to determine sequentially all the functions* a_{pr} $(p = r + 1 , \ldots , m; r = 1 , \ldots , m)$ *and the spectral measures* $\hat{\sigma}_r$ $(r = 2 , \ldots , m)$.

It remains to determine the matrix $c(\lambda) = a(\lambda)^{-1}$. It is not difficult to verify that $c_{pp}(\lambda) = 1$, $c_{pq}(\lambda) = 0$ at $q > p$ *and that the other elements of the matrix* $c(\lambda)$ *are given by the formulae*

$$c_{p,p-1}(\lambda) = -a_{p,p-1}(\lambda) \quad (p = 2 , \ldots , m) ,$$

$$c_{pq}(\lambda) = -a_{pq}(\lambda) - \sum_{r=q+1}^{p-1} a_{pr}(\lambda)c_{rp}(\lambda) \quad \begin{pmatrix} q = 1 , \ldots , p - 2 \\ p = 3 , \ldots , m \end{pmatrix} .$$

$$(4.8.89)$$

It is easy to understand that the obtained solution of Eqs. (4.8.89) is not unique. To be certain in that fact it is sufficient to notice that at any change of the enumeration of the measures $\hat{Z}_1 , \ldots , \hat{Z}_m$ and $Z_1 , \ldots , Z_m$ the outlined method will give another solution. ◁

R e m a r k. In the case of the existence of the spectral density $s_x(\lambda)$ while determining the matrices $a(\lambda)$ and $c(\lambda)$ the vector white noise $V(\lambda)$ with the correlated components is substituted by the white noise $\hat{V}(\lambda)$ with the uncorrelated components by means of a standard procedure of the orthogonalization.

E x a m p l e 4.8.3. Find an approximate canonical expansion of the vector stationary random function $X(t) = [X_1(t) , \ldots , X_n(t)]^T$ of a scalar argument with the matrix of the spectral densities $s^x(\lambda) = \{s^x_{lh}(\lambda)\} , (l, h = 1, 2 , \ldots , n)$ at the time interval $|t| < T$. Analogously to Example 4.8.2 for the component $X_h(t)$ $(h = 1 , \ldots , n)$ we have:

$$X_h(t) = m_h^x + \sum_{r=1}^{n} \int_{-\infty}^{\infty} V_r(\lambda)a_{rh}(\lambda)e^{i\lambda t}d\lambda = m_h^x + \int_{-\infty}^{\infty} U_h(\lambda)e^{i\lambda t}d\lambda \quad \text{(I)}$$

$$X_{hN}(t) = m_h^x + \sum_{\nu=-N}^{N} \sum_{r=1}^{n} V_\nu^{(r)}a_{rh}(\lambda_\nu)e^{i\lambda_\nu t} = m_h^x + \sum_{\nu=-N}^{N} U_{h\nu} e^{i\lambda_\nu t} , \quad \text{(II)}$$

$$X_h(t) - X_{hN}(t) = \int_{-\infty}^{-\Omega} U_h(\lambda)e^{i\lambda t}d\lambda + \int_{\Omega}^{\infty} U_h(\lambda)e^{i\lambda t}d\lambda$$

$$+ \sum_{\nu=-N}^{N} \sum_{r=1}^{n} \int_{\lambda_\nu-\alpha}^{\lambda_\nu+\alpha} V_r(\lambda) \left[a_{rh}(\lambda)e^{i\lambda t} - a_{rh}(\lambda_\nu)e^{i\lambda_\nu t} \right] d\lambda . \quad \text{(III)}$$

It is denoted here

$$V_\nu^{(r)} = \int\limits_{\lambda_\nu - \alpha}^{\lambda_\nu + \alpha} V_r(\lambda)d\lambda \quad (r = 1, \ldots, n; \ \nu = 0 \pm 1, \ldots, \pm N);$$

$$D_\nu^{(r)} = \int\limits_{\lambda_\nu - \alpha}^{\lambda_\nu + \alpha} \nu_r(\lambda)d\lambda \quad (r = 1, \ldots, n; \ \nu = 0 \pm 1, \ldots, \pm N);$$

$$\nu_r(\lambda) = \frac{1}{a_{rh}(\lambda)} \sum_{l=1}^{n} \left[\overline{a_{lr}^-(\lambda)s_{hl}^x(\lambda)} \right] ;$$

where $a_{lr}^-(\lambda)$ is an element of the inverse to matrix $a_{rh}(\lambda)$. As the items in the right-hand side of (III) are uncorrelated and the expectation of the difference $X_h - X_{hN}$ is equal to zero then the expectation of the modulus square of the difference $X_h - X_{hN}$ is equal to the sum of the variances of the items. Therefore

$$E\,|X_h(t) - X_{hN}(t)|^2 = \int\limits_{\infty}^{-\Omega} s_{hh}^x(\lambda)d\lambda + \int\limits_{\Omega}^{\infty} s_{hh}^x(\lambda)d\lambda$$

$$+ \sum_{\nu=-N}^{N} \sum_{r=1}^{n} \int\limits_{\lambda_\nu - \alpha}^{\lambda_\nu + \alpha} \nu_r(\lambda)\,|a_{rh}(\lambda)|^2 \left|1 - \frac{a_{rh}(\lambda_\nu)}{a_{rh}(\lambda)}e^{i(\lambda_\nu - \lambda)t}\right|^2 d\lambda. \qquad (\mathrm{IV})$$

Accounting that

$$s_{hl}^x(\lambda) = \sum_{r=1}^{n} \nu_r(\lambda)a_{rh}(\lambda)\overline{a_{rl}(\lambda)},$$

and choosing ε from the condition

$$\left|1 - \frac{a_{rh}(\lambda_\nu)}{a_{rh}(\lambda)}e^{i(\lambda_\nu - \lambda)t}\right|^2 < \varepsilon, \quad |t| < T,$$

we get from (IV) at sufficiently small α the following estimate:

$$\sum_{\nu=-N}^{N} \int\limits_{\lambda_\nu - \alpha}^{\lambda_\nu + \alpha} s_{hh}^x(\lambda) \left|1 - \frac{a_{rh}(i\lambda_\nu)}{a_{rh}(\lambda)}e^{i(\lambda_\nu - \lambda)t}\right|^2 d\lambda < \varepsilon \int\limits_{-\Omega}^{\Omega} s_{hh}^x(\lambda)d\lambda \le \varepsilon D_h^x .$$

Thus the first two integrals in (IV) as well as the sum of the integrals in (IV) may be made as small as possible by sufficiently small interval of the frequencies 2α.

P r o b l e m s

4.1. Show that the Cauchy distribution may serve as the one-dimensional distribution of a process with independent increments keeping in mind that the characteristic function $g_1(\lambda; t) = e^{-\alpha t |\lambda|}$ of the Cauchy distribution with the density $f_1(x; t) = \alpha t / \pi(\alpha^2 t^2 + x^2)$ satisfies condition (4.6.10): $h(\lambda; t_1, t_2) = g_1(\lambda; t_2)/g_1(\lambda; t_1) = e^{-\alpha(t_2 - t_1)|\lambda|}$ at $t_2 > t_1$. But such a process has neither the expectation nor the second order moment and consequently, cannot be a process with uncorrelated increments.

4.2. Let us consider a process with uncorrelated but dependent increments

$$X(t) = \left(\frac{\alpha}{\Lambda} A + U\right) \sin \Lambda t + \left(\frac{\alpha}{\Lambda} U - Z\right) (1 - \cos \Lambda t),$$

where Z, U, Λ are independent random variables, Z and U are normally distributed with zero expectations and the same variance D, and Λ is distributed according to Cauchy law with the density $f(\lambda) = \alpha/\pi(\alpha^2 + \lambda^2)$.

S o l u t i o n. In order to prove that the process $X(t)$ is a process with uncorrelated increments we shall find its covariance function. For this purpose we first evaluate the conditional covariance function of the process $X(t)$ given the value λ of Λ:

$$K_x(t_1, t_2 \mid \lambda) = D \frac{\alpha^2 + \lambda^2}{\lambda^2} \left[1 + \cos \Lambda(t_1 - t_2) - \cos \lambda t_1 - \cos \lambda t_2\right].$$

After that according to the formula of total expectation we find the covariance function of the process $X(t)$:

$$K_x(t_1, t_2) = \int\limits_0^\infty K_x(t_1, t_2 \mid \lambda) f_1(\lambda) d\lambda$$

$$= \frac{2D\alpha}{\pi} \left[\int\limits_0^\infty \frac{1 - \cos \lambda t_1}{\lambda^2} d\lambda \right.$$

$$\left. + \int\limits_0^\infty \frac{1 - \cos \lambda t_2}{\lambda^2} d\lambda - \int\limits_0^\infty \frac{1 - \cos \lambda(t_1 - t_2)}{\lambda^2} d\lambda \right].$$

Using the formula

$$\int\limits_0^\infty \frac{1 - \cos \beta \lambda}{\lambda^2} d\lambda = \frac{\pi}{2} |\beta| ,$$

we get $K_x(t_1, t_2) = 2D\alpha \min(t_1, t_2)$. Hence it is clear that $X(t)$ represents a process with uncorrelated increments.

It is clear by intuition that $X(t)$ is not a process with independent increments. To prove this let us find the joint distribution of the random variables

$$X = X(t) = \left(\frac{\alpha}{\Lambda}Z + U\right)\sin \Lambda t + \left(\frac{\alpha}{\Lambda}U - Z\right)(1 - \cos \Lambda t),$$

$$Y = X(s) - X(t) = \left(\frac{\alpha}{\Lambda}Z + U\right)(\sin \Lambda s - \sin \Lambda t)$$

$$+ \left(\frac{\alpha}{\Lambda}U - Z\right)(\cos \Lambda t - \cos \Lambda s).$$

t and s are fixed. The equations

$$x = \left(\frac{\alpha}{\lambda}z + u\right)\sin \lambda t + \left(\frac{\alpha}{\lambda}u - z\right)(1 - \cos \lambda t),$$

$$y = \left(\frac{\alpha}{\lambda}z + u\right)(\sin \lambda s - \sin \lambda t) + \left(\frac{\alpha}{\lambda}u - z\right)(\cos \lambda t - \cos \lambda s)$$

have unique solution relative to z and u at $\lambda \neq 0$. Using the known formula for the density of a composite function of random variables for this case we find the joint density of the random variables X and Y:

$$f(x, y) = \frac{\alpha}{2\pi^2 D} \int\limits_{-\infty}^{\infty} \exp\left\{-\lambda^2\left[1 - \cos \lambda(s - t)\right]x^2\right.$$

$$+ \left[1 - \cos \lambda(s - t) + \cos \lambda s - \cos \lambda t\right]xy + (1 - \cos \lambda t)y^2\right\}$$

$$\times D^{-1}(\alpha^2 + \lambda^2)^{-1}\left[\sin \lambda(s - t) - \sin \lambda s + \sin \lambda t\right]^{-2}$$

$$\times \frac{\lambda^2 d\lambda}{(\alpha^2 + \lambda^2)^2 \,|\sin \lambda(s - t) - \sin \lambda s + \sin \lambda t\,|}.$$

Hence, it is clear that the joint density of the random variables X and Y cannot be represented as the product of their densities. Consequently, X and Y are dependent what proves our statement.

4.3. A step random process $P(t)$ which increases by unit jumps at random moments forming a Poisson stream is called *a Poisson process*. As the numbers of the jumps of a Poisson process in nonintersecting time intervals are independent according to the definition, so a Poisson process is a process with independent increments. We denote by $\nu(t)$ the intensity of a Poisson stream of the jumps of the process $P(t)$, a function of time in general. Show that its first and second moments and one- and multi-dimensional characteristic functions are defined by formulae

$$\mu = EP(t) = DP(t) = \int\limits_{t_0}^{t} \nu(\tau)d\tau. \qquad \text{(I)}$$

$$g_1(\lambda; t) = \exp\left\{(e^{i\lambda} - 1)\int_0^t \nu(\tau)d\tau\right\}.\tag{II}$$

$$g_n(\lambda_1, \ldots, \lambda_n; t_1, \ldots, t_n) = \exp\left\{(e^{i(\lambda_1+\cdots+\lambda_n)})\int_0^{t_1}\nu(\tau)d\tau\right.$$

$$\left. +e^{i(\lambda_2+\cdots+\lambda_n)}\int_{t_1}^{t_2}\nu(\tau)d\tau + \cdots + e^{i\lambda_n}\int_{t_{n-1}}^{t_n}\nu(\tau)d\tau - \int_0^{t_n}\nu(\tau)d\tau\right\}.\tag{III}$$

4.4. A scalar or vector random step process $X(t)$ with independent identically distributed jumps at random time moments forming a Poisson stream is called *a general Poisson process*. Find its moments and characteristic functions.

S o l u t i o n. As the sums of independent random variables which do not contain common items are independent, a general Poisson process represents a process with independent increments. If the jumps have finite expectation m_0 and covariance matrix (the variance in the case of a scalar process $X(t)$) K_0, then it is also a process with uncorrelated increments. Therefore, taking into account that $X(t_0) = 0$ we come to the conclusion that the conditional expectation and the conditional covariance matrix of the value of the process $X(t)$ at the moments t (when the number of jumps in the time interval (t_0, t) is m) are equal to mm_0 and mK_0 respectively. Consequently, its conditional second order moment is equal to $m(K_0 + m_0 m_0^T)$. Taking into consideration that the number of jumps in the time interval (t_0, t) represents a random variable with the Poisson distribution we find the probability of m jumps in the interval (t_0, t)

$$p_m = \frac{\mu^m}{m!}e^{-\mu}, \quad \mu = \int_{t_0}^t \nu_0(\tau)d\tau,\tag{I}$$

where $\nu_0(\tau)$ is the intensity of the stream of jumps. After that, according to the total expectation formula we find the unconditional expectation and second order moments of the process $X(t)$ at the instant t:

$$m_x(t) = EX(t) = m_0 \sum_{m=1}^{\infty} m\frac{\mu^m}{m!}e^{-\mu} = m_0\mu,\tag{II}$$

$$\Gamma_x(t) = (K_0 + m_0 m_0^T) \sum_{m=1}^{\infty} m\frac{\mu^m}{m!}e^{-\nu} = (K_0 + m_0 m_0^T)\mu.\tag{III}$$

Consequently, the covariance matrix of the value of the process $X(t)$ at the instant t is determined by the formula

$$k(t) = \Gamma_x(t) - m_x(t)m_x(t)^T = K_0\mu = K_0 \int_{t_0}^{t} \nu_0(\tau)d\tau. \qquad \text{(IV)}$$

Thus the process $X(t)$ has the finite second order moment and consequently, is a process with uncorrelated increments.

In order to find the one-dimensional characteristic functin of a general Poisson process $X(t)$,

$$X(t) = \sum_{k=1}^{P(t)} X_k, \qquad \text{(V)}$$

where $\{X_k\}$ is a sequence of independent identically distributed jumps $X_1, X_2, \ldots$ of the process $X(t)$ and $P(t)$ is a simple Poisson process generated by the same stream of the intensity $\nu_0(t)$, we denote by $g(\lambda)$ the characteristic function of the jumps $X_1, X_2, \ldots$. Then taking into account that the characteristic function of the sum of independent random variables is equal to the product of the characteristic functions of the items we find the conditional characteristic function of the process $X(t)$ at $P(t) = m$:

$$g_{P(t)=m}(\lambda; t) = E \exp\left\{ i\lambda^T \sum_{k=1}^{m} X_k \right\} = g^m(\lambda). \qquad \text{(VI)}$$

According to the formula of the total expectation we find

$$g_1(\lambda; t) = \sum_{m=0}^{\infty} g^m(\lambda)\frac{\mu^m}{m!}e^{-\mu} = e^{\mu[g(\lambda)-1]}, \quad \mu = \int_{t_0}^{t} \nu_0(\tau)d\tau. \qquad \text{(VII)}$$

After determining the one-dimensional characteristic function of a general Poisson process $X(t)$ according to formula (4.6.11) we calculate its multi-dimensional characteristic functions at $n = 2, 3, \ldots$,

$$g_n(\lambda_1, \ldots, \lambda_n; t_1, \ldots, t_n) = \exp\left\{ g(\lambda_1 + \cdots + \lambda_n) \int_{0}^{t_1} \nu_0(\tau)d\tau \right.$$

$$\left. +g(\lambda_2 + \cdots + \lambda_n) \int_{t_1}^{t_2} \nu_0(\tau)d\tau + \cdots + g(\lambda_n) \int_{t_{n-1}}^{t_n} \nu_0(\tau)d\tau - \int_{0}^{t_n} \nu_0(\tau)d\tau \right\}.$$

$$\text{(VIII)}$$

4.5. Under the conditions of Problems 4.4 consider a general Poisson process $X(t)$ whose distribution of the independent jump changes depends on time moments during which these changes occur. Find the distribution of the increment of the process $X(t)$ on the infinitesimal time interval $(t, s]$.

Sol u t i o n. It is sufficient to notice for this purpose that at the infinitesimal $s - t$ the probability of the fact that the number of the jumps at the interval $(t, s]$ will be more than 1 has the order $o(s - t)$. Then accounting that at the absence of the jumps in the interval $(t, s]$ a conditional characteristic function of the increment is equal to unit, and at one jump is equal to the characteristic function $g(\lambda; t)$ of the jump value, we have

$$g(\lambda; t, s) = 1 \cdot e^{-\nu_0(t)(s-t)} + g(\lambda; t)\nu_0(t)(s - t) + o(s - t)$$

$$= 1 + [g(\lambda; t) - 1]\nu_0(t)(s - t) + o(s - t). \tag{I}$$

Besides that find a joint characteristic function of the increments of a general Poisson process $X(t)$ and of the simple Poisson process which generates it at the infinitesimal time interval $(t, s]$. According to the definition we have

$$g(\lambda_1, \lambda_2; t, s) = E \exp\left\{i\lambda_1^T \Delta X + i\lambda_2 \Delta P\right\}$$

$$= e^{-\nu_0(t)(s-t)} + e^{i\lambda_2} g(\lambda_1; t)\nu_0(t)(s - t) + o(s - t)$$

$$= 1 + \left[e^{i\lambda_2} g(\lambda_1; t) - 1\right]\nu_0(t)(s - t) + o(s - t). \tag{II}$$

4.6. Show that the Cauchy process is presentable in the form of (4.6.47), i.e.

$$X(t) = \int_{-1}^{1} x P^0(t, dx) + \int_{|x| > 1} x P(t, dx), \quad \mu(t, x) = \alpha t/\pi. \tag{I}$$

Sol u t i o n. Using formulae we have

$$\frac{1}{\pi} \int_{-\infty}^{\infty} \frac{\cos \lambda x - 1}{x^2} dx = -|\lambda|, \quad \int_{-1}^{1} \frac{\sin \lambda x - \lambda x}{x^2} dx = 0, \quad \int_{|x| > 1} \frac{\sin \lambda x}{x^2} dx = 0,$$

So by virtue of (4.6.48) we get

$$\ln g_1^0(\lambda; t) = -\alpha t |\lambda| = \frac{\alpha t}{\pi} \int_{-1}^{1} (e^{i\lambda x} - 1 - i\lambda x)\frac{dx}{x^2} + \frac{\alpha t}{\pi} \int_{|x| > 1} (e^{i\lambda x} - 1)\frac{dx}{x^2}.$$

$$\tag{II}$$

4.7. Let us consider the θ-integrals

$$Y_{\theta_1} = \int_0^T W(\tau)d_{\theta_1}W(\tau)\,, \quad Y_{\theta_2} = \int_0^T W(\tau)d_{\theta_2}W(\tau)\,, \tag{I}$$

corresponding to two different values θ_1 and θ_2 of the parameter θ, where $W(t)$ is a standard Wiener process. Show that the stochastic integrals corresponding to different values of the parameter θ are different.

S o l u t i o n. By virtue of definition (4.6.41) taking $t_p^{(n)} = pT/n$ ($p = 0, 1,\,, \ldots, n$),

$$\Delta Y = Y_{\theta_2} - Y_{\theta_1}$$

$$= (\theta_2 - \theta_1)\,\underset{n\to\infty}{\text{l.i.m.}}\ \int_{n\to\infty} \sum_{p=1}^{n}\left[W\left(\frac{pT}{n}\right) - W\left(\frac{(p-1)T}{n}\right)\right]^2. \tag{II}$$

Let us find the expectation and the variance of the random variable ΔY. As the random variable $W(pT/n) - W((p-1)T/n)$ is normally distributed with zero expectation and the variance equal to T/n we have

$$E\Delta Y = (\theta_2 - \theta_1)\lim_{n\to\infty}\sum_{p=1}^{n}\frac{T}{n} = (\theta_2 - \theta_1)T\,, \tag{III}$$

$$E(\Delta Y)^2 = (\theta_2 - \theta_1)^2\lim_{n\to\infty}\left\{\sum_{p=1}^{n}E\left[W(pT/n) - W((p-1)T/n)\right]^4\right.$$

$$+ \sum_{p\neq q}E\left[W(pT/n) - W((p-1)T/n)\right]^2$$

$$\left.\times\left[W(qT/n) - W((q-1)T/n)\right]^2\right\}, \tag{IV}$$

where the second sum is extended over all $p, q, p \neq q$. But a central fourth order moment of a normally distributed random variable is equal to the triple square of its variance. Consequently, $E[W(pT/n) - W((p-1)T/n)]^4 = 3(T/n)^2$ and

$$E(\Delta Y)^2 = (\theta_2 - \theta_1)^2\lim_{n\to\infty}\left\{3\sum_{p=1}^{n}\frac{T^2}{n^2} + \sum_{p\neq q}\frac{T^2}{n^2}\right\}$$

$$= (\theta_2 - \theta_1)^2\lim_{n\to\infty}\left\{2\sum_{p=1}^{n}\frac{T^2}{n^2} + \sum_{p,q=1}^{n}\frac{T^2}{n^2}\right\}$$

$$= (\theta_2 - \theta_1)^2 \lim_{n \to \infty} \left(2\frac{T^2}{n} + T^2 \right) = (\theta_2 - \theta_1)^2 T^2 . \qquad \text{(V)}$$

Now we may find the variance of the random variable ΔY, $D\Delta Y = E(\Delta Y)^2 - (E\Delta Y)^2 = 0$. Thus the variable ΔY is nonrandom, and consequently, coincides with its expectation, $\Delta Y = (\theta_2 - \theta_1)T$. This proves that the stochastic integrals corresponding to different values of the parameter θ are different.

4.8. Prove that the Wiener measure in the space of all scalar functions of the scalar argument $t \in [0, \infty)$ is determined on the semialgebra of measurable rectangles of finite product of measurable spaces $(X^T, \mathcal{A}^T)$

$$X^T = \prod_{t \in T} X_t, \quad \mathcal{A}^T = \prod_{t \in T} \mathcal{A}_t, \quad X_t = R, \quad \mathcal{A}_t = \mathcal{B}, \quad T = [0, \infty), \quad \text{(I)}$$

i.e. the space of all the functions $x(t)$ with the domain $T = [0, \infty)$ by the following formula:

$$\mu_W(A_1 \times \cdots \times A_n) = \frac{1}{\sqrt{(2\pi)^n t_1(t_2 - t_1)\ldots(t_n - t_{n-1})}}$$

$$\times \int_{A_1} \cdots \int_{A_n} \exp\left\{ -\frac{x_1^2}{2t_1} - \frac{1}{2}\sum_{k=2}^{n} \frac{(x_k - x_{k-1})^2}{t_k - t_{k-1}} \right\} dx_1 \cdots dx_n , \qquad \text{(I)}$$

where $A_1, \ldots, A_n$ are the intervals of the real axis, $t_1, \ldots, t_n \in [0, \infty)$, $t_1 < t_2 < \cdots < t_n$, $(n = 1, 2, \ldots)$, $\mathcal{B}$ is a σ-algebra of Borel sets on R. It is evident that the measure μ_W is additive on the semialgebra of the rectangles of the space X^T. Notice that the measure μ_W may be also considered as a measure on any subspace $(X^{[a,b]}, \mathcal{A}^{[a,b]})$ of the space $(X^T, \mathcal{A}^T)$. The Wiener measure may serve as an example of a measure in an infinite-dimensional space.

4.9. Prove that the Wiener measure (Problem 4.8) is finite and its value in the whole space X^T is equal to 1, $\mu_W(X^T) = 1$.

4.10. Prove that the Wiener measure μ_W for the set of functions from X^T for which its increment on the interval $(t, t + \tau)$ is no less than $\varepsilon > 0$ tends to zero at $\tau \to 0$ at any fixed t.

S o l u t i o n. For calculation of μ_W of the set of functions for which $|x(t + \tau) - x(t)| \geq \varepsilon$ we use the functions

$$\Phi(u) = \frac{1}{\sqrt{\pi}} \int_0^u e^{-x^2/2} \, dx, \quad \Phi(\infty) = \frac{1}{2}, \quad \Phi(-u) = -\Phi(u) .$$

As a result we get

$$\mu_W(\{x(t) : |x(t + \tau) - x(t)| \geq \varepsilon\})$$

$$= \frac{1}{2\pi\sqrt{t\tau}} \int_{-\infty}^{\infty} dx_1 \int_{|x_2-x_1|\geq\varepsilon} \exp\left\{-\frac{x_1^2}{2t} - \frac{(x_2-x_1)^2}{2\tau}\right\} dx_2$$

$$= \frac{2}{\pi\sqrt{t\tau}} \int_{0}^{\infty} \exp\left\{-\frac{x_1^2}{2t}\right\} dx_1 \int_{\varepsilon}^{\infty} \exp\left\{-\frac{x^2}{2\tau}\right\} dx$$

$$= 1 - 2\Phi(\varepsilon/\sqrt{\tau}) \to 0 \quad \text{at} \quad \tau \to 0 \quad \forall \varepsilon > 0. \tag{I}$$

Thus for any sequence $\{\tau_n\}$, $\tau_n > 0$ that converges to zero the sequence of functionals form function $x(t) \in X^T$ and $f_n(x) = x(t + \tau_n) - x(t)$ ($n = 1, 2, \ldots$) converges to zero by the measure μ_W.

4.11. On the basis of the result of Problem 4.10 prove that the Wiener measure μ_W may be considered as concentrated on the subspace C of continuous functions from the space X^T of all scalar functions of the variable $t \in [0, \infty)$.

S o l u t i o n. Denote by S the set of binary-rational numbers $m2^{-p}$ ($m = 0, 1, 2, \ldots$; $p = 1, 2, \ldots$). Then applying formula (I) of Problem 4.10 and the inequality:

$$1 - 2\Phi(u) < k!2^{2k+1}u^{-2k-1}/\sqrt{2\pi}, (k = 1, 2, \ldots), \tag{I}$$

valid for any natural number k, we get

$$\mu_W\left(\{x(t) : | x((m+1)2^{-p}) - x(m2^{-p}) | \geq \varepsilon\}\right)$$

$$< k!2^{2k+1}(\varepsilon2^{p/2})^{-2k-1}/\sqrt{2\pi} \quad \forall p, m. \tag{II}$$

Let us put

$$z_p = \sup_{m} | x((m+1)2^{-p}) - x(m2^{-p}) |, \tag{III}$$

where the upper bound is taken over all m satisfying the condition "$m2^{-p} < a, a > 0$ is an arbitrary number". Then we have

$$\{x(t) : z_p \geq \varepsilon\} \subset \bigcup_{m=0}^{[a2^p]} \{x(t) : | x((m+1)2^{-p}) - x(m2^{-p}) | \geq \varepsilon\}. \tag{IV}$$

Putting $\varepsilon = A2^{-p\alpha}$, $0 < \alpha < 1/2$, $A > 0$, $c = (a+1)k!2^{2k+1}/\sqrt{2\pi}A^{2k+1}$ we receive from inequality (II)

$$\mu_W\left(\{x(t) : z_p \geq A2^{-p\alpha}\}\right) < c2^{-p(\beta-1)}, \tag{V}$$

where $\beta = (1/2 - \alpha)(2k + 1)$. Define the sets

$$A_p = \{x(t) : z_p \geq A2^{-p\alpha}\}, \quad B_n = \bigcup_{p=n}^{\infty} A_p, \quad N = \bigcap_{n=1}^{\infty} B_n . \qquad \text{(VI)}$$

Using the obvious inequality $\mu_W(N) \leq \mu_W(B_n) \, \forall n$ and inequality (V), we have

$$\mu_W(B_n) \leq \sum_{p=n}^{\infty} \mu_W(A_p) < c \sum_{p=n}^{\infty} 2^{-p(\beta-1)} . \qquad \text{(VII)}$$

After choosing k in such a way that $\beta > 1$ we get from (V)–(VII)

$$\mu_W(B_n) < c2^{-n(\beta-1)}/(1 - 2^{-\beta+1}) \quad \forall n , \qquad \text{(VIII)}$$

$$\mu_W(N) \leq \mu_W(B_n) < c'2^{-n(\beta-1)} , \qquad \text{(IX)}$$

where $c' = c/(1 - 2^{-\beta+1})$. As inequality (IX) is valid for all n then $\mu_W(N) = 0$. Let $\bar{N} = \bigcup_{n=1}^{\infty} \bar{B}_n$. Any function $x(t) \in \bar{N}$ belongs to set $\bar{B}_n$, and consequently, to all sets $\bar{A}_p, p \geq n$. So,

$$z_p = \sup_m \mid x((m + 1)2^{-p}) - x(m2^{-p}) \mid < A2^{-p\alpha} \qquad \text{(X)}$$

for all sufficiently large p. Let $s, s' \in S$ be arbitrary binary-rational numbers, q be the least natural number satisfying the condition $\mid s' - s \mid < 2^{-q+1}$. In this case there exists a point $s_0 = m2^{-q}$ between the points s and s' for some m and

$$s - s_0 = \pm \sum_{p=1}^{l} k_p 2^{-q-p} , \qquad \text{(XI)}$$

where the numbers k_l are equal to zero or unit. If we realize the transfer from the point s_0 to the point s sequentially by the steps of the length $k_1 2^{-q-1}$, $\ldots , k_l 2^{-q-l}$ then due to (X) we get

$$\mid x(s) - x(s_0) \mid < A \sum_{p=1}^{l} 2^{-(q-p)\alpha} < A2^{-q\alpha} \sum_{p=1}^{\infty} 2^{-p\alpha} = \frac{A2^{-(q+1)\alpha}}{1 - 2^{-\alpha}} , \qquad \text{(XII)}$$

$$\mid x(s') - x(s) \mid < \frac{2A2^{-(q+1)\alpha}}{1 - 2^{-\alpha}} ,$$

$$\sup_{\substack{s,\,s' \in S \\ |\,s' - s\,| < 2^{-q}}} |\,x(s') - x(s)\,| < \frac{2A2^{-(q+1)\alpha}}{1 - 2^{-\alpha}}. \qquad \text{(XIII)}$$

It is clear that $\sup |\,x(s') - x(s)\,| \to 0$ at $s' \to s$, $s,\, s' \in S \subset [0,\, a]$. This proves that all the functions $x(t) \in X^T$ excluding the set N of zero measure μ_W are continuous on the set S of binary-rational numbers for any finite interval $[0,\, a]$. Now let us take an arbitrary $t \in [0,\, a)$ and the sequence $\{s_n\}$ of binary-rational numbers converging to t and $|\,s_p - t\,| < 2^{-p}$. Define the sets

$$A_p = \left\{ x(t') :|\, x(t) - x(s_p)\,| \geq A2^{-p\alpha} \right\},$$

$$B_n = \bigcup_{p=n}^{\infty} A_p, \quad N = \bigcap_{n=1}^{\infty} B_n. \qquad \text{(XIV)}$$

Analogously we prove that $\mu_W(N) = 0$ and that the sequence $\{x(s_p)\}$ converges to $x(t)$ beyond the set N. Thus at any fixed $t \in [0,\, a)$ almost all relative to the measure μ_W functions from the space X^T are continuous at the point t for any a. This result make possible to extend the measure μ_W on the σ-algebra $\mathcal{A}^T C = \{B\, :\, B = AC,\, C \in \mathcal{A}^T\}$ of the subspace $C \subset X^T$ of continuous functions from the space of all functions X^T putting $\mu_W(B) = \mu_W(A),\, B \in \mathcal{A}^T C$.

As it is known any function from the space X^T without changing the values of measure on rectangle sets from the space X^T and also on all sets of an induced σ-algebra may be replaced by some so called *separable* function. Such a function is almostly defined for all t by its values on some countable set called a *separant*. In our case separant is the set of all binary-rational numbers and separable function is the set of all continuous functions.

Notice that the Lebesgue extention of the measure μ_W initially defined on the sets of the rectangles of all functions X^T may be determined on the space C of all continuous functions.

4.12. Prove that the σ-algebra $\mathcal{A}^T C$ in Problem 4.11 coincides with the σ-algebra $\mathcal{C}$ of the space $C = C[0, \infty)$ of continuous functions with supnorm induced by the set of all balls.

4.13. Prove that almost all relative to the Wiener measure μ_W functions of the space $C \subset X^T$ are equal to zero at $t = 0$.

4.14. Let us consider the function $f(x) = \varphi_1(s)x^2(t_1) + \varphi_2(s)x^2(t_2)$ where $\varphi_1(s), \varphi_2(s)$ are the continuous functions on the bounded closed interval S, i.e. the elements of the B-space $C(S)$, and $x(t)$ is a scalar function belonging to the space X^T with the σ-algebra $\mathcal{A}^T$, $T = [0, \infty)$ of Problem 4.8. Prove that the function $f(x)$ maps the measurable space $(X^T, \mathcal{A}^T)$ into the separable B-space $C(S)$.

S o l u t i o n. This function is $(\mathcal{A}^T, \mathcal{B})$-measurable as the inverse image of any ball of the space $C(S)$

$$f^{-1}(S_r(\psi)) = \left\{ x(t) : \psi(s) - r < \varphi_1(s)\,x^2(t_1) + \varphi_2(s)\,x^2(t_2) \right.$$

$$< \psi(s) + r, \ \forall s \in S \}$$

may be presented as a countable intersection of the cylinders in X^T with two-dimensional bases $\psi(s) - r < \varphi_1(s)\, x_1^2 + \varphi_2(s)\, x_2^2 < \psi(s) + r$ in the product of the spaces $X_{t_1} \times X_{t_2}$ correspondent to all rational $s \in S$.

Analogously the $(\mathcal{A}^T, \mathcal{B})$-measurability of the function

$$f(x) = \sum_{k=1}^{n} \varphi_k(s)\, x^2(t_k),$$

mapping the space X^T of Problem 4.8 in the separable B-space $C(S)$ is proved.

4.15. Prove the measurability of the function

$$f(x) = f_x(s) = \int_a^b \varphi(s, t)\, x^2(t)\, dt,$$

where $\varphi(s,t)$ is a continuous function of s, t, $s \in S$, S is a bounded closed set.

S o l u t i o n. This function maps the space of all functions $\mathcal{A}^T$ of the variable $t \in T = [0, \infty)$ into the separable B-space $C(S)$. In accordance with the result of Problem 2.3.8 the function $f(x)$ is determined almost everywhere in X^T relative to the Wiener measure μ_W. Therefore almost for all functions $x(t) \in X^T$ there exists the limit

$$f(x) = \lim_{n \to \infty} f_n(x) = \lim_{n \to \infty} \frac{b-a}{n-1} \sum_{k=0}^{n-1} \varphi(s, t_k)\, x^2(t_k),$$

$t_k = a + k\,(b - a)/(n - 1)$. But all the functions $f_n(x)$ are $(\mathcal{A}^T, \mathcal{B})$-measurable according to the result of Problem 4.14. Consequently, $f(x)$ represents the limit almost everywhere relative to the measure μ_W of the convergent sequence of $(\mathcal{A}^T, \mathcal{B})$-measurable functions. So the function $f(x)$ is $(\mathcal{A}_W^T, \mathcal{B})$-measurable where $\mathcal{A}_W^T$ is the complement of the σ-algebra $\mathcal{A}^T$ relative to the measure μ_W, i.e. μ_W is measurable.

4.16.   Calculate the integral of the function $f(x) = \varphi_1(s)x^2(t_1) + \varphi_2(s)x^2(t_2)$ over the whole space of the functions X^T on the Wiener measure μ_W (Problem 4.8) if $\varphi_1(s)$ and $\varphi_2(s)$ are continuous functions on bounded closed interval S of the real axis.

S o l u t i o n. In a given case the argument of the function $f(x)$ is the function $x(t)$ from the space X^S of scalar functions of $t \in S$ on which the Wiener measure μ_W is determined. The space of the values of the function $f(x)$ is the B-space of continuous functions $C(S)$. As the function $f(x)$ depends only on the values of the function $x(t)$ in two fixed points t_1, t_2 then for calculating the integral $\int f(x)\mu_W(dx)$ we may use the expression of Problem 4.8 of the measure μ_W on rectangular sets at $n = 2$. As a result we receive

$$\int f(x)\mu_W(dx) = \frac{1}{2\pi\sqrt{t_1(t_2 - t_1)}} \int\limits_{-\infty}^{\infty}\int\limits_{-\infty}^{\infty} [\varphi_1(s)x_1^2$$

$$+\varphi_2(s)x_2^2]\exp\left\{-\frac{x_1^2}{2t_1} - \frac{(x_2 - x_1)^2}{2(t_2 - t_1)}\right\}dx_1 dx_2 . \tag{I}$$

For calculating the integrals over x_1, x_2 we use formulae

$$\int\limits_{-\infty}^{\infty} e^{\eta x - cx^2/2}dx = \sqrt{\frac{2\pi}{c}}e^{\eta^2/2c}, \quad \int\limits_{-\infty}^{\infty} e^{-cx^2/2}dx = \sqrt{\frac{2\pi}{c}},$$

$$\int\limits_{-\infty}^{\infty} xe^{-cx^2/2}dx = 0, \quad \int\limits_{-\infty}^{\infty} x^2 e^{-cx^2/2}dx = \frac{1}{c}\sqrt{\frac{2\pi}{c}}. \tag{II}$$

So we find

$$\int f(x)\mu_W(dx) = \varphi_1(s)t_1 + \varphi_2(s)t_2 . \tag{III}$$

4.17. Using Problem 4.16 calculate the integral of the function

$$f(x) = \sum_{k=1}^{n} \varphi_k(s)x^2(t_k), \quad \varphi_k \in C(S), \quad (k = 1, 2, \ldots, n)$$

over the whole space of functions X^S with the Wiener measure μ_W.

4.18. Show that the integral of the function

$$f(x) = \sum_{k=1}^{n} \varphi_k(s)x^m(t_k)$$

over the whole space of functions X^T relative to the Wiener measure is equal to zero at odd $m = 2r + 1$. At even $m = 2r$

$$\int f(x)\mu_W(dx) = (2r - 1)!! \sum_{k=1}^{n} \varphi_k(s)t_k^r .$$

4.19. Calculate the integral relative to the Wiener measure of the function

$$f(x) = \varphi_1(s)x(t_1) + \varphi_2(s)x(t_2), \quad \varphi_1, \varphi_2 \in C(S),$$

over the rectangle set of space X^T with the base $A = (a_1, b_1) \times (a_2, b_2)$ in the product of the spaces $X_{t_1} \times X_{t_2}$.

4.20. Show that the following formula is valid

$$\int f(x)\mu_W(dx) = \exp\left\{ \sum_{k,l=1}^{n} \varphi_k(s)\varphi_l(s)\min(t_k, t_l) \right\} \tag{I}$$

for the integral over the whole space X^T relative to the Wiener measure of the function

$$f(x) = \exp\left\{ \sum_{k=1}^{n} \varphi_k(s)x(t_k) \right\}. \tag{II}$$

4.21. Calculate the integral

$$\int f(x)\mu_W(dx), \quad f(x) = \int_a^b \varphi(s,t)x^2(t)\, dt. \tag{I}$$

In Problem 4.15 it was evaluated that the function $f(x)$ is $(\mathcal{A}_W^T, \mathcal{B})$-measurable, and in Problem 4.16 was calculated the integral for the function obtained from $f(x)$ by the substitution of the integral by the sum. On the basis of these results, putting as in Problem 4.15

$$f_n(x) = \frac{b-a}{n-1} \sum_{k=0}^{n-1} \varphi(s, t_k)x^2(t_k), \tag{II}$$

we get

$$\int f_n(x)\mu_W(dx) = \frac{b-a}{n-1} \sum_{k=0}^{n-1} \varphi(s, t_k)t_k, \tag{III}$$

$$\lim_{n\to\infty} \int f_n(x)\mu_W(dx) = \int_a^b \varphi(s,t)t\, dt. \tag{IV}$$

It remains to prove the possibility of the limit passage under the integral sign. For this purpose we notice that

$$\| f_n(x) \| \le g_n(x) = \frac{b-a}{n-1} \sum_{k=1}^{n-1} \sup_{s\in S} |\varphi(s, t_k)|x^2(t_k). \tag{V}$$

Putting

$$g(x) = \lim_{n \to \infty} g_n(x) = \int_a^b \sup_{s \in S} |\varphi(s,t)| x^2(t)\, dt, \qquad \text{(VI)}$$

we obtain similarly as earlier in (III)–(VI)

$$\lim_{n \to \infty} \int g_n(x)\mu_W(dx) = \int_a^b \sup_{s \in S} |\varphi(s,t)| t\, dt. \qquad \text{(VII)}$$

But in accordance with the Fatou lemma

$$\int g(x)\mu_W(dx) \le \lim_{n \to \infty} \int g_n(x)\mu_W(dx). \qquad \text{(VIII)}$$

Thus the function $g(x)$ is μ_W-integrable and the functions f_n are bounded in the measure by μ_W-integrable function $g(x) + c$ at some $c > 0$. As $f_n(x) \to f(x)$ almost everywhere relative to the measure μ_W we get

$$\int f(x)\mu_W(dx) = \lim_{n \to \infty} \int f_n(x)\mu_W(dx) = \int_a^b \varphi(s,t)t\, dt. \qquad \text{(IX)}$$

4.22.   Using the procedure of Problem 4.21 prove that the integral of the function

$$f(x) = \int_a^b \varphi(s,t)x^{2r}(t)\, dt \qquad \text{(I)}$$

over the Wiener measure μ_W extended on the whole space X^T is determined by

$$\int f(x)\mu_W(dx) = (2r-1)!! \int_a^b \varphi(s,t)t^r\, dt. \qquad \text{(II)}$$

4.23. Prove that the integral of the function

$$f(x) = \exp\left\{ \int_a^b \varphi(s,t)x(t)\, dt \right\} \qquad \text{(I)}$$

over the Wiener measure μ_W extended on the whole space X^T is determined by

$$\int f(x)\mu_W(dx) = \exp\left\{ \int_a^b\int_a^b \varphi(s,t_1)\varphi(s,t_2) \min(t_1,t_2)\, dt_1\, dt_2 \right\}. \qquad \text{(II)}$$

S o l u t i o n. Introducing the functions

$$f_n(x) = \exp\left\{\frac{b-a}{n-1}\sum_{k=0}^{n-1}\varphi(s,t_k)x(t_k)\right\}$$

and using the result of Problem 4.16 we find similarly as in Problem 4.21

$$\lim_{n\to\infty}\int f_n(x)\mu_W(dx) = \exp\left\{\int_a^b\int_a^b \varphi(s,t_1)\varphi(s,t_2)\min(t_1,t_2)\,dt_1\,dt_2\right\}.$$

$$(\text{III})$$

It remains to prove the possibility of the limit passage under the integral sign. We notice for this purpose that according to the Fatou lemma which may be used as a result of nonnegativity of all the functions $f_n(x)$ at each $s \in S$,

$$\int f(x)\mu_W(dx) = \int \lim f_n(x)\mu_W(dx) \le \lim \int f_n(x)\mu_W(dx).$$

From here and from (III) it is clear that the function $f(x)$ is μ_W-integrable and its norm is equal to

$$\| f(x) \| = \sup_{s\in S}\exp\left\{\int_a^b \varphi(s,t)x(t)\,dt\right\}.$$

And as the sequence $\{f_n(x)\}$ converges to $f(x)$ almost everywhere relative to the measure μ_W then

$$\| f_n(x) \| = \sup_{s\in S}\exp\left\{\frac{b-a}{n-1}\sum_{k=0}^{n-1}\varphi(s,t_k)x(t_k)\right\} \le C \| f(x) \|$$

at some $C > 1$. Thus the sequence of the functions $\{f_n(x)\}$ convergent almost everywhere relative to μ_W is bounded in the norm by μ_W-integrable function. Hence according to the Lebesgue theorem it follows that

$$\int f(x)\mu_W(dx) = \lim_{n\to\infty}\int f_n(x)\mu_W(dx),$$

and we obtain the required formula (II).

The formulae obtained as a result of solution of Problems 4.22 and 4.23 and the conditions of the existence of the Bochner integral show that the functions

$$f_1(x) = \int_0^\infty \varphi(s,t)x^{2r}(t)\,dt,$$

$$f_2(x) = \exp\left\{\int\limits_0^\infty \varphi(s,t)x(t)\,dt\right\}$$

are μ_W-integrable if and only if the improper Riemann integrals

$$\int\limits_0^\infty \sup_{s\in S}|\varphi(s,t)|t\,dt, \qquad \int\limits_0^\infty\int\limits_0^\infty \sup_{s\in S}|\varphi(s,t_1)\varphi(s,t_2)|\min(t_1,t_2)\,dt_1\,dt_2$$

converge. Notice that in Problem 4.16 and in Problems 4.17–4.20, 4.22 and 4.23 the conditions of the existence of the Bochner integral (the integrability of the norm of the functions $f(x)$ considered as the functions of s, $s \in S$ from the space $C(S)$) are fulfilled.

4.24. Calculate the integrals relative to the Wiener measure μ_W extended over the whole space X^T of the following functions:

$$f_1(x) = \sin \int\limits_a^b \varphi(s,t)\,x(t)\,dt,$$

$$f_2(x) = \cos \int\limits_a^b \varphi(s,t)\,x(t)\,dt,$$

$$f_3(x) = \sinh \int\limits_a^b \varphi(s,t)\,x(t)\,dt,$$

$$f_4(x) = \cosh \int\limits_a^b \varphi(s,t)\,x(t)\,dt,$$

$$f_5(x) = \exp\left\{\int\limits_a^b \varphi_1(s,t)\,x(t)\,dt\right\}\sin \int\limits_a^b \varphi_2(s,t)\,x(t)\,dt,$$

$$f_6(x) = \exp\left\{\int\limits_a^b \varphi_1(s,t)\,x(t)\,dt\right\}\cos \int\limits_a^b \varphi_2(s,t)\,x(t)\,dt,$$

$$f_7(x) = \exp\left\{\int\limits_a^b \varphi_1(s,t)\,x(t)\,dt\right\}\sinh \int\limits_a^b \varphi_2(s,t)\,x(t)\,dt,$$

$$f_8(x) = \exp\left\{\int\limits_a^b \varphi_1(s,t)\,x(t)\,dt\right\}\cosh \int\limits_a^b \varphi_2(s,t)\,x(t)\,dt,$$

$$f_9(x) = \psi(x)\exp\left\{\int\limits_a^b \varphi(s,t)\,x(t)\,dt\right\}, \quad \psi(x)$$

$$= \int\limits_a^b \ldots \int\limits_a^b \varphi(s,t_1)\ldots\varphi(s,t_m)x(t_1)\ldots x(t_m)\,dt_1\ldots dt_m,$$

$$f_{10}(x) = \psi(x)\sin \int\limits_a^b \varphi(s,t)\,x(t)\,dt,$$

$$f_{11}(x) = \psi(x)\cos \int\limits_a^b \varphi(s,t)\,x(t)\,dt,$$

$$f_{12}(x) = \psi(x)\,\mathrm{sh}\int_a^b \varphi(s,t)\,x(t)\,dt,$$

$$f_{13}(x) = \psi(x)\,\mathrm{ch}\int_a^b \varphi(s,t)\,x(t)\,dt.$$

I n s t r u c t i o n. For calculating the integrals of f_9–f_{13} replace in the formulae of Problem 4.23 and in the formulae for the integrals of the functions $f_1,\ldots,f_4$ the function $\varphi(s,t)$ by the function $\alpha\varphi(s,t)$, differentiate the obtained formulae m times with respect to α and put $\alpha = 1$ after that.

4.25. Calculate the integral relative to the Wiener measure μ_W extended over the whole space X^T of the function

$$f(x) = \int_a^b \cdots \int_a^b \varphi(s,t_1,\ldots,t_m)x(t_1)\ldots x(t_m)\,dt_1\ldots dt_m.$$

If the functions φ in all previous formulae do not depend on s then all the integrals in Problem 4.16 and in Problems 4.17–4.20, 4.23 and 4.24 represent abstract Lebesgue integrals.

4.26. Consider the spaces $L_p(X^T, \mathcal{A}_W^T, \mu_W)$. The measure μ_W may be extended to the space of the continuous functions $C \subset X^T$ with the σ-algebra $C\mathcal{A}^T$. Setting in the space C an ordinary norm $\| x \| = \sup | x(t) |$ we get the space $C(T)$. Show that the σ-algebra in the space $C(T)$ induced by a set of all open balls enters into the σ-algebra $C(T)\mathcal{A}^T$.

S o l u t i o n. For this purpose we notice that any open ball in $C(T)$ represents the limit of decreasing sequence of the rectangles:

$$\{x(t) \,:\, | x(t) - \varphi(t) | < r \quad \forall t \subset T\}$$

$$= \lim_{n\to\infty}\{x(t) \,:\, | x(s_k) - \varphi(s_k) | < r \quad (k = 1, \ldots, n)\},$$

where $\{s_n\}$ is a set of rational numbers of the interval $[0, \infty)$. Consequently, any ball in $C(T)$ belongs to the σ-algebra $C(T)\mathcal{A}^T$. It means that the whole σ-algebra C induced by a set of the balls in $C(T)$ enters into $C(T)\mathcal{A}^T$. But the σ-algebra C in $C(T)$ is induced by a countable set of the balls of rational radii with the centres at the points of a dense countable set (the space $C(T)$ is separable as a countable set of continuous functions with rational values in rational points is dense in $C(T)$). Consequently, all the spaces $L_p(C[a,\,b],\, C,\, \mu_W)$ are separable. In particular, the H-space of scalar functions $L_2(C[a,\,b],\, C,\, \mu_W)$ is separable.

4.27. Whether the function $f(x)$ of Problem 4.23 belongs to the space $L_p(C(R),\, C,\, \mu_W)$ at any $p \geq 1$ and whether the sequence of the functions $\{f_n(x)\}$ converges to $f(x)$ in p-mean?

4.28. Prove that the value of the measure μ_W on any set $C \in \mathcal{A}^{[t_1,\infty)}$ is expressed by the formula

$$\mu_W(C) = \frac{1}{\sqrt{2\pi t_1}} \int \exp\{-\frac{x^2}{2t_1}\}\, dx \int \mathbf{1}_C(x,y)\nu_{t_1}(dy;x) \qquad (\mathrm{I})$$

at any fixed t_1. At all possible values $t_1 \in [0,\infty)$ formula (I) determines the measure μ_W on all the sets $C \in \mathcal{A}^T$.

4.29. Let $(X^T, \mathcal{A}^T)$ be an infinite product of the measurable spaces

$$X^T = \prod_{t \in T} X_t, \quad \mathcal{A}^T = \prod_{t \in T} \mathcal{A}_t, \quad X_t = R^m, \quad \mathcal{A}_t = \mathcal{B}^m, \quad T = [0,\infty),$$

where $\mathcal{B}^m$ is the σ-algebra of Borel sets of the space R^m. It is evident that X^T represents a space of all m-dimensional vector functions with the domain T. We shall determine on the semialgebra of the measurable rectangles of the space $(X^T, \mathcal{A}^T)$ the measure by formula

$$\mu(A_1 \times \cdots \times A_n) = \frac{1}{(2\pi)^m \sqrt{|K|}} \int_{A_1} \cdots \int_{A_n} \exp\{-x^T K^{-1} x/2\}\, dx_1\, dx_2$$

$$\times \lambda_{t_3;t_1,t_2}(dx_3; x_1, x_2) \ldots \lambda_{t_n;t_1,\ldots,t_{n-1}}(dx_n; x_1,\ldots,x_n), \qquad (\mathrm{I})$$

where

$$x = \begin{bmatrix} x_1 \\ x_2 \end{bmatrix}, \quad K = \begin{bmatrix} K(t_1,t_1) & K(t_1,t_2) \\ K(t_2,t_1) & K(t_2,t_2) \end{bmatrix},$$

$K(s_1, s_2)$ is a continuous matrix function of the dimension $m \times m$ satisfying the following conditions:

(i) $K(s_2, s_1) = K(s_1, s_2)^T$;

(ii) $\sum_{i,j=1}^{N} u_i^T K(s_i, s_j) u_j > 0$ at any N, $s_1, \ldots, s_N$ and m-dimensional vectors $u_1, \ldots, u_N$;

(iii) $\operatorname{tr} \sigma(t_1, t_2) < B|t_1 - t_2|^\gamma$ at some $B, \gamma > 0$, where

$$\sigma(t_1, t_2) = K(t_1,t_1) - K(t_1,t_2) - K(t_2,t_1) + K(t_2,t_2),$$

$\lambda_{t_k;t_1,\ldots,t_{k-1}}(B; x_1,\ldots,x_{k-1})$ $(k = 3,\ldots,n)$ are finite nonnegative measures in the correspondent spaces $(X_{t_k}, \mathcal{A}_{t_k})$ representing at each $B \in \mathcal{A}_{t_k}$ $(\mathcal{A}_{t_1} \times \cdots \times \mathcal{A}_{t_{k-1}}, \mathcal{B})$-measurable functions $x_1, \ldots, x_{k-1}$. The measure μ is σ-additive on the semialgebra $\mathcal{C}$ of rectangles of the space X^T and may be uniquely extended on the σ-algebra $\mathcal{A}_\mu^T$ obtained by the supplement of $\mathcal{A}^T$ relative to μ.

Prove that the measure μ in (I) is finite.

I n s t r u c t i o n. Apply the formula

$$\mu(\{x(t) : |x_k(t_2) - x_k(t_1)| \geq \eta\}$$

$$= \frac{1}{2\pi\sqrt{|K|}} \underset{|x_{2k}-x_{1k}|>\eta}{\int\int} \exp\{-x^T K^{-1} x/2\}\, dx_1\, dx_2$$

$$= 1 - 2\Phi(\eta/\sqrt{\sigma_{kk}(t_1,t_2)}) \quad (k=1,\ldots,m),$$

where $\sigma_{kk}(t_1,t_2)$ are diagonal elements of the matrix $\sigma(t_1,t_2)$, an inequality of (I) from Problem 4.11 and take into account that

$$|x(t_2) - x(t_1)| = \sqrt{\sum_{k=1}^{m}[x_k(t_2) - x_k(t_1)]^2} \leq \sum_{k=1}^{m}|x_k(t_2) - x_k(t_1)|;$$

as a result

$$\{x(t) : |x(t_2) - x(t_1)| \geq \varepsilon\} \subset \bigcup_{k=1}^{m}\{x(t) : |x_k(t_2) - x_k(t_1)| \geq \varepsilon/m\},$$

and take α of Problem 4.11 in the interval $(0, \gamma/2)$.

4.30. Prove that the integral over the measure μ of Problem 4.29 of the function

$$f(x) = \sum_{h,l=1}^{n} \varphi_{hl}(t_1,\ldots,t_n)x(t_h)x(t_l)^T, \tag{I}$$

where φ_{hl} $(h,l = 1,\ldots,n)$ are continuous matrix functions of the dimension $m \times m$ with the values in a separable B-space extended to the whole space C is determined by formula

$$\int f(x)\mu(dx) = \sum_{h,l=1}^{n} \varphi_{hl}(t_1,\ldots,t_n)K(t_h,t_l). \tag{II}$$

4.31. Under the conditions of Problem 4.29 determine the measure

$$\mu(A_1 \times \cdots \times A_n) = \frac{1}{(2\pi)^{2m}\sqrt{|K|}} \int_{A_1} \cdots \int_{A_n} \exp\{-x^T K^{-1} x/2\}\, dx_1$$

$$\ldots dx_4 \lambda_{t_5;t_1,\ldots,t_4}(dx_5; x_1, \ldots, x_4) \ldots \lambda_{t_n;t_1,\ldots,t_{n-1}}(dx_n; x_1, \ldots, x_{n-1}), \tag{I}$$

where

$$x = \begin{bmatrix} x_1 \\ x_2 \\ x_3 \\ x_4 \end{bmatrix}, \quad K = \begin{bmatrix} K(t_1,t_1) & K(t_1,t_2) & K(t_1,t_3) & K(t_1,t_4) \\ K(t_2,t_1) & K(t_2,t_2) & K(t_2,t_3) & K(t_2,t_4) \\ K(t_3,t_1) & K(t_3,t_2) & K(t_3,t_3) & K(t_3,t_4) \\ K(t_4,t_1) & K(t_4,t_2) & K(t_4,t_3) & K(t_4,t_4) \end{bmatrix},$$

$K(t_1,t_2)$ is a continuous together with its derivatives till the second order inclusively the matrix function of the dimension $m \times m$ possessing properties (i), (ii) from Problem 4.29 and the additional property

$$\operatorname{tr} \sigma(t_1,t_2,t_3,t_4) < B\tau^\gamma$$

at some $B > 0,\ \gamma > 2,\ \tau = \max_{i,j} |t_i - t_j|$,

$$\sigma(t_1,t_2,t_3,t_4) = K(t_1,t_1) + K(t_2,t_2) + K(t_3,t_3) + K(t_4,t_4)$$

$$-K(t_1,t_2) - K(t_2,t_1) - K(t_1,t_3) - K(t_3,t_1) + K(t_1,t_4) + K(t_4,t_1)$$

$$+K(t_2,t_3) + K(t_3,t_2) - K(t_2,t_4) - K(t_4,t_2) - K(t_3,t_4) - K(t_4,t_3);$$

the measures $\lambda_{t_k;t_1,\ldots,t_{k-1}}$ $(k = 5,\ldots,n)$ are the same as in Problem 3.8.6. Prove that almost all the functions of the space X^T relative to the measure μ are continuous and have continuous first derivatives.

I n s t r u c t i o n. Use formula

$$\frac{1}{(2\pi)^{2m}\sqrt{|K|}} \underset{|x_{4k}-x_{3k}-x_{2k}+x_{1k}|\geq\eta}{\int\int\int\int} \exp\{-x^T K^{-1} x/2\}\, dx_1\, dx_2\, dx_3\, dx_4$$

$$= 1 - 2\Phi(\eta/\sqrt{\sigma_{kk}(t_1,t_2,t_3,t_4)}\,),$$

where $\sigma_{kk}(t_1,t_2,t_3,t_4)$ are the diagonal elements of the matrix $\sigma(t_1,t_2,t_3,t_4)$, use from inequality (I) Problem 4.11, take α in the interval $(1,\gamma/2)$.

CHAPTER 5
GENERAL THEORY OF STOCHASTIC SYSTEMS AND ITS APPLICATIONS

Chapter 5 is dedicated to the general theory of stochastic systems described by integral, differential, integrodifferential, difference and etc stochastic equations in the finite- and infinite-dimensional spaces. Section 5.1 is devoted to the stochastic Itô differentials, the Stratonovich differentials and the θ-differentials. The Itô formula and its various generalizations are proved. Basic definitions concerning the stochastic differential equations, the change of variables and sufficient conditions of the existence and the uniqueness of the solutions are given in Sections 5.2 and 5.3. Practical problems connected with the transformation of the system equations into the stochastic differential equations based on the shaping filters are considered in Section 5.4. Numerical methods and algorithms for the finite-dimensional stochastic differential equations solution are presented in Section 5.5. Section 5.6 contains the general theory of the multi-dimensional distributions in the continuous stochastic systems described by the stochastic integral and the differential equations. The multi-dimensional distributions in the stochastic discrete and continuous-discrete systems are considered in Section 5.7. Structural theory of the distributions in the stochastic composed systems is given in Section 5.8. Section 5.9 is devoted to the theory of the multi-dimensional distributions in the stochastic infinite-dimensional systems. The last Section 5.10 contains applications to filtering based on the online data processing.

5.1. Stochastic Differentials

5.1.1. Stochastic Differential of Itô Process

A random process $Z(t)$ is called *the Itô process* if it is representable in the form

$$Z(t) = Z(t_0) + \int\limits_{t_0}^{t} X(\tau)d\tau + \int\limits_{t_0}^{t} Y_1(\tau)dW(\tau)$$

$$+ \int\limits_{t_0}^{t}\int\limits_{R_0^q} Y_2(\tau, u)P^0(d\tau, du).$$

$$(5.1.1)$$

Here $t_0 > 0$, $W(t)$ is the Wiener process, $P^0(\Delta, A)$ is the independent of $W(t)$ centered Poisson measure, the first integral is the m.s. integral, the second integral is the Itô stochastic integral, the third integral is the stochastic integral with respect to the Poisson measure $P(\Delta, A)$, $X(t)$, $Y_1(t)$, $Y_2(t, u)$ are random functions satisfying the conditions of the existence of the respective integrals. If $Z(t)$ and $W(t)$ are the vector random processes, then $X(t)$ and $Y_2(t, u)$ represent the vector random functions of the same dimensions as $Z(t)$ and $Y_1(t)$ represent the matrix random function.

The formal expression

$$dZ = X(t)dt + Y_1(t)dW(t) + \int_{R_0^q} Y_2(t, u)P^0(dt, du). \qquad (5.1.2)$$

is called *the stochastic differential* of the Itô process $Z(t)$.

R e m a r k. It should be emphasized that expression (5.1.2) has no usual sense of differential since it involves the infinitesimal quantities of different orders. Really, replacing the integrals by the respective integral sums, the increment of $Z(t)$ on a small interval $(t, t + \Delta t]$ may be expressed as

$$\Delta Z = X(t)\Delta t + Y_1(t)\Delta W(t) + \int_{R_0^q} Y_2(t, u)P^0((t, t + \Delta t], du), \quad (5.1.3)$$

where $\Delta W = W(t + \Delta t) - W(t)$. Since

$$E(\Delta W)^2 = E\left[W(t + \Delta t) - W(t)\right]^2 = \int_t^{t+\Delta t} \nu(\tau)d\tau = O(\Delta t),$$

the second item in the right-hand side of (5.1.3) has the order $\sqrt{\Delta t}$.

In the special case where $Y_2(t, u) \equiv 0$ and $Y_1(t) = Y(t)$ with probability 1 the Itô stochastic differential of the Itô process

$$Z(t) = Z(t_0) + \int_{t_0}^t X(\tau)dt + \int_{t_0}^t Y(\tau)dW(\tau) \qquad (5.1.4)$$

is given by

$$dZ = X\,dt + Y\,dW, \qquad (5.1.5)$$

where the argument t is omitted for brevity.

5.1.2. Itô Formula

Theorem 5.1.1. *Let $\varphi(z,t)$ be a scalar real function, $Z(t)$ be an Itô process given by Eq. (5.1.4). If the scalar function $\varphi = \varphi(z,t)$ is continuous and bounded together with its first and second derivatives, its second derivatives satisfy the Lipschitz condition*

$$\left| \frac{\partial^2 \varphi(z',t)}{\partial z_p \partial z_q} - \frac{\partial^2 \varphi(z,t)}{\partial z_p \partial z_q} \right| < b \sum_{r=1}^{n} |z_r' - z_r| \, , \qquad (5.1.6)$$

n being the dimensional of the vector z, the random function $Y(t)$ has the finite absolute fourth order moments then $U(t) = \varphi(Z(t),t)$ is the Itô process and its stochastic differential is given by

$$dU = \left\{ \varphi_t(Z,t) + \varphi_z(Z,t)^T X + \frac{1}{2} \mathrm{tr}\left[\varphi_{zz}(Z,t) Y \nu Y^T\right] \right\} dt$$

$$+ \varphi_z(Z,t)^T Y \, dW, \qquad (5.1.7)$$

where $\varphi_t(z,t) = \partial\varphi(z,t)/\partial t$; $\varphi_z = \partial\varphi(z,t)/\partial z$ is the matrix-column of the derivatives with respect to the components of vector z; $\varphi_{zz}(z,t) = (\partial/\partial z)(\partial/\partial z)^T \varphi(z,t)$ is the square matrix of the second derivatives of $\varphi(z,t)$ with respect to the components of the vector z and $\nu(t)$ is the intensity of the Wiener process $W(t)$.

R e m a r k. For linear function $\varphi(z,t)$ formula (5.1.7) coincides with usual formula for the differentiation of a composite function.

$\triangleright$ Since

$$\varphi\left(Z(t_0) + \sum_{k=1}^{N} \Delta Z^k, t\right) - \varphi(Z(t_0), t_0)$$

$$= \sum_{k=1}^{N} \left[\varphi(Z^k + \Delta Z^k, t_k) - \varphi(Z^k, t_{k-1})\right],$$

where $t_k = t_0 + kh$, $h = (t - t_0)/N$, $Z^k = Z(t_{k-1})$, $\Delta Z^k = X^k h + Y^k \Delta W^k$; $X^k = X(t_{k-1})$, $Y^k = Y(t_{k-1})$, $\Delta W^k = W(t_k) - W(t_{k-1})$ we get using the Taylor formula with the residual in the Lagrange form and taking into account that ΔZ^k is of order $\sqrt{h}$,

$$\varphi\left(Z(t_0) + \sum_{k=1}^{N} \Delta Z^k, t\right) - \varphi(Z(t_0), t_0) = \sum_{k=1}^{N} \{\varphi_t^k h + \varphi_z^{kT} \Delta Z^k\}$$

$$+\frac{1}{2}\operatorname{tr}\left[\varphi_{zz\theta}^{k}Y^{k}\Delta W^{kT}Y^{kT}\right]+o(h)\Big\}$$

$$=\sum_{k=1}^{N}\left\{\varphi_{t}^{k}+\varphi_{z}^{kT}X^{k}+\frac{1}{2}\operatorname{tr}\left[\varphi_{zz}^{k}Y^{k}\nu^{k}Y^{kT}\right]\right\}h+\sum_{k=1}^{N}\varphi_{z}^{kT}Y^{k}\Delta W^{k}$$

$$+\frac{1}{2}\sum_{k=1}^{N}\operatorname{tr}\left[\varphi_{zz}^{k}Y^{k}(\Delta W^{k}\Delta W^{kT}-\nu^{k}h)Y^{kT}\right]$$

$$+\frac{1}{2}\sum_{k=1}^{N}\operatorname{tr}\left[(\varphi_{zz\theta}^{k}-\varphi_{zz}^{k})Y^{k}\Delta W^{k}\Delta W^{kT}Y^{kT}\right]+O(h)\,,\qquad(5.1.8)$$

where the superscript k indicates the values of the respective functions at $Z = Z(t_{k-1})$, $t = t_{k-1}$; $\varphi_{zz\theta}^{k}$ means the value of φ_{zz} at $Z = Z(t_{k-1}) + \theta\Delta Z^{k}$, $t = t_{k-1} + \theta h$, $\theta \in (0,1]$. Now we estimate the last two sums in (5.1.8).

At first we evaluate the expectation of the product of two different items of the third sum in (5.1.8). Owing to independence X^{l}, Y^{l}, X^{k}, Y^{k}, ΔW^{l} of ΔW^{k} at $l < k$

$$E\operatorname{tr}[\varphi_{zz}^{k}Y^{k}(\Delta W^{k}\Delta W^{kT}-\nu^{k}h)Y^{kT}]\operatorname{tr}[\varphi_{zz}^{l}Y^{l}(\Delta W^{l}\Delta W^{lT}-\nu^{l}h)Y^{lT}]$$

$$= E\operatorname{tr}\left[\varphi_{zz}^{k}Y^{k}(E\Delta W^{k}\Delta W^{kT}-\nu^{k}h)Y^{kT}\right]$$

$$\times\operatorname{tr}\left[\varphi_{zz}^{l}Y^{l}(\Delta W^{l}\Delta W^{lT}-\nu^{l}h)Y^{lT}\right]=0\,,\qquad(5.1.9)$$

since $E\Delta W\Delta W^{T}=\nu\Delta t=\nu h$. As for the expectation of the square of an item in the third sum (5.1.8) we have the following estimate:

$$E\left\{\operatorname{tr}\left[\varphi_{zz}^{k}Y^{k}(\Delta W^{k}\Delta W^{kT}-\nu^{k}h)Y^{kT}\right]\right\}^{2}$$

$$< a^{2}\sum_{j_{1},h_{1},j_{2},h_{2}=1}^{n}\sum_{p_{1},q_{1},p_{2},q_{2}=1}^{m}E\left|Y_{j_{1}p_{1}}^{k}Y_{h_{1}q_{1}}^{k}Y_{j_{2}p_{2}}^{k}Y_{h_{2}q_{2}}^{k}\right|$$

$$\times E\left|\Delta W_{p_{1}}^{k}\Delta W_{q_{1}}^{k}-\nu_{p_{1}q_{1}}^{k}h\right|\left|\Delta W_{p_{2}}^{k}\Delta W_{q_{2}}^{k}-\nu_{p_{2}q_{2}}h\right|\,,$$

where a is the upper bound of the moduli of the function φ and its first and the second derivatives, n and m are the dimensions of the vectors Z and W respectively. Now we note that the random variable ΔW is normally distributed, hence its fourth order moments are of the order h^{2}. Taking into account that the fourth order moments of random functions $|Y_{jp}(t)|$ are finite, we have

$$E\left\{\operatorname{tr}\left[\varphi_{zz}^{k}Y^{k}(\Delta W^{k}\Delta W^{kT}-\nu^{k}h)Y^{kT}\right]\right\}^{2}=o(h)\,.\qquad(5.1.10)$$

Using (5.1.9) and (5.1.10) we get

$$E\left\{\sum_{k=1}^{N}\operatorname{tr}\left[\varphi_{zz}^{k}Y^{k}(\Delta W^{k}\Delta W^{kT}-\nu^{k}h)Y^{kT}\right]\right\}^{2}$$

$$=\sum_{k=1}^{N}E\left\{\operatorname{tr}\left[\varphi_{zz}^{k}Y^{k}(\Delta W^{k}\Delta W^{kT}-\nu^{k}h)Y^{kT}\right]\right\}^{2}$$

$$=No(h)=No(N^{-1}). \tag{5.1.11}$$

This result shows that the means square of the third in (5.1.8) vanishes as $N\to\infty$.

In the same way using (5.1.6) we find

$$\left|\operatorname{tr}\left[(\varphi_{zz\theta}^{k}-\varphi_{zz}^{k})Y^{k}\Delta W^{k}\Delta W^{kT}Y^{kT}\right]\right|$$

$$=\left|\sum_{j,h=1}^{n}\sum_{p,q=1}^{m}\left(\frac{\partial^{2}\varphi_{\theta}^{k}}{\partial z_{j}\partial z_{h}}-\frac{\partial^{2}\varphi^{k}}{\partial z_{j}\partial z_{h}}\right)Y_{jp}^{k}Y_{hq}^{k}\Delta W_{p}^{k}\Delta W_{q}^{k}\right|$$

$$<b\sum_{j,h=1}^{n}\sum_{p,q,r=1}^{m}\left|Y_{jp}^{k}Y_{hq}^{k}\right|\left|\Delta W_{p}^{k}\Delta W_{q}^{k}\Delta W_{r}^{k}\right|.$$

By the independence Y^{k} of ΔW^{k}

$$E\left|\operatorname{tr}\left[(\varphi_{zz\theta}^{k}-\varphi_{zz}^{k})Y^{k}\Delta W^{k}\Delta W^{kT}Y^{kT}\right]\right|$$

$$<b\sum_{j,h=1}^{n}\sum_{p,q,r=1}^{m}E\left|Y_{jp}^{k}Y_{hq}^{k}\right|E\left|\Delta W_{p}^{k}\Delta W_{q}^{k}\Delta W_{r}^{k}\right|.$$

But by the Cauchy–Buniakowski–Schwarz inequality, taking into account that ΔW is normally distributed we have

$$E\left|\Delta W_{p}^{k}\Delta W_{q}^{k}\Delta W_{r}^{k}\right|\leq\sqrt{E(\Delta W_{p}^{k}\Delta W_{q}^{k}\Delta W_{r}^{k})^{2}}=O(h^{3/2})=o(h),$$

$$E\left|\operatorname{tr}\left[(\varphi_{zz\theta}^{k}-\varphi_{zz}^{k})Y^{k}\Delta W^{k}\Delta W^{kT}Y^{kT}\right]\right|=o(h),$$

$$E\left|\sum_{k=1}^{N}\operatorname{tr}\left[(\varphi_{zz\theta}^{k}-\varphi_{zz}^{k})Y^{k}\Delta W^{k}\Delta W^{kT}Y^{kT}\right]\right|=No(h)=No(N^{-1}).$$

So the last sum in (5.1.8) tends to zero in probability as $N\to\infty$.

By virtue of the results obtained formula (5.1.8) may be rewritten as

$$\varphi\left(Z(t_0) + \sum_{k=1}^{N} \Delta Z^k, t\right) - \varphi(Z(t_0), t_0)$$

$$= \sum_{k=1}^{N} \left\{\varphi_t^k + \varphi_z^{kT} X^k + \frac{1}{2}\operatorname{tr}\left[\varphi_{zz}^k Y^k \nu^k Y^{kT}\right]\right\} h$$

$$+ \sum_{k=1}^{N} \varphi_z^{kT} Y^k \Delta W^k + O(h).$$

Passing to the limit as $N \to \infty$, $h = (t - t_0)/N \to 0$ we get

$$\varphi(Z(t), t) = \varphi(Z(t_0), t_0) + \int_{t_0}^{t} \left\{\varphi_t(Z_\tau, \tau) + \varphi_z(Z_\tau, \tau)^T X(\tau)\right.$$

$$\left. + \frac{1}{2}\operatorname{tr}\left[\varphi_{zz}(Z_\tau, \tau)Y(\tau)\nu(\tau)Y(\tau)^T\right]\right\}d\tau + \int_{t_0}^{t} \varphi_z(Z_\tau, \tau)^T Y(\tau)dW(\tau).$$

$$(5.1.12)$$

Since all the derivatives here are bounded, the random function $X(\tau)$ has finite second order moments and the random function $Y(t)$ has finite fourth order moments the both integrals in (5.1.12) exist proving that $U(t) = \varphi(Z(t), t)$ is an Itô process and its stochastic differential is given by formula (5.1.7). ◁

In the special case when the function $\varphi(z, t)$ represents the product of two first components of the vector z, $U(t) = Z_1(t)Z_2(t)$ formula (5.1.7) yields the formula for the stochastic differential of the product of two random processes determined by formula (5.1.4)

$$d(Z_1 Z_2) = Z_1 dZ_2 + Z_2 dZ_1 + \frac{1}{2}\left(Y_1 \nu Y_2^T + Y_2 \nu Y_1^T\right) dt$$

where Y_1 and Y_2 are the first and the second rows of the matrix Y respectively. But for any matrix-rows a and b of the same dimension and for any symmetric square matrix c, $c = c^T$ we have $acb^T = bca^T$. Consequently, $Y_1 \nu Y_2^T = Y_2 \nu Y_1^T$ and we obtain

$$d(Z_1 Z_2) = Z_1\, dZ_2 + Z_2\, dZ_1 + Y_1 \nu Y_2^T\, dt. \qquad (5.1.13)$$

$\triangleright$ Formula (5.1.7) may also be written for a vector function $\varphi(z,t) = [\varphi_1(z,t) \ldots \varphi_n(z,t)]^T$. For this purpose we denote by $\varphi_{zz} : A$ the matrix column whose elements are the traces of the products of the matrices of the second derivatives with respect to the components t of the corresponding elements $\varphi_p(z,t)$ of the matrix-column $\varphi_2(z,t)$ by the matrix A:

$$\varphi_{zz} : A = [\operatorname{tr}(\varphi_{1zz}A) \ldots \operatorname{tr}(\varphi_{nzz}A)]^T. \qquad (5.1.14)$$

Using notation (5.1.14) we obtain from (5.1.7) the following formula for the stochastic differential of a vector random function $U(t) = \varphi(Z(t), t)$:

$$dU = \left\{ \varphi_t(Z,t) + \varphi_z(Z,t)^T X + \frac{1}{2}\varphi_{zz}(Z,t) : Y\nu Y^T \right\} dt +$$

$$+\varphi_z(Z,t)^T Y \, dW, \qquad (5.1.15)$$

where $\varphi_z(z,t) = (\partial/\partial z)\varphi(z,t)^T$, $\varphi_{zz} = (\partial/\partial z)(\partial^T/\partial z)\varphi(z,t)$. $\triangleleft$

So the following theorem is valid.

Theorem 5.1.2. *If the components of the vector function $U(t) = \varphi(Z(t),t)$ satisfy conditions of* Theorem 5.1.1 *then the stochastic differential is given by formula* (5.1.15).

The stochastic Itô differentials of some typical composite functions of the standard Wiener process and of the multi-dimensional Wiener process are given in Appendix 1.

E x a m p l e 5.1.1. The stochastic Itô differential of the square of the standard Wiener process $W^2(t)$ in accordance with (5.1.7) is equal to $dW^2(t) = dt + 2W(t)\,dW(t)$, as in this case $X(t) = 0$, $Y(t) = 1$, $Z(t) = W(t)$, $\varphi(Z,t) = W^2$. Integrating this formula from 0 to t and taking into account that $W(0) = 0$ we find

$$W^2(t) = t + 2\int_0^t W(\tau)\,dW(\tau), \quad \int_0^t W(\tau)\,dW(\tau) = \frac{1}{2}[W^2(t) - t].$$

Thus the Itô integral of the standard Wiener process with respect to this process cannot be evaluated with the aid of the usual formula of integration of powers of a variable.

5.1.3. Generalized Itô Formulae

Theorem 5.1.3. *Let $\varphi(z,t)$ be a scalar real function, $Z(t)$ be a general Itô process given by (5.1.1). If the function $\varphi = \varphi(z,t)$ is continuous and bounded with its first and second derivatives, its second derivatives satisfy Lipschitz condition (5.1.6) and the random function $Y_1(t)$ has the finite absolute fourth order moments then $U(t) = \varphi(Z(t),t)$ is the Itô process and its stochastic differential is given by:*

$$dU = \left\{ \varphi_t(Z,t) + \varphi_z(Z,t)^T X + \frac{1}{2}\operatorname{tr}\left[\varphi_{zz}(Z,t)Y_1 \nu Y_1^T\right]\right\} dt$$

$$+ \int_{R_0^q} \left[\varphi(Z+Y_2,t)^T - \varphi(Z,t) - \varphi_z(Z,t)^T Y_2\right] \mu_P(dt,du)$$

$$+\varphi_z(Z,t)^T Y_1 dW + \int_{R_0^q}\left[\varphi(Z+Y_2,t) - \varphi(Z,t)\right]P^0(dt,du). \qquad (5.1.16)$$

▷ At first we shall consider the case where the integration with respect to u in Eq. (5.1.1) is extended over the ball $0 < |u| < R$ with the origin pricked out. Replacing in this case the integrals in Eq. (5.1.1) by respective integral sums we write

$$I = \varphi\left(Z(t_0) + \sum_{k=1}^{N}(\Delta Z_1^k + \Delta Z_2^k), t\right) - \varphi(Z(t_0),t_0)$$

$$= \sum_{k=1}^{N}\left[\varphi(Z^k + \Delta Z_1^k + \Delta Z_2^k, t_k) - \varphi(Z^k + \Delta Z_2^k, t_{k-1})\right]$$

$$+ \sum_{k=1}^{N}\left[\varphi(Z^k + \Delta Z_2^k, t_{k-1}) - \varphi(Z^k, t_{k-1})\right], \qquad (5.1.17)$$

where $\Delta Z_1^k = X^k h + Y_1^k \Delta W^k$, $\Delta Z_2^k = \sum\limits_{s=1}^{M_1} Y_2^{ks} P^{ks}$, $Y_2^{ks} = Y_2(t_{k-1}, u_s)$; $P^{ks} = P^0((t_{k-1}, t_k], A_s)$. The first sum in (5.1.17) is transformed exactly in the same way as in Subsection 5.1.2. To transform the second sum we consider the random variable

$$V_1 = \varphi\left(z + \sum_{s=1}^{M_1} Y_2^{ks} P^{ks}, t\right) - \varphi(z,t) - \sum_{s=1}^{M_1}\left[\varphi(z + Y_2^{ks}, t)\right.$$

$$\left. -\varphi(z,t) - \varphi_z(z,t)^T Y_2^{ks}\right]\mu_P^{ks} - \sum_{s=1}^{M_1}\left[\varphi(z + Y_2^{ks}, t) - \varphi(z,t)\right]P^{ks},$$

where $\mu_P^{ks} = \mu_P((t_{k-1}, t_k], A_s)$. The value V_1 corresponding to $P((t_{k-1}, t_k],$ $A_s) = 0 \; \forall s$ and therefore to $P^{ks} = -\mu_P^{ks} \; \forall s$ is equal to

$$V_1 = \varphi\left(z + \sum_{s=1}^{M_1} Y_2^{ks} \mu_P^{ks}, t\right) - \varphi(z,t) - \sum_{s=1}^{M_1}\left[\varphi(z + Y_2^{ks}, t)\right.$$

$$-\varphi(z,t) - \varphi_z(z,t)^T Y_2^{ks}\right]\mu_P^{ks} + \sum_{s=1}^{M_1}\left[\varphi(z + Y_2^{ks}, t) - \varphi(z,t)\right]\mu_P^{ks}$$

$$= \varphi\left(z - \sum_{s=1}^{M_1} Y_2^{ks}\mu_P^{ks}, t\right) - \varphi(z,t) + \sum_{s=1}^{M_1}\varphi_z(z,t)^T Y_2^{ks}\mu_P^{ks} = o(\mu_P^{ks}) = o(h)$$

and its probability is equal to $\prod_{s=1}^{M_1} e^{-\mu_P^{ks}} = 1 - \sum_{s=1}^{M_1}\mu_P^{ks} + o(h)$ by virtue of the

independence of the random variables P^{ks}. The value V_1 corresponding to one of the variables $P((t_{k-1}, t_k], A_s)$ say $P((t_{k-1}, t_k], A_\sigma)$, equal to 1 and others zero, $P^{k\sigma} = 1 - \mu_P^{k\sigma}$, $P^{ks} = -\mu_P^{ks}$ at $s \neq \sigma$, is equal to

$$V_1 = \varphi\left(z + Y_2^{k\sigma} - \sum_{s=1}^{M_1} Y_2^{ks}\mu_P^{ks}, t\right) - \varphi(z,t) - \sum_{s=1}^{M_1}\left[\varphi(z + Y_2^{ks}, t)\right.$$

$$-\varphi(z,t) - \varphi_z(z,t)^T Y_2^{ks}\right]\mu_P^{ks} - \left[\varphi(z + Y_2^{k\sigma}, t) - \varphi(z,t)\right]$$

$$+ \sum_{s=1}^{M_1}\left[\varphi(z + Y_2^{ks}, t) - \varphi(z,t)\right]\mu_P^{ks} = \varphi\left(z + Y_2^{k\sigma} - \sum_{s=1}^{M_1} Y_2^{ks}\mu_P^{ks}, t\right)$$

$$-\varphi(z + Y_2^{k\sigma}, t) + \sum_{s=1}^{M_1}\varphi_z(z,t)^T Y_2^{ks}\mu_P^{ks} = o(\mu_P^{ks}) = o(h).$$

Its probability is equal to $\mu_P^{k\sigma} \prod_{s=1}^{M_1} e^{-\mu_P^{ks}} = o(\mu_P^{k\sigma}) = o(h)$. The values of V_1

corresponding to all other values P^{ks} are bounded and their probabilities are equal

to $\prod_{s=1}^{M_1} \frac{(\mu_P^{ks})^{m_s}}{m_s!} e^{-\mu_P^{ks}}$, $m_1 + \cdots + m_{M_1} > 1$. So we have the following estimate

for the expectation of $|V_1|$:

$$E\,|V_1| < o(h) + C \sum_{m_1 + \cdots + m_{M_1} > 1} \prod_{s=1}^{M_1} \frac{(\mu_P^{ks})^{m_s}}{m_s!} e^{-\mu_P^{ks}}$$

$$= o(h) + C\left(1 - \exp\left\{-\sum_{s=1}^{M_1}\mu_P^{ks}\right\} - \sum_{s=1}^{M_1}\mu_P^{ks}\exp\left\{-\sum_{s=1}^{M_1}\mu_P^{ks}\right\}\right) = o(h)\,.$$

Hence

$$E\left|\sum_{k=1}^{N}\left[\varphi(Z^k + \Delta Z_2^k, t_{k-1}) - \varphi(Z^k, t_{k-1})\right]\right.$$

$$-\sum_{k=1}^{N}\left[\varphi(Z^k + Y_2^{ks}, t_{k-1}) - \varphi(Z^k, t_{k-1}) - \varphi_z(Z^k, t_{k-1})^T Y_2^{ks}\right]\mu_P^{ks}$$

$$\left.-\sum_{k=1}^{N}\left[\varphi(Z^k + Y_2^{ks}, t_{k-1}) - \varphi(Z^k, t_{k-1})\right]P^{ks}\right| < No(h) = No(N^{-1})\,.$$

This proves together with the estimates of Subsection 5.1.2 that (5.1.17), may be rewritten as

$$I = \varphi\left(Z(t_0) + \sum_{k=1}^{N}(\Delta Z_1^k + \Delta Z_2^k), t\right) - \varphi(Z(t_0), t_0)$$

$$= \sum_{k=1}^{N}\left\{\varphi_t(Z^k + \Delta Z_2^k, t_{k-1}) + \varphi_z(Z^k + \Delta Z_2^k, t_{k-1})^T X(t_{k-1})\right.$$

$$+ \frac{1}{2}\operatorname{tr}\left[\varphi_{zz}(Z^k + \Delta Z_2^k, t_{k-1})Y_1(t_{k-1})\nu(t_{k-1})Y_1(t_{k-1})^T\right]\right\}h$$

$$+ \sum_{k=1}^{N}\varphi_z(Z^k + \Delta Z_2^k, t_{k-1})^T Y_1(t_{k-1})\Delta W^k$$

$$+ \sum_{k=1}^{N}\sum_{s=1}^{M_1}\left[\varphi(Z^k + Y_2^{ks}, t_{k-1}) - \varphi(Z^k, t_{k-1}) - \varphi_z(Z^k, t_{k-1})^T Y_2^{ks}\right]\mu_P^{ks}$$

$$+ \sum_{k=1}^{N}\sum_{s=1}^{M_1}\left[\varphi(Z^k + Y_2^{ks}, t_{k-1}) - \varphi(Z^k, t_{k-1})\right]P^{ks} + o(h)\,. \qquad (5.1.18)$$

Finally we notice that due to the boundedness of the function $\varphi(z, t)$ and its first and second derivatives and to (5.1.6) the first factors in the first two sums in (5.1.18) differ from their values at $\Delta Z_2^k = 0$ by infinitesimals $o(h)$. Therefore ΔZ_2^k in these sums may be omitted giving

$$\varphi\left(Z(t_0) + \sum_{k=1}^{N}(\Delta Z_1^k + \Delta Z_2^k), t\right) - \varphi(Z(t_0), t_0)$$

$$= \sum_{k=1}^{N} \left\{ \varphi_t(Z^k, t_{k-1}) + \varphi_z(Z^k, t_{k-1})^T X(t_{k-1}) \right.$$

$$\left. + \frac{1}{2} \operatorname{tr} \left[\varphi_{zz}(Z^k, t_{k-1}) Y_1(t_{k-1}) \nu(t_{k-1}) Y_1(t_{k-1})^T \right] \right\} h$$

$$+ \sum_{k=1}^{N} \varphi_z(Z^k, t_{k-1})^T Y_1(t_{k-1}) \Delta W^k + \sum_{k=1}^{N} \sum_{s=1}^{M_1} \left[\varphi(Z^k + Y_2^{ks}, t_{k-1}) \right.$$

$$\left. - \varphi(Z^k, t_{k-1}) - \varphi_z(Z^k, t_{k-1})^T Y_2^{ks} \right] \mu_P^{ks}$$

$$+ \sum_{k=1}^{N} \sum_{s=1}^{M_1} \left[\varphi(Z^k + Y_2^{ks}, t_{k-1}) - \varphi(Z^k, t_{k-1}) \right] P^{ks} + O(h) .$$

Passing to the limit when h and maximal diameter of the sets A_s tend to zero we find

$$\varphi(Z(t), t) = \varphi(Z(t_0), t_0) + \int_{t_0}^{t} \left\{ \varphi_t(Z_\tau, \tau) + \varphi_z(Z_\tau, \tau)^T X(\tau) \right.$$

$$\left. + \frac{1}{2} \operatorname{tr} \left[\varphi_{zz}(Z_\tau, \tau) Y_1(\tau) \nu(\tau) Y_1(\tau)^T \right] \right\} d\tau + \int_{t_0}^{t} \varphi_z(Z_\tau, \tau)^T Y_1(\tau) dW(\tau)$$

$$+ \int_{t_0}^{t} \int_{0 < |u| < R} [\varphi(Z_\tau + Y_2(\tau, u), \tau) - \varphi(Z_\tau, \tau) - \varphi_z(Z_\tau, \tau)^T Y_2(\tau, u)] \mu_P(d\tau, du)$$

$$+ \int_{t_0}^{t} \int_{0 < |u| < R} [\varphi(Z_\tau + Y_2(\tau, u), \tau) - \varphi(Z_\tau, \tau)] P^0(d\tau, du) ,$$

where $Z_\tau = Z(\tau)$. Thus the theorem is proved for the case of integration with respect to u over the set $0 < |u| < R$. So passing to the limit as $R \to \infty$ we get

$$\varphi(Z(t), t) = \varphi(Z(t_0), t_0) + \int_{t_0}^{t} \left\{ \varphi_t(Z_\tau, \tau) + \varphi_z(Z_\tau, \tau)^T X(\tau) \right.$$

$$\left. + \frac{1}{2} \operatorname{tr} \left[\varphi_{zz}(Z_\tau, \tau) Y_1(\tau) \nu(\tau) Y_1(\tau)^T \right] \right\} d\tau$$

$$+ \int\limits_{t_0}^{t} \varphi_z(Z_\tau, \tau)^T Y_1(\tau) dW(\tau) + \int\limits_{t_0}^{t} \int\limits_{R_0^q} [\varphi(Z_\tau + Y_2(\tau, u), \tau) - \varphi(Z_\tau, \tau)$$

$$-\varphi_z(Z_\tau, \tau)^T Y_2(\tau, u)] \mu_P(d\tau, du)$$

$$+ \int\limits_{t_0}^{t} \int\limits_{R_0^q} [\varphi(Z_\tau + Y_2(\tau, u), \tau) - \varphi(Z_\tau, \tau)] P^0(d\tau, du). \qquad (5.1.19)$$

This accomplishes the proof. ◁

In the special case of a centered Poisson process $W(t) = P(t) - EP(t)$, $EP(t) = \mu = \int\limits_0^t \nu(\tau) d\tau$, then Eq. (5.1.1) takes the form

$$dZ = X\,dt + Y\,dW. \qquad (5.1.20)$$

So we get the following statment.

Corollary 5.1.3. *Under the conditions of* Theorem 5.1.3 *and for a centred Poisson process in* Eq. (5.1.1) *the following Itô formula is valid:*

$$dU = \{\varphi_t(Z, t) + \varphi_z(Z, t)^T (X - Y\nu) +$$

$$+ [\varphi(Z + Y, t) - \varphi(Z, t)]\nu\} dt + [\varphi(Z + Y, t) - \varphi(Z, t)] dW. \qquad (5.1.21)$$

This formula of differentiation of a composite function is quite unlike the usual formula of mathematical analysis. And only in the case of the function $\varphi(z, t)$ linear relative to z, it does coincide with the usual formula. Formula (5.1.21) is also valid for a vector random process $U(t) = \varphi(Z(t), t)$.

E x a m p l e 5.1.2. The stochastic Itô differential of the square of the Poisson process $U(t) = P^2(t)$ according to (5.1.21) is determined by the formula

$$dU = [(P + 1)^2 - P^2](\nu\,dt + dW) = (2P + 1)\,dP,$$

since $Z(t) = P(t)$, $\varphi(Z, t) = Z^2$, $dZ = dP = \nu\,dt + dW$ in this case and consequently, $X(t) = \nu(t)$, $Y(t) = 1$. Thus in the case of a centered Poisson process $W(t)$ the usual formula for the differentials of powers of a variable is not applicable, too.

Now we pass to the case where a process with the independent increments in (5.1.4) is determined by formula (Subsection 4.6.3)

$$W(t) = W_0(t) + \int\limits_{R_0^q} c(x) P^0(t, dx),\qquad(5.1.22)$$

where $W_0(t)$ being a Wiener process. Using formula (5.1.16) we get for scalar function $U(t) = \varphi(Z(t), t)$ the following expression:

$$dU = \{\varphi_t(Z, t) + \varphi_z(Z, t)^T X(t)$$

$$+\frac{1}{2}\mathrm{tr}\,[\,\varphi_{zz}(Z, t) Y(t) \nu_0(t) Y(t)^T\,] + \int\limits_{R_0^q} [\,\varphi(Z + Y(t) c(x), t) - \varphi(Z, t)$$

$$-\varphi_z(Z, t)^T Y(t) c(x)\,]\nu_P(t, x)\, dx\} \, dt + \varphi_z(Z, t)^T Y(t)\, dW_0$$

$$+ \int\limits_{R_0^q} [\,\varphi(Z + Y(t) c(x), t) - \varphi(Z, t)\,] \, P^0(dt, dx)\,.\qquad(5.1.23)$$

For vector function $U(t) = \varphi(Z(t), t)$ we have in a similar way

$$dU = \{\varphi_t(Z, t) + \varphi_z(Z, t)^T X(t)$$

$$+\frac{1}{2}\varphi_{zz}(Z, t)\,:\,Y(t) \nu_0(t) Y(t)^T + \int\limits_{R_0^q} [\,\varphi(Z + Y(t) c(x), t) - \varphi(Z, t)$$

$$-\varphi_z(Z, t)^T Y(t) c(x)\,]\nu_P(t, x)\, dx\} \, dt + \varphi_z(Z, t)^T Y(t)\, dW_0$$

$$+ \int\limits_{R_0^q} [\,\varphi(Z + Y(t) c(x), t) - \varphi(Z, t)\,] \, P^0(dt, dx).\qquad(5.1.24)$$

So the following statement are valid.

Theorem 5.1.4. *Under the conditions of* Theorem 5.1.3 *and for the process* $W(t)$ *with the independet increments* (5.1.22) *the scalar generalized Itô formula is given by* (5.1.23).

Theorem 5.1.5. *If the components of vector function* $U(t) = \varphi(Z(t), t)$ *satisfy conditions of* Theorem 5.1.3 *and for a process* $W(t)$ *with the independent increments* (5.1.22) *the vector generalized Itô formula is expressed by* (5.1.24).

5.1.4. Other Types of Stochastic Differentials

In conformity with the variety of possible interpretations of the stochastic integral in Eq. (5.1.1) one may introduce various types of stochastic differentials. So if the stochastic integral in Eq. (5.1.1) represents the θ-integral then the formula

$$d_\theta Z = X\,dt + Y\,d_\theta W \qquad (5.1.25)$$

similar to (5.1.5) determines the stochastic θ-differential of the process $Z(t)$.

If the same process $Z(t)$ may be represented by formula (5.1.4) for two different interpretations of the stochastic integral,

$$Z(t) = Z_0 + \int_{t_0}^{t} X_{\theta_1}(\tau)\,d\tau + \int_{t_0}^{t} Y_{\theta_1}(\tau)d_{\theta_1}W(\tau)$$

$$= Z_0 + \int_{t_0}^{t} X_{\theta_2}(\tau)\,d\tau + \int_{t_0}^{t} Y_{\theta_2}(\tau)d_{\theta_2}W(\tau), \qquad (5.1.26)$$

then the corresponding stochastic differentials $d_{\theta_1}Z$ and $d_{\theta_2}Z$ are equal to each other, $d_{\theta_1}Z = d_{\theta_2}Z$, or

$$X_{\theta_1}\,dt + Y_{\theta_1}\,d_{\theta_1}W = X_{\theta_2}\,dt + Y_{\theta_2}\,d_{\theta_2}W\,. \qquad (5.1.27)$$

Thus *the various stochastic differentials of the same random process differ only by the form of representation of this process by formula* (5.1.4).

Formulae (5.1.4) and (5.1.7) permit the relations between various types of the stochastic differentials in the special case (where the scalar random process $U(t)$ represents a function of a Wiener process) $U(t) = \varphi(W(t), t)$ to be established latter in accordance with (5.1.7)

So the following statements is true.

Theorem 5.1.6. *The stochastic Itô differential of a sclar composite function* $U(t) = \varphi(W(t), t)$ *where* $W(t)$ *being the Wiener process is difined by*

$$dU = \left\{ \varphi_t(W, t) + \frac{1}{2}\mathrm{tr}\left[\varphi_{ww}(W, t)\nu(t) \right] \right\} dt + \varphi_w(W, t)^T dW\,. \qquad (5.1.28)$$

$\triangleright$ In order to transform the Itô differential dU to the form of the θ-differential we rewrite this formula in accordance with the definition of the θ-integral in Subsection 4.6.9 in the form

$$dU = \left\{ \varphi_t(W,t) + \frac{1}{2}\operatorname{tr}\left[\varphi_{ww}(W,t)\nu(t)\right] \right\} dt$$

$$+\left[(1-\theta)\varphi_w(W(t),t)^T + \theta\varphi_w(W(t+dt),t+dt)^T\right]dW$$

$$-\theta\left[\varphi_w(W(t+dt),t+dt)^T - \varphi_w(W(t),t)^T\right]dW. \qquad (5.1.29)$$

It remains to represent the components of the difference $\varphi_w(W(t+dt),t+dt)$ $-\varphi_w(W(t),t)$ in the last item by formula (5.1.7). Here we have to retain only the term containing dW as the other terms will give the infinitesimals of higher orders in the expression of dU. As a result the components of the difference $d\varphi_w(W,t)$ $= \varphi_w(W(t+dt),t+dt) - \varphi_w(W(t),t)$ will be expressed by the formula

$$\frac{\partial}{\partial w_k}\varphi_w(W(t+dt),t+dt) - \frac{\partial}{\partial w_k}\varphi_w(W(t),t)$$

$$= \sum_{h=1}^{m} \frac{\partial^2}{\partial w_k\,\partial w_h}\varphi(W(t),t)\,dW_h.$$

Consequently,

$$\left[\varphi_w(W(t+dt),t+dt)^T - \varphi_w(W(t),t)^T\right]dW$$

$$= \sum_{k=1}^{m}\left[\frac{\partial}{\partial w_k}\varphi(W(t+dt),t+dt) - \frac{\partial}{\partial w_k}\varphi(W(t),t)\right]dW_k$$

$$= \sum_{k,h=1}^{m}\frac{\partial^2}{\partial w_k\,\partial w_h}\varphi(W(t),t)\,dW_k\,dW_h = \operatorname{tr}\left[\varphi_{ww}(W,t)\,dW\,dW^T\right].$$

In the same way as in Subsection 5.1.2 we make sure that only the expectation of the random variable $dW\,dW^T$ should be retained as the mean square value of the random part of $dW\,dW^T$ which is an infinitesimal of higher order than dW. The expectation of $dW\,dW^T$ (i.e. the covariance matrix of the random vector dW) is equal to $\nu(t)\,dt$. As a result the formula for dU takes the form

$$dU = \left\{ \varphi_t(W,t) + \left(\frac{1}{2}-\theta\right)\operatorname{tr}\left[\varphi_{ww}(W,t)\nu(t)\right] \right\} dt$$

$$+\left[(1-\theta)\varphi_w(W(t),t)^T + \theta\varphi_w(W(t+dt),t+dt)^T\right]dW$$

or

$$d_\theta U = \left\{ \varphi_t(W, t) + \left(\frac{1}{2} - \theta \right) \operatorname{tr} \left[\varphi_{ww}(W, t)\nu(t) \right] \right\} dt + \varphi_w(W, t)^T d_\theta W. \; \triangleleft$$

$$(5.1.30)$$

Thus we get the statement.

Theorem 5.1.7. *The stochastic θ-differential of a scalar composite function $U(t) = \varphi(W(t), t)$ where $W(t)$ being the Wiener process is given by formula* (5.1.30).

Corollary 5.1.7. *Under the conditions of* Theorem 5.1.7 *stochastic Strotonovich symmetrized differential ($\theta = 1/2$) is expressed by formula*

$$d_{1/2}U = \varphi_t(W, t)\, dt + \varphi_w(W, t)^T d_{1/2}W\,. \qquad (5.1.31)$$

R e m a r k 1. The usual formula of the differentiation of a composite function is valid for the symmetrized differentials of the functions of a Wiener process. This is some advantage of the symmetrized stochastic integrals and the corresponding symmetrized differentials over the other types of the stochastic integrals and the differentials.

R e m a r k 2. But we do not manage to find a formula for the θ-differentials of functions of more complex processes with the independent increments, and in particular, of functions of a Poisson process, although as we have seen above it does not offer any difficulties for the Itô differential. This is one of the numerous advantages of the stochastic Itô integrals and the differentials over all the other types of the stochastic integrals and differentials. Further we shall meet with other, considerably more essential advantages.

$\triangleright$ From formula (5.1.30) for the θ-differential of the function $\varphi(W, t)$ it follows

$$\varphi(W, t) = \varphi(W(t_0), t_0) + \int_{t_0}^{t} \left\{ \varphi_t(W_\tau, \tau) + \left(\frac{1}{2} - \theta \right) \right.$$

$$\left. \times \operatorname{tr} \left[\varphi_{ww}(W_\tau, \tau)\nu(\tau) \right] \right\} d\tau + \int_{t_0}^{t} \varphi_w(W_\tau, \tau)^T d_\theta W(\tau)\,. \qquad (5.1.32)$$

Comparing formula (5.1.32) with the Itô integral formula containing the Itô integral,

$$\varphi(W, t) = \varphi(W(t_0), t_0) + \int_{t_0}^{t} \left\{ \varphi_t(W_\tau, \tau) + \frac{1}{2} \operatorname{tr} \left[\varphi_{ww}(W_\tau, \tau)\nu(\tau) \right] \right\} d\tau$$

$$+ \int\limits_{t_0}^{t} \varphi_w(W_\tau, \tau)^T \, dW(\tau), \tag{5.1.33}$$

we obtain the relation between the θ-integral and the corresponding Itô integral of the function $\varphi_w(W_\tau, \tau)^T$

$$\int\limits_{t_0}^{t} \varphi_w(W_\tau, \tau)^T d_\theta W(\tau)$$

$$= \int\limits_{t_0}^{t} \varphi_w(W_\tau, \tau)^T dW(\tau) + \theta \int\limits_{t_0}^{t} \mathrm{tr}\left[\varphi_{ww}(W_\tau, \tau)\nu(\tau)\right] d\tau. \tag{5.1.34}$$

As $\varphi_w(w, t)$ may be any function satisfying the conditions of existence of both integrals in the right-hand side of this formula $\varphi_w(w, t)$ may be replaced in this formula by any differentiable function $\psi(w, t)$ satisfying these conditions. ◁

Thus we proved the following statement.

Theorem 5.1.8. *For any matrix-row $\psi(w, t)$ the following relation between the θ-integral and the Itô integral exists:*

$$\int\limits_{t_0}^{t} \psi(W_\tau, \tau)d_\theta W(\tau)$$

$$= \int\limits_{t_0}^{t} \psi(W_\tau, \tau)\, dW(\tau) + \theta \int\limits_{t_0}^{t} \mathrm{tr}\left[\frac{\partial\psi(W_\tau, \tau)}{\partial w}\nu(\tau)\right] d\tau. \tag{5.1.35}$$

E x a m p l e 5.1.6. For the function $\psi(W(t), t) = W^n(t)$ of the standard Wiener process $W(t)$ we get

$$\int\limits_{0}^{t} W^n(\tau)d_\theta W(\tau) = \int\limits_{0}^{t} W^n(\tau)dW(\tau) + \theta n \int\limits_{0}^{t} W^{n-1}(\tau)d\tau. \tag{I}$$

But from the formula for the Itô differential of the function $W^{n+1}(t)$

$$dW^{n+1}(t) = \frac{1}{2}(n+1)nW^{n-1}(t)dt + (n+1)W^n(t)dW(t), \tag{II}$$

it follows that

$$\int\limits_0^T W^n(\tau)dW(\tau) = \frac{W^{n+1}(t)}{n+1} - \frac{n}{2}\int\limits_0^T W^{n-1}(\tau)d\tau \qquad \text{(III)}$$

(remember that $W(0) = 0$). Substituting expression (III) into formula (II) we obtain

$$\int\limits_0^T W^n(\tau)d_\theta W(\tau) = \frac{W^{n+1}(T)}{n+1} + \left(\theta - \frac{1}{2}\right)n\int\limits_0^T W^{n-1}(\tau)d\tau. \qquad \text{(IV)}$$

At $\theta = 1/2$ this formula coincides with the usual formula of integration of a power of a variable.

5.2. Stochastic Differential Equations

5.2.1. Definitions

The differential equation

$$\dot{X} = a(X,t) + b(X,t)V \qquad (5.2.1)$$

is called *a stochastic differential equation* if the random function (generalized) $V = V(t)$ represents a white noise in the strict sense (Subsection 4.6.4). Let X_0 be a random vector of the same dimension as the random function $X(t)$. Eq. (5.2.1) with the initial condition $X(t_0) = X_0$ determines the random process $X(t)$.

In order to give an exact sense to Eq. (5.2.1) and to the above statement we shall integrate formally Eq. (5.2.1) in the limits from t_0 to t with the initial condition $X(t_0) = X_0$. As a result we obtain

$$X(t) = X_0 + \int\limits_{t_0}^t a(X(\tau),\tau)d\tau + \int\limits_{t_0}^t b(X(\tau),\tau)V(\tau)d\tau,$$

where the first integral represents a m.s. integral. Introducing the process with independent increments $W(t)$ whose derivative is a white noise $V(t)$ we rewrite the previous equation in the form

$$X(t) = X_0 + \int\limits_{t_0}^t a(X(\tau),\tau)d\tau + \int\limits_{t_0}^t b(X(\tau),\tau)dW(\tau). \qquad (5.2.2)$$

Eq. (5.2.2) has an exact sense. Eq. (5.2.1) or the equivalent equation

$$dX = a(X,t)dt + b(X,t)dW \qquad (5.2.3)$$

with the initial condition $X(t_0) = X_0$ represents a concise form of Eq. (5.2.2).

Eq. (5.2.2) in which the second integral represents a stochastic Itô integral is called *a stochastic Itô integral equation* and the corresponding differential Eq. (5.2.1) or (5.2.3) is called *a stochastic Itô differential equation*.

A random process $X(t)$ satisfying Eq. (5.2.2) in which the integrals represent the a m.s. limits of the corresponding integral sums is called *a mean square* or shortly, a *m.s. solution* of stochastic integral Eq. (5.2.2) and of the corresponding stochastic differential Eq. (5.2.1) or Eq. (5.2.3) with the initial condition $X(t_0) = X_0$.

If the integrals in Eq. (5.2.2) exist for every realization of the processes $W(t)$ and $X(t)$ and equality in (5.2.2) holds for every realization then the random process $X(t)$ is called *a solution in the realizations* of Eq. (5.2.2) and of the corresponding stochastic differential Eq. (5.2.1) or Eq. (5.2.3) with the initial condition $X(t_0) = X_0$.

Certainly the above definitions include as a special case linear stochastic differential equations where the function $a(x,t)$ is linear relative to x, and $b(x,t)$ is independent of x. For nonlinear stochastic differential equations we have to restrict ourselves to the case of a white noise in the strict sense $V(t)$, i.e. to the case where the process $W(t)$ is a process with independent increments (Subsection 4.6.3).

A more general stochastic Itô differential equation in which a and b depend on the process $W = W(t)$ is reduced to the equation of (5.2.3) form:

$$dX = a(X,W,t)dt + b(X,W,t)dW . \qquad (5.2.4)$$

In this case putting $Y(t) = W(t)$ we reduce equation (5.2.4) to the set of two equatons:

$$dX = a(X,Y,t)dt + b(X,Y,t)dW , \quad dY = dW . \qquad (5.2.5)$$

While introducing a composite vector process $\left[X(t)^T Y(t)^T \right]^T$ we get from Eq. (5.2.5) a stochastic Itô differential equation of Eq. (5.2.3) form

$$d\begin{bmatrix} X \\ Y \end{bmatrix} = \begin{bmatrix} a(X,Y,t) \\ 0 \end{bmatrix} + \begin{bmatrix} b(X,Y,t) \\ I \end{bmatrix} dW . \qquad (5.2.6)$$

Eq. (5.2.6) with the initial conditions $X(t_0) = X_0$, $Y(t_0) = Y_0$ (here Y_0 is a random variable independent of X_0 whose distribution is the one-dimensional distribution of the process $W(t)$ at $t = t_0$) is equivalent to the former stochastic differential equation with the initial condition $X(t_0) = X_0$.

5.2.2. Stochastic Differential Equation Determines Markov Process

Theorem 5.2.1. *Stochastic Itô differential* Eq. (5.2.1) *or* Eq. (5.2.3) *with the initial condition* $X(t_0) = X_0$, *where* X_0 *is a random variable independent of the future values of the white noise* $V(s)$, $s > t_0$, *(the future increments* $W(s) - W(t)$, $s > t \geq t_0$, *of the process* W) *determines the Markov random process.*

    ▷ In order to understand this it is sufficient to notice that the value of the process $X(t)$ determined by Eq. (5.2.1) at any t is completely and uniquely determined by its value $X(\tau)$ at some instant $\tau \in [t_0, t)$ and by the values of the white noise V in the interval $[\tau, t)$. In consequence of this and by virtue of the independence of the initial value $X(t_0) = X_0$ of the future values of the white noise V the aggregate of the random variables $X_{\tau_1}, \ldots, X_{\tau_N}, X_\tau$ at any N, $\tau_1 < \cdots < \tau_N < \tau$, $\tau_1 \geq t_0$ is independent of the aggregate of the values of the white noise V in the interval $[\tau, t)$ (more exactly of the aggregate of the increments of the process W on any nonintersecting subintervals of the interval $[\tau, t)$). Hence the conditional distribution of the value of the process $X(t)$ at any instant $t > \tau$ relative to its values $X_{\tau_1}, \ldots, X_{\tau_N}, X_\tau$ at the inatants $\tau_1, \ldots, \tau_N, \tau$ depends only on X_τ and does not depend on $X_{\tau_1}, \ldots, X_{\tau_N}$. It means that $X(t)$ is a Markov random process (Subsection 2.7.4). ◁

5.2.3. Change of Variables in Stochastic Differential Equation

While solving problems connected with stochastic differential equations as well as problems with ordinary differential equations it is often expedient to simplify the equations by a suitable change of variables. But in the general case the change of variables in stochastic differential equations differs from the change of variables in the ordinary differential equations owing to the necessity of using the Itô formula or generalized Itô formula instead of the ordinary formula for differentiation of a composite function.

Suppose that it is necessary to fulfill the invertible change of variables $Y = \varphi(X, t)$ in stochastic differential Eq. (5.2.1). For this purpose it is necessary in the general case of a nonlinear function φ to find the

Itô stochastic differential of the process $Y(t)$ using Itô formulae (Subsections 5.1.2 and 5.1.3) and then to replace X in the formula thus obtained by its expression from the equation $Y = \varphi(X, t)$. Certainly this equation must have the unique solution $X = \varphi^{-1}(Y, t)$. Otherwise the change of variables gives nothing. As a result a stochastic differential equation for the process $Y(t)$ is obtained.

In the special case of a linear change of variables, the function φ being linear in X, the change of variables may be performed in the same way as in ordinary differential equations.

E x a m p l e 5.2.1. To integrate the linear stochastic differential equation

$$dX = (aX + a_0)dt + bdW , \tag{I}$$

we make the change of variables $X = u(t, t_0)Y$ where $u(t, t_0)$ is the solution of the homogeneous equation $\dot{u} = au$ with the initial condition $u(t_0, t_0) = I$. As the matrix $u(t, t_0)$ is invertible (Subsection 1.2.1) it follows from $X = u(t, t_0)Y$ that $Y = u(t, t_0)^{-1}X$. Differentiating this formula we obtain

$$dY = du(t, t_0)^{-1}X + u(t, t_0)^{-1}dX ,$$

or taking into account that $du(t, t_0)^{-1} = -u(t, t_0)^{-1}adt$

$$dY = -u(t, t_0)^{-1}aX\,dt + u(t, t_0)^{-1}(aX + a_0)dt + u(t, t_0)^{-1}bdW$$

$$= u(t, t_0)^{-1}a_0\,dt + u(t, t_0)^{-1}bdW .$$

The right-hand side of this formula does not depend on Y. Therefore integrating the obtained equation we find

$$Y = Y_0 + \int_{t_0}^{t} u(\tau, t_0)^{-1}a_0(\tau)d\tau + \int_{t_0}^{t} u(t, t_0)^{-1}b(\tau)dW(\tau) . \tag{II}$$

Thus formula (II) is valid for any processes with independent increments $W(t)$ including such processes which do not have second-order moments, for instance, for the Cauchy process.

E x a m p l e 5.2.2. Let us consider the scalar equation

$$dX = (aX + a_0)dt + (bX + b_0)dW , \tag{I}$$

where $W(t)$ is a scalar Wiener process. We shall make the change of variables in this equation $Y = Xe^{-Z}$ where

$$Z(t) = \int_{t_0}^{t} \left[a(\tau) - \frac{\nu}{2}b^2(\tau) \right] d\tau + \int_{t_0}^{t} b(\tau)dW(\tau).$$

In this case $Y = \varphi(X, Z) = Xe^{-Z}$ and consequently

$$\varphi_t(X, Z) = 0, \qquad \varphi_x(X, Z) = e^{-Z}, \qquad \varphi_z(X, Z) = -Xe^{-Z},$$
$$\varphi_{xx}(X, Z) = 0, \quad \varphi_{xz}(X, Z) = -e^{-Z}, \quad \varphi_{zz}(X, Z) = Xe^{-Z}.$$

Using Itô formula (5.1.7) we find

$$dY = \left\{ e^{-Z}(aX + a_0) - Xe^{-Z}\left(a - \frac{\nu}{2}b^2\right) \right.$$

$$\left. + \frac{1}{2}\,\mathrm{tr}\begin{bmatrix} 0 & -e^{-Z} \\ -e^{-Z} & Xe^{-Z} \end{bmatrix}\begin{bmatrix} bX + b_0 \\ b \end{bmatrix}\nu\,[\,bx + b_0 b\,] \right\}dt$$

$$+ e^{-Z}(bX + b_0)dW - Xe^{-Z}bdW$$

$$= \left\{ e^{-Z}\left(a_0 + \frac{\nu}{2}b^2 X\right) - \frac{1}{2}e^{-Z}b^2\nu X - e^{-Z}bb_0\nu \right\}dt$$

$$+ e^{-Z}b_0 dW = e^{-Z}\left[(a_0 - \nu bb_0)dt + b_0 dW \right]. \tag{II}$$

The right-hand side of this equation does not contain Y. Therefore integrating Eqs. (I) and (II) we obtain

$$Y = Y_0 + \int_{t_0}^{t} e^{-Z(\tau)}\left[a_0(\tau) - \nu(\tau)b(\tau)b_0(\tau) \right]d\tau + \int_{t_0}^{t} e^{-Z(\tau)}b_0(\tau)dW(\tau),$$

$$\tag{III}$$

$$X(t) = X_0 e^{Z(t)} + \int_{t_0}^{t} e^{Z(t)-Z(\tau)}\left[a_0(\tau) - \nu(\tau)b(\tau)b_0(\tau) \right]d\tau$$

$$+ \int_{t_0}^{t} e^{Z(t)-Z(\tau)}b_0(\tau)dW(\tau). \tag{IV}$$

5.2.4. Other Types of Stochastic Differential Equations

If we understand the second integral in Eq. (5.2.2) as a stochastic θ-integral then Eq. (5.2.1) or Eq. (5.2.3) will be a stochastic differential equation of the other kind. We shall call Eq. (5.2.1) or Eq. (5.2.3) *a stochastic differential equation with the θ-differential*.

We shall write a stochastic differential equation with a θ-differential corresponding to the case where the second integral in Eq. (5.2.2) represents a θ-integral in the form

$$\frac{d_\theta X}{dt} = a(X,t) + b(X,t)V \qquad (5.2.7)$$

or

$$d_\theta X = a(X,t)dt + b(X,t)d_\theta W . \qquad (5.2.8)$$

If $\theta = 1/2$, Eq. (5.2.7) and Eq. (5.2.8) represent the symmetrized stochastic differential equations.

R e m a r k. By virtue of the discrepancy of stochastic integrals of different types Eqs. (5.2.1), Eq. (5.2.7) at different θ with the same functions $a(x,t)$ and $b(x,t)$ determine different random processes. It is necessary to remember this and always indicate in what sense a stochastic differential equation is understood. Later on, without special mentioning, we shall always use the Itô equations. And only in the exceptional cases of using stochastic differential equations of other types we shall indicate in what sense they are understood.

5.2.5. Transformation of Stochastic Differential Equation into Itô Equation

In order to replace a stochastic differential equation with the θ-differential determining a given random process by the Itô equation which determines the same process (certainly with other function $a(x,t)$; the function $b(x,t)$ as we shall see remains invariant). We shall solve this problem only for the case of the Wiener process $W(t)$.

▷ In order to find the way of transformation of an equation with a θ-differential into an Itô equation we solve first the inverse problem of finding the equation with θ-differential for a given Itô equation. Suppose that the process $X(t)$ is determined by Itô Eq. (5.2.3). This equation determines the Itô differential of the process $X(t)$. To transform it into the form of a θ-differential we rewrite this equation remembering the definition of a θ-integral in the form

$$dX = a(X,t)dt + \left[(1-\theta)b(X(t),t) + \theta b(X(t+dt),t+dt)\right]dW$$

$$-\theta\left[b(X(t+dt),t+dt)-b(X(t),t)\right]dW,$$

or neglecting infinitesimals of higher orders

$$dX = a(X,t)dt + \left[(1-\theta)b(X(t),t)+\theta b(X(t+dt),t+dt)\right]dW$$

$$-\theta db(X,t)dW, \tag{5.2.9}$$

where $db(X,t)$ is an Itô differential of the process $U(t) = b(X(t),t)$. The Itô differentials of the elements of the matrix $U(t)$ are determined by formulae of Subsection 5.1.2. Here the process $X(t)$ plays the role of the process $Z(t)$; $a(X,t)$ and $b(X,t)$ play the role of the random functions $X(t)$ and $Y(t)$ respectively. Substituting the expressions of the differentials of the elements of the matrix $U(t)$ into Eq. (5.2.9) we have to take into account only the terms with $dW(t)$ which will give in Eq. (5.2.9) the items of order dt. All the other terms will give the infinitesimal items of higher orders. As a result for the components of the vector $db(X,t)dW$ we obtain

$$(db(X,t)dW)_r = \sum_{s=1}^{q} db_{rs}(X,t)dW_s$$

$$= \sum_{s=1}^{q}\sum_{h=1}^{p}\sum_{l=1}^{q} \frac{\partial b_{rs}(X,t)}{\partial X_h} b_{hl}(X,t)dW_l dW_s.$$

It is easy to see that the right-hand side of this equality represents the rth element of the matrix-column $\left[(\partial/\partial\xi)^T b(X,t)dW dW^T b(\xi,t)^T\right]_{\xi=X}^{T}$. Consequently $db(X,t)dW = \left[(\partial/\partial\xi)^T b(X,t)dW dW^T b(\xi,t)^T\right]_{\xi=X}^{T}$. In the same way as in Subsection 5.1.2 we come to the conclusion that it is sufficient to retain only the expectation of $db(X,t)dW$ which has the order dt. The random part of this variable which has the same order dt is infinitesimal of higher order in comparison with dW. As a result formula (5.2.9) takes the form

$$dX = \left\{a(X,t) - \theta\left[(\partial/\partial\xi)^T b(X,t)\nu(t)b(\xi,t)^T\right]_{\xi=X}^{T}\right\} dt$$

$$+ \left[(1-\theta)b(X(t),t)+\theta b(X(t+dt),t+dt)\right]dW$$

or

$$d_\theta X = \left\{a(X,t) - \theta\left[(\partial/\partial\xi)^T b(X,t)\nu(t)b(\xi,t)^T\right]_{\xi=X}^{T}\right\} dt + b(X,t)d_\theta W.$$

$$\tag{5.2.10}$$

Thus the stochastic differential equation with the θ-differential determining the same random process as Itô Eq. (5.2.3) has the form of Eq. (5.2.10). ◁

Thus we have the following statement.

Theorem 5.2.2. *If the random process $X(t)$ is determined by the stochastic differential equation with the θ-differential*

$$d_\theta X = a_1(X,t)dt + b(X,t)d_\theta W \,, \qquad (5.2.11)$$

then the Itô equation determining the same process $X(t)$ has the form of Eq. (5.2.3) where

$$a(X,t) = a_1(X,t) + \theta \left[(\partial/\partial\xi)^T b(X,t)\nu(t)b(\xi,t)^T \right]_{\xi=X}^T . \qquad (5.2.12)$$

R e m a r k 1. The statement proved concerns also Eq. (5.2.1) and the equation

$$d_\theta X/dt = a_1(X,t) + b(X,t)V \qquad (5.2.13)$$

of the form Eq. (5.2.7) corresponding to Eq. (5.2.11).

R e m a r k 2. Note that if the function $b(x,t)$ is independent of x the second item in (5.2.12) is equal to zero and $a(x,t) = a_1(x,t)$. Therefore, in the case where the coefficient at the white noise is independent of an unknown function X all the types of stochastic differential equations with a Wiener process $W(t)$ (with a normally distributed white noise $V(t)$) determining the same process $X(t)$ coincide. The transformation of the equation with a θ-differential into an Itô equation and vice versa is possible only in the case of the function $b(x,t)$ differentiable with respect to x. If the function $b(x,t)$ has no first derivatives with respect to the components of the vector x (either continuous or discontinuous) then the Itô equation determining the same process as a given equation with a θ-differential does not exist.

E x a m p l e 5.2.3. Transform the equation with a θ-differential

$$d_\theta X/dt = -X^3 + XV \qquad (I)$$

into an Itô equation if $V(t)$ is a normally distributed white noise with unit intensity. In this case $a_1(X,t) = -X^3$, $b(X,t) = X$ and formula (5.2.12) gives $a(X,t) = -X^3 + \theta X$. The corresponding Itô equation has the form

$$dX/dt = -X^3 + \theta X + XV \,. \qquad (II)$$

5.2.6. Sufficient Conditions of Existence and Uniqueness of m.s. Solution of Stochastic Differential Equation

Theorem 5.2.3. *Consider the stochastic integral equation*

$$X(t) = \Phi(t) + \int_{t_0}^{t} a(X(\tau), \tau, t)d\tau + \int_{t_0}^{t} b(X(\tau), \tau, t)Z(d\tau) \qquad (5.2.14)$$

where $\Phi(t)$ is a m.s. continuous random function independent of the stochastic measure Z. Suppose that

(i) for any m.s. continuous random functions $Y(\tau)$ and $Y_1(\tau)$ the random function $a(Y(\tau), \tau, t)$ as a function of τ, t is m.s. continuous on the closed set $[t_0, t] \times [t_0, t_0 + T]$ and satisfies the inequality

$$\| a(Y_1(\tau), \tau, t) - a(Y(\tau), \tau, t) \| < k \| Y_1(\tau) - Y(\tau) \| , \qquad (5.2.15)$$

where the norm $\| \cdot \|$ represents the norm of the element in B-space $L_2(\Omega, S, P, R^n)$ of the n-dimensional vector functions of the variable ω;

(ii) for any m.s. continuous random functions $Y(\tau)$ and $Y_1(\tau)$ whose values at each given τ are the functions of the values of the stochastic measure Z on Borel sets of the interval $[t_0, \tau)$ the function $b(y, \tau, t)$ satisfies inequalities

$$\left\| \int_{t_0}^{t} b(Y(\tau), \tau, t)Z(d\tau) \right\|^2 < c_1^2(t - t_0), \qquad (5.2.16)$$

$$\left\| \int_{t_0}^{t} [b(Y_1(\tau), \tau, t) - b(Y(\tau), \tau, t)] Z/d\tau \right\|^2 < k_1^2 \int_{t_o}^{t} \| Y_1(\tau) - Y(\tau) \|^2 d\tau .$$

$$(5.2.17)$$

Conditions (i) and (ii) are sufficient for the existence and the uniqueness of the m.s. solution of Eq. (5.2.14).

▷ For proving the existence of m.s. solution of Eqs. (5.2.14) we construct the sequence of the random functions for $p = 0, 1, 2 \ldots$

$$Y_0(t) = \Phi(t) ,$$

$$Y_{p+1}(t) = \Phi(t) + \int_{t_0}^{t} a(Y_p(\tau), \tau, t)d\tau + \int_{t_0}^{t} b(Y_p(\tau), \tau, t)Z(d\tau) . \qquad (5.2.18)$$

We estimate the differences

$$Y_1(t) - Y_0(t) = \Phi(t) + \int_{t_0}^{t} a(Y_0(\tau), \tau, t)d\tau + \int_{t_0}^{t} b(X_0(\tau), \tau, t)Z(d\tau),$$

$$(5.2.19)$$

$$Y_{p+1}(t) - Y_p(t) = \Phi(t) + \int_{t_0}^{t} \left[a(Y_p(\tau), \tau, t) - a(Y_{p-1}(\tau), \tau, t) \right] d\tau$$

$$+ \int_{t_0}^{t} \left[b(Y_p(\tau), \tau, t) - b(Y_{p-1}(\tau), \tau, t) \right] Z(d\tau). \qquad (5.2.20)$$

As $a(Y_0(\tau), \tau, t)$ is m.s. continuous then there exists such constant c that $\| a(Y_0(\tau), \tau, t) \| < c$ at $\tau \in [t_0, t]$, $t \in [t_0, t_0 + T]$. From here and from (5.2.16) we obtain

$$\| Y_1(t) - Y_0(t) \| = \left\| \int_{t_0}^{t} a(Y_0(\tau), \tau, t)d\tau \right\|$$

$$+ \left\| \int_{t_0}^{t} b(Y_0(\tau), \tau, t)Z(d\tau) \right\| < c(t - t_0) + c_1(t - t_0)^{1/2}. \qquad (5.2.21)$$

And as $t - t_0 \leq T$ at $t \in [t_0, t_0 + T]$ then from (5.2.12) we have

$$\| Y_1(t) - Y_0(t) \|_2 \leq (c\sqrt{T} + c_1)(t - t_0)^{1/2}. \qquad (5.2.22)$$

Using estimate (5.2.22) and inequality (5.2.17) we get by the induction

$$\| Y_{p+1}(t) - Y_p(t) \|_2 < (c\sqrt{T} + c_1) \prod_{q=1}^{p} \left(\frac{2k\sqrt{T}}{q+2} + \frac{k_1}{\sqrt{q+1}} \right) (t - t_0)^{(p+1)/2}. \qquad (5.2.23)$$

The series

$$\sum_{p=1}^{\infty} \alpha_p (t - t_0)^{(p+1)/2} = \sum_{p=1}^{\infty} \prod_{q=1}^{p} \left(\frac{2k\sqrt{T}}{q+2} + \frac{k_1}{\sqrt{q+1}} \right) (t - t_0)^{(p+1)/2}$$

$$(5.2.24)$$

converges for all t as at $p \to \infty$

$$\frac{\alpha_p(t-t_0)^{(p+2)/2}}{\alpha_{p-1}(t-t_0)^{p/2}} = \left(\frac{2k\sqrt{T}}{p+2} + \frac{k_1}{\sqrt{p+1}}\right)(t-t_o)^{1/2} \to 0. \qquad (5.2.25)$$

By virtue of (5.2.24) and (5.2.25) it follows that the sequence $\{Y_p(t)\}$ m.s. converges to the limit random function $Y(t)$ and the sequence of the integrals $\left\{\int_{t_0}^{t} a(Y_p(\tau),\tau,t)d\tau\right\}$ m.s converges to the integrals $\int_{t_0}^{t} a(Y(\tau),\tau,t)d\tau$.
Besides that by virtue of (5.2.17) we have the estimate

$$\left\|\int_{t_0}^{t} b(Y_p(\tau),\tau,t)Z(d\tau) - \int_{t_0}^{t} b(Y(\tau),\tau,t)Z(d\tau)\right\|^2$$

$$< k_1^2 \int_{t_0}^{t} \|Y_p(\tau) - Y(\tau)\|^2 \, d\tau. \qquad (5.2.26)$$

As at $p \to \infty$

$$\|Y_p(\tau) - Y(\tau)\|^2 = \lim_{q\to\infty} \|Y_p(\tau) - Y_q(\tau)\|^2$$

$$< (c\sqrt{T} + c_1)^2 \left[\sum_{r=p}^{\infty} \alpha_r(t-t_0)^{(r+1)/2}\right]^2 \to 0, \qquad (5.2.27)$$

then the right-hand side of inequality (5.2.26) at $p \to \infty$ tends to zero. This fact proves the convergence of the sequence of the stochastic integrals $\left\{\int_{t_0}^{t} b(Y_p(\tau),\tau,t)Z(d\tau)\right\}$ to the stochastic integral $\left\{\int_{t_0}^{t} b(Y(\tau),\tau,t)Z(d\tau)\right\}$.
Thus passing to the limit in (5.2.18) at $p \to \infty$ we make sure that the limit function $Y(t)$ satisfies Eq. (5.2.14), and consequently, (5.2.2) also. ◁

Now we shall prove the uniqueness of the solution.
▷ Let $Y(t)$ be any solution of Eq. (5.2.14). After substituting it into equation (5.2.14) and subtracting sequence (5.2.18) from it we have (5.2.27) at $X(\tau) = Y(\tau)$. Passing to the limit we get $\|X(t) - Y(t)\| = 0$. This proves the uniqueness of the m.s. solution of Eq. (5.2.14). ◁

5.3. Differential Equations Containing Random Functions

5.3.1. Definitions

Let $(\Omega, \mathcal{S}, P)$ be a complete probability space, $F(X,t) = f(x,t,\omega)$ be the n-dimensional vector random function of the n-dimensional vector X and the scalar variable t. Let us consider a differential equation of the following form:

$$\dot{X} = F(X,t), \quad X(0) = X_0, \tag{5.3.1}$$

where the differentiation over t is denoted by a point. If there exists the random function $X(t)$ satisfying in some sense Eq. (5.3.1) then it is called *a solution* or an *integral* of differential Eq. (5.3.1). Depending on the fact what sense is contributed of the expression "satisfies the equation" different types of the integrals of the differential equations which contain random functions are obtained. In the application problems certainly we are interested in such integral $X(t)$ of Eq. (5.3.1) whose each realization satisfies the correspondent realization of Eq. (5.3.1). But it is often expedient to consider the other types of the integrals of the differential equations.

The random function $X(t) = x(t,\omega)$ is called *an integral* of Eq. (5.3.1) in the realizations or shortly, *an integral* of Eq. (5.3.1) *at the interval* T if almost for all ω its realization $x = x(t) = x(t,\omega)$ satisfies the equation

$$\dot{x} = f(x,t,\omega). \tag{5.3.2}$$

The random function $X(t) = x(t,\omega)$ is called *an integral of* Eq. (5.3.1) *almost sure* (an integral a.s.) at the interval T if at each $t \in T$ the function $x(t,\omega)$ satisfies Eq. (5.3.2) almost at all ω.

A set of the exceptional values of ω at which $x(t,\omega)$ does not satisfy Eq. (5.3.2) at a given $t \in T$ in the general case depends on t in consequence of which we denote it by N_t, $N_t \subset N_t^0 \in \mathcal{S}$, $P(N_t^0) = 0$. The union of such exceptional sets $\bigcup\limits_{t \in T} N_t$ correspondent to all $t \in T$ may be not zero set. Therefore a.s. integral of Eq. (5.3.1) may have a nontrivial set of the realizations which do not satisfy Eq. (5.3.2) almost at all $t \in T$. Moreover $X(t)$ may have none absolutely continuous realization satisfying Eq. (5.3.2) almost at all $t \in T$. In such cases a.s. integral of Eq. (5.3.1) will be not an integral of this equation in the realizations.

The integral of Eq. (5.3.1) in the realizations is always its a.s. integral but not vice versa, a.s. integral may be not an integral in the realizations.

Now we suppose that the random function $F(x,t)$ in (5.3.1) possesses such a property that for any m.s. continuous function $Y(t)$ the random function $G(t) = F(Y(t),t)$ is m.s. integrable at some interval T. Then the random function

$$Z(t) = \int_{t_0}^{t} F(Y(\tau),\tau)d\tau, \quad t, t_0 \in T,$$

is m.s. continuous and m.s. differentiable on T and $\dot{Z}(t) = F(Y(t),t)$. Therefore Eq. (5.3.1) may be replaced by integral equation

$$X(t) = X_0 + \int_{t_0}^{t} F(X(\tau),\tau)d\tau, \tag{5.3.3}$$

where the integral is assumed in m.s. sense. The random function $X(t)$ is called *an integral of* Eq. (5.3.1) *in mean square (m.s. integral) at the interval* T if it satisfies Eq. (5.3.3) at t_0, $t \in T$. In other words the random function $X(t)$ is called a m.s. integral of Eq. (5.3.1) at the interval T if it is m.s. differentiable and its m.s. derivative $\dot{X}(t)$ at each $t \in T$ a.s. coincides with $F(X(t),t)$. Zero set N_t of the exceptional values ω at which $\dot{x}(t,\omega) \neq f(x(t,\omega),t,\omega)$ at given $t \in T$ in the general case depends on t therefore $\bigcup\limits_{t \in T} N_t$ may be not zero set. Consequently, m.s. integral of Eq. (5.3.1) may be not its integral in the realizations. Moreover m.s. integral of Eq. (5.3.1) may have none differentiable realization.

E x a m p l e 5.3.1. Let the probability space $(\Omega, \mathcal{S}, P)$ represent the real axis with a normal distribution (on σ-algebra of Borel sets). The random function $X(t) = x(t,\omega) = x_0 e^{\omega t}$, $\omega \in \Omega$ is simultaneously an integral in the realizations, an a.s. integral and m.s. integral of the differential equation $\dot{X} = \omega X$ satisfying the initial condition $X(0) = x_0$.

E x a m p l e 5.3.2. Under the conditions of Example 5.3.1 the random function

$$Y(t) = \begin{cases} x_0 e^{\omega t} & \text{at} \quad t \neq \omega/n, \quad n = \pm 1, \pm 2, \ldots, \\ 0 & \text{at} \quad t = \omega/n \end{cases}$$

is an a.s. and m.s. integral of the differential equation of Example 5.3.1 at the initial condition $Y(0) = x_0$ as at any t only the countable set N_t of the realizations correspondent to $\omega = nt$ $(n = \pm 1, \pm 2, \ldots)$ does not satisfy this equation. But the random function $Y(t)$ is not an integral in the realizations as at the infinitesimal interval $[0, T)$ each its realization has a countable set of the discontinuity points at which it is not differentiable.

E x a m p l e   5.3.3. Let the probability space $(\Omega, \mathcal{S}, P)$ be the real axis with Cauchy distribution which is characterized by the density $\alpha/\pi(\alpha^2 + \omega^2)$. The random function $X(t) = x_0 e^{\omega t}$ represents an integral in the realizations and a.s. integral of the equation $\dot{X} = \omega X$. But it is not a m.s. integral of this equation as it has no finite second order moment at none $t \neq 0$.

E x a m p l e   5.3.4. Under the conditions of Example 5.3.3 the random function $Y(t)$ of Example 5.3.2 represents an a.s. integral of the equation $\dot{Y} = \omega Y$. But it is neither an integral in the realizations nor m.s. integral.

5.3.2. Sufficient Conditions of Existence and Uniqueness of m.s. Solutions of Differential Equations Containing Random Functions

The following theorem gives the sufficient condition for the a.s. or m.s. integral of Eq. (5.3.1) to be an integral of this equation in the realizations.

Theorem 5.3.1. *If the function $f(x, t, \omega)$ in Eq. (5.3.2) is continuous as the function of x and t at almost all $\omega \in \Omega$ and almost all the realizations of a.s. integral (m.s. integral) of Eq. (5.3.1) have continuous derivatives then an a.s. integral (m.s. integral) is at the same time an integral of Eq. (5.3.1) in the realizations.*

$\triangleright$ It follows from the continuity of the function $f(x, t, \omega)$ on x and t that it may be represented almost at all ω as the limit of the sequence of the functions

$$f_n(x, t, \omega) = \sum f(x_p^n, t_q^n, \omega) \mathbf{1}_{X_p^n}(x) \mathbf{1}_{T_q^n}(t) \tag{5.3.4}$$

where $\bigcup_p X_p^n = X$, $\bigcup_q T_q^n = T$ are some sequences of the partitions of the n-dimensional space X and the real axis T into mutually exclusive rectangles and the intervals respectively. It follows from the measurability of the sections of the function $f(x, t, \omega)$ at any x, t that each of the functions of (5.3.4) of the variable ω may be represented as the limit of the sequence of the simple functions:

$$f(x_p^n, t_q^n, \omega) = \lim \sum_{m=1}^{N_{pq}^n} f_{pq}^{nm} \mathbf{1}_{A_{pq}^{nm}}(\omega). \tag{5.3.5}$$

Thus by virtue of (5.3.5) the function $f(x, t, \omega)$ is representable as the limit of the sequence of the step functions of the form

$$f_n^m(x, t, \omega) = \sum_{p,q,m} f_{pq}^{nm} \mathbf{1}_{X_p^n}(x) \mathbf{1}_{T_q^n}(t) \mathbf{1}_{A_{pq}^{nm}}(\omega). \tag{5.3.6}$$

As the sets of the partitions X_{pq}^n, T_q^n, A_{pq}^{nm} are measurable relatively to σ-algebra $\mathcal{B}^n$ of Borel sets of the n-dimensional space, σ-algebra $\mathcal{B}$ of Borel sets of the real axis and σ-algebra $\mathcal{S}$ of probability space $(\Omega, \mathcal{S}, P)$ correspondingly then the functions $f_n^m(x, t, \omega)$ are measurable functions of x, t, ω, i.e. elementary functions of a point in the product of the correspondent spaces. Consequently, the function $f(x, t, \omega)$ is a measurable functions of x, t, ω. Let $X(t) = x(t, \omega)$ be a.s. integral of Eq. (5.3.1) at the interval T whose almost all the realizations have continuous derivatives. Then at any t the function $f(x(t, \omega), t, \omega)$ is a measurable function of ω. Thus $Y(t) = F(X(t), t)$ as a random function of time t has almost all continuous realizations by virtue of the continuity of f and almost all realizations of $X(t)$.

Further from the definition of a.s. integral it follows that $\dot{X}(t)$ and $Y(t)$ are equivalent, i.e. at any fixed t coincide with provability 1 on any countable set of the points. And as any continuous function of a finite-dimensional argument is completely determined by its values on the countable everywhere dense set of the points, for instance, with the rational coordinates then almost all realizations of $\dot{X}(t)$ and $Y(t) = F(X(t), t)$ coincide everywhere on T. And it means that at almost all ω the realizations $x(t) = x(t, \omega)$ of the random function $X(t)$ satisfy Eq. (5.3.1), i.e. $X(t)$ represents integral of Eq. (5.3.1) in the realizations.

If $X(t)$ is m.s. integral (5.3.1) whose almost all realizations have continuous derivatives then m.s. derivative $\dot{X} = F(X(t), t)$ is equivalent to the derivative in the realizations which serves at the same time as a.s. derivative. Consequently, m.s. integral $X(t)$ is also a.s. integral of Eq. (5.3.1). But as it was already proved $X(t)$ represents integral of Eq. (5.3.1) in the realizations. ◁

Now we prove the main theorem of the existence and the uniqueness of m.s. integrals in conformity to more general than Eq. (5.3.3) integral equation of the form

$$X(t) = \Phi(t) + \int_{t_0}^{t} F(X(\tau), \tau, t)d\tau \tag{5.3.7}$$

where $\Phi(t)$ is a m.s. continuous random function.

Suppose that for any m.s. continuous random functions $Y(\tau)$, $Y_1(\tau)$ the random function $F(Y(\tau), \tau, t)$ is m.s. continuous on the closed sets of $\tau \in [t_0, t]$ and $t \in [t_0, t_0 + T]$ and satisfies on these sets the condition

$$E\,\|F(Y_1(\tau), \tau, t) - F(Y(\tau), \tau, t)\|^2 < k^2 E\,\|Y_1(\tau) - Y(\tau)\|^2, \tag{5.3.8}$$

where the norm of the vector is assumed as the Euclidean norm. Then at the condition of (5.3.8) Eq. (5.3.7) has the unique solution at the interval $[t_0, t_0 + T]$.

▷ We shall cite the proof in connection with the random functions with the values in B-space $L_2(\Omega, \mathcal{S}, P, R^n)$ of n-dimensional vector functions of the variable ω. As it is known from functional analysis the norm of the value of the random function $X(t) = x(t, \omega)$ at each t is determined by formula

$$\|X(t)\|_2^2 = \int \|x(t, \omega)\|^2 P(d\omega) = E\|X(t)\|^2. \qquad (5.3.9)$$

In other words, the square of the norm $\|X(t)\|^2$ of the value $X(t)$ at a given t in B-space $L_2(\Omega, \mathcal{S}, P, R^n)$ represents an expectation of the square of its Euclidean norm $\|X(t)\|$. Then Eqs. (5.3.1), (5.3.3) and (5.3.7) may be considered as the equations in B-space $L_2(\Omega, \mathcal{S}, P, R^n)$ and condition (5.3.8) will be written in the form of the Lipchitz condition

$$\|F_1(Y_1(\tau), \tau, t) - F(Y(\tau), \tau, t)\| < k\|Y_1(\tau) - Y(\tau)\|. \qquad (5.3.10)$$

Therefore for the completion of the proof it is sufficient to use the known from the theory of the ordinary differential equations the Picard method of the successive approximations.

We determine the sequence of m.s. continuous random functions for $p = 0, 1, 2, \ldots$:

$$X_0(t) = \Phi(t), \quad X_{p+1}(t) = \Phi(t) + \int_{t_0}^{t} F(X_p(\tau), \tau, t)d\tau, \qquad (5.3.11)$$

and estimate the norms of the differences

$$X_1(t) - X_0(t) = \int_{t_0}^{t} F(X_0(\tau), \tau, t)d\tau, \qquad (5.3.12)$$

$$X_{p+1}(t) - X_p(t) = \int_{t_0}^{t} [F(X_p(\tau), \tau, t) - F(X_{p-1}(\tau), \tau, t)]d\tau. \qquad (5.3.13)$$

As according to the condition the random function $F(X_0(\tau), \tau, t)$ of the variables τ, t is m.s. continuous in consequence of m.s. continuity of $\Phi(\tau)$ then its second order moment $E\|F(X_0(\tau), \tau, t)\|^2 = \|F(X_0(\tau), \tau, t)\|^2$ is continuous and finite at all τ, t, and consequently, achieves its upper bound on the finite closed interval

$\tau \in [t_0, t]$, $t \in [t_0, t_0 + T]$. Therefore $\|F(X_0(\tau), \tau, t)\|^2 < c$ at some $c < \infty$. Hence from (5.3.12) we find

$$\|X_1(t) - X_0(t)\| \leq \int_{t_0}^{t} \|F(X_0(\tau), \tau, t)\| \, d\tau < c(t - t_0). \qquad (5.3.14)$$

After that from (5.3.13) and (5.3.10) by the induction we get

$$\|X_{p+1}(t) - X_p(t)\| < c \frac{k^p (t - t_0)^{m+1}}{(p+1)!}. \qquad (5.3.15)$$

At any p and m from (5.3.15) we have

$$\|X_{p+m}(t) - X_p(t)\| < \sum_{q=p}^{p+m-1} c \frac{k^q (t - t_0)^{q+1}}{(q+1)!} \to 0. \qquad (5.3.16)$$

It proves that the sequence $\{X_p(t)\}$ at any t is fundamental. And as B-space $L_2(\Omega, \mathcal{S}, P, R^n)$ is complete then there exists m.s. limit of the sequence. Therefore it follows from (5.3.10), (5.3.15) and (5.3.16) that at $p \to \infty$, $F(X_p(\tau), \tau, t) \to F(X((\tau), \tau, t)$,

$$\|F(X_p(\tau), \tau, t)\|_2 \leq \|F(X_0(\tau), \tau, t)\|_2$$

$$+ \sum_{q=1}^{p} \|F(X_q(\tau), \tau, t) - F(X_{q-1}(\tau), \tau, t)\| < c e^{k(\tau - t_0)}. \qquad (5.3.17)$$

Thus at all τ the sequence $F(X(\tau), \tau, t)$ is bounded on the norm by the integrable function of τ. Consequently, on the basis of Lebesgue theorem which is known from functional analysis the function $F(X(\tau), \tau, t)$ is integrable over τ and we may pass in (5.3.11) to the limit at $p \to \infty$ under the sign of the integral. It means that the limit function $X(t)$ satisfies Eq. (5.3.7) what proves the existence of the m.s. solution of Eq. (5.3.7). $\triangleleft$

Now we shall prove the uniqueness of the m.s. solution.

$\triangleright$ Let $Y(t)$ be any solution of Eq. (5.3.7). After substituting it into (5.3.7) and subtracting it from (5.3.11) we obtain

$$X_0(t) - Y(t) = - \int_{t_0}^{t} F(Y(\tau), \tau, t) d\tau,$$

$$X_{p+1}(t) - Y(t) = \int_{t_0}^{t} [F(X_p(\tau), \tau, t) - F(Y((\tau), \tau, t)] d\tau,$$

$$\|X_{p+1}(t) - Y(t)\| = c\frac{k^{p+1}(t-t_0)^{p+2}}{(p+2)!}.$$

Hence passing to the limit at $p \rightarrow \infty$ we shall have $\|X(t) - Y(t)\| = 0$. It proves the uniqueness of the m.s. solution of Eq. (5.3.7), and consequently, also of Eq. (5.3.1) (with the accuracy till the equivalence). ◁

5.4. Transformation of System Equations into Stochastic Equations

5.4.1. Principal Possibility of Replacing Random Function in Differential Equation by White Noise

All the random functions in the equations which describe real phenomena differ from a white noise. Nevertheless it is highly alluring to try to use the well developed methods of the theory of random processes determined by stochastic differential equations for studying various real systems. In other words, it is expedient to use stochastic differential equations for constructing suitable mathematical models of real systems.

It is clear that we may speak about the direct replacement of the random function in a differential equation by a white noise only in a case where this random function enters linearly into the equation. Therefore we shall consider first a system whose behaviour is described by the equation:

$$\dot{Y} = a(Y,t) + b(Y,t)X, \qquad (5.4.1)$$

where Y is the state vector of the system, $a(y,t)$ and $b(y,t)$ are known functions of the vector y and of time t, and $X(t)$ is a random function with zero expectation. The function $a(y,t)$ represents a vector of the same dimension p as the state vector Y, and $b(y,t)$ is a $p \times m$ matrix, m being the dimension of the vector random function $X(t)$. If the correlation interval of the random function $X(t)$ is sufficiently small then we may consider $X(t)$ practically as a white noise (Subsection 3.3.3). The question arises in what sense should we understand the stochastic differential equation obtained by the formal replacement of the random function $X(t)$ by a white noise. It is clear that we cannot treat this equation in the general case as the Itô equation. To understand this we notice that on the basis of the definitions of the Itô integral and the Itô equation the value of the coefficient in front of the white noise in an Itô equation at any time instant t is independent of the value of the white noise at the same instant (more exactly, the random variables $b(Y(t),t)$ and $dW(t)$ are independent at any t). Consequently, $Eb(Y(t),t)V(t) = Eb(Y(t),t)EV(t) = 0$. At the same time if the random function $X(t)$ in (5.4.1) is not a white noise, then in the general case $Eb(Y(t),t)X(t) \neq 0$. It means that treating the obtained equation as an Itô equation we introduce inevitably an error in all the calculations. Intuitively we may come to the conclusion that the stochastic differential equation obtained by replacing the random function $X(t)$ in (5.4.1) by a white noise should be considered as a symmetrized stochastic differential equation. The question in what sense stochastic differential equation, obtained by replacing the random function $X(t)$ by a white noise should be understood, may be solved in practical problems only simultaneously with constructing the mathematical model of a system with the account of all adopted assumptions. But it is often impossible in practical problems even to enumerate all neglected delays while constructing the mathematical model of a system and *a fortiori* to estimate their magnitudes. Therefore, it is easy to come to the conclusion that the simple replacement of the random function in a differential equation by a white noise cannot be recommended. Consequently, one should search for another method for the transformation of the differential equations of a system into stochastic differential equations. Such a method will be considered in the following section.

5.4.2. Method of Shaping Filters

Another method of transforming differential equations describing the system state into the stochastic differential equations consists of replacing the random function (in the general case a vector one) in a differential equation by some other random function which may be represented as the result of a memoryless transformation of the solution of some stochastic differential equation. This method is applicable also in those cases where a random function enters nonlinearly into the differential equation of a system. Therefore we shall consider the differential equation describing the evolution of a system state in the general form:

$$\dot{Y} = f(Y, X, t), \quad Y(t_0) = Y_0, \tag{5.4.2}$$

where $X = X(t)$ is a random function. If the random function $X(t)$ may be represented as $X(t) = \omega(U(t), t)$ where $\omega(u, t)$ is some function of the vector u and of time t, and $U(t)$ is the random process determined by the stochastic differential equation

$$\dot{U} = \varphi(U, t) + \psi(U, t)V \tag{5.4.3}$$

with the initial condition $U(t_0) = U_0$, then the vector random process $Y_1(t) = \left[\, Y(t)^T U(t)^T \,\right]^T$ represents the solution of the set of two differential equations:

$$\dot{Y} = f(Y, \omega(U,t), t), \quad \dot{U} = \varphi(U,t) + \psi(U,t)V \qquad (5.4.4)$$

with the initial condition $Y(t_0) = Y_0$, $U(t_0) = U_0$. This set of equations may be written in the form of one stochastic differential equation for Y_1:

$$\dot{Y}_1 = a(Y_1, t) + b(Y_1, t)V, \qquad (5.4.5)$$

where

$$Y_1 = \begin{bmatrix} Y \\ U \end{bmatrix}, \quad a(Y_1, t) = \begin{bmatrix} f(Y, \omega(U,t), t) \\ \varphi(U,t) \end{bmatrix}, \quad b(Y_1, t) = \begin{bmatrix} 0 \\ \psi(U,t) \end{bmatrix}.$$

This stochastic differential equation must always be understood in the same sense as Eq. (5.4.3) which determines the process $U(t)$. In particular, if Eq. (5.4.3) represents the Itô equation, then the equation for the process $Y_1(t) = \left[\, Y(t)^T U(t)^T \,\right]^T$ is the Itô equation too.

The equation $X(t) = \omega(U(t), t)$ and Eq. (5.4.3) may be considered as a model of some system shaping the random function $X(t)$ from a white noise V, i.e. as a system which gives as the output the random function $X(t)$, receiving at the input the white noise V. Such a system is called *a shaping filter* for the random function $X(t)$ (Subsection 4.8.4). In consequence of this the outlined method of transforming a system equation into a stochastic differential equation is called *the method of shaping filters*. Methods of linear shaping filters are given in Subsections 4.8.4 and 6.5.1–6.5.4.

5.4.3. Equations of Stochastic Differential Systems. Central Problem of Stochastic Systems Theory

At first we suppose that the evolution of the state (the expanded state, in general) of the stochastic system is described by the stochastic Itô differential equation of the form Eq. (5.2.1)

$$\dot{Y} = a(Y,t) + b(Y,t)V, \qquad (5.4.6)$$

where $a(y,t)$ and $b(y,t)$ are the known functions of y and t of $(p \times 1)$ and $(p \times m)$ dimension correspondingly. The white noise V in Eq. (5.4.6) we shall assume as m-dimensional vector white noise in the strict sense (Subsection 4.6.4). The initial value $Y(t_0) = Y_0$ of the state vector we shall always consider as independent of the white noise $V(t)$, $t > t_0$.

The system which is described by Eq. (5.4.6) for the normally distributed white noise $V = V(t)$ with zero expectation $EV = 0$ and with the matrix of the intensities $\nu = \nu(t)$ we shall call *the normal stochastic differential system*.

For some composite stochastic differential systems in the finite-dimensional space the stochastic differential equatin of the form is used

$$dY = a(Y,t)dt + b(Y,t)dW_0 + \int\limits_{R_0^q} c(Y,t,u)P^0(t,du). \qquad (5.4.7)$$

Here $a = a(y,t)$ and $b = b(y,t)$ are the known $(p \times 1)$-dimensional and $(p \times m)$-dimensional functions of the vector Y and time t; $W_0 = W_0(t)$ is a m-dimensional Wiener random process of the intensity $\nu_0 = \nu_0(t)$; $c(y,t,u)$ is $(p \times 1)$-dimensional function of y, t and of the auxiliary $(q \times 1)$-dimensional parameter u; $\int\limits_{\Delta} dP^0(t,A)$ is a centered Poisson measure:

$$\int\limits_{\Delta} dP^0(t,A) = \int\limits_{\Delta} dP(t,A) - \int\limits_{\Delta} \nu_P(t,A)dt, \qquad (5.4.8)$$

where $\int\limits_{\Delta} dP(t,A)$ is a number of the jumps of a Poisson process at time interval Δ; $\nu_P(t,A)$ is the intensity of the Poisson process $P(t,A)$; A is some Borel set of the space R_0^q with the pricked origin of the coordinates. The integral in Eq. (5.4.7) is extended on the whole space R_0^q with the pricked origin of the coordinates. The initial value Y_0 of the vector Y represents a random variable which does not depend on the increments of Wiener process $W_0(t)$ and of the Poisson process $P(t,A)$ at time intervals $\Delta = (t_1, t_2]$ which follows t_0, $t_0 \leq t_1 \leq t_2$ for any set A.

R e m a r k. In the case when the integrand $c(y,t,u)$ in Eq. (5.4.7) admits the presentation $c(y,t,u) = b(y,t)c'(u)$ Eq. (5.4.7) is reduced to the form of (5.4.6) if we assume

$$W(t) = W_0(t) + \int\limits_{R_0^q} c'(u)P^0(t,du).$$

In some practical cases to stochastic differential Eq. (5.4.7) may be reduced the stochastic integrodifferential equations of the form of Eq. (5.4.9)

$$dY_t = a(Y_{t_0}^t, t)dt + b(Y_{t_0}^t, t)dW_0 + \int\limits_{R_0^q} c(Y_{t_0}^t, t, u)P^0(t, du). \qquad (5.4.9)$$

Here $a = a(y_{t_0}^t, t)$, $b = b(y_{t_0}^t, t)$, $c = c(y_{t_0}^t, t, u)$ at each instant depend not only on the current value of the process $Y(t)$ but on all its values $Y(\tau)$ at the interval $[t_0, t)$, i.e. the components of the vector functions a, c and the elements of the matrix b are the functionals of the process $Y(\tau)$, $\tau \in [t_0, t)$.

In problems of practice for calculating the probabilities of the events connected with the random functions it is sufficient to know the multi-dimensional distributions (Section 2.7). Therefore *the central problem of the theory of the continuous stochastic systems* (Subsection 1.6.5) is the problem of probability analysis of the multi-dimensional distributions of the processes which satisfy the stochastic differential equations of the form Eq. (5.4.6) or Eq. (5.4.7) at the correspondent initial conditions.

In the theory of continuous stochastic systems two principally different approaches to distributions calculation are recognized. The first general approach is based on the statistical simulation, i.e. on the direct numerical solution of stochastic differential Eqs. (5.4.6), (5.4.7) with the corresponding statistical data processing (Section 5.5). The second general approach is based on the theory of continuous Markov processes and supposes the analytical simulation, i.e. the solution of equations in the functional spaces for the one-dimensional and the multi-dimensional distributions (Section 5.6).

Along with the general methods of the stochastic systems theory the special methods are distinguished which are oriented on the linear stochastic systems (Chapter 6) and on the nonlinear stochastic systems (Chapter 7).

5.5. Numerical Integration of Stochastic Differential Equation

5.5.1. Introductory Remarks

Integration of stochastic differential equations has some peculiarities. The fact is that all numerical integration methods of the ordinary differential equations besides the simplest Euler method are based on the calculation of the increments of the sought functions on each step by means of using the integral theorem about the mean value. In the accordance with this fact the right-hand sides of the equations (the derivatives of the sought functions) are taken in the mean points of the intervals. Various methods of the numerical integration differ in essence from one another only by the way of approximate finding the mean values of the right-hand sides of the equations.

The theorem about the mean value is not applicable to the stochastic integrals. But some analog of the theorem about the mean value is valid for the stochastic integrals with respect to the nonrandom functions. It consists in the fact that the best approximation of the stochastic integral with respect to the continuous nonrandom function gives the product of the value of a given function by the increment of the process over which the integration is performed. Therefore, for instance, all the methods of the numerical integration of the ordinary differential equations may be used for the stochastic differential equations (5.4.6) whose coefficient at the white noise is a determinate time functions t, i.e. $b(Y, t) = b(t)$. If the function $b(Y, t)$ in Eq. (5.4.6) depends on Y then the numerical integration method of such equations is required to be chosen depending on the fact in what sense the stochastic integral is assumed. In the case of the Itô equation the increment of the process $Y(t)$ at the interval $[t, t + \Delta t)$ should be determined by formula $\Delta Y = a(Y_t, t)\Delta t + b(Y_t, t)\Delta W$ what corresponds to the Euler method of the numerical integration. Thus the Itô equations may be integrated by Euler method. But for better accuracy of the calculation of the first item we may take the value of the function $a(Y_t, t)$ in some mean point of the integration interval $[t, t + \Delta t)$. Here we may use any method of the numerical integration, for instance, the widely known Adams method and the Runge–Kutta method. But the value of the function $b(Y_t, t)$ should be always taken at the initial point of the integration interval. In the case of the equation in the θ-differential the value of this equation should be taken at the point $t + \theta \Delta t$. But for its approximate finding we have to elaborate new methods of the numerical integration.

While simulating the stochastic differential equations by means of the analog computers for the functions $b(Y, t)$ which depends on Y it is necessary beforehand to reduce the Itô equations to the symmetrized (Stratonovich) form. It may be explained by the fact that at such simulating a white noise should be substituted by a process with a small correlation interval but such one which is different from zero as a white noise is physically nonrealizable. In this case a simulating process will be close to the solution of a stochastic differential equation if and only if a given equation is assumed as a symmetrized stochastic differential equation. While simulating the processes which are determined by Eq. (5.4.6) it is necessary to simulate the random variables or random processes. At numerical integration of the stochastic differential equations on digital computers it is required to simulate the increment ΔW of the random process $W(t)$ at each step. While simulating by means of the analog computer it is necessary to generate the wide-band random processes.

5.5.2. Existence of Exact Difference Equation Correspondent to Given Stochastic Differential Equation

Let us consider the Itô stochastic differential equation (5.4.7). Suppose that it is required to substitute this equation by a difference equation for the values of the process $Y(t)$ in a given discrete series of the equidistant points $\{t_k\}$, $t_k = kh$ where h is an interval between adjacent points t_k, $h = t_{k+1} - t_k$. In principal the problem is solved exactly as the values of Markov process $Y(t)$ at the points t_k form Markov random sequence $\{\bar{Y}_k\}$, $\bar{Y}_k = Y(kh)$, and any Markov sequence is determined by some stochastic difference equation. But to compose this exact difference equation by means of a given differential equation is practically impossible. First of all for its composing it is necessary to find a transient distribution of Markov process $Y(t)$ determined by Eq. (5.4.7) and then we may compose a difference equation for the sequence $\{\bar{Y}_n\}$ by means of the found transient distribution. But the exact evaluation of the transient distribution of the process $Y(t)$ is possible only in some specials cases. In the general case we are satisfied with an approximate evaluation of a transient distribution of the process $Y(t)$. As a result by this transient distribution we may obtain only an approximate difference equation. The latter makes it expedient the application of very complicated algorithm derivation of an exact difference equation for the sequence $\{\bar{Y}_n\}$. As we are satisfied only with an approximate difference equation even at the application of a given algorithm then it is expedient to make it by more simple ways.

At first we substitute an integral over the variable u in Eq. (5.4.7) by the correspondent integral sum. As a result Eq. (5.4.7) will be replaced by the following:

$$dY = a(Y,t)dt + b(Y,t)dW_0 + \sum_{i=1}^{N} c_i(Y,t)dP_i^0, \qquad (5.5.1)$$

where $c_i(y,t)$ are the p-dimensional vector functions which represent the values of the function $c(y,t,u)$ in some mean point u_i correspondent to the elements A_i of the partition of q-dimensional sphere of sufficiently large radius, $u_i \in A_i$ ($i = 1,\ldots,N$), and $P_i^0(t)$ are the centered simple Poisson processes (Subsection 4.6.6):

$$P_i^0(t) = P^0([0,t), A_i) - \mu([0,t), A_i), \quad i = 1, 2, \ldots, N. \qquad (5.5.2)$$

The intensities of these processes are determined in terms of the expectation $\mu(\Delta, A)$ of Poisson measure $P(\Delta, A)$ by formula

$$\nu_i(t) = d\mu([0,t), A_i)/dt. \qquad (5.5.3)$$

The replacement of Eq. (5.5.1) by the correspondent difference equation is performed by the Euler method.

5.5.3. Euler Method

The simplest way of the replacement of Eq. (5.5.1) by a difference equation consists in the replacement of all the differentials by the elements of the integral sums:

$$Y((n+1)h) - Y(nh) = a(Y(nh), nh)h + b(Y(nh, nh))[W_0((n+1)h)$$

$$-W_0(nh)] + \sum_{i=1}^{N} c_i(Y(nh), nh)[P_i^0((n+1)h) - P_i^0(nh)].$$

Putting

$$\bar{Y}_n = Y(nh), \quad \varphi_n(\bar{Y}_n) = Y(nh) + a(Y(nh), nh), \quad \psi_{1n}(\bar{Y}_n) = b(Y(nh), nh),$$

$$\psi_{in}(\bar{Y}_n) = c_{i-1}(Y(nh), nh), \quad V_{1n} = W_0((n+1)h) - W_0(nh), \quad (5.5.4)$$

$$V_{in} = P_{i-1}^0((n+1)h) - P_{i-1}^0(nh), \quad i = 2, \ldots, N+1,$$

we get a stochastic difference equation

$$\bar{Y}_{n+1} = \varphi_n(\bar{Y}_n) + \sum_{i=1}^{N+1} \psi_{in}(\bar{Y}_n)V_{in}.$$

Introducing the $p \times (m + N)$ block matrix

$$\psi_n(\bar{Y}_n) = [\psi_{1n}(\bar{Y}_n) \ldots \psi_{N+1,n}(\bar{Y}_n)] \qquad (5.5.5)$$

and $(m + N)$-dimensional random vector

$$V_n = [V_{1n}^T \ V_{2n} \ \cdots \ V_{N+1,n}]^T, \qquad (5.5.6)$$

we may write the obtained difference equation in short form:

$$\bar{Y}_{n+1} = \varphi_n(\bar{Y}_n) + \psi_n(\bar{Y}_n)V_n. \qquad (5.5.7)$$

As the Wiener and the Poisson processes are the processes with the independent increments (Subsection 4.6.3) then the random vectors V_n form the sequence of the independent random vectors $\{V_n\}$ and the blocks V_{1n} of the vectors V_n have the normal distribution $N(0, \bar{G}_n)$ where

$$\bar{G}_n = \int\limits_{nh}^{(n+1)h} \nu_0(\tau)d\tau \cong \nu_0(nh)h; \qquad (5.5.8)$$

$\nu_0(t)$ is the intensity of the Wiener process $W_0(t)$, the scalar blocks V_{in} ($i = 2, \ldots, N+1$) have the Poisson distributions with the parameters

$$\mu_{in} = \int\limits_{nh}^{(n+1)h} \nu_i(\tau)d\tau \cong \nu_i(nh)h. \qquad (5.5.9)$$

These distributions determine completely the distributions of the random vectors V_n. The covariance matrix G_n of the vector V_n represents a block-diagonal matrix

$$G_n = \begin{bmatrix} \bar{G}_n & 0 & 0 & \ldots & 0 \\ 0 & \mu_{2n} & 0 & \ldots & 0 \\ \ldots & \ldots & \ldots & \ldots & \ldots \\ 0 & 0 & 0 & \ldots & \mu_{N+1,n} \end{bmatrix}. \qquad (5.5.10)$$

Eq. (5.5.7) determines $\bar{Y}_{n+1}$ at a given $\bar{Y}_n$ with the accuracy till h in the determinate item $\varphi_n(\bar{Y}_n)$ and with the accuracy till $\sqrt{h}$ in the random item $\psi_n(\bar{Y}_n)V_n$.

R e m a r k. The stated method of the substitution of the stochastic differential equation by a difference one in essence does not differ from the Euler method of the numerical integration of the ordinary differential equations.

5.5.4. More Exact Equations

For deriving more accurate difference equations than Eq. (5.5.7) we substitute (5.5.1) by the correspondent integral equation

$$\Delta Y_n = \int\limits_{nh}^{(n+1)h} a(Y_\tau, \tau)d\tau + \int\limits_{nh}^{(n+1)h} b(Y_\tau, \tau)dW_0(\tau)$$

$$+ \sum_{i=1}^{N} \int\limits_{nh}^{(n+1)h} c_i(Y_\tau, \tau)dP_i^0(d\tau). \qquad (5.5.11)$$

For the purpose of the approximate calculation of the integrals we determine Y_τ by means of the linear interpolation of the random function $Y(t)$ at the interval $(nh, (n+1)h)$. Then setting as usual $\Delta Y_n = Y((n+1)h) - Y(nh)$ we have

$$a(Y_\tau, \tau) \cong a\left(\bar{Y}_n + \frac{\tau - nh}{h}\Delta Y_n, \tau\right), \quad b(Y_\tau, \tau) \cong b\left(\bar{Y}_n + \frac{\tau - nh}{h}\Delta Y_n, \tau\right),$$

$$\tag{5.5.12}$$

$$c_i(Y_\tau, \tau) \cong c_i\left(\bar{Y}_n + \frac{\tau - nh}{h}\Delta Y_n, \tau\right), \quad i = 1, \dots, N.$$

For calculating the right-hand sides in these formulae we use the generalized Itô formula (5.1.24). So we have

$$\varphi(Y + dY, t + dt) = \varphi(Y, t) + \{\varphi_t(Y, t) + \varphi_y(Y, t)^T a(Y, t)$$

$$+ \frac{1}{2}\varphi_{yy}(Y, t) : \sigma(Y, t) + \sum_{i=1}^{N}[\varphi(Y + c_i(Y, t), t) - \varphi(Y, t)$$

$$- \varphi_y(Y, t)^T c_i(Y, t)]\nu_i(t)\}dt + \varphi_y(Y, t)^T b(X, t)dW_0$$

$$+ \sum_{i=1}^{N}[\varphi(Y + c_i(Y, t), t) - \varphi(Y, t)]dP_i^0, \tag{5.5.13}$$

where $\varphi_t(y, t)$ is a partial derivative of the function $\varphi(y, t)$ over time t; $\varphi_y(y, t)$ is a matrix whose rows represent the partial derivatives of the matrix-row $\varphi(y, t)^T$ with respect to the components of the vector y; $\varphi_{yy}(y, t) : \sigma(y, t)$ is a vector whose components are the traces of the products of the matrices of the second derivatives of the correspondent components of the vector function $\varphi(y, t)$ over the components of the vector y on the matrix $\sigma(y, t)$:

$$[\varphi_{yy}(y, t) : \sigma(y, t)]_k = \text{tr}[\varphi_{kyy}(y, t)\sigma(y, t)], \tag{5.5.14}$$

$$\sigma(y, t) = b(y, t)\nu_0(t)b(y, t)^T. \tag{5.5.15}$$

It is convenient to transform formula (5.5.13) in such a way that namely the Poisson processes $P_i(t)$ would enter into it directly, not the centered processes $P_i^0(t)$. Taking into account that

$$P_i^0(t) = P_i(t) - \int_0^t \nu_i(\tau)d\tau, \tag{5.5.16}$$

we may rewrite formula (5.5.13) in the form

$$\varphi(Y + dY, t + dt) = \varphi(Y, t) + \{\varphi_t(Y, t) + \varphi_y(Y, t)^T [a(Y, t)$$

$$- \sum_{i=1}^{N} c_i(Y, t)\nu_i(t)] + \frac{1}{2}\varphi_{yy}(Y, t) : \sigma(Y, t)\}dt + \varphi_y(Y, t)^T$$

$$\times b(Y, t)dW_0 + \sum_{i=1}^{N}[\varphi(Y + c_i(Y, t), t) - \varphi(Y, t)]dP_i. \tag{5.5.17}$$

For extending formulae (5.5.13) and (5.5.17) over the matrix functions φ it is necessary to modify the notation of some its members in such a way that the expressions $\varphi_y(y, t)$ in the context of the matrices algebra will have also the sense for the matrix fucntion φ. Taking into account that in the case of the scalar or the vector function φ

$$\varphi_y(y, t)^T u = \sum_{i=1}^{n} \frac{\partial\varphi(y, t)}{\partial y_i} u_i = u^T \frac{\partial}{\partial y}\varphi(y, t)$$

for any p-dimensional vector u we may substitute the items $\varphi_y(Y, t)^T$ $\times\ a(Y, t)$, $\varphi_y(Y, t)^T b(Y, t)dW_0$, $\varphi_y(Y, t)^T c_i(Y, t)$ by the items $a(Y, t)^T$ $\times\ (\partial/\partial y)^T \varphi(Y, t)$, $dW_0^T b(Y, t)(\partial/\partial y)\varphi(Y, t)$, $c_i(Y, t)^T (\partial/\partial y)\varphi(Y, t)$ correspondingly. Therefore formulae (5.5.13) and (5.5.17) will be also valid for the matrix functions φ as $u^T (\partial/\partial y)$ represents a scalar differential operator for any p-dimensional vector u and its application to the vector or the matrix function φ means its application to all components of the vector φ or to all elements of the matrix φ. The variable $\varphi_{yy}(Y, t) : \sigma(Y, t)$ in the case of the matrix function φ represents a matrix whose elements are the traces of the products by the matrix $\sigma(y, t)$ of the matrices of the second derivatives of the correspondent elements of the matrix $\varphi(y, t)$ with respect to the components of the vector y:

$$[\varphi_{yy}(y, t) : \sigma(y, t)]_{kl} = \text{tr}[\varphi_{klyy}(y, t)\sigma(y, t)]. \tag{5.5.18}$$

Using formula (5.5.17) and its modification suitable for the matrix function φ and accounting also (5.5.15) and the fact that according to (5.4.7) and (5.5.16)

$$\frac{\tau - nh}{h}\Delta Y_n = \frac{\tau - nh}{h}a(\bar{Y}_n, nh)h + \frac{\tau - nh}{h}b(\bar{Y}_n, nh)\Delta W_n$$

$$+ \sum_{i=1}^{N} \frac{\tau - nh}{h}c_i(\bar{Y}_n, nh)(\Delta P_{in} - \nu_{in}h), \tag{5.5.19}$$

(where as earlier $\Delta W_n = W_0((n+1)h) - W_0(nh)$, $\Delta P_{in} = P_i((n+1)h) - P_i(nh)$) we shall have with the accuracy till infinitesimal of higher order relatively to h the following equalities:

$$a\left(\bar{Y}_n + \frac{\tau - nh}{h}\Delta Y_n, \tau\right) = a + a_t(\tau - nh) + \frac{\tau - nh}{h}a_y^T$$

$$\times \left(a - \sum_{i=1}^{N} c_i \nu_{in}\right)h + \frac{1}{2}\left(\frac{\tau - nh}{h}\right)^2 (a_{yy} : \sigma)h + \frac{\tau - nh}{h}a_y^T b\Delta W_n$$

$$+ \sum_{i=1}^{N}\left[a\left(\bar{Y}_n + \frac{\tau - nh}{h}c_i, nh\right) - a\right]\Delta P_{in}, \qquad (5.5.20)$$

$$b\left(\bar{Y}_n + \frac{\tau - nh}{h}\Delta Y_n, \tau\right) = b + b_t(\tau - nh) + \frac{\tau - nh}{h}$$

$$\times \left(a^T - \sum_{i=1}^{N} c_i^T \nu_{in}\right)\frac{\partial}{\partial y}bh + \frac{1}{2}\left(\frac{\tau - nh}{h}\right)^2 (b_{yy} : \sigma)h + \frac{\tau - nh}{h}$$

$$\times \Delta W_n^T b^T \frac{\partial}{\partial y}b + \sum_{i=1}^{N}\left[b\left(\bar{Y}_n + \frac{\tau - nh}{h}c_i, nh\right) - b\right]\Delta P_{in}, \qquad (5.5.21)$$

$$c_i\left(\bar{Y}_n + \frac{\tau - nh}{h}\Delta Y_n, \tau\right) = c_i + c_{it}(\tau - nh) + \frac{\tau - nh}{h}c_{iy}^T$$

$$\times \left(a - \sum_{j=1}^{N} c_j \nu_{jn}\right)h + \frac{1}{2}\left(\frac{\tau - nh}{h}\right)^2 (c_{iyy} : \sigma)h + \frac{\tau - nh}{h}c_{iy}^T b\Delta W_n$$

$$+ \sum_{j=1}^{N}\left[c_i\left(\bar{Y}_n + \frac{\tau - nh}{h}c_j, nh\right) - c_i\right]\Delta P_{in}. \qquad (5.5.22)$$

Here the arguments $\bar{Y}_n$, nh at all functions which depend on are omitted them for brevity.

For increasing the accuracy of the calculations, in particular, of the covariance matrix of a normally distributed random vector $\Delta W_n = W_0((n+1)h) - W_0(nh)$and the parameters of Poisson distributions of the random variables $\Delta P_{in} = P_i((n+1)h) - P_i(nh)$ we may take the values of the intensities $\nu_0(t)$ of the process $W_0(t)$ in formula (5.5.15) and $\nu_i(t)$ of Poisson streams which generate the processes $P_i(t)$ at the mean point $nh + h/2$ at the interval $(nh, (n+1)h)$:

$$\sigma(\bar{Y}_n, nh) = b(\bar{Y}_n, nh)\nu_0\left(nh + \frac{h}{2}\right) b(\bar{Y}_n, nh)^T,$$

$$\nu_{in} = \nu_i\left(nh + \frac{h}{2}\right), \quad i = 1, \ldots, N. \tag{5.5.23}$$

To facilitate the further calculations we find the increments of the functions a, b, c_i in the sums of equation (5.5.19) by means of the linear interpolation at small interval $(nh, (n+1)h)$:

$$a\left(\bar{Y}_n + \frac{\tau - nh}{h}c_i, nh\right) - a \cong \frac{\tau - nh}{h}\Delta_i a_n, \quad b\left(\bar{Y}_n + \frac{\tau - nh}{h}c_i, nh\right) - b$$

$$\cong \frac{\tau - nh}{h}\Delta_i b_n, \quad c_i\left(\bar{Y}_n + \frac{\tau - nh}{h}c_j, nh\right) - c_i \cong \frac{\tau - nh}{h}\Delta_j c_{in},$$

$$\tag{5.5.24}$$

$$\Delta_i a_n = a(\bar{Y}_n + c_i, nh) - a(\bar{Y}_n, nh), \quad \Delta_i b_n = b(\bar{Y}_n + c_i, nh) - b(\bar{Y}_n, nh),$$
$$\Delta_j c_{in} = c_i(\bar{Y}_n + c_j, nh) - c_i(\bar{Y}_n, nh).$$

$$\tag{5.5.25}$$

Using formulae (5.5.20)–(5.5.24) we find the following approximate expressions of the integrals in (5.5.11)

$$\int\limits_{nh}^{(n+1)h} a(Y_\tau, \tau)d\tau = ah + \frac{1}{2}\left[a_t + a_y^T\left(a - \sum_{i=1}^{N}c_i\nu_{in}\right) + \frac{1}{3}a_{yy} : \sigma\right]h^2$$

$$+\frac{1}{2}\left(a_y^T b\Delta W_n + \sum_{i=1}^{N}\Delta_i a_n\Delta P_{in}\right)h, \tag{5.5.26}$$

$$\int\limits_{nh}^{(n+1)h} b(Y_\tau, \tau)dW_0(\tau) = b\Delta W_n + \left\{\left[b_t + \left(a^T - \sum_{i=1}^{N}c_i^T\nu_{in}\right)\frac{\partial}{\partial y}b\right]h\right.$$

$$\left. +\Delta W_n^T b^T\frac{\partial}{\partial y}b + \sum_{i=1}^{N}\Delta_i b_n\Delta P_{in}\right\}\int\limits_{nh}^{(n+1)h}\frac{\tau - nh}{h}dW_0(\tau)$$

$$+\frac{1}{2}(b_{yy} : \sigma)\int\limits_{nh}^{(n+1)h}\left(\frac{\tau - nh}{h}\right)^2 dW_0(\tau), \tag{5.5.27}$$

$$\int\limits_{nh}^{(n+1)h} c_i(Y_\tau,\tau)dP_i^0(\tau) = c_i\Delta P_{in} + \left\{\left[c_{it} + c_{iy}^T\left(a - \sum_{j=1}^{N} c_j\nu_{jn}\right)\right]h\right.$$

$$\left. + c_{iy}^T b\Delta W_n + \sum_{j=1}^{N}\Delta_j c_{in}\Delta P_{jn}\right\}\int\limits_{nh}^{(n+1)h}\frac{\tau - nh}{h}dP_i^0(\tau)$$

$$+\frac{1}{2}(c_{iyy}:\sigma)\int\limits_{nh}^{(n+1)h}\left(\frac{\tau - nh}{h}\right)^2 dP_i^0(\tau), \qquad (5.5.28)$$

$$\int\limits_{nh}^{(n+1)h} c_i(Y_\tau,\tau)\nu_i(\tau)d\tau = \left\{c_i h + \frac{1}{2}\left[c_{it} + c_{iy}^T\left(a - \sum_{j=1}^{N} c_j\nu_{jn}\right)\right.\right.$$

$$\left.\left. +\frac{1}{3}c_{iyy}:\sigma\right]h^2 + \frac{1}{2}\left(c_{iy}^T b\Delta W_n + \sum_{j=1}^{N}\Delta_j c_{in}\Delta P_{jn}\right)h\right\}\nu_{in}. \quad (5.5.29)$$

Further on substituting the obtained approximate expressions of the integrals in to (5.5.11) and taking into considerations that $\Delta Y_n = Y((n+1)h) - Y_i(nh) = \bar{Y}_{n+1} - \bar{Y}_n$ we come to the difference equation

$$\bar{Y}_{n+1} = \varphi_n(\bar{Y}_n) + \sum_{i=1}^{N+1}[\psi_{in}(\bar{Y}_n)V_{in} + \psi'_{in}(\bar{Y}_n, V_{in}^{(1)})V'_{in} + \psi''_{in}(\bar{Y}_n)V''_{in}].$$

$$(5.5.30)$$

Here the following denotions are introduced:

$$\varphi_n(\bar{Y}_n) = \bar{Y}_n + \left[a(\bar{Y}_n, nh) - \sum_{j=1}^{N} c_j(\bar{Y}_n, nh)\nu_{jn}\right]h+$$

$$+\frac{1}{2}\left\{a_t(\bar{Y}_n, nh) - \sum_{j=1}^{N} c_{jt}(\bar{Y}_n, nh)\nu_{jn} + \left[a_x(\bar{Y}_n, nh)^T\right.\right.$$

$$\left.- \sum_{j=1}^{N} c_{jy}(\bar{Y}_n, hn)^T \nu_{jn}\right]\left[a(\bar{Y}_n, nh) - \sum_{j=1}^{N} c_j(\bar{Y}_n, hn)\nu_{jn}t\right]$$

$$+\frac{1}{3}\left[a_{yy}(\bar{Y}_n, nh) - \sum_{j=1}^{N} c_{jyy}(\bar{Y}_n, nh)\nu_{jn}\right]:\sigma(\bar{Y}_n, nh)\right\}h^2, \quad (5.5.31)$$

$$\psi_{1n}(\bar{Y}_n) = b(\bar{Y}_n, nh) + \frac{1}{2}\left[a_y(\bar{Y}_{nh}, nh)^T - \sum_{j=1}^{N} c_j(\bar{Y}_n, hn)^T \nu_{jn} \right] b(\bar{Y}_n, nh),$$

$$(5.5.32)$$

$$\psi_{in}(\bar{Y}_n) = c_{i-1}(\bar{Y}_n, nh) + \left[\Delta_{i-1} a_n - \sum_{j=1}^{N} \Delta_{i-1} c_{jn} \nu_{jn} \right] h, \qquad (5.5.33)$$

$$\psi'_{1n}(Y_n, V_n^{(1)}) = \left\{ b(\bar{Y}_n, nh) + \left[a(\bar{Y}_n, nh)^T \right.\right.$$

$$\left. - \sum_{j=1}^{N} c_j(\bar{Y}_n, hn)^T \nu_{jn} \right] \frac{\partial}{\partial y} b(\bar{Y}_n, nh)$$

$$+ V_{1n}^T b(\bar{Y}_n, nh)^T \frac{\partial}{\partial y} b(\bar{Y}_n, nh) \Bigg\} h + \sum_{j=1}^{N} \Delta_j b_n V_{j+1,n}, \qquad (5.5.34)$$

$$\psi'_{in}(\bar{Y}_n, V_n^{(1)}) = \left\{ c_{i-1,t}(\bar{Y}_n, nh) + c_{i-1,y}(\bar{Y}_n, nh)^T \right.$$

$$\times \left[a(\bar{Y}_n, nh) - \sum_{j=1}^{N} c_j(\bar{Y}_n, hn)\nu_{jn} \right] \Bigg\} h$$

$$+ c_{i-1,y}(\bar{Y}_n, nh)^T b(\bar{Y}_n, nh) V_{1,n} + \sum_{j=1}^{N} \Delta_j c_{j-1,n} V_{j+1,n}, \qquad (5.5.35)$$

$$\psi''_{1n}(\bar{Y}_n) = \frac{1}{2}[b_{yy}(\bar{Y}_n, nh) : \sigma(\bar{Y}_n, nh)], \qquad (5.5.36)$$

$$\psi''_{in}(\bar{Y}_n) = \frac{1}{2}[c_{i-1,yy}(\bar{Y}_n, nh) : \sigma(\bar{Y}_n, nh)], \qquad (5.5.37)$$

$$V_{1n} = \Delta W_n = W_0((n+1)h) - W_0(nh), \qquad (5.5.38)$$

$$V_{in} = \Delta P_{i-1,n} = P_{i-1}((n+1)h) - P_{i-1}(nh), \quad i = 2, \ldots, N, \qquad (5.5.39)$$

$$V'_{1n} = \int_{nh}^{(n+1)h} \frac{\tau - nh}{h} dW_0(\tau), \quad V''_{1n} = \int_{nh}^{(n+1)h} \left(\frac{\tau - nh}{h} \right)^2 dW_0(\tau),$$

$$(5.5.40)$$

$$V'_{in} = \int_{nh}^{(n+1)h} \frac{\tau - nh}{h} dP^0_{i-1}(\tau), \tag{5.5.41}$$

$$V''_{in} = \int_{nh}^{(n+1)h} \left(\frac{\tau - nh}{h}\right)^2 dP^0_{i-1}(\tau), \quad i = 2, \ldots, N+1. \tag{5.5.42}$$

Introducing the vector $V^{(1)}_n = [\, V^T_{1n} \; V_{2n} \; \ldots \; V_{n+1} \,]^T$, the block matrix

$$\psi_n(\bar{Y}_n, V^{(1)}_n) = [\, \psi_{1n}(\bar{Y}_n) \; \psi'_{1n}(\bar{Y}_n, V^{(1)}_n) \; \psi''_{1n}(\bar{Y}_n)$$

$$\ldots \psi_{N+1,n}(\bar{Y}_n) \; \psi'_{N+1,n}(\bar{Y}_n, V^{(1)}_n) \; \psi''_{N+1,n}(\bar{Y}_n)\,]$$

and the block random vector

$$V_n = [\, V^T_{1n} \; V'^T_{1n} \; V''^T_{1n} \; V^T_{2n} \; V'^T_{2n} \; V''^T_{2n} \; \ldots, \; V_{N+1,n} \; V'_{N+1,n} \; V''_{N+1,n} \,]^T, \tag{5.5.43}$$

we may write Eq. (5.5.30) in short form

$$\bar{Y}_{n+1} = \varphi_n(\bar{Y}_n) + \psi_n(\bar{Y}_n, V^{(1)}_n)V_n. \tag{5.5.44}$$

The variables which enter into Eqs. (5.5.44) are determined by formulae (5.5.31)–(5.5.43).

5.5.5. Distribution of Random Vector V_n

We find a distribution of the random vector V_n. It is clear that the expectations of the random variables V'_{1n}, V''_{1n} are equal to zero and that the scalar vector $[\, V^T_{1n} \; V'^T_{1n} \; V''^T_{1n} \,]$ has the normal distribution, and the scalar variables V_{2n}, $\ldots$, $V_{N+1,n}$ have the Poisson distributions with the parameters

$$\mu_{in} = \int_{nh}^{(n+1)h} \nu_{i-1}(\tau)d\tau = \nu_{i-1}\left(nh + \frac{1}{2}h\right)h, \quad i = 2, \ldots, N+1. \tag{5.5.45}$$

It is also evident that the random vectors $[\, V^T_{1n}, \; V'^T_{1n}, \; V''^T_{1n} \,]^T, \ldots, [\, V^T_{N+1,n}, \; V'^T_{N+1,n}, \; V''^T_{N+1,n} \,]^T$ are independent by virtue of the independence of the processes $W_0(t)$, $P_1(t), \ldots, P_N(t)$ and that at different n the variables V_n are independent. But at any given i, n the vectors V_{in}, V'_{in}, V''_{in} are dependent.

For complete definition of a distribution of the random vector V_n in Eq. (5.5.44) it is sufficient to find the covariance matrix of the normally distributed random vector $\begin{bmatrix} V_{in}^T & V_{in}'^T & V_{in}''^T \end{bmatrix}^T$. Using the known formulae for the covariance and cross-covariance matrices of the stochastic integrals we find the blocks of the covariance matrix K_{1n} of the random vector $\begin{bmatrix} V_{in}^T & V_{in}'^T & V_{in}''^T \end{bmatrix}^T$:

$$K_{1n,11} = EV_{1n}V_{1n}^T = \int_{nh}^{(n+1)h} \nu_0(\tau)d\tau \cong \nu_0(nh + h/2)h,$$

$$K_{1n,12} = EV_{1n}V_{1n}'^T = \int_{nh}^{(n+1)h} \frac{\tau - nh}{h}\nu_0(\tau)d\tau \cong \frac{1}{2}\nu_0(nh + h/2)h,$$

$$K_{1n,13} = EV_{1n}V_{1n}''^T = \int_{nh}^{(n+1)h} \left(\frac{\tau - nh}{h}\right)^2 \nu_0(\tau)d\tau \cong \frac{1}{3}\nu_0(nh + h/2)h,$$

$$K_{1n,21} = K_{1n,12}, \quad K_{1n,31} = K_{1n,13},$$

$$K_{1n,23} = EV_{1n}V_{1n}''^T = \int_{nh}^{(n+1)h} \left(\frac{\tau - nh}{h}\right)^3 \nu_0(\tau)d\tau \cong \frac{1}{4}\nu_0(nh + h/2)h,$$

$$K_{1n,31} = K_{1n,13}, \quad K_{1n,32} = K_{1n,23},$$

$$K_{1n,33} = EV_{1n}''V_{1n}''^T = \int_{nh}^{(n+1)h} \left(\frac{\tau - nh}{h}\right)^4 \nu_0(\tau)d\tau \cong \frac{1}{5}\nu_0(nh + h/2)h.$$

$$(5.5.46)$$

In practice it is expedient to approximate the stochastic integrals V_{in}', V_{in}'' in (5.5.41), (5.5.42) ($i = 2, \ldots, N+1$) by means of the following analog of the integral theorem about the mean for the stochastic integrals:

$$V_{in}' = \int_{nh}^{(n+1)h} \frac{\tau - nh}{h}dP_{i-1}^0(\tau) \cong \left[\frac{\int_{nh}^{(n+1)h} \left(\frac{\tau-nh}{h}\right)\nu_{i-1}(\tau)d\tau}{\int_{nh}^{(n+1)h} \nu_{i-1}(\tau)d\tau} \right]$$

$$\times \Delta P_{i-1,n} \cong \frac{1}{2}\Delta P_{i-1,n} = \frac{1}{2}V_{in}, \qquad (5.5.47)$$

$$V_{in}'' = \int\limits_{nh}^{(n+1)h} \left(\frac{\tau - nh}{h}\right)^2 dP_{i-1}^0(\tau) \cong \left[\frac{\int\limits_{nh}^{(n+1)h} \left(\frac{\tau-nh}{h}\right)^2 \nu_{i-1}(\tau)d\tau}{\int\limits_{nh}^{(n+1)h} \nu_{i-1}(\tau)d\tau}\right]$$

$$\times \Delta P_{i-1,n} \cong \frac{1}{3}\Delta P_{i-1,n} = \frac{1}{3}V_{in}, \quad i = 2,\ldots,N+1. \qquad (5.5.48)$$

5.5.6. Analysis of More Precise Equations

At the statistical simulation of a system by means of Eq. (5.5.44) it is also not difficult to simulate the random variables distributed according to the normal and the Poisson laws. It is easy to see that the right-hand side of difference equation (5.5.44) is determined with the accuracy till h^2 in the determinate (at a given $\bar{Y}_n$) item $\varphi_n(\bar{Y}_n)$ and with the accuracy till $h^{3/2}$ in the random item $\psi_n(\bar{Y}_n, V_n^{(1)})V_n$.

While deriving Eq. (5.5.44) two small errors took place. Firstly, while replacing $\bar{Y}_\tau$ by the variable $\bar{Y}_n + (\tau - nh)\Delta Y_n/h$ we substituted the random functions $b(Y_\tau,\tau)$ and $c_i(Y_\tau,\tau)$ independent of $dW_0(\tau)$ and $dP_i^0(\tau)$ by the nonrandom functions depending on the random parameter ΔY_n which depends on the values $dW_0(\tau)$ and $dP_i^0(\tau)$ at the interval $(nh,(n+1)h)$. Secondly, at $c_i(y,t) \neq 0$ even at one i the realizations of the random process $Y(t)$ have the discontinuities of the first kind in the random points despite of m.s. continuity of $Y(t)$. Therefore strictly speaking we cannot perform the linear interpolation of a given process. The first error may be removed by two ways. The first one consists in the replacement of the interpolation of the process $Y(t)$ by its extrapolation what is equivalent to the substitution of ΔY_n by the variable ΔY_{n-1} in the obtained expression for Y_τ. But this way will lead to the appearance of the variables $\bar{Y}_{n-1}$ and $V_{n-1}^{(1)}$ in the right-hand side of a difference equation, i.e. to the substitution of the equation of the first order by the difference equation of the second order. The second way consists in the abandonment of the interpolation of the process $Y(t)$ at the interval and in the direct expression of the increments of the functions $a(Y_\tau,\tau)$, $b(Y_\tau,\tau)$, $c_i(Y_\tau,\tau)$ at small interval $(nh,(n+1)h)$ according to the generalized Itô formula with the substitution the differentials by the increments. At this method the second error is also removed. But the difference equation obtained in such a way will be more complicated. The random variables which represent two-fold integrals over the components of the Wiener process $W(t) = W_0(t)$ and over the Poisson processes will be included into this equation:

$$\int\limits_{nh}^{(n+1)h}\int\limits_{nh}^{\tau} dW_j(\sigma)dW_j(\tau), \qquad \int\limits_{nh}^{(n+1)h}\int\limits_{nh}^{\tau} dP_i(\sigma)dP_j^0(\tau),$$

$$\int\limits_{nh}^{(n+1)h}\int\limits_{nh}^{\tau} dP_i^0(\sigma)dW_j(\tau), \qquad \int\limits_{nh}^{(n+1)h}\int\limits_{nh}^{\tau} dP_j^0(\sigma)dW_i(\tau). \qquad (5.5.49)$$

It is very difficult to find the distributions of these random variables and only the first two of them are easily calculated at $j = i$:

$$\int\limits_{nh}^{(n+1)h}\int\limits_{nh}^{\tau} dW_i(\sigma)dW_i(\tau) = \frac{[\Delta W_{in}]^2 - \nu_{ii}(nh + h/2)h}{2},$$

$$\int\limits_{nh}^{(n+1)h}\int\limits_{nh}^{\tau} dP_i^0(\sigma)dP_i^0(\tau) = \frac{[\Delta P_{in}]^2 - \Delta P_{in}}{2}. \qquad (5.5.50)$$

As regards to the second error it cannot essentially influence on the result as the probability of the occurrence of a jump of the Poisson process on sufficiently small interval $(nh, (n + 1)h)$ is infinitesimal.

The accuracy of the approximation of a stochastic differential equation by a difference one may be increased later on. In particular, in one of the ways it is sufficient to express $a(Y_\tau, \tau)$, $b(Y_\tau, \tau)$, $c_i(Y_\tau, \tau)$ at the interval $(nh, (n + 1)h)$ by the Itô integral formula:

$$a(Y_\tau, \tau) = a(\bar Y_n, nh) + \int\limits_{nh}^{\tau} a_t(Y_s, s) + a_x(Y_s, s)^T \left[a(Y_s, s) \right.$$

$$-\sum_{i=1}^{N} c_i(Y_s, \nu_i(s)) + \frac{1}{2}a_{yy}(Y_s, s) : \sigma(Y_s, s) \bigg] ds + \int\limits_{nh}^{\tau} a_y(Y_s, s)^T$$

$$\times b(Y_s, s)dW_0(s) + \sum_{i=1}^{N}\int\limits_{nh}^{\tau} [a(Y_s + c_i(Y_s, s), s) - a(Y_s, s)]dP_i^0(s). \qquad (5.5.51)$$

Formulae for $b(Y_\tau, \tau)$, $c_i(Y_\tau, \tau)$ are analogous. To the integrals obtained we may then to apply the same method which was used for the calculation of the integral in (5.5.11). As a result we shall obtain the right-hand side of the difference equation with the accuracy till h^3 in the determinate (at a given $\bar Y_n$) item and $h^{5/2}$ in the random item. The process of the more precise definition of the difference equation correspondent to a given stochastic differential equation may be continued. But each new correction requires the existence of the derivatives of the functions a, b, c_i of more higher orders.

We may use another method for precising the difference equation. Namely, the integrands in (5.5.51) and in analogous formulae for $b(Y_\tau, \tau)$, $c_i(Y_\tau, \tau)$ we may express by means of the generalized Itô formula substituting in it the differentials by the increments. Here the three-fold integrals over the components of the Wiener process $W_0(t)$ and the Poisson processes $P_i(t)$ will enter into the difference equation. For further correction of the approximation the stochastic differential equation by the difference one in this case the integrands in (5.5.51) and in the correspondent formulae for $b(Y_\tau, \tau)$, $c_i(Y_\tau, \tau)$ in its turn should be presented by the integral Itô formula and later on may be used the differential Itô formula with the change of the differentials by the increments. This process may be continued further and as a result it will lead to the presentation of the process $Y(t)$ at the interval $(nh, (n+1)h)$ by the stochastic analogs of Taylor formula. In this case the multiple stochastic integrals over the Wiener process $W_0(t)$ and over the Poisson processes $P_i(t)$ will enter into the difference equation. The finding of a distribution of these integrals represents insuperable difficulty. And only the integrals of any multiplicity over one and the same component of the Wiener process $W_0(t)$ or over one and the same Poisson process $P_i(t)$ are calculated very simply.

In order to avoid the calculations of the derivatives of the functions $a(Y_\tau, \tau)$, $b(Y_\tau, \tau)$, $c_i(Y_\tau, \tau)$ while using two stated ways of the approximation of a stochastic differential equation by a difference one, we may recommend to substitute them by the retios of the finite increments, for instance, at the interval $(nh, (n+1)h)$ over the variable t and at the intervals $(\bar{Y}_{nk}, \bar{Y}_{nk} + a_k(\bar{Y}_n, nh)h)$ over the components of the vector y.

5.5.7. Strong and Weak Approximations of Nonlinear Stochastic Differential Equations

The obtained in Subsections 5.5.2–5.5.6 difference equations may be used both for the theoretical research and for the numerical integration of the stochastic differential equations. In both cases we must know the distribution of all random variables which enter into the difference equations. In a given case the difference equations will represent the so-called *a strong approximation* of the stochastic differential equations. While numerical integrating such an approximation is necessary if is required to obtain the realizations of the process $Y(t)$. But often there is no need to obtain the realizations of the process but it is sufficient only to have the estimates of the moments or of the expectations of some functions of the value at some instant of the random process $Y(t)$. In such cases we may give up the use of the exact distributions entering into the difference equations of the random variables and substitute them by some more simple distributions with the same moment characteristics. At the substitution of the random variables by the variables with other distributions the difference equation will represent *a weak approximation* of a stochastic differential equation.

For the stationary and nonstationary stochastic linear differential equations standard numerical methods of the ordinary differential equations may be used.

Systematic presentation of the modern numerical methods for the nonlinear stochastic differential equations integration based on a strong and a weak approximation and different classes of the numerical schemes are considered for, example, in (Artemiev and Averina 1997, Kloeden and Platen 1994, 1997, 1999).

5.6. Multi-Dimensional Distributions in Stochastic Differential Systems

5.6.1. One-Dimensional Characteristic Function

Let us consider the system whose state vector (in the general case the extended state vector) is described by the Itô stochastic differential Eq. (5.4.6) where V is a white noise in the strict sense (Subsection 4.6.4). The problem is to find all the multi-dimensional distributions of the state vector of the system $Y(t)$, supposing that the one-dimensional distribution (and consequently, all the multi-dimensional distributions (Subsection 4.6.3) of the process with the independent increments

$$W(t) = W(t_0) + \int\limits_{t_0}^{t} V(\tau)d\tau \qquad (5.6.1)$$

is known.

▷ In order to find the one-dimensional characteristic function of the random process $Y(t)$ let us consider two time instants t and $t + \Delta t$. According to the definition of a characteristic function (Subsection 3.5.1) the values of the one- dimensional characteristic fucntion $g_1(\lambda; t)$ of the process $Y(t)$ at the instants t and $t + \Delta t$ are expressed by

$$g_1(\lambda; t) = Ee^{i\lambda^T Y(t)}, \quad g_1(\lambda; t + \Delta t) = Ee^{i\lambda^T Y(t+\Delta)}. \qquad (5.6.2)$$

Subtracting the first formula (5.6.2) from the second one termwise we shall have

$$g_1(\lambda; t + \Delta t) - g_1(\lambda; t)$$

$$= E\left[e^{i\lambda^T Y(t+\Delta)} - e^{i\lambda^T Y(t)}\right] = E\left[e^{i\lambda^T \Delta Y(t)} - 1\right]e^{i\lambda^T Y(t)}. \qquad (5.6.3)$$

But it follows from Eq. (5.4.6) and Eq. (5.6.1) that with accuracy up to the infinite-simals of higher order

$$\Delta Y(t) = Y(t + \Delta) - Y(t) = a(Y(t), t)\Delta t + b((Y(t), t)\Delta W, \qquad (5.6.4)$$

or omitting the argument t of the function $Y(t)$ in (5.6.4) we get

$$\Delta Y(t) = a(Y, t)\Delta t + b(Y, t)\Delta W, \ \ \Delta W = W(t + \Delta) - W(t). \qquad (5.6.5)$$

Substituting (5.6.5) into (5.6.3) we find

$$g_1(\lambda; t + \Delta t) - g_1(\lambda; t) = E\left\{ e^{i\lambda^T [a(Y,t)\Delta t + b(Y,t)\Delta W]} - 1 \right\} e^{i\lambda^T Y}$$

$$= E\left\{ e^{i\lambda^T [a(Y,t)\Delta t + b(Y,t)\Delta W]} - e^{i\lambda^T b(Y,t)\Delta W} \right.$$

$$\left. + e^{i\lambda^T b(Y,t)\Delta W} - 1 \right\} e^{i\lambda^T Y(t)},$$

or with accuracy up to the infinitesimals of higher order relative to Δt

$$g_1(\lambda; t + \Delta t) - g_1(\lambda; t)$$

$$= E\left\{ e^{i\lambda^T b(Y,t)\Delta W} \left(e^{i\lambda^T a(Y,t)\Delta t} - 1 \right) + e^{i\lambda^T b(Y,t)\Delta W} - 1 \right\} e^{i\lambda^T Y}$$

$$= E\left\{ e^{i\lambda^T b(Y,t)\Delta W} i\lambda^T a(Y, t)\Delta t + e^{i\lambda^T b(Y,t)\Delta W} - 1 \right\} e^{i\lambda^T Y}.$$

$$(5.6.6)$$

For calculating the expectation in (5.6.6) let us use the formula of total expectation (Subsection 3.4.2) taking first the conditional expectation at fixed value y of the random variable Y, and then taking the expectation with the account of the randomness of Y. As in the Itô Eq. (5.4.6) the random variables $Y = Y(t)$ and $\Delta W = W(t + \Delta t) - W(t)$ are independent, the conditional expectation of any function of the random variable ΔW at a given y does not depend on Y and is equal to the unconditional expectation of this function. Consequently,

$$E e^{i\lambda^T b(Y,t)\Delta W} i\lambda^T a(Y, t) e^{i\lambda^T Y} = E\left[e^{i\lambda^T b(Y,t)\Delta W} i\lambda^T a(Y, t) e^{i\lambda^T Y} \mid Y \right]$$

$$= E\left\{ i\lambda^T a(Y, t) e^{i\lambda^T Y} E\left[e^{i\lambda^T b(Y,t)\Delta W} \mid Y \right] \right\}$$

$$= E\left[i\lambda^T a(Y, t) e^{i\lambda^T Y} E e^{i\lambda^T b(Y,t)\Delta W} \right] \qquad (5.6.7)$$

and similarly

$$E e^{i\lambda^T b(Y,t)\Delta W + i\lambda^T Y} = E\left[e^{i\lambda^T Y} E e^{i\lambda^T b(Y,t)\Delta W} \right], \qquad (5.6.8)$$

the expectation in brackets being taken at fixed Y. But $Ee^{i\mu^T \Delta W}$ represents the characteristic function of the increment ΔW of the process $W(t)$ which we shall denote by $h(\mu; t, t + \Delta t)$:

$$Ee^{i\mu^T \Delta W} = h(\mu; t, t + \Delta t). \tag{5.6.9}$$

Therefore

$$Ee^{i\lambda^T b(Y,t)\Delta W} = h(b(Y,t)^T \lambda; t, t + \Delta t). \tag{5.6.10}$$

and formulae (5.6.7) and (5.6.8) take the form

$$Ee^{i\lambda^T b(Y,t)\Delta W} i\lambda^T a(Y,t)e^{i\lambda^T Y} = Ei\lambda^T a(Y,t)h(b(Y,t)^T \lambda; t, t + \Delta t)e^{i\lambda^T Y}, \tag{5.6.11}$$

$$Ee^{i\lambda^T b(Y,t)\Delta W + i\lambda^T Y} = Eh(b(Y,t)^T \lambda; t, t + \Delta t)e^{i\lambda^T Y}. \tag{5.6.12}$$

Substituting expressions (5.6.11) and (5.6.12) into (5.6.6) we obtain

$$g_1(\lambda; t + \Delta t) - g_1(\lambda; t) = E\{i\lambda^T h(b(Y,t)^T \lambda; t, t + \Delta t)\Delta t$$

$$+ h(b(Y,t)^T \lambda; t, t + \Delta t) - 1\}e^{i\lambda^T Y}. \tag{5.6.13}$$

Now we use formula (4.6.10) determining the characteristic function of the increment of a process with independent increments in terms of its one-dimensional characteristic function. Then denoting by $h_1(\mu; t)$ the one-dimensional characteristic function of the process $W(t)$ we shall have

$$h(\mu; t, t + \Delta t) - 1 = \frac{h(\mu; t + \Delta t)}{h_1(\mu; t)} - 1 = \frac{h_1(\mu; t + \Delta t) - h_1(\mu; t)}{h_1(\mu; t)}. \tag{5.6.14}$$

Supposing that $h_1(\mu; t)$ has the continuous time derivative and using the Lagrange finite increments formula we find from (5.6.14)

$$h_1(\mu; t + \Delta t) - h_1(\mu; t) = \frac{\partial h_1(\mu; \tau)}{\partial t} \tag{5.6.15}$$

where $\tau \in (t, t + \Delta t)$. Consequently, (5.6.14) is equal to

$$h(\mu; t + \Delta t) - 1 = \frac{1}{h_1(\mu; t)} \frac{\partial h_1(\mu; \tau)}{\partial t} \Delta t,$$

or with the accuracy up to the infinitesimals of higher orders relative to Δt

$$h(\mu; t + \Delta t) - 1 = \frac{1}{h_1(\mu; t)} \frac{\partial h_1(\mu; t)}{\partial t} \Delta t, \tag{5.6.16}$$

Putting

$$\chi(\mu;t) = \frac{1}{h_1(\mu;t)} \frac{\partial h_1(\mu;t)}{\partial t} \qquad (5.6.17)$$

we find for (5.6.16)

$$h(\mu;t+\Delta t) - 1 = \chi(\mu;t)\Delta t. \qquad (5.6.18)$$

Substituting (5.6.18) into (5.6.13) we get with the accuracy up to the infinitesimals of higher order relative to Δt

$$g_1(\lambda;t+\Delta t) - g_1(\lambda;t) = E\left\{i\lambda^T a(Y,t) + \chi(b(Y,t)^T\lambda;t)\right\} e^{i\lambda^T Y} \Delta t. \qquad (5.6.19)$$

Finally, dividing both parts of this formula by Δt and passing to the limit at $\Delta t \to 0$ we obtain the equation which determines the one-dimensional characteristic function $g_1(\lambda;t)$ of the state vector $Y = Y(t)$:

$$\frac{\partial g_1(\lambda;t)}{\partial t} = E\left\{i\lambda^T a(Y,t) + \chi(b(Y,t)^T\lambda;t)\right\} e^{i\lambda^T Y}. \vartriangleleft \qquad (5.6.20)$$

The expectation in the right-hand side of (5.6.20) is determined by the one-dimensional distribution of the process $Y(t)$ which in turn is completely determined by its one-dimensional characteristic function $g_1(\lambda;t)$. Consequently, the right-hand side of formula (5.6.20) at a given t represents a functional of the characteristic function $g_1(\lambda;t)$. Therefore (5.6.20) represents the equation determining $g_1(\lambda;t)$.

Let $Y_0 = Y(t_0)$ be the initial value of the state vector of the system at the instant t_0 representing a random variable independent of the values of the white noise $V(t)$ at $t \geq t_0$. Denote the characteristic function of the random variable Y_0 by $g_0(\lambda)$. Then the initial condition for Eq. (5.6.20) will be

$$g_1(\lambda;t_0) = g_0(\lambda). \qquad (5.6.21)$$

Eq. (5.6.20) *and the initial condition (5.6.21) determine completely and uniquely the one-dimensional characteristic function $g_1(\lambda;t)$ of the state vector of the system (5.4.6) at any time instant $t \geq t_0$.*

Eq. (5.6.20) was first obtained at the begining of the 1940s by Pugachev (Pugachev 1944) who apparently was the first who studied stochastic differential equations with an arbitrary process with the independent increments. Bernstein, Itô and Gukhman studied only the stochastic differential equations with the Wiener processes (Bernstein 1934, 1938, Itô 1951a, b, Gikhman 1947, 1950a, b).

R e m a r k 1. It goes without saying that Eq. (5.6.20) is valid only if the expectation in the right-hand side exists. It follows from our derivation of Eq. (5.6.20) that in this case $g_1(\lambda; t)$ is differentiable with respect to t and satisfies Eq. (5.6.20). Some sufficient conditions of the existence of the one-dimensional distributions are given in Subsection 5.6.12.

R e m a r k 2. We emphasize that Eq. (5.6.20) is valid only if Eq. (5.4.6) is an Itô equation. Only under this condition the instantaneous value of the state vector of the system $Y(t)$ is independent of the increment $\Delta W = W(t + \Delta t) - W(t)$ of the process $W(t)$, and the conditional expectation of the random variable $e^{i\mu^T \Delta W}$ relative to Y coincides with its unconditional expectation. We used this fact while deriving formula (5.6.13).

5.6.2. Multi-Dimensional Characteristic Functions

Analogously the equations determining other multi-dimensional characteristic functions of the state vector of a system are derived.

▷ Let

$$g_n(\lambda_1, \ldots, \lambda_n; t_1, \ldots, t_n) = E \exp\left\{i\sum_{k=1}^{n} \lambda_k^T Y(t_k)\right\}, \qquad (5.6.22)$$

be the n-dimensional characteristic function of the state vector Y of a system. Taking two values t_n and $t_n + \Delta t$ of the last argument t_n at fixed values of all preceding arguments $t_1 < t_2 < \cdots < t_{n-1} < t_n$ and repeating the arguments of Subsection 5.6.1 we obtain the following equation for the n-dimensional characteristic function of the process $Y(t)$:

$$\frac{\partial g_n(\lambda_1, \ldots, \lambda_n; t_1, \ldots, t_n)}{\partial t_n}$$

$$= E\left\{\left[i\lambda_n^T a(Y_{t_n}, t_n) + \chi(\lambda_n; Y_{t_n}, t_n)\right] \exp\left[i\sum_{k=1}^{n} \lambda_k^T Y_{t_k}\right]\right\} \qquad (5.6.23)$$

$$(n = 1, 2, \ldots).$$

The right-hand side of this equality is determined by the n-dimensional distribution of the process $Y(t)$, i.e. depends on $g_n(\lambda_1, \ldots, \lambda_n; t_1, \ldots, t_n)$. Consequently, (5.6.23) represents an equation relative to $g_n(\lambda_1, \ldots, \lambda_n; t_1, \ldots, t_n)$. In order to find the initial condition for Eq. (5.6.23) we put in (5.6.22) $t_n = t_{n-1}$:

$$g_n(\lambda_1, \ldots, \lambda_n; t_1, \ldots, t_{n-1}, t_{n-1})$$

$$= E \exp \left\{ i(\lambda_1^T Y_{t_1} + \cdots + \lambda_{n-1}^T Y_{t_{n-1}} + \lambda_n^T Y_{t_{n-1}}) \right\}$$

$$= g_{n-1}(\lambda_1, \ldots, \lambda_{n-1} + \lambda_n; t_1, \ldots, t_{n-1}). \qquad (5.6.24)$$

The right-hand side of (5.6.23) represents the $(n-1)$-dimensional characteristic function of the process $Y(t)$ with λ_{n-1} replaced by the sum $\lambda_{n-1} + \lambda_n$. Consequently, (5.6.24) is equal to

$$g_n(\lambda_1, \ldots, \lambda_n; t_1, \ldots, t_{n-1}, t_{n-1})$$

$$= g_{n-1}(\lambda_1, \ldots, \lambda_{n-2}, \lambda_{n-1} + \lambda_n; t_1, \ldots, t_{n-1}) \ (n = 2, 3, \ldots).$$
$$(5.6.25)$$

This equality serves as the initial condition for Eq. (5.6.23) at $t_1 < t_2 < \cdots < t_{n-1} < t_n$.

In order to find $g_n(\lambda_1, \ldots, \lambda_n; t_1, \ldots, t_n)$ at any $t_1, \ldots, t_n$ suppose that for some permutations $(s_1, \ldots, s_n)$ of the numbers $(1, \ldots, n)$ $t_{s_1} < t_{s_2} < \cdots < t_{s_{n-1}} < t_{s_n}$ (such a permutations always exists). Then by virtue of the condition of symmetry of multi-dimensional distributions

$$g_n(\lambda_1, \ldots, \lambda_n; t_1, \ldots, t_n) = g_n(\lambda_{s_1}, \ldots, \lambda_{s_n}; t_{s_1}, \ldots, t_{s_n}).$$
$$(5.6.26)$$

This relation determines $g_n(\lambda_1, \ldots, \lambda_n; t_1, \ldots, t_n)$ at all $\lambda_1, \ldots, \lambda_n;$ $t_1, \ldots, t_n$, if this function is known for any ordered set of time instants. ◁

Thus Eq. (5.6.20) *with initial condition* (5.6.21), *Eq.* (5.6.23) *with initial condition* (5.6.25) *and condition of symmetry of multi-dimensional distributions* (5.6.26) *at* $n = 2, 3, \ldots$ *determine successively all the multi-dimensional distributions of the state vector* $Y(t)$ *of the system* (5.4.6).

R e m a r k 1. Sufficient conditions of the existence of the multi-dimensional distributions are given in Subsection 5.6.12.

R e m a r k 2. According to the Kolmogorov theorem (Subsection 2.6.2) the multi-dimensional distributions of a random function uniquely determine its distribution in the corresponding function space. Thus the equations for the multi-dimensional characteristic function given here determine uniquely the random function distribution in functional space too. These equations represent generally ordinary linear homogeneous differential equations with unbounded operators in Banach spaces of continuous bounded functions of the variables λ and $\lambda_1, \ldots, \lambda_n$ respectively.

5.6.3. Specific Form of Equation for Characteristic Functions

Suppose that the one-dimensional density $f_1(y; t)$ of the state vector of a system Y exists. Then the expectation in Eq. (5.6.20) is easily calculated. Using formula (5.6.17) in the form

$$\chi(\mu; t) = \chi(b(y, t)^T \lambda; t) \qquad (5.6.27)$$

we have in Eq. (5.6.20)

$$\frac{\partial g_1(\lambda;t)}{\partial t} = \int_{-\infty}^{\infty} \left[i\lambda^T a(y,t) + \chi(b(y,t)^T \lambda;t) \right] e^{i\lambda^T y} f_1(y;t) dy. \quad (5.6.28)$$

But

$$f_1(y;t) = \frac{1}{(2\pi)^p} \int_{-\infty}^{\infty} e^{-i\mu^T y} g_1(\mu;t) d\mu, \quad (5.6.29)$$

where p is the dimension of the state vector Y, and the integral with respect to every component of the p-dimensional vector μ is assumed as the principal value of the integral in the sense of Cauchy if $g_1(\mu;t)$ is not absolutely integrable. Eqs. (5.6.28), (5.6.29) represent a set of two equations relative to $g_1(\lambda;t)$ and $f_1(y;t)$. Substituting $f_1(y;t)$ from Eq. (5.6.29) into Eq. (5.6.28) we obtain the linear integrodifferential equation for $g_1(\lambda;t)$:

$$\frac{\partial g_1(\lambda;t)}{\partial t} = \frac{1}{(2\pi)^p} \int_{-\infty}^{\infty} \int_{-\infty}^{\infty} [i\lambda^T a(y,t) + \chi(b(y,t)^T \lambda;t)]$$

$$\times e^{i(\lambda^T - \mu^T)y} g_1(\mu;t) d\mu dy. \quad (5.6.30)$$

Analogously supposing that all the multi-dimensional densities of the process $Y(t)$ exist we transform Eq. (5.6.23) into the form of a linear integrodifferential equation relative to $g_n(\lambda_1, \ldots, \lambda_n; t_1, \ldots, t_n)$:

$$\frac{\partial}{\partial t_n} g_n(\lambda_1, \ldots, \lambda_n; t_1, \ldots, t_n)$$

$$= \frac{1}{(2\pi)^{np}} \int_{-\infty}^{\infty} \cdots \int_{-\infty}^{\infty} \left[i\lambda^T a(y_n, t_n) + \chi(b(y_n, t_n)^T \lambda_n; t_n) \right]$$

$$\times \exp\left\{ i \sum_{k=1}^{n} (\lambda_k^T - \mu_k^T) y_k \right\} g_n(\mu_1, \ldots, \mu_n; t_1, \ldots, t_n)$$

$$\times d\mu_1 \ldots d\mu_n dy_1 \ldots dy_n. \quad (5.6.31)$$

5.6.4. Equations for Multi-Dimensional Densities

Analogously replacing in Eq. (5.6.28) the variable of integration y by η, multiplying this equation by $(2\pi)^{-p}e^{-i\lambda^T y}$ and integrating it with respect to λ we obtain the integrodifferential equation for the one-dimensional density $f_1(y;t)$:

$$\frac{\partial f_1(y;t)}{\partial t} = \frac{1}{(2\pi)^p} \int\limits_{-\infty}^{\infty} \int\limits_{-\infty}^{\infty} [i\lambda^T a(\eta,t) + \chi(b(\eta,t)^T\lambda;t)]e^{i\lambda^T(\eta-y)} f_1(\eta;t)\,d\eta\,d\lambda.$$

(5.6.32)

In exactly the same way from Eq. (5.6.31) in which the expectation is expressed in terms of the n-dimensional density we obtain:

$$\frac{\partial}{\partial t_n} f_n(y_1,\ldots,y_n;t_1,\ldots,t_n)$$

$$= \frac{1}{(2\pi)^{np}} \int\limits_{-\infty}^{\infty} \cdots \int\limits_{-\infty}^{\infty} \left[i\lambda_n^T a(\eta_n,t_n) + \chi(b(\eta_n,t_n)^T\lambda_n;t_n) \right]$$

$$\times \exp\left\{ i\sum_{k=1}^{n} \lambda_k^T(\eta_k - y_k) \right\} f_n(\eta_1,\ldots,\eta_n;t_1,\ldots,t_n)$$

$$\times d\eta_1\ldots d\eta_n\,d\lambda_1\ldots d\lambda_n.$$

(5.6.33)

The initial condition for Eq. (5.6.23) has the form

$$f_1(y;t_0) = f_0(y),$$

(5.6.34)

where $f_0(y)$ is the density of the initial value Y_0 of the state vector. The initial condition for Eq. (5.6.33) is

$$f_n(y_1,\ldots,y_{n-1},y_n;t_1,\ldots,t_{n-1},t_{n-1})$$

$$= f_{n-1}(y_1,\ldots,y_{n-1};t_1,\ldots,t_{n-1})\delta(y_n - y_{n-1}),$$

(5.6.35)

(δ-function in the right-hand side is explained by the fact that at $t_n = t_{n-1}$ the random variable $Y(t_n)$ coincides with the variable $Y(t_{n-1})$ with probability 1).

5.6.5. Formulae for Function $\chi(\mu; t)$

A specific form of the function $\chi = \chi(\mu; t)$ in the obtained equations for the characteristic functions is determined by the character of the process with the independent increments $W(t)$.

1^0 If $W(t)$ is the Wiener process, then its one-dimensional characteristic function $h_1(\mu; t)$ in accordance with Subsection 4.6.5 is determined by the formula

$$h_1(\mu; t) = \exp\left\{ -\frac{1}{2}\mu^T \int_0^t \nu(\tau)d\tau\mu \right\}. \tag{5.6.36}$$

According to (5.6.17) the function $\chi(\mu; t)$ represents the logarithmic derivative of the characteristic function $h_1(\mu; t)$ with respect to t: $\chi(\mu; t) = \partial/\partial t \ln h_1(\mu; t)$. Substituting here the expression of the function $h_1(\mu; t)$ we find

$$\chi(\mu; t) = -\mu^T \nu(t)\mu/2. \tag{5.6.37}$$

2^0 If $W(t)$ represents the general Poisson process then according to the result of Subsection 4.6.6

$$h_1(\mu; t) = \exp\left\{ [g(\mu) - 1] \int_0^t \nu(\tau)d\tau \right\}, \tag{5.6.38}$$

where $g(\mu)$ is the characteristic function of the jumps, and $\nu(\tau)$ is the intensity of the jumps stream $W(t)$. After taking the logarithmic derivative of the function $\chi(\mu; t)$ we find the function $\chi(\mu; t)$ for the general Poisson process

$$\chi(\mu; t) = [g(\mu) - 1]\nu(t). \tag{5.6.39}$$

In the special case of the simple Poisson process with the unit jumps we have $g(\mu) = e^{i\mu}$.

3^0 If the process $W(t)$ represents the linear combination of the independent Wiener and Poisson processes:

$$W(t) = W_0(t) + \sum_{k=1}^{N} c_k P_k(t), \tag{5.6.40}$$

then the function $\chi(\mu; t)$ is determined by

$$\chi(\mu; t) = -\frac{1}{2}\mu^T \nu_0(t)\mu + \sum_{k=1}^{N} \left[g_k(c_k^T \mu) - 1 \right] \nu_k(t), \tag{5.6.41}$$

where $\nu_0(t)$ is the intensity of the Wiener process $W_0(t)$, $g_k(\lambda)$ is the characteristic function of the jumps of the general Poisson process $P_k(t)$ and $\nu_k(t)$ is the intensity of its jumps stream.

R e m a r k 1. While deriving formulae (5.6.39) and (5.6.41) there is no need to assume that the expectation of the process $W(t)$ may be different from zero since Eqs. (5.6.20) and (5.6.23) are valid in this case. If the expectation of $W(t)$ is zero then all the Poisson processes involved should be replaced by the corresponding centered ones, and the functions $g(\mu)$ and $g_k(\rho)$ in (5.6.41) should be replaced by $g(\mu)-i\mu^T m_0$ and $g_k(\rho)-i\rho^T m_k$ respectively, m_0 and m_k being the expectations of the jumps of the processes $W(t)$ and $P_k(t)$.

R e m a r k 2. The Poisson processes especially simple one are of special importance for the queueing theory and for the reliability theory of technical systems connected with it. Using the approximate description of the queueing processes and of the processes of failure and restoration of failed elements by means of stochastic differential equations one should use formulae (5.6.39) or (5.6.41) for the function $\chi(\mu;t)$ at $\nu_0 = 0$. Here as a rule there will be $g(\mu) = g_k(\mu) = e^{i\mu}$.

4^0 If the vector $W(t)$ consists of N independent subvectors, i.e. $W(t) = [\,W_1(t)^T \ldots W_N(t)^T\,]^T$ then while partitioning the vector μ into the correspondent blocks $\mu = [\,\mu_1^T \ldots \mu_N^T\,]^T$ we have

$$h_1(\mu;t) = [\,h_{11}(\mu_1;t) \ldots h_{1N}(\mu_N;t)\,]\,, \qquad (5.6.42)$$

$$\chi(\mu;t) = \sum_{k=1}^{N} \chi_k(\mu_k;t)\,, \qquad (5.6.43)$$

where $h_{1k}(\mu_k;t)$ and $\chi_k(\mu_k;t)$ are the one-dimensional characteristic function and the correspondent function χ of the process $W_k(t)$. Hence partitioning the matrix $b(y,t)$ correspondingly, $b(y,t) = [b_1(y,t) \ldots b_N(y,t)]$, we find

$$\chi(b(y,t)^T\lambda;t) = \sum_{k=1}^{N} \chi_k(b_k(y,t)^T\lambda;t)\,. \qquad (5.6.44)$$

5^0 In the general case the process with the independent increments $W(t)$ is determined by formula (Subsection 4.6.3)

$$W(t) = W_0(t) + \int_{R_0^q} c(x)P(t,dx)\,. \qquad (5.6.45)$$

Differentiating over t the expression

$$\ln h_1(\lambda;t) = -\frac{1}{2}\lambda^T \nu_0(t)\lambda + \int\limits_{R_0^q} \left[e^{i\lambda^T c(x)} - 1 - i\lambda^T c(x) \right] \mu(t,x)dx$$

$$(5.6.46)$$

for the one-dimensional characteristic function of the process $W(t)$ and accounting the formula for an expectation of the Poisson process and the formula for a covariance function of the Wiener process we get

$$\chi(\rho;t) = -\frac{1}{2}\rho^T \nu_0(t)\rho + \int\limits_{R_0^q} \left[e^{i\rho^T c(x)} - 1 - i\rho^T c(x) \right] \nu_P(t,x)dx \,.$$

$$(5.6.47)$$

Here $\nu_0(t)$ is the intensity of the Wiener process $W_0(t)$ and $\nu_P(t,x)dx = [\partial\mu(t,x)/\partial t]\,dx$ is the intensity of the Poisson stream of the jumps equal to $c(x)$ of the process $W(t)$ (Subsection 4.6.3).

5.6.6. *Equations for Multi-Dimensional Densities in Case of Wiener Process*

$\triangleright$ Let us consider now the special case of the Wiener process $W(t)$. In this case in accordance with (5.6.37) we have

$$\chi(b(\eta,t)^T\lambda;t) = -\frac{1}{2}\lambda^T b(\eta,t)\nu(t)b(\eta,t)^T \lambda \,,\qquad (5.6.48)$$

and Eq. (5.6.32) takes the form

$$\frac{\partial f_1(y;t)}{\partial t} = \frac{1}{(2\pi)^p} \int\limits_{-\infty}^{\infty} \int\limits_{-\infty}^{\infty} \left[i\lambda^T a(\eta,t) - \frac{1}{2}\lambda^T b(\eta,t)\nu(t)b(\eta;t)^T \lambda \right]$$

$$\times e^{i\lambda^T (\eta,y)} f_1(\eta;t)d\eta d\lambda \,. \qquad (5.6.49)$$

or taking into consideration that $u^T A u = \mathrm{tr}\,(uu^T A)$ for any n-vector u and any $n \times n$-matrix A we get from (5.6.49)

$$\frac{\partial f_1(y;t)}{\partial t} = \frac{1}{(2\pi)^p} \int\limits_{-\infty}^{\infty} \int\limits_{-\infty}^{\infty} \left\{ i\lambda^T a(\eta,t) - \frac{1}{2}\mathrm{tr}\,\left[\lambda\lambda^T b(\eta;t)\nu(t)b(\eta,t)^T \right] \right\}$$

$$\times e^{i\lambda^T (\eta-y)} f_1(\eta;t)d\eta d\lambda \,. \qquad (5.6.50)$$

Now we use formula for the integral representation of δ-function

$$\frac{1}{(2\pi)^p} \int\limits_{-\infty}^{\infty} e^{i\lambda^T(\eta-y)} d\lambda = \delta(\eta-y)\,. \tag{5.6.51}$$

Differentiating (5.6.51) with respect to η we get

$$\frac{1}{(2\pi)^p} \int\limits_{-\infty}^{\infty} i\lambda e^{i\lambda^T(\eta-y)} d\lambda = \frac{\partial}{\partial\eta}\delta(\eta-y) = \delta'(\eta-y), \tag{5.6.52}$$

$$-\frac{1}{(2\pi)^p} \int\limits_{-\infty}^{\infty} \lambda\lambda^T e^{i\lambda^T(\eta-y)} d\lambda = \frac{\partial}{\partial\eta}\frac{\partial^T}{\partial\eta}\delta(\eta-y) = \delta''(\eta-y). \tag{5.6.53}$$

Here λ and η are the p-dimensional vectors and in accordance with this $\partial/\partial\eta = [\partial/\partial\eta_1 \ldots \partial/\partial\eta_p]^T$ is the vector of gradient in the p-dimensional space, and $(\partial/\partial\eta)\times(\partial^T/\partial\eta)$ is the square matrix of the operators of double differentiation with respect to the components of the vector η. In other words, $\delta'(\eta-y)$ represents a matrix column whose elements are the partial derivatives of the δ-function with respect to the components of the vector argument, and $\delta''(\eta-y)$ represents square matrix whose elements are the second derivatives ot the δ-function with respect to the components of the vector argument. On the basis of formulae (5.6.51)–(5.6.53) we write

$$\frac{1}{(2\pi)^p} \int\limits_{-\infty}^{\infty}\int\limits_{-\infty}^{\infty} i\lambda^T a(\eta,t) e^{i\lambda^T(\eta-y)} f_1(\eta;t) d\eta d\lambda$$

$$= \int\limits_{-\infty}^{\infty} \delta'(\eta-y)^T a(\eta,t) f_1(\eta;t) d\eta = -\frac{\partial^T}{\partial y}\left[a(y,t)f_1(y,t)\right]\,, \tag{5.6.54}$$

$$-\frac{1}{(2\pi)^p} \int\limits_{-\infty}^{\infty}\int\limits_{-\infty}^{\infty} \lambda\lambda^T b(\eta,t)\nu(t)b(\eta,t)^T e^{i\lambda^T(\eta-y)} f_1(\eta;t) d\eta d\lambda$$

$$= \int\limits_{-\infty}^{\infty} \delta''(\eta-y)b(\eta,t)\nu(t)b(\eta,t)^T f_1(\eta;t) d\eta$$

$$= \frac{\partial}{\partial y}\frac{\partial^T}{\partial y}\left[b(y,t)\nu(t)b(y,t)^T f_1(y,t)\right]\,. \tag{5.6.55}$$

Using formulae (5.6.54) and (5.6.55) we represent Eq. (5.6.50) in the form

$$\frac{\partial f_1(y;t)}{\partial t} = -\frac{\partial^T}{\partial y}\left[\,a(y,t)f_1(y,t)\,\right]+$$

$$+\frac{1}{2}\,\mathrm{tr}\left\{\frac{\partial}{\partial y}\frac{\partial^T}{\partial y}\left[\,b(y,t)\nu(t)b(y,t)^T f_1(y,t)\,\right]\right\}. \tag{5.6.56}$$

Eq. (5.6.56) is also valid for the n-dimensional density $f_n(y_1,\ldots,y_n;\,t_1,\ldots,t_n)$ if the differentiation with respect to t and y is considered as the differentiation with respect to t_n and y_n. To be sure of that it is sufficient to repeat the previous evaluations for Eq. (5.6.33) with the function $\chi(\mu;t)$ determined by formula (5.6.37). ◁

Thus we obtained the parabolic linear homogeneous equation of the second order in partial derivatives. This equation, firstly in various special cases of a scalar Y, and then for a vector Y was obtained at the beginining of the century by Fokker, Einstein, Smoluchovsky and others while studying the Brownian motion and the diffusion (see, for instance, Einstein and Smoluchovsky 1936). Therefore Eq. (5.6.56) is usually called the Fokker–Planck equation. The first rigorous derivation of Eq. (5.6.56) was given at the end of the 1920s and the begining of the 1930s by Kolmogorov. This is the reason why it is sometimes called the Fokker–Planck–Kolmogorov equation.

The first application of Eq. (5.6.56) for studying dynamic systems was given in (Andronov, Witt and Pontryagin 1933) in which the problem of studing the fluctuations of the phase portrait of a system with one degree of freedom was stated, and the steady state distribution of the representative point on the phase plane for the stationary system with one degree of freedom was found.

E x a m p l e 5.6.1. Using the formula for differentiation of a composite function of a Wiener process (5.1.7) we make sure that the equation

$$\dot{Y} = 1 + 2\sqrt{Y}V, \ V = \dot{W}, \tag{I}$$

where V is a normally distributed white noise of the unit intensity has the solution $Y(t) = W^2(t)$. As the one-dimensional distribution of the standard Wiener process is normal with zero expectation and the variance t, the one-dimensional characteristic function and the density of the process Y are determined by the formulae

$$g_1(\lambda;t) = (1 - 2i\lambda t)^{-1/2}, \tag{II}$$

$$f_1(y;t) = (2\pi ty)^{-1/2} e^{-y/2t} 1(y)\,. \tag{III}$$

Eqs. (5.6.20) and (5.6.56) have in this case the form

$$\frac{\partial g_1(\lambda;t)}{\partial t} = i\lambda g_1(\lambda;t) - 2\lambda^2 EY e^{i\lambda Y}\,, \tag{IV}$$

$$\frac{\partial f_1(y;t)}{\partial t} = -\frac{\partial f_1(y;t)}{\partial y} + 2\frac{\partial^2}{\partial y^2}[yf_1(y;t)]\,. \tag{V}$$

It is easy to see that the functions obtained $g_1(\lambda;t)$ and $f_1(y;t)$ satisfy Eqs. (I) and (V), and the expectation $Ee^{i\lambda Y}$ exists in Eq. (IV) since $EY = EW^2 = DW = t$.

E x a m p l e 5.6.2. Using the formula of differentiation of a composite function of a Wiener process (5.1.7) we make sure that the equation

$$\dot{Y} = Y^3 - Y^2 V,\ \ V = \dot{W}, \tag{I}$$

where V is a normally distributed white noise of unit intensity has the solution $Y(t) = 1/W(t)$. The one-dimensional density and the one-dimensional characteristic function of the process $Y(t)$ are determined by the formulae

$$f_1(y;t) = \frac{1}{y^2\sqrt{2\pi t}} e^{-1/2ty^2}\,, \tag{II}$$

$$g_1(\lambda;t) = Ee^{i\lambda Y} = Ee^{i\lambda/W} = \frac{1}{\sqrt{2\pi t}}\int_{-\infty}^{\infty} e^{i\lambda/w - w^2/2t}\,dw\,. \tag{III}$$

Eqs. (5.6.20) and (5.6.56) have in this case the form

$$\frac{\partial g_1(\lambda;t)}{\partial t} = E\left[i\lambda Y^3 - \frac{1}{2}\lambda^2 Y^4\right] e^{i\lambda Y}\,, \tag{IV}$$

$$\frac{\partial f_1(y;t)}{\partial t} = -\frac{\partial}{\partial y}[y^3 f_1(y;t)] + \frac{1}{2}\frac{\partial^2}{\partial y^2}[y^4 f_1(y;t)]\,. \tag{V}$$

E x a m p l e 5.6.3. For the linear scalar equation

$$\dot{Y} + \alpha Y = \gamma V\,,\ \ \gamma^2 = 2D\alpha\,, \tag{I}$$

Eqs. (5.6.20) and (5.6.56) have the form:

$$\frac{\partial g_1}{\partial t} = -\alpha\lambda\frac{\partial g_1}{\partial \lambda} - \gamma^2\lambda^2 g_1\,,\ \ \frac{\partial f_1}{\partial t} = \alpha\frac{\partial f_1}{\partial y} + \frac{\gamma^2}{2}\frac{\partial^2 f_1}{\partial y^2}\,. \tag{II}$$

5.6.7. Equation for Transition Density

In particular, writing Eq. (5.6.56) for the two-dimensional density $f_2(y_1, y_2; t_1, t_2)$ and replacing $f_2(y_1, y_2; t_1, t_2)$ by its expression in terms of the one-dimensional and the transition density, $f_2(y_1, y_2; t_1, t_2)$ $f_1(y_1; t_1)f(y_2; t_2|y_1; t_1)$, we obtain after cancelling out $f_1(y_1; t_1)$:

$$\frac{\partial f(y_2; t_2|y_1; t_1)}{\partial t_2} = -\frac{\partial^T}{\partial y_2}\left[a(y_2, t_2)f(y_2; t_2|y_1; t_1)\right]$$

$$+\frac{1}{2}\operatorname{tr}\left\{\frac{\partial}{\partial y_2}\frac{\partial^T}{\partial y_2}\left[b(y_2, t_2)\nu(t_2)b(y_2, t_2)^T f(y_2; t_2|y_1; t_1)\right]\right\}. \qquad (5.6.57)$$

This equation is of the same form as Eq. (5.6.56).

Thus if *the transition density $f(y, t|\eta; \tau)$ of the Markov process determined by stochastic differential* Eq. (5.4.6) *is considered as a function of two first arguments y and t, $t > \tau$ given η, τ, then it is determined by* Eq. (5.6.56) *with the initial condition for the transition density having evidently the form*

$$f(y; \tau|\eta; \tau) = \delta(y - \eta). \qquad (5.6.58)$$

R e m a r k. We obtained Eq. (5.6.56) as a special case of Eq. (5.6.32) where $W(t)$ is a Wiener process. Eqs. (5.6.32) and even Eq. (5.6.20) are valid in more general situations, for any process with the independent increments $W(t)$. But Eq. (5.6.56) becomes valid in some cases where Eq. (5.6.20) fails owing to the fact that the expectation in its right-hand side does not exist.

After determining the one-dimensional density $f_1(y; t)$ and the transition density $f(y; t|\eta; \tau)$ of the Markov process $Y(t)$ by the integration of Eq. (5.6.56) with the corresponding initial conditions we may find all the multi-dimensional densities of the process $Y(t)$ by the formula

$$f_n(y_1, \ldots, y_n; t_1, \ldots, t_n) = f_1(y_1; t_1)f(y_2; t_2|y_1; t_1)\ldots$$

$$\ldots f(y_n; t_n|y_{n-1}; t_{n-1}) \text{ at } t_1 \leq t_2 \leq \cdots \leq t_n. \qquad (5.6.59)$$

If $t_{s_1} \leq t_{s_2} \leq \cdots \leq t_{s_n}$ for some permutation $(s_1, \ldots, s_n)$ of the numbers $1, \ldots, n$ then the n-dimensional density $f_n(y_1, \ldots, y_n; t_1, \ldots, t_n)$ is expressed by the same formula with the change of all the subscripts $1, \ldots, n$ by the corresponding subscripts $s_1, \ldots, s_n$.

5.6.8. Case of $a(y,t)$ Polynomial in y and $b(y,t)$ Independent on y

If the function $a(y,t)$ in Eq. (5.4.6) represents a polynomial in y, and the white noise coefficient $b(y,t)$ does not depend on y, $b(y,t) = b(t)$, then Eqs. (5.6.20) and (5.6.23) are reduced to the same linear equation in the partial derivatives whose order is equal to the degree of the polynomial $a(y,t)$.

$\triangleright$ To prove this we first note that if $b(y,t) = b(t)$ is independent of y, then the Eq. (5.6.20) may be rewritten in the form

$$\frac{\partial g_1(\lambda;t)}{\partial t} = Ei\lambda^T a(Y,t)e^{i\lambda^T Y} + \chi(b(t)^T\lambda;t)g_1(\lambda;t). \qquad (5.6.60)$$

Now we notice that the differentiation of the characteristic function $g_1(\lambda;t) = Ee^{i\lambda^T Y(t)}$ with respect to λ leads to the appearance of the factor $iY(t)$ under the sign of expectation $\partial g_1(\lambda;t)/\partial\lambda = iEY(t)e^{i\lambda^T Y(t)}$, where $\partial/\partial\lambda = [\partial/\partial\lambda_1 \ldots \partial/\partial\lambda_p]^T$ is the operator of gradient in the p-dimensional space. In other words, the differentiation of the function $g_1(\lambda;t)$ with respect to the component λ_k of the vector λ leads to the appearance of the factor iY_k under the sign of expectation. Hence it follows that

$$\begin{aligned}
EY_1^{k_1}\ldots Y_p^{k_p}e^{i\lambda^T Y} &= \frac{\partial^{k_1+\cdots+k_p}}{\partial(i\lambda_1)^{k_1}\ldots\partial(i\lambda_p)^{k_p}}Ee^{i\lambda^T Y}\\
&= \frac{\partial^{k_1+\cdots+k_p}g_1(\lambda;t)}{\partial(i\lambda_1)^{k_1}\ldots\partial(i\lambda_p)^{k_p}}
\end{aligned} \qquad (5.6.61)$$

at any $k_1, \ldots, k_p \geq 0$. Thus, every term of a polynomial with the components of the vector Y under the sign of expectation in Eq. (5.6.60) corresponds to exactly the same term with the components of the vector of gradient $\partial/\partial(i\lambda)$. Hence, it follows that if the function $a(y,t)$ represents a polynomial relative to y, then the vector Y in Eq. (5.6.60) under the sign of expectation may be replaced throughout by the vector $\partial/\partial(i\lambda)$ outside the sign of expectations. Then remembering that $Ee^{i\lambda^T Y} = g_1(\lambda;t)$ we may rewrite Eq. (5.6.20) in the form

$$\frac{\partial g_1(\lambda;t)}{\partial t} = \left[i\lambda^T a\left(\frac{\partial}{i\partial\lambda},t\right) + \chi(b(t)^T\lambda;t)\right]g_1(\lambda;t). \qquad (5.6.62)$$

This equation in partial derivatives is written in the operator form. In order to write it in the explicit form of an equation in partial derivatives it is necessary to replace the vector y in the polynomial $a(y,t)$ by the operator of differentiation $\partial/i\partial\lambda$ and to apply the obtained linear differential operator to the function $g_1(\lambda; t)$, considered as a function of λ. In other words, each term $y_1^{k_1} \ldots y_p^{k_p}$ of the polynomial $a(y, t)$ should be replaced by the corresponding operator $(\partial/i\partial\lambda_1)^{k_1} \ldots (\partial/i\partial\lambda_p)^{k_p}$.

Eq. (5.6.62) represents a linear homogeneous equation in partial derivatives relative to g_1 whose order is equal to the degree of the polynomial $a(y, t)$. ◁

Similarly, the equation for the n-dimensional characteristic function may be rewritten in this case in the form

$$\frac{\partial g_n}{\partial t_n} = \left[i\lambda_n^T a\left(\frac{\partial}{i\partial\lambda_n}, t_n\right) + \chi(b(t)^T \lambda_n; t_n) \right] g_n . \qquad (5.6.63)$$

This equation differs from Eq. (5.6.62) only in notations.

5.6.9. Case of $a(y, t)$ Polynomial and Normal White Noise

Eq. (5.6.20) determining the one-dimensional characteristic function of the process $Y(t)$ may be also reduced to a linear equation in partial derivatives in the case where the functions $a(y, t)$ and $b(y, t)$ in Eq. (5.4.6) both represent the polynomials in y if the white noise V is normally distributed.

▷ In this case Eq. (5.6.20) due to formula (5.6.37) for the function $\chi(\mu; t)$ has the form

$$\frac{\partial g_1(\lambda; t)}{\partial t} = E\left\{ i\lambda^T a(Y, t) - \frac{1}{2}\lambda^T b(Y, t)\nu(t)b(Y, t)^T \lambda \right\} e^{i\lambda^T Y} . \qquad (5.6.64)$$

If $a(y, t)$ and $b(y, t)$ are polynomials in y then as shown in Subsection 5.6.8 the vector Y under the sign of expectation may be replaced throughout by the operator $\partial/\partial(i\lambda)$ outside the sign of expectation. As a result we obtain

$$\frac{\partial g_1(\lambda; t)}{\partial t} = \left[i\lambda^T a\left(\frac{\partial}{i\partial\lambda}, t\right) \right.$$

$$\left. -\frac{1}{2}\lambda^T b\left(\frac{\partial}{i\partial\lambda}, t\right) \nu(t)b\left(\frac{\partial}{i\partial\lambda}, t\right)^T \lambda \right] g_1(\lambda; t). ◁ \qquad (5.6.65)$$

Eq. (5.6.23) for the n-dimensional characteristic function g_n with the change of g_1, λ and t by g_n, λ_n and t_n respectively is also reduced to this equation in such a case.

It is clear that Eq. (5.6.65) represents a linear homogeneous equation in partial derivatives of order $\max(P, 2Q)$, P being the degree of the polynomial $a(y, t)$, and Q the degree of the polynomial $b(y, t)$.

E x a m p l e 5.6.4. In the case of the equation

$$\dot{Y} = -Y^3 + YV \tag{I}$$

Eq. (5.6.65) has the form

$$\frac{\partial g_1}{\partial t} = \lambda \frac{\partial^3 g_1}{\partial \lambda^3} + \frac{1}{2}\lambda^2 \nu \frac{\partial^2 g_1}{\partial \lambda^2} \, . \tag{II}$$

E x a m p l e 5.6.5. For the equations

$$\dot{Y}_1 = -Y_1^2 Y_2 \, , \quad \dot{Y}_2 = Y_1 Y_2 + Y_1 V \tag{I}$$

Eq. (5.6.65) has the form

$$\frac{\partial g_1}{\partial t} = \lambda_1 \frac{\partial^3 g_1}{\partial \lambda_1^2 \partial \lambda_2} - i\lambda_2 \frac{\partial^2 g_1}{\partial \lambda_1 \partial \lambda_2} + \frac{1}{2}\lambda_2^2 \nu \frac{\partial^2 g_1}{\partial \lambda^2} \, . \tag{II}$$

The integration of Eqs. (5.6.62) and (5.6.65) generally represents a difficult problem. In practice we have to solve it usually by approximate methods of numerical integration. And only in the special case of linear Eq. (5.4.6). Eqs. (5.6.62) and (5.6.65) represent the equations of the first order which are easily integrated by a standard method (Section 6.2).

R e m a r k. Eqs. (5.6.20) and (5.6.23) in the case of the normally distributed white noise may be also reduced to the linear equations in the partial derivatives, Eq. (5.6.65) if the function $a(y, t)$ represents a polynomial in y, and $b(y, t)$ is not a polynomial but the function $b(y, t)\nu b(y, t)^T$ represents a polynomial in y. The equations in which $a(y, t)$ and $\chi(\lambda; y, t)$ are the polynomials relatively to y are of special interest.

E x a m p l e 5.6.6. Let us consider scalar nonlinear stochastic differential Eq. (5.4.6) at $a(y, t) = ay + a_0$, $b(y, t) = b\sqrt{y}$. Eq. (5.6.65) will have the following form:

$$\frac{\partial g_1}{\partial t} = \lambda \left(a + \frac{1}{2}ib^2\nu\lambda \right) \frac{\partial g_1}{\partial \lambda} + ia_0\lambda g_1 \, .$$

This is the linear equation in the partial derivatives of the first order. Here $b(y, t)$ is not a polynomials over y but $\sigma(y, t) = b(y, t)\nu \, b(y, t) = \nu b^2 y$ is a polynomial of the first degree.

5.6.10. Multi-Dimensional Distributions in Stochastic Continuous Composite Systems

At first let us obtain an equation for the one-dimensional characteristic function of the random process $Y(t)$ described by the nonlinear stochastic differential Eq. (5.4.7).

$\triangleright$ For obtaining an equation for the one-dimensional characteristic function $g_1(\lambda; t) = E e^{i\lambda^T Y(t)}$ we calculate at first the stochastic differential of the random process $e^{i\lambda^T Y(t)}$ by means of the generalized Itô formula (5.1.23) for a stochastic differential of the scalar function $\varphi(Y, t)$:

$$d\varphi(Y(t), t) = \left\{ \varphi_t(Y, t) + \varphi_y(Y, t)^T a(Y, t) \right.$$

$$+ \frac{1}{2} \, \mathrm{tr} \, \left[\varphi_{yy}(Y, t) b(Y, t) \nu_0(t) b(Y, t)^T \right] \Big\} dt + \int_{R_0^q} [\varphi(Y + c(Y, t, u), t)$$

$$- \varphi(Y, t) - \varphi_y(Y, t)^T c(Y, t, u)] \mu_P(dt, du) + \varphi_y(Y, t)^T b(Y, t) dW_0$$

$$+ \int_{R_0^q} [\varphi(Y + c(Y, t, u), t) - \varphi(Y, t)] \, P^0(dt, du) \,. \tag{5.6.66}$$

Here in the correspondence with the definition (Subsection 4.6.6) of the Poisson measure the random function $P(\Delta, A)$ where $\Delta = (t_1, t_2]$ represents for any set A the simple Poisson process; $\mu_P(\Delta, A)$ is its expectation; $P^0(\Delta, A) = P(\Delta, A) - \mu_P(\Delta, A)$. Denoting the intensity of the correspondent Poisson stream of the events in terms of $\nu_P(\Delta, A)$ it is not difficult to present the expression for $\mu_P(\Delta, A)$ in the form

$$\mu_P(\Delta, A) = EP(\Delta, A) = \int_\Delta \nu_P(\tau, A) d\tau \,. \tag{5.6.67}$$

By virtue of (5.6.66) we have

$$de^{i\lambda^T Y(t)} = \left\{ i\lambda^T a(Y, t) - \frac{1}{2} \lambda^T b(Y, t) \nu_0(t) b(Y, t)^T \lambda \right\} e^{i\lambda^T Y(t)} dt$$

$$+ \int_{R_0^q} \left\{ e^{i\lambda^T [Y + c(Y, t, u)]} - e^{i\lambda^T Y} - i\lambda^T c(Y, t, u) e^{i\lambda^T Y} \right\} \mu_P(t, du)$$

$$+ i\lambda^T b(Y, t) e^{i\lambda^T Y} dW_0 + \int_{R_0^q} \left\{ e^{i\lambda^T [Y + c(Y, t, u)]} - e^{i\lambda^T Y} \right\} P^0(t, du) \,.$$

$$\tag{5.6.68}$$

Using expression (5.6.67) for $\mu_P(\Delta, A)$ and taking the operation of an expectation in the left and right-hand sides (5.6.68) we come to the sought equation for the one-dimensional characteristic function:

$$\frac{\partial g_1(\lambda; t)}{\partial t} = E\left\{\left[i\lambda^T a(Y, t) + \chi(\lambda; Y, t)\right] e^{i\lambda^T Y}\right\}, \qquad (5.6.69)$$

where

$$\chi(\lambda; Y, t) = -\frac{1}{2}\lambda^T b(Y, t)\nu_0(t)b(Y, t)^T \lambda$$

$$+ \int_{R_0^q} \left\{e^{i\lambda^T c(Y, t, u)} - 1 - i\lambda^T c(Y, t, u)\right\} \nu_P(t, du). \lhd \qquad (5.6.70)$$

Condition (5.6.21) serves for Eq. (5.6.69) as the initial condition.

The multi-dimensional characteristic functions (5.6.22) are found from Eq. (5.6.23), the initial conditions (5.6.25) and with function $\chi(\lambda; Y, t)$ determined by (5.6.70).

Some specific cases of Eqs. (5.6.20), (5.6.30), (5.6.32), (5.6.69) for the one-dimensional densities are given in Problems 5.24–5.27.

5.6.11. Stationary Processes in Stochastic Differential Systems

The solution of Eqs. (5.6.20), (5.6.69) and (5.6.23) in every specific case depends on the initial one-dimensional characteristic function $g_0(\lambda)$. At arbitrary t_0 and $g_0(\lambda)$ the process in the system described by Eqs. (5.4.6) or (5.4.7) will be nonstationary in the general case.

The problem of the existence of the stationary regimes in the stationary differential systems, i.e. when the process $Y(t)$ in the system is stationary, is of great importance. Let us consider stationary differential system (5.4.7) under the action of the stationary white noise V. In this case the functions $a(y, t)$, $b(y, t)$, $c(y, t, u)$, $\chi(\mu; y, t)$ do not explicitly depend on time, $a(y, t) = a(y)$, $b(y, t) = b(y)$, $c(y, t, u) = c(y, u)$, $\chi(\mu; y, t) = \chi(\mu; y)$. If in such a system the process $Y(t)$ is stationary relatively to the one-dimensional distribution then its one-dimensional characteristic function $g_1(\lambda; t)$ does not depend on time too, $g_1(\lambda; t) = g_1(\lambda)$, $\partial g_1(\lambda)/\partial t = 0$ and Eq. (5.2.69) takes the form

$$E\left\{\left[i\lambda^T a(Y) + \chi(\lambda; Y, t))\right] e^{i\lambda^T Y}\right\} = 0. \qquad (5.6.71)$$

If this equation has a nontrivial solution $g_1(\lambda) \neq 0$, $g_1(0) = 1$, then the stationary relatively to the one-dimensional distribution process is possible in the system. We shall prove that this process is the strictly stationary process.

▷ Suppose that the process $Y(t)$ is stationary in the $(n-1)$-dimensional distribution. In this case

$$g_{n-1}(\lambda_1 , \ldots , \lambda_{n-1}; t_1 , \ldots , t_{n-1})$$

$$= g_{n-1}(\lambda_1 , \ldots , \lambda_{n-1}; \tau_1 , \ldots , \tau_{n-2}), \qquad (5.6.72)$$

where $\tau_1 = t_2 - t_1 , \ldots , \tau_{n-2} = t_{n-1} - t_1$. Let us consider Eq. (5.6.23) for $g_n(\lambda_1 , \ldots , \lambda_n; t_1 , \ldots , t_n)$. By means of the change of variables $t_n = t_1 + \tau_{n-1}$ we transform this equation into the form

$$\partial g_n(\lambda_1 , \ldots , \lambda_n; t_1 , \ldots , t_{n-1}, t_1 + \tau_{n-1})/\partial \tau_{n-1}$$

$$= E\left\{i\lambda_n^T a(Y_{t_n}) + \chi(\lambda_n; Y_{t_n})\right\} \exp\left\{i\lambda_1^T Y_{t_1} + \cdots + i\lambda_n^T Y_{t_n}\right\}. \qquad (5.6.73)$$

The initial condition for this equation has the form

$$g_n(\lambda_1 , \ldots , \lambda_n; t_1 , \ldots , t_{n-1}, t_{n-1})$$

$$= g_{n-1}(\lambda_1 , \ldots , \lambda_{n-1} + \lambda_n; \tau_1 , \ldots , \tau_{n-2}). \qquad (5.6.74)$$

As the right-hand sides of the equation and of the initial condition do not explicitly depend on $t_1 , \ldots , t_n$ the solution of the equation depends only on $\tau_1, \ldots, \tau_{n-1}$:

$$g_n(\lambda_1 , \ldots , \lambda_n; t_1 , \ldots , t_1 + \tau_{n-1}) = g_n(\lambda_1 , \ldots , \lambda_n; \tau_1 , \ldots , \tau_{n-1}).$$
$$(5.6.75)$$

Consequently, the process $Y(t)$ is also stationary in the n-dimensional distribution. And as the process $Y(t)$ is stationary in the one-dimensional distribution it is also stationary in all the multi-dimensional distributions. ◁

Thus, *if in a stationary stochastic differential system there exists a stationary process, then it is the strictly stationary process. The multi-dimensional distributions of this process are determined by* Eqs. (5.6.33), (5.6.25) *at* $g_1(\lambda; t) = g_1(\lambda)$, *where* $g_1(\lambda)$ *is determined by* Eq. (5.6.69) *at* $\partial g_1/\partial t = 0$.

If the equation which determines the one-dimensional characteristic function $g_1(\lambda)$ of the stationary process $Y(t)$ has several solutions then every one of these solutions determines the possible stationary process in the system.

In the case of the normally distributed white noise the one-dimensional distribution of the stationary process in the stationary system may be obtained by the integration of Eq. (5.6.56) putting $\partial f_1 / \partial t = 0$,

$$-\frac{\partial^T}{\partial y} \left[a(y) f_1(y) \right] + \frac{1}{2} \operatorname{tr} \left\{ \frac{\partial}{\partial y} \frac{\partial^T}{\partial y} \left[b(y) \nu b(y)^T f_1(y) \right] \right\} = 0 . \qquad (5.6.76)$$

E x a m p l e 5.6.7. For scalar nonlinear system

$$\dot{Y} = a(Y) + b(Y) V , \quad Y(t_0) = Y_0 , \qquad (I)$$

where V is the stationary normal m-dimensional white noise of the intensity ν, $a(y)$ is the scalar nonlinear function, $b(y)$ is m-row of the nonlinear functions, Eq. (5.6.56) has the form

$$\frac{\partial f_1(y, t)}{\partial t} = -\frac{\partial}{\partial y} \left[a(y) f_1(y; t) \right] + \frac{1}{2} \frac{\partial^2}{\partial y^2} \left[\sigma(y) f_1(y; t) \right] , \qquad (II)$$

where $\sigma(y) = b(y) \nu b(y)^T$. Hence for a stationary in the strict sense random process the one-dimensional density $f_1(y)$ satisfies an ordinary differential equation

$$\frac{d^2}{dy^2} \left[\sigma(y) f_1(y) \right] - 2 \frac{d}{dy} \left[a(y) f_1(y) \right] = 0 . \qquad (III)$$

After integration we find

$$f_1(y) = \frac{c}{\sigma(y)} \exp \left\{ 2\varphi(y) \right\} , \quad \sigma(y) = b(y) \nu b(y)^T , \qquad (IV)$$

$$\varphi(y) = \int_0^y \frac{a(x)}{\sigma(x)} dx , \quad c^{-1} = \int_{-\infty}^{\infty} \exp \left\{ 2\varphi(z) \right\} \frac{dz}{\sigma(z)} < \infty . \qquad (V)$$

In particular, at $b(y) = 1$, $\sigma(y) = \nu$ we get from (V)

$$f_1(y) = \frac{c}{\nu} \exp \left\{ \frac{2}{\nu} \int_0^y a(x) dx \right\} . \qquad (VI)$$

Hence at $a(y) = -l \operatorname{sgn} y$ we have the following expressions for the density $f_1(y)$ correspondent to the characteristic function $g_1(\lambda)$:

$$f_1(y) = \gamma e^{-\gamma |\eta|} , \quad \gamma = \frac{2l}{\nu} , \quad g_1(\lambda) = \frac{2\gamma^2}{\lambda^2 + \gamma^2} . \qquad (VII)$$

For stationary system (I) with the parametric noises when $a(y) = a_0 + a_1 y$, $b(y) = b_0 + b_1 y$ (b_0 and b_1 are the matrices-rows) the one-dimensional density $f_1(y)$ of the stationary process by virtue of (III) is determined by the known Pearson's curves equation

$$\frac{df_1}{dy} = \frac{c_0 + c_1 y}{c_2 + c_3 y + c_4 y^2} f_1 \,, \qquad \text{(VIII)}$$

where $c_0 = c_0 - b_0 \nu b_1^T$, $c_1 = a_1 - b_1 \nu b_1^T$, $2c_2 = b_0 \nu b_0^T$, $c_3 = b_0 \nu b_1^T$, $2c_4 = b_1 \nu b_1^T$.

The invariant measure methods for finite-dimensional Eqs. (5.6.56) and (5.6.76) are considered, for example, in (Sinitsyn and Moschyk 1991, 1992a, b, Soize 1994). See also Problems 5.15–5.21.

5.6.12. Sufficient Conditions of Existence of Multi-Dimensional Distributions

Theorem 5.6.1. *If W in Eq. (5.4.6) represents a second order process with the independent increments whose expectation $m(t)$ and corariance function $k(t)$ are given by*

(i) $m(t) = m(t_0) + \int\limits_{t_0}^{t} m'(\tau)d\tau,$

(ii) $k(t) = k(t_0) + \int\limits_{t_0}^{t} \nu(\tau)d\tau$

and whose one-dimensional characteristic function $h_1(\mu;t)$ is differentiable with respect to t and the functions $a(y,t)$ and $b(y,t)$ satisfy the conditions,

(iii) $E\,|a(Y,t)|^2 < k_1 + k_2 E\,|Y|^2,$

(iv) $E\,\mathrm{tr}\,b(Y,t)\nu(t)b(Y,t)^T < k_3 + k_4 E\,|Y|^2,$

(v) $E\,|a(y',t) - a(Y,t)|^2 < c_1^2 E\,|Y' - Y|^2,$

(vi) $E\,\mathrm{tr}\,[\,b(Y',t) - b(Y,t)\,]\nu(t)\,[\,b(Y',t)^T - b(Y,t)^T\,] < c_2^2 E|Y' - Y|^2$ *for some k_1, k_2, k_3, k_4, c_1, c_2 and for any random variables Y, Y' with the finite second moments then the multi-dimensional characteristic functions $g_n(\lambda_1, \ldots, \lambda_n; t_1, \ldots, t_n)$ $(n = 1, 2, \ldots)$ of the process Y determined by Eq. (5.4.6) and the initial condition $Y(t_0) = Y_0$, Y_0 being a random variable with the finite second moment satisfy Eqs. (5.6.20), (5.6.23) and initial conditions (5.6.21), (5.6.25) where $\chi(\mu;t)$ represents the logarithmic derivative of the one-dimensional characteristic function $h_1(\mu;t)$ of the process W with respect to t and is defined by (5.6.17).*

▷ Fist of all we observe that under the conditions of the theorem Eq. (5.4.6) has a unique strong solution $Y(t)$ satisfying $Y(t_0) = Y_0$ which represents a second order random process (Subsection 5.2.7). We have then for any $t_1 < \cdots t_n < s$,

$$g_n(\lambda_1, \ldots, \lambda_n; t_1, \ldots, t_{n-1}, s) - g_n(\lambda_1, \ldots, \lambda_n; t_1, \ldots, t_n)$$

$$= E\left[\exp\left\{i\sum_{k=1}^{n-1}\lambda_k^T Y_{t_k} + i\lambda_k^T Y_s\right\} - \exp\left\{i\sum_{k=1}^{n}\lambda_k^T Y_s\right\}\right]$$

$$= E\left\{\exp\left\{i\lambda_n^T(Y_s - Y_{t_k}) - 1\right\}\right\}\exp\left\{i\sum_{k=1}^{n}\lambda_k^T Y_{t_k}\right\}. \tag{5.6.77}$$

Now we notice that by Eq. (5.4.6) at $s > t_n$,

$$Y_s - Y_{t_n} = a(Y_{t_n}, t_n)(s - t_n) + b(Y_{t_n}, t_n)(W_s - W_{t_n})$$

$$+ \int_{t_n}^{s} [b(Y_\tau, \tau) - b(Y_{t_n}, t_n)]\, dW_\tau + o_p(s - t_n). \tag{5.6.78}$$

Substituting (5.6.78) into (5.6.77) gives

$$g_n(\lambda_1, \ldots, \lambda_n; t_1, \ldots, t_{n-1}, s) - g_n(\lambda_1, \ldots, \lambda_n; t_1, \ldots, t_{n-1}, t_n)$$

$$= E\left[\exp\left\{i\lambda_n^T b(Y_{t_n}, t_n)(W_s - W_{t_n})\right\}\left\{i\lambda_n^T a(Y_{t_n}, t_n)(s - t_n)\right.\right.$$

$$\left. + i\lambda_n^T \int_{t_n}^{t} [b(Y_\tau, \tau) - b(Y_{t_n}, t_n)]\, dW_\tau + o_p(s - t_n)\right\}$$

$$+ \exp\left\{i\lambda_n^T b(Y_{t_n}, t_n)(W_s - W_{t_n})\right\} - 1\bigg] \exp\left\{i\sum_{k=1}^{n}\lambda_k^T Y_{t_k}\right\}. \tag{5.6.79}$$

But

$$\left| E\left[\exp\left\{i\lambda_n^T b(Y_{t_n}, t_n)(W_s - W_{t_n})\right\} - 1\right] \int_{t_n}^{s} [bY_\tau, \tau) - b(Y_{t_n}, t_n)]\, dW_\tau \right.$$

$$\left. \times \exp\left\{i\sum_{k=1}^{n}\lambda_k^T Y_{t_k}\right\}\right|^2 \le E\left|\exp\left\{i\lambda_n^T b(Y_{t_n}, t_n)(W_s - W_{t_n})\right\} - 1\right|^2$$

$$\times E\left|\int_{t_n}^{s}\left[b(Y_\tau,\tau)-b(Y_{t_n},t_n)\right]dW_\tau\right|^2$$

$$=\left[2-E\exp\left\{i\lambda_n^T b(Y_{t_n},t_n)(W_s-W_{t_n})\right\}\right.$$

$$\left.-E\exp\left\{-i\lambda_n^T b(Y_{t_n},t_n)(W_s-W_{t_n})\right\}\right]\int_{t_n}^{s}\operatorname{tr}E\left[b(Y_\tau,\tau)\right.$$

$$\left.-b(Y_{t_n},t_n)\right]\nu_\tau\left[b(Y_\tau,\tau)^T-b(Y_{t_n},t_n)^T\right]d\tau. \tag{5.6.80}$$

Averaging at first with respect to $W_s-W_{t_n}$ for fixed Y_{t_n} and taking into account that $W_s-W_{t_n}$ is independent of Y_{t_n} we get

$$E\exp\left\{i\lambda_n^T b(Y_{t_n},t_n)(W_s-W_{t_n})\right\}=Eh(b(Y_{t_n},t_n)^T\lambda_n;t_n,s) \tag{5.6.81}$$

where $h(\mu;t,s)$ is the characteristic function of the random variable W_s-W_t. Remembering that (Subsection 4.6.5)

$$h(\mu;t,s)=\frac{h_1(\mu;s)}{h_1(\mu;t)}=1+\frac{1}{h_1(\mu;t)}\frac{\partial h_1(\mu;t)}{\partial t}(s-t)+o(s-t)$$

$$=1+\chi(\mu;t)(s-t)+o(s-t) \tag{5.6.82}$$

we obtain from (5.6.81)

$$E\exp\left\{i\lambda_n^T b(Y_{t_n},t_n)(W_s-W_{t_n})\right\}$$

$$=1+\chi(b(Y_{t_n},t_n)^T\lambda_n;t_n)(s-t_n)+o(s-t_n). \tag{5.6.83}$$

Now we consider the integral in the last term of (5.6.80). By (vi)

$$\int_{t_n}^{s}\operatorname{tr}E\left[b(Y_\tau,\tau)-b(Y_{t_n},t_n)\right]\nu_\tau\left[b(Y_\tau,\tau)^T-b(Y_{t_n},t_n)^T\right]d\tau$$

$$<c_2^2\int_{t_n}^{s}E\left|Y_\tau-Y_{t_n}\right|^2d\tau=o(s-t_n) \tag{5.6.84}$$

since $E\left|Y_\tau-Y_{t_n}\right|^2$ is a continuous function of τ and is equal to zero at $\tau=t_n$. Using (5.6.83), (5.6.84) and (i), (ii) we obtain from (5.6.80)

$$E\exp\left\{i\lambda_n^T b(Y_{t_n},t_n)(W_s-W_t)+i\sum_{k=1}^{n}\lambda_k^T Y_{t_k}\right\}\int_{t_n}^{s}\left[b(Y_\tau,\tau)\right.$$

$$-b(Y_{t_n}, t_n)]\, dW_\tau = E \exp\left\{ i \sum_{k=1}^{n} \lambda_k^T Y_{t_k} \right\} \int_{t_n}^{s} [\, b(Y_\tau, \tau) - b(Y_{t_n}, t_n)\,]\, dW_\tau$$

$$+o(s - t_n) = \int_{t_n}^{s} E \exp\left\{ i \sum_{k=1}^{n} \lambda_k^T Y_{t_k} \right\} [\, b(Y_\tau, \tau) - b(Y_{t_n}, t_n)\,]\, m_\tau'\, d\tau$$

$$+o(s - t_n) = o(s - t_n) \tag{5.6.85}$$

since $b(Y_\tau, \tau) \to b(Y_{t_n}, t_n)$ in probability by virtue of (vi). Using (5.6.85) Eq. (5.6.79) becomes

$$g_n(\lambda_1, \ldots, \lambda_n; t_1, \ldots, t_{n-1}, s) - g_n(\lambda_1, \ldots, \lambda_n; t_1, \ldots, t_{n-1}, t_n)$$

$$= E\big[\exp\left\{ i\lambda_n^T b(Y_{t_n}, t_n)(W_s - W_{t_n}) \right\} i\lambda_n^T a(Y_{t_n}, t_n)(s - t_n)$$

$$+ \exp\left\{ i\lambda_n^T b(Y_{t_n}, t_n)(W_s - W_{t_n}) \right\} - 1\big] \exp\left\{ i \sum_{k=1}^{n} \lambda_k^T Y_{t_k} \right\} + o(s - t_n). \tag{5.6.86}$$

Averaging with respect to $W_s - W_{t_n}$ for fixed $Y_{t_1}, \ldots, Y_{t_n}$ and taking into account that $W_s - W_{t_n}$ is independent of $Y_{t_1}, \ldots, Y_{t_n}$ we obtain

$$g_n(\lambda_1, \ldots, \lambda_n; t_1, \ldots, t_{n-1}, s) - g_n(\lambda_1, \ldots, \lambda_n; t_1, \ldots, t_{n-1}, t_n)$$

$$= E\big[i\lambda_n^T a(Y_{t_n}, t_n) h(Y_{t_n}, t_n)^T \lambda_n, t_n, s)$$

$$+ h(b(Y_{t_n}, t_n)^T \lambda_n; t_n, s) - 1\big] \exp\left\{ i \sum_{k=1}^{n} \lambda_k^T Y_{t_k} \right\} + o(s - t_n). \tag{5.6.87}$$

Observing that by virtue of (5.6.17)

$$h(\mu; t, s) - 1 = \frac{h_1(\mu; s) - h_1(\mu; t)}{h_1(\mu; t)} = \frac{1}{h_1(\mu; t)} \frac{\partial h_1(\mu; t)}{\partial t}(s - t) + o(s - t)$$

$$= \chi(\mu; t)(s - t) + o(s - t) \tag{5.6.88}$$

we get

$$g_n(\lambda_1, \ldots, \lambda_n; t_1, \ldots, t_{n-1}, s) - g_n(\lambda_1, \ldots, \lambda_n; t_1, \ldots, t_{n-1}, t_n)$$

$$= E\big\{ i\lambda_n^T a(Y_{t_n}, t_n) h(Y_{t_n}, t_n)^T \lambda_n, t_n, s)$$

$$+\chi(b(Y_{t_n}, t_n)^T \lambda_n; t_n)\} \exp\left\{ i \sum_{k=1}^{n} \lambda_k^T Y_{t_k} \right\} (s - t_n) + o(s - t_n). \quad (5.6.89)$$

The expectation in (5.6.89) exists according to (iii) and (iv) since $|\chi(\mu; t)| < c\,|\mu|^2$ for some $c > 0$. Dividing (5.6.89) by $(s - t_n)$ and putting then $s \downarrow t_n$ yields Eq. (5.6.23). The initial condition (5.6.21) for Eq. (5.6.23) with $n = 1$ is obvious. The initial condition (5.6.25) for $g_n(\lambda_1, \ldots, \lambda_n; t_1, \ldots, t_n)$, $n \geq 2$, follows from the obvious relation (5.6.24) proving the theorem. $\triangleleft$

Now we shall extend Theorem 5.6.1 to the case of the discontinuous process with the independent increments W.

$\triangleright$ Let $W = W(t)$ in Eq. (5.4.6) be the process with the independent increments with random jumps W_k (certainly independent of one another) at fixed points τ_k $(k = 1, 2, \ldots)$. The one-dimensional characteristic function $h_1(\mu; t)$ of the process $W(t)$ has in this case the jumps $[g_k(\mu) - 1] h_1(\mu; \tau_k)$ at the points τ_k, $g_k(\mu)$ being the characteristic function of W_k $(k = 1, 2, \ldots)$. Hence the derivative of $h_1(\mu; t)$ with respect to t can exist only as a generalized function containing the linear combination of δ-functions $\delta(t - \tau_k)$ with the coefficient $[g_k(\mu) - 1] h_1(\mu; \tau_k)$ and respectively the function $\chi(\mu; t)$ determined by (5.6.17) can exist only as a generalized function containing the linear combination of δ-functions $\delta(t - \tau_k)$ with the coefficients $g_k(\mu) - 1$. Now we observe that the solution $Y(t)$ of Eq. (5.4.6) has in this case the random jumps $b(Y_{\tau_k}, \tau_k) W_k$ at the points τ_k. Hence the n-dimensional characteristic function $g_n(\lambda_1, \ldots, \lambda_n; t_1, \ldots, t_n)$ of the process $Y(t)$ considered as a function of t_n for fixed $t_1, \ldots, t_{n-1}$ has the jumps $\left[g_k(b(Y_{t_k}, t_k)^T \lambda_n) - 1 \right] g_n(\lambda_1, \ldots, \lambda_n; t_1, \ldots, t_{n-1}, \tau_k)$ at the points τ_k $(k = 1, 2, \ldots)$. These jumps evidently are determined by Eq. (5.6.23) with $\chi(\mu; t)$ containing the linear combination of δ-functions $\delta(t - \tau_k)$ with the coefficients $g_k(\mu) - 1$. $\triangleleft$

Thus we have the following statement.

Theorem 5.6.2. *If W in Eq. (5.4.6) is the second order process with the independent increments with the random jumps at the fixed sequence of points $\{\tau_k\}$ whose expectation $m(t)$ and covariance function $k(t)$ are representable in the form* (i) *and* (ii) *where $\nu(\tau)$ contains the linear combination of δ-functions $\delta(t - \tau_k)$ and whose one-dimensional characteristic function $h_1(\mu; t)$ is differentiable with respect to t at all t different from τ_k and the functions $a(y, t)$ and $b(y, t)$ satisfy conditions* (iii)–(vi) *for all t different from τ_k then the multi-dimensional characteristic functions of the process $Y(t)$ determined by Eq. (5.4.6) with* *the initial condition* $Y(t_0) = Y_0$ *satisfy* Eqs. (5.6.20), (5.6.21), (5.6.23), (5.6.25) *with function* $\chi(\mu;t)$ *containing the respective linear combination of* δ*-functions* $\delta(t - \tau_k)$ $(k = 1, 2, \ldots)$.

For stochastic differential systems described by Eq. (5.4.7) the corresponding sufficient conditions (i)–(vi) must be complemented by the following ones:

(vii) $\int\limits_{R_0^q} |c(y,t,\tau,u)|^2 \, \nu_P(\tau,du) < k_5 + k_6 \, |y|^2$,

(viii) $\int\limits_{R_0^q} |c(y,t,\tau,u) - c(y,t,\tau,u)|^2 \, \nu_P(\tau,du) < c_3 \, |y - y|^2$.

5.7. Multi-Dimensional Distributions in Stochastic Discrete and Continuous-Discrete Systems

5.7.1. Multi-Dimensional Distributions in Stochastic Discrete Systems

Let us consider the nonlinear stochastic discrete systems described by stochastic difference equations

$$Y_{k+1} = \omega_k(Y_k, V_k) \quad (k = 1, 2, \ldots), \tag{5.7.1}$$

$$Y_{k+1} = a_k(Y_k) + b_k(Y_k)V_k \quad (k = 1, 2, \ldots), \tag{5.7.2}$$

where the functions $\omega_k(y_k, v_k)$, $a_k(y_k)$ and $b_k(y_k)$ have the dimensions $(p \times 1)$, $(p \times 1)$ and $(p \times m)$ correspondingly. The characteristic functions $h_k = h_k(\rho)$ of the vector random variables V_k and the correspondent densities $\eta_k = \eta_k(v)$ we assume to be known.

By means of direct calculation on the basis of Eq. (5.7.1) similarly as in Subsections 5.6.1 and 5.6.2 we come to the following recurrent formulae which determine the one- and multi-dimensional densities and the characteristic functions:

$$f_k(y) = \frac{1}{(2\pi)^p} \int\limits_{-\infty}^{\infty} e^{-i\lambda^T y} g_k(\lambda)d\lambda, \tag{5.7.3}$$

$$g_k(\lambda) = E \exp\left\{i\lambda^T Y_k\right\}, \tag{5.7.4}$$

$$f_{k_1,\ldots,k_n}(y_1, \ldots, y_n) = \frac{1}{(2\pi)^{np}} \int\limits_{-\infty}^{\infty} \exp\left\{-i\sum_{h=1}^{n}\lambda_h^T y_h\right\}$$

$$\times g_{k_1,\ldots,k_n}(\lambda_1, \ldots, \lambda_n)d\lambda_1\ldots d\lambda_n, \tag{5.7.5}$$

$$g_{k_1,\ldots,k_n}(\lambda_1,\ldots,\lambda_n) = E \exp\left\{i \sum_{l=1}^{n} \lambda_l^T Y_{k_l}\right\}, \tag{5.7.6}$$

$$g_{k+1}(\lambda) = E \exp\left\{i\lambda^T \omega_k(Y_k, V_k)\right\}$$

$$= \int_{-\infty}^{\infty} \int_{-\infty}^{\infty} e^{i\lambda^T \omega_k(y,v)} f_k(y) h_k(v)\, dy\, dv, \tag{5.7.7}$$

$$g_{k_1,\ldots,k_n}(\lambda_1,\ldots,\lambda_n) = E \exp\left\{i \sum_{l=1}^{n-1} \lambda_l^T Y_{k_l} + i\lambda_n^T \omega_{k_n}(Y_{k_n}, V_{k_n})\right\}$$

$$= \int_{-\infty}^{\infty} \cdots \int_{-\infty}^{\infty} \int_{-\infty}^{\infty} \exp\left\{i \sum_{h=1}^{n-1} \lambda_h^T y_h + i\lambda_n^T \omega_{k_n}(y_n, v_n)\right\}$$

$$\times f_{k_1,\ldots,k_n}(y_1,\ldots,y_n)\, dy_1 \ldots dy_n\, dv_n, \tag{5.7.8}$$

and

$$g_{k_1,\ldots,k_{n-1},k_{n-1}}(\lambda_1,\ldots,\lambda_n) = g_{k_1,\ldots,k_{n-1}}(\lambda_1,\ldots,\lambda_{n-1}+\lambda_n), \tag{5.7.9}$$

$$g_{k_1,\ldots,k_n}(\lambda_1,\ldots,\lambda_n) = g_{k_{s_1},\ldots,k_{s_n}}(\lambda_{s_1},\ldots,\lambda_{s_n}),$$

where $(s_1,\ldots,s_n)$ is any permutation of the numbers $(1,\ldots,n)$ such that $k_{s_1} < k_{s_2} < \cdots < k_{s_n}$.

Thus *the one- and the multi-dimensional characteristic functions and the densities of the state vector in stochastic nonlinear discrete system (5.7.1) are determined by recurrent formulae (5.7.3)–(5.7.9).*

The general recurrent formulae for the distributions in stochastic discrete system (5.7.2) immediately follow from correspondent formulae (5.7.3)–(5.7.9).

5.7.2. Distributions in Stochastic Continuous-Discrete Systems

Let us consider the continuous-discrete system whose state vector Y is determined by the joint system of stochastic equations of the form Eqs. (5.4.6) and (5.7.1):

$$\dot{Y}' = a(Y,t) + b(Y,t)V, \tag{5.7.10}$$

$$Y'' = \sum_{k=0}^{\infty} Y_k'' \mathbf{1}_{A_k}(t), \quad Y_{k+1}'' = \omega_k(Y_k, V_k), \tag{5.7.11}$$

where Y_k is the value of $Y(t)$ at $t = t^{(k)}$; $Y_k = \left[Y_k'^T Y_k''^T \right]^T = Y(t^{(k)})$ $(k = 0, 1, 2, \ldots)$; a, b, ω_k are the functions of the indicated arguments; $\mathbf{1}_{A_k}(t)$ is an indicator of an interval $A_k = [t^{(k)}, t^{(k+1)})$ $(k = 0, 1, 2, \ldots)$; V is a white noise in the strict sense; $\{V_k\}$ is the sequence of the independent vector random variables independent of the white noise V.

The one-dimensional characteristic function $h_1(\mu; t)$ of the process with the independent increments $W(t)$ whose weak m.s. derivative is the white noise V, and the distributions of the random variables V_k we shall consider as the known ones. We introduce the random processes

$$Y'''(t) = \sum_{k=0}^{\infty} Y_k' \mathbf{1}_{A_k}(t), \quad \bar{Y}(t) = \left[Y'(t)^T Y''(t)^T Y'''(t)^T \right]^T. \quad (5.7.12)$$

For the one-dimensional characteristic function of the process $\bar{Y}(t)$,

$$g_1(\lambda; t) = E e^{i\lambda^T \bar{Y}(t)} = E \exp\left\{ i\lambda'^T Y'(t) + i\lambda''^T Y''(t) + i\lambda'''^T Y'''(t) \right\}$$

$$= E \exp\left\{ i\lambda'^T Y_k' + i\lambda''^T Y_k'' + i\lambda'''^T Y_k' \right\}$$

$$(5.7.13)$$

we obtain analogously as in Subsection 5.6.1 the following equation:

$$\frac{\partial g_1(\lambda; t)}{\partial t} = E \exp\left\{ \left[i\lambda'^T a(Y, t) + \chi(b(Y, t)^T \lambda'; t) \right] e^{i\lambda^T \bar{Y}} \right\}. \quad (5.7.14)$$

Assuming as the initial instant $t_0 = t^{(0)}$ we have the initial condition for Eq. (5.7.14)

$$g_1(\lambda; t_0) = E \exp\left\{ i(\lambda'^T + \lambda'''^T)Y_0' + i\lambda''^T Y_0'' \right\}$$

$$= g_0\left(\left[\lambda'^T + \lambda'''^T, \lambda''^T \right]^T \right), \quad (5.7.15)$$

where $g_0(\rho)$ is the characteristic function of the initial value $Y_0 = Y(t_0)$ of the process $Y(t)$.

At the instant $t = t^{(k)}$ the value $g_1(\lambda; t)$ evidently, is equal to $E \exp\{i(\lambda'^T + \lambda'''^T)Y_k' + i\lambda''^T Y_k''\}$, i.e. is equal to the value $g_k([\lambda'^T + \lambda'''^T, \lambda''^T]^T)$ of the characteristic function $g_k(\rho)$ of the random variable $Y_k = \left[Y_k'^T Y_k''^T \right]^T$. If the function $\chi(\mu; t)$ is the continuous function of t at any μ then $g_1(\lambda; t)$ tends to $E \exp\{i(\lambda'^T Y_{k+1}' + \lambda''^T Y_k'' + i\lambda'''^T Y_k'\}$

at $t \to t^{(k+1)}$, i.e. to the joint characteristic function $g'_k(\lambda', \lambda'', \lambda''')$ of the random variables Y'_{k+1}, Y''_k, Y'_k,

$$g_1(\lambda; t^{(k+1)} - 0) = \lim_{t \to t^{(k+1)}} g_1(\lambda; t) = g'_k(\lambda', \lambda'', \lambda''').$$

At the instant $t^{(k+1)}$ the function $g_1(\lambda; t)$ changes its value by a jump and becomes equal to

$$E \exp \left\{ i(\lambda'^T + \lambda'''^T) Y'_{k+1} + i\lambda''^T Y''_{k+1} \right\} = g_{k+1}\left(\left[\lambda'^T + \lambda'''^T, \lambda''^T \right]^T \right).$$
$$(5.7.16)$$

For calculating this value we substitute here the expression for Y''_{k+1} from Eq. (5.7.11). Then we get with the account of (5.7.11)

$$g_1(\lambda; t^{(k+1)}) = E \exp \left\{ i(\lambda'^T + \lambda'''^T) Y'_{k+1} + i\lambda''^T \omega_k(Y_k, V_k) \right\}. \quad (5.7.17)$$

In the consequence of the independence of the sequences $\{V_k\}$ of the white noise V, V_k of V_0, V_1, ... , V_{k-1} the random variables Y_k and Y'_{k+1} do not depend on V_k. Consequently, the expectation in the right-hand side of Eq. (5.7.17) is completely determined by the known distribution of the random variable V_k and the joint characteristic function $g'_k(\lambda', \lambda'', \lambda''')$ of the random variables Y_{k+1}, Y''_k, Y'_k, i.e. by the function $g_1(\lambda; t^{(k+1)} - 0)$.

Thus Eq. (5.7.14) *with initial condition (5.7.15) and formula (5.7.17) determine the evolution of the one-dimensional characteristic function $g_1(\lambda; t)$ of the process $\bar{Y}(t) = \left[Y'(t)^T Y''(t)^T Y'''(t)^T \right]^T$ and its step-wise increments at the instants $t^{(k)}$ $(k = 1, 2, \ldots)$ of the stochastic continuous-discrete system described by Eqs. (5.7.10) and (5.7.11).*

The equation for the n-dimensional characteristic function $g_n(\lambda_1, \ldots, \lambda_n; t_1, \ldots, t_n)$ of the random process $\bar{Y}(t) = [Y'(t)^T Y''(t)^T \times Y'''(t)^T]^T$ determined by (5.7.10) and (5.7.11) have the form

$$\partial g_n(\lambda_1, \ldots, \lambda_n; t_1, \ldots, t_n)/\partial t_n$$

$$= E \left\{ i{\lambda'_n}^T a(Y(t_n), t_n) + \chi(b(Y(t_n), t_n)^T \lambda'_n; t_n) \right\}$$

$$\times \exp \left\{ i\lambda_1^T \bar{Y}(t_1) + \cdots + i\lambda_n^T \bar{Y}(t_n) \right\}. \quad (5.7.18)$$

Here the value of the characteristic function $g_n(\lambda_1, \ldots, \lambda_n; t_1, \ldots, t_n)$ at $t_n = t^{(k+1)} \geq t_{n-1} > \cdots > t_1$ is equal

$$g_n(\lambda_1, \ldots, \lambda_n; t_1, \ldots, t_{n-1}, t^{(k+1)}) = E \exp\{i\lambda_1^T \bar{Y}(t_1) +$$

$$\cdots + i\lambda_{n-1}^T \bar{Y}(t_{n-1}) + i({\lambda'_n}^T + {\lambda'''_n}^T) Y'_{k+1} + i{\lambda''_n}^T \omega_k(Y_k, V_k)\}. \quad (5.7.19)$$

At the point $t_n = t^{(k+1)}$ the characteristic function $g_n(\lambda_1, \ldots, \lambda_n; t_1, \ldots, t_n)$ changes step-wise from

$$g_n(\lambda_1, \ldots, \lambda_n; t_1, \ldots, t_{n-1}, t^{(k-1)} - 0) = E\exp\{i\lambda_1^T \bar{Y}(t_1) +$$

$$\cdots + i\lambda_{n-1}^T \bar{Y}(t_{n-1}) + i{\lambda_n'}^T Y_{k+1}' + i{\lambda_n''}^T Y_k'' + i{\lambda_n'''}^T Y_k'\}$$

till the value $g_n(\lambda_1, \ldots, \lambda_n; t_1, \ldots, t_{n-1}, t^{(k+1)})$ determined by formula (5.7.19).

The right-hand side of Eq. (5.7.19) is completely determined by the known distribution of the random variable V_k and the joint characteristic function $g_n(\lambda_1, \ldots, \lambda_n; t_1, \ldots, t_{n-1}, t^{(k+1)} - 0)$ of the random vectors $\bar{Y}(t_1), \ldots, \bar{Y}(t_{n-1}), Y_{k+1}', Y_k'', Y_k'$.

Thus Eq. (5.7.18) *with initial condition* (5.7.17) *and formula* (5.7.19) *determine the evolution and the step-wise increments of* $g_n(\lambda_1, \ldots, \lambda_n; t_1, \ldots, t_n)$ *at the point* $t^{(k+1)}$ *at the increase of* t_n *beginning the value* t_{n-1} *for the stochastic continuous-discrete system described by* Eqs. (5.7.10) *and* (5.7.11).

5.8. Distributions in Stochastic Composed Systems

5.8.1. Structural Theory of Stochastic Composed Systems

Using main theorems of Sections 2.3–2.6 we get the following results concerning the conditional probability measures (CPM) for typical connections of stochastic systems (Subsection 1.6.2).

▷ CPM's μ_A and μ_B of the stochastic systems A and B connected in parallel (Fig.1.6.1) determine a conditional probability measure in the direct product of the spaces $Y \times Z$ of the outputs of these systems. This measure is expressed by formula

$$\mu_{A \oplus B}(\mathcal{E} \mid x) = \iint_{\mathcal{E}} \mu_A(dy \mid x)\mu_B(dz \mid x) = \int_Y \mu_B(\mathcal{E}_y \mid x)\mu_A(dy \mid x),$$

$$(5.8.1)$$

where $\mathcal{E} \in \mathcal{B} \times \mathcal{C}$, $\mathcal{E}_y = \{z : (y, z) \in \mathcal{E}\}$ is the section of the set $\mathcal{E}$ at the point y, $\mathcal{B}, \mathcal{C}$ are σ-algebras of the sets in the spaces Y and Z correspondingly. The integration over y in the latter integral is performed practically over the projection $\mathcal{E}$ on the set Y as $\mathcal{E}_y \neq \varnothing$ only in the case when y belongs to this projection. ◁

Theorem 5.8.1. *The parallel connection of the stochastic systems* A *and* B *represents the stochastic system* $A \oplus B$ *whose CMP's* $\mu_{A \oplus B}$ *is determined by formula* (5.8.1).

▷ Analogously CMP's μ_A, μ_B of the systems A, B connected sequentially (Fig.1.6.2) are determined. The conditional probability measure is determined in the direct product $Y \times Z$. This measure is given by formula

$$\mu_{A\otimes B}(\mathcal{E} \mid x) = \int_Y \mu_B(\mathcal{E}_y \mid x, y)\mu_A(dy \mid x), \quad \mathcal{E} \in \mathcal{B} \times \mathcal{C}. \qquad (5.8.2)$$

A special case of the sequential connection (Fig.1.6.2) presented on Fig.1.6.3 is also stochastic system (we denote it by C) whose CPM is obtained from (5.8.3) at $\mathcal{E} = Y \times F$, $F \in \mathcal{C}$:

$$\mu_C(F \mid x) = \int_Y \mu_B(F \mid x, y)\mu_A(dy \mid x), \quad F \in \mathcal{C}. \triangleleft \qquad (5.8.3)$$

Theorem 5.8.2. *The sequential connection of the stochastic systems A and B represents the stochastic system C whose CPM μ_C is determined by formula* (5.8.3).

▷ Let us consider the closure $[A]$ of the stochastic system by elementary feedback (Fig.1.6.4a,b). Let $\mu_A(G \mid x, y)$, $G \in \mathcal{B}$ be CPM of the system A. Suppose that there exists such measurable mapping $T_y : Y \to Y$ of the space Y into itself that the measure $\mu_A(T_y^{-1}G \mid x, y)$ does not depend on y. This latter measure if it exists we shall denote by $\mu(G \mid x)$ then the identity is valid

$$\mu(G \mid x) = \mu_A(T_y^{-1}G \mid x, y), \quad G \in \mathcal{B}, \ (x, y) \in X \times Y. \qquad (5.8.4)$$

We put here $y = y_0$. If T_{y_0} is a convertible transformation then

$$\mu(T_{y_0}G \mid x) = \mu_A(G \mid x, y_0), \quad G \in \mathcal{B},$$

where $\mu_A(G \mid x, y_0)$ is a conditional probability measure of the output Y of the system $[A]$ at a given $x \in X$. Therefore CPM of the system $[A]$ may be naturally determined by formula

$$\mu_{[A]}(G \mid x) = \mu(T_{y_0}G \mid x), \quad G \in \mathcal{B}. \qquad (5.8.5)$$

This formula is reduced also to the form

$$\mu_{[A]}(G \mid x) = \mu_A(T_y^{-1}T_{y_0}G \mid x, y), \quad G \in \mathcal{B}, \qquad (5.8.6)$$

in consequence of (5.8.4). On the basis of formula

$$\mu_A(T_y^{-1}G \mid x, y) = \mu_A(T_{y_0}^{-1}G \mid x, y_0) \tag{5.8.7}$$

or equivalently

$$\mu_A(G \mid x, y) = \mu_A(T_{y_0}^{-1}T_y G \mid x, y_0), \tag{5.8.8}$$

which follows from (5.8.4) the right-hand side in (5.8.6) does not depend on the choice of $y_0 \in Y$. ◁

Theorem 5.8.3. *The closure of the stochastic system A by an elementary feedback represents the stochastic system $[A]$ whose CPM $\mu_{[A]}$ is determined by formula* (5.8.6).

The stated theorems establish the completeness of the class of the stochastic systems relatively to all possible typical connections (Subsection 1.6.2). Formulae (5.8.1), (5.8.2) and (5.8.6) give the principal opportunity to find CPM of any connections of the stochastic systems by given CPM's of the connected systems. The established properties of a class of the stochastic systems permit to use the structural methods for study of the composed stochastic systems. After partitioning a composite system into the subsystems we may study the characteristics of the subsystems and further to study the behaviour of the system as a whole. The structural methods for the linear and the nonlinear stochastic systems will be considered in Chapters 6 and 7. For the stochastic composed systems reducible to the systems considered in Sections 5.6 and 5.7 their CPM's are determined by the multi-dimensional distributions.

5.8.2. Distributions in Stochastic Systems with Random Structure

Let us consider a system which is described by stochastic differential equation:

$$\dot{Y} = a(Y, S(t), t) + b(Y, S(t), t)V \ (k = 1, \ldots, N), \tag{5.8.9}$$

where $S = S(t)$ be some step random process $S(t)$ which describes the random changes of a system structure. Let $s_1, \ldots, s_N$ be possible values of $S(t)$. At $S(t) = s_k$ the system is described by the Itô equation

$$\dot{Y} = a(Y, s_k, t) + b(Y, s_k, t)V \ (k = 1, \ldots, N). \tag{5.8.10}$$

We consider a case when the transitions of a system from one structure to another form the Poisson stream of the events. We denote the Poisson process of the transitions into the kth structure by $P_k(t)$ and its intensity by $\nu_k(Y, S, t)$ taking into account that it depends on the random process S and may also depend on the system state at the transition moment. It easy to see that the process $S(t)$ is determined by the following Itô stochastic differential equation:

$$dS = \sum_{k=1}^{N}(s_k - S)dP_k \,. \qquad (5.8.11)$$

Supposing that while changing the structure of the system its state vector may receive a random increment by a jump we write the sought stochastic differential equation of a system with randomly changed structure in the form

$$dY = a(Y,S,t)dt + b(Y,S,t)dW + \sum_{k=1}^{N} dQ_k, \qquad (5.8.12)$$

where $Q_k(t)$ is the general Poisson process with variable distribution of the jumps generated by simple Poisson process $P_k(t)$ $(k = 1\,,\ \ldots\,,\ N)$. The distribution of each jump of the process $Q_k(t)$ may depend not only on time but on the state vector of the system Y and its structure at the transition moment into the kth structure. Thus after adding to the state vector of the system Y the value of the process S determining its structure we shall get a system of Itô stochastic differential equations which describes the behaviour of a system with randomly changed structure. Naturally, the vector $\bar{Y} = \left[\, S\ Y^T\,\right]^T$ will be its state vector.

The equations of a system with randomly changing structure (5.8.10) and (5.8.11) at Poisson stream of the changes of the structure may be written in the general form of the Itô equations

$$d\bar{Y} = \bar{a}(\bar{Y},t)dt + \bar{b}(\bar{Y},t)d\bar{W} \,. \qquad (5.8.13)$$

Here $\bar{Y} = \left[\, S\ Y^T\,\right]^T$ is the state vector of a system; $\bar{W}(t)$ is a process with independent increments which consists of $N+1$ independent blocks $W(t)$, $\left[\, P_1 Q_1^T\,\right]^T$, $\ldots,\ \left[\, P_N Q_N^T\,\right]^T$; $\bar{a}(\bar{y},t)$, $\bar{b}(\bar{y},t)$ are block matrices equal to

$$\bar{a}(\bar{y},t) = \begin{bmatrix} 0 \\ a(y,s,t) \end{bmatrix},$$

$$\bar{b}(\bar{y},t) = \left[\begin{array}{c|cc|c|cc} 0 & s_1 - s & 0 & & s_N - s & 0 \\ b(y,s,t) & 0 & I & \cdots & 0 & I \end{array}\right].$$

Using Eq. (5.8.12) we set up Eq. (5.6.7) for an one- dimensional characteristic function of the state vector for the system with the randomly changing structure.

For this purpose we find at first the function $\bar{\chi}(\bar{\mu};t)$ correspondent to the process $\bar{W}(t)$, $\bar{\mu} = [\mu\,\mu_1 \ldots \mu_N]^T$. In accordance with the mentioned above in Subsection 5.6.3 the function $\bar{\chi}(\bar{\mu};t)$ in this case represents the sum of the functions χ correspondent to the independent blocks $W(t)$, $[P_1(t)Q_1(t)^T]^T \ldots [P_N(t)Q_N(t)^T]^T$ which compose the process $\bar{W}(t)$:

$$\bar{\chi}(\bar{\mu};Y,S,t) = \chi(\mu;t) + \sum_{k=1}^{N} \chi_k(\bar{\mu}_k;Y,S,t)\,, \tag{5.8.14}$$

where $\chi(\mu;t)$, $\chi_k(\bar{\mu}_k;Y,S,t)$ are the functions χ of the processes $W(t)$, $[P_k(t) Q_k(t)^T]^T$ correspondingly $(k = 1,\ldots,N)$; $\bar{\mu} = [\mu^T\,\bar{\mu}_1^T \ldots \bar{\mu}_N^T]^T$ is the partition of the vector $\bar{\mu}$ into the correspondent blocks; $\bar{\mu}_k = [\mu_{k0}\,\mu_k^T]^T$ is the expansion of the vector $\bar{\mu}_k$ into the blocks correspondent to the processes $P_k(t)$ and $Q_k(t)$. Here we took into consideration that the intensity of the streams of the jumps of the processes $P_k(t)$ and the distributions of the jumps of the processes $Q_k(t)$ may depend on the state vector of the system $\bar{Y} = [S\,Y^T]^T$. Using formula (5.6.8) for the functions χ and the formula for the characteristic function of the increments of the Poisson process and of the general Poisson process which is generated by it at infinitesimal interval $(t,s]$ according to which

$$h_k(\bar{\mu}_k;Y,S,t)$$

$$= 1 + \left[e^{i\mu_{k0}} g_k(\mu_k;Y,S,t) - 1\right] \nu_k(Y,S,t)(s-t) + o(s-t)\,. \tag{5.8.15}$$

We find from (5.8.15)

$$\chi_k(\bar{\mu}_k;Y,S,t) = \left[e^{i\mu_{k0}} g_k(\mu_k;Y,S,t) - 1\right] \nu_k(Y,S,t)\,, \tag{5.8.16}$$

where $g_k(\mu_k;y,s,t)$ is a characteristic function of the jumps of the process $Q_k(t)$ at the instant t. Thus (5.8.14) takes the form

$$\bar{\chi}(\bar{\mu};Y,S,t) = \chi(\mu;t) + \sum_{k=1}^{N} \left[e^{i\mu_{k0}} g_k(\mu_k;Y,S,t) - 1\right] \nu_k(Y,S,t)\,. \tag{5.8.17}$$

Therefore by virtue of (5.8.17) we have

$$\bar{\chi}(\bar{b}(Y,S,t)^T\bar{\lambda};Y,S,t) = \chi(b(Y,S,t)^T\lambda;t)$$

$$+ \sum_{k=1}^{N} \left[e^{i(s_k-S)\lambda_0} g_k(\lambda;Y,S,t) - 1\right] \nu_k(Y,S,t)\,, \tag{5.8.18}$$

where $\bar{\lambda} = [\lambda_0 \lambda^T]^T$ is the partition of the vector $\bar{\lambda}$ into the blocks correspondent to the blocks S, Y of the state vector of the system $\bar{Y}$. Substituting (5.8.18) into (5.6.7) we obtain the equation for the one-dimensional characteristic function of the process in the system with the random changes of the structure:

$$\frac{\partial g_1(\bar{\lambda}; t)}{\partial t} = E\left\{ i\lambda^T a(Y, S, t) + \chi(b(Y, S, t)^T \lambda; t) \right.$$

$$+ \sum_{k=1}^{N} \left[e^{i(s_k - S)\lambda_0} g_k(\lambda; Y, S, t) - 1 \right] \nu_k(Y, S, t) \right\} e^{i\lambda_0 S + i\lambda^T Y} . \qquad (5.8.19)$$

Basing on Eqs. (5.8.19) for the one-dimensional characteristic function we may derive the equations for the conditional characteristic functions in the different structures and for the probabilities of these structures. Taking into account that $S(t)$ is a discrete random variable with possible values $s_1, \ldots, s_N$ and the probabilities of these values $p_1(t), \ldots, p_N(t)$ at each t we find by formula of the total expectation

$$g_1(\bar{\lambda}; t) = E e^{i\lambda_0 S(t) + i\lambda^T Y(t)} = \sum_{k=1}^{N} p_l e^{i\lambda_0 s_l} g_1(\lambda; t \mid s_l), \qquad (5.8.20)$$

where $g_1(\lambda; t \mid s_l)$ is a conditional characteristic function of the state vector of the system Y in the lth structure. Analogously an expectation in the right-hand of Eq. (5.8.19) for $g_1(\bar{\lambda}; t)$ is calculated. As a result the equation will take the form

$$\sum_{l=1}^{N} e^{i\lambda_0 s_l} \frac{\partial p_l g_1(\lambda; t \mid s_l)}{\partial t}$$

$$= \sum_{l=1}^{N} p_l e^{i\lambda_0 s_l} E\left[\left\{ i\lambda^T a(Y, s_l, t) + \chi(b(Y, s_l, t)^T \lambda; t) \right\} e^{i\lambda^T Y} \mid s_l \right]$$

$$+ \sum_{l=1}^{N} p_l e^{i\lambda_0 s_l} \sum_{k=1}^{N}{}^{(l)} \left[e^{i\lambda_0 (s_k - s_l)} E g_k(\lambda; Y, s_l, t) \nu_k(Y, s_l, t) e^{i\lambda^T Y} \mid s_l \right]$$

$$- \sum_{l=1}^{N} p_l e^{i\lambda_0 s_l} E\left[\sum_{k=1}^{N}{}^{(l)} \nu_k(Y, s_l, t) e^{i\lambda^T Y} \mid s_l \right] . \qquad (5.8.21)$$

Here the index (l) at the sums over k shows that the item correspondent to $k = l$ does not enter into the sum. It does not enter into the sum as the processes $P_l(t)$

and $Q_l(t)$ maintain the constant values at the absence of the change of the structure as a result of which $g_l(\lambda; Y, s_l, t) = 1$. After changing the order of the summation in the first double sum (5.8.21) we get

$$\sum_{k=1}^{N} e^{i\lambda_0 s_k} \sum_{l=1}^{N}{}^{(k)} p_l E\left[g_k(\lambda; Y, s_l, t) \nu_k(Y, s_l, t) e^{i\lambda^T Y} \mid s_l \right].
\qquad (5.8.22)$$

After replacing in (5.8.22) the indexes k and l (the double sum will not change in this case) and substituting the obtained expression into Eq. (5.8.21) we shall have

$$\sum_{l=1}^{N} e^{i\lambda_0 s_l} \frac{\partial p_l g_1(\lambda; t \mid s_l)}{\partial t}$$

$$= \sum_{l=1}^{N} p_l e^{i\lambda_0 s_l} E\left[\left\{ i\lambda^T a(Y, s_l, t) + \chi(b(Y, s_l, t)^T \lambda; t) \right\} e^{i\lambda^T Y} \mid s_l \right]$$

$$+ \sum_{l=1}^{N} p_l e^{i\lambda_0 s_l} \sum_{k=1}^{N}{}^{(l)} p_k E\left[g_l(\lambda; Y, s_k, t) \nu_l(Y, s_k, t) e^{i\lambda^T Y} \mid s_k \right]$$

$$- \sum_{l=1}^{N} p_l e^{i\lambda_0 s_l} E\left[\nu_l(Y, t) e^{i\lambda^T Y} \mid s_l \right],
\qquad (5.8.23)$$

where for brevity is assumed $\nu_l(Y, t) = \sum_{k=1}^{N}{}^{(l)} \nu_l(Y, s_k, t)$. Comparing the coefficients at similar exponential functions and denoting by $f_1(y; t \mid s_l)$ a conditional one-dimensional density of the process $Y(t)$ in the lth structure we set up the system of the equations from Eq. (5.8.23)

$$\frac{\partial p_l g_1(\lambda; t \mid s_l)}{\partial t} = p_l \int_{-\infty}^{\infty} \{ i\lambda^T a(y, s_l, t)$$

$$+ \chi(b(y, s_l, t)^T \lambda; t) e^{i\lambda^T y} \} f_1(y; t \mid s_l) dy$$

$$+ \sum_{k=1}^{N}{}^{(l)} p_k \int_{-\infty}^{\infty} g_l(\lambda; y, s_k, t) \nu_l(y, s_k, t) e^{i\lambda^T y} f_1(y; t \mid s_k) dy$$

$$- p_l \int_{-\infty}^{\infty} \nu_l(y, t) e^{i\lambda^T y} f_1(y; t \mid s_l) dy \quad (l = 1, \dots, N).
\qquad (5.8.24)$$

Putting in Eq. (5.8.24) $\lambda = 0$ and taking into consideration that $g_1(0; t \mid s_l)$ $= 1$, $\chi(0; t) = 0$ we get the equations for the probabilities of the structures ($l = 1, \ldots, N$) of the system at the moment t:

$$\dot{p}_l = \sum_{k=1}^{N}{}^{(l)} p_k \int_{-\infty}^{\infty} \nu_l(y, s_k, t) f_1(y; t \mid s_k) dy - p_l \int_{-\infty}^{\infty} \nu_l(y, t) f_1(y; t \mid s_l) dy.$$

$$(5.8.25)$$

Using Eqs. (5.8.24) we derive the equations for the conditional densities from the equations for the conditional characteristic functions in the case of the Wiener process $W(t)$. For this purpose following Subsection 5.6.6 we change the denotion of the integration variable y by η, multiply the equation for $p_l g_1(\lambda; t \mid s_l)$ by $e^{i\lambda^T y}/(2\pi)^p$ and integrate it over λ. The first integral is transformed similarly as in Subsection 5.6.6 and gives as a result an item which is analogous to the right-hand side of Eq. (5.6.42). The second integral as a result of formula

$$\frac{1}{(2\pi)^p} \int_{-\infty}^{\infty} g_l(\lambda; \eta, s_k, t) e^{i\lambda^T(\eta - y)} d\lambda = q_l(y - \eta; \eta, s_k, t), \qquad (5.8.26)$$

where $q_l(y; \eta, s_k, t)$ is the density of the jumps of the process $Q_k(t)$ at given η, s_k and t is transformed to the form

$$\int_{-\infty}^{\infty} q_l(y - \eta; \eta, s_k, t) \nu_l(\eta, s_k, t) f_1(\eta; t \mid s_k) d\eta. \qquad (5.8.27)$$

The third integral in Eq. (5.8.23) as a result of (5.6.37) takes the form $\nu_l(y, t) f_1(y; t \mid s_l)$. As a result for $l = 1, \ldots, N$ we have by virtue of (5.8.27)

$$\frac{\partial p_l f_1(y; t \mid s_l)}{\partial t} = -p_l \frac{\partial^T}{\partial y} [a(y, s_l, t) f_1(y; t \mid s_l)]$$

$$+ \frac{p_l}{2} \operatorname{tr} \left\{ \frac{\partial}{\partial y} \frac{\partial^T}{\partial y} [b(y, s_l, t) \nu(t) b(y, s_l, t)^T f_1(y; t \mid s_l)] \right\}$$

$$+ \sum_{k=1}^{N}{}^{(l)} p_k \int_{-\infty}^{\infty} q_l(y - \eta; \eta, s_k, t) \nu_l(\eta, s_k, t) f_1(\eta; t \mid s_k) d\eta - p_l \nu_l(y; t) f_1(y; t \mid s_l).$$

$$(5.8.28)$$

In important special case where at the transition from one structure to another the state vector of the system Y does not receive a random increment the items with dQ_k in the equation for Y are absent $(Q_k(t) \equiv 0)$. Here in the equation for the one-dimensional characteristic function $g_1(\lambda; t)$ and the equations for the conditional one-dimensional characteristic functions $g_1(\lambda; t \mid s_l)$ at different structures we have $g_l(\lambda; y, s_k, t) \equiv 1$. Correspondingly in the equations for the conditional one-dimensional densities $f_1(y; t \mid s_l)$ at the different structures we get $q_l(y - \eta; \eta, s_k, t) = \delta(y - \eta)$ and the equations for $f_1(y; t \mid s_l)$ $(l = 1, \ldots, N)$ take the form

$$\frac{\partial p_l f_1(y; t \mid s_l)}{\partial t} = -p_l \frac{\partial^T}{\partial y} \left[a(y, s_l, t) f_1(y; t \mid s_l) \right]$$

$$+ \frac{p_l}{2} \operatorname{tr} \left\{ \frac{\partial}{\partial y} \frac{\partial^T}{\partial y} \left[b(y, s_l, t) \nu(t) b(y, s_l, t)^T f_1(y; t \mid s_l) \right] \right\}$$

$$+ \sum_{k=1}^{N} {}^{(l)} \nu_l(y, s_k, t) \left[p_k f_1(y; t \mid s_k) - p_l f_1(y; t \mid s_l) \right] \quad (l = 1, \ldots, N).$$

$$(5.8.29)$$

5.9. Distributions in Stochastic Infinite-Dimensional Systems

5.9.1. Distributions in Stochastic Continuous Hilbert Systems

At first let us consider the random process $Y = Y(t)$ with the values in separable H-space $\mathcal{Y}$ which is determined by the Itô stochastic differential equation of the form:

$$dY = a(Y, t)dt + b(Y, t)dW, \quad Y(t_0) = Y_0. \tag{5.9.1}$$

Here $W = W(t)$ is a random process with the independent increments in some other H-space from $\mathcal{W}$ to $\mathcal{Y}$, $a(Y, t)$ is an operator mapping $\mathcal{Y} \times R$ into $\mathcal{Y}$, $b(Y, t)$ is an operator-valued function mapping $\mathcal{Y} \times R$ into the space of the linear operators which act from $\mathcal{W}$ to $\mathcal{Y}$, Y_0 is a random variable in the space $\mathcal{Y}$ independent of the future increments of the process W. The central problem of stochastic systems theory consists in finding the multi-dimensional functionals of the process Y:

$$g_n(\lambda_1, \ldots, \lambda_n; t_1, \ldots, t_n) = E \exp \left[i \sum_{k=1}^{n} (\lambda_k, Y(t_k)) \right] \quad (n = 1, 2, \ldots). \tag{5.9.2}$$

Here $\lambda_1, \ldots, \lambda_n$ are the elements of the adjoint space $\mathcal{Y}^*$, (λ_k, Y_{t_k}) is a scalar product of the elements λ_k and Y_{t_k} in H-space $\mathcal{Y}$. The multi-dimensional functionals $g_n = g_n(\lambda_1, \ldots, \lambda_n; t_1, \ldots, t_n)$ completely determine at the known conditions a distribution of a process (Section 3.6). The stated problem is easily solved while performing the following conditions.

(i) The space $\mathcal{Y}$ of the values of the process $Y(t)$ at each t is a separable H-space.

(ii) The random process $W(t)$ represents the Wiener process with the values in the separable H-space $\mathcal{W}$ whose covariance operator of the value at each given t is determined by formula

$$K_w(t) = \int_{t_0}^{t} \nu(\tau)d\tau, \qquad (5.9.3)$$

where $\nu(t)$ is the intensity of the process $W(t)$ which represents at each t the trace-class adjoint operator.

(iii) The function $a(Y, t)$ with the values in the H-space $\mathcal{Y}$ is determined by formula

$$a(Y, t) = a_1 Y + a'(Y, t). \qquad (5.9.4)$$

Here a_1 being a closed linear operator which is an infinitesimally generating operator of strongly continuous semigroup of the restricted operators $\{u(t)\}$ satisfying the condition $\|u(t)\| \le ce^{-\gamma t}$ at some $c, \gamma > 0$ and equation

$$du/dt = a_1 u, \quad u(0) = I, \qquad (5.9.5)$$

where I is an unit operator, $a'(Y, t)$ is the strongly differentiable function over Y and t with the values in the H-space $\mathcal{Y}$.

(iv) The operator-valued function $b(Y, t)$ at each Y and t represents such compact linear operator that the self-adjoint operator

$$v(Y, t, \tau) = u(t - \tau)\sigma(Y, t)u(t - \tau)^*, \quad \sigma(Y, t) = b(Y, t)\nu(t)b(Y, t)^*, \qquad (5.9.6)$$

is trace-class and at each t the following conditions is fulfilled:

$$\int_{t_0}^{t} \operatorname{tr} v(Y, t, \tau)d\tau < \infty. \qquad (5.9.7)$$

The representability of the function $a(Y, t)$ in Eq. (5.9.1) in the form of (5.9.4) with the restrictions on a_1, $a'(Y, t)$ and condition (5.9.7) are

sufficient for the existence of the solution of the Itô stochastic integral
equation

$$Y(t) = u(t)Y_0 + \int_{t_0}^{t} u(t-\tau)a'(Y(\tau),\tau)d\tau + \int_{t_0}^{t} u(t-\tau)b(Y(\tau),\tau)dW ,$$

$$(5.9.8)$$

and for the existence and the integrability of the operator of the second
order moment. The property of the operators $\nu(t)$, $\sigma(Y,t)$, $v(Y,t,\tau)$ to be
trace-class is necessary and sufficient for $\nu(t)\Delta t$, $\sigma(Y,t)\Delta t$, $v(Y,t,\tau)\Delta\tau$
be the covariance operators of the normally distributed random variab-
les ΔW, $b(Y,t)\Delta W(t)$ and $u(t-\tau)b(Y,\tau)\Delta W(\tau)$ for the infinitesimals
Δt and $\Delta\tau$.

Let us show that while performing conditions (ii)–(iv) the multi-
dimensional characteristic functionals (5.9.2) are determined by equations

$$\frac{\partial g_n(\lambda_1,\ldots,\lambda_n;t_1,\ldots,t_n)}{\partial t_n} = E\left\{\left[i(\lambda_n,a(Y_{t_n},t_n))\right.\right.$$

$$\left.\left.-\frac{1}{2}(\lambda_n,\sigma(Y_{t_n},t_n))\lambda_n)\right]\exp\left[i\sum_{k=1}^{n}(\lambda_k,Y_{t_k})\right]\right\} \quad (n=1,2,,\ldots,)$$

$$(5.9.9)$$

and the initial conditions

$$g_1(\lambda_1;0) = g_0(\lambda_1),\qquad\qquad (5.9.10)$$

$$g_n(\lambda_1,\ldots,\lambda_n;t_1,\ldots,t_{n-1},t_{n-1}) = g_{n-1}(\lambda_1,\ldots,\lambda_{n-1}+\lambda_n;t_1,\ldots,t_{n-1}),$$

$$(5.9.11)$$

where $g_0(\lambda_1)$ is the characteristic functional of the random variable Y_0.

▷ In fact it is sufficient to use Itô formula (5.1.15) at a given n for the calculation
of the stochastic differential of the composite function $\exp\left[i\sum_{k=1}^{n}(\lambda_k,Y_{t_k})\right]$ which
is considered as the function of a scalar random process $(\lambda_n,Y(t_n))$ at the fixed
$\lambda_1,\ldots,\lambda_{n-1},Y_{t_1},\ldots,Y_{t_{n-1}}$ and to use the formula of the transformation of
a covariance operator in linear spaces (Subsection 3.3.2). ◁

In the stationary case when in Eq.(5.9.1) $a'(Y,t) = a(Y)$, $b(Y,t)$
$= b(Y)$ and $\nu(t) = \nu$ do not depend on time there exists a stable stationary process at conditions (i)–(iv). The one-dimensional distribution
of the stationary process $Y(t)$ is determined by (5.9.9) at $n = 1$ and

$\partial g_1/\partial t = 0$. As it is known this process is stable if $\| u(t) \| < e^{-\gamma t}$ for some $\gamma > 0$.

E x a m p l e 5.9.1. Let us consider a stochastic system described by a nonlinear wave stochastic differential equation

$$\frac{\partial^2 U}{\partial t^2} = c^2(1 + V_1)\frac{\partial^2 U}{\partial x^2} + 2h_1\frac{\partial U}{\partial t} - \frac{2}{3}h_2\left(\frac{\partial U}{\partial t}\right)^3 + V_2 \qquad \text{(I)}$$

with the boundary conditions $U(0,t) = U(2l,t) = 0$. Here $U = U(x,t)$; $V_{1,2} = V_{1,2}(x,t)$ are the independent normally distributed white noises with the intensities $\nu_{1,2} = \nu_{1,2}(x_1, x_2, t)$; c, $h_{1,2}$ are constant coefficients. After putting

$$Y_1 = U , \quad Y_2 = \partial U/\partial t , \quad Y = [Y_1\, Y_2]^T ,$$

$$a(Y,t) = \left[\begin{array}{c} Y_2 \\ c^2\partial^2 Y_1/\partial x^2 + 2h_1 Y_2 - (2/3)\,h_2 Y_2^3 \end{array}\right] , \qquad \text{(II)}$$

$$b(Y,t) = \left[\begin{array}{cc} 0 & 0 \\ c^2\partial^2 Y_1/\partial x^2 & 1 \end{array}\right] ,$$

we reduce Eq. (I) to the form of Eq. (5.9.1). Eq. (5.9.9) for the one-dimensional characteristic functional $g_1 = g_1(\lambda_1, \lambda_2; t)$ will take the following form:

$$\frac{\partial g_1(\lambda_1, \lambda_2; t)}{\partial t} = E\left\{\left[i(\lambda_1, Y_2) + i\left(\lambda_2, c^2\frac{\partial^2 Y_1}{\partial x^2} + 2h_1 Y_2 - \frac{2}{3}h_2 Y_2^3\right)\right.\right.$$

$$\left.\left. -\frac{c^4}{2}\left(\lambda_2, \left(\frac{\partial^2 Y_1}{\partial x^2}\right)\nu_1\left(\frac{\partial^2 Y_1}{\partial x^2}\right)^* \lambda_2\right) - \frac{1}{2}(\lambda_2, \nu_2\lambda_2)\right] e^{i(\lambda_1, Y_1)+i(\lambda_2, Y_2)}\right\} ,$$

$$\text{(III)}$$

where a scalar product is given by the formula $(\varphi, \psi) = \int\limits_0^{2l} \varphi(x)\psi(x)dx$.

Along with Eq. (5.9.1) there are the stochastic differential equations in the separable H-space $\mathcal{Y}$ of the following form:

$$dY = a(Y,t)dt + b(Y,t)dW_0 + \int\limits_{R_0^q} c(Y,t,u)P^0(dt, du) . \qquad \text{(5.9.12)}$$

Here $W_0 = W_0(t)$ is the Wiener process with the values in the H-space $\mathcal{W}_0$, $P^0(\Delta, \mathcal{E})$ is the centered Poisson measure, $\Delta = (t_1, t_2)$ is some time interval, $\mathcal{E}$ is some Borel set of the space R_0^q representing the q-dimensional space with pricked origin, $a = a(Y, t)$ and $c = c(Y, t, u)$ are the functions mapping $\mathcal{Y} \times R$ and $\mathcal{Y} \times R \times R_0^q$ into $\mathcal{Y}$, $b = b(Y, t)$ is an operator-valued function mapping $\mathcal{Y} \times R$ into the space of the linear operators which act from $\mathcal{W}_0$ to $\mathcal{Y}$. We shall assume that the initial value Y_0 is an element of the space $\mathcal{Y}$ and does not depend on the increments of the Wiener process W_0 and the values of the Poisson measure $P^0(\Delta, \mathcal{E})$ at the time interval $\Delta = (t_1, t_2]$, $t_0 \leq t_1 < t_2$ for every $\mathcal{E}$.

In this case similarly as in the previous case it is shown that multidimensional characteristic functionals (5.9.2) satisfy the equations

$$\frac{\partial g_n(\lambda_1, \ldots, \lambda_n; t_1, \ldots, t_n)}{\partial t_n} = E\bigg\{ [i(\lambda_n, a(Y_{t_n}, t_n))$$

$$+\chi(\lambda_n, Y_{t_n}; t_n)] \exp\left[i \sum_{k=1}^{n} (\lambda_k, Y_{t_k}) \right] \bigg\} \quad (n = 1, 2, \ldots) \qquad (5.9.13)$$

with initial conditions (5.9.10) and (5.9.11) where

$$\chi(\lambda, y; t) = -\frac{1}{2}(\lambda, \sigma_0(y, t))$$

$$+ \int_{R_0^q} \left\{ e^{i(\lambda, c(y, t, u))} - 1 - i(\lambda, c(y, t, u)) \right\} \nu_P(t, du), \qquad (5.9.14)$$

$\sigma_0(y, t) = b(y, t)\nu_0(t)b(y, t)^*$, ν_0 is the intensity of the process W_0 which represents the trace-class operator, $b(y, t)^*$ is an operator adjoint with $b(y, t)$, $\nu_P(t, du)$ is the intensity of the simple Poisson process.

The obtained results may be generalized on the stochastic continuous systems in B-spaces with a basis (Pugachev and Sinitsyn 2000).

5.9.2. Distributions in Markov Stochastic Discrete Systems

Let us obtain the equations of the discrete Markov systems which are important in applications. For this purpose we consider the random sequence $\{Y_k\}$ with the values in some measurable space $(\mathcal{Y}, \mathcal{B})$ which satisfies the following stochastic difference equation:

$$Y_{k+1} = \omega_k(Y_k, V_k), \quad k = 1, 2, \ldots, \qquad (5.9.15)$$

where $\{V_k\}$ is the sequence of the independent random variables with the values in other measurable space $(\mathcal{V}, \mathcal{E})$; $\omega_k(y_k, v_k)$ are $(\mathcal{B} \times \mathcal{E}, \mathcal{B})$-

measurable functions mapping $\mathcal{Y} \times \mathcal{V}$ into $\mathcal{Y}$. Let the initial value Y_1 be independent of the sequence $\{V_k\}$; $\mu_1(B)$, $B \in \mathcal{B}$ is the distribution of Y_1; $\nu_k(E)$, $E \in \mathcal{E}$ is the distribution of V_k. In this case $\{Y_k\}$ is a Markov sequence. Consequently (Subsection 2.7.9), its multi-dimensional distributions are completely determined by the initial distribution $\mu_1(B)$, $B \in \mathcal{B}$ and by the transition distribution $\mu_{k+1}(B \mid Y_k)$, $B \in \mathcal{B}$ and

$$\mu_{k+1}(B \mid y) = \nu_k(\omega_k^{-1}(y, B)), \qquad (5.9.16)$$

where $\omega_k^{-1}(y, B) \in \mathcal{V}$ is the inverse image of the set $B \in \mathcal{Y}$ which is determined by the function $\omega_k(y, v)$ at the fixed y. Recurrent Eq. (5.9.15) at $k = 1, 2, \ldots$ and given initial distribution $\mu_1(B)$ permit to find the distributions of the sequence $\{Y_k\}$.

5.10. Applications. Optimal Online Filtering

5.10.1. Introductory Remarks

Problems of determining the state of a system using the results of some measurements often arise in practice. As the measurements are always accompanied by random errors one should speak not about the determination of the system state but about its estimation (filtering, extrapolation, interpolation) by statistical processing of the results of measurements.

At practice a filtering problem is technically solved by means of passing the measured signal $Y(t)$ through the device which is called *a filter* for "filtering off" the noise and for obtaining a signal at the output which reproduces the required process $X(t)$ with possibly great accuracy.

Let us consider the continuous stochastic system described by Eq. (5.4.6)

$$\dot{X} = \varphi(X, t) + \psi(X, t)V \qquad (5.10.1)$$

where X is the n-dimensional state vector of the system, V is the r-dimensional vector white normal noise, and $\varphi(x, t)$, $\psi(x, t)$ are known functions of the system state and time. The values of the function $\varphi(x, t)$ are evidently the n-dimensional vectors, and the values of the function $\psi(x, t)$ are the $n \times r$ matrices.

If the state vector of the system X is continuously measured then the n-dimensional random process $Y(t) = X(t) + U(t)$ will be the result of measurements where $U(t)$ is the measurement error which represents usually a random function of time. However, it is not the components of the state vector, but some functions of the state vector are usually measured among which may be some of the components

of the state vector. The result of the measurements is determined in the general case
by the formula

$$Y = Y(t) = \varphi_0(X, U, t) \tag{5.10.2}$$

where Y is the n_1-dimensional vector, U is the measurement error representing a
vector random function of time of dimension $r \geq n_1$, and $\varphi_0(x, u, t)$ is the known
function of the system state, measurement error and time. In the general case the
function φ_0 is nonlinear both relative to the state vector and to the measurement
error. The dimension of the error vector U in problems of practice cannot be smaller
than the number of measured variables r_1 as each measurement is accompanied by
an error and the errors of various instruments are usually independent (or at least
contain independent items).

In some cases the measurements are performed by instruments possessing inertia
or the results of measurements may be filtered by various filters. So the general model
of measurements which are performed in a system may be described by the differential
equation

$$\dot{Y} = \varphi_1(Y, X, U, t). \tag{5.10.3}$$

The result of measurements represent the random process $Y(t)$. We come thus to
the problem of filtering of the state vector of a system X at each instant $t > t_0$ using
the results of continuous measurement of the process Y determined by Eq. (5.10.3)
in the time interval $[t_0, t]$.

R e m a r k. We emphasize that such a statement of the filtering problem is
adequate only if online data processing, i.e. data processing in the real time basis
is possible. If online data processing is impossible and one has to be content with
post data processing when the measurement is finished, then the above statement of
the filtering problem has no sense. The estimates will always be better if the whole
measurement is used to estimate the state vector of a system at any instant t during
the experiment. Therefore, other techniques for solving filtering problem must be
used in such cases (Subsection 1.6.5).

Thus we come to the exact statement of the filtering problem.

A vector random process $\left[Y^T X^T \right]^T$ *is determined by the Itô stochastic differential equations*

$$dY = \varphi_1(Y, X, t)dt + \psi_1(Y, X, t)dW, \tag{5.10.4}$$

$$dX = \varphi(Y, X, t)dt + \psi(Y, X, t)dW, \tag{5.10.5}$$

where Y *is a* n_1*-dimensional random process,* X *is a* n*-dimensional process,* W
is a n*-dimensional process,* $\psi_1(y, x, t)$ *and* $\varphi(y, x, t)$ *are known vector functions
mapping the space* $R^{n_1} \times R^n \times R$ *into the spaces* R^{n_1} *and* R^n *respectively, and*
$\varphi(y, x, t)$ *and* $\psi(y, x, t)$ *are the known matrix functions mapping* $R^{n_1} \times R^n \times R$

into $R^{n_1 r}$ and R^{nr} respectively. It is required to filter the process $\big($the state vector of the system$\big)$ at any instant $t > t_0$ using the results of continuous measurement of the process Y in the time interval $[t_0, t]$.

5.10.2. General Formula for Optimal Filtering

It is natural to require that the filtering of the process $Y(t)$ be optimal in some sense. The mean square error (m.s.e.) criterion $E \big| \hat{X}_t - X_t \big|^2 = \min$ serves as a natural criterion of optimality in many problems of mathematical statistics. If we assume this criterion, then the general solution of the filtering problem follows directly from the known property of the second order moments: the smallest of all second order moments of a scalar random variable is its variance. Hence it follows that the best approximation of a random variable by a nonrandom variable from the m.s.e. criterion point of view is given by its expectation. In particular, the best approximation of a random variable using the results of measurement is given by the conditional expectation relative to the results of measurements.

We denote by $Y_{t_0}^t$ the aggregate of the values of the measured process in the time interval $[t_0, t]$, $Y_{t_0}^t = \{Y(\tau) : \tau \in [t_0 . t]\}$. Then the optimal estimate of the vector $X_u = X(u)$, which gives the solution of problem at $u = t$ is determined by the formula

$$\hat{X}_u = E\left[X_u \mid Y_{t_0}^t \right] . \tag{5.10.6}$$

This formula determines the optimal estimate of the value X_u of any random function $X(u)$ using the results of measurement of other random function $Y(t)$ in the interval $[t_0, t]$. It is also valid for the case of a vector argument t and the measurement of the random function $Y(t)$ on any set T of values of t.

In the general case it is necessary for practical application of formula (5.10.6) to find the conditional distribution of X_u. This problem is very difficult and in the general case is not yet solved. In the particular case of the processes $Y(t)$ and $X(t)$ determined by Eqs. (5.10.4), (5.10.65) it may be solved under some additional restrictions. But even in these cases the practical application of formula (5.10.6) is a matter of great and often insuperable difficulty. The reason is that the determination of the conditional distribution always requires very cumbersome and time-consuming calculations which can be performed only after measurement. The statement of problems in Subsection 5.10.1 implies that the estimates must be calculated in the real time while the results of measurement appear. But the difficulties of application of formula (5.10.6) do not reduce the importance of the optimal estimation. This theory is necessary for studying the potential accuracy of the estimates, i.e. maximal achievable accuracy of the estimation.

5.10.3. Auxiliary Problem

The general formula for the stochastic differential of the optimal estimate of a given function of the state vector of a system underlies the optimal filtration theory. Let $f(X_t, t)$ be some scalar function of the n-dimensional state vector of a system and of time. Its optimal estimate using the results of observation $Y_{t_0}^t$ according to (5.10.6) is determined by the formula

$$\hat{f}(t) = E\left[f(X_t, t) \mid Y_{t_o}^t \right]. \tag{5.10.7}$$

This estimate represents a functional of the random process $Y(t)$ in the variable time interval $[t_0, t]$, and consequently, is itself a random function of t. We put the auxiliary mathematical problem: to find the Itô stochastic differential of this random process.

This problem may be solved under the condition that $W(t)$ in Eqs. (5.10.4), (5.10.5) represents the Wiener process whose dimension r is not smaller than the dimension n_1 of the measurement process $Y(t)$, and that the function φ_1 in Eq. (5.10.4) does not depend on X. Eqs. (5.10.4), (5.10.5) in this case have the form

$$dY = \varphi_1(Y, X, t)dt + \psi_1(Y, t)dW, \tag{5.10.8}$$
$$dX = \varphi(Y, X, t)dt + \psi(Y, X, t)dW. \tag{5.10.9}$$

Lemma 5.10.1. *If $\sigma_1(y, t) = \psi_1(y, t)\nu((t)\psi_1(y, t)^T$ where $\nu(t)$ being the intensity of the Wiener process then Eqs. (5.10.8), (5.10.9) may be transformed to the following form:*

$$dY = \varphi_1(Y, X, t)dt + \psi_1''(Y, t)dW_2,$$
$$dX = \varphi(Y, X, t)dt + \psi_1'(Y, X, t)dW_1 + \psi_1''(Y, X)dW_2. \tag{5.10.10}$$

Here $W_1 = W_1(t)$ and $W_2(t)$ are two independent Wiener processes formed by the first $r_1 = r - n_1$ components and by the last components of the process $W' = W'(t)$ with independent components:

$$W'(t) = \int\limits_0^t \cdot \Omega(y, \tau)\nu^{-1/2}(\tau)dW(\tau)$$

where ψ, ψ_1', ψ_1'' be some matrices and $\Omega(y, t)$ be any orthogonal real matrix possibly depending on y and time t.

▷ As it is known the power of a symmetrical nonnegative definite matrix ν is determined by the formula $\nu^s = A\Lambda^s A^T$, where $\Lambda = \mathrm{diag}\{\lambda_1, \ldots, \lambda_q\}$

is the diagonal matrix to which ν is reduced by the orthogonal transformation A, and $\Lambda^s = \text{diag}\left\{\lambda_1^s, \ldots, \lambda_q^s\right\}$. Then the random process defined by (5.10.11) represents a Wiener process with independent components. Really, the covariance matrix of the value of the process $W'(t)$ at a given t is determined by the formula

$$EW'(t)W'(t)^T = \int_0^t \Omega(y,\tau)\nu^{-1/2}(\tau)\nu(\tau)\nu^{-1/2}(\tau)\Omega(y,\tau)^T \, d\tau = It,$$

since $\Omega(y,\tau)\Omega(y,\tau)^T \equiv I$ by virtue of the orthogonality of the matrix $\Omega(y,\tau)$. This formula shows that the components of the process $W'(t)$ are uncorrelated, and consequently, are independent as a Wiener process is normally distributed. Since the intensity of the process $W'(t)$ is equal to the identity matrix, each component of the process $W(t)$ represents a standard Wiener process.

Let us pass from the process $W(t)$ to the process $W'(t)$ in Eqs. (5.10.8), (5.10.9). Then the differential dW will be replaced by $\nu^{1/2}(t)\Omega(y,t)^T dW'$. Now we choose the orthogonal matrix $\Omega(y,t)$ in such a way that the matrix $b_1(y,t)$ be reduced to a block form $\left[0 \ \psi_1''(y,t)\right]$ where the first block represents the $n_1 \times (r-n_1)$ matrix whose elements are all equal to zero, and the second block represents an $n_1 \times n_1$ matrix. It is possible under some conditions, as the orthogonal $r \times r$ matrix Ω has $r(r-1)/2$ arbitrary elements which must satisfy $n_1(r-n_1)$ conditions, and $r(r-1)/2 \geq r^2/4$ at $r \geq 2$, $n_1(r-n_1) \leq r^2/4$. As a result of such a transformation

$$\begin{aligned}
\begin{bmatrix} \psi_1(Y,t) \\ \psi(Y,X,t) \end{bmatrix} dW &= \begin{bmatrix} \psi_1(Y,t) \\ \psi(Y,X,t) \end{bmatrix} \nu^{1/2}(t)\Omega(Y,t)^T dW' \\
&= \begin{bmatrix} \psi_1(Y,t)\nu^{1/2}(t)\Omega(Y,t)^T \\ \psi(Y,X,t)\nu^{1/2}(t)\Omega(Y,t) \end{bmatrix} dW' \\
&= \begin{bmatrix} 0 & \psi_1''(Y,t) \\ \psi'(Y,X,t) & \psi''(Y,X,t) \end{bmatrix} dW' \\
&= \begin{bmatrix} \psi_1''(Y,t)dW_2 \\ \psi'(Y,X,t)dW_1 + \psi''(Y,X,t)dW_2 \end{bmatrix}.
\end{aligned}$$

To find the matrices ψ', ψ'' and ψ_1'' and establish the conditions under which the transformation of Eqs. (5.10.8), (5.10.9) into the form of Eqs. (5.10.10) is possible we evaluate the conditional covariance matrix of the random vector

$$\begin{bmatrix} \psi_1(Y,t) \\ \psi(Y,X,t) \end{bmatrix} \Delta W = \begin{bmatrix} 0 & \psi_1''(Y,t) \\ \psi'(Y,X,t) & \psi''(Y,X,t) \end{bmatrix} \Delta W'$$

relative to the random variables Y, X where $\Delta W = W(t+\Delta t) - W(t)$, $\Delta W' = W'(t+\Delta t) - W'(t)$. Taking into account that ΔW and $\Delta W'$ are independent of Y_t, X_t and their covariance matrices are equal to $\nu(t)\Delta t + o(\Delta t)$ and $I\Delta t$ respectively, using the formula of transformation of the covariance matrix at a linear transformation of a random vector and omitting for brevity the arguments of the functions ψ, ψ_1, ψ', ψ'', ψ_1'' we have

$$\begin{bmatrix} \psi_1 \\ \psi \end{bmatrix} \nu \begin{bmatrix} \psi_1^T \psi^T \end{bmatrix} \Delta t + o(\Delta t) = \begin{bmatrix} 0 & \psi_1'' \\ \psi' & \psi'' \end{bmatrix} \begin{bmatrix} 0 & \psi'^T \\ \psi_1''^T & \psi''^T \end{bmatrix} \Delta t.$$

Hence, dividing both parts by Δt we obtain as $\Delta t \to 0$

$$\begin{bmatrix} \psi_1 \nu \psi_1^T & \psi_1 \nu \psi^T \\ \psi \nu \psi_1^T & \psi \nu \psi^T \end{bmatrix} = \begin{bmatrix} \psi_1'' \psi_1''^T & \psi_1'' \psi''^T \\ \psi'' \psi_1''^T & \psi' \psi^T + \psi'' \psi''^T \end{bmatrix},$$

or

$$\psi_1'' \psi_1''^T = \psi_1 \nu \psi_1^T, \quad \psi'' \psi_1''^T = \psi \nu \psi_1^T, \quad \psi' \psi'^T + \psi'' \psi''^T = \psi \nu \psi^T.$$

$$(5.10.11)$$

It remains to solve these equations successively for ψ_1'', ψ'' and ψ'.

The matrix ψ_1'' determined by formula (5.10.11) is independent of X, as ψ_1 is independent of X. The second Eq. (5.10.11) has the solution if and only if the matrix $\psi_1''(y,t)$ is invertible at all y, t. It is necessary for this that the matrix $\psi_1 \nu \psi_1^T$ be invertible at all y, t since on the basis of the first Eq. (5.10.11) $|\psi_1''|^2 = |\psi_1 \nu \psi_1^T|$. In this case the second Eq. (5.10.11) gives $\psi'' = \psi \nu \psi_1^T (\psi_1''^{-1})^T$. After that the third Eq. (5.10.11) is reduced to the form

$$\psi' \psi'^T = \psi \nu \psi^T - \psi \nu \psi_1^T (\psi_1''^{-1})^T (\psi_1''^{-1}) \psi_1 \nu \psi^T,$$

or taking into account that by virtue of the first Eq. (5.10.11) $(\psi_1''^{-1})^T (\psi_1''^{-1}) = (\psi_1 \nu \psi_1^T)^{-1}$,

$$\psi' \psi'^T = \psi \nu \psi^T - \psi \nu \psi_1^T (\psi_1 \nu \psi_1^T)^{-1} \psi_1 \nu \psi^T.$$

The right-hand side of this equation represents a symmetrical matrix. Let us prove that it is nonnegative definite. For this purpose we notice that the matrix

$$A = \begin{bmatrix} \psi_1 \nu \psi^T & \psi_1 \nu \psi^T \\ \psi \nu \psi_1^T & \psi \nu \psi^T \end{bmatrix},$$

being proportional to the covariance matrix of the random vector $\begin{bmatrix} \psi_1^T \psi^T \end{bmatrix}^T dW$ is nonnegative definite in consequence of which at any vector $u = \begin{bmatrix} u_1^T u_2^T \end{bmatrix}^T$

$$u^T A u = u_1^T \psi_1 \nu \psi_1^T u_1 + u_2^T \psi \nu \psi_1^T u_1 + u_1^T \psi_1 \nu \psi^T u_2 + u_2^T \psi \nu \psi^T u_2 \geq 0.$$

Putting here $u_1 = -(\psi_1\nu\psi_1^T)^{-1}\psi_1\nu\psi^T u_2$ we shall have

$$u_2^T\psi\nu\psi_1^T(\psi_1\nu\psi_1^T)^{-1}\psi_1\nu\psi^T u_2 - u_2^T\psi\nu\psi_1^T(\psi_1\nu\psi_1^T)^{-1}\psi_1\nu\psi^T u_2$$

$$-u_2^T\psi\nu\psi_1^T(\psi_1\nu\psi_1^T)^{-1}\psi_1\nu\psi^T u_2 + u_2^T\psi\nu\psi^T u_2 \geq 0,$$

or after cancellations

$$u_2^T\left[\psi\nu\psi^T - \psi\nu\psi_1^T(\psi_1\nu\psi_1^T)^{-1}\psi_1\nu\psi^T\right]u_2 \geq 0.$$

The validity of this inequality for all n-dimensional vectors u_2 proves our statement. After this the solution of the last Eq. (5.10.11) is given by the explicit formula $\psi' = \psi\left[\nu - \nu\psi_1^T(\psi_1\nu\psi_1^T)^{-1}\psi_1\nu\right]^{1/2}$. It is evident that multiplying the matrices ψ_1'' and ψ' on the right by arbitrary orthogonal matrices of the corresponding dimensions we obtain another solution of Eqs. (5.10.11).

Thus Eqs. (5.10.11) have an infinite set of solutions, and consequently, Eqs. (5.10.8), (5.10.9) are reducible to the form of Eqs. (5.10.10) if the matrix $\sigma_1(y,t) = \psi_1(y,t)\nu(t)\psi_1(y,t)^T$ is invertible at all y, t. ◁

Lemma 5.10.2. *The stochastic differential of the optimal estimate of the random variable $f(X_t,t)$ for Eqs. (5.10.4), (5.10.5) is given by formula:*

$$d\hat{f} = E[f_t(X,t) + f_x(X,t)^T\varphi(Y,X,t)$$

$$+\frac{1}{2}\mathrm{tr}\left\{f_{xx}(X,t)(\psi\nu\psi^T)(Y,X,t)\right\}\,|\,Y_{t_0}^t]dt + E[f(X,t)\left\{\varphi_1(Y,X,t)^T - \hat{\varphi}_1^T\right\}$$

$$+f_x(X,t)^T(\psi\nu\psi_1^T)(Y,X,t)\,|\,Y_{t_0}^t](\psi_1\nu\psi_1^T)^{-1}(Y,t)(dY - \hat{\varphi}_1 dt), \quad (5.10.12)$$

where

$$(\psi\nu\psi^T)(y,x,t) = \psi(y,x,t)\nu(t)\psi(y,x,t)^T,$$

$$(\psi\nu\psi_1^T)(y,x,t) = \psi(y,x,t)\nu(t)\psi_1(y,t)^T,$$

$$(\psi_1\nu\psi_1^T)^{-1}(y,t) = [\psi_1(y,t)\nu(t)\psi_1(y,t)^T]^{-1}, \quad (5.10.13)$$

$$\hat{\varphi}_1 = \int_{-\infty}^{\infty} \varphi_1 p_t(x)dx = E[\varphi_1(X_t,Y_t,t\,|\,Y_{t_0}^t), \quad (5.10.14)$$

$p_t(x)$ being the conditional density of X_t relative to $Y_{t_0}^t$; the derivatives f_t, f_x, f_{xx} and all the conditional expectations in its right-hand side are supposed to exist.

▷ Let us consider two close instants t and $t + \Delta t$, $\Delta t > 0$, and write

$$\Delta\hat{f}(t) = \hat{f}(t + \Delta t) - \hat{f}(t)$$

$$= E\left[f(X_{t+\Delta t}, t+\Delta t) \mid Y_{t_0}^{t+\Delta t}\right] - E\left[f(X_t, t) \mid Y_{t_0}^t\right]$$

$$= E\left[f(X_{t+\Delta t}, t+\Delta t) - f(X_t, t) \mid Y_{t_0}^t\right]$$

$$+ E\left[f(X_{t+\Delta t}, t+\Delta t) \mid Y_{t_0}^{t+\Delta t}\right] - E\left[f(X_{t+\Delta t}, t+\Delta) \mid Y_{t_0}^t\right]. \quad (5.10.15)$$

We evaluate every item in the right-hand side separately. Using the differentiation formula of a composite function (5.1.7) by virtue of the second Eqs. (5.10.10) we get omitting for brevity the arguments of the functions

$$f(X_{t+\Delta t}, t+\Delta t) - f(X_t, t)$$

$$= \left\{ f_t(X_t, t) + f_x(X_t, t)^T \varphi + \frac{1}{2} \mathrm{tr}\left(f_{xx}(X_t, t)[\psi'\psi''] \begin{bmatrix} \psi'^T \\ \psi''^T \end{bmatrix} \right) \right\} \Delta t$$

$$+ f_x(X_t, t)^T (\psi' \Delta W_1 + \psi'' \Delta W_2) = \left\{ f_t(X_t, t) + f_x(X_t, t)^T \varphi \right.$$

$$+ \frac{1}{2} \mathrm{tr}\left[f_{xx}(X_t, t)(\psi'\psi'^T + \psi''\psi''^T) \right] \right\} \Delta t + f_x(X_t, t)^T (\psi' \Delta W_1 + \psi'' \Delta W_2).$$

Substituting here the expression of ΔW_2 from the first Eqs. (5.10.10), $\Delta W_2 = \psi_1''^{-1}(\Delta Y - \varphi_1 \Delta t)$, taking the conditional expectation of the obtained expression relative to $Y_{t_0}^t$, using the third Eqs. (5.10.11) and taking into account the independence of ΔW_1, ΔW_2, ΔY we obtain at fixed Δt

$$E\left[f(X_{t+\Delta t}, t+\Delta t) - f(X_t, t) \mid Y_{t_0}^t\right]$$

$$E\left[f_t(X_t, t) + + f_x(X_t, t)^T \varphi + \frac{1}{2} \mathrm{tr}\left\{ f_{xx}(X_t, t)\psi\nu\psi^T \right\} \mid Y_{t_0}^t \right] \Delta t$$

$$E\left[f_x(X_t, t)^T \psi'' \psi_1''^{-1}(\Delta Y - \varphi_1 \Delta t) \mid Y_{t_0}^t \right].$$

Further we find from the first two equations of Eqs. (5.10.11)

$$\psi'' = \psi\nu\psi_1^T(\psi_1''^{-1})^T, \quad \psi_1''^{-1} = \psi_1''^{T}(\psi_1\nu\psi_1^T)^{-1}.$$

Substituting these expressions into the previous formula and taking into account that ψ_1 does not depend on X_t we shall have

$$E\left[f(X_{t+\Delta t} + \Delta t) - f(X_t, t) \mid Y_{t_0}^t \right]$$

$$= E\left[f_t(Y_t, t) + f_x(X_t, t)^T \varphi + \frac{1}{2} \mathrm{tr}\left\{ f_{xx}(X_t, t)\psi\nu\psi^T \right\} \mid Y_{t_0}^t \right] \Delta t$$

$$+E\left[\,f_x(X_t,t)^T\psi\nu\psi_1^T(\psi_1\nu\psi_1^T)^{-1}(\Delta Y-\varphi_1\Delta t)\,|\,Y_{t_0}^t\,\right]. \qquad (5.10.16)$$

For evaluating the second item in (5.10.5) we shall find the conditional density of the random vector X_t, relative to $Y_{t_0}^{t+\Delta t}$. Let $p_t(x)$ be the conditional density of the random vector X_t relative to $Y_{t_0}^t$, $q_t(\eta\mid y,x)$ the conditional density of the random variable ΔY at given values y, x of the random vectors Y_t, X_t at a fixed value x of the random vector X_t. As the first equation of (5.10.10) determines the Markov process Y_t, the density $q_t(\eta\mid y,x)$ coincides with the conditional density of the random variable ΔY at given values $y_{t_0}^t$ and x of the random variables $Y_{t_0}^t$ and X_t. Consequently, on the basis of the multiplication theorem of the densities the conditional density $p_t'(x)$ of the random vector X_t relative to $Y_{t_o}^{t+\Delta t}$ or, what is the same, relative to $Y_{t_0}^t$ and ΔY is determined by the formula

$$p_t'(x)=\frac{p_t(x)q_t(\Delta Y\mid Y_t,x)}{\displaystyle\int\limits_{-\infty}^{\infty}p_t(x)q_t(\Delta Y\mid Y_t,x)dx}. \qquad (5.10.17)$$

Here the numerator represents the joint conditional density of the random variables X_t and ΔY relative to $Y_{t_0}^t$, and the integral in the denominator represents the conditional density of the random variable ΔY relative to $Y_{t_0}^t$. For determining $q_t(\Delta y\mid Y_t,x)$ we notice that by the first equation of (5.10.10) at $X=x$

$$\Delta Y=\varphi_1(Y_t,x,t)\Delta t+\psi_1''(Y_t,t)\Delta W_2.$$

Hence it is clear that at given values of Y_t and $X_t=x$ the variable ΔY represents the linear function of the normally distributed random variable ΔW_2. Consequently, the conditional distribution of ΔY at given values of Y_t and $X_t=x$ is normal and for determining $q_t(\Delta\mid Y_t,x)$ it is sufficient to find the corresponding conditional expectation and covariance matrix of the random vector ΔY. Using the known formulae for the expectation and the covariance matrix of a linear function of a random vector, taking into account that the covariance matrix of the random vector ΔW_2 is equal to $I\Delta t$, and omitting the arguments of the functions φ_1 and ψ_1'' we obtain

$$E\left[\Delta Y\mid Y_t,x\right]=\varphi_1\Delta t,\quad K_{\Delta Y}=\psi_1''\psi_1''^T\Delta t,$$

or, by virtue of the first Eq. (5.10.11)

$$E\left[\Delta Y\mid Y_t,x\right]=\varphi_1\Delta t,\quad K_{\Delta Y}=\psi_1''\nu\psi_1''^T\Delta t,$$

Consequently,

$$q_t(\Delta Y\mid Y_t,x)=c\exp\left\{-\frac{1}{2\Delta t}(\Delta Y^T-\varphi_1^T\Delta t)(\psi_1\nu\psi_1^T)^{-1}(\Delta Y-\varphi_1\Delta t)\right\},$$

c being the normalizing constant. Taking into account that the quantities Δt and ΔY are infinitesimal we may represent the exponential function by the Taylor formula with the residual of order $o(\Delta t)$. Then we obtain

$$q_t(\Delta Y \mid Y_t, x) = c \exp\left\{-\frac{1}{2\Delta t}\Delta Y^T (\psi_1 \nu \psi_1^T 1)^{-1}\Delta Y\right.$$

$$\left. +\varphi_1^T (\psi_1 \nu \psi_1^T)^{-1}\Delta Y - \frac{1}{2}\varphi_1^T (\psi_1 \nu \psi_1^T)^{-1}\varphi_1 \Delta t\right\}$$

$$= c \exp\left\{-\frac{1}{2\Delta t}\Delta Y^T (\psi_1 \nu \psi_1^T 1)^{-1}\Delta Y\right\}\left[1 + \varphi_1^T (\psi_1 \nu \psi_1^T)^{-1}\Delta Y\right.$$

$$-\frac{1}{2}\varphi_1^T (\psi_1 \nu \psi_1^T)^{-1}\varphi_t \Delta t + \frac{1}{2}\varphi_1^T (\psi_1 \nu \psi_1^T)^{-1}\Delta Y \Delta Y^T (\psi_1 \nu \psi_1^T)^{-1}\varphi_1].$$

Retaining only the expectation of the random variable $\Delta Y \Delta Y^T$,

$$E\Delta Y \Delta Y^T = \varphi_1^2 \Delta t^2 + \psi_1'' \psi_1'' \Delta t = \psi_1 \nu \psi_1^T \Delta t + o(\Delta t),$$

we obtain

$$q_t(\Delta Y \mid Y_t, x)$$

$$= c \exp\left\{-\frac{1}{2\Delta t}\Delta Y^T (\psi_1 \nu \psi_1^T)^{-1}\Delta Y\right\}\left[1 + \varphi_1^T (\psi_1 \nu \psi_1^T)^{-1}\Delta Y\right],$$

and, as ψ_1 is independent of x,

$$\int\limits_{-\infty}^{\infty} p_t(x)q_t(\Delta Y \mid Y_t, x)dx$$

$$= c \exp\left\{-\frac{1}{2\Delta t}\Delta Y^T (\psi_1 \nu \psi_1^T)^{-1}\Delta Y\right\}\left[1 + \hat{\varphi}_1^T (\psi_1 \nu \psi_1^T)^{-1}\Delta Y\right],$$

where $\hat{\varphi}_1$ is defined by (5.10.14). Substituting the obtained expressions of $q_t(\Delta Y \mid \Delta Y_t, x)$ and of the integral into (5.10.17) we shall have after the cancellation

$$p_t'(x) = p_t(x)\frac{1 + \varphi_1^T (\psi_1 \nu \psi_1^T)^{-1}\Delta Y}{1 + \hat{\varphi}_1^T (\psi_1 \nu \psi_1^T)^{-1}\Delta Y}.$$

Representing $1/[1 + \hat{\varphi}_1^T (\psi_1 \nu \psi_1^T)^{-1}\Delta Y]$ by the Taylor formula with the residual of order $o(\Delta t)$, performing the multiplication and retaining only the expectation of the random variable $\Delta Y \Delta Y^T$ as before we get

$$p_t'(x) = p_t(x)[1 + \varphi_1^T (\psi_1 \nu \psi_1^T)^{-1}\Delta Y][1 - \hat{\varphi}_1^T (\psi_1 \nu \psi_1^T)^{-1}\Delta Y$$

$$+\hat{\varphi}_1^T(\psi_1\nu\psi_1^T)^{-1}\Delta Y\Delta Y^T(\psi_1\nu\psi_1^T)^{-1}\hat{\varphi}_1]$$

$$= p_t(x)\{1 + (\varphi_1^T - \hat{\varphi}_1^T)(\psi_1\nu\psi_1^T)^{-1}[\Delta Y - \Delta Y\Delta Y^T(\psi_1\nu\psi_1^T)^{-1}\hat{\varphi}_1^T]\}$$

$$= p_t(x)[1 + (\varphi_1^T - \hat{\varphi}_1^T)(\psi_1\nu\psi_1^T)^{-1}(\Delta Y - \hat{\varphi}_1\Delta t)].$$

Now we may evaluate the second item in the right-hand side of formula (5.10.5). Taking into account that according to (5.1.7)

$$f(X_{t+\Delta t}, t + \Delta t)$$

$$= f(X_t, t) + \{f_t(X_t, t) + f_x(X_t, t)^T\varphi + \frac{1}{2}\text{tr}[f_{xx}(X_t, t)\psi\nu\psi^T]\}\Delta t$$

$$+ f_x(X_t, t)^T(\psi'\Delta W_1 + \psi''\Delta W_2)$$

$$= f(X_t, t) + f_1\Delta t + f_x(X_t, t)^T(\psi'\Delta W_1 + \psi''\Delta W_2)$$

where ψ' and ψ'' depend only on Y_t, X_t, t, we find

$$E\left[f(X_{t+\Delta t}, t + \Delta t) \mid Y_{t_0}^{t+\Delta t}\right] - E\left[f(X_{t+\Delta t}, t + \Delta t) \mid Y_{t_0}^{t}\right]$$

$$= \int_{-\infty}^{\infty}\left[f(x, t) + f_1\Delta t + f_x(x, t)^T(\psi'\Delta W_1 + \psi''\Delta W_2)\right]$$

$$\times[p_t'(x) - p_t(x)]\,dx = \int_{-\infty}^{\infty}[f(x, t) + f_1\Delta t + f_x(x, t)^T(\psi'\Delta W_1$$

$$+ \psi''\Delta W_2)](\varphi_1^T - \hat{\varphi}_1^T)^{-1}(\psi_1\nu\psi_1^T)^{-1}(\Delta Y - \hat{\varphi}_1\Delta t)p_t(x)\,dx.$$

Using the first Eq. (5.10.10) and retaining here only the items of order not higher than Δt we get

$$E\left[f(X_{t+\Delta t}, t + \Delta t) \mid Y_{t_0}^{t+\Delta t}\right] - E\left[f(X_{t+\Delta t}, t + \Delta t) \mid Y_{t_0}^{t}\right]$$

$$= \int_{-\infty}^{\infty} f(x, t) + (\varphi_1^T - \hat{\varphi}_1^T)p_t(x)\,dx(\psi_1\nu\psi_1^T)^{-1}(\Delta Y - \hat{\varphi}_1\Delta t)$$

$$+ \int_{-\infty}^{\infty} f_x(x, t)^T(\psi'\Delta W_1\Delta W_2^T\psi_1''^T + \psi''\Delta W_2\Delta W_2^T\psi_1''^T)$$

$$\times(\psi_1\nu\psi_1^T)^{-1}(\varphi_1 - \hat{\varphi}_1)p_t(x)\,dx = E[f(X_t, t)(\varphi_1^T - \hat{\varphi}_1^T) \mid Y_{t_0}^{t}]$$

$$\times(\psi_1 \nu \psi_1^T)^{-1}(\Delta Y - \hat{\varphi}_1 dt) + \int\limits_{-\infty}^{\infty} f_x(x,t)^T(\psi' \Delta W_1 \Delta W_2^T \psi_1''^T$$

$$+\psi'' \Delta W_2 \Delta W_2^T \psi_1''^T)(\psi_1 \nu \psi_1^T)^{-1}(\varphi_1 - \hat{\varphi}_1)p_t(x)dx. \qquad (5.10.18)$$

Finally, retaining only the expectations of the random variables $\Delta W_1 \Delta W_2^T$ and $\Delta W_2 \Delta W_2^T$ taking into consideration that $E\Delta W_1 \Delta W_2^T = 0$ in consequence of the independence of W_2 and W_1 and the second Eq. (5.10.11) $\psi'' \psi_1''^T = \psi \nu \psi_1^T$ we obtain from (5.10.18)

$$E\left[f(X_{t+\Delta t}, t+\Delta t) \mid Y_{t_0}^{t+\Delta t}\right] - E\left[f(X_{t+\Delta t}, t+\Delta t) \mid Y_{t_0}^t\right]$$

$$= E\left[f(X_t,t)(\varphi_1^T - \hat{\varphi}_1^T) \mid Y_{t_0}^t\right](\psi_1 \nu \psi_1^T)^{-1}(\Delta Y - \hat{\varphi}_1 \Delta t)$$

$$+E\left[f_x(X_t,t)^T \psi \nu \psi_1^T(\psi_1 \nu \psi_1^T)^{-1}(\varphi_1 - \hat{\varphi}_1) \mid Y_{t_0}^t\right]\Delta t. \qquad (5.10.19)$$

Substituting expressions (5.10.16) and (5.10.19) into (5.10.15) and replacing the increments by the differentials we obtain the final formula (5.10.12). ◁

5.10.4. Equation for Conditional Characteristic Function

Putting in (5.10.12) $f(x,t) = e^{i\lambda^T x}$ let us evaluate the stochastic equation for the conditional characteristic function of the random vector X_t:

$$g_t(\lambda) = E\left[e^{i\lambda^T X_t} \mid Y_{t_0}^t\right]. \qquad (5.10.20)$$

▷ Noting that in this case

$$f_t = 0, \ f_x = i\lambda e^{i\lambda^T x}, \ f_{xx} = -\lambda\lambda^T e^{i\lambda^T x},$$

$$\mathrm{tr}\left\{\lambda\lambda^T(\psi\nu\psi^T)(y,x,t)\right\} = \lambda^T(\psi\nu\psi^T)(y,x,t)\lambda$$

we obtain from Eq. (5.10.12)

$$dg_t(\lambda) = E\left[\left\{i\lambda^T \varphi(Y,X,t) - \frac{1}{2}\lambda^T(\psi\nu\psi^T)(Y,X,t)\lambda\right\}e^{i\lambda^T X} \mid Y_{t_0}^t\right]dt$$

$$+E[\{\varphi_1(Y,X,t)^T - \hat{\varphi}_1^T + i\lambda^T(\psi\nu\psi_1^T)(Y,X,t))\}$$

$$\times e^{i\lambda^T X} \mid Y_{t_0}^t](\psi_1 \nu \psi_1^T)^{-1}(Y,t)(dY - \hat{\varphi}_1 dt). \ \triangleleft \qquad (5.10.21)$$

Here the right-hand side represents a functional of the characteristic function $g_t(\lambda)$ considered as a function of λ. In fact, the conditional distribution of the random vector X_t is completely and uniquely determined by this characteristic function. Therefore, Eq. (5.10.21) represents a stochastic differential equation for the conditional characteristic function $g_t(\lambda)$. This equation is nonlinear as $\hat{\varphi}_1 = E[\varphi_1(Y, X, t) \mid Y_{t_0}^t]$ is also a functional of $g_t(\lambda)$.

At the initial moment t_0 the function $g_{t_0}(\lambda)$ represents the conditional characteristic function of the random variable X_0 relative to Y_0. This serves as the initial condition for Eq. (5.10.21).

After solving Eq. (5.10.21) we may easily evaluate the optimal estimate $\hat{X}_t$ of the state vector X_t determined by formula (5.10.6). It is sufficient for this purpose to remember the expression of the expectation in terms of the characteristic function. Then we obtain

$$\hat{X}_t = E\left[X_t \mid Y_{t_0}^t\right] = \left[\frac{\partial g_t(\lambda)}{\partial \lambda}\right]_{\lambda=0}. \tag{5.10.22}$$

5.10.5. Equation for Conditional Density

The stochastic equation for the conditional density $p_t(x)$ of the random vector X_t is derived in exactly the same way as in Subsection 5.6.6:

$$dp_t(x) = -\frac{\partial^T}{\partial x}\left[\varphi(Y, x, t)p_t(x)\right]dt$$

$$+\frac{1}{2}\mathrm{tr}\left\{\frac{\partial}{\partial x}\frac{\partial^T}{\partial x}\left[(\psi\nu\psi^T)(Y, x, t)p_t(x)\right]\right\}dt + \left\{\left[\varphi_1(Y, x, t)^T - \hat{\varphi}_1^T\right]p_t(x)\right.$$

$$\left.-\frac{\partial}{\partial x}\left[(\psi\nu\psi_1^T)(Y, x, t)p_t(x)\right]\right\}(\psi_1\nu\psi_1^T)^{-1}(Y, t)(dY - \hat{\varphi}_1 dt) \tag{5.10.23}$$

or

$$dp_t(x) = L^* p_t(x)dt + \left\{\left[\psi_1(Y, x, t)^T - \varphi_1^T\right]p_t(x)\right.$$

$$\left.-\frac{\partial^T}{\partial x}\left[(\psi\nu\psi^T)(Y, x, t)p_t(x)\right]\right\}(\psi_1\nu\psi_1^T)^{-1}(Y, t)(dY - \hat{\varphi}_1 dt) \tag{5.10.24}$$

where L^* being the adjoint operator for the operator

$$L = \varphi(Y, x, t)^T\frac{\partial}{\partial x} + \frac{1}{2}\mathrm{tr}\left[(\psi\nu\psi^T)(Y, x, t)\frac{\partial}{\partial x}\frac{\partial^T}{\partial x}\right]. \tag{5.10.25}$$

Noting (5.10.14) we conclude that Eq. (5.10.23) represents a nonlinear stochastic integrodifferential equation relative conditional density $p_t(x)$.

At the initial moment t_0 the function $p_{t_0}(x)$ represents the conditional density of the random variable X_0 relative to Y_0. Thus, $p_{t_0}(x)$ serves as the initial condition for Eq. (5.10.23). After solving Eq. (5.10.23) we may find according to formula (5.10.6) the optimal estimate $\hat{X}_t$ of the state vector X_t of the system:

$$\hat{X}_t = E\left[X_t \mid Y_{t_0}^t \right] = \int_{-\infty}^{\infty} x p_t(x)\,dx. \qquad (5.10.26)$$

But it is very difficult to realize this possibility as well as to evaluate $\hat{X}_t$ by formula (5.10.22). Both Eqs. (5.10.21) and (5.10.23) may be solved only after obtaining the results of measurements $Y_{t_0}^t$, and therefore must be solved in real time basis during the system functioning.

As formula (5.10.8) determines the stochastic Itô differential of the random process $\hat{f}(t)$ Eqs. (5.10.21) and (5.10.23) represent the Itô stochastic equations.

Eq. (5.10.23) was first obtained in another form and under more rigid restrictions in (Stratonovich 1961, 1966) namely in the form of the Stratonovich stochastic equation. At the same time, the equation for p_t in the Itô form was obtained in (Kushner 1964a,b, 1967) also under more rigid restrictions. Therefore, it is usually called the Stratonovich–Kushner equation.

5.10.6. Equations for Nonnormalized Conditional Density and Characteristic Function

Let us introduce a nonnormalized density

$$\tilde{p}_t(x) = \mu p_t(x) \qquad (5.10.27)$$

where μ does not depend on x (but certainly may depend on Y_t, $Y_{t_0}^t$). We define μ by the following stochastic differential equation:

$$d\mu = \tilde{\varphi}\,dt + \tilde{\psi}\,dY \qquad (5.10.28)$$

where φ and ψ are the functions of Y_t and Y_{t_0} which we shall choose later on. In this case we shall suppose the existence of continuous first derivatives of the functions φ and ψ_1 and the second derivatives of the functions of ψ over the components of the vector x.

According to the Itô formula for a stochastic differential of the product of two processes determined by stochastic differential equations (5.1.13) we ascertain that the coefficients at dW in Eqs. (5.10.23) and (5.10.28) are equal respectively

$$\left\{ (\varphi_1^T - \hat{\varphi}_1^T)p_t - \frac{\partial}{\partial x}\left[\psi \nu \psi_1^T p_t \right] \right\} (\psi_1 \nu \psi_1^T)^{-1} \psi_1$$

and $\psi\psi_1$. So we have

$$d\tilde{p}_t = \mu dp_t$$

$$+p_t d\mu + \left\{ (\varphi_1^T - \hat{\varphi}_1^T)p_t - \frac{\partial}{\partial x}\left[\psi\nu\psi_1^T\right] \right\} (\psi_1\nu\psi_1^T)^{-1}(\psi_1\nu\psi_1^T)\tilde{\psi}^T dt.$$

$$(5.10.29)$$

After substituting the expressions for dp_t and $d\mu$ in (5.10.29) from Eqs. (5.10.23) and (5.10.28) and taking into consideration that μ does not depend on x we get

$$d\tilde{p}_t = L^*\mu p_t dt - \{(\varphi_1^T - \hat{\varphi}_1^T)\mu p_t - \frac{\partial}{\partial x}[\psi\nu\psi_1^T \mu p_t](\psi_1\nu\psi_1^T)^{-1}\hat{\varphi}_1\}dt$$

$$+\{(\varphi_1^T - \hat{\varphi}_1^T)\mu p_t - \frac{\partial}{\partial x}[\psi\nu\psi_1^T \mu p_t]\}(\psi_1\nu\psi_1^T)^{-1}dY + \tilde{\varphi}p_t dt$$

$$+\tilde{\psi}p_t dY + \{\varphi_1^T - \hat{\varphi}_1^T)p_t - \frac{\partial^T}{\partial x}[\psi\nu\psi_1^T p_t]\}\tilde{\psi}^T dt. \qquad (5.10.30)$$

It is clear from (5.10.30) that if we put $\tilde{\varphi} \equiv 0$, $\tilde{\psi} = \mu\hat{\varphi}_1^T(\psi_1\nu\psi_1^T)^{-1}$ then accounting (5.10.27) we obtain a linear stochastic differential equation in partial derivatives with the parametric noises (Zakai 1969) for $\tilde{p}_t(x)$:

$$d\tilde{p}_t(x) = L^*\tilde{p}_t(x) + \{\varphi_1(y, x, t)^T\tilde{p}_t(x)$$

$$-\frac{\partial^T}{\partial x}[(\psi\nu\psi_1)^T(Y, x, t)\tilde{p}_t(x)]\}(\psi_1\nu\psi_1)^{-1}(Y, t)dY. \qquad (5.10.31)$$

After the integration of Eq. (5.10.31) at the initial condition $\tilde{p}_t(x) = p_{t_0}(x) = p_0(x \mid Y_0)$ the coefficient μ in (5.10.27) will be determined by formula

$$\mu = \int_{-\infty}^{\infty} \tilde{p}_t(x)dx \qquad (5.10.32)$$

and formula (5.10.26) for the optimal estimate will take the form

$$\hat{X}_t = \mu^{-1}\int_{-\infty}^{\infty} x\tilde{p}_t(x)dx. \qquad (5.10.33)$$

Let us obtain the equation for nonnormalized conditional characteristic function. We denote by $\tilde{g}_t(\lambda)$ a nonnormalized characteristic function responding to the nonnormalized density $\tilde{p}_t(x)$. According to the definition we have

$$\tilde{g}_t(\lambda) = \int_{-\infty}^{\infty} e^{i\lambda^T x}\mu p_t(x)dx = \mu g_t(\lambda), \qquad (5.10.34)$$

where $g_t(\lambda)$ is the conditional characteristic function of X_t.

Show that the equation for $\tilde{g}_t(\lambda)$ has the following form:

$$d\tilde{g}_t(\lambda) = \int_{-\infty}^{\infty} i\lambda^T \varphi(Y, x, t) - \frac{1}{2}\lambda^T (\psi\nu\psi^T)(Y, x, t)\lambda\} e^{i\lambda^T x}$$

$$\times \tilde{p}_t(x)dx\,dt + \int_{-\infty}^{\infty} \{\varphi_1(Y, x, t)^T$$

$$+ i\lambda^T (\psi\nu\psi_1^T)(Y, x, t)\} e^{i\lambda^T x}\tilde{p}_t(x)dx(\psi_1\nu\psi_1^T)^{-1}(y, t)dY. \qquad (5.10.35)$$

$\triangleright$ Using formula (5.1.13) for a differential of the product and accounting that the coefficients at dW in Eqs. (5.10.23) and (5.10.28) are equal respectively

$$\int_{-\infty}^{\infty} (\varphi_1^T - \hat{\varphi}_1^T + i\lambda\psi\nu\psi_1^T) e^{i\lambda^T x} p_t(x)dx(\psi_1\nu\psi_1^T)^{-1}\psi_1$$

and $\tilde{\psi}\psi_1 = \mu\hat{\varphi}_1^T(\psi_1\nu\psi_1^T)^{-1}\psi_1$, we find

$$d\tilde{g}_t(\lambda) = \mu dg_t(\lambda) + g_t(\lambda)d\mu$$

$$+ \int_{-\infty}^{\infty} (\varphi_1^T - \hat{\varphi}_1^T + i\lambda^T \psi\nu\psi_1^T) e^{i\lambda^T x} p_t(x)dx(\psi_1\nu\psi_1^T)^{-1}\mu\hat{\varphi}_1\,dt. \qquad (5.10.36)$$

Substituting in Eq. (5.10.28) the expressions for $dg_t(\lambda)$ from (5.10.21), $d\mu$ from Eq. (5.10.28) we get taking into the considerations formulae (5.10.27), (5.10.34) and the equalities $\tilde{\varphi} \equiv 0$, $\tilde{\psi} = \mu\hat{\varphi}_1^T(\psi_1\nu\psi_1^T)^{-1}$:

$$d\tilde{g}_t(\lambda) = \int_{-\infty}^{\infty} (i\lambda^T \varphi - \frac{1}{2}\lambda^T \psi\nu\psi^T \lambda) e^{i\lambda^T x}\tilde{p}_t(x)dx\,dt$$

$$+ \int_{-\infty}^{\infty} (\varphi_1^T - \hat{\varphi}_1^T + i\lambda^T \psi\nu\psi_1^T) e^{i\lambda^T x}\tilde{p}_t(x)(\psi_1\nu\psi_1^T)^{-1}(dY - \hat{\varphi}_1 dt)$$

$$+ \tilde{g}_t(\lambda)\hat{a}_1^T(\psi_1\nu\psi_1^T)^{-1}dY$$

$$+ \int\limits_{-\infty}^{\infty} (\varphi_1^T - \hat\varphi_1^T + i\lambda^T \psi\nu\psi_1^T)e^{i\lambda^T x}\tilde{p}_t(x)dx(\psi_1\nu\psi_1^{-1})^T\hat\varphi_1 dt. \quad (5.10.37)$$

Here after reductions we come to Eq. (5.10.27). ◁

5.10.7. Stochastic Differential of Conditional Expectation

Formula (5.10.6) determined the optimal estimate as the conditional expectation $\hat{X}$ of the corresponding random variable X. The accuracy of the optimal estimate at given results of measurement is characterized by the conditional covariance matrix R of the estimated random variable. These formulae are easily derived from general formula (5.10.12). As formula (5.10.12) determines the stochastic differential of a scalar function of the system state it is necessary to apply it for each element of matrices $\hat{X}$ and R separately.

▷ Putting in (5.10.12) $f(X,t) = X_l$ we shall have $f_t = 0$, $f_x = [0\ldots1\ldots0]^T$, $f_{xx} = 0$, and formula (5.10.12) will give

$$d\hat{X}_l = \hat\varphi_l dt + E[X_l(\varphi_1^T - \hat\varphi_1^T)$$

$$+(\psi\nu\psi_1^T)_l \mid Y_{t_0}^t](\psi_1\nu\psi_1^T)^{-1}(dY - \hat\varphi_1 dt) \ (l = 1, \ldots, n) \quad (5.10.38)$$

where in accordance with (5.10.14) $\hat\varphi_l = E[\varphi_l(Y, X, t) \mid Y_{t_0}^t]$, $(\psi\nu\psi_1^T)_l$ being the lth row of the matrix $\psi\nu\psi_1^T$ and the arguments of the functions φ_1, $\psi\nu\psi_1^T$ and $(\psi_1\nu\psi_1^T)^{-1}$ are omitted for brevity. It follows from (5.10.30) the matrix form for the stochastic differential of the optimal estimate $\hat{X}$ of the state vector X of the system:

$$d\hat{X} = \hat\varphi dt + E[X\{\varphi_1(Y, X, t)^T - \hat\varphi_1^T\}$$

$$+(\psi\nu\psi_1^T)(Y, X, t) \mid Y_{t_0}^t](\psi_1\nu\psi_1^T)^{-1}(Y, t)(dY - \hat\varphi_1 dt). \ ◁$$

5.10.8. Stochastic Differential of Conditional Second Order Moment

Let us find the stochastic differential of the conditional second order moment of the state vector of the system.

▷ Putting in (5.10.12) $f(X, t) = X_k X_l$ we shall have at $k < l$

$$f_t = 0, \ f_x = [0 \ldots X_l \ldots X_k \ldots 0]^T,$$

$$
f_{xx} = \begin{array}{cc} & \begin{array}{ccc} k & & l \end{array} \\ \begin{bmatrix} 0 & \cdots & 0 & \cdots & 0 & \cdots & 0 \\ \cdot & \cdot\cdot & \cdot & \cdot\cdot & \cdot & \cdot\cdot & \cdot \\ 0 & \cdots & 0 & \cdots & 1 & \cdots & 0 \\ \cdot & \cdot\cdot & \cdot & \cdot\cdot & \cdot & \cdot\cdot & \cdot \\ 0 & \cdots & 1 & \cdots & 0 & \cdots & 0 \\ \cdot & \cdot\cdot & \cdot & \cdot\cdot & \cdot & \cdot\cdot & \cdot \\ 0 & \cdots & 0 & \cdots & 0 & \cdots & 0 \end{bmatrix} & \begin{array}{c} \\ \\ k \\ \\ l \\ \\ \end{array} \end{array},
$$

and formula (5.10.12) will give

$$
d\Gamma_{kl} = E\left[\, X_k\varphi_l + X_l\varphi_k + (\psi\nu\psi^T)_{kl} \mid Y_{t_0}^t \,\right] dt
$$

$$
+ E[X_k X_l(\varphi_1^T - \hat{\varphi}_1^T) + X_k(\psi\nu\psi_1^T)_l
$$

$$
+ X_l(\psi\nu\psi_1^T)_k \mid Y_{t_0}^t](\psi_1\nu\psi_1^T)^{-1}(dY - \hat{\varphi}_1 dt) \quad (k,l = 1,\ldots,n) \quad (5.10.39)
$$

where in addition to the previous notations $d\Gamma_{kl} = E\left[\, X_k X_l\varphi_k \mid Y_{t_0}^t \,\right]$, and $(\psi\nu\psi^T)_{kl}$ is the corresponding element of the matrix $\psi\nu\psi^T$.

To derive the corresponding matrix formula we rewrite (5.10.39) in the form

$$
d\Gamma_{kl} = E\left[\, X_k\varphi_l + X_l\varphi_k + (\psi\nu\psi^T)_{kl} \mid Y_{t_0}^t \,\right] dt
$$

$$
+ \sum_{\rho=1}^{r} E[X_k X_l a_\rho + X_k b_{lk} + X_l b_{k\rho} \mid Y_{t_0}^t](dY_\rho - \hat{\varphi}_{1\rho} dt)
$$

where a_ρ is the ρth element of the matrix-row $(\varphi_1^T - \hat{\varphi}_1^T)(\psi_1\nu\psi_1^T)^{-1}$, and $b_{k\rho}$ is the element of the kth row and of the ρth column of the matrix $\psi\nu\psi_1^T(\psi_1\nu\psi_1^T)^{-1}$. Then, denoting by b_ρ the ρth column of the matrix $\psi\nu\psi_1^T(\psi_1\nu\psi_1^T)^{-1}$, $b_\rho = [b_{1\rho}\ldots b_{p\rho}]^T$ $(\rho = 1,\ldots,r)$, we obtain the following formula for the stochastic differential of the conditional second order moment Γ of the state vector of the system:

$$
d\Gamma = E[X\varphi(Y,X,t)^T + \varphi(Y,X,t)X^T
$$

$$
+ (\psi\nu\psi^T)(Y,X,t) \mid Y_{t_0}^t]dt + \sum_{\rho=1}^{r} E[XX^T a_\rho(Y,X,t)
$$

$$
+ Xb_\rho(Y,X,t)^T + b_\rho(Y,X,t)X^T \mid Y_{t_0}^t](dY_\rho - \hat{\varphi}_{1\rho} dt). \quad \triangleleft \qquad (5.10.40)
$$

5.10.9. Stochastic Differential of Conditional Covariance Matrix

To find the stochastic differential of the conditional covariance matrix R of the state vector of the system let us use the known formula which connects the expectation, the second order moment and the covariance matrix of a random vector $R = \Gamma - \hat{X}\hat{X}^T$, or, in the scalar form $R_{kl} = \Gamma_{kl} - \hat{X}_k\hat{X}_l$.

▷ From this formula follows $dR_{kl} = d\Gamma_{kl} - d(\hat{X}_k\hat{X}_l)$. For finding $d(\hat{X}_k\hat{X}_l)$ we use formula (5.1.13) for the stochastic differential of the product of two random processes in the case of a Wiener process W. Taking into consideration that due to (5.10.30)

$$E\left[\, X_k(\varphi_1^T - \hat{\varphi}_1^T) + (\psi\nu\psi_1^T)_k \mid Y_{t_0}^t \,\right] (\psi_1\nu\psi_1^T)^{-1}\psi_1,$$

$$E\left[\, X_l(\varphi_1^T - \hat{\varphi}_1^T) + (\psi\nu\psi_1^T)_l \mid Y_{t_0}^t \,\right] (\psi_1\nu\psi_1^T)^{-1}\psi_1,$$

play the role of the matrix-rows Y_1 and Y_2 in this case, we find

$$d(\hat{X}_k\hat{X}_l) = \hat{X}_k d\hat{X}_l + \hat{X}_l d\hat{X}_k$$

$$+E\left[\, X_k(\varphi_1^T - \hat{\varphi}_1^T) + (\psi\nu\psi_1^T)_k \mid Y_{t_0}^t \,\right] (\psi_1\nu\psi_1^T)^{-1}\psi_1\nu\psi_1^T(\psi_1\nu\psi_1^T)^{-1}$$

$$\times E\left[\, X_l(\varphi_1^T - \hat{\varphi}_1^T) + (\psi_1\nu\psi^T)_l \mid Y_{t_0}^t \,\right] dt.$$

Substituting here the expressions of $d\hat{X}_k$ and $d\hat{X}_l$ from (5.10.30) we shall have

$$d(\hat{X}_k\hat{X}_l) = \{\hat{X}_k\hat{\varphi}_l + \hat{X}_l\hat{\varphi}_k + E[X_k(\varphi_1^T - \hat{\varphi}_1^T)$$

$$+(\psi\nu\psi_1^T)_k \mid Y_{t_0}^t](\psi_1\nu\psi_1^T)^{-1}E[X_l(\varphi_1 - \hat{\varphi}_1) + (\psi\nu\psi_1^T)_l^T \mid Y_{t_0}^t]\}dt$$

$$+E[(\hat{X}_kX_l + \hat{X}_lX_k)(\varphi_1^T - \hat{\varphi}_1^T) + \hat{X}_k(\psi\nu\psi_1^T)_l$$

$$+\hat{X}_l(\psi\nu\psi_1^T)_k \mid Y_{t_0}^t](\psi_1\nu\psi_1^T)^{-1}(dY - \hat{\varphi}_l dt).$$

Subtracting this formula from (5.10.32) and adding a zero item

$$E[\hat{X}_k\hat{X}_l(\varphi_1^T - \hat{\varphi}_1^T) \mid Y_{t_0}^t] = \hat{X}_k\hat{X}_l(\hat{\varphi}_1^T - \hat{\varphi}_1^T) = 0$$

we get

$$dR_{kl} = \{E[(X_k - \hat{X}_k)\varphi_l + (X_l - \hat{X}_l)\varphi_k + (\psi\nu\psi_1^T)_{kl} \mid Y_{t_0}^t]$$

$$-E[X_k(\varphi_1^T - \hat{\varphi}_1^T) + (\psi\nu\psi_1^T)_k \mid Y_{t_0}^t](\psi_1\nu\psi_11^T)^{-1}E[X_l(\varphi_1 - \hat{\varphi}_1)$$

$$+(\psi\nu\psi_1^T)_l^T \mid Y_{t_0}^t]\}dt + E[(X_k - \hat{X}_k)(X_l - \hat{X}_l)(\varphi_1^T - \hat{\varphi}_1^T)$$

$$+(X_k - \hat{X}_k)(\psi\nu\psi_1^T)_l + (X_l - \hat{X}_l)(\psi\nu\psi_1^T)_k \mid Y_{t_0}^t](\psi_1\nu\psi_1^T)^{-1}(dY - \hat{\varphi}_1 dt)$$

$$(5.10.41)$$

$$(k, l = 1, \ldots, n).$$

Transforming the last item in (5.10.41) in exactly the same way as in (5.10.32) we obtain the matrix formula for the stochastic differential of the conditional covariance matrix R of the state vector (5.10.41) of the system:

$$dR = \{E[(X - \hat{X})\varphi(Y, X, t)^T + \varphi(Y, X, t)(X^T - \hat{X}^T)$$

$$+(\psi\nu\psi^T)(Y, X, t) \mid Y_{t_0}^t] - E[X\{\varphi_1(Y, X, t)^T - \hat{\varphi}_1^T\}$$

$$+(\psi\nu\psi_1^T)(Y, X, t) \mid Y_{t_0}^t](\psi_1\nu\psi_1^T)^{-1}(Y, t)E[\{\varphi_1(Y, X, t) - \hat{\varphi}_1\}X^T$$

$$+(\psi_1\nu\psi^T)(Y, X, t) \mid Y_{t_0}^t]\}dt$$

$$+\sum_{\rho=1}^{r} E[(X - \hat{X})(X^T - \hat{X}^T)a_\rho(Y, X, t) + (X - \hat{X})b_\rho(Y, X, t)^T$$

$$+b_\rho(Y, X, t)(X - \hat{X})^T \mid Y_{t_0}^t](dY_\rho - \hat{\varphi}_{1\rho}dt). \vartriangleleft \qquad (5.10.42)$$

The mean square of the error of the optimal estimate $\hat{X}$ given the result of observation $Y_{t_0}^t$, i.e. the conditional expectation of the squared modulus of the error is evidently equal to the trace of the conditional covariance matrix of the state vector of the system

$$E[|X - \hat{X}|^2 \mid Y_{t_0}^t] = \operatorname{tr} R.$$

It is clear that (5.10.31) and (5.10.33) or (5.10.31) and (5.10.42) are not stochastic differential equations relative to m and Γ or m and R in the general case, as their right-hand sides depend on the unknown conditional distribution. Nevertheless, as will be shown later, formulae (5.10.31), (5.10.33) and (5.10.42) may serve as the basis for various approximate methods of filtering.

5.10.10. Optimal Filtering in Stochastic Discrete Systems

Basic formulae of Subsection 5.10.2 for discrete m.s.e. filtering for composite random vector $[Y_l^T X_l^T]^T$ defined by stochastic difference equation of the form of Eqs. (5.7.2)

$$X_{l+1} = \varphi_l(X_l) + \psi_l(X_l)V_l, \qquad (5.10.43)$$

$$Y_l = \varphi_l'(X_l) + \psi_l'(X_l)V_l', \qquad (5.10.44)$$

for conditional densities $p(x_l \mid y^l)$, $y^l = \{y_0, \ldots, y_l\}$ and $p(y_{l+1} \mid y_l)$ satisfy the following recurrent equations:

$$p(x_{l+1} \mid y^{l+1}) = \frac{p(y_{l+1} \mid x_{l+1})}{p(y_{l+1} \mid y^l)} \int\limits_{R^n} p(x_l \mid y^l) p(x_{l+1} \mid x_l) dx_l, \quad (5.10.45)$$

$$p(y_{l+1} \mid y^l) = \int\limits_{R^n} \int\limits_{R^n} p(x_l \mid y^l) p(x_{l+1} \mid x_l) p(y_{l+1} \mid x_{l+1}) dx_l dx_{l+1}.$$

$$(5.10.46)$$

Problems

5.1. Find the multi-dimensional distributions of the random function $X(t)$ representing the fluctuations of the output voltage of a network subject to the action of charged particles emitted by an electronic device (the so-called *shot effect*).

S o l u t i o n. The stream of charged particles initiating voltage fluctuation in a network may be considered with sufficient accuracy as a Poisson stream. We shall consider the intensity of stream of particles (the mean number of particles per unit of time) as a known function of time $\mu(t)$. Electric impulses received by a network from different particles we shall consider as independent, identically distributed random variables independent of random instants of action of particles. By $w(t, \tau)$ we denote the weighting function of the network (supposed to be linear), i.e. the output voltage of the network at the instant t in response to the unit electric impulse acting at the instant τ. Then the output random voltage at the moment t will be determined by

$$X(t) = \sum_{k=1}^{N} A_k w(t, T_k), \qquad (I)$$

where N is the random number of impulses acting on the network in the time interval (t_0, t); $T_1, \ldots, T_N$ the random instants of action of the impulses, and $A_1, \ldots, A_N$ the random magnitudes of the impulses acting on the input of the circuit.

The simplest way to find the n-dimensional distribution of the random function $X(t)$ is to evaluate the joint characteristic function of its values at the instants $t_1, \ldots, t_n$. Denoting by t the largest of the numbers $t_1, \ldots, t_n$, $t = \max(t_1, \ldots, t_n)$, and taking into account that $w(t, T_k) = 0$ at $t < T_k$ (the network is certainly non-anticipative) in consequence of which $X(t_1), \ldots, X(t_n)$

may all be evaluated by formula (I) at the same N, representing the number of the impulses in the time interval (t_0, t), we obtain

$$g_{t_1,\ldots,t_n}(\lambda_1,\ldots,\lambda_n) = E \exp\left\{ i \sum_{p=1}^{n} \lambda_p X(t_p) \right\}$$

$$= E \exp\left\{ i \sum_{k=1}^{N} A_k \sum_{p=1}^{n} \lambda_p w(t_p, T_k) \right\}. \tag{II}$$

We evaluate the expectation on this formula of total expectation calculating first the conditional expectation at a given value m of the random number of impulses N, and then taking the expectation of the obtained conditional expectation with the account of the randomness of the number of impulses. Let σ be the expectation of the number of impulses which act on the network in the time interval (t_0, t) and B_m the event consisting in the appearance of exactly m impulses $(m = 0, 1, 2, \ldots)$ during the time interval (t_0, t). Then the probability of the events B_m will be determined by the formula

$$P(B_m) = \frac{\sigma^m}{m!} e^{-\sigma}, \quad \sigma = \int_{t_0}^{t} \mu(\tau) d\tau \ (m = 0, 1, 2, \ldots). \tag{III}$$

To evaluate the conditional expectation relative to the event B_m in (II) we shall use again the formula of total expectation taking first the conditional expectation relative to the random variables $T_1, \ldots, T_m$. Then taking into account the independence of the random variables $A_1, \ldots, A_m$ and their independence of the random variables $T_1, \ldots, T_m$ we obtain

$$E\left[\exp\left\{ i \sum_{k=1}^{N} A_k \sum_{p=1}^{n} \lambda_p w(t_p, T_k) \right\} | B_m \right]$$

$$= E \exp\left\{ i \sum_{k=1}^{m} A_k \sum_{p=1}^{n} \lambda_p w(t_p, T_k) \right\}$$

$$= E\left[E\left[\exp\left\{ i \sum_{k=1}^{m} A_k \sum_{p=1}^{n} \lambda_p w(t_p, T_k) \right\} | T_1, \ldots, T_m \right] \right]$$

$$= E \prod_{k=1}^{m} g_a \left(\sum_{p=1}^{n} \lambda_p w(t_p, T_k) \right), \tag{IV}$$

where $g_a(\lambda)$ is the characteristic function of the random variable A_k $(k = 1, 2, \ldots)$. For evaluating the last expectation we shall find the joint conditional distribution of the instants of action of the impulses $T_1, \ldots, T_m$ relative to the event B_m. Let us take arbitrary instants of time $\tau_1, \ldots, \tau_m$, $\tau_1 < \cdots < \tau_m$, in the interval (t_0, t). To evaluate the distribution function $F(\tau_1, \ldots, \tau_m)$ at a given value m of the number of impulses N in the time interval (t_0, t) it is sufficient to calculate the sum of probabilities of all possible distributions of the impulses $T_1, \ldots, T_m$, $T_1 \leq T_2 \leq \cdots \leq T_m$, in the intervals $(t_0, \tau_1), (\tau_1, \tau_2), \ldots, (\tau_{m-1}, \tau_m)$ satisfying the conditions $T_1 < \tau_1, \ldots, T_m < \tau_m$, and to divide this probability by the probability of occurrence of m impulses in the interval (t_0, t), i.e. by the probability of the even B_m. After taking the mixed derivative with respect to $\tau_1, \ldots, \tau_m$ of the conditional distribution function $F(\tau_1, \ldots, \tau_m)$ found in such a way we shall find the conditional density $f(\tau_1, \ldots, \tau_m)$ of the random instants of action of the impulses $T_1, \ldots, T_m$ at $N = m$ (i.e. under the supposition that m impulses appear in the interval (t_0, t)). Then only one item depending on all the variables $\tau_1, \ldots, \tau_m$, which is proportional to the probability of occurrence of one impulse in everyone of the intervals $(t_0, \tau_1), (\tau_1, \tau_2), \ldots, (\tau_{m-1}, \tau_m)$:

$$\frac{m!}{\sigma^m} \int_{t_0}^{\tau_1} \mu(\tau)d\tau \int_{\tau_1}^{\tau_2} \mu(\tau)d\tau \ldots \int_{\tau_{m-1}}^{\tau_m} \mu(\tau)d\tau, \tag{V}$$

will give the result different from zero. As a result we obtain $f(\tau_1, \ldots, \tau_m) = \frac{m!}{\sigma^m}\mu(\tau_1)\ldots\mu(\tau_m)$ at $t_0 < \tau_1 \leq \tau_2 \leq \cdots \leq \tau_{m-1} < t$ and $f(\tau_1, \ldots, \tau_m) = 0$ when the condition $t_0 < \tau_1 \leq \tau_2 \leq \cdots \leq \tau_{m-1} < t$ is not fulfilled. After determining the conditional density of the moments of action of the impulses $T_1, \ldots, T_m$ we find the expectation in (V):

$$E\left[\exp\left\{i\sum_{k=1}^{N} A_k \sum_{p=1}^{n} \lambda_p w(t_p, T_k)\right\} | B_m\right]$$

$$= \frac{m!}{\sigma^m} \int_{t_0}^{t} \mu(\tau_m)d\tau_m \int_{\tau_{m-1}}^{\tau_m} \mu(\tau_{m-1})d\tau_{m-1}$$

$$\ldots \int_{t_0}^{\tau_2} \prod_{k=1}^{m} g_a\left(\sum_{p=1}^{n}\lambda_p w(t_p, \tau_k)\right) \mu(\tau_1)d\tau_1. \tag{VI}$$

The integrand here is symmetric relative to the variables $\tau_1, \ldots, \tau_m$. Consequently, any permutation of the variables $\tau_1, \ldots, \tau_m$ does not change the integral in

(VI). Therefore instead of taking one integral in (VI) multiplied by $m!$ one may take the sum of $m!$ integrals obtained by all possible permutations of the variables $\tau_1, \ldots, \tau_m$. But the sum of all such integrals is equal to the integral over the m-dimensional cube with the side (t_0, t). Consequently,

$$E\left[\exp\left\{i\sum_{k=1}^{N} A_k \sum_{p=1}^{n} \lambda_p w(t_p, T_k)\right\} | B_m\right]$$

$$= \frac{1}{\sigma^m} \int_{t_0}^{t} \cdots \int_{t_0}^{t} \prod_{k=1}^{m}\left\{g_a\left(\sum_{p=1}^{n}\lambda_p w(t_p, \tau_k)\right)\mu(\tau_k)\right\} d\tau_1 \ldots d\tau_m$$

$$= \frac{1}{\sigma^m}\left[\int_{t_0}^{t} g_a\left(\sum_{p=1}^{n}\lambda_p w(t_p, \tau)\right)\mu(\tau)d\tau\right]^m. \qquad \text{(VII)}$$

To evaluate the expectation in (II) it is now sufficient to multiply the right-hand side of the latter formula by the probability of the event B_m and to sum up the obtained expressions corresponding to all possible values of m from 0 to ∞. As a result we obtain

$$g_{t_1,\ldots,t_n}(\lambda_1,\ldots,\lambda_n) = \sum_{m=0}^{\infty}\frac{1}{m!}\left[\int_{t_0}^{t} g_a\left(\sum_{p=1}^{n}\lambda_p w(t_p, \tau)\right)\mu(\tau)d\tau\right]^m e^{-\sigma}$$

$$= \exp\left\{\int_{t_0}^{t}\left[g_a\left(\sum_{p=1}^{n}\lambda_p w(t_p, \tau)\right) - 1\right]\mu(\tau)d\tau\right\}. \qquad \text{(VIII)}$$

Thus the multi-dimensional characteristic functions of the random function $X(t)$ are determined by

$$g_{t_1,\ldots,t_n}(\lambda_1,\ldots,\lambda_n)$$

$$= \exp\left\{\int_{t_0}^{\max(t_1,\ldots,t_n)}\left[g_a\left(\sum_{p=1}^{n}\lambda_p w(t_p, \tau)\right) - 1\right]\mu(\tau)d\tau\right\}. \qquad \text{(IX)}$$

$$(n = 1, 2, \ldots).$$

5.2. Derive the formulae for the stochastic Itô differentials of the functions of the standard Wiener process given in Table A.1.1. Obtain analogous formulae for the stochastic θ-differentials.

5.3. Derive the formulae for the stochastic Itô differentials of the functions of a vector Wiener process given in Table A.1.2. Obtain analogous formulae for the stochastic θ-differentials.

5.4. Prove that the stochastic Itô differential of a scalar function $U = \varphi(P)$ of a Poisson process P is determined by the formula $dU = [\varphi(P+1) - \varphi(P)]dP$. Evaluate by this formula the stochastic differentials of the functions given in Table A.1.1.

5.5. Show that the scalar stochastic differential equations of the first order with the θ-differential

(i) $\dot{Y} = -\varepsilon(1 + V_2)Y + kV_1$,
(ii) $\dot{Y} = (1 + V_2)\varphi(Y) + kV_1$,
(iii) $\dot{Y} = \varphi_0(Y, t) + \sum_{l=1}^{n} \varphi_l(Y, t)V_l$

may be transformed into the following Itô equations:

(i) $\dot{Y} = -\varepsilon(1 - 2\theta\varepsilon\nu_{22} + V_2)Y + k(-2\theta\varepsilon\nu_{21} + V_1)$,
(ii) $\dot{Y} = [1 + 2\theta\varepsilon\nu_{22}\varphi'(Y) + V_2]\varphi(Y) + k[2\theta\nu_{12}\varphi'(Y) + V_1]$,
(iii) $\dot{Y} = \varphi_0(Y, t) + \theta \sum_{p,q=1}^{n} \nu_{pq}\varphi_p(Y, t)\varphi_q'(Y, t) + \sum_{l=1}^{n} \varphi_l(Y, t)V_l$,

where $\varphi' = \partial\varphi/\partial y$, and $V = [V_1 \ \ldots \ V_n]^T$ is the normally distributed white noise with the intensity ν. Under what conditions to the Itô equations coincide in this case with the initial equations?

5.6. Let $Y(t)$ be such a m.s. $n - 1$ times differentiable random process that its $(n-1)^{\text{th}}$ m.s. derivative has the Itô differential

$$dY^{(n-1)} = \varphi(Y, \dot{Y}, \ldots, Y^{(n-1)}, t)dt + \psi(Y, \dot{Y}, \ldots, Y^{(n-1)}, t)dW. \quad \text{(I)}$$

This relation is called *the stochastic Itô differential equation of the nth order relative to the process* $Y(t)$. This equation may also be written in the form

$$Y^{(n)} = \varphi(Y, \dot{Y}, \ldots, Y^{(n-1)}, t) + \psi(Y, \dot{Y}, \ldots, Y^{(n-1)}, t)V \quad \text{(II)}$$

where $V = dW/dt$ is the white noise in the strict sense. Prove that Eq. (II) is transformed into a standard form of the stochastic Itô differential equation for some n-dimensional vector random process. In the case of the Wiener process $W(t)$ derive the formulae of the transition from the Itô equation of the nth order to the equation in the θ-differential of the Nth order and vice versa.

5.7. Show that the scalar stochastic differential equations of the second order with the θ-differential:

(i) $\ddot{Y} + 2\xi\omega_0(1 + V_3)\dot{Y} + \omega_0^2(1 + V_2)Y = kV_1$,
(ii) $\ddot{Y} = \varphi_0(Y, \dot{Y}, t) + \sum_{l=1}^{n} \varphi_l(Y, \dot{Y}, t)V_l$

correspond to the following Itô equations:

(i) $\ddot{Y} + 2\xi\omega_0(1 - 2\theta\nu_{33}\xi\omega_0 + V_3)\dot{Y} + \omega_0^2(1 - 2\theta\nu_{23}\xi\omega_0 + V_2)Y$
$= k(-2\theta\xi\nu_{13}\omega_0 + V_1),$

(ii) $\ddot{Y} = \varphi_0(Y, \dot{Y}, t) + \theta \sum\limits_{l,h=1}^{n} \nu_{lh} \varphi_h'(Y, \dot{Y}, t) \varphi_l(Y, \dot{Y}, t)$

$+ \sum\limits_{l=1}^{n} \varphi_l(Y, \dot{Y}, t)V_l,$ where $\varphi_h'(Y, \dot{Y}, t) = \partial\varphi_h(Y, \dot{Y}, t)/\partial\dot{Y}$, and V
$= [V_1 \ldots V_n]^T$ is the normally distributed white noise of the intensity ν. Find
the conditions under which the Itô equations coincide with the initial equations.

5.8. Show that the n-dimensional characteristic function g_n and the distribution
density f_n at the random initial condition Y_0 and at the absence of the noises in Eq.
(5.4.6) when $b(Y, t) \equiv 0$ are determined by the following formula:

$$g_n = g_n(\lambda_1, \ldots, \lambda_n; t_1, \ldots, t_n)$$

$$= \int\limits_{-\infty}^{\infty} e^{i\lambda_1^T \varphi(t_1,\eta)+\cdots+i\lambda_n^T \varphi(t_n\eta)} f_0(\eta)d\eta \quad (n = 1, 2, \ldots); \qquad (I)$$

$$f_n = f_n(y_1, \ldots, y_n; t_1, \ldots, t_n)$$

$$= f_0(\varphi^{-1}(t_1, y_1))|\varphi_y^{-1}(t_1, y_1)|\delta(y_2 - \varphi(t_2, \varphi^{-1}(t_1, (y_1))))$$

$$\ldots \delta(y_n - \varphi(t_n, \varphi^{-1}(t_1, (y_1)))) \quad (n = 2, 3, \ldots), \qquad (II)$$

where $Y(t) = \varphi(t, Y_0)$ is the solution of a system which represents the function of
the time and the random initial value Y_0.

5.9. Show that for the one-dimensional nonlinear system $\dot{Y} = -\gamma Y^{1+r}$ (r
> -1) with the random initial value Y_0 its n-dimensional density is determined at
$|y_l| < (r\gamma t_l)^{-1/r}$ by the following formula:

$$f_n(y_1, \ldots, y_n; t_1, \ldots, t_n) = f_0(y_1(1 - r\gamma t_1 y_1^r)^{1/r}(1 - r\gamma t_1 y_1^r)^{-(1+r)/n})$$

$$\times \prod\limits_{l=2}^{n} \delta(y_l - y_1(1 - r\gamma t_1 y_1^r)^{-(1+r)/r})[1 + r\gamma t_l y_1^r(1 - r\gamma t_1 y_1^r)^{-(1+r)}]^{-1/r},$$

where $f_0(y_0)$ is the density of the random initial value Y_0.

5.10. Find the one-dimensional distribution of a stationary process in the conservative mechanical system:

$$\dot{q} = \partial H/\partial p, \quad \dot{p} = -\partial H/\partial q,$$

with the random initial conditions where $q = [\, q_1 \, \ldots \, q_n \,]^T$ is the vector of the generalized coordinates; $p = A\dot{q} = [\, p_1 \, \ldots \, p_n \,]^T$ is the vector of the generalized impulses; A is the generalized mass (constant); H is the Hamilton function (the total energy of a system) equal to $H = H(q, p) = p^T A^{-1} p/2 + \Pi(q)$ where $\Pi(q)$ is the potential energy of a system. Show that at $\Pi(q) = q^T C q/2$ the stationary distributions of the coordinates and the impulses are determined by the densities $\varphi_1(q^T C q)$ and $\varphi_2(p^T A^{-1} p)$ where φ_1 and φ_2 to a great extent are the arbitrary functions (more exactly they both are expressed in terms of one arbitrary function). Show that the distributions of the doubled potential energy $Y_1 = q^T C q$ and the doubled kinetic energy $Y_2 = p^T A^{-1} p$ are determined by $f_1(y_1) = c_1 y_1^{(n-2)/2} \varphi_1(y_1) \mathbf{1}(y_1)$, $f_2(y_2) = c_2 y_2^{(n-2)/2} \varphi_2(y_2) \mathbf{1}(y_2)$, where c_1 and c_2 are the normalized constants. In the special case of the exponential functions φ_1 and φ_2, $\varphi_1(x) = \varphi_2(x) = e^{-x/2}$, the distributions of thedouble potential and the kinetic energy present χ^2-distribution with n degrees of freedom. Find the distribution of the total energy $Y = (Y_1 + Y_2)/2$.

5.11. Check that for the nonlinear system described by the second order equation

$$\ddot{Y} + 2\varepsilon \dot{Y} + \varphi'(Y) = V \quad (\varepsilon > 0), \tag{I}$$

the one-dimensional density and the one-dimensional characteristic function of the strictly stationary process are determined by the formulae:

$$f_1(y, \dot{y}) = c \exp\{-h^2[\varphi(y) + \dot{y}^2/2]\} \quad (h^2 = 2\varepsilon\nu), \tag{II}$$

$$g_1(\lambda', \lambda'') = \frac{\sqrt{2\pi}}{h} c e^{-\frac{\lambda''^2}{2h^2}} \int_{-\infty}^{\infty} e^{i\lambda' y - h^2 \varphi(y)} dy, \tag{III}$$

where c is the normalizing constant, V is the stationary normally distributed white noise of the intensity ν, $\varphi(y)$ is the nonlinear function, $\varphi'(y) = \partial \varphi(y)/\partial y$. Consider the special cases: (a) $\varphi'(y) = a_1 y + a_3 y^3$; (b) $\varphi'(y) = a \operatorname{sgn} y$; (c) $\varphi'(y) = a \sin y$.

5.12. Show that the one-dimensional density of the strictly stationary process in the nonlinear system

$$\dot{Y} = \psi(Y) - 2\varepsilon a Y + bV, \quad \varepsilon > 0, \tag{I}$$

is determined by the formula:

$$f_1(y) = \sqrt{\alpha^n/\pi^n} \, e^{-\alpha y^T y}, \quad \alpha = 4\varepsilon/\nu. \tag{II}$$

Here $\psi(y) = [\, \psi_1(y) \, \ldots \, \psi_n(y) \,]^T$, $\partial \psi_k(y)/\partial y_k = 0$ $(k = 1, \ldots, n)$, $y^T \psi(y) = \sum_{k=1}^{n} y_k \psi_k = 0$; $V = [\, V_1, \ldots, V_m \,]^T$ is the normally distributed stationary white noise with the independent components of the same intensity ν, b is the constant $n \times n$-matrix, a is the constant $n \times m$-matrix satisfying the condition $a + a^T = 2bb^T$.

5.13. Show that the one-dimensional density of the strictly stationary process in the system

$$\dot{Y} = a\varphi'(Y) + bV, \tag{I}$$

is determined by the formula: $f_1(y) = ce^{2\varphi(y)/\nu}$, c being the normalization constant, $\varphi(y)$ is the nonlinear function, $\varphi'(y) = \partial\varphi(y)/\partial y$, b is the constant (rectangular in general) matrix; a is the constant matrix satisfying the condition $a + a^T = 2bb^T$, V is the stationary normally distributed white noise with the independent components of the same intensity ν.

5.14. Show that the one-dimensional distribution of the strictly stationary process $Y(t) = [Y_1(t)\, Y_2(t)]^T$ in the nonlinear stochastic system

$$\dot{Y}_1 = \alpha\,\mathrm{sgn}(Y_2 - \beta Y_1), \quad \dot{Y}_2 = -Y_1 + V, \tag{I}$$

is determined by:

$$f_1(y_1, y_2) = c\exp\{-\beta(y_1^2 + 2\alpha|y_2 - \beta y_1|)/\nu\}, \tag{II}$$

where c is the normalizing constant V, is the stationary normally distributed white noise of the intensity ν. Find the corresponding characteristic function.

5.15. Show that the one-dimensional distribution of the stationary process in the nonlinear mechanical system with n degrees of freedom whose behaviour is described by the equations in canonical variables

$$\dot{q} = \partial H/\partial p, \quad \dot{p} = -\partial H/\partial q - 2\varepsilon p + b(q)V, \quad b(q) = A(q)^{1/2} \tag{I}$$

is determined by formula: 1985):

$$f_1(q, p) = ce^{-\alpha H}, \quad \alpha = 4\varepsilon/\nu, \tag{II}$$

where c is the normalization constant. Here $q = [q_1 \ldots q_n]^T$ is the vector of generalized coordinates, $p = A(q)\dot{q} = [p_1 \ldots p_n]^T$ is the vector of generalized impulses, $A(q)$ is the generalized mass which represents symmetric positive definite matrix depending in general on q, H is Hamiltonian function (the total energy of the system) which is determined by the formula $H = H(q, p) = p^T A(q)^{-1}p/2 + \Pi(q)$, $\Pi(q)$ being the potential energy of the system, 2ε is a specific coefficient of viscous friction, $V = [V_1 \ldots V_n]^T$ is the vector of generalized forces whose components represent independent normally distributed stationary white noises of intensity ν. Find the known Gibbs formula (Gibbs 1928) at $b(q) = I$.

5.16. Show that the function $f_1 = f_1(y; t)$ will be the solution of Eq. (5.6.56) with the normal white noise V if and only if the vector function $a = a(y, t)$ assumes such a presentation

$$a(y, t) = a_1^1(y, t) + a_2^1(y, t), \tag{I}$$

that the function f_1 is the density of the invariant measure of the ordinary differential equation

$$\dot{Y} = a_1^1(Y, t), \tag{II}$$

i.e. satisfies the condition

$$\frac{\partial f_1}{\partial t} + \frac{\partial^T}{\partial y}(a_1^1 f_1) = 0, \tag{III}$$

and the correspondent function $a_2^1 = a_2^1(y, t)$ is determined by formula

$$a_2^1(y, t) = \frac{1}{2}\left[\sigma\frac{\partial \ln f_1}{\partial y} + \left(\frac{\partial^T}{\partial y}\sigma\right)^T\right], \quad \sigma = b\nu b^T. \tag{IV}$$

5.17. Suppose that for Eq. (5.4.6) at a normal white noise V the presentation (I) of Problem 5.16 a density $\mu_1 = \mu_1(y, t)$ of integral invariant (II) are known. Besides let

(i) the matrix σ be invertible, $\det |\sigma^{-1}| \neq 0$,

(ii) the vector function $\gamma_1 = \gamma_1(y, t)$ where

$$\gamma_1(y, t) = \sigma^{-1}\left\{a_2^1 - \frac{1}{f_1}\left[\frac{\partial}{\partial y}(\sigma f_1)\right]^T\right\}$$

satisfies the condition of the curl absence: $\operatorname{rot}\gamma_1 = 0$ or

$$\partial\gamma_{1i}/\partial y_j = \partial\gamma_{1j}/\partial y_i \quad (i, j = 1, \ldots, p),$$

(iii) the scalar function

$$F_1(y, t) = \int \gamma_1^T(y, t)dy,$$

is the first integral of Eq. (II),

(iv) the condition of the normalization is fulfilled

$$\int \mu_1(y, t)\exp F_1(y, t)dy = 1.$$

Show that there exists the solution of Eq. (5.4.6) $Y = Y(t)$ for which the one-dimensional density $f_1(y; t)$ is determined by formula

$$f_1(y; t) = \mu_1(y; t) \exp F_1(y, t). \tag{I}$$

5.18. Show that the function $\mu_1(y, t)$ of Problem 5.17 is equal to $f_1(y; t)$ if and only if there exist such a matrix function $A_1 = A_1(y, t)$ that

$$a_2^1 \mu_1 = (\partial A_1 / \partial y)^T, \quad A_1 + A_1^T = \sigma \mu_1, \tag{I}$$

where $a_2^1 = a - a_1^1$. Find the correspondent formulae for the transition function and the multi-dimensional densities.

5.19. Consider Problem 5.17 for the system with the singular total matrix $\sigma = b\nu b^T$,

$$\dot{Y}' = Q(Y', Y'', t), \quad \dot{Y}'' = P(Y', Y'', t) + b_0(Y', Y'', t)V_0, \tag{I}$$

where Y' and Y'' are the s-dimensional vectors, $2s = p$, V_0 is the r-dimensional normal white noise of the intensity $\nu_0 = \nu_0(t)$, p is the dimension of the vector Y, $\sigma_0 = b_0(Y', Y'', t)\nu_0(t)b_0(Y', Y'', t)^T$ is the nonsingular matrix. Show that the one-dimensional distribution is given by the following formulae:

$$f_1(y', y'', t) = \mu_1(y', y'', t) \exp F_1(y', y'', t), \tag{II}$$

where

$$F_1(y', y'', t) = \int \gamma_1^T \, dy' dy'', \tag{III}$$

$$\gamma_1(y', y'', t) = \sigma_0^{-1} P_2^1, \quad \partial \gamma_{1i} / \partial y_j = \partial \gamma_{1j} / \partial y_i, \tag{IV}$$

$$\frac{\partial \mu_1}{\partial t} + \frac{\partial^T}{\partial y'}(Q \mu_1) + \frac{\partial^T}{\partial y''}(P_1^1 \mu_1) = 0, \tag{V}$$

$$P_2^1 = \frac{\sigma_0}{2} \frac{\partial \ln f_1}{\partial y''}, \quad P = P_1^1 + P_1^1. \tag{VI}$$

5.20. Show that the solution of Problem 5.18 for the system with the singular total matrix σ exist if and only if the conditions are fulfilled:

$$P_2^1 \mu_1 = (\partial B_1 / \partial y'')^T, \quad B_1 + B_1^T = \sigma_0 \mu_1, \tag{I}$$

where $P_2^1 = P - P_1^1$, μ_1 is the density of the finite integral invariant of the equations $\dot{Y}' = Q$, $\dot{Y}'' = P_1^1$. Show that $f_1(y', y'', t) = \mu_1(y', y'', y)$.

5.21. Show that for any vectors q, p and vector functions $Q(q,p)$, $P(q,p)$ of the same dimensions as q, p if there exists a function $N(q)$ satisfying the condition

$$\frac{\partial^T}{\partial q}[N(q)Q(q,p)] + N(q)\frac{\partial^T}{\partial p}P(p,q) = 0,$$

then the one-dimensional distribution of the solution stochastic system

$$\dot{q} = Q(q,p), \quad \dot{p} = P(q,p) - 2\varepsilon(H)a(q)\frac{\partial H}{\partial p} + b(q)V, \qquad (\mathrm{I})$$

may be presented by formula:

$$f_1(q,p) = cN(q)e^{-\beta\psi(H)}, \quad \psi(H) = \int\limits_0^H \varepsilon(\eta)d\eta, \quad \beta = 4/\nu. \qquad (\mathrm{II})$$

Here $b(q)$ is any rectangular matrix, $a(q)$ is any square matrix satisfying the condition $a(q) + a(q)^T = 2b(q)b(q)^T$, $H = H(q,p)$ is the first integral of Eq. (I) at $a(q) = b(q) = 0$, $2\varepsilon(H)$ is coefficient of viscous friction, $V = V(t)$ is the normal white noise with independent components and equal intensities ν.

Find in special case $\varepsilon = \mathrm{const}$ the following generalized Gibbs formula: $f_1(q,p) = cN(q)e^{-\alpha H(q,p)}$, $\alpha = 4\varepsilon/\nu$.

5.22. Show that for the scalar nonstationary stochastic system

$$\dot{Y} = a_0(t) + a(t)Y + b(Y,t)V, \quad b(Y,t) = \sqrt{c(t)Y}, \quad Y(t_0) = Y_0, \qquad (\mathrm{I})$$

the solution of Eq. (5.6.65) for the one-dimensional characteristic function has the form

$$g_1(\lambda;t) = g_0(\varphi(t,\lambda)^{-1})\exp\left[i\int\limits_{t_0}^t \psi(\tau,\varphi(t,\lambda))a_0(\tau)d\tau\right]. \qquad (\mathrm{II})$$

Here V is the normal white noise of the intensity $\nu(t)$,

$$\varphi(t,\lambda) = \frac{1}{\lambda}\left\{\exp\left[-\int\limits_{t_0}^t a(\tau)d\tau\right]\right.$$

$$\left. -\frac{i\lambda}{2}\int\limits_{t_0}^t \nu(\tau)c(\tau)\exp\left[-\int\limits_{t_0}^t a(\sigma)d\sigma\right]d\tau\right\},$$

$$\psi(\tau, \varphi(t, \lambda)) = \frac{\exp\left[-\int_{t_0}^{\tau} a(\tau_1)d\tau_1\right]}{i/2 \int_{t_0}^{\tau} \nu(\tau_1)c(\tau_1) \exp\left[-\int_{t_0}^{\tau_1} a(\sigma)d\sigma\right] d\tau_1 + \varphi(t, \tau)}.$$

At the constant fucntions, a_0, a, c, $\nu = 1$ and the initial condition $g_0(\lambda) = (1 - 2iD_0)^{-1/2}$ we obtain

$$g_1(\lambda; t) = [\exp(at) - ic\lambda(\exp(at) - 1)/2a]^{-2aa_0}$$

$$\times [1 - 2iD_0 r(t)]^{-1/2} \exp\left(2\frac{aa_0}{c}t\right), \tag{III}$$

where $r(t) = \lambda[\exp(at) - ic\lambda(\exp(at) - 1)/2a]^{-1}$. Give the generalization for the case of vector Eq. (I) when $b(y, t) = \mathrm{diag}\,[b_1(y, t) \ldots b_p(y, t)]$, $b_j^2(y, t) = c_j(t)^T y$.

    5.23. Show that under the conditions of the previous problem the multi-dimensional characteristic function is determined by the recurrent formula:

$$g_n(\lambda_1, \ldots, \lambda_n; t_1, \ldots, t_n) = g_{n-1}(\lambda_1, \ldots, \lambda_{n-1})$$

$$+\varphi^{-1}(t_n, \lambda_n); t_1, \ldots, t_{n-1}) \exp\left[i \int_{t_{n-1}}^{t_n} \psi(\tau, \varphi(t_n, \lambda_n))a_0(\tau)d\tau\right], \tag{I}$$

where $n = 2, 3, \ldots$,

$$\varphi(t_n, \lambda_n) = \frac{1}{\lambda_n}\left\{\exp\left[-\int_{t_0}^{t} a(\tau)d\tau\right]\right.$$

$$\left. -\frac{i\lambda_n}{2} \int_{t_{n-1}}^{t_n} \nu(\tau)c(\tau) \exp\left[-\int_{t_{n-1}}^{t_n} a(\sigma)d\sigma\right] d\tau\right\}, \tag{II}$$

$$\psi(\tau, \varphi(t_n, \lambda_n)) = \frac{\exp\left[-\int_{t_{n-1}}^{t_n} a(\sigma)d\sigma\right]}{i/2 \int_{t_{n-1}}^{\tau} \nu(\xi)c(\xi) \exp\left[-\int_{t_{n-1}}^{\xi} a(\sigma)d\sigma\right] d\xi + \varphi(t_n, \lambda_n)}.$$

$$\tag{III}$$

5.24. Show that under the conditions of Example 5.6.6 at δ-impulse Poisson actions with the values of the jumps c_k Eq. (5.6.32) has the form

$$\frac{\partial f_1(y;t)}{\partial t} = \alpha \frac{\partial^T}{\partial y}\left[yf_1 y;t)\right] + \sum_{k=1}^{N} \nu_k(t)\left[f_1(y - \gamma c_k;t) - f_1(y;t)\right].$$

5.25. Using formula (5.6.39) and Eq. (5.6.32) show that in the case of the general Poisson process $W(t)$ and unit matrix $b(y,t) = I$ the one-dimensional density is defined by the following integrodifferential Kolmogorov–Feller equation:

$$\frac{\partial f_1(y;t)}{\partial t} = -\frac{\partial^T}{\partial y}\left[a(y,t)f_1(y;t)\right] + \nu(t)\left[(f * f_1)(y,t) - f_1(y;t)\right], \quad \text{(I)}$$

where $f * f_1$ being the convolution of the density $f(y)$ of the jumps magnitude of the Poisson process with the density $f_1(y;t)$:

$$(f * f_1)(y,t) = \int_{-\infty}^{\infty} f(y - \eta)f_1(\eta;t)d\eta. \quad \text{(II)}$$

Write a similar equation for $f_n(y_1, \ldots, y_n; t_1, \ldots, t_n)$. Obtain from Eq. (I) the corresponding equation for the case where $W(t)$ represents the product of a constant h by the simple Poisson process.

5.26. Using formula (5.6.41) for the function $\chi(\rho;t)$ show that Eq. (5.6.32) is reduced to the following integrodinfferential generalized Kolmogorov–Feller:

$$\frac{\partial f_1(y;t)}{\partial t} = -\frac{\partial^T}{\partial y}\left[a(y,t)f_1 y;t)\right]$$

$$+ \frac{1}{2}\operatorname{tr}\left\{\frac{\partial}{\partial y}\frac{\partial^T}{\partial y}\left[b(y,t)\nu_0(t)b(y,t)^T f_1(y;t)\right]\right\}$$

$$+ \sum_{k=1}^{N} \nu_k(t)\left[\int_{-\infty}^{\infty} \varphi_k(y - \eta;\eta,t)f_1(\eta;t)d\eta - f_1(y;t)\right],$$

where $\varphi_k(\xi,\eta,t)$ is the density corresponding to the characteristic function $g_k(c_k^T b(y,t)^T \lambda)$, i.e. the density of the random vector representing the **product of the matrix $(b(y,t)\, c_k)$ by the random vector Y_k which is the random jump magnitude of the general Poisson process $P_k(t)$**.

5.27. Show that in the case where $\chi(\rho; t)$ is given by (5.6.47) Eq. (5.6.32) is reduced to

$$\frac{\partial f_1(y; t)}{\partial t} = -\frac{\partial}{\partial y}[a(y, t)f_1(y; t)] + \frac{1}{2}\text{tr}\left\{\frac{\partial}{\partial y}\frac{\partial^T}{\partial y}[b(y, t)\nu_0(t)b(y, t)^T f_1(y; t)]\right\}$$

$$+ \int_{R_0^q}\left\{f_1(y - b(y, t)c(x); t) - f_1(y; t) + \frac{\partial^T}{\partial y}[b(y, t)c(x)f_1(y; t)]\right\}\nu_P(t, x)dx .$$

CHAPTER 6
METHODS OF LINEAR STOCHASTIC SYSTEMS THEORY AND THEIR APPLICATIONS

Chapter 6 is devoted to the methods of the linear stochastic systems described by the linear stochastic integral, differential, difference and other operator equations in the finite-dimensional and infinite-dimensional spaces. In Section 6.1 the spectral and correlation methods are stated. Section 6.2 contains the account of the methods of general stochastic systems theory based on the multi-dimensional distributions. In Sections 6.3 and 6.4 methods based the canonical representations are considered. Sections 6.5 and 6.6 present applications of linear stochastic systems theory to the linear shaping filters design, analytical modeling and optimal linear filtering based the online data processing.

6.1. Spectral and Correlation Methods

6.1.1. Representation of State Vector

Let us consider a linear system. In this case Eq. (5.4.6) has the form

$$\dot{Y} = aY + a_0 + bV, \qquad (6.1.1)$$

where $a = a(t)$, $a_0 = a_0(t)$, $b = b(t)$ may in the general case be the functions of time t, and V is a white noise whose intensity ν may be a function of time t.

R e m a r k. For finding the first and the second order moments of the random process $Y(t)$ determined by linear differential Eq. (6.1.1) there is no necessity that the white noise V be the white noise in the strict sense, i.e. the derivative of a process with the independent increments. It is sufficient that it be simply the white noise (i.e. the derivative of the process with the uncorrelated increments) as well as in the theory of Subsection 3.2.3. Therefore we shall consider V as an arbitrary white noise in this section. In other sections studying the multi-dimensional distributions of the process $Y(t)$ we shall always consider V as the white noise in the strict sense.

Using formula (1.2.8) determining the solution of Eq. (6.1.1) we get *representation on the system state vector Y by the formula*

$$Y(t) = u(t,t_0)Y_0 + \int_{t_0}^{t} u(t,\tau)b(\tau)V(\tau)d\tau + \int_{t_0}^{t} u(t,\tau)a_0(\tau)d\tau, \quad (6.1.2)$$

where $u(t,\tau)$ is the matrix determined as a function of t by the homogeneous differential equation $du/dt = a(t)u$ and the initial condition $u(\tau,\tau) = I$.

6.1.2. Representations of First and Second Order Moments

Taking into account that the expectation of a white noise is equal to zero we find by virtue of (6.1.2) the following *formula for the expectation of the system state vector $Y(t)$*

$$m(t) = u(t,t_0)m_0 + \int_{t_0}^{t} u(t,\tau)a_0(\tau)d\tau. \qquad (6.1.3)$$

where m_0 is the expectation of the initial value Y_0 of the state vector Y.

The covariance function of the state vector Y is determined by the formula

$$K(t_1,t_2) = u(t_1,t_0)K_0 u(t_2,t_0)^* +$$

$$+ \int_{t_0}^{\min(t_1,t_2)} u(t_1,\tau)b(\tau)\nu(\tau)b(\tau)^T u(t_2,\tau)^* d\tau, \qquad (6.1.4)$$

where K_0 is the covariance matrix of the initial value Y_0 of the state vector Y.

While deriving this formula we remembered that the initial state of the system Y_0 is independent of the white noise $V(t)$ at $t > t_0$ and that $u(t,\tau) = 0$ at $\tau > t$. Due to this the upper limit of integration is equal to $\min(t_1,t_2)$. Besides that we took into account that the coefficients of the equations of real systems are always real while the elements of the matrix $u(t,\tau)$ may be complex in this case.

After determining the expectation and the covariance function of the state vector Y we may find its second order moment which in this case of real-valued functions $m(t)$ and $K(t_1,t_2)$ gives

$$\Gamma(t_1,t_2) = K(t_1,t_2) + m(t_1)m(t_2)^T. \qquad (6.1.5)$$

6.1.3. Differential Equation for Expectation

In practical problems it is usually sufficient to find probabilistic characteristics of the state vector of a system Y at every given instant

t (determined by the one-dimensional distribution), i.e. only the values $K(t,t) = K(t)$ and $\Gamma(t,t) = \Gamma(t)$ of the covariance function and the second order moment. In other words, it is sufficient to find the expectation, the covariance matrix and the second order moment of the state vector Y at every instant t. It goes without saying the all these variables for a linear system may be determined by formulae (6.1.3)–(6.1.5) at $t_1 = t_2 = t$. But in the case of a linear system they may be calculated more simply, namely by the integration of the corresponding linear differential equations.

$\triangleright$ In order to derive the differential equation for the expectation of the vector Y we differentiate formula (6.1.3):

$$\dot{m}(t) = u_t(t, t_0)m_0 + \int_{t_0}^{t} u_t(t, \tau)a_0(\tau)d\tau + a_0(t)$$

$$= a(t)\left[u(t, t_0)m_0 + \int_{t_0}^{t} u(t, \tau)a_0(\tau)d\tau \right] + a_0(t).$$

But the expression in the brackets is equal to $m = m(t)$ by virtue of (6.1.3). Consequently,

$$\dot{m} = am + a_0. \quad \triangleleft \tag{6.1.6}$$

Thus we obtained the differential equation for the expectation of the vector Y. *Integrating* Eq. *(6.1.6) with the initial condition* $m(t_0)$ $= m_0$ *we may calculate the expectation of the random vector* Y *in stochastic linear system (6.1.1).*

6.1.4. Differential Equation for Covariance Matrix

$\triangleright$ In order to derive the equation for the covariance matrix $K(t)$ of the vector Y we put in (6.1.4) $t_1 = t_2 = t$:

$$K(t) = K(t, t) = u(t, t_0)K_0 u(t, t_0)^*$$

$$+ \int_{t_0}^{t} u(t, \tau)b(\tau)\nu(\tau)b(\tau)^T u(t, \tau)^* d\tau. \tag{6.1.7}$$

Differentiating this formula with respect to t we obtain

$$\dot{K}(t) = u_t(t, t_0) K_0 u(t, t_0)^*$$

$$+ \int_{t_0}^{t} u_t(t, \tau) b(\tau) \nu(\tau) b(\tau)^T u(t, \tau)^* d\tau + u(t, t_0) K_0 u_t(t, t_0)^*$$

$$+ \int_{t_0}^{t} u(t, \tau) b(\tau) \nu(\tau) b(\tau)^T u_t(t, \tau) d\tau + b(t) \nu(t) b(t)^T ,$$

or as $u_t(t, \tau) = a(t) u(t, \tau)$, $u_t(t, \tau)^* = u(t, \tau)^* a(t)^T$ (Subsection 1.2.2),

$$\dot{K}(t) = a(t) \left[u(t, t_0) K_0 u(t, t_0)^* \right.$$

$$+ \int_{t_0}^{t} u(t, \tau) b(\tau) \nu(\tau) b(\tau)^T u(t, \tau)^* d\tau \left. \right] + \left[u(t, t_0) K_0 u(t, t_0)^* \right.$$

$$+ \int_{t_0}^{t} u(t, \tau) b(\tau) \nu(\tau) b(\tau)^T u(t, \tau)^* d\tau \left. \right] a(t)^T + b(t) \nu(t) b(t)^T .$$

But the expression in the brackets by virtue of (6.1.7) is equal to $K = K(t)$. Consequently,

$$\dot{K} = a K + K a^T + b \nu b^T . \triangleleft \tag{6.1.8}$$

Thus we obtained the differential equation for the covariance matrix of the value of the random vector Y at a given t. *Integrating* Eq. (6.1.8) *with the initial condition* $K(t_0) = K_0$ *we may calculate the covariance matrix of the random vector Y in stochastic linear system* (6.1.1).

6.1.5. Differential Equation for Second Order Moment

▷ For deriving the differential equation for the initial second order moment $\Gamma(t)$ of the vector Y we put in (6.1.5) $t_1 = t_2 = t$. Then we shall have $\Gamma(t) = K(t) + m(t) m(t)^T$ or, omitting the argument t

$$\Gamma = K + m m^T . \tag{6.1.9}$$

Differentiating this formula we find

$$\dot{\Gamma} = \dot{K} + \dot{m}m^T + m\dot{m}^T .$$

Substituting here the expressions $\dot{m}$ and $\dot{K}$ from Eqs. (6.1.6), (6.1.8) we obtain

$$\dot{\Gamma} = aK + Ka^T + b\nu b^T + amm^T + a_0 m^T + mm^T a^T + ma_0^T ,$$

or using formula (6.1.9)

$$\dot{\Gamma} = a\Gamma + \Gamma a^T + b\nu b^T + a_0 m^T + ma_0^T . \triangleleft \qquad (6.1.10)$$

Integrating Eq. (6.1.10) *after* Eq. (6.1.6) *which determines the expectation* m *with the initial condition* $\Gamma(t_0) = \Gamma_0 = K_0 + m_0 m_0^T$ *we may calculate the initial second order moment of the random vector* Y *in stochastic linear system* (6.1.1).

R e m a r k. It is evident that the direct numerical integration of Eqs. (6.1.6), (6.1.8) or Eqs. (6.1.6), (6.1.10) is much simpler than the calculation of m and K or m and Γ by formulae (6.1.3)–(6.1.5) with preceding multiple integration of the homogeneous equation $\dot{u} = au$ with the initial condition $u(\tau, \tau) = I$ at various τ necessary for finding $u(t, \tau)$ as a function of two variables.

6.1.6. Differential Equation for Covariance Function

$\triangleright$ Starting from formula (6.1.4) we shall derive the equation for the covariance function $K(t_1, t_2)$ of the random process $Y(t)$ considered as a function t_2 at any fixed t_1. For this purpose we rewrite formula (6.1.4) for the case $t_1 < t_2$:

$$K(t_1, t_2) = u(t_1, t_0)K_0 u(t_2, t_0)^* + \int_{t_0}^{t_1} u(t_1, \tau)b(\tau)\nu(\tau)b^T(\tau)u(t_2, \tau)^* d\tau .$$

Differentiating the is formula with respect to t_2 we find

$$\frac{\partial K(t_1, t_2)}{\partial t_2} = u(t_1, t_0)K_0 u_{t_2}(t_2, t_0)^*$$

$$+ \int_{t_0}^{t_1} u(t_1, \tau)b(\tau)\nu(\tau)b(\tau)^T u_{t_2}(t_2, \tau)^* d\tau = u(t_1, t_0)K_0 u(t_2, t_0)^* a^T(t_2)$$

$$+ \int_{t_0}^{t_1} u(t_1, \tau) b(\tau) \nu(\tau) b(\tau)^T u(t_2, \tau)^* a^T(t_2) d\tau \,,$$

or

$$\partial K(t_1, t_2)/\partial t_2 = K(t_1, t_2) a(t_2)^T \,, \quad t_1 < t_2 \,. \tag{6.1.11}$$

The initial condition for this equation has the form $K(t_1, t_1) = K(t_1)$. ◁

Integrating Eq. (6.1.11) *at various values of* t_1 *we shall obtain a number of sections of the covariance function* $K(t_1, t_2)$ *at* $t_2 > t_1$. *To obtain* $K(t_1, t_2)$ *at* $t_2 < t_1$ *we use*

$$K(t_1, t_2) = K(t_2, t_1)^T \,. \tag{6.1.12}$$

Summarizing the results in Subsections 6.1.1–6.1.6 we obtain.

Theorem 6.1.1. Eqs. (6.1.6), (6.1.8), (6.1.10)–(6.1.12) *with the corresponding initial conditions determine completely the expectation, the second order moment and the covariance function of the process* $Y(t)$ *in stochastic linear differential system defined by* Eq. (6.1.1).

R e m a r k. Theorem 6.1.1 is basic for different correlation methods of linear stochastic differential systems. The white noise in Eq. (6.1.1) is understood in the wie sense (Subsection 3.2.3).

E x a m p l e 6.1.1. For the random process $Y(t)$ determined by the equation

$$\dot{Y} + \alpha Y = \sqrt{2D\alpha} V \,, \tag{I}$$

V being a white noise of unit intensity, Eqs. (6.1.6), (6.1.8), (6.1.10), (6.1.11) have the form

$$\dot{m}_y = -\alpha m_y \,, \quad \dot{D}_y = -2\alpha D_y + 2D\alpha \,, \quad K_y = D_y \,,$$
$$\dot{\Gamma}_y = -2\alpha \Gamma_y + 2D\alpha \,, \quad \partial K_y(t_1, t_2)/\partial t_2 = -\alpha K_y(t_1, t_2) \,. \tag{II}$$

The condition $K_x(t_1, T_1) = D_x(t_1)$ will serve as the initial condition for the last equation.

E x a m p l e 6.1.2. For the vector process which is determined by the differential equations

$$\dot{Y}_1 = Y_2 + q_1 V \,, \quad \dot{Y}_2 = -b^2 Y_1 - 2a Y_2 + q_2 V \,, \tag{I}$$

where V is a white noise of unit intensity, Eqs. (6.1.6), (6.1.8), (6.1.10) and (6.1.11) have the form

$$\dot{m}_1 = m_2 , \quad \dot{m}_2 = -b^2 m_1 - 2a m_2 , \tag{II}$$

$$\dot{K} = \begin{bmatrix} 0 & 1 \\ -b^2 & -2a \end{bmatrix} K + K \begin{bmatrix} 0 & -b^2 \\ 1 & -2a \end{bmatrix} + \begin{bmatrix} q_1 \\ q_2 \end{bmatrix} [q_1 q_2] , \tag{III}$$

$$\dot{\Gamma} = \begin{bmatrix} 0 & 1 \\ -b^2 & -2a \end{bmatrix} \Gamma + \Gamma \begin{bmatrix} 0 & -b^2 \\ 1 & -2a \end{bmatrix} + \begin{bmatrix} q_1 \\ q_2 \end{bmatrix} [q_1 q_2] , \tag{IV}$$

$$\frac{\partial K_x(t_1, t_2)}{\partial t_2} = K(t_1, t_2) \begin{bmatrix} 0 & -b^2 \\ 1 & -2a \end{bmatrix} . \tag{V}$$

E x a m p l e 6.1.3. For the random process $Y(t)$ determined by the equation

$$\dot{Y} = (\mu + 2at)Y - (\mu + 2at)(a + bt) + b + be^{\mu t} V , \tag{I}$$

where V is a white noise of the intensity $\nu = 2t$ Eqs. (6.1.6), (6.1.8), (6.1.10), (6.1.11) have the form

$$\dot{m}_y = (\mu + 2at)m_y - (\mu + 2at)(a + bt) + b , \tag{II}$$

$$\dot{D}_y = 2(\mu + 2at)D_y + 2b^2 t e^{2\mu t} , \quad K_y = D_y , \tag{III}$$

$$\dot{\Gamma}_y = 2(\mu + 2at)\Gamma_y + 2b^2 t e^{2\mu t} - 2(\mu + 2at)(a + bt)m_y + 2b m_y , \tag{IV}$$

$$\partial K_y(t_1, t_2)/\partial t_2 = (\mu + 2at)K_y(t_1, t_2) . \tag{V}$$

6.1.7. Stationary Processes in Stationary Linear Systems

We consider a stable stationary linear system (6.1.1) under the influence of a stationary white noise with constant intensity ν. In this case a, a_0, b are constant, the function $u(t, \tau)$ depends only on the difference of the arguments, $u(t, \tau) = w(t - \tau)$, and formulae (6.1.3), (6.1.7) and (6.1.4) at $t_0 \to -\infty$ take the form

$$m = \int_0^\infty w(\xi)a_0 d\xi , \tag{6.1.13}$$

$$K = \int_0^\infty w(\xi)b\nu b^T w(\xi)^* d\xi , \tag{6.1.14}$$

$$k(\tau) = K(t + \tau, t) = \int_0^\infty w(\xi + \tau)b\nu b^T w(\xi)^* d\xi \quad (\tau \geq 0) . \tag{6.1.15}$$

It is clear from these formulae that under the influence of a stationary white noise the wide sense stationary process settles in course of time in a stable stationary linear differential system. It is also clear from these formulae with the account of (1.1.14) that *the condition of stability of a system is not only sufficient but also necessary for the existence of a stationary process.*

As m and K are constant in a steady state we get linear algebraic equations for m and K,

$$am + a_0 = 0, \quad aK + Ka^T + b\nu b^T = 0 \qquad (6.1.16)$$

putting in Eqs. (6.1.6), (6.1.8) $\dot{m} = 0$, $\dot{K} = 0$.

If the initial values m_0 and K_0 and K satisfy these equations, then Eqs. (6.1.6), (6.1.8) have the evident solution: $m = m_0$, $K = K_0$. In this case at any t_0 the process $Y(t)$ is wide sense stationary.

To find the covariance function $k(\tau)$ of the stationary process $Y(t)$ in a linear system we transpose Eq. (6.1.11) after changing the order of the variables t_1 and t_2 in it. Then remembering that the matrix $a(t) = a$ is constant and that $K(t_2,t_1)^T = K(t_1,t_2)$ we shall have at $t_1 > t_2$ $\partial K_x(t_1,t_2)/\partial t_2 = aK(t_1,t_2)$. Putting here $K(t_1,t_2) = k(\tau)$, $\tau = t_1 - t_2$, $t_2 = t$ and performing the change of variables $t_1 = t + \tau$ we obtain the equation

$$dk(\tau)/d\tau = ak(\tau), \quad \tau > 0. \qquad (6.1.17)$$

This ordinary linear differential equation with the initial condition $k(0) = K$ determines $k(\tau)$ at $\tau > 0$. At $\tau < 0$ according to the property of the covariance function of a stationary random process it is determined by the formula $k(\tau) = k(-\tau)^T$.

Corollary 6.1.1. *In asymptotically stable linear stationary system defined by Eq. (6.1.1) under the influence of the stationary white noise the wide sense stationary process settles with the expectation, the covariance matrix and the covariance function given by* Eqs. (6.1.16), (6.1.17).

E x a m p l e 6.1.4. Under the conditions of Example 6.1.1 the variance D_y and the covariance function $k_y(\tau)$ of the stationary process $Y(t)$ are determined by the equations

$$-\alpha D_y + \alpha D = 0, \quad dk_y(\tau)/d\tau = -\alpha k_y(\tau), \quad k_y(0) = D, \quad \tau \geq 0.$$

For the existence of a stationary process in the system the condition $\alpha > 0$ is necessary and sufficient.

E x a m p l e 6.1.5. For Example 6.1.2 at $a > 0$ the covariance matrix K and the covariance function $k(\tau)$ of the stationary process are defined by the equations

$$\begin{bmatrix} 0 & 1 \\ -b^2 & -2a \end{bmatrix} K + K \begin{bmatrix} 0 & -b^2 \\ 1 & -2a \end{bmatrix} + \begin{bmatrix} q_1 \\ q_2 \end{bmatrix} [q_1 q_2] = 0 \,,$$

$$\frac{dk(\tau)}{d\tau} = \begin{bmatrix} 0 & 1 \\ -b^2 & -2a \end{bmatrix} k(\tau) \,, \quad \tau > 0 \,, \quad k(0) = K \,.$$

Notice that the obtained results are easily extended to nonstationary linear systems (6.1.1) at constant a, b and ν and an arbitrary time function $a_0(t)$. In this case the equations obtained for K and $k(\tau)$ remain valid, and m represents a function of time determined by Eq. (6.1.6). The process $Y(t)$ is covariance stationary in the system for which m is determined by Eq. (6.1.6) at any initial condition, and K and $k(\tau)$ are determined by the method outlined.

6.1.8. Spectral Methods

Let us consider a stable stationary linear system with the transfer functions $\Phi(s)$ whose input is a stationary random process $X(t)$ with the spectral density $s_x(\omega)$. We shall prove that the output $Y(t)$ of this system is a steady state, i.e. under the infinitely long action of the input represents a stationary random function.

$\triangleright$ We represent the random process $X(t)$ by the spectral representation (Subsection 4.8.6)

$$X(t) = m_x(t) + \int_{-\infty}^{\infty} e^{i\omega t} V(\omega) d\omega \,. \tag{6.1.18}$$

On the basis of the superposition principle to this spectral representation it corresponds the spectral representation of the output

$$Y(t) = m_y(t) + \int_{-\infty}^{\infty} e^{i\omega t} \Phi(i\omega) V(\omega) d\omega \,. \tag{6.1.19}$$

We find the covariance function of the random function $Y(t)$:

$$K_y(t_1, t_2) = \int_{-\infty}^{\infty} e^{i\omega t_1} \overline{e^{i\omega t_2}} \Phi(i\omega) s_x(\omega) \Phi(i\omega)^* d\omega$$

$$= \int_{-\infty}^{\infty} e^{i\omega(t_1-t_2)} \Phi(i\omega) s_x(\omega) \Phi(i\omega)^* \, d\omega \,. \tag{6.1.20}$$

Hence it is clear that the output of the system in a steady state represents a stationary random function of time. Let us find its spectral density. Using formulae of Subsection 4.8.8 at $\omega = \lambda$:

$$k_y(\tau) = \int_{-\infty}^{\infty} s_y(\omega) e^{i\omega\tau} \, d\omega \,, \quad s_y(\omega) = \frac{1}{2\pi} \int_{-\infty}^{\infty} k_y(\tau) e^{-i\omega\tau} \, d\tau \,, \tag{6.1.21}$$

we come to the conclusion that the spectral density of the output of the system $Y(t)$ is determined by

$$s_y(\omega) = \Phi(i\omega) s_x(\omega) \Phi(i\omega)^* \,. \lhd \tag{6.1.22}$$

Now we shall find the cross-covariance function of the input and the output of the system. According to (6.1.8), (6.1.9) we find

$$K_{xy}(t_1,t_2) = \int_{-\infty}^{\infty} e^{i\omega(t_1-t_2)} s_x(\omega) \Phi(i\omega)^* \, d\omega \,. \tag{6.1.23}$$

Hence it is clear that the input and the output in a steady state are cross-stationary and the cross-spectral density of the input and the output of the system is determined by

$$s_{xy}(\omega) = s_x(\omega) \Phi(i\omega)^* \,. \tag{6.1.24}$$

R e m a r k. Formulae (6.1.22)–(6.1.24) are valid only for stable stationary systems functioning in a steady state. Practically the formulae are applicable when the time of input action exceeds the time of the transient process. In should be noticed that if a system is described by a differential equation and consequently, its transfer function is rational, then the output may be also a stationary random function for any time of system functioning with special initial conditions. Namely, the random initial value of the output $Y(t_0) = Y_0$ should be chosen in such a way that the covariance matrix of the random vector $Z(t) = \left[X(t)^T Y(t)^T \right]^T$ be independent of t. For this it is sufficient to take the random initial value Y_0 whose covariance matrix and cross-covariance matrix are equal to

$$K_{y_0} = k_y(0) = \int_{-\infty}^{\infty} \Phi(i\omega) s_x(\omega) \Phi(i\omega)^* \, d\omega \,,$$

$$\tag{6.1.25}$$

$$K_{x_0 y_0} = k_{xy}(0) = \int_{-\infty}^{\infty} s_x(\omega) \Phi(i\omega)^* \, d\omega \,.$$

But the appearance of these initial conditions in real systems is practically negligible. So we proved the following proposition.

Theorem 6.1.2. *The output $Y(t)$ of asympotically stable stationary linear system with transfer function $\Phi(s)$ under the influence of the wide sense stationary input $X(t)$ in a steady state is the wide sense stationary process with the spectral and correlation characteristics given by Eqs.* (6.1.21)–(6.1.24).

$\triangleright$ For calculating the variances and the covariances of the output components of a stable stationary system functioning in a steady state under the action of a stationary input at every given instant t it is sufficient to put in formula (6.1.20) $t_1 = t_2 = t$. Then we get the following formula for the covariance matrix of the value of the output at the instant t:

$$K_y = k_y(0) = \int\limits_{-\infty}^{\infty} \Phi(i\omega)s_x(\omega)\Phi(i\omega)^* \, d\omega \, . \tag{6.1.26}$$

In practical problems, to find the variances and the covariances of the output components by formula (6.1.26) we have to calculate the integrals of rational functions. For calculating such integrals it is convenient to use the formula

$$\int\limits_{-\infty}^{\infty} \frac{b_0(i\omega)^{2n-2} + b_1(i\omega)^{2n-3} + \cdots + b_{2n-1}(i\omega) + b_{2n-2}}{\left| a_0(i\omega)^n + a_1(i\omega)^{n-1} + \cdots + a_{n-1}(i\omega) + a_n \right|^2} \, d\omega$$

$$= (-1)^{n+1} \frac{\pi}{a_0} \frac{D_n}{\Delta_n} \, . \tag{6.1.27}$$

Here

$$\Delta_n = \begin{vmatrix} c_{11} & c_{12} & \cdots & c_{1n} \\ c_{21} & c_{22} & \cdots & c_{2n} \\ \cdots & \cdots & \cdots & \cdots \\ c_{n1} & c_{n2} & \cdots & c_{nn} \end{vmatrix} \, , \quad D_n = \begin{vmatrix} b_0 & c_{12} & \cdots & c_{1n} \\ b_2 & c_{22} & \cdots & c_{2n} \\ \cdots & \cdots & \cdots & \cdots \\ b_{2n-2} & c_{n2} & \cdots & c_{nn} \end{vmatrix} \, ,$$

where $c_{pq} = a_{2p-q}$ at $0 \le 2p - q \le n$; $c_{pq} = 0$ at $2p - q < 0$ or $2p - q > n$.

For using formula (6.1.27) it is necessary to represent the numerator in (6.1.26) in the form of a polynomial relative to $i\omega$ and the denominator as the square of the modulus of the polynomial relative to $i\omega$ with positive coefficients $a_0, a_1, \ldots, a_n$.

We may also use formula (6.1.27) for calculating the covariances of the input components with the output components according to the formula

$$K_{xy} = k_{xy}(0) = \int\limits_{-\infty}^{\infty} s_x(\omega)\Phi(i\omega)^* \, d\omega. \, \triangleleft \tag{6.1.28}$$

R e m a r k. The solution of the linear differential equation for the covariance function $k(\tau)$ of a stable system represents a linear combination of functions containing damped exponents. To each such function it corresponds a rational spectral density. Thus the stationary process in the stationary linear system (6.1.1) possesses the rational spectral density. We may also come to this conclusion if we use formula (1.2.57) at $s = i\omega$: $\Phi(i\omega) = -(a - i\omega I)^{-1}b$.

E x a m p l e 6.1.6. Find the variance of a random function whose spectral density is determined by the formula $s_y(\omega) = D\alpha/\pi(\alpha^2 + \omega^2)$. According to formula (6.1.26) we find

$$D_y = \frac{D\alpha}{\pi} \int\limits_{-\infty}^{\infty} \frac{d\omega}{\alpha^2 + \omega^2} = \frac{D\alpha}{\pi} \int\limits_{-\infty}^{\infty} \frac{d\omega}{|i\omega + \alpha|^2}. \tag{I}$$

In this case $n = 1$, $a_0 = 1$, $a_1 = \alpha$, $b_0 = 1$, $c_{11} = a_1 = \alpha$, $\Delta_1 = c_{11} = \alpha$, $D_1 = b_0 = 1$ and formula (6.1.27) gives

$$\int\limits_{-\infty}^{\infty} \frac{d\omega}{|i\omega + \alpha|^2} = \frac{\pi}{\alpha}. \tag{II}$$

Substituting (II) into (I) we obtain as expected $D_y = D$.

E x a m p l e 6.1.7. Calculate the variance of a random function whose spectral density is determined by the formula

$$s_y(\omega) = \frac{D}{\pi} \frac{(\alpha + \gamma\omega_0)\beta^2 + (\alpha - \gamma\omega_0)\omega^2}{\beta^4 + 2(\alpha^2 - \omega_0^2)\omega^2 + \omega^4}, \quad \beta^2 = \alpha^2 + \omega_0^2. \tag{I}$$

Taking into account that

$$\beta^4 + 2(\alpha^2 - \omega_0^2)\omega^2 + \omega^4 = \left[\alpha^2 + (\omega - \omega_0)^2\right]\left[\alpha^2 + (\omega + \omega_0)^2\right]$$

$$= |\alpha + i(\omega - \omega_0)|^2 \, |\alpha + i(\omega + \omega_0^2|^2 = |\alpha^2 + i\alpha(\omega + \omega_0)$$

$$+ i\alpha(\omega - \omega_0) + i^2(\omega - \omega_0)(\omega + \omega_0)|^2 = |\alpha^2 + 2\alpha(i\omega)$$

$$+ i^2(\omega^2 - \omega_0^2)|^2 = \left|(i\omega)^2 + 2\alpha(i\omega) + \alpha^2 + \omega_0^2\right|^2 = \left|(i\omega)^2 + 2\alpha(i\omega) + \beta^2\right|^2,$$

$$(\alpha + \gamma\omega_0)\beta^2 + (\alpha - \gamma\omega_0)\omega^2 = (\alpha + \gamma\omega_0)\beta^2 - (\alpha - \gamma\omega_0)(i\omega)^2,$$

we have

$$D_y = \frac{D}{\pi} \int\limits_{-\infty}^{\infty} \frac{(\alpha + \gamma\omega_0)\beta^2 + (\alpha - \gamma\omega_0)\omega^2}{\beta^4 + 2(\alpha^2 - \omega_0^2)\omega^2 + \omega^4} d\omega$$

$$= \frac{D}{\pi} \int\limits_{-\infty}^{\infty} \frac{-(\alpha - \gamma\omega_0)(i\omega_0)^2 + (\alpha + \gamma\omega_0)\beta^2}{|(i\omega)^2 + 2\alpha(i\omega) + \beta^2|^2} d\omega \,.$$

$$(\mathrm{II})$$

In this case $n = 2$, $a_0 = 1$, $a_1 = \alpha$, $a_2 = \beta^2$, $b_0 = -\alpha + \gamma\omega_0$, $b_1 = (\alpha + \gamma\omega_0)\beta^2$, $c_{11} = a_1 = 2\alpha$, $c_{12} = a_0 = 1$, $c_{21} = 0$, $c_{22} = a_2 = \beta^2$,

$$\Delta_2 = \begin{vmatrix} 2\alpha & 1 \\ 0 & \beta^2 \end{vmatrix} = 2\alpha\beta^2 \,, \quad D_2 = \begin{vmatrix} -\alpha + \gamma\omega_0 & 1 \\ (\alpha + \gamma\omega_0)\beta^2 & \beta^2 \end{vmatrix} = -2\alpha\beta^2 \,.$$

Consequently, according to (6.1.27) we get from (II)

$$\int\limits_{-\infty}^{\infty} \frac{-(\alpha - \gamma\omega_0)(i\omega_0)^2 + (\alpha + \gamma\omega_0)\beta^2}{|(i\omega)^2 + 2\alpha(i\omega) + \beta^2|^2} d\omega = -\frac{\pi}{1} \cdot \frac{-2\alpha\beta^2}{2\alpha\beta^2} = \pi \,.$$

Substituting this expression into the formula for D_y we obtain as expected $D_y = D$.

6.1.9. Spectral and Correlation Methods for Stochastic Discrete Linear Systems

Let us consider at first a stochastic discrete linear system which is described by the following difference equation of the first order:

$$Y_{l+1} = a_l Y_l + b_l V_l + a_{0l} \quad (l = 1, 2, \ldots) \,, \tag{6.1.29}$$

where Y_l is a state vector of the dimension p; V_l are independent vector random variables of the dimension m with zero expectations and the covariance matrix G_l; a_l, b_l and a_{0l} are the matrix coefficients of the dimension $(p \times p)$, $(p \times 1)$ and $(p \times m)$ respectively.

Analogously to stated in Subsections 6.1.2–6.1.4 the following recurrent equations for the expectation $m_l = EY_l$ and the covariance matrix $K_l = E(Y_l - m_l)(Y_l - m_l)^T$ are derived:

$$m_{l+1} = a_l m_l + a_{0l} \,, \tag{6.1.30}$$

$$K_{l+1} = a_l K_l a_l^T + b_l G_l b_l^T \,. \tag{6.1.31}$$

The initial conditions for these equations serve $m_1 = EY_1$, $K_1 = EY_1 Y_1^T$.

For the covariance function $K(l, h) = E(Y_l - m_l)(Y_h - m_h)^T$ we have recurrent equation

$$K(l, h+1) = K(l, h)a_h^T , \quad l < h \tag{6.1.32}$$

and the initial condition $K(l, l) = K_l$. At $l > h$ we have $K(l, h) = K(h, l)^T$.

In the case of stationary discrete linear system (6.1.29) the matrices b_l and the vector a_{0l} do not depend on the discrete time l: $a_l = a$, $b_l = b$, $a_{0l} = a$ and Eq. (6.1.29) takes the form

$$Y_{l+1} = aY_l + a_0 + bV_l . \tag{6.1.33}$$

The variables V_l have one and the same distribution, and consequently one and the same covariance matrix G. If there exists the stationary discrete random process $\{Y_l\}$ in Eq. (6.1.33) then its expectation $m = m_l = m_{l+1}$ and covariance matrix $K = K_l = K_{l+1}$ are constant. By virtue of Eqs. (6.1.30) and (6.1.31) they satisfy the algebraic equations:

$$m = am + a_0 , \quad K = aKa^T + bGb^T . \tag{6.1.34}$$

It is evident that these equations have the unique solution if the matrix $(I - a)$ is convertible.

Thus we proved the following statements.

Theorem 6.1.3. *Recurrent Eqs. (6.1.30)–(6.1.32) with the corresponding initial conditions determine completely the expectation, the covariance matrix and the covariance function of the process Y_l in stochastic discrete linear system (6.1.29).*

Corollary 6.1.3. *In asympotically stable linear stationary system defined by Eq. (6.1.33) under the influence of the stationary white noise the wide sense stationary process settles with the expectation, the covariance matrix and the covariance function given by Eqs. (6.1.34).*

Now we consider the more general equation:

$$A(\nabla)Y_k = c + B(\nabla)V_k , \tag{6.1.35}$$

where $A(\nabla)$ and $B(\nabla)$ are the matrix polynomials over the shift operator ∇:

$$A(\nabla) = I\nabla^p - \sum_{r=0}^{p-1} a_r \nabla^r , \quad B(\nabla) = \sum_{s=-N}^{p} b_s \nabla^s$$

(the polynomials $B(\nabla)$ contain also the negative degrees of ∇ if $N > 0$). According to the spectral theory and under the conditions of asymptotically stable linear system the spectral density $s_y(\omega)$ for the stationary sequence $\{Y_l\}$ is expressed in terms of the spectral density $s_v(\omega)$ of the sequence $\{V_l\}$ and the transfer function $\Phi(z)$ of Eq. (6.1.35):

$$s_y(\omega) = \Phi(e^{i\omega})s_v(\omega)\Phi(e^{i\omega})^* . \tag{6.1.36}$$

Consequently,

$$K = \int\limits_0^{2\pi} s_y(\omega)d\omega = \int\limits_0^{2\pi} \Phi(e^{i\omega})s_v(\omega)\Phi(e^{i\omega})^* d\omega . \tag{6.1.37}$$

As the variables V_l are independent then the spectral density $s_v(\omega)$ of the sequence $\{V_l\}$ is constant: $s_v(\omega) = G/2\pi$ where G is the covariance matrix of the variables V_l. Substituting this expression into (6.1.37) we get

$$K = \frac{1}{2\pi} \int\limits_0^{2\pi} \Phi(e^{i\omega})G\Phi(e^{i\omega})^* d\omega . \tag{6.1.38}$$

After the change of the variables $z = e^{i\omega}$ formula (6.1.38) will take the form

$$K = \frac{1}{2\pi i} \oint \Phi(z)G\Phi(e^z)^* \frac{dz}{z} , \tag{6.1.39}$$

where the unit circle $|z| = 1$ with the curve counterlock-wise serves as the integration counter. Later on it is expedient to pass to a new variable for the calculation

$$\lambda = \frac{z - 1}{i(z + 1)} = \mathrm{tg}\, \frac{\omega}{2} .$$

Then the unit circle of the plane of the variable z turns into the real axis of the variable λ and formula (6.1.39) will take the form

$$K = \frac{1}{\pi} \int\limits_{-\infty}^{\infty} \Phi\left(\frac{1 + i\lambda}{1 - i\lambda}\right) G\Phi\left(\frac{1 + i\lambda}{1 - i\lambda}\right)^* \frac{d\lambda}{1 + \lambda^2} . \tag{6.1.40}$$

As the transfer function $\Phi(z)$ of Eq. (6.1.35) is the rational function of z then the integrand function in (6.1.40) will be the rational function

λ. This fact gives the opportunity to use for the calculations of the co-variance matrix K the known formulae for the integrals over the rational functions (Subsection 6.1.8).

In some cases it is also expedient to find the covariance function

$$k(l) = E(Y_h - m_h)\left(Y_{h+l}^T - m_{h+l}^T\right) \tag{6.1.41}$$

of the stationary sequence $\{Y_l\}$. For this purpose we use the formulae of the connection between the spectral-correlation characteristics (Subsection 4.8.6):

$$k(l) = \frac{1}{2\pi} \int_0^{2\pi} \Phi(e^{i\omega})G\Phi(e^{i\omega})^* e^{il\omega}\, d\omega\,. \tag{6.1.42}$$

Then the correspondent formulae with the variables of the integration z and λ have the form

$$k_y(l) = \frac{1}{2\pi} \int_0^{2\pi} \Phi(e^{i\omega})G\Phi(e^{i\omega})^* e^{il\omega}\, d\omega\,, \tag{6.1.43}$$

$$k_y(l) = \frac{1}{\pi} \int_{-\infty}^{\infty} \Phi\left(\frac{1+i\lambda}{1-i\lambda}\right) G\Phi\left(\frac{1+i\lambda}{1-i\lambda}\right) \frac{(1+i\lambda)^{2l}}{(1+\lambda^2)^{l+1}}\, d\lambda\,. \tag{6.1.44}$$

Formula (6.1.44) is more convenient for the calculations than formulae (6.1.42) and (6.1.43) as the integrand in it is rational in consequence of which we may use the known formula (6.1.27) for the integrals over the rational functions for the calculation of the elements of the matrix $k(l)$.

R e m a r k. As regards to the constant expectation of the stationary sequence $\{Y_k\}$, it is defined by the trivial equation: $A(1)m_y = c$. This equation has always the unique solution $m_y = A(1)^{-1}c$ as 1 cannot be the root of the determinant of the polynomial $|A(z)|$ in the case of the asymptotically stable system (6.1.35).

Thus we have the statement.

Theorem 6.1.4. *In asympotically stable discrete linear system (6.1.35) under the influence of the stationary white noise the wide sense stationary process settles with the spectral and the correlation characteristics defined by Eqs. (6.1.36), (6.1.38)–(6.1.40), (6.1.42)–(6.1.44).*

6.1.10. Correlation Theory of Linear Transformations

Let the expectation $m_x(t)$ and the covariance function $K_x(t, t')$ of the input $X(t)$ be given. Let us find $m_y(s)$ and $K_y(s, s')$ of the output $Y(s)$ for the linear transformation

$$Y(s) = AX(t), \tag{6.1.45}$$

where A is an arbitrary linear operator. This problem is easily solved if we assume that the operation of the expectation and the operator A are commutative. It is often valid for linear operators in the problems of practice.

According to the properties of the first and the second order moments we have

$$EY(s) = E[AX(t)] = AE(X(t),$$
$$E\left[Y^0(s)\overline{Y^0(s')}\right] = E\left[A_t X^0(t)\overline{A_{t'} X^0(t')}\right] \tag{6.1.46}$$
$$= E\left[A_t \overline{A_{t'}} X^0(t)\overline{X^0(t')}\right] = A_t \overline{A_{t'}} E\left[X^0(t)\overline{X^0(t')}\right]$$

or

$$m_y(s) = Am_x(t), \tag{6.1.47}$$
$$\cdot K_y(s, s') = A_t \overline{A_{t'}} K_x(t, t') = \overline{A_{t'}} A_t K_x(t, t'). \tag{6.1.48}$$

Here the index at the operator A indicates that this operator acts upon the function of a given argument at the fixed values of all other variables.

It is evident that formula (6.1.46) is also valid for the initial moment of the second order: $\Gamma_y(s, s') = A_t \overline{A_{t'}} \Gamma_x(t, t') = \overline{A_{t'}} A_t \Gamma_x(t, t')$.

Thus *formulae (6.1.44) and (6.1.48) are at the basis of the correlation theory of the linear transformations of random functions.* Hence the correspondent formulae of Subsection 6.1.1–6.1.9 are obtained as the special case for the concrete operators A.

The general solution of this problem is given in Sections 6.3 and 6.4.

6.1.11. Spectral and Correlation Methods for Stochastic Infinite-Dimensional Linear Systems

Let us consider a linear stochastic differential system in a separable H-space $\mathcal{Y}$ (Subsection 5.9.1):

$$dY = (aY + a_0)at + bdW. \tag{6.1.49}$$

Using formulae of Subsection 6.1.10 we get for expectation m, $m = EY^0$, covariance operator K, $K\lambda = E\left[Y_t^0(\lambda, Y_t^0)\right]$, $\lambda \in \mathcal{Y}$ where $Y_t^0 = Y_t - m$ of the process $Y_t = Y(t)$ at a given instant t the following equations:

$$\dot{m} = am + a_0\,, \tag{6.1.50}$$

$$\dot{K}\lambda = aK\lambda + K\lambda a^* + \sigma\lambda \ (\sigma = b\nu b^*)\,. \tag{6.1.51}$$

For covariance operator $K(t_1, t_2)\lambda_2 = E\left[Y_{t_1}^0(\lambda_2, Y_{t_2}^0)\right]$ at t_1 and t_2 we have

$$\frac{\partial K(t_1, t_2)}{\partial t_2}\lambda_2 = E\left[Y_{t_1}^0\left(\lambda_2, aY_{t_2}^0\right)\right]\,, \quad t_2 > t_1\,. \tag{6.1.52}$$

The following conditions serve as the initial conditions for $m(t)$, $K(t)\lambda$ and $K(t_1, t_2)\lambda_2$ in operator equations (6.1.50)–(6.1.51):

$$m_0 = m(t_0)\,, \quad K(t_0)\lambda = E\left[Y_{t_0}^0(\lambda, Y_{t_0}^0)\right]\,, \quad K(t, t)\lambda_2 = E\left[Y_t^0(\lambda_2, Y_t^0)\right]\,. \tag{6.1.53}$$

For determining the processes stationary in the wide sense in (6.1.49) at a, a_0, b and σ independent of time it is sufficient to put in Eqs. (6.1.50) and (6.1.51) $\dot{m} = 0$ and $\dot{K} = 0$. As a result we obtain the equations:

$$am + a_0 = 0\,, \quad aK\lambda + K\lambda a^* + \sigma\lambda = 0\,. \tag{6.1.54}$$

For the covariance operator of the stationary process

$$K(t_1, t_2)\lambda_2 = E\left[Y_{t_1}^0(\lambda_2, Y_{t_2}^0)\right] = k(\tau)\lambda_2 \ (t_1 - t_2 = \tau)$$

Eq. (6.1.52) takes the form

$$\frac{dk(\tau)}{d\tau} = ak(\tau)\,. \tag{6.1.55}$$

In the correspondence with the formulae of Subsection 4.8.6 the following equation for the spectral operator $s_y(\omega)$ holds:

$$s_y(\omega)\lambda_2 = \Phi(i\omega)s_v(\omega)\Phi(i\omega)^*\,, \tag{6.1.56}$$

where $s_v = s_v(\omega)$ is a spectral operator of the white noise V, $\Phi(i\omega) = -(a - i\omega I)^{-1}b$.

Thus we get the following propositions.

Theorem 6.1.5. *Eqs. (6.1.50)–(6.1.52) with initial conditions (6.1.53) determine completely the expectation and covariance operators in stochastic linear differential system defined by Eq. (6.1.40).*

Corolary 6.1.5. *In asymptotically stable stochastic linear differential system (6.1.49) under the influence stationary white noise the wide sense stationary process settles with the spectral and the correlation operators defined by Eqs. (6.1.54)–(6.1.56).*

E x a m p l e 6.1.8. Under the conditions of Example 5.9.1 Eqs. (6.1.50) and (6.1.51) at constant c, h, ν for $m_i = m_i(x_1, x_2, t)$, $K_{ij} = K_{ij}(x_1, x_2, t)$ have the following form:

$$\frac{\partial m_1}{\partial t} = m_2, \quad \frac{\partial m_1}{\partial t} = c^2 \frac{\partial^2 m_1}{\partial x^2} + 2hm_2, \tag{I}$$

$$\frac{\partial K_{11}}{\partial t} = K_{21} + K_{12}, \quad \frac{\partial K_{12}}{\partial t} = K_{22} + c^2 \frac{\partial^2 K_{11}}{\partial x_2^2} + hK_{12},$$

$$\frac{\partial K_{21}}{\partial t} = K_{22} + c^2 \frac{\partial^2 K_{11}}{\partial x_1^2} + 2hK_{21},$$

$$\frac{\partial K_{22}}{\partial t} = c^2 \left(\frac{\partial^2 K_{12}}{\partial x_1^2} + \frac{\partial^2 K_{21}}{\partial x_2^2} \right) + 4hK_{22} + \nu. \tag{II}$$

6.2. Methods of General Linear Stochastic Systems Theory

6.2.1. Equations for Characteristic Functions

In the case of a linear system (6.1.1) where V being the white noise (in the strict sense) Eqs. (5.6.48), (5.6.49) as $a(\partial/i\partial\lambda, t) = -ia(t)\partial/\partial\lambda + a_0(t)$ take the form

$$\frac{\partial g_1}{\partial t} = \lambda^T a \frac{\partial g_1}{\partial \lambda} + \left[i\lambda^T a_0 + \chi(b^T \lambda; t) \right] g_1, \tag{6.2.1}$$

$$\frac{\partial g_n}{\partial t_n} = \lambda_n^T a(t_n) \frac{\partial g_n}{\partial \lambda_n} + \left[i\lambda_n^T a_0(t_n) + \chi(b(t_n)^T \lambda_n; t_n) \right] g_n. \tag{6.2.2}$$

Thus, after integrating Eq. (6.2.1) with the corresponding initial conditions (5.6.6) and (5.6.12)

$$g_1(\lambda; t_0) = g_0,$$

$$g_n(\lambda_1, \ldots, \lambda_n; t_1, \ldots, t_{n-1}, t_{n-1})$$

$$= g_{n-1}(\lambda_1, \ldots, \lambda_{n-2}, \lambda_{n-1} + \lambda_n; t_1, \ldots, t_{n-1}), \quad n = 2, 3, \ldots \quad (6.2.3)$$

we shall find all the multi-dimensional characteristic functions of the process $Y(t)$.

6.2.2. Integration of Equations for Characteristic Functions

In order to integrate Eq. (6.2.1) we rewrite the first item in its right-hand side in scalar form omitting subscript 1:

$$\frac{\partial g}{\partial t} - \sum_{k=1}^{p} \left(\sum_{l=1}^{p} a_{lk} \lambda_l \right) \frac{\partial g}{\partial \lambda_k} = \left[i\lambda^T a_0 + \chi(b^T \lambda; t) \right] g . \qquad (6.2.4)$$

The standard algorithm for integrating equations of the first order in partial derivatives linear relative to the derivatives consists of the following operations.

(i) The integration of the corresponding set of ordinary differential equations which has in this case the form

$$\frac{dt}{1} = \frac{d\lambda_1}{-\sum\limits_{l=1}^{p} a_{l1} \lambda_l} = \cdots = \frac{d\lambda_p}{-\sum\limits_{l=1}^{p} a_{lp} \lambda_l} = \frac{dg}{\left[i\lambda^T a_0 + \chi(b^T \lambda; t) \right] g} . \qquad (6.2.5)$$

This set of equations is obtained from the original equation in partial derivatives in the following way: the differential of each of the variables t, $\lambda_1, \ldots, \lambda_p$ is divided by the coefficient of the derivative of the function g with respect to this variable, the differential of the function is divided by the right-hand side of the equation and the ratios thus obtained are equalized to one another.

(ii) The general solution of Eqs. (6.2.5) is represented in the form

$$\varphi_k(t, \lambda_1, \ldots, \lambda_p, g) = c_k \quad (k = 1, \ldots, p+1), \qquad (6.2.6)$$

where $c_1, \ldots, c_p, c_{p+1}$ are arbitrary constants (i.e. in the form of $p+1$ independent first integrals of these equations).

(iii) An arbitrary function $f(x_1, \ldots, x_p)$ of p variables is taken and the equation is written

$$\varphi_{p+1}(t, \lambda_1, \ldots, \lambda_p, g)$$

$$= f(\varphi_1(t, \lambda_1, \ldots, \lambda_p, g), \ldots, \varphi_p(t, \lambda_1, \ldots, \lambda_p, g)). \qquad (6.2.7)$$

The solution of this equation relative to g depending on the arbitrary function f represents a complete integral of the equation in partial derivatives.

$\triangleright$ Let us apply this algorithm for the solution of Eq. (6.2.1). In this case the coefficients of the derivatives are linear function of the variables $\lambda_1, \ldots, \lambda_p$ independent of the unknown function g, and the right-hand side of the equation is linear relative to g.

Rewriting the set of Eqs. (6.2.5) in the Cauchy form we shall have

$$\frac{d\lambda_k}{dt} = -\sum_{l=1}^{p} a_{lk}\lambda_l \ (k = 1, \ldots, p), \qquad (6.2.8)$$

$$\frac{dg}{dt} = \left[i\lambda^T a_0 + \chi(b^T\lambda; t) \right] g. \qquad (6.2.9)$$

The first p of these equations form a set of linear homogeneous equations relative to $\lambda_1, \ldots, \lambda_p$ whose matrix of coefficients represents the transposed matrix of the coefficients of the original Eq. (6.1.1) with the sign reversed. In matrix form this set of equations is

$$d\lambda/dt = -a^T\lambda. \qquad (6.2.10)$$

In order to find the general solution of this equation we denote by $u(t, \tau)$ the solution of the homogeneous equation $du/dt = au$ with the initial condition $u(\tau, \tau) = I$, $t \geq \tau \geq t_0$. Then the solution $v(t, \tau)$ of the adjoint equation $dv/dt = -a^T v$ with the initial condition $v(\tau, \tau) = I$ is determined by formula $v(t, \tau) = u(t, \tau)^{-1T}$ (Subsection 1.2.2). Consequently, the general solution of Eq. (6.2.10) is determined by formula

$$\lambda = u(t, \alpha)^{-1T} c, \qquad (6.2.11)$$

where c is an arbitrary constant p-dimensional vector, $c = \left[c_1 \ldots c_p \right]^T$, and α is an arbitrary time instant within the interval $[t_0, t)$. Substituting expression (6.2.11) into Eq. (6.2.9) we transform it to the form

$$dg/dt = \left[ic^T u(t, \alpha)^{-1} a_0(t) + \chi(b(t)^T u(t, \alpha)^{-1T} c; t) \right] g. \qquad (6.2.12)$$

The general solution of Eq. (6.2.12) has the form

$$g = c_{p+1} \exp\left\{ \int_{\alpha}^{t} \left[ic^T u(\tau, \alpha)^{-1} a_0(\tau) + \chi(b(\tau)^T u(\tau, \alpha)^{-1T} c; \tau) \right] d\tau \right\},$$

where c_{p+1} is an arbitrary scalar constant. Substituting into this equation the expression of c from (6.2.11) $c = u(t, \alpha)^T \lambda$, and solving the obtained equation relative to c_{p+1} we get

$$c_{p+1} = g \exp \left\{ - \int_\alpha^t [i\lambda^T u(t, \alpha) u(\tau, \alpha)^{-1} a_0(\tau) \right.$$
$$\left. + \chi(b(\tau)^T u(\tau, \alpha)^{-1T} u(t, \alpha)^T \lambda; \tau)] d\tau \right\}. \tag{6.2.13}$$

But according to Subsection 1.2.2 for any $\alpha < \tau < t$

$$u(t, \alpha) = u(t, \tau) u(\tau, \alpha). \tag{6.2.14}$$

Using formula (6.2.14) we reduce the previous formula to the form

$$c_{p+1} = g \exp \left\{ -i\lambda^T \int_\alpha^t u(t, \tau) a_0(\tau) d\tau - \int_\alpha^t \chi(b(\tau)^T u(\tau, \tau)^T \lambda; \tau) d\tau \right\}. \tag{6.2.15}$$

The right-hand side of this formula and the components of the vector $u(t, \alpha)^T \lambda$ represent respectively the functions $\varphi_{p+1}, \varphi_1, \ldots, \varphi_p$ of point (iii) of the algorithm. Consequently, taking an arbitrary function of the p-dimensional vector $f(x)$ we get Eq. (6.2.7) which has in this case the form

$$g \exp \left\{ -i\lambda^T \int_\alpha^t u(t, \tau) a_0(\tau) d\tau - \int_\alpha^t \chi(b(\tau)^T u(t, \tau)^T \lambda; \tau) d\tau \right\}$$
$$= f(u(t, \alpha)^T \lambda). \tag{6.2.16}$$

After solving this equation relative to g we find the complete integral of the equation in partial derivatives (6.2.1):

$$g(\lambda; t) = f(u(t, \alpha)^T \lambda) \exp \left\{ -i\lambda^T \int_\alpha^t u(t, \tau) a_0(\tau) d\tau \right.$$
$$\left. + \int_\alpha^t \chi(b(\tau)^T u(t, \tau)^T \lambda; \tau) d\tau \right\}. \tag{6.2.17}$$

The arbitrary function $f(x)$ in this formula is determined from the corresponding initial condition at $t = \alpha$. Therefore, using formula (6.2.17) and initial conditions (6.2.13) we may determine successively all the multi-dimensional characteristic functions of the process $Y(t)$. ◁

6.2.3. Explicit Formulae for Multi-Dimensional Characteristic Functions

Now let us use formula (6.2.17) to derive explicit expressions for all the multi-dimensional characteristic functions of the state vector of a linear system (6.1.1).

▷ Putting in (6.2.17) $\alpha = t_0$, $g(\lambda; t) = g_1(\lambda; t)$ and using the initial condition (6.2.3) $g_1(\lambda; t_0) = g_0(\lambda)$ we obtain $f(u(t_0, t_0)^T \lambda) = g_0(\lambda)$ or $f(\lambda) = g_0(\lambda)$ since $u(\tau, \tau) = I$ at any τ. This formula completely determines the function $f(x)$ in (6.2.17) for the case of the one-dimensional characteristic function $g_1(\lambda, t)$. Thus, the one-dimensional characteristic function $g_1(\lambda; t)$of the process $Y(t)$ is determined by the formula

$$
\begin{aligned}
g_1(\lambda; t) = {} & g_0(u(t, t_0)^T \lambda) \exp \left\{ i\lambda^T \int_{t_0}^{t} u(t, \tau) a_0(\tau) d\tau \right. \\
& \left. + \int_{t_0}^{t} \chi(b(\tau)^T u(t, \tau)^T \lambda; \tau) d\tau \right\}.
\end{aligned}
$$
(6.2.18)

For determining the two-dimensional characteristic function $g_2(\lambda_1, \lambda_2; t_1, t_2)$ at $t_1 < t_2$ we put $g(\lambda; t) = g_2(\lambda_1, \lambda_2; t_1, t_2)$ in (6.2.17) in accordance with intial condition (6.2.3) at $n = 2$, $\alpha = t_1$, $\lambda = \lambda_2$, $t = t_2$. Then we obtain

$$
\begin{aligned}
g_2(\lambda_1 \lambda_2; t_1, t_2) = {} & f(u(t_2, t_1)^T \lambda_2) \exp \left\{ i\lambda_2^T \int_{t_1}^{t_2} u(t_2, \tau) a_0(\tau) d\tau \right. \\
& \left. + \int_{t_1}^{t_2} \chi(b(\tau)^T u(t_2, \tau)^T \lambda_2; \tau) d\tau \right\}.
\end{aligned}
$$
(6.2.19)

To find the function f we use the initial condition (6.2.3). Putting $t_2 = t_1$ we get on the basis of (6.2.3) and (6.2.18)

$$
f(\lambda_2) = g_1(\lambda_1 + \lambda_2; t_1) = g_0(u(t_1, t_0)^T (\lambda_1 + \lambda_2))
$$

$$\times \exp\left\{ i(\lambda_1^T + \lambda_2^T) \int_{t_0}^{t_1} u(t_1,\tau)a_0(\tau)d\tau + \int_{t_0}^{t_1} \chi(b(\tau)^T u(t_1,\tau)^T(\lambda_1+\lambda_2);\tau)d\tau \right\}.$$

Consequently,

$$f(u(t_2,t_1)^T \lambda_2) = g_0(u(t_1,t_0)^T \lambda_1 + u(t_1,t_0)^T u(t_2,t_1)^T \lambda_2)$$

$$\times \exp\left\{ i\left[\lambda_1^T + \lambda_2^T u(t_2,t_1)\right] \int_{t_0}^{t_1} u(t_1,\tau)a_0(\tau)d\tau \right.$$

$$\left. + \int_{t_0}^{t_1} \chi(b(\tau)^T \left[u(t_1,\tau)^T \lambda_1 + u(t_1,\tau)^T u(t_2,t_1)^T \lambda_2 \right];\tau)d\tau \right\}.$$

$$(6.2.20)$$

But according to (6.2.14)

$$u(t_2,t_1)u(t_1,\tau) = u(t_2,\tau), \quad u(t_1,t_0)^T u(t_2,t_1)^T = u(t_2,t_0)^T,$$
$$u(t_1,\tau)^T u(t_2,t_1)^T = u(t_2,\tau)^T.$$

Hence formula (6.2.20) becomes

$$f(u(t_2,t_1)^T \lambda_2) = g_0(u(t_1,t_0)^T \lambda_1 + u(t_2,t_0)^T \lambda_2)$$

$$\times \exp\left\{ i\lambda_1^T \int_{t_0}^{t_1} u(t_1,\tau)a_0(\tau)d\tau + i\lambda_2^T \int_{t_0}^{t_1} u(t_2,\tau)a_0(\tau)d\tau \right.$$

$$\left. + \int_{t_0}^{t_1} \chi(b(\tau)^T \left[u(t_1,\tau)^T \lambda_1 + u(t_2,\tau)^T \lambda_2 \right];\tau)d\tau \right\}.$$

$$(6.2.21)$$

Substituting expression (6.2.21) into (6.2.19) we find the two-dimensional characteristic function $g_2(\lambda_1, \lambda_2; t_1, t_2)$:

$$g_2(\lambda_1, \lambda_2; t_1, t_2) = g_0(u(t_1,t_0)^T \lambda_1 + u(t_2,t_0)^T \lambda_2)$$

$$\times \exp\left\{ i\lambda_1^T \int_{t_0}^{t_1} u(t_1,\tau)a_0(\tau)d\tau + i\lambda_2^T \int_{t_0}^{t_2} u(t_2,\tau)a_0(\tau)d\tau \right.$$

$$+ \int_{t_0}^{t_2} \chi(b(\tau)^T \left[u(t_1,\tau)^T \lambda_1 + u(t_2,\tau)^T \lambda_2) \right];\tau)d\tau$$

$$\left. + \int_{t_1}^{t_2} \chi(b(\tau)^T u(t_2,\tau)^T \lambda_2; \tau)d\tau \right\}. \qquad (6.2.22)$$

The easiest way to find the other characteristic functions is to use the induction method. Observing the regularities in obtaining $g_2(\lambda_1, \lambda_2; t_1, t_2)$ from $g_1(\lambda_1; t_1)$ we may write the following expression for the n-dimensional characteristic function $g_n(\lambda_1, \ldots, \lambda_n; t_1, \ldots, t_n)$ at any n, $t_1 < t_2 < \cdots < t_n$:

$$g_n(\lambda_1, \ldots, \lambda_n; t_1, \ldots, t_n)$$

$$= g_0\left(\sum_{k=1}^{n} u(t_k, t_0)^T \lambda_k\right) \exp\left\{i \sum_{k=1}^{n} \lambda_k^T \int_{t_0}^{t_k} u(t_k, \tau) a_0(\tau) d\tau\right.$$

$$\left. + \sum_{k=1}^{n} \int_{t_{k-1}}^{t_k} \chi(b(\tau)^T \sum_{l=k}^{n} u(t_l, \tau)^T \lambda_l; \tau) d\tau\right\} \quad (n = 1, 2, \ldots). \qquad (6.2.23)$$

Supposing that formula (6.2.23) is valid for some n and using formula (6.2.17) with the initial condition (6.2.3), we make sure in the same way as formula (6.2.22) was derived that formula (6.2.23) is also valid for $n + 1$. Since it was proved for $n = 1$ and $n = 2$ it is also valid for any n. ◁

Thus we get the following statement.

Theorem 6.2.1. *Formula* (6.2.23) *gives the explicit expressions for all the multi-dimensional characteristic functions of the random process* $Y(t)$ *which is defined by* Eq. (6.1.1).

R e m a r k. Using this formula one may find all the multi-dimensional distributions of the state vector of a linear system. To find the multi-dimensional characteristic functions of the state vector of a system in the case where $Y(t)$ represents its extended state vector including all the additional variables, the state vectors and the outputs of all the shaping filters involved while transforming the equations of the system into stochastic differential equations, it is necessary to equate to zero in (6.2.23) the components of all the vectors $\lambda_1, \ldots, \lambda_n$ corresponding to all the components of the vector Y which are of no interest. After that we may find all the multi-dimensional distributions of the output of the system using the formula for the characteristic function of a linear function of a random variable. Thus formula (6.2.23) gives a complete solution of the problem of determining the distribution of the state vector and of the output of any linear system whose behaviour is described by a stochastic linear differential equation.

E x a m p l e 6.2.1. The output Y of the system is determined by the equation

$$\dot{Y} + \alpha Y = \gamma V, \qquad (I)$$

where $\gamma = \sqrt{2D\alpha}$, and V is the white noise of unit intensity.

In this case $a = -\alpha$, $a_0 = 0$, $b = \gamma$, $u(t,\tau) = e^{-\alpha(t-\tau)}$ and formula (6.2.23) gives for $t_0 = 0$, $t_1 \le t_2 \le \cdots \le t_n$

$$g_n(\lambda_1, \ldots, \lambda_n; t_1, \ldots, t_n)$$

$$= g_0 \left(\sum_{k=1}^{n} e^{-\alpha t_k} \lambda_k \right) \exp\left\{ \sum_{k=1}^{n} \int_{t_{k-1}}^{t_k} \chi \left(\gamma \sum_{l=k}^{n} e^{-\alpha(t_l-\tau)} \lambda_l; \tau \right) d\tau \right\}. \quad \text{(II)}$$

In the special case of a normally distributed white noise $V(t)$ the function $\chi(\mu; t)$ is determined by (5.6.23), and the obtained formula assumes the form

$$g_n(\lambda_1, \ldots, \lambda_n; t_1, \ldots, t_n) = g_0 \left(\sum_{k=1}^{n} e^{-\alpha t_k} \lambda_k \right)$$

$$\times \exp\left\{ -\frac{1}{2}\gamma^2 \sum_{k=1}^{n} \int_{t_{k-1}}^{t_k} \sum_{l=k}^{n} e^{-\alpha(t_h+t_l-2\tau)} \lambda_h \lambda_l d\tau \right\} = g_0 \left(\sum_{k=1}^{n} e^{-\alpha t_k} \lambda_k \right)$$

$$\times \exp\left\{ -\frac{\gamma^2}{2\alpha} \sum_{k=1}^{n} \left(e^{2\alpha t_k} - e^{2\alpha t_{k-1}} \right) \sum_{h,l=k}^{n} e^{-\alpha(t_h+t_l)} \lambda_h \lambda_l \right\}. \quad \text{(III)}$$

Changing the order of summation and taking into account that $k \le \min(h, l)$ in all the items we obtain

$$g_n(\lambda_1, \ldots, \lambda_n; t_1, \ldots, t_n) = g_0 \left(\sum_{k=1}^{n} e^{-\alpha t_k} \lambda_k \right)$$

$$\times \exp\left\{ -\frac{\gamma^2}{2\alpha} \sum_{h,l=k}^{n} e^{-\alpha(t_h+t_l)} \left(e^{2\alpha \min(t_h,t_l)} - 1 \right) \lambda_h \lambda_l \right\}$$

$$= g_0 \left(\sum_{k=1}^{n} e^{-\alpha t_k} \lambda_k \right) \exp\left\{ -\frac{D}{2} \sum_{h,l=k}^{n} \left(e^{-\alpha|t_h-t_l|} - e^{-\alpha(t_h+t_l)} \right) \lambda_h \lambda_l \right\}$$

$$\text{(IV)}$$

$$n = 1, 2, \ldots.$$

As this formula is symmetric in pairs $(\lambda_1, t_1), \ldots, (\lambda_n, t_n)$ it is valid at all $t_1, \ldots, t_n$.

If V is a weak mean square derivative of a Poisson process (a Poisson stream of δ-functions with independent equally distributed random coefficients), then according to (5.6.25) $\chi(\mu; t) = [g(\mu) - 1]\nu_0$, where $g(\mu)$ is the characteristic function of jumps of the general Poisson process, and ν is the intensity of jumps stream. The formula obtained in this case takes the form

$$g_n(\lambda_1, \ldots, \lambda_n; t_1, \ldots, t_n) = g_0\left(\sum_{k=1}^{n} e^{-\alpha t_k}\lambda_k\right)$$

$$\times \exp\left\{\nu \sum_{k=1}^{n} \int_{t_{k-1}}^{t_k} \left[g\left(\gamma\sum_{l=k}^{n} e^{-\alpha(t_l-\tau)}\lambda_l\right) - 1\right] d\tau\right\}. \tag{V}$$

E x a m p l e 6.2.2. Let us consider a servo-system which consists of an integrator and an amplifier with the negative feedback. The differential equation of this system has the form

$$\dot{Y} + \beta Y = \beta X(t), \tag{I}$$

where $X(t)$ is a stationary random function with the expectation $m_x(t) = a + bt$ and the covariance function $k_x(\tau) = De^{-\alpha|\tau|}$ described by the differential equation of Example 6.2.1.. Thus, the considered servo system is described by the linear stochastic differential equations

$$\dot{Y} = -\beta Y + \beta Y_1 + \beta(a+bt), \quad \dot{Y}_1 = -\alpha Y_1 + \sqrt{2D\alpha}V, \quad Y_1 = X^0(t). \tag{II}$$

In matrix form these equations are

$$\frac{d}{dt}\begin{bmatrix} Y \\ Y_1 \end{bmatrix} = \begin{bmatrix} -\beta & \beta \\ 0 & -\alpha \end{bmatrix}\begin{bmatrix} Y \\ Y_1 \end{bmatrix} + \begin{bmatrix} \beta(a+bt) \\ 0 \end{bmatrix} + \begin{bmatrix} 0 \\ \sqrt{2D\alpha} \end{bmatrix}V_1. \tag{III}$$

Thus we have in this case

$$\lambda_k = \begin{bmatrix} \lambda'_k \\ \lambda''_k \end{bmatrix}, \quad a = \begin{bmatrix} -\beta & \beta \\ 0 & \alpha \end{bmatrix}, \quad a_0 = \begin{bmatrix} \beta(a+bt) \\ 0 \end{bmatrix}, \quad b = \begin{bmatrix} 0 \\ \sqrt{2D\alpha} \end{bmatrix},$$

$$u(t, \tau) = \begin{bmatrix} e^{-\beta(t-\tau)} & \frac{\beta}{\beta-\alpha}\{e^{-\alpha(t-\tau)} - e^{-\beta(t-\tau)}\} \\ 0 & e^{-\alpha(t-\tau)} \end{bmatrix}. \tag{IV}$$

Suppose that the white noise V is normally distributed. Then taking into account that $\nu = 1$ we find by (5.6.23) $\chi(\mu; t) = \mu^2/2$. Substituting the expressions of λ, a, a_0, b, $u(t, \tau)$ and $\chi(\mu; t)$ into (6.2.23) at $t_0 = 0$, after elementary but rather cumbersome evaluations we find the multi-dimensional characteristic functions of the vector random function $[\,Y(t)Y_1(t)\,]^T$:

$$g_n(\lambda_1', \lambda_1'', \ldots, \lambda_n', \lambda_n''; t_1, \ldots, t_n)$$

$$= g_0\left(\sum_{k=1}^n e^{-\beta t_k}\lambda_k', \frac{\beta}{\beta-\alpha}\sum_{k=1}^n \left[\left(e^{-\alpha t_k}-e^{-\beta t_k}\right)\lambda_k' + e^{-\alpha t_k}\lambda_k''\right]\right)$$

$$\times \exp\left\{i\sum_{k=1}^n \lambda_k\left[a+b\left(t_k-\frac{1}{\beta}\right)-\left(a-\frac{b}{\beta}\right)e^{-\beta t_k}\right]\right.$$

$$-\frac{D\beta}{2(\alpha-\beta)^2}\sum_{l,h=1}^n \left\{\beta\left[e^{-\alpha|t_l-t_h|}-e^{-\alpha(t_l+t_h)}\right]\right.$$

$$+\alpha\left[e^{-\beta|t_l-t_h|}-e^{-\beta(t_l+t_h)}\right]-\frac{2\alpha\beta}{\alpha+\beta}\left(e^{-\alpha|t_l-t_h|}\right.$$

$$+e^{-\beta|t_l-t_h|}-2e^{-\alpha t_l-\beta t_h}\right)\right\}\lambda_l'\lambda_h' + \frac{D\beta}{\alpha-\beta}\sum_{l,h=1}^n\left\{e^{-\alpha|t_l-t_h|}-e^{-\alpha(t_l+t_h)}\right.$$

$$-\frac{2\alpha}{\alpha+\beta}\left(e^{-\beta|t_l-t_h|}-e^{-\alpha t_l-\beta t_h}\right)\right\}\lambda_h'\lambda_l''$$

$$-\frac{D}{2}\sum_{l,h=1}^n\left[e^{-\alpha|t_l-t_h|}-e^{-\alpha(t_l+t_h)}\right]\lambda_l''\lambda_h''\right\}\quad (n=1,2,\ldots).\quad\text{(V)}$$

This formula being symmetric in triplets $(\lambda_1', \lambda_1'', t_1), \ldots, (\lambda_n', \lambda_n'', t_n)$ is valid at all $t_1, \ldots, t_n$, despite the fact that formula (6.2.23) is valid in the general case only at $t_1 \leq t_2 \leq \cdots \leq t_n$.

In order to determine the multi-dimensional distributions of the output of the system Y it is sufficient to put in the obtained formula $\lambda_k' = \lambda_k$, $\lambda_k'' = 0$ ($k = 1, \ldots, n$). Then the multi-dimensional characteristic functions $k_n(\lambda_1, \ldots, \lambda_n; t_1, \ldots, t_n)$ of the output of the system $Y(t)$ are determined by

$$k_n(\lambda_1, , \ldots, \lambda_n; t_1, \ldots, t_n) = g_0\left(\sum_{k=1}^n e^{-\beta t_k}\lambda_k, \sum_{k=1}^n \left(e^{-\alpha t_k}-e^{-\beta t_k}\right)\lambda_k\right)$$

$$\times \exp\left\{i\sum_{k=1}^n \lambda_k\left[a+b\left(t_k-\frac{1}{\beta}\right)-\left(a-\frac{b}{\beta}\right)e^{-\beta t_k}\right]\right.$$

$$-\frac{D\beta}{2(\alpha-\beta)^2} \sum_{l,h=1}^{n} \left\{ \beta \left[e^{-\alpha|t_l-t_h|} - e^{-\alpha(t_l+t_h)} \right] + \alpha \left[e^{-\beta|t_l-t_h|} - e^{-\beta(t_l+t_h)} \right] \right.$$

$$\left. -\frac{2\alpha\beta}{\alpha+\beta} \left(e^{-\alpha|t_l-t_h|} + e^{-\beta|t_l-t_h|} - 2e^{-\alpha t_l-\beta t_h} \right) \right\} \lambda_l \lambda_h \right\} \quad (n=1,2,\ldots).$$

$$(\text{VI})$$

6.2.4. Case of Normal Distribution of System State

▷ In the special case of a normally distributed white noise V in Eq. (6.2.11) the function $\chi(\mu;t)$ is determined by formula (5.6.23): $\chi(\mu;t) = -\mu^T \nu(t)\mu/2$. Therefore

$$\chi\left(b(\tau)^T \sum_{l=k}^{n} u(t_l,\tau)^T \lambda_l; \tau \right)$$

$$= -\frac{1}{2} \sum_{l,h=k}^{n} \lambda_l^T u(t_l,\tau) b(\tau) \nu(\tau) b(\tau)^T u(t_h,\tau)^T \lambda_h , \qquad (6.2.24)$$

and formula (6.2.23) gives

$$g_n(\lambda_1, , \ldots, \lambda_n; t_1, \ldots, t_n)$$

$$= g_0\left(\sum_{k=1}^{n} u(t_k,t_0)^T \lambda_k \right) \exp\left\{ i \sum_{k=1}^{l} \lambda_k^T \int_{t_0}^{t_k} u(t_k,\tau) a_0(\tau) d\tau \right.$$

$$-\frac{1}{2} \sum_{k=1}^{n} \sum_{h=k}^{n} \lambda_l^T \int_{t_0}^{t_k} u(t_l,\tau) b(\tau) \nu(\tau) b(\tau)^T u(t_h,\tau)^T d\tau \lambda_h \right\} . \qquad (6.2.25)$$

We change in (6.2.25) the order of summation in the last sum, first we shall sum up with respect to k, and after that with respect to l and h. Observing that $1 \le k \le l$, h we get at $t_1 \le t_2 \le \cdots \le t_n$

$$g_n(\lambda_1, , \ldots, \lambda_n; t_1, \ldots, t_n)$$

$$= g_0\left(\sum_{k=1}^{n} u(t_k,t_0)^T \lambda_k \right) \exp\left\{ i \sum_{k=1}^{l} \lambda_k^T \int_{t_0}^{t_k} u(t_k,\tau) a_0(\tau) d\tau \right.$$

$$-\frac{1}{2}\sum_{l,h=1}^{n}\lambda_l^T \sum_{k=1}^{\min(l,h)}\int_{t_{k-1}}^{t_k} u(t_l,\tau)b(\tau)\nu(\tau)b(\tau)^T u(t_h,\tau)^T d\tau\,\lambda_h\Bigg\} \qquad (6.2.26)$$

$$(n=1,2,\ldots)\,.$$

But

$$\sum_{k=1}^{\min(l,h)}\int_{t_{k-1}}^{t_k} u(t_l,\tau)b(\tau)\nu(\tau)b(\tau)^T u(t_h,\tau)^T d\tau$$

$$=\int_{t_0}^{\min(t_l,t_h)} u(t_l,\tau)b(\tau)\nu(\tau)b(\tau)^T u(t_h,\tau)^T d\tau\,. \qquad (6.2.27)$$

Consequently,

$$g_n(\lambda_1,\,,\ldots,\lambda_n;t_1,\ldots,t_n)$$

$$=g_0\left(\sum_{k=1}^{n}u(t_k,t_0)^T\lambda_k\right)\exp\Big\{i\sum_{k=1}^{n}\lambda_k^T\int_{t_0}^{t_k}u(t_k,\tau)a_0(\tau)d\tau$$

$$-\frac{1}{2}\sum_{l,h=1}^{n}\lambda_l^T\int_{t_0}^{\min(t_l,t_h)}u(t_l,\tau)b(\tau)\nu(\tau)b(\tau)^T u(t_h,\tau)^T d\tau\,\lambda_h\Big\}\quad (n=1,2,\ldots).$$

$$(6.2.28)$$

This formula is symmetric relative to the pairs $(\lambda_1,t_1),\ldots,(\lambda_n,t_n)$. Therefore, it is valid at all $t_1,\ldots,t_n$ though it was derived for $t_1\le\cdots\le t_n$.

If the initial distribution is normal, then

$$g_0(\lambda)=\exp\left\{i\lambda^T m_0-\frac{1}{2}\lambda^T K_0\lambda\right\} \qquad (6.2.29)$$

and formula (6.2.28) takes the form

$$g_n(\lambda_1,\,,\ldots,\lambda_n;t_1,\ldots,t_n)=\exp\Big\{i\sum_{k=1}^{n}\lambda_k^T\Big[u(t_k,t_0)m_0$$

$$+\int_{t_0}^{t_k}u(t_k,\tau)a_0(\tau)\Big]-\frac{1}{2}\sum_{l,h=1}^{n}\lambda_l^T\Big[u(t_l,t_0)K_0 u(t_h,t_0)^T$$

$$+ \int\limits_{t_0}^{\min(t_l,t_h)} u(t_l,\tau)b(\tau)\nu(\tau)b(\tau)^T u(t_h,\tau)^T\,d\tau\Bigg]\lambda_h\Bigg\} \quad (n=1,2,\ldots)\,.$$

$$(6.2.30)$$

Formula (6.2.30) shows that in this case all the multi-dimensional distributions of the random process $Y(t)$ are normal. Consequently, the process $Y(t)$ represents a normally distributed random function with the expectation

$$m(t) = u(t,t_0)m_0 + \int\limits_{t_0}^{t} u(t,\tau)a_0(\tau)d\tau \qquad (6.2.31)$$

and the covariance function

$$K(t_1,t_2) = u(t_1,t_0)K_0 u(t_2,t_0)^* + \int\limits_{t_0}^{\min(t_1,t_2)} u(t_1,\tau)b(\tau)\nu(\tau)b(\tau)^T u(t_2,\tau)^*\,d\tau$$

$$(6.2.32)$$

in complete conformity with formulae of Section 6.1. But these formulae were obtained for the considerably more general case of any distribution of the initial state and any white noise V.

If the system is stable, when $u(t,t_0) \to 0$, then $t \to \infty$ at any t_0. Therefore

$$g_0\left(\sum_{k=1}^{n} u(t_h,t_0)^T\lambda_k\right) \to g_0(0) = 1$$

when $\min(t_1,\ldots,t_n) \to \infty$. It follows from (6.2.28) that in this case all the multi-dimensional distributions of the process $Y(t)$ tend to the corresponding normal distributions when $\min(t_1,\ldots t_n) \to \infty$ independently of the initial condition $g_0(\lambda)$. ◁

So we proved the following propositions.

Corollary 6.2.1. *If the initial state of the system Y_0 and the white noise V in Eq. (6.1.1) are normally distributed, then the state of the system considered as a function of time represents the normally distributed random process.*

Corollary 6.2.2. *For a stable linear system the distribution of the process $Y(t)$ is asymptotically normal if the white noise V in Eq. (6.1.1.) is normally distributed.*

6.2.5. Strictly Stationary Processes in Stationary Linear Systems

Let us consider a stable stationary linear system (6.1.1) under the action of a strictly stationary white noise. In this case a, a_0 and b are constant, the function $u(t,\tau)$ depends only on the difference of the arguments, $u(t,\tau) = w(t-\tau)$, and the function χ does not depend on time $\chi(\mu;t) = \chi(\mu)$. Let find the one-dimensional and the multi-dimensional characteristic functions.

$\triangleright$ Putting in formula (6.2.18) $t_0 = -\infty$, $u(t,\tau) = w(\sigma)$, $\sigma = t - \tau$, $u(t,t_0) = w(\infty) = 0$ and remembering the known properties of characteristic functions we come to the following explicit formula for the one-dimensional characteristic function of the process $Y(t)$:

$$g_1(\lambda) = \exp\left\{ i\lambda^T \int_0^\infty w(\sigma)a_0 d\sigma + \int_0^\infty \chi(b^T w(\sigma)^T \lambda)d\sigma \right\}. \qquad (6.2.33)$$

Hence, it is clear that the one-dimensional characteristic function of the state vector of a stable stationary linear differential system is independent of time t at $t_0 = -\infty$ (this is the reason why we denoted it simply by $g_1(\lambda)$). Consequently, the stationary in the one-dimensional distribution process settles in course of time in such a system. In accordance with the theorem of Subsection 5.6.10 this process is strictly stationary. The one-dimensional characteristic function of this process certainly is given by (6.2.1) at $\partial g_1/\partial t = 0$. Formula (6.2.33) shows that in the case of a normally distributed white noise where $\chi(\mu) = -\mu^T \nu \mu/2$ the one-dimensional distribution of the stationary process in the system is normal and the expectation m and the covariance matrix K of the value of the process $Y(t)$ at any t are determined by the formulae of Subsection 6.1.7.

Putting in (6.2.23) $t_0 = -\infty$, $u(t,\tau) = w(\sigma)$, $\sigma = t - \tau$ we find at $t_1 \leq \cdots \leq t_n$

$$g_n(\lambda_1, , \ldots, \lambda_n; t_1, \ldots, t_n)$$

$$= \exp\left\{ i\sum_{k=1}^n \lambda_k^T \int_0^\infty w(\sigma)a_0 d\sigma + \int_0^\infty \chi\left(b^T \sum_{l=1}^n w(t_l - t_1 + \sigma)^T \lambda_l \right) d\sigma \right.$$

$$\left. + \sum_{k=2}^n \int_0^{t_k - t_{k-1}} \chi\left(b^T \int_{l=k}^n w(t_l - t_k + \sigma)^T \lambda_l \right) d\sigma \right\} \quad (n = 1, 2, \ldots). \triangleleft$$

$$(6.2.34)$$

Theorem 6.2.2. *Formulae* (6.2.33) *and* (6.2.34) *determine all the multi-dimensional distributions of the strictly stationary random process in a stable stationary linear differential system.*

R e m a r k. At an arbitrary initial instant t_0 the process in the system will be strictly stationary if its initial distribution is determined by the characteristic function $g_0(\lambda) = g_1(\lambda)$ defined in (6.2.33). In order to prove this statement it is sufficient to substitute this expression into (6.2.23) at $u(t, \tau) = w(t - \tau)$ replacing λ by the sum $\sum_{k=1}^{n} w(t_k - t_0)^T \lambda_k$, and to recall that according to property (6.2.14) of the function $u(t, \tau)$ we have $w(t_k - t_0)w(\sigma) = w(t_k - t_0 + \sigma)$ at all t_k, t_0, σ.

In the special case of a normally distributed white noise V and a normally distributed initial value of the process $Y(t)$ it is clear directly from the results of Subsection 6.1.7. In this case the multi-dimensional characteristic functions of the strictly stationary process are determined by the formula which follows from (6.2.28)

$$g_n(\lambda_1, , \ldots, \lambda_n; t_1, \ldots, t_n) = \exp\left\{ i \sum_{k=1}^{n} \lambda_k^T \int_0^\infty w(\sigma)a_0 d\sigma \right.$$

$$\left. -\frac{1}{2} \sum_{k,l=1}^{n} \lambda_k^T \int_0^\infty w(t_l - t_k + \sigma)b\nu b^T w(\sigma)^T d\sigma \lambda_l \right\} \quad (n = 1, 2, \ldots). \quad (6.2.35)$$

So we get the following proposition.

Corollary 6.2.2. *At the normally distributed white noise and the initial conditions in stable stationary linear system* (6.1.1) *there exist the stationary (in the strict sence) process with the normal mutli-dimensional distributions* (6.2.35).

E x a m p l e 6.2.3. Under the conditions of Example 6.2.1 the multi-dimensional distributions of the strictly stationary process $Y(t)$ $(n = 1, 2, \ldots)$ in the case of the normally distributed white noise V are determined by the formula

$$g_n(\lambda_1, , \ldots, \lambda_n; t_1, \ldots, t_n) = \exp\left\{ -\frac{D}{2} \sum_{k,l=1}^{n} e^{-\alpha|t_k - t_l|} \lambda_k \lambda_l \right\}. \quad (I)$$

E x a m p l e 6.2.4. Under the conditions of Example 6.2.2 the multi-dimensional characteristic functions of the strictly stationary output $Y(t)$ $(n = 1, 2, \ldots)$ are determined by the formula

$$g_n(\lambda_1, , \ldots, \lambda_n; t_1, \ldots, t_n)$$

$$= \exp\left\{ -\frac{D\beta}{2(\beta^2 - \alpha^2)} \sum_{h,l=1}^{n} \left(\beta e^{-\alpha |t_h - t_l|} - \alpha e^{-\beta |t_h - t_l|} \right) \lambda_h \lambda_l \right\}. \qquad \text{(I)}$$

6.2.6. Methods of General Theory of Stochastic Discrete Linear Systems

We present Eq. (5.3.76) for discrete linear system (6.1.36) with the account of the independence of Y_k and V_k in the form

$$g_{k+1}(\lambda) = \exp\left\{ i\lambda^T a_{0k} \right\} h_k(b_k^T \lambda) g_k(a_k^T \lambda), \qquad (6.2.36)$$

While solving the equation we obtain the explicit formula for an one-dimensional characteristic function:

$$g_{k+1}(\lambda) = g_1(a_1^T \ldots a_k^T \lambda) \exp\left\{ i\lambda^T a_{0k} \right\} h_k(b_k^T \lambda)$$

$$\times \prod_{s=1}^{k-1} \exp\left\{ i\lambda^T a_k \ldots a_{s+1} a_{0s} \right\} h_s(b_s^T a_{s+1}^T \ldots a_k^T \lambda). \qquad (6.2.37)$$

Analogously we find a difference equation of the n-dimensional characteristic function:

$$g_{k_1, \ldots, k_n+1}(\lambda_1, \ldots, \lambda_n)$$

$$= \exp\left\{ i\lambda_n^T a_{0k_n} \right\} h_{k_n}(b_{k_n}^T \lambda_n) g_{k_1, \ldots, k_n}(\lambda_1, \ldots, \lambda_{n-1}, a_{k_n}^T \lambda_n). \qquad (6.2.38)$$

As solving Eq. (6.2.38) at the initial conditions (5.3.78) we get the recurrent formula for the multi-dimensional distributions:

$$g_{k_1, \ldots, k_n}(\lambda_1, \ldots, \lambda_n) = g_{k_1, \ldots, k_{n-1}}(\lambda_1, \ldots, \lambda_{n-1} + a_{k_{n-1}}^T, a_{k_1}^T \ldots a_{k_n}^T \lambda_n)$$

$$\times \exp\left\{ i\lambda_n^T a_{0k_n} \right\} h_{k_n}(b_{k_n}^T \lambda_n)$$

$$\times \prod_{s=1}^{k_n-1} \exp\left\{ i\lambda_n^T a_{k_n} \ldots a_{s+1} a_{0s} \right\} h_s(b_s^T a_{s+1}^T \ldots a_{k_n}^T \lambda_n). \qquad (6.2.39)$$

Thus we have the following statements.

Theorem 6.2.3. *Finite formula (6.2.37) and recurrent formula (6.2.39) give the explicit expressions for all multi-dimensional characteristic functions of discrete linear system (6.1.29).*

Corollary 6.2.3. *At the normal distributions of the variables V_k and Y_1 Eqs. (6.1.37)–(6.1.39) completely determine all multi-dimensional normal distributions of the sequence $\{Y_k\}$.*

Theorem 6.2.4. *If there exists a stationary process in a stationary discrete system then it is stationary in the strict sense. The one-dimensional distribution is determined by Eqs. (5.3.76) at $g_1(\lambda; t)$ $= g_1(\lambda)$.*

6.2.7. Methods of Stochastic Infinite-Dimensional Linear Systems Theory

For the stochastic linear differential system in the separable H-space $\mathcal{Y}$:

$$dY = (aY + a_0)dt + bdW \qquad (6.2.40)$$

equation for the one-dimensional characteristic functional of $Y = Y_t$

$$g = g_1(\lambda; t) = E \exp\left[i\lambda, Y(t)\right] \qquad (6.2.41)$$

may be presented in the following form:

$$\frac{\partial g}{\partial t} = (\lambda, aD_\lambda g) + \left[i(\lambda, a_0) - \frac{1}{2}(\lambda, \sigma\lambda) \right] g. \qquad (6.2.42)$$

Recall that in (6.2.40)–(6.2.42) λ is an element of the space $\mathcal{Y}$, $W = W(t)$ is the Wiener process with the values in the separable H-space $\mathcal{W}$, the covariance operator whose values at each t are determined by formula (5.9.3) and its intensity ν at each t is the trace-class self-adjoint operator; $a = A + f_1$ where A is the closed linear operator which satisfies conditions (iii) of Subsection 5.9.1, $f_1 = f_1(t)$ is the bounded linear operator at each t; $a_0 = a_0(t)$ is some element of the H-space $\mathcal{Y}$; (λ, a_0) is the scalar product of the elements λ and a_0 from $\mathcal{Y}$; $b = b(t)$ is the operator mapping $\mathcal{W}$ into $\mathcal{Y}$; $\sigma = b\nu b^*$ is the operator which satisfies conditions (iv) of Subsection 5.9.1. Condition (5.9.10) serves as the initial condition for Eq. (6.2.42).

In the case when the operators a_0, a_1 and σ do not depend on time supposing $\partial g/\partial t = 0$ we come to the following equation for determining the stationary process in Eq. (6.2.40):

$$(\lambda, aD_\lambda g) + \left[i(\lambda, a_0) - \frac{1}{2}(\lambda, \sigma\lambda) \right] g = 0. \qquad (6.2.43)$$

Theorem 6.2.4. *Under the conditions (i)–(iv) of Subsection 5.9.1 the one-dimensional characteristic functional (6.2.41) of the process Y_t in stochastic linear differential system (6.2.40) is defined by Eq. (6.2.42).*

Corollary 6.2.4. *Under conditions of* Theorem 6.2.4 *in stationary system* (6.2.40) *the one-dimensional characteristic functional of the strictly stationary process is defined by* Eq. (6.2.43).

E x a m p l e 6.2.5. Let us consider a linear stochastic differential system which is described by the following stochastic wave equation relative to $U = U(x, t)$:

$$\frac{\partial^2 U}{\partial t^2} = c^2 \frac{\partial^2 U}{\partial x^2} + 2h \frac{\partial U}{\partial t} + V \tag{I}$$

with the boundary conditions $U(0, t) = U(2l, t) = 0$. Here $V = V(x, t)$ is a normally distributed white noise of the intensity $\nu = \nu(x_1, x_2, t)$; c, h are the constant coefficients. After putting $Y_1 = U$, $Y_2 = (\partial/\partial t)U$, $Y = [Y_1 Y_2]^T$,

$$aY = \begin{bmatrix} Y_2 \\ c^2 \partial Y_1/\partial x_2 + 2hY_2 \end{bmatrix}, \quad b = \begin{bmatrix} 0 \\ 1 \end{bmatrix}, \tag{II}$$

we present Eq. (6.2.42) for the one-dimensional characteristic fucntional $g = g(\lambda_1, \lambda_2; t) = \exp\{i(\lambda_1, Y_1) + i(\lambda_2, Y_2)\}$ in the form

$$\frac{\partial g}{\partial t} = -\frac{1}{2}(\lambda_2, \nu\lambda_2)g + iE\left\{\left[(\lambda_1, Y_2) + \left(\lambda_2, c^2\frac{\partial^2 Y_1}{\partial x^2}\right)\right.\right.$$

$$\left.\left. + 2h(\lambda_2, Y_2)\right]e^{i(\lambda_1, Y_1) + i(\lambda_2, Y_2)}\right\}, \tag{III}$$

where the scalar product is given by the formula $(\varphi, \psi) = \int\limits_0^{2l} \varphi(x)\psi(x)dx$.

As it follows from Eq. (5.9.9) the multi-dimensional characteristic functionals g_n which are determined by (5.9.2) $(n = 2, 3, \ldots)$ satisfy Eq. (6.2.42). Here (5.9.10) and (5.9.11) will be the initial conditions.

6.3. Methods of Linear Stochastic Systems Theory based on Canonical Expansions

6.3.1. Introductory Remarks

In elementary theory of the random processes the method of canonical expansions is used generally for the correlation analysis and simulation of the nonstationary processes which are obtained as a result of linear (integral, differential, discrete and etc) transformations. Hence we distinguish two methods.

According to the first one the canonical expansions of the random input and weighting (transfer) functions of a system are supposed to be given. Let the random input $X(t)$ and its covariance function $K_x(t,t')$ of an one-dimensional system be given by a canonical expansion of the form (3.9.6) and (3.9.7):

$$X(t) = m_x(t) + \sum_{\nu=1}^{\infty} V_\nu x_\nu(t), \quad K_x(t,t') = \sum_{\nu=1}^{\infty} D_\nu x_\nu(t)\overline{x_\nu(t')}. \quad (6.3.1)$$

For linear integral transformation (1.1.13)

$$Y(t) = \int_{t_0}^{t} g(t,\tau)X(\tau)d\tau, \quad (6.3.2)$$

where $g(t,\tau)$ is a weighting function of a system the output $Y(t)$ and its covariance function will be expressed in terms of the canonical expansions:

$$Y(t) = m_y(t) + \sum_{\nu=1}^{\infty} V_\nu y_\nu(t), \quad K_y(t,t') = \sum_{\nu=1}^{\infty} D_\nu y_\nu(t)\overline{y_\nu(t')}, \quad (6.3.3)$$

and

$$m_y(t) = \int_{t_0}^{t} g(t,\tau)m_x(\tau)d\tau, \quad y_\nu(t) = \int_{t_0}^{t} g(t,\tau)x_\nu(\tau)d\tau \quad (\nu = 1,2\ldots)$$

$$(6.3.4)$$

the second method supposes the knowledge about the canonical expansions of a random input (6.3.1) and the differential difference equation which describe the system. In this case it is necessary to set up and solve the correspondent linear differential or difference equations for the expectations and the coordinate functions and later to use formulae (6.3.3).

E x a m p l e 6.3.1. Find the variance of the output in the system

$$T\dot{Y} + Y = X, \quad (I)$$

on whose the white noise input acts with constant intensity ν beginning with $t_0 = 0$. The system was at rest till the instant $t_0 = 0$. The weighting function of the system $g(t,\tau)$ was determined in Example 1.1.6. Therefore we have the following exact solution for the variance:

$$D_y^{\text{ES}}(t) = \frac{\nu}{T^2} \int_0^t \int_0^t e^{-(2t-\tau-\tau')/T} \delta(\tau-\tau')d\tau d\tau' = \frac{\nu}{2T}\left(1 - e^{-2t/T}\right).$$

$$\text{(II)}$$

Hence at $t \to \infty$ we find $D_y^{\text{ES}}(\infty) = \nu/2T$.

For applying the first method we use a canonical expansion of a white noise at the interval T_1 obtained in Example 3.9.10. Therefore we have for the coordinate functions according to formula (6.3.4):

$$y_\nu(t) = \frac{1}{TT_1}\int_0^t e^{-(t-\tau)/T+i\omega_\nu\tau}d\tau = \frac{e^{i\omega_\nu t} - e^{-t/T}}{T_1(1+iT\omega_\nu)}. \qquad \text{(III)}$$

Substituting this expression of the coordinate functions into formula (6.3.3) at $t = t'$ and accounting that $D_\nu = \nu T_1$ we get

$$D_y(t) = \frac{\nu}{T_1}\sum_{n=-\infty}^{\infty}\frac{\left|e^{i\omega_n t} - e^{-t/T}\right|}{1+T^2\omega_n^2}. \qquad \text{(IV)}$$

The second methods gives for the coordinate functions a linear differential equation

$$T\dot{y}_n + y_n = \frac{1}{T_1}e^{i\omega_n t}. \qquad \text{(V)}$$

While integrating Eq. (V) at zero initial condition we shall obtain formula (III). For the estimate of the accuracy of the presentation of the variance Y by a finite segment of series (IV) we notice that at $t \gg T$ formula (IV) may be approximately substituted by the following one:

$$D_y(t) \approx D_y(\infty) = \frac{\nu}{T_1}\sum_{n=-\infty}^{\infty}\frac{1}{1+(T\omega_n)^2}$$

$$= \frac{\nu}{T_1}\left(1 + 2\sum_{n=1}^{\infty}\frac{T_1^2}{T_1^2 + 4\pi^2 n^2 T^2}\right). \qquad \text{(VI)}$$

In practice we may use formula (VI) at $t \geq 3T$. In particular, at $T_1/T_0 = 4$ while restricting ourselves to the first six terms of the series in formula (VI) we find

$$D_y(t) \approx \frac{\nu}{T_1}\left(1 + \sum_{n=1}^{6}\frac{8}{4+\pi^2 n^2}\right) \approx \frac{1,94}{T_1}\nu. \qquad \text{(VII)}$$

Thus at restricting oneselves to the first six terms in (VI) what corresponds to the restriction in the canonical expansion of a white noise by the first 13 terms (i.e. by 6 harmonics) we assume an error about 3%.

R e m a r k. Example 6.3.1 convinces us in the fact that for the analysis of linear dynamic systems we may practically use the canonical expansions of input random disturbances restricting ourselves by moderate number of the terms even in those cases when it is necessary to take very large number of the terms of these canonical expansions for the presentation of the disturbances themselves with sufficient accuracy. Usually knowing the cut frequency (Subsection 1.1.7) it is possible to estimate approximately what coordinate functions in the canonical expansion of the output variable should be known and what functions can be negleced.

6.3.2. Theory of General Linear Transfomations

Theorem 6.3.1. *Let T be a linear operator mapping the phase space X of the random variable X into some linear space Y, continuous in the weak topologies of the spaces X and Y determined by the sets of the linear functionals F and G respectively and measurable relatively to the σ-algebras of $\mathcal{A}$ and $\mathcal{B}$ generated by the classes of the sets $\{x : (f_1 x , \ldots , f_n x) \in A\}$ and $\{y : (g_1 y , \ldots , g_n y) \in B\}$ at all finite sets $f_1 , \ldots , f_n \in F, g_1 , \ldots , g_n \in G$ and Borel sets A, B of the correspondent spaces. Then the operator T will be $(\mathcal{A}, \mathcal{B})$-measurable*

$\triangleright$ For proving the $(\mathcal{A}, \mathcal{B})$-measurability of the operator T we notice that by virtue of its continuity the mapping gT at any $g \in G$ represents a continuous linear functional on X. As in the weak topology only those linear functionals by which this topology is determined are continuous then $gT \in F$.Consequently, in this case there exists the adjoint operator T^* determined on the whole space G which maps G into F, $T^* g = gT \in F$. Therefore the set $\{y : (g_1 y , \ldots , g_n y) \in B\} \in \mathcal{B}$ serves as the inverse image of the set $\{x : ((T^* g_1)x , \ldots , (T^* g_n)x) \in B\} \in \mathcal{A}$. As the inverse set $\mathcal{A}_T$ of the σ-algebra $\mathcal{B}$ is the minimal σ-algebra which contains all the sets $\{x : ((T^* g_1)x , \ldots , (T^* g_n)x) \in B\}$ then $\mathcal{A}_T \subset \mathcal{A}$ what proves $(\mathcal{A}, \mathcal{B})$-measurability of the operator T. $\triangleleft$

Let us consider a linear transformation of the following form:

$$Y = TX + b , \qquad (6.3.5)$$

where b is some vector of the space Y. Usually it is required to find the expectation m_y and the covariance operator K_y of the random variable Y assuming the expectation m_x and the covariance operator K_x of the random variable X to be given. If the auxiliary H-space (Subsection 3.9.2) is separable then for finding the covariance operator of the random variable Y we may use any canonical expansion of the random variable X:

$$X = m_x + \sum_{\nu=1}^{\infty} V_\nu x_\nu .$$

(6.3.6)

This expansion m.s. converges in the weak topology of the space X and its residual is estimated by formula (Subsection 3.9.3)

$$E \left| f X^0 - \sum_{\nu=1}^{n} V_\nu f x_\nu \right|^2 = E \left| f X^0 \right|^2 - \sum_{\nu=1}^{n} D_\nu \left| f x_\nu \right|^2$$

$$= f K_x f - \sum_{\nu=1}^{n} D_\nu \left| f x_\nu \right|^2 , \quad f \in F ,$$

(6.3.7)

where D_ν is the variance of the random variable V_ν.

It follows from the comparison of (6.3.7) with formula (3.9.18) that the expectation of the modulus square of the residual of the canonical expansion of the variable fX at any $f \in F$ is equal to the correspondent residual of the canonical expansion of the variance of the variable fX.

Substituting formally expansion (6.3.6) into (6.3.5) we get

$$Y = T m_x + \sum_{\nu=1}^{\infty} V_\nu T x_\nu + b$$

(6.3.8)

Here it goes without saying that the expectation m_x and the coordinate vectors x_ν must belong to the domain D_T of the operator T. Using the linearity property of the operator of the expectation and taking into account that the expectations of the variables V_ν are equal to zero we find $m_y = T m_x + b$. Taking into the consideration this fact and putting $y_\nu = T x_\nu$ we rewrite formula (6.3.7) in the form

$$Y = m_y + \sum_{\nu=1}^{\infty} V_\nu y_\nu .$$

(6.3.9)

After obtaining canonical expansion (6.3.5) of the random variable Y we may find its covariance operator by formula (3.9.18).

Theorem 6.3.2. *If the linear operator T is continuous in the weak topologies of the spaces X and Y then canonical expansion (6.3.9) of the* *random variable $Y = TX + b$ m.s. converges in the weak topology of the space Y.*

$\triangleright$ According to (6.3.3) for any $g \in G$ the formula is valid

$$E \left| gY^0 - \sum_{\nu=1}^{n} V_\nu g y_\nu \right|^2 = E \left| gY^0 \right|^2 - \sum_{\nu=1}^{n} D_\nu \left| g y_\nu \right|^2$$

$$= E \left| gTX^0 \right|^2 - \sum_{\nu=1}^{n} D_\nu \left| gT x_\nu \right|^2 .$$

By virtue of the continuity of the operator T there exists the continuous adjoint operator T^* determined on the whole space G and $gTx = (T^* g)x$ for any $x \in X$ and here $T^* g \in F$. Thus using again formula (6.3.7) we get

$$E \left| gY^0 - \sum_{\nu=1}^{n} V_\nu g y_\nu \right|^2 = E \left| (T^* g)X^0 \right|^2 - \sum_{\nu=1}^{n} D_\nu \left| (T^* g)x_\nu \right|^2$$

$$= E \left| (T^* g)X^0 - \sum_{\nu=1}^{n} V_\nu (T^* g)x_\nu \right|^2 .$$

The latter expression tends to zero at $n \to \infty$ because of the weak m.s. convergence of the canonical expansion of the random variable X what proves the weak m.s. convergence of obtained canonical expansion (6.3.7) of the random variable Y. $\triangleleft$

R e m a r k. The proved theorem with some modifications is extended over more large class of the linear operators. Namely as it is easy to see all our arguments remain valid for any linear operator T which has the adjoint operator T^* except that the canonical expansion of the random variables gY will converge not for any functional $g \in G$ but for those functionals $g \in G$ which belong to the domain D_{T^*} of the adjoint operator T^* and are mapped by the operator T^* at the points of the space F. Thus canonical expansion (6.3.9) of the random variable Y in this case m.s. converges only in more weak topology which is determined in the space Y by the set of the linear functionals G' belonging to the intersection of the space G with such part of the domain of the adjoint operator T^* which is mapped by this operator into F. Here the operator T is measurable relatively to the σ-algebra $\mathcal{A}$ of the space X and the σ-algebra $\mathcal{B}'$ of the space Y generated by the class of the sets $\{y : (g_1 y, \ldots, g_n y) \in B\}$ at all finite sets $g_1, \ldots, g_n \in G'$ and all Borel sets B of the correspondent finite-dimensional spaces. It is proved similarly as the $(\mathcal{A}, \mathcal{B})$-measurability of the continuous operator T if we account that $T^* g \in F$ for any $g \in G'$. It goes without saying that this generalization has the sense only in that case when the space G' is sufficiently complete for the fact that from $gy = 0$ for any $g \in G'$ will follow $y = 0$.

E x a m p l e 6.3.2. Let X be a space of the finite-dimensional vector functions of the scalar variable t, $t \in T_1$ which are continuous together with their derivatives till the order N inclusively, F be a space of the linear functionals on X of the form

$$fx = \sum_{p=0}^{N} \sum_{k=1}^{m} f_{pk}^T x^{(p)}(t_k) \,. \tag{I}$$

Let us consider the operator of the differentiation $D = d/dt$. The operator D^l maps the space X into the space Y of the functions which are continuous together with their derivatives till the order $N - l$ inclusively $(l = 1, \ldots, N)$. We assume the set of the linear functionals on Y of the form

$$gy = \sum_{q=0}^{N-l} \sum_{k=1}^{m} g_{qk}^T y^{(q)}(t_k) \tag{II}$$

as the space G. Now we consider the random variable $X(t)$ with the realizations in X. The operator D^l transforms it into the random function $Y(t) = D^l X(t) = X^{(l)}(t)$ with the realizations in the space Y. As

$$gD^l x = \sum_{q=0}^{N-l} \sum_{k=1}^{m} g_{qk}^T x^{(l+q)}(t_k) \,, \tag{III}$$

then the functional gD^l at any $g \in G$ belongs to F. Consequently, the operator D^l is continuous in the weak topologies of the spaces X and Y, and we may use Theorem 6.3.2. The adjoint operator $\left(D^l\right)^*$ in a given case is determined on the whole space G and maps G on the subspace F' of the space F:

$$F' = \left\{ f : fx = \sum_{p=l+1}^{N} \sum_{k=1}^{m} f_{pk}^T x^{(p)}(t_k) \right\} \,. \tag{IV}$$

Therefore after expressing the random function $X(t)$ by any canonical expansion we shall obtain for its derivative of the order l the canonical expansion

$$X^{(l)}(t) = m_x^{(l)}(t) + \sum_{\nu=1}^{\infty} V_\nu x_\nu^{(l)}(t) \,, \quad t \in T_1 \,. \tag{V}$$

E x a m p l e 6.3.3. Let $X(t)$, $t \in T_1$ be a vector random function of the finite-dimensional vector t, X be its phase space, F be a set of all linear functionals on X representable in the form

$$fx = \int f(t)^T x(t)dt \tag{I}$$

for almost all the functions $x(t) \in X$ (relatively to the probability measure μ_x of the random function $X(t)$). Let us consider the linear integral operator $Lx = \int l(s,t)x(t)dt$ which maps X into the space Y of the m-dimensional vector functions of the finite-dimensional vector s, $s \in S$. Here we shall assume that at any fixed $s \in S$ each row of the matrix $l(s,t)$ determines some linear functional from F. We determine on Y a set of the linear functionals G of the form $gy = \sum_{k=1}^{N} g_k^T y(s_k)$.

Then for any $g \in G$ we shall have $gLx = \int \sum_{k=1}^{N} g_k^T l(s_k,t)x(t)dt$. Putting

$$f(t) = \sum_{k=1}^{N} l(s_k,t)^T g_k \,, \tag{II}$$

we reduce this formula to the form (I). Thus in a given case the operator L is continuous in the weak topologies of the spaces X and Y formed by the sets of the linear functionals F and G correspondingly To any functional $g \in G$, i.e. to any finite set of the pairs $\{s_k, g_k\}$ the adjoint operator L^* sets up the functional $f \in F$ which is determined by formulae (I) and (II). Thus to any canonical expansion of the random function $X(t)$ the corresponds canonical expansion

$$Y(s) = m_y(s) + \sum_{\nu=1}^{\infty} V_\nu y_\nu(s) \tag{III}$$

of the random function $Y(s) = LX(t) = \int l(s,t)X(t)dt$ and here

$$m_y(s) = \int l(s,t)m_x(t)dt\,, \quad y_\nu(s) = \int l(s,t)x_\nu(t)dt\,. \tag{IV}$$

Expansion (III) m.s. converges at all $s \in S$. After obtaining canonical expansion (III) of the random function $Y(s)$ we find its covariance function $K_y(s_1, s_2)$.

E x a m p l e 6.3.4. Let us consider a special case of the previous example where $X(t)$ represents a white noise of the intensity $\nu(t)$. In this case we take the space of the main functions as F and use the canonical expansion of the white noise of Example 3.9.10. As it was shown in this case any orthonormal system of the functions $\{\varphi_n(t)\}$ with the weight $\nu(t)$, $\int \nu(t)\varphi_p(t)^T\overline{\varphi_q(t)}dt = \delta_{pq}$ gives the canonical expansion of the white noise $X(t) = \sum\limits_{p=1}^{\infty} V_p\varphi_p(t)$ which weakly m.s. converges in the sense that for any main function $f(t) \in F$ the series $fX = \sum\limits_{p=1}^{\infty} V_p \int f(t)^T \varphi_p(t)dt$ m.s. converges. The expectation of a white noise as usual we assume equal to zero.

Let us consider the random function

$$Y(s) = LX(t) = \int l(s,t)X(t)dt, \tag{I}$$

where each row of the matrix $l(s,t)$ at any fixed $s \in S$ forms the vector function from the space of the main functions. In this case the operator L is continuous in the weak topologies of the spaces X and Y and the stated method gives a canonical expansion of the random function $Y(s)$,

$$Y(s) = \sum_{p=1}^{\infty} V_p \int l(s,t)\varphi_p(t)dt, \tag{II}$$

which m.s. converges at any $s \in S$.

E x a m p l e 6.3.5. Let $X(t)$ be a finite-dimensional vector random function which represents a random variable with the phase space $L_p(T_1, \mathcal{J}, \mu), p \geq 1$. We assume as F the adjoint space $L_q(T_1, \mathcal{J}, \mu), q = p/(p-1)$. Let us consider the linear integral operator

$$y(s) = Lx = \int l(s,t)x(t)\mu(dt), \tag{I}$$

where $l(s,t)$ satisfies the following conditions:

(i) at each $s \in S$ $l(s,t)$ belongs to the space $L_q(T_1, \mathcal{J}, \mu)$;

(ii) at each $t \in T_1$ $l(s,t)$ belongs to the space of the finite-dimensional vector functions $L_m(S, \mathcal{S}, \nu)$ $m \geq 1$;

(iii) for any function $x(t) \in L_p(T_1, \mathcal{J}, \mu)$ the function $y(s) = \int l(s,t)x(t) \times \mu(dt)$ belongs to the space $L_m(S, \mathcal{S}, \nu)$;

(iv) for any function $g(s)$ which belongs to the space $L_n(S, \mathcal{S}, \nu)$, $n = m/(m-1)$ the function $f(t) = \int l(s,t)^T g(s)\nu(ds)$ belongs to the space $L_q(T_1, \mathcal{J}, \mu)$.

In this case for any $x(t) \in L_p(T_1, \mathcal{J}, \mu)$ and any $g(s) \in L_n(S, \mathcal{S}, \nu)$

$$gy = \int g(s)^T y(s)\nu(ds) = \int g(s)^T \nu(ds) \int l(s,t)x(t)\mu(dt)$$

$$= \int \left\{ \int g(s)^T l(s,t)\nu(ds) \right\} x(t)\mu(dt) = \int f(t)^T x(t)\mu(dt). \qquad \text{(II)}$$

It is clear from (I) and (II) that $gL \in F$, and consequently, the operator L from $L_p(T_1, \mathcal{J}, \mu)$ into $L_m(S, \mathcal{S}, \nu)$ is continuous in the weak topologies determined by the correspondent dual spaces $F = L_q(T_1, \mathcal{J}, \mu)$ and $G = L_n(S, \mathcal{S}, \nu)$. Consequently, any canonical expansion of the random function $X(t)$ which is weakly m.s. convergent in $L_p(T_1, \mathcal{J}, \mu)$ gives canonical expansion (III) of Example 6.3.4 of the random function which weakly m.s. converges in $L_m(S, \mathcal{S}, \nu)$,

$$Y(s) = LX = \int l(s,t)X(t)\mu(dt), \qquad \text{(III)}$$

where $m_y(s) = \int l(s,t)m(t)\mu(dt)$, $y_\nu(s) = \int l(s,t)x_\nu(t)\mu(dt)$.

6.3.3. Assignment of Probability Measures by Canonical Expansions

As we have already seen in Subsection 3.9.2 to each vector f from the H-space H_x connected with a given random variable X corresponds a scalar random variable (more exactly a class of the equivalent random variables) fX. The measurability of the function fx of the variable x for f belonging to the factor-space F/N is evident. The measurability of fx for $f \in H_x$, $f \bar{\in} F/N$ follows from the fact that fx as the limit of the sequence $\{f_n x\}$, $f_n \in F/N$ of the space $L_2(X, \mathcal{A}, \mu_x)$ also belongs to $L_2(X, \mathcal{A}, \mu_x)$ by virtue of the completeness of $L_2(X, \mathcal{A}, \mu_x)$. Therefore *the probability measure $V = \{V_\nu\}$ of the coefficients $V_\nu = f_\nu X^0$ of canonical expansion (6.3.6) of the random variable X is completely determined by the probability measure μ_v of the random vector V on the cylindrical sets of its phase space,*

$$\mu_v(\{v : (v_{\nu_1}, \ldots, v_{\nu_n}) \in B\})$$

$$= \mu_x(\{x : (f_{\nu_1} x^0, \ldots, f_{\nu_n} x^0) \in B\}), \quad x^0 = x - m_x. \qquad (6.3.10)$$

Knowing the measure μ_v on the cylindrical sets we may continue it uniquely by the theorem about the measure extension over the σ-algebra $\mathcal{B}^N$ of the phase set V. Here the σ-algebra of Borel sets of the complex plane is designated as $\mathcal{B}$, and as N is the set of positive integers..

Our interest is in the inverse problem: *knowing the probability measure μ_v of the random vector V find the probability measure μ_x of the random variable X.*

▷ For solving this problem at first we establish the measurability of the linear functional $\varphi_f v = \sum_{\nu=1}^{\infty} v_\nu(f x_\nu)$ relatively to the σ-algebra $\mathcal{B}^N$ at any $f \in H_x$. But the measurability follows immediately from m.s. convergence of the series

$$\varphi_f V = \sum_{\nu=1}^{\infty} V_\nu(f x_\nu) = f X^0 \tag{6.3.11}$$

at any $f \in H_x$ (Subsection 3.9.3). In reality from m.s. convergence of this series follows the existence of such subsequence $\{n_p\}$ that the sequence of the random variables $\left\{\varphi_f^p V\right\}$ where $\varphi_f^p V = \sum_{\nu=1}^{n_p} V_\nu(f x_\nu)$, $(p = 1, 2, \ldots)$ a.s. converges to $\varphi_f V$. In other words, the sequence of measurable linear functionals $\left\{\varphi_f^p v\right\}$ converges almost everywhere (relatively to the measure μ_v) to the linear functional $\varphi_f v$. Consequently, the functional $\varphi_f v$ is measurable. ◁

▷ Formula (6.3.6) which determines a canonical expansion of the random variable X establishes the mapping of the phase space of the random vector V into the space X. As the inverse image of the set $\{x : (g_1 x, \ldots, g_n x) \in B\}$ at any $g_1, \ldots, g_n \in H_x$ and any Borel set B of the n-dimensional space at this mapping by virtue of (6.3.11) serves the set $\{v : (g_1 m_x + \varphi_{g_1} v, \ldots, g_n m_x + \varphi_{g_n} v) \in B\}$. This set belongs to the σ-algebra $\mathcal{B}^N$ because of the measurability of the linear functionals $\varphi_{g_1}, \ldots, \varphi_{g_n}$. Knowing the probability measure μ_v of the random vector V we may determine the probability measure μ_x of the random variable X on all the sets of the type $\{x : (g_1 x, \ldots, g_n x) \in B\}$:

$$\mu_x(\{x : (g_1 x, \ldots, g_n x) \in B\})$$

$$= \mu_v(\{v : (g_1 m_x + \varphi_{g_1} v, \ldots, g_n m_x + \varphi_{g_n} v) \in B\}). \tag{6.3.12}$$

The measure μ_x is uniquely expended over the σ-algebra $\mathcal{A}$ generated by the class of all such sets ◁

Thus the following statement is valid.

Theorem 6.3.3. *In the case of the separable space H_x the probability measure of the random variable X is uniquely determined by the probability measure of the coefficients vector V of its canonical expansion.*

E x a m p l e 6.3.6. Let X be a normally distributed random variable with the characteristic functional (Subsection 3.7.1)

$$g_x(f) = \exp\left\{ifm_x - \frac{1}{2}fK_x f\right\}. \tag{I}$$

In the phase space of the random vector $V = \{V_\nu\}$, where $V_\nu = f_\nu X^0$ we determine a set of the linear functionals H of the form $hv = \sum_{\nu=1}^{\infty} h_\nu v_\nu$ and $\sum_{\nu=1}^{\infty} D_\nu h_\nu^2 < \infty$. Then the characteristic functional of the random vector V will be determined for any $h \in H$ by formula

$$
\begin{aligned}
g_v(h) &= E e^{ihV} = E e^{\,i \sum_{\nu=1}^{\infty} h_\nu V_\nu} \\
&= E e^{\,i \left(\sum_{\nu=1}^{\infty} h_\nu f_\nu \right) X^0} = e^{\,-i \left(\sum_{\nu=1}^{\infty} h_\nu f_\nu \right) m_x} \cdot g_x \left(\sum_{\nu=1}^{\infty} h_\nu f_\nu \right) .
\end{aligned}
\tag{II}
$$

After substituting expression (I) for g_x into (II) and taking into consideration that $f_\nu K_x f_\mu = (f_\nu, f_\mu) = D_\nu \delta_{\nu\mu}$ (Subsection 3.9.3) we get

$$
g_v(h) = \exp \left\{ -\frac{1}{2} \sum_{\nu=1}^{\infty} D_\nu h_\nu^2 \right\} .
\tag{III}
$$

The formula shows that the random vector V is normally distributed and its components are independent.

E x a m p l e 6.3.7. Suppose that the random vector V of the coefficients of canonical expansion (6.3.5) of the random variables X is normally distributed (and consequently, its components are independent). Its probability measure μ_v on the measurable cylindrical sets is expressed by formula

$$
\mu_v(\{v : \{v_{\nu_1}, \ldots, v_{\nu_n}\} \in B\})
$$

$$
= [(2\pi)^n D_{\nu_1}, \ldots, D_{\nu_n}]^{-1/2} \int_{B} \overset{\cdots}{\int} \exp \left\{ -\frac{1}{2} \sum_{p=1}^{\infty} \frac{z_p^2}{D_{\nu_p}} \right\} dz_1, \ldots, dz_n .
\tag{I}
$$

On the sets of the type $\left\{ v : (h^{(1)}v, \ldots, h^{(n)}v) \in B \right\}$ at any linearly independent $h^{(1)}, \ldots, h^{(n)} \in H$ the measure μ_v is determined by formula

$$
\mu_v(\left\{ v : \left\{ h^{(1)}v, \ldots, h^{(n)}v \right\} \in B \right\})
$$

$$
= [(2\pi)^n \, |C|\,]^{-1/2} \int_{B} \exp \left\{ -\frac{1}{2} z^T C^{-1} z \right\} dz ,
\tag{II}
$$

where C is a matrix with the elements $c_{pq} = \sum\limits_{\nu=1}^{\infty} D_\nu h_\nu^{(p)} h_\nu^{(q)}$ $(p, q = 1, \ldots, n)$.

Using formula (6.3.12) we find the probability measure of the random variable X:

$$\mu_x(\{x \ : \ (g_1 x, \ldots, g_n x) \in B\})$$

$$= [(2\pi)^n \ |D| \,]^{-1/2} \int\limits_{B-a} \exp\left\{-\frac{1}{2} z^T D^{-1} z\right\} dz, \qquad \text{(III)}$$

where a is a vector with the components $a_p = g_p m_x$ $(p = 1, \ldots, n)$; D is a matrix whose elements with the account that $(\varphi_{g_p})_\nu = g_p x_\nu$ are determined by formula $d_{pq} = \sum\limits_{\nu=1}^{\infty} D_\nu (g_p x_\nu) g_q x_\nu$ $(p, q = 1, \ldots, n)$. After determining the measure μ_x on the sets of such a type it is uniquely extended over the σ-algebra $\mathcal{A}$ generated by these sets.

R e m a r k. In a given case the measure μ_x may be determined in another way. Namely, knowing the characteristic functional of the random vector V we may find the characteristic functional of the random variable X:

$$g_x(f) = E e^{ifX} = e^{ifm_x} E e^{ifX^0} = e^{ifm_x} E \exp\left\{i \sum\limits_{\nu=1}^{\infty} V_\nu(f x_\nu)\right\}. \qquad \text{(IV)}$$

The latter expectation represents the value of the characteristic functional of the random vector V at $h_\nu = f x_\nu$ $(\nu = 1, 2, \ldots)$. As a result we obtain

$$g_x(f) = e^{ifm_x} \exp\left\{-\frac{1}{2} \sum\limits_{\nu=1}^{\infty} D_\nu(f x_\nu)^2\right\}. \qquad \text{(V)}$$

E x a m p l e 6.3.8. Let $X(t)$, $t \in T_1$ be the r-dimensional normally distributed vector random function, F be a set of all linear functionals of the form $fx = \sum\limits_{h=1}^{N} f_h^T x(t_h)$, $t_1, \ldots, t_N \in T_1$. According to the results obtained in the previous example the multi-dimensional distributions of the random function $X(t)$ are determined by formula

$$\mu_{t_1, \ldots, t_n}(\{x_1, \ldots, x_n \ : \ (x_1, \ldots, x_n) \in B\})$$

$$= [(2\pi)^{nr} \ |D| \,]^{-1/2} \int\limits_{B-a} \exp\left\{-\frac{1}{2} z^T Dz\right\} dz. \qquad \text{(I)}$$

Here a is the nr-dimensional vector with the components $a_p = m_{p-[(p-1)/r]r}$ $\times \left(t_{[(p-1)/r]+1}\right)$ $(p = 1, \ldots, nr)$, D is the matrix whose elements are expressed in terms of the coordinate functions $x_\nu(t)$ of the canonical expansion of the random functions $X(t)$ by formula

$$d_{pq} = \sum_{\nu=1}^{\infty} D_\nu a_\nu^p a_\nu^q \,, \quad a_\nu^s = x_{\nu, s - [(s-1)/r]r}\left(t_{[(s-1)/r]+1}\right) \quad (s = 1, \ldots, nr).$$

(II)

Formula (V) of Example 6.3.7 gives the following expression of the characteristic functional of the random function $X(t)$:

$$g_x(f) = \exp\left\{ i \int f(t)^T m_x(t) dt - \frac{1}{2} \sum_{\nu=1}^{\infty} D_\nu \left[\int f(t)^T x_\nu(t) dt \right]^2 \right\}. \quad \text{(III)}$$

Putting here, in particular, $f(t) = \sum\limits_{l=1}^{n} \sum\limits_{m=1}^{r} d_{(l-1)r+m} e_m \delta(t - t_l)$, where e_1, $\ldots, e_r$ are the unit vectors of coordinate axes of the r-dimensional space we det the n-dimensional characteristic function of the random function $X(t)$.

6.3.4. Solution of Equations by Canonical Expansions

The canonical expansions allow to find the solutions of the linear operator equations of the type

$$LK_x = \Phi, \qquad (6.3.13)$$

where L is an unknown linear operator, K_x is a covariance operator of some random variable X, Φ is known linearly-adjoint operator. In particular, the problem of finding the regression for the random variables X and Y with the normal joint distribution is reduced to this equation. In the latter case $\Phi = K_{yx}$.

Let X be a phase space of the random variable X, F is a space of the linear functionals by means of which the weak topology and the σ-algebra $\mathcal{A}$ in the space X are determined, Y is a space in which the operator Φ maps F, G is a space of the linear functionals by means of which the weak topology and the σ-algebra $\mathcal{B}$ in the space Y are determined. Usually in practice it is required to find not any solution of Eq. (6.3.13) (it may have not unique solution) but only such solution L that the random variable gLX will have the finite second order moment at any $g \in G$. Besides that it is usually required that the operator L would be permutable with the operator of an expectation.

Theorem 6.3.4. *If the auxiliary H-space H_x (Subsection 3.9.2) connected with the random variable X is separable then any canonical* *expansion of the variable X used for the representation of the solution Eq. (6.3.13) in all the cases when it exists.*

▷ Firstly, we define the necessary conditions for the existance of the solution Eq. (6.3.13). Let L be a required solution of Eq. (6.3.13). It represents the $(\mathcal{A}, \mathcal{B})$-measurable linear operator as only in this case LX will be a random variable in $(Y, \mathcal{B})$. Let $\{f_\nu\}$ be the orthogonal basis in H_x, $\| f_\nu \|^2 = D_\nu$. Then the random variable X may be presented by canonical expansion (6.3.6) where $V_\nu = f_\nu X^0$ are uncorrelated random variables with zero expectation and the variances which are equal to the correspondent numbers D_ν, and x_ν are the coordinate vectors satisfying the conditions (Subsection 3.9.3):

$$x_\nu = D_\nu^{-1} K_x f_\nu , \quad f_\nu x_\mu = \delta_{\nu\mu} . \tag{6.3.14}$$

From the fact that the random variable gLX has the finite second order moment at any $g \in G$ and the operator L is permutable with the operator of an expectation follows the convergence of the series

$$\sum_{\nu=1}^{\infty} D_\nu \, |gLx_\nu|^2 \tag{6.3.15}$$

at any $g \in G$. As by virtue of the first formula of (6.3.14)

$$E\left[gLX^0 - \sum_{\nu=1}^{n} V_\nu(gLx_\nu) \right] \overline{V}_\mu = gLEX^0\overline{V}_\mu - D_\mu(gLx_\mu)$$

$$= gLEX^0(\overline{f_\mu X^0}) - D_\mu(gLx_\mu) = gLK_x f_\mu - D_\mu gLx_\mu = 0 \ (\mu = 1, \ldots, n),$$

then $E\left| gLX^0 - \sum_{\nu=1}^{n} V_\nu(gLx_\nu) \right|^2 = E\left| gLX^0 \right|^2 - \sum_{\nu=1}^{n} D_\nu \, |gLx_\nu|^2$. Hence it follows that series (6.3.15) converges at any $g \in G$. Further, it follows from the canonical expansion of the covariance operator: $K_x f = \sum_{\nu=1}^{\infty} D_\nu x_\nu (\overline{f x_\nu})$, $f \in H_x$, and from the convergence of series (6.3.15) that $LK_x f = \sum_{\nu=1}^{\infty} D_\nu (Lx_\nu)(\overline{f x_\nu})$, where the series weakly converges at any $f \in H_x$ in the weak topology of the space Y. Substituting this expression into (6.3.13) we get $\sum_{\nu=1}^{\infty} D_\nu (Lx_\nu)(\overline{f x_\nu}) = \Phi f$, $f \in H_x$. From here and from the convergence of series (6.3.15) it follows that Eq. (6.3.15) may have the required solution only in that case when the operator Φ is representable by the expansion

$$\Phi = \sum_{\nu=1}^{\infty} D_\nu y_\nu (\overline{\cdot x_\nu}) , \tag{6.3.16}$$

where y_ν are the vectors of the space Y which satisfy the condition

$$\sum_{\nu=1}^{\infty} D_\nu \left|g y_\nu\right|^2 < \infty \quad \text{at any} \quad g \in G. \tag{6.3.17}$$

For determining the coefficients y_ν we replace the index of the summation ν in (6.3.16) by the index μ and apply the operator Φ to the vector f_ν. Then by virtue of (6.3.14) we obtain $D_\nu y_\nu = \Phi f_\nu$. Consequently,

$$y_\nu = D_\nu^{-1} \Phi f_\nu \quad (\nu = 1, 2, \ldots). \vartriangleleft \tag{6.3.18}$$

$\vartriangleright$ The operator L in (6.3.13) is determined only at the domain Δ_{K_x} of the covariance operator K_x. Therefore the variable LX has the sense only at $m_x \in \Delta_{K_x}$. In this case there exists such a vector $f_0 \in H_x$ that $m_x = K_x f_0$. And as $\{f_\nu\}$ is the basis in H_x then $f_0 = \sum c_\mu f_\mu$ and by virtue of (6.3.14) $m_x = \sum_{\mu=1}^{\infty} D_\mu c_\mu x_\mu$ and this series converges in the weak topology of the space X. In order to obtain the explicit expression for the coefficients c_ν we apply to the previous equality for the functional f_ν. Then taking into accounting the second relation of (6.3.14) we get $D_\nu c_\nu = f_\nu m_x$, $(\nu = 1, 2, \ldots)$ and consequently,

$$m_x = \sum_{\mu=1}^{\infty} (f_\nu m_x) x_\nu. \tag{6.3.19}$$

Finally, noticing that on the basis of the first relation of (6.3.14) $f_\nu m_x = f_\nu K_x f_0 = \overline{f_0 K_x f_\nu} = D_\nu \overline{f_0 x_\nu}$ and accounting the convergence of the series $\sum D_\nu \left|f x_\nu\right|^2$ at any $f \in H_x$ we come to the conclusion that the following condition must be filfilled:

$$\sum_{\nu=1}^{\infty} D_\nu^{-1} \left|f_\nu m_x\right|^2 < \infty. \vartriangleleft \tag{6.3.20}$$

Theorem 6.3.5. *For the existence of the linear operator L which satisfies Eq. (6.3.1), is permutable with the operator of the expectation and gives the finite second order moment of the random variable gLX, conditions* (6.3.16), (6.3.17), (6.3.19, (6.3.20) *are necessary.*

We prove that these conditions are also sufficient for the existence of the solution of Eq. (6.3.13).

$\triangleright$ For this purpose we shall find the operator L in the form of the series

$$L = \sum_{\mu=1}^{\infty} \eta_\mu f_\mu \,, \tag{6.3.21}$$

where η_μ are yet unknown vectors. Then taking into the consideration (6.3.7) for any $f \in H_x$ we get $LK_x f = \sum_{\mu=1}^{\infty} \eta_\mu f_\mu K_x f = \sum_{\mu=1}^{\infty} \eta_\mu \overline{f K_x f_\mu}$
$= \sum_{\mu=1}^{\infty} D_\mu \eta_\mu (\overline{f x_\mu})$. In order the operator L would satisfy Eq. (6.3.13) it is neces-
sary that at any $f \in H_x$ the following equality will be fulfilled: Φf
$= \sum_{\mu=1}^{\infty} D_\mu \eta_\mu (\overline{f x_\mu})$. After comparing this expansion with (6.3.16) we obtain

$$\sum_{\mu=1}^{\infty} D_\mu \eta_\mu (\overline{f x_\mu}) = \sum_{\mu=1}^{\infty} D_\mu y_\mu (\overline{f x_\mu}) \text{ at any } f \in H_x. \text{ Putting here } f = f_\nu$$

we find $\eta_\nu = y_\nu \ (\nu = 1, 2, \ldots)$. Substituting this expression into (6.3.21) and accounting (6.3.18) we obtain a fomal solution of Eq. (6.3.13) in the form of series

$$L = \sum_{\nu=1}^{\infty} y_\nu f_\nu = \sum_{\nu=1}^{\infty} D_\nu^{-1} (\Phi f_\nu) f_\nu \,. \tag{6.3.22}$$

Let us take a look in what sense expansion (6.3.22) determines the solution of Eq. (6.3.13) which possesses the required properties. For this purpose we consider the random variable $Y = LX = \sum_{\nu=1}^{\infty} y_\nu f_\nu X = \sum_{\nu=1}^{\infty} y_\nu (f_\nu m_x + V_\nu)$. This expansion weakly m.s. converges as it follows from (6.3.17) and (6.3.20) that at any $\varepsilon > 0$, $g \in G$, for all p and at sufficiently large n

$$E \left| \sum_{\nu=n}^{n+p} (g y_\nu)(f_\nu m_x + V_\nu) \right|^2 = \left| \sum_{\nu=n}^{n+p} (g y_\nu)(f_\nu m_x) \right|^2 + \sum_{\nu=n}^{n+p} D_\nu \, |g y_\nu|^2$$

$$\leq \sum_{\nu=n}^{n+p} D_\nu \, |g y_\nu|^2 \cdot \sum_{\nu=n}^{n+p} D_\nu^{-1} \, |f_\nu m_x|^2 + \sum_{\nu=n}^{n+p} D_\nu \, |g y_\nu|^2 < \varepsilon \,.$$

Hence and from (6.3.22) on the basis of stated in Subsection 3.9.2 it follows that at any $g \in G, f \in H_x$

$$E(gY) = E(gLX) = \lim_{n \to \infty} \sum_{\nu=1}^{n} (g y_\nu)(f_\nu m_x) = g \sum_{\nu=1}^{\infty} y_\nu f_\nu m_x = g L m_x \,,$$

$$E(gLX)(\overline{fX^0}) = \lim_{n \to \infty} \sum_{\nu=1}^{n} (gy_\nu)EV_\nu(\overline{fX^0})) = \sum_{\nu=1}^{\infty} (gy_\nu)E(f_\nu X^0)(\overline{fX^0})$$

$$= g \sum_{\nu=1}^{\infty} y_\nu f_\nu K_x f = gLK_x f = gLEX^0(\overline{fX^0}) = gLEX(\overline{fX^0}),$$

$$E\,|gY|^2 = E\,|gLX|^2 = \lim_{n \to \infty} \left\{ \left| \sum_{\nu=1}^{n} (gy_\nu)(f_\nu m_x) \right|^2 + \sum_{\nu=1}^{n} D_\nu\,|gy_\nu|^2 \right\}$$

$$= \left| \sum_{\nu=1}^{\infty} (gy_\nu)(f_\nu m_x) \right|^2 + \sum_{\nu=1}^{\infty} D_\nu\,|gy_\nu|^2$$

$$\leq \sum_{\nu=1}^{\infty} D_\nu\,|gy_\nu|^2 \left\{ \sum_{\nu=1}^{\infty} D_\nu^{-1}\,|f_\nu m_x|^2 + 1 \right\} < \infty.$$

Thus $E(LX) = Lm_x$, $E(LX)(\overline{fX^0}) = LEX(\overline{fX^0})$, i.e. the operator L is permutable with the operator of an expectation and the random variable $gY = gLX$ has the finite second order moment at any $g \in G$. $\triangleleft$

While fulfilling the conditions of Theorem 6.3.4 formula (6.3.22) determines the solution of Eq. (6.3.13) which satisfies the required conditions in such a sense that the finite segments of series (6.3.22) $L_n = \sum\limits_{\nu=1}^{n} y_\nu f_\nu = \sum\limits_{\nu=1}^{n} D_\nu^{-1}(\Phi f_\nu)f_\nu$ form the sequence of the random variables $\{L_n X\}$ weakly m.s. convergent to the random variable $Y = LX$ which has the covariance operator K_y determined on the whole space G.

It may be said that if conditions (6.3.16) and (6.3.17) are fulfilled and the condition $m_x \in \Delta_{K_x}$ and conditions (6.3.19) and (6.3.20) which follow from it are not fulfilled then the statemnts of Theorem 6.3.5 are valid for centered random variables X^0, $Y^0 = LX^0$ and we have the following statement.

Theorem 6.3.6. *Conditions (6.3.15) and (6.3.17) are necessary and sufficient for the existence of such a solution of* Eq. *(6.3.13) that the random variable $Y^0 = LX^0$ has the covariance operator K_y determined on the whole space G and the operator L is permutable with the operator of the expectation: $E(LX^0)(\overline{fX^0}) = LEX^0(\overline{fX^0})$ for any $f \in H_x$.*

In particular, if X and Y are two random variables with a normal joint distribution then putting in (6.3.22) $\Phi = K_{yx}$ we shall get the following expression for an operator of the regression of the random variable Y over X:

$$L = \sum_{\nu=1}^{\infty} D_{\nu}^{-1}(K_{yx}f_{\nu})f_{\nu} \, . \tag{6.3.23}$$

After putting this expression into the first formula of (3.7.9) with the change of x by the random variable X we shall obtain the following expression for the conditional expectation of the random variable Y relatively to X:

$$m_{y|X} = E(Y|X) = m_y + LX^0$$

$$= m_y + \sum_{\nu=1}^{\infty} D_{\nu}^{-1}(K_{yx}f_{\nu})f_{\nu}X^0 = m_y + \sum_{\nu=1}^{\infty} D_{\nu}^{-1}V_{\nu}(K_{yx}f_{\nu}) \, .$$

$$\tag{6.3.24}$$

This formula gives a canonical expansion of the random variable $m_{y|X}$ (regression Y on X) with the coordinate functions $y_{\nu} = D_{\nu}^{-1}K_{yx}f_{\nu}$.

E x a m p l e 6.3.9. Let $X(t)$, $t \in T_1$ be n-dimensional vector random function of the finite-dimensional vector t; Y be a space of m-dimensional vector functions of the variable $s \in S$. Let us consider an operator equation

$$L_t K_x(t,\tau) = \varphi(s,\tau) \quad (\tau \in T_1, s \in S), \tag{I}$$

where L_t is an unknown operator which acts over the functions of the variable t; $\varphi(s,\tau)$ is $m \times n$-matrix whose each column at any fixed $\tau \in T$ represents a vector function which belongs to the space Y. In a given case the space H_x is a set of all vector functions $f(t)$ (including also the generalized functions) for which (Subsection 3.9.3) $\iint f(t_1)^T K_x(t_1,t_2)\overline{f(t_2)}dt_1 dt_2 < \infty$. This space is separable and any orthogonal basis in it $\{f_{\nu}(t)\}$ determines a canonical expansion of the random function $X(t)$. As usual we assume as the space G the set of all linear functionals on Y of the form $gy = \sum_{p=1}^{N} g_p^T y_p(s)$. Formulae (6.3.16) and (6.3.18) in a given case give:

$$\Phi f = \int \varphi(s,\tau)\overline{f(\tau)}d\tau = \sum_{\nu=1}^{\infty} D_{\nu}y_{\nu}(s)\int x_{\nu}(\tau)^*\overline{f(\tau)}d\tau \tag{II}$$

for any function $f(t) \in H_x$ and $y_{\nu}(s) = \frac{1}{D_{\nu}}\int \varphi(s,\tau)\overline{f_{\nu}(\tau)}d\tau$ $(\nu = 1,2,\ldots)$. The operator Φ as we remember is a linear-adjoint one therefore Eq. (I) has a solution for which the random function $Y^0(s) = L_t X^0(t)$ of the variable s contains a finite covariance function at all $s \in S$ if and only if the function $\varphi(s,\tau)$ satisfies condition

(II) for any function $f(t) \in H_x$ and the series $\sum D_\nu |y_\nu(s)|^2$ converges at any $s \in S$. If these conditions are fulfilled then the operator L is determined by the expansion

$$L_t x = \sum_{\nu=1}^{\infty} y_\nu(s) \int f_\nu(t)^T x(t) dt$$

$$= \sum_{\nu=1}^{\infty} \frac{1}{D_\nu} \int \varphi(s,\tau)\overline{f_\nu(\tau)}d\tau \int f_\nu(t)^T x(t) dt. \tag{III}$$

This operator possesses such a property that the series $L_t X^0 = \sum_{\nu=1}^{\infty} y_\nu(s)$

$\times \int f_\nu(t)^T X^0(t) dt = \sum_{\nu=1}^{\infty} V_\nu y_\nu(s)$ m.s. converges and determines the random

function $Y^0(s)$ with the finite covariance function at all $s \in S$.

Notice that it follows in noway from (II) that L_t represents a linear integral operator as from m.s. convergence of the series $L_t X^0(t)$ determined from (III) by the change of $x(t)$ by $X^0(t)$ does not follow the convergence of the series

$$\sum_{\nu=1}^{\infty} \frac{1}{D_\nu} \left\{ \int \varphi(s,\tau)\overline{f_\nu(\tau)}d\tau \right\} f_\nu(t)^T . \tag{IV}$$

But any finite segment of series (II) is a linear integral operator of the form $L_n x$

$= \int l_n(s,t)x(t) dt = \int \left\{ \sum_{\nu=1}^{n} y_\nu(s) f_\nu(t)^T \right\} x(t) dt.$ Therefore Eq. (I) may be

formally regarded as linear integral equation

$$\int l(s,t) K_x(t,\tau) = \varphi(s,\tau) \quad (\tau \in T_1, s \in S), \tag{V}$$

and its solution may be written on the basis of (II) in the form of series

$$l(s,t) = \sum_{\nu=1}^{\infty} y_\nu(s) f_\nu(t)^T = \sum_{\nu=1}^{\infty} \frac{1}{D_\nu} \left\{ \int \varphi(s,\tau)\overline{f_\nu(\tau)}d\tau \right\} f_\nu(t)^T . \tag{VI}$$

At the same moment it should be remembered that this series converges only in such a sense that the correspondent series for m.s. integral $\int l(s,t) X^0(t) dt$

$= \sum_{\nu=1}^{\infty} V_\nu y_\nu(s)$ m.s. converges at any $s \in S$ to the random variable with the

finite covariance function. In particular, the regression of the random function $Y(s)$ over $X(t)$ at their normal joint distribution is determined according to (6.3.24) by the following formula:

$$m_{y|X} = E(Y|X) = m_y(s) + \sum_{\nu=1}^{\infty} D_\nu^{-1} V_\nu \int K_{yx}(s,\tau) f_\nu(\tau) d\tau. \qquad \text{(VII)}$$

In the majority of the problems of the practice it is necessary to applythe operator L_t to the random function $X(t)$ but not only to the centered random function $X^0(t)$. As we have already seen for the existence of such an operator it is necessary and sufficient the fulfilment of conditions (6.3.19) and (6.3.20) besides conditions (6.3.16) and (6.3.17). In a given case these conditions take the form

$$m_x(t) = \sum_{\nu=1}^{\infty} x_\nu(t) \int f_\nu(\tau)^T m_x(\tau) d\tau, \quad \sum_{\nu=1}^{\infty} D_\nu^{-1} \left| \int f_\nu((\tau)^T m_x(\tau) d\tau \right|^2 < \infty.$$

$$\text{(VIII)}$$

Therefore formula (III) determines the required solution of Eq. (I) and the series $LX = \sum_{\nu=1}^{\infty} y_\nu(s) \left\{ \int f_\nu(t)^T m_x(t) dt + V_\nu \right\}$ m.s. converges at each $s \in S$ to the random variable with the finite matrix of the second order moments.

E x a m p l e 6.3.10. In the special case of the finite set $T_1 = \{t_1, \ldots, t_N\}$ all the functions $f(t) \in H_x$ are the linear combinations of the δ-functions with the peculiarities at the points $t_1, \ldots, t_N$, the canonical expansion of the random function represents the finite sum which contains no more than nN items (Example 3.9.8), and the formulae of the previous example give the solution of Eq. (I) of the previous example in the form of the finite sum. The operator L_t in this case is a matrix. At large N the solutionof Eq. (I) of the previous example which is presented by the formulae of the previous example is considerably simpler for the calculation than the solution which is obtained by the inversion of the block matrix $\{K_x(t_p, t_q)\}$.

E x a m p l e 6.3.11. Example 6.3.9 is easily generalized on the case when the spaces X and Y represent the spaces of the functions with the values in any linear spaces with the weak topologies which are determined by the sets of the linear functionals Λ_x and Λ_y respectively. In this case in all previous formulae $f(t)'$ is a function with the values in Λ_x at any t and the variables $g_1', \ldots, g_N'$ are the vectors of the space Λ_y. The convergence of expansion (III) of Example 6.3.9 at the replacement of $x(t)$ by $X^0(t)$ or $X(t)$ in a given case should be considered as m.s. convergence at each $s \in S$ in the weak topology of the space Y of the values of the functions $y(s) \in Y$.

The stated methods may be also used for the calculation of the Radon–Nikodym derivatives of the output measure μ_y by input measure μ_x, $d\mu_y/d\mu_x$ (Pugachev and Sinitsyn 2000).

6.4. Methods of Linear Stochastic Systems Theory based on Integral Canonical Representations

6.4.1. Introductory Remarks

The integral canonical expansions (Section 4.8) along with the canonical expansions (Sections 3.9, 6.3) are widely used for the analysis and the simulation of the random processes in the dynamic systems with random disturbances. The methods of the application of the integral canonical representations are analogous stated in Subsection 6.3.1.

According to the first method if an one-dimensional linear stochastic system is given by a linear integral transformation (6.3.2), and an input and its covariance function are determined by some integral canonical representation (Section 4.8):

$$X(t) = m_x(t) + \sum_{r=1}^{N} \int_{\Lambda_r} V_r(\lambda) x_r(t, \lambda) d\lambda \,,$$

$$K_x(t, t') = \sum_{r=1}^{N} \int_{\Lambda_r} \nu_r(\lambda) x_r(t, \lambda) \overline{x_r(t', \lambda)} d\lambda \,,$$

$$(6.4.1)$$

then the output $Y(t)$ by virtue of (6.3.2) will be expressed by the integral canonical representation:

$$Y(t) = m_y(t) + \sum_{r=1}^{N} \int_{\Lambda_r} V_r(\lambda) y_r(t, \lambda) d\lambda \,,$$

$$K_y(t, t') = \sum_{r=1}^{N} \int_{\Lambda_r} \nu_r(\lambda) y_r(t, \lambda) \overline{y_r(t', \lambda)} d\lambda \,,$$

$$(6.4.2)$$

and

$$m_y(t) = \int_{t_0}^{t} g(t, \tau) m_x(\tau) d\tau,$$

$$(6.4.3)$$

$$y_\nu(t, \lambda) = \int_{t_0}^{t} g(t, \tau) x_\nu(\tau, \lambda) d\tau \quad (\nu = 1, \ldots, N).$$

The formulae for the finite-dimensional linear system and the vector inputs are written analogously.

Now we consider the second method. Let the equations be given which describe the behaviour of a linear stochastic system. Then for the correlation analysis we may use the first method after determining before-hand the weighting functions of a system. For obtaining the equations which for the expectations all random inputs and outputs should be substituted by their correspondent expectations in the equations of the considered system. While deriving the equations which determine the coordinate functions it is necessary to replace all random inputs and outputs by the correspondent coordinate functions in the equations of the system.

E x a m p l e 6.4.1. Find a variance of the output of the stationary oscillating link

$$\ddot{Y} + a_1\dot{Y} + a_0 Y = X \tag{I}$$

for $m_x = 0$, $k_x(\tau) = D_x e^{-\alpha|\tau|}$. At $N = 1$, $\lambda = \omega$ we have $x(t,\omega) = e^{i\omega t}$ and

$$\ddot{y}(t,\omega) + a_1\dot{y}(t,\omega) + a_0 y(t,\omega) = e^{i\omega t}, \quad y(0,\omega) = 0. \tag{II}$$

Hence after integrating we find

$$y(t,\omega) = \frac{e^{i\omega t}}{a_0 + a_1 i\omega - \omega^2} + \frac{(\lambda_2 - i\omega)e^{\lambda_1 t} - (\lambda_1 - i\omega)e^{\lambda_2 t}}{(\lambda_1 - \lambda_2)(a_0 + a_1 i\omega - \omega^2)}, \tag{III}$$

where $\lambda_{1,2} = -\beta \pm i\gamma$ are the roots of the characteristic equation $\lambda^2 + a_1\lambda + a_0 = 0$. Consequently,

$$D_y(t) = \frac{\alpha D_x}{\pi}$$

$$\times \int_{-\infty}^{\infty} \frac{(\cos\omega t - \frac{\beta}{\gamma}e^{-\beta t}\sin\gamma t - e^{-\beta t}\cos\gamma t)^2 + (\sin\omega t - \frac{\omega}{\gamma}e^{-\beta t}\sin\gamma t)^2}{(\alpha^2 + \omega^2)\,|a_0 + a_1 i\omega - \omega^2|^2}\,d\omega. \tag{IV}$$

At $t \to \infty$ for the asymptotically stable system $(\beta > 0)$ we find the stationary value of the variance from (IV)

$$D_y(\infty) = \frac{D_x(\alpha + a_1)}{a_0 a_1(\alpha^2 + a_1\alpha + a_0)}. \tag{V}$$

We may come to the same result if we account that

$$D_y(\infty) = \int_0^{\infty} s_y(\omega)\,d\omega, \quad s_y(\omega) = |\Phi(i\omega)|^2 s_x(\omega),$$

$$s_x(\omega) = \frac{D_x}{\pi}\frac{\alpha}{\alpha^2+\omega^2}\,,\quad \Phi(i\omega) = \frac{1}{a_0 + a_1 i\omega - \omega^2}\,.$$

6.4.2. General Theory of Integral Canonical Representations of Linear Transformations

▷ After expressing the random variable X by the integral canonical representation (Section 4.8)

$$X = m_x + \sum_{r=1}^{\infty}\int x_r(\lambda)Z_r(d\lambda)\,, \tag{6.4.4}$$

we obtain for the random variable (6.3.5) the analogous canonical representation

$$Y = m_y + \sum_{r=1}^{\infty}\int y_r(\lambda)Z_r(d\lambda)\,, \tag{6.4.5}$$

where $m_y = Tm_x + b$, $y_r(\lambda) = Tx_r(\lambda)$ $(\lambda \in \Lambda, r = 1, 2, \ldots)$. If the operator T is continuous in the weak topologies of the spaces X and Y then representation (6.4.5) similarly as in (6.4.4) represents weakly m.s. convergent series of the weak stochastic integrals. In other words, for any linear functional $g \in G$ the random variable $gY = gTX$ is expressed by m.s. convergent series of the stochastic integrals:

$$gY = gm_y + \sum_{r=1}^{\infty}\int gy_r(\lambda)Z_r(d\lambda)\,. \tag{6.4.6}$$

After obtaining canonical representation (6.4.5) of the random variable Y we may find its covariance operator by formulae of Subsection 4.8.2. ◁

Thus the following general theorem is valid.

Theorem 6.4.1. *If the linear operator T is continuous in the weak topologies of the spaces X and Y then integral canonical representation (6.4.5) of the random variable Y m.s. converges in the weak topology of the space Y.*

R e m a r k. If the operator T is not continuous in the weak topologies of the spaces X and Y but has the adjoint operator T^* then the same is valid not for all $g \in G$ but only for those which belong to the subspace G of the space G formed by the intersection of G with the part of the domain D_{T^*} of the operator T^* which is mapped by it into F. If the subspace G is sufficiently rich in our ordinary sense then formula (6.4.5) gives the opportunity to solve the stated problem.

E x a m p l e   6.4.2. After expressing the random function $X(t)$ by the integral canonical representation in Example 6.3.2

$$X(t) = m_x(t) + \sum_{r=1}^{\infty} \int x_r(t, \lambda) Z_r(d\lambda) , \qquad \text{(I)}$$

we get for its lth derivative the integral representation

$$X^{(l)}(t) = m_x^{(l)}(t) + \sum_{r=1}^{\infty} \int x_r^{(l)}(t, \lambda) Z_r(d\lambda) , \qquad \text{(II)}$$

where $x_r^{(l)}(t, \lambda)$ is the lth derivative of the function $x(t, \lambda)$ over t.

E x a m p l e   6.4.3. In Example 6.3.3 after expressing the random function $X(t)$ by representation (I) we get for the random function $Y(s)$ an analogous integral canonical representation

$$Y(s) = m_y(s) + \sum_{r=1}^{\infty} \int y_r(s, \lambda) Z_r(d\lambda) , \qquad \text{(I)}$$

where

$$y_r(s, \lambda) = \int l(s, t) x_r(t, \lambda) dt . \qquad \text{(II)}$$

After obtaining representation (I) of the random function $Y(s)$ we may find its covariance function $K_y(s_1, s_2)$ by means of the known formulae.

E x a m p l e   6.4.4. In the special case when $X(t)$ is a stationary scalar random function of the finite-dimensional argument t it may be expressed by spectral presentation (Subsection 4.8.6):

$$X(t) = m_x + \int e^{i\lambda^T t} Z(d\lambda) . \qquad \text{(I)}$$

Then we get for $Y(s)$ the integral canonical representation

$$Y(s) = m_y(s) + \int y(s, \lambda) Z(d\lambda) , \qquad \text{(II)}$$

where it is denoted

$$m_y(s) = \int l(s, t) m_x \, dt , \quad y(s, \lambda) = \int l(s, t) e^{i\lambda^T t} dt . \qquad \text{(III)}$$

In particular, when $t = s$ is time and $Y(t)$ represents an output of asymptotically stable stationary linear system which performs in the stationary regime then $y(t, \lambda) = \Phi(i\lambda)e^{i\lambda t}$, where $\Phi(i\lambda)$ is a frequency characteristic of a system. In this case the formula for the covariance function of the stationary random function $Y(t)$ gives an integral representation

$$k_y(\tau) = \int\limits_{-\infty}^{\infty} e^{i\lambda\tau}\Phi(i\lambda)\Phi(i\lambda)^*\sigma(d\lambda)\,, \qquad \text{(IV)}$$

where σ is the spectral measure of the random function $X(t)$. Finally when there exists the spectral density of the random function $X(t)$, and $Y(t)$ is the scalar random function then the obtained formula gives the known expression for the spectral density of the random function $Y(t)$: $s_y(\lambda) = |\Phi(i\lambda)|^2 s_x(\lambda)$.

E x a m p l e 6.4.5. In the other special case when $X(t)$ is a stationary m-dimensional vector random function it may be expressed in terms of the integral canonical representation (Subsection 4.8.7):

$$X(t) = m_x + \sum_{p=1}^{m} \int a_p(\lambda)e^{i\lambda^T t}\hat{Z}_p(d\lambda)\,. \qquad \text{(I)}$$

Then we get for the random function $Y(s)$ the integral canonical representation

$$Y(s) = m_y(s) + \sum_{p=1}^{m} \int y_p(s, \lambda)\hat{Z}_p(d\lambda)\,, \qquad \text{(II)}$$

where $m_y(s)$ is determined by the formula as in the previous example and

$$y_p(s, \lambda) = \int l(s, t)a_p(\lambda)e^{i\lambda^T t}dt\,. \qquad \text{(III)}$$

If $t = s$ is time and $Y(s)$ represents the output of the asymptotically stationary linear system which performs in stationary regime then $y_p(t, \lambda) = \Phi(i\lambda)a_p(\lambda)e^{i\lambda t}$ where $\Phi(i\lambda)$ is the frequency characteristic of the system. In this case for the covariance function of the stationary random function $Y(t)$ the following integral representation takes place:

$$k_y(\tau) = \sum_{p=1}^{m} \int e^{i\lambda\tau}\Phi(i\lambda)a_p(\lambda)a_p(\lambda)^*\Phi(i\lambda)^*\hat{\sigma}_p(d\lambda)\,, \qquad \text{(IV)}$$

where $\hat{\sigma}_p$ is a covariance function of the stochastic measure $\hat{Z}_p$. According to the formula for the matrix $\sigma(B)$ of the spectral and cross-spectral measures of the components of the vector random function $X(t)$ (Subsection 4.8.3) we have

$$\sum_{p=1}^{m}\int_B a_p(\lambda)a_p(\lambda)^* \hat{\sigma}_p(d\lambda) = \sigma(B)\,, \tag{V}$$

and consequently, we may present formula for $k_y(\tau)$ in the form

$$k_y(\tau) = \int e^{i\lambda\tau}\Phi(i\lambda)\sigma(d\lambda)\Phi(i\lambda)^* \,. \tag{VI}$$

In the case when there exists the matrix of the spectral and the cross-spectral densities of the components of the vector random function $X(t)$ the obtained formula gives the following expression for the matrix of the spectral and the cross-spectral densities of the components of the vector random function $Y(t)$:

$$s_y(\lambda) = \Phi(i\lambda)s_x(\lambda)\Phi(i\lambda)^* \,. \tag{VII}$$

E x a m p l e 6.4.6. Interpreting the random function $X(t)$ of Example 6.3.5 as representation (I) of Example 6.4.2, we obtain for the function $Y(s)$ representation (I) of Example 6.4.3 where

$$y_r(s,\lambda) = \int l(s,t)x_r(t,\lambda)\mu(dt)\,.$$

The correspondent integral canonical representation of the covariance function $K_y(s_1, s_2)$ of the random function $Y(s)$ is found by the formulae of Subsection 4.8.2.

6.4.3. Solution of Equations by Integral Canonical Representations Method

Analogously we may solve Eq. (6.3.13) by means of the integral canonical representation of the random variable X. As we have already seen in Section 4.8 any sequence of the expansion of unit $\{\mathcal{E}_\nu\}$ and the orthogonal vectors $\{\gamma_\nu\}$ such that the set of the vectors of all $\mathcal{E}_\nu(B)\gamma_\nu$ is dense in H_x give the following integral canonical representation of the random variable X,

$$X = m_x + \sum_{r=1}^{\infty}\int x_r(\lambda)Z_r(d\lambda)\,. \tag{6.4.7}$$

Here $Z_r = \mathcal{E}_r \gamma_r X^0$ are uncorrelated stochastic measures, and the functions $x_r(\lambda)$ satisfy on the basis of Subsection 4.8.2 the following conditions:

$$\int_B x_r(\lambda)\sigma_r(d\lambda) = K_x \mathcal{E}_e(B)\gamma_r , \qquad (6.4.8)$$

$$\mathcal{E}_r(B)\gamma_r x_\mu(\lambda) = \delta_{r\mu}\mathbf{1}_B(\lambda) \qquad (6.4.9)$$

for any set B $(B \subset \Lambda, B \in \mathcal{S})$ where σ_r is a covariance function of the stochastic measure Z_r.

$\triangleright$ If L is a solution of Eq. (6.3.13) which possesses the required properties then by virtue of (6.4.8) for any $g \in G$ and $B \in \mathcal{S}$

$$E\left[gLX^0 - \sum_{r=1}^n \int gLx_r(\lambda)Z_r(d\lambda)\right]\overline{Z_\mu(B)}$$

$$= gLEX^0\overline{Z_\mu(B)} - \int_B gLx_\mu(\lambda)\sigma_\mu(d\lambda)$$

$$= gLEX^0\left[\overline{\mathcal{E}_\mu(B)\gamma_\mu X^0}\right] - gLK_x\mathcal{E}_\mu(B)\gamma_\mu = 0, \quad (\mu = 1, \ldots, n).$$

Consequently, according to (6.4.8) we have the equality

$$E\left|gLX^0 - \sum_{r=1}^n \int gLx_r(\lambda)Z_r(d\lambda)\right|^2$$

$$= E\left|gLX^0\right|^2 - \sum_{r=1}^n \int gLx_r(\lambda)E\left[\overline{gLX^0}Z_r(d\lambda)\right]$$

$$= E\left|gLX^0\right|^2 - \sum_{r=1}^n \int gLx_r(\lambda)E\left[\overline{gLK_x\mathcal{E}_r(d\lambda)\gamma_r}\right]$$

$$= E\left|gLX^0\right|^2 - \sum_{r=1}^n \int |gLx_r(\lambda)|^2 \sigma_r(d\lambda) .$$

Hence it follows that the condition is fulfilled

$$\sum_{r=1}^n \int |gLx_r(\lambda)|^2 \sigma_r(d\lambda) < \infty . \qquad (6.4.10)$$

Further it follows from the canonical representation of the covariance operator $K_x f$
$= \sum\limits_{r=1}^{\infty} \int x_r(\lambda)\overline{fx_r(\lambda)}\sigma_r(d\lambda)$ that at any $f \in H_x$ by virtue of condition (6.4.10)
the series

$$LK_x f = \sum_{r=1}^{\infty} \int Lx_r(\lambda)\overline{fx_r(\lambda)}\sigma_r(d\lambda)$$

converges in the weak topology of the space Y. It follows from the obtained expression
$LK_x g$ that Eq. (6.3.13) may have the required solution only in that case when the
operator Φ is representable by

$$\Phi = \sum_{r=1}^{\infty} \int y_r(\lambda)\left[\overline{\cdot x_r(\lambda)}\right]\sigma_r(d\lambda), \qquad (6.4.11)$$

and if we account (6.4.10) then for any $g \in G$ the following condition takes place:

$$\sum_{r=1}^{\infty} \int |gy_r(\lambda)|^2 \sigma_r(d\lambda) < \infty. \qquad (6.4.12)$$

For determining the functions $y_r(\lambda)$ we substitute in (6.4.11) the index of the summation ν by μ and apply the operator Φ to the vector $\mathcal{E}_r(B)\gamma_r$. Then taking into
consideration (6.4.9) we get $\Phi\mathcal{E}_r(B)\gamma_r = \int_B y_r(\lambda)\sigma_r(d\lambda)$. Hence it follows that
for any $g \in G$ the function $gy_r(\lambda)$ represents the Radon–Nikodym derivative of the
scalar measure $g\Phi\mathcal{E}_r(B)\gamma_r$ over the measure σ_r. This result may be written in the
form

$$y_r(\lambda) = \frac{d\Phi\mathcal{E}_r\gamma_r}{d\sigma_r}(\lambda). \qquad (6.4.13)$$

If the operator L should be applied to the random variable X but not only
to the centered random variable X^0 then one more condition is necessary, namely
$m_x \in \Delta_{K_x}$ which on the basis of Subsection 4.8.2 with account that $u_r = fx_r(\lambda)$
at any $f \in H_x$ gives

$$m_x = K_x f_0 = K_x \sum_{r=1}^{\infty} \int f_0 x_r(\lambda)\mathcal{E}_r(d\lambda)\gamma_r = \sum_{r=1}^{\infty} \int \overline{f_0 x_r(\lambda)}K_x\mathcal{E}_r(d\lambda)\gamma_r \ .$$

Hence taking into the consideration (6.4.8) we get $m_x = \sum\limits_{r=1}^{\infty} \int x_r(\lambda)\overline{f_0 x_r(\lambda)}$
$\times \sigma_r(d\lambda)$. Therefore the expectation must be presentable by the expansion

$$m_x = \sum_{r=1}^{\infty} \int \zeta_r(\lambda)x_r(\lambda)\sigma_r(d\lambda), \qquad (6.4.14)$$

and even

$$\sum_{r=1}^{\infty} \int |\zeta_r(\lambda)|^2 \, \sigma_r(d\lambda) = \sum_{r=1}^{\infty} \int |f_0 x_r(\lambda)|^2 \, \sigma_r(d\lambda) = f_0 K_x f_0 < \infty . \lhd$$

$$(6.4.15)$$

Theorem 6.4.2. *For the existence of the operator L satisfying Eq. (6.3.13) and possessing the required properties conditions (6.4.11), (6.4.12), (6.4.14) and (6.4.15) are necessary.*

These conditions are also sufficient for the existence of the required solution of Eq. (6.3.13). We shall prove it.

$\triangleright$ We shall find the solution of Eq. (6.3.13) in the form

$$L = \sum_{r=1}^{\infty} \int \eta_r(\lambda)\mathcal{E}_r(d\lambda)\gamma_r .$$

$$(6.4.16)$$

Then we have for any $f \in H_x$

$$LK_x f = \sum_{r=1}^{\infty} \int \eta_r(\lambda)\mathcal{E}_r(d\lambda)\gamma_r K_x f = \sum_{r=1}^{\infty} \int \eta_r(\lambda)\overline{f K_x \mathcal{E}_r(d\lambda)\gamma_r} ,$$

or accounting (6.4.8)

$$LK_x f = \sum_{r=1}^{\infty} \int \eta_r(\lambda)\overline{f x_r(\lambda)}\sigma_r(d\lambda) .$$

In order the operator L will satisfy Eq. (6.3.13) it is necessary at any $f \in H_x$ the equality $\Phi f = \sum_{r=1}^{\infty} \int \eta_r(\lambda)\overline{f x_r(\lambda)}\sigma_r(d\lambda)$ holds. After comparing this equality with (6.4.11) we shall have at any $f \in H_x$

$$\sum_{\mu=1}^{\infty} \int \eta_\mu(\lambda)\overline{f x_\mu(\lambda)}\sigma_\mu(d\lambda) = \sum_{\mu=1}^{\infty} \int y_\mu(\lambda)\overline{f x_\mu(\lambda)}\sigma_\mu(d\lambda) .$$

Putting $f = \mathcal{E}_r(B)\gamma_r$ by virtue of (6.4.9) we get $\int_B \eta_r(\lambda)\sigma_r(d\lambda) = \int_B y_r(\lambda)$
$\times \sigma_r(d\lambda)$. As this equality is valid for any set $B \in \mathcal{S}$ then $\eta_r(\lambda) = y_r(\lambda)$ almost

at all λ relatively to the measure σ_r. Substituting the latter integral expression into (6.4.16) and with account of (6.4.13) we find that

$$L = \sum_{r=1}^{\infty} \int y_r(\lambda)\mathcal{E}_r(d\lambda)\gamma_r = \sum_{r=1}^{\infty} \int \frac{d\Phi\mathcal{E}_r\gamma_r}{d\sigma_r}(\lambda)\mathcal{E}_r(d\lambda)\gamma_r \,. \quad \lhd \qquad (6.4.17)$$

Similarly as under conditions of Subsection 6.3.1 the following statement is proved.

Theorem 6.4.3. *In the case of the fulfilment of conditions (6.4.11) and (6.4.12) the random variable $Y^0 = LX^0$ has the covariance operator K_y determined on the whole space G and the operator L is permutable with the operator of the expectation. And if besides (6.4.11) and (6.4.12) conditions (6.4.14) and (6.4.15) are also fulfilled then there exists the random variable $Y = LX$ which has the expectation $m_y = Lm_x$ and the operator of the second order moment determined on the whole space G.*

Let be exist the measures μ_p with the values in H_x, the nonnegative measures σ_p and the functions $x_p(\lambda)$ with the values in X satisfying the following conditions (Subsection 4.8.4) in some measurable space $(\Lambda, \mathcal{S})$:

$$K_x\mu_p(B) = \int_B x_p(\lambda)\sigma_p(d\lambda), \quad B \in \mathcal{S}, \qquad (6.4.18)$$

$$\mu_p(B)x_q(\lambda) = \delta_{pq}\mathbf{1}_B(\lambda). \qquad (6.4.19)$$

Then $\sum_{p=1}^{N} \int fx_p(\lambda)\mu_p(d\lambda) = f$ at any $f \in H_x$ and we may take $\mathcal{E}_1(B)$

$= \cdots = \mathcal{E}_N(B) = \sum_{p=1}^{N} \int \mathbf{1}_B(\lambda)[\cdot x_p(\lambda)]\mu_p(d\lambda), \ \gamma_r = \mu_r(\Lambda) \ (r = 1,$

$\dots, N)$. In this case formula (6.4.17) takes the form

$$L = \sum_{p=1}^{N} \int y_p(\lambda)\mu_p(d\lambda) = \sum_{p=1}^{N} \int \frac{d\Phi\mu_p}{d\sigma_p}(\lambda)\mu_p(d\lambda). \qquad (6.4.20)$$

E x a m p l e 6.4.7. For solving the equation of Example 6.3.9 by the integral canonical representations method it is required to find some number N of the vector measures $\mu_p(t, B)$ (of the same dimension that the vector random function $X(t)$), N nonnegative measures $\sigma_p(B)$ and N sets of the vector functions $x_p(t, \lambda)$ (of the same dimension that $X(t)$) determined on some measurable space $(\Lambda, \mathcal{S})$ and satisfying the conditions of Subsection 4.8.5. Then formula (6.4.20) will give the following expression of the linear operator L_t which satisfies Eq. (I) of Example 6.3.9:

$$L_t x = \sum_{p=1}^{N} \int l(s,t)x(t)dt \, , \tag{I}$$

where

$$l(s,t) = \sum_{p=1}^{N} \int \left\{ \int \varphi(s,\tau) \frac{\overline{d\mu_p(\tau,\lambda)}}{d\sigma_p} d\tau \right\} \mu_p(t,d\lambda)^T \, . \tag{II}$$

Here in order the operator L_t will be permutable with the operator of an expectation and the random function $Y^0(s) = L_t X^0(t)$ will have the finite covariance function at all $s \in S$ it is necessary and sufficient that the function $\varphi(s,\tau)$ may be represented as $\varphi(s,\tau) = \sum_{p=1}^{N} \int y_p(s,\lambda)x_p(\tau,\lambda)^* \sigma_p(d\lambda)$, where $y_p(s,\lambda)$ $= \int \varphi(s,\tau) \frac{d\mu_p(\tau,\lambda)}{d\sigma_p} d\tau$ $(p=1,\ldots,N)$ and at all $s \in S$ the condition of the form (6.4.12) will be fulfilled: $\sum_{p=1}^{\infty} \int |y_p(s,\lambda)|^2 \sigma_p(d\lambda) < \infty.$

In the special case of the finite-dimensional space Λ and absolutely continuous measures $\mu_p(t,B)$ and $\sigma_p(B)$ formula (II) takes the form

$$l(s,t) = \sum_{p=1}^{N} \int \frac{1}{\nu_p(\lambda)} \left\{ \int \varphi(s,\tau)\overline{a_p(\tau,\lambda)}d\tau \right\} a_p(t,\lambda)^T d\lambda \, , \tag{III}$$

where $a_p(t,\lambda)$ and $\nu_p(\lambda)$ are the Radon–Nikodym derivatives of the measures $\mu_p(t,B)$ and $\sigma_p(B)$ over the Lebesgue measure respectively.

E x a m p l e 6.4.8. Let us consider a special case of the previous example when the random function $X(t)$ is real and its forming filter is known (Subsection 4.8.4). Let $w(t,\tau)$ be a weighting function of a forming filter, $w^-(t,\tau)$ be a weighting function of an inverse system. Then we may assume as N the number of the components of the random function $X(t)$ and as the matrices-columns $x_p(t,\lambda)$ and $a_p(t,\lambda)$ $(p=1,\ldots,N)$ the correspondent columns of the matrices $x(t,\lambda) = w(t,\lambda)$ and $a(t,\lambda) = w^-(\lambda,t)^T$. Considering equal to unit the intensities of the components of a white noise which is transformed by a forming filter into the random function $X(t)$ we get from (III) the following formula:

$$l(s,t) = \int \left\{ \int \varphi(s,\tau)w^-(\lambda,\tau)^T d\tau \right\} w^-(\lambda,t)d\lambda \, . \tag{IV}$$

From this result as the special cases the Wiener formula known from the theory of optimal linear systems and its generalizations on the finite-dimensional linear systems are derived.

6.5. Applications. Shaping Filters Design and Modeling

6.5.1. Shaping Filters for Scalar Stationary Processes

Let us show how we may design the shaping filter for a scalar stationary random function $X(t)$ with rational spectral density.

$\triangleright$ Suppose that the spectral density of a real scalar stationary random function $X(t)$ with zero expectation is determined by the formula

$$s_x(\omega) = P(\omega)/Q(\omega), \qquad (6.5.1)$$

where $P(\omega)$ and $Q(\omega)$ are the polynomials. As $s_x(\omega)$ is an even function, $P(\omega)$ and $Q(\omega)$ contain only even degrees of ω. In practical problems the degree $2m$ of the numerator $P(\omega)$ is always smaller then the degree $2n$ of the denominator $Q(\omega)$, as only in this case the variance of the random function $X(t)$ equal to the integral of the spectral density is finite. Besides that it is necessary for the convergence of the integral of the spectral density that the denominator $Q(\omega)$ of the spectral density be different from zero at any real value of ω. Finally, since a spectral density is essentially positive at all (real) values of the frequency ω, all the coefficients of the polynomials $P(\omega)$ and $Q(\omega)$ are real. It follows from these properties of the polynomials $P(\omega)$ and $Q(\omega)$ that the roots of each of them are pair-wise conjugate complex numbers and to each root corresponds the root of the same polynomial of opposite sign. In other words, the roots of the polynomials $P(\omega)$ and $Q(\omega)$ are disposed on the plane of the complex variable ω symmetrically relative to the real and imaginary axes.

Let us factorize the polynomials $P(\omega)$ and $Q(\omega)$ and select the factors corresponding to the roots disposed in the upper half-plane of the complex variables ω. If $P(\omega)$ has real roots then all these roots are multiple roots of even multiplicity as each of such roots is the point where the plot of the spectral density is tangent to the real axis being disposed wholly above it. The half of the factors corresponding to each of such roots should be considered as belonging to the upper half-plane and the other half as belonging to the lower half-plane. Taking all the factors in the factorized polynomial $P(\omega)$, corresponding to the upper half-plane with the arithmetic square root of the coefficient p_{2m} of ω^{2m} in this polynomial and i^m as additional factors we shall obtain the polynomial of degree m in ω whose all the roots are disposed in the upper half-plane and on the real axis. We denote this polynomial by $H(i\omega)$. It is easy to see that all the coefficients of the polynomial $H(i\omega)$ considered as a polynomial in $i\omega$ are positive. Now let us take some root of the polynomial $H(i\omega)$. Let this root be $\alpha + i\beta$, $\beta \geq 0$. If $\alpha \neq 0$, then according to the property of the polynomials $P(\omega)$ and $Q(\omega)$ proved above the polynomial $H(i\omega)$ has also the root $-\alpha + i\beta$. Consequently, the polynomial $H(i\omega)$ represents the product of the positive number $\sqrt{p_{2m}}$ by the factors of the form $i(\omega - \alpha - i\beta)i(\omega + \alpha - i\beta)$ and $i(\omega - i\beta)$, where $\beta \geq 0$. But

$$i(\omega - \alpha - i\beta)i(\omega + \alpha - i\beta) = (i\omega)^2 + 2\beta(i\omega) + \alpha^2 + \beta^2 \,,$$

$$i(\omega - i\beta) = i\omega + \beta \,,$$

Consequently, the product of any number of such factors represents the polynomial in $i\omega$ with positive coefficients.

The remaining m factors in the factorized polynomial $P(\omega)$ corresponding to the roots disposed in the lower half-plane multiplied by $\sqrt{p_{2m}}$ and by $(-i)^m$ form the polynomial which is obtained from $H(i\omega)$ by the change of sign at $i\omega$, i.e. $H(-i\omega)$. Therefore, each root $\alpha + i\beta$ of the polynomial $P(\omega)$ in the upper half-plane corresponds to the root $\alpha - i\beta$ in the lower half-plane. Therefore, each factor of the form $i(\omega - \alpha - i\beta)i(\omega + \alpha - i\beta) = (i\omega)^2 + 2\beta(i\omega) + \alpha^2 + \beta^2$ or $i(\omega - i\beta) = i\omega + \beta$ of the polynomial $H(i\omega)$ corresponds to the remaining factor of the form $-i(\omega - \alpha + i\beta)(-i)(\omega + \alpha + i\beta) = (-i\omega)^2 + 2\beta(-i\omega) + \alpha^2 + \beta^2$ or respectively $-i(\omega + i\beta) = -i\omega + \beta$.

Thus selecting in the factorized polynomial $P(\omega)$ the factors corresponding to the roots disposed in the upper half-plane and adding the factor $i^m \sqrt{p_{2m}}$ we represent the polynomial $P(\omega)$ at real values of ω in the form

$$P(\omega) = H(i\omega)H(-i\omega) = |H(i\omega)|^2 \,, \qquad (6.5.2)$$

where $H(i\omega)$ is a polynomial in $i\omega$ with positive coefficients. In exactly the same way selecting in the factorized polynomial $Q(\omega)$ the factors corresponding to the roots disposed in the upper half-plane, adding to them the factor $i^n \sqrt{q_{2n}}$, q_{2n} being the coefficient of ω^{2n} in the polynomial $Q(\omega)$, and denoting the resulting polynomial by $F(i\omega)$ we represent $Q(\omega)$ at real values of ω in the form

$$Q(\omega) = F(i\omega)F(-i\omega) = |F(i\omega)|^2 \,, \qquad (6.5.3)$$

where $F(i\omega)$ is the polynomial in $i\omega$ with positive coefficients. Representing the numerator and the denominator of the spectral density $s_x(\omega)$ by formulae (6.5.2), (6.5.3) we obtain the following expression for the spectral density $s_x(\omega)$:

$$s_x(\omega) = |H(i\omega)/F(i\omega)|^2 \,. \qquad (6.5.4)$$

Such a representation of the spectral density is called *factorization*.

Notice now that the function

$$\Phi(s) = H(s)/F(s) \qquad (6.5.5)$$

represents the transfer function of some stationary linear system. Since the multiplication of a complex number by the imaginary unit i represents the counter-clock wise rotation of the vector representing this number by the angle $\pi/2$ the left half-plane of the complex variables $s = i\omega$ corresponds to the upper half-plane of the complex variable ω. Consequently, all the roots of the polynomials $H(s)$ and $F(s)$ are disposed in the left half-plane of the variables s, and the function $\Phi(s)$ determined by formula (6.5.5) represents the transfer function of a stable stationary linear system.

Recalling that the stationary random function with constant spectral density s_0 represents a white noise of intensity $2\pi s_0$ we come to the conclusion that the stationary random function $X(t)$ with rational spectral density $s_x(\omega)$ may be considered as the response of the stable stationary linear system with the transfer function $\Phi(s)$ determined by formula (6.5.5) to the input representing a white noise with unit spectral density. Consequently, the system with the transfer function $\Phi(s) = H(s)/F(s)$ represents the shaping filter for the random function $X(t)$.

On the basis of the results of Subsections 1.1.6 and 1.2.5 the stationary linear system with the transfer function $\Phi(s) = H(s)/F(s)$ is described by the linear differential equation with the operator $F(D)$, $D = d/dt$, in the left-hand side (acting on the output) and the operator $H(D)$ in the right-hand side (acting on the input).

Taking into consideration that the input of the shaping filter is the white noise $V(t)$ with unit spectral density whose intensity is equal to 2π, and the output of the shaping filter is the random function $X(t)$, we obtain the differential equation of the shaping filter: $F(D)X = H(D)V$ or denoting the coefficient of s^k in the polynomial $F(s)$ by a_k $(k = 0, 1, \ldots, n)$, and the coefficient of s^k in the polynomial $H(s)$ by b_k $(k = 0, 1, \ldots, m)$,

$$\sum_{k=0}^{n} a_k X^{(k)} = \sum_{k=0}^{m} b_k V^{(k)}.$$

Thus the shaping filter for the random function $X(t)$ represents a system described by a linear stochastic differential equation of the nth order. Replacing this equation by the corresponding set of differential equations of the first order (Subsection 1.2.3) we obtain the equations of the shaping filter in the form

$$\begin{aligned}
\dot{X}_{1k} &= X_{1,k+1} \quad (k = 1, \ldots, n-m-1), \\
\dot{X}_{1k} &= X_{1,k+1} + q_k V \quad (k = n-m, \ldots, n-1), \\
\dot{X}_{1n} &= -a_n^{-1} \sum_{k=1}^{n} a_{k-1} X_{1k} + q_n V,
\end{aligned}$$

$$(6.5.6)$$

and $X(t) = \omega(X_1(t), t) = X_{11}(t)$, $X_1(t) = [X_{11}(t) \ldots X_{1n}(t)]^T$ where

$q_{n-m}, \ldots, q_n$ are constant coefficients determined by formulae:

$$q_{n-m} = a_n^{-1} b_m \,,$$

$$q_k = a_n^{-1} \left(b_{n-k} - \sum_{h=n-m}^{k-1} a_{n-k+h} q_h \right) \quad (k = n - m + 1 , \ldots , n) . \triangleleft$$

$$(6.5.7)$$

R e m a r k. In some cases it is desirable to form a random function $X(t)$ of a white noise of unit intensity. The spectral density of such a white noise is equal to $1/2\pi$. Consequently, representing the spectral density $s_x(\omega)$ of the random function $X(t)$ in the form (6.5.4) it is necessary in such cases to separate from $|H(i\omega)/F(i\omega)|^2$ the factor $1/2\pi$. We may attain it by including into the polynomial $H(i\omega)$ the additional factor $\sqrt{2\pi}$ or by including into the polynomial $F(i\omega)$ the additional factor $1/\sqrt{2\pi}$. As a result instead of (6.5.4) we get $s_x(\omega) = |H(i\omega)/F(i\omega)|^2 (1/2\pi)$.

While applying formula (6.5.4) or more general formulae (6.5.8) two conditions must be satisfied: the condition of stability of a system and the condition of stationarity of its output. The first condition is satisfied by virtue of our method of constructing the shaping filter. This filter is stable since all the poles of its transfer function (6.5.5) are disposed in the left half-plane of the variables s. The second condition may be satisfied by the choice of initial conditions for Eqs. (6.5.6). This is possible because the shaping filter is a mere theoretical tool for reducing system differential equations to stochastic equations due to which the initial conditions for the equations of the shaping filter may be taken arbitrarily. We shall now show how we may do it

$$s_y(\omega = \Phi(i\omega)s_x(\omega)\Phi(i\omega)^* \,. \qquad (6.5.8)$$

$\triangleright$ As it is known, a stationary random process has a constant covariance matrix (the variance in the case of a scalar process). In other words, the covariance matrix of the value of the stationary random process $X_1(t)$ at any time moment t is independent of t and is equal to the value of the covariance function $k_{x_1}(\tau)$ of this process at $\tau = 0$, $k_{x_1}(0)$,

$$k_{x_1}(0) = \int_{-\infty}^{\infty} s_{x_1}(\omega)d\omega \,. \qquad (6.5.9)$$

Therefore, for providing the stationarity of the process $X_1(t)$ at the output of the shaping filter one should take the random initial value $X_1(t_0) = X_{10}$ at the instant

t_0 with the covariance matrix $k_{x_1}(0)$ determined by formula (6.5.9). It remains to find the spectral density $s_{x_1}(\omega)$ of the process $X_1(t)$ determined by Eq. (6.5.6).

Using formula (6.5.8) we find

$$s_{x_1}(\omega) = \Phi_1(i\omega)\Phi_1(-i\omega)^T s_0 , \qquad (6.5.10)$$

where s_0 is the spectral density of the white noise V equal to 1 if $s_x(\omega)$ is represented by formula (6.5.4) or $1/2\pi$ if $s_x(\omega)$ is represented by formula (6.5.8). For determining the transfer function $\Phi_1(s)$ of that part of the shaping filter which gives the vector random process $X_1(t) = [X_{11}(t)\ldots X_{1n}(t)]^T$ at the output we notice that the transfer function of the whole shaping filter whose output represents the process $X(t) = X_{11}(t)$ was found earlier. It is determined by formula (6.5.5), $\Phi_{11}(s) = \Phi(s) = H(s)/F(s)$. Therefore, the remaining elements of the matrix-column $\Phi_1(s)$ may be found from the first $n-1$ by replacing V by the function e^{st}, X_{1k} by the function $\Phi_{1k}(s)e^{st}$ $(k = 1, \ldots, n)$, and the operator d/dt by the variable s (Subsection 1.2.5). As a result we obtain the equations

$$s\Phi_{1k}(s) = \Phi_{1,k+1}(s) \quad (k = 1, \ldots, n - m - 1),$$
$$s\Phi_{1k}(s) = \Phi_{1,k+1}(s) + q_k \quad (k = n - m, \ldots, n - 1).$$

Solving these equations in succession and putting in the first of them $\Phi_{11}(s) = \Phi(s) = H(s)/F(s)$ we find the elements of the matrix-column $\Phi_1(s)$:

$$\Phi_{1k}(s) = s^{k-1}\Phi(s) = s^{k-1}H(s)/F(s) \quad (k = 1, \ldots, n - m),$$
$$\Phi_{1k}(s) = s^{k-1}\Phi(s) - \sum_{l=0}^{k-1-n+m} q_{k-l-1}s^l$$
$$= s^{k-1}H(s)/F(s) - \sum_{l=0}^{k-1-n+m} q_{k-l-1}s^l \quad (k = n - m + 1, \ldots, n).$$

Reducing the expressions of $\Phi_{1k}(s)$ $(k = n - m + 1, \ldots, n)$ to a common denominator and using formula (6.5.7) we obtain after simple but rather cumbersome evaluations

$$\Phi_{11}(s) = H(s)/F(s) , \quad \Phi_{1k}(s) = H_k(s)/F(s) \ (k = 2, \ldots, n) . \quad (6.5.11)$$

Here

$$H_k(s) = s^{k-1}H(s) \quad (k = 2, \ldots, n - m) , \qquad (6.5.12)$$

$$H_k(s) = \sum_{r=0}^{n-1} c_{kr}s^r \quad (k = n - m + 1, \ldots, n) , \qquad (6.5.13)$$

where

$$c_{kr} = -\sum_{l=k-1-r}^{k-1} a_{l-k+r+1} q_l \quad (r = 0, \ldots, m - n + k - 1),$$

$$c_{kr} = -\sum_{l=n-m}^{k-1} a_{l-k+r+1} q_l \quad (r = m - n + k, \ldots, k - 2)$$

(only if $n - m \geq 2$),

$$c_{kr} = \sum_{l=k}^{k-1+n-r} a_{l-k+r+1} q_l \quad (r = k - 1, \ldots, n - 1). \qquad (6.5.14)$$

For calculating the elements of the matrix $k_{x_1}(0)$ determined by formula (6.5.9) we may use formula (6.1.27). ◁

E x a m p l e 6.5.1. Design the shaping filter for a stationary random function $X(t)$ with the spectral density

$$s_x(\omega) = \frac{D}{\pi} \frac{\alpha}{\alpha^2 + \omega^2}. \qquad (I)$$

In this case the numerator $P(\omega)$ represents a constant and the denominator has two pure imaginary roots $\pm i\alpha$. Consequently, we may assume $H(i\omega) = \sqrt{D\alpha/\pi}$, $F(i\omega) = i\omega + \alpha$. Then we obtain

$$s_x(\omega) = |\Phi(i\omega)|^2, \quad \Phi(s) = \sqrt{\frac{D\alpha}{\pi}} \frac{1}{s + \alpha}. \qquad (II)$$

Thus the shaping filter represents in this case the first order filter with the time constant $T = 1/\alpha$ and the gain $k = \sqrt{D/\pi\alpha}$. The differential equation of this shaping filter has the form

$$\dot{X} + \alpha X = \sqrt{D\alpha/\pi} V, \qquad (III)$$

V being the white noise of the intensity $\nu = 2\pi$.

Putting $H(i\omega) = \sqrt{2D\alpha}$, $F(i\omega) = i\omega + \alpha$, we have

$$s_x(\omega) = \frac{1}{2\pi} |\Phi(i\omega)|^2, \quad \Phi(s) = \frac{\sqrt{2D\alpha}}{s + \alpha}. \qquad (IV)$$

The differential equation of the shaping filter will in this case take the form

$$\dot{X} + \alpha X = \sqrt{2D\alpha}\,V\,,\tag{V}$$

V being a white noise of unit intensity. Taking in both cases the random initial value $X_0 = X(t_0)$ with the variance

$$D_{x_0} = \int\limits_{-\infty}^{\infty} s_x(\omega)d\omega = \frac{D\alpha}{\pi}\int\limits_{-\infty}^{\infty}\frac{d\omega}{\alpha^2 + \omega^2} = D\,,\tag{VI}$$

we shall obtain at the output of the shaping filter the stationary random process $X(t)$ with given spectral density $s_x(\omega)$.

E x a m p l e 6.5.2. Design the shaping filter for a stationary random function with spectral density

$$s_x(\omega) = \frac{D}{\pi}\frac{(a + \gamma\omega_0)b^2 + (a - \gamma\omega_0)\omega^2}{b^4 + 2(a^2 - \omega_0^2)\omega^2 + \omega^4}\,,\quad \omega_0^2 = b^2 - a^2\,.\tag{I}$$

In this case the numerator $P(\omega)$ has two roots $\omega = \pm ib_1$, $b_1^2 = b^2(a + \gamma\omega_0)(a - \gamma\omega_0)^{-1}$, and the denominator $Q(\omega)$ has four roots $\pm(\omega_0 \pm ia)$. Selecting the roots disposed in the upper half-plane according to the outlined method we assume

$$H(i\omega) = \sqrt{2D(a - \gamma\omega_0)}\,i(\omega - ib_1) = \sqrt{2D(a - \gamma\omega_0)}(i\omega + b_1)\,,$$
$$F(i\omega) = i(\omega - \omega_0 - ia)\,i(\omega + \omega_0 - ia) = (i\omega)^2 + 2a(i\omega) + b^2\,.$$

Then we shall have

$$s_x(\omega) = \frac{1}{2\pi}\,|\Phi(i\omega)|^2\,,\quad \Phi(s) = \frac{H(s)}{F(s)} = \frac{\sqrt{2D(a - \gamma\omega_0)}(s + b_1)}{s^2 + 2as + b^2}\,.\tag{II}$$

The differential equation of this shaping filter has the form

$$\ddot{X} + 2a\dot{X} + b^2 X = \sqrt{2D(a - \gamma\omega_0)}(\dot{V} + b_1 V)\,,\tag{III}$$

V being a white noise of unit intensity. To this equation corresponds the set of two first-order Eqs. (6.5.6) for the vector random function $X_1(t) = \big[X(t)\ \dot{X}(t) - q_1 V(t)\big]^T$:

$$\dot{X}_{11} = X_{12} + q_1 V\,,\quad \dot{X}_{12} = -b^2 X_{11} - 2a X_{12} + q_2 V\,,\tag{IV}$$

where according to formulae (6.5.7) $q_1 = \sqrt{2D(a - \gamma\omega_0)}$, $q_2 = \sqrt{2D(a - \gamma\omega_0)}$ $\times (b_1 - 2a)$.

For determining the covariance matrix of the initial value $X_{10} = X_1(t_0)$ we find by formula (6.5.11) the elements of the matrix-column $\Phi_1(s)$:

$$\Phi_{11}(s) = \Phi(s) = \frac{q_1(s + b_1)}{s^2 + 2as + b^2}, \quad \Phi_{12}(s) = \frac{q_2 s - q_1 b^2}{s^2 + 2as + b^2}. \quad \text{(V)}$$

After this we find by formulae (6.5.10), (6.5.4), (6.5.8)

$$k_{x_{11}}(0) = \frac{q_1^2}{2\pi} \int_{-\infty}^{\infty} \frac{-(i\omega)^2 + b_1^2}{|(i\omega)^2 + 2a(i\omega) + b^2|^2} d\omega = D, \quad \text{(VI)}$$

$$k_{x_{11}x_{12}}(0) = -\frac{q_1}{2\pi} \int_{-\infty}^{\infty} \frac{(i\omega + b_1)(q_2 i\omega + b^2 q_1)}{|(i\omega)^2 + 2a(i\omega) + b^2|^2} d\omega = -D(a - \gamma\omega_0), \quad \text{(VII)}$$

$$k_{x_{12}}(0) = \frac{1}{2\pi} \int_{-\infty}^{\infty} \frac{-q_2^2(i\omega)^2 + b^4 q_1^2}{|(i\omega)^2 + 2ai\omega + b^2|^2} d\omega$$

$$= D\left[b^2 - b\sqrt{a^2 - \gamma^2\omega_0^2} + 2a(a - \gamma\omega_0) \right]. \quad \text{(VIII)}$$

Taking the random initial value $X_{10} = X_1(t_0)$ with the covariance matrix whose elements are determined by the derived formulae we shall get at the first output of the shaping filter the stationary random function $X(t)$ with given spectral density $s_x(\omega)$.

6.5.2. Shaping Filter for Vector Stationary Process

Analogously the problem of finding the shaping filter for a stationary vector random function $X(t)$ with a rational spectral density $s_x(\omega)$ is solved. In this case all the elements of the matrix $s_x(\omega)$ represent the ratios of the polynomials in ω. In accordance with (6.1.22) for finding the shaping filter for $X(t)$ it is sufficient to represent the spectral density $s_x(\omega)$ by the formula

$$s_x(\omega) = F(i\omega)^{-1} H(i\omega) H(i\omega)^* F(i\omega)^{-1*}, \quad \text{(6.5.15)}$$

where $F(s)$ and $H(s)$ are the matrices whose elements are all polynomials in s, and all the roots of the determinants of these matrices $F(s)$ and $H(s)$ are disposed in the left half-plane of the complex variable s. The representation of the matrix of spectral and cross-spectral densities $s_x(\omega)$ of the components of a stationary vector random function in such a form is called *a factorization* of the matrix $s_x(\omega)$. After factorizing the spectral density $s_x(\omega)$ we may write the differential equation of the shaping filter, transform it into a set of the first order differential Eqs. (6.5.6) and find the initial conditions for them providing the stationarity of the process at the output of the filter in the same way as in Subsection 6.5.1.

E x a m p l e 6.5.3. The spectral density of two-dimensional stationary vector random function $X(t)$ is determined by the formula

$$
s_x(\omega) = \begin{bmatrix} \dfrac{k_1^2}{\alpha_1^2 + \omega^2} & \dfrac{k_1 k_2}{\alpha_1 \alpha_2 - (\alpha_1 - \alpha_2)i\omega + \omega^2} \\[3mm] \dfrac{k_1 k_2}{\alpha_1 \alpha_2 + (\alpha_1 - \alpha_2)i\omega + \omega^2} & \dfrac{k_2^2}{\alpha_2^2 + \omega^2} \end{bmatrix} .
\tag{I}
$$

Find the shaping filter for $X(t)$. In this case

$$
\begin{aligned}
s_x(\omega) = {} & \begin{bmatrix} (\alpha_1 + i\omega)^{-1} & 0 \\ 0 & (\alpha_2 + i\omega)^{-1} \end{bmatrix} \begin{bmatrix} k_1 \\ k_2 \end{bmatrix} \begin{bmatrix} k_1 k_2 \end{bmatrix} \\
& \times \begin{bmatrix} (\alpha_1 - i\omega)^{-1} & 0 \\ 0 & (\alpha_2 - i\omega)^{-1} \end{bmatrix} ,
\end{aligned}
\tag{II}
$$

$$
F(s) = \begin{bmatrix} \alpha_1 + s & 0 \\ 0 & \alpha_2 + s \end{bmatrix} , \quad H(s) = \begin{bmatrix} k_1 \\ k_2 \end{bmatrix} ,
\tag{III}
$$

and the set of differential equations of the shaping filter has the form

$$
\dot{X}_1 + \alpha_1 X_1 = k_1 V , \quad \dot{X}_2 + \alpha_2 X_2 = k_2 V ,
\tag{IV}
$$

V being a white noise of the intensity 2π. Taking the random initial value of the vector $X_0 = X(t_0)$ with the covariance matrix

$$
k_x(0) = \int_{-\infty}^{\infty} s_x(\omega)\,d\omega = \begin{bmatrix} \pi k_1^2/\alpha_1 & 2\pi k_1 k_2/(\alpha_1 + \alpha_2) \\ 2\pi k_1 k_2/(\alpha_1 + \alpha_2) & \pi k_2^2/\alpha_2 \end{bmatrix} ,
\tag{V}
$$

we shall obtain at the output of the filter the two-dimensional stationary vector random process with given spectral density $s_x(\omega)$.

6.5.3. Shaping Filter for Process Reducible to Stationary One

The method outlined in Subsections 6.5.1 and 6.5.2 may also be used to design filters for random processes reducible to stationary ones. The case of the process reducible to stationary by the transformation of the process itself is trivial. If

$X(t) = \psi(X_1(t), t)$ where $X_1(t)$ is a stationary random function, then substituting this expression into differential Eq. (5.4.2) we obtain an equation with a stationary random function in the right-hand side. Therefore it remains to consider the case of a process $X(t)$ reducible to stationary by the transformation of the argument.

$\triangleright$ Let $X(t) = \psi(X_1(\varphi(t)), t)$ where $X_1(s)$ is a stationary random function with zero expectation and a rational spectral density, and $s = \varphi(t)$ being a monotonous increasing differentiable function. After constructing the shaping filter for the random function $X_1(s)$ we write the differential equation of this filter and the corresponding initial condition in the form

$$dX_1 = aX_1 ds + b dW_1, \quad X_1(\varphi(t_0)) = X_0 \qquad (6.5.16)$$

where a and b are constant matrices of the corresponding dimensions, and $W_1 = W_1(s)$ is a process with uncorrelated increments with the covariance function

$$K_{w_1}(s_1, s_2) = k_1(\min(s_1, s_2)), \quad k_1(s) = k_1(s_0) + \nu_1(s - s_0), \quad (6.5.17)$$

ν_1 being the constant intensity of the stationary white noise $V_1(s) = \dot{W}_1(s)$, and $s_0 = \varphi(t_0)$. Let us consider the random process $W(t) = W_1(\varphi(t))$. It is evident that $W(t)$ is a process with uncorrelated increments as the monotonous increasing function $s = \varphi(t)$ maps any nonintersecting intervals on the s-axis on nonintersecting intervals of the t-axis. The covariance function of the process $W(t)$ is determined by the evident formula

$$K_w(t_1, t_2) = K_{w_1}(\varphi(t_1), \varphi(t_2)) = k(\min(\varphi(t_1), \varphi(t_2))), \qquad (6.5.18)$$

where

$$k(t) = k_1(\varphi(t)) = k_1(\varphi(t_0)) + \nu_1[\varphi(t) - \varphi(t_0)] = k(t_0) + \int_{t_0}^{t} \nu_1 \dot{\varphi}(\tau) d\tau.$$

Hence, it is clear that the intensity of the white noise $V(t) = \dot{W}(t)$ is equal to $\nu(t) = \nu_1 \dot{\varphi}(t)$. The change of variables $s = \varphi(t)$ in Eq. (6.5.16) yields

$$dX_1(\varphi(t)) = a\dot{\varphi}(t)X_1(\varphi(t))dt + b dW, \quad X_1(\varphi(t_0)) = X_0. \qquad (6.5.19)$$

Introducing the random process $X_2(t) = X_1(\varphi(t))$, we represent Eq. (6.5.19) in the form

$$dX_2 = a\dot{\varphi}(t)X_2 dt + b dW, \quad X_2(t_0) = X_0. \qquad (6.5.20)$$

Thus the shaping filter in this case is described by Eq. (6.5.20) and by the formula $X(t) = \psi(X_2(t), t)$. ◁

Note the fact that after writing the equation of the shaping filter for the random function $X_1(s)$ in the form of an ordinary differential equation with a white noise in the right-hand side and performing the change of variables $s = \varphi(t)$ in this equation we shall get in the coefficient in front of the white noise in Eq. (6.5.20) the additional factor $\dot{\varphi}(t)$. But such a derivation of Eq. (6.5.20) is false. This illustrates the fact that ordinary rules of the change of independent variable are not applicable when there is a white noise in a differential equation. It is necessary to remember this when we deal with stochastic differential equations.

The covariance function of the initial value of the process $X_1(s)$, $X_0 = X_1(\varphi(t_0))$ is determined by Eqs. (6.5.9)–(6.5.11).

E x a m p l e 6.5.4. The random function $X(t)$ with the expectation $m_x(t) = c_0 + c_1 t$ and the covariance function: $K_x(t_1, t_2) = De^{\mu(t_1+t_2)-\alpha|t_1^2-t_2^2|}$ is reduced to a stationary one by the transformations: $X(t) = c_0 + c_1 t + e^{\mu t} X_1(s)$, $s = t^2$. Here $X_1(s)$ is a stationary random function with zero expectation and the covariance function $k_{x_1}(\sigma) = De^{-\alpha|\sigma|}$, $\sigma = s_1 - s_2$. This covariance function corresponds to the spectral density $s_{x_1}(\omega) = D\alpha/\pi(\alpha^2 + \omega^2)$. Using the method of Subsection 6.5.1 we find the differential equation of the random function $X_1(s)$:

$$dX_1/ds = -\alpha X_1 + \sqrt{2D\alpha}V_1 , \qquad (\mathrm{I})$$

V_1 being a white noise of unit intensity. Passing to the variable t we replace Eq. (I) by Eq. (6.5.20) for the random function $X_2(t) = X_1(t^2)$ which has in this case the form

$$\dot{X}_2 = -2\alpha t X_2 + \sqrt{2D\alpha}V , \qquad (\mathrm{II})$$

where V is a white noise of the intensity $2t$. The initial condition for this equation in accordance with Eq. (6.5.20) has the form: $X_2(t_0) = X_0$, where X_0 is a random variable with zero expectation and the variance

$$D_{x_0} = \int\limits_{-\infty}^{\infty} s_{x_1}(\omega)d\omega = \frac{D\alpha}{\pi} \int\limits_{-\infty}^{\infty} \frac{d\omega}{\alpha^2 + \omega^2} = D . \qquad (\mathrm{III})$$

6.5.4. Shaping Filters for Stationary Functions of Vector Argument

Many problems of practice lead to the partial differential equations which contain stationary or reducible to stationary random functions of a vector argument.

For the application of the methods of the linear stochastic systems theory to such problems it is expedient to present the random functions entering into the equations as the solutions of the linear stochastic differential equations. Here we give a general method of finding the linear stochastic differential equations in the correspondent H-spaces for a wide class of the stationary functions of a vector argument.

Let $X(t, x)$ be a scalar stationary random function of the scalar variable t which we shall assume as the time and N-dimensional vector variable x; $k(t_1 - t_2, x_1 - x_2)$ be its covariance function which is equal to $k(t_1 - t_2, 0)$ at $x_1 = x_2 = x$ and $k(0, x_1 - x_2)$ at $t_1 = t_2 = t$. We shall consider here normally distributed random functions $X(t, x)$ satisfying two conditions:

(i) at a given x the random function $X_x(t)$ has a rational spectral density independent of x

$$s(\omega) = \frac{1}{2\pi} \int_{-\infty}^{\infty} k(\tau, 0) e^{-i\omega\tau} d\tau = \frac{Q(\omega^2)}{P(\omega^2)}, \qquad (6.5.21)$$

where $P(\omega^2)$ and $Q(\omega^2)$ are the polynomials and $P(\omega^2)$ has no zeros on the real axis;

(ii) at a given t the covariance function of the random function $X_t(x)$ at some $C > 0$ and $\alpha > 0$ satisfies the inequality

$$|k(0, x_1 - x_2)| < Ce^{-\alpha|x_1 - x_2|} \qquad (6.5.22)$$

For obtaining a stochastic differential equation for the random function $X = X_x(t)$ considered as a random function of the time t at a given x we may apply the same method of shaping filters which was used for the scalar random process (Subsection 6.5.1). According to this method it is necessary to present the spectral density $s_x(\omega)$ in the form

$$s_x(\omega) = \frac{\nu_0}{2\pi} \left| \frac{H(i\omega)}{F(i\omega)} \right|^2, \qquad (6.5.23)$$

where $F(s)$ and $H(s)$ are the polynoms whose roots at $s = i\omega$ be in the left half-plane; ν_0 is the constant which is usually obtained for the sake of convenience to write the equation for $X(t, x)$. Formula (6.5.23) allows to treat $X_x(t)$ as the result of the passage of the white noise of the intensity ν_0 through a stationary linear system with the transfer function $H(s)/F(s)$. This fact will give for $X_x(t)$ a linear stochastic differential equation

$$F(\partial/\partial t)X = H(\partial/\partial t)V, \qquad (6.5.24)$$

where $V = V(t, x) = V_x(t)$ is the stationary normally distributed white noise which depends on time t at a given x whose intensity is equal to ν_0.

If we have

$$F(s) = \sum_{k=0}^{n} a_k s^k \; , \;\; H(s) = \sum_{k=0}^{m} b_k s^k \; , \;\; m < n \, , \tag{6.5.25}$$

then Eq. (6.5.24) represents a stochastic linear differential equation of the nth order which should be assumed as the Itô standard equation:

$$dZ = AZdt + qdW. \tag{6.5.26}$$

Here Z is the vector with the components $Z_1 = X$, Z_2, ..., Z_n, A is the matrix with the elements $a_{ij} = \delta_{i,i+1}$, $i = 1$, ..., $n-1$, $a_{ni} = -a_n^{-1} a_{i-1}$, $i = 1$, ..., n, q is the vector with the components $q_1 = \cdots = q_{n-m-1} = 0$, $q_{n-m} = a_n^{-1} b_m$,

$$q_k = a_n^{-1} \left(b_{n-k} - \sum_{h=n-m}^{k-1} a_{n-k+h} q_h \right) \, , \;\; k = n - m + 1 \, , \ldots , \, n;$$

$W = W(t, x)$ is the scalar Wiener process with the values in the space of the functions of the variable x whose weak m.s. derivative is the white noise $V = V_x(t)$ of the intensity ν:

$$W = W(t, x) = \int_0^t V_x(\tau)d\tau. \tag{6.5.27}$$

Eq. (6.5.2) is reduced to Eq. (6.5.26) by a formal change of the variables (6.5.6):

$$Z_1 = X \, , \;\; Z_{k+1} = \frac{\partial Z_k}{\partial t} \, , \;\; k = 1 \, , \ldots , \, n - m - 1 \, ,$$

$$Z_{k+1} = \frac{\partial Z_k}{\partial t} - q_k V \, , \;\; k = n - m \, , \ldots , \, n - 1 \, , \tag{6.5.28}$$

where q_k are determined by formulae (6.5.7) and $q\nu q^T = qq^T \nu$.

We shall consider the random function $X(t, x)$ as a random process with the values $X_t(\cdot)$ in the H-space $L_2(D)$ of the functions of the variable x, $D \subset R^N$. Then as a covariance operator of the value of this process at a given t will be an integral operator with the square-integrable kernel $k_x(0, x - \xi)$:

$$K_x \varphi = \int_D k_x(0, x - \xi)\varphi(\xi)d\xi \, . \tag{6.5.29}$$

The analogous formula takes the place for the covariance operator K_{z_t} of the vector process $Z_t(x)$:

$$K_{z_t}\varphi = \int_D k_{z_t}(x - \xi)\varphi(\xi)d\xi \,,$$

where the kernel $k_{z_t}(x_1 - x_2) = k_z(0, x_1 - x_2)$ is a covariance function of the process $Z_t(x)$ at a given t. For finding the kernel $\tilde{\nu}(x - \xi)$ of an operator of the intensity of the Wiener process W we use an equation which connects the covariance matrix k_{z_t} of the value of the process $Z = Z(t, x)$ at a given t with the intensity ν of the Wiener process W. This equation has the form

$$\frac{\partial k_{z_t}}{\partial t} = A k_{z_t} + k_{z_t} A^T + qq^T \nu \,. \tag{6.5.30}$$

As the process Z is stationary then the covariance matrix k_{z_t} of its value at a given t does not depend on t and is expressed in terms of its covariance function $k_z(t_1 - t_2, x_1 - x_2)$ by the formula $k_{z_t} = k_z(0, x_1 - x_2)$. Therefore Eq. (6.5.30) has in a given case the form

$$0 = A k_z(0, x_1 - x_2) + k_z(0, x_1 - x_2)A^T + qq^T \nu(x_1 - x_2) \,. \tag{6.5.31}$$

Matrix Eq. (6.5.31) represents a system of $n(n + 1)/2$ scalar linear algebraic equations with $n(n + 1)/2$ unknowns, namely $n(n + 1)/2 - 1$ of the unknown elements of the matrix $k_z(0, x_1 - x_2)$ and the unknown $\nu(x_1 - x_2)$. One element of the matrix $k_z(0, x_1 - x_2)$ is known: $k_{z_1,z_1}(0, x_1 - x_2) = k_x(0, x_1 - x_2)$ as $Z_1(t, x) = X(t, x)$. After excluding from (6.5.29) the unknown elements of the matrix $k_z(0, x_1 - x_2)$ we shall find the unknown function $\tilde{\nu}(x_1 - x_2)$:

$$\tilde{\nu}(x_1 - x_2) = c k_x(0, x_1 - x_2) \,, \tag{6.5.32}$$

where c is some constant. At $x_1 = x_2 = x$ this formula gives the intensity ν_0 of the process $W(t, x)$ at a given x: $\nu_0 = c k_x(0, 0)$.

From (5.9.3) it follows the formula for the kernels of the operators:

$$k_w(t, x_1 - x_2) = \int_0^t \tilde{\nu}(x_1 - x_2)d\tau = c k_x(0, x_1 - x_2)t \,. \tag{6.5.33}$$

Here k_w is the kernel of the operator $K_w(t)$ in (5.9.3) at a given t. Condition (6.5.22) provides the trace-type of the operators k_w and ν as in this case by virtue of the square integralability of the kernel k_x for any orthonormal basis $\{\varphi_k(x)\}$ in $L_2(D)$:

$$\sum_{k=1}^{\infty} \int\limits_{D}\int\limits_{D} \tilde{\nu}(x-\xi)\varphi_k(x)\varphi_k(\xi)dxd\xi$$

$$= c\sum_{k=1}^{\infty} \int\limits_{D}\int\limits_{D} k_x(0, x-\xi)\varphi_k(x)\varphi_k(\xi)dxd\xi < \infty. \qquad (6.5.34)$$

R e m a r k. For the rectangular region D it is easy to check out this fact by the direct integration after using estimate (6.5.22) as for the basis $\{\varphi_k(x)\}$ in a given case may be taken a trigonometrical basis correspondent to the Fourier series expansion of the function $f(x) \in L_2(D)$.

Thus the one-dimensional normal distribution of the Wiener process W is concentrated on the H-space $L_2(D)$. In this case the Wiener process with the values in $L_2(D)$ possesses at some additional condition the main properties of the finite-dimensional Wiener processes: the continuity over t and nondifferentiability over t in none of the points of its realizations.

Eq. (6.5.26) represents a linear stochastic differential equation in the H-space $L_2^n(D)$ of the n-dimensional vector functions of the variable x. The stated method permits also to find the linear stochastic differential equations in the H-space for the random functions $X = X(t, x)$ reducible to the stationary ones. The case of the function reducible to the stationary one by the transformation of the function itself $X(t, x) = \psi(X_1(t, x), t, x)$ where $X_1(t, x)$ is a stationary random function is trivial. There it remains to consider the case when the random function $X(t, x)$ is reduced to a stationary function by the transformation both a function and an argument.

▷ Let $X(t, x) = \psi(X_1(\varphi(t), x), t, x)$ where $X_1(s, x)$ is a stationary random function, $\varphi(t)$ is monotonously increasing differentiable function; $\varphi(0) = 0$. After writing Eq. (6.5.26) for the random function $X_1(s, x)$,

$$dZ = aZ_1ds + bdW_1, \qquad (6.5.35)$$

where $Z_1(s, x) = X_1(s, x)$, and $W_1(s, x)$ is a Wiener process with the kernel of the covariance operator $\nu(x_1 - x_2)s$ and after changing the variables $s = \varphi(t)$ we obtain a linear stochastic differentiable equation in the H-space $L_2^n(D)$:

$$dZ = A\dot{\varphi}Zdt + qdW. \qquad (6.5.36)$$

Here $W = W(t, x) = W_1(s, x)$ is a Wiener process with the kernel of the covariance operator of its value at a given t equal to $k(t, x_1 - x_2) = \tilde{\nu}_1(x_1 - x_2)\varphi(t)$

where $\tilde{\nu}_1(x_1 - x_2)$ is the kernel of an operator of the intensity of the Wiener process $W_1(s)$. ◁

6.5.5. Software for Analytical Modeling Problems

At practice analytical modeling problems for the linear stochastic systems are usually solved by means of the universal or the specialized computer libraries and the dialog packages. The corresponding algorithms are based on equations given in Sections 6.1–6.3, 6.5.

Let us consider some examples of analytical modeling solved on the specialized dialog packages "StS-Analysis" (Versions 1.0 and 2.0) based on equations of Sections 6.1 and 6.5.

Working with any version of the "StS-Analysis" package, an user must only type on the display with the aid of the keyboard the original stochastic equations in the natural mathematical form without any special programmers tricks, introduce the corresponding initial conditions and indicate which characteristics must be calculated. Once the original stochastic equations are introduced the user may correct them, i.e. to remove some equations, to replace some of them by others, to correct errors and so on. When all the data are introduced and checked the user has to start computations. Then the package "StS-Analysis" automatically derives the equations for the corresponding distribution parameters and solves them. The user may observe visually the behaviour of a system under study in the process of calculations by tracing the graphs or the tables of current values of various characteristics on the display and get the copies of these graphs and tables from the file after the calculations are finished. So the use of any of the versions of the package "StS-Analysis" does not require that the user know mathematics and programming.

Any version of the "StS-Analysis" package automatically derives the equations for the distribution parameters from the original stochastic equations introduced by the user in the natural mathematical form and solves these equations. It should be remarked that the "StS-Analysis" package does not simulate a stochastic system and does not use the statistical simulation (Monte-Carlo) method but performs the theoretical calculation of the necessary characteristics of a system behaviour by the linear stochastic systems theory methods.

E x a m p l e 6.5.5. For the stochastic two-dimensional linear discrete system

$$Y_1(l+1) = aY_1(l) + bY_2(l) + V_1(l), \qquad\qquad \text{(I)}$$
$$Y_2(l+1) = -bY_1(l) + aY_2(l) \qquad\qquad \text{(II)}$$

(a, b being known constants) the exact equations for the expectations, the variances and the covariances (Subsection 6.1.9) have the following form:

$$m_1(l+1) = am_1(l) + bm_2(l), \qquad\qquad \text{(III)}$$

$$m_2(l+1) = -bm_1(l) + am_2(l), \qquad\qquad\text{(IV)}$$
$$D_1(l+1) = a^2 D_1(l) + b^2 D_2(l) + 2abK_{12}(l) + 1, \qquad\text{(V)}$$
$$D_2(l+1) = b^2 D_1(l) + a^2 D_2(l) - 2abK_{12}(l), \qquad\text{(VI)}$$
$$K_{12}(l+1) = -abD_1(l) + abD_2(l) + (a^2 - b^2)K_{12}(l). \qquad\text{(VI)}$$

Graphs of $m_1(l)$, $m_2(l)$, $D_1(l)$, $D_2(l)$ and $K_{12}(l)$ at initial conditions: $EY_2(0) = 1$, $EY_1(0) = DY_1(0) = DY_2(0) = EY_1(0)Y_2(0) = 0$ are given on Fig.6.5.1.

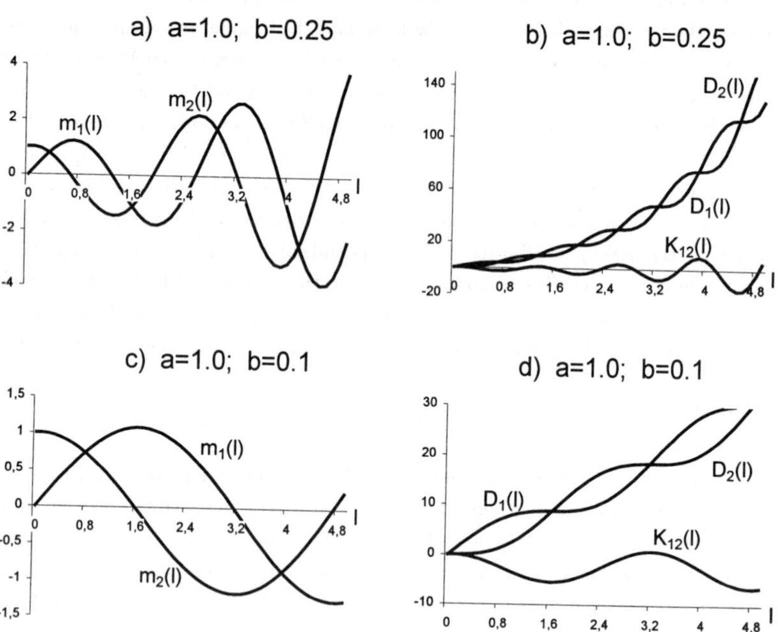

Fig. 6.5.1

6.6. Applications. Optimal Online Linear Filtering

6.6.1. Equations of Linear Filtering

The optimal filtering problem may be solved completely in the case of linear Eqs. (5.10.4) and (5.10.5)

$$dY = (bY + b_1 X + b_0)dt + \psi_1 dW, \tag{6.6.1}$$
$$dX = (aY + a_1 X + a_0)dt + \psi dW, \tag{6.6.2}$$

where the coefficients a, a_1, a_0, b, b_1, b_0, ψ and ψ_1 in the general case depend on time t. In this case the distribution of the process $\left[Y(t)^T X(t)^T\right]^T$ is normal if the distribution of its initial value $\left[Y_0^T X_0^T\right]^T$ is normal. Consequently, the conditional distribution of the state vector of the system relative to the process $Y_{t_0}^t$ is also normal and for its determining it is sufficient to find the conditional expectation $\hat{X}$ and the conditional covariance matrix R. For this purpose we use formulae (5.10.35) and (5.10.31) for the stochastic differentials of the random variables $\hat{X}$ and R.

$\triangleright$ Substituting into (5.10.31) and (5.10.35) the expressions $\varphi(y, x, t) = ay + a_1 x + a_0$, $\varphi_1(y, x, t) = by + b_1 x + b_0$ we obtain

$$\begin{aligned} d\hat{X} = {}& (aY + a_1 \hat{X} + a_0)dt + \{E[X(X^T - \hat{X}^T) \mid Y_{t_0}^t]b_1^T \\ & + \psi\nu\psi_1^T\}(\psi_1\nu\psi_1^T)^{-1}[dY - (bY + b_1\hat{X} + b_0)dt], \end{aligned} \tag{6.6.3}$$

$$\begin{aligned} dR = {}& E[(X - \hat{X})(Y^T a^T + X^T a_1^T + a_0^T) \\ & + (aY + a_1 X + a_0)(X^T - \hat{X}^T) + \psi\nu\psi^T \mid Y_{t_0}^t]dt \\ & - E[(X - \hat{X})(Y^T b^T + X^T b_1^T + b_0^T) + \psi\nu\psi_1^T \mid Y_{t_0}^t] \\ & \times (\psi_1\nu\psi_1^T)^{-1} E[(bY + b_1 X + b_0)(X^T - \hat{X}^T) + \psi_1\nu\psi^T \mid Y_{t_0}^t]dt \\ & + \sum_{l=1}^{n_1}\left\{\sum_{h=1}^{n} \alpha_{hl} E[(X_h - \hat{X}_h)(X - \hat{X})(X^T - \hat{X}^T) \mid Y_{t_0}^T]\right. \\ & \left. + E[X - \hat{X} \mid Y_{t_0}^t]\beta_l^T + \beta_l E[X^T - \hat{X}^T \mid Y_{t_0}^t]\right\} \\ & \times \left[dY_l - \left(\sum_{k=1}^{n_1} b_{lk} Y_k + \sum_{\rho=1}^{n} b_{1l\rho}\hat{X}_\rho + b_{0l}\right)dt\right] \end{aligned} \tag{6.6.4}$$

where α_{hr} are the elements of the matrix $\alpha = b_1^T(\psi_1\nu\psi_1^T)^{-1}$ and β_r is the rth column of the matrix $\psi\nu\psi_1^T(\psi_1\nu\psi_1^T)^{-1}$. Taking into account that

$$\begin{aligned} E[X(X^T - \hat{X}^T) \mid Y_{t_0}^t] &= E[(X - \hat{X})X^T \mid Y_{t_0}^t] \\ &= E[(X - \hat{X})(X^T - \hat{X}^T) \mid Y_{t_0}^t] = R, \end{aligned}$$

$$E[X - \hat{X} \mid Y_{t_0}^t] = \hat{X} - \hat{X} = 0,$$

$$E[(X - \hat{X})Y^T \mid Y_{t_0}^t] = E[X - \hat{X} \mid Y_{t_0}^t]Y^T = 0$$

and that the conditional central third order moments are equal to zero in consequence of the normality of the conditional distribution we transform formulae (6.6.3), (6.6.4) into the form

$$
\begin{aligned}
d\hat{X} &= (aY + a_1 X + a_0)dt \\
&\quad + (Rb_1^T + \psi\nu\psi_1^T)(\psi_1\nu\psi_1^T)^{-1}[dY - (bY + b_1\hat{X} + b_0)dt], \quad (6.6.5) \\
dR &= [a_1 R + Ra_1^T + \psi\nu\psi^T \\
&\quad - (Rb_1^T + \psi\nu\psi_1^T(\psi_1\nu\psi_1^T)^{-1}(bR + \psi_1\nu\psi^T)]dt. \triangleleft \qquad (6.6.6)
\end{aligned}
$$

Thus, in the case of linear Eqs. (6.6.1), (6.6.2) formulae (5.10.31) and (5.10.32) give the set of equations determining the optimal estimate $\hat{X}$ of the state vector of the system X and its conditional covariance matrix R characterizing the accuracy of the estimate $\hat{X}$.

R e m a r k 1. Now pay attention to two circumstances. Firstly, the matrix Riccati equation (6.6.6) for the conditional covariance matrix R does not contain $\hat{X}$, and consequently, may be integrated separately. After determining R we may calculate the quantity

$$\beta = (Rb_1^T + \psi\nu\psi_1^T)(\psi_1\nu\psi_1^T)^{-1} \qquad (6.6.7)$$

and determine the estimate $\hat{X}$ of the state vector of the system X by the integration of Eq.(6.6.5) which may be rewritten in the form

$$d\hat{X} = (aY + a_1\hat{X} + a_0)dt + \beta[dY - (bY + b_1\hat{X} + b_0)dt]. \qquad (6.6.8)$$

R e m a r k 2. Eq. (6.6.6) does not contain the results of measurements. Therefore the conditional covariance matrix of the state vector X coincides with the conditional covariance matrix of the error vector $\tilde{X} = \hat{X} - X$ of the estimate $\hat{X}$. This gives the opportunity to determine R and to calculate the coefficient β in Eq. (6.6.8) beforehand when the measurements are not yet performed. Then the estimate $\hat{X}$ will be determined by the integration of Eq. (6.6.8) using the current results of the measurements. Eq. (6.6.8) is sufficiently simple and may be integrated in on-line data processing.

Eqs. (6.6.6) and (6.6.8) with the initial conditions

$$\hat{X}(t_0) = \hat{X}_0 = E[X_0 \mid Y_0], \quad R(t_0) = R_0 = E[(X_0 - \hat{X}_0)(X_0^T - \hat{X}_0^T) \mid Y_0]$$
$$(6.6.9)$$

determine the optimal estimate $\hat{X}$ of the vector X and the covariance matrix R of the error $\tilde{X} = \hat{X} - X$ at all $t > t_0$. This optimal estimate $\hat{X}$ is unbiased at all $t > t_0$, as by virtue of the formula of total expectation and formula (5.10.6)

$$E(\hat{X}_t - X_t) = E\{E(\hat{X}_t - X_t) \mid Y_{t_0}^t]\} = E(\hat{X}_t - \hat{X}_t) = 0.$$

Eqs. (6.6.6)–(6.6.8) solve completely and exactly the problem of optimal linear filtering of the state of a linear system defined by Eqs. (6.6.1), (6.6.2).

6.6.2. Kalman–Bucy Filter

Eqs. (6.6.6) and (6.6.8) were first obtained by Kalman and Bucy in 1960s in the case of $a = b = 0$. This case corresponds to the practical problem of filtering of the signal X determined by (6.6.2) at $a = 0$ when the signal $b_1 X + b_0$ is measured with an additive disturbance representing a white noise. In this case Eq. (6.6.1), (6.6.2) may be rewritten in the form

$$\dot{X} = a_1 X + a_0 + \psi V, \quad Z = b_1 X + b_0 + \psi_1 V. \tag{6.6.10}$$

V being a white noise, i.e. the derivative of a Wiener process $W(t)$. The equation of the optimal filter (6.6.8) has in this case the form

$$\dot{\hat{X}} = a_1 \hat{X} + a_0 + \beta(Z - b_1 \hat{X} - b_0). \tag{6.6.11}$$

Eq. (6.6.11) determines the structure of the optimal filter. Namely, the optimal filter may be obtained from the given system (the system generating the signal X; Fig.6.6.1a by the installation of the amplifier with the gain β before its input and by applying negative feedback containing the amplifier with the gain b_1 to the system obtained (Fig.6.6.1b). Receiving the measured process Z at the input with the function of time b_0, subtracted from it the filter obtained in this way will give the optimal estimate $\hat{X}$ of the vector X at the output.

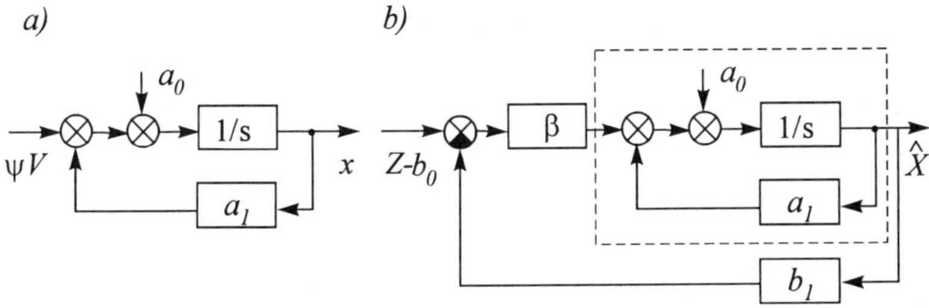

Fig. 6.6.1

As the process $Y(t)$ representing the integral of the measured process $Z(t)$ is not used in the optimal filter, the conditional expectation of the vector X_0, $\hat{X}_0 = EX_0$ should be assumed as the initial value of the estimate $\hat{X}$ at $t = t_0$, and the conditional covariance matrix of the random vector X_0, $R_0 = E(X_0 - EX_0)(X_0^T - EX_0^T)$ should be assumed as the initial value of the matrix R while integrating Eq. (6.6.6) and calculating the gain β. But these conditional characteristics are usually unknown. This compels one to take arbitrary initial of $\hat{X}$ and R. Certainly, $\hat{X}$ will not be the optimal estimate of the vector X in this case and may be only an asymptotically optimal estimate if only the first equation of (6.6.10) and Eq. (6.6.6) describe a stable process (i.e. if the system described by the first equation of (6.6.10) and Eq. (6.6.6) is stable).

It goes without saying that the gains β and b_1 represent in the general case the matrices. So we consider here the amplifiers of vector signals performing linear transformations of the vectors received at the input which are determined by the corresponding matrices, i.e. by matrix gains.

Optimal linear filters constructed by the method outlined are usually called *Kalman–Bucy filters*.

For the linear stochastic systems the problems of Kalman–Bucy filter design are usually solved by means of the universal or the specialized computer libraries and dialog packages.

E x a m p l e 6.6.1. Find the optimal filter for filtering a sinusoid signal of a given frequency ω_0 which is measured with an additive normally distributed white noise independent of the signal.

The sinusoid signal $X_1(t)$ and its derivative $X_2(t) = \dot{X}_1(t)$ may be considered as the components of the state vector of a system described by the differential equation

$$\frac{d}{dt}\begin{bmatrix} X_1 \\ X_2 \end{bmatrix} = \begin{bmatrix} 0 & 1 \\ -\omega_0^2 & 0 \end{bmatrix}\begin{bmatrix} X_1 \\ X_2 \end{bmatrix}. \tag{I}$$

The measured process is determined by the formula

$$\dot{Y} = Z = X_1 + V = [1\ 0]X + V. \tag{II}$$

Consequently, in this case $a = a_0 = b = b_0 = \psi = 0$, $b_1 = [1\ 0]$, $\psi_1 = 1$,

$$a_1 = \begin{bmatrix} 0 & 1 \\ -\omega_0^2 & 0 \end{bmatrix}.$$

Eq. (6.6.6) represents the set of equations

$$\dot{R}_{11} = 2R_{12} - \nu^{-1}R_{11}^2, \quad \dot{R}_{12} = \omega_0^2 R_{11} + R_{22} - \nu^{-1}R_{11}R_{12},$$

$$\dot{R}_{22} = -2\omega_0^2 R_{12} - \nu^{-1} R_{12}^2. \tag{III}$$

After determining R_{11}, R_{12} and R_{22} by the integration of these equations with the initial conditions $R_{11}(t_0) = EX_{01}^{02}$, $R_{12}(t_0) = EX_{01}^0 X_{02}^0$, $R_{22}(t_0) = EX_{02}^{02}$ we find m.s.e. of the filtering of the signal X_1, $R_{11} = E[(X_1 - X_1)^2 \mid Y_{t_0}^t]$ and of its derivative X_2, $R_{22} = E[(X_2 - \hat{X}_2)^2 \mid Y_{t_0}^t]$, and the gain

$$\beta = \nu^{-1} R b_1^T = \nu^{-1} \begin{bmatrix} R_{11} & R_{12} \\ R_{12} & R_{22} \end{bmatrix} \begin{bmatrix} 1 \\ 0 \end{bmatrix} = \nu^{-1} \begin{bmatrix} R_{11} \\ R_{12} \end{bmatrix} = \begin{bmatrix} \beta_1 \\ \beta_2 \end{bmatrix}. \tag{IV}$$

Eq. (6.6.8) for optimal estimates $\hat{X}_1$, $\hat{X}_2$ has in this case the form

$$\frac{d}{dt} \begin{bmatrix} \hat{X}_1 \\ \hat{X}_2 \end{bmatrix} = \begin{bmatrix} 0 & 1 \\ -\omega_0^2 & 0 \end{bmatrix} \begin{bmatrix} \hat{X}_1 \\ \hat{X}_2 \end{bmatrix} + \nu^{-1} \begin{bmatrix} R_{11} \\ R_{12} \end{bmatrix} (Z - \hat{X}_1). \tag{V}$$

The block diagram of the optimal filter found is shown in Fig.6.6.2.

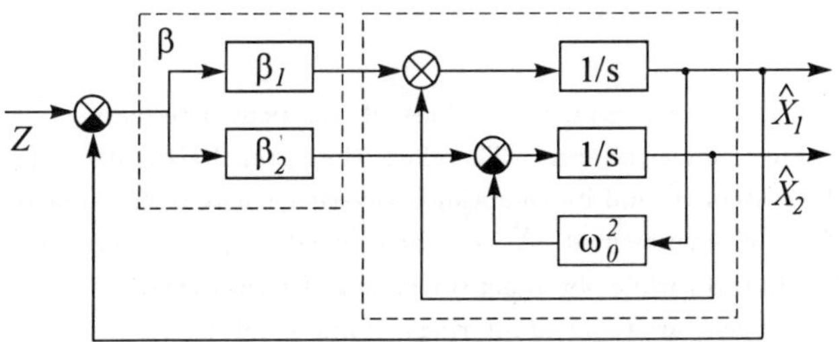

Fig. 6.6.2

6.6.3. Case of Equations Linear in State Vector

Let us now consider the more general case of Eqs. (5.10.9) linear only in the state vector of the system

$$dY = [b_1(Y,t)X + b_0(Y,t)]dt + \psi_1(Y,t)dW, \tag{6.6.12}$$
$$dX = [a_1(Y,t)X + a_0(Y,t)]dt + \psi(Y,t)dW. \tag{6.6.13}$$

In this case for any realization $y(t)$ of the measurement process $Y(t)$ Eq. (6.6.13) is linear in X. Therefore, we may make the conclusion that the conditional distribution of the random vector X_t is normal for any realization $y_{t_0}^t = \{y_\tau; \tau \in [t_0, t]\}$ of the measurements $Y_{t_0}^t$ at all $t > t_0$ if the initial distribution of X is normal.

▷ Assuming the conditional distribution of X_t relative to $Y_{t_0}^t$ as normal we write for this case formulae (5.10.31) and (5.10.35) for the stochastic differentials of the optimal estimate $\hat{X}$ and the covariance matrix R of the state vector of the system. Substituting into (5.10.31) and (5.10.35) the expressions

$$\varphi(y, x, t) = a_1(y, t)x + a_0(y, t), \quad \varphi_1(y, x, t) = b_1(y, t)x + b_0(y, t), \quad (6.6.14)$$

taking into account the independence of ψ on X and recalling that all the central moments of the third order are equal to zero for a normal distribution, we shall obtain in the same way as in Subsection 6.6.1

$$d\hat{X} = [a_1(Y, t)\hat{X} + a_0(Y, t)]dt + [Rb_1(Y, t)^T + (\psi\nu\psi_1^T)(Y, t)]$$
$$\times (\psi_1\nu\psi_1^T)^{-1}(Y, t)\{dY - [b_1(Y, t)\hat{X} + b_0(Y, t)]dt\}, \quad (6.6.15)$$

$$dR = \{a_1(Y, t)R + Ra_1(Y, t)^T + (\psi\nu\psi^T)(Y, t) - [Rb_1(Y, t)^T$$
$$+ (\psi\nu\psi_1^T)(Y, t)](\psi_1\nu\psi_1^T)^{-1}(Y, t)[b_1(Y, t)R + (\psi\nu\psi_1^T)(Y, t)]\}dt. \ \triangleleft$$
$$(6.6.16)$$

These equations represent a closed set of equations determining $\hat{X}$ and R as well in the case of linear filtering. Therefore, the optimal estimate $\hat{X}$ of the state vector of the system X and its conditional covariance matrix R characterizing the accuracy of the optimal estimate $\hat{X}$, may be calculated by the integration of the set Eqs. (6.6.15), (6.6.16) while obtaining the results of measurements.

Notice that contrary to linear filtering it is impossible in this case to calculate R beforehand when the results of observations are not yet received, as the coefficients of Eq. (6.6.16) depend on the results of measurements. Therefore the optimal filter in this case must perform the simultaneous integration of both Eqs. (6.6.15), (6.6.16). It leads to the essential increase of the order of the optimal filter.

▷ Now we shall prove that the conditional distribution of the state vector of the system is normal in this case. For this purpose it is sufficient to show that Eq. (5.10.21) for the conditional characteristic function $g_t(\lambda)$ of the vector X_t has the solution

$$g_t(\lambda) = \exp\left\{i\lambda^T \hat{X} - \frac{1}{2}\lambda^T R\lambda\right\} \quad (6.6.17)$$

where $\hat{X}$ and R are determined by Eqs. (6.6.15), (6.6.16).

Substituting into Eq. (5.10.31) expressions (6.6.14) of the functions φ and φ_1 we reduce it to the form

$$dg_t(\lambda) = E[\{i\lambda^T(a_1 X + a_0) - \frac{1}{2}\lambda^T \psi\nu\psi^T\}e^{i\lambda^T X} \mid Y_{t_0}^t]dt$$

$$+E[\{(X^T-\hat{X}^T)b_1^T+i\lambda^T\psi\nu\psi_1^T\}e^{i\lambda^TX}\mid Y_{t_0}^t](\psi_1\nu\psi_1^T)^{-1}[dY-(b_1\hat{X}+b_0)dt]$$
$$(6.6.18)$$

where the arguments of the functions a_1, a_0, b_1, b_0, ψ and ψ_1 are omitted for brevity. Taking into account that

$$E[iXe^{i\lambda^TX}\mid Y_{t_0}^t]=\frac{\partial g_t(\lambda)}{\partial\lambda} \qquad (6.6.19)$$

we get from Eq. (6.6.18)

$$dg_t(\lambda)=\left[\lambda^Ta_1\frac{\partial g_t(\lambda)}{\partial\lambda}+\left(i\lambda^Ta_0-\frac{1}{2}\lambda^T\psi\nu\psi^T\right)g_t(\lambda)\right]dt$$

$$-\left[i\frac{\partial^Tg_t(\lambda)}{\partial\lambda}b_1^T+(\hat{X}^Tb_1^T-i\lambda^T\psi\nu\psi_1^T)g_t(\lambda)\right]$$

$$\times(\psi_1\nu\psi_1^T)^{-1}\left[dY-(b_1\hat{X}+b_0)dt\right]. \qquad (6.6.20)$$

We evaluate separately the left- and right-hand sides of this equation for the function $g_t(\lambda)$ determined by formula (6.6.17). Using differentiation formula (5.1.15) of a composite function in the case of a Wiener process W, taking into consideration that the stochastic differential of the process $\hat{X}(t)$ is determined by formula (6.6.15) and that $\left(Rb^T+\psi\nu\psi_1^T\right)(\psi_1\nu\psi_1^T)^{-1}\psi_1$ plays the role of the matrix Y in (5.1.15) in this case, we find the stochastic differential of the process $g_t(\lambda)$ determined by formula (6.6.17):

$$dg_t(\lambda)=g_t(\lambda)\left\{i\lambda^Td\hat{X}-\frac{1}{2}\mathrm{tr}[\lambda\lambda^T(Rb_1^T+\psi\nu\psi_1^T)\right.$$

$$\left.\times(\psi_1\nu\psi_1^T)^{-1}\psi_1\nu\psi_1^T(\psi_1\nu\psi_1^T)^{-1}(b_1R+\psi_1\nu\psi^T)]dt\right\}$$

$$-\frac{1}{2}\lambda^TdR\lambda g_t(\lambda)=g_t(\lambda)\left\{\left[i\lambda^T(a_1\hat{X}+a_0)\right.\right.$$

$$\left.-\frac{1}{2}\lambda^T(Rb_1^T+\psi\nu\psi_1^T)(\psi_1\nu\psi_1^T)^{-1}(b_1R+\psi_1\nu\psi^T)\lambda\right]dt$$

$$-\frac{1}{2}\lambda^T[a_1R+Ra_1^T+\psi\nu\psi^T-(Rb_1^T+\psi\nu\psi_1^T)$$

$$\left.\times(\psi_1\nu\psi_1^T)^{-1}(b_1R+\psi_1\nu\psi^T)]\lambda\right\}dt+g_t(\lambda)i\lambda^T(Rb_1^T$$

$$+\psi\nu\psi_1^T)(\psi_1\nu\psi_1^T)^{-1}[dY - (b_1\hat{X} + b_0)dt]$$

$$= g_t(\lambda)\left[i\lambda^T(a_1\hat{X} + a_0) - \frac{1}{2}\lambda^T(a_1R + Ra_1^T + \psi\nu\psi^T)\lambda\right]dt$$

$$+g_t(\lambda)i\lambda^T(Rb_1^T + \psi\nu\psi_1^T)(\psi_1\nu\psi_1^T)^{-1}\left[dY - (b_1\hat{X} + b_0)dt\right]. \quad (6.6.21)$$

As $\lambda^T Ra_1^T = \lambda^T a_1 R\lambda$, we get from (6.6.21)

$$dg_t(\lambda) = g_t(\lambda\left[i\lambda^T(a_1\hat{X} + a_0) - \lambda^T a_1 R\lambda - \frac{1}{2}\lambda^T\psi\nu\psi^T\lambda\right]dt$$

$$+g_t(\lambda)i\lambda^T(Rb_1^T + \psi\nu\psi_1^T)(\psi_1\nu\psi_1^T)^{-1}\left[dY - (b_1\hat{X} + b_0)dt\right]. \quad (6.6.22)$$

For evaluating the right-hand side of Eq. (6.6.20) we notice that (6.6.17) yields

$$\frac{\partial g_t(\lambda)}{\partial\lambda} = g_t(\lambda)(i\hat{X} - R\lambda).$$

Using this formula we find

$$\left[\lambda^T a_1\frac{\partial g_t(\lambda)}{\partial\lambda} + \left(i\lambda^T a_0 - \frac{1}{2}\lambda^T\psi\nu\psi^T\lambda\right)g_t(\lambda)\right]dt$$

$$-\left[i\frac{\partial^T g_t(\lambda)}{\partial\lambda}b_1^T + (\hat{X}^T b_1^T - i\lambda^T\psi\nu\psi_1^T)g_t(\lambda)\right]$$

$$\times(\psi_1\nu\psi_1^T)^{-1}\left[dY - (b_1\hat{X} + b_0)dt\right]$$

$$= g_t(\lambda)\left[i\lambda^T(a_1\hat{X} + a_0) - \lambda^T a_1 R\lambda - \frac{1}{2}\lambda^T\psi\nu\psi^T\lambda\right]dt$$

$$+g_t(\lambda)(i\lambda^T Rb_1^T + i\lambda^T\psi\nu\psi_1^T)(\psi_1\nu\psi_1^T)^{-1}[dY - (b_1\hat{X} + b_0)dt]. \quad (6.6.23)$$

Eq. (6.6.23) is identical to (6.6.22). Consequently, the normal conditional characteristic function determined by formula (6.6.17) with the parameters $\hat{X}$ and R determined by Eqs. (6.6.15), (6.6.16) satisfies Eq. (6.6.20). This proves the normality of the conditional distribution of the random vector X_t and the validity of Eqs. (6.6.15), (6.6.16) which determine the parameters $\hat{X}_t$ and R_t of this distribution if the initial values of $\hat{X}$ and R at the moment t_0 represent the conditional expectation and the conditional covariance matrix of the random vector $X_0 = X_{t_0}$ relative to $Y_0 = Y_{t_0}$ respectively. ◁

The outlined generalization of the linear optimal filtering theory for the case of Eq. (5.10.9) linear only in the state vector X was given in (Liptser and Shiryaev 1968, 1977).

R e m a r k. The method outlined also allows to estimate unknown parameter θ in Eqs. (6.6.12), (6.6.13) provided that these equation are linear in X, θ. In this case the usual approach reduces the problem to equations of the form of Eqs. (6.6.12), (6.6.13) for the extended state vector $X' = \left[\, X^T \theta^T \,\right]^T$.

E x a m p l e 6.6.2. The process $Y(t)$ is determined by the Itô stochastic differential equation

$$dY = \left[\, b_1(Y,t)\theta + b_0(Y,t) \,\right] dt \psi_1(y,t)dW \tag{I}$$

where θ is the unknown multi-dimensional vector parameter. Find the optimal estimate of θ at each $t > t_0$ using the measurements results of the process Y in time interval $[t_0, t]$.

Replacing the vector θ by the random process $\Theta(t)$ which is determined by the differential equation $d\Theta(t) = 0$, and assuming Θ as the state vector X of the corresponding system we obtain the differential equation of form Eqs. (6.6.12), (6.6.13) at $a_1(Y,t) = 0$, $a_0(Y,t) = 0$, $\psi(Y,t) = 0$. Eqs. (6.6.15), (6.6.16) take in this case the form

$$d\hat{\Theta} = Rb_1(Y,t)^T(\psi_1\nu\psi_1^T)^{-1}(Y,t)\left\{dY - [b_1(Y,t)\hat{\Theta} + b_0(Y,t)]dt\right\}, \tag{II}$$

$$dR = -Rb_1(Y,t)^T(\psi_1\nu\psi_1^T)^{-1}(Y,t)b_1(Y,t)Rdt. \tag{III}$$

As the parameter θ is unknown and may be nonrandom the initial values of $\hat{\Theta}$ and R are taken arbitrarily. The equations obtained will give in this case the estimate $\hat{\Theta}$ which will be optimal assuming that the parameter θ is random and has the normal distribution which the expectation $\hat{\Theta}_0$ and the covariance matrix R_0.

P r o b l e m s

6.1. Show that for the stationary random processes with covariance functions $De^{-\alpha|\tau|}(1 + \alpha|\tau|)$ and $De^{-\alpha|\tau|}\left(1 + \alpha|\tau| + \frac{\alpha^2}{3}\tau^2\right)$ the intensities of the equivalent white noises are equal to $4D/\alpha$ and $16D/3\alpha$ respectively. Calculate the correlation intervals.

6.2. Show that for the process reducible to a stationary one determined by formula $Y(t) = b_0(t) + b_1(t)X(t)$, where $X(t)$ being the stationary function the intensity of the equivalent white noise and the correlation interval are determined by

$$\nu(t) = b_1(t) \int\limits_{-\infty}^{\infty} b_1(t+\tau)k_x(\tau)d\tau, \quad \tau_c = \frac{1}{2}\max_t \frac{\nu(t)}{b_1^2(t)k_x(0)}.$$

Calculate $\nu(t)$ and τ_c for the process determined by formula $Y(t) = b_0(t) + b_1(t)X(\varphi(t))$, where $\varphi(t)$ being monotone increasing function.

6.3. Show that for the two-dimensional stationary system of Problem 1.3 with uncorrelated inputs $X_1(t)$ and $X_2(t)$ with spectral densities $s_1(\omega)$ and $s_2(\omega)$ the elements of the spectral density matrix $s_y(\omega)$ of the process $Y(t) = [\,Y_1(t)\,Y_2(t)\,]^T$ are determined by

$$s_{11}(\omega) = s_1(\omega)|\Phi_{11}(i\omega)|^2 + s_2(\omega)|\Phi_{12}(i\omega)|^2,$$

$$s_{12}(\omega) = s_1(\omega)\Phi_{11}(i\omega)\overline{\Phi_{21}(i\omega)} + s_2(\omega)\Phi_{12}(i\omega)\overline{\Phi_{22}(i\omega)},$$

$$s_{21}(\omega) = s_1(\omega)\overline{\Phi_{11}(i\omega)}\Phi_{21}(i\omega) + s_2(\omega)\overline{\Phi_{12}(i\omega)}\Phi_{22}(i\omega),$$

$$s_{22}(\omega) = s_1(\omega)|\Phi_{21}(i\omega)|^2 + s_2(\omega)|\Phi_{22}(i\omega)|^2.$$

Find the cross-spectral density of the processes $X(t) = [X_1(t)X_2(t)]^T$ and $Y(t)$.

6.4. Check that for the two-dimensional stationary system of Problem 1.4 with a stationary input $X(t) = [X_1(t)X_2(t)]^T$ with spectral density $s(\omega)$ the elements of the matrices $s_y(\omega)$ and $s_{xy}(\omega)$ are determined by the formulae

$$s_{y_1y_1} = s_{11}(\omega)\,|\psi_1(\omega)|^2 + s_{22}(\omega)\,|\psi_2(\omega)|^2 + s_{12}(\omega)[\psi_1(\omega)\overline{\psi_2(\omega)} + \psi_2(\omega)\overline{\psi_1(\omega)}],$$

$$s_{y_1y_2} = -s_{11}(\omega)\psi_1(\omega)\overline{\psi_2(\omega)} + s_{22}(\omega)\psi_2(\omega)\overline{\psi_1(\omega)} + s_{12}(\omega)(\,|\psi_1(\omega)|^2 - |\psi_2(\omega)|^2),$$

$$s_{y_2y_1} = -s_{11}(\omega)\psi_2(\omega)\overline{\psi_2(\omega)} + s_{12}(\omega)(\,|\psi_1(\omega)|^2 - |\psi_2(\omega)|^2),$$

$$s_{y_2y_2} = s_{11}(\omega)\,|\psi_2(\omega)|^2 + s_{22}(\omega)\,|\psi_1(\omega)|^2 - s_{12}(\omega)[\overline{\psi_1(\omega)}\psi_2(\omega) + \psi_1(\omega)\overline{\psi_2(\omega)}],$$

$$s_{x_1y_1} = s_{11}(\omega)\overline{\psi_1(\omega)} + s_{12}(\omega)\overline{\psi_2(\omega)},$$

$$s_{x_1y_2} = -s_{11}(\omega)\overline{\psi_2(\omega)} + s_{12}(\omega)\overline{\psi_1(\omega)},$$

$$s_{x_2y_1} = s_{12}(\omega)\overline{\psi_1(\omega)} + s_{22}(\omega)\overline{\psi_2(\omega)},$$

$$s_{x_2y_2} = -s_{12}(\omega)\overline{\psi_2(\omega)} + s_{22}(\omega)\overline{\psi_1(\omega)},$$

where

$$\psi_1(\omega) = \left[\varepsilon(\omega^2 + \varepsilon^2 + \omega^2) + i\omega(\omega^2 - \varepsilon^2 - \omega^2)\right] c^2(\omega),$$
$$\psi_2(\omega) = \left[(\omega^2 + \varepsilon^2 - \omega^2) - 2\varepsilon i\omega\right] c^2(\omega)\omega,$$
$$c^{-2}(\omega) = (\omega^2 + \varepsilon^2 - \omega^2)^2 + 4\varepsilon^2\omega^2).$$

6.5. Show that for the system of the third order of Problem 1.9 at $a = -2\varepsilon I$, $b = I$, $s_x(\omega) = sI$, I being the 3×3 unit matrix, the spectral density of the output is equal to $s_y(\omega) = sI/(\omega^2 + 4\varepsilon^2)$.

6.6. Show that under the conditions of Problem 1.10 at $X = [0\,0\,\sqrt{2D\alpha}]^T V$, V being a white noise of unit intensity, the spectral density of the output is equal to

$$s_y(\omega) = \frac{2D\alpha}{(2\pi)^3}
\begin{bmatrix}
|\Phi_{13}(i\omega)|^2 & \Phi_{13}(i\omega)\overline{\Phi_{23}(i\omega)} & \Phi_{13}(i\omega)\overline{\Phi_{33}(i\omega)} \\[2mm]
\Phi_{23}(i\omega)\overline{\Phi_{13}(i\omega)} & |\Phi_{23}(i\omega)|^2 & \Phi_{23}(i\omega)\overline{\Phi_{33}(i\omega)} \\[2mm]
\Phi_{33}(i\omega)\overline{\Phi_{13}(i\omega)} & \Phi_{33}(i\omega)\overline{\Phi_{23}(i\omega)} & |\Phi_{33}(i\omega)|^2
\end{bmatrix}.$$

6.7. For the system with two degrees of freedom of Problem 1.11 with the vector of the generalized forces representing a stationary process with spectral density $s_Q(\omega)$ show that the elements of the matrices $s_q(\omega)$ and $s_{Qq}(\omega)$ are determined by the formulae

$$s_{q_h q_l}(\omega) = s_{11}(\omega)\Phi_{h1}(i\omega)\overline{\Phi_{l1}(i\omega)} + s_{22}(\omega)\Phi_{h2}(i\omega)\overline{\Phi_{l2}(i\omega)}$$

$$+ s_{12}(\omega)\left[\Phi_{h1}(i\omega)\overline{\Phi_{l2}(i\omega)} + \Phi_{h2}(i\omega)\overline{\Phi_{l1}(i\omega)}\right],$$
$$s_{Q_1 q_h}(\omega) = s_{11}(\omega)\overline{\Phi_{h1}(i\omega)} + s_{12}(\omega)\overline{\Phi_{h2}(i\omega)},$$
$$s_{Q_2 q_h}(\omega) = s_{12}(\omega)\overline{\Phi_{h1}(i\omega)} + s_{22}(\omega)\overline{\Phi_{h2}(i\omega)} \quad (h = 1, 2).$$

6.8. Find the spectral density $s_y(\omega)$ and the covariance matrix K_y of the stationary vector process $Y(t)$ in the stationary system

$$\begin{pmatrix} \dot{Y}_1 \\ \dot{Y}_2 \\ \dot{Y}_3 \\ \dot{Y}_4 \end{pmatrix} = \begin{pmatrix} 0 & 1 & 0 & 0 \\ -\omega_0^2 & -2\varepsilon & 1 & 0 \\ 0 & 0 & 0 & 1 \\ 0 & 0 & -\omega_1^2 & -2\varepsilon_1 \end{pmatrix} \begin{pmatrix} Y_1 \\ Y_2 \\ Y_3 \\ Y_4 \end{pmatrix} + \begin{pmatrix} 0 \\ 0 \\ q_1 \\ q_2 \end{pmatrix} V,$$

where V is a white noise of unit intensity, and $q_1, q_2, \omega_0, \omega_1, \varepsilon, \varepsilon_1, (\varepsilon, \varepsilon_1 > 0)$ are some constant coefficients.

6.9. Find the spectral density $s_y(\omega)$ and the variance D_y of the vertical oscillations of the body of a car in linear approximation (Problem 1.15) assuming that the spectral density of the road microprofile is known, $s_l(\omega) = s(\rho)/v$, $\rho = 2\pi/\sigma$.

6.10. Find the spectral density $s_x(\omega)$ and the variance D_x of the stationary random oscillations of an object with dynamic damper of oscillations in Problem 1.16 under the action of the random disturbing force with the spectral density $s_F(\omega)$.

6.11. For the stationary linear system of Problem 1.12 with three degrees of freedom at $A = I$, $C = \omega_0^2 I$, $B = 2\varepsilon I$, $B' = C' = 0$, $s_Q(\omega) = (2\pi)^3 s I$ (I being the unit 3×3-matrix) derive the formulae for the spectral density $s_y(\omega)$ of the stationary process $Y(t) = [\, q_1(t) \, q_2(t) \, q_3(t) \, \dot{q}_1(t) \, \dot{q}_2(t) \, \dot{q}_3(t) \,]^T$. Find also the covariance matrix K_y.

6.12. Derive the formulae for the spectral and the cross-spectral densities $s_x(\omega)$ and $s_{xy}(\omega)$ for the system of Problem 1.13 at $B = 2\varepsilon A$ and $B = \eta C$ assuming that the spectral density $s_x(\omega)$ of the input is known.

6.13. Show that for the linear stationary system with the transfer function $\Phi(s) = (d_1 s + d_0)(c_3 s^3 + c_2 s^2 + c_1 s + c_0)^{-1}$ the input being a white noise with the intensity ν, the variance of the output is equal to $D_y = \dfrac{\nu(c_0 d_1^2 + c_2 d_0^2)}{2c_0(c_1 c_2 - c_0 c_3)}$. Calculate also the covariance k_{xy}.

6.14. Show that the covariance function $k_y(\tau)$ and the spectral density $s_y(\omega)$ of the random function $Y(t) = X_1(t)X_2(t)$, where $X_1(t)$ and $X_2(t)$ are the components of the two-dimensional real normally distributed random process with the known covariance function and spectral density

$$k_x(\tau) = \begin{bmatrix} k_{11}(\tau) & k_{12}(\tau) \\ k_{21}(\tau) & k_{22}(\tau) \end{bmatrix}, \quad s_x(\omega) = \begin{bmatrix} s_{11}(\omega) & s_{12}(\omega) \\ s_{21}(\omega) & s_{22}(\omega) \end{bmatrix}$$

are calculated by the formulae

$$k_y(\tau) = k_1(\tau)k_2(\tau) + k_{12}(\tau)k_{21}(\tau) + m_1^2 k_1(\tau) + m_2^2 k_1(\tau),$$

$$s_y(\omega) = \int\limits_{-\infty}^{\infty} s_1(\omega - \zeta)s_2(\zeta)d\zeta + \int\limits_{-\infty}^{\infty} s_{12}(\omega - \zeta)s_{21}(\zeta)d\zeta$$

$$+ m_1^2 s_2(\omega) + m_2^2 s_1(\omega).$$

Give the generalization to the case of $Y(t) = X_1^T(t)X_2(t)$ where $X_1(t)$ and $X_2(t)$ are the n-dimensional vectors.

6.15. Find the variances and the covariances of Y_1 and Y_2 in Problem 1.6 at the input X being the white noise with the intensity ν.

6.16. Find the variance of Y_1 in Problem 1.7 for X being the white noise with the intensity ν.

6.17. Find the variance and the covariance function of Y_1 in Problem 1.7 for X being the white noise with intensity ν.

6.18. Deduce formulae for the expectations of the number of intersections Y of a constant level a and of the number of stationary points Y_1 during the time T for a stationary normally distributed random process $X(t)$ with the expectation m_x and the covariance function $k_x(\tau)$:

$$m_y = \frac{T}{\pi}\sqrt{-\frac{k_x''(0)}{k_x(0)}}\exp-\left[(m_x-a)^2/k_x(0)\right], \quad m_{y1} = \frac{T}{\pi}\sqrt{-\frac{k_x^{IV}(0)}{k_x''(0)}}$$

6.19. Show that for the system

$$\dot{Y}_1 = Y_2, \quad \dot{Y}_2 = a_0 + V, \tag{I}$$

where a_0 is the constant, V is the white noise of the constant intensity ν, the expectations, the variance, the covariance and the cross-covariance functions of the processes $Y_1(t)$ and $Y_2(t)$ at zero initial conditions are determined by the formulae

$$m_1 = a_0 t^2/2, \quad m_2 = a_0 t, \quad k_{11} = \nu t^3/3, \quad k_{12} = \nu t^2/2, \quad k_{22} = \nu t, \tag{II}$$

$$K_{11}(t_1,t_2) = \nu t_1 t_2 \min(t_1,t_2), \quad K_{12}(t_1,t_2) = \nu t_1/2\min(t_1,t_2),$$
$$K_{21}(t_1,t_2) = \nu t_2/2\min(t_1,t_2), \quad K_{22}(t_1,t_2) = \nu\min(t_1,t_2). \tag{III}$$

6.20. Prove that for the Langevin equations

$$\dot{Y}_1 = Y_2, \quad \dot{Y}_2 = -2\varepsilon Y_2 + V, \tag{I}$$

where $\varepsilon > 0$ is the constant, V is the white noise of constant intensity ν, the variances and the covariance of the processes $Y_1(t)$ and $Y_2(t)$ at zero initial conditions are determined by the formulae:

$$k_{11} = \nu(4\varepsilon t - 3 + 4e^{-2\varepsilon t} - e^{-4\varepsilon t})/16\varepsilon^3,$$

$$k_{12} = \nu(1 - 2e^{-2\varepsilon t} + e^{-4\varepsilon t})/8\varepsilon^3, \quad k_{22} = \nu(1 - e^{-4\varepsilon t})/4\varepsilon. \tag{II}$$

6.21. Show that under the conditions of Problem 1.5, the generalized force Q being the white noise of intensity ν, Eqs. (6.1.8) and (6.1.11) have the form

$$\dot{k}_{11} = 2k_{12}/A, \quad \dot{k}_{12} = -Ck_{11} - (B/A)k_{12} + (1/A)k_{22},$$
$$\dot{k}_{22} = -2Ck_{12} - (2B/A)k_{22} + \nu, \tag{I}$$
$$\partial K_{11}(t_1,t_2)/\partial t_2 = K_{12}(t_1,t_2)/A,$$
$$\partial K_{12}(t_1,t_2)/\partial t_2 = -CK_{11}(t_1,t_2) - BK_{12}(t_1,t_2)/A,$$
$$\partial K_{21}(t_1,t_2)/\partial t_2 = K_{22}(t_1,t_2)/A,$$
$$\partial K_{22}(t_1,t_2)/\partial t_2 = -CK_{21}(t_1,t_2) - BK_{22}(t_1,t_2)/A. \tag{II}$$

Write the explicit formulae for the solution of Eqs. (I), (II).

6.22. Prove that the covariance and cross-covariance functions of the stationary and cross-stationary processes $Y_1(t)$ and $Y_2(t)$ of Example 1.1.9, the input $X(t)$ being a white noise of the intensity ν, are determined by the formulae

$$k_{11}(\tau) = \frac{\nu}{4\zeta\omega_0^3}e^{-\zeta\omega_0|\tau|}(\cos\omega_c\tau + \gamma\sin\omega_c|\tau|),$$

$$k_{22}(\tau) = \frac{\nu}{4\zeta\omega_0}e^{-\zeta\omega_0|\tau|}(\cos\omega_c\tau - \gamma\sin\omega_c|\tau|),$$

$$k_{12}(\tau) = -\frac{\nu}{4\zeta\omega_c\omega_0}e^{-\zeta\omega_0|\tau|}\sin\omega_c|\tau| \ (\gamma = \zeta/\sqrt{1-\zeta^2}, \ \omega_c = \omega_0\sqrt{1-\zeta^2}).$$

6.23. Show that under the conditions of Problem 6.21 the variance of the coordinate and of the impulse in the regime of stationary random oscillations, i.e. when the oscillations represent a two-dimensional stationary random process, are determined by the formulae $k_{11} = \nu/2BC$, $k_{12} = 0$, $k_{22} = A\nu/2B$. Show that in the case of heavy damping $(B^2 > 4AC)$, the process of stationary oscillations settling consists of two stages: first, the oscillations of the impulse settle, then follows the slow settling of the oscillations of the coordinate.

6.24. Show that the set of Eqs. (6.1.8) for the system of Problem 1.3 at $x_h = \sum_{l=1}^{n} b_{hl}V_l$ $(h = 1, 2)$, where $V = [V_1 \ldots V_n]^T$ is a white noise of intensity ν, and b_{hl} are the coefficients depending on time has the form

$$\dot{k}_{11} = 2(a_{11}k_{11} + a_{12}k_{12}) + \sum_{r,h=1}^{n} \nu_{rh}b_{1r}b_{1h},$$

$$\dot{k}_{12} = a_{21}k_{11} + (a_{11} + a_{22})k_{12} + a_{12}k_{22} + \sum_{r,h=1}^{n} \nu_{rh}b_{1r}b_{2h},$$

$$\dot{k}_{22} = 2(a_{21}k_{12} + a_{22}k_{22}) + \sum_{r,h=1}^{n} \nu_{rh}b_{2r}b_{2h}.$$

Obtain from these equations the corresponding equations for the system of Problem 1.4 at $X_1 = V_1$, $X_2 = V_2$. Find the stationary solutions in the case of constant intensity ν. Consider the process of settling of the stationary solution.

6.25. Show that for the linear system with one degree of freedom of Problem 1.5 in the case where the generalized force Q represents a stationary random process with the spectral density $s_x(\omega) = D\alpha/\pi(\alpha^2 + \omega^2)$ the set of Eq. (6.1.8) for the components $Y_1 = q$, $Y_2 = p$, $Y_3 = Q/A$ of the state vector has the form

$$\dot{k}_{11} = 2k_{12}, \quad \dot{k}_{12} = k_{23} - \alpha k_{13},$$

$$\dot{k}_{12} = k_{22} - \omega_0^2 k_{11} - 2\varepsilon k_{12} + k_{13}, \quad \dot{k}_{22} = -2(\omega_0^2 k_{12} + 2\varepsilon k_{22} - k_{23}),$$

$$\dot{k}_{23} = -\omega_0^2 k_{13} - (\alpha + 2\varepsilon)k_{23} + k_{33}, \quad \dot{k}_{33} = -2\alpha(k_{33} - D).$$

6.26. Under the conditions of Problem 1.15 show that if a car is moving with the speed v and the covariance function of the road profile is exponential, $k_q(\sigma) = De^{-\alpha|\sigma|}$, then the state variables $Y_1 = Y$, $Y_2 = \dot{Y}$, $Y_3 = q$ (components of the expended state vector) satisfy Eq. (6.1.1)

$$a = \begin{pmatrix} 0 & 1 & 0 \\ -\omega_0^2 & -2\varepsilon & \omega_1^2 \\ 0 & 0 & -\alpha v \end{pmatrix}, \quad a_0 = 0, \quad b = \begin{pmatrix} 0 \\ b_2 \\ b_3 \end{pmatrix}$$

with the white noise of unit intensity. Verify the validity of the following formulae for the variances and the covariances in the stationary regime:

$$k_{11} = \frac{D}{\omega_0^2}\left[2\varepsilon\alpha v + \left(1 + \frac{\alpha v}{2\varepsilon}\right)\frac{\omega_1^2 \mu}{\varepsilon}\right], \quad k_{12} = 0, \quad k_{13} = \mu D,$$

$$k_{22} = 2D\alpha v\left(\varepsilon + \frac{\mu\omega_1^2}{4\varepsilon}\right), \quad k_{33} = D, \quad k_{23} = \alpha v k_{13} = \alpha v \mu D$$

$$\left(\mu = \frac{\omega_0^2 + 2\varepsilon\alpha v}{\omega_0^2 + 2\varepsilon\alpha v + \alpha^2 v^2}\right).$$

6.27. Show that for the system with two degrees of freedom of Problem 1.12 at $Q = V = [V_1\ V_2]^T$ where V is a white noise of the intensity ν Eqs. (6.1.8) have the form

$$\dot{k}_{11} = 2(A_{22}^- k_{13} - A_{12}^- k_{14}),$$

$$\dot{k}_{12} = -A_{12}^- k_{13} + A_{11}^- k_{14} + A_{22}^- k_{23} - A_{12}^- k_{24},$$

$$\dot{k}_{13} = -C_{11}k_{11} - (C_{12} + C_{12}')k_{12} - [B_{11}A_{22}^- - (B_{12} + B_{12}'A_{12}^-)]k_{13}$$
$$- [(B_{12} + B_{12}')A_{11}^- - B_{11}A_{12}^-]k_{14} + A_{22}^- k_{33} - A_{12}^- k_{34},$$

$$\dot{k}_{14} = -(C_{12} - C_{12}')k_{11} - C_{22}k_{12} - [(B_{12} - B_{12}')A_{22}^- - B_{22}A_{12}^-]k_{13}$$
$$- [B_{22}A_{11}^- - (B_{12} + B_{12}')A_{12}^-]k_{14} + A_{22}^- k_{34} - A_{12}^- k_{44},$$

$$\dot{k}_{22} = 2(-A_{12}^- k_{23} + A_{11}^- k_{24}),$$

$$\dot{k}_{23} = -C_{11}k_{12} - (C_{12} + C_{12}')k_{22} - [B_{11}A_{22}^- - (B_{12} + B_{12}')A_{12}^-]k_{23}$$
$$- [(B_{12} + B_{12}')A_{11}^- - B_{11}A_{12}^-]k_{24} - A_{12}^- k_{33} + A_{11}^- k_{34},$$

$$\dot{k}_{24} = -(C_{12} - C_{12}')k_{12} - C_{22}k_{22} - [(B_{12} - B_{12}')A_{22}^- - B_{22}A_{12}^-]k_{23}$$

$$-[B_{22}A_{11}^- - (B_{12} + B_{12}')A_{12}^-]k_{24} - A_{12}^-k_{34} + A_{11}^-k_{44},$$

$$\dot{k}_{33} = -2C_{11}k_{14} - 2(C_{12} + C_{12}')k_{23} - 2[B_{11}A_{22}^- - (B_{12} + B_{12}')A_{12}^-]k_{33}$$

$$-2[(B_{12} + B_{12}')A_{11}^- - B_{11}A_{12}^-]k_{34} + \nu_{11},$$

$$\dot{k}_{34} = -C_{11}k_{14} - (C_{12} - C_{12}')k_{13} - C_{22}k_{23} - (C_{12} + C_{12}')k_{24}$$

$$-[B_{11}A_{22}^- - B_{22}A_{11}^- - 2B_{12}A_{12}^-]k_{34} - [(B_{12} + B_{12}')A_{11}^- - B_{11}A_{12}^-]k_{44}$$

$$-[(B_{12} - B_{12}')A_{22}^- - B_{22}A_{12}^-]k_{33} + \nu_{12},$$

$$\dot{k}_{44} = -2(C_{12} - C_{12}')k_{14} - 2C_{22}k_{24} - 2[(B_{12} - B_{12}')A_{22}^- - B_{22}A_{12}^-]k_{34}$$

$$+2[B_{22}A_{11}^- - (B_{12} - B_{12}')A_{12}^-]k_{44} + \nu_{22},$$

where A_{ij}^- being the elements of the inverse matrix A^{-1}.

6.28. Write the equations for the variances and the covariances of the coordinates of the linear system with the dynamic damper of Problem 1.16. Show that at $b = 0$, $(c_d = 0)$ variance of the process $X(t)$ in the regime of stationary random oscillations is determined by: $D_x = \nu[\mu_d^2 c + (\mu_d + \mu)b_d^2]/2c^2b_d\mu_d^2$.

6.29. Using the equations of Problem 6.27 write the equations for the variances and the covariance of the processes $Y_1(t)$ and $Y_2(t)$ at $A_{11} = A_{22} = A$; $A_{12} = 0$; $B_{11} = B_{22} = B$; $B_{12} = 0$; $B_{12}' = H$; $\nu_{11} = \nu_{22} = \nu$; $\nu_{12} = 0$. Verify that the solutions of these equations with zero initial conditions are determined by the formulae:

(a) At $C_{11} = C_{12} = C_{22} = C_{12}' = 0$ and $\Lambda = \nu/(H^2 + B^2)$:

$$k_{11} = k_{22} = \Lambda t, \quad k_{13} = k_{24} = \Lambda/2, \quad k_{12} = k_{34} = 0$$

$$k_{14} = -k_{23} = H\Lambda/2B, \quad k_{33} = k_{44} = \nu/2AB;$$

(b) At $C_{11} = C_{12} = C_{22} = 0$, $C_{12}' = K(BH > AK)$ and $\kappa = \nu/\delta$, $\delta = BH - AK$:

$$k_{11} = k_{22} = B\kappa/2K, \quad k_{33} = k_{44} = \kappa H/A,$$

$$k_{12} = k_{13} = k_{24} = k_{34} = 0, \quad k_{14} = -k_{23} = \kappa/2.$$

Using formulae (5.1.15), (5.1.16), (5.1.24) calculate the derivative of

$$H = (p^T A^{-1}p + q^T Cq)/2.$$

Show at the stationary random oscillation that the mean power EH is wasted on damping $(B' \neq 0)$ and position nonconservative $(C' \neq 0)$ forces.

6.30. Show that Eqs. (6.1.8) for stationary linear system of Problem 1.11 at $Q = lV$ where l is a constant $m \times n$-matrix, V is an n-dimensional white noise of intensity ν represents the set of equations:

$$\dot{K}_{11} = A^{-1}K_{21} + K_{21}A^{-1},$$

$$\dot{K}_{12} = A^{-1}K_{22} - K_{11}(C - C') - K_{12}A^{-1}(B - B'),$$

$$\dot{K}_{21} = K_{22}A^{-1} - (C + C')K_{11} - (B + B')A^{-1}K_{21},$$

$$\dot{K}_{22} = -(C + C')K_{12} - K_{21}(C - C') - (B + B')A^{-1}K_{22} -$$

$$-K_{22}A^{-1}(B - B') + l\nu l^{T},$$

where K_{11} and K_{22} are the covariance matrices of the vectors of canonical variables $Y_1 = q$, $Y_2 = p$, and the blocks K_{12} and K_{21} are the cross-covariance matrices of Y_1 and Y_2. Consider the special cases: (a) $C = C' = 0$, $B = 2\varepsilon A$; (b) $C = C' = 0$, $B = \eta C$; (c) $|C| < 0$, $B2 = \varepsilon A$.

6.31. Write Eqs. (6.1.11) for the blocks of the covariance and the cross-covariance functions of the canonical variables of the system of Problem 6.30 for the regime of stationary random oscillations.

6.32. Show that under the conditions of Problem 6.19 the one-dimensional characteristic function of the process $Y(t) = [\,Y_1(t)\,Y_2(t)\,]^{T}$ is determined by the formula

$$g_1(\lambda', \lambda''; t) = g_0(\lambda', \lambda'') \exp\left\{ ia_0 t(\lambda'' + \lambda' t/2) \right.$$

$$\left. + \int_0^t \chi((t - \tau)\lambda' + \lambda''; \tau)d\tau \right\}.$$

6.33. Show that under the conditions of Problem 6.20 for a stationary white noise with any function $\chi(\mu)$ the one-dimensional stationary distribution of the process $Y_2(t)$ is determined by the characteristic function:

$$g_1(\lambda) = \exp\left\{ \frac{1}{2\varepsilon} \int_0^\lambda \frac{\chi(\mu)}{\mu} d\mu \right\}.$$

6.34. Under the conditions of Problem 6.25 write the formula for the one-dimensional characteristic function of the strictly stationary process $Y_1(t)$ in the case of an arbitrary function $\chi(\mu)$. Consider also the case of $\chi(\mu) = -\mu^{T}\nu\mu/2$.

6.35. The three-dimensional Brownian motion of a particle under the action of the disturbing force representing the stationary white noise is described by the vector Langevin equation

$$\ddot{Y} + 2\varepsilon\dot{Y} = V, \tag{I}$$

Y being the state vector of the particle, V is the stationary white noise with the known function $\chi(\mu)$. Show that the one-dimensional characteristic function of the velocity vector $U = \dot{Y}$ of the particle in the stationary regime is determined by the formula

$$g_1(\lambda'') = \exp\left\{ \int_0^\infty \chi(e^{-2\varepsilon\tau}\lambda'')d\tau \right\} \quad (\lambda'' = [\,\lambda''_x\,\lambda''_y\,\lambda''_z\,]). \tag{II}$$

Consider the case of the normally distributed white noise.

6.36. Show that the multi-dimensional distributions of the random process $Y(t)$ determined by the differential equation

$$\dot{Y} = a_0 + aY + bX(t), \tag{I}$$

where $X(t)$ is the random function representing the solution of the stochastic differential equation

$$\dot{X} = \alpha_0 + \alpha X + \beta V \tag{II}$$

with independent initial values Y_0 and X_0 are determined by the formula:

$$g_n^y(\lambda_1, \ldots, \lambda_n; t_1, \ldots, t_n) = g_0^y\left(\sum_{k=1}^n u_{11}(t_k, t_0)^T \lambda \right) g_0^x\left(\sum_{k=1}^n u_{12}(t_k, t_0)^T \lambda \right)$$

$$\times \exp\left\{ -i\sum_{k=1}^n \lambda_k^T \left[\int_{t_0}^{t_k} u_{11}(t_k, \tau)\alpha_0(\tau)d\tau + \int_{t_0}^{t_k} u_{12}(t_k, \tau)a_0(\tau)d\tau \right] \right.$$

$$\left. + \sum_{k=1}^n \int_{t_{k-1}}^{t_k} \chi\left(\beta(\tau)^T \sum_{l=k}^n u_{12}(t_l, \tau)^T \lambda_l; \tau \right) d\tau \right\}, \quad \lambda = (\lambda_1 \ldots \lambda_n)^T,$$

$$\tag{III}$$

where $g_0^y(\rho)$, $g_0^x(\sigma)$ are the characteristic functions of the initial values Y_0, X_0 of the process $Y(t)$, $X(t)$; $u_{11}(t, \tau)$ is the fundamental matrix of solutions of the homogeneous equation $du_{11}/dt = au_{11}$, $u_{11}(\tau, \tau) = I_p$; $u_{12}(t, \tau)$ is the solution of the equation $du_{12}/dt = au_{12} + bu_{22}$ with zero initial condition, $u_{12}(\tau, \tau) = 0$;

$u_{22}(t, \tau)$ is the fundamental matrix of solutions of the equation $du_{22}/dt = \alpha u_{22}$; $u_{22}(\tau, \tau) = I_q$ (I_q is the unit matrix of dimension q).

6.37. Show that the stationary variance D_y of the output of the system consisting of the impulse device which generates the rectangular impulses of the duration T_d (with the period repetition T_r) and the linear link with the transfer function $\Phi(s)$ under the action of the impulse white noise with the variance D of each impulse are determined by the formulae:

1) $D_y = kD/(k + 2)$ at $\Phi(s) = k/s$,

2) $D_y = k^2 D\rho(z_1)$ at $\Phi(s) = k/(Ts + 1)$.

Here $\rho(z_1) = z_1^2 (z_1^{-\gamma} - 1)^2/(1 - z_1^2)$, $z_1 = e^{-T_r/T}$, $\gamma = T_d/T_r$.

6.38. Find the expressions for the spectral densities and the variances of the output for typical linear systems with distributed parameters (Table A.5.4).

6.39. Consider Example 6.3.1 for the case when $T = T_1$, $m_x(t) = a + bt_0$, $K_x(t, t') = De^{\beta(t+t') - \alpha|t-t'|}$.

S o l u t i o n. According to the first method if we use the result of Example 3.9.12 at $\nu(t) = \sqrt{D}e^{\beta t}$, choosing the coordinate functions in the form

$$x_n(t) = \frac{2\sqrt{D}e^{\beta t}}{T + \lambda_n} \sin\left[\omega_n\left(t - \frac{T}{2}\right) + \frac{\pi n}{2}\right] \quad (n = 1, 2, \ldots), \qquad \text{(I)}$$

we obtain respectively

$$m_y = \frac{1}{T}\int_0^t e^{-(t-\tau)/T}(a + b\tau)d\tau = a + b(t - T) - (a - bT)e^{-t/T}, \qquad \text{(II)}$$

$$y_n(t) = \frac{2\sqrt{D}}{(T + \lambda_n)\sqrt{(1 + \beta T)^2 + \omega_n^2 T^2}}\left[e^{\beta t}\sin(\omega_n t + \varphi_n) - e^{-t/T}\sin\varphi_n\right], \qquad \text{(III)}$$

$$D_y(t) = 2D\sum_{n=1}^{\infty} \frac{\lambda_n\left[e^{\beta t}\sin(\omega_n t + \varphi_n) - e^{-t/T}\sin\varphi_n\right]}{(T + \lambda_n)\left[(1 + \beta T)^2 + \omega_n^2 T^2\right]}, \qquad \text{(IV)}$$

$$\left(\varphi_n = \frac{n\pi}{2} - \frac{\omega_n T}{2} - \operatorname{arctg}\frac{\omega_n T}{1 + \beta T}\right).$$

Here the exact solution of te variance $D_y^{\text{ES}}(t)$ is determined by formula

$$D_y^{\text{ES}}(t) = \frac{D}{T^2}\int_0^t\int_0^t e^{-(2t-\tau-\tau')/T_0 + \beta(\tau+\tau') - \alpha|\tau-\tau'|}d\tau d\tau'$$

$$= \frac{D\left(e^{2\beta t} - e^{-2t/T}\right)}{(1 + \beta T)\left[1 + (\alpha + \beta)T\right]} - \frac{2D\left(e^{\beta_1 t} - 1\right)}{(1 + \beta T)^2 + \alpha^2 T^2} \tag{V}$$

$$(\beta_1 = \beta - \alpha + 1/T).$$

While using the second method the following equations are solved

$$T\dot{m}_y + m_y = a + bt\,,\ \ T\ddot{y}_n + y_n = \frac{2\sqrt{D}e^{\beta t}}{T + \lambda_n}\sin\left[\omega_n\left(t - \frac{T}{2}\right) + \frac{\pi n}{2}\right],$$

$$\tag{VI}$$

their integrals which vanish at $t = 0$ are expressed by formulae (I) and (III).

CHAPTER 7
METHODS OF NONLINEAR STOCHASTIC SYSTEMS THEORY AND THEIR APPLICATIONS

The last Chapter is dedicated to the basic methods of the nonlinear stochastic systems described by the stochastic integral, differential, integrodifferential, difference and other nonlinear operator equations. Section 7.1 contains the derivation of the formulae for the derivatives with respect to the moments of different orders. Special attention is paid to the derivation of the infinite system of the equations for the moments. The equations for the moments in the stochastic linear systems with the parametric noise are composed. In Section 7.2 the methods of the normal approximation and the statistical linearization for determining the one- and the multi-dimensional distributions are stated. The methods of the equivalent linearization and the general methods of the normalization which form the basis of the structural methods of nonlinear stochastic systems theory have a significant place in Section 7.2. Section 7.3 contains the brief account of the methods of general nonlinear stochastic systems and the statement of the general problem of the parametrization of the one- and the multi-dimensional distributions. Methods of the reduction of the number of the equations for the distributions parameters are considered. The moments methods, the semiinvariants methods, and the moment-semiinvariants methods for defining the one- and the multi-dimensional distributions are stated in Section 7.4. Section 7.5 is dedicated to the methods of the orthogonal expansions and the quasi-moments. Method of structural parametrization based on the ellipsoidal approximation of the one- and the multi-dimensional distributions is given in Section 7.6. Section 7.7 is devoted to the methods based on the canonical expansions and the integral canonical representations. Sections 7.8 and 7.9 contain the applications of the nonlinear stochastic systems theory methods to the analysis, modeling and conditionally optimal filtering.

7.1. Moments of State Vector

7.1.1. Formula for Derivative of Expectation

$\triangleright$ The simplest way to find the formula for the time derivative of the expectation of the state vector of a system is to pass to the expectations directly in Eq. (5.4.6) for the normal white noise with the matrix intensity $\nu = \nu(t)$. Then taking into consideration that in the Itô equation the value of the process $Y(t)$ at any t is independent of the value of the white noise V at the same instant we obtain the following *formula for the time derivative of the expectation* $m(t) = EY(t)$ *of the* *process* $Y(t)$:

$$\dot{m} = Ea(Y,t) . \triangleleft \tag{7.1.1}$$

R e m a r k.   This formula is not a closed equation which determines the expectation of the process $Y(t)$, as the expectation in the right-hand side depends on the unknown one-dimensional distribution of the process $Y(t)$.

7.1.2. Formula for Derivative of Second Order Moment

▷ Let us denote by $\gamma_{kj} = EY_k(t)Y_j(t)$ the components of the matrix of moments of the second order $\Gamma = EY(t)Y(t)^T$. Using formula (5.1.7), we evaluate a stochastic differential of the product $Y_k(t)Y_j(t)$. As a result we obtain the following expression for a stochastic differential of the product:

$$dY_k(t)Y_j(t) = \left\{ Y_j a_k(Y,t) + Y_k a_j(Y,t) + \sum_{r=1}^{m}\sum_{s=1}^{m} b_{kr}(Y,t)b_{js}(Y,t)\nu_{rs}(t) \right\} dt$$

$$+ \sum_{r=1}^{m} \left[Y_k b_{kr}(Y,t) + Y_j b_{jr}(Y,t) \right] dW_r . \tag{7.1.2}$$

While evaluating the expectation of both parts of (7.1.2) and accounting that $\Gamma = [\gamma_{kj}]\ (k, j = 1 , \ldots , p)$ we come to the sought formula for the derivative of the second order moment

$$\dot{\Gamma} = E\left\{ a(Y,t)Y^T + Ya(Y,t)^T + \sigma(Y,t) \right\} . \tag{7.1.3}$$

Here

$$\sigma(Y,t) = b(Y,t)\nu(t)b(Y,t)^T , \triangleleft \tag{7.1.4}$$

R e m a r k.   Formulae (7.1.1) and (7.1.3) are not closed equations for m and Γ in the general case, as their right-hand sides depend on the one-dimensional distribution of the process $Y(t)$, i.e. not only on m and Γ. And only in some special cases the right-hand sides of formulae (7.1.1) and (7.1.3) may be the functions of m and Γ independent of other characteristics of the one-dimensional distribution of the process $Y(t)$. In such cases (7.1.1) and (7.1.3) will be ordinary differential equations which determine the moments m and Γ. So, for instance, in the case of linear functions a and b formulae (7.1.1) and (7.1.3) represent the equations which determine the expectation m and the second order moment Γ of the process $Y(t)$ separately. These equations certainly coincide with equations of Subsection 6.1.1 obtained for the more general case.

7.1.3. Formula for Derivative of Covariance Matrix

The formula for the time derivative of the covariance matrix K of the vector Y is easily derived from formulae (7.1.1) and (7.1.3).

$\triangleright$ In order to obtain this formula it is sufficient to differentiate the relation $K = \Gamma - mm^T$ with respect to t and substitute into the obtained formula the expressions of $\dot{m}$ and $\dot{\Gamma}$ from (7.1.1) and (7.1.3). As a result we get

$$\dot{K} = E\left\{a(Y,t)(Y^T - m^T) + (Y - m)a(Y,t)^T + \sigma(Y,t)\right\} . \triangleleft \quad (7.1.5)$$

So the results obtained in Subsections 7.1.1 and 7.1.2 may be summarized in the following statement.

Theorem 7.1.1. *If the random process $Y(t)$ defined by Eq. (5.4.6) with the normal white noise $V = V(t)$ has the finite expectation m, the second order moment Γ and the covariance matrix K then formulae (7.1.1), (7.1.3), (7.1.5) are valid.*

For formulae for the derivatives of expectation and second order moments in the case of Eqs. (5.4.7) and (5.4.6) for V being the nongaussian white noise see Problems 7.1 and 7.2.

E x a m p l e 7.1.1. For the system described by the equation

$$\dot{Y} = -Y^3 + YV , \qquad (I)$$

formulae (7.1.1), (7.1.3) and (7.1.5) have the form

$$\dot{m} = -EY^3 , \quad \dot{\alpha}_2 = -2EY^4 + \nu\alpha_2 ,$$
$$\dot{D} = -2EY^4 + 2mEY^3 + \nu(m^2 + D) . \qquad (II)$$

In this case the process $Y(t)$ is scalar, and consequently, Γ and K represent respectively the initial second order moment $\alpha_2(t)$ and the variance $D(t)$ of the value of the process $Y(t)$ at the instant t. The right-hand sides of formulae (II) depend on the third and the fourth order moments of the process $Y(t)$, and consequently, formulae (II) are not the equations determining m, α_2 and D. The attempts to close these equations by adding the equations for the third and the fourth order moments will lead to the appearance of the moments of the fifth and sixth orders, and so on. Consequently, such attempts will not lead to a closed set of finite number of equations.

E x a m p l e 7.1.2. For the system

$$\dot{Y} = aY + a_0 + b\sqrt{Y}V \tag{I}$$

formulae (7.1.1), (7.1.3) and (7.1.6) give

$$\dot{m} = am + a_0, \quad \dot{\alpha}_2 = 2(a\alpha_2 + a_0 m) + b^2\nu m, \quad \dot{D} = 2aD + b^2\nu m. \tag{II}$$

These ordinary differential equations with given initial conditions: $m(t_0) = m_0$, $\alpha_2(t_0) = \alpha_{20}$, $D(t_0) = D_0 = \alpha_{20} - m_0^2$ determine consecutively the moments m, α_2 and D.

7.1.4. Formulae for Derivatives of Second Order Moment and Covariance Function

Later on we shall also need the formulae for the partial derivatives of the second order moment $\Gamma(t_1, t_2)$ and the covariance function $K(t_1, t_2)$ of the process $Y(t)$ defined by Eqs. (5.4.6) with respect to the second argument t_2 at $t_1 < t_2$.

▷ Using

$$\begin{aligned} d_{t_2}\Gamma(t_1, t_2) &= EY(t_1)dY(t_2)^T, \\ d_{t_2}K(t_1, t_2) &= E\left[Y(t_1) - m(t_1)\right]dY(t_2)^T \end{aligned} \tag{7.1.6}$$

and using Eqs. (5.4.6) we shall get at $t_1 < t_2$ the following formulae:

$$\partial\Gamma(t_1, t_2)/\partial t_2 = EY_{t_1}a(Y_{t_2}, t_2)^T, \tag{7.1.7}$$

$$\partial K(t_1, t_2)/\partial t_2 = E(Y_{t_1} - m_{t_1})a(Y_{t_2}, t_2)^T. \triangleleft \tag{7.1.8}$$

Theorem 7.1.2. *If the random process $Y(t)$ defined by Eq. (5.4.6) with the normal white noise $V = V(t)$ has the finite first and the second order moments for instants t_1 and t_2 then the matrix of the initial moments and the covariance matrix satisfy formulae (7.1.7) and (7.1.8).*

E x a m p l e 7.1.3. For the system of Example 7.1.1 formulae (7.1.7) and (7.1.8) have the form

$$\partial\Gamma(t_1, t_2)/\partial t_2 = -EY_{t_1}Y_{t_2}^3, \quad \partial K(t_1, t_2)/\partial t_2 = -E(Y_{t_1} - m_{t_1})Y_{t_2}^3.$$

E x a m p l e 7.1.4. For the system of Example 7.1.2 formulae (7.1.7) and (7.1.8) have the form

$$\partial\Gamma(t_1,t_2)/\partial t_2 = a(t_2)\Gamma(t_1,t_2)\,, \quad \partial K(t_1,t_2)/\partial t_2 = a(t_2)K(t_1,t_2)\,. \quad \text{(I)}$$

These formulae represent the ordinary differential equations at any t_1. Together with the initial conditions: $\Gamma(t_1,t_1) = \alpha_2(t_1)$, $K(t_1,t_1) = D(t_1)$ they determine the second order moment $\Gamma(t_1,t_2)$ and the covariance function $K(t_1,t_2)$ of the process $Y(t)$ at $t_1 < t_2$:

$$\Gamma(t_1,t_2) = \alpha_2(t_1)\exp\left\{\int_{t_1}^{t_2} a(\tau)d\tau\right\}, \quad K(t_1,t_2) = D(t_1)\exp\left\{\int_{t_1}^{t_2} a(\tau)d\tau\right\}.$$

$$\text{(II)}$$

7.1.5. Infinite Set of Equations for Moments

In the case of Subsection 5.6.9 in which the equations for the characteristic functions were reduced to linear equations in partial derivatives one may obtain the infinite set of ordinary linear differential equations determining all the moments of the one-dimensional distribution of the state vector. Let us consider equations in the case of Subsection 5.6.9 where functions $a(y,t)$ and $\sigma(y,t) = b(y,t)\nu(t)b(y,t)^T$ represent the polynomials in y and the white noise V is normally distributed.

▷ Representing the components of the vector $a(y,t)$ and the elements of the matrix $\sigma(y,t) = b(y,t)\nu(t)b(y,t)^T$ in the explicit form of the polynomials

$$a_r(y,t) = \sum_{h_1,\ldots,h_p=0}^{N} a_{r,h_1,\ldots,h_p} y_1^{h_1}\cdots y_p^{h_p} \quad (r=1,\ldots,p), \quad (7.1.9)$$

$$\sigma_{rs}(y,t) = \sum_{h_1,\ldots,h_p=0}^{N} \sigma_{rs,h_1,\ldots,h_p} y_1^{h_1}\cdots y_p^{h_p} \quad (r,s=1,\ldots,p) \quad (7.1.10)$$

with the coefficients depending on time t in the general case, we rewrite Eq. (5.6.64) in partial derivatives in the form

$$\frac{\partial g_1}{\partial t} = \sum_{r=1}^{p} i\lambda_r \sum_{h_1,\ldots,h_p=0}^{N} a_{r,h_1,\ldots,h_p} \frac{\partial^{h_1+\cdots+h_p} g_1}{\partial(i\lambda_p)^{h_1}\cdots\partial(i\lambda_p)^{h_p}}$$

$$+\frac{1}{2}\sum_{r,s=1}^{p}(i\lambda_r)(i\lambda_p)\sum_{h_1,\ldots,h_p=1}^{N}\sigma_{rs,h_1,\ldots,h_p}\frac{\partial^{h_1+\cdots+h_p}g_1}{\partial(i\lambda_1)^{h_1}\cdots\partial(i\lambda_p)^{h_p}}\,.$$

$$(7.1.11)$$

Differentiating Eq. (7.1.11) k_1 times with respect to $i\lambda_1$, k_2 times with respect to $i\lambda_2$, $\ldots$, k_p times with respect to $i\lambda_p$ and putting after that $\lambda=0$ we get

$$\dot{\alpha}_{k_1,\ldots,k_p}=\sum_{r=1}^{p}k_r\sum_{h_1,\ldots,h_p=0}^{N}a_{r,h_1,\ldots,h_p}\alpha_{h_1+k_1,\ldots,h_r+k_r-1,\ldots,h_p+k_p}$$

$$+\frac{1}{2}\sum_{r=1}^{p}k_r(k_r-1)\sum_{h_1,\ldots,h_p=0}^{N}\sigma_{rr,h_1,\ldots,h_p}\alpha_{h_1+k_1,\ldots,h_r+k_r-2,\ldots,h_p+k+p}$$

$$+\sum_{r=1}^{p}\sum_{s=1}^{r-1}k_rk_s\sum_{h_1,\ldots,h_p=0}^{N}\sigma_{rs,h_1,\ldots,h_p}$$

$$\times\alpha_{h_1+k_1,\ldots,h_s+k_s-1,\ldots,h_r+k_r-1,\ldots,h_p+k_p}\qquad(7.1.12)$$

$$(k_1,\ldots,k_p=0,1,2,\ldots;\ |k|=k_1+\cdots+k_p=1,2,\ldots)\,.$$

Eqs. (7.1.12) may also be rewritten in an abbreviated form using the multi-indexes at $k_1,\ldots,k_p=0,1,2,\ldots;\ |k|=k_1+\cdots+k_p=1,2,\ldots$:

$$\dot{\alpha}_k=\sum_{r=1}^{p}k_r\sum_{h_1,\ldots,h_p=0}^{N}a_{r,h}\alpha_{h+k-e_r}+$$

$$+\frac{1}{2}\sum_{r=1}^{p}k_r(k_r-1)\sum_{h_1,\ldots,h_p=0}^{N}\sigma_{rr,h}\alpha_{h+k-2e_r}+$$

$$+\sum_{r=1}^{p}\sum_{s=1}^{r-1}k_rk_s\sum_{h_1,\ldots,h_p=0}^{N}\sigma_{rs,h}\alpha_{h+k-e_r-e_s}\,.\ \triangleleft\qquad(7.1.13)$$

In Eqs. (7.1.12) α_s is equal to zero if at least one of the components of the multi-index s is negative, and is equal to unity if all the components of the multi-index s are equal to zero, $\alpha_0=1$.

So we get the following statement.

Theorem 7.1.3. *If the random process $Y(t)$ defined by* Eq. (5.4.6) *with the normal white noise $V=V(t)$ has the finite initial moments α_k then α_k satisfy the infinite set of* Eqs. (7.1.12) *or* (7.1.13).

E x a m p l e 7.1.5. For the system of Example (7.1.1) with the normally distributed white noise V Eqs. (7.1.13) have the form

$$\dot{\alpha}_k = -k\alpha_{k+2} + (\nu/2)k(k-1)\alpha_k \quad (k = 1, 2, \ldots).$$

E x a m p l e 7.1.6. For the system

$$\dot{Y}_1 = -Y_1 Y_2 , \quad Y_2 = -aY_2 + hV \tag{I}$$

with a normally distributed white noise V Eqs. (7.1.13) have the form

$$\dot{\alpha}_{rs} = -r\alpha_{r,s+1} - sa\alpha_{rs} + h^2 \nu s(s-1)\alpha_{r,s-2}/2 \quad (r, s = 0, 1, 2, \ldots). \tag{II}$$

Eqs. (7.1.12), (7.1.13) for stochastic system (5.4.6) where V being the nongaussian white noise are given in Problem 7.1.3.

7.1.6. Linear System with Parametric Noises

The state vector of a linear system with parametric white noises is determined by the equation (Subsection 1.4.3)

$$\dot{Y} = aY + a_0 + \left(b_0 + \sum_{h=1}^{p} b_h Y_h \right) V . \tag{7.1.14}$$

▷ Eq. (7.1.1) has in this case the form

$$\dot{m} = am + a_0 . \tag{7.1.15}$$

Eq. (7.1.15) with the corresponding initial condition $m(t_0) = m_0$ completely determines the expectation m of the vector Y. Eq. (7.1.3) has in this case the form

$$\dot{\Gamma} = aEYY^T + a_0 EY^T + EYY^T a^T$$
$$+ EY a_0^T + E\left(b_0 + \sum_{h=1}^{p} b_h Y_h \right) \nu \left(b_0^T + \sum_{h=1}^{p} b_h^T Y_h \right) . \tag{7.1.16}$$

Opening in Eq. (7.1.16) the brackets we get

$$\dot{\Gamma} = a\Gamma + \Gamma a^T + a_0 m^T + m a_0^T + b_0 \nu b_0^T$$

$$+ \sum_{h=1}^{p} (b_h \nu b_0^T + b_0 \nu b_h^T) EY_h + \sum_{h,l=1}^{p} b_h \nu b_l^T EY_h Y_l$$

$$= a\Gamma + \Gamma a^T + a_0 m^T + m a_0^T + b_0 \nu b_0^T$$

$$+ \sum_{h=1}^{p} (b_h \nu b_0^T + b_0 \nu b_h^T) m_h + \sum_{h,l=1}^{p} b_h \nu b_l^T \gamma_{hl} . \tag{7.1.17}$$

Eq. (7.1.17) does not contain any characteristics of the random vector Y except in its expectation and second moments (the elements of the matrix Γ). Consequently, after integrating Eq. (7.1.15) which determines the expectation m of the vector Y, Eq. (7.1.17) with the initial condition $\Gamma(t_0) = \Gamma_0$ (and consequently, $\gamma_{hl}(t_0) = \gamma_{hl}^0$) completely determines the second order moment $\Gamma(t)$ of the vector $Y(t)$.

In exactly the same way we reduce Eq. (7.1.5) to the form

$$\dot{K} = aK + Ka^T + b_0 \nu b_0^T + \sum_{h=1}^{p} (b_h \nu b_0^T + b_0 \nu b_h^T) m_h$$

$$+ \sum_{h,l=1}^{p} b_h \nu b_l^T (m_h m_l + k_{hl}) , \quad K = [k_{hl}] , \tag{7.1.18}$$

where k_{hl} is the covariance of the components Y_h and Y_l of the vector Y $(h, l = 1, \ldots, p)$. Eq. (7.1.18) with the initial conditions $K(t_0) = K_0$ $(k_{lh} = k_{lh}^0)$ completely determines the covariance matrix $K(t)$ of the vector $Y(t)$ at any time instant t after finding its expectation m. ◁

Thus if we confine ourselves to the moments of the first and the second order of the state vector and of the output while studying a linear system with parametric white noises (7.1.14) *then these moments may be exactly determined by subsequent integration of* Eqs. (7.1.15). (7.1.17) *or* Eqs. (7.1.15), (7.1.18) *as well as in the case of a linear system.*

For a linear system with parametric white noises (7.1.14) formula (7.1.8) gives the following equation for the covariance function of the process $Y(t)$ at $t_2 > t_1$:

$$\partial K(t_1, t_2)/\partial t_2 = K(t_1, t_2) a(t_2)^T . \tag{7.1.19}$$

The initial condition for this equation has the form $K(t_1, t_1) = K(t_1)$.

At last it is easy to conclude that *the moments of the state vector* Y *of a system* (7.1.14) *are determined by the infinite set of* Eqs. (7.1.13) *which in this case is*

decomposed into independent sets of equations for the moments of each given order k
$(k_1, \ldots, k_p = 0, 1, 2, \ldots; \ |k| = k_1 + \cdots + k_p = 1, 2, \ldots)$:

$$
\dot{\alpha}_k = \sum_{r=1}^{p} k_r \left(a_{r,0} \alpha_{k-e_r} + \sum_{q=1}^{p} a_{r,e_q} \alpha_{k+e_q-e_r} \right)
$$

$$
+ \frac{1}{2} \sum_{r=1}^{p} k_r (k_r - 1) \left(\sigma_{rr,0} \alpha_{k-e_r} + \sum_{q=1}^{p} \sigma_{rr,e_q} \alpha_{k+e_q-e_r} \right.
$$

$$
+ \sum_{q,u=1}^{p} \sigma_{rr,e_q+e_u} \alpha_{k+e_q+e_u-2e_r} \left. \right) + \sum_{r=2}^{p} \sum_{s=1}^{p-1} k_r k_s \left(\sigma_{rs,0} \alpha_{k-e_r-e_s} \right.
$$

$$
+ \sum_{q=1}^{p} \sigma_{rs,e_q} \alpha_{k+e_q-e_r-e_s} + \sum_{q,u=1}^{p} \sigma_{rs,e_q+e_u} \alpha_{k+e_q+e_u-e_r-e_s} \left. \right) .
$$

$$
(7.1.20)
$$

Here as well as in Eqs. (7.1.13) α_s is equal to zero if at least one of the components of the multi-index s is negative, and is equal to unity if all the components of the multi-index s are equal to zero.

E x a m p l e 7.1.7. In the case of scalar Y and V Eq. (7.1.14) has the form

$$
\dot{Y} = aY + a_0 + (b_0 + b_1 Y)V . \tag{I}
$$

Eq. (7.1.15) has in this case the same form as in the general case of vector Y, V. Eqs. (7.1.17), (7.1.18) for the second initial moment $\alpha_2 = \Gamma$ and the variance $D = K$ of the process Y take the form

$$
\dot{\alpha}_2 = (2a + b_1^2 \nu)\alpha_2 + 2(a_0 + b_0 b_1 \nu)m + b_0^2 \nu ,
$$

$$
\dot{D} = (2a - b_1^2 \nu)D + 2b_0 b_1 \nu m + b_1^2 \nu m^2 + b_0^2 \nu . \tag{II}
$$

In the case of a normally distributed white noise V we may also obtain the exact equations for the moments of higher orders. These equations have the form

$$
\dot{\alpha}_k = k \left[a + \frac{1}{2}(k-1)b_1^2 \nu \right] \alpha_k + k \left[a_0 + (k-1)b_0 b_1 \nu \right] \alpha_{k-1}
$$

$$
+ \frac{1}{2} k(k-1)b_0^2 \nu \alpha_{k-2} \quad (k = 3, 4, \ldots) . \tag{III}
$$

Let us consider the stationary linear system (7.1.14) with stationary parametric white noises. In this case a, a_0, b_0, b_h and ν are constant. Therefore, putting in

Eqs. (7.1.15), (7.1.18) $\dot{m} = 0$ and $\dot{K} = 0$ we obtain algebraic equations for the expectation and the elements of the covariance matrix of the value of the stationary process in the system at any instant t:

$$am + a_0 = 0, \qquad (7.1.21)$$

$$aK + Ka^T + b_0\nu b_0^T + \sum_{h=1}^{p}(b_h\nu b_0^T + b_0\nu b_h^T)m_h + \sum_{h,l=1}^{p} b_h\nu b_l^T(m_h m_l + k_{hl}) = 0.$$

$$(7.1.22)$$

For finding the covariance function of the stationary process $k(\tau)$, $\tau = t_1 - t_2$, we write the general equation for the covariance function

$$\partial K(t_1, t_2)/\partial t_2 = K(t_1, t_2)a^T \text{ at } t_2 > t_1. \qquad (7.1.23)$$

Hence, using the property of the covariance function $K(t_1, t_2) = K(t_2, t_1)^T$ and transposing Eq. (7.1.23) we shall have

$$\partial K(t_2, t_1)/\partial t_2 = aK(t_2, t_1) \text{ at } t_1 < t_2. \qquad (7.1.24)$$

Changing the places of t_1 and t_2 in this equation, putting after this $t_2 = t_1$, $t_1 = t + \tau$ and fixing t we shall come to the following equation for the covariance function of the stationary process in the system:

$$dk(\tau)/d\tau = ak(\tau). \qquad (7.1.25)$$

Eq. (7.1.25) *with the initial condition* $k(0) = K$ *determines the covariance function of the stationary process at* $\tau > 0$. *At* $\tau < 0$ *covariance function is determined by the formula* $k(\tau) = k(-\tau)^T$.

We may also determine the moments of higher orders of the one-dimensional distribution of the stationary process in the system. For this purpose one should put

$$\dot{\alpha}_k = 0 \ (k_1, \ldots, k_p = 0, 1, 2, \ldots; \ |k| = k_1 + \cdots + k_p = 1, 2, \ldots). \qquad (7.1.26)$$

R e m a r k. It should be noticed that the wide sense stationary process in a linear system with parametric white noises may physically exist only in the case where the systems described by deterministic Eq. (7.1.15), (7.1.18) are stable. But this condition is generally not sufficient for the existence of the strictly stationary process in the system.

7.1.7. Formulae for Moments in Stochastic Discrete Nonlinear Systems

The formulae for the first and the second order moments in conformity to discrete system (5.7.1) are obtained on the basis of Eq. (5.7.7) and have the following form

$$m_{l+1} = E\omega_l(Y_l, V_l),\qquad\qquad (7.1.27)$$
$$K_{k+1} = E\left[\omega_l(Y_l, V_l) - m_{l+1}\right]\left[\omega_l(Y_l, V_l)^T - m_{l+1}^T\right].\quad (7.1.28)$$

Analogously on the basis of Eqs. (5.7.14), (5.7.15) and (5.7.17) the formulae for the moments are derived also for the case of continuously-discrete systems (5.7.10) and (5.7.11). For the nonlinear systems formulae (7.1.27) and (7.1.28) are not closed equations.

7.1.8. Equations for Moments of Discrete Linear System with Parametric Noises

As an example of direct derivation of the equations of Subsection 7.1.7 for the first and the second order moments let us consider the following discrete linear system with the parametric noises:

$$Y_{l+1} = a_l Y_l + a_{0l} + \left(b_{0l} + \sum_{j=1}^{p} b_{lj} Y_{lj}\right) V_l,\qquad (7.1.29)$$

where $b_{0l}, b_{1l}, \ldots, b_{pl}$ are the matrices of the same dimension as the matrix a_{0l} and $Y_{l1}, \ldots, Y_{lp}$ are the components of the vector Y_l.

$\triangleright$ After taking the expectations of both parts of Eq. (7.1.29) and accounting the independence of Y_l and V_l and also the equality $EV_l = 0$ on the basis of the theorem of the expectations multiplication for the independent variables we get

$$m_{l+1} = a_l m_l + a_{0l}.\qquad (7.1.30)$$

Similarly as for an ordinary linear system while deriving the equations for the covariance matrix $K_l = E(Y_l - m_l)(Y_l - m_l)^T$ of the process $\{Y_l\}$ we subtract from Eq. (7.1.29) term-wise Eq. (7.1.30) and multiply the result by the correspondent equality for the transpose matrix. After taking the expectation of both parts of the obtained equality we come to the equation

$$E(Y_{l+1} - m_{l+1})(Y_{l+1}^T - m_{l+1}^T) = a_l E(Y_l - m_l)(Y_l^T - m_l^T)a_l^T$$

$$+ E \left(b_{0l} + \sum_{j=1}^{p} b_{jl} Y_{jl} \right) V_l V_l^T \left(b_{0l} + \sum_{j=1}^{p} b_{jl}^T Y_{lj} \right) . \tag{7.1.31}$$

But by virtue of the independence of Y_l and V_l and according to formula of the total expectation we have

$$E \left(b_{0l} + \sum_{j=1}^{p} b_{jl} Y_{lj} \right) V_l V_l^T \left(b_{0l}^T + \sum_{j=1}^{p} b_{jl} Y_{lj} \right)$$

$$= E \left(b_{0l} + \sum_{j=1}^{p} b_{jl} Y_{lj} \right) E \left[V_l V_l^T \mid Y_l \right] \left(b_{0l}^T + \sum_{j=1}^{p} b_{jl}^T Y_{lj} \right)$$

$$= E \left(b_{0l} + \sum_{j=1}^{p} b_{jl} Y_{lj} \right) E \left[V_l V_l^T \right] \left(b_{0l}^T + \sum_{j=1}^{p} b_{jl}^T Y_{lj} \right)$$

$$= E \left(b_{0l} + \sum_{j=1}^{p} b_{jl} Y_{lj} \right) G_l \left(b_{0l}^T + \sum_{j=1}^{p} b_{jl}^T Y_{lj} \right)$$

$$= b_{0l} G_l b_{0l}^T + \sum_{j=1}^{p} \left(b_{0l} G_l b_{jl}^T + b_{jl} G_l b_{0l}^T \right) m_{lj}$$

$$+ \sum_{j=1}^{p} \sum_{h=1}^{p} b_{jl} G_l b_{jl}^T \left(m_{lj} m_{lh} + k_{ljh} \right) , \tag{7.1.32}$$

where m_{lj} are the components of the vector m_l; k_{ljh} are the elements of the matrix K_l; G_l is the covariance matrix of the variables V_l. Substituting expression (7.1.32) into Eq. (7.1.31) we get the sought difference equation for the matrix K_l:

$$K_{l+1} = a_l K_l a_l^T + b_{0l} G_l b_{0l}^T + \sum_{j=1}^{p} (b_{0l} G_l b_{jl}^T$$

$$+ b_{jl} G_l b_{0l}^T) m_{jl} + \sum_{j=1}^{p} \sum_{h=1}^{p} b_{jl} G_l b_{hl}^T (m_{lj} m_{lh} + k_{ljh}) . \triangleleft \tag{7.1.33}$$

Thus Eq. (7.1.30) *with the initial condition* $m_1 = EY_1$ *determines completely the expectation* $\{m_l\}$ *of the process* $\{Y_l\}$, *and Eq.* (7.1.32) *with the initial value* $K_1 = E(Y_1 - m_1)(Y_1 - m_1)^T$ *determines the covariance matrix* $\{K_l\}$ *of the process* $\{Y_l\}$.

For finding the covariance function $K(j,h)$ *of the process* $\{Y_l\}$ *we evaluate by means of Eqs.* (7.1.30) *and* (7.1.33) $E(Y_j - m_j)(Y_{h+1}^T - m_{h+1}^T)$ *at* $h \geq j$. *Accounting the independence of* Y_h, Y_j *and* V_l *at* $j \leq h$ *and the inequality* $EV_l = 0$ *we obtain*

$$K(j, h+1) = K(j, h)a_h^T . \lhd \qquad (7.1.34)$$

Eq. (7.1.34) *with the initial condition* $K(j,j) = K_j$ *determines completely* $K(j,h)$ *at* $h > j$. *At* $h < j$ *we have* $K(j,h) = K(h,j)^T$.

R e m a r k. The distribution of the process $\{Y_l\}$ will be not normal at the normal distributions V_l and Y_1 in consequence of the nonlinearity of Eq. (7.1.29) relatively to Y_l and V_l.

7.2. Methods of Normal Approximation and Statistical Linearization

7.2.1. Normal Approximations Method. One-Dimensional Distribution

In the general case the exact solution of the equations (Section 5.6) determining the multi-dimensional distributions of the state vector of a system is impossible. The simplest approximate method of finding the multi-dimensional distributions of the state vector of a nonlinear system is the method of approximation of these distributions by normal distributions. It is clear that the closer is the system to a linear one the more precise are the calculations by the normal approximation method. But as is known by experience of using this method it may also give good results for essentially nonlinear systems.

$\rhd$ Approximating the one-dimensional distribution of the random process $Y(t)$ in normal stochastic differential system (5.4.6) by a normal one we shall have

$$g_1(\lambda; t) \approx \exp\left\{ i\lambda^T m - \frac{1}{2}\lambda^T K\lambda \right\}, \qquad (7.2.1)$$

$$f_1(y; t) \approx [(2\pi)^p \ |K|]^{-1/2} \exp\left\{ -\frac{1}{2}(y^T - m^T)K^{-1}(y - m) \right\}, \qquad (7.2.2)$$

where m and K are the unknown expectation and covariance matrix of the state vector of the system Y.

After calculating the expectations in (7.1.1) and (7.1.5) for the normal distribution $N(m, K)$ we obtain the ordinary differential equations approximately determining m and K:

$$\dot{m} = \varphi_1(m, K, t), \quad m(t_0) = m_0, \tag{7.2.3}$$

$$\dot{K} = \varphi_2(m, K, t), \quad K(t_0) = K_0, \tag{7.2.4}$$

where

$$\varphi_1(m, K, t) = E_N a(Y, t), \tag{7.2.5}$$

$$\varphi_2(m, K, t) = \varphi_{21}(m, K, t) + \varphi_{21}(m, K, t)^T + \varphi_{22}(m, K, t), \tag{7.2.6}$$

$$\varphi_{21}(m, K, t) = E_N a(Y, t)(Y^T - m^T), \tag{7.2.7}$$

$$\sigma(Y, t) = b(Y, t)\nu(t)b(Y, t)^T, \tag{7.2.8}$$

$$\varphi_{22}(m, K, t) = E_N \sigma(Y, t) = E_N b(Y, t)\nu(t)b(Y, t)^T \tag{7.2.9}$$

and the subscript N denotes that the expectation is calculated for the normal distribution $N(m, K)$ of the random variable Y:

$$E_N(\cdot) = \frac{1}{(2\pi)^p} \int\limits_{-\infty}^{\infty} (\cdot) \exp\left[-\frac{1}{2}(y^T - m^T)K^{-1}(y - m) \right] dy. \tag{7.2.10}$$

The number of equations for m and K by the normal approximation method (NAM) is equal to $Q_{\text{NAM}} = p(p+3)/2$. ◁

Thus, *the NAM for the one-dimensional distributions in the normal stochastic differential system is based on* Eqs. (7.2.3), (7.2.4) *at* (7.2.5)–(7.2.10) *and corresponding initial conditions.*

7.2.2. Statistical Linearization Method

In the special case of the unit matrix $b(y, t)$, $b(y, t) = I$, $\varphi_{22}(m, K, t) = \nu(t)$. In this case NAM gives the same equations as the Kazakov statistical linearization method (SLM).

▷ Really, SLM is based on the approximate formula

$$\varphi(Y, t) \approx \varphi_0 + k_1(Y - m), \tag{7.2.11}$$

where φ_0 and k_1 are determined by minimizing the mean square error under the assumption of normality of the distribution of Y. Here φ_0 and k_1 are given by

the formulae $\varphi_0 = \varphi_1(m, K, t)$ and $k_1 = k_1(m, K, t) = \varphi_{21}(m, K, t)K^{-1}$.
Replacing the function $a(Y, t)$ by the obtained linear function of Y we reduce Eq.
(5.4.6) in the case of $b(Y, t) = I$ to a linear stochastic differential equation

$$\dot{Y} = \varphi_0 + k_1(Y - m) + V . \tag{7.2.12}$$

Using methods of Section 6.1 we get in this case the following approximate equations
for m and K:

$$\dot{m} = \varphi_0 , \tag{7.2.13}$$

$$\dot{K} = k_1 K + K k_1^T + \nu . \tag{7.2.14}$$

Substituting here the obtained expressions of φ_0 and k_1 we see that these equations
coincide with Eqs. (7.2.3), (7.2.4) at $\varphi_{22}(m, K, t) = \nu$. ◁

For SLM practical application the tables of formulae were compiled
for φ_0 and $k_1 = \left[\partial/\partial m)\varphi_0^T\right]^T$ for typical scalar and vector composite
functions. In Appendix 3 the formulae for φ_0 are given for some typical
composite functions. These formulae may be used for the determination
of the function $\varphi_{21}(m, K, t) = k_1 K = \left[(\partial/\partial m)\varphi_0^T\right]^T K$, while setting
up Eqs. (7.2.3), (7.2.4) of the normal approximation method. We may
also use these formulae for determining the function $\varphi_{22}(m, K, t)$, as
according to (7.2.9) it represents the first item in the formula of the form
of Eq. (7.2.11) for the statistically linearized function $b(Y, t)\nu(t)b(Y, t)^T$.

For NAM equations in case of Eqs. (5.4.7) and (5.4.6) at the nongaussian white noise see Problems 7.9 and 7.10.

E x a m p l e 7.2.1. Let us consider the system described by the stochastic
differential equation of Example 7.1.1. In this case $a(y, t) = -y^3$, $b(y, t) = y$ and
formulae (7.2.5), (7.2.7), (7.2.9) give

$$\varphi_1(m, D, t) = -\frac{1}{\sqrt{2\pi D}} \int_{-\infty}^{\infty} y^3 e^{-(y-m)^2/2D}\, dy, \tag{I}$$

$$\varphi_{21}(m, D, t) = -\frac{1}{\sqrt{2\pi D}} \int_{-\infty}^{\infty} y^3(y - m)e^{-(y-m)^2/2D}\, dy, \tag{II}$$

$$\varphi_{22}(m, D, t) = \frac{\nu}{\sqrt{2\pi D}} \int_{-\infty}^{\infty} y^2 e^{-(y-m)^2/2D}\, dy. \tag{III}$$

Taking into account

$$y^2 = m^2 + 2m(y - m) + (y - m)^2 \,,$$
$$y^3 = m^3 + 3m^2(y - m) + 3m(y - m)^2 + (y - m)^3, \qquad \text{(IV)}$$

that for the normal distribution the central moments of the odd orders are equal to zero, and the central fourth order moment is equal to $3D^2$ we obtain from Eqs. (I)–(IV)

$$\varphi_1(m, D, t) = -m(m^2 + 3D)\,, \quad \varphi_{21}(m, D, t) = -3D(m^2 + D)\,,$$
$$\varphi_{22}(m, D, t) = \nu(m^2 + D)\,,$$
$$\varphi_2(m, D, t) = -6D(m^2 + D) + \nu(m^2 + D) = (\nu - 6D)(m^2 + D)\,.$$

Consequently, Eqs. (7.2.3), (7.2.4) have in this case the form

$$\dot{m} = -m(m^2 + 3D)\,, \quad \dot{D} = (\nu - 6D)(m^2 + D)\,. \qquad \text{(V)}$$

After integrating Eqs. (V) with the initial conditions $m(t_0) = m_0$, $D(t_0) = D_0$ we determine completely the approximate normal one-dimensional distribution of the process $Y(t)$. The obtained equations may also be derived from the first two equations of Example 7.1.5 by substituting into them the expressions of the third and the fourth moments in terms of the expectation and the variance for the normal distribution.

E x a m p l e 7.2.2. For the system described by the equation

$$\dot{Y} = -\varphi(Y) + V\,, \qquad \text{(I)}$$

where $\varphi(y)$ is the nonlinear characteristic of the limiter shown in Fig.7.2.1a formulae (7.2.5), (7.2.7), (7.2.9) give

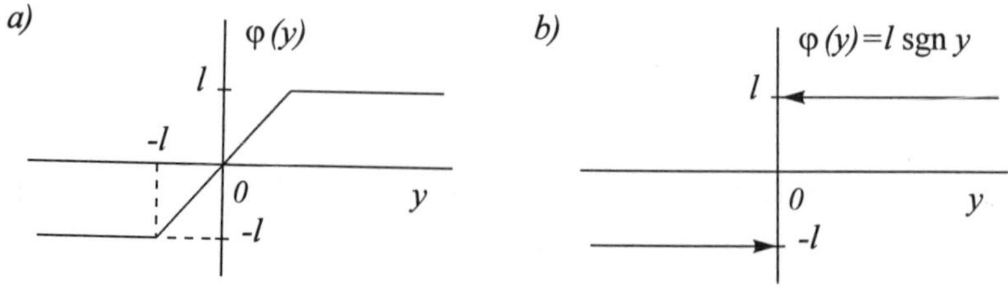

Fig. 7.2.1

$$\varphi_1(m, D, t) = -\frac{1}{\sqrt{2\pi D}}\left[-l\int_{-\infty}^{-l} e^{-(y-m)^2/2D}\,dy + \int_{-l}^{l} ye^{-(y-m)^2/2D}\,dy\right.$$

$$\left. +l\int_{l}^{\infty} e^{-(y-m)^2/2D}\,dy\right], \quad \varphi_{22}(m, D, t) = \nu, \qquad \text{(II)}$$

$$\varphi_{21}(m, D, t) = -\frac{1}{\sqrt{2\pi D}}\left[-l\int_{-\infty}^{-l} (y-m)e^{-(y-m)^2/2D}\,dy\right.$$

$$\left. +\int_{-l}^{l} y(y-m)e^{-(y-m)^2/2D}\,dy + l\int_{i}^{\infty} (y-m)e^{-(y-m)^2/2D}\,dy\right]. \qquad \text{(III)}$$

For performing the integration we notice that the integrals of a normal density representing the probabilities of the occurrence of a normally distributed random variable in the corresponding intervals are expressed in terms of the Laplace function $\Phi(x) = -1/\sqrt{2\pi}\int_0^x e^{-u^2/2}\,du$ by the formula

$$\frac{1}{\sqrt{2\pi D}}\int_\alpha^\beta e^{-(y-m)^2/2D}\,dy = \Phi\left(\frac{\beta - m}{\sqrt{D}}\right) - \Phi\left(\frac{\alpha - m}{\sqrt{D}}\right). \qquad \text{(IV)}$$

The other integrals are reduced to the integrals

$$\int_\alpha^\beta ue^{-u^2/2}\,du = -\left(e^{-\beta^2/2} - e^{-\alpha^2/2}\right), \qquad \text{(V)}$$

$$\int_\alpha^\beta u^2 e^{-u^2/2}\,du = -\int_\alpha^\beta ude^{-u^2/2} = -\left(\beta e^{-\beta^2/2} - \alpha e^{-\alpha^2/2}\right)$$

$$+\int_\alpha^\beta e^{-u^2/2}\,du = -\left(\beta e^{-\beta^2/2} - \alpha e^{-\alpha^2/2}\right) + \sqrt{2\pi}\,[\Phi(\beta) - \Phi(\alpha)]. \qquad \text{(VI)}$$

Using formulae (IV)–(VI) and taking into account that $\Phi(-x) = -\Phi(x)$, $\Phi(\infty) = 1/2$ we find

$$\varphi_1(m, D, t) = -\left[(l+m)\Phi\left(\frac{l+m}{\sqrt{D}}\right) - (l-m)\Phi\left(\frac{l-m}{\sqrt{D}}\right)\right]$$
$$+ \sqrt{\frac{D}{2\pi}}\left\{\exp\left[-\frac{(l-m)^2}{2D}\right] - \exp\left[-\frac{(l+m)^2}{2D}\right]\right\}, \quad \text{(VII)}$$

$$\varphi_{21}(m, D, t) = -D\left[\Phi\left(\frac{l+m}{\sqrt{D}}\right) + \Phi\left(\frac{l-m}{\sqrt{D}}\right)\right]. \quad \text{(VIII)}$$

Eqs. (7.2.3), (7.2.4) have in this case the form

$$\dot{m} = -\left[(l+m)\Phi\left(\frac{l+m}{\sqrt{D}}\right) - (l-m)\Phi\left(\frac{l-m}{\sqrt{D}}\right)\right]$$
$$+ \sqrt{\frac{D}{2\pi}}\left\{\exp\left[-\frac{(l-m)^2}{2D}\right] - \exp\left[-\frac{(l+m)^2}{2D}\right]\right\}, \quad \text{(IX)}$$

$$\dot{D} = -2D\left[\Phi\left(\frac{l+m}{\sqrt{D}}\right) + \Phi\left(\frac{l-m}{\sqrt{D}}\right)\right] + \nu. \quad \text{(X)}$$

Analogously in the case when $\varphi(y) = l\,\mathrm{sgn}\,y$ (Fig.7.2.1b), we obtain

$$\dot{m} = -\varphi_1(m, D), \quad \varphi_1 = 2l\Phi\left(\frac{m}{\sqrt{D}}\right), \quad \text{(XI)}$$

$$\dot{D} = -2Dk_1(m, D) + \nu, \quad k_1(m, D) = l\sqrt{\frac{2}{\pi D}}e^{-m^2/2D}. \quad \text{(XII)}$$

E x a m p l e 7.2.3. For the system with the two-dimensional state-space by Eq. (I) of Example 7.1.6 the matrices $a(Y, t)$ and $b(Y, t)$ are determined by the formulae

$$a(Y, t) = -\begin{bmatrix} Y_1 & Y_2 \\ a & Y_2 \end{bmatrix}, \quad b(Y, t) = \begin{bmatrix} 0 \\ h \end{bmatrix}, \quad \text{(I)}$$

and formulae (7.2.5), (7.2.7), (7.2.9) give

$$\varphi_1(m, K, t) = -E_N\begin{bmatrix} Y_1 & Y_2 \\ a & Y_2 \end{bmatrix} = -\begin{bmatrix} k_{12} \\ am_2 \end{bmatrix}, \quad \text{(II)}$$

$$\varphi_{21}(m, K, t) = -E_N\begin{bmatrix} Y_1 & Y_2 \\ a & Y_2 \end{bmatrix}[Y_1 - m_1 \ Y_2 - m_2]$$

$$= -E_N\begin{bmatrix} Y_1Y_2Y_1^0 & Y_1Y_2Y_2^0 \\ aY_2Y_1^0 & aY_2Y_2^0 \end{bmatrix}, \quad \text{(III)}$$

$$\varphi_{22}(m, K, t) = E_N\begin{bmatrix} 0 \\ h \end{bmatrix}\nu[0 \ h] = \begin{bmatrix} 0 & 0 \\ 0 & \nu h^2 \end{bmatrix}. \quad \text{(IV)}$$

Taking into account that for the normal distribution all the central moments of the odd orders are equal to zero we find from (I)–(IV)

$$E_N Y_1 Y_2 Y_1^0 = E_N \left\{ m_1 m_2 Y_1^0 + m_1 Y_2^0 Y_1^0 + m_2 Y_1^{02} + Y_1^{02} Y_2^0 \right\}$$

$$= m_1 k_{12} + m_2 k_{11}, \tag{V}$$

$$E_N Y_1 Y_2 Y_2^0 = m_1 k_{22} + m_2 k_{12}, \quad E_N Y_2 Y_1^0 = k_{12}, \quad E_N Y_2 Y_2^0 = k_{22}, \tag{VI}$$

$$\varphi_{21}(m, K, t) = - \begin{bmatrix} m_2 k_{11} + m_1 k_{12} & m_2 k_{12} + m_1 k_{22} \\ a k_{12} & a k_{22} \end{bmatrix}, \tag{VII}$$

$$\varphi_2(m, K, t) = \varphi_{21}(m, K, t) + \varphi_{21}(m, K, t)^T + \varphi_{22}(m, K, t)$$

$$= \begin{bmatrix} -2(m_2 k_{11} + m_1 k_{12}) & -(m_2 + \alpha) k_{12} - m_1 k_{22} \\ -(m_2 + a) k_{12} - m_1 k_{22} & \nu h^2 - 2 a k_{22} \end{bmatrix}. \tag{VIII}$$

Thus Eqs. (7.2.3), (7.2.4) using (V)–(VIII) have in this case the form

$$\dot{m}_1 = -m_1 m_2 - k_{12}, \quad \dot{m}_2 = -a m_2, \tag{IX}$$

$$\dot{k}_{11} = -2(m_2 k_{11} + m_1 k_{12}), \quad \dot{k}_{12} = -(m_2 + a) k_{12} - m_1 k_{22},$$

$$\dot{k}_{22} = \nu h^2 - 2 a k_{22}. \tag{X}$$

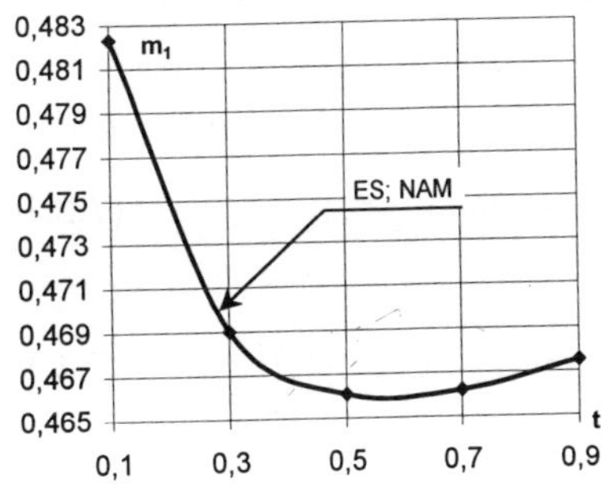

Fig. 7.2.2

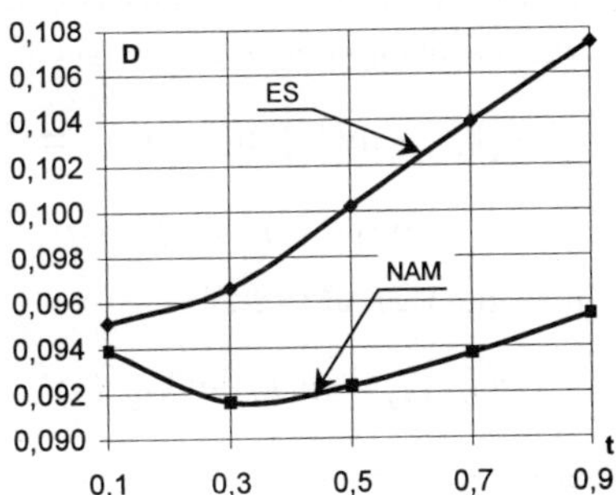

Fig. 7.2.3

The results of calculations of the solution curves Eqs. (IX), (X) are shown in Fig.7.2.2 and 7.2.3 at $m_1(0) = m_2(0) = 0,5$; $k_{11}(0) = 0,1$; $k_{12} = (0)$; $k_{22}(0) = 1$. The exact distribution curve is given for comparison. So the expectation coincides with the exact one. The accuracy of NAM for $\sigma = \sqrt{D}$, $D = k_{11}$ is about 3%.

7.2.3. Normal Approximation Method. Multi-Dimensional Distributions

In exactly the same way the normal approximations of all other multi-dimensional distributions of the process $Y(t)$ may be found. As the normal distribution is completely determined by the moments of the first and the second orders, and the latter are determined by the two-dimensional distribution, it is sufficient to find the normal approximation of the two-dimensional characteristic function $g_2(\lambda_1, \lambda_2; t_1, t_2)$. For this purpose it is sufficient to find the covariance function $K(t_1, t_2)$ of the process $Y(t)$.

▷ In order to obtain an equation for $K(t_1, t_2)$ in the case of system (5.4.6) it is sufficient to evaluate approximately the expectation in (7.1.8) replacing the unknown two-dimensional distribution of the process $Y(t)$ by a normal one. As a result we obtain

$$\partial K(t_1, t_2)/\partial t_2 = E_N \left[Y_{t_1} - m(t_1) \right] a(Y_{t_2}, t_2)^T \tag{7.2.15}$$

where the subscript N at the expectation means that it is calculated for a normal joint distribution of the random variables Y_{t_1} and Y_{t_2}. For calculating this expectation we use the formula of total expectation: first, we evaluate the conditional expectation relative to Y_{t_2}, and after that we evaluate the expectation of the obtained function of the random variable Y_{t_2}. Then using the known formula for the conditional expectation of the part of the components of a normally distributed random vector relative to the remaining components we get

$$E_N[Y_{t_1} - m(t_1)]a(Y_{t_2}, t_2)^T = E_N\{E_N[Y_{t_1}|Y_{t_2}] - m(t_2)\}a(Y_{t_2}, t_2)^T$$

$$= K(t_1, t_2)K(t_2)^{-1}E_N[Y_{t_2} - m(t_2)]a(Y_{t_2}, t_2)^T . \tag{7.2.16}$$

But on the basis of (7.2.7)

$$E_N[Y_{t_2} - m(t_2)]a(Y_{t_2}, t_2)^T = \varphi_{21}(m(t_2), K(t_2), t_2)^T . \tag{7.2.17}$$

Consequently, from (7.2.10) and (7.2.17) we get

$$E_N[Y_{t_1} - m(t_1)]a(Y_{t_2}, t_2)^T = K(t_1, t_2)K^{-1}(t_2)\varphi_{21}(m(t_2), K(t_2), t_2)^T . \tag{7.2.18}$$

Substituting expression (7.2.18) into (7.2.15) we obtain the approximate equation for the covariance function of the process $Y(t)$:

$$\partial K(t_1, t_2)/\partial t_2 = K(t_1, t_2)K(t_2)^{-1}\varphi_{21}(m(t_2), K(t_2), t_2)^T . \tag{7.2.19}$$

This equation with the initial condition $K(t_1, t_1) = K(t_1)$ represents at any fixed t_1 an ordinary linear differential equation determining $K(t_1, t_2)$ as the function of $t_2, t_1 < t_2$. ◁

Thus Eqs. (7.2.3), (7.2.4), (7.2.19) define consecutively the expectation $m(t)$, the covariance matrix $K(t,t) = K(t)$ and the covariance function $K(t_1, t_2)$ of the process $Y(t)$ according to NAM in the normal stochastic differential system described by. After this we may determine all the multi-dimensional distributions (approximately normal) of the process $Y(t)$ by formulae of (Subsection 2.7.5):

$$g_{t_1, \ldots, t_n}(\lambda_1, \ldots, \lambda_n) = \exp\left\{ i\bar{\lambda}^T \bar{m}_n - \frac{1}{2}\bar{\lambda}^T \bar{K}_n \bar{\lambda} \right\} \quad (n = 1, 2, \ldots),$$

$$(7.2.20)$$

$$\bar{\lambda} = \left[\lambda_1^T \lambda_2^T \ldots \lambda_n^T\right]^T, \quad \bar{m}_n = \left[m_y(t_1)^T m_y(t_2)^T \ldots m_y(t_n)^T\right]^T,$$

$$\bar{K}_n = \begin{bmatrix} K(t_1, t_1) & K(t_1, t_2) & \ldots & K(t_1, t_n) \\ K(t_2, t_1) & K(t_2, t_2) & \ldots & K(t_2, t_n) \\ \ldots & \ldots & \ldots & \ldots \\ K(t_n, t_1) & K(t_n, t_2) & \ldots & K(t_n, t_n) \end{bmatrix}$$

$$f_n(y_1, \ldots, y_n; t_1, \ldots, t_n)$$

$$\approx \left[(2\pi)^n |\bar{K}_n|\right]^{-1/2} \exp\left\{ -\frac{1}{2}(\bar{y}_n^T - \bar{m}_n^T)\bar{K}_n^{-1}(\bar{y}_n - \bar{m}_n) \right\} \quad (n = 1, 2, \ldots),$$

$$(7.2.21)$$

where $\bar{y}_n = [y_1^T y_2^T \ldots y_n^T]^T$.

E x a m p l e 7.2.4. Under the conditions of Example 7.2.1 Eq. (7.2.19) has the form

$$\partial K(t_1, t_2)/\partial t_2 = -3 \left[m^2(t_2) + D(t_2) \right] K(t_1, t_2). \qquad (I)$$

The solution of this equation with the initial condition $K(t_1, t_1) = D(t_1)$ is expressed by the formula

$$K(t_1, t_2) = D(t_1) \exp\left[-3 \int_{t_1}^{t_2} \left\{ m^2(\tau) + D(\tau) \right\} d\tau \right]. \qquad (II)$$

According to this formula we may calculate the covariance function of the process $Y(t)$ after the integration of the equations determining its expectation $m(t)$ and variance $D(t)$.

E x a m p l e 7.2.5. Under the conditions of Example 7.2.2 (Fig.7.2.1a) Eq. (7.2.19) has the form

$$\frac{\partial K(t_1, t_2)}{\partial t_2} = -\left[\Phi\left(\frac{l + m(t_2)}{\sqrt{D(t_2)}} \right) + \Phi\left(\frac{l - m(t_2)}{\sqrt{D(t_2)}} \right) \right] K(t_1, t_2). \qquad (I)$$

The solution of this equation with the initial condition $K(t_1, t_2) = D(t_1)$ is expressed by the formula

$$K(t_1, t_2) = D(t_1) \exp \left[- \int_{t_1}^{t_2} \left\{ \Phi \left(\frac{l + m(\tau)}{\sqrt{D(\tau)}} \right) + \Phi \left(\frac{l - m(\tau)}{\sqrt{D(\tau)}} \right) \right\} d\tau \right] .$$

$$(II)$$

For the special case (Fig.7.2.1b) in formulae (XI) and (XII) of Example 7.2.1 we need to put

$$k_1(m(t), D(t)) = l \sqrt{\frac{2}{\pi D(t)}} \exp \left[- \frac{m(t)^2}{2D(t)} \right] .$$

$$(III)$$

E x a m p l e 7.2.6. To obtain by NAM the equation for the covariance function

$$K(t_1, t_2) = \begin{bmatrix} K_{11}(t_1, t_2) & K_{12}(t_1, t_2) \\ K_{21}(t_1, t_2) & K_{22}(t_1, t_2) \end{bmatrix}$$

of the process $Y(t) = [Y_1(t) Y_2(t)]^T$ under the conditions of Example 7.2.3 we represent the formula obtained for $\varphi_{21}(m, K, t)$ in the form

$$\varphi_{21}(m, K, t) = - \begin{bmatrix} m_2 & m_1 \\ 0 & a \end{bmatrix} K .$$

Substituting this expression at $t = t_2$ into Eq. (7.2.19) we obtain

$$\frac{\partial K(t_1, t_2)}{\partial t_2} = -K(t_1, t_2) \begin{bmatrix} m_2(t_2) & 0 \\ m_1(t_2) & a \end{bmatrix}$$

or in scalar form

$$\partial K_{11}(t_1, t_2)/\partial t_2 = -m_2(t_2) K_{11}(t_1, t_2) - m_1(t_2) K_{12}(t_1, t_2) ,$$
$$\partial K_{12}(t_1, t_2)/\partial t_2 = -a K_{12}(t_1, t_2) , \quad \partial K_{22}(t_1, t_2)/\partial t_2 = -a K_{22}(t_1, t_2) ,$$
$$\partial K_{21}(t_1, t_2)/\partial t_2 = -m_2(t_2) K_{21}(t_1, t_2) - m_1(t_2) K_{22}(t_1, t_2) .$$

$$(I)$$

The initial conditions for Eqs. (I) have the form $K_{hl}(t_1, t_1) = k_{hl}(t_1)$ $(h, l = 1, 2)$. The second and the fourth equations are easily integrated. Their solutions are expressed by the formulae

$$K_{12}(t_1, t_2) = k_{12}(t_1) e^{-a(t_2 - t_1)} , \quad t_2 > t_1 ,$$
$$K_{22}(t_1, t_2) = k_{22}(t_1) e^{-a(t_2 - t_1)} , \quad t_2 > t_1 .$$

$$(II)$$

Substituting these expressions into the first and the third Eqs. (I) respectively we find $K_{11}(t_1, t_2)$ and $K_{21}(t_1, t_2)$.

7.2.4. Approximate Determination of Stationary Processes

NAM may also be used to determine approximately the characteristics of the strictly stationary process in the normal nonlinear stochastic differential system with the stationary normal white noise V. In this case $a(y,t) = a(y)$, $b(y,t) = b(y)$, $\nu(t) = \nu$ do not depend on time t, and consequently, the functions φ_1 and φ_2 determined by formulae (7.2.5) and (7.2.6) are also independent of t.

For finding the expectation and the covariance matrix of the value of the stationary process at any t we should put in Eqs. (7.2.3), (7.2.4) $\dot{m} = 0$, $\dot{K} = 0$. As a result we obtain the equations

$$\varphi_1(m, K) = 0 \,, \tag{7.2.22}$$

$$\varphi_2(m, K) = 0 \,. \tag{7.2.23}$$

If the constant vector m and the nonnegative definite matrix K exist satisfying these equations, and this solution of Eqs. (7.2.3), (7.2.4) is stable according to Liapunov then we may suppose that this solution characterizes the stationary process in the system. In this case for determining the covariance function $k(\tau)$ of the stationary process we should use the same method as in Subsection 7.2.3. As a result we obtain

$$dk(\tau)/d\tau = \varphi_{21}(m, K)K^{-1}k(\tau) \,. \tag{7.2.24}$$

Thus Eq. (7.2.24) with *the initial condition $k(0) = K$ determines the covariance function of the stationary process in the system at $\tau > 0$. At $\tau < 0$ we use $k(\tau) = k(-\tau)^T$.*

After finding m, K and $k(\tau)$ all approximately normal multi-dimensional distributions of the strictly stationary process in the system are determined by formula (7.2.20) or (7.2.21).

R e m a r k. But it should be remembered that Eqs. (7.2.3), (7.2.4) are approximate in consequence of which the equations obtained from Eqs. (7.2.3), (7.2.4) at $\dot{m} = 0$, $\dot{K} = 0$ may have the required solutions also in the case where the strictly stationary process in the system does not exist. Therefore, using NAM for the determination of the stationary processes in the nonlinear stochastic differential systems one should be careful and check each obtained stationary solution by the methods of Section 7.4 with the account of the moments of higher orders.

E x a m p l e 7.2.7. The nonlinear system of Examples 7.1.1 and 7.2.1 may serve as an example of such a situation. It is not difficult to check that NAM gives in this case two stationary process: $m = 0$, $D = 0$ and $m = 0$, $D = \nu/6$. But using the method of the moments with the account of the moments up to the fourth order we make sure that the second stationary process does not exist and only the first solution corresponds to the strictly stationary process in the system.

So the results in Subsections 7.2.1–7.2.4 may be summarized in the form of the following statements.

Theorem 7.2.1. *If the random process $Y(t)$ in normal stochastic differential system (5.4.6) has the finite the first and second order moments then its one- and multi-dimensional distributions may be approximated by normal ones (7.2.1), (7.2.2), (7.2.20), (7.2.21) with $m(t)$, $K(t)$ and $K(t_1, t_2)$ defined by Eqs. (7.2.3), (7.2.4), (7.2.19).*

Corollary 7.2.1. *If the stationary (in the strict sense) process in stationary normal stochastic differential system (5.4.6) has the finite the first and the second order moments then its stationary one- and multi-dimensional distributions may be approximately expressed by normal ones with m, K and $k(\tau)$ defined by Eqs. (7.2.22), (7.2.23) and (7.2.24).*

E x a m p l e   7.2.8.   Under the conditions of Example 7.2.2 at $\varphi(Y) = l\,\mathrm{sgn}\,y$ the variance of the stationary process Y which was found by SLM is equal $D_{\mathrm{SML}} = \pi\nu^2/8l^2$. The exact solution of the variance with the account of the results of Example 5.6.9 is $D_{\mathrm{ES}} = 1/2l^2$. The relative error of the calculation $\sigma = \sqrt{D}$ at $\nu = 1$ is $1 - \sqrt{\pi/4} \approx 11\%$. The covariance function $k_{\mathrm{SLM}}(\tau)$ according to Eq. (7.2.31) satisfies the equation

$$dk_{\mathrm{SLM}}(\tau)/d\tau = -\alpha k_{\mathrm{SLM}}(\tau)\,, \quad \alpha = 4l^2/\pi\nu\,.$$

Hence we find $k_{\mathrm{SLM}}(\tau) = D_{\mathrm{SLM}}e^{-\alpha|\tau|}$.

7.2.5. Normal Approximation Method for Stochastic Discrete Nonlinear Systems

Analogously Subsections 7.2.1–7.2.4 we get the following statements.

Theorem 7.2.2. *If the discrete random process in stochastic normal discrete system (5.7.1) has the finite the first and the second order moments then its one- and multi-dimensional distributions may be approximately expressed by normal ones*

$$g_l(\lambda) \approx \exp\left\{ i\lambda^T m_l - \frac{1}{2}\lambda^T K_l \lambda \right\}, \tag{7.2.25}$$

$$f_l(y_l) \approx \left[(2\pi)^p\, |K_l|\, \right]^{-1/2} \exp\left\{ -\frac{1}{2}(y_l^T - m_l^T)K_l^{-1}(y_l - m_l) \right\}, \tag{7.2.26}$$

$$g_{l_1,\ldots,l_n}(\lambda_1,\ldots,\lambda_n) \approx \exp\left\{i\bar{\lambda}^T\bar{m}_n - \frac{1}{2}\bar{\lambda}^T\bar{K}_n\bar{\lambda}\right\}$$

$$\left(\bar{\lambda} = [\lambda_1^T \ldots \lambda_n^T]^T, \quad \bar{m}_n = [m_{l_1}^T \ldots m_{l_n}^T]^T\right), \qquad (7.2.27)$$

$$\bar{K}_n = \begin{bmatrix} K(l_1,l_1) & K(l_1,l_2) & \ldots & K(l_1,l_n) \\ K(l_2,l_1) & K(l_2,l_2) & \ldots & K(l_2,l_n) \\ \ldots & \ldots & \ldots & \ldots \\ K(l_n,l_1) & K(l_n,l_2) & \ldots & K(l_n,l_n) \end{bmatrix},$$

$$f_{l_1,\ldots,l_n}(y_{l_1},\ldots,y_n) \approx \left[(2\pi)^n\,|\bar{K}_n|\right]^{-1/2}$$

$$\times \exp\left\{-\frac{1}{2}(\bar{y}_n^T - \bar{m}_n^T)\bar{K}_n^{-1}(\bar{y}_n - \bar{m}_m)\right\} \quad \left(\bar{y}_n = [y_{l_1}^T y_{l_2}^T \ldots y_{l_n}^T]^T\right)$$

$$(7.2.28)$$

with m_l, K_l and $K(l,h)$ defined by difference equations:

$$m_{l+1} = E_N\omega_l(Y_l,V_l), \quad m_1 = EY_1, \quad l = 1,2,\ldots, \qquad (7.2.29)$$

$$K_{l+1} = E_N\omega_l(Y_l,V_l)\omega_l(Y_l,V_l)^T - E_N\omega_l(Y_l,V_l)E\omega_l(Y_l,V_l)^T,$$
$$K_1 = E(Y_1 - m_1)(Y_1 - m_1)^T, \quad l = 1,2,\ldots, \qquad (7.2.30)$$

$$K(l,h) = E_N Y_l\omega_h(Y_h,V_h)^T - m_l E_N\omega_h(Y_h,V_h)^T,$$
$$K(l,l) = K_l \text{ at } l < h, \quad l,h = 1,2,\ldots, \qquad (7.2.31)$$
$$K(l,h) = K(h,l) = K(h,l)^T \text{ at } l > h,$$

where $E_{Nl}(\cdot) = \displaystyle\int_{-\infty}^{\infty}\int_{-\infty}^{\infty}(\cdot)f_{Nl}(y)\eta_{Nl}(v)dydv$; $f_{Nl}(y)$ and $\eta_{Nl}(v)$ being the normal densities of random values Y_l and V_l.

7.2.6. Normal Approximation Method for Stochastic Infinite-Dimensional Nonlinear Systems

In the general case the exact solution of the equations of Section 5.9 for the multi-dimensional functionals g_n of the normal stochastic systems is impossible. After taking the derivatives of the first and the second order with respect to both parts of Eq. (5.9.1) at $n = 1$ we obtain the formulae of the time derivatives of an operator of the expectation m and the covariance operator K of the process Y_t:

$$\dot{m} = Ea(Y,t), \qquad (7.2.32)$$

$$\dot{K}\lambda = E[a(Y,t)(\lambda,(Y - m)) + (Y - m)(\lambda,a(Y,t))$$
$$+ \sigma(Y,t)\lambda], \quad \sigma(Y,t) = b(Y,t)\nu(t)b(Y,t)^*. \qquad (7.2.33)$$

If we calculate here the expectations replacing the unknown true one-dimensional distribution by a normal one (3.7.00):

$$g_N(\lambda;t) = \exp\left[i(\lambda,m) - \frac{1}{2}(K\lambda,\lambda)\right] = \exp\left[i(\lambda,m) - \frac{1}{2}(\lambda,K\lambda)\right],$$

$$(7.2.34)$$

then Eq. (7.2.32), (7.2.33) at $E = E_N$ together with the correspondent initial conditions will represent a closed system of the ordinary differential equations in a separable H-space $\mathcal{Y}$ which approximately determines the operators m and K.

Similarly after taking the mixed second strong derivative over λ_1 and λ_2 with respect to both parts of Eq. (5.9.9) at $n = 2$ multiplied by $[-i(\lambda_1, m_{t_1}) - i(\lambda_1, m_{t_2})]$ and evaluating the expectation for the two-dimensional normal distribution we get

$$\frac{\partial K(t_1,t_2)}{\partial t_2}\lambda_2 = E_N(Y_{t_1} - m_{t_1})(\lambda_2, a(Y_{t_2} - m_{t_2})), \quad t_2 > t_1 \qquad (7.2.35)$$

for the covariance operator

$$K(t_1,t_2)\lambda_2 = E_N((Y_{t_1} - m_{t_1}), \lambda_2(Y_{t_2} - m_{t_2})). \qquad (7.2.36)$$

Eq. (7.2.35) at the fixed t_1 is an ordinary differential equation in a H-space $\mathcal{Y}$ for the operator $K(t_1,t_2)$ as the function t_2 at the initial condition

$$K(t_1,t_1) = K(t_1). \qquad (7.2.37)$$

For approximate finding the stationary process in (5.4.10) at the stationary white noise $V = W$ we may also use the normal approximation method. Accounting that in this case $a(Y,t) = a(Y)$, $b(y,t) = b(Y)$, $\nu(t) = \nu$, $\sigma(Y,t) = \sigma(Y)$ do not depend on time and putting in Eqs. (7.2.32) and (7.2.33) $\dot{m} = 0$ and $\dot{K} = 0$, $E = E_N$ we shall have

$$E_N a(Y) = 0, \qquad (7.2.38)$$

$$E_N\left[a(Y)(\lambda, (Y - m)) + (Y - m)(\lambda, a(Y)) + \sigma(Y)\lambda\right] = 0. \qquad (7.2.39)$$

The obtained results may be summarized as follows.

Theorem 7.2.3. *If the random process Y_t in the normal stochastic differential system defined by Eq. (5.4.9) in a separable H-space has the bounded the first order operator moment m and the trace-type covariance operator K then its one-dimensional characteristic functional may be* *approximated by normal ones (7.2.34) with m and K satisfing Eqs. (7.2.32), (7.2.33).*

Corollary 7.2.3. *If the stationary (in the strict sense) process Y_t in stationary normal stochastic differential system (5.4.9) has the trace-type covariance operator then its approximate characteristic functional is given by (7.2.34) with m and K defined by Eqs . (7.2.38), (7.2.39).*

E x a m p l e 7.2.9. Under the conditions of Example 5.9.1 Eqs. (7.2.32), (7.2.33) for $m_i(x,t)$ and $K_{ij} = K_{ij}(x_1, x_2, t)$, $i,j = 1,2$ give

$$\frac{\partial m_1}{\partial t} = m_2 \,, \quad \frac{\partial m_2}{\partial t} = c^2 \frac{\partial^2 m_1}{\partial x^2} + 2h_1 m_2 - \frac{2}{3} h_2 m_2 (m_2^2 + 3K_{22}) \,, \quad \text{(I)}$$

$$
\begin{cases}
\dfrac{\partial K_{11}}{\partial t} = K_{21} + K_{12} \,, \quad \dfrac{\partial K_{12}}{\partial t} = K_{22} + c^2 \dfrac{\partial^2 K_{11}}{\partial x^2} + h(x_2, t) K_{12} \,, \\[3mm]
\dfrac{\partial K_{21}}{\partial t} = K_{22} + c^2 \dfrac{\partial^2 K_{11}}{\partial x_1^2} + 2h(x_1, t) K_{21} \,, \\[3mm]
\dfrac{\partial K_{22}}{\partial t} = c^2 \left(\dfrac{\partial^2 K_{12}}{\partial x_1^2} + \dfrac{\partial^2 K_{21}}{\partial x_2^2} \right) + 4h(x_1, t) K_{22} + \\[3mm]
\qquad + c^4 \nu_1 \left(\dfrac{\partial^4 K_{11}}{\partial x_1^2 \partial x_2^2} + \dfrac{\partial^2 m_1}{\partial x_1^2} \cdot \dfrac{\partial^2 m_2}{\partial x_2^2} \right) + \nu_2 \,,
\end{cases}
\quad \text{(II)}
$$

where $h(x,t) = h_1 - h_2 \left[m_2^2(x,t) + K_{22}(x,x,t) \right]$.

7.2.7. Methods of Equivalent Linearization and Structural Methods

Let $f(u,y)$ be the joint density of the random vectors U and Y, m_y and K_y be the expectation and the covariance matrix of the random vector Y, and $|K_y| \neq 0$. From the known equations of the m.s. linear regression (Section 3.4):

$$g K_y = K_{uy} \,, \quad a = m_y - g m_u \tag{7.2.40}$$

or

$$g \Gamma_y = \Gamma_{uy} \tag{7.2.41}$$

we get

$$g = K_{uy} K_y^{-1} = \int\limits_{-\infty}^{\infty} \int\limits_{-\infty}^{\infty} (u - m_u)(y - m_y)^T K_y^{-1} f(u,y) du\, dy$$

$$= \int\limits_{-\infty}^{\infty} [\, m_u(y) - m_u \,](y - m_y)^T K_y^{-1} f_1(y) dy \,, \qquad (7.2.42)$$

where $f_1(y)$ is the density of the random vector Y. This formula together with the approximate formula

$$m_u(Y) \approx m_u + g(Y - m_y) \qquad (7.2.43)$$

yields *the Kazakov statistical linearization* of the regression $m_u(Y)$ (Subsection 7.2.2). Analogously, in case of (7.2.41) we have

$$g = \Gamma_{uy}\Gamma_y^{-1} = \int\limits_{-\infty}^{\infty} \int\limits_{-\infty}^{\infty} uy^T \Gamma_y^{-1} f(u, y) du dy = \int\limits_{-\infty}^{\infty} m_u(y) y^T \Gamma_y^{-1} f_1(y) dy \,.$$

$$(7.2.44)$$

Formula (7.2.44) together with the approximate formula

$$m_u(Y) \approx gY \qquad (7.2.45)$$

yields *the Booton statistical linearization* of the regression $m_u(Y)$.

The regression $m_u(y)$ represents deterministic regression model for $U = \varphi(Y)$. To obtain a stochastic regression model it is sufficient to represent $U = U(Y)$ in the form $U = m_u(y) + U'$ where U' being some random variable. For finding a deterministic linear stochastic model it is sufficient to know the expectations m_u, m_y and the covariance matrices K_y, K_{uy}. But for finding a stochastic linear model it is also necessary to know the distribution of the variable U for any y or at least expectation $m_u(y)$ (the regression) and the covariance matrix $K_u(y)$ (coinciding with the covariance matrix $K_{u'}(y)$ of the random variable U').

A more general problem of the best approximation of the regression by the finite linear combination of given functions $\varphi_1(y), \ldots, \varphi_N(y)$ is reduced to the problem of the best approximation of the regression, as any linear combination of the functions $\varphi_1(y), \ldots, \varphi_N(y)$ represents a linear function of the variables $u_1 = \varphi_1(y), \ldots, u_N = \varphi_N(y)$.

E x a m p l e 7.2.10. Design the optimal polynomial regression model in the case of scalar Y and U. Putting

$$u_1 = y, \quad u_2 = y_1^2, \ldots, u_N = y^N,$$

$$u = [\, 1 \, u_1 \, u_2 \ldots u_N \,]^T, \quad g = [\, ag_1 \ldots g_N \,] \qquad (I)$$

we reduce the problem to the finding model

$$u(y) = g(u). \tag{II}$$

Eq. (7.2.41) in this case takes the form

$$[\,ag_1 \ldots g_N\,] = \begin{bmatrix} 1 & \alpha_1 & \cdots & \alpha_N \\ \alpha_1 & \alpha_2 & \cdots & \alpha_{N+1} \\ \cdots & \cdots & \cdots & \cdots \\ \alpha_N & \alpha_{N+1} & \cdots & \alpha_{2N} \end{bmatrix} = [\,m_u\gamma_1 \ldots \gamma_N\,] \tag{III}$$

where $\alpha_p = EY^p$ is the pth moment of Y, and $\gamma_p = EUY^p$ is the $(p+1)^{\text{th}}$ moment of $[\,Y\,U\,]^T$ containing U in the first degree. Eqs. (II), (III) determine the polynomial m.s. regression of U on Y of the degree N which serves as the optimal polynomial regression model.

The equations of the composed nonlinear stochastic systems after the equivalent linearization of the nonlinearities entering into the equations may be studied by the methods of Chapter 6 and first of all by the structural ones. Hence usually for the description of the composed stochastic systems the different describing functions are taken, for instance, the equivalent weighting functions $g = g(t, \tau; m, K)$ and the transfer function $\Phi(iw, m, K)$.

The implementation of the spectral and the correlation methods (Subsection 6.1.8) for Eq. (5.4.6) assumed as symmetrized Stratonovich equation allows to substitute the joint solution of Eqs. (7.2.22)–(7.2.24) by the solution of Eq. (7.2.22) and the corresponding the equations for the matrix s_y of the spectral densities and the covariance matrix K:

$$s_y = s_y(\omega; m, K) = \Phi(i\omega; m, K)s_v(\omega)\Phi(i\omega; m, K)^*, \tag{7.2.46}$$

$$K = \int_{-\infty}^{\infty} s_y(\omega; m, K)d\omega. \tag{7.2.47}$$

Here $s_v(\omega) = \nu/2\pi$ is the spectral density of the stationary white noise V; $\Phi(i\omega; m, K)$ is the equivalent transfer function of the statistically linearized system for Y^0,

$$Y^0 = \Phi(s; m, K)V, \quad Y^0 = Y - m, \quad s = d/dt, \tag{7.2.48}$$

equal to

$$\Phi(s; m, K) = -[\bar{a}(m, K) - sI]^{-1}\bar{b}(m, K), \tag{7.2.49}$$

where $\bar{a}(m, K) \;=\; E_N a(Y, t) Y^{0T} \;=\; \varphi_{21}(m, K) K^{-1}; \quad \bar{b}(m, K) = E_N b(Y, t).$

7.2.8. General Problem of Stochastic Systems Normalization

In the class of the stochastic systems there are the simplest systems which permit comparatively simple mathematical description and at the connections form the systems of the same type. These are the normal systems. *A normal system* we shall call such a stochastic system whose the joint distribution of the input and the output is normal at the normal distribution of the input. It goes without saying that at the distribution of the input different from the normal one the joint distribution of the input and the output may be nonnormal even a normal system.

The opportunity of applying the well developed methods of the linear systems theory (Chapter 6) to the normal systems and the relative simplicity of the mathematical description of the normal system cause the natural quest for extending the methods of the research of the normal systems as the approximate ones over sufficiently the large class of the stochastic systems different from normal. Thus *the problem of the normalization* of the stochastic nonlinear systems arises, i.e. the problem of finding a suitable normal model for a given nonlinear stochastic system.

Let us prove the following theorem.

Theorem 7.2.4. *For the normalization of the given stochastic nonlinear system it is sufficient to linearize the regression of the output on the input and to average over all possible realizations of the input the conditional covariance operator of the output at the given input.*

▷ In order to linearize the regression we may try to approximate it by the linear dependence on the criterion of the minimum of the mean square error (m.s.e.) at any instant (to use the method of least squares). Then the problem will be reduced to the definition of the linear operator L and the variable a from the condition

$$E|a + LX - E[Y|X]|^2 = \min, \qquad (7.2.50)$$

where $E[Y|x]$ is a conditional expectation of the output Y at the given realization x of the input X (Section 3.4). We apply to the random variable $Z = a + LX - Y$ the known relation:

$$E|Z|^2 = E|E[Z|X]|^2 + ED[Z|X], \qquad (7.2.51)$$

which is valid for any random variables X, Z where $D[\,Z|x\,]$ is a conditional variance of the variable Z at a given realization x of the variable X. As a result we obtain from (7.2.50)

$$E|a + LX - Y|^2 = E|a + LX - E[\,Y|X\,]|^2 + ED[\,Y|X\,]. \qquad (7.2.52)$$

Hence it is clear that the problem of the optimal linear approximation of the regression of the random variable Y on X is equivalent to the problem of the optimal linear approximation of the random variable Y itself on the m.s.e. criterion. $\triangleleft$

7.3. Methods of General Nonlinear Stochastic Systems Theory. Parametrization of Multi-Dimensional Distributions

7.3.1. Introductory Remarks

Methods of the general stochastic systems theory (Sections 5.6–5.9) practically are used only for the scalar nonlinear systems and the specific classes of the finite- and infinite-dimensional stochastic systems.

Among the various quantative analytical methods of the stochastic finite-dimensional and infinite-dimensional systems theory the methods based on the parametrization of the distributions are widely used. These methods permit to design practically simple software when it is possible to have the definite information about analytical properties of the stochastic system under research.

NAM generalization leads for Eqs. (5.4.6), (5.4.7), (5.7.1), (5.7.2), (5.9.1) to the various approximate methods based on the parametrization of the distributions. In order to explain the general idea of these methods we note that the normal approximation of the one-dimensional distribution represents essentially the approximation of the characteristic function $g_1(\lambda; t)$ and the corresponding density $f_1(y; t)$ by the normal characteristic function and the density depending on the finite set of parameters, i.e. the expectations, the variances and the covariances of the components of the state vector. These parameters naturally depend on instant t, and NAM gives the ordinary differential equations for these parameters. The normal approximation of the n-dimensional distribution represents the replacing the corresponding characteristic function and the density by normal ones.

The set of the parameters on which depends this normal distribution consists of the expectations, the variances and the covariances of the components of the state vector at given the instants $t_1, \ldots, t_n$ found while determining the one-dimensional distribution and besides that of the covariances of the components of the state vectors at different instants. Each of these covariances depends on two instants, and consequently, is determined by the two-dimensional distribution. In accordance with this the normal approximation method gives the ordinary differential equation for the covariance function of the state vector considered as a function of time t_2 at fixed instants $t_1 < t_2$.

Thus NAM is based on replacing of the unknown distributions by the known ones (namely normal) depending on the finite set of unknown parameters and on the derivation of the ordinary differential equations for these parameters.

7.3.2. General Problem of Multi-Dimensional Distributions Parametrization

Approximating the one-dimensional characteristic function $g_1(\lambda; t)$ and the corresponding density $f_1(y; t)$ by some known functions $g_1^*(\lambda; \vartheta)$, $f_1^*(y; \vartheta)$ depending on the finite-dimensional vector parameter ϑ, we reduce the problem of approximate determination of the one-dimensional distribution to the derivation from equations for the one-dimensional characteristic functions (Section 5.6) of the ordinary differential equations determining ϑ as a function of time t. In the same way the problem of approximation of other multi-dimensional distributions may be reduced to the derivation from equations of Section 5.6 the ordinary differential equations determining the finite set of unknown parameters. While approximating the multi-dimensional distributions it is expedient to choose such sequences of functions $\{g_n^*(\lambda_1, \ldots, \lambda_n; \vartheta_n)\}$ and $\{f_n^*(y_1, \ldots, y_n; \vartheta_n)\}$ that at any n the set of components of the vector ϑ_n include as a subset the set of components of the vector ϑ_{n-1}. Then approximating the n-dimensional distribution we have to determine only those components of the vector ϑ_n which were not determined earlier when the functions $g_1, f_1, \ldots, g_{n-1}, f_{n-1}$ were approximated.

R e m a r k. Notice that in the general case the functions $g_n^*(\lambda_1, \ldots, \lambda_n; \vartheta_n)$ and $f_n^*(y_1, \ldots, y_n; \vartheta_n)$ approximating the functions $g_1(\lambda_1, \ldots, \lambda_n; t_1, \ldots, t_n)$ and $f_n(y_1, \ldots, y_n; t_1, \ldots, t_n)$ may not be characteristic functions and the corresponding densities themselves. They may give a good approximation of $g_n(\lambda_1, \ldots, \lambda_n; t_1, \ldots, t_n)$ and $f_n(y_1, \ldots, y_n; t_1, \ldots, t_n)$ if they are not the characteristic functions and densities. So, for instance, we may obtain a good approximation of the densities $f_n(y_1, \ldots, y_n; t_1, \ldots, t_n)$ by the truncated orthogonal expansions which are not densities themselves (Section 3.8). The corresponding truncated expansions for the characteristic functions $g_n(\lambda_1, \ldots, \lambda_n; t_1, \ldots, t_n)$ are not characteristic functions in this case.

Various approximate methods for the solution of the equations of Section 5.6 determining the multi-dimensional distributions of the state vector of a system $Y(t)$ may be derived depending on the nature of the parameters on which depend the functions $f_n^*(y_1, \ldots, y_n; \vartheta_n)$ and $g_n^*(\lambda_1, \ldots, \lambda_n; \vartheta_n)$ approximating the unknown multi-dimensional densities $f_n(y_1, \ldots, y_n; t_1, \ldots, t_n)$ and characteristic functions $g_n(\lambda_1, \ldots, \lambda_n; t_1, \ldots, t_n)$.

As it was mentioned in Section 5.6 equations for the multi-dimensional characteristic functions are the linear differential equations in B-space. For general methods of the numerical analysis of linear differential equations in abstract spaces see, for example, (Pugachev and Sinitsyn 1999).

In Chapter 7 we shall outline the approximate methods for the various types of stochastic systems based on using the moments and the semiinvariants (Section 7.4) up to the given order or the coefficients of the given truncated orthogonal expansions of densities (Section 7.5) as parameters.

7.3.3. Reducing Number of Equations for Distribution Parameters

While deriving equations for the parameters of the one-dimensional distribution in specific problems it is useful to remember that the number N_q^r of the moments of the rth order of the q-dimensional random vector and the total number P_q^r of the moments of orders not exceeding N of the q-dimensional random vector are determined by formulae

$$N_q^r = C_{q+r-1}^r = \frac{(q+r-1)!}{r!(q-1)!}, \tag{7.3.1}$$

$$P_q^N = \sum_{r=1}^{N} N_q^r = C_{N+q}^N - 1 = \frac{(N+q)!}{N!q!} - 1. \tag{7.3.2}$$

The main difficulty in practical applications of the methods based on the parametrization of the distributions is the rapid increase of the number of equations for the moments, the semiinvariants or the coefficients of the truncated orthogonal expansions on the dimension p of the state vector (extended in the general case) Y and the maximal order N of the moments involved. Table 7.3.1 illustrates the dependence of the number of equations for the parameters of the one- dimensional distribution on p and N. But in practice one often has to deal with the cases where p amounts to 100 or even to 200. The number of equations for the parameters becomes exceedingly large in such cases. For instance, at $p = 100$, $N = 6$ the number of equations for the parameters amounts to $1.7 \cdot 10^9$.

Table 7.3.1

Moments Max Order N	State Vector Dimension p										
	1	2	3	4	5	6	7	8	9	10	20
2	2	5	9	14	20	27	35	44	54	65	230
4	4	14	34	69	125	209	329	494	714	1000	10625
6	6	27	83	209	461	923	1715	3002	5004	8007	
8	8	44	164	494	1286	3002	6434				
10	10	65	1000	3002	8007						

To reduce the number of equations for the distribution parameters, one may use such an approximation of the distribution which involves the mixed moments or semiinvariants only of the second order and is independent of the mixed moments or semiinvariants of higher orders. Table 7.3.2 shows the dependence of the number of equations for the parameters of the one-dimensional distribution on the dimension p of the state vector Y and the highest order N of the moments of every component of the vector Y involved into such an approximation. The line corresponding to $N = 2$ shows the number of equations for the first and the second order moments for the normal approximation method.

Table 7.3.2

Moments Max Order N	State Vector Dimension p							
	10	20	30	40	50	100	150	200
2	65	230	495	860	1325	5150	11475	20300
4	85	270	555	840	1425	5350	11475	20700
6	105	310	615	1020	1525	5550	12075	21100
8	125	350	675	1100	1625	5750	12375	21500
10	145	390	735	1180	1725	5950	12675	21900

Comparing Tables 7.3.1 and 7.3.2 we see that the number of equations for the distribution parameters may be greatly reduced by using the approximations of distributions involving mixed moments or semiinvariants only of the second order. In particular, in the above example of $p = 100$, $N = 6$ the number of equations is reduced from $1.7 \cdot 10^9$ to 5500, i.e. to approximately $1/(3 \cdot 10^5)$ of its primary value. At $p = 10$, $N = 6$ the number of equations is reduced from 8007 to 105, i.e. to approximately $1/80$ of its primary value.

Basic specific methods of reducing the number of equations for the distribution parameters will be presented in Sections 7.4 and 7.5 for the stochastic continuous and the discrete systems.

7.4. Methods of Moments and Semiinvariants

7.4.1. One-Dimensional Distribution. Initial Moments

Suppose that the parameter ϑ on which depend the functions $g_1^*(\lambda; \vartheta)$, $f_1^*(y; \vartheta)$ approximating the one-dimensional characteristic function $g_1(\lambda; t)$ and the corresponding density $f_1(y; \vartheta)$ represents the set of the moments of the vector Y up to a given order N.

As the function $f_1^*(y; \vartheta)$ which approximates the density $f_1(y; t)$ a truncated orthogonal expansion of $f_1(y; t)$ of the form (3.8.9) may be used,

$$f_1(y; t) \approx f_1^*(y; \vartheta) = w_1(y) \left[1 + \sum_{l=3}^{N} \sum_{|\nu|=l} c_\nu p_\nu(y) \right]. \tag{7.4.1}$$

The coefficients c_ν represent here linear combinations (3.8.5) of the moments of the random vector $Y(t)$ up to the Nth order

$$c_\nu = q_\nu(\alpha) \ (\nu_1, \ldots, \nu_p = 0, 1, \ldots, N; \ |\nu| = \nu_1 + \cdots + \nu_p = 3, \ldots, N). \tag{7.4.2}$$

In Eqs. (7.4.2) $q_\nu(\alpha)$ represents the result of replacing all the monomials $y_1^{r_1} \ldots y_p^{r_p}$ in the expression of the polynomial $q_\nu(y)$ by the respective moments $\alpha_{r_1, \ldots, r_p}$. In the general case the coefficients of the polynomials $p_\nu(y)$ and $q_\nu(y)$ depend on the first and the second order moments of the vector $Y(t)$, since the density $w_1(y)$ in Eq. (7.4.1) has the same moments of the first and the second order as $f_1(y; t)$.

▷ For the normal stochastic differential system (5.4.6) the equations for α_r are easily derived by means of the calculation the stochastic differential of the product $Y_1^{r_1} \ldots Y_p^{r_p}$ by formula (5.1.7) with the account of (3.8.5) and (3.8.11). As a result we get the following set of the equations for the moments:

$$\dot{\alpha}_r = \varphi_{r,0}(m, K, t) + \sum_{k=3}^{N} \sum_{|\nu|=k} \varphi_{r,\nu}(m, K, t) q_\nu(\alpha) \tag{7.4.3}$$

$$(r_1, \ldots, r_p) = 0, 1, \ldots, N; \ |r| = 1, \ldots, N),$$

where

$$\varphi_{r,\nu}(m, K, t) = \sum_{s=1}^{p} \nu_s \int y_1^\nu \ldots y_s^{\nu_s - 1} \ldots y_p^{\nu_p} a_s(y, t) w_1(t) p_r(y) dy$$

$$+ \frac{1}{2} \sum_{s=1}^{p} \nu_s(\nu_s - 1) \int_{-\infty}^{\infty} y_1^{\nu_1} \ldots y_s^{\nu_s - 2} \ldots y_p^{\nu_p} \sigma_{ss}(y, t) w_1(y) p_r(y) dy$$

$$+ \sum_{s=2}^{p} \sum_{l=1}^{s-1} \nu_s \nu_l \int_{-\infty}^{\infty} y_1^{\nu_1} \ldots y_l^{\nu_l - 1} \ldots \ldots y_s^{\nu_s - 1} \ldots y_p^{\nu_p} \sigma_{sl}(y, t) w_1 p_l(y) dy.$$

$$\tag{7.4.4}$$

Here the functions $\varphi_{r,0}(m, K, t)$ are obtained from the latter formula at $p_r(y) = p_0 = 1$. ◁

Integrating the set of Eqs. (7.4.3) at the initial conditions

$$\alpha_r(t_0) = \alpha_r^0 \quad (r_1, \ldots, r_p = 0, 1, \ldots, N; \ |r| = 1, \ldots, N), \tag{7.4.5}$$

we find all the coordinates of the vector ϑ as the function of time t (α_r^0 are the moments of the initial value Y_0 of the vector $Y(t)$ at $t = t_0$).

It goes without saying that at $|r| = 1$ and $|r| = 2$ Eqs. (7.4.3) for the first and the second order moments coincide with the equations obtained from (7.1.1) and (7.1.3) by the substitution of the density $f_1(y; t)$ in the expressions of the expectations by its approximating function $f_1^*(y; \vartheta)$.

Eqs. (7.4.3) are evidently linear relative to the moments α_r of orders higher than the second, $|r| = 3, \ldots, N$, and are nonlinear relative to the first and the second order moments, since the density $w_1(y)$ and the coefficients of the polynomials $p_\nu(y)$ and $q_\nu(y)$ depend on the first and the second order moments, yielding the dependence of the coefficients $\varphi_{r,0}, \varphi_{r,\nu}$ in Eqs. (7.4.3) on the first and the second order moments.

Expansion (7.4.1) may in particular be a Hermite polynomial expansion of the density $f_1(y; t)$. One may also use a truncated Edgeworth series for the approximation of $f_1(y; t)$ (Subsection 3.8.4). In this case the number of items in the sum (7.4.1) with account of the moments up to the Nth order increases up to $3N - 6$ in consequence of which we may expect greater accuracy of the approximation.

Thus *the initial moments method for determining one-dimensional distribution (7.4.1) in normal nonlinear stochastic differential system (5.4.6) is based on* Eqs. (7.4.2), (7.4.3) *and initial conditions* (7.4.5).

7.4.2. One-Dimensional Distribution. Central Moments

Since the initial and the central moments of a random variable are connected with each other by the one-to-one relation (Subsection 3.8.3) we may use the equations for the central moments instead of those for the initial ones.

▷ While deriving the equations for μ_r we calculate the stochastic differential of the product $[Y_1(t) - m_1(t)]^{r_1} \ldots [Y_p(t) - m_p(t)]^{r_p}$ by means of Itô formula. But at first we notice that formulae (7.1.1) and (7.1.5) at approximation (7.4.1) may be written in the following form:

$$\dot{m} = \varphi_{1,0}(m, K, t) + \sum_{l=3}^{N} \sum_{|\nu|=l} \varphi_{1,\nu}(m, K, t) q_\nu(\alpha), \qquad (7.4.6)$$

$$\dot{K} = \varphi_{2,0}(m, K, t) + \sum_{l=3}^{N} \sum_{|\nu|=l} \varphi_{2,\nu}(m, K, t) q_\nu(\alpha), \qquad (7.4.7)$$

where

$$\varphi_{1,\nu}(m, K, t) = \int_{-\infty}^{\infty} a(y, t) w_1(y) p_\nu(y) dy, \qquad (7.4.8)$$

$$\varphi_{2,\nu}(m, K, t) = \int_{-\infty}^{\infty} \{ a(y, t)(y^T - m^T) + (y - m)a(y, t)^T$$

$$+ \sigma(y, t)\} w_1(y) p_\nu(y) dy, \qquad (7.4.9)$$

$$\sigma(y, t) = b(y, t)\nu(t)b(y, t)^T. \qquad (7.4.10)$$

Here the coefficients $\varphi_{1,0}(m, K, t)$ and $\varphi_{2,0}(m, K, t)$ are determined by formulae (7.4.8) and (7.4.9) at $p_\nu = p_0(y) = 1$ with the change ν by 0.

Evaluating directly by formula (5.1.7) the stochastic differential of the product $[Y_1(t) - m_1(t)]^{r_1} \ldots [Y_p(t) - m_p(t)]^{r_p}$ we get with the account of (7.4.8) and (7.4.9)

$$\dot{\mu}_r = \varphi_{r,0}(m, K, t) + \sum_{l=3}^{N} \sum_{|\nu|=l} \varphi_{r,\nu}(m, K, t) q_\nu(\alpha)$$

$$- \sum_{s=1}^{p} r_s \mu_{r-e_s} \dot{m}_s \ (|r| = 3, \ldots, N), \qquad (7.4.11)$$

where the functions $\varphi_{r,\nu}(m, K, t)$ are determined according to (7.4.4) by means of the change of the products $y_1 \ldots y_p$ by $(y_1 - m_1) \ldots (y_p - m_p)$, and e_s is a vector whose all the components are equal to 0 besides the sth component equal to 1. ◁

Integrating Eqs. (7.4.6), (7.4.11) with the respective initial conditions

$$m(t_0) = m_0, \ \mu_r(t_0) = \mu_r^0 \ (r_1, \ldots, r_p = 0, 1, \ldots, N; \ |r| = 2, \ldots, N), \qquad (7.4.12)$$

we find approximately all the components of the vector ϑ as a function of time t.

It is clear that Eqs. (7.4.6) coincide with those obtained from Eqs. (7.1.1) by replacing the density $f_1(y; t)$ by its approximating function $f_1^*(y; \vartheta)$, and equations Eqs. (7.4.11) for the second order moments, $|r| = 2$, coincide with those obtained in the same way from Eqs. (7.1.5).

The initial moments in $q_\nu(\alpha) = G_\nu(\mu)$ must be expressed in terms of the central ones. It is evident that Eqs. (7.4.6) and (7.4.11) are always nonlinear owing to the presence of the items of the form $\mu_{r-e_h} q_\nu(\alpha)$.

Notice that while using orthogonal Hermite polynomial expansions we have $q_\nu(\alpha) = G_\nu(\mu)$ (Subsection 3.8.2) and, consequently, $q_\nu(\alpha)$ are automatically obtained as the linear combinations of the central moments. Approximating $f_1(y; t)$ by a truncated Edgeworth series (Subsection 3.8.4) the semiinvariants in the coefficients of Edgeworth series must be replaced by their expressions in terms of the central moments.

Thus *in the basis of the central moments method of determining one-dimensional distribution* (7.4.1) *for normal nonlinear stochastic differential system* (5.4.6) *lay Eqs.* (7.4.6), (7.4.7), (7.4.11) *at initial conditions* (7.4.12).

The equations for the moments of stochastic systems (5.4.7) and (5.4.6) in the case of nongaussian white noise are given in Problems 7.23–7.24.

E x a m p l e 7.3.1. Under the conditions of Example 7.1.1 supposing that V is the normal white noise we find the approximation of the one-dimensional distribution of the process $Y(t)$ by a fragment of the Hermite polynomial expansion of the density $f_1(y; t)$ (Subsection 3.8.3):

$$f_1(y; t) \approx f_1^*(y; \vartheta)$$

$$= (2\pi D)^{-1/2} e^{-(y-m)^2/2D} \{1 + c_3 H_3(y - m)/3! + c_4 H_4(y - m)/4!\}. \tag{I}$$

According to formulae of Subsection 3.8.3 the quasimoments $c_3 = G_3(\mu)$ and $c_4 = G_4(\mu)$ represent the linear functions of the central moments μ_3 and μ_4 respectively also depending on the expectation m and the variance D of the process $Y(t)$. Thus the vector ϑ represents the four-dimensional vector with the components m, D, μ_3 and μ_4.

On the basis of formulae (40 and (5) of Appendix 2 the polynomials $H_\nu(x)$ and $G_\nu(x)$ are determined in this case by the formulae

$$H_\nu(x) = (-1)^\nu e^{x^2/2D} \frac{d^\nu}{dx^\nu} e^{-x^2/2D}, \tag{II}$$

$$G_\nu(x) = (-1)^\nu e^{x^2/2D} \left[\frac{d^\nu}{dy^\nu} e^{-Dy^2/2} \right]_{y=x/D}. \tag{III}$$

These formulae give

$$\begin{aligned}
&H_0(x) = 1, \ H_1(x) = x/D, \ H_2(x) = (x^2 - D)/D^2, \\
&H_3(x) = (x^3 - 3Dx)/D^3, \ H_4(x) = (x^4 - 6Dx^2 + 3D^2)/D^4, \\
&H_5(x) = (x^5 - 10Dx^3 + 15D^2x)/D^5, \\
&H_6(x) = (x^6 - 15Dx^4 + 45D^2x^2 - 15D^3)/D^6,
\end{aligned} \tag{IV}$$

$$\begin{aligned}
&G_0(x) = 1, \ G_1(x) = x, \ G_2(x) = x^2 - D, \ G_3(x) = x^3 - 3Dx, \\
&G_4(x) = x^4 - 6Dx^2 + 3D^2, \ G_5(x) = x^5 - 10Dx^3 + 15D^2x, \\
&G_6(x) = x^6 - 15Dx^4 + 45D^2x^2 - 15D^3.
\end{aligned} \tag{V}$$

Eqs. (7.4.3) have in this case the form

$$\begin{aligned}
&\dot\alpha_1 = -\alpha_3, \ \dot\alpha_2 = \nu\alpha_2 - 2\alpha_4, \\
&\dot\alpha_3 = 3\nu\alpha_3 - 30(\alpha_2 - 2\alpha_1^2)\alpha_3 - 15\alpha_1\alpha_4 + 90\alpha_1\alpha_2^2 - 180\alpha^3\alpha_2 + 72\alpha_1^5, \\
&\dot\alpha_4 = 160\alpha_1^3\alpha_3 + 6\nu\alpha_4 - 60\alpha_2\alpha_4 + 120\alpha_2^3 - 480\alpha_1^4\alpha_2 + 256\alpha_1^6.
\end{aligned} \tag{VI}$$

The easiest way to obtain these equations is to substitute into the first four equations of the infinite set of Example 7.1.5 the expressions of α_5 and α_6 following from the equations

$$q_5(\alpha) = G_5(\mu) = \mu_5 - 10D\mu_3 = 0\,,$$
$$q_6(\alpha) = G_6(\mu) = \mu_6 - 15D\mu_4 + 30D^3 = 0\,, \qquad \text{(VII)}$$

for the moments of the truncated orthogonal expansion of the density.

Eqs. (7.4.6), (7.4.7), (7.4.11) for the expectation, the variance, the third and the fourth order central moments have in this case the form:

$$\dot m = -(m^2+3D)m-\mu_3\,,\quad \dot D = \nu(m^2+D)-6m^2D-6m\mu_3-2\mu_4, \quad \text{(VIII)}$$

$$\dot\mu_3 = 3mD(2\nu+3D)+3(\nu - 3m^2 - 9D)\mu_3 - 9m\mu_4\,,$$
$$\dot\mu_4 = 6D(\nu m^2+20D^2)+12(\nu - 9D)m\mu_3+4\mu_3^2+6(\nu - 2m^2 - 10D)\mu_4.$$
$$\text{(IX)}$$

E x a m p l e 7.4.2. We find the approximation of the one-dimensional density of the two-dimensional process $\left[Y_1(t)Y_2(t)\right]^T$ by the truncated Hermite polynomial expansion with the account of the moments up to the fourth order

$$f_1(y;t) \approx f_1^*(y;\vartheta)$$

$$= \left[(2\pi)^2(k_{11}k_{22} - k_{12}^2)\right]^{-1/2}\exp\{-[k_{22}(y_1 - m_1)^2$$
$$-2k_{12}(y_1 - m_1)(y_2 - m_2) + k_{11}(y_2 - m_2)^2]/2(k_{11}k_{22} - k_{12}^2)\}$$

$$\times\left\{1+\sum_{k=3}^{4}\sum_{\nu_1+\nu_2=k}\frac{c_{\nu_1\nu_2}}{\nu_1!\nu_2!}H_{\nu_1\nu_2}(y - m)\right\}, \qquad \text{(I)}$$

where $c_{\nu_1\nu_2} = G_{\nu_1\nu_2}(\nu)$ $(\nu_1, \nu_2 = 0, 1, 2, 3, 4; \nu_1 + \nu_2 = 3, 4)$. Hermite polynomials $H_{pq}(x)$ and $G_{pq}(x)$ are determined by formulae (4) and (5) of Appendix 2 which have in this case the form

$$H_{p,q}(x) = (-1)^{p+q}e^{(c_{11}x_1^2+2c_{12}x_1x_2+c_{22}x_2^2)/2}\frac{\partial^{p+q}}{\partial x_1^p\partial x_2^q}$$
$$\times\, e^{-(c_{11}x_1^2+2c_{12}x_1x_2+c_{22}x_2^2)/2}, \qquad \text{(II)}$$

$$G_{p,q}(x) = (-1)^{p+q}e^{(c_{11}x_1^2+2c_{12}x_1x_2+c_{22}x_2^2)/2}$$
$$\times\left[\frac{\partial^{p+q}}{\partial y_1^p\partial y_2^q}e^{-(k_{11}y_1^2+2k_{12}y_1y_2+k_{22}y_2^2)/2}\right]_{y=Cx}, \qquad \text{(III)}$$

where

$$K = \begin{bmatrix} k_{11} & k_{12} \\ k_{12} & k_{22} \end{bmatrix}\,,\quad K^{-1} = C = \begin{bmatrix} c_{11} & c_{12} \\ c_{12} & c_{22} \end{bmatrix}. \qquad \text{(IV)}$$

Formulae (I)–(IV) give

$$H_{0,0}(x) = 1\,, \quad H_{1,0}(x) = c_{11}x_1 + c_{12}x_2\,, \quad H_{0,1}(x) = c_{12}x_1 + c_{22}x_2, \quad \text{(V)}$$
$$H_{2,0}(x) = (c_{11}x_1 + c_{12}x_2)^2 - c_{11}\,,$$
$$H_{1,1}(x) = (c_{11}x_1 + c_{12}x_2)(c_{12}x_1 + c_{22}x_2) - c_{12}\,,$$
$$H_{0,2}(x) = (c_{12}x_1 + c_{22}x_2)^2 - c_{22}, \quad \text{(VI)}$$

$$H_{3,0}(x) = (c_{11}x_1 + c_{12}x_2)^2 - 3c_{11}(c_{11}x_1 + c_{12}x_2),$$
$$H_{2,1}(x) = (c_{11}x_1 + c_{12}x_2)^2(c_{12}x_1 + c_{22}x_2) - 2c_{12}(c_{11}x_1 + c_{12}x_2) -$$
$$- c_{11}(c_{12}x_1 + c_{22}x_2)\,,$$
$$H_{1,2}(x) = (c_{11}x_1 + c_{12}x_2)(c_{12}x_1 + c_{22}x_2)^2 - c_{22}(c_{11}x_1 + c_{12}x_2) -$$
$$- 2c_{12}(c_{12}x_1 + c_{22}x_2)\,,$$
$$H_{0,3}(x) = (c_{12}x_1 + c_{22}x_2)^3 - 3c_{22}(c_{12}x_1 + c_{22}x_2), \quad \text{(VII)}$$

$$H_{4,0}(x) = (c_{11}x_1 + c_{12}x_2)^4 - 6c_{11}(c_{11}x_1 + c_{12}x_2) + 3c_{11}^2\,,$$
$$H_{3,1}(x) = (c_{11}x_1 + c_{12}x_2)^3(c_{12}x_1 + c_{22}x_2) - 3c_{12}(c_{11}x_1 + c_{12}x_2)^2$$
$$- 3(c_{11}x_1 + c_{12}x_2)(c_{12}x_1 + c_{22}x_2) + 3c_{11}c_{12}\,,$$
$$H_{2,2}(x) = (c_{11}x_1 + c_{12}x_2)^2(c_{12}x_1 + c_{22}x_2)^2 - c_{22}(c_{11}x_1 + c_{12}x_2)^2$$
$$- 4c_{12}(c_{11}x_1 + c_{12}x_2)(c_{12}x_2 + c_{22}x_2)$$
$$- c_{11}(c_{12}x_1 + c_{22}x_2)^2 + c_{11}c_{22} + 2c_{12}^2\,,$$
$$H_{1,3}(x) = (c_{11}x_1 + c_{12}x_2)(c_{12}x_1 + c_{22}x_2)^3 - 3c_{22}(c_{11}x_1 + c_{12}x_2)$$
$$\times (c_{12}x_1 + c_{22}x_2) - 3c_{12}(c_{12}x_1 + c_{22}x_2)^2 + 3c_{12}c_{22}\,,$$
$$H_{0,4}(x) = (c_{12}x_1 + c_{22}x_2)^4 - 6c_{22}(c_{12}x_1 + c_{22}x_2)^2 + 3c_{22}^2\,, \quad \text{(VIII)}$$

$$G_{0,0}(x) = 1, \quad G_{1,0}(x) = x_1, \quad G_{0,1}(x) = x_2, \quad \text{(IX)}$$

$$G_{2,0}(x) = x_1^2 - k_{11}, \quad G_{1,1}(x) = x_1x_2 - k_{12}, \quad G_{0,2}(x) = x_2^2 - k_{22}, \quad \text{(X)}$$

$$G_{3,0}(x) = x_1^3 - 3k_{11}x_1\,, \quad G_{1,2}(x) = x_1x_2^2 - k_{22}x_1 - 2k_{12}x_2\,,$$
$$G_{2,1}(x) = x_1^2x_2 - 2k_{12}x_1 - k_{11}x_2\,,$$
$$G_{0,3}(x) = x_2^3 - 3k_{22}x_2\,, \quad \text{(XI)}$$

$$G_{4,0}(x) = x_1^4 - 6k_{11}x_1^2 + 3k_{11}^2,$$
$$G_{3,1}(x) = x_1^3 x_2 - 3k_{12}x_1^2 - 3k_{11}x_1 x_2 + 3k_{11}k_{12},$$
$$G_{2,2}(x) = x_1^2 x_2^2 - k_{22}x_1^2 - 4k_{12}x_1 x_2 - k_{11}x_2^2 + k_{11}k_{22} + 2k_{12}^2,$$
$$G_{1,3}(x) = x_1 x_2^3 - 3k_{22}x_1 x_2 - 3k_{12}x_2^2 + 3k_{12}k_{22},$$
$$G_{0,4}(x) = x_2^4 - 6k_{22}x_2^2 + 3k_{22}^2, \tag{XII}$$

$$G_{5,0}(x) = x_1^5 - 10k_{11}x_1^3 + 15k_{11}^2 x_1,$$
$$G_{4,1}(x) = x_1^4 x_2 - 4k_{12}x_1^3 - 6k_{11}x_1^2 x_2 + 12k_{11}k_{12}x_1 + 3k_{11}^2 x_2,$$
$$G_{3,2}(x) = x_1^3 x_2^2 - k_{22}x_1^2 - 6k_{12}x_1^2 x_2 - 3k_{11}x_1 x_2^2$$
$$+ 3(k_{11}k_{22} + 2k_{12}^2)x_1 + 6k_{11}k_{12}x_2,$$
$$G_{2,3}(x) = x_1^2 x_2^3 - 3k_{22}x_1^2 x_2 - 6k_{12}x_1 x_2^2 - k_{11}x_2^3$$
$$+ 6k_{12}k_{22}x_1 + 3(k_{11}k_{22} + 2k_{12}^2)x_2,$$
$$G_{1,4}(x) = x_1 x_2^4 - 6k_{22}x_1 x_2^2 - 4k_{12}x_2^3 + 3k_{22}^2 x_1 + 12k_{12}k_{22}x_2,$$
$$G_{0,5}(x) = x_2^5 - 10k_{22}x_2^3 + 15k_{22}^2 x_2. \tag{XIII}$$

Eqs. (7.4.3) have in this case the following form:

$$\dot{\alpha}_{10} = -\alpha_{11}, \, \dot{\alpha}_{01} = -a\alpha_{01}, \tag{XIV}$$

$$\dot{\alpha}_{20} = -2\alpha_{21}, \; \dot{\alpha}_{11} = -\alpha_{12} - a\alpha_{11}, \; \dot{\alpha}_{02} = -2a\alpha_{02} + h^2\nu, \tag{XV}$$

$$\dot{\alpha}_{30} = -3\alpha_{31}, \; \dot{\alpha}_{21} = -2\alpha_{22} - a\alpha_{21},$$
$$\dot{\alpha}_{12} = -\alpha_{13} - 2a\alpha_{12} + h^2\nu\alpha_{10}, \; \dot{\alpha}_{03} = -3a\alpha_{03} + 3h^2\nu\alpha_{01}, \tag{XVI}$$

$$\dot{\alpha}_{40} = -4[\alpha_{01}\alpha_{40} + 4\alpha_{10}\alpha_{31} + 4(\alpha_{11} - 2\alpha_{10}\alpha_{01})(\alpha_{30} - 3\alpha_{10}\alpha_{20})$$
$$+ 6(\alpha_{20} + 2\alpha_{10}^2)(\alpha_{21} - \alpha_{01}\alpha_{20} - 2\alpha_{10}\alpha_{11}) - 24\alpha_{10}^4\alpha_{01}],$$
$$\dot{\alpha}_{31} = -3[2\alpha_{11}\alpha_{31} + 3\alpha_{10}\alpha_{22} + (\alpha_{02} - 2\alpha_{01}^2)(\alpha_{30} - 3\alpha_{10}\alpha_{20})$$
$$+ 6(\alpha_{11} - 2\alpha_{10}\alpha_{01})(\alpha_{21} - \alpha_{20}\alpha_{01} - \alpha_{10}\alpha_{11})$$
$$+ 3(\alpha_{20} - 2\alpha_{10}^2)(\alpha_{12} - \alpha_{10}\alpha_{02} - 2\alpha_{01}\alpha_{11}) - 24\alpha_{10}^2\alpha_{01}^2] - a\alpha_{31},$$
$$\dot{\alpha}_{22} = -2[3\alpha_{01}\alpha_{22} + 2\alpha_{10}\alpha_{13} + 3(\alpha_{02} - 2\alpha_{01}^2)(\alpha_{21} - \alpha_{01}\alpha_{20} - 2\alpha_{10}\alpha_{11})$$
$$+ 6(\alpha_{11} - 2\alpha_{10}\alpha_{01})(\alpha_{12} - \alpha_{10}\alpha_{02} - 2\alpha_{01}\alpha_{11})$$
$$+ (\alpha_{20} - 2\alpha_{10}^2)(\alpha_{03} - 3\alpha_{01}\alpha_{02}) - 24\alpha_{10}^2\alpha_{01}^3] - 2a\alpha_{22},$$
$$\dot{\alpha}_{13} = [4\alpha_{01}\alpha_{13} + \alpha_{10}\alpha_{04} + 6(\alpha_{02} - 2\alpha_{01}^2)(\alpha_{12} - \alpha_{10}\alpha_{02} - 2\alpha_{01}\alpha_{11})$$
$$+ 4(\alpha_{11} - 2\alpha_{10}\alpha_{01})(\alpha_{03} - 3\alpha_{01}\alpha_{02}) - 24\alpha_{10}\alpha_{01}^4] - 3a\alpha_{13} + 3h^2\nu\alpha_{11},$$
$$\dot{\alpha}_{04} = -4a\alpha_{04} + 6h^2\nu\alpha_{02}. \tag{XVII}$$

The simplest way to obtain these equations is to substitute into the first 14 equations of the infinite set of equations of Example 7.1.6 (corresponding to $r_1 + r_2 = 1, 2, 3, 4$) the expressions of the fifth order moments in terms of the moments of lower orders from the relations $G_{4,1}(\mu) = G_{3,2}(\mu) = G_{1,4}(\mu) = 0$ for the moments of the function $f_1^*(y; \vartheta)$. These relations have in this case the form

$$\mu_{41} - 4\mu_{30}k_{12} - 6k_{11}\mu_{21} = 0 \,, \quad \mu_{32} - k_{22}\mu_{30} - 6k_{12}\mu_{21} - 3k_{11}\mu_{12} = 0 \,,$$
$$\mu_{23} - 3k_{22}\mu_{21} - 6k_{12}\mu_{12} - k_{11}\mu_{03} = 0 \,, \quad \mu_{14} - 6k_{22}\mu_{12} - 4k_{12}\mu_{03} = 0 \,.$$
$$\text{(XVIII)}$$

Expressing here the central moments in terms of the initial ones, solving these equations relative to α_{41}, α_{32}, α_{23} and α_{14} and substituting the expressions obtained into the equations of Example 7.1.6 corresponding to $r_1 = 4$, $r_2 = 1$; $r_1 = 3$, $r_2 = 2$; $r_1 = 2$, $r_2 = 3$; $r_1 = 1$, $r_2 = 4$ we get the above equations.

Eqs. (7.4.6), (7.4.7), (7.4.11) for the expectations and the central moments have in this case the form:

$$\dot{m}_1 = -m_1 m_2 - k_{12} \,, \quad \dot{m}_2 = -am_2 \,, \qquad \text{(XIX)}$$

$$\dot{k}_{11} = -2(m_2 k_{11} + m_1 k_{12} + \mu_{21}) \,, \quad \dot{k}_{12} = -(m_2 + a)k_{12} - m_1 k_{22} - \mu_{12} \,,$$
$$\dot{k}_{22} = -2ak_{22} + h^2\nu \,, \qquad \text{(XX)}$$

$$\dot{\mu}_{30} = 3(k_{11}k_{12} - m_2\mu_{30} - m_1\mu_{21} - \mu_{31}) \,,$$
$$\dot{\mu}_{21} = 2[k_{12}^2 - (m_2 + a/2)\mu_{21} - m_1\mu_2 - \mu_{22}] \,,$$
$$\dot{\mu}_{12} = k_{12}k_{22} - (m_2 + 2a)\mu_{12} - m_1\mu_{03} - \mu_{13} \,, \quad \dot{\mu}_{03} = -3a\mu_{03}, \text{(XXI)}$$

$$\dot{\mu}_{40} = -4(3k_{12}\mu_{30} + 6k_{11}\mu_{21} + m_2\mu_{40} + m_1\mu_{31}) \,,$$
$$\dot{\mu}_{31} = -3(k_{22}\mu_{30} + 5k_{12}\mu_{21} + 3k_{11}\mu_{12}) - (3m_2 + a)\mu_{31} - 3m_1\mu_{22} \,,$$
$$\dot{\mu}_{22} = -2[3k_{22}\mu_{21} + 5k_{12}\mu_{12} + k_{11}\mu_{03} + (m_2 + a)\mu_{22} + m_1\mu_{13}] + h^2\nu k_{11} \,,$$
$$\dot{\mu}_{04} = -4a\mu_{04} + 6h^2\nu k_{22} \,,$$
$$\dot{\mu}_{13} = -3(2k_{22}\mu_{12} + k_{12}\mu_{03}) - (m_2 + 3a)\mu_{13} - m_1\mu_{04} + 3h^2\nu k_{12} \,. \text{(XXII)}$$

The simplest way to obtain Eqs. (XIX)–(XXII) is to differentiate the random functions $[Y_1(t) - m_1(t)]^r [Y_2(t) - m_2(t)]^s$ $(r, s = 1, 2, 3, 4; r + s = 1, 2, 3, 4)$ using Itô formula (5.1.7), to take the expectations and to replace the moments μ_{41}, μ_{32}, μ_{23}, μ_{14} in the equations obtained by their expressions from Eqs. (XVIII).

7.4.3. Multi-Dimensional Distributions. Initial Moments

In exactly the same way the approximate equations for the moments of the multi-dimensional distributions may be evaluated in the case where the moments are assumed as parameters of distributions.

▷ The multi-dimensional approximation of the n-dimensional distribution of the random process $Y(t)$, expressed in terms of moments

$$\alpha_{r_{11},\,\ldots,\,r_{1p},\,\ldots,\,r_{n1},\,\ldots,\,r_{np}}(t_1,\,\ldots,\,t_n)$$

$$= EY_1^{r_{11}}(t_1)\ldots Y_p^{r_{1p}}(t_1)\ldots Y_1^{r_{n1}}(t_n)\ldots Y_p^{r_{np}}(t_n)\,, \qquad (7.4.13)$$

is determined by the formula (3.8.45):

$$f_n(y_1,\,\ldots,\,y_n;t_1,\,\ldots,\,t_n)\approx f_n^*(y_1,\,\ldots,\,y_n;\vartheta_n)=w_n(y_1,\,\ldots,\,y_n)$$

$$\times\left\{1+\sum_{l=1}^N\sum_{|\nu_1|+\cdots+|\nu_n|=l}c_{\nu_1,\,\ldots,\,\nu_n}p_{\nu_1,\,\ldots,\,\nu_n}(y_1,\,\ldots,\,y_n)\right\}\,, \qquad (7.4.14)$$

$$\partial\alpha_{\kappa_1,\,\ldots,\,\kappa_n}/\partial t_n=\varphi_{\kappa_1,\,\ldots,\,\kappa_n,0}(\bar{m}_n,\bar{K}_n,t_n)$$

$$+\sum_{l=3}^N\sum_{|\nu_1|+\cdots+|\nu_n|=l}q_{\nu_1,\,\ldots,\,\nu_n}(\alpha_1,\,\ldots,\,\alpha_n)$$

$$\times\varphi_{\kappa_1,\,\ldots,\,\kappa_n,\nu_1,\,\ldots,\,\nu_n}(\bar{m},\bar{K}_n,t_n)\,,$$

$$(|\kappa_1|,\,\ldots,\,|\kappa_n|=1,\,\ldots,\,N,\ |\kappa_1|+\cdots+|\kappa_n|=n,\,\ldots,\,N)\,, \qquad (7.4.15)$$

where

$$\varphi_{\kappa_1,\,\ldots,\,\kappa_n;\nu_1,\,\ldots,\,\nu_n}(\bar{m}_n\bar{K}_n,t_n)$$

$$=\sum_{s=1}^p\nu_{ns}\int_{-\infty}^\infty\cdots\int_{-\infty}^\infty y_{11}^{\nu_{11}}\ldots y_{1p}^{\nu_{1p}}\ldots y_{n-1,1}^{\nu_{n-1,1}}$$

$$\ldots y_{n-1,p}^{\nu_{n-1,p}}y_{n1}^{\nu_{n1}}\ldots y_{ns}^{\nu_{ns}-1}\ldots y_{np}^{\nu_{np}}a_s(y_n,t_n)$$

$$\times w_n(y_1,\,\ldots,\,y_n)p_{\nu_1,\,\ldots,\,\nu_n}(y_1,\,\ldots,\,y_n)dy_1\ldots dy_n$$

$$+\frac{1}{2}\sum_{s=1}^p\nu_{ns}(\nu_{ns}-1)\int_{-\infty}^\infty\cdots\int_{-\infty}^\infty y_{11}^{\nu_{11}}\ldots y_{1p}^{\nu_{1p}}$$

$$\ldots y_{n-1,1}^{\nu_{n-1,1}}\ldots y_{n-1,p}^{\nu_{n-1,p}}y_{n1}^{\nu_{n1}}\ldots y_{ns}^{\nu_{ns}-2}\ldots y_{np}^{\nu_{np}}\sigma_{ss}(y_n,t_n)$$

$$\times w_n(y_1, \ldots, y_n) p_{\nu_1, \ldots, \nu_n}(y_1, \ldots, y_n) dy_1 \ldots dy_n$$

$$+ \sum_{s=2}^{p} \sum_{n=1}^{s-1} \nu_{ns} \nu_{nr} \int_{-\infty}^{\infty} \cdots \int_{-\infty}^{\infty} y_{11}^{\nu_{11}} \cdots y_{1p}^{\nu_{1p}}$$

$$\cdots y_{n-1,1}^{\nu_{n-1,1}} \cdots y_{n-1,p}^{\nu_{n-1,p}} y_{n1}^{\nu_{n1}} \cdots y_{nr}^{\nu_{nr}-1} \cdots y_{ns}^{\nu_{ns}-1} \cdots y_{np}^{\nu_{np}} \sigma_{sr}(y_n, t_n)$$

$$\times w_n(y_1, \ldots, y_n) p_{\nu_1, \ldots, \nu_n}(y_1, \ldots, y_n) dy_1 \ldots dy_n \tag{7.4.16}$$

$$(\,|\kappa_1| + \cdots + |\kappa_n| = 3, \ldots, N; \ |\nu_1| + \cdots + |\nu_n| = 0, 1, \ldots, N;$$

$$|\nu_1| + \cdots + |\nu_n| = 3, \ldots, N)\,.$$

Here the functions $\varphi_{\kappa_1, \ldots, \kappa_n, 0}$ in (7.4.15) are determined by (7.4.16) at $p_{\nu_1, \ldots, \nu_n}(y_1, \ldots, y_n) = p_{0, \ldots, 0}(y_n) = 1$ and $\nu_1, \ldots, \nu_n = 0$. The initial conditions for (7.4.15) have the form

$$\alpha_{r_1, \ldots, r_n}(t_1, \ldots, t_{n-1}, t_{n-1}) = \alpha_{r_1, \ldots, r_{n-1}+r_n}(t_1, \ldots, t_{n-1})\,, \tag{7.4.17}$$

as

$$EY_1^{r_{11}}(t_1) \ldots Y_p^{r_{1p}}(t_1) \ldots Y_1^{r_{n-1,1}}(t_{n-1}) \ldots Y_p^{r_{n-1,p}}(t_{n-1}) \ldots Y_1^{r_{n1}}(t_{n-1})$$

$$\ldots Y_p^{r_{np}}(t_{n-1}) = MY_1^{r_{11}}(t_1) \ldots Y_p^{r_{1,p}}(t_1) \ldots Y_1^{r_{n-1,1}+r_{n1}}(t_{n-1})$$

$$\ldots Y_p^{r_{n-1,p}+r_{np}}(t_{n-1})\,. \ \triangleleft$$

While setting up the equations for the moments of the n-dimensional distribution we have restrict ourselves only to those moments which depend on all n variables $t_1, \ldots, t_n$, i.e. for which none of the sums $|r_1|, \ldots, |r_n|$ is equal to zero as the moments which depend only on the part of the variables $t_1, \ldots, t_n$, for example, on $t_{r+1}, \ldots, t_n$ represent the moments of $(n-k)$-dimensional distribution and consequently, are determined earlier at the approximation of $(n-k)$-dimensional distribution. It is natural that Eqs. (7.4.15) at $|r_1| = \cdots = |r_n| = 0$ coincide with the correspondent equations for the moments of $(n-k)$-dimensional distribution. If at all n approximating function $f_n^*(y_1, \ldots, y_n; \vartheta_n)$ depends on the moments no higher than N^{th} order then at $n = N$ we have to set up Eqs. (7.4.15) only for the moments correspondent to $|r_1| = \cdots = |r_n| = 0$ and after determining the N-dimensional distribution all other multi-dimensional distributions will be determined uniquely. In particular, at the approximation of all multi-dimensional densities by the orthogonal expansions, for example, bu the consistent orthogonal expansions (Subsection 3.8.7) of the form (7.4.5) with the account of the moments no higher than the Nth order, all the multi-dimensional distributions of the process $Y(t)$ will be determined uniquely after finding the moments of N-dimensional distribution. Eqs. (7.4.15) at $n > 2$ are the linear ordinary differential equations.

Thus *the initial moments method for determining multi-dimensional distributions (7.4.14) in normal nonlinear stochastic differential system (5.4.6) is based on* Eqs. (7.4.15).

7.4.4. Multi-Dimensional Distributions. Central Moments

The approximate equations for the central moments of the n-dimensional distribution of the random process $Y(t)$ at $t_1 < \cdots < t_n$ may be derived in exactly the same way as the equations for the central moments of the one-dimensional distribution were obtained in Subsection 7.4.2):

$$\partial \mu_{r_1,\ldots,r_n}(t_1,\ldots,t_n)/\partial t_n = \varphi_{r_1,\ldots,r_n,0}(\bar{m}_n,\bar{K}_n,t_n)$$

$$+ \sum_{k=3}^{N} \sum_{|\nu_1|+\cdots+|\nu_n|=k} q_{\nu_1,\ldots,\nu_n}(\alpha_1,\ldots,\alpha_n)$$

$$\times \varphi_{r_1,\ldots,r_n,\nu_1,\ldots,\nu_n}(\bar{m}_n,\bar{K}_n,t_n) - \sum_{s=1}^{p} \nu_{ns}\mu_{r_1,\ldots,r_n-e_s}\dot{m}_s(t_n)$$

$$(|r_1|,\ldots,|r_n| = 1,\ldots,N; \; |r_1| + \cdots |r_n| = \max(3,n),\ldots,N).$$
$$(7.4.18)$$

Here the functions $\varphi_{r_1,\ldots,r_n,\nu_1,\ldots,\nu_n}(\bar{m}_m,\bar{K}_n,t_n)$ are determined by formulae (7.4.16) with the substitution the degrees of $y_{11},\ldots,y_{1p}$, $\ldots,y_{n1},\ldots,y_{np}$ by the degrees of $y_{11} - m_1(t_1),\ldots,y_{1p} - m_p(t_1)$, $\ldots,y_{n1} - m_1(t_n),\ldots,y_{np} - m_p(t_n)$, and the initial moments in $q_{\nu_1,\ldots,\nu_n}(\alpha_1,\ldots,\alpha_n)$ are replaced by their expressions in terms of the central moments. The initial conditions (7.4.18) have the form

$$\mu_{r_1,\ldots,r_n}(t_1,\ldots,t_{n-1},t_{n-1}) = \mu_{r_1,\ldots,r_{n-1}+r_n}(t_1,\ldots,t_{n-1}).$$
$$(7.4.19)$$

For the two-dimensional distribution ($n = 2$) it is necessary to add to Eqs. (7.4.18) the following equation for the covariance function:

$$\partial K(t_1,t_2)/\partial t_2 = \varphi_{30}(\bar{m}_2,\bar{K}_2,t_2)$$

$$+\sum_{l=3}^{N}\sum_{|\nu_1|+|\nu_2|=l} q_{\nu_1,\nu_2}(\alpha_1,\alpha_2)\varphi_{3,\nu_1,\nu_2}(\bar{m}_2,\bar{K}_2,t_2),\qquad (7.4.20)$$

where

$$\varphi_{3,\nu_1,\nu_2}(\bar{m}_2,\bar{K}_2,t_2) = \int_{-\infty}^{\infty}\cdots\int_{-\infty}^{\infty}[y_1 - m(t_1)]\,a(y_2,t_2)^T$$

$$\times w_2(y_1,y_2)p_{\nu_1,\nu_2}(y_1,y_2)dy_1dy_2 \qquad (7.4.21)$$

$$(|\nu_1|,|\nu_2| = 1,\ldots,N;\ |\nu_1|+|\nu_2| = 3,\ldots,N).$$

Here the function φ_{30} is determined by (7.4.21) at $p_{\nu_1,\nu_2}(y_1,y_2)$ $= p_{0,0}(y_1,y_2) = 1$. Eqs. (7.4.18), (7.4.20) for $n = 2$ with the initial conditions $K(t_1,t_1) = K(t_1)$, $\mu_{r_1,r_2}(t_1,t_1) = \mu_{r_1+r_2}(t)$ determine the sought two-dimensional distribution.

Thus *as the basis of the central moments method for determining multi-dimensional distributions* (7.4.14) *in normal nonlinear stochastic differential system* (5.4.6) *serve* Eqs. (7.4.18) *at initial conditions* (7.1.19).

E x a m p l e 7.4.3. For the system of Example 7.4.1 the two-dimensional distribution is determined approximately by formula

$$f_2(y_1,y_2;t_1,t_2) \approx \left[2\pi\sqrt{k_{11}k_{22}-k_{12}^2}\right]^{-1}\exp\{-[k_{22}(y_1-m_1)^2$$

$$-2k_{12}(y_1-m_1)(y_2-m_2)+k_{11}(y_2-m_2)^2]/2(k_{11}k_{22}-k_{12}^2)\}$$

$$\times\left[1+\sum_{k=3}^{4}\sum_{\nu_1+\nu_2=k}\frac{c_{\nu_1\nu_2}}{\nu_1!\nu_2!}H_{\nu_1\nu_2}(y_1-m_1,y_2-m_2)\right],\qquad (I)$$

where $m_1 = m(t_1)$; $m_2 = m(t_2)$; $k_{11} = D(t_1)$; $k_{22} = D(t_2)$; $c_{\nu_1\nu_2}$ $= G_{\nu_1\nu_2}(\mu)$; $\mu_{30} = \mu_3(t_1)$, $\mu_{03} = \mu_3(t_2)$; $\mu_{40} = \mu_4(t_1)$; $\mu_{04} = \mu_4(t_2)$, and the moments $k_{12} = K(t_1,t_2)$, $\mu_{21} = \mu_{21}(t_1,t_2)$; $\mu_{12} = \mu_{12}(t_1,t_2)$; μ_{31} $= \mu_{31}(t_1,t_2)$; $\mu_{22} = \mu_{22}(t_1,t_2)$ and $\mu_{13} = \mu_{13}(t_1,t_2)$ are determined by Eqs. (7.4.20) which have in this case the form:

$$\partial K(t_1,t_2)/\partial t_2 = -3m^2(t_2)K(t_1,t_2) - 3m(t_2)\mu_{12}(t_1,t_2) - \mu_{13}(t_1,t_2),$$
$$(II)$$

$$\partial\mu_{21}(t_1,t_2)/\partial t_2 = -3\left[m^2(t_2)+D(t_2)\right]\mu_{21}(t_1,t_2)$$
$$-6K(t_1,t_2)\mu_{12}(t_1,t_2) - 3m(t_2)\mu_{22}(t_1,t_2) + 3m(t_2)D(t_1)D(t_2),(III)$$

$$\partial\mu_{12}(t_1,t_2)/\partial t_2 = 2\left[\nu m(t_2) + m(t_2)D(t_2) - 6\mu_3(t_2)\right]K(t_1,t_2)$$
$$+ \left[\nu - 6m^2(t_2) - 12D(t_2)\right]\mu_{12}(t_1,t_2) - 6m(t_2)\mu_{13}(t_1,t_2), \quad \text{(IV)}$$

$$\partial\mu_{31}(t_1,t_2)/\partial t_2 = 6\left[3D(t_1)D(t_2) + 2K^2(t_1,t_2)\right]K(t_1,t_2)$$
$$- 18m(t_2)K(t_1,t_2)\mu_{21}(t_1,t_2) - 9m(t_2)D(t_1)\mu_{12}(t_1,t_2)$$
$$- 3[m^2(t_2) + D(t_2)]\mu_{31}(t_1,t_2) - 3D(t_1)\mu_{13}(t_1,t_2) + \mu_3(t_1)\mu_3(t_2), \quad \text{(V)}$$

$$\partial\mu_{22}(t_1,t_2)/\partial t_2 = 12D(t_2)\left[D(t_1)D(t_2) + 2K^2(t_1,t_2)\right]$$
$$+ 2\left[\nu m(t_2) - 8m(t_2)D(t_2) + \mu_3(t_2)\right]\mu_{21}(t_1,t_2)$$
$$- 36m(t_2)K(t_1,t_2)\mu_{12}(t_1,t_2) + \left[\nu - 6m^2(t_2) - 12D(t_2)\right]$$
$$\times \mu_{22}(t_1,t_2) - 16K(t_1,t_2)\mu_{13}(t_1,t_2) + D(t_1)[\nu m^2(t_2)$$
$$- 6m(t_2)\mu_3(t_2) - 2\mu_4(t_2)], \quad \text{(VI)}$$

$$\partial\mu_{13}(t_1,t_2)/\partial t_2 = 3\left[\nu m^2(t_2) + 30D^2(t_2) - 12m(t_2)\mu_3(t_2) - 5\mu_4(t_2)\right]$$
$$\times K(t_1,t_2) + 3\left[2\nu m(t_2) - 15m(t_2)D(t_2) + \mu_3(t_2)\right]\mu_{12}(t_1,t_2)$$
$$+ 3\left[\nu - 3m^2(t_2) - 10D(t_2)\right]\mu_{13}(t_1,t_2). \quad \text{(VII)}$$

The simplest way to obtain these equations is to differentiate the random functions $[Y(t_1) - m(t_1)]^p [Y(t_2) - m(t_2)]^q$ with respect to t_2 according to the rule of the differentiation of a composite function of the process determined by the stochastic differential equation in the case of the Wiener process $W(t)$, to take the expectations and to substitute into the equations obtained the expressions of the fifth and the sixth moments in terms of the moments of lower orders from the equations

$$G_{2,3}(\mu) = G_{1,4}(\mu) = G_{3,2}(\mu) = G_{3,3}(\mu) = G_{2,4}(\mu) = G_{1,5}(\mu) = 0,$$
$$\text{(VIII)}$$

which yield in this case

$$\mu_{23}(t_1,t_2) = D(t_1)\mu_3(t_2) + 3D(t_2)\mu_{21}(t_1,t_2) + 6K(t_1,t_2)\mu_{12}(t_1,t_2),$$
$$\mu_{14}(t_1,t_2) = 4\mu_3(t_2)K(t_1,t_2) + 6D\mu_{12}(t_1,t_2),$$
$$\mu_{32}(t_1,t_2) = D(t_2)\mu_3(t_1) + 3D(t_1)\mu_{12}(t_1,t_2) + 6K(t_1,t_2)\mu_{21}(t_1,t_2),$$
$$\mu_{33}(t_1,t_2) = 3D(t_2)\mu_{31}(t_1,t_2) + 9K(t_1,t_2)\mu_{22}(t_1,t_2)$$
$$+ 3D(t_1)\mu_{13}(t_1,t_2) - 6\left[3D(t_1)D(t_2) + 2K^2(t_1,t_2)\right]K(t_1,t_2),$$
$$\mu_{24}(t_1,t_2) = 6D(t_2)\mu_{22}(t_1,t_2) + 8K(t_1,t_2)\mu_{13}(t_1,t_2)$$
$$- 24D(t_2)K^2(t_1,t_2) + D(t_1)\left[\mu_4(t_2) - 6D^2(t_2)\right],$$
$$\mu_{15}(t_1,t_2) = 5\mu_4(t_2)K(t_1,t_2) + 10D(t_2)\mu_{13}(t_1,t_2).$$
$$\text{(IX)}$$

Integrating the obtained ordinary (at fixed t_1) differential equations with the initial conditions

$$K(t_1,t_1) = D(t_1), \quad \mu_{21}(t_1,t_1) = \mu_{12}(t_1,t_1) = \mu_3(t_1), \qquad \text{(X)}$$

$$\mu_{31}(t_1,t_1) = \mu_{22}(t_1,t_1) = \mu_{13}(t_1,t_1) = \mu_4(t_1), \qquad \text{(XI)}$$

we find all the moments up to the fourth order of the two-dimensional distribution of the process $Y(t)$.

The other multi-dimensional distributions of the process $Y(t)$ are determined by the approximate formula

$$f_n(y_1, \ldots, y_n; t_1, \ldots, t_n) \approx$$

$$\approx \left[(2\pi)^n \, |K|\,\right]^{-1/2} \exp\left\{-\frac{1}{2}(y^T - m^T)K^{-1}(y - m)\right\}$$

$$\times \left[1 + \sum_{k=3}^{4} \sum_{|\nu|=k} \frac{c_\nu}{\nu_1!\ldots\nu_n!} H_\nu(y - m)\right],$$

$$\text{(XII)}$$

where $y = [\,y_1 \ldots y_n\,]^T$, $m = [\,m(t_1)\ldots m(t_n)\,]^T$,

$$K = \begin{bmatrix} D(t_1) & K(t_1,t_2) & \ldots & K(t_1,t_n) \\ K(t_1,t_2) & D(t_2) & \ldots & K(t_2,t_n) \\ \ldots & \ldots & \ldots & \ldots \\ K(t_1,t_n) & K(t_2,t_n) & \ldots & D(t_n) \end{bmatrix}, \quad c_\nu = G_\nu(\mu). \qquad \text{(XIII)}$$

Here the moments $\mu_{\nu_1,\ldots,\nu_n} = \mu_{\nu_1,\ldots,\nu_n}(t_1,\ldots,t_n)$, with some of the subscripts $\nu_1, \ldots, \nu_n$ equal to zero, coincide with the moments of the distributions of smaller dimensions, these are obtained by the deletion of all zero subscripts together with the corresponding arguments, for instance, $\mu_{004} = \mu_4(t_3)$, $\mu_{0112} = \mu_{112}(t_2,t_3,t_4)$, $\mu_{01030} = \mu_{13}(t_2,t_4)$.

The moments $\mu_{111}(t_1,t_2,t_3)$, $\mu_{211}(t_1,t_2,t_3)$, $\mu_{121}(t_1,t_2,t_3)$ and $\mu_{0112}(t_1,t_2,t_3)$ of the three-dimensional distribution depending on all three arguments t_1, t_2, t_3 are determined by the following equations:

$$\partial\mu_{111}(t_1,t_2,t_3)/\partial t_3 = -3\left[m^2(t_3) + D(t_3)\right]\mu_{111}(t_1,t_2,t_3)$$
$$- 3m(t_3)\mu_{112}(t_1,t_2,t_3) + 3m(t_3)D(t_3)K(t_1,t_2)$$
$$- 3K(t_1,t_3)\mu_{12}(t_2,t_3) - 3K(t_2,t_3)\mu_{12}(t_1,t_2), \qquad \text{(XIV)}$$

$$\partial\mu_{211}(t_1,t_2,t_3)/\partial t_3 = -12m(t_3)K(t_1,t_3)\mu_{111}(t_1,t_2,t_3)$$

$$- 3 \left[m^2(t_3) + D(t_3) \right] \mu_{211}(t_1, t_2, t_3) - 6K(t_1, t_3)\mu_{112}(t_1, t_2, t_3)$$
$$+ 6D(t_1)D(t_3)K(t_2, t_3) + 12K(t_1, t_3)[K(t_1, t_3)K(t_2, t_3)$$
$$+ D(t_3)K(t_1, t_2)] + \mu_3(t_3)\mu_{21}(t_1, t_2) - 6m(t_3)K(t_2, t_3)\mu_{21}(t_1, t_2)$$
$$- 3m(t_3) \left[D(t_1)\mu_{12}(t_2, t_3) + 2K(t_1, t_2)\mu_{12}(t_1, t_3) \right]$$
$$- 3K(t_2, t_3)\mu_{22}(t_1, t_3) - 2K(t_1, t_2)\mu_{13}(t_1, t_3)$$
$$- D(t_1)\mu_{13}(t_2, t_3), \tag{XV}$$

$$\partial\mu_{121}(t_1, t_2, t_3)/\partial t_3 = -12m(t_3)K(t_2, t_3)\mu_{111}(t_1 1, t_2, t_3)$$
$$- 3 \left[m^2(t_3) + D(t_3) \right] \mu_{121}(t_1, t_2, t_3) - 6K(t_2, t_3)\mu_{112}(t_1, t_2, t_3)$$
$$+ 6D(t_3) \left[D(t_2)K(t_1, t_3) + 2K(t_1, t_2)K(t_2, t_3) \right] + 12K(t_1, t_3)K^2(t_2, t_3)$$
$$- 6m(t_3)K(t_1, t_3)\mu_{21}(t_2, t_3) + \mu_3(t_2)\mu_{12}(t_1, t_2)$$
$$- 3m(t_3)D(t_2)\mu_{12}(t_1, t_3) - 6m(t_3)K(t_1, t_2)\mu_{12}(t_2, t_3)$$
$$- 3K(t_1, t_3)\mu_{22}(t_2, t_3) - D(t_2)\mu_{13}(t_1, t_3) - 2K(t_1, t_2)\mu_{13}(t_2, t_3), \tag{XVI}$$

$$\partial\mu_{112}(t_1, t_2, t_3)/\partial t_3 = 2 \left[\nu m(t_3) - 6m(t_3)D(t_3) + \mu_3(t_3) \right] \mu_{111}(t_1, t_2, t_3)$$
$$+ \left[\nu - 6m^2(t_3) - 12D(t_3) \right] \mu_{112}(t_1, t_2, t_3) + \nu m^2(t_3)K(t_1, t_2)$$
$$+ 12 \left[K(t_1, t_2)D(t_3) + 4K(t_1, t_3)K(t_2, t_3) \right] D(t_3) - 2K(t_1, t_2)\mu_4(t_3)$$
$$- 6m(t_3)[3K(t_2, t_3)\mu_{12}(t_1, t_3) + 3K(t_1, t_3)\mu_{12}(t_2, t_3)$$
$$+ K(t_1, t_2)\mu_3(t_3)] - 8 \left[K(t_1, t_3)\mu_{13}(t_2, t_3) + K(t_2, t_3)\mu_{13}(t_1, t_3) \right] \tag{XVII}$$

and the initial conditions

$$\mu_{111}(t_1, t_2, t_3) = \mu_{12}(t_1, t_2), \quad \mu_{211}(t_1, t_2, t_3) = \mu_{22}(t_1, t_2)$$
$$\mu_{121}(t_1, t_2, t_3) = \mu_{112}(t_1, t_2, t_3) = \mu_{13}(t_1, t_2). \tag{XVIII}$$

The unique fourth order moment depending on four variables t_1, t_2, t_3, t_4 is determined by the corresponding equation

$$\partial\mu_{1111}(t_1, t_2, t_3, t_4)/\partial t_4 = -3 \left[m^2(t_4) + D(t_4) \right] \mu_{111}(t_1, t_2, t_3, t_4)$$
$$+ 6 \left[K(t_1, t_2)K(t_3, t_4) + K(t_1, t_3)K(t_2, t_4) + K(t_1, t_4)K(t_2 t_3) \right] D(t_4)$$
$$+ 12K(t_1, t_4)K(t_2, t_4)K(t_3, t_4) - 3m(t_4)[K(t_1, t_2)\mu_{12}(t_3, t_4)$$
$$+ K(t_1, t_3)\mu_{12}(t_2, t_4) + K(t_2, t_3)\mu_{12}(t_1, t_4)] - K(t_1, t_2)\mu_{13}(t_3, t_4)$$
$$- K(t_1, t_3)\mu_{13}(t_2, t_4) - K(t_2, t_3)\mu_{13}(t_1, t_4) + \mu_3(t_4)\mu_{111}(t_1, t_2, t_3)$$
$$- 6m(t_4)[K(t_3, t_4)\mu_{111}(t_1, t_2, t_4) + K(t_2, t_4)\mu_{111}(t_1, t_3, t_4)$$
$$+ K(t_1, t_4)\mu_{111}(t_2, t_3, t_4)] - 3[K(t_3, t_4)\mu_{112}(t_1, t_2, t_4)$$
$$+ K(t_2, t_4)\mu_{112}(t_1, t_3, t_4) + K(t_1, t_4)\mu_{112}(t_2, t_3, t_4)] \tag{XIX}$$

with the initial condition $\mu_{1111}(t_1, t_2, t_3, t_4) = \mu_{112}(t_1, t_2, t_3)$. Since moments of the order no higher than the fourth cannot depend on more than four of the arguments $t_1, \ldots, t_n$, after integrating Eq. (XIX) approximate formula (XII) will determine all the multi-dimensional distributions of the process $Y(t)$. The above equations for the moments of the three- and the four-dimensional distributions are derived in the same way as the equations for the moments of the two-dimensional distribution.

7.4.5. Approximate Determination of Stationary Processes

For approximate determination by the method of moments of the one-dimensional distribution of the strictly stationary process in a stationary nonlinear stochastic differential system one should put in Eqs. (7.4.3) $\dot\alpha_r = 0$ or in Eqs. (7.4.11) $\dot\mu_r = 0$. If the equations obtained in such a way have the solution which may serve as a set of the moments of some random variable then we may suppose that the strictly stationary process in the system exists. In this case for the determination of the other multi-dimensional distributions of this stationary process one should replace in Eqs. (7.4.15) or Eqs. (7.4.18) the derivatives with respect to t_n by the derivative with respect to $\tau_{n-1} = t_n - t_1$, and write the initial conditions in the form

$$\alpha_{r_1, \ldots, r_n}(\tau_1, \ldots, \tau_{n-2}, \tau_{n-2}) = \alpha_{r_1, \ldots, r_{n-1}+r_n}(\tau_1, \ldots, \tau_{n-2})$$
$$(7.4.22)$$

or respectively in the form

$$\mu_{r_1, \ldots, r_n}(\tau_1, \ldots, \tau_{n-2}, \tau_{n-2}) = \mu_{r_1, \ldots, r_{n-1}+r_n}(\tau_1, \ldots, \tau_{n-2}).$$
$$(7.4.23)$$

However, one should take into account the possibility of appearance of "superfluous" of the obtained equations in consequence of the approximate character of original equations.

So the results of Subsections 7.4.1–7.4.5 may be summarized in the following statements.

Theorem 7.4.1. *If the random process $Y(t)$ in normal stochastic differential system (5.4.6) has the finite initial moments then its one- and multi-dimensional distributions may be approximated by expansions (7.4.1), (7.4.14) with initial moments defined by Eqs. (7.4.3), (7.4.15) at initial conditions (7.4.5), (7.4.17).*

Theorem 7.4.2. *If the random process $Y(t)$ in normal stochastic differential system (5.4.6) has finite central moments then its one- and multi-dimensional distributions may be approximated by expansions*

(7.4.1), (7.4.14) *with central moments defined by Eqs.* (7.4.6), (7.4.7), (7.4.11), (7.4.18) *at initial conditions* (7.4.12), (7.4.19).

Corollary 7.4.1. *If the stationary (in the strict sense) process in stationary normal stochastic differential system* (5.4.6) *exists and has the finite initial central moments then its one- and multi-dimensional distributions may be approximately expressed by expansions* (7.4.1), (7.4.14) *with initial or central moments defined by* (7.4.22) *or* (7.4.23).

The problems 7.26–7.29 concern the corresponding equations in cases of Eqs. (5.4.7) and (5.4.6) with the nongaussian white noise.

7.4.6. Moments Method for Stochastic Discrete Nonlinear Systems

For the stochastic nonlinear discrete system (5.7.1) corresponding Eqs. (7.4.15) and (7.4.18) for the initial and the central moments

$$\alpha_{l_1,\dots,l_n}^{r_1,\dots,r_n} = EY_{l_11}^{r_{11}}\dots Y_{l_1p}^{r_{np}}\dots Y_{l_n1}^{r_{n1}}\dots Y_{l_np}^{r_{np}}, \tag{7.4.24}$$

$$\mu_{l_1,\dots,l_n}^{r_1,\dots,r_n} = E(Y_{l_11} - m_{l_11})^{r_{11}}\dots(Y_{l_1p} - m_{l_1p})^{r_{1p}}$$
$$\dots(Y_{l_n1} - m_{l_n1})^{r_{n1}}\dots(Y_{l_np} - m_{l_np})^{r_{np}}, \tag{7.4.25}$$

are simplified and take the form

$$\alpha_{l+1}^{r_1} = E\omega_{l1}^{r_{11}}(Y_l, V_l)\dots\omega_{lp}^{r_{1p}}(Y_l, V_l), \tag{7.4.26}$$

$$\alpha_{l_1,\dots,l_{n-1},l_n+1}^{r_1,\dots,r_n} = EY_{l_11}^{r_{11}}\dots Y_{l_n-1,p}^{r_n-1,p}\omega_{l_n1}^{r_{n1}}(Y_{l_n}, V_{l_n})$$
$$\dots\omega_{l_np}^{r_{np}}(Y_{l_n}, V_{l_n}), \tag{7.4.27}$$

$$\mu_{l+1}^{r_1} = E\left[\omega_{l1}(Y_l, V_l) - E\omega_{l1}(Y_l, V_l)\right]^{r_{11}} \times$$
$$\cdots \times \left[\omega_{lp}(Y_l, V_l) - E\omega_{lp}(Y_l, V_l)\right]^{r_{1p}}, \tag{7.4.28}$$

$$\mu_{l_1,\dots,l_{n-1},l_n+1}^{r_1,\dots,r_n} = E(Y_{l_1} - m_{l_1})^{r_{11}}\dots(Y_{l_{n-1}p} - m_{l_{n-1}p})^{r_{n-1,p}}$$
$$\times \left[\omega_{l_n1}(Y_{l_n}, V_{l_n}) - E\omega_{l_n1}(Y_{l_n}, V_{l_n})\right]^{r_{n1}} \times$$
$$\cdots \times \left[\omega_{l_np}(Y_{l_n}, V_{l_n}) - E\omega_{l_np}(Y_{l_n}, V_{l_n})\right]^{r_{np}}, \tag{7.4.29}$$

where

$$l_1 < \cdots < l_{n-1} \le l_n; \quad r_{11},\dots,r_{np} = 0,1,\dots,N-n+1;$$

$$r_{11},\dots,r_{1p} = 0,1,\dots,N; \quad |r_1| = r_{11}+\cdots+r_{1p} = 1,\dots,N;$$

$$|r_1|, \ldots, |r_n| = 1, \ldots, N - n + 1;$$

$$|r_1| + \cdots + |r_n| = n, \ldots, N; \quad n = 2, \ldots, N.$$

Substituting the unknown distributions in formulae (7.4.26)–(7.4.29) by their approximating functions we get a closed system of the difference equations which determines sequentially all the multidimensional distributions of the process $\{V_l\}$. Eqs. (7.4.26) and (7.4.28) at the initial α_1^r and μ_1^r determine α_l^r and μ_l^r for all $l > 1$. Eqs. (7.4.27) and (7.4.29) at the initial conditions

$$\alpha_{l_1,\ldots,l_{n-1},l_{n-1}}^{r_1,\ldots,r_n} = \alpha_{l_1,\ldots,l_{n-1}}^{r_1,\ldots,r_{n-1}+r_n} \quad (n = 2, \ldots, N), \tag{7.4.30}$$

$$\mu_{l_1,\ldots,l_{n-1},l_{n-1}}^{r_1,\ldots,r_n} = \mu_{l_1,\ldots,l_{n-1}}^{r_1,\ldots,r_{n-1}+r_n} \quad (n = 2, \ldots, N) \tag{7.4.31}$$

determine the correspondent moments of all $l_n > l_{n-1}$.

7.4.7. Moments Method for Stochastic Infinite-Dimensional Systems

It is clear from expression (5.9.2) that the operators of the initial moments $\alpha_{r_1,\ldots,r_n}$ and of the central moments $\mu_{r_1,\ldots,r_n}$ are expressed as the values of the correspondent strong derivatives of the characteristic functional in zero:

$$\alpha_{r_1,\ldots,r_n} = \left[\frac{D^{|r|} g_n}{D_{\lambda_1}^{r_1} \ldots D_{\lambda_n}^{r_n}} \right]_{\lambda=0}, \tag{7.4.32}$$

$$\mu_{r_1,\ldots,r_n} = \left[\frac{D^{|r|} \tilde{g}_n}{D_{\lambda_1}^{r_1} \ldots D_{\lambda_n}^{r_n}} \right]_{\lambda=0}, \tag{7.4.33}$$

where $|r| = r_1 + \cdots + r_n = 1, 2, \ldots; r_1, \ldots, r_n = 0, 1, 2 \ldots,$

$$\tilde{g}_n = g_n \exp\left\{ -i \sum_{k=1}^{n} (\lambda_k, m_{t_k}) \right\}. \tag{7.4.34}$$

After taking the strong derivatives with respect to the left and to the right part of Eq. (5.9.9) till the order n inclusively at $n = 1, 2, \ldots$ we obtain the equations for all operators of the initial moments till the order N inclusively. If we multiply the equations by $\exp\left\{ -i \sum_{k=1}^{n} (\lambda_k, m_{t_k}) \right\}$ before the differentiation then we shall have the analogous equations for the operators of the central moments.

After calculating the expectations in the equations at the approximation of the probability measures and the correspondent characteristic functionals by the known expressions which depend on the operators of the moments till the order N inclusively we shall obtain a closed system of the ordinary differential equations in H-space $\mathcal{Y}$ for all necessary operators of the moments.

It should be stressed that the equations obtained for the operators of the moments may be used also for the approximate determination of the stationary processes.

7.4.8. Semiinvariant and Moment-Semiinvariant Methods

Sometimes the semiinvariants instead of the moments are assumed as parameters of distributions (Subsection 3.8.3). This is motivated by the fact that contrary to the moments the semiinvariants do not increase while the order increases. So, for instance, for the normal distribution all the semiinvariants of the third and higher orders are equal to zero. Therefore for the distributions close to normal the semiinvariants of the third, the fourth and higher orders are small. This gives the opportunity to approximate the distributions neglecting the semiinvariants of orders higher than the prescribed number.

$\triangleright$ In order to derive the differential equations for the semiinvariants of the one-dimensional distribution of the process $Y(t)$ we recall that according to the definition the semiinvariants $\kappa_{r_1,\,\dots,\,r_p}$ are expressed in terms of the characteristic function $g_1(\lambda;t)$ by the formula (Subsection 3.8.3)

$$\kappa_r = \kappa_{r_1,\,\dots,\,r_p} = \left[\frac{\partial^{|r|} \ln g_1(\lambda;t)}{\partial(i\lambda_1)^{r_1} \dots (i\lambda_p)^{r_p}} \right]_{\lambda=0} . \qquad (7.4.35)$$

Differentiating this formula with respect to time t, substituting into the obtained formula the expression of $\partial g_1(\lambda;t)/\partial t$ from Eq. (5.6.28) and replacing the unknown density $f_1(y;t)$ by its approximating function $f_1^*(y;\vartheta)$ we shall have

$$\dot\kappa_r = \int\limits_{-\infty}^{\infty} \left[\frac{\partial^{|r|}}{\partial(i\lambda_1)^{r_1} \dots (i\lambda_p)^{r_p}} \{ i\lambda^T a(y,t) + \chi(b(y,t)^T \lambda;t) \} \right.$$

$$\left. \times \exp\{ i\lambda^T y - \ln g_1(\lambda;t) \} \right]_{\lambda=0} f_1^*(y;\vartheta)dy \qquad (7.4.36)$$

$$(r_1,\,\dots,\,r_p = 0,1,\,\dots,\,N; \ |r| = 1,\,\dots,\,N),$$

where ϑ is the vector whose components are the semiinvariants κ_r up to the Nth order. The corresponding semiinvariants of the random variable $Y_0 = Y(t_0)$ should be taken as the initial values of the semiinvariants κ_r at $t = t_0$. ◁

In problems of practice the semiinvariant methods possess the following advantages (Malachov 1978, Stratonovich 1961). Firstly, the semiinvariants have independent applied statistical sense and may be signed in the definite degree independently of each other. Secondly, the semiinvariants of the higher order for the typical nonganssian random variables and process are calculated rather simple. Thirdly to the finite set of the semiinvariants it corresponds usually a simple approximation of the distribution. But we notice that in the general case the equations for the semiinvariants are more complex than the equations for the initial and central moments.

For the algorithmization of the semiinvariants calculation in the physical applications the technique of the semiinvariant (cumulant) brackets and diagrams is used (Malachov 1978).

E x a m p l e 7.4.4. For the system of Example 7.1.1 using the approximation of $f_1(y; t)$ by the truncated Hermite polynomial expansion with the account of the moments up to the fourth order, Eqs. (7.4.36) have the form $\left(m = \kappa_1, D = \kappa_2 \right)$

$$\dot{m} = -m(m^2 + 3D) - \kappa_3 , \tag{I}$$

$$\dot{D} = (\nu - 6D)(m^2 + D) - 6m\kappa_3 - 2\kappa_4 , \tag{II}$$

$$\dot{\kappa}_3 = 6(\nu - 3D)mD + 3(\nu - 3m^2 - 9D)\kappa_3 - 9m\kappa_4 ,$$

$$\dot{\kappa}_4 = 6(\nu + 2m^2)D^2 + 36D^3 + 4\kappa_3^2 + 12(\nu - 6D)m\kappa_3$$

$$+6(\nu - 2m^2 - 8D)\kappa_4 . \tag{III}$$

While approximating $f_1(y; t)$ by the truncated Edgeworth series with the account of the semiinvariants up to the fourth order the additional item $10\kappa_3^2$ in the right-hand side of the last Eq. (III) appears.

Setting up the differential equations for the semiinvariants is considerably more difficult than setting up the equations for the moments, especially in the case of the polynomial functions $a(y, t)$ and $b(y, t)$ and the Wiener process $W(t)$. In the latter case for setting up the equations for the moments it is sufficient to find stochastic differentials of the products $Y_1^{r_1}(t) \ldots Y_p^{r_p}(t)$ or $[Y_1(t) - m_1(t)]^{r_1} \ldots [Y_p(t) - m_p(t)]^{r_p}$ using the formula of the differentiation of a composite function (5.1.7), to take the expectations and to express in the obtained equalities the moments higher than the Nth order in accordance with the adopted approximation of the density $f_1(y;t)$. In exactly the same way the equations are derived for the moments of the multi-dimensional distributions. But it is very difficult to derive the equations for the semiinvariants even in this simplest case. This circumstance leads to the idea to use a combined method based on setting up the equations for the moments and closing these equations by the relations obtained by equalizing to zero all the semiinvariants of orders higher than a given number N (Dashevskii 1975). Equalizing to zero the semiinvariants of orders higher than N and expressing them in terms of the central moments we obtain the relations from which we may determine the moments of orders higher than N involved into the obtained equations in terms of the moments of orders $2,\ldots,N$:

$$\mu_{r_1,\ldots,r_s} = \sum_{h=2}^{[(r_1+\cdots+r_s)/2]} \frac{(-1)^h}{h} r_1!\ldots r_s!$$

$$\times \sum_{\rho_{11}+\cdots+\rho_{1h}=r_1} \cdots \sum_{\rho_{s1}+\cdots+\rho_{sh}=r_s} \prod_{l=1}^{h} \frac{\mu_{1l},\ldots,\mu_{sl}}{\rho_{1l}!\ldots\rho_{sl}!}, \qquad (7.4.37)$$

where the summation with respect to $\rho_{11},\ldots,\rho_{1h},\ldots,\rho_{s1},\ldots,\rho_{sh}$ extends to all $\rho_{11},\ldots,\rho_{1h},\ldots,\rho_{s1},\ldots,\rho_{sh}$ for which $2 \le \rho_{1l}+\cdots+\rho_{sl} < |r|$ $(l=1,\ldots,h)$, $s=np$.

The moment-semiinvariant method (MSM) of setting up the equations for the moments is based on the approximation of the logarithm of the characteristic function by the polynomial of the Nth degree:

$$\ln g_1(\lambda;t) \approx \ln g_1^*(\lambda;\vartheta) = i\lambda^T m + \sum_{k=2}^{N} \sum_{|r|=k} \frac{(i\lambda_1)^{r_1}\ldots(i\lambda_p)^{r_p}}{r_1!\ldots r_p!}\kappa_p\,.$$

$$(7.4.38)$$

It is evident that such an approximation of a distribution differs in the general case from the approximation of the density by a finite segment of an orthogonal expansion or of the Edgeworth series. It is therefore natural that the equations for the moments obtained by the moment-semiinvariant method do not coincide in the general case with the equations which the approximation of the density by fragment of an orthogonal expansion or of the Edgeworth series gives.

It goes without saying that the approximation (7.4.38) of a distribution is practically applicable only if $W(t)$ in Eq. (5.4.6) represents the Wiener process and $a(y,t)$ and $b(y,t)\nu(t)b(y,t)^T$ are the polynomials in y. In other cases the evaluation of expectations in the equations obtained offers insuperable difficulties.

E x a m p l e 7.4.5. Under the conditions of Example 7.4.1 while equating to zero the semiinvariants of the fifth and the sixth orders to zero we obtain the relations: $\kappa_5 = \mu_5 - 10\mu_3 = 0$, $\kappa_6 = \mu_6 - 15D\mu_4 - 10\mu_3^2 + 30D^3 = 0$. Hence we find $\mu_5 = 10D\mu_3$, $\mu_6 = 15D\mu_4 + 10\mu_3^2 - 30D^3$. The first of these equalities coincides with the equality of Example 7.4.1 obtained at the condition $G_5(\mu) = 0$, the second one differs from the equality obtained in Example 7.4.1 when $G_1(\mu) = 0$ by an additional item $10\mu_3^2$ in the right-hand side. In consequence of this fact the first three equations of Example 7.4.1 for the central moments will be also valid in this case, and in the right-hand side of the fourth equation the item $40\mu_3^2$ will be added similarly as at the approximation of the density by the segment of the Edgeworth series with the account of the fourth order moments. This coincidence is explained by the fact that the equality $G_6(\mu) = 10\kappa_3^2$ obtained while applying the Edgeworth series is equivalent to the equality $G_6(\mu) = 10\mu_3^2$ obtained by equating κ_6 to zero.

E x a m p l e 7.4.6. Under the conditions of Example 7.4.2 the equations for the central moments obtained by the MSM with the account of the semiinvariants till the fourth order coincide with the equations obtained in Example 7.4.2 by means of the approximation of the density $f_1(y; t)$ by the segment Hermite expansion in terms of the polynomials or by the segment of the Edgeworth series with the account of the moments till the fourth order. It is explained by the fact that into the equations for the moments of the fourth order enter the moments not higher than the fifth order, and semiinvariants of the fifth order κ_{pq}, $p + q = 5$ coincide with the correspondent quasimoments $c_{pq} = G_{pq}(\mu)$. Therefore the equalities $\kappa_{pq} = 0$ and $G_{pq} = 0$ at $p + q = 5$ are equivalent. It certainly does not mean that the considered approximations are equivalent. In order to be sure in the stated it is sufficient to account that (Appendix 2) $\kappa_{pq} \neq G_{pq}(\mu)$ at $p + q \geq 6$ in consequence of which the considered approximations of the distribution will give different equations for the moments higher than the fourth order. On Fig. 7.4.1–7.4.5 the results of the calculations of the moments m_{10}, $D_{11} = D$, α_{40}, α_{60}, α_{80} of the random variable Y_1 in accordance with time t at $a = 5$, $h = \nu = 1$, $t \in [0, 1]$ and at the independent normal initial distribution of $[Y_1 Y_2]^T$ with the parameters $m_{10,0} = m_{20,0} = 0, 5$, $D_{11,0} = 0, 1$, $D_{22,0} = 1$ are given accounting the semiinvariants till the fourth and the tenth order. The accuracy of MSM while calculating α_{60} and α_{80} for $N = 4$ is 10% and 30% respectively and is smaller than 1% at $N = 10$.

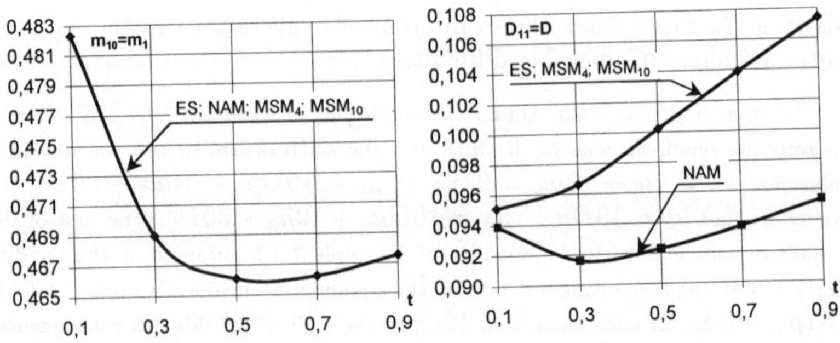

Fig. 7.4.1　　　　　　　　　　　　　　　Fig. 7.4.2

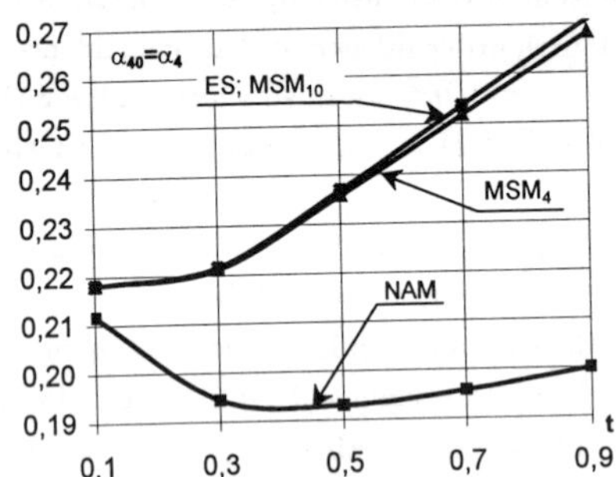

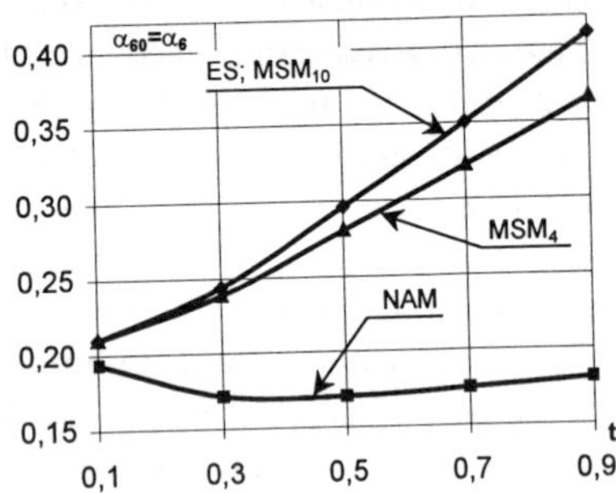

Fig. 7.4.3　　　　　　　　　　　　　　　Fig. 7.4.4

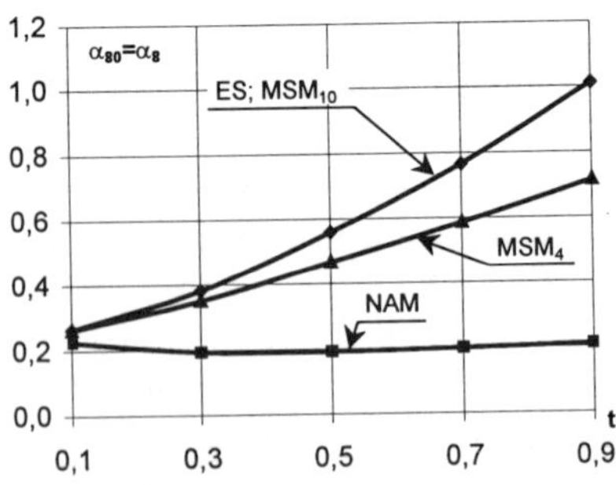

Fig. 7.4.5

7.5. Methods based on Orthogonal Expansions and Quasimoments

7.5.1. Orthogonal Expansion of One-Dimensional Distribution

While approximating the distributions by a truncated orthogonal expansion it is natural to assume as distribution parameters the expectation, the covariance matrix (on which the basic distributions w_n and consequently the polynomials $p_\nu(y)$, $q_\nu(y)$ depend) and the coefficients of the expansion.

▷ In order to derive the equations for the expectation and the covariance matrix for normal stochastic system (5.4.6) we substitute into the expressions of the expectations in (7.1.1) and (7.1.5) of the approximating function

$$f_1^*(y; \vartheta) = w_1(y) \left[1 + \sum_{l=3}^{N} \sum_{|\nu|=l} c_\nu p_\nu(y) \right]. \qquad (7.5.1)$$

instead of $f_1(y; \vartheta)$. Then we obtain

$$\dot{m} = \int_{-\infty}^{\infty} a(y,t) w_1(y) \left[1 + \sum_{l=3}^{N} \sum_{|\nu|=l} c_\nu p_\nu(y) \right] dy, \qquad (7.5.2)$$

$$\dot{K} = \int_{-\infty}^{\infty} [a(y,t)(y^T - m^T) + (y - m)a(y,t)^T +$$

$$+ b(y,t)\nu(t)b(y,t)^T] w_1(y) \left[1 + \sum_{l=3}^{N} \sum_{|\nu|=l} c_\nu p_\nu(y) \right] dy. \qquad (7.5.3)$$

Introducing the notations

$$\varphi_{10}(m, K, t) = \int_{-\infty}^{\infty} a(y,t) w_1(y) dy, \qquad (7.5.4)$$

$$\varphi_{1\nu}(m, K, t) = \int_{-\infty}^{\infty} a(y,t) p_\nu(y) w_1(y) dy, \qquad (7.5.5)$$

$$\varphi_{20}(m, K, t) = \int\limits_{-\infty}^{\infty} [a(y,t)(y^T - m^T) +$$

$$+ (y-m)a(y,t)^T + b(y,t)\nu(t)b(y,t)^T]w_1(y)dy, \qquad (7.5.6)$$

$$\varphi_{2\nu}(m, K, t) = \int\limits_{-\infty}^{\infty} [a(y,t)(y^T - m^T) + (y-m)a(y,t)^T$$

$$+ b(y,t)\nu(t)b(y,t)^T]p_\nu(y)w_1(y)dy, \qquad (7.5.7)$$

we represent these equations as

$$\dot{m} = \varphi_{10}(m, K, t) + \sum_{l=3}^{N} \sum_{|\nu|=l} \varphi_{1\nu}(m, K, t)c_\nu, \qquad (7.5.8)$$

$$\dot{K} = \varphi_{20}(m, K, t) + \sum_{l=3}^{N} \sum_{|\nu|=l} \varphi_{2\nu}(m, K, t)c_\nu \quad \triangleleft. \qquad (7.5.9)$$

Now let derive the equations for the coefficients c_ν in (7.5.1).

▷ We use formula (3.8.7) changing index ν on κ:

$$c_\kappa = \left[q_\kappa \left(\frac{\partial}{i\partial\lambda} \right) g_1(\lambda; t) \right]_{\lambda=0}. \qquad (7.5.10)$$

Differentiating this formula with respect to time t and taking into consideration that the polynomial $q_\kappa(y)$ depends on the expectation m and the covariance matrix K of the random vector Y_t which are the functions of time we find

$$\dot{c}_\kappa = \left[q_\kappa \left(\frac{\partial}{i\partial\lambda} \right) \frac{\partial g_1(\lambda; t)}{\partial t} \right]_{\lambda=0} + \left[q_\kappa^m \left(\frac{\partial}{i\partial\lambda} \right)^T g_1(\lambda; t) \right]_{\lambda=0} \cdot \dot{m}$$

$$+ \operatorname{tr} \left\{ \left[q_\kappa^K \left(\frac{\partial}{i\partial\lambda} \right) g_1(\lambda; t) \right]_{\lambda=0} \cdot \dot{K} \right\}, \qquad (7.5.11)$$

where $q_\kappa^m(y)$ is the matrix-column of the derivatives of the polynomial $q_\kappa(y)$ with respect to the components of the vector m, and q_κ^K is the square matrix of the derivatives of the polynomial $q_\kappa(y)$ with respect to the elements of the matrix K.

Substituting into this formula the expression $\partial g_1(\lambda;t)/\partial t$ from Eq. (5.6.20) and replacing the density $f_1(y;t)$ by its truncated orthogonal expansion (7.5.1) we obtain

$$
\dot{c}_\kappa = \int\limits_{-\infty}^{\infty} \left\{ q_\kappa \left(\frac{\partial}{i\partial\lambda} \right) \left[i\lambda^T a(y,t) + \chi(b(y,t)^T\lambda;t) \right] e^{i\lambda^T y} \right\}_{\lambda=0}
$$

$$
\times \left[1 + \sum_{l=3}^{N} \sum_{|\nu|=l} c_\nu p_\nu(y) \right] w_1(y) dy + q_\kappa^m(\alpha)\dot{m} + \mathrm{tr}\left[q_\kappa^K(\alpha)\dot{K} \right] , \quad (7.5.12)
$$

where $\chi(\mu;t)$ for the normal white noise is determined by (5.6.37), $q_\kappa^m(\alpha)$ and $q_\kappa^K(\alpha)$ represent as usual the result of replacing the monomials $y_1^{r_1} \ldots y_p^{r_p}$ in the expressions of the polynomials $q_\kappa^m(y)$ and $q_\kappa^K(y)$ by the corresponding moments $\alpha_{r_1}, \ldots, \alpha_{r_p}$. Putting

$$
\varphi_{\kappa 0}(m, K, t) = \int\limits_{-\infty}^{\infty} \left\{ q_\kappa \left(\frac{\partial}{i\partial\lambda} \right) \right.
$$

$$
\left. \times \left[i\lambda^T a(y,t) + \chi(b(y,t)^T\lambda;t) \right] e^{i\lambda^T y} \right\}_{\lambda=0} w_1(y) dy ,
$$
$$
(7.5.13)
$$

$$
\varphi_{\kappa \nu}(m, K, t) = \int\limits_{-\infty}^{\infty} \left\{ q_\kappa \left(\frac{\partial}{i\partial\lambda} \right) \left[i\lambda^T a(y,t) \right. \right.
$$
$$
(7.5.14)
$$
$$
\left. \left. + \chi(b(y,t)^T\lambda;t) \right] e^{i\lambda^T y} \right\}_{\lambda=0} p_\nu(y) w_1(y) dy
$$

and substituting the expressions of $\dot{m}$ and $\dot{K}$ from (7.5.8), (7.5.9) we obtain the equations

$$
\dot{c}_\kappa = \varphi_{\kappa 0}(m, K, t) + \varphi_{10}(m, K, t)^T q_\kappa^m(\alpha) + \mathrm{tr}\left[\varphi_{20}(m, K, t) q_\kappa^K(\alpha) \right]
$$

$$
+ \sum_{l=3}^{N} \sum_{|\nu|=l} \left\{ \varphi_{\kappa \nu}(m, K, t) + \varphi_{1\nu}(m, K, t)^T q_\kappa^m(\alpha) + \mathrm{tr}\left[\varphi_{2\nu}(m, K, t) q_\kappa^K(\alpha) \right] \right.
$$

$$
+ \sum_{l=3}^{N} \sum_{|\nu|=l} \varphi_{2\nu}(m, K, t) \} c_\nu \quad (|\kappa| = 3, \ldots, N) . \quad (7.5.15)
$$

Here the moments $\alpha_{r_1}, \ldots, \alpha_{r_p}$ in $q_\kappa^m(\alpha)$ and $q_\kappa^K(\alpha)$ must be replaced by their expressions in terms of the coefficients c_ν in accordance with (3.8.14). ◁

In Appendix 4 the formulae for the evaluation of the integrands in (7.5.13) and (7.5.14) are given.

Eqs. (7.5.8), (7.5.9), (7.5.15) *form the set of equations which determine approximately all the parameters* m, K, c_ν ($|\nu| = 3, \ldots, N$) *of the truncated orthogonal expansion* $f_1^*(y; \vartheta)$ *of the density* $f_1(y; t)$. *As the initial values of* m, K, c_ν ($|\nu| = 3, \ldots, N$) *at* $t = t_0$ *the corresponding parameters of the truncated orthogonal expansion of the density* $f_0(y)$ *of the random variable* Y_0 *should be taken:*

$$m(t_0) = m_0, \quad K(t_0) = K_0, \quad c_\nu(t_0) = c_{\nu 0}. \qquad (7.5.16)$$

R e m a r k 1. Eqs. (7.5.8), (7.5.9) and (7.5.15) are also valid in the case when only $q_\nu(y)$ represent the polynomials, and $p_\nu(y)$ are not the polynomials. In particular these equations are valid only in that case when the right hand side of formula (7.5.1) is a segment of the expansion $f_1(y; t)$ in terms of the derivatives of the density $w_1(y)$. In this case $p_\nu(y) = w_1^{(\nu)}(y)/w_1(y)$, and $q_\nu(y)$ are the polynomials (Subsection 3.8.1). Such expansion was used for approximating $f_1(y; t)$ in (Ahlbehrendt and Kempe 1984).

R e m a r k 2. If $q_\nu(y)$ are not the polynomials then the previous evaluations in the part of the derivation of Eqs. (7.5.15) are inapplicable. In this case for deriving the equations for the coefficients c_ν the stochastic differential of the function $q_\nu(Y, t)$ of the process $Y(t)$ by formula (5.1.7) should be evaluated and the expectation of the obtained expression should be taken. As a result at the normal white noise the equations analogous to Eqs. (7.5.15) are obtained.

R e m a r k 3. Eqs. (7.5.15) are nonlinear relatively to the coefficients c_ν. If we reject out the requirement of the coincidence of the first and the second order moments of the distribution $w_1(y)$ and assign the moments for $w_1(y)$ a priori then the linear equations for c_ν are obtained. But in this case we have to take a great deal of the numbers of the expansions.

The equations of the method of the orthogonal expansions for stochastic nonlinear systems described by Eq. (5.4.6) at the nongaussian white noise and for Eq. (5.4.7) are given in Problems 7.11 and 7.12.

7.5.2. Method of Quasimoments

If the density $f_1(y; t)$ is approximated by a truncated Hermite polynomial expansion (Appendix 2) then $p_\nu(y) = H_\nu(y - m)(\nu_1! \ldots \nu_p!)^{-1}$, $q_\nu(y) = G_\nu(y - m)$ and consequently, according (3.8.25)

$$c_\nu = q_\nu(\alpha) = G_\nu(\mu). \qquad (7.5.17)$$

$\triangleright$ In order to find the vector $q_\kappa^m(\alpha)$ and the matrix $q_\kappa^K(\alpha)$ we use formulae (9), (12) and (13) of Appendix 2. Then we obtain

$$q_{\kappa r}^m = \frac{\partial}{\partial m_r} G_\kappa(y-m) = -\frac{\partial}{\partial y_r} G_\kappa(y-m) = -\kappa_r G_{\kappa-e_r}(y-m), \quad (7.5.18)$$

$$q_{\kappa r}^K = \frac{\partial}{\partial K_{rr}} G_\kappa(y-m) = -\frac{1}{2}\kappa_r(\kappa_r-1)G_{\kappa-2e_r}(y-m), \quad (7.5.19)$$

$$q_{\kappa r s}^K = \frac{\partial}{\partial K_{rs}} G_\kappa(y-m) = -\kappa_r\kappa_s G_{\kappa-e_r-e_s}(y-m), \quad (7.5.20)$$

where e_r is the vector whose components are all equal to zero except the rth component which is equal to 1. Replacing here the monomials of the form $(y_1 - m_1)^{r_1}\ldots(y_p - m_p)^{r_p}$ by the corresponding moments $\mu_{r_1,\ldots,r_p}$ and remembering that $G_\nu(\mu) = c_\nu$ and that $c_\nu = 0$ at $|\nu| < 3$ we find the elements of the matrix-column $q_\kappa^m(\alpha)$ and of the square matrix $q_\kappa^K(\alpha)$:

$$q_{\kappa r}^m(\alpha) = -\kappa_r c_{\kappa-e_r} \quad (r = 1,\ldots,p; \ |\kappa| = 4,\ldots,N); \quad (7.5.21)$$

$$q_{\kappa r r}^K(\alpha) = -\frac{1}{2}\kappa_r(\kappa_r-1)c_{\kappa-2e_r}, \quad q_{\kappa r s}^K(\alpha) = -\kappa_r\kappa_s c_{\kappa-e_r-e_s}, \quad (7.5.22)$$

$$(r,s = 1,\ldots,p; \ s > r; \ |\kappa| = 5,\ldots,N).$$

At $|\kappa| = 3$ we have $q_\kappa^m(\alpha) = 0$ and $q_\kappa^K(\alpha) = 0$. $\triangleleft$

Thus, *while approximating the density $f_1(y;t)$ by the truncated Hermite polynomial expansion the elements of the matrices $q_\kappa^m(\alpha)$ and $q_\kappa^K(\alpha)$ are proportional to the corresponding quasimoments.* Eqs. (7.5.8), (7.5.9), (7.5.15) *determine in this case the expectation m, the covariance matrix K and the quasimoments $c_\nu(|\nu| = 3,\ldots,N)$ for normal stochastic system* (5.4.6).

The formulae for evaluating the integrands in the quasimoments method are given in Appendix 4.

E x a m p l e 7.5.1. In the stochastic system of Example 7.4.4 equations for the quasimoments are identical with the corresponding equations for the semiinvariants as $\kappa_l = c_l$ $(l = 1,\ldots,5)$.

E x a m p l e 7.5.2. Under the conditions of Example 7.1.6 approximating $f_1(y;t)$ by the truncated Hermite polynomial expansion with the account of the polynomials of up to the fourth degree Eqs. (7.5.8), (7.5.9) and (7.5.15) have the form (certainly they coincide with the corresponding equations for the semiinvariants, the

difference between them will be only when the Hermite polynomials of no lower than
the sixth degree are involved):

$$\dot{m}_1 = -m_1 m_2 - k_{12}, \quad \dot{m}_2 = -a m_2, \tag{I}$$

$$\dot{k}_{11} = -2(m_2 k_{11} + m_1 k_{12} + c_{21}),$$

$$\dot{k}_{12} = -(m_2 + a)k_{12} - m_1 k_{22} - c_{12}, \quad \dot{k}_{22} = -2a k_{22} + h^2 \nu \tag{II}$$

$$\dot{c}_{30} = -3(2k_{11}k_{12} + m_2 c_{30} + m_1 c_{21} + c_{31}),$$

$$\dot{c}_{21} = -2\left[k_{11}k_{22} + k_{12}^2 + (m_2 + a)c_{21} + m_1 c_{12} + c_{22}\right],$$

$$\dot{c}_{12} = -2k_{12}k_{22} - (m_2 + 2a)c_{12} - m_1 c_{03} - c_{13}, \quad \dot{c}_{03} = -3a c_{03},$$

$$\dot{c}_{40} = -4(3k_{12}c_{30} + 3k_{11}c_{21} + m_2 c_{40} + m_1 c_{31}),$$

$$\dot{c}_{31} = -3(k_{22}c_{30} + 3k_{12}c_{21} + 2k_{11}c_{12}) - (3m_2 + a)c_{31} - 3m_1 c_{22},$$

$$\dot{c}_{22} = -2\left[2k_{12}c_{21} + 3k_{12}c_{12} + k_{11}c_{03} + (m_2 + a)c_{22} + m_1 c_{13}\right],$$

$$\dot{c}_{13} = -3(k_{22}c_{12} + k_{12}c_{03}) - (m_2 + 3a)c_{13} - m_1 c_{04}, \quad \dot{c}_{04} = -4a c_{04}. \tag{III}$$

7.5.3. Consistent Orthogonal Expansions of Multi-Dimensional Distributions

One may obtain an approximate solution of Eqs. (5.6.20), (5.6.23)
for the multi-dimensional distributions of the random process determined
by Eq. (5.4.6) for the normal white noise in the form of truncated
consistent orthogonal expansions of the multi-dimensional densities f_n
$= f_n(y_1, \ldots, y_n; t_1, \ldots, t_n)$:

$$f_n = f_n(y_1, \ldots, y_n; t_1, \ldots, t_n) \approx f_n^*(y_1, \ldots, y_n; \vartheta_n) = w_n(y_1, \ldots, y_n)$$

$$\times \left[1 + \sum_{l=3}^{N} \sum_{|\nu_1| + \cdots + |\nu_n| = l} c_{\nu_1, \ldots, \nu_n} p_{\nu_1, \ldots, \nu_n}(y_1, \ldots, y_n)\right] \tag{7.5.23}$$

$$(n = 1, 2, \ldots),$$

where $w_n = \{w_n(y_1, \ldots, y_n)\}$ $(n = 1, 2, \ldots)$ is the consistent sequence
of basic densities which have the same moments of the first and the
second order as the corresponding densities f_n, in consequence of which
each of the densities w_n depends on the values of the expectation $m(t)$
and of the covariance function $K(t, t')$ of the process $Y(t)$ at t, t'
$= t_1, \ldots, t_n$ $(n = 1, 2, \ldots)$ as parameters.

After replacing in the expression of the expectation in Eq. (7.1.8) for the covariance function of the process $Y(t)$ the two-dimensional density $f_2(y_1, y_2; t_1, t_2)$ by its approximating segment of the orthogonal expansion (7.5.23) we get

$$\partial K(t_1, t_2)/\partial t_2$$

$$= \varphi_{20}(m, K, t_1, t_2) + \sum_{l=3}^{N} \sum_{|\nu_1|+|\nu_2|=l} \varphi_{2\nu_1\nu_2}(m, K, t_1, t_2) c_{\nu_1\nu_2}(t_1, t_2),$$

$$(7.5.24)$$

where

$$\varphi_{20}(m, K, t_1, t_2) = \int_{-\infty}^{\infty} \int_{-\infty}^{\infty} [y_1 - m(t_1)] \, a(y_2, t_2)^T w_2(y_1, y_2) dy_1 \, dy_2 \,,$$

$$(7.5.25)$$

$$\varphi_{2\nu_1\nu_2}(m, K, t_1, t_2)$$

$$= \int_{-\infty}^{\infty} \int_{-\infty}^{\infty} [y_1 - m(t_1)] \, a(y_2, t_2)^T p_{\nu_1\nu_2}(y_1, y_2) w_2(y_1, y_2) dy_1 \, dy_2 \,. \quad (7.5.26)$$

The functions φ_{20} and $\varphi_{2\nu_1\nu_2}$ depend on the values m_{t_1}, m_{t_2}, K_{t_1}, K_{t_2} of the expectation $m(t)$, the covariance matrix $K(t)$ and on the value of the covariance function $K(t_1, t_2)$ of the process $Y(t)$.

▷ In order to derive the equations for the coefficients of the consistent orthogonal expansions of the multi-dimensional densities of the process $Y(t)$,we notice that according to (3.8.46)

$$c_{\kappa_1, \dots, \kappa_n}(t_1, \dots, t_n) = \left[q_{\kappa_1, \dots, \kappa_n} \left(\frac{\partial}{i\partial\lambda_1}, \dots, \frac{\partial}{i\partial\lambda_n} \right) \right.$$

$$\left. \times g_n(\lambda_1, \dots, \lambda_n; t_1, \dots, t_n) \right]_{\lambda_1 = \dots = \lambda_n = 0} . \qquad (7.5.27)$$

Differentiating this formula with respect to t_n we shall have

$$\partial c_{\kappa_1, \dots, \kappa_n}(t_1, \dots, t_n)/\partial t_n$$

$$= \left[q_{\kappa_1, \dots, \kappa_n} \left(\frac{\partial}{i\partial\lambda_1}, \dots, \frac{\partial}{i\partial\lambda_n} \right) \frac{\partial g_n}{\partial t} \right]_{\lambda_1 = \dots = \lambda_n = 0}$$

$$+ \left[q_{\kappa_1, \dots, \kappa_n}^m \left(\frac{\partial}{i\partial\lambda_1}, \dots, \frac{\partial}{i\partial\lambda_n} \right)^T g_n \right]_{\lambda_1 = \dots = \lambda_n = 0} \cdot \dot{m}_{t_n}$$

$$+\operatorname{tr}\left\{\sum_{h=1}^{n-1}\left[q_{\kappa_1,\ldots,\kappa_n}^{K_h}\left(\frac{\partial}{i\partial\lambda_1},\ldots,\frac{\partial}{i\partial\lambda_n}\right)g_n\right]_{\lambda_1=\cdots=\lambda_n=0}\cdot\frac{\partial K(t_h,t_n)}{\partial t_n}\right\}$$

$$+\operatorname{tr}\left\{\left[q_{\kappa_1,\ldots,\kappa_n}^{K_n}\left(\frac{\partial}{i\partial\lambda_1},\ldots,\frac{\partial}{i\partial\lambda_n}\right)g_n\right]_{\lambda_1=\cdots=\lambda_n=0}\cdot\dot{K}_{t_n}\right\}$$

$$=\left[q_{\kappa_1,\ldots,\kappa_n}\left(\frac{\partial}{i\partial\lambda_1},\ldots,\frac{\partial}{i\partial\lambda_n}\right)\frac{\partial g_n}{\partial t}\right]_{\lambda_1=\cdots=\lambda_n=0}+q_{\kappa_1,\ldots,\kappa_n}^m(\alpha)^T\dot{m}_{t_n}$$

$$+\operatorname{tr}\left[\sum_{h=1}^{n-1}q_{\kappa_1,\ldots,\kappa_n}^{K_h}(\alpha)\frac{\partial K(t_h,t_n)}{\partial t_n}\right]+\operatorname{tr}\left[q_{\kappa_1,\ldots,\kappa_n}^{K_n}(\alpha)\dot{K}_{t_n}\right]\,,\quad(7.5.28)$$

where $q_{\kappa_1,\ldots,\kappa_n}^m(y_1,\ldots,y_n)$ is the matrix-column of the derivatives of the polynomial $q_{\kappa_1,\ldots,\kappa_n}(y_1,\ldots,y_n)$ with respect to the components of the vector m_{t_n} $=m(t_n)$; $q_{\kappa_1,\ldots,\kappa_n}^{K_h}(y_1,\ldots,y_n)$ is the square matrix of the derivatives of the polynomial $q_{\kappa_1,\ldots,\kappa_n}(y_1,\ldots,y_n)$ with respect to the elements of the matrix $K(t_h,t_n)$ $(h=1,\ldots,n)$, $q_{\kappa_1,\ldots,\kappa_n}^m(\alpha)$; $q_{\kappa_1,\ldots,\kappa_n}^{K_h}(\alpha)$ is the result of the replacement of the monomials of the form $y_{11}^{r_{11}}\ldots y_{1p}^{r_{1p}}\ldots y_{n1}^{r_{n1}}\ldots y_{np}^{r_{np}}$ by the corresponding moments $\alpha_{r_{11},\ldots,r_{1p},\ldots,r_{n1},\ldots,r_{np}}$ in the expressions of the polynomials $q_{\kappa_1,\ldots,\kappa_n}^m(\alpha)$, $q_{\kappa_1,\ldots,\kappa_n}^{K_h}(\alpha)$. Substituting the expression into the previous formula from (5.6.23), replacing the density $f_n(y_1,\ldots,y_n;t_1,\ldots,t_n)$ by its approximating function (7.5.23) and taking into account that $(\partial/i\partial\lambda_k)e^{i\lambda_k^T y_k}=y_k e^{i\lambda^T y}$ we obtain from (7.5.28) at $t_1<\cdots<t_n$

$$\partial c_{\kappa_1,\ldots,\kappa_n}(t_1,\ldots,t_n)/\partial t_n$$

$$=\int_{-\infty}^{\infty}\cdots\int_{-\infty}^{\infty}\left\{q_{\kappa_1,\ldots,\kappa_n}\left(y_1,\ldots,y_{n-1},\frac{\partial}{i\partial\lambda_n}\right)[i\lambda_n^T a(y_n,t_n)\right.$$

$$\left.+\chi(b(y_n,t_n)^T\lambda_n;t_n)]e^{i\lambda_n^T y_n}\right\}_{\lambda_n=0}f_n^*(y_1,\ldots,y_n;\vartheta_n)dy_1\ldots dy_n$$

$$+q_{\kappa_1,\ldots,\kappa_n}^m(\alpha)^T\dot{m}_{t_n}+\sum_{h=1}^{n-1}\operatorname{tr}\left[q_{\kappa_1,\ldots,\kappa_n}^{K_h}(\alpha)\frac{\partial K(t_h,t_n)}{\partial t_n}\right]$$

$$+\operatorname{tr}\left[q_{\kappa_1,\ldots,\kappa_n}^{K_n}(\alpha)\dot{K}_{t_n}\right]\,.\qquad(7.5.29)$$

Replacing here $\dot{m}_{t_n}$, $\dot{K}_{t_n}$ and $\partial K(t_h,t_n)/\partial t_n$ by their expressions from (7.5.8), (7.5.9) and (7.5.24) we reduce these equations to

$$\partial c_{\kappa_1,\ldots,\kappa_n}(t_1,\ldots,t_n)/\partial t_n=\varphi_{\kappa_1,\ldots,\kappa_n,0}(m,K,t_1,\ldots,t_n)$$

$$+\left\{\varphi_{10}(m_{t_n}, K_{t_n}, t_n)^T + \sum_{l=3}^{N} \sum_{|\nu|=l} c_\nu(t_n)\varphi_{1\nu}(m_{t_n}, K_{t_n}, t_n)^T\right\} q^m_{\kappa_1,\dots,\kappa_n}(\alpha)$$

$$+\operatorname{tr}\left[\left\{\varphi_{20}(m_{t_n}, K_{t_n}, t_n) + \sum_{l=3}^{N} \sum_{|\nu|=l} c_\nu(t_n)\varphi_{2\nu}(m_{t_n}, K_{t_n}, t_n)\right\}\right.$$

$$\times q^{K_n}_{\kappa_1,\dots,\kappa_n}(\alpha)\Bigg] + \sum_{h=1}^{n-1} \operatorname{tr}\left[\left\{\varphi_2(m_{t_n}, K_{t_n}, t_n)^T\right.\right.$$

$$\left.\left.+\sum_{l=3}^{N} \sum_{|\nu_1|+|\nu_2|=l} c_{\nu_1\nu_2}(t_h,t_n)\varphi_{2\nu_1\nu_2}(m, K, t_h, t_n)^T\right\} q^{K_h}_{\kappa_1,\dots,\kappa_n}(\alpha)\right]$$

$$+\sum_{l=3}^{N} \sum_{|\nu_1|+\dots+|\nu_n|=l} \varphi_{\kappa_1,\dots,\kappa_n,\nu_1,\dots,\nu_n}(m, K, t_1, \dots, t_n) c_{\nu_1,\dots,\nu_n}(t_1, \dots, t_n),$$

$$(7.5.30)$$

where $|\kappa_1|, \dots, |\kappa_n| = 1, \dots, N-n+1$; $|\kappa_1|+\dots+|\kappa_n| = n, \dots, N$.
Here in addition to the previous notations

$$\varphi_{\kappa_1,\dots,\kappa_n,0}(m, K, t_1, \dots, t_n)$$

$$= \int_{-\infty}^{\infty} \cdots \int_{-\infty}^{\infty} \left\{ q_{\kappa_1,\dots,\kappa_n}\left(y_1, \dots, y_{n-1}, \frac{\partial}{i\partial\lambda_n}\right)[i\lambda_n^T a(y_n, t_n)\right.$$

$$\left.+ \chi(b(y_n, t_n)^T \lambda_n; t_n)]e^{i\lambda_n^T y_n}\right\}_{\lambda_n=0} w_n(y_1, \dots, y_n)dy_1, \dots, dy_n,$$

$$(7.5.31)$$

$$\varphi_{\kappa_1,\dots,\kappa_n,\nu_1,\dots,\nu_n}(m, K, t_1, \dots, t_n)$$

$$= \int_{-\infty}^{\infty} \cdots \int_{-\infty}^{\infty} \left\{ q_{\kappa_1,\dots,\kappa_n}\left(y_1, \dots, y_{n-1}, \frac{\partial}{i\partial\lambda_n}\right)[i\lambda_n^T a(y_n, t_n)\right.$$

$$\left.+ \chi(b(y_n, t_n)^T \lambda_n; t_n)]e^{i\lambda_n^T y_n}\right\}_{\lambda_n=0} p_{\nu_1,\dots,\nu_n}(y_1, \dots, y_n)$$

$$\times w_n(y_1, \dots, y_n)dy_1, \dots, dy_n. \qquad (7.5.32)$$

These functions depend on the values $m_{t_h} = m(t_h)$ and $K(t_h, t_l)$ $(h, l = 1, \dots, n)$ of the expectation $m(t)$ and the covariance function $K(t, t')$ of the process $Y(t)$, that is shown by the letters m, K, t_1, $\dots$, t_n as the arguments.

On the basis of property (3.8.51) of the coefficients of the consistent orthogonal expansions initial conditions for Eqs. (7.6.30) have the form

$$c_{\kappa_1,\ldots,\kappa_n}(t_1,\ldots,t_{n-1},t_{n-1}) = c_{\kappa_1,\ldots,\kappa_{n-1}+\kappa_n}(t_1,\ldots,t_{n-1}). \triangleleft$$

$$(7.5.33)$$

R e m a r k. Notice that at every n only those coefficients $c_{\nu_1,\ldots,\nu_n}$ in Eq. (7.5.30) (also in Eq. (7.5.24) at $n = 2$) are unknown which depend on all n arguments $t_1,\ldots,t_n$. By virtue of property (3.8.47) of the coefficients of consistent orthogonal expansions only those coefficients $c_{\nu_1,\ldots,\nu_n}$ depend on all n arguments of $t_1,\ldots,t_n$ for which none of the vector subscripts $\nu_1,\ldots,\nu_n$ is equal to zero, i.e. none of the sums $|\nu_1|,\ldots,|\nu_n|$ is equal to zero. The coefficients $c_{\nu_1,\ldots,\nu_n}$ with some subscripts $\nu_1,\ldots,\nu_n$ equal to zero depend only on those of the arguments $t_1,\ldots,t_n$ which correspond to the non zero subscripts. These coefficients are determined before while finding the coefficients of the expansions of the distributions of lower dimensions.

Thus, *after integrating Eqs. (7.5.8), (7.5.9), (7.5.15) we determine approximately all the parameters m, K, c_ν ($|\nu| = 3,\ldots,N$) of the truncated expansion of the one-dimensional density $f_1(y;t)$ as the function of time t. After integrating Eqs. (7.5.30) with initial conditions (7.5.33) at $n = 2$, $t_2 > t_1$ and Eq. (7.5.24) with the initial condition $K(t_1,t_1) = K(t_1)$ we determine approximately the covariance function $K(t_1,t_2)$ of the process $Y(t)$ and all the coefficients $c_{\nu_1\nu_2}$ of truncated expansion (7.5.23) of the two-dimensional density $f_2(y_1,y_2;t_1,t_2)$ which remained unknown. Further, subsequently integrating Eq. (7.5.30) with initial conditions (7.5.33) at $t_1 < \cdots < t_n$ we determine approximately the coefficients $c_{\nu_1,\ldots,\nu_n}$ of the truncated expansions (7.5.23) of the densities $f_n(y_1,\ldots,y_n;t_1,\ldots,t_n)$ $(n = 3,\ldots,N)$. After that the coefficients of truncated expansions (7.5.23) of all other multi-dimensional densities of the process $Y(t)$ are determined by formula (3.8.47) as at $n > N$ at least $n - N$ subscripts are equal to zero for any of the coefficients $c_{\nu_1,\ldots,\nu_n}$ in (7.5.23). As a result all the multi-dimensional distributions of the process $Y(t)$ will be approximately determined.*

7.5.4. Consistent Hermite Polynomial Expansions

In the same way as in Subsection 7.5.2 it is proved that while using the consistent Hermite polynomial expansions the elements of the matrices $q^m_{\kappa_1,\ldots,\kappa_n}(\alpha)$, $q^{K_h}_{\kappa_1,\ldots,\kappa_n}(\alpha)$ are proportional to the corresponding quasimoments

$$q^m_{\kappa_1,\ldots,\kappa_n,r}(\alpha) = -\kappa_{nr}c_{\kappa_1,\ldots,\kappa_n-e_r} \tag{7.5.34}$$

$$(r = 1,\ldots,p;\ |\kappa_1| + \cdots + |\kappa_n| = 4,\ldots,N),$$

$$q^{K_h}_{\kappa_1,\ldots,\kappa_n,rs}(\alpha) = -\kappa_{hr}\kappa_{ns}c_{\kappa_1,\ldots,\kappa_h-e_r,\ldots,\kappa_n-e_s} \tag{7.5.35}$$

$$(r,s = 1,\ldots,p;\ h = 1,\ldots,n-1;\ |\kappa_1| + \cdots + |\kappa_n| = 5,\ldots,N),$$

$$q^{K_n}_{\kappa_1,\ldots,\kappa_n,rr}(\alpha) = -\frac{1}{2}\kappa_{nr}(\kappa_{nr}-1)c_{\kappa_1,\ldots,\kappa_n-2e_r}, \tag{7.5.36}$$

$$q^{K_n}_{\kappa_1,\ldots,\kappa_n,rs}(\alpha) = -\kappa_{nr}\kappa_{ns}c_{\kappa_1,\ldots,\kappa_n-e_r-e_s} \tag{7.5.37}$$

$$(r,s = 1,\ldots,p;\ s \neq r;\ |\kappa_1| + \cdots + |\kappa_n| = 5,\ldots,N),$$

$$\begin{cases} q^m_{\kappa_1,\ldots,\kappa_n}(\alpha) = 0 \text{ at } |\kappa_1| + \cdots + |\kappa_n| = 3; \\ q^{K_h}_{\kappa_1,\ldots,\kappa_n}(\alpha) = 0 \text{ at } |\kappa_1| + \cdots + |\kappa_n| = 3 \text{ and } |\kappa_1| + \cdots + |\kappa_n| = 4; \end{cases} \tag{7.5.38}$$

In the case of the normal white noise V in Eq. (5.4.6) we have

$$\left\{ q_{\kappa_1,\ldots,\kappa_n}(y_1,\ldots,y_{n-1},\partial/i\partial\lambda_n)[i\lambda_n^T a(y_n,t_n) \right.$$

$$\left. +\chi(b(y_n,t_n)^T\lambda_n;t_n)]e^{i\lambda_n^T y_n} \right\}_{\lambda_n=0}$$

$$= \left\{ G_{\kappa_1,\ldots,\kappa_n}(y_1 - m_{t_1},\ldots,y_{n-1} - m_{t_{n-1}},\partial/i\partial\lambda_n)[i\lambda_n^T a(y_n,t_n) \right.$$

$$\left. +\chi(b(y_n,t_n)^T\lambda_n;t_n)]\exp\{i\lambda_n^T(y_n - m_{t_n})\} \right\}_{\lambda_n=0}$$

$$= \sum_{s=1}^{p} \kappa_{ns}a_s(y_n,t_n)G_{\kappa_1,\ldots,\kappa_n-e_s}(y_1 - m_{t_1},\ldots,y_n - m_{t_n})$$

$$+\frac{1}{2}\sum_{s=1}^{p} \kappa_{ns}(\kappa_{ns}-1)\sigma_{ss}(y_n,t_n)G_{\kappa_1,\ldots,\kappa_n-2e_s}(y_1 - m_{t_1},\ldots,y_n - m_{t_n})$$

$$+\sum_{q=2}^{p}\sum_{s=1}^{q-1} \kappa_{ns}\kappa_{nq}\sigma_{sq}(y_n,t_n)G_{\kappa_1,\ldots,\kappa_n-e_s-e_q}(y_1 - m_{t_1},\ldots,y_n - m_{t_n}). \tag{7.5.39}$$

As well as the equations for the moments (Section 7.4) Eqs. (7.5.29), (7.5.30) for the coefficients of the orthogonal expansions of the multi-dimensional densities of the process $Y(t)$ may be derived in problems of practice by differentiation of the random functions $q_{\kappa_1,\ldots,\kappa_n}(Y(t_1)$ $\ldots Y(t_n))$ with respect to t_n using Itô formula (5.1.7) of differentiation of a composite function (5.1.7) and the evaluation of the expectations of the obtained expressions by means of the approximate representation of the corresponding densities by truncated orthogonal expansions (7.5.23).

E x a m p l e   7.5.3. In the problem of Example 7.1.1 using the approximation of the distributions by the truncated Hermite polynomial expansions with the account of the polynomials of up to the fourth degree the covariance function $K(t_1,t_2)$ and the quasimoments $c_{21}(t_1,t_2)$, $c_{12}(t_1,t_2)$, $c_{31}(t_1,t_2)$, $c_{22}(t_1,t_2)$, $c_{13}(t_1,t_2)$ of the two-dimensional distribution of the process Y are determined by the equations:

$$\partial K(t_1,t_2)/\partial t_2 = -3[m^2(t_2) + D(t_2)]K(t_1,t_2)$$

$$-3m(t_2)c_{12}(t_1,t_2) - c_{13}(t_1,t_2), \tag{I}$$

$$\partial c_{21}(t_1,t_2)/\partial t_2 = -3m(t_2)K^2(t_1,t_2) - 3[m^2(t_2) + D(t_2)]c_{21}(t_1,t_2)$$

$$-6K(t_1,t_2)c_{12}(t_1,t_2) - 3m(t_2)c_{22}(t_1,t_2), \tag{II}$$

$$\partial c_{12}(t_1,t_2)/\partial t_2 = 2[\nu m(t_2) - 8m(t_2)D(t_2) - 6c_3(t_2)]K(t_1,t_2)$$

$$+[\nu - 6m^2(t_2) - 12D(t_2)]c_{12}(t_1,t_2) - 6m(t_2)c_{13}(t_1,t_2), \tag{III}$$

$$\partial c_{31}(t_1,t_2)/\partial t_2 = -6K^3(t_1,t_2) - 18m(t_2)K(t_1,t_2)c_{21}(t_1,t_2)$$

$$-3[m^2(t_2) + D(t_2)]c_{31}(t_1,t_2) - 9K(t_1,t_2)c_{22}(t_1,t_2) + c_3(t_1)c_3(t_2), \tag{IV}$$

$$\partial c_{22}(t_1,t_2)/\partial t_2 = 2[\nu - 6D(t_2)]K^2(t_1,t_2) + 2[\nu m(t_2)$$

$$-8m(t_2)D(t_2) + c_3(t_2)]c_{21}(t_1,t_2) - 24m(t_2)K(t_1,t_2)c_{12}(t_1,t_2)$$

$$+[\nu - 6m^2(t_2) - 12D(t_2)]c_{22}(t_1,t_2) - 12K(t_1,t_2)c_{13}(t_1,t_2) + 6D(t_1)D^2(t_2), \tag{V}$$

$$\partial c_{13}(t_1,t_2)/\partial t_2 = 3\{D(t_2)[2\nu + 3D(t_2)] - 6m(t_2)c_3(t_2) - 3c_4(t_2)\}K(t_1,t_2)$$

$$+3[2\nu m(t_2) - 12m(t_2)D(t_2) + c_3(t_2)]c_{12}(t_1,t_2)$$

$$+3[\nu - 3m^2(t_2) - 9D(t_2)]c_{13}(t_1,t_2) \tag{VI}$$

with the initial conditions

$$K(t_1,t_2) = D(t_1); \quad c_{21}(t_1,t_1) = c_{12}(t_1,t_1) = c_3(t_1),$$

$$c_{31}(t_1,t_1) = c_{22}(t_1,t_1) = c_{13}(t_1,t_1) = c_4(t_1). \tag{VII}$$

The quasimoments $c_{111}(t_1, t_2, t_3)$, $c_{211}(t_1, t_2, t_3)$, $c_{121}(t_1, t_2, t_3)$, $c_{112}(t_1, t_2, t_3)$ of the three-dimensional distribution are determined by the equations

$$\partial c_{111}(t_1, t_2, t_3)/\partial t_3 = -3[m^2(t_3) + D(t_3)]c_{111}(t_1, t_2, t_3)$$

$$-3m(t_3)c_{112}(t_1, t_2, t_3) - 6m(t_3)K(t_1, t_3)K(t_2, t_3)$$

$$-3K(t_1, t_3)c_{12}(t_2, t_3) - 3K(t_2, t_3)c_{12}(t_1, t_3), \qquad \text{(VIII)}$$

$$\partial c_{211}(t_1, t_2, t_3)/\partial t_3 = -12m(t_3)K(t_1, t_3)c_{111}(t_1, t_2, t_3)$$

$$-3[m^2(t_3) + D(t_3)]c_{211}(t_1, t_2, t_3) - 6K(t_1, t_3)c_{112}(t_1, t_2, t_3)$$

$$-6K^2(t_1, t_3)K(t_2, t_3) + c_3(t_3)c_{21}(t_1, t_3)$$

$$-6m(t_3)K(t_2, t_3)c_{21}(t_1, t_3) - 3K(t_2, t_3)c_{22}(t_1, t_3), \qquad \text{(IX)}$$

$$\partial c_{121}(t_1, t_2, t_3)/\partial t_3 = -12m(t_3)K(t_2, t_3)c_{111}(t_1, t_2, t_3)$$

$$-3[m^2(t_3) + D(t_3)]c_{121}(t_1, t_2, t_3) - 6K(t_1, t_3)c_{112}(t_1, t_2, t_3)$$

$$-6K(t_1, t_3)K^2(t_2, t_3) - 6m(t_3)K(t_1, t_3)c_{21}(t_2, t_3)$$

$$+c_3(t_3)c_{12}(t_1, t_2) - 3K(t_1, t_3)c_{22}(t_2, t_3), \qquad \text{(X)}$$

$$\partial c_{112}(t_1, t_2, t_3)/\partial t_3 = 2[\nu m(t_3) - 6m(t_3)D(t_3) + c_3(t_3)]c_{111}(t_1, t_2, t_3)$$

$$+[\nu - 6m^2(t_3) - 12D(t_3)]c_{112}(t_1, t_2, t_3)$$

$$+2[\nu - 6D(t_3)]K(t_1, t_3)K(t_2, t_3) - 12m(t_3)[K(t_2, t_3)c_{12}(t_1, t_3)$$

$$+K(t_1, t_3)c_{12}(t_2, t_3)] - 6[K(t_2, t_3)c_{13}(t_1, t_3) + K(t_1, t_3)c_{13}(t_2, t_3)] \quad \text{(XI)}$$

with the initial conditions: $c_{111}(t_1, t_2, t_3) = c_{12}(t_1, t_2)$; $c_{211}(t_1, t_2, t_3) = c_{22}(t_1, t_2)$; $c_{121}(t_1, t_2, t_3) = c_{112}(t_1, t_2, t_3) = c_{13}(t_1, t_2)$. The only quasimoment of the four-dimensional distribution $c_{1111}(t_1, t_2, t_3, t_4)$ depending on four variables t_1, t_2, t_3, t_4 is determined by the equation

$$\partial c_{1111}(t_1, t_2, t_3, t_4)/\partial t_4 = -3[m^2(t_4) + D(t_4)]c_{1111}(t_1, t_2, t_3, t_4)$$

$$-6K(t_1, t_4)K(t_2, t_4)K(t_3, t_4) + c_3(t_4)c_{111}(t_1, t_2, t_3)$$

$$-6m(t_4)[K(t_3, t_4)c_{111}(t_1, t_2, t_4) + K(t_2, t_4)c_{111}(t_1, t_3, t_4)$$

$$+K(t_1, t_4)c_{111}(t_2, t_3, t_4)] - 3[K(t_3, t_4)c_{112}(t_1, t_2, t_4)$$

$$+K(t_2, t_4)c_{112}(t_1, t_3, t_4) + K(t_1, t_4)c_{112}(t_2, t_3, t_4)] \qquad \text{(XII)}$$

with the initial condition $c_{1111}(t_1, t_2, t_3, t_4) = c_{112}(t_1, t_2, t_3)$.

7.5.5. Approximate Determination of Stationary Processes

For the approximate determination of the one-dimensional distribution of the strictly stationary process in stochastic stationary nonlinear differential system (5.4.6) by the methods based on the orthogonal expansions one should put in Eqs. (7.5.8), (7.5.9), (7.5.15)

$$\dot{m} = 0, \quad \dot{K} = 0, \quad \dot{c}_\kappa = 0. \tag{7.5.40}$$

If the equations obtained in such a way have the solution which may serve as the parameters of the corresponding truncated orthogonal expansion of the one-dimensional distribution, then we may suppose that the strictly stationary process exists in a system. In this case for determining other multi-dimensional distributions of this stationary process one should replace the derivatives with respect to t_n in Eqs. (7.5.24), (7.5.30) by the derivatives with respect to $\tau_{n-1} = t_n - t_1$ and write the initial conditions (7.5.33) in the form

$$c_{\kappa_1,\ldots,\kappa_n}(\tau_1, \ldots, \tau_{n-2}, \tau_{n-2}) = c_{\kappa_1,\ldots,\kappa_{n-1}+\kappa_n}(\tau_1, \ldots, \tau_{n-2}). \tag{7.5.41}$$

We summarize the results obtained in Subsections 7.5.1–7.5.5 as the following statements.

Theorem 7.5.1. *If the random process $Y(t)$ in normal stochastic differential system (5.4.6) has the finite coefficients of the consistent orthogonal expansions then its one- and multi-dimensional distributions may be approximated by expansions (7.5.1), (7.5.23) with $m(t)$, $K(t)$, $c_\nu(t)$, $K(t_1,t_2)$, $c_{\nu_1,\ldots,\nu_n}(t_1, \ldots, t_n)$ defined by Eqs. (7.5.8), (7.5.9), (7.5.15), (7.5.24), (7.5.30) at initial conditions (7.5.16), (7.5.33).*

Corollary 7.5.1. *If the stationary (in the strict sense) process in stationary normal stochastic differential system (5.4.6) exists and has the finite coefficients of the consistent orthogonal expansions then its one- and multi-dimensional distributions may be approximately expressed by (7.5.1), (7.5.23) with m, K, c_ν, $K(\tau)$, $c_{\nu_1,\ldots,\nu_n}(\tau_1, \ldots, \tau_{n-2}, \tau_{n-2})$ defined by Eqs. (7.5.40), (7.5.41).*

The equations of the orthogonal expansions method for Eqs. (5.4.6) at the nongaussian white noise and Eqs. (5.4.7) are given in Problems 7.34–7.35.

7.5.6. Reducing Number of Equations

The main difficulty of practical use of the stated methods especially for the multi-dimensional systems is a fast increase the number of the equations for the coefficients of the segments of the orthogonal expansions with the increase of the dimension p of the state vector y and the maximal order N of coefficients used. The dependence of the equations number for the moments cited in Tables 7.3.1 and 7.3.2 remains valid for the coefficients of orthogonal expansions, too.

While comparing the ciphers ν, $\left[\nu_1^T \ldots \nu_n^T\right]^T$ in Tables 7.3.1 and 7.3.2 we see that the number of the equations of the orthogonal expansion may be considerably reduced by means of the approximations of the distributions which include the mixed coefficients of the second order only. Such an approximation of the distribution may be obtained by different ways. For instance, we may assume that all mixed coefficients of the orthogonal expansions of higher than the second order are equal to zero. Another way consists of the use of the known expressions of the mixed central moments in terms of the variances and the covariances of the components of the random vector for the normal distributions (7.4.35). Both these approaches give the opportunity to use any approximations of the distributions and the correspondent equations for the coefficients of the orthogonal expansions on which the approximating function $f_1^*(y; \vartheta)$ or $f_n^*(y_1, \ldots, y_n; \vartheta_n)$ depends in Eqs. (5.4.6). Thus while using the quasimoments method it is natural to neglect all the terms of the expansions containing the coefficients c_ν, $c_{\nu_1, \ldots, \nu_n}$ with more than one component of the multi-indexes ν, $[\nu_1^T \ldots \nu_n^T]^T$ different from zero.

The approximation of a distribution (Subsection 7.4.4) may be obtained in various ways. For instance, one may assume that all mixed semiinvariants of orders higher than the second are equal to zero. Then one obtains the recursive formula (7.4.37) determining the mixed moments of the third, the fourth and higher orders as the function of the moments of the components of the random vector and the mixed second order moments (Kashkarova and Shin 1986). Another way is to use the known expressions for the mixed central moments in terms of the variances and the covariances of the components of the random vector for the normal distributions

$$\mu_{r_1, \ldots, r_s} = \frac{r_1! \ldots r_s!}{2^m m!} \sum k_{h_1 l_1} \ldots k_{h_m l_m} \text{ if } r_1 + \cdots + r_s = 2m,$$

$$\mu_{r_1, \ldots, r_s} = 0, \text{ if } r_1 + \cdots + r_s = 2m + 1 \ (m = 1, 2 \ldots). \qquad (7.5.42)$$

Both these approaches allow one to use any of the distributions of Sections 7.4 and 7.5 and to write the corresponding equations for moments, the semiinvariants and the coefficients of orthogonal expansions on which the approximating density depends. If the method of orthogonal expansions (in particular the method of quasimoments) is used, then it is natural to neglect all the terms of the expansions containing coefficients c_ν, $c_{\nu_1,\ldots,\nu_n}$ with more than one component of the multi-indexes ν, $\left[\nu_1^T \ldots \nu_n^T\right]^T$ different from zero.

E x a m p l e 7.5.4. For the system of Examples 7.4.1 and 7.4.3 the method of this subsection using formula (7.5.42) gives

$$\mu_{12}(t_1,t_2) = \mu_{21}(t_1,t_2) = 0\,, \quad \mu_{13}(t_1,t_2) = 3D(t_2)K(t_1,t_2)\,,$$
$$\mu_{22}(t_1,t_2) = D(t_1)D(t_2) + 2K^2(t_1,t_2)\,, \tag{I}$$
$$\mu_{31}(t_1,t_2) = 3D(t_1)K(t_1,t_2)\,,$$

$$\mu_{111}(t_1,t_2,t_3) = 0\,,$$
$$\mu_{112}(t_1,t_2,t_3) = D(t_3)K(t_1,t_2) + 2K(t_1,t_3)K(t_2,t_3)\,,$$
$$\mu_{121}(t_1,t_2,t_3) = D(t_2)K(t_1,t_3) + 2K(t_1,t_2)K(t_2,t_3)\,, \tag{II}$$
$$\mu_{211}(t_1,t_2,t_3) = D(t_1)K(t_2,t_3) + 2K(t_1,t_2)K(t_1,t_3)\,,$$

$$\mu_{1111}(t_1,t_2,t_3,t_4) = K(t_1,t_2)K(t_3,t_4)$$
$$+ K(t_1,t_3)K(t_2,t_4) + K(t_1,t_4)K(t_2,t_3)\,. \tag{III}$$

Substituting the obtained expressions of $\mu_{12}(t_1,t_2)$ and $\mu_{13}(t_1,t_2)$ into the equation of Example 7.4.3 for the covariance function we get

$$\partial K(t_1,t_2)/\partial t_2 = -3\left[m^2(t_2) + D(t_2)\right]K(t_1,t_2)\,, \quad t_2 > t_1\,. \tag{IV}$$

Integrating this approximate equation with the initial condition $K(t_1,t_2) = D(t_1)$ we find all the mixed fourth order moments using the above formulae. Then all the multi-dimensional distributions of the process $Y(t)$ will be determined in the same way as in Examples 7.4.1 and 7.4.3.

The same equation for the covariance function yields the method of quasimoments if in formulae (I) and (II) of Example 7.4.3 we put $c_\nu = 0$ if more than one component of the multi-index ν is different from zero.

E x a m p l e 7.5.5. For the system of Example 7.4.2 formulae (7.4.37) give

$$\mu_{12} = \mu_{21} = 0\,, \quad \mu_{13} = 3k_{12}k_{22}\,, \quad \mu_{22} = k_{11}k_{22} + 2k_{12}^2\,, \quad \mu_{31} = 3k_{11}k_{12}\,,$$
$$\tag{I}$$

and the equations of Example 7.4.2 for the moments of the one-dimensional distribution are reduced to:

$$\dot{m}_1 = -m_1 m_2 - k_{12}\,, \quad \dot{m}_2 = -a m_2\,, \tag{II}$$

$$\dot{k}_{11} = -2(m_2 k_{11} + m_1 k_{12})\,, \quad \dot{k}_{12} = -(m_2 + a)k_{12} - m_1 k_{22}\,,$$

$$\dot{k}_{22} = -2a k_{22} + h^2 \nu\,, \tag{III}$$

$$\dot{\mu}_{30} = -3(m_2 \mu_{30} + 2k_{11}k_{12})\,, \quad \dot{\mu}_{03} = -3a\mu_{03}\,,$$

$$\dot{\mu}_{40} = -4(3k_{12}\mu_{30} + m_2\mu_{40} + 3m_1 k_{11}k_{12})\,, \quad \dot{\mu}_{04} = -4a\mu_{04} + 6h^2\nu k_{22}. \tag{IV}$$

Equalizing to zero the coefficients $c_{12}, c_{21}, c_{13}, c_{22}, c_{31}$ in the method of quasimoments yields the above equations for $m_1, m_2, k_{11}, k_{12}, k_{22}$, but the equations for $\mu_{03}, \mu_{30}, \mu_{40}$ and μ_{04} are replaced by

$$\dot{c}_{30} = -3(2k_{11}k_{12} + m_2 c_{30})\,, \quad \dot{c}_{03} = -3a c_{03}\,,$$

$$\dot{c}_{40} = -4(3k_{12}c_{30} + m_2 c_{40})\,, \quad \dot{c}_{04} = -4a c_{04}\,. \tag{V}$$

As for the multi-dimensional distributions, all the approximations of this subsection yield the equations

$$\partial K_{11}(t_1, t_2)/\partial t_2 = -m_2(t_2)K_{11}(t_1, t_2) - m_1(t_2)K_{12}(t_1, t_2)\,,$$
$$\partial K_{12}(t_1, t_2)/\partial t_2 = -a K_{12}(t_1, t_2)\,,$$
$$\partial K_{21}(t_1, t_2)/\partial t_2 = -m_2(t_2)K_{21}(t_1, t_2) - m_1(t_2)K_{22}(t_1, t_2)\,,$$
$$\partial K_{22}(t_1, t_2)/\partial t_2 = -a K_{22}(t_1, t_2)\,, \quad t_2 > t_1\,, \tag{VI}$$

for the covariance function of the process $Y(t) = [Y_1(t_1)Y(t_2)]^T$. Integrating at first the equations for the moments or quasimoments of the one-dimensional distribution, and then the above equations for the covariance functions with the initial conditions: $K_{11}(t_1, t_1) = k_{11}(t_1)$, $K_{12}(t_1, t_1) = K_{21}(t_1, t_1) = k_{12}(t_1)$, $K_{22}(t_1, t_1) = k_{22}(t_1)$ we may find all the multi-dimensional distributions of the process $Y(t) = [Y(t_1)Y(t_2)]^T$ using the corresponding fragments of the Hermite polynomial expansions or of the Edgeworth series.

Examples 7.5.4 and 7.5.5 illustrate essential simplifications which give the approximations of distributions considered in this subsection. In the problem of Example 7.4.2 the number of equations for the complete set of moments of the first four orders amounts to 115 instead of 13 obtained in Example 7.5.5.

7.5.7. Methods based on Orthogonal Expansions for Discrete Nonlinear Systems

As applied to stochastic discrete nonlinear systems (5.7.1) we obtain the following equations of the orthogonal expansions method:

$$f_n^*(y_1, \ldots, y_n; \vartheta^{(n)}) = w(y_1, \ldots, y_n, m, K)$$

$$\times \left\{ 1 + \sum_{h=3}^{N} \sum_{|\nu_1| + \cdots + |\nu_n| = h} c_{\nu_1, \ldots, \nu_n}^{l_1, \ldots, l_n} p_{\nu_1, \ldots, \nu_n}(y_1, \ldots, y_n) \right\}, \quad (7.5.43)$$

$$c_\nu^{l+1} = E q_\nu(\omega_l(Y_l, V_l)) = \left[q_\nu \left(\frac{\partial}{i\partial\lambda} \right) E \exp\left\{ i\lambda^T \omega_l(Y_l, V_l) \right\} \right]_{\lambda=0},$$
$$(7.5.44)$$

$$c_{\nu_1, \ldots, \nu_n}^{l_1, \ldots, l_n+1} = E q_{\nu_1, \ldots, \nu_n}\left(Y_{l_1}, \ldots, Y_{l_{n-1}}, \omega_{l_n}(Y_{l_n}, V_{l_n}) \right). \quad (7.5.45)$$

Here

$$m = [m_{l_1}^T \ldots m_{l_n}^T]^T, \quad K = \begin{bmatrix} K_{l_1} & K_{l_1 l_2} & \ldots & K_{l_1 l_n} \\ K_{l_1 l_2}^T & K_{l_2} & \ldots & K_{l_2 l_n} \\ \cdots & \cdots & \cdots & \cdots \\ K_{l_1 l_n}^T & K_{l_2 l_n}^T & \ldots & K_{l_n} \end{bmatrix}, \quad (7.5.46)$$

$$c_{\nu_1, \ldots, \nu_n}^{l_1, \ldots, l_n}$$

$$= \left[q_{\nu_1, \ldots, \nu_n} \left(\frac{\partial}{i\partial\lambda_1} \cdots \frac{\partial}{i\partial\lambda_n} \right) g_{l_1, \ldots, l_n}(\lambda_1, \ldots, \lambda_n) \right]_{\lambda_1 = \cdots = \lambda_n = 0},$$
$$(7.5.47)$$

$\{p_{\nu_1, \ldots, \nu_n}, q_{\nu_1, \ldots, \nu_n}\}$ being the biorthogonal polynomials with

$$\int_{-\infty}^{\infty} \cdots \int_{-\infty}^{\infty} w(y_1, \ldots, y_n; m, K) p_{\nu_1, \ldots, \nu_n}(y_1, \ldots, y_n)$$

$$\times q_{\mu_1, \ldots, \mu_n}(y_1, \ldots, y_n) dy_1 \ldots dy_n = \delta_{\nu_1 \mu_1} \ldots \delta_{\nu_n \mu_n}, \quad (7.5.48)$$

where $\nu_l = [\nu_{l_1} \ldots \nu_{l_p}]^T$, $l = 1, 2, \ldots$, $\nu_{l_1}, \ldots, \nu_{l_n} = 0, 1, \ldots, N$; $|\nu_1|, \ldots, |\nu_n| = 1, \ldots, N - n + 1$, $|\nu_l| = |\nu_{l_1}| + \cdots + |\nu_{l_p}| = n$, $\ldots, N$.

Thus Eqs. (7.2.29), (7.2.30), (7.5.44), (7.2.31), (7.5.45) *at* $n = 2$ *and* Eq. (7.5.45) *at* $n > 2$ *determine recurrently all the multi-dimensional distributions* $\{Y_l\}$ *in stochastic discrete nonlinear system* (5.7.1).

For the continuous-discrete stochastic systems the corresponding equations are given in (Pugachev and Sinitsyn 2000)

7.5.9. Other Methods of Distributions Parametrization

The distributions parametrization methods stated above are based on the consistent biorthogonal expansions of the densities and lead to an infinite system of the nonlinear ordinary differential equations for the coefficients of the expansion. For obtaining the approximate solutions this system is substituted by a finite one. The estimating of the influence of the rejected terms in the general case encounters with difficulties. The Fourier series method (Moschuk 1994) is based only on the orthogonal expansions of the densities and is reduced to the solution of an infinite system of the linear ordinary differential equations for the coefficients of the expansion. Here "the linearity" of the reduced system allows to estimate the accuracy of the Fourier series method. The main advantage of a given method over the known ones is in the availability of the accuracy estimates and in the possibility to construct analytical expressions for the coefficients of the expansion of an one-dimensional density and transfer densities. The corresponding algorithms are simpler than in the moments methods, the semiinvariants and quasimoments methods. The method is easily generalized on the stochastic discrete and continuous-discrete nonlinear systems and on the stochastic composite systems as well.

The method of the orthogonal expansions may be successfully used in those cases when the additional restrictions over the structure of a distribution are known (Section 7.6).

7.6. Structural Parametrization of Distributions. Ellipsoidal Approximation Method

7.6.1. Introductory Remarks

For structural parametrization and/or approximation of the probability densities of the random vectors we shall apply the ellipsoidal densities, i.e. the densities whose planes of the levels of equal probability are similar concentric ellipsoids (the ellipses for two-dimensional vectors, the ellipsoids for three-dimensional vectors, the hyperellipsoids for the vectors of the dimension more than three). In particular, a normal distribution in any finite-dimensional space has an ellipsoidal structure. The distinctive characteristics of such distributions consists in the fact that their probability densities are the functions of positively determined quadratic form $u = u(y) = (y^T - m^T)C(y - m)$ where m is an expectation of the random vector Y, C is some positively determined matrix.

▷ Let us find ellipsoidal approximation (EA) for the density of r-dimensional random vector by means of the truncated expansion based on biorthogonal polynomials $\{p_{r,\nu}(u(y)), q_{r,\nu}(u(y))\}$, depending only on the quadratic form $u = u(y)$ for which some probability density of the ellipsoidal structure $w(u(y))$ serves as the weight:

$$\int_{-\infty}^{\infty} w(u(y))p_{r,\nu}(u(y))q_{r,\mu}(u(y))dy = \delta_{\nu\mu}. \qquad (7.6.1)$$

The indexes ν and μ at the polynomials mean their degrees relatively to the variable u. The concrete form and the properties of the polynomials are determined further. But without the loss of generality we may assume that $q_{r,0}(u) = p_{r,0}(u) = 1$. Then the probability density of the vector Y may be approximately presented by the expression of the form

$$f(y) \approx f^*(u) = w(u)\sum_{\nu=0}^{N} c_{r,\nu}p_{r,\nu}(u). \qquad (7.6.2)$$

For determining the coefficients $c_{r,\nu}$ in (7.6.2) we multiply term-wise equality (7.6.2) by $q_{r,\mu}(u)$ and integrate over the whole space R^r. Then by virtue of condition (7.6.1) for $\mu = 1, \ldots, N$ we obtain

$$\int_{-\infty}^{\infty} f(y)q_{r,\mu}(u)dy = \int_{-\infty}^{\infty} w(u)\sum_{\nu=0}^{N} c_{r,\nu}p_{r,\nu}(u)q_{r,\mu}(u)dy = c_{r,\mu}.$$

Thus the coefficients $c_{r,\nu}$ are determined by formula

$$c_{r,\nu} = \int_{-\infty}^{\infty} f(y)q_{r,\nu}(u)dy = Eq_{r,\nu}(U), \quad (\nu = 1, \ldots, N). \qquad (7.6.3)$$

As $p_{r,0}(u)$ and $q_{r,0}(u)$ are reciprocal constants (the polynomials of zero degree) then always $c_{r,0}p_{r,0} = 1$ and we come to the result:

$$f(y) \approx f^*(u) = w(u)\left[1 + \sum_{\nu=2}^{N} c_{r,\nu}p_{r,\nu}(u)\right]. \triangleleft \qquad (7.6.4)$$

Formula (7.6.4) expresses the essence of the ellipsoidal approxima-tion of the probability density of the random vector Y.

$\triangleright$ For constructing the system of the polynomials $\{p_{r,\nu}(u), q_{r,\mu}(u)\}$ satisfying the condition of the biorthonormality relatively to the quadratic form $u = (y - m)^T C(y - m)$ we transform condition (7.6.1). The change of variables $x = y - m = C^{-1/2}\zeta$ gives $u = \zeta^T\zeta = |\zeta|^2$ and

$$\int_{-\infty}^{\infty} w(u)p_{r,\nu}(u)q_{r,\mu}(u)dy = |C|^{-1/2}\int_{-\infty}^{\infty} w(|\zeta|^2)p_{r,\nu}(|\zeta|^2)q_{r,\mu}(|\zeta|^2)d\zeta.$$

$$(7.6.5)$$

We make the change of variables $\zeta_1 = \alpha_1\sqrt{u}, \ldots, \zeta_r = \alpha_r\sqrt{u}$. Hence we take into account that $\zeta_1^2 + \cdots + \zeta_r^2 = \zeta^T\zeta = u$ and consequently, $\alpha_1^2 + \cdots + \alpha_r^2 = 1$.

After the transformations (7.6.5) is reduced to the following form:

$$\int_{-\infty}^{\infty} p_{r,\nu}(u)q_{r,\mu}(u)w(u)dy$$

$$= \frac{|C|^{-1/2}\pi^{r/2}}{\Gamma(r/2)}\int_{0}^{\infty} u^{r/2-1}p_{r,\nu}(u)q_{r,\mu}(u)w(u)du. \qquad (7.6.6)$$

For the applications the case when the normal distribution is chosen as the distribution $w(u)$ is of great importance

$$w(u) = w(x^T Cx) = \frac{1}{\sqrt{(2\pi)^r|K|}}\exp(-x^T K^{-1}x/2); \qquad (7.6.7)$$

accounting that $C = K^{-1}$ we reduce the condition of the biorthonormality (7.6.6) to the form

$$\frac{1}{2^{r/2}\Gamma(r/2)}\int_{0}^{\infty} p_{r,\nu}(u)q_{r,\mu}(u)u^{r/2-1}e^{-u/2}du = \delta_{\nu\mu}. \quad \triangleleft \qquad (7.6.8)$$

Thus *the problem of the choosing of the polynomials system $\{p_{r,\nu}(u), q_{r,\mu}(u)\}$ which is used at the ellipsoidal approximation of the densities (7.6.4) and (7.6.3) is reduced to finding a biorthonormal system of the polynomials for which the χ^2-distribution with r degrees of the freedom serves as the weight.*

Taking into consideration that χ^2-distribution represents a special case of the gamma-distribution (Appendices 2, 7)

$$\omega(u) = \frac{k^{\alpha+1}}{\Gamma(\alpha+1)} u^{\alpha} e^{-ku}, \tag{7.6.9}$$

we consider a system of the polynomials $S_{\nu}^{\alpha}(u)$,

$$S_{\nu}^{\alpha}(u) = (-k)^{-\nu} u^{-\alpha} e^{ku} \frac{d^{\nu}}{du^{\nu}} \left(u^{\alpha+\nu} e^{-ku} \right), \quad \alpha > 1, u \geq 0. \tag{7.6.10}$$

At $\alpha = r/2 - 1$ and $k = 1/2$ we obtain a system of the polynomials which are orthogonal relatively to χ^2-distribution with r degrees of the freedom:

$$S_{r,\nu}(u) = S_{\nu}^{r/2-1}(u) = \sum_{\mu=0}^{\nu} (-1)^{\nu+\mu} C_{\nu}^{\mu} \frac{(r + 2\nu - 2)!!}{(r + 2\mu - 2)!!} u^{\mu}. \tag{7.6.11}$$

The main properties of the polynomials $S_{\nu}^{\alpha}(u)$ and $S_{r,\nu}(u)$ are given in Appendix 2. Between the polynomials $S_{r,\nu}(u)$ and the system of the polynomials $\{p_{r,\nu}(u), q_{r,\mu}(u)\}$ the following relations exist:

$$p_{r,\nu}(u) = S_{r,\nu}(u), \tag{7.6.12}$$

$$q_{r,\nu}(u) = \frac{(r - 2)!!}{(r + 2\nu - 2)!!(2\nu)!!} S_{r,\nu}(u), \quad r \geq 2. \tag{7.6.13}$$

Now we consider the expansion of the densities of nonnegative random variables with respect to the polynomials $S_{\nu}^{\alpha}(u)$ and $S_{r,\nu}(u)$. Let $L_2([0,\infty))$ be a H-space of the functions integrable with the modulus square. After test we make sure that *the functions*

$$\varphi_{\nu}(u) = \sqrt{\omega(u)} S_{\nu}^{\alpha}(u) = \sqrt{\frac{k^{\alpha+1}}{\Gamma(\alpha+1)}} u^{\alpha/2} e^{-u/4} S_{\nu}^{\alpha}(u) \tag{7.6.14}$$

form the orthogonal basis in $L_2([0,\infty))$. Therefore any function $F(u) \in L_2([0,\infty))$ may be presented as m.s. convergent (in the norm of the space) expansion

$$F(u) = \sqrt{\omega(u)} \sum_{\nu=0}^{\infty} c_{\nu} S_{\nu}^{\alpha}(u), \tag{7.6.15}$$

whose coefficients are determined by the relation

$$c_\nu = \int\limits_0^\infty F(u)S_\nu^\alpha(u)du \Big/ \frac{\nu!\Gamma(\alpha+\nu+1)}{k^{2\nu}\Gamma(\alpha+1)}. \qquad (7.6.16)$$

In particular any probability density $\varphi(u)$ of the nonnegative random variable satisfying the condition $\left\{\varphi(u)/\sqrt{\omega(u)}\right\} \in L_2([0,\infty))$,

$$\int\limits_0^\infty \frac{\varphi^2(u)}{\omega(u)}du = \int\limits_0^\infty \frac{\Gamma(\alpha+1)}{k^{\alpha+1}}u^\alpha e^{u/2}\varphi^2(u)du < \infty, \qquad (7.6.17)$$

may be presented by the expansion

$$\varphi(u) = \omega(u) \sum_{\nu=0}^\infty c_\nu S_\nu^\alpha(u). \qquad (7.6.18)$$

Now we consider the H-space $L_2(R^r)$ and the orthogonal system of the functions in them where the polynomials $S_{r,\nu}(u)$ are given by formula (7.6.11), and $w(u)$ is a normal distribution of the r-dimensional random vector (7.6.7). It is evident that this system is not complete in $L_2(R^r)$. But the expansion of the probability density $f(u) = f((y^T - m^T)C(y - m))$ of the random vector Y which has an ellipsoidal structure over the polynomials $p_{r,\nu}(u) = S_{r,\nu}(u)$,

$$f(u) = w(u) \sum_{\nu=0}^\infty c_{r,\nu} p_{r,\nu}(u), \qquad (7.6.19)$$

m.s. converges to the function $f(u)$ itself. The coefficients of the expansion in this case are determined by relation (7.6.16) which by virtue of (7.6.12) and (7.6.13) will take the form

$$c_{r,\nu} = \int\limits_{-\infty}^\infty f(u)p_{r,\nu}(u)dy \Big/ \frac{(2\nu)!!(r+2\nu-2)!!}{(r-2)!!}. \qquad (7.6.20)$$

Thus *the system of the functions $\{\sqrt{w(u)}S_{r,\nu}(u)\}$ forms the basis in the subspace of the space $L_2(R^r)$ generated by the functions $f(u)$ of the quadratic form $u = (y-m)^T C(y-m)$.*

R e m a r k. In practice the weak convergence of the probability measures generated by the segments of the density expansion to the probability measure generated by the density itself is more important than m.s. convergence of the segments of the density expansion over the polynomials $S_{r,\nu}(u)$ to the density namely,

$$\int\limits_{A} w(u) \sum_{\nu=0}^{N} c_{r,\nu} p_{r,\nu}(u) dy \to \int\limits_{A} f(u) dy \qquad (7.6.21)$$

uniformly relative to A at $N \to \infty$ on the σ-algebra of Borel sets of the space R^r. Thus the partial sums of series (7.6.19) give the approximation of the distribution, i.e. the probability of any event A determined by the density $f(u)$ with any degree of the accuracy. The finite segment of this expansion may be practically used for an approximate presentation of $f(u)$ with any degree of the accuracy even in those cases when $f(u)/\sqrt{w(u)}$ does not belong to $L_2(R^r)$. In this case it is sufficient to substitute $f(u)$ by the truncated density (Subsection 3.8.1). Expansion (7.6.19) is valid only for the densities which have the ellipsoidal structure. It is impossible in principal to approximate with any degree of the accuracy by means of the ellipsoidal approximation (7.6.4) the densities which arbitrarily depend on the vector y.

It is not difficult to make sure in the fact that *at the expansion over the polynomial $S_{r,\nu}(u)$ the probability densities of the random vector Y and all its possible projections are consistent.* In other words, at integrating the expansions over the polynomials $S_{h+l,\nu}(u)$, $h + l = r$, of the probability densities of the r-dimensional vector Y,

$$f(y) = \frac{1}{\sqrt{(2\pi)^{h+l}|K|}} e^{-u/2} \left[1 + \sum_{\nu=2}^{N} c_{h+l,\nu} S_{h+l,\nu}(u) \right],$$

$$u = (y - m)^T K^{-1}(y - m), \quad y = [y'^T \ y''^T]^T, \qquad (7.6.22)$$

on all the components of the l-dimensional vector y'' we obtain the expansion over the polynomials $S_{h,\nu}(u_1)$ of the probability density of the h-dimensional vector Y' with the same coefficients,

$$f(y') = \frac{1}{\sqrt{(2\pi)^h|K_{11}|}} e^{-u_1/2} \left[1 + \sum_{\nu=2}^{N} c_{h,\nu} S_{h,\nu}(u_1) \right],$$

$$u_1 = (y' - m')^T K_{11}^{-1}(y' - m'), \quad c_{h,\nu} = c_{h+l,\nu}, \qquad (7.6.23)$$

where K_{11} is a covariance matrix of the vector Y'.

But approximation (7.6.23) the probability density of h-dimensional random vector Y' obtained by the integration of expansion (7.6.22) the density of $(h + l)$-dimensional vector is not optimal ellipsoidal approximation of the density.

For the random r-dimensional vector with an arbitrary distribution the ellipsoidal approximation (7.6.4) of its distribution determines exactly the moments till the Nth order inclusively of the quadratic form $U = (Y - m)^T K^{-1}(Y - m)$, i.e.

$$EU^\mu = E^{\text{EA}}U^\mu, \quad \mu \le N. \tag{7.6.24}$$

(E^{EA} stands for expectation relative to ellipsoidally approximated distribution).

The ellipsoidal approximation of any distribution determines exactly the expectation of the polynomials $p_{r,0}(u), q_{r,0}(u), \ldots, p_{r,N}(u), q_{r,N}(u)$. In this case the initial moments of the order s, $s = s_1 + \cdots + s_r$ of the random vector Y at the approximation (7.6.4) determined by formula

$$\alpha_{s_1,\ldots,s_r} = \alpha_s = EY_1^{s_1} \ldots Y_r^{s_r}$$

$$\approx \int\limits_{-\infty}^{\infty} y_1^{s_1} \ldots y_r^{s_r} w(u) dy + \sum_{\nu=2}^{N} c_{r,\nu} \int\limits_{-\infty}^{\infty} y_1^{s_1} \ldots y_r^{s_r} p_{r,\nu}(u) w(u) dy. \tag{7.6.25}$$

Thus at the ellipsoidal approximation of the distribution of the random vector its moments are combined as the sums of the correspondent moments of the normal distribution and the expectations of the products of the polynomials $p_{r,\nu}(u)$ by the degrees of the components of the vector Y at the normal density $w(u)$.

7.6.2. Ellipsoidal Approximation of One-Dimensional Distribution

We use the ellipsoidal approximation method (EAM) for finding the one-dimensional probability density $f_1(y; t)$ of the p-dimensional random process $Y(t)$ which is determined by the Eq. (5.4.6) at the normally distributed white noise. Suppose that we know a distribution of the initial value $Y_0 = Y(t_0)$ of the random process $Y(t)$. Following the idea of EAM (Subsection 7.6.1) we present the one-dimensional density in the form of a segment of the orthogonal expansion (7.6.4) in terms of the polynomials dependent on the quadratic form $u = (y^T - m^T)C(y - m)$

where m and $K = C^{-1}$ are the expectation and the covariance matrix
of the random process $Y(t)$:

$$f_1(y;t) \cong f_1^{\mathrm{EAM}}(u) = w_1(u)\left[1 + \sum_{\nu=2}^{N} c_{p,\nu} p_{p,\nu}(u)\right]. \qquad (7.6.26)$$

Here $w_1(u)$ is the normal density of the p-dimensional random vector
which is chosen in correspondence with the requirement $c_{p,1} = 0$. The
optimal coefficients of the expansion $c_{p,\nu}$ are determined by relation
(7.6.3) which in our case will take the form

$$c_{p,\nu} = \int_{-\infty}^{\infty} f_1(y;t) q_{p,\nu}(u)\,dy = E q_{p,\nu}(U), \quad (\nu = 1,\dots,N). \qquad (7.6.27)$$

The set of the polynomials $\{p_{p,\nu}(u), q_{p,\nu}(u)\}$ is constructed on the
base of the orthogonal set of the polynomials $\{S_{p,\nu}(u)\}$ (Subsection 7.6.1)
according to the following rule which provides the biorthonormality of
system (7.6.1) at $p \geq 2$:

$$p_{p,\nu}(u) = S_{p,\nu}(u), \quad q_{p,\nu}(u) = \frac{(p-2)!!}{(p+2\nu-2)!!(2\nu)!!} S_{p,\nu}(u), \qquad (7.6.28)$$

where the polynomial $S_{p,\nu}(u)$ is given by formula (7.6.11). Thus the
solution of the problem of finding the one-dimensional probability density by EAM is reduced to finding the expectation m and the covariance
matrix K of the state vector of normal system (5.4.6) and the coefficients
of the correspondent expansion $c_{p,\nu}$ also.

$\triangleright$ For deriving the equations for the expectation $m(t) = EY(t)$ and the
covariance matrix $K(t) = E(Y(t) - m(t))(Y(t) - m(t))^T$ for normal system
we use Itô formula (5.1.15). As a result after the evaluation of the expectations
accounting (7.6.26) we get the equations

$$\dot{m} = \varphi_{10}(m, K, t) + \sum_{\nu=2}^{N} c_{p,\nu}\varphi_{1\nu}(m, K, t), \qquad (7.6.29)$$

$$\dot{K} = \varphi_{20}(m, K, t) + \sum_{\nu=2}^{N} c_{p,\nu}\varphi_{2\nu}(m, K, t), \qquad (7.6.30)$$

where the following indications are introduced

$$\varphi_{10}(m, K, t) = \int\limits_{-\infty}^{\infty} a(y, t) w_1(u) dy, \qquad (7.6.31)$$

$$\varphi_{1\nu}(m, K, t) = \int\limits_{-\infty}^{\infty} a(y, t) p_{p,\nu}(u) w_1(u) dy, \qquad (7.6.32)$$

$$\varphi_{20}(m, K, t) = \int\limits_{-\infty}^{\infty} [\, a(y, t)(y^T - m^T) + (y - m)a(y, t)^T + \sigma(y, t)\,] w_1(u) dy,$$

$$(7.6.33)$$

$$\varphi_{2\nu}(m, K, t) = \int\limits_{-\infty}^{\infty} [\, a(y, t)(y^T - m^T) + (y - m)a(y, t)^T$$

$$+ \sigma(y, t)\,] p_{p,\nu}(u) w_1(u) dy, \qquad (7.6.34)$$

where

$$\sigma(y, t) = b(y, t)\nu_0(t)b(y, t)^T. \quad \triangleleft \qquad (7.6.35)$$

Let us show that in the case of the nonsingular initial distribution
the differential equations for the coefficients of the expansion $c_{p,\kappa}$ determined in (7.6.27) by virtue of (5.4.6) satisfy the following equations:

$$\dot{c}_{p,\kappa} = - \left(\frac{c_{p,\kappa-1}}{2p} - \frac{\kappa c_{p,\kappa}}{p} \right)$$

$$\times \mathrm{tr} \left\{ K^{-1}\varphi_{20}(m, K, t) + P^{-1} \sum_{\nu=2}^{N} c_{p,\nu}\varphi_{2\nu}(m, K, t) \right\}$$

$$+ \psi_{\kappa 0}(m, K, t) + \sum_{\nu=2}^{N} c_{p,\nu}\psi_{\kappa\nu}(m, K, t), \quad \kappa = 2, \dots, N, \qquad (7.6.36)$$

where

$$\psi_{\kappa 0}(m, K, t) = \int\limits_{-\infty}^{\infty} [\, q'_{p,\kappa}(u)(2(y - m)^T K^{-1} a(y, t) + \mathrm{tr}\, K^{-1}\sigma(y, t))$$

$$+2q_\kappa''(u)(y-m)^T K^{-1}\sigma(y,t)(y-m), \tag{7.6.37}$$

$$\psi_{\kappa\nu}(m,K,t) = \int\limits_{-\infty}^{\infty}[\,q_{p,\kappa}'(u)[\,(2(y-m)^T K^{-1}a(y,t) + \mathrm{tr}\,K^{-1}\sigma(y,t)]$$

$$+2q_\kappa''(u)(y-m)^T K^{-1}\sigma(y,t)K^{-1}(y-m). \tag{7.6.38}$$

Eqs. (7.6.29), (7.6.30), (7.6.36) at the initial conditions

$$m(t_0) = m_0, \quad K(t_0) = K_0, \quad c_{p,\kappa}(t_0) = c_{p,\kappa}^0 \quad (\kappa = 2,\ldots,N) \tag{7.6.39}$$

determine m, K, $c_{p,2},\ldots,c_{p,N}$ as time functions. For finding the variables $c_{p,\kappa}^0$ the density of the initial value Y_0 of the state vector of a system should be approximated by formula (7.6.26).

$\triangleright$ In fact in Itô formula (5.1.15) for the nonlinear function $\varphi(Y(t),t)$ $= q_{p,k}(U)$ the items which enter in it sequentially have the following form. The first item is

$$\varphi_t(Y,t) = \frac{\partial q_{p,k}(U)}{\partial t} = q_{p,k}'(U)\frac{\partial U}{\partial t}, \tag{7.6.40}$$

where

$$\frac{\partial U}{\partial t} = -2(Y^T - m^T)C\dot{m} + (Y^T - m^T)\dot{C}(Y - m) \tag{7.6.41}$$

If we differentiate the relation $KC = I$ and find from it $\dot{C}$

$$\dot{K}C + K\dot{C} = 0, \quad \dot{C} = -K^{-1}\dot{K}C = -C\dot{K}C, \tag{7.6.42}$$

then (7.6.40) gives

$$\frac{\partial q_{p,\kappa}(U)}{\partial t} = q_{p,\kappa}'(U)(-2(Y-m)^T C\dot{m} - (Y-m)^T C\dot{K}C(Y-m)). \tag{7.6.43}$$

The expression $\varphi_y(Y,t)$ in the second item of the Itô formula with the account of the value of the derivative $\partial U/\partial Y = 2C(Y-m)$ will take the form

$$\varphi_y(Y,t) = 2q_{p,\kappa}'(U)C(Y-m). \tag{7.6.44}$$

Therefore the second item will be completely equal to

$$\varphi_y(Y,t)^T a = 2q_{p,\kappa}'(U)(Y-m)^T Ca(Y,t). \tag{7.6.45}$$

For calculating the third item we shall find at first the expression $\varphi_{yy}(Y,t)$:

$$\varphi_{yy}(Y,t) = \frac{\partial \varphi_y(Y,t)^T}{\partial Y} = \frac{2\partial q'_{p,\kappa}(U)(Y-m)^T C}{\partial Y}$$

$$= 2q''_{p,\kappa}(U)\frac{\partial U}{\partial Y}(Y-m)^T C + 2q'_{p,\kappa}(U)\frac{\partial (Y-m)^T}{\partial Y}C$$

$$= 4q''_{p,\kappa}(U)C(Y-m)(Y-m)^T C + 2q_{p,\kappa}(U)C. \tag{7.6.46}$$

The third item with the account of the indication of (7.6.35) will be written as

$$\frac{1}{2}\mathrm{tr}[\,\varphi_{yy}(Y,t)c\nu_0 c^T\,] = \frac{1}{2}\mathrm{tr}\{[\,4q''_{p,\kappa}(U)C(Y-m)(Y-m)^T C$$

$$+2q_{p,\kappa}(U)\,]\sigma(Y,t)\}. \tag{7.6.47}$$

Finally after denoting $b = b(y,t)$ we have

$$\varphi_y(Y,t)^T b\,dW = 2q'_{p,k}(U)(Y-m)^T Cb\,dW. \tag{7.6.48}$$

At the following stage we shall find the expectations of the obtained items. For the first item of (7.6.43) we have

$$Eq'_{p,\kappa}(U)(Y^T - m^T)C = \int\limits_{-\infty}^{\infty} q'_{p,\kappa}(u)(y^T - m^T)Cf_1^{\mathrm{EA}}(u)dy.$$

We evaluate the expectation in this formula. While introducing a new vector variable $\bar{y} = C^{1/2}(y-m)$ and noticing that $u = (y^T - m^T)C^{1/2}C^{1/2}(y-m) = |\bar{y}|^2$ we obtain

$$Eq'_{p,\kappa}(U)(Y^T - m^T)C = \int\limits_{-\infty}^{\infty} q'_{p,\kappa}(|\bar{y}|^2)\bar{Y}^T C^{1/2}f_1^{\mathrm{EA}}(|\bar{y}|^2)|C|^{1/2}d\bar{y} = 0,$$

$$\tag{7.6.49}$$

as the integrand is odd relatively to each of the coordinates of the vector $\bar{y}$ and the integral with respect to odd function in symmetrical limits is always equal to zero. For evaluating the expectation in the second item of (7.6.43) we use formula $\xi^T A\xi = \mathrm{tr}[\,A\xi\xi^T\,]$ which is valid for any vectors ξ and the quadratic matrices A. We obtain

$$(Y^T - m^T)C\dot{K}C(Y-m) = \mathrm{tr}[\,\dot{K}C(Y-m)(Y-m)^T C. \tag{7.6.50}$$

Thus it is necessary to find an expectation

$$Eq'_{p,\kappa}(U)C(Y-m)(Y-m)^T C = E_{w_1} q'_{p,\kappa}(U)C(Y-m)(Y-m)^T C$$

$$+\sum_{\nu=2}^{N} c_{p,\nu} E_{w_1} q'_{p,\kappa}(U)p_{p,\nu}(U)C(Y-m)(Y-m)^T C.$$

Here and further on E_{w_1} means an expectation at the density $w_1(u)$. It is evident that the finding of this expectation is reduced to the evaluation of the expectations of the form $E_{w_1}\{U^k C(Y-m)(Y-m)^T C\}$ as $q'_{p,\kappa}(u)$ and $p_{p,\nu}(u)$ are the polynomials of the variable u. After performing the change of the variables $\bar{y} = C^{1/2}(y-m) = C^{1/2}x$ in the integral which expresses E_{w_1} we may write

$$E_{w_1}\{U^k CXX^T C\} = E_{w_1}\{U^k C^{1/2}\bar{Y}\bar{Y}^T C^{1/2}\} = C^{1/2}E_{w_1}\{U^k \bar{Y}\bar{Y}^T\}C^{1/2}.$$

At first we calculate the diagonal elements of the matrix $E_{w_1} U^k \bar{Y}\bar{Y}^T$. As the distribution of the random vector $\bar{Y} = C^{1/2}X$ is spherically symmetric then $E_{w_1} U^k \bar{Y}_1^2 = \cdots = E_{w_1} U^k \bar{Y}_p^2$, i.e. all the diagonal elements of the matrix $E_{w_1} U^k \bar{Y}\bar{Y}^T$ are identical. Therefore we may put $E_{w_1}\{U^k \bar{Y}_i^2\} = \alpha$. After summarizing over i and taking into account that $U = \sum_{i=1}^{p} \bar{Y}_i^2 = \bar{Y}^T \bar{Y}$ we have $E_{w_1}\{U^k \bar{Y}^T \bar{Y}\} = E_{w_1} U^{k+1} = p\alpha$.

As far as the nondiagonal elements of this matrix concerned they turn out to be zero for the same reason that in formula (7.6.49). As a result we obtain $C^{1/2}E_{w_1} U^k \bar{Y}\bar{Y}^T C^{1/2} = (E_{w_1}U^{k+1}/p)C$. Consequently, $E_w U^k CXX^T C = (E_{w_1}U^{k+1}/p)C$. Thus

$$Eq'_{p,\kappa}(U)CXX^T C = \frac{1}{p}E_{w_1}\{q'_{p,\kappa}(U)UC\}$$

$$+\sum_{\nu=2}^{N} \frac{c_{p,\nu}}{p}E_{w_1}\{q'_{p,\kappa}(U)p_{p,\nu}(U)UC\}. \tag{7.6.51}$$

For calculating the expectations in (7.6.51) it is sufficient to know the probability the density $w(u)$ of the variable U correspondent to the normal probability density $w_1(u)$:

$$\omega(u) = \frac{1}{2^{p/2}\Gamma(p/2)}u^{p/2-1}e^{-u/2}. \tag{7.6.52}$$

Now we find

$$Eq'_{p,\kappa}(U)CXX^T C = \left\{ \frac{1}{2^{p/2}\Gamma(p/2)p} \int_0^\infty q'_{p,\kappa}(u)u^{p/2}e^{-u/2}du \right.$$

$$+ \sum_{\nu=2}^{N} c_{p,\nu} \frac{1}{2^{p/2}\Gamma(p/2)p} \int_0^\infty q'_{p,\kappa}(u)p_{p,\nu}(u)u^{p/2}e^{-u/2}du \Bigg\} C. \qquad (7.6.53)$$

For the sake of the convenience we introduce the denotions

$$\gamma_{\kappa 0} = \frac{1}{2^{p/2}\Gamma(p/2)p} \int_0^\infty u^{p/2}q'_{p,\kappa}(u)e^{-u/2}du, \qquad (7.6.54)$$

$$\gamma_{\kappa\nu} = \frac{1}{2^{p/2}\Gamma(p/2)p} \int_0^\infty u^{p/2}q'_{p,\kappa}(u)p_{p,\nu}(u)e^{-u/2}du. \qquad (7.6.55)$$

Notice that in the consequence of the orthogonality of $p_{p,\nu}(u)$ to all the functions u^λ at $\lambda > \nu$ the variable $\gamma_{\kappa\nu}$ vanishes at $\nu > \kappa$. Therefore

$$Eq'_{p,\kappa}(U)CXX^TC = \left[\gamma_{\kappa 0} + \sum_{\nu=2}^{\kappa} c_{p,\nu}\gamma_{\kappa\nu} \right] C.$$

We evaluate the variables $\gamma_{\kappa 0}$ and $\gamma_{\kappa\nu}$, determined by formulae (7.6.54) and (7.6.55). At first we find $\gamma_{\kappa\kappa}$. But virtue of the orthogonality of u^λ to the polynomial $p_{p,\kappa}(u)$ at $\lambda < \kappa$

$$\gamma_{\kappa\kappa} = \frac{1}{p2^{p/2}\Gamma(p/2)} \int_0^\infty \frac{(p+2\kappa-2)!!(2\kappa)!!}{(p-2)!!}\kappa u^{\kappa-1}p_{p,\kappa}(u)u^{p/2}e^{-u/2}du.$$

$$(7.6.56)$$

For the same reason formula for $\gamma_{\kappa\kappa}$ may be written in accordance with formula (7.6.3) and Appendix 7 in the form

$$\gamma_{\kappa\kappa} = \frac{\kappa}{p2^{p/2}\Gamma(p/2)} \cdot \frac{(p+2\kappa-2)!!(2\kappa)!!}{(p-2)!!} \int_0^\infty p^2_{p,\kappa}(u)u^{p/2-1}e^{-u/2}du = \frac{\kappa}{p}.$$

$$(7.6.57)$$

Now we evaluate $\gamma_{\kappa\nu}$ for arbitrary $\nu < \kappa$. While integrating by parts we obtain

$$\gamma_{\kappa\nu} = \frac{1}{p2^{p/2}\Gamma(p/2)} \int_0^\infty q'_{p,\kappa}(u)p_{p,\nu}(u)u^{p/2}e^{-u/2}du$$

$$= \frac{1}{p2^{p/2}\Gamma(p/2)} \left[q_{p,\kappa}(u)p_{p,\nu}(u)u^{p/2}e^{-u/2}\big|_0^\infty \right.$$

$$-\frac{p}{2}\int_0^\infty q_{p,\kappa}(u)p_{p,\nu}(u)u^{p/2-1}e^{-u/2}du - \int_0^\infty q_{p,\kappa}(u)p'_{p,\nu}(u)u^{p/2}e^{-u/2}du$$

$$\left. + \frac{1}{2}\int_0^\infty q_{p,\kappa}(u)p_{p,\nu}(u)u^{p/2}e^{-u/2}du \right]. \qquad (7.6.58)$$

It is evident that the first item is equal to zero. The first integral in (7.6.58) at $\nu < \kappa$ vanishes by virtue of the condition of the biorthonormality (7.6.8). The equality to zero of the second integral is shown by means of the same way as in (7.6.57). Analogously the third integral is equal to zero at $\nu + 1 < \kappa$ or correspondingly at $\nu < \kappa - 1$. At $\nu = \kappa - 1$ it is different from zero. We shall determine it using the same line of reasoning as at deriving (7.6.56) and (7.6.57):

$$\gamma_{\kappa,\kappa-1} = \frac{1}{p2^{(p/2)+1}\Gamma(p/2)}\int_0^\infty q_{p,\kappa}(u)p_{p,\kappa-1}(u)u^{p/2}e^{-u/2}du$$

$$= \frac{1}{p2^{(p/2)+1}\Gamma(p/2)}\int_0^\infty q_{p,\kappa}(u)p_{p,\kappa-1}(u)uu^{(p/2)-1}e^{-u/2}du$$

$$= \frac{1}{p2^{(p/2)+1}\Gamma(p/2)}\int_0^\infty q_{p,\kappa}(u)u^\kappa u^{(p/2)-1}e^{-u/2}du$$

$$= \frac{1}{p2^{(p/2)+1}\Gamma(p/2)}\int_0^\infty q_{p,\kappa}(u)q_{p,\kappa}(u)u^{p/2-1}e^{-u/2}du = \frac{1}{2p}.$$

Combining the results of the evaluation of (7.6.57) and (7.6.58) we write

$$\gamma_{\kappa\nu} = 0 \quad \nu < \kappa - 1, \quad \gamma_{\kappa,\kappa-1} = 1/2p, \quad \gamma_{\kappa\kappa} = \kappa/p. \qquad (7.6.59)$$

And finally, expression (7.6.40) transformed in the correspondence with (7.6.50) has the form

$$E\varphi_t(Y,t) = -Eq'_{p,\kappa}(U)\mathrm{tr}[\dot{K}C(Y-m)(Y-m)^T C]$$

$$= -\mathrm{tr}[\dot{K}EC(Y-m)(Y-m)^T Cq'_{p,\kappa}(U)]$$

$$= -\mathrm{tr}\left[\dot{K}\left(\frac{c_{p,\kappa-1}}{2p} + \frac{\kappa c_{p,\kappa}}{p}\right)C\right] = -\left(\frac{c_{p,\kappa-1}}{2p} + \frac{\kappa c_{p,\kappa}}{p}\right)\mathrm{tr}(\dot{K}C). \qquad (7.6.60)$$

The first and the second items of (7.6.48) have the form

$$E \operatorname{tr} \left\{ [\, 2q_\kappa''(U)C(Y-m)(Y-m)^T C + q_\kappa'(U)C\,]\sigma(Y,t)\right\}$$

$$= E\{2q_\kappa''(U)C(Y-m)^T C\sigma(Y,t)C(Y-m) + q_\kappa'(U)\operatorname{tr}C\sigma(Y,t)\}. \quad (7.6.61)$$

The expectations of the items in (7.6.48) vanish by virtue of the equality to zero of the expectations of Wiener process $W(t)$.

After performed transformations the expectation of the stochastic differential of the polynomial $q_{p,\kappa}(U)$ will be equal to

$$Edq_{p,\kappa}(U)$$

$$= -\left\{ \left(\frac{c_{p,\kappa-1}}{2p} + \frac{\kappa c_{p,\kappa}}{p} \right) \operatorname{tr}(C\dot{K}) + 2Eq_{p,\kappa}'(U)(Y-m)^T Ca(Y,t) \right.$$

$$\left. +2Eq_{p,\kappa}''(U)(Y-m)^T C\sigma(Y,t)C(Y-m) + Eq_{p,\kappa}'(U)\operatorname{tr}(C\sigma(Y,t)) \right\} dt.$$

$$(7.6.62)$$

From (7.6.62) by virtue of (7.6.27) the differential equation for the coefficient of the expansion $c_{p,\kappa}$ takes the form

$$\dot{c}_{p,\kappa} = -\left\{ \left(\frac{c_{p,\kappa-1}}{2p} + \frac{\kappa c_{p,\kappa}}{p} \right) \operatorname{tr}(C\dot{K}) + Eq_{p,\kappa}'(U)[\, 2(Y-m)^T Ca(Y,t) \right.$$

$$\left. +\operatorname{tr} C\sigma(Y,t)\,] + 2Eq_{p,\kappa}''(U)(Y-m)^T C\sigma(Y,t)C(Y-m). \quad (7.6.63)$$

After excluding $\dot{K}$ by means of (7.6.30) and changing C by K^{-1} we come to the sought Eqs (7.6.36). ◁

Thus Eqs. (7.6.29), (7.6.30), (7.6.36) *form the basis of the ellipsoidal approximation of the one-dimensional distribution in normal nonlinear stochastic system* (5.4.6).

R e m a r k 1. The ellipsoidal approximation method of the distributions is not applicable to the case of the singular initial distribution as in this case the matrix $K_0 = K(t_0)$ is irreversible and the matrix C does not exist. The properties of the consistency (7.6.23) allow to avoid this difficulty. Let the rank of the covariance matrix K_0 be equal to h, $h < p$. We partition the vector into two blocks, $X = [\, X'^T \ X''^T \,]^T$ where $X' = Y' - m'$ is the h-dimensional vector; $X'' = Y'' - m''$ is the $(p-h)$-dimensional vector, and X' consists of linearly independent components of the vector X, then $X'' = AX'$ for some matrix A. After using the ellipsoidal approximation of the distribution of the vector Y' we obtain the following approximate expression for the probability density of the initial value Y_0 of the vector Y:

$$f(y_0) \approx \frac{e^{-u_1/2}}{\sqrt{(2\pi)^n |K_{11}|}} \left[1 + \sum_{\nu=2}^{N} c_{h,\nu}^0 p_{h,\nu}(u_1) \right]$$

$$\times \delta(y'' - m_0'' - A(y' - m_0')), \quad u_1 = x'^T K_{11}^{-1} x', \qquad (7.6.64)$$

where K_{11} is the covariance matrix of the vector Y'. Supposing that the distribution has the same structure over a small time interval $[t_0, t_0 + \tau_1]$,

$$f_1(y; t) \approx \frac{e^{-u_1/2}}{\sqrt{(2\pi)^n |K_{11}|}} \left[1 + \sum_{\nu=2}^{N} c_{h,\nu} p_{h,\nu}(u_1) \right]$$

$$\times \delta(y'' - m_0 - A(y' - m_0')), \qquad (7.6.65)$$

the equations for the coefficients of the expansion $c_{h,\nu}$ only for the random vector Y' should be made up and they should be integrated till the order $t_0 + \tau_1$ when the matrix K ceases to be singular jointly with the equations for m, K at the approximation of the density (7.6.65). Then beginning with the instant $t_0 + \tau_1$ we may integrate Eqs. (7.6.29), (7.6.30 and (7.6.36) putting $c_{p,\nu}(t_0 + \tau_1) = c_{h,\nu}(t_1 + \tau_1)$ on the basis of (7.6.23). We may recommend to increase h at each step of the numerical integration of Eqs. (7.6.29), (7.6.30), (7.6.36) in accordance with the increasing the rank of the matrix K.

R e m a r k 2. At the nonrandom initial conditions $Y_0 = y_0$ the probability density of the initial state represents $\delta(y - y_0)$. In this case we integrate the equations for m, K at $f_1(y; t) = \delta(y - y_0)$ from the moment t_0 till the instant $t_0 + \tau_1$ when the matrix K ceases to be singular, and beginning with the moment $t_0 + \tau_1$ we pass to the integration of Eqs. (7.6.29), (7.6.30), (7.6.36) at the ellipsoidal approximation of the probability density of the vector Y. At increasing the rank h of the matrix K at each step of the numerical integration at the interval $[t_1, t_0 + \tau_1]$ we may also pass to the integration of the expansion composed for the h-dimensional vector Y' at the approximation (7.6.65).

7.6.3. Ellipsoidal Approximation of Multi-Dimensional Distributions

According to EAM we present approximately the n-dimensional probability density of the process $Y(t)$ in the form

$$f_n(y_1, \ldots, y_n; t_1, \ldots, t_n) \approx f_n^{\text{EAM}}(u)$$

$$= w_n(u) \left[1 + \sum_{\nu=2}^{N} c_{np,\nu} p_{np,\nu}(u) \right], \quad c_{np,\nu} = E q_{np,\nu}(U). \qquad (7.6.66)$$

Here the following denotions are used:

$$U = (\tilde{Y}_n - \tilde{m}_n)^T C_n (\tilde{Y}_n - \tilde{m}_n); \quad C_n = K_n^{-1};$$

$$\tilde{Y}_n = [\, Y_1^T \ Y_2^T \ \ldots \ Y_n^T \,]; \quad Y_k = Y(t_k); \tag{7.6.67}$$

$$\tilde{m}_n = [\, m_1^T \ m_2^T \ \ldots \ m_n^T \,]; \quad m_k = m(t_k);$$

$$K_n = \begin{bmatrix} K(t_1,t_1) & K(t_1,t_2) & \ldots & K(t_1,t_n) \\ K(t_2,t_1) & K(t_2,t_2) & \ldots & K(t_2,t_n) \\ \ldots & \ldots & \ldots & \ldots \\ K(t_n,t_1) & K(t_n,t_2) & \ldots & K(t_n,t_n) \end{bmatrix}, \tag{7.6.68}$$

where $K(t_i,t_j)$ is the covariance matrix of the process $Y(t)$;

$$C_n = \begin{bmatrix} C_{11}^{(n)} & C_{12}^{(n)} & \ldots & C_{1n}^{(n)} \\ C_{21}^{(n)} & C_{22}^{(n)} & \ldots & C_{2n}^{(n)} \\ \ldots & \ldots & \ldots & \ldots \\ C_{n1}^{(n)} & C_{n2}^{(n)} & \ldots & C_{nn}^{(n)} \end{bmatrix} \tag{7.6.69}$$

is the matrix which consists of the block $C_{kl}^{(n)}$ of the dimension $p \times p$; $\bar{C}_k = [\, C_{k1}^{(n)} \ \ldots \ C_{kn}^{(n)} \,]$ is its the kth block.

As $w_n(u)$ we take the normal probability density

$$w_n(u) = [(2\pi)^n |K_n|]^{-1/2} e^{-u/2}. \tag{7.6.70}$$

The polynomials $p_{np,\nu}(u)$ and $q_{np,\nu}(u)$ in the correspondence with (7.6.12) and (7.6.13) we determine by formulae

$$p_{np,\nu}(u) = S_{np,\nu}(u) = \sum_{\mu=0}^{\nu} (-1)^{\nu-\mu} C_\nu^\mu \frac{(np + 2\nu - 2)!!}{(np + 2\mu - 2)!!} u^\mu, \tag{7.6.71}$$

$$q_{np,\nu}(u) = \frac{(np - 2)!!}{(np + 2\nu - 2)!!(2\nu)!!} p_{np,\nu}(u). \tag{7.6.72}$$

The symbol n in the index of the coefficients of the expansion and the polynomials means the belonging to the n-dimensional distribution.

$\triangleright$ For composing the differential equations for the coefficients of the expansion $c_{np,\kappa}$ we find by Itô formula (5.1.15) the stochastic differential of the polynomial $q_{np,\kappa}(U)$ which is considered as the function of the random process $Y(t_n)$ at fixed $t_1, \ldots, t_{n-1}, 0 \le t_1 \le \ldots \le t_{n-1} \le t_n$ and take the expectation of the obtained result.

Beforehand we present the quadratic form in the following way:

$$U = U_n = (\tilde{Y}_n - \tilde{m}_n)^T C_n (\tilde{Y}_n - \tilde{m}_n) = \sum_{k,l=1}^{n} (Y_k - m_k)^T C_{kl}^{(n)} (Y_l - m_l)$$

$$= \sum_{k,l=1}^{n-1} (Y_k - m_k)^T C_{kl}^{(n)} (Y_l - m_l) + 2 \sum_{k=1}^{n-1} (Y_n - m_n)^T C_{nk}^{(n)} (Y_k - m_k)$$

$$+ (Y_n - m_n)^T C_{nn}^{(n)} (Y_n - m_n). \tag{7.6.73}$$

While finding the items in the Itô formula we use the results of Subsection 7.6.2. Similarly as in the previous case we evaluate sequentially each of the items. From (7.6.13) we have

$$\varphi_t(Y,t) = \frac{\partial q_{np,\kappa}(U)}{\partial t_n} = q'_{np,\kappa}(U) \frac{\partial U}{\partial t_n}$$

$$= q'_{np,\kappa}(U)[(\tilde{Y}_n - \tilde{m}_n)^T \frac{\partial C_n}{\partial t_n}(\tilde{Y}_n - \tilde{m}_n) - 2 m_n^T \bar{C}_n (\tilde{Y}_n - \tilde{m}_n). \tag{7.6.74}$$

Analogously as in Subsection 7.6.2

$$\varphi_y(Y,t) = \frac{\partial q_{np,\kappa}(U)}{\partial Y_n} = 2 q'_{np,\kappa}(U) \bar{C}_n (\tilde{Y}_n - \tilde{m}_n), \tag{7.6.75}$$

$$\varphi_{yy}(Y,t) = \frac{\partial^2 q_{np,\kappa}(U)}{\partial Y_n \partial Y_n^T}$$

$$= 4 q''_{np,\kappa}(U) \bar{C}_n (\tilde{Y}_n - \tilde{m}_n)(\tilde{Y}_n - \tilde{m}_n)^T \bar{C}_n^T + 2 q'_{np,\kappa}(U) C_{nn}^{(n)}, \tag{7.6.76}$$

$$\frac{1}{2} \text{tr}\{\varphi_{yy}(Y_n,t) c\nu c^T\} = \frac{1}{2} \text{tr}\left\{\frac{\partial^2 q_{np,\kappa}(U)}{\partial Y_n \partial Y_n^T} \sigma(Y_n, t_n)\right\}$$

$$= 2 q''_{np,\kappa}(U)(\tilde{Y}_n - \tilde{m}_n)^T \bar{C}_n^T \sigma(Y_n, t_n) \bar{C}_n (\tilde{Y}_n - \tilde{m}_n)$$

$$+ q'_{np,\kappa}(U) \text{tr}[C_{nn}^{(n)} \sigma(Y_n, t_n)]. \tag{7.6.77}$$

Finally the Itô differential of the polynomial $q_{np,\kappa}(U)$ will take the form

$$dq_{np,\kappa}(U) = \left\{ q'_{np,\kappa}(U) \left((\tilde{Y}_n - \tilde{m}_n)^T \frac{\partial C_n}{\partial t_n}(\tilde{Y}_n - \tilde{m}_n) - 2 m_n^T \bar{C}_n (\tilde{Y}_n - \tilde{m}_n) \right) \right.$$

$$+ 2 q'_{np,\kappa}(U)(\tilde{Y}_n - \tilde{m}_n) \bar{C}_n^T a(Y_n, t_n) + 2 q''_{np,\kappa}(U)(\tilde{Y}_n - \tilde{m}_n)^T \bar{C}_n^T$$

$$\times \sigma(Y_n, t_n)\bar{C}_n(\tilde{Y}_n - \tilde{m}_n) + q'_{np,\kappa}(U)\mathrm{tr}[\,C_{nn}^{(n)}\sigma(Y_n, t_n)\,]\Big\}dt_n, \qquad (7.6.78)$$

where

$$\bar{U} = \sum_{k,l=1}^{n-1} (Y_k - m_k)^T C_{kl}^{(n)}(Y_k - m_l)$$

$$+2\sum_{k=1}^{n-1}(Y_n - m_n)^T C_{nk}^{(n)}(Y_k - m_k) + (Y_n - m_n)^T C_{nn}^{(n)}(Y_n - m_n). \qquad (7.6.79)$$

We evaluate the expectation of this differential. The expectation of the latter item of (7.6.78) vanishes by virtue of the equality to zero of the expectations of Wiener process $W(t)$. We consider in detail the finding of the expectation of each term in the first item in (7.6.78).

For the first term we have the following expression

$$Eq'_{np,\kappa}(U)(\tilde{Y}_n - \tilde{m}_n)^T \frac{\partial C_n}{\partial t_n}(\tilde{Y}_n - \tilde{m}_n). \qquad (7.6.80)$$

After differentiating over t_n the relation $C_n K_n = I$ we find

$$\frac{\partial C_n}{\partial t_n} = -C_n \frac{\partial K_n}{\partial t_n} C_n. \qquad (7.6.81)$$

Substituting (7.6.81) into (7.6.80) we get

$$-Eq'_{np,\kappa}(U)(\tilde{Y}_n - \tilde{m}_n)^T C_n \frac{\partial K_n}{\partial t_n} C_n(\tilde{Y}_n - \tilde{m}_n)$$

$$= -Eq'_{np,\kappa}(U)\mathrm{tr}\left[\frac{\partial K_n}{\partial t_n}C_n(\tilde{Y}_n - \tilde{m}_n)(\tilde{Y}_n - \tilde{m}_n)^T C_n\right]. \qquad (7.6.82)$$

After extracting from (7.6.82) the expression $Eq'_{np,\kappa}(U)C_n(\tilde{Y}_n - \tilde{m}_n)(\tilde{Y}_n - \tilde{m}_n)^T C_n$ we come to

$$Eq'_{np,\kappa}(U)C_n(\tilde{Y}_n - \tilde{m}_n)(\tilde{Y}_n - \tilde{m}_n)^T C_n$$

$$= \left\{\frac{1}{np2^{np/2}\Gamma(np/2)}\int_0^\infty q'_{np,\kappa}(u)u^{np/2}e^{-u/2}du\right.$$

$$+\sum_{\nu=2}^N \frac{c_{np,\nu}}{np2^{np/2}\Gamma(np/2)}\int_0^\infty q'_{np,\kappa}(u)p_{np,\nu}(u)u^{np/2}e^{-u/2}du\left.\right\}C_n. \qquad (7.6.83)$$

The expression in the braces may be evaluated. In the correspondence with (7.6.59) it turns to be equal to $c_{np,\kappa-1}/2np + \kappa c_{np,\kappa}/np$. And finally, we have

$$Eq'_{np,\kappa}(U)(\tilde{Y}_n - \tilde{m}_n)^T \frac{\partial C_n}{\partial t_n}(\tilde{Y}_n - \tilde{m}_n)$$

$$= -\left(\frac{c_{np,\kappa-1}}{2np} + \frac{\kappa c_{np,\kappa}}{np}\right) \mathrm{tr}\left[\frac{\partial K_n}{\partial t_n}C_n\right]. \tag{7.6.84}$$

The equality to zero of the expectation $-2m_n^T \bar{C}_n(\tilde{Y}_n - \tilde{m}_n)$ is proved in the same way as it was made while obtaining formulae (7.6.49). After performing the transformations the equation for the coefficients of the expansion $c_{np,\kappa}$ will take the form

$$\frac{\partial c_{np,\kappa}}{\partial t_n} = -\left(\frac{c_{np,\kappa-1}}{2np} + \frac{\kappa c_{np,\kappa}}{np}\right) \mathrm{tr}\left[\frac{\partial K_n}{\partial t_n}C_n\right]$$

$$+ \int\limits_{-\infty}^{\infty} \cdots \int\limits_{-\infty}^{\infty} \{2q'_{np,\kappa}(u)(\tilde{y}_n - \tilde{m}_n)^T \bar{C}_n^T a(y_n, t_n)$$

$$+2q''_{np,\kappa}(u)(\tilde{y}_n - \tilde{m}_n)^T \bar{C}_n^T \sigma(y_n, t_n)\bar{C}_n(\tilde{y}_n - \tilde{m}_n)$$

$$+q'_{np,\kappa}(u)\mathrm{tr}[C_{nn}^{(n)}\sigma(y_n, t_n)]\}. \tag{7.6.85}$$

Eqs. (7.6.85) is an ordinary differential equation but not an equation in partial derivatives as the variables $t_1, \ldots, t_{n-1}$, $t_1 \le t_2 \le \cdots \le t_{n-1}$ at the function $c_{np,\kappa}$ are fixed.

After putting in (7.6.23) $h + l = np$, $l = p$, $h = (n-1)p$ and determining the polynomials $p_{np,\nu}(u)$ and $q_{np,\nu}(u)$ by formulae (7.6.71) and (7.6.72) we get

$$\int\limits_{-\infty}^{\infty} w_n(u)\left[1 + \sum_{\nu=2}^{N} c_{np,\nu}(t_1, \ldots, t_n)p_{np,\nu}(u)\right] dy_n$$

$$= w_{n-1}(u)\left[1 + \sum_{\nu=2}^{N} c_{np,\nu}(t_1, \ldots, t_n)p_{(n-1)p,\nu}(u)\right]. \tag{7.6.86}$$

Hence it is clear that it is impossible to speak about the complete consistency of n- and $(n-1)$-dimensional distributions in the case of the ellipsoidal approximation because of the dependence of the coefficients in the expansion of $(n-1)$-dimensional density on t_n. Actually such dependence is missing. Therefore we may achieve the consistency only at some fixed t_n. It is natural to attain the consistency at $t_n = t_{n-1}$. But putting t_n equal to t_{n-1} we come to the singularity of the matrix

K_n and to the indefiniteness of the matrix C_n. It makes impossible the integration of the equations for the expectation, the covariance matrix and the coefficients of the expansion. Therefore we shall require the consistency of n- and $(n-1)$-dimensional distributions at $t_n = t_{n-1} + \tau_n$ where τ_n is a small quantity which is sufficient for providing the existence of the matrix C_n. Then the initial conditions for Eqs. (7.6.85) which determined the coefficients of the expansion may be written in the form

$$c_{np,\nu}(t_1, t_2, \ldots, t_{n-1}, t_{n-1} + \tau_n) = c_{(n-1)p,\nu}(t_1, \ldots, t_{n-1}). \qquad (7.6.87)$$

In the case of the two-dimensional distribution, $n = 2$, the equation for the covariance function of the process $Y(t)$ should be added to the equations for the coefficients of expansion (7.6.85). After evaluating the expectation in the known equality

$$\frac{\partial K(t_1, t_2)}{\partial t_2} = E(Y(t_1) - m(t_1))a(Y(t_2), t_2)^T, \qquad (7.6.88)$$

and taking

$$w_2(u) = \left[(2\pi)^2 |K_n|\right]^{-1/2} \exp\{-(\tilde{y}_n - \tilde{m}_n)^T K_n^{-1}(\tilde{y}_n - \tilde{m}_n)\}, \qquad (7.6.89)$$

$$p_{2p,\nu}(u) = S_{2p,\nu}(u) = \sum_{\mu=0}^{\nu}(-1)^{\nu-\mu}C_\nu^\mu \frac{(2p + 2\nu - 2)!!}{(2p + 2\mu - 2)!!}u^\mu \qquad (7.6.90)$$

at the ellipsoidal approximation of the two-dimensional density

$$f_2^{\mathrm{EAM}}(u) = w_2(u)\left[1 + \sum_{\nu=2}^{N} c_{2p,\nu}(t_1, t_2)p_{2p,\nu}(u)\right], \qquad (7.6.91)$$

we obtain

$$\frac{\partial K(t_1, t_2)}{\partial t_2} = \int_{-\infty}^{\infty} \int_{-\infty}^{\infty} (y_1 - m_1)a(y_2, t_2)^T f_2^{\mathrm{EAM}}(u)dy_1 dy_2. \qquad (7.6.92)$$

The initial condition should be added to this equation,

$$K(t_1, t_1) = K(t_1). \qquad (7.6.93)$$

Because of the singularity of the two-dimensional distribution at $t_2 = t_1$ the equation for the covariance function till the instant $t_1 + \tau_2$ is integrated at the singular distribution

$$f_2^{\mathrm{EAM}}(u) \approx w_1(u)\left[1 + \sum_{\nu=2}^{N} c_{1p,\nu}(t_1)p_{1p,\nu}(u)\right]\delta(y_1 - y_2), \qquad (7.6.94)$$

and beginning with $t_1 + \tau_2$ it is integrated jointly with Eqs. (7.6.85). The variable τ_2 is determined while integrating when the matrix K_2 becomes nonsingular. ◁

Thus *in the basis of the ellipsoidal approximation of the multi-dimensional densities in normal nonlinear stochastic system (5.4.6) lay* Eqs. (7.6.29), (7.6.30), (7.6.36), (7.6.85), (7.6.92) *with the correspondent initial conditions.*

For the approximate evaluation of the one-dimensional distribution of the stationary process (in the stricted sense) in the stationary nonlinear stochastic system (5.4.6) by the methods based on the ellipsoidal approximation of the distributions it is worth to put in Eqs. (7.6.29), (7.6.30), (7.6.36)

$$\dot{m} = 0, \quad \dot{K} = 0, \quad \dot{c}_{p,\kappa} = 0, \quad \kappa = 1, \ldots, N. \tag{7.6.95}$$

If the equations obtained in such a way have the solution which may serve as the vector of the parameters of the correspondent ellipsoidal approximation of the one-dimensional distribution then we may suppose that the stationary process exists in the system. In this case for determining other multi-dimensional distributions of this stationary process we must substitute in Eqs. (7.6.85) and (7.6.92) the derivatives over t_n by the derivatives over $\tau_{n-1} = t_n - t_1$ and the initial conditions (7.6.87) take in the form

$$c_{np,\kappa}(\tau_1, \ldots, \tau_{n-2}, \tau_{n-2} + \Delta) = c_{(n-1)p,\kappa}(\tau_1, \ldots, \tau_{n-2}). \tag{7.6.96}$$

where Δ is a small quantity which provides the nonsingularity of the matrix C in the equations for the parameters of the distribution.

7.6.4. Moments of State Vector

Taking into account approximate presentation (7.6.26) of one-dimensional density $f_1(y; t)$ and formula (7.6.25) for the moments of the random vector at the ellipsoidal approximation of its distribution we obtain for the initial moments of the vector Y the following approximate formula:

$$\alpha_{r_1, \ldots, r_p} = \alpha_r = EY_1^{r_1} \ldots Y_p^{r_p} = \int\limits_{-\infty}^{\infty} y_1^{r_1} \ldots y_p^{r_p} f_1(y; t) dy$$

$$\approx \int\limits_{-\infty}^{\infty} y_1^{r_1} \ldots y_p^{r_p} w_1(u) dy + \sum_{\nu=2}^{N} c_{p,\nu} \int\limits_{-\infty}^{\infty} y_1^{r_1} \ldots y_p^{r_p} p_{p,\nu}(u) w_1(u) dy,$$

$$\tag{7.6.97}$$

or

$$EY_1^{r_1} \ldots Y_p^{r_p} = E_{w_1} Y_1^{r_1} \ldots Y_p^{r_p}$$

$$+ \sum_{\nu=2}^{N} c_{p,\nu} E_{w_1} p_{p,\nu}(U) Y_1^{r_1} \ldots Y_p^{r_p}, \quad r = [r_1 \ldots r_p]. \qquad (7.6.98)$$

7.6.5. EAM Accuracy

The accuracy of approximation a function by expansion (7.6.4) may be performed by different ways but the most natural one is the way of the estimate of the accuracy of the distribution approximation in the comparison of the probability characteristics calculated by means of the known density and its approximate expression. The most complete estimate of the accuracy of the approximation may be obtained by the comparison of the probability occurence on the sets of some given class. Besides that taking into consideration that the probability density is usually approximated by a finite segment of its orthogonal expansion for instance, over Hermite polynomials or by a finite segment of the Edgeworth series (Section 3.8) which contain the moments till the fourth order the accuracy may be characterized by the accuracy of the definition of the moments of the random vector or its separate components, in particular, of the fourth order moments.

At first we compare the accuracy of EAM on the probabilities of the occurence on the sets. Let $\mathcal{A}$ be class of the sets A for which it is required to estimate the accuracy of the definition of the occurence probability. The error of the definition of the occurence probability on Borel set A at the approximation of the density of the r-dimensional random vector (7.6.4) is equal to

$$\int_A f^*(y)dy - \int_A f(y)dy. \qquad (7.6.99)$$

Therefore as the measure of the accuracy of the approximation we assume the variable

$$\varepsilon = \sup_{A \subset \mathcal{A}} \left| \int_A f^*(y)dy - \int_A f(y)dy \right|. \qquad (7.6.100)$$

Practically we may evaluate the values ε only in the case when the class of the sets $\mathcal{A}$ contains a small number of the sets. So, for example, if as a class of the sets $\mathcal{A}$ the σ-algebra of Borel sets is chosen then the finding of ε is practically impossible. Therefore it is expedient to assume the finite systems of the typical regions as a class of the sets $\mathcal{A}$. It is natural to choose as such regions the ellipsoids in the space R^r, the ellipsoidal cylinders with the bases in the subspace of the space R^r, the layers formed by the hyperplanes of the dimension $r-1$ for the EAM distributions. These layers may be considered as the ellipsoidal cylinders with the one-dimensional bases. For calculating the probabilities of the occurences into the ellipsoidal cylinders with the bases in the subspaces R^h of the space R^r we have to use the ellipsoidal approximation of the distribution of the correspondent h components of the vector Y.

Let E_0 be the ellipsoid $(y-m)^T C(y-m) < u_0$, $C = K^{-1}$. The probability of the occurence in the ellipsoid E_0 at the ellipsoidal approximation (7.6.4) is equal to

$$p_0 = P(E_0) = \int_{E_0} f^{\text{EAM}}(y)dy = \int_{E_0} w(u)\left[1 + \sum_{\nu=2}^{N} c_{r,\nu}p_{r,\nu}(u)\right]dy$$

$$= \frac{|K|^{1/2}\pi^{1/2}}{\Gamma(r/2)}\int_0^{u_0} u^{r/2-1}w(u)\left[1 + \sum_{\nu=2}^{N} c_{r,\nu}p_{r,\nu}(u)\right]du$$

$$= \int_0^{u_0} \frac{u^{r/2-1}e^{-u/2}}{\Gamma(r/2)2^{r/2}}du + \sum_{\nu=2}^{N} c_{r,\nu}\int_0^{u_0}\frac{u^{r/2-1}e^{-u/2}p_{r,\nu}(u)}{\Gamma(r/2)2^{r/2}}du$$

$$= \frac{1}{\Gamma(r/2)}\left[\gamma(r/2, u_0/2) - \frac{u_0^{r/2}e^{-u_0/2}}{2^{r/2-1}}\sum_{\nu=2}^{N} c_{r,\nu}p_{r+1,\nu-1}(u_0)\right], \quad (7.6.101)$$

where $\gamma(x,y)$ is not complete γ function (Appendices 2, 7).

For illustrating the ellipsoidal approximation of the distributions, in particular, the approximations of the densities of one components or joint densities of two or three components we may use the ellipsoids in the correspondent subspaces R^r whose probabilities of the occurence are given.

For finding the ellipsoid E_0 whose probability of the occurence $A = P(E_0)$ is given it is necessary to solve Eq. (7.6.101) relatively to u_0. The root of this equation we shall find by the Newton method known from numerical analysis which gives the following recurrent formula for the sequential approximations for the equation $A(E_0) = \varphi(u_0)\, u_0$ the root u_0:

$$u_0^{n+1} = u_0^n + \varphi'(u_0^n)^{-1}[A(E_0) - \varphi(u_0^n)]$$

$$= u_0^n + 2^{r/2} \left(\frac{\Gamma(r/2)A(E_0) - \gamma(r/2, u_0^n/2)}{(u_0^n)^{r/2-1} e^{-u_0^n/2} \left[1 + \sum\limits_{\nu=2}^{N} c_{r,\nu} p_{r,\nu}(u_0^n) \right]} \right)$$

$$+ 2u_0^n \frac{\sum\limits_{\nu=2}^{N} c_{r,p} p_{r+1,\nu-1}(u_0^n)}{1 + \sum\limits_{\nu=2}^{N} c_{r,\nu} p_{r,\nu}(u_0^n)}. \tag{7.6.102}$$

It is expedient to choose the value found from the equation $\gamma(r/2, u_0^0/2) = A(E_0)\Gamma(r/2)$ as zero approximation of the variable u_0.

Now we consider the second way of EAM estimation of the accuracy. It consists in the comparison of the values of the moments calculated by means of the known distribution and its ellipsoidal approximation. So, for example, the error of the definition of the initial moment of the order d at the ellipsoidal approximation (7.6.4) of the probability density of the r-dimensional random vector will be equal to

$$\delta_1(d_1, \ldots, d_r) = \int\limits_{-\infty}^{\infty} y_1^{d_1} \ldots y_r^{d_r} f^{\text{EAM}}(y) dy - \int\limits_{-\infty}^{\infty} y_1^{d_1} \ldots y_r^{d_r} f(y) dy$$

$$= E_{\text{EAM}} Y_1^{d_1} \ldots Y_r^{d_r} - E Y_1^{d_1} \ldots Y_r^{d_r}, \quad d_1 + \cdots + d_r = d, \tag{7.6.103}$$

$$\delta_2(d_1, \ldots, d_r) = \frac{\delta_1(d_1, \ldots, d_r)}{E Y_1^{d_1} \ldots Y_r^{d_r}}. \tag{7.6.104}$$

The variable $\delta_1(d_1, \ldots, d_r)$ represents an absolute error of the definition of the initial moment $E Y_1^{d_1} \ldots Y_r^{d_r}$; $\delta_2(d_1, \ldots, d_r)$ represents a relative error.

Let the density $f(y)$ of the r-dimensional random vector with an arbitrary covariance matrix and the moments which exist till the order d inclusively be given. By means of the orthogonal transformation Ω we reduce the covariance matrix $K = [K_{ij}]$ to the diagonal form $\tilde{K} = \text{diag}(\lambda_1, \ldots, \lambda_r) = \Omega K \Omega^T$. Then $C = K^{-1} = [C_{ij}]$ will turn into the diagonal matrix $\tilde{C} = \tilde{K}^{-1} = \text{diag}(1/\lambda_1, \ldots, 1/\lambda_r)$. Hence the quadratic form u will take the form

$$u = x^T K^{-1} x = \tilde{x}^T \tilde{K}^{-1} \tilde{x} = \sum_{i=1}^{r} \tilde{x}_i^2/\lambda_i, \quad x = y - m, \quad \tilde{x} = \Omega^T x,$$

$$\tag{7.6.105}$$

where m is an expectation of the vector Y. The probability density of the vector $\tilde{X}$ is expressed in terms of the density $f(y)$ of the vector Y by the formula

$$\tilde{f}(\tilde{x}) = \frac{1}{|\Omega^T|} f(\Omega^{-1T}(\tilde{x} + \Omega^T m)). \qquad (7.6.106)$$

The estimation of the accuracy of the approximation we shall perform by the comparison of the values of the moments of the vector $\tilde{X}$. The detailed derivation and the calculations we shall perform for the moments of the fourth order ($d = 4$). Nevertheless the method which is described further is also applicable for the calculation of the moments of any order. While calculating the moments of an arbitrary component for instance, the first one of the r-dimensional vector $\tilde{X}$ by formula (7.6.25) and substituting density (7.6.106) by its ellipsoidal approximation (7.6.4) we have

$$\int\limits_{-\infty}^{\infty} \cdots \int\limits_{-\infty}^{\infty} \tilde{x}_1^4 \tilde{f}(\tilde{x}) d\tilde{x}_1 \ldots d\tilde{x}_r \cong \int\limits_{-\infty}^{\infty} \cdots \int\limits_{-\infty}^{\infty} \tilde{x}_1^4 w(u) d\tilde{x}_1 \ldots d\tilde{x}_r$$

$$+ \sum_{\nu=2}^{N} c_{r,\nu} \int\limits_{-\infty}^{\infty} \cdots \int\limits_{-\infty}^{\infty} \tilde{x}_1^4 p_{r,\nu}(u) w(u) d\tilde{x}_1 \ldots d\tilde{x}_r, \qquad (7.6.107)$$

where $w(u)$ is the r-dimensional normal density. We integrate (7.6.107) over $\tilde{x}_2, \ldots, \tilde{x}_r$ using (7.6.23). As a result we obtain

$$\int\limits_{-\infty}^{\infty} \cdots \int\limits_{-\infty}^{\infty} \tilde{x}_1^4 \tilde{f}(\tilde{x}) d\tilde{x}_1 \ldots d\tilde{x}_r \cong \int\limits_{-\infty}^{\infty} \tilde{x}_1^4 w(u_1) d\tilde{x}_1$$

$$+ \sum_{\nu=2}^{N} c_{r,\nu} \int\limits_{-\infty}^{\infty} \tilde{x}_1^4 p_{1,\nu}(u_1) w(u_1) d\tilde{x}_1, \qquad (7.6.108)$$

where $u_1 = \tilde{x}_1^2/\lambda_1$, $w(u_1)$ is one-dimensional density. We express $\tilde{x}_1$ in terms of u_1 in the form $\tilde{x}_1^4 = \lambda_1^2 u_1^2$. Then the second item in (7.6.108) will be

$$\sum_{\nu=2}^{N} c_{r,\nu} \int\limits_{-\infty}^{\infty} \lambda_1^2 u_1^2 p_{1,\nu}(u_1) w(u_1) d\tilde{x}_1. \qquad (7.6.109)$$

For evaluating this integral we use property (7.6.1) of the orthogonality of the polynomial $p_{r,\nu}(u)$ to all degrees of u^λ at $\lambda < \nu$. According to (7.6.1), u_1^2 will be orthogonal to all polynomials $p_{1,\nu}(u_1)$ beginning with ν which is equal to 3. As a result the sum in (7.6.109) will contain only one item correspondent to ν and equal to 2. After calculating the fourth moment of the normal distribution Eq. (7.6.107) will be written as

$$\int\limits_{-\infty}^{\infty} \cdots \int\limits_{-\infty}^{\infty} \tilde{x}_1^4 \tilde{f}(\tilde{x})d\tilde{x}_1 \ldots d\tilde{x}_r \cong 3D_1^2 + c_{r,2}\int\limits_{-\infty}^{\infty} \tilde{x}_1^4 p_{1,2}(u_1)w(u_1)d\tilde{x}_1.$$

$$(7.6.110)$$

We shall determine the coefficients $c_{r,\nu}$ from relation (7.6.3). For further evaluations it is convenient to rewrite expression (7.6.13) for the polynomial $q_{r,\nu}(u)$ in other form:

$$q_{r,\nu}(u) = \frac{(r-2)!!}{(r+2\nu-2)!!(2\nu)!!}S_{r,\nu}(u)$$

$$= \frac{(r-2)!!}{(r+2\nu-2)!!2^\nu\nu!}\sum_{\mu=0}^{\nu}\frac{(-1)^{\nu+\mu}\nu!(r+2\nu-2)!!}{\mu!(\nu-\mu)!(r+2\mu-2)!!}u^\mu$$

$$= \frac{1}{2^\nu}\sum_{\mu=0}^{\nu}\frac{(-1)^{\nu+\mu}u^\mu}{\mu!(\nu-\mu)!r(r+2)\cdots(r+2\mu-2)}. \qquad (7.6.111)$$

It is adopted in the latter formula that the product $r(r+2)\cdots(r+2\mu-2)$ obtained at the division of $(r-2)!!$ by $(r+2\mu-2)!!$ is equal to 1 at $\mu = 0$.

Now we shall find the μth degrees of the quadratic form u written in the form of (7.6.105):

$$u^\mu = \sum_{\mu_1+\cdots+\mu_r=\mu}\frac{\mu!}{\mu_1!\cdots\mu_r!}\frac{\tilde{x}_1^{2\mu_1}\ldots\tilde{x}_r^{2\mu_r}}{\lambda_1^{\mu_1}\ldots\lambda_r^{\mu_r}}. \qquad (7.6.112)$$

Here we used the formula for raising the polynomial to entire positive degree. We determine the expectation of expression (7.6.112):

$$EU^\mu = \sum_{\mu_1+\cdots+\mu_r=\mu}\frac{\mu!}{\mu_1!\cdots\mu_r!}\frac{E\tilde{X}_1^{2\mu_1}\ldots\tilde{X}_r^{2\mu_r}}{\lambda_1^{\mu_1}\ldots\lambda_r^{\mu_r}}.$$

Then the expectations of the degrees U^μ and the whole polynomial $q_{r,\nu}(U)$ in new denotions will be correspondingly equal to

$$EU^\mu = \sum_{\mu_1+\cdots+\mu_r=\mu}\mu!\gamma_{\mu_1\ldots\mu_r}, \qquad (7.6.113)$$

$$c_{r,\nu} = Eq_{r,\nu}(U) = \frac{1}{2^\nu} \sum_{\mu=0}^{\nu} (-1)^{\mu+\nu} \frac{\sum\limits_{\mu_1+\cdots+\mu_r=\mu} \gamma_{\mu_1\ldots\mu_r}}{(\nu-\mu)! r(r+2)\cdots(r+2\mu-2)},$$

$$(7.6.114)$$

where

$$\gamma_{\mu_1\ldots\mu_r} = \frac{E\tilde{X}_1^{2\mu_1}\cdots\tilde{X}_r^{2\mu_r}}{\mu_1!\cdots\mu_r!\,\lambda_1^{\mu_1}\cdots\lambda_r^{\mu_r}}, \quad \gamma_{0\ldots0} = 1. \qquad (7.6.115)$$

As it was shown above for finding the fourth order moment of a separate component it is necessary to know only the value $c_{r,2}$. We evaluate separately each item in (7.6.114) putting $\nu = 2$. At $\mu = 0$ the numerator is equal to 1 and the first item is equal to $1/2$. As $D\tilde{X}_k = \lambda_k$ then at $\mu = 1$ we have

$$\frac{\sum\limits_{\mu_1+\cdots+\mu_r=1} \gamma_{\mu_1\ldots\mu_r}}{1!\,r} = \frac{1}{r}\sum_{k=1}^{r} \gamma_{0\ldots\underset{k}{1},\ldots0} = \frac{1}{r}\sum_{k=1}^{r} \frac{E\tilde{X}_k^2}{\lambda_k} = 1. \qquad (7.6.116)$$

And finally, at $\mu = 2$ only two types of the expressions $\gamma_{\mu_1\ldots\mu_r}$ are possible. The first one appears when only one of μ_i is equal to 2, the others are zeros, the second is μ_i and μ_j are equal to 1, the others are zeros. As a result we obtain

$$c_{r,2} = \frac{1}{4}\left[-\frac{1}{2} + \frac{\sum\limits_{i=1}^{r} \gamma_{0\ldots\underset{r}{2}\ldots0} + \sum\limits_{i=1}^{r-1}\sum\limits_{j=i+1}^{r} \gamma_{0\,\ldots\,\underset{i}{1}0\ldots0\underset{j}{1}0\ldots0}}{r(r+2)} \right], \qquad (7.6.117)$$

where

$$\gamma_{0\ldots\underset{i}{2}\ldots0} = E\tilde{X}_i^4/\lambda_i^2, \quad \gamma_{0\ldots\underset{i}{1}\ldots\underset{j}{1}\ldots0} = E\tilde{X}_i^2\tilde{X}_j^2/\lambda_i\lambda_j. \qquad (7.6.118)$$

For the second item in (7.6.110) we have

$$E_{w_1}p_{1,2}(U_1)\tilde{X}_1^4 = 2^2\cdot2!R_{4,2} = 8R_{4,2} = 24D_1^2, \qquad (7.6.119)$$

as

$$R_{4,2} = \sum_{|h|=4} C_4^h \mu_h^w \alpha_{4-h}^w = C_4^4 \mu_4^w \alpha_0^w = \mu_4^w = 3D_1^2,$$

where μ_h^w and α_{4-h}^w are the central and the initial moments of the normal distribution $N(0,\tilde{K})$ respectively.

Thus the ellipsoidal approximation of the r-dimensional distribution gives for the fourth moment of the first and the ith coordinates

$$E_{\text{EA}}\tilde{X}_1^4 = 3D_1^2 + 24c_{r,2}D_1^2, \quad E_{\text{EA}}\tilde{X}_i^4 = 3D_i^2 + 24c_{r,2}D_i^2. \qquad (7.6.120)$$

By virtue of (7.6.103) and (7.6.104) we have

$$\delta_1(4,0,\dots,0) = E_{EA}\tilde{X}_1^4 - E\tilde{X}_1^4, \qquad (7.6.121)$$

$$\delta_2(4,0,\dots,0) = \frac{\delta_1(4,0,\dots,0)}{E\tilde{X}_1^4}. \qquad (7.6.122)$$

The fourth order moments of the vector may be also given by the expressions of the type $E\tilde{X}_i^2\tilde{X}_j^2$, $E\tilde{X}_i\tilde{X}_j^3$, $E\tilde{X}_i\tilde{X}_j\tilde{X}_k\tilde{X}_l$. We evaluate $E\tilde{X}_i^2\tilde{X}_j^2$. The remained two types of the moments are determined analogously. After choosing, for example, $i = 1$ and $j = 2$, we write

$$\int\limits_{-\infty}^{\infty} \dots \int\limits_{-\infty}^{\infty} \tilde{x}_1^2\tilde{x}_2^2\tilde{f}(\tilde{x})d\tilde{x}_1\dots d\tilde{x}_r \cong \int\limits_{-\infty}^{\infty} \dots \int\limits_{-\infty}^{\infty} \tilde{x}_1^2\tilde{x}_2^2 w(u)d\tilde{x}_1\dots d\tilde{x}_r$$

$$+ \sum_{\nu=2}^{N} c_{r,\nu} \int\limits_{-\infty}^{\infty} \dots \int\limits_{-\infty}^{\infty} \tilde{x}_1^2\tilde{x}_2^2 p_{r,\nu}(u)w(u)d\tilde{x}_1\dots d\tilde{x}_r. \qquad (7.6.123)$$

Let us integrate over $\tilde{x}_3,\dots,\tilde{x}_r$ using (7.6.23). As a result we obtain

$$\int\limits_{-\infty}^{\infty} \dots \int\limits_{-\infty}^{\infty} \tilde{x}_1^2\tilde{x}_2^2\tilde{f}(\tilde{x})d\tilde{x}_1\, d\tilde{x}_2 \cong \int\limits_{-\infty}^{\infty} \int\limits_{-\infty}^{\infty} \tilde{x}_1^2\tilde{x}_2^2 w(u_2)d\tilde{x}_1\, d\tilde{x}_2$$

$$+ \sum_{\nu=2}^{N} c_{r,\nu} \int\limits_{-\infty}^{\infty} \int\limits_{-\infty}^{\infty} \tilde{x}_1^2\tilde{x}_2^2 p_{2,\nu}(u_2)w(u_2)d\tilde{x}_1\, d\tilde{x}_2, \qquad (7.6.124)$$

where $u_2 = \tilde{x}_1^2/\lambda_1 + \tilde{x}_2^2/\lambda_2$, $w(u_2)$ is two-dimensional normal density. We rewrite (7.6.124) in the form of the expectations and substitute the values of the normal moments

$$E_{\text{EA}}\tilde{X}_1\tilde{X}_2 = D_1D_2 + \sum_{\nu=2}^{N} c_{r,\nu} E_w\tilde{X}_1^2\tilde{X}_2^2 p_{2,\nu}(U_2). \qquad (7.6.125)$$

We evaluate the expectation in (7.6.125):

$$E_{w_1} p_{2,\nu}(U_2)\tilde{X}_1^2 \tilde{X}_2^2 = E_w p_{2,2}(U_2)\tilde{X}_1^2 \tilde{X}_2^2$$

$$= 2^2 \cdot 2! R_{2,2,2} = 8R_{2,2,2} = 8D_1 D_2 \qquad (7.6.126)$$

by virtue of the fact that at $\nu = 2$ it turns out that $E_w p_{2,\nu}(U_2)\tilde{X}_1^2 \tilde{X}_2^2 = 0$, and at $h_i \leq r_i$

$$R_{2,2,2} = \sum_{|h|=4} C_2^{h_1} C_2^{h_2} \mu_{h_1 h_2}^w \alpha_{2-h_1,2-h_2}^w = C_2^2 C_2^2 \mu_{2,2}^w \alpha_{0,0}^w = D_1 D_2.$$

According to (7.6.126) the ellipsoidal approximation (7.6.125) of the mixed fourth order moment will take the form

$$E\tilde{X}_1^2 \tilde{X}_2^2 = D_1 D_2 + 8c_{r,2}D_1 D_2. \qquad (7.6.127)$$

Therefore finally we get:

$$\delta_1(2,2,0,\dots,0) = E_{\mathrm{EA}}\tilde{X}_1^2 \tilde{X}_2^2 - E\tilde{X}_1^2 \tilde{X}_2^2, \qquad (7.6.128)$$

$$\delta_2(2,2,0,\dots,0) = \delta_1(2,2,0,\dots,0)/E\tilde{X}_1^2 \tilde{X}_2^2. \qquad (7.6.129)$$

E x a m p l e   7.6.1.   Under conditions of Example 7.5.5 Eqs. (7.6.29), (7.6.30), (7.6.36) have the form:

$$\dot{m}_1 = -m_1 m_2 - k_{12}, \quad \dot{m}_2 = -am_2; \qquad (\mathrm{I})$$

$$\dot{k}_{11} = -2(m_2 k_{11} + m_1 k_{12}), \quad \dot{k}_{12} = -2(m_2 + a)k_{12} - m_1 k_{22},$$

$$\dot{k}_{22} = -2ak_{22} + h^2 \nu; \qquad (\mathrm{II})$$

$$\dot{c}_{2,2} = 4h^2 \nu C_{22} + c_{2,2}\{(m_2 + a)(6 + 8C_{12}k_{12}) + 8m_1 C_{12}k_{12} - 3h^2 \nu C_{22}\},$$
$$(\mathrm{III})$$

$$\dot{c}_{2,3} = -24h^2 \nu C_{22} + 3c_{2,2}\{(m_2 + \alpha)(C_{12}k_{12} - 5) + m_1 C_{12}k_{22} + 4,5h^2 \nu C_{22}\}$$

$$+ 3c_{2,3}\{6(m_2 + a)C_{12}k_{12} + 6m_1 C_{12}k_{12} - 3h^2 \nu C_{22}\}, \qquad (\mathrm{IV})$$

$$\dot{c}_{2,4} = 192h^2 \nu C_{22} + 24h^2 \nu C_{22}c_{2,2} + 4c_{2,3}\{(m_2 + a)(C_{12}k_{12} - 7) + m_1 C_{12}k_{22}$$

$$+ 6,5h^2 \nu C_{22}\} + 2c_{2,4}\{(m_2 + a)(14 + 16C_{12}k_{12}) + 16m_1 C_{12}k_{22} - 7h^2 \nu C_{22}\},$$

$$(\mathrm{V})$$

$$\dot{c}_{2,5} = -1920h^2\nu C_{22} - 240h^2\nu C_{22}c_{2,2} + 40h^2\nu C_{22}c_{2,3}$$

$$+5c_{2,4}\{(m_2 + a)(C_{12}k_{12} - 9) + m_1 C_{12}k_{22} - 8,5h^2\nu C_{22}\}$$

$$+5c_{2,5}\{(m_2 + a)(9 + C_{12}k_{12}) + 10m_1 C_{12}k_{22} - 4,5h^2\nu C_{22}\}. \qquad \text{(VI)}$$

Here C_{ij} are the elements of the matrix C; $C = K^{-1}$.

The peculiarity of these equations is the fact that Eq. (I) for the expectation and Eq. (II) for the covariance matrix are separated from Eqs. (III)–(VI) for the coefficients of the expansion. Here Eq. (I), (II) coincide with the correspondent equations of NAM (Example 7.2.3). If we wish to restrict our consideration to the moments till the fourth order we may put $c_{2,3} = c_{2,4} = c_{2,5} = 0$ in the obtained equations. By the same procedure the equations in the case of accounting the moments till the sixth or the eighth order are obtained.

On Figs. 7.6.1–7.6.5 the accuracy of EAM while calculating the intial moments $\alpha_1 = m_1$, $D = k_{11}$, α_2, α_4, α_6, α_8 in accordance with time t at the interval $[0, 1]$: $\alpha = 5$, $h = 1$ and at the initial conditions: $m_1(0) = 0,5$, $m_2(0) = 0,5$, $k_{11}(0) = D_1(0) = 0,1$; $k_{22}(0) = D_2(0) = 0,1$; $k_{12}(0) = 0$ is shown . The accuracy of α_4, α_6, α_8 ranges about -2%, 8%, 20% relative to the exact solution respectively. Hence we see that EAM gives a good accuracy similar as MSM (Subsection 7.4.7). In this case EAM with the account of the moments will the 10th order gives 9 equations instend of 65 ones at MSM.

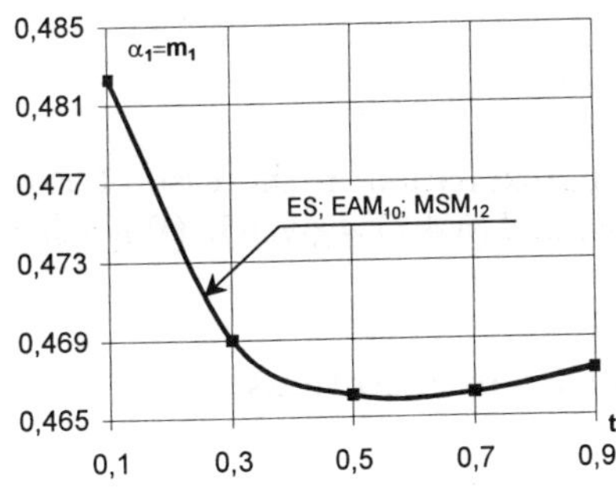

Fig. 7.6.1

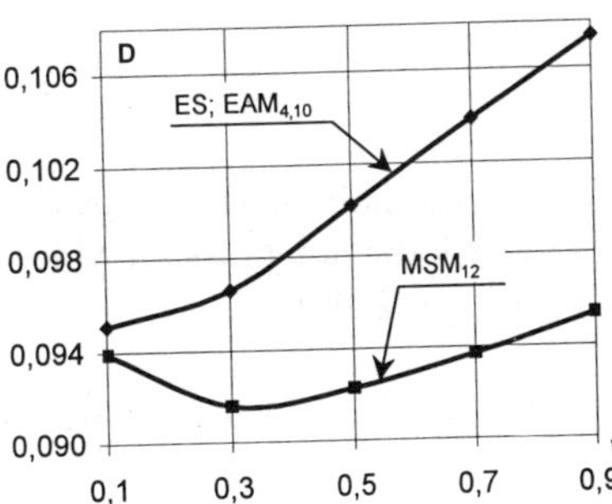

Fig. 7.6.2

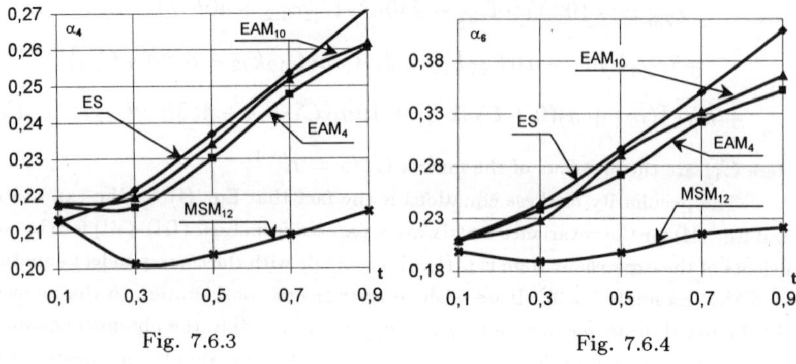

Fig. 7.6.3   Fig. 7.6.4

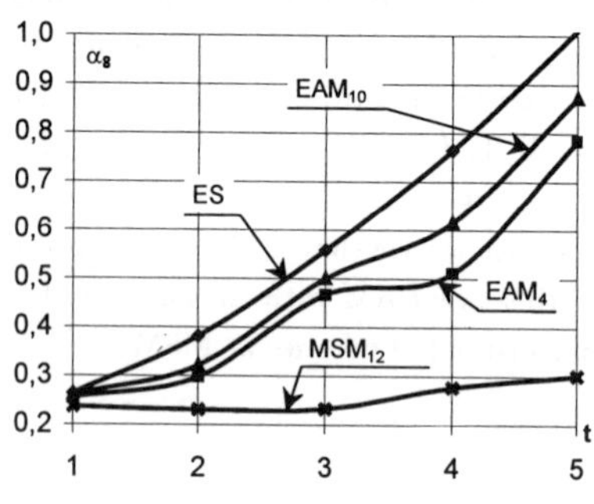

Fig. 7.6.5

The method of the ellipsoidal approximation in the case of Eqs. (5.4.6) with the nongaussian white noise and (5.4.7) is given in (Pugachev and Sinitsyn 2000).

7.7. Methods based on Canonical Representations

7.7.1. Linearization by Canonical Expansions

The linearization of the operator equations from the point of view of general theory of the stochastic systems transformation (Section 6.5) may be used in two variants. Firstly, a given dependence between random functions may be directly linearized and by such a way the nonlinear equations connecting the random functions may be substituted by linear equations. Secondly, the canonical expansions method which leads to the replacement of the operations over the random functions by the operations over the ordinary random variables may be used. After that we may use an ordinary in the applications method of linearization of the functional dependences between random variables.

The method of direct linearization of the nonlinear transformation of random functions consists in the replacement of all given equations connecting the random functions by approximate linear equations which reflect adequately the real dependence between the random functions in the domain practically possible realizations of the random functions. As the expectations of the random variables are the mean values near which their possible realizations are considered then in practice it is more convenient to perform the linearization of the relations between the random functions relatively their deviations from the expectations, i.e. the centered random functions. In this case all the functions which enter into the given equations should be expanded into the Taylor series over the centered random functions and rejected the terms of these series higher than the first degree. The degree of the accuracy of the approximation obtained in such a way may be estimated by maximal possible value of the rejected terms in the region of practically possible realizations of random functions. After substituting the given equations connecting the random functions by the approximate linear equations we may use the methods of Section 6.3 for the approximate determination of the expectations and the covariance functions found as a result of the considered nonlinear transformation.

Let us pass to the application of the canonical expansions method to the approximate analysis of the nonlinear transformations of the random functions. Let the scalar random function $Y(t)$ be obtained as a result of the transformation of the scalar random functions $X(t)$ by means of the nonlinear operator A:

$$Y(t) = AX(t). \tag{7.7.1}$$

Substituting here instead of the random function $X(t)$ any its canonical expansion we find

$$Y(t) = A\left\{ m_x(t) + \sum_\nu V_\nu x_\nu(t) \right\} = \varphi(V_1, V_2, \ldots, t). \tag{7.7.2}$$

Equality (7.7.2) represents the random function $Y(t)$ as some function, in general case, nonlinear of the random variables V_ν into which the argument t enters as a parameter. While linearizing this function and taking into consideration that $EV_\nu = 0$, $\nu = 1, 2, \ldots$ we obtain the following approximate expansion of the function $Y(t)$ and its covariance function $K_y(t, t')$:

$$Y(t) \approx Am_x(t) + \sum_\nu V_\nu y_\nu(t), \quad K_y(t, t') \approx \sum_\nu D_\nu y_\nu(t)\overline{y_\nu(t')}. \quad (7.7.3)$$

Here

$$y_\nu(t) = \left[\frac{\partial \varphi(V_1, V_2, \ldots, t)}{\partial V_\nu} \right]_0 = \left[\frac{\partial}{\partial V_\nu} A \left\{ m_x(t) + \sum_\mu V_\mu x_\mu(t) \right\} \right]_0,$$
$$(7.7.4)$$

where zero below indicates that after the calculation of the derivative all the random variables should be equated to zero, $V_\nu = 0$.

We apply the stated arguments to the nonlinear differential system

$$\dot{Y}_k = f_k(t, Y_1, \ldots, Y_m, X_1, \ldots, X_n) \quad (k = 1, \ldots, m). \quad (7.7.5)$$

Here f_k are the known functions and the random disturbances $X_s = X_s(t, Y_1, \ldots, Y_m)$ allow the joint canonical expansion

$$X_s = m_s^x + \sum_\nu V_\nu x_{\nu s} \quad (s = 1, \ldots, n), \quad (7.7.6)$$

where $m_s^x = m_s^x(t, Y_1, \ldots, Y_m)$; $x_{\nu s} = x_{\nu s}(t, Y_1, \ldots, Y_m)$; V_ν are the uncorrelated random coefficients with zero expectations and the variances D_ν. If we substitute (7.7.6) into (7.7.5) then the right-hand sides of Eqs. (7.7.5) will depend on the random coefficients V_ν. Here they will be known functions of the variables t, $Y_1, \ldots, Y_m$ and V_ν. As a result of the integration of these equations we obtain $Y_k = \varphi_k(t, V_1, V_2, \ldots)$. If we consider φ_k as the functions of V_ν sufficiently close to the linear ones in the region of practically possible values of V_ν then we obtain in the linear approximation

$$Y_k \approx [Y_k] + \sum_\nu V_\nu \left[\frac{\partial Y_k}{\partial V_\nu} \right]_0, \quad (7.7.7)$$

where by zero below is indicated that after the calculation all the derivatives should be taken at $V_\nu = 0$.

Hence using the formulae for the canonical expansions of the vector random functions (Subsection 3.9.3) we find the approximate canonical expansion Y_k and the covariance functions:

$$Y_k \approx m_k^y + \sum_\nu V_\nu y_{\nu k},$$

$$K^y_{pq}(t,t') \approx \sum_\nu D_\nu y_{\nu p}(t)\overline{y_{\nu q}(t)} \quad (p,q,k=1,\dots,m).\tag{7.7.8}$$

where $m^y_k \approx [Y_k]_0$, $y_{\nu k}(t) = [\partial Y_k/\partial V_\nu]_0$. From (7.7.5) after the differentiation m^y_k, $y_{\nu k}$ over time and accounting that

$$\frac{d}{dt}\left(\frac{\partial Y_k}{\partial V_\nu}\right) = \sum_{p=1}^{m}\frac{\partial f_k}{\partial Y_p}\frac{\partial Y_p}{\partial V_\nu} + \sum_{q=1}^{n}\frac{\partial f_k}{\partial X_q}\frac{\partial X_q}{\partial V_\nu} \quad (k=1,\dots,m),$$

$$\frac{\partial X_q}{\partial V_\nu} = x_{\nu q} + \sum_{p=1}^{m}\frac{\partial m^x_q}{\partial Y_p}\frac{\partial Y_p}{\partial V_\nu} + \sum_\mu\sum_{p=1}^{m} V_\mu \frac{\partial x_{\nu p}}{\partial V_p}\frac{\partial Y_p}{\partial V_\nu},$$

we find

$$\dot{m}^y_k = \frac{d}{dt}[Y_k]_0 = [f_k]_0,\tag{7.7.9}$$

$$\dot{y}_{\nu k}(t) = \frac{d}{dt}\left[\frac{\partial Y_k}{\partial V_\nu}\right]_0 = \sum_{p=1}^{m}\left(\frac{\partial f_k}{\partial m^y_p} + \sum_{q=1}^{n}\frac{\partial f_k}{\partial m^x_q}\frac{\partial m^x_q}{\partial m^y_p}\right)$$

$$\times \left[\frac{\partial Y_p}{\partial V_\nu}\right]_0 + \sum_{q=1}^{n}\frac{\partial f_k}{\partial m^x_q}x_{\nu q}.\tag{7.7.10}$$

Consequently, (7.7.9) and (7.7.10) lead to the following ordinary differential equations for m^y_k and $y_{\nu k}$:

$$\dot{m}^y_k = f_k(t, m^y_1, \dots, m^y_m, m^x_1, \dots, m^x_n),\tag{7.7.11}$$

$$\dot{y}_{\nu k} = \sum_{p=1}^{m} a_{kp}(t)y_{\nu p} + \sum_{q=1}^{n} b_{kq}(t)c_{\nu q}(t),\tag{7.7.12}$$

where

$$m^x_s = m^x_s(t, m^y_1, \dots, m^y_m),$$

$$a_{kp}(t) = \frac{\partial f_k}{\partial m^y_p} + \sum_{q=1}^{n}\frac{\partial f_k}{\partial m^x_q}\frac{\partial m^x_q}{\partial m^y_p} \quad (k,p=1,\dots,m),$$

$$b_{kq}(t) = \frac{\partial f_k}{\partial m^x_q} \quad (k=1,\dots,m;\ p=1,\dots,n),\tag{7.7.13}$$

$$c_{\nu q}(t) = x_{\nu q}(t, m^y_1, \dots, m^y_m) \quad (q=1,\dots,n).$$

Thus *after presenting the vector random function* $X = [X_1 \dots X_n]^T$ *by canonical expansion (7.7.6) we get a system of ordinary nonlinear* *differential Eqs. (7.7.11) for an approximate determination of the expectation of the random function $Y = [Y_1 \ldots Y_m]^T$ and a system of the linear ordinary differential equations with the one and the same matrix at unknown set of Eqs. (7.7.12), for determination of the coordinate functions. After integrating Eqs. (7.7.11) and (7.7.12) we shall find by formulae (7.7.8) the approximate canonical expansion Y and its covariance function $K^y(t,t')$.*

It remains to solve the problem about the initial conditions for Eqs. (7.7.11) and (7.7.12). If the initial conditions are nonrandom, i.e. $t = t_0$, $Y_k = y_{k0}$ ($k = 1, \ldots, m$) then it is necessary to integrate the equations at the initial conditions $t = t_0$, $m_k^y = y_{k0}$, $y_{\nu k} = 0$. For the random initial conditions $t = t_0$, $Y_k = Y_{k0}$ ($k = 1, \ldots, m$) Eqs. (7.7.11) and (7.7.12) are integrated at the initial conditions $t = t_0$, $m_k^y = m_{k0}^y$, $y_{\nu k} = y_{\nu k0}$. In the latter case m_{k0} and $y_{\nu k0}$ are found from the joint canonical expansion of $(n+m)$-dimensional random function $\left[X^T Y_0^T \right]^T$.

Analogously Eqs. (7.7.1) and (7.7.12) are written for the equations unsolved relatively to the derivatives, $F(Y, \dot{Y}, \ldots, Y^{(l)}, X, \dot{X}, \ldots, X^{(r)}, t) = 0$; $X = X(Y,t)$ ($l, r = 1, 2, \ldots$).

E x a m p l e   7.7.1. Consider the scalar nonlinear system

$$\dot{Y} = 1 - Y^2 + X \,, \quad m_y(0) = 1 \,, \tag{I}$$

where $X = X(t)$ is a stationary random process with zero expectation and the known spectral density $s_x(\omega)$. We use a segment of the canonical expansion of the stationary random function $X(t)$ at the interval of the duration $2T$ (Example 4.8.5):

$$X(t) = \sum_{\nu=-N}^{N} V_\nu e^{i\omega_\nu t} \,, \quad 0 < t < 2T \,. \tag{II}$$

Here the variances V_ν are equal to $D[V_\nu] = D_\nu = \int\limits_{\omega_\nu - \alpha}^{\omega_\nu + \alpha} s_x(\omega)d\omega$, $\nu = 0, \pm1,$ $\pm2, \ldots$, where 2α is the length of each of $(2N+1)$ segments of the interval of the frequences $(-\Omega, \Omega)$. In a given case Eq. (7.7.11) has the form $\dot{m}_y = 1 - m_y^2$, $m_y(0) = 1$. It allows the integral $m_y(t) \equiv 1$. Eq. (7.7.12) approximately determining the coordinate functions $\varphi_\nu = y_\nu(t, \omega_\nu)$ with account $m_y(t) \equiv 1$ has the form $\dot{y}_\nu = -2y_\nu + e^{i\omega_\nu t}$. Its integral vanishing at $t = 0$ is equal

$$y_\nu(t, \omega_\nu) = e^{-2t} \int\limits_0^t e^{2\tau + i\omega_\nu \tau} d\tau = \frac{e^{i\omega_\nu t} - e^{-2t}}{2 + i\omega_\nu} \,. \tag{III}$$

Consequently,

$$D_y(t) \approx \sum_{\nu=-N}^{N} D_\nu \frac{\left| e^{i\omega_\nu t} - e^{2t} \right|^2}{4 + i\omega_\nu^2} = \frac{D_0}{4} + 2\sum_{\nu=1}^{N} D_\nu \frac{1 - 2e^{-2t}\cos\omega_\nu t + e^{-4t}}{4 + \omega_\nu^2}.$$

$$(IV)$$

E x a m p l e 7.7.2. Solve Example 7.7.1 supposing that the input random disturbance $X = X(t, Y)$ is a real stationary random function with the spectral density $s_x(\omega, \lambda)$. It is evident that similarly as in Example 7.7.1, $m_y(t) \equiv 1$. For determining the variances of the random function Y we use a segment of its canonical expansion in the region $0 < t < 2T$, $|Y| < S$:

$$X(t, Y) \approx \sum_{\nu=-N}^{N} \sum_{\mu=-N}^{N} V_{\nu\mu} e^{i(\omega_\nu t + \lambda_\mu Y)}.$$

$$(I)$$

Here the variances $V_{\nu\mu}$ are calculated by formula

$$D[V_{\nu\mu}] = D_{\nu\mu} = \int_{\omega_\nu-\alpha}^{\omega_\nu+\alpha} d\omega \int_{\lambda_\nu-\beta}^{\lambda_\mu-+\beta} s_x(\omega, \lambda) d\lambda,$$

$$(II)$$

where 2α and 2β are the lengths of each of $(2N+1)$ segments of the intervals of the frequencies $(-\Omega, \Omega)$ and $(-\Lambda, \Lambda)$. The coordinate function $\varphi_{\nu\mu} = y_{\nu\mu}(t, \omega_\nu, \lambda_\mu)$ satisfies the equation $\dot{y}_{\nu\mu} = -2y_{\nu\mu} + e^{i(\omega_\nu t + \lambda_\mu)}$ and the integral of this equation which vanishes at $t = 0$ is determined by formula

$$y_{\nu\mu}(t, \omega_\nu, \lambda_\mu) = e^{i\lambda_\mu} \frac{e^{i\omega_\nu t} - e^{-2t}}{2 + i\omega_\nu}.$$

$$(III)$$

Consequently, acconting (I)–(III) we shall find

$$D_y(t) = \sum_{\nu=-N}^{N} \sum_{\mu=-N}^{N} D_{\nu\mu} \frac{\left| e^{i\omega_\nu t} - e^{-2t} \right|^2}{4 + \omega_\nu^2} = \sum_{\nu=-N}^{N} \tilde{D}_\nu \frac{\left| e^{i\omega_\nu t} - e^{-2t} \right|^2}{4 + \omega_\nu^2}$$

$$= \frac{\tilde{D}_0}{4} + 2\sum_{\nu=1}^{N} \tilde{D}_\nu \frac{1 - e^{-2t}\cos\omega_\nu t + e^{-4t}}{4 + \omega_\nu^2},$$

$$(IV)$$

where $\tilde{D}_\nu = \tilde{D}_{-\nu} = \sum_{\mu=-N}^{N} D_{\nu\mu} \approx \int_{\omega_\nu-\alpha}^{\omega_\nu+\alpha} d\omega \int_{-\infty}^{\infty} s_x(\omega, \lambda) d\lambda.$

7.7.2. Linearization by Integral Canonical Representations

Let us consider system (7.7.5). We present the random disturbances $X_s = X_s(t, Y)$ by the joint integral canonical representation (Subsection 4.8.2)

$$X_s(Y, t) = m_s^x(Y, t) + \varepsilon \sum_{r=1}^{N} \int_{\Lambda_r} V_r(\lambda) x_{rs}(Y, t, \lambda) d\lambda \quad (s = 1, \ldots, n),$$

(7.7.14)

where an arbitrary parameter ε is temporarily introduced. If we substitute (7.7.14) into (7.7.5) then the right-hand sides of Eqs. (7.7.5) will be the functions of Y_k, t and ε. Therefore as a result of the integration of Eqs. (7.7.5) we have $Y_k = \varphi_k(t, \varepsilon)$ ($k = 1, \ldots, m$). Linearizing these functions by any ordinary way relatively to the parameter ε we get an approximate formula:

$$Y_k \approx [Y_k]_0 + \left[\frac{\partial Y}{\partial \varepsilon} \right]_0,$$

(7.7.15)

where by zero below the values of the correspondent variables at $\varepsilon = 0$ are indicated. For determining $[Y_k]_0$ it is necessary to put $\varepsilon = 0$ after the substitution of expressions (7.7.14) into (7.7.5). Then assuming $[Y_k]_0 = [m_1^y \ldots m_m^y]$ we get for them Eqs. (7.7.7). To find $[\partial Y_k / \partial \varepsilon]_0$ (7.7.5) should be differentiated over ε. We have

$$\frac{d}{dt} \frac{\partial Y_k}{\partial \varepsilon} = \sum_{p=1}^{m} \left\{ \frac{\partial f_k}{\partial Y_p} + \sum_{q=1}^{n} \frac{\partial f_k}{\partial X_q} \left[\frac{\partial m_q^x}{\partial Y_p} + \varepsilon \sum_{r=1}^{N} \int_{\Lambda_r} V_r(\lambda) \frac{\partial x_{rq}}{\partial Y_p} d\lambda \right] \right\} \frac{\partial Y_q}{\partial \varepsilon}$$

$$+ \sum_{q=1}^{n} \frac{\partial f_k}{\partial X_q} \sum_{r=1}^{N} \int_{\Lambda_r} V_r(\lambda) x_{rq} d\lambda.$$

(7.7.16)

Putting here $\varepsilon = 0$ and taking into account that X_s in this case turn into m_s^x and Y_k turn into $[Y_k]_0 = m_k^y$ we get

$$\frac{d}{dt} \left[\frac{\partial Y_k}{\partial \varepsilon} \right]_0 = \sum_{p=1}^{m} \left(\frac{\partial f_k}{\partial m_p^y} + \sum_{q=1}^{n} \frac{\partial f_k}{\partial m_q^x} \frac{\partial m_q^x}{\partial m_q^y} \right)$$

$$\times \left[\frac{\partial Y_p}{\partial \varepsilon} \right]_0 + \sum_{q=1}^{n} \frac{\partial f_k}{\partial m_q^x} \sum_{r=1}^{N} \int_{\Lambda_r} V_r(\lambda) x_{rq}(m^y, t, \lambda) d\lambda,$$

(7.7.17)

where m_q^x and f_k are determined by (7.7.13). Introducing into (7.7.17) the functions $a_{kp}(t)$, $b_{kq}(t)$ and the functions $c_{rq}(t,\lambda)$ according to (7.7.13) we may write equations (7.7.17) in the form

$$\frac{d}{dt}\left[\frac{\partial Y_k}{\partial \varepsilon}\right]_0 = \sum_{p=1}^{m} a_{kp}(t)\left[\frac{\partial Y_k}{\partial \varepsilon}\right]_0 + \sum_{q=1}^{n} b_{kq}(t)\sum_{r=1}^{N}\int_{\Lambda_r} V_r(\lambda)c_{rq}(t,\lambda)d\lambda\,.$$

$$(7.7.18)$$

Let $y_{rk} = y_{rk}(t,\lambda)$ be the integral integrals of equations

$$\dot{y}_{rk} = \sum_{p=1}^{m} a_{kp}(t)y_{rp} + \sum_{q=1}^{n} b_{kq}(t)c_{rq}(t,\lambda)\,. \qquad (7.7.19)$$

Then the integral of linear equations (7.7.18) will be expressed by formula

$$\left[\frac{\partial Y_k}{\partial \varepsilon}\right]_0 = \sum_{r=1}^{N}\int_{\Lambda_r} V_r(\lambda)y_{rk}(t,\lambda)d\lambda\,. \qquad (7.7.20)$$

Substituting (7.7.20) into (7.7.15) and putting $\varepsilon = 1$ we obtain an approximate integral canonical representation

$$Y_k(t) \approx m_k^y(t) + \sum_{r=1}^{N}\int_{\Lambda_r} V_r(\lambda)y_{rk}(t,\lambda)d\lambda \quad (k = 1,\,\ldots,\,m)\,. \qquad (7.7.21)$$

Here the covariance functions are approximately determined by the following formula:

$$K_{pq}^y(t,t') \approx \sum_{r=1}^{N}\int_{\Lambda_r} \nu_r(\lambda)y_{rp}(t,\lambda)\overline{y_{rq}(t',\lambda)}d\lambda \quad (p,q = 1,\,\ldots,\,m),$$

$$(7.7.22)$$

where $\nu_r(\lambda)$ being the intensity of the white noise $V_r(\lambda)$ $(r = 1,\,\ldots,\,N)$.

Thus *to integral canonical representation of random input X (7.7.14) at $\varepsilon = 1$ corresponds approximate integral canonical representation Y (7.7.21) with coordinate functions y_{rk} defined by Eqs. (7.7.19) as function of time t at fixed λ.*

7.7.3. Canonical Representations of Nonlinear Transformations Reducible to Linear Ones

In many practical problems the nonlinear transformations of the random variables which are called reducible to linear ones are of great importance.

Under conditions of Subsection 6.3.2 let f_τ be a linear functional on the phase space of all random variables X which depends on the parameter τ and $f_\tau \in F$ at all $\tau \in T$; $\varphi(z_1, \ldots, z_n)$ is a function of the scalar variables $z_1, \ldots, z_n$ with the values in the linear space U with the weak topology determined by a set $\mathcal{K}$ of the linear functionals and the correspondent σ-algebra $\mathcal{U}$ (formed similarly as the σ-algebras $\mathcal{A}$ and $\mathcal{B}$ in the spaces X and Y of Section 4.8). Suppose that $\varphi(z_1, \ldots, z_n)$ is measurable relatively to the Borel σ-algebra of the n-dimensional space and the σ-algebra $\mathcal{U}$. Then the function

$$U(\tau_1, \ldots, \tau_n) = \varphi(f_{\tau_1} X, \ldots, f_{\tau_n} X) \tag{7.7.23}$$

will be a random function of the variables $\tau_1, \ldots, \tau_n \in T$ with the values in $(U, \mathcal{U})$. Its expectation $m_u(\tau_1, \ldots, \tau_n)$ and the second order moment $\Gamma(\tau_1, \ldots, \tau_n, \sigma_1, \ldots, \sigma_n)$ are determined by the evident formulae

$$m_u(\tau_1, \ldots, \tau_n) = \int \varphi(f_{\tau_1} x, \ldots, f_{\tau_n} x) \mu_x(dx), \tag{7.7.24}$$

$$\Gamma_u(\tau_1, \ldots, \tau_n, \sigma_1, \ldots, \sigma_n)k$$

$$\int \varphi(f_{\tau_1} x, \ldots, f_{\tau_n} x) \left[\overline{k\varphi(f_{\tau_1} x, \ldots, f_{\tau_n} x)} \right] \mu_x(dx), \quad k \in \mathcal{K}. \tag{7.7.25}$$

After finding m_u and Γ_u the covariance function of the random function U will be determined by formula (3.3.5):

$$K_u(\tau_1, \ldots, \tau_n, \sigma_1, \ldots, \sigma_n)k = \Gamma_u(\tau_1, \ldots, \tau_n, \sigma_1, \ldots, \sigma_n)k$$

$$-m_u(\tau_1, \ldots, \tau_n) \left[\overline{km_u(\sigma_1, \ldots, \sigma_n)} \right]. \tag{7.7.26}$$

It is evident that there is no need to know the probability measure μ_x of the random variable X for the calculations by formulae (7.7.24) and (7.7.25) but it is sufficient to know only its contraction on the σ-algebra $\mathcal{A}_{f_{\tau_1}, \ldots, f_{\tau_{2n}}}$ at any fixed $f_{\tau_1}, \ldots, f_{\tau_{2n}} \in F$.

Let L be a linear operator mapping the phase functional space of the random variable $U = U(\tau_1, \ldots, \tau_n)$, $\tau_1, \ldots, \tau_n \in T$ into the linear space Y (Subsection 6.3.2) which measurable relatively to the σ-algebras

$$\mathcal{U}^{T^n} = \prod_{\tau_1, \ldots, \tau_n \in T} \mathcal{U}_{\tau_1, \ldots, \tau_n} \qquad (7.7.27)$$

and $\mathcal{B}$. Then $Y = LU$ will be a random variable in $(Y, \mathcal{B})$. The random variable Y represents the result of the linear transformation of the random function $U(\tau_1, \ldots, \tau_n)$ and at the same time the result of the nonlinear in the general case transformation of the random variable X:

$$Y = L\varphi(f_{\tau_1} X, \ldots, f_{\tau_n} X). \qquad (7.7.28)$$

It is natural to name the nonlinear transformations of such a type by *the transformations reducible to linear ones*.

For determining the expectation m_y and the covariance operator K_y of the random variable Y it is sufficient to find any canonical expansion of the random function $U(\tau_1, \ldots, \tau_n)$ and then use the method stated in Subsection 6.3.2. It is sufficient for this purpose that the operator L will be continuous in the weak topologies of the phase space of the random function $U(\tau_1, \ldots, \tau_n)$ and in the weak topology of the space Y. The topology of the phase space of U is determined by the set Λ of the linear functionals of the form

$$\lambda u = \sum_{p=1}^{N} k_p u(\tau_1^{(p)}, \ldots, \tau_n^{(p)}), \quad k_1, \ldots, k_N \in \mathcal{K}. \qquad (7.7.29)$$

Instead of L-continuity it is sufficient that L will have at least the adjoint will have the adjoint operator L^* whose intersection with G of the part of its domain D_{L^*} mapped into Λ contains a sufficient set of the linear functionals.

It is easy to see that all mentioned remains valid in the case when f_τ is assumed as any finite set $f_\tau^{(1)}, \ldots, f_\tau^{(r)} \in F$, $\tau \in T$ of the linear functionals dependent on τ. In this case for the calculation by formulae (7.7.24) and (7.7.25) it is necessary to know the contraction of the probability measure μ_x of the random variable X on the σ-algebra $\mathcal{A}_{f_1, \ldots, f_{2rn}}$ at any fixed $f_1, \ldots, f_{2rn} \in F$.

E x a m p l e 7.7.3. Let $X(t)$ be the r-dimensional vector random function $t \in T$; $\varphi(s, z_1, \ldots, z_n)$ be the m-dimensional vector function of the variable

$s \in S$ and the r-dimensional vector variables $z_1, \ldots, z_n$. In a given case X represents a space of the r-dimensional vector functions of the variable $t \in T$, and U be a space of the m-dimensional vector functions of the variable $s \in S$. Taking $f_\tau x = x(\tau)$, i.e. $f_\tau^{(p)} x = x_p(\tau)$, $(p = 1, \ldots, r)$ we obtain

$$U = U(s, \tau_1, \ldots, \tau_n) = \varphi(s, X(\tau_1), \ldots, X(\tau_n)). \qquad (\mathrm{I})$$

The contraction of the probability measure of the random function $X(t)$ on the σ-algebra $\mathcal{A}_{f_{\tau_1}^{i_1}, \ldots, f_{\tau_n}^{i_r}}$ in a given case represents the n-dimensional distribution $\mu_{\tau_1, \ldots, \tau_n}$ of the random function $X(t)$, and its contraction on the σ-algebra $\mathcal{A}_{f_{\tau_1}^{i_1}, \ldots, f_{\tau_n}^{i_r}, f_{\sigma_1}^{i_1}, \ldots, f_{\sigma_n}^{i_r}}$ be the $2n$-dimensional distribution $\mu_{\tau_1, \ldots, \tau_n, \sigma_1, \ldots, \sigma_n}$ of the random function $X(t)$. Formulae (7.7.24) and (7.7.25) give in this case

$$m_u(s, \tau_1, \ldots, \tau_n) = \int \varphi(s, x_1, \ldots, x_n) \mu_{\tau_1, \ldots, \tau_n}(dx), \qquad (\mathrm{II})$$

$$\Gamma_u(s_1, \tau_1, \ldots, \tau_n, s_2, \sigma_1, \ldots, \sigma_n)$$
$$= \int \varphi(s_1, x_1, \ldots, x_n)\overline{\varphi(s_2, x_{n+1}, \ldots, x_{2n})} \mu_{\tau_1, \ldots, \tau_n, \sigma_1, \ldots, \sigma_n}(dx),$$
$$(\mathrm{III})$$

where the integration in formula (II) is performed over the variables $x_1, \ldots, x_n$, and in (III) over the variables $x_1, \ldots, x_{2n}$. The covariance function of the random function $U(s, \tau_1, \ldots, \tau_n)$ according to (7.7.26) is determined by formula

$$K_u(s_1, \tau_1, \ldots, \tau_n, s_2, \sigma_1, \ldots, \sigma_n) = \Gamma_u(s_1, \tau_1, \ldots, \tau_n, s_2, \sigma_1, \ldots, \sigma_n)$$

$$-m_u(s_1, \tau_1, \ldots, \tau_n)\overline{m_u(s_2, \sigma_1, \ldots, \sigma_n)}. \qquad (\mathrm{IV})$$

Let us consider the function

$$Y(s) = \int \cdots \int l(s, \tau_1, \ldots, \tau_n)\varphi(s, X(\tau_1), \ldots, X(\tau_n))\nu_1(d\tau_1) \ldots \nu_n(d\tau_n). \qquad (\mathrm{V})$$

As in a given case

$$Y(s) = \int \cdots \int l(s, \tau_1, \ldots, \tau_n)U(s, \tau_1, \ldots, \tau_n)\nu_1(d\tau_1) \ldots \nu_n(d\tau_n), \qquad (\mathrm{VI})$$

then the considered transformation of the random function $X(t)$ refers to the class of the nonlinear transformations reduced to the linear ones. If the function $l(s, \tau_1, \ldots, \tau_n)$ satisfies the conditions of one of Examples 6.3.1–6.3.9 then for determining the covariance function of the random function $Y(s)$ we may use the canonical expansions method (under the conditions of Examples 6.3.2 and 6.3.6 we restrict ourselves with the case of the finitedimensional argument t). After expressing the random function $U(s, \tau_1, \ldots, \tau_n)$, $s \in S$, $\tau_1, \ldots, \tau_n \in T$ by some canonical expansion

$$U(s, \tau_1, \ldots, \tau_n) = m_u(s, \tau_1, \ldots, \tau_n) + \sum_{r=1}^{\infty} V_r u_r(s, \tau_1, \ldots, \tau_n) \quad \text{(VII)}$$

or by the integral canonical representation

$$U(s, \tau_1, \ldots, \tau_n) = m_u(s, \tau_1, \ldots, \tau_n) + \sum_{r=1}^{\infty} \int u_r(s, \tau_1, \ldots, \tau_n, \lambda) Z_r(d\lambda),$$
$$\text{(VIII)}$$

we get for the random function $Y(s)$ correspondent canonical expansion (III) of Example 6.3.3 or integral canonical representation (I)–(III) of Example 6.6.3 where

$$m_y(s) = \int \ldots \int l(s, \tau_1, \ldots, \tau_n) m_u(s, \tau_1, \ldots, \tau_n) \nu_1(d\tau_1) \ldots \nu_n(d\tau_n),$$
$$\text{(IX)}$$

$$y_r(s) = \int \ldots \int l(s, \tau_1, \ldots, \tau_n) u_r(s, \tau_1, \ldots, \tau_n) \nu_1(d\tau_1) \ldots \nu_n(d\tau_n),$$
$$\text{(X)}$$

$$y_r(s, \lambda) = \int \ldots \int l(s, \tau_1, \ldots, \tau_n) u_r(s, \tau_1, \ldots, \tau_n, \lambda) \nu_1(d\tau_1) \ldots r_n(d\tau_n).$$
$$\text{(XI)}$$

E x a m p l e 7.7.4. Consider a special case of the real scalar random function $X(t)$ and the n-dimensional vector function $\varphi(s, z_1, \ldots, z_n)$ with the components $\varphi_1(s, z_1, \ldots, z_n) = z_1$, $\varphi_p(s, z_1, \ldots, z_n) = z_1 \ldots z_p$ $(p = 2, \ldots, n)$, $t \in T$. In this case $U(s, z_1, \ldots, z_n)$ does not depend on s and represents the vector random function of $\tau_1, \ldots, \tau_n$ with the components $U_p(\tau_1, \ldots, \tau_n) = X(\tau_1) \ldots X(\tau_p)$, $(p = 1, \ldots, n)$. If t is a finite-dimensional variable, and $\nu_1, \ldots, \nu_n$ coincide with the Lebesgue measure in the space T then formula(V) represents the random function $Y(s)$ by the finite segment of the Volterra series:

$$Y(s) = \sum_{p=1}^{n} \int \ldots \int l_p(s, \tau_1, \ldots, \tau_p) X(\tau_1) \ldots X(\tau_p) d\tau_1 \ldots d\tau_p. \quad \text{(XII)}$$

The expectation of the random function $U(\tau_1, \ldots, \tau_n)$ is a vector with the components $m_{u_p}(\tau_1, \ldots, \tau_n) = \alpha_p(\tau_1, \ldots, \tau_n)$, α_p being p^{th} moment of $X(t)$, and its second order moment is the matrix with the elements $\Gamma_{u_p u_q}(\tau, \sigma) = \alpha_{p+q}(\tau, \sigma)$,

$p, q = 1, \ldots, n$, where $\tau = (\tau_1, \ldots, \tau_n)$, $\sigma = (\sigma_1, \ldots, \sigma_n)$. Thus for finding the canonical expansion of the random function $U(\tau_1, \ldots, \tau_n)$ it is sufficient in this case to know the moments of the random function $X(t)$ till the order $2n$ inclusively.

The generalization of the considered problem is the problem of finding the expectation and the covariance function of the random function $Y(s)$ which is expressed by the finite segment of the Volterra series (I) in Example 7.7.3 terms of the h-dimensional vector random function $X(t)$ of the finite-dimensional argument t. Here the function $\varphi(s, z_1, \ldots, z_n)$ represents a vector function with the components $z_1^{(k_1)} \ldots z_p^{(k_p)}$ $(k_1, \ldots, k_p = 1, \ldots, h; p = 1, \ldots, n)$, and the random function $U(\tau_1, \ldots, \tau_n)$ represents a vector random function with the components $X_{k_1}(\tau_1) \ldots X_{k_p}(\tau_p)$ $(k_1, \ldots, k_p = 1, \ldots, h; p = 1, \ldots, n)$. For finding the canonical expansion of the random function $U(\tau_1, \ldots, \tau_n)$ it is sufficient to know the moments of the random function $X(t)$ till the order $2n$ inclusively.

7.7.4. General Methods of Nonlinear Stochastic Systems Theory based on Canonical Representations

In Subsection 7.7.1 the case was considered when the function φ in (7.7.2) with sufficient accuracy may be assumed linear in the region of practically possible values V_ν. If this condition is not fulfilled then we may pose the question about the approximation of the function Y by the polynomials relatively to V_ν. In order such an approximation be possible it is sufficient that the function determined by (7.7.2) relatively to V_ν will be continuous. But practically the approximation Y by the polynomials is obtained simpler in the case when as the polynomials the segments of the Maclaurin series are taken. In this case besides the continuity the existence of its derivatives till the definite order is necessary what will require the correspondent restrictions on the nonlinear operator A. After expanding Y as the function of the random variables V_ν determined by (7.7.2) into the Maclaurin series we shall obtain in the general case the noncanonical expansion

$$Y(s) = Am_x(s) + \sum_{\nu_1, \ldots, \nu_n} V_{\nu_1} \ldots V_{\nu_n} y_{\nu_1, \ldots, \nu_n}(s), \qquad (7.7.30)$$

where

$$y_{\nu_1, \ldots, \nu_n}(s) = \frac{1}{n!} \left[\frac{\partial^n}{\partial V_{\nu_1} \ldots V_{\nu_n}} A \left\{ m_x(t) + \sum_\mu V_\mu x_\mu(t) \right\} \right]. \qquad (7.7.31)$$

Formula (7.7.31) give the formal expansion of $Y(s)$ in powers of the random variables V_ν. Therefore if the moments of the random variables V_ν of all necessary orders are assigned then it is easy to determine the moments of the random function $Z(s) = Y(s) - Am_x(t)$.

Let us introduce the denotion $\mu_{\nu_1,\,\ldots,\,\nu_n} = EV_{\nu_1}V_{\nu_2}\ldots V_{\nu_n}$ for the moments of the random variables. While applying to expansion (7.7.30) the operation of an expectation we find the first order moment of the random $Z(s)$:

$$m_z(s) = \alpha_1^z(s) = m_y(s) - Am_x(s) = \sum_{n=1}^{\infty}\ \sum_{\nu_1,\,\ldots,\,\nu_n} \mu_{\nu_1,\,\ldots,\,\nu_n} y_{\nu_1,\,\ldots,\,\nu_n}(s)\,.$$

$$(7.7.32)$$

Analogously the initial moments of the kth order are found. We have

$$Z(s_1)\ldots Z(s_k)$$

$$= \sum_{n_1,\,\ldots,\,n_k=1}\ \sum_{\nu_1^{(1)},\,\ldots,\,\nu_{n_1}^{(1)}}\ \cdots\ \sum_{\nu_1^{(k)},\,\ldots,\,\nu_n^{(k)}} V_{\nu_1^{(1)}}\ldots V_{\nu_{n_1}^{(1)}}\ldots V_{\nu_1^{(k)}}$$

$$\ldots V_{\nu_{n_k}^{(k)}} y_{\nu_1^{(1)},\,\ldots,\,\nu_n^{(1)}}(s_1)\ldots y_{\nu_1^{(k)},\,\ldots,\,\nu_{n_k}^{(k)}}(s_k)\,.\qquad (7.7.33)$$

Hence applying the operation of the expectation to (7.7.33) we get a formal expansion of the kth order moment by the moments of the random variables V_ν:

$$\alpha_k^z(s_1,\,\ldots,\,s_k)$$

$$= \sum_{n_1,\,\ldots,\,n_k=1}^{\infty}\ \sum_{\nu_1^{(1)},\,\ldots,\,\nu_{n_1}^{(1)}}\ \cdots\ \sum_{\nu_1^{(k)},\,\ldots,\,\nu_{n_k}^{(k)}} \mu_{\nu_1^{(1)},\,\ldots,\,\nu_{n_1}^{(1)},\,\ldots,\,\nu_1^{(k)},\,\ldots,\,\nu_{n_k}^{(k)}}$$

$$\times y_{\nu_1^{(1)},\,\ldots,\,\nu_{n_1}^{(1)}}(s_1)\ldots y_{\nu_1^{(k)},\,\ldots,\,\nu_{n_k}^{(k)}}(s_k)\,.\qquad (7.7.34)$$

In practice while using expansion (7.7.34) it is necessary to restrict ourselves to the finite segments of the expansions. Therefore the question arises inevitably about the error estimate of the expansions obtained in such a way and about the convergence of these expansions. From practical point of view the question about the estimate of an expectation of the modulus square of the residual of (7.7.30) is cardinal as the smallness of this variable in comparison with the variance of the sum of accounted terms of expansion (7.7.30) will serve as the criterion of the accuracy of the definition of practically most important moments of the first $\alpha_1^z(s)$

and the second $\alpha_2^z(s)$ orders of the random function $Z(s)$ by means of the correspondent segments of (7.7.34).

E x a m p l e 7.7.5. Under the conditions of Example 7.7.1 let us calculate the expectation m_y of the random function Y with the accuracy till the second order moments relatively to the random variables V_ν and the variance D_y with the accuracy till the fourth order moments relatively to V_ν supposing that the random function is normally distributed. After substituting canonical expansion (II) for the functions $X(t)$ of Example 7.7.1 into initial Eq. (I) and expanding the integral of this equation in powers of V_ν we shall get with the accuracy till the terms of the third power

$$
\begin{aligned}
Y(t) \approx 1 + &\sum_{\nu=-\infty}^{\infty} V_\nu y_\nu(t) + \sum_{\nu_1=-\infty}^{\infty} \sum_{\nu_2=-\infty}^{\infty} V_{\nu_1} V_{\nu_2} y_{\nu_1 \nu_2}(t) \\
+ &\sum_{\nu_1=-\infty}^{\infty} \sum_{\nu_2=-\infty}^{\infty} \sum_{\nu_3=-\infty}^{\infty} V_{\nu_1} V_{\nu_2} V_{\nu_3} y_{\nu_1 \nu_2 \nu_3}(t) .
\end{aligned}
\tag{I}
$$

Here the functions $y_\nu(t)$ are determined by formulae (III) of Example 7.7.1. For determining the functions $y_{\nu_1 \nu_2}(t)$ we differentiate Eq. (I) of Example 7.7.1 once over V_{ν_1} and once over V_{ν_2}. As a result we have

$$
\frac{d}{dt}\left[\frac{\partial^2 Y}{\partial V_{\nu_1} \partial V_{\nu_2}} \right] = -2Y \frac{\partial^2 Y}{\partial V_{\nu_1} \partial V_{\nu_2}} - 2 \frac{\partial Y}{\partial V_{\nu_1}} \frac{\partial Y}{\partial V_{\nu_2}} .
\tag{II}
$$

After dividing Eq. (II) by 2 and assuming $V_\nu = 0$ and accounting that Y in this case is equal to 1, $\partial Y / \partial V_{\nu_1}$ turns into y_ν, and $(1/2)\partial^2 Y / \partial V_{\nu_1} \partial V_{\nu_2}$ into $y_{\nu_1 \nu_2}$ we get

$$
\dot{y}_{\nu_1 \nu_2} = -2 y_{\nu_1 \nu_2} - y_{\nu_1} y_{\nu_2} .
\tag{III}
$$

Differentiating Eq. (II) over V_{ν_2} and putting after this $V_\nu = 0$ we get

$$
\dot{y}_{\nu_1 \nu_2 \nu_3} = -2 y_{\nu_1 \nu_2 \nu_3} - \frac{1}{3}\left(y_{\nu_1} y_{\nu_2 \nu_3} + y_{\nu_2} y_{\nu_1 \nu_3} + y_{\nu_3} y_{\nu_1 \nu_2} \right) .
\tag{IV}
$$

As by the condition $Y(0) = 1$ then the integrals of Eqs. (III) and (IV) at $t = 0$ vanish, $y_{\nu_1 \nu_2}(0) = 0$, $y_{\nu_1 \nu_2 \nu_3}(0) = 0$. After substituting expression (III) of Example 7.7.1 for y_ν into (II) and integrating this equation we find $y_{\nu_1 \nu_2}(t)$. Analogously substituting y_ν and $y_{\nu_1 \nu_2}(t)$ into (IV) we find $y_{\nu_1 \nu_2 \nu_3}(t)$.

By virtue of the fact that the random function $X(t)$ is normally distributed all V_ν are normally distributed, and consequently, V_ν are not only uncorrelated but are independent as well. Therefore all their third order moments are equal to zero and between the fourth order moments only those are different from zero whose indexes coincide pairwise: $\mu_{\nu\nu\nu\nu} = EV_\nu^2 = 3D_\nu^2$, $\mu_{\nu_1\nu_1\nu_2\nu_2} = EV_{\nu_1}^2 EV_{\nu_2}^2 = D_{\nu_1}D_{\nu_2}$. Consequently, after calculating with the stipulated accuracy finally we get

$$
m_y(t) = 1 + \sum_{\nu=-\infty}^{\infty} D_\nu y_{\nu\nu}(t)\,, \quad D_y(t) = \sum_{\nu=-\infty}^{\infty} D_\nu y_\nu^2(t)
$$

$$
+ 2 \sum_{\nu_1=-\infty}^{\infty} \sum_{\nu_2=-\infty}^{\infty} D_{\nu_1} D_{\nu_2} \left[y_{\nu_1\nu_2}^2(t) + 3y_{\nu_1}(t)y_{\nu_1\nu_2\nu_3}(t) \right]\,.
$$

$$(V)$$

Expressing the random function $Y(s)$ in terms of V_ν we have the principal possibility of finding the n-dimensional distributions of the random function at any n. In the general case the determination of the distributions of the random functions by means of the canonical expansions is connected with great calculations difficulties. But for some classes of the nonlinear stochastic systems the canonical expansions method may be sucessfully used together with the distributions parametrization methods stated in Sections 7.2–7.6. Hence the essential advantages connected with the calculation of the integrals with respect to the nonlinear functions in the equations for the parameters of the nongaussian distributions V_ν appear.

E x a m p l e 7.7.6. Let us consider the nonlinear differential system of the form

$$
F(t, D)Y = \varphi(Y) + X(t)\,, \quad X(t) = m_x(t) + \sum_\nu V_\nu x_\nu(t)\,, \qquad (I)
$$

where $F(t, D)$ is the polynomial relatively to the differentiation operator $D = d/dt$ with the coefficients dependent on time t; $\varphi = \varphi(Y)$ is the nonlinear, for instance, the discontinuous function, and $\varphi(0) = 0$; $X(t)$ is the normally distributed random function. Performing by the formulae of Subsection 7.2.2 the statistical linearization of the nonlinear function φ we substitute initial Eqs. (I) by linear one:

$$
F(t, D)Y = k_0(m_y, D_y)m_y + k_1(m_y, D_y)(Y - m_y) + X(t)\,. \qquad (II)
$$

The functions $m_y(t)$ and $y_\nu(t)$ which determine the canonical expansion $Y(t)$ and $K_y(t, t')$ in the form

$$
Y = m_y(t) + \sum_\nu V_\nu y_\nu(t)\,, \quad K_y(t, t') = \sum_\nu D_\nu y_\nu(t)\overline{y_\nu(t)}\,, \qquad (III)
$$

are defined by the following equations, see (II):

$$F(t, D)m_y = k_0(m_y, D_y)m_y + m_x \,,$$
$$F(t, D)y_\nu = k_1(m_y, D_y)y_\nu + x_\nu \,, \tag{IV}$$

where $D_y = \sum\limits_\nu D_y \, |y_\nu(t)|^2$. The joint integration of nonlinear ordinary Eqs. (IV) allows to find $m_y(t)$, $y_\nu(t)$ and $K_y(t, t')$. In the case when the function $X(t)$ has the nongaussian distribution Eqs. (IV) will have the following form:

$$F(t, D)m_y = k_0(m_y, D^{(\mu)})m_y + m_x \,;$$
$$F(t, D)y_\nu = k_{1\nu}(m_y, D^{(\mu)})y_\nu + x_\nu \,, \tag{V}$$

where the coefficients $k_0(m_y, D^{(\mu)})$ and $k_{1\nu}(m_y, D^{(\mu)})$ will depend on the random variables V_μ. Here the coefficients $k_{1\nu}$ will be different for various V_ν.

The canoncial expansions method forms the basis of:
- the methods of equivalent disturbances (Dostupov 1970),
- the methods of the stochastic functional series (Adamian 1983),
- the statistical linearization and the normalization methods based on the canonical expansions (Sinitsyn 1974, Pugachev and Sinitsyn 1978),
- the modifications of the methods of Sections 7.4–7.6 based on the canonical expansions.

7.8. Applications. Analysis of Stochastic Systems and Analytical Modeling

7.8.1. Accuracy Analysis of Single Loop Nonlinear Control Systems

At first let us consider the one-dimensional stationary control system (Fig.7.8.1) consisting of the linear stationary system with the transfer function $\Phi(s)$ and the nonlinear element whose input and output variables are connected by the single-valued nonlinear dependence $U = \varphi(Y)$. The input X we shall assume as the stationary random function with the constant expectation m_x and the spectral density $s_x(\omega)$. Using the statistical linerization of the nonlinear element in the form

$$U = \varphi(Y) \approx \varphi_0(m_y, D_y) + k_1(m_y, D_y)Y^0 \,, \quad Y^0 = Y - m_y \,, \tag{7.8.1}$$

we shall obtain the following dependences between the expectations m_u and m_y and the centred random functions $U^0 = U - m_u$ and Y^0:

$$m_u = k_0(m_y, D_y)m_y \,, \quad U^0 = k_1(m_y, D_y)Y^0 \,, \tag{7.8.2}$$

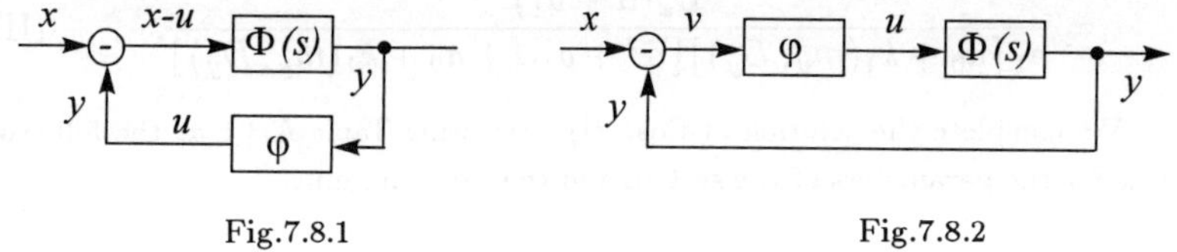

Fig.7.8.1   Fig.7.8.2

where

$$k_0(m_y, D_y) = \varphi_0(m_y, D_y)m_y^{-1}.$$

From the relations (Fig.7.8.1) $m_y = \Phi(0)(m_x - m_u)$, $Y^0 = \Phi(s)(X^0 - U^0)$ accounting (7.8.1), (7.8.2) we have the following statistically linearized operator equations:

$$m_y = \Phi_y^{(0)}(0; m_y, D_y)m_x , \ \Phi_y^{(0)}(0; m_y, D_y) = \frac{\Phi(0)}{1 + k_0(m_y, D_y)\Phi(0)} ,$$

$$Y^0 = \Phi_y^{(1)}(s; m_y, D_y)X^0 , \ \Phi_y^{(1)}(s; m_y, D_y) = \frac{\Phi(s)}{1 + k_1(m_y, D_y)\Phi(s)} .$$
$$\tag{7.8.3}$$

So from Eq. (7.8.3) we get the corresponding equation for the variance

$$D_y = \int_{-\infty}^{\infty} \left| \frac{\Phi(i\omega)}{1 + k_1(m_y, D_y)\Phi(i\omega)} \right|^2 s_x(\omega)d\omega . \tag{7.8.4}$$

Solving numerically Eqs. (7.8.3) and (7.8.4) jointly we find the expectation m_y and the variance D_y.

E x a m p l e 7.8.1. Find the expectation and the variance of the output of the system which consists of the stationary oscillatory link, with the transfer function $\Phi(s) = 1/(s^2 + a_1 s + sa_0)$, and the negative feedback across the relay element arranged in Fig.7.8.1. Assume that the input X is the stationary random process,

$$m_x = \text{const}, \ k_x(\tau) = D_x e^{-\alpha|\tau|}, \ s_x(\omega) = \frac{D_x}{\pi}\frac{\alpha}{\alpha^2 + \omega^2}. \tag{I}$$

From Eqs. (7.8.3) and (7.8.4) we have

$$m_y = \frac{m_x}{a_0 + k_0(m_y, D_y)},$$

$$D_y = \frac{\alpha D_x}{\pi} \int_{-\infty}^{\infty} \frac{d\omega}{(a_1 + \omega^2) \left| a_0 - \omega^2 + a_1 i\omega + k_1(m_y, D_y) \right|^2} \tag{II}$$

$$= \frac{D_x(\alpha + a_1)}{a_1\left[a_0 + k_1(m_y, D_y)\right]\left[\alpha^2 + a_1\alpha + a_0 + k_1(m_y, D_y)\right]}. \tag{III}$$

We complete the solution of Eqs. (I), (II) using Table A.3.1 at the following values for the parameters of the system and the random input:

$$a_0 = 1, \ \alpha^2 = 0.2, \ a_1 = 1/\alpha, \ l = 1, \ m_x = 1. \tag{IV}$$

Substituting values (IV) into Eq. (II) at

$$k_0 = \frac{2l}{m}\Phi\left(\frac{m}{\sqrt{D}}\right) \tag{V}$$

Eq. (II) becomes

$$m_y = \frac{m_y}{m_y + 2\Phi\left(\dfrac{m_y}{\sqrt{D_y}}\right)}. \tag{VI}$$

It is clear that Eq. (VI) is equivalent to the following one:

$$m_y = 1 - 2\Phi\left(\frac{m_y}{\sqrt{D_y}}\right). \tag{VII}$$

From Eq. (III) at values (IV) and relations

$$k_1 = \frac{l}{\sqrt{D}}\Phi'\left(\frac{m}{\sqrt{D}}\right), \quad \Phi'\left(\frac{m}{\sqrt{D}}\right) = \frac{1}{\sqrt{2\pi}}e^{-m^2/2D} \tag{VIII}$$

we get implicit equation for D_y

$$D_y = \frac{1.2D_x}{\left[1 + \dfrac{1}{\sqrt{D_y}}\Phi\left(\dfrac{m_y}{\sqrt{D_y}}\right)\right]\left[2.2 + \dfrac{1}{\sqrt{D_y}}\Phi\left(\dfrac{m_y}{\sqrt{D_y}}\right)\right]} \tag{IX}$$

The results of the numerical solution of Eqs. (VI), (IX) are given in Table 7.8.1.

Table 7.8.1.

D_x	0.1	0.25	0.5	1	2	3
m_y	0.1	0.2	0.2	0.3	0.4	0.5
$\sqrt{D_y}$	0.1	0.1	0.2	0.3	0.5	0.8

For the nonlinear control system illustrated on Fig.7.8.2 we have the following expressions of the transfer functions of the statistically linearized system:

$$\Phi_y^{(0)}(s, m_v, D_v) = \frac{k_0(m_v, D_v)\Phi(s)}{1 + k_0(m_v, D_v)\Phi(s)}, \qquad (7.8.5)$$

$$\Phi_y^{(1)}(s, m_v, D_v) = \frac{k_1(m_v, D_v)\Phi(s)}{1 + k_1(m_v, D_v)\Phi(s)}, \qquad (7.8.6)$$

$$\Phi_v^{(0)}(s; m_v, D_v) = \frac{1}{1 + k_0(m_v, D_v)\Phi(s)}, \qquad (7.8.7)$$

$$\Phi_v^{(1)}(s; m_v, D_v) = \frac{1}{1 + k_1(m_v, D_v)\Phi(s)}. \qquad (7.8.8)$$

Consequently,

$$m_y = \frac{k_0(m_v, D_v)\Phi(0)}{1 + k_0(m_v, D_v)\Phi(0)} m_x, \quad m_v = m_x - m_y = \frac{m_x}{1 + k_0(m_v, D_v)\Phi(0)}, \qquad (7.8.9)$$

$$D_y = k_1^2(m_v, D_v) \int_{-\infty}^{\infty} s_x(\omega) \left| \frac{\Phi(i\omega)}{1 + k_1(m_v, D_v)\Phi(i\omega)} \right|^2 d\omega,$$

$$D_v = \int_{-\infty}^{\infty} \frac{s_x(\omega)d(\omega)}{|1 + k_1(m_v, D_v)\Phi(i\omega)|^2}. \qquad (7.8.10)$$

Solving numerically Eqs. (7.8.9) and (7.8.10) jointly, we find at first m_v and D_v, and further m_y and D_y.

E x a m p l e 7.8.2. Find the accuracy characteristics of the control system (Fig.7.8.2) consisting of the linear drive with the transfer function $\Phi(s) = a/s(Ts + 1)$, the device for comparing the input and the output, and the relay element emitting drive control signals of the constant absolute magnitude. Analogously to the previous example we have at first

$$\Phi_y^{(0)} = \frac{k_0 a}{T^2 s^2 + s + k_0 a}, \qquad (I)$$

and $m_y = m_x$. Then

$$\Phi_v^{(1)} = \frac{s(Ts + 1)}{T^2 s^2 + s + k_1 a}, \qquad (II)$$

$$\Phi_y^{(1)} = \frac{k_1 a}{T^2 s^2 + s + k_1 a}, \qquad (III)$$

$$D_y = \frac{\alpha D_x k_1^2 a^2}{\pi} \int_{-\infty}^{\infty} \frac{d\omega}{(\alpha^2 + \omega^2)\, |k_1 a - T\omega^2 + i\omega|^2}$$

$$= \frac{D_x k_1 a(1 + \alpha T)}{\alpha(1 + \alpha T) + k_1 a}, \tag{IV}$$

$$D_v = \frac{\alpha D_x}{\pi} \int_{-\infty}^{\infty} \frac{\omega^2 (T^2 \omega^2 + 1)\, d\omega}{(\alpha^2 + \omega^2)\, |k_1 a - T\omega^2 + i\omega|^2}$$

$$= \alpha D_x \frac{1 + \alpha T + k_1 a T}{\alpha(1 + \alpha T) + k_1 a}. \tag{V}$$

Here k_0 and k_1 are defined by formulae (V), (VIII) of Example 7.8.1. The results of numerical solution of Eqs. (IV), (V) at $l = T = 1$, $\alpha = 0.5$, $a = 5$ are given in Table 7.8.2.

Table 7.8.2.

$\sqrt{D_x}$	1	1.5	2	2.5	3	3.5	4
$\sqrt{D_v}$	0.7	1.1	1.5	2.0	2.4	2.9	3.3
$\sqrt{D_y}$	1.2	1.7	2.2	2.7	3.1	3.5	4.0
k_0	1.1	0.7	0.5	0.4	0.3	0.3	0.2
k_1	1.2	0.8	0.6	0.5	0.4	0.3	0.3

Table 7.8.2 demonstrates the "choking" effect of the relay element by large fluctuations.

7.8.2. Solid Dynamics in Stochastic Media

Let us consider the dynamics of the solid with one fixed point (Example 1.1.5) in the stochastic media using the following models of the stochastic media:

$$1^0 \quad \begin{aligned} M_x &= -\varphi_0 I_x \omega_x + V_x', \\ M_y &= -\varphi_0 I_y \omega_y + V_y', \\ M_z &= -\varphi_0 I_z \omega_z + V_z', \end{aligned} \tag{7.8.11}$$

V_x', V_y', V_z' being the normal white noises with the equal intensities, $\nu_x' = \nu_y' = \nu_z' = \nu_0$, φ_0 being the viscous coefficient;

$$2^0 \qquad M_x = -\varphi_0 I_x \omega_x + V_x'',$$
$$M_y = -\varphi_0 I_y \omega_y + V_y'', \qquad (7.8.12)$$
$$M_z = -\varphi_0 I_z \omega_z + V_z'',$$

V_x'', V_y'', V_z'' being the normal white noises with intensities satisfying the conditions:
$\nu_x'' = \nu_0 I_x$, $\nu_y'' = \nu_0 I_y$, $\nu_z'' = \nu_0 I_z$;

$$3^0 \qquad M_x = -\varphi_1 I_x \omega_x + V_x''',$$
$$M_y = -\varphi_2 I_y \omega_y + V_y''', \qquad (7.8.13)$$
$$M_z = -\varphi_3 I_z \omega_z + V_z''',$$

V_x''', V_y''', V_z''' being the independent normal white noises with the intensities ν_x''', ν_y''', ν_z''', φ_1, φ_2, φ_3 being the viscous coefficients.

In case 1^0 the Itô–Euler dynamical equations have the following form:

$$I_x \dot{\omega}_x + (I_z - I_y)\omega_y \omega_z = -\varphi_0 I_x \omega_x + V_x',$$
$$I_y \dot{\omega}_y + (I_x - I_z)\omega_z \omega_x = -\varphi_0 I_y \omega_y + V_y', \qquad (7.8.14)$$
$$I_z \dot{\omega}_z + (I_y - I_x)\omega_x \omega_y = -\varphi_0 I_x z \omega_z + V_z'.$$

Eqs. (7.8.14) have the stationary regime of fluctuations with the following one-dimensional density:

$$f_1(\omega_x, \omega_y, \omega_z) = c \exp\left[-(\varphi_0 \eta / \nu_0)\right], \qquad (7.8.15)$$

where

$$\eta = I_x^2 \omega_x^2 + I_y^2 \omega_y^2 + I_z^2 \omega_z^2, \quad c = \left[\varphi_0 / (\pi \nu_0)\right]^{3/2} I_x I_y I_z.$$

More than that. In the nonstationary regime of fluctuations Eqs. (7.8.14) may be transformed into the following set of the independent linear equations

$$\frac{dL_i}{dt} = -\varphi_0 L_i + V_i \quad (i = 1, 2, 3). \qquad (7.8.16)$$

From Eq. (7.8.16) using the results of Subsection 6.2.3 we get the explicit formula for the multi-dimensional characteristic fucntions for the variables L_i $(i = 1, 2, 3)$,

$$g_n(\lambda_1, \ldots, \lambda_n; t_1, \ldots, t_n) = g_0 \left(\sum_{k=1}^{n} e^{-\varphi_0 t_k} \lambda_k \right)$$

$$\times \exp\left\{ -\frac{\nu_0}{4\varphi_0} \sum_{h=1}^{n} \sum_{l=1}^{n} \left(e^{-\varphi_0 |t_h - t_l|} - e^{-\varphi_0 (t_h + t_l)} \right) \lambda_h \lambda_l \right\}, \qquad (7.8.17)$$

where $g_0(\lambda)$ being the characteristic function of the initial conditions: $L_i(t_0)$. Hence the density of the variable $\eta = L_1^2 + L_2^2 + L_3^2 \geq 0$ and its first and second moments are

$$f_\eta(\eta) = (\varphi_0/\nu_0)\Gamma(3/2)\sqrt{\eta}\exp(-\varphi_0\eta/\nu_0), \qquad (7.8.18)$$

(Γ being the gamma-function, see Appendix 7),

$$E\eta = 3\nu_0/(2\varphi_0), \quad D\eta = 3\nu_0/(4\varphi_0). \qquad (7.8.19)$$

In case 2^0 Eqs. (7.8.13) have the stationary regime of fluctuations with the following one-dimensinal density:

$$f_1(\omega_x, \omega_y, \omega_z) = c\exp\left(-\frac{2\varphi_0}{\nu_0}T\right), \qquad (7.8.20)$$

where

$$2T = I_x\omega_x^2 + I_y\omega_y^2 + I_z\omega_z^2, \quad c = \sqrt{I_x I_y I_z}\left(\frac{\varphi_0}{\pi\nu_0}\right)^{3/2}.$$

Using Itô formula (5.1.7) we get the first order nonlinear stochastic differential equation for the variable T,

$$\dot{T} = -2\varphi_0 T + 3\nu_0/2 + \sqrt{2T}\, V. \qquad (7.8.21)$$

Here V being the normmal white noise of the intensity ν_0. The equations for the multi-dimensional characteristic function $g_n'(\lambda_1, \ldots, \lambda_n; t_1, \ldots, t_n)$ for $T(t)$ defined by Eq. (7.8.21) are as follows:

$$\frac{\partial g_n'}{\partial t_n} = \lambda_n(-2\varphi_0 + i\nu_0\lambda_n)\frac{\partial g_n'}{\partial \lambda_n} + \frac{3}{2}i\nu_0\lambda_n g_n'. \qquad (7.8.22)$$

The solution of Eq. (7.8.22) for $n = 1, 2, \ldots$ according to Subsection 6.2.2 at the intial conditions

$$\begin{aligned}
&g_1'(\lambda; t_0) = g_0'(\lambda), \\
&g_n'(\lambda_1, \ldots, \lambda_n; t_1, \ldots, t_n, t_{n-1}, t_{n-1}) \\
&\quad = g_{n-1}'(\lambda_1, \ldots, \lambda_{n-2}, \lambda_{n-1} + \lambda_n; t_1, \ldots, t_{n-1})
\end{aligned} \qquad (7.8.23)$$

may be presented in the following explicit form:

$$g_1'(\lambda; t) = g_0'(\lambda e^{-2\varphi_0(t-t_0)}\psi^{-1}(\lambda, t - t_0)\psi^{-3/2}(\lambda, t - t_0)), \qquad (7.8.24)$$

$$g_n'(\lambda_1, \ldots, \lambda_n; t_1, \ldots, t_n) = g_{n-1}(\lambda_1, \ldots, \lambda_{n-2}, \lambda_{n-1} + \lambda_n e^{-\varphi_0(t_n - t_{n-1})}$$

$$\psi^{-1}(\lambda_n, t_n - t_{n-1}); t_1, \ldots, t_{n-1})\psi^{-3/2}(\lambda_n, t_n - t_{n-1}) \quad (n = 2, 3, \ldots),$$
$$(7.8.25)$$

where

$$\psi(\lambda, \tau) = 1 - i\lambda\nu_0(1 - e^{-2\varphi_0\tau})/(2\varphi_0). \qquad (7.8.26)$$

In special case $g_0(\lambda) = (1 - i\lambda\delta)^{-3/2}$ from formula (7.8.24) we have

$$g_1(\lambda; t) = \left[1 - i\lambda\frac{\nu_0}{2\varphi_0} + i\lambda\left(\frac{\nu_0}{2\varphi_0} - \delta\right)e^{-2\varphi_0(t-t_0)}\right]^{-3/2}, \qquad (7.8.27)$$

$$g_1(\lambda; \infty) = (1 - i\lambda\nu_0/2\varphi_0)^{-3/2}. \qquad (7.8.28)$$

Formula (7.8.28) corresponds to the χ^2-distrubution with three degree of freedom.

From Eq. (7.8.21) we have the linear equations for the expectation $m' = ET$ and the variance $D' = DT$,

$$\frac{dm'}{dt} = -2\varphi_0 m' + \frac{3\nu_0}{2}, \quad \frac{dD'}{dt} = -4\varphi_0 D' + 2\nu_0 m'. \qquad (7.8.29)$$

In case 3^0 we have the following Itô–Euler dynamical equations

$$\dot{\omega}_x = e_1\omega_y\omega_z - \varphi_1\omega_x + V_1,$$
$$\dot{\omega}_y = e_2\omega_z\omega_x - \varphi_2\omega_y + V_2, \qquad (7.8.30)$$
$$\dot{\omega}_z = e_3\omega_x\omega_y - \varphi_3\omega_z + V_3,$$

where

$$e_1 = \frac{I_y - I_z}{I_x}, \quad e_2 = \frac{I_z - I_x}{I_y}, \quad e_3 = \frac{I_x - I_y}{I_z},$$

$$V_1 = \frac{V_x'''}{I_x}, \quad V_2 = \frac{V_y'''}{I_y}, \quad V_3 = \frac{V_z'''}{I_z}.$$

From Eqs. (7.8.30) using the moments method (Subsectio 7.4.1) we get the infinite set of equations for the initial moments $\alpha_{nmk} = E\omega_x^n\omega_y^m\omega_z^k$,

$$\dot{\alpha}_{nmk} = -(n\varphi_1 + m\varphi_2 + k\varphi_3)\alpha_{nmk} + ne_1\alpha_{n-1,m+1,k+1} + me_2\alpha_{n+1,m-1,k+1}$$

$$+ke_3\alpha_{n+1,m+1,k-1} + \frac{1}{2}\nu_1 n(n-1)\alpha_{n-2,m,k} + \frac{1}{2}\nu_2 m(m-1)\alpha_{n,m-2,k}$$

$$+\frac{1}{2}\nu_3 k(k-1)\alpha_{n,m,k-2}. \qquad (7.8.31)$$

With the accuracy up to the moments of the fourth order we get from Eq. (7.8.31):

$$\alpha_{100} = \alpha_{010} = \alpha_{110} = \alpha_{101} = \alpha_{011} = 0,$$
$$\alpha_{200} = \nu_1/(2\varphi_1), \quad \alpha_{020} = \nu_2/(2\varphi_2), \quad \alpha_{002} = \nu_3/(2\varphi_3). \tag{7.8.32}$$

R e m a r k 1. The nonlinear stochatic differential equations for the solid in the stochastic media admit the exact normal stationary solution

$$f(\omega) = c \exp(-\omega^T \Lambda \omega). \tag{7.8.33}$$

Here

$$\omega = [\omega_x \omega_y \omega_z]^T, \quad \Lambda = \lambda I + \mu I^2 = I\nu^{-1}\varphi, \quad c = (2\pi)^{-3/2} |\Lambda|^{1/2},$$

$$I = \begin{bmatrix} I_x & 0 & 0 \\ 0 & I_y & 0 \\ 0 & 0 & I_z \end{bmatrix}, \quad E_3 = \begin{bmatrix} 1 & 0 & 0 \\ 0 & 1 & 0 \\ 0 & 0 & 1 \end{bmatrix},$$

$$\nu = \begin{bmatrix} \nu_x & 0 & 0 \\ 0 & \nu_y & 0 \\ 0 & 0 & \nu_z \end{bmatrix}, \quad \varphi = \begin{bmatrix} \varphi_1 & 0 & 0 \\ 0 & \varphi_2 & 0 \\ 0 & 0 & \varphi_3 \end{bmatrix},$$

where λ and μ are some real coefficients.

7.8.3. Stochastic Hereditary Systems

Many problems of the theory of hereditary systemms (for instance, the systems which contain the elements described by hereditary mechanics, physics, bilogy, chemistry) under the conditions of random disturbances lead to the study of the following stochastic integrodifferential system:

$$\dot{X} = a(X,t) + \int_{t_0}^{t} a_1(X(\tau),\tau,t)d\tau$$

$$+ \left[b(X,t) + \int_{t_0}^{t} b_1(X(\tau),\tau,t)d\tau \right] \dot{W}, \quad X(t_0) = X_0, \tag{7.8.34}$$

which is considered in the Itô sense. Here $X \in R^q$ is the state vector; $\dot{W} \in R^q$ is the vector white noise of the intensity $\nu = \nu(t)$ independent of the initial condition X_0; $W = W(t)$ is the random second order process with the independent increments;

$$a = a(X,t), a \; : \; R^q \times R \to R^p;$$
$$a_1 = a_1(X(\tau),\tau,t), a_1 \; : \; R^p \times R \times R \to R^p;$$
$$b = a(X,t), b \; : \; R^q \times R \to R^{pq};$$
$$b_1 = b_1(X(\tau),\tau,t), b_1 \; : \; R^p \times R \times R \to R^{pq}.$$

In applications the functions a_1 and b_1 are represented in the form

$$a_1 = A(t,\tau)\varphi(X(\tau),\tau), \quad b_1 = \sum_{h=1}^{N} B_h(t,\tau)\psi_h(X((\tau),\tau), \qquad (7.8.35)$$

where hereditary kernels $A(t,\tau)$ and $B_h(t,\tau)$ are $(p \times p)$- and $(p \times q)$-martix functions, $\psi_h(X(\tau),\tau)$ is the $(q \times q)$-matrix function, $\varphi(X(\tau),\tau)$ is the p-dimensional vector fucntion. Let us suppose that the hereditary kernels $A(t,\tau) = \{A_{ij}(t,\tau)\}$, $i,\,j = \overline{1,p}$ and $B_h(t,\tau) = B_{hij}(t,\tau)$, $h = \overline{1,N}$, $i = \overline{1,p}$, $j = \overline{1,q}$ satisffy the conditions of the nonanticipativeness and the conditions of dying hereditary (or assymptotical stability)

$$A_{ij}(t,\tau) = 0, \quad B_{hij}(t,\tau) = 0, \quad \forall \tau > t, \qquad (7.8.36)$$

$$\int_{-\infty}^{\infty} |A_{ij}(t,\tau)|\,d\tau < \infty, \quad \int_{-\infty}^{\infty} |B_{hij}(t,\tau)|\,d\tau < \infty. \qquad (7.8.37)$$

In the case where the hereditary kernels satisfy the conditions

$$A_{ij}(t,\tau) = \tilde{A}_{ij}(\xi), \quad B_{hij}(t,\tau) = \tilde{B}_{hij}(\xi), \quad \xi = t - \tau, \qquad (7.8.38)$$

one say about the invariance of the stationary of the hereditary. The important class of hereditary kernels are singular kernels

$$A_{ij}(t,\tau) = A_{ij}^{+}(t)A_{ij}^{-}(\tau), \quad B_{hij}(t,\tau) = B_{hij}^{+}(t)B_{hij}^{-}(\tau). \qquad (7.8.39)$$

The simplest exmaples of the functions satisfying conditions (7.8.36)–(7.8.39) are the functions $\exp(-\alpha\,|\xi|)\mathbf{1}(\xi)$, $\exp(-\alpha\,|\xi|)[\cos\omega\xi + \gamma\sin\omega\,|\xi|]\mathbf{1}(\xi)$, $\mathbf{1}(\xi)$ is the unit step function.

One of the most effective ways of obtaining the statistical dynamics equations for Eq. (7.8.34) is their reduction to the stochastic differential equations by means of the approximation of the real hereditary kernels by the functions satisfying some ordinary differential equations.

Let us consider Eq. (7.8.34) under condition (7.8.35)

$$\dot{X} = a(X,t) + \int_{t_0}^{t} A(t,\tau)\varphi(X(\tau),\tau)d\tau$$

$$+ \left[b(X,t) + \sum_{h=1}^{N} \int_{t_0}^{t} B_h(t,\tau)\psi_h(X(\tau),\tau)d\tau \right] \dot{W}, \quad X(t_0) = X_0. \quad (7.8.40)$$

We shall determine a class of admissible kernels by conditions (7.8.36), (7.8.37) and by the order and the form of linear ordinary equations to which the kernels as the functions of t and τ must satisfy.

Let the kernels $A_{ij}(t,\tau)$ and $B_{hij}(t,\tau)$ in (7.8.40) satisfy (7.8.36), (7.8.37) and at fixed τ be the solutions of linear differential equations

$$\sum_{i=1}^{p} F_{ri}^t A_{ij}(t,\tau) = H_{rj}^t \delta(t-\tau); \ r,j = \overline{1,p},$$

$$\sum_{i=1}^{p} \Phi_{hri}^t B_{hij}(t,\tau) = \Psi_{hrj}^t \delta(t-\tau), \ r = \overline{1,p}, \ j = \overline{1,q},$$

$$(7.8.41)$$

at a fixed t the kernels are determined by the formulae

$$A_{lj}(t,\tau) = \sum_{s=1}^{p} H_{sj}^{*\vartheta} A_{is}'(t,\tau), \ l,j = \overline{1,p},$$

$$B_{hij}(t,\tau) = \sum_{s=1}^{p} \Psi_{hsj}^{*\tau} B_{his}'(t,\tau), \ l = \overline{1,p}, \ j = \overline{1,q},$$

$$(7.8.42)$$

and the functions $A_{is}'(t,\tau)$ and $B_{his}'(t,\tau)$ entering into (7.8.42) at a fixed t are the solutions of the linear differential equations

$$\sum_{s=1}^{p} F_{sr}^{*\tau} A_{ls}'(t,\tau) = \delta_{rl} \delta(t-\tau), \ l,r = \overline{1,p},$$

$$\sum_{s=1}^{p} \Phi_{hsr}^{*\tau} B_{hls}'(t,\tau) = \delta_{rl} \delta(t-\tau), \ r = \overline{1,p}, \ l = \overline{1,q}.$$

$$(7.8.43)$$

Here $F_{rl} = F_{rl}(t,D)$, $H_{rl} = H_{rl}(t,D)$, $\Phi_{hrl} = \Phi_{hrl}(t,D)$, $\Psi_{hrl} = \Psi_{hrl}(t,D)$ are some linear differential operators, the order of the operators H_{rl} and Ψ_{hrl} denoted by m, the order of F_{rl} and Ψ_{hrl} is denoted by n, $n \geq m$, the superscript t at the operator indicates that the operator acts on the function considered as the function of t at a fixed τ.

As it is known from the linear differential systems theory (Section 1.2) the integral terms which enter into (7.8.40)

$$Y_i = \int_{t_0}^{t} \sum_{k=1}^{p} A_{ik}(t,\tau)\varphi_k(X(\tau),\tau)d\tau, \ i = \overline{1,p},$$

$$U_{hij} = \int_{t_0}^{t} \sum_{k=1}^{a} B_{hik}(t,\tau)\psi_{hkj}(X(\tau),\tau)d\tau, \ i = \overline{1,p}, \ j = \overline{1,q},$$

$$(7.8.44)$$

represent the solutions of the following equations:

$$\sum_{k=1}^{p} F_{rk} Y_k = \sum_{k=1}^{p} H_{rk}\varphi_k, \quad r = \overline{1,p},$$

$$\sum_{l=1}^{q}\sum_{k=1}^{p} \Phi_{hrk} U_{hkl} = \sum_{l=1}^{q}\sum_{k=1}^{q} \Psi_{hrk}\psi_{hkl}, \quad r = \overline{1,p}. \tag{7.8.45}$$

of their kernels satisfy (7.8.41)–(7.8.43).

It is clear that here the nonlinear functions $\varphi_k = \varphi_k(X,t)$ and $\psi_{hkl}(X,t)$ must be differentiable not less than m times.

It is evident that if we introduce the vectors Z' and ε' of the dimensions $p(1 + qN) \times 1$ and $(p + q^2 N) \times 1$ and the vectors U and ε of dimensions $pqN \times 1$ and $q^2 N \times 1$ assuming

$$Z' = \left[Y^T U^T \right]^T, \quad Y = \left[Y_1 \ldots Y_p \right]^T,$$

$$U = \left[U_1^T \ldots U_N^T \right]^T, \quad U_h = \left[U_{h1}^T \ldots U_{hp}^T \right]^T,$$

$$h = \overline{1,N}, \quad \varepsilon' = \left[\varphi^T \varepsilon^T \right]^T, \quad \varphi = \left[\varphi_1 \ldots \varphi_p \right]^T, \varepsilon = \left[\psi_1^T \ldots \psi_N^T \right]^T. \tag{7.8.46}$$

Then the scalar Eqs. (7.8.45) may be written in the form

$$F' Z' = H' \varepsilon'. \tag{7.8.47}$$

Here the matrix differential operators F' and H' are

$$F' = \{F'_{rl}(t, D)\} = \sum_{k=0}^{n} \alpha_k D^k, \quad r, l = \overline{1, p(1 + qN)},$$

$$H' = \{H'_{rl}(t, D)\} = \sum_{k=0}^{m} \beta_k D^k, \quad r, l = \overline{1, (p + q^2 N)}, \tag{7.8.48}$$

where $\alpha_k = \alpha_k(t)$ and $\beta_k = \beta_k(t)$ are the $(p(1 + qN) \times p(1 + qN))$ and $(p + q^2 N) \times (p + q^2 N)$ matrix coefficients. Further we shall replace (7.8.47) by the set of equations in the Cauchy for according to the known rules of (Subsection 1.2.3):

$$\dot{Z}''_k = Z''_{k+1} + q_k \varepsilon', \quad k = \overline{1, n-1}, \quad \dot{Z}''_n = -\sum_{l=1}^{n} \alpha_n^{-1}\alpha_{l-1} Z''_l + q_n \varepsilon'. \tag{7.8.49}$$

Here Z_h'' are the $p(1+qN)$-dimensional vectors, $h = \overline{1,n}$, $q_k = q_k(t)$ being the $p(1+qN) \times q^2 N$ matrix determined by the formulae

$$q_k = \alpha_k^{-1}\left[\beta_{n-1} - \sum_{h=0}^{k-1}\sum_{l=0}^{k-h} C_{n-k-l}^{n-k}\alpha_{n-k+h+l}q_h^{(l)}\right], \quad k = \overline{1,n-1},$$

$$q_n = \alpha_n^{-1}\left[\beta_0 - \sum_{h=0}^{n-1}\sum_{l=0}^{n-h} \alpha_{h+l}q_h^{(l)}\right]. \tag{7.8.50}$$

Here the dependence between Z' and ε' will exist. So

$$Z' = Z_1'' + q_0\varepsilon', \quad q_0 = \alpha_n^{-1}\beta_n. \tag{7.8.51}$$

By virtue of (7.8.46) and (7.8.51), integral terms (7.8.44) take the form

$$Y = \int_{t_0}^t A(t,\tau)\varphi(X(\tau),\tau)d\tau = a'(Z_1'' + q_0\varepsilon'),$$

$$\sum_{h=1}^N U_h = \sum_{h=1}^N \int_{t_0}^t B_h(t,\tau)\psi_h(X(\tau),\tau)d\tau = \left(\sum_{h=1}^N b_h'\right)(Z_1'' + q_0\varepsilon'),$$

$$\tag{7.8.52}$$

where the following notations are introduces:

$$a' = [I_p\ 0], \quad b_h' = \Lambda_h[0\ I_{pqN}], \quad \Lambda_h\begin{bmatrix} & & 1 & & \\ 0 & \cdots & & \cdots & 0 \\ & & n & & \end{bmatrix}, \tag{7.8.53}$$

and I_p is the $(p \times p)$ unit matrix. Consequently, Eq. (7.8.40) together with Eqs. (7.8.49), (7.8.52) for the extended state vector of the dimension $np(1+qN)$,

$$Z = \left[X^T Z^T\right]^T, \quad Z'' = \left[Z_1''^T \ldots Z_n''^T\right]^T$$

is reduced to

$$\dot{Z} = c(Z,t) + l(Z,t)\dot{W}, \quad Z(t_0) = Z_0, \tag{7.8.54}$$

where

$$c(Z,t) = \begin{bmatrix} a(X,t) + a'\left[Z_1'' + q_0(t)\varepsilon'(X,t)\right] \\ \gamma(t)Z'' + \gamma'(t)\varepsilon'(X,t) \end{bmatrix},$$

$$l(Z,t) = \begin{bmatrix} b(X,t) + \left(\sum_{h=1}^N b_h'\right)\left[Z_1'' + q_0(t)\varepsilon'(X,t)\right] \\ 0 \end{bmatrix},$$

$$Z_0 = \left[X_0^T 0 \right]^T, \quad \gamma'(t) = \text{diag} \left[q_1(t) \ldots q_n(t) \right],$$

$$\gamma(t) = \begin{bmatrix} I & 0 & \ldots & 0 \\ 0 & I & \ldots & 0 \\ \ldots & \ldots & \ldots & \ldots \\ 0 & 0 & \ldots & I \\ -\alpha_n^{-1}\alpha_0 & -\alpha_n^{-1}\alpha_1 & \ldots & -\alpha_n^{-1}\alpha_{n-1} \end{bmatrix}, \quad I = I_{p(1+qN)}.$$

$$(7.8.55)$$

The matrices a', b_h' which enter into (7.8.51) are determined by (7.8.53); q_l, $l = \overline{0,n}$ are determined by (7.8.50), (7.8.51); ε' is determined by (7.8.46). The dimensions of $c(Z,t)$ and $l(Z,t)$ are equal to $np(2+qN)$ and $np(2+qN) \times q$. Eqs. (7.8.50) are linear in Z'' and nonlinear in X.

Let us consider now the class of kernels satisfying the conditions (7.8.36) and the condition of sationarity (7.8.38). As it is known from the theory of linear stationary systems (Subsection 1.2.5) in this case it is sufficient to suppose instead of (7.8.37) that the Laplace transforms of hereditary kernels $\tilde{A}(\xi)$ and $\tilde{B}_h(\xi)$ are rational functions of the scalar variable s, i.e. admit the representation of the form

$$\int_0^\infty \tilde{A}(\xi)e^{-s\xi}d\xi = F(s)^{-1}H(s), \quad \int_0^\infty \tilde{B}_h(\xi)e^{-s\xi}d\xi = \Phi_h(s)^{-1}\Psi(s).$$

$$(7.8.56)$$

Here the order of the matrix polynomials $H(s) = \{H_{rl}(s)\}$ and $\Psi_h(s) = \{\Psi_{hrl}(s)\}$ is equal to m and the order of the polynomials $F(s) = \{F_{rl}(s)\}$ and $\Phi_h(s) = \{\Phi_{hrl}(s)\}$ is equal to n, $n \geq m$. So Eq. (7.8.40) by means of extending the state vector up to dimension $np(2+qN)$ may be reduced to Eq. (7.8.54).

In conclusion we consider the class of the hereditary kernels satisfying the conditions (7.8.36), (7.8.37) and (7.8.39). In such a case the order of the operators F_{ri} and Φ_{hri} is equal to one, $n = 1$, the order of the operators H_{rj} and Ψ_{hrj} is equal to 0, $m = 0$. In this special case Eqs. (7.8.44), (7.8.45) take the following form:

$$\int_{t_0}^t A(t,\tau)\varphi(X(\tau),\tau)d\tau = A^+Y; \quad Y = [Y_1 \ldots Y_p]^T,$$

$$(7.8.57)$$

$$\int_{t_0}^t B_h(t,\tau)\psi_h(X(\tau),\tau)d\tau = B^+U_h; \quad U_h = \left[U_{h1}^T \ldots U_{hq}^T \right]^T,$$

$$\dot{Y} = A^-Y, \quad Y(t_0) = 0, \quad \dot{U}_h = B_h^-U_h, \quad U_h(t_0) = 0. \qquad (7.8.58)$$

Here $A^\pm = \left\{ A_{ij}^\pm(t) \right\}$ and $B_h^\pm = \left\{ B_{hij}^\pm(t) \right\}$ are $(p \times p)$- and $(p \times q)$-matrices, Y and $U_{hi} = [U_{hi}, \ldots, U_{hip}]^T$ are the p-dimensional vectors. As a result

Eq. (7.8.40) is reduced to Eq. (7.8.54) for the extended $p(2 + qN)$-dimensional state vector $Z = \left[X^T Y^T U^T \right]^T$ if we introduce (pqN)-dimensional vector $U = \left[U_1^T \ldots U_N^T \right]^T$, the $p(2 + qN)$-dimensional vector function $c(Z, t)$, the $(p(2 + qN) \times q)$-dimensional matrix-function $l(Z, t)$ and the initial $p(2 + qN)$-dimensional vector Z_0 assuming

$$c(Z, t) = \begin{bmatrix} a(X, t) + A^+ Y \\ A^- Y \\ B^- U \end{bmatrix}, \quad l(Z, t) = \begin{bmatrix} l(X, t) + B^+ U \\ 0 \\ 0 \end{bmatrix},$$

$$B^\pm = \left[B_1^\pm \ldots B_N^\pm \right], \quad Z_0 \left[X_0^T \ 0 \ 0 \right]^T. \tag{7.8.59}$$

The functions $c(Z, t)$ and $l(Z, t)$ are linear in Y and U and nonlinear in X. We emphasize that in the case of singular kernels the differentiability of φ and ψ_h is not obligatory. So Eq. (7.8.40) by means of extending the state vector up to the $p(2 + qN)$-dimension may be reduced to (7.8.54) under condition (7.8.59).

E x a m p l e 7.8.3. The scalar linear stochastic integrodifferential system

$$\dot{X} = a_0 + a_1 X + A \int_0^t e^{-\alpha(t-\tau)} X(\tau) d\tau + b \dot{W}, \quad X_0 = 0, \tag{I}$$

$(a_0, a_1, A, b, \alpha$ are constant) is reduced to the following linear two-dimensional stochastic differential system:

$$\dot{X} = a_0 + a_1 X + (A/\alpha) Y + b \dot{W}, \quad \dot{Y} = \alpha(X - Y), \quad X_0 = Y_0 = 0. \tag{II}$$

E x a m p l e 7.8.4. The scalar system with the parametric noise

$$\dot{X} = a_0 + a_1 X + \left[b + B \int_0^t e^{-\beta(t-\tau)} X(\tau) d\tau \right] \dot{W}, \quad X_0 = 0, \tag{I}$$

$(a_0, a_1, b, B, \beta$ are constant) is reduced to the following two-dimensional stochastic differential with the parametric noise:

$$\dot{X} = a_0 + a_1 X + [b + (B/\beta) Y] \dot{W}, \quad \dot{Y} = \beta(X - Y), \quad X_0 = Y_0 = 0. \tag{II}$$

E x a m p l e 7.8.5. The scalar linear nonlinear stochastic system

$$\dot{X} = a_0 + a_1 X + \int_0^t e^{-\alpha(t-\tau)} \varphi(X(\tau)) d\tau + b \dot{W}, \quad X_0 = 0, \tag{I}$$

(a_0, a_1, α, b are constant) is reduced to the following nonlinear two-dimensional stochastic differential system:

$$\dot{X} = a_0 + a_1 X + (1/\alpha)Y + b\dot{W}, \quad \dot{Y} = \alpha\left[\varphi(X) - Y\right], \quad X_0 = Y_0 = 0. \quad \text{(II)}$$

7.8.4. Normalization of Risk and Profit Financial Transactions

In stochastic finance as it is known (Shiryaev 1999) there are widely used continuous and discrete stochastic models defined by the stochastic differential and difference equations.

Normalization of such stochastic models by NAM gives the possibility to design the specialized algorithms for various problem of the financial analysis and modeling. The results of the statistical simulation given in (Artemiev et al 1996) for the following types of the stochastic differential equation

$$\dot{Y} = \alpha_0(t)\left[Y^*(t) - Y\right] + \beta_0(t)Y^\gamma V, \qquad (7.8.60)$$

$$\begin{aligned}
\dot{Y}_1 &= \alpha_{01}(t)\left[Y_1^*(t) - Y_1\right] + \beta_{01}(t)Y_1^{\gamma_1} V_1, \\
\dot{Y}_2 &= \alpha_{02}(t)\left[Y_2^*(t) - Y_2\right] + \beta_{02}(t)Y_2^{\delta_2} Y_3^{\gamma_2} V_2, \\
\dot{Y}_3 &= \alpha_{03}(t)(Y_3 - Y_4) + \beta_{03}(t)Y_3^{\gamma_3} V_3, \\
\dot{Y}_4 &= \alpha_{04}(t)(Y_3 - Y_4)
\end{aligned} \qquad (7.8.61)$$

($\alpha_0(t)$, $\alpha_{0i}(t)$, $\beta_{0i}(t)$, $Y_i^*(t)$ being some known time functions) confirm practically good NAM accuracy (5-10%) at γ_1, δ_2, γ_2, $\gamma_3 \leq 1$.

7.8.5. Normalization of Queueing Systems

An important class of the nonlinear stochastic systems is the class of the queueing systems. For normalization of the queueing systems it is expedient to decompose the queueing system into the typical subsystems and use the normalization method to each subsystem separately. Because of the completeness of the class of the normal systems relatively to the connections the normal model of the queueing system is obtained as the result.

As it is known any queueing system may be considered as a system consisting of the following typical elements: the queueing channels, the waiting places in queue, the distributor $\mathcal{D}$ which determines the direction of the motion of every client in queue depending on the availability of the free channels or the places in queue with the account of the priority. The stream of the clients which is directed into the given channel or at the given place in queue serves as the input of each channel and each place in queue. The stream of clients served by the channel and the signal fixing the channel (is it free or occupied) are the outputs of the channel. The streams of the clients and the signals of the feedback from the channels and the places in queue carrying the information about the state of each channel and each place in queue serve as the inputs of the distributor. The streams of the clients directed into the channels and in queue and the stream of the clients who received the refusals in service will be the outputs of the distributor $\mathcal{D}$. Example of scheme of the queueing system with n channels and without queue is illustrated on Fig.7.8.3.

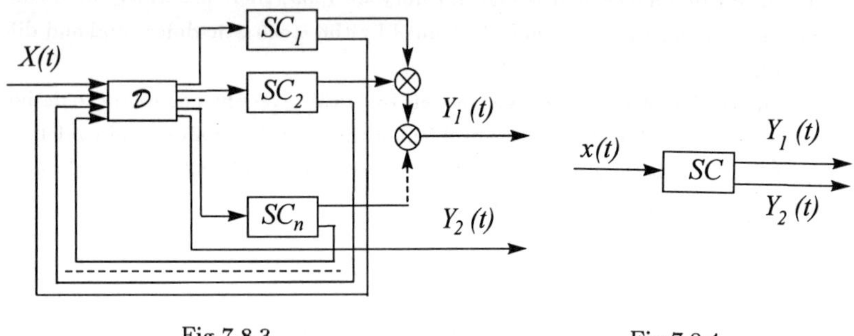

Fig.7.8.3   Fig.7.8.4

We show how we may normalize the service channel, main element of the queueing system (Fig.7.8.4). Let $\varphi(\sigma)$ be a probability density of the queueing time T. The input of a channel, i.e. the stream of the clients entering into the given channel, may be presented in the form of the sequence of unit impulses correspondent to the instants of the clients arrivals. It is also convenient to present the output stream of the clients in the form of the sequence of unit impulses which appear at the instants when the clients leave the channel. The other output of the channel which carries the information about its state may be presented as the binary signal equal to zero in the channel is free and is equal to 1 if the channel is occupied. At such agreement the channel will response on unit input impulse $\delta(t - \tau)$ which acts at the definite instant τ by unit impulse $\delta(t - \tau - T)$ at the random instant $(\tau + T)$ on one input and by the signal equal to 1 inside of the random interval $(\tau, \tau + T)$ and equal to zero outside this interval on another output:

$$x(t) = \delta(t - \tau), \ \ Y_1(t) = \delta(t - \tau - T), \ \ Y_2(t) = 1(t - \tau)1(T + \tau - t).$$
$$(7.8.62)$$

Taking into consideration that the response of a linear system on unit impulse represents its weighting function we shall consider queueing channel as a linear system with the following random weighting functions:

$$G_1(t, \tau) = \delta(t - \tau - T), \ \ G_2(t, \tau) = 1(t - \tau)1(T + \tau - t). \quad (7.8.63)$$

Then its response on any determinate input $x(t)$ will be expressed by formulae

$$Y_1(t) = \int_{t_0}^{t} \delta(t - \tau - T)x(\tau)d\tau, \quad Y_2(t) = \int_{t_0}^{t} 1(T + \tau - t)x(\tau)d\tau. \quad (7.8.64)$$

Now for the linearization of the regression it is sufficient to find the conditional expectations of the outputs $Y_1(t)$ and $Y_2(t)$ at a given realization $x(t)$ of the random signal $X(t)$. As a result we obtain

$$E[Y_1(t)|x] = \int_{t_0}^{t} x(\tau)d\tau \int_{0}^{\infty} \delta(t - \tau - \sigma)\varphi(\sigma)d\sigma = \int_{t_0}^{t} \varphi(t - \tau)x(\tau)d\tau,$$

$$(7.8.65)$$

$$E[Y_2(t)|x] = \int_{t_0}^{t} x(\tau)d\tau \int_{0}^{\infty} 1(\sigma + \tau - t)\varphi(\sigma)d\sigma = \int_{t_0}^{t} \psi(t - \tau)x(\tau)d\tau,$$

$$(7.8.66)$$

where $\psi(t) = \int_{t}^{\infty} \varphi(\sigma)d\sigma = 1 - F(t)$ is the probability of the fact that queueing time T will exceed t. Formulae (7.8.65) and (7.8.66) show that the queuing channel may be considered as a linear system with the weighting functions $\varphi(\zeta)$ and $\psi(\zeta)$ and with the additive noises on its outputs (Fig.7.8.5).

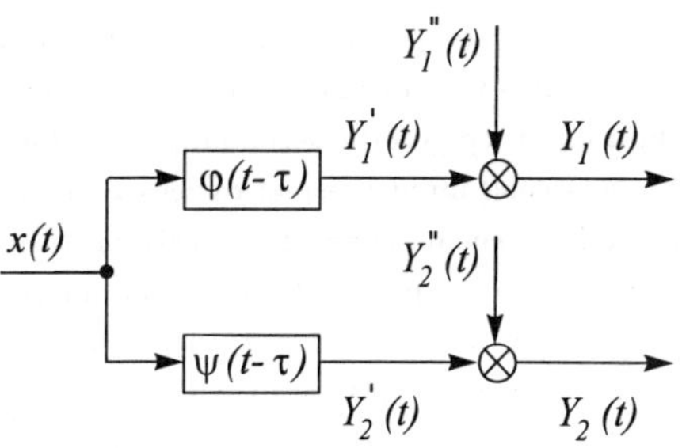

Fig. 7.8.5

At the exponential distribution of queuing time T when $\varphi(\zeta) = \mu e^{-\mu\zeta}$, $\psi(\zeta) = e^{-\mu\zeta}$ the channel may be considered as the parallel connection of two aperiodic links with one and the same time constant $1/\mu$ and with the gains 1 and $1/\mu$ with the additive noises on the outputs.

In case of determinate queueing time T when $\varphi(\zeta) = \delta(\zeta - T)$, $\psi(\zeta) = 1(\zeta)1(T - \zeta)$ the channel may be considered as the parallel connection of the ideal lagged link and the integrator with the finite memory T on whose outputs additive noises act.

Now it is sufficient for the normalization of the channel to average on all possible realizations $x(t)$ of the normally distributed input $X(t)$ the covariance function of the vector output at the given $x(t)$. From (7.8.65) and (7.8.66) after averaging we obtain

$$\hat{K}_{y_1|x}(t,t') = \int_{t_0}^{t} \varphi(\min\{t,t'\} - \tau)\Gamma_x(\tau, \tau + |t - \tau'|)d\tau$$

$$- \int_{t_0}^{t} \int_{t_0}^{t'} \varphi(t - \tau)\varphi(t' - \tau')\Gamma_x(\tau, \tau')d\tau d\tau', \qquad (7.8.67)$$

$$\hat{K}_{y_1y_2|x}(t,t') = \int_{t_0}^{t'} \int_{t_0}^{t} \varphi(t-\tau)[\,\mathbf{1}(t-t'-\tau+\tau') - \psi(t'-\tau')\,]\Gamma_x(\tau, \tau')d\tau d\tau',$$

$$(7.8.68)$$

$$\hat{K}_{y_2|x}(t,t') = \int_{t_0}^{t} \int_{t_0}^{t'} [\,\psi(\max\{t-\tau; t'-\tau'\}) - \psi(t-\tau)\psi(t'-\tau)\,]\Gamma_x(\tau, \tau')d\tau,$$

$$(7.8.69)$$

where $\Gamma_x(t,t')$ is the second initial moment of the input.

Now we show how we may construct the normal model of the distributor $\mathcal{D}$ in the one-dimensional queueing system without queue. In this case the distributor $\mathcal{D}$ represents a determinate system with two inputs and two outputs. The stream of the clients enters on the first input, and on the second input – a binary signal about the state of a channel. If the second input is equal to zero then the distributor directs the stream of the clients in queueing channel. In this case the first output of the distributor coincides with the first input, and the second one is equal to zero. If the second input is equal to 1 then the distributor gives the clients a refusal and directs an input stream from the system. In this case the first output of the distributor is equal to zero, and the second one coincides with the first input. Thus the dependence between the inputs $x_1(t)$, $x_2(t)$ and the outputs $y_1(t)$, $y_2(t)$ of the distributor with the one-channel queueing system without quence has the form

$$y_1(t) = x_1(t)[\,1 - x_2(t)\,], \quad y_2(t) = x_1(t)x_2(t). \qquad (7.8.70)$$

These formulae show that the distributor $\mathcal{D}$ represents the parallel connection of two elementary determinate nonlinear inertialess links. For the normalization of the distributor it is sufficient to linearize dependences (7.8.70). We may do it, for example, by the Taylor formula expansion of $y_1(t)$ and $y_2(t)$ in the vicinity of the expectation of the vector of random inputs. As a result we get

$$Y_1(t) = [\,1 - m_{x_2}(t)\,]X_1(t) - m_{x_1}(t)X_2^0(t),$$

$$Y_2(t) = m_{x_2}(t)X_1(t) + m_{x_1}(t)X_2^0(t). \qquad (7.8.71)$$

E x a m p l e 7.8.6. For testing the accuracy of the approximation of the queueing systems by the normal models we use the stated method to the simplest one-channel queueing system without queue. Supposing the distribution of time queueing as an exponential one we obtain a structural scheme of the considered system (Fig.7.8.6). Here $\mathcal{D}$ is an distributor which determines the movement direction of each next client; $\mu/(\mu + s)$ and $1/(\mu + s)$ are the transfer functions describing the distribution of time queueing; $Y_1(t)$ and $Y_2(t)$ are the input and the output. The equations which describe the behaviour of the system have the form:

$$\dot{Y}_1' + (m_x + \mu)Y_1' = m_x m_{y_1} - \mu m_x U'' + (\mu - m_{y_1})X, \qquad (I)$$

$$Y_1 = Y_1' + Y_1'', \quad Y_2 = (m_{y_1}/\mu)X + (m_x/\mu)(Y_1' - m_{y_1}) + m_x U'', \quad (II)$$

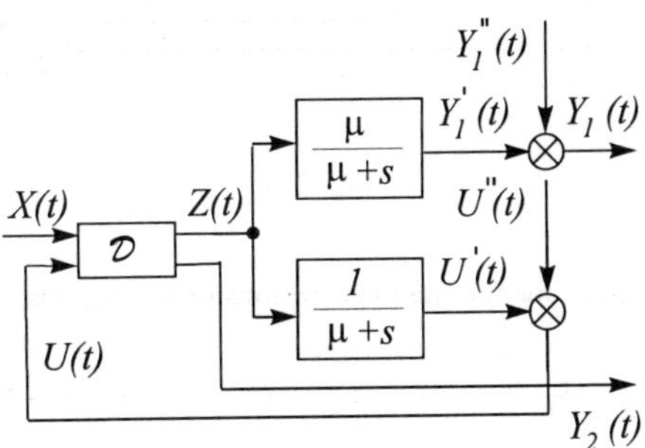

Fig. 7.8.6

where U'' and Y_1'' are the normal noises with zero expectations. Hence we obtain the equations for the expectations of the signals:

$$\dot{m}_{y_1} + (m_x + \mu)m_{y_1} = \mu m_x, \quad m_{y_2} = (m_{y_1}/\mu)m_x. \qquad (III)$$

The stationary (steady state) values of the expectations of the outputs are determined by formulae

$$m_{y_1} = \mu m_x/(m_x + \mu), \quad m_{y_2} = m_x^2/(m_x + \mu). \qquad (IV)$$

In the special case of the Poisson input stream the expectation m_x is equal to the intensity of the stream λ and the obtained formulae give the known exact solution.

For the determinate time queueing T the structural scheme of a system is represented in Fig.7.8.7. Consequently, the equations of the system have the form:

$$Y_1 = Y_1' + Y_1'', \quad Y_1'(t) = Z(t - T), \quad Z = (1 - m_u)X - m_x U^0,$$

$$U(t) = \int_{t-T}^{t} Z(\tau)d\tau + U''(t), \quad Y_2 = m_u X + m_x U^0. \tag{V}$$

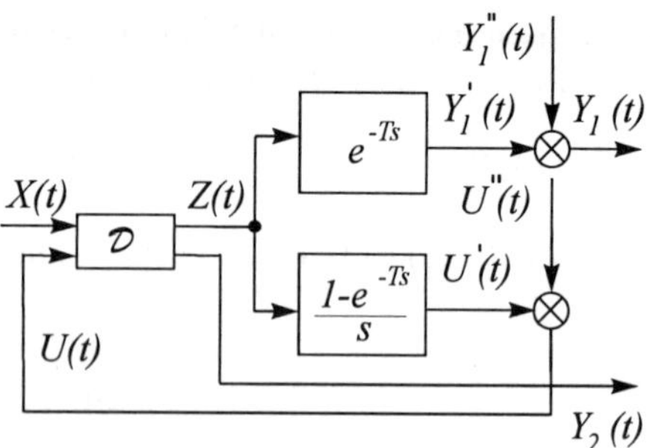

Fig. 7.8.7

Hence we get the equations for the expectations of the signals

$$m_{y_1}(t) = m_z(t - T), \quad m_z = (1 - m_u)m_x,$$

$$m_u(t) = \int_{t-T}^{t} m_z(\tau)d\tau, \quad m_{y_2} = m_u m_x. \tag{VI}$$

In the steady state Eqs. (VI) give the opportunity to derive the following expressions of the constants m_{y_1}, m_{y_2}, m_z and m_u:

$$m_{y_1} = m_z = m_u/T = m_x/(1 + Tm_x), \quad m_{y_2} = Tm_x^2/(1 + Tm_x). \tag{VII}$$

Solution (VII) also coincides with exact one in the case of the Poisson input stream. Now we notice that at the exponential distribution of time queueing the mean time

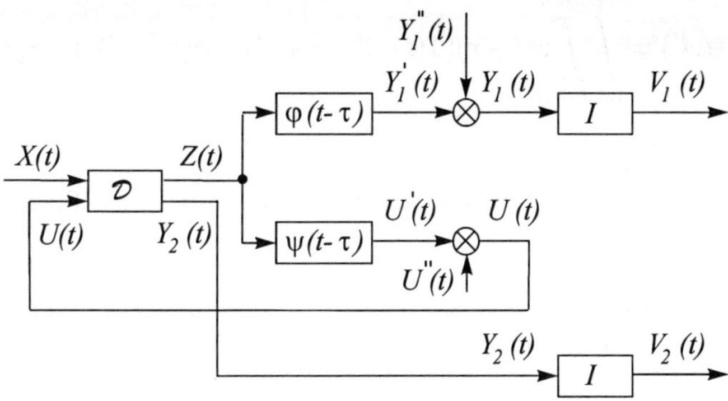

Fig. 7.8.8

queueing is equal to $1/\mu$. Therefore the formulae obtained for the case of the determinate time queueing coincide with the formulae at the exponential distribution of time queueing.

E x a m p l e   7.8.7. Let us estimate the mean number of the served clients who received the refusal and the variance of the clients number on the outputs of the one-channel queuing system without queue (Fig.7.8.8). Here D is the distributor which determines the direction of the movement of each next client in dependence of the fact whether the channel is free or occupied; $\varphi(t - \tau)$ and $\psi(t - \tau)$ are the weighting functions connected with the probability density $\varphi(\sigma)$ of time queueing T by the relation $\psi(\sigma) = \int\limits_{\sigma}^{\infty} \varphi(\zeta)d\zeta$; $X(t)$ is the input stream of the clients into the system; $Y_1(t)$ is the stream of the served clients; $Y_2(t)$ is the stream of the clients who received the refusal and I denote the integrators whose outputs represent the numbers of the served clients and the clients who received refusals beginning with the instant t_0 till the instant t; $Y_1''(t)$ and $U''(t)$ are the inner noises of the system which are determined by formulae (7.8.67)–(7.8.69) at $X = Z$:

$$K_{y_1''}(t,t') = \int\limits_{t_0}^{\min\{t,t'\}} \varphi(\min\{t,t'\} - \tau)\Gamma_z(\tau, \tau + |t - t'|)d\tau$$

$$- \int\limits_{t_0}^{t} \int\limits_{t_0}^{t'} \varphi(t - \tau)\varphi(t' - \tau')\Gamma_z(\tau, \tau')d\tau d\tau', \tag{I}$$

$$K_{y_1''u''}(t,t') = \int_{t_0}^{t}\int_{t_0}^{t'} \varphi(t-\tau)[\,\mathbf{1}(t-t'-\tau+\tau') - \psi(t'-\tau')\,]\Gamma_z(\tau,\tau')d\tau d\tau',$$

$$(\mathrm{II})$$

$$K_{u''}(t,t') = \int_{t_0}^{t}\int_{t_0}^{t'} [\,\psi(\max\{t-\tau, t'-\tau\}) - \psi(t-\tau)\psi(t'-\tau')\,]\Gamma_z(\tau,\tau')d\tau d\tau'.$$

$$(\mathrm{III})$$

Thus the set of the equations which describe the performance of the queueing system has the following form:

$$Z(t) = [\,1 - m_u(t)\,]X(t) - m_x(t)U^0(t),\qquad (\mathrm{IV})$$

$$Y_2(t) = m_u(t)X(t) + m_x(t)U^0(t),\qquad (\mathrm{V})$$

$$Y_1(t) = \int_{t_0}^{t} \varphi(t-\tau)Z(\tau)d\tau + Y_1''(t),\qquad (\mathrm{VI})$$

$$U(t) = \int_{t_0}^{t} \psi(t-\tau)Z(\tau)d\tau + U''(t),\qquad (\mathrm{VII})$$

where $U^0(t)$ is a centered random function; $m_x(t)$ and $m_u(t)$ are the expectations of $X(t)$ and $U(t)$. Accounting that $m_{u''} = m_{y_1''} = 0$, we get from Eqs. (IV)–(VII) the set of the equations for the expectations:

$$m_z(t) = [\,1 - m_u(t)\,]m_x(t), \quad m_{y_2}(t) = m_u(t)m_x(t),$$

$$m_{y_1}(t) = \int_{t_0}^{t} \varphi(t-\tau)m_z(\tau)d\tau, \quad m_u(t) = \int_{t_0}^{t} \psi(t-\tau)m_z(\tau)d\tau. \quad (\mathrm{VIII})$$

Hence in the steady state for the Poisson stream $m_x(t) = \lambda$ putting $t_0 = -\infty$ we find

$$m_z = \epsilon\lambda, \quad m_u = \epsilon_1\lambda, \quad \epsilon_1 = \lambda \int_{-\infty}^{t} \psi(t-\tau)d\tau, \quad \epsilon^{-1} = 1 + \epsilon_1. \qquad (\mathrm{IX})$$

Taking into consideration the relations

$$\int_{-\infty}^{t} \varphi(t-\tau)d\tau = \int_{0}^{\infty} \varphi(\sigma)d\sigma = 1,$$

$$\int\limits_{-\infty}^{t} \psi(t-\tau)d\tau = \int\limits_{0}^{\infty} \psi(\sigma)d\sigma = \int\limits_{0}^{\infty} [1-\Psi(\sigma)]d\sigma$$

$$= \int\limits_{0}^{\infty} \sigma\varphi(\sigma) = \mu^{-1}, \quad \Psi(\sigma) = \int\limits_{-\infty}^{\sigma} \varphi(\zeta)d\zeta, \tag{X}$$

from (IX) finally we obtain

$$m_z = m_{y_1} = \lambda\mu/(\lambda+\mu), \ \ m_u = \lambda/(\lambda+\mu), \ \ m_{y_2} = \lambda^2/(\lambda+\mu). \tag{XI}$$

Thus the normalization method for the steady state gives the exact values for the expectations of the outputs at any distribution of time queueing $\varphi(\sigma)$.

At last we give an approximate estimate of the variances of the outputs for the steady state of the performance. In this case it is convenient to present Eqs. (IV)–(VII) for $Y_1(t)$ and $Y_2(t)$ in the following form:

$$Y_1(t) = Y_1'(t) + Y_1''(t) - Y_1'''(t), \quad Y_2(t) = Y_2'(t) + Y_2''(t), \tag{XII}$$

$$\left.\begin{array}{ll} Y_1'(t) = F(D)X(t), & Y_1'''(t) = R(D)U''(t), \\ Y_2'(t) = P(D)X(t), & Y_2''(t) = Q(D)U''(t). \end{array}\right\} \tag{XIII}$$

Here

$$\left.\begin{array}{ll} F(D) = \dfrac{(1-m_u)\Phi_1(D)}{1+m_x\Phi_2(D)}, & R(D) = \dfrac{m_x\Phi_1(D)}{1+m_x\Phi_2(D)}, \\[3mm] P(D) = \dfrac{m_u+m_x\Phi_2(D)}{1+m_x\Phi_2(D)}, & Q(D) = \dfrac{m_x}{1+m_x\Phi_2(D)}, \end{array}\right\} \tag{XIV}$$

where $D = d/dt$; $\Phi_1(D)$ and $\Phi_2(D)$ are the transfer functions correspondent to the weighting functions $\varphi(t-\tau)$ and $\psi(t-\tau)$ respectively.

It follows from the properties of a normal stochastic system that the input $X(t)$ is uncorrelated with the inner noises $Y_1''(t)$ and $U''(t)$ and it means that $Y_1'(t)$ is also uncorrelated with $Y_1''(t)$ and with $Y_1'''(t)$. Analogously $Y_2'(t)$ is uncorrelated with $Y_2''(t)$. Consequently,

$$K_{y_1}(t,t') = K_{y_1'}(t,t') + K_{y_1''}(t,t')$$

$$+K_{y_1'''}(t,t') - K_{y_1'''y_1''}(t,t') - K_{y_1'''y_1''}(t,t'), \tag{XV}$$

$$K_{y_2}(t,t') = K_{y_2'}(t,t') + K_{y_2''}(t,t''). \tag{XVI}$$

The variances of the total numbers of the clients $V_1(t)$ and $V_2(t)$ are determined by formulae

$$D_{v_i}(t) = \int_{t_0}^{t} \int_{t_0}^{t} K_{y_i}(\tau, \tau') d\tau d\tau' \quad (i = 1, 2). \tag{XVII}$$

In the special case when $X(t)$ is a white noise of the intensity λ, $K_x(t, t') = \lambda \delta(t - t')$ applying the formulae of the linear theory of the stationary systems (Subsection 6.1.8) to Eqs. (XIII) we obtain

$$K_{y_1}(t, t') = \frac{\lambda \mu^4}{2(\lambda + \mu)^3} e^{-(\lambda+\mu)|t-t'|}, \tag{XVIII}$$

$$K_{y_2'}(t, t') = \frac{\lambda^3}{(\lambda + \mu)^2} \delta(t - t') + \frac{\lambda^3 \mu(2\lambda + 3\mu)}{2(\lambda + \mu)^3} e^{-(\lambda+\mu)|t-t'|}, \tag{XIX}$$

$$K_{y_1''}(t, t') = \frac{\lambda \mu^2}{(\lambda + \mu)^2} \delta(t - t') - \frac{\lambda \mu^2}{2(\lambda + \mu)} e^{-(\lambda+\mu)|t-t'|}, \tag{XX}$$

$$K_{u''}(t, t') = \frac{\lambda^2}{(\lambda + \mu)^2} + \frac{\lambda \mu^2}{2(\lambda + \mu)^3} e^{-(\lambda+\mu)|t-t'|}, \tag{XXI}$$

$$K_{y_1'''}(t, t') = \frac{\lambda^4 \mu^2}{(\lambda + \mu)^4} + \frac{\lambda^3 \mu^4}{4(\lambda + \mu)^5} e^{-(\lambda+\mu)|t-t'|}$$
$$+ \frac{\lambda^3 \mu^4}{4(\lambda + \mu)^4} |t - t'| e^{-(\lambda+\mu)|t-t'|}, \tag{XXII}$$

$$K_{y_2''}(t, t') = \frac{\lambda^4 \mu^2}{(\lambda + \mu)^4} + \frac{\lambda^3 \mu^2 [(\lambda + \mu)^2 + \mu^2]}{4(\lambda + \mu)^5} e^{-(\lambda+\mu)|t-t'|}$$
$$- \frac{\lambda^4 \mu^2 (\lambda + 2\mu)}{4(\lambda + \mu)^4} |t - t'| e^{-(\lambda+\mu)|t-t'|}, \tag{XXIII}$$

$$K_{y_1''' y_1''}(t, t') = -\frac{\lambda^2 \mu^3}{4(\lambda + \mu)^3} e^{-(\lambda+\mu)|t-t'|}$$
$$+ \frac{\lambda^2 \mu^3}{2(\lambda + \mu)^2} |t - t'| e^{-(\lambda+\mu)|t-t'|}. \tag{XXIV}$$

Accounting (XV), (XVI) and (XVIII)–(XXIV) after evaluations we find from (XVII)

$$D_{v_1}(t) = \frac{\lambda^4 \mu^2}{(\lambda + \mu)^4} t^2 + \frac{\lambda \mu^2 (\mu^3 + \lambda \mu^2 - \lambda^3)}{(\lambda + \mu)^6} t$$

$$+\frac{\lambda^2\mu^2(2\lambda^3 + 14\lambda^2\mu + 19\lambda\mu^2 + 10\mu^3)}{2(\lambda + \mu)^7}$$

$$-\frac{\lambda^2\mu^3(4\lambda^2 + 7\lambda\mu + 4\mu^2)}{2(\lambda + \mu)^6}te^{-(\lambda+\mu)t}$$

$$-\frac{\lambda^2\mu^2(2\lambda^3 + 14\lambda^2\mu + 19\lambda\mu^2 + 10\mu^3)}{2(\lambda + \mu)^7}e^{-(\lambda+\mu)t}, \qquad \text{(XXV)}$$

$$D_{v_2}(t) = \frac{\lambda^4\mu^2}{(\lambda + \mu)^4}t^2 + \frac{\lambda^3(\lambda^4 + 6\lambda^3\mu + 13\lambda^2\mu^2 + 12\lambda\mu^3 + 5\mu^4)}{(\lambda + \mu)^6}t$$

$$-\frac{\lambda^4\mu^2(\lambda + 2\mu)}{2(\lambda + \mu)^6}te^{-(\lambda+\mu)t} - \frac{\lambda^3\mu(4\lambda^3 + 13\lambda^2\mu + 14\lambda\mu^2 + 8\mu^3)}{2(\lambda + \mu)^7}$$

$$+\frac{\lambda^3\mu(4\lambda^3 + 13\lambda^2\mu + 14\lambda\mu^2 + 8\mu^3)}{2(\lambda + \mu)^7}e^{-(\lambda+\mu)t}. \qquad \text{(XXVI)}$$

From (XXV) at $t \to \infty$ for $D_{v_1}(t)$ we obtain $D_{v_1}(t) = D_{v_2}(t) = \lambda^4\mu^2/(\lambda+\mu)^4$, i.e. the variances of the number those clients who are served during time unit and those who are received the refusal are equal and do not depend on time.

7.8.6. Software for Nonlinear Stochastic Systems Analytical Modeling

We see that the theory of stochastic systems given in Chapters 1, 5–7 is rather complicated and requires considerable mathematical background for studying it. To solve any practical problem of stochastic systems analysis one has to derive the respective equations determining distribution parameters from original stochastic equations and then to produce a program for solving these equations. This is a tedious and time-consuming job which can be done only by those who have an extensive knowledge of mathematics and some skill in programming. At practice two main approaches to the software design for the distributions parametrization methods in nonlinear stochastic systems are usually used:

- the creation of the specialized libraries for different programming languages;
- the creation of the specialized dialogue software packages.

The library approach is effective for the analytical modeling and the automatic derivation of the equations for the distributions parameters and for the statistical simulation of linear stochastic systems also. The choice of one or another method is determined by available information about numerical model of a real system and random disturbances, required accuracy of the calculations of system efficiency and by the information technologies of research also. For the nonlinear stochastic systems the derivation of equations for the distribution parameters is rather cumbersome and accounts about 90% of the whole process. In this situation it is necessary to create the dialogue packages which realize the main operations of the nonlinear analysis.

In recent years actual prerequisites for further improvement of the software appeared on the basis of using modern highly effective software of data processing and information storage. One of the directions of the development is connected with the implementation of the mathematical packages which possess the wide possibilities of analytical derivation such as "Mathematica" what allows to expand the classes of used functions. The other direction of the development is the implementation of the modern object oriented control systems of the data storage of equations of the intermediate results.

The package "StS–Analysis, version 1.0" is designed for studying the stochastic differential systems described by Eq. (5.4.6) with the standard Wiener process $W(t)$ by NAM for the polynomial in y function $a(y,t)$ of not higher than the third degree and the function $b(y,t)$ independent of y or linear in y. The package "StS–Analysis, version 1.0" has been widely used in Russia since 1987. The package "StS–Analysis, version 2.0" is destined for studying by NAM the stochastic systems described by the difference equations.

Working with any version of the "StS–Analysis" package, the user must only type on the display with the aid of the keyboard the original stochastic equations in the natural mathematical form without any special programmers tricks, introduce the corresponding initial conditions and fix the statistical characteristics must be calculated. Once the original stochastic equations are introduced the user may correct them, i.e. to remove some equations, to replace some of them by others, to correct errors and so on. When all the data are introduced and checked the user has to start computations. Then the package "StS–Analysis" automatically derives the NAM equations for the corresponding distribution parameters and solves them. The user may observe visually the behaviour of a system under study in the process of calculations by tracing the graphs or the tables of current values of various statistical characteristics on the display and get the copies of these graphs and tables from the file after the calculations are finished. So the use of any of the versions of the package "StS–Analysis" does not require that the user know mathematics and programming.

Any version of the "StS–Analysis" package automatically derives the NAM equations for the distribution parameters from original stochastic equations introduced by the user in the natural mathematical form and solves these equations.

It should be remarked that the "StS–Analysis" package does not simulate a stochastic system and does not use the statistical simulation (Monte-Carlo) method but performs the theoretical determination of necessary probability characteristics of a system behaviour by the normal approximation method.

E x a m p l e 7.8.8. Let us consider the scalar stochastic nonlinear discrete system

$$Y_1(l+1) = -aY_1^3(l) + Y_1(l) + bV_1(l). \tag{I}$$

Formulae for the expectation and the variace have the form

$$m_1(l+1) = -aEY_1^3(l) + m_1(l), \tag{II}$$

$$D_1(l+1) = a^2\big(EY_1^6(l) - EY_1^3(l)EY_1^3(l)\big)$$
$$- 2a\big(EY_1^4(l) - m_1(l)EY^3(l)\big) + D_1(l) + b^2. \tag{III}$$

The higher moments approximately are expressed in terms of the first and the second moments:

$$EY_1^2(l) = D_1(l) + m_1^2(l), \tag{IV}$$

$$EY_1^3(l) = m_1(l)EY_1^2(l) + 2m_1(l)D_1(l), \tag{V}$$

$$EY_1^4(l) = m_1(l)EY_1^3(l) + 3D_1(l)EY_1^2(l), \tag{VI}$$

$$EY_1^5(l) = m_1(l)EY_1^4(l) + 4D_1(l)EY_1^3(l), \tag{VII}$$

$$EY_1^6(l) = m_1(l)EY_1^5(l) + 5D_1(l)EY_1^4(l). \tag{VIII}$$

On Fig 7.8.9 graphs of m_1 and D_1 for $a = 0.05$, $b = 0.05$ and $a = 0.1$, $b = 0.05$ at the initial conditions $m_1(0) = 1$, $D_1(0) = 0.5$ obtained by NAM are given.

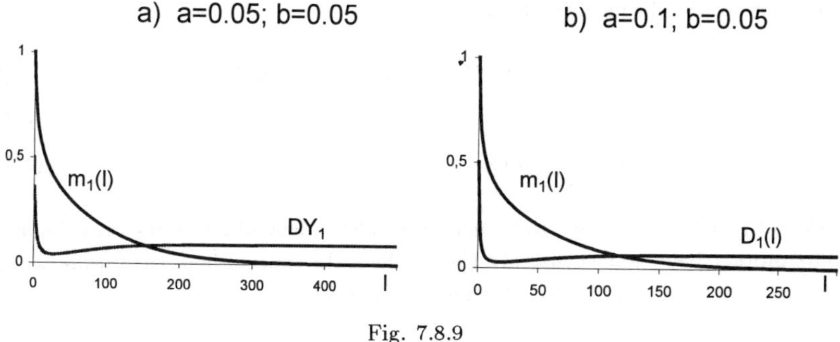

Fig. 7.8.9

E x a m p l e 7.8.9. Stochastic three-dimensional nonlinear discrete system is described by the following equatons:

$$Y_1(l+1) = Y_1(l) + aY_1(l), \tag{I}$$

$$Y_2(l+1) = Y_2(l) + 2aY_3(l), \tag{II}$$

$$Y_3(l+1) = Y_3(l) - aY_2(l) - 2aY_3(l)$$
$$- aY_2^3(l) + 3aY_1(l)V_1(l) \tag{III}$$

at the initial conditions

$$EY_1(0) = EY_2(0) = 0.5, \quad EY_3(0) = 0.8$$
$$DY_1(0) = DY_2(0) = 1, \quad EY_1(0)Y_2(0) = 0.4. \tag{IV}$$

Other variances and covariances are assumed equal to zero. Graphs of the expectations and the variances obtained by NAM are given on Fig.7.8.10.

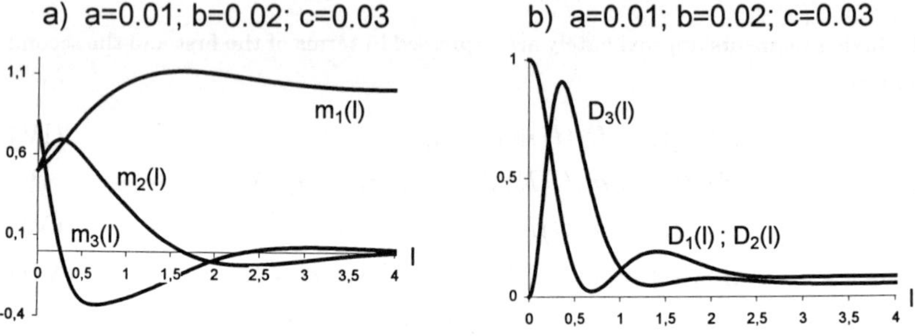

Fig. 7.8.10

E x a m p l e 7.8.10. Let us consider the stochastic nonlinear differential equations of the chemical reactor:

$$\dot{Y_1} = 1 + Y_1^2 Y_2 - (1 + Y_3)Y_1 + 0.5V_1, \tag{I}$$
$$\dot{Y_2} = Y_1 Y_3 - Y_1^2 Y_3 + V_2, \tag{II}$$
$$\dot{Y_3} = -Y_1 Y_3 + c, \tag{III}$$

where V_1 and V_2 are the normal mutually independent white noises with the unit intensities, c being some constant. The graphs of the expectations, the variances and the covariances at $c = 1.3$ and the initial conditions

$$EY_1(0) = 0.5, \quad EY_2(0) = 3.3, \quad EY_3(0) = 3.0,$$
$$DY_1(0) = 0.01, \quad DY_2(0) = 0.5. \ DY_3(0) = 0 \tag{IV}$$

obtained by NAM are given on Fig.7.8.11.

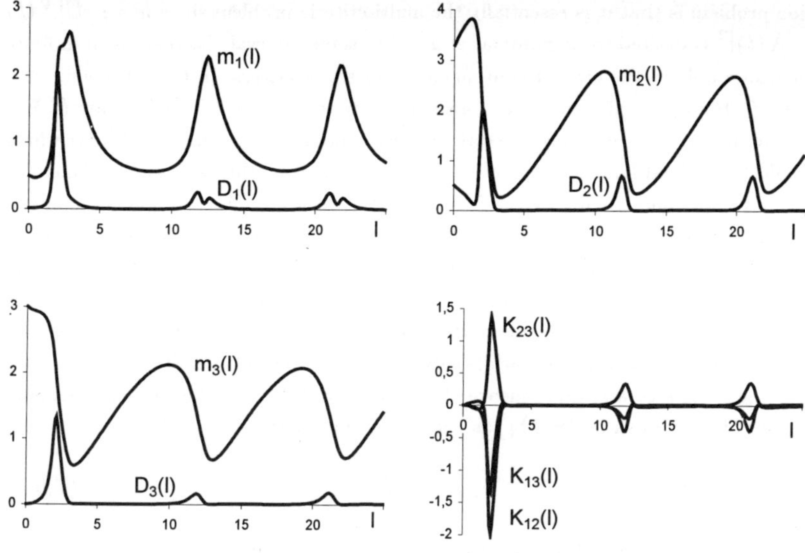

Fig. 7.8.11

7.9. Online Conditionally Optimal Nonlinear Filtering

7.9.1. Preliminary Remarks

The simplicity of the Kalman and the Kalman–Bucy filters (Subsection 6.6.2) leads to the idea of conditionally optimal filtering. This idea is to reject the absolute (unconditional) optimality and to restrict oneself with finding the optimal estimate in some bounded class of admissible estimates satisfying some rather simple (differential or difference) equations which may be solved online while receiving the results of the measurements. The main problem of applying the conditionally optimal filtering theory is to define a class of admissible filters. It is recommended usually to define the class of admissible filters as the set of all filters governed by the differential or the difference equations of the prescribed form with some undetermined coefficients. The problem of optimization in this case is reduced to finding optimal values of all undetermined coefficients which appear generally time functions.

The first characteristic feature of the nonlinear conditionally optimal estimation problem is that it is essentially the multi-criteria problem since m.s.e. $E|\hat{X}(t) - X(t)|^2$ is desired to be minimal at any t in some interval. So there is an infinite (noncountable in the case of continuous time) set of criteria in this problem. Formula (5.10.6) gives the absolutely optimal estimate which is optimal at any t. But restricting the class of admissible estimates we cannot hope to find an estimate which would be optimal for any t. Only in the linear problems this is possible. Thus one has to restrict oneself only by the Pareto optimal decision. The second characteristic feature of the conditionally optimal filtering is that the optimal values of the coefficients in the equations of a filter are determined only by prior data without using measurements. So the optimal filter may be designed beforehand quite in the same way as the Kalman or the Kalman–Bucy filters. The measurements need to be used only in the process of filtering while solving the equations of the filter.

Let $\{X_l\}$ be a state sequence of random variables. Another measurement sequence of random variables $\{Y_l\}$ is defined by the equation

$$Y_l = \omega_l'(X_l, V_l), \tag{7.9.1}$$

$\{V_l\}$ being a sequence of independent random variables with known distributions. In problems of practice the random variables V_l cause the errors. This is the reason why they are usually called *measurement errors*. The problem is to find an estimate $\hat{X}_l$ at each l of the random variable X_l using the measurements of the previous random variables $Y_1 , \ldots , Y_{l-1}$ in the class of admissible estimates defined by the formula $\hat{X}_l = AU_l$, A being some matrix of the rank n, and the difference equation

$$U_{l+1} = \delta_l \zeta_l(U_l, Y_l) + \gamma_l \tag{7.9.2}$$

with a given sequence of the vector structural functions $\zeta_l(u, y)$ and all possible values of the matrices of coefficients δ_l and the coefficients vectors γ_l. Any given sequence of the structural functions $\zeta_l(u, y)$ determines the class of admissible filters. Any choice of $\{\delta_l\}$, $\{\gamma_l\}$ and the given functions $\zeta_l(u, y)$ determines the admissible filter. The dimension n_1 of the vectors Y_l is supposed less than the dimension n of the vectors X_l, $n_1 < n$.

Certainly such an estimate $\hat{X}_l$ is required which minimizes in some sense the mean square error (m.s.e.)

$$E \left| \hat{X}_l - X_l \right|^2 \tag{7.9.3}$$

at any l. As this problem represents the multi-criteria one the question arises, in what sense $\{\delta_l\}$, $\{\gamma_l\}$ may be optimal. We shall call the filter which minimizes (7.9.3) at each l when the values of δ_h, γ_h at all the preceding steps, $h < l$, are known *the conditionally optimal filter*.

As for sequence $\{X_l\}$, it may be defined also by the difference equation of the form

$$X_{l+1} = \omega_l(X_l, V_l). \tag{7.9.4}$$

The structural functions $\zeta_l(u, y)$ in (7.9.2) may be taken arbitrarily in principle. But the accuracy of filtering depends essentially upon the choice of these functions. We may recommend to choose $\zeta_l(u, y)$ in such a way that the respective class of the admissible filters would contain some known suboptimal filter. Then the conditionally optimal filter will certainly always be better (at least no worse) than this suboptimal filter.

7.9.2. Equations Determining Optimal Filter Coefficients

We shall solve the following problem: *given Eq. (7.9.4) defining the Markov sequence of the random variables $\{X_l\}$ and the distributions of independent random variables V_l find the optimal values of δ_l, γ_l in Eq. (7.9.2) minimizing the mean square of the error (7.9.3) at each step (each l) supposing that all coefficients δ_l, γ_l, $h < l$, are determined at previous steps.*

Eq. (7.9.2) determines the mode of using the information contained in Y_1, $\ldots, Y_{l-1}$ (i.e. the information received at previous steps). The information is used only via U_l, and only the last received measurement Y_l enters explicitly in (7.9.2) and therefore is used directly. Writing (7.9.2) in the form

$$\hat{X}_{l+1} = A\delta_l\zeta_l(U_l, Y_l) + A\gamma_l \tag{7.9.5}$$

we see that $\hat{X}_{l+1}$ is the optimal estimate of X_{l+1} if and only if the right-hand side of (7.9.5) represents the linear mean square regression of X_{l+1} on the random variable $\zeta_l(U_l, Y_l)$. It follows from linear regression theory that the right hand-side of (7.9.5) represents the linear mean square regression of X_{l+1} on $\zeta_l(U_l, Y_l)$ if and only if the estimate is unbiased and the error $\hat{X}_{l+1} - X_{l+1}$ is uncorrelated with $\zeta_l(U_l, Y_l)$. Thus we have the following equations for the optimal values of δ_l, γ_l:

$$A\delta_l E\zeta_l(U_l, Y_l) + A\gamma_l = EX_{l+1}, \tag{7.9.6}$$

$$E(AU_{l+1} - X_{l+1})\zeta_l(U_l, Y_l)^T = 0. \tag{7.9.7}$$

Putting in (7.9.6)

$$m_{l+1} = EX_{l+1}, \quad L_l = E\zeta_l(U_l, Y_l) \tag{7.9.8}$$

we get

$$A\gamma_l = m_{l+1} - A\delta_l L_l. \tag{7.9.9}$$

Substituting into (7.9.7) expression (7.9.2) of U_{l+1} we obtain the equation for δ_l:

$$A\delta_l E\left[\zeta_l(U_l, Y_l) - L_l\right]\zeta_l(U_l, Y_l)^T - E(X_{l+1} - m_{l+1})\zeta_l(U_l, Y_l)^T = 0.$$

$$(7.9.10)$$

Denoting

$$B_l = E\left[\zeta_l(U_l, Y_l) - L_l\right]\zeta_l(U_l, Y_l)^T, \quad D_l = E(X_{l+1} - m_{l+1})\zeta_l(U_l, Y_l)^T$$

$$(7.9.11)$$

we find from (7.9.10)

$$A\delta_l = D_l B_l^{-1}.$$

$$(7.9.12)$$

Certainly this formula determines $A\delta_l$ only if the matrix B_l is invertible. This is always the case if the components of the vector random variables $\zeta_l(U_l, Y_l)$ are linearly independent.

It should be remarked that the number of columns of the matrix A usually exceeds the number ofthe raws due to which Eqs. (7.9.9), (7.9.12) do not determine δ_l, γ_l uniquely. Therefore some the elements of the matrix δ_l and matrix-column γ_l may be chosen arbitrarily.

The problem of Subsections 7.9.1 cannot be considered as solved till the expectations in formulae (7.9.8) and (7.9.11) are determined. To find these expectations we substitute in (7.9.8) and (7.9.11) the expressions of X_{l+1} and X_l from Eqs. (7.9.4) and (7.9.33). Then we obtain

$$m_{l+1} = E\omega_l(X_l, V_l), \quad L_l = E\zeta_l(U_l, \omega_l'(X_l, V_l)), \qquad (7.9.13)$$

$$B_l = E\left[\zeta_l(U_l, \omega_l'(X_l, V_l)) - L_l\right]\zeta_l(U_l, \omega_l'(X_l, V_l))^T,$$

$$D_l = E\left[\zeta_l(U_l, \omega_l'(X_l, V_l)) - m_{l+1}\right]\zeta_l(U_l, \omega_l'(X_l, V_l))^T. \qquad (7.9.14)$$

These formulae show that m_{l+1}, L_l, B_l, D_l may be calculated in the general case if the joint distribution of the random variables X_l, U_l and the distribution of the random variables V_l are known. To find the joint distribution of X_l, U_l which represents evidently the one-dimensional distribution of the sequence of composed random vectors $\{[X_l^T U_l^T]^T\}$ we deduce the difference equation for the characteristic function

$$g_{1,l}(\lambda, \mu) = E\exp\left\{i\lambda^T X_l + i\mu^T U_l\right\}.$$

Using (7.9.4), (7.9.3) and (7.9.2) we get

$$g_{1,l+1}(\lambda, \mu) = E\exp\left\{i\lambda^T\omega_l(X_l, V_l) \right.$$
$$\left. + i\mu^T\left[\delta_l\zeta_l(U_l, \omega_l'(X_l, V_l)) + \gamma_l\right]\right\}. \qquad (7.9.15)$$

The expectation here is determined by the joint distribution of X_l, U_l, i.e. by the characteristic function $g_{1,l}(\lambda, \mu)$, and the known distribution of the random variable

V_l which is independent of X_l, U_l by (7.9.4), (7.9.3), (7.9.2). So (7.9.15) represents the difference equation determining together with the initial characteristic function $g_{1,1}(\lambda, \mu) = E \exp\left\{i\lambda^T X_1 + i\mu^T U_1\right\}$, which is supposed known, the sequence of characteristic functions $\{g_{1,l}(\lambda, \mu)\}$. Eqs. (7.9.9), (7.9.12) with (7.9.13), (7.9.14) and (7.9.15) represent the set of equations determining γ_l, δ_l and $g_{1,l}(\lambda, \mu)$ at each step (each l). To solve these equations one may use any of the approximate methods of Sections 7.2–7.6.

7.9.3. Estimation of Filtering Accuracy

The knowledge of the joint distribution of the random variables X_l, U_l obtained while solving Eqs. (7.9.9), (7.9.2), (7.9.15) enables one to estimate the accuracy of the conditionally optimal filter a priori in the process of designing the filter. Really knowing the distribution of X_l, U_l one may find the joint distribution of X_l, $\hat{X}_l$ and then estimate the mean square of the error of filtering (7.9.3) and find confidence regions for X_l at each step. Eq. (7.9.15) determines also the joint distribution of X_l, U_l and therefore of X_l, $\hat{X}_l$ for any admissible filter defined by (7.9.2) at arbitrary δ_l, γ_l. So one may estimate the accuracy of filtering characterized either by m.s.e. or by confidential regions for X_l for any admissible filter. Thus the theory of conditionally optimal filtering allows the accuracy prior estimation of any admissible filter, in particular, of conditionally filter while designing it.

7.9.4. Generalized Kalman Filter

Let us consider the special case of Subsection 7.9.3 in which $A = I$ and Eq. (7.9.4) has the form of Eq. (7.1.29)

$$X_{l+1} = a_l X_l + a_{0l} + \left(c_{0l} + \sum_{r=1}^{n} c_{rl} X_{lr} \right) V_l \qquad (7.9.16)$$

where X_{lr} is the rth component of the vector X_l. This case corresponds to the linear systems with parametric noises. Eq. (7.9.1) we assume in similar form

$$Y_l = b_l Y_l + b_{0l} + \left(d_{0l} + \sum_{r=1}^{n} d_{rl} Y_{lr} \right) V_l. \qquad (7.9.17)$$

As a class of admissible filters we assume the class of linear filters so that (7.9.2) becomes

$$\hat{X}_{l+1} = \delta_{1l}\hat{X}_l + \delta_{2l}Y_l + \gamma_l. \qquad (7.9.18)$$

In (7.9.16), (7.9.18) a_l, c_{0l}, c_{rl}, b_l, d_{0l}, d_{rl}, δ_{1l}, δ_{2l} represent the matrices and a_{0l}, b_{0l}, γ_l the vectors of the respective dimensions.

We see $\delta_l = [\delta_{1l}\delta_{2l}]$, $\hat{X}_l = U_l$, $\zeta_l(\hat{X}_l, Y_l) = \left[\hat{X}_l^T Y_l^T\right]^T$ in this case. So taking into account that $E\hat{X}_l = EX_l = m_l$ and X_l, $\hat{X}_l$ are independent of V_l we find by formulae (7.9.13), (7.9.14)

$$m_{l+1} = a_l m_l + a_{0l}, \quad L_l = \left[m_l^T \; m_l^T b_l^T + b_{0l}^T\right]^T, \tag{7.9.19}$$

$$B_l = \begin{bmatrix} K_{\hat{x}_l} & K_{\hat{x}_l x_l} b_l^T \\ b_l K_{x_l \hat{x}_l} & b_l K_{x_l} b_l^T + H_{l22} \end{bmatrix},$$

$$D_l = \left[a_l K_{x_l \hat{x}_l} \quad b_l K_{x_l} b_l^T + H_{l22}\right] \tag{7.9.20}$$

where K_y is the covariance matrix of a random vector Y, K_{yu} the cross-covariance matrix of random vectors Y and U,

$$H_{l11} = c_{0l} G_l c_{0l}^T + \sum_{r=1}^{n} m_{lr} (c_{0l} G_l c_{rl}^T + c_{rl} G_l c_{0l}^T)$$

$$+ \sum_{r,s=1}^{n} (m_{lr} m_{ls} + k_{lrs}) c_{rl} G_l c_{sl}^T,$$

$$H_{l12} = c_{0l} G_l d_{0l}^T + \sum_{r=1}^{n} m_{lr} (c_{0l} G_l d_{rl}^T + c_{rl} G_l d_{0l}^T)$$

$$+ \sum_{r,s=1}^{n} (m_{lr} m_{ls} + k_{lrs}) c_{rl} G_l d_{sl}^T, \tag{7.9.21}$$

$$H_{l22} = d_{0l} G_l d_{0l}^T + \sum_{r=1}^{n} m_{lr} (d_{0l} G_l d_{rl}^T + d_{rl} G_l d_{0l}^T)$$

$$+ \sum_{r,s=1}^{n} (m_{lr} m_{ls} + k_{lrs}) d_{rl} G_l d_{sl}^T,$$

G_l being the covariance matrix of the random vector V_l, m_{l1}, $\ldots$, m_{lr} the components of the vector $m_l = EX_l$, k_{lrs} $(r, s = 1, \ldots, n)$ the elements of the covariance matrix $K_l = K_{x_l}$ of the random vector X_l. While deriving (7.9.9), (7.9.20) we supposed without loss of generality that the expectations of all the random variables V_l are equal to zero.

Now we find from (7.9.7)

$$E(\hat{X}_{l+1} - X_{l+1})\hat{X}_l^T = 0, \quad E(\hat{X}_{l+1} - X_{l+1})Y_l^T = 0.$$

Multiplying the first of these equalities from the right by δ_{1l}^T and the second by δ_{2l}^T, adding the resulting equalities and adding to the result the evident equality $E(\hat{X}_{l+1} - X_{l+1})\gamma_l^T = 0$ due to the unbiasedness of the estimate $\hat{x}_{l+1}$ we get by (7.9.18) $E(\hat{X}_{l+1} - X_{l+1})\hat{X}_{l+1}^T = 0$. As this equality holds for any l we may rewrite it as $E(X_l - X_l)\hat{X}_l = 0$ yielding, in consequence of the unbiasedness of $\hat{X}_l$, $K_{x_l} = K_{x_l \hat{x}_l}$ and due to the symmetry of the real covariance matrices, $K_{\hat{x}_l x_l} = K_{x_l \hat{x}_l}^T = K_{\hat{x}_l}$. Using this fact and (7.9.20) we get the following equations for the matrices δ_{1l} and δ_{2l}:

$$\delta_{1l} K_{\hat{x}_l} + \delta_{2l} b_l K_{\hat{x}_l} = a_l K_{\hat{x}_l}, \tag{7.9.22}$$

$$\delta_{1l} K_{\hat{x}_l} b_l^T + \delta_{2l}(b_l K_{x_l} b_l^T + H_{l22}) = a_l K_{x_l} b_l^T + H_{l12}. \tag{7.9.23}$$

Multiplying Eq. (7.9.22) from the right by b_l^T and subtracting from Eq. (7.9.23) we obtain

$$\delta_{2l}[b_l(K_{x_l} - K_{\hat{x}_l})b_l^T + H_{l22}] = a_l(K_{x_l} - K_{\hat{x}_l})b_l^T + H_{l12}. \tag{7.9.24}$$

Finally we notice that due to $K_{x_l \hat{x}_l} = K_{\hat{x}_l x_l} = K_{\hat{x}_l}$ and the unbiasedness of $\hat{X}_l$

$$K_{x_l} - K_{\hat{x}_l} = K_{x_l} - K_{x_l \hat{x}_l} - K_{\hat{x}_l x_l} + K_{\hat{x}_l}$$
$$= E(\hat{X}_l - X_l)(\hat{X}_l^T - X_l^T)$$

represents the covariance matrix of the error $\tilde{X}_l = \hat{X}_l - X_l$ of filtering at the lth step. So denoting it by R_l Eq. (7.9.24) becomes

$$\delta_{2l}(b_l R_l b_l^T + H_{l22}) = a_l R_l b_l^T + H_{l12}. \tag{7.9.25}$$

Solving Eq. (7.9.25) we find

$$\delta_{2l} = (a_l R_l b_l^T + H_{l12})(b_l R_l b_l^T + H_{l22})^{-1}. \tag{7.9.26}$$

After that Eq. (7.9.22) gives

$$\delta_{1l} = a_l - \delta_{2l} b_l. \tag{7.9.27}$$

Certainly formula (7.9.26) and (7.9.27) are valid only if the matrices $b_l R_l b_l^T + H_{l22}$ and $K_{\hat{x}_l}$ are invertible. This is generally the case as the components of the vector error $\tilde{X}_l = \hat{X}_l - X_l$ and of the vector $\hat{X}_l$ are usually linearly independent. Once δ_{1l}, δ_{2l} determined we find γ_l from (7.9.9). Substituting (7.9.19) and the expression of $\delta_l = [\delta_{1l} \delta_{2l}] = [a_l - \delta_{2l} b_l \ \delta_{2l}]$ following from (7.9.27) into (7.9.9) we get

$$\gamma_l = a_{0l} - \delta_{2l} b_{0l}. \tag{7.9.28}$$

Substituting (7.9.27) and (7.9.28) into (7.9.18) we represent the equation of the conditionally optimal filter in the form

$$\hat{X}_{l+1} = a_l \hat{X}_l + a_{0l} + \delta_{2l}(Y_l - b_l \hat{X}_l - b_{0l}). \qquad (7.9.29)$$

This is the usual Kalman filter equation (Subsection 6.6.2). But in our more general problem where Eqs. (7.9.16) and (7.9.17) are nonlinear the matrix coefficient δ_{2l} is determined by formulae (7.9.26) and (7.9.21) different from the corresponding formula of Kalman linear filter theory.

To accomplish the solution of the problem it is necessary to indicate how the expectation m_l and the covariance matrix K_l of the random vector X_l and the covariance matrix R_l of the error of filtering may be found at each step. For this purpose we deduce the difference equation for the error of filtering $\tilde{X}_l = \hat{X}_l - X_l$. Subtracting Eq. (7.9.16) from (7.9.29) and using (7.9.17) we get

$$\tilde{X}_{l+1} = (a_l - \delta_{2l} b_l)\tilde{X}_l$$
$$+ \left[\delta_{2l} d_{0l} - c_{0l} + \sum_{r=1}^{n} (\delta_{2l} d_{rl} - c_{rl}) X_{lr} \right] V_l. \qquad (7.9.30)$$

Now remembering that $m_l = EX_l$, $K_l = E(X_l - m_l)X_l^T$, $R_l = E\tilde{X}_l \tilde{X}_l^T$ ($E\tilde{X}_l = 0$) and taking into account that X_l, $\tilde{X}_l$ are independent of V_l we find from (7.9.16) and (7.9.30) the difference equations for m_l, K_l, R_l,

$$m_{l+1} = a_l m_l + a_{0l}, \qquad (7.9.31)$$

$$K_{l+1} = a_l K_l a_l^T + H_{l11}, \qquad (7.9.32)$$

$$R_{l+1} = a_l R_l a_l^T - \delta_{2l}(b_l R_l a_l^T + H_{l12}^T) + H_{l11}, \qquad (7.9.33)$$

where H_{l11}, H_{l12} are determined by (7.9.21).

Eqs. (7.9.31)–(7.9.33) and formula (7.9.26) determine m_l, K_l, R_l, δ_{2l} at each step. Eqs. (7.9.34), (7.9.32) are linear difference equations determining successively the sequence $\{m_l\}$, $\{K_l\}$ while Eq. (7.9.33) is nonlinear difference equation with respect to R_l due to the dependence of δ_{2l} on R_l. As always in the conditionally optimal filtering and in Kalman linear filtering the sequences m_l, K_l, R_l, δ_{2l} may be calculated beforehand in the process of designing the filter, as Eqs. (7.9.31), (7.9.30) and formula (7.9.26) do not contain observations.

R e m a r k. The linear filter (7.9.29) of the special kind (X_l, ... , X_{l-1} are incorporated into (7.9.29) only via $\hat{X}_l$) happens to be the optimal filter among all the linear filters. The fact that the conditionally optimal filter coincides in our case with the optimal linear filter at any l is the natural consequence of the fact that the latter satisfies Eq. (7.9.18), i.e. is contained in the class of admissible filters. It is interesting also to notice that in our special case only the second order moments of X_l, $\hat{X}_l$ were needed to determine the equation of the conditionally optimal filter. This is the reason why Eq. (7.9.15) was not used. But if one wants to characterize the accuracy of filtering not only by the mean square of the error R_l but also by confidence regions for X_l, then Eq. (7.9.15) should be used.

7.9.5. Software for Conditionally Optimal Discrete Filter Design

The solution of Eqs. (7.9.9), (7.9.12), (7.9.15) is very cumbersome and time-consuming. By this reason the problem of conditionally optimal filter design is difficult and practical applications of the conditionally optimal filtering theory are severely restricted. The only way to provide the wide applications of the conditionally optimal filters is to create such specialized software for the c conditionally optimal filter design. Aiming this goal the package "StS–Filter" was produced in the Institute of Informatics Problems of Russian Academy of Sciences. This package is destined for the automatic solution of Eqs. (7.9.9), (7.9.12), (7.9.15) for the case of the polynomial functions $\omega_l(x, v)$, $\omega_l'(x, v)$, $\zeta_l(x, v)$ in x of not higher than the third degree and linear in v, y, the coefficients of v, y being independent of x or linear in x. For solving Eq. (7.9.15) the normal approximation method is used (Section 7.2).

Eqs. (7.9.1) and (7.9.4) presented in the autoregression form

$$X(l + 1) = \varphi(l, X(l) + \psi(l, X(l))V(l), \tag{7.9.34}$$

$$Y(l) = \varphi'(l, X(l) + \psi(l, X(l))W(l), \tag{7.9.35}$$

are the basis of the algorithms of software package "StS–Filter". In Eqs. (7.9.34), (7.9.35) the following notations are introduced: $X(l)$ and $Y(l)$ being the state and the measurements vectors of the dimensions n and n_1, $V(l)$ and $W(l)$ being the discrete normal white noises with variances $G(l)$ and $Q(l)$, the initial values $X(0)$, $Y(0)$, $V(0)$ and $W(0)$ being the mutually independent normal variables.

The structure of the conditionally optimal filter is chosen in the following form:

$$\hat{X}(l + 1) = \alpha(l)\xi(l, \hat{X}(l)) + \beta(l)\eta(l, \hat{X}(l))Y(l) + \gamma(l). \tag{7.9.36}$$

Here $\xi = \xi(l, \hat{X}(l))$ and $\eta = \eta(l, \hat{X}(l))$ being the known structural filter functions, $\alpha = \alpha(l)$, $\beta = \beta(l)$, $\gamma = \gamma(l)$ being the unknown filter gains determined from m.s.e. condition

$$\min_{\alpha,\beta,\gamma} E\tilde{X}(l + 1)^T \tilde{X}(l + 1), \quad \tilde{X}(l + 1) = \hat{X}(l + 1) - X(l) \tag{7.9.37}$$

at known values $\alpha = \alpha(j)$, $\beta = \beta(j)$, $\gamma = \gamma(j)$ for $j < l$, $j = 1, 2\ldots$).

Optimal values of filter coefficients $\alpha = \alpha(l)$, $\beta = \beta(l)$, $\gamma = \gamma(l)$ by virtue of (7.9.37) determined from the solution of the set of linear equations

$$\alpha\kappa_{11} + \beta\kappa_{21} = \kappa_{01}, \tag{7.9.38}$$

$$\alpha\kappa_{12} + \beta\kappa_{22} = \kappa_{02}, \tag{7.9.39}$$

$$\gamma = \mu_0 - \alpha\mu_1 - \beta\mu_2, \tag{7.9.40}$$

where

$$\kappa_{01} = E\left[\varphi(l, X(l)) - \mu_0(l)\right]\xi(l, \hat{X}(l))^T, \tag{7.9.41}$$

$$\kappa_{02} = E\left[\varphi(l, X(l)) - \mu_0(l)\right]\varphi'(l, X(l))^T \eta(l, \hat{X}(l))^T$$
$$+ E\left[\psi(l, X(l))G(l)\psi'(l, X(l))^T \eta(l, \hat{X}(l))^T\right] \tag{7.9.42}$$

$$\kappa_{11} = E\left[\xi(l, \hat{X}(l)) - \mu_1(l)\right]\xi(l, \hat{X}(l))^T, \tag{7.9.43}$$

$$\kappa_{12} = E\left[\xi(l, \hat{X}(l)) - \mu_1(l)\right]\varphi(l, X(l))^T \eta(l, \hat{X}(l))^T, \tag{7.9.44}$$

$$\kappa_{22} = E\left[\eta(l, \hat{X}(l))\varphi'(l, X(l)) - \mu_2(l)\right]\varphi'(l, X(l))^T \eta(l, \hat{X}(l))^T$$

$$+ E\left[\eta(l, \hat{X}(l))\psi'(l, X(l))G(l)\psi'(l, X(l))^T \eta(l, \hat{X}(l))^T\right], \tag{7.9.45}$$

$$\mu_0(l) = E\varphi(l, X(l)), \quad \mu_1(l) = E\xi(l, \hat{X}(l)),$$
$$\mu_2 = E\eta(l, \hat{X}(l))\varphi'(l, X(l)). \tag{7.9.46}$$

In the case of linear Eqs. (7.9.34)–(7.9.36) conditionally optimal filter coincides with the Kalman–Bucy filter. For nonlinear Eqs. (7.9.34)–(7.9.36) with the polynomial functions φ, φ', ψ, ψ' the expectations in (7.9.41)–(7.9.46) represent the moments of the random vector $\left[X(l)^T \hat{X}(l)^T\right]^T$ and are determined by the normal approximation method (Subsection 7.2.1).

Software package "StS–Filter" while designing the conditionally optimal filter performs:
- calculation of the filter coefficients $\alpha = \alpha(l)$, $\beta = \beta(l)$, $\gamma = \gamma(l)$,
- calculation of the accuracy characteristics

$$R(l) = E\tilde{X}(l)\tilde{X}(l)^T, \quad \tilde{X}(l) = \hat{X}(l) - X(l),$$

$$m_x(l) = EX(l), \quad m_{\hat{x}} = E\hat{X}(l),$$

$$K_x(l) = E(X(l) - m_x(l))(X(l) - m_x(l))^T,$$

$$K_{x\hat{x}}(l) = E(X(l) - m_x(l))(\hat{X}(l) - m_{\hat{x}}(l))^T,$$

$$K_{\hat{x}}(l) = E(\hat{X}(l) - m_{\hat{x}}(l))(\hat{X}(l) - m_{\hat{x}}(l))^T,$$

- estimates the performance by the statistical simulation.

Theory of the nonlinear conditionally optimal estimation and control and its applications are detailly considered, for example, in (Pugachev and Sinitsyn 1987, 1993, 1996).

E x a m p l e 7.9.1. Let us design the conditionally optimal filter for the follwing stochastic discrete nonlinear system

$$X_1(l+1) = X_1(l) + hX_2(l), \tag{I}$$
$$X_2(l+1) = X_2(l) + ha_0 + ha_2 X_2^2(l) + bV_1(l), \tag{II}$$

where V_1 being the descrete normal white noise with the zero expectation and the unit intensity, a_0, a_2 and b being constant coefficients. Measurements equation we take in the form

$$Y_1(l) = X_1(l) + cV_2(l), \tag{III}$$

where $V_2(l)$ being the discrete white noise with zero expectation and the unit intensity. The noises $V_1(l)$, $V_2(l)$ are mutually independent.

Eq. (7.9.36) for the nonlinear conditionally optimal filter may be expanded as

$$\hat{X}_1(l+1) = \alpha_{11}(l)\hat{X}_1(l) + \alpha_{12}(l)\hat{X}_2(l)$$
$$+ \beta_1(l)Y_1(l) + \gamma_1(l), \tag{IV}$$
$$\hat{X}_2(l+1) = \alpha_{21}(l)\hat{X}_2(l) + \alpha_{22}(l)\hat{X}_2^2(l)$$
$$+ \beta_2(l)Y_1(l) + \gamma_2(l). \tag{VI}$$

The results of the conditionally optimal filter design at

$$a_0 = 0, \quad h = 0.1, \quad b = 1, \quad c = 200 \tag{V}$$

and initial conditions

$$E\hat{X}_1(0) = EX_1(0) = 4000,$$
$$E\hat{X}_2(0) = E\hat{X}_2(0) = -20,$$
$$DX_1(0) = 160000, \quad D\hat{X}_1 = 10000, \tag{VII}$$
$$DX_2(0) = D\hat{X}_2(0) = 1$$

are given on Figs.7.9.1-7.9.3. The expections, the variances and the covariances of the processes $X_i(l)$ $(i = 1, 2)$ are given on Fig.7.9.1. The variances and the covariances of $\hat{X}_i(l)$ are practicaly identical. Graphs of $\alpha_{11}(l)$, $\alpha_{12}(l)$, $\alpha_{21}(l)$, $\alpha_{22}(l)$, $\beta_1(l)$, $\beta_2(l)$, $\gamma_1(l)$, $\gamma_2(l)$ are shown on Fig.7.9.2. The results of statistical simulation the processes $X_i(l)$, $\hat{X}_i(l)$ $(i = 1, 2)$ are presented on Fig.7.9.3. Maximal m.s.e. is does not exceed 1%. For the linear Kalman filter (at $\alpha_{22} \equiv 0$) maximal m.s.e. is equal to 5%.

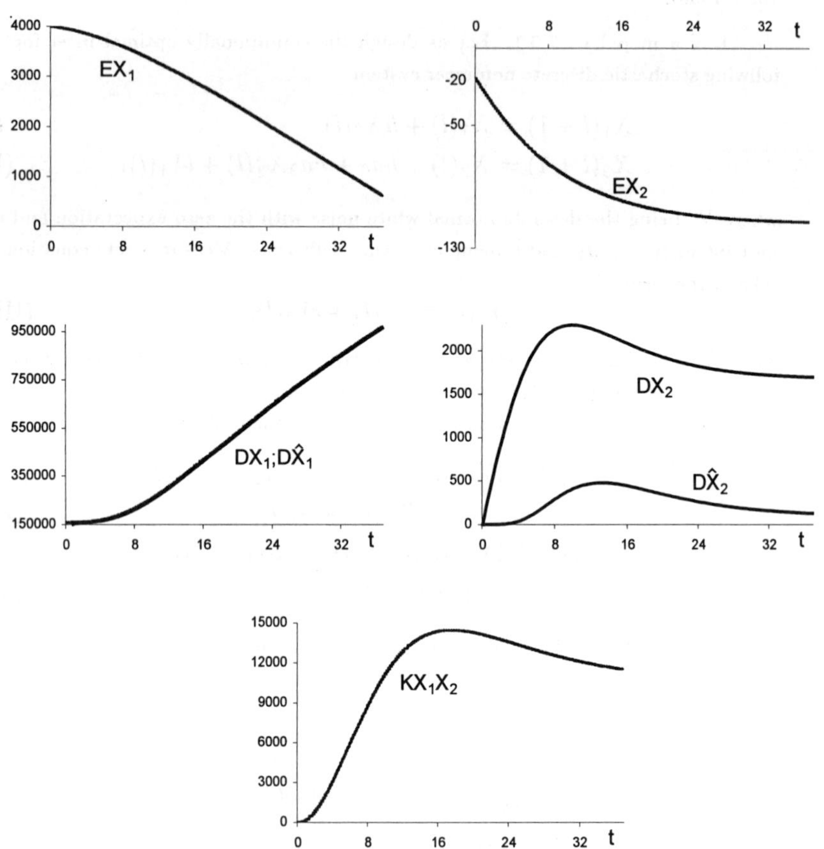
4000
3000
2000
1000
0
EX₁
0   8   16   24   32   t
0   8   16   24   32   t
-20
-50
-130
EX₂
950000
750000
550000
350000
150000
DX₁;DX̂₁
0   8   16   24   32   t
2000
1500
1000
500
0
DX₂
DX̂₂
0   8   16   24   32   t
15000
12000
9000
6000
3000
0
KX₁X₂
0   8   16   24   32   t

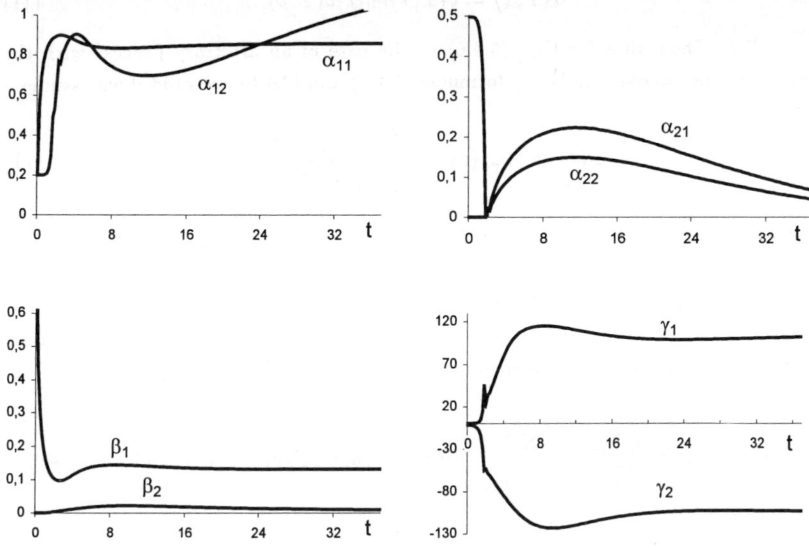

Fig. 7.9.2

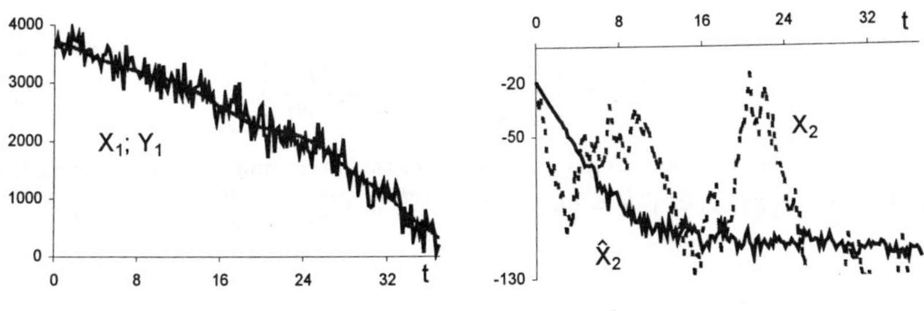

Fig. 7.9.3

Problems

7.1. Using formula (5.1.16) show that in the case (5.4.7) the function $\sigma(Y, t)$ in formulae (7.1.3), (7.1.5) must be replaced by $\bar{\sigma}((Y, t)$,

$$\bar{\sigma}(Y, t) = \sigma(Y, t) + \int\limits_{R_0^q} c(Y, t, u) c(Y, t, u)^T \nu_P(t, du), \qquad (I)$$

$$\sigma(Y,t) = b(Y,t)\nu_0(t)b(Y,t)^T. \tag{II}$$

7.2. Show that for Eq. (5.4.6) in the case of an arbitrary process with the independent uncrements $W(t)$ formulae (7.1.3) and (7.1.5) are valid if we assume

$$\nu(t) = [\nu_{rs}(t)] = \left[\frac{\partial^2 \chi(\lambda;t)}{\partial(i\lambda_1)\partial(i\lambda_s)}\right], \tag{I}$$

where $\chi(\lambda;t) = \partial/\partial t \ln h_1(\lambda;t)$, $h_1(\lambda;t)$ be the one-dimensional characteristic functions of $W(t)$.

7.3. Show that the infinite set of equations for the moments in the case Eq. (5.4.6) where $a(y,t)$ is defined by (7.4.1), $b(y,t) = b(t)$, $V = \dot{W}$, W being an arbitrary process with the independent increments has the following form:

$$\dot{\alpha}_{k_1,\ldots,k_p} = \sum_{r=1}^{p} k_r \sum_{h_1,\ldots,h_p=0}^{N} a_{r,h_1,\ldots,h_p}\alpha_{h_1+k_1,\ldots,h_r+k_r-1,\ldots,h_p+k_p}$$

$$+ \sum_{h_1=0}^{k_1}\cdots\sum_{h_p=0}^{k_p} C_{k_1}^{h_1}\ldots C_{k_p}^{h_p}\chi_{h_1,\ldots,h_p}\alpha_{k_1-h_1,\ldots,k_p-h_p} \tag{I}$$

$$(k_1,\ldots,k_p = 0,1,2,\ldots; k_1+\cdots k_p = 1,2,\ldots),$$

where

$$\chi_{h_1,\ldots,h_p} = \left[\frac{\partial^{h_1,\ldots,h_p}\chi(\lambda;t)}{\partial(i\lambda_1)^{h_1}\ldots\partial(i\lambda_p)^{h_p}}\right]_{\lambda=0}. \tag{II}$$

Also show that Eqs. (I) may be written in the more compact form using multi-indexes $k = [k_1\ldots k_p]$, $h = [h_1\ldots h_p]$, $C_k^h = C_{k_1}^{h_1}\ldots C_{k_p}^{h_p}$:

$$\dot{\alpha}_k = \sum_{r=1}^{p} k_r \sum_{h=0}^{N} a_{r,h}\alpha_{h+k-e_r} + \sum_{h=0}^{k} C_k^h \chi_h \alpha_{k-h} \tag{III}$$

$$(k_1,\ldots,k_p = 0,1,2,\ldots; |k| = k_1+\cdots+k_p = 1,2,\ldots),$$

where e_r is the vector whose components are equal to zero except the rth component which is equal to the unity, $e_r = \left[\begin{smallmatrix}0\ldots0\,1\,0\ldots0\\ r\end{smallmatrix}\right]^T$,

$$\alpha_k = \alpha_{k_1,\ldots,k_p}, \quad a_{r,h} = a_{r,h_1,\ldots,h_p}, \quad \chi_h = \chi_{h_1,\ldots,h_p}, \tag{IV}$$

$\chi_h = 0$ at $|h| = h_1+\cdots+h_p < 2$ and $\chi_h = \nu_{rs}$ at $h = e_r + e_s$. In Eqs. (III) α_s is equal to zero if at least one of the components of the multi-index s is negative, and is equal to unity if all the components of the multi-index s are equal to zero, $\alpha_0 = 1$.

7.4. For the linear system (1.4.5) with the parametric noises $N(t)$ considered as the system with the θ-differentials write out the correspondent Itô equations assuming that the white noise is normally distributed.

7.5. Consider the linear system with parametric noises

$$A\ddot{q} + (1 + V_3)(B + B')\dot{q} + (1 + V_2)(C + C')q = (1 + V_1)Q^*, \qquad \text{(I)}$$

where $q = [\,q_1 \ldots q_n\,]^T$ and $Q^* = [\,Q_1^* \ldots Q_n^*\,]^T$ are the n-dimensional vectors; A, B, C are the symmetric matrices; B', C' are the antisymmetric $n \times n$-matrices; $V = [\,V_1\,V_2\,V_3\,]^T$ is the normally distributed white noise of the intensity ν. Show that to Eq. (I) corresponds the following Itô equation:

$$A\ddot{q} + (1 + V_3)(B + B')\dot{q} + (1 + V_2)(C + C')q$$

$$= (1 + V_1)Q^* + \theta[\,-\nu_{13}Q^* A^{-1}(B - B')$$

$$+\nu_{23}(C + C')q A^{-1}(B - B') + \nu_{33}(B + B')\dot{q} A^{-1}(B - B')\,]. \qquad \text{(II)}$$

7.6. Considering the equations of Problem 1.14 at $\Pi = \psi V$, $Q = \varphi V$ where $\psi = \psi(q, t)$, $\varphi = \varphi(q, \dot{q}, t)$ are matrix functions of the stated variables, V is the normally distributed vector white noise of the intensity ν as the equations with the θ-differential, reduce them to the Itô equations for the canonical variables.

7.7. Show that for the linear stochastic system with the parametric noise of Problem 5.5 (i) equations for the expectation, the variance and the covariance function have the form

$$\dot{m} = -\varepsilon(1 - 2\theta\varepsilon\nu_{22})m - 2\varepsilon\theta k\nu_{12}, \qquad \text{(I)}$$

$$\dot{D} = -2\varepsilon(1 - 2\theta\varepsilon\nu_{22})D + k^2\nu_{11} + \varepsilon^2\nu_{22}(m^2 + D) - 2\varepsilon k m\nu_{12}, \qquad \text{(II)}$$

$$\partial K(t_1, t_2)/\partial t_2 = -\varepsilon(1 - 2\theta\varepsilon\nu_{22})K(t_1, t_2). \qquad \text{(III)}$$

Find the stationary values of m and D and the conditions of the stationary regime existence.

7.8. Show that for the linear stochastic system with parametric noises of Problem 5.7 (i) equations for m and K have the form

$$\dot{m}_1 = m_2, \quad \dot{m}_2 = -m_1\omega_0^2(1 - 2\theta\zeta\omega_0\nu_{23})$$

$$-2m_2\zeta\omega_0(1 - 2\theta\zeta\omega_0\nu_{33}) - 2\theta k\zeta\omega_0\nu_{13}; \qquad \text{(I)}$$

$$\dot{k}_{11} = 2k_{12}, \quad \dot{k}_{12} = -\omega_0^2(1 - 2\theta\zeta\omega_0\nu_{23})k_{11} - 2\zeta\omega_0(1 - 2\theta\zeta\omega_0\nu_{33})k_{12} + k_{22},$$

$$\dot{k}_{22} = -2\omega_0^2(1 - 2\theta\zeta\omega_0\nu_{23})k_{12} - 4\zeta\omega_0(1 - 2\theta\zeta\omega_0\nu_{33})k_{22} + k^2\nu_{11}$$
$$-2m_1 k\omega_0^2\nu_{12} - 4m_2\zeta\omega_0 k\nu_{13} + \omega_0^4\nu_{22}(m_1^2 + k_{11})$$
$$+4\zeta\omega_0^3\nu_{23}(m_1 m_2 + k_{12}) + 4\zeta^2\omega_0^2\, nu_{33}(m_2^2 + k_{22}). \tag{II}$$

7.9. Using formula (5.1.16) show that Eqs. (7.2.3), (7.2.4) are valid for Eq. (5.4.7) at functions φ_1, φ_2, φ_{21} defined by (7.2.5)–(7.2.7), at $\varphi_{22} = E_N\bar{\sigma}$ and $\bar{\sigma}$ and σ defined by formula (I) of Problem 7.1.

7.10. Sow that NAM equations for stochastic differential system (5.4.6) in the case of an arbitrary process with the independent increments W coincide with the equations of Subsection 7.2.1 under condition (I) of Problem 7.2.

7.11. Derive NAM equations for nonlinear stochastic discrete system (5.7.2).

7.12. Derive NAM equations for the continuous-discrete system defined by Eqs. (5.7.10), (5.7.11).

7.13. For the typical composite functions derive formulae for φ_0 for NAM given in Appendix 3.

7.14.   Show that NAM gives the exponential covariance function $k(\tau)$ $= De^{-\alpha|\tau|}$ for the stationary process in the system of Example 7.2.1 where
(i) $D = (\nu/6b_3)^{1/2}$, $\alpha = (3/2\nu b_3)^{1/2}$ at $\varphi(y) = b_3 y^3$;
(ii) $D = (\sqrt{2\pi\nu}/8b_2)^{1/2}$, $\alpha = (8/\pi)^{1/2}(\sqrt{2\pi\nu}/8b_6)^{1/3}$ at $\varphi(y)$ $= b_2 y^2 \, \mathrm{sgn}\, y$.

7.15. Using the equations of Problems 7.14 in the case (i) deduce the formula for the standard deviation $\sigma = \sqrt{D}$ for an arbitrary instant.

7.16. Show that the expectation m and the variance D of the state vector of the system
$$T\dot{Y} + Y = b(X - \alpha Y^3), \quad X = m_x + V, \tag{I}$$

(V being the stationary white noise of the intensity ν) obtained by NAM are determined by the equations

$$T\dot{m} = bm_x - m[1 + \alpha b(m^2 + 3D)], \quad T^2\dot{D} = b^2\nu - 2DT[1 + 3\alpha b(m_D^2)]. \tag{II}$$

Find the moments method values of m and D for the regime of stationary oscillations.

Using MM with the account of the moments of up to the fourth order verify whether or not this method rejects the obtained m and D as a superfluous solution.

7.17. Using NAM show that the variances and the covariance of the state variables in the stochastic nonlinear system

$$T\dot{Y}_1 + Y_1 = b\,\mathrm{sgn}(Y_2 - Y_1), \quad \dot{Y}_2 + \alpha Y_2 = \sqrt{2D\alpha}V, \tag{I}$$

(the expectations of the initial values of Y_1 and Y_2 being equal to zero, V the white noise of the unit intensity) are determined by the equations

$$T\dot{k}_{11} = 2\beta k_{12} + 2(1 + \beta)k_{11}, \quad T\dot{k}_{12} = \beta k_{22} - (1 + \alpha T + \beta)k_{12},$$

$$k_{22} = 2\alpha(D - k_{22}), \quad \beta = b[\,(k_{11} + k_{22} - 2k_{12})\pi/2\,]^{-1/2}. \tag{II}$$

Find k_{11}, k_{12}, k_{22} for the stationary regime. Using MM with the account of the moments of up to the fourth order verify whether or not this method rejects the obtained variance k_{11} for the stationary regime as a superfluous solution. Solve this problem also in the case of the nonzero expectations of the initial values of Y_1 and Y_2.

7.18. Using NAM show that for the system

$$T\ddot{Y} + \dot{Y} = b\,\mathrm{sgn}\,(X - Y), \quad \dot{X} + \alpha X = \sqrt{2D\alpha}V, \tag{I}$$

V being the normal white noise of the unit intensity, NAM equations for the variances and the covariances of the random variables $Y_1 = Y$, $Y_2 = \dot{Y}$, $Y_3 = X$ have the form

$$\dot{k}_{11} = 2k_{12}, \quad T\dot{k}_{12} = Tk_{22} - k_{12} + \beta(k_{13} - k_{11}),$$

$$\dot{k}_{13} = k_{23} - \alpha k_{13}, \quad \dot{k}_{33} = 2\alpha(D - k_{33}),$$

$$T\dot{k}_{22} = 2k_{22} + 2\beta(k_{23} - k_{12}), \quad T\dot{k}_{23} = -(1 + \alpha T)k_{23} + \beta(k_{33} - k_{13}),$$

$$\beta = b[\,(k_{11} + k_{33} - 2k_{13})\pi/2\,]^{-1/2}, \tag{II}$$

if the expectations of the initial values of Y_1, Y_2, Y_3 are equal to zero. Find the stationary values of the variances and the covariances.

7.19. For the stochastic nonlinear oscillator

$$\ddot{Y} + \varphi(Y, \dot{Y}) + \omega_0^2 Y = V, \tag{I}$$

where V is the white noise of the intensity ν show that the variances and the covariances of the random variables $Y_1 = Y$, $Y_2 = \dot{Y}$ are determined by the following NAM equations:

$$\dot{k}_{11} = 2k_{12}, \quad \dot{k}_{12} = k_{22} - (\omega_0^2 + k_1)k_{11} - k_2 k_{12},$$

$$\dot{k}_{22} = -2(\omega_0^2 + k_1)k_{12} - 2k_2 k_{22} + \nu, \tag{II}$$

where k_1 and k_2 are the statistical linearization coefficients of the functions $\varphi(y_1, y_2)$, the expectations of the initial values of Y_1, Y_2 being equal to zero.

7.20. Using NAM and the results of Problem 7.19 verify the validity of the following approximate formulae for the variances and the covariances of the stationary process in the systems:

(i) the oscillator with dry friction $\varphi(y_1, y_2) = b_0\,\mathrm{sgn}\,y_2$,

$$k_{11} = \pi\nu^2/8\omega_0^2 b_0^2, \quad k_{12} = 0, \quad k_{22} = \omega_0^2 k_{11};$$

(ii) the Rayleigh oscillator $\varphi(y_1, y_2) = -b_1 y_2 + b_3 y_2^3$,

$$k_{11} = (1/6\omega_0^2 b_3)(b_1 + \sqrt{b_1^2 + 6\nu b_3}), \quad k_{12} = 0, \quad k_{22} = \omega_0^2 k_{11};$$

(iii) the Van der Paul oscillator $\varphi(y_1, y_2) = b(y_1^2 - 1)y_2$,

$$k_{11} = (1 + \sqrt{1 + 2\nu/\omega_0^2 b})/2, \quad k_{12} = 0, \quad k_{22} = \omega_0^2 k_{11}.$$

7.21. Using formula (5.1.16) show that the equations for the initial moments in the case of Eq. (5.4.7) may be presented in the form (7.4.3) if we add to function $\varphi_{r,\nu}$ defined by (7.4.4) the following item:

$$\Delta\varphi_{r,\nu} = \int\limits_{-\infty}^{\infty} \int\limits_{R_0^q} \left\{ [y_1 + c_1(y, t, u)]^{\nu_1} \ldots [Y - p + c_p(y, t, u)]^{\nu_p} \right.$$

$$\left. -y_1^{\nu_1} \ldots y_p^{\nu_p} - \sum_{s=1}^{p} \nu_c y_1^{\nu_1} \ldots y_s^{\nu_s-1} \ldots y_s^{\nu_p} c_s(y, t, u) \right\}$$

$$\times \nu_P(t, du) w_1(y) p_r(y) dy. \tag{I}$$

7.22. Show that the equations for the initial moments in the case of Eq. (5.4.6) for the nongaussian white noise have the form (7.4.3) at

$$\varphi_{r,0} = \int\limits_{-\infty}^{\infty} \left\{ \frac{\partial^{|r|}}{\partial(i\lambda_1)^{r_1} \ldots \partial(i\lambda_p)^{r_p}} \right.$$

$$\times \left. \left[i\lambda^T a(y, t) + \chi(b(y, t)^T \lambda; t) \right] e^{i\lambda^T y} \right\}_{\lambda=0} w_1(y) dy, \tag{I}$$

$$\varphi_{r,\nu} = \int\limits_{-\infty}^{\infty} \left\{ \frac{\partial^{|r|}}{\partial(i\lambda_1)^{r_1} \ldots \partial(i\lambda_p)^{r_p}} \right.$$

$$\times \left. \left[i\lambda^T a(y, t) + \chi(b(y, t)^T \lambda; t) \right] e^{i\lambda^T y} \right\}_{\lambda=0} p_\nu(y) w_1(y) dy. \tag{II}$$

7.23. Using formula (5.1.16) show that the equations for the central moments in the case of Eq. (5.4.7) have the form (7.4.6), (7.4.7), (7.4.11), where $\varphi_{1,\nu}$ is defined by (7.4.8), $\varphi_{2,\nu}$ is defined by (7.4.9) at σ, substituted by $\bar{\sigma}$ of Problem 7.1 and $\varphi_{r,\nu}$ is defined in Problem 7.21.

7.24. Show that the equations for the central moments in the case of Eq. (5.4.6) with the nongaussian white noise have the form

$$\dot{m}_{1,0,h} = \varphi_{1,0,h}(m,K,t) + \sum_{k=3}^{N} \sum_{|\nu|=k} \varphi_{1,\nu,h}(m,K,t)q_\nu(\alpha) \ (h=1,\ldots,p),$$

(I)

$$\dot{\mu}_r = \varphi_{r,0}(m,K,t) - \sum_{h=1}^{p} r_h \varphi_{1,0,h}(m,K,t)\mu_{r-e_h}$$

$$+ \sum_{l=3}^{N} \sum_{|\nu|=l} \left[\varphi_{r,\nu}(m,K,t) - \sum_{h=1}^{p} r_h \varphi_{1,\nu,h}(m,K,t)\mu_{r-e_h} \right] q_\nu(\alpha) \quad \text{(II)}$$

$$(r_1,\ldots,r_p = 0,1,\ldots,N; \ |r| = 2,\ldots,N),$$

where

$$\varphi_{1,0,h}(m,K,t) = \int_{-\infty}^{\infty} a_h(y,t)w_1(y)dy,$$

(III)

$$\varphi_{1,\nu,h}(m,K,t) = \int_{-\infty}^{\infty} a_h(y,t)p_\nu(y)w_1(y)dy,$$

(IV)

$$\varphi_{r,0}(m,K,t) = \int_{-\infty}^{\infty} \left\{ \frac{\partial^{|r|}}{\partial(i\lambda_1)^{r_1}\ldots\partial(i\lambda_p)^{r_p}} [i\lambda^T a(y,t) \right.$$

$$\left. + \chi(b(y,t)^T \lambda; t)]e^{i\lambda^T(y-m)} \right\}_{\lambda=0} w_1(y)dy,$$

(V)

$$\varphi_{r,\nu}(m,K,t) = \int_{-\infty}^{\infty} \left\{ \frac{\partial^{|r|}}{\partial(i\lambda_1)^{r_1}\ldots\partial(i\lambda_p)^{r_p}} [i\lambda^T a(y,t) \right.$$

$$\left. + \chi(b(y,t)^T \lambda; t)]e^{i\lambda^T(y-m)} \right\}_{\lambda=0} p_\nu(y)w_1(y)dy.$$

(VI)

7.25. Derive the equations of Subsections 7.4.1, 7.4.2 for the initial and central moments in conformity to stochastic discrete system (5.7.2).

7.26. Show that the equations for the multi-dimensional initial moments in the case of Eq. (5.4.7) have the form (7.4.15) with $\varphi_{\kappa_1,\ldots,\kappa_n,\nu_1,\ldots,\nu_n}(\bar{m},\bar{K}_n,t_n)$ containing the following additional term:

$$\Delta\varphi_{\kappa_1,\ldots,\kappa_n;\nu_1,\ldots,\nu_n}(\bar{m}_n \bar{K}_n, t_n)$$

$$= \int\limits_{-\infty}^{\infty} \cdots \int\limits_{-\infty}^{\infty} \int\limits_{R_0^q} y_{11}^{\nu_{11}} \cdots y_{1p}^{\nu_{1p}} \cdots y_{n-1,1}^{\nu_{n-1,1}}$$

$$\cdots y_{\nu-1,p}^{\nu_{n-1,p}} \left\{ \left[y_{n1} + c_1(y_n, t_n, u) \right]^{\nu_{n1}} \right.$$

$$\cdots \left[y_{np} + c_p(y_n, t_n, u) \right]^{\nu_{np}} - y_{n1}^{\nu_{n1}} \cdots y_{np}^{\nu_{np}}$$

$$\left. - \sum_{s=1}^{p} \nu_{ns} y_{n1}^{\nu_{n1}} \cdots y_{ns}^{\nu_{ns}-1} \cdots y_{np}^{\nu_{np}} c_s(y_n, t_n, u) \right\} \nu_P(t_n, du)$$

$$\times w_n(y_1, \ldots, y_n) p_{\nu_1, \ldots, \nu_n}(y_1, \ldots, y_n) dy_1 \ldots dy_n \qquad \text{(I)}$$

7.27. Show that the equations for the multi-dimensional moments in the case of Eq. (5.4.6) with the nongaussian white noise have the form

$$\partial \alpha_{r_1, \ldots, r_n}(t_1, \ldots, t_n)/\partial t_n = \int\limits_{-\infty}^{\infty} \cdots \int\limits_{-\infty}^{\infty} \left\{ \frac{\partial^{|r|}}{\partial(i\lambda_{n1})^{r_{n1}} \ldots \partial(i\lambda_{np})^{r_{np}}} \right.$$

$$\left. \times \left[i\lambda_n^T a(y_n, t_n) + \chi(b(y_n, t_n)^T \lambda_n; t_n) \right] e^{i\lambda_n^T y_n} \right\}_{\lambda_n = 0}$$

$$\times y_{11}^{r_{11}} \cdots y_{1p}^{r_{1p}} \cdots y_{n-1,1}^{r_{n-1,1}} \cdots y_{n-1,p}^{r_{n-1,p}} f_n^*(y_1, \ldots, Y_n; \theta_n) dy_1 \ldots dy_n \qquad \text{(I)}$$

7.28. Derive the equations for the multi-dimensional moments in the case of Eq. (5.4.6) for the nongaussian white noise.

7.29. Derive the equation for the multi-dimensional central moments in the case of Eq. (5.4.7).

7.30. For the system of Example 1.3.3 in the case when the vector of random disturbances $N(t) = [\, N_1(t) \, N_2(t) \,]^T$ is the normal two-dimensional white noise with the matrix of the intensity ν show that the moments $\alpha_{r,s}$ of the process $Y = [\, Y_1 \, Y_2 \,]^T$ are determined by an infinite set of equations for the moments:

$$\dot{\alpha}_{r,s} = r\alpha_{r-1,s+1} - s\omega_0^2 \alpha_{r+1,s-1} - 2\varepsilon s \alpha_{r,s} + \nu_{12}\omega_0^4 r(r-1)\alpha_{r-1,s}$$

$$+ \frac{1}{2}\omega_0^4 s(s-1)(\nu_{11}\alpha_{r,s-2} + 2\nu_{12}\alpha_{r+1,s-2} + \nu_{22}\alpha_{r+2,s-2})$$

$$(r, s = 0, 1, 2, \ldots, |r| + |s| = 1, 2, \ldots).$$

7.31. Under condition of Example 7.2.0 at $\varphi(y) = y^3$ show that the equations for the first three semiinvariants m, D and κ_3 have the form

$$\dot{m} = -\kappa_3 - 3mD - m^3, \qquad \text{(I)}$$

$$\dot{D} = -6m\kappa_3 - 6m^3 D - 6D^2 + 1 \,,$$
$$\dot{\kappa}_3 = -9\kappa_3 m^2 - 18m^3 D - 27D\kappa_3 \,. \qquad \text{(II)}$$

7.32. Prove that for the stochastic two-dimensional system

$$\dot{Y}_1 = a_0 - aY_1 + h_1 V_1 \,, \quad \dot{Y}_2 = -Y_1 Y_2 + h_2 V_2 \qquad \text{(I)}$$

(a_0, a, h_1, $h_2 > 0$, V_1 and V_2 are the independent normal white noises with the unit intensities) the equations of the semiinvariant method with the accuracy till the first four semiinvariants κ_{ij} (i, $j = 0, 1, 2, 3, 4$) have the form

$$\dot{\kappa}_{01} = -a\kappa_{10} + a_0 \,, \quad \dot{\kappa}_{10} = \kappa_{10}\kappa_{01} \,, \quad \dot{\kappa}_{01} = -2a\kappa_{20} + h_1^2 \,, \qquad \text{(I)}$$

$$\dot{\kappa}_{11} = -a\kappa_{11} - \kappa_{01}\kappa_{20} - \kappa_{10}\kappa_{11} - \kappa_{21} \,,$$

$$\dot{\kappa}_{20} = -2(\kappa_{11}\kappa_{01} + \kappa_{10}\kappa_{02} + \kappa_{12}) + h_2^2 \,, \qquad \text{(II)}$$

$$\dot{\kappa}_{30} = -3a\kappa_{30} \,, \quad \dot{\kappa}_{40} = -4a\kappa_{40} \,,$$

$$\dot{\kappa}_{21} = -2a\kappa_{21} - \kappa_{01}\kappa_{30} - 2\kappa_{20}\kappa_{11} - \kappa_{10}\kappa_{21} - \kappa_{31} \,,$$

$$\dot{\kappa}_{12} = -a\kappa_{12} - 2(\kappa_{01}\kappa_{21} + \kappa_{20}\kappa_{02} + \kappa_{11}^2 + \kappa_{10}\kappa_{12} + \kappa_{22}) \,,$$

$$\dot{\kappa}_{03} = -3(\kappa_{01}\kappa_{12} + 2\kappa_{11}\kappa_{02} + \kappa_{10}\kappa_{03} + 3\kappa_{13}) \,, \qquad \text{(III)}$$

$$\dot{\kappa}_{31} = -3a\kappa_{31} - \kappa_{01}\kappa_{40} - 3\kappa_{30}\kappa_{11} - 3\kappa_{20}\kappa_{21} - \kappa_{10}\kappa_{31} \,,$$

$$\dot{\kappa}_{22} = -2a\kappa_{22} - 2(\kappa_{01}\kappa_{31} + \kappa_{30}\kappa_{02} + 3\kappa_{21}\kappa_{11} + 2\kappa_{20}\kappa_{12} + \kappa_{10}\kappa_{22}) \,,$$

$$\dot{\kappa}_{13} = -a\kappa_{13} - 3(\kappa_{01}\kappa_{22} + 2\kappa_{21}\kappa_{02} + \kappa_{20}\kappa_{03} + 3\kappa_{11}\kappa_{12} + \kappa_{10}\kappa_{13}) \,,$$

$$\dot{\kappa}_{04} = -4(\kappa_{01}\kappa_{13} + 3\kappa_{12}\kappa_{02} + \kappa_{10}\kappa_{04} + 3\kappa_{11}\kappa_{03}) \,. \qquad \text{(IV)}$$

Show that at $a_0 \gg h_1^2 a$ disappears the difference between the values $\kappa_{02}^{(2)}$, $\kappa_{02}^{(3)}$, $\kappa_{02}^{(4)}$ calculated with the accuracy till the semiinvariants of the second fourth orders.

7.33. Derive the equations for the first two moments of Problems 5.23 and 5.24. Show that the solution of these equations at the initial conditions $m_0 = EY_0$, $D_0 = DY_0^2$ has the form

$$\alpha_1(t) = \alpha_{10} + \alpha_{11} \exp(-at) \,,$$
$$\alpha_2(t) = \alpha_{20} + \alpha_{21} \exp(-at) + \alpha_{22} \exp(-2at) \,,$$

where $\alpha_{10} = a_0/a$; $\alpha_{11} = m_0 - \alpha_{10}$; $\alpha_{20} = (1 + 2a_0 a_1)a_0/2a_1 a^2$; $\alpha_{21} = \alpha_{11}(1 + 2a_0 a_1)/aa_1$; $\alpha_{22} = D_0 - \alpha_{20} - \alpha_{21}$. Find the stationary solutions.

7.34. Show that the equations for the coefficients of the orthogonal expansions in the case of Eq. (5.4.7) have the form of Eqs. (7.5.8), (7.5.9) at

$$\varphi_{20}(m, K, t) = \int_{-\infty}^{\infty} [a(y,t)(y^T - m^T) + (y - m)a(y,t)^T + \bar{\sigma}(y,t)]w_1(y)dy \,,$$

$$\text{(I)}$$

$$\varphi_{2\nu}(m, K, t) = \int_{-\infty}^{\infty} [a(y)(y^T - m^T) + (y - m)a(y, t)^T + \bar{\sigma}(y, t)]p_\nu w_1(y)(y)dy,$$

$$\text{(II)}$$

$$\varphi_{\kappa\nu}(m, P, t) = \int_{-\infty}^{\infty} \left\{ \frac{\partial^T q_\kappa(y)}{\partial y} a(y, t) + \frac{1}{2}\, \text{tr} \left[\frac{\partial}{\partial y}\frac{\partial^T}{\partial y} q_\kappa(y)\sigma(y, t) \right] \right.$$

$$+ \int_{R_0^q} \left[q_\kappa(y) + c(y, t, u)) - q_\kappa(y) - \frac{\partial^T q_\kappa(y)}{\partial y}c(y, t, u) \right] \qquad \text{(III)}$$

$$\times \left. \nu_P(t, du) \right\} w_1(y)p_\nu(y)dy \ (\,|\kappa|\,,\,|\nu| = 3\,,\,\ldots\,,\,N);$$

and further,

$$\dot{c}_\kappa = \varphi_{\kappa 0}(m, K, t) + \psi_{\kappa 0}^m(m, K, t)^T \dot{m} + \frac{1}{2}\, \text{tr} \left[\psi_{\kappa 0}^K(m, K, t)\dot{K} \right]$$

$$+ \sum_{l=3}^{N} \sum_{|\nu|=l} c_\nu \left[\varphi_{\kappa\nu}(m, K, t) + \psi_{\kappa\nu}^m(m, K, t)^T \dot{m} \right]$$

$$+ \frac{1}{2}\, \text{tr} \left[\psi_{\kappa\nu}^K(m, K, t)\dot{K} \right] \ (\,|\kappa| = 3\,,\,\ldots\,,\,N)\,. \qquad \text{(IV)}$$

Here $\varphi_{\kappa\nu}(m, K, t)$ is defined by (III), and

$$\psi_{\kappa\nu}^m(m, K, t) = \int_{-\infty}^{\infty} q_\kappa^m(y)w_1(y)p_\nu(y)dy \ (\,|\kappa|\,,\,|\nu| = 3\,,\,\ldots\,,\,N)\,, \ \text{(V)}$$

$$\psi_{\kappa\nu}^K(m, K, t) = \int_{-\infty}^{\infty} q_\kappa^K(y)w_1(y)p_\nu(y)dy \ (\,|\kappa|\,,\,|\nu| = 3\,,\,\ldots\,,\,N)\,, \text{(VI)}$$

functions $\varphi_{\kappa\nu}(m, K, t)$, $\psi_{\kappa 0}^m(m, K, t)$, $\psi_{\kappa 0}^K(m, K, t)$ are defined by formulae of Appendix 4 at $p_\nu(y) = p_0(y) = 1$.

7.35. Show that the equations for the multi-dimensional coefficients of the orthogonal expansions in the case of Eq. (5.4.7) are as follows:

$$\partial c_{\kappa_1, \ldots, \kappa_n}(t_1, \ldots, t_n)/\partial t_n = \varphi_{\kappa_1, \ldots, \kappa_n, 0}(\bar{m}_n, \bar{K}_n, t_n)$$

$$+ \psi_{\kappa_1, \ldots, \kappa_n, 0}^m(\bar{m}_n, \bar{K}_n, t_n)^T \dot{m}(t_n) + \text{tr} \left[\psi_{\kappa_1, \ldots, \kappa_n, 0}^{K_h}(\bar{m}_n, \bar{K}_n, t_n)\dot{K}(t_n) \right]$$

$$+ \sum_{h=1}^{n-1} \mathrm{tr}\left[\psi_{\kappa_1,\ldots,\kappa_n,0}^{K_n}(\bar{m}_n, \bar{K}_n, t_n) \frac{\partial K(t_n, t_h)}{\partial t_n} \right]$$

$$+ \sum_{l=3}^{N} \sum_{|\nu_1|+\cdots+|\nu_n|=l} c_{\kappa_1,\ldots,\kappa_n,0}\Bigg\{ \varphi_{\kappa_1,\ldots,\kappa_n,\nu_1,\ldots,\nu_n}(\bar{m}_n, \bar{K}_n, t_n)$$

$$+ \psi_{\kappa_1,\ldots,\kappa_n,\nu_1,\ldots,\nu_n}^{m}(\bar{m}_n, \bar{K}_n, t_n)\dot{m}(t_n)$$

$$+ \mathrm{tr}\left[\psi_{\kappa_1,\ldots,\kappa_n,\nu_1,\ldots,\nu_n}^{K}(\bar{m}_n, \bar{K}_n, t_n)\dot{K}(t_n) \right]$$

$$+ \sum_{h=1}^{n-1} \mathrm{tr}\left[\psi_{\kappa_1,\ldots,\kappa_n,\nu_1,\ldots,\nu_n}^{K_h}(\bar{m}_n, \bar{K}_n, t_n)\frac{\partial K(t_n, t_h)}{\partial t_n} \right]\Bigg\} \qquad \text{(I)}$$

$$(\,|\kappa_1|\,,\ldots,\,|\kappa_n|=1,\ldots,N;\,|\kappa_1|+\cdots+|\kappa_n|=\max n,\ldots,N\,),$$

where

$$\varphi_{\kappa_1,\ldots,\kappa_n,\nu_1,\ldots,\nu_n}(\bar{m}_n, \bar{K}_n, t_n)$$

$$= \int_{-\infty}^{\infty} \cdots \int_{-\infty}^{\infty} \Bigg\{ \frac{\partial^T q_{\kappa_1,\ldots,\kappa_n}(y_1,\ldots,y_n)}{\partial y_n} a(y_n, t_n)$$

$$+ \frac{1}{2}\mathrm{tr}\left[\frac{\partial}{\partial y_n}\frac{\partial^T}{\partial y_n} q_{\kappa_1,\ldots,\kappa_n}(y_1,\ldots,y_n)\sigma(y_n, t_n) \right]$$

$$+ \int_{R_0^q}\Bigg[q_{\kappa_1,\ldots,\kappa_n}(y_1,\ldots,y_{n-1},y_n+c(y_n,t_n,u))$$

$$- q_{\kappa_1,\ldots,\kappa_n}(y_1,\ldots,y_n) - \frac{\partial^T q_{\kappa_1,\ldots,\kappa_n}(y_1,\ldots,y_n)}{\partial y_n}c(y_n,t_n,u) \Bigg]$$

$$\times \nu_P(t_n, du)\Bigg\} w_n(y_1,\ldots,y_n)p_{\nu_1,\ldots,\nu_n}(y_1,\ldots,y_n)dy_1\ldots dy_n, \qquad \text{(II)}$$

$$\psi_{\kappa_1,\ldots,\kappa_n,\nu_1,\ldots,\nu_n}^{m}(\bar{m}_n, \bar{K}_n, t_n) = \int_{-\infty}^{\infty} \cdots \int_{-\infty}^{\infty} q_{\kappa_1,\ldots,\kappa_n}^{m}(y_1,\ldots,y_n)$$

$$\times w_n(y_1,\ldots,y_n)p_{\nu_1,\ldots,\nu_n}(y_1,\ldots,y_n)dy_1\ldots dy_n, \qquad \text{(III)}$$

$$\psi_{\kappa_1,\ldots,\kappa_n,\nu_1,\ldots,\nu_n}^{K}(\bar{m}_n, \bar{K}_n, t_n) = \int_{-\infty}^{\infty} \cdots \int_{-\infty}^{\infty} q_{\kappa_1,\ldots,\kappa_n}^{K}(y_1,\ldots,y_n)$$

$$\times w_n(y_1 , \ldots , y_n)p_{\nu_1 , \ldots , \nu_n}(y_1 , \ldots , y_n)dy_1 \ldots dy_n , \qquad \text{(IV)}$$

$$\psi^{K_h}_{\kappa_1 , \ldots , \kappa_n, \nu_1 , \ldots , \nu_n}(\bar{m}_n , \bar{K}_n , t_n) = \int\limits_{-\infty}^{\infty} \cdots \int\limits_{-\infty}^{\infty} q^{K_h}_{\kappa_1 , \ldots , \kappa_n}(y_1 , \ldots , y_n)$$

$$\times w_n(y_1 , \ldots , y_n)p_{\nu_1 , \ldots , \nu_n}(y_1 , \ldots , y_n)dy_1 \ldots dy_n , \quad (h = 1 , \ldots , n{-}1).$$
$$\text{(V)}$$

Functions $\varphi_{\kappa_1 , \ldots , \kappa_n, 0}$, $\psi_{\kappa_1 , \ldots , \kappa_n, 0}$, $\psi^{K}_{\kappa_1 , \ldots , \kappa_n, 0}$, $\psi^{K_h}_{\kappa_1 , \ldots , \kappa_n, 0}$ are defined by formulae (II)–(V) and $p_{\nu_1 , \ldots , \nu_n}(y_1 , \ldots , y_n) = p_{0, \ldots, 0}(y_1 , \ldots , y_n) = 1$ at $\nu_1 , \ldots , \nu_n = 0$.

7.36. Show that the parameters of the one-dimensional distribution of the stationary random process in the system of Example 5.6.8 are determined as functions of the first four moments by the equations

$$nc_2\alpha_{n-1} - [(n+1)c_3 + c_0]\alpha_n + [(n+2)c_4 + c_1]\alpha_{n+1} = 0, \quad n = 0,1,2,3,$$

where α_n is the initial moment of the nth order, $\alpha_0 = 1$.

7.37. Using the parametrization of the distribution by means of the approximate representation of the density by the fragment of the Fourier series,

$$f_1(y;t) \approx f_1^*(t;a) = \sum_{\nu_1, \ldots, \nu_r = -s}^{s} a_{\nu_1, \ldots, \nu_r} e^{-i\pi(\nu_1 y_1 + \cdots + \nu_r y_r)/\Delta}$$

in the r-dimensional cube $(-\Delta, \Delta)^r$ and equalizing it to zero outside this cube, derive the approximate ordinary differential equations for the Fourier coefficients $a_{\nu_1, \ldots, \nu_r}$ for the one-dimensional density of the state vector of the nonlinear system.

7.38. Show that EAM equations for the stochastic nonlinear system:

$$\dot{Y}_1 = Y_2, \ \dot{Y}_2 = -\varepsilon Y_2 - \omega_0^2 Y_1 - \alpha Y_1^3 + F\cos\omega t + V(t) \qquad \text{(I)}$$

while accounting the moments till the fourth order have the following form:

$$\dot{m}_1 = m_2, \ \dot{m}_2 = -\varepsilon m_2 - \omega_0^2 m_1 - \alpha(m_1^3 + 3m_1 k_{11}) + F\cos\omega t, \qquad \text{(II)}$$

$$\dot{k}_{11} = 2k_{12}, \ \dot{k}_{12} = k_{22} - \varepsilon k_{12} - \omega_0^2 k_{11} - 3\alpha k_{11}(m_1^2 + k_{11}) - 24\alpha c_{2,2}k_{11}^2,$$

$$\dot{k}_{22} = -\varepsilon k_{22} - 2\omega_0^2 k_{12} - 6\alpha k_{12}(m_1^2 + k_{11}) - 48\alpha c_{2,2}k_{11}k_{12} + \nu, \qquad \text{(III)}$$

$$\dot{c}_{2,2} = \frac{c_{2,2}}{|K|}[6\alpha k_{11}^2(k_{12} + m_1 m_2) - k_{11}\nu] + \frac{3\alpha k_{11}^2}{8|K|}(k_{12} + m_1 m_2), \qquad \text{(IV)}$$

where ε, α, ω_0, F and ω_0 are some positive constants; ν is the intensity of the normal white noise $V(t)$; $|K| = k_{11}k_{22} - k_{12}^2$.

APPENDICES

1. Stochastic Itô Differentials of Typical Composite Functions

Functions of Standard Wiener Process
Table A.1.1

Function	Stochastic Differential
1. $U = W^a$	$dU = aW^{a-1}dW + 1/2a(a-1)W^{a-2}dt$
2. $U = e^{aW}$	$dU = e^{aW}dW + 1/2a^2 e^{aW}dt$
3. $U = a^W$	$dU = a^W \ln a\, dW + 1/2a^W \ln^2 a\, dt$
4. $U = \log_b aW$	$dU = \dfrac{a}{W \ln b}dW - \dfrac{a^2}{2W^2 \ln b}dt$
5. $U = \sin aW$	$dU = a\cos aW\, dW - 1/2a^2 \sin aW\, dt$
6. $U = \cos aW$	$dU = -a\sin aW\, dW - 1/2a^2 \cos aW\, dt$
7. $U = \operatorname{tg} aW$	$dU = \dfrac{a}{\cos^2 aW}dW + \dfrac{a^2 \sin aW}{\cos^3 aW}dt$
8. $U = \operatorname{ctg} aW$	$dU = -\dfrac{a}{\sin^2 aW}dW + \dfrac{a^2 \cos aW}{\sin^3 aW}dt$
9. $U = \sec aW$	$dU = \dfrac{a\sin aW}{\cos^2 aW}dW + \dfrac{a^2(1+\sin^2 aW)}{2\cos^3 aW}dt$
10. $U = \operatorname{cosec} aW$	$dU = -\dfrac{a\cos aW}{\sin^2 aW}dW + \dfrac{a^2(1+\cos^2 aW)}{2\sin^3 aW}dt$
11. $U = \arcsin aW$	$dU = \dfrac{a}{\sqrt{1-a^2W^2}}dW + \dfrac{a^3 W}{2\sqrt{(1-a^2W^2)^3}}dt$
12. $U = \arccos aW$	$dU = \dfrac{-a}{\sqrt{1-a^2W^2}}dW - \dfrac{a^3 W}{2\sqrt{(1-a^2W^2)^3}}dt$
13. $U = \operatorname{arctg} aW$	$dU = \dfrac{a}{1+a^2W^2}dW - \dfrac{a^3 W}{(1+a^2W^2)^2}dt$
14. $U = \operatorname{arcctg} aW$	$dU = \dfrac{-a}{1+a^2W^2}dW + \dfrac{a^3 W}{(1+a^2W^2)^2}dt$
15. $U = \operatorname{sh} aW$	$dU = a\operatorname{ch} aW\, dW + 1/2a^2 \operatorname{sh} aW\, dt$
16. $U = \operatorname{ch} aW$	$dU = a\operatorname{sh} aW\, dW + 1/2a^2 \operatorname{ch} aW\, dt$

Table A.1.1 (continued)

Function	Stochastic Differential
17. $U = \operatorname{th} aW$	$dU = \dfrac{a}{\operatorname{ch}^2 aW}dW - \dfrac{a^2 \operatorname{sh} aW}{2\operatorname{ch}^3 aW}dt$
18. $U = \operatorname{cth} aW$	$dU = \dfrac{-a}{\operatorname{sh}^2 aW}dW + \dfrac{a^2 \operatorname{ch} aW}{2\operatorname{sh}^3 aW}dt$
19. $U = \operatorname{arsh} aW$	$dU = \dfrac{a}{\sqrt{1 + a^2 W^2}}dW - \dfrac{a^3 W}{2\sqrt{(1 + a^2 W^2)^3}}dt$
20. $U = \operatorname{arch} aW$	$dU = \dfrac{a}{\sqrt{a^2 W^2 - 1}}dW - \dfrac{a^3 W}{2\sqrt{(a^2 W^2 - 1)^3}}dt$
21. $U = \operatorname{arth} aW$	$dU = \dfrac{a}{1 - a^2 W^2}dW - \dfrac{a^3 W}{(1 - a^2 W^2)^2}dt$
22. $U = \operatorname{arcth} aW$	$dU = \dfrac{a}{1 - a^2 W^2}dW - \dfrac{a^3 W}{(1 - a^2 W^2)^2}dt$

Functions of Multi-Dimensional Wiener Process

Table A.1.2

Function	Stochastic Differential
1. $U = W_1 W_2$	$dU = W_2 dW_1 + W_1 dW_2 + \nu_{12}dt$
2. $U = W_1 W_2 W_3$	$dU = W_2 W_3 dW_1 + W_1 W_3 dW_2 + W_1 W_2 dW_3 +$ $+(\nu_{12}W_3 + \nu_{13}W_2 + \nu_{23}W_1)dt$
3. $U = W_1 W_2^{-1}$	$dU = W_2^{-1}dW_1 + W_1 W_2^{-2}dW_2 +$ $+(\nu_{22}W_1 W_2^{-3} - \nu_{12}W_2^{-2})dt$
4. $U = (W_1^2 + W_2^2)^{1/2}$	$dU = (W_1^2 + W_2^2)^{-1/2}(W_1 dW_1 + W_2 dW_2) +$ $+1/2(W_1^2 + W_2^2)^{-3/2}(\nu_{11}W_2^2 +$ $+\nu_{22}W_1^2 - 2\nu_{12}W_1 W_2)dt$
5. $U = (W_1^2 + W_2^2 +$ $+W_3^2)^{1/2}$	$dU = (W_1^2 + W_2^2 + W_3^2)^{-1/2}(W_1 dW_1 +$ $+W_2 dW_2 + W_3 dW_3) +$ $+1/2(W_1^2 + W_2^2 + W_3^2)^{-3/2}[\nu_{11}(W_2^2 + W_3^2) +$ $+\nu_{22}(W_1^2 + W_3^2) + \nu_{33}(W_1^2 + W_2^2) -$ $-2\nu_{12}W_1 W_2 - 2\nu_{13}W_1 W_3 - 2\nu_{23}W_2 W_3]dt$
6. $U = W^T aW$	$dU = W^T(a + a^T)dW + [W^T \dot{a}W +$ $+1/2\operatorname{tr}(a + a^T)\nu]dt$
7. $U = e^{a^T W}$	$dU = e^{a^T W}a^T dW + [e^{a^T W}\dot{a}^T W +$ $+1/2e^{a^T W}\operatorname{tr}(aa^T \nu)]dt$

2. Hermite Polynomials. Polynomials Orthogonal to Gamma and χ^2-Distibutions

2.1. Hermite Polynomials

Let K be a symmetric positive definite real $n \times n$ matrix. Hermite polynomials of the n-dimensional vector x are determined as the coefficients of the expansion of the generating functions

$$\varphi(u) = \exp\left\{ x^T K^{-1} u - \frac{1}{2} u^T K^{-1} u \right\}, \quad \psi(u) = \exp\left\{ x^T u - \frac{1}{2} u^T K u \right\}$$

into a power series in the variables $u_1, \ldots, u_n$ which form the vector $u = [u_1 \ldots u_n]^T$:

$$\exp\left\{ x^T K^{-1} u \frac{1}{2} U^T K^{-1} u \right\} = \sum \frac{u_1^{m_1} \ldots u_n^{m_n}}{m_1! \ldots m_n!} H_m(x) \tag{1}$$
$$m = [m_1 \ldots m_n]^T,$$

$$\exp\left\{ x^T u - \frac{1}{2} u^T K u \right\} = \sum \frac{u_1^{m_1} \ldots u_n^{m_n}}{m_1! \ldots m_n!} G_m(x), \quad m = [m_1 \ldots m_n]^T, \tag{2}$$

where the summation is performed with respect to all nonnegative integers m_1, $\ldots, m_n$.

It is easy to prove that $G_m(x)$ represents a polynomial of the degree m_1 in the variable x_1, of the degree m_2 in $x_2, \ldots$, of the degreee m_n while $H_m(x)$ represents a polynomial of the degree m_1 in $c_{11} x_1 + \cdots + c_{1n} x_n$, of the degree m_2 in $c_{21} x_1 + \cdots + c_{2n} x_n, \ldots$, of the degree m_n in $c_{n1} x_1 + \cdots + c_{nn} x_n$, $c_{pq} = c_{qp}$ being the elements of the matrix $C = K^{-1}$. The common degree of the polynomials $H_m(x)$ and $G_m(x)$ is equal to the sum of the components of the multi-index $m = [m_1 \ldots m_n]^T$, $|m| = m_1 + \cdots + m_n$.

Taking into consideration that

$$x^T K^{-1} u - \frac{1}{2} u^T K^{-1} u = \frac{1}{2} x^T K^{-1} x - \frac{1}{2}(x^T - u^T) K^{-1}(x - u),$$

$$x^T u - \frac{1}{2} u^T K u = \frac{1}{2} x^T K^{-1} x - \frac{1}{2}(x^T K^{-1} - u^T) K(K^{-1} x - u), \tag{3}$$

we conclude that $e^{-x^T K^{-1} x/2} H_m(x)$ represent the coefficients of the expansion of the function $e^{-v^T K^{-1} v/2}$, $v = -x + u$ into the Taylor series in the vicinity of the point $v = -x$, and $e^{-x^T K^{-1} x/2} G_m(x)$ are the coefficients of the expansions of the function $e^{-v^T K v/2}$, $v = -K^{-1}x + u$ into the Taylor series in the vicinity of the point $v = -K^{-1}x$. Consequently,

$$e^{-x^T K^{-1}x/2} H_m(x) = \left[\frac{\partial^{|m|}}{\partial v_1^{m_1} \ldots \partial v_n^{m_n}} e^{-v^T K^{-1}v/2} \right]_{v=-x},$$

$$e^{-x^T K^{-1}x/2} G_m(x) = \left[\frac{\partial^{|m|}}{\partial v_1^{m_1} \ldots \partial v_n^{m_n}} e^{-v^T K^{-1}v/2} \right]_{v=-K^{-1}x}.$$

Hence it follows that:

$$H_m(x) = (-1)^{|m|} e^{x^T K^{-1}x/2} \frac{\partial^{|m|}}{\partial x_1^{m_1} \ldots \partial x_n^{m_n}} e^{-x^T P^{-1}x/2}, \qquad (4)$$

$$G_m(x) = (-1)^{|m|} e^{x^T K^{-1}x/2} \left[\frac{\partial^{|m|}}{\partial y_1^{m_1} \ldots \partial y_n^{m_n}} e^{-y^T K y/2} \right]_{y=-K^{-1}x}. \qquad (5)$$

In the special case of $n = 1$, $K = 1/2$ the definition of the Hermite polynomials of a scalar variable used in the mathematical analysis is obtained from (4) to (5):

$$H_m(x) = (-1)^m e^{x^2} \frac{d^m}{dx^m} e^{-x^2}, \quad G_m(x) = 2^{-m} H_m(x).$$

In the other special case of $n = 1$, $K = 1$ the definition of the Hermite polynomials of a scalar variable used in the probability theory is obtained from (4) and (5):

$$H_m(x) = G_m(x) = (-1)^m e^{x^2/2} \frac{d^m}{dx^m} e^{-x^2/2}.$$

These polynomials are sometimes denoted by $He_m(x)$, in contradistinction to other Hermite polynomials (Abramowitz and Stegun 1964).

Let us prove that *the polynomials $H_m(x)$ and $G_m(x)$ possess the property of biorthogonality with the weight $w(x) = [(2\pi)^n |K|]^{-1/2} e^{-x^T K^{-1}x/2}$*. For this purpose we derive the formula

$$\frac{1}{\sqrt{(2\pi)^n |K|}} \int_{-\infty}^{\infty} e^{-x^T K^{-1}x/2} H_m(x) G_l(x)\, dx = m_1! \ldots m_n! \delta_{ml}, \qquad (6)$$

where $\delta_{ml} = 1$ at $l = m$ and $\delta_{ml} = 0$ if at least one of the components of the vector subscript l does not coincide with the corresponding component of the subscript m. The weight $w(x)$ represents the n-dimensional normal density.

▷ To prove formula(6) we multiply term-wise the equalities

$$\exp\left\{-\frac{1}{2}(x^T - u^T)K^{-1}(x-u)\right\} = \sum \frac{u_1^{m_1}\cdots u_n^{m_n}}{m_1!\ldots m_n!}e^{-x^T K^{-1}x}H_m(x),$$

$$\exp\left\{x^T u - \frac{1}{2}u^T K u\right\} = \sum \frac{u_1^{l_1}\cdots u_n^{l_n}}{l_1!\ldots l_n!}G_l(x), \tag{7}$$

which follow from (1)–(3). As a result we get

$$\exp\left\{-\frac{1}{2}u^T K u + x^T u - \frac{1}{2}(x^T - u^T)K^{-1}(x-u)\right\}$$

$$= \sum \frac{u_1^{m_1+l_1}\cdots u_n^{m_n+l_n}}{m_1!l_1!\ldots m_n!l_n!}e^{-x^T K^{-1}x/2}H_m(x)G_l(x).$$

Integrating this equality with respect to x we obtain

$$e^{-u^T K u/2 + |u|^2}\int_{-\infty}^{\infty}e^{u^T(x-u)-(x^T-u^T)K^{-1}(x-u)/2}dx$$

$$= \sum \frac{u_1^{m_1+l_1}\cdots u_n^{m_n+l_n}}{m_1!l_1!\ldots m_n!l_n!}\int_{-\infty}^{\infty}e^{-x^T K^{-1}x/2}H_m(x)G_l(x)dx,$$

where $|u|^2 = u^T u = u_1^2 + \cdots + u_n^2$. Hence, taking into consideration that

$$\int_{-\infty}^{\infty}e^{u^T v - v^T K^{-1}v/2}dv = \sqrt{(2\pi)^n|K|}e^{-u^T K u/2},$$

we deduce the equality

$$\sqrt{(2\pi)^n|K|}e^{u_1^2+\cdots+u_n^2} = \sum \frac{u_1^{m_1+l_1}\cdots u_n^{m_n+l_n}}{m_1!l_1!\ldots m_n!l_n!}$$

$$\times \int_{-\infty}^{\infty}e^{-x^T K^{-1}x/2}H_m(x)G_l(x)dx.$$

Expanding the exponential functions $e^{u_1^2},\ldots,e^{u_n^2}$ into powers series in $u_1^2,\ldots,u_n^2$ we come to the identity

$$\sqrt{(2\pi)^n|K|}\sum \frac{u_1^{2m_1}\cdots u_n^{2m_n}}{m_1!\ldots m_n!}$$

$$= \sum \frac{u_1^{m_1+l_1}\ldots u_n^{m_n+l_n}}{m_1!l_1!\ldots m_n!l_n!} \int\limits_{-\infty}^{\infty} e^{-x^T K^{-1}x/2} H_m(x)G_l(x)dx.$$

Comparing the coefficients of the same degrees of the variables $u_1,\ldots,u_n$ in the left- and right-hand sides of the obtained equality we get

$$\int\limits_{-\infty}^{\infty} e^{-x^T K^{-1}x/2} H_m(x)G_l(x)dx = 0, \ l \neq m,$$

$$\frac{1}{m_1!\ldots m_n!} \int\limits_{-\infty}^{\infty} e^{-x^T K^{-1}x/2} H_m(x)G_m(x)dx = \sqrt{(2\pi)^n|K|}.\lhd$$

Now let us derive the formulae for the derivatives of the polynomials $G_m(x)$ with respect to the components of the vector x and with respect to the elements of the matrix K.

$\rhd$ Differentiating formula (2) with respect to x_p we get

$$\sum \frac{u_1^{m_1}\ldots u_n^{m_n}}{m_1!\ldots m_n!} \frac{\partial}{\partial x_p} G_{m_1,\ldots,m_n}(x) = u_p \exp\left\{x^T u - \frac{1}{2}u^T K u\right\}$$

$$= \sum \frac{u_1^{m_1}\ldots u_p^{m_p+1}\ldots u_n^{m_n}}{m_1!\ldots m_n!} G_{m_1,\ldots,m_n}(x)$$

$$= \sum \frac{u_1^{m_1}\ldots u_n^{m_n}}{m_1!\ldots m_n!} m_p G_{m_1,\ldots m_{p-1},\ldots,m_n}(x). \tag{8}$$

Hence, it follows that

$$\frac{\partial}{\partial x_p} G_{m_1,\ldots,m_n}(x) = m_p G_{m_1,\ldots,m_{p-1},\ldots,m_n}(x). \tag{9}$$

Thus *the differentiation of the polynomial $G_{m_1,\ldots,m_n}(x)$ with respect to x_p is reduced to the decrease by unity of the subscript m_p and to the multiplication of the result by m_p.* Using this rule for the repeated differentiation of formula (9) we get

$$\frac{\partial^2}{\partial x_p^2} G_{m_1,\ldots,m_n}(x) = m_p(m_p-1)G_{m_1,\ldots,m_{p-2},\ldots,m_n}(x), \tag{10}$$

$$\frac{\partial^2}{\partial x_p \partial x_q} G_{m_1,\ldots,m_n}(x) = m_p m_q G_{m_1,\ldots,m_{p-1},\ldots,m_{q-1},\ldots,m_n}(x), \ \text{at } p < q. \tag{11}$$

Analogously differentiating formula (2) with respect to k_{pp} and to k_{pq} we obtain

$$\sum \frac{u_1^{m_1}\dots u_n^{m_n}}{m_1!\dots m_n!}\frac{\partial}{\partial k_{pp}}G_{m_1,\dots,m_n}(x) = -\frac{1}{2}u_p^2\exp\left\{x^T u - \frac{1}{2}u^T K u\right\}$$

$$= -\frac{1}{2}\sum \frac{u_1^{m_1}\dots u_n^{m_n}}{m_1!\dots m_n!}m_p(m_p-1)G_{m_1,\dots,m_{p-2},\dots,m_n}(x),$$

$$\sum \frac{u_1^{m_1}\dots u_n^{m_n}}{m_1!\dots m_n!}\frac{\partial}{\partial k_{pq}}G_{m_1,\dots,m_n}(x) = -u_p u_q^2\exp\left\{x^T u - \frac{1}{2}u^T K u\right\}$$

$$= -\sum \frac{u_1^{m_1}\dots u_n^{m_n}}{m_1!\dots m_n!}m_p m_q G_{m_1,\dots,m_{p-1},\dots,m_{q-1},\dots,m_n}(x).$$

Hence follow the formulae

$$\frac{\partial}{\partial k_{pp}}G_{m_1,\dots,m_n}(x) = -\frac{1}{2}m_p(m_p-1)G_{m_1,\dots,m_{p-2},\dots,m_n}(x)$$

$$= -\frac{1}{2}\frac{\partial^2}{\partial x_p^2}G_{m_1,\dots,m_n}(x), \tag{12}$$

$$\frac{\partial}{\partial k_{pq}}G_{m_1,\dots,m_n}(x) = -m_p m_q G_{m_1,\dots,m_{p-1},\dots,m_{q-1},\dots,m_n}(x)$$

$$= -\frac{\partial^2}{\partial x_p\,\partial x_q}G_{m_1,\dots,m_n}(x).\triangleleft \tag{13}$$

In exactly the same way the formulae for the derivatives of the polynomials $H_m(x)$ are derived.

Let us consider now the sequence of the invertible symmetrical positive definite matrices $\{K_n\}$

$$K_n = \begin{bmatrix} K_{11} & K_{12} & \dots & K_{1n} \\ K_{12} & K_{22} & \dots & K_{2n} \\ \vdots & \vdots & \dots & \vdots \\ K_{1n} & K_{2n} & \dots & K_{nn} \end{bmatrix}.$$

For each n we determine the corresponding system of Hermite polynomials $H_m(x^{(n)}) = H_{m_1,\dots,m_n}(x_1,\dots,x_n)$, $G_m(x^{(n)}) = G_{m_1,\dots,m_n}(x_1,\dots,x_n)$ where $x^{(n)} = [x_1\dots x_n]^T$, $m = [m_1\dots m_n]^T$. We state some theorems determining the relations between Hermite polynomials corresponding to different values of n.

Theorem 1. At any $m_1, \ldots, m_{n-1}, m_n$

$$\frac{1}{\sqrt{(2\pi)^n |K_n|}} \int_{-\infty}^{\infty} e^{-x^T K_n^{-1} x/2} H_{m_1,\ldots,m_n}(x_1, \ldots, x_{n-1}, x_n)\, dx_n$$

$$= \frac{1}{\sqrt{(2\pi)^{n-1}|K_{n-1}|}} e^{-x'^T K_{n-1}^{-1} x'/2} H_{m_1,\ldots,m_{n-1}}(x_1, \ldots, x_{n-1}) \delta_{0,m_n}, \tag{14}$$

where $\delta_{00} = 1$, $\delta_{0k} = 0$ at $k \neq 0$, $x' = [\,x_1 \ldots x_{n-1}\,]^T$.

Theorem 2. At any $m_1, \ldots, m_{n-1}$

$$G_{m_1,\ldots,m_{n-1},0}(x_1, \ldots, x_n) = G_{m_1,\ldots,m_{n-1}}(x_1, \ldots, x_{n-1}). \tag{15}$$

The following two theorems give the relations between the Hermite polynomials corresponding to different n in the limiting case where $k_{pn} = k_{p,n-1}$ ($p = 1,$ $\ldots, n-1$), $k_{nn} = k_{n-1,n-1}$, i.e. in the case of the singular normal distribution $N(0, K_n)$. In this case the components X_{n-1} and X_n of the random vector $X = [\,X_1 \ldots X_{n-1} X_n\,]$ coincide with the probability 1, $X_n = X_{n-1}$. Hence, the density

$$w_n(x) = [\,(2\pi)^n |K_n|\,]^{-1/2} \exp\{-x^T K_n^{-1} x/2\}$$

is replaced by the corresponding singular normal density

$$w'_n(x) = [\,(2\pi)^{n-1}|K_{n-1}|\,]^{-1/2} \exp\{-x'^T K_{n-1}^{-1} x'/2\} \delta(x_n - x_{n-1}).$$

Theorem 3. At $k_{pn} = k_{p,n-1}$ ($p = 1, \ldots, n-1$), $k_{nn} = k_{n-1,n-1}$

$$\sum_{k=0}^{m_{n-1}} C_{m_{n-1}}^k H_{m_1,\ldots,m_{n-1}-k,k}(x_1, \ldots, x_{n-1}, x_{n-1})$$

$$= H_{m_1,\ldots,m_{n-1}}(x_1, \ldots, x_{n-1}). \tag{16}$$

Theorem 4. At $k_{pn} = k_{p,n-1}$ ($p = 1, \ldots, n-1$), $k_{nn} = k_{n-1,n-1}$

$$G_{m_1,\ldots,m_{n-1},m_n}(x_1, \ldots, x_{n-1}, x_{n-1}) = G_{m_1,\ldots,m_{n-1}+m_n}(x_1, \ldots, x_{n-1}). \tag{17}$$

To study other properties of the Hermite polynomials the reader may use the book (Bateman and Erdélyi 1953).

2.2. Polynomials Orthogonal to Gamma and χ^2-Distributions

It is possible to use the orthogonal system of the polynomials $S_\nu^\alpha(u)$:

$$S_\nu^\alpha(u) = (-k)^{-\nu} u^{-\alpha} e^{ku} \frac{d^\nu}{du^\nu}(u^{\alpha+\nu} e^{-ku}), \quad \alpha > -1, u \geq 0. \qquad (18)$$

The polynomials $S_\nu^\alpha(u)$ are connected with the generalized Laguerre polynomials $L_\nu^\alpha(x)$ (Appendix 7) by the relation

$$S_\nu^\alpha(u) = (-k)^{-\nu} \nu! L_\nu^\alpha(ku).$$

After differentialting in (18) we get

$$S_\nu^\alpha(u) = (-k)^{-\nu} \sum_{\mu=0}^{\nu} C_\nu^\mu \frac{\Gamma(\alpha+\nu+1)}{\Gamma(\alpha+\mu+1)}(-ku)^\mu. \qquad (19)$$

As a generating function of these polynomials serves

$$\varphi(s) = \frac{e^{us/(1+s/k)}}{(1+s/k)^{\alpha+1}} = \sum_{\nu=0}^{\infty} \frac{s^\nu}{\nu!} S_\nu^\alpha(u).$$

The system of the functions $u^{\alpha/2} e^{-ku/2} S_\nu^\alpha(u)$ $(\nu = 0, 1, 2, \ldots)$ is complete in the space $L_2([0, \infty))$. Consequently, any function $f(u)$, in particular, any density of a random variable satisfying the condition

$$\int_0^\infty \frac{f^2(u)}{u^{\alpha/2} e^{-ku/2}} \, du < \infty, \qquad (20)$$

may be presented by the expansion

$$f(u) = u^\alpha e^{-ku} \sum_{\nu=0}^{\infty} c_\nu S_\nu^\alpha(u), \qquad (21)$$

where

$$c_\nu = \frac{k^{2\nu} \Gamma(\alpha+1)}{\nu! \Gamma(\alpha+\nu+1)} \int_0^\infty f(u) S_\nu^\alpha(u) du.$$

Any density $f(u)$ not satisfying condition (20) may be also presented by a finite segment of expansion (21) with any degree of the accuracy. In the special case at

$\alpha = p/2 - 1$, $k = 1/2$ we obtain a system of the polynomials which is orthogonal to the χ^2-distribution with p degrees of freedom:

$$S_{p\nu}(u) = S_\nu^{p/2-1}(u) = \sum_{\mu=0}^{\nu}(-1)^{\nu+\mu}C_\nu^\mu \frac{(p+2\nu-2)!!}{(p+2\mu-2)!!}\, u^\mu . \qquad (22)$$

Let us consider the main properties of the polynomials $S_\alpha^\nu(u)$ orthogonal to the gamma-distribution.

Property 1:

$$S_\nu^{\alpha-1}(u) = S_\nu^\alpha(u) + \frac{\nu}{k}S_{\nu-1}^\alpha(u). \qquad (23)$$

$\triangleright$ Let use the formula (19) which determines the polynomials $S_\nu^\alpha(u)$. Then for $S_{\nu-1}^\alpha(u)$ and $S_\nu^{\alpha-1}(u)$ we obtain the following presentations:

$$S_{\nu-1}^\alpha(u) = (-k)^{-\nu+1}\sum_{\mu=0}^{\nu-1}C_{\nu-1}^\mu \frac{\Gamma(\alpha+\nu)}{\Gamma(\alpha+\mu+1)}(-ku)^\mu,$$

$$S_\nu^{\alpha-1}(u) = (-k)^{-\nu}\sum_{\mu=0}^{\nu}C_\nu^\mu \frac{\Gamma(\alpha+\nu)}{\Gamma(\alpha+\mu)}(-ku)^\mu.$$

We calculate the sum in the right-hand side of equality (23) and show that it is equal to the left-hand side, i.e. to the polynomial $S_\nu^{\alpha-1}(u)$. We have

$$S_\nu^\alpha(u) + \frac{\nu}{k}S_{\nu-1}^\alpha(u) = (-k)^{-\nu}\sum_{\mu=0}^{\nu-1}\frac{\Gamma(\alpha+\nu+1)}{\Gamma(\alpha+\mu+1)}C_\nu^\mu(-ku)^\mu$$

$$+(-k)^{-\nu}C_\nu^\nu \frac{\Gamma(\alpha+\nu+1)}{\Gamma(\alpha+\nu+1)}(-ku)^\nu$$

$$+\frac{\nu}{k}(-k)^{-\nu+1}\sum_{\mu=0}^{\nu-1}C_{\nu-1}^\mu \frac{\Gamma(\alpha+\nu)}{\Gamma(\alpha+\mu+1)}(-ku)^\mu$$

$$= (-k)^{-\nu}\sum_{\mu=0}^{\nu-1}\frac{\Gamma(\alpha+\nu)}{\Gamma(\alpha+\mu+1)}\left[C_\nu^\mu(\alpha+\nu)-\nu C_{\nu-1}^\mu\right](-ku)^\mu + u^\nu .$$

Using the property of the gamma-function $\Gamma(n+1) = n\Gamma(n)$ we find

$$S_\nu^\alpha(u) + \frac{\nu}{k}S_{\nu-1}^\alpha(u) = (-k)^{-\nu}\sum_{\mu=0}^{\nu-1}\frac{\Gamma(\alpha+\nu)}{\Gamma(\alpha+\mu+1)}$$

$$\times \left[(\alpha + \nu)\frac{\nu!}{\mu!(\nu - \mu)!} - \nu\frac{(\nu - 1)!}{\mu!(\nu - 1 - \mu)!} \right](-ku)^\mu + u^\nu$$

$$= (-k)^{-\nu} \sum_{\mu=0}^{\nu-1} \frac{\Gamma(\alpha + \nu)}{\Gamma(\alpha + \mu + 1)} \frac{\nu!}{\mu!(\nu - \mu - 1)!} \left[\frac{\alpha + \nu}{\nu - \mu} - 1 \right](-ku)^\mu + u^\nu$$

$$= (-k)^{-\nu} \sum_{\mu=0}^{\nu-1} \frac{\Gamma(\alpha + \nu)}{\Gamma(\alpha + \mu + 1)} \frac{\nu!}{\mu!(\nu - \mu - 1)!} \left[\frac{\alpha + \mu}{\nu - \mu} \right](-ku)^\mu + u^\nu$$

$$= (-k)^{-\nu} \sum_{\mu=0}^{\nu-1} \frac{\Gamma(\alpha + \nu)}{(\alpha + \mu)\Gamma(\alpha + \mu)} \frac{\nu!}{\mu!(\nu - \mu)!}(\alpha + \mu)(-ku)^\mu + u^\nu$$

$$= (-k)^{-\nu} \sum_{\mu=0}^{\nu-1} C_\nu^\mu \frac{\Gamma(\alpha + \nu)}{\Gamma(\alpha + \mu)}(-ku)^\mu + u^\nu$$

$$= (-k)^{-\nu} \sum_{\mu=0}^{\nu} C_\nu^\mu \frac{\Gamma(\alpha + \nu)}{\Gamma(\alpha + \mu)}(-ku)^\mu = S_\nu^{\alpha-1}. \triangleleft$$

Property 2.

$$\frac{d}{du}\left[e^{-ku}u^\alpha S_\nu^\alpha(u) \right] = (-k)e^{-ku}u^{\alpha-1}S_{\nu+1}^{\alpha-1}(u). \qquad (24)$$

$\triangleright$ After differentiating in the left-hand side and substituting expression (19) for the polynomial $S_\nu^\alpha(u)$ we use the property of the gamma-function $\Gamma(n + 1) = n\Gamma(n)$. As a result we obtain

$$\frac{d}{du}\left[e^{-ku}u^\alpha S_\nu^\alpha(u) \right] = e^{-ku}\left[\sum_{\mu=1}^{\nu+1} C_\nu^{\mu-1}\frac{\Gamma(\alpha + \nu + 1)}{\Gamma(\alpha + \mu)}(-k)^{-\nu+\mu}u^{\alpha+\mu-1} \right.$$

$$\left. + \sum_{\mu=0}^{\nu} C_\nu^\mu \frac{\Gamma(\alpha + \nu + 1)}{\Gamma(\alpha + \mu)}(-k)^{-\nu+\mu}u^{\alpha+\mu-1} \right].$$

Let us unite both sums using the equality $C_{\nu+1}^\mu = C_\nu^\mu + C_\nu^{\mu-1}$. After the transformation we get

$$\frac{d}{du}\left[e^{-ku}u^\alpha S_\nu^\alpha(u) \right] = e^{-ku}\left[\sum_{\mu=0}^{\nu+1} C_{\nu+1}^\mu \frac{\Gamma(\alpha + \nu + 1)}{\Gamma(\alpha + \mu)}(-k)^{-\nu+\mu}u^{\alpha+\mu-1} \right].$$

It is not difficult to notice that after taking out the sign of the sum the variable $(-k)^{-\nu}u^{\alpha-1}$ the residual expression represents the polynomial $S_{\nu+1}^{\alpha-1}$ multiplied by the number $(-k)$. As a result we obtain

$$\frac{d}{du}\left[e^{-ku}u^{\alpha}S_{\nu}^{\alpha}(u)\right] = (-k)e^{-ku}u^{\alpha-1}S_{\nu+1}^{\alpha-1}(u).$$

From Property 1 we may derive the following relation which establishes the integral dependence between the polynomials $S_{\nu}^{\alpha}(u)$, $\nu \geq 1$. ◁

Property 3:

$$\int_0^{u_0} u^{\alpha}e^{-ku}S_{\nu}^{\alpha}(u)du = -\frac{1}{k}u_0^{\alpha+1}e^{-ku_0}S_{\nu-1}^{\alpha+1}(u_0). \tag{25}$$

Relation (25) gives as new property of incomplete gamma-function:

$$\gamma(x,y) = \int_0^y e^{-t}t^{x-1}dt. \tag{26}$$

We shall formulate it as one of the properties of the polynomials $S_{\nu}^{\alpha}(u)$.

Property 4: For the incomplete gamma-functions and any values of the parameters α and k the following equality is true

$$(-k)^{-\nu-\alpha}\sum_{\mu=0}^{\nu}(-1)^{\mu-\nu}C_{\nu}^{\mu}\frac{\Gamma(\alpha+\nu+1)}{\Gamma(\alpha+\mu+1)}\gamma(\alpha+\mu+1,ku_0)$$

$$= -u_0^{\alpha+1}e^{-ku_0}S_{\nu-1}^{\alpha+1}(u_0). \tag{27}$$

▷ After substituting (19) into (25) we shall have

$$\int_0^{u_0} u^{\alpha}e^{-ku}S_{\nu}^{\alpha}(u)du = -(k)^{-\nu}\int_0^{u_0} u^{\alpha}e^{-ku}\sum_{\mu=0}^{\nu}C_{\nu}^{\mu}\frac{\Gamma(\alpha+\nu+1)}{\Gamma(\alpha+\mu+1)}(-ku)^{\mu}du$$

$$= k^{-\nu}\int_0^{u_0}\frac{(ku)^{\alpha}}{k^{\alpha}}e^{-ku}\sum_{\mu=0}^{\nu}(-1)^{\mu-\nu}C_{\nu}^{\mu}\frac{\Gamma(\alpha+\nu+1)}{\Gamma(\alpha+\mu+1)}(ku)^{\mu}du. \tag{28}$$

Let us make the change of variable $t = ku$ in (28) then

$$\int_0^{u_0} u^\alpha e^{-ku} S_\nu^\alpha(u)\,du$$

$$= k^{-\alpha-1-\nu} \int_0^{ku_0} e^{-t} \sum_{\mu=0}^{\nu} C_\nu^\mu \frac{\Gamma(\alpha+\nu+1)}{\Gamma(\alpha+\mu+1)} (-1)^{\mu-\nu} t^{\alpha+\mu}\,dt. \qquad (29)$$

Applying formula (26) to the right-hand side of (29) we find

$$\int_0^{u_0} u^\alpha e^{-ku} S_\nu^\alpha(u)\,du$$

$$= k^{-\nu-1-\alpha} \sum_{\mu=0}^{\nu} (-1)^{\mu-\nu} C_\nu^\mu \frac{\Gamma(\alpha+\nu+1)}{\Gamma(\alpha+\mu+1)} \gamma(\alpha ha + \mu + 1, ku_0).$$

But according to (25) we have

$$k^{-\nu-1-\alpha} \sum_{\mu=0}^{\nu} (-1)^{\mu-\nu} C_\nu^\mu \frac{\Gamma(\alpha+\nu+1)}{\Gamma(\alpha+\mu+1)} \gamma(\alpha ha + \mu + 1, ku_0)$$

$$= -\frac{1}{k} u_0^{\alpha+1} e^{-ku_0} S_{\nu-1}^{\alpha+1}(u_0).$$

After multiplying the both parts of the latter equality by k we come to (27). ◁

Property 5. The system of the polynomials $\{S_\nu^\alpha(u)\}$ at a given $\alpha > -1$ represents an orthogonal system of the polynomials for which the gamma-distribution $\omega(u) = k^{\alpha+1} u^\alpha e^{-ku}/\Gamma(\alpha+1)$ serves as the weight:

$$\int_0^{\infty} \frac{k^{\alpha+1} u^\alpha}{\Gamma(\alpha+1)} e^{-ku} S_\nu^\alpha(u) S_\mu^\alpha(u)\,du = \begin{cases} 0, & \text{at } \nu \neq \mu, \\ \dfrac{\nu!\,\Gamma(\alpha+\nu+1)}{k^{2\nu}\Gamma(\alpha+1)} & \text{at } \nu = \mu. \end{cases} \qquad (30)$$

▷ For proving the orthogonality of the polynomials $S_\nu^\alpha(u)$ and $S_\mu^\alpha(u)$ it is sufficient to show the orthogonality of $S_\nu^\alpha(u)$ to the degrees u^λ, $\lambda < \nu$. We have

$$\int_0^{\infty} \frac{k^{\alpha+1} u^\alpha}{\Gamma(\alpha+1)} e^{-ku} S_\nu^\alpha(u)(ku)^\lambda\,du$$

$$= (-k)^{-\nu} \sum_{\mu=0}^{\nu} C_\nu^\mu \frac{\Gamma(\alpha+\nu+1)}{\Gamma(\alpha+\mu+1)} (-1)^\mu \frac{\Gamma(\alpha+\nu+\mu+1)}{\Gamma(\alpha+1)}.$$

Let us partition this sum into two ones taking into account that

$$C_\nu^\mu = C_{\nu-1}^\mu + C_{\nu-1}^{\mu-1}, \;\; 0 < \mu < \nu, \;\; C_\nu^0 = C_{\nu-1}^0 = C_\nu^\nu = C_{\nu-1}^{\nu-1} = 1.$$

Then

$$\int_0^\infty \frac{k^{\alpha+1} u^\alpha}{\Gamma(\alpha+1)} e^{-ku} S_\nu^\alpha(u)(ku)^\lambda \, du$$

$$= (-k)^{-\nu} \sum_{\mu=0}^{\nu-1} C_{\nu-1}^\mu \frac{\Gamma(\alpha+\nu+1)\Gamma(\alpha+\lambda+\mu+1)}{\Gamma(\alpha+\mu+1)\Gamma(\alpha+1)}$$

$$+ (-k)^\nu \sum_{\mu=0}^{\nu-1} (-1)^{\mu+1} C_{\nu-1}^\mu \frac{\Gamma(\alpha+\nu+1)\Gamma(\alpha+\lambda+\mu+2)}{\Gamma(\alpha+\mu+2)\Gamma(\alpha+1)}.$$

Here $\mu - 1$ is taken as μ in the second sum what involves the change of the lower limit of the summation by zero, and the upper limit by the number $\nu - 1$. Thus we have

$$\int_0^\infty \frac{k^{\alpha+1} u^\alpha}{\Gamma(\alpha+1)} e^{-ku} S_\nu^\alpha(u)(ku)^\lambda \, du$$

$$= (-k)^{-\nu} \sum_{\mu=0}^{\nu-1} (-1)^\mu C_{\nu-1}^\mu \frac{\Gamma(\alpha+\nu+1)\Gamma(\alpha+\lambda+\mu+1)}{\Gamma(\alpha+\mu+1)\Gamma(\alpha+1)} \left(\frac{\alpha+\lambda+\mu+1}{\alpha+\mu+1} \right)$$

$$= (-k)^{-\nu} \lambda \sum_{\mu=0}^{\nu-1} (-1)^{\mu-1} C_{\nu-1}^\mu \frac{\Gamma(\alpha+\nu+1)\Gamma(\alpha+\lambda+\mu+1)}{\Gamma(\alpha+\mu+1)\Gamma(\alpha+1)}.$$

The obtained sum is analogous to the initial sum but has one less item. Accounting that $C_{\nu-1}^\mu = C_{\nu-2}^\mu + C_{\nu-2}^{\mu-1}$ we partition once more the sum into two ones and substitute $\mu - 1$ by μ, then

$$\int_0^\infty \frac{k^{\alpha+1} u^\alpha}{\Gamma(\alpha+1)} e^{-ku} S_\nu^\alpha(u)(ku)^\lambda \, du$$

$$= (-k)^{-\nu} \lambda(\lambda-1) \sum_{\mu=0}^{\nu-2} (-1)^{\mu-2} C_{\nu-2}^\mu \frac{\Gamma(\alpha+\nu+1)\Gamma(\alpha+\lambda+\mu+1)}{\Gamma(\alpha+\mu+3)\Gamma(\alpha+1)}.$$

After performing this transformation λ times we shall have

$$\int_0^\infty \frac{k^{\alpha+1} u^\alpha}{\Gamma(\alpha+1)} e^{-ku} S_\nu^\alpha(u)(ku)^\lambda \, du$$

$$= (-k)^{-\nu} \lambda! \sum_{\mu=0}^{\nu-\lambda} (-1)^{\mu-\lambda} C_{\nu-\lambda}^\mu \frac{\Gamma(\alpha+\nu+1)}{\Gamma(\alpha+1)}$$

$$= (-k)^{-\nu} \lambda! \frac{\Gamma(\alpha+\nu+1)}{\Gamma(\alpha+1)} (-1)^\lambda \sum_{\mu=0}^{\nu-\lambda} C_{\nu-\lambda}^\mu (-1)^\mu = 0,$$

as the sum represents a binomial expansion of the variable $(1-1)^{\nu-\lambda}$. At $\lambda = \nu$ the same formula gives

$$\int_0^\infty \frac{k^{\alpha+1} u^\alpha}{\Gamma(\alpha+1)} e^{-ku} S_\nu^\alpha(u)(ku)^\nu \, du = \frac{k^{-\nu} \nu! \Gamma(\alpha+\nu+1)}{\Gamma(\alpha+1)}.$$

Consequently,

$$\int_0^\infty \frac{k^{\alpha+1} u^\alpha}{\Gamma(\alpha+1)} e^{-ku} \left[S_\nu^\alpha(u) \right]^2 du$$

$$= \int_0^\infty \frac{k^{\alpha+1} u^\alpha}{\Gamma(\alpha+1)} e^{-ku} S_\nu^\alpha(u) u^\nu \, du = \frac{\nu! \Gamma(\alpha+\nu+1)}{k^{2\nu} \Gamma(\alpha+1)}. \triangleleft$$

Now we shall consider the properties of the polynomials $S_{r,\nu}(u)$, which are orthogonal to the χ^2-distribution with r of freedom (22).

Property 1.

$$S_{r-1,\nu}(u) = S_{r,\nu}(u) + 2\nu S_{r,\nu-1}(u). \tag{31}$$

$\triangleright$ For proving this statement it is sufficient to put in formula (23) $\alpha = r/2 - 1$ and $k = 1/2$. $\triangleleft$

Property 2. The polynomials $S_{r,\nu}(u)$ are consistent in such a sense that as a result of the integration of the polynomial $S_{h+l,\nu}(u_1 + u_2)$ correspondent to χ^2-distribution with $h + l$ degrees of freedom over the χ^2-distribution with l degrees of freedom the polynomial $S_{h,\nu}(u_1)$ correspondent to the χ^2-distribution with h degrees of freedom is obtained:

$$\int_0^\infty S_{h+l,\nu}(u_1 + u_2)\frac{u_2^{l/2-1}e^{-u_2/2}}{2^{l/2}\Gamma(l/2)}\,du_2 = S_{h,\nu}(u_1). \qquad (32)$$

$\triangleright$ It is evident that the integral in the left-hand side represents the polynomial of νth degree relatively to u_1. This polynomial is orthogonal to the χ^2-distribution with h degrees of freedom at all $\lambda > \nu$. Really, we have

$$\int_0^\infty u_1^\lambda \frac{u_1^{h/2-1}e^{-u_1/2}}{2^{h/2}\Gamma(h/2)}\,du_1 \int_0^\infty S_{h+l,\nu}(u_1 + u_2)\frac{u_2^{l/2-1}e^{-u_2/2}}{2^{l/2}\Gamma(l/2)}\,du_2$$

$$= (2^{(h+l)/2}\Gamma(l/2)\Gamma(h/2))^{-1}$$

$$\times \int_0^\infty du_1 \int_{u_1}^\infty S_{h+l,\nu}(u)u_1^{\lambda+h/2-1}e^{-u_1/2}(u - u_1)^{l/2-1}e^{-(u-u_1)/2}\,du$$

$$= (2^{(h+l)/2}\Gamma(l/2)\Gamma(h/2))^{-1}$$

$$\times \int_0^\infty S_{h+l,\nu}(u)e^{-u/2}\,du \int_0^u u_1^{\lambda+h/2-1}(u - u_1)^{l/2-1}\,du_1. \qquad (33)$$

Here we made the change of variables $u = u_1 + u_2$ and after that changed the order of the integration. Further after performing the change of variables in the integral over u_1 we get

$$\int_0^u u_1^{\lambda+h/2-1}(u - u_1)^{l/2-1}\,du_1 = \int_0^1 (ut)^{\lambda+h/2-1}(u - ut)^{l/2-1}u\,dt$$

$$= u^{\lambda+h/2+l/2-1}\int_0^1 t^{\lambda+h/2-1}(1 - t)^{l/2-1}\,dt.$$

The latter integral represents the beta-function $B(\lambda + h/2, l/2)$. Using the known expression of beta-function $B(p, q)$ in terms of gamma-function $B(p, q) = \Gamma(p)\Gamma(q)/\Gamma(p + q)$ we shall find

$$\int_0^u u_1^{\lambda+h/2-1}(u - u_1)^{l/2-1}\,du_1 = u^{\lambda+h/2+l/2-1}\frac{\Gamma\left(\lambda + \frac{h}{2}\right)\Gamma\left(\frac{l}{2}\right)}{\Gamma\left(\lambda + \frac{h}{2} + \frac{l}{2}\right)}.$$

We substitute the obtained result into (33):

$$(2^{(h+l)/2}\Gamma(h/2)\Gamma(l/2))^{-1}$$

$$\times \int_0^\infty S_{h+l,\nu}(u)e^{-u/2}u^\lambda u^{h/2+l/2-1}du\,\frac{\Gamma\left(\lambda+\frac{h}{2}\right)\Gamma\left(\frac{l}{2}\right)}{\Gamma\left(\lambda+\frac{h}{2}+\frac{l}{2}\right)} = 0,$$

as the polynomial $S_{h+l,\nu}(u)$ is orthogonal to u^λ at all $\lambda > \nu$ relative to χ^2-distribution with $h+l$ degrees of freedom. But the polynomial of the νth degree orthogonal to all u^λ, $\lambda = 0, 1, 2, \ldots, \nu-1$ is determined uniquely with the accuracy till a constant multiplier. Consequently, the integral in (32) may differ from the polynomial $S_{h,\nu}(u_1)$ only by a constant multiplier which as it is easy to notice is equal to 1. ◁

Let us indicate one more property of $S_{r,\nu}(u)$ which follows from the properties of the generalized Laguerre polynomials. Accounting the relations between the polynomials $L_\nu^\alpha(u)$ and $S_\nu^\alpha(u)$, $S_\nu^\alpha(u) = (-k)^\nu \nu!\, L_\nu^\alpha(ku)$ we write

$$S_\nu^\alpha(u) = -\frac{2\nu+\alpha-1-ku}{k}S_{\nu-1}^\alpha(u) - \frac{(\nu-1)(\alpha+\nu-1)}{k^2}S_{\nu-2}^\alpha(u). \quad (34)$$

Property 3. Relation (34) in the case of the χ^2-distribution takes the following form:

$$S_{r,\nu}(u) = -(4\nu+r-4-u)S_{r,\nu-1}(u) - 2(\nu-1)(r-4+2\nu)S_{r,\nu-2}(u). \quad (35)$$

3. Statistical Linearization of Typical Composite Functions

3.1. Scalar Argument Functions (Table A.3.1)

$$\varphi(X) \approx \varphi_0(m, D) + k_1(m, D)X^0,$$

where

$$X^0 = X - m, \quad m = EX, \quad D = E\left|X^0\right|^2,$$

$$k_1(m, D) = \partial\varphi_0(m, D)/\partial m,$$

$$\Phi(z) = \frac{1}{\sqrt{2\pi}}\int_0^z e^{-t^2/2}dt, \quad \Phi'(z) = \frac{1}{\sqrt{2\pi}}e^{-z^2/2},$$

$$\zeta = m/\sqrt{D}, \quad \zeta_a^+ = (m+a)/\sqrt{D}, \quad \zeta_a^- = (m-a)/\sqrt{D}.$$

Table A.3.1

$\varphi(X)$	$\varphi_0(m, D)$
1. $1(X)$	$0,5 + \Phi(\zeta)$
2. $1(X - a)$	$0,5 + \Phi(\zeta_a^-)$
3. $\operatorname{sgn} X$	$2\Phi(\zeta)$
4. X^2	$\alpha_2 = m^2 + D$
5. $X^2 \operatorname{sgn} X$	$2\left[(m^2 + D)\Phi(\zeta) + m\sqrt{D}\Phi'(\zeta)\right]$
6. X^3	$\alpha_3 = m(m^2 + 3D)$
7. $X^n \ (n = 2, 3, \ldots)$	$\alpha_n = m\alpha_{n-1} + (n - 1)D\alpha_{n-2},$ $\alpha_0 = 1, \alpha_1 = m$
8. e^{aX}	$\beta_0 = e^{am + a^2 D/2}$
9. Xe^{aX}	$\beta_1 = (m + aD)e^{am + a^2 D/2}$
10. $X^n e^{aX} \ (n = 2, 3, \ldots)$	$\beta_n = (m + aD)\beta_{n-1} + (n - 1)D\beta_{n-2}$
11. $\sin aX$	$\gamma_0^s = e^{-a^2 D/2}\sin am$
12. $\cos aX$	$\gamma_0^c = e^{-a^2 D/2}\cos am$
13. $X \sin aX$	$\gamma_1^s = e^{-a^2 D/2}(aD\cos am + m\sin am)$
14. $X \cos aX$	$\gamma_1^c = e^{-a^2 D/2}(m\cos am - aD\sin am)$
15. $X^n \sin aX \ (n = 2, 3, \ldots)$	$\gamma_n^s = aD\gamma_{n-1}^c + m\gamma_{n-1}^s + (n - 1)D\gamma_{n-2}^s$
16. $X^n \cos aX \ (n = 2, 3, \ldots)$	$\gamma_n^c = m\gamma_{n-1}^s - aD\gamma_{n-1}^s + (n - 1)D\gamma_{n-2}^c$
17. 1 at $X > a$ 0 at $\|X\| \le a$ -1 at $X < -a$	$\Phi(\zeta_a^+) - \Phi(\zeta_a^-)$
18. l at $X > a$ lX/a at $\|X\| \le a$ $-l$ at $X < -a$	$(l/a)\{(m + a)\Phi(\zeta_a^+) - (m - a)\Phi(\zeta_a^-) +$ $+\sqrt{D}[\Phi'(\zeta_a^+) - \Phi'(\zeta_a^-)]\}$
19. $X1(X)$	$m[0,5 + \Phi(\zeta)] + \sqrt{D}\Phi'(\zeta)$

3.2. Vector Argument Functions (Table A.3.2):

$$\varphi(X) \approx \varphi_0(m, K) + k_1(m, K)X^0,$$

where

$$X^0 = X - m, \quad X = [X_1 \ldots X_n]^T, \quad m = EX = [m_1 \ldots m_n]^T,$$

$$K = EX^0 X^{0T}, \quad k_1(m, K) = \left[(\partial/\partial m)\varphi_0(m, K)^T\right]^T,$$

$$\nu = [\nu_1 \ldots \nu_n], \quad |\nu| = \nu_1 + \cdot + \nu_n,$$

$$e_p = \begin{bmatrix} 0 \ldots \underset{p}{1} \ldots 0 \end{bmatrix}, \quad k_p = [k_{p1} \ldots k_{pn}] \quad (p = 1, \ldots, n).$$

Here k_{pq} $(p, q = 1, \ldots, n)$ are the elements of the covariance matrix $K = EX^0 X^{0T}$.

Table A.3.2

$\varphi(X)$	$\varphi_0(m, K)$
1. $X_1 X_2$	$m_1 m_2 + k_{12}$
2. $X_1 X_2 X_3$	$m_1 m_2 m_3 + m_1 k_{23} + m_3 k_{12}$
3. $X_1^{\nu_1} \ldots X_n^{\nu_n}$, $\nu_1, \ldots, \nu_n$ are nonnegative integers $|\nu| > 1$	$\alpha_\nu = m_p \alpha_{\nu - e_p} + \sum_{r=1}^{n} \nu_r k_{pr} \alpha_{\nu - e_p - e_r} - k_{pp} \alpha_{\nu - 2e_p}, \quad \alpha_0 = 1, \quad \alpha_{e_p} = m_p$
4. $e^{a^T X}$	$\beta_0 = e^{a^T m + a^T K a/2}$
5. $X_p e^{a^T X}$	$\beta_{e_p} = (m_p + k_p a) e^{a^T m + a^T K a/2}$
6. $X_1^{\nu_1} \ldots X_n^{\nu_n} e^{a^T X}$, $\nu_1, \ldots, \nu_n$ are nonnegative integers $|\nu| > 1$	$\beta_\nu = (m_p + k_p a) \beta_{\nu - e_p} + \sum_{r=1}^{n} \nu_r k_{pr} \beta_{\nu - e_p - e_r} - k_{pp} \beta_{\nu - 2e_p}$
7. $\sin a^T X$	$\gamma_0^s = e^{-a^T K a/2} \sin a^T m$
8. $\cos a^T X$	$\gamma_0^c = e^{-a^T K a/2} \cos a^T m$
9. $X_p \sin a^T X$	$\gamma_{e_p}^s = e^{-a^T K a/2}(k_p a \cos a^T m + m_p \sin a^T m)$
10. $X_p \cos a^T X$	$\gamma_{e_p}^c = e^{-a^T K a/2}(m_p \cos a^T m + K_p a \sin a^T m)$
11. $X_1^{\nu_1} \ldots X_n^{\nu_n} \sin a^T X$, $\nu_1, \ldots, \nu_n$ are nonnegative integers $|\nu| > 1$	$\gamma_\nu^s = k_p a \gamma_{\nu - e_p}^c + m_p \gamma_{\nu - e_p}^s + \sum_{r=1}^{n} \nu_r k_{pr} \gamma_{\nu - e_p - e_r}^s - k_{pp} \gamma_{\nu - 2e_p}^s$

$\varphi(X)$	$\varphi_0(m, K)$
12. $X_1^{\nu_1}\ldots X_n^{\nu_n}\cos a^T X,$ $\nu_1,\ldots,\nu_n$ – nonnegative integers $\lvert\nu\rvert > 1$	$\gamma_\nu^c = m_p\gamma_{\nu-e_p}^c - k_p a\gamma_{\nu-e_p}^s$ $+\sum_{r=1}^{n}\nu_r k_{pr}\gamma_{\nu-e_p-e_r}^c - k_{pp}\gamma_{\nu-2e_p}^c$
13. $X^T a X$	$\operatorname{tr}\left\{a(K + mm^T)\right\}$
14. $\operatorname{sgn}\left(X_1 - a\operatorname{sgn} X_2\right)$	$[1 - 2\Phi(\eta)$ $-2k_{12}m_1\Phi'(\eta)/\sqrt{k_{11}(k_{11}k_{22} - k_{12}^2)}]$ $\times\Phi(\eta_a^+) + [1 + 2\Phi(\eta)$ $+2k_{12}m_1\Phi'(\eta)/\sqrt{k_{11}(k_{11}k_{22} - k_{12}^2)}]$ $\times\Phi(\eta_a^-) + 2k_{12}\Phi'(\eta)[\Phi'(\eta_a^-)$ $-\Phi'(\eta_a^+)]/\sqrt{k_{11}(k_{11}k_{22} - k_{12}^2)},$ $\eta_a^+ = (m_1 + a)/\sqrt{k_{11}},$ $\eta_a^- = (m_1 + a)/\sqrt{k_{11}}.$ (Approximate formula for small l, $l = k_{12}a/\sqrt{k_{11}(k_{11}k_{22} - k_{12}^2)})$

4. Integrands Evaluation in Equations of Methods based on Parametrization of Distributions

4.1. Moments Methods

Owing to the known expressions of the semiinvariants in terms of the characteristic function (Section 3.8)

$$\kappa_{s_1,\ldots,s_q}(t) = \left[\frac{\partial^{\lvert s\rvert}\chi(\mu;t)}{\partial(i\mu_1)^{s_1}\ldots\partial(i\mu_q)^{s_q}}\right]_{\mu=0} = \kappa_{s_1,\ldots,s_q}^w(t), \qquad (1)$$

$\kappa_{s_1,\ldots,s_q}^w(t)$ $(s_1,\ldots,s_q = 0,1,\ldots,N;\ \lvert s\rvert = 2,\ldots,N)$ being the semi-invariants of the process $W(t)$ ($\kappa_{s_1,\ldots,s_q}^w = 0$ when $\lvert s\rvert = 1$ since these semiin-variants represent the expectations of the components of the process $W(t)$ which are all equal to zero. Hence, by the Taylor formula

$$\chi(\mu;t) = \sum_{k=2}^{N}\sum_{\lvert s\rvert=k}\frac{(i\mu_1)^{s_1}\ldots(i\mu_q)^{s_q}}{s_1!\ldots s_q!}\chi_{s_1,\ldots,s_q}(t) + \rho_N \qquad (2)$$

where ρ_N is the residual term. Substituting here the expression $\mu = b(y,t)^T \lambda$ and collecting the items of the same degrees in λ_1 , ... , λ_p we obtain

$$\chi(b(y,t)^T \lambda;t) = \sum_{k=2}^{N} \sum_{|h|=k} \frac{(i\lambda_1)^{h_1}\dots(i\lambda_p)^{h_p}}{h_1!\dots h_p!}\omega_{h_1,\dots,h_p}(y,t) + \rho_N , \quad (3)$$

where

$$\omega_{h_1,\dots,h_p}(y,t)$$

$$= h_1!\dots h_p! \sum_{\rho_{11}+\dots+\rho_{q_1}=h_1} \dots \sum_{\rho_{1p}+\dots\rho_{qp}=h_p} \frac{\kappa_{\rho_{11}+\dots+\rho_{1p},\dots,\rho_{q_1}+\dots+\rho_{qp}}(t)}{\rho_{11}!\dots\rho_{1p}!\dots\rho_{q1}!\dots\rho_{qp}!}$$

$$\times b_{11}^{\rho_{11}}(y,t)\dots b_{pq}^{\rho_{pq}}(y,t) \qquad (4)$$

$$(h_1,\dots,h_p = 0,1,\dots,N; \quad |h| = 2,\dots,N).$$

If in particular $h_r = h_s = 1$, $h_u = 0$ at $u \neq r, s$, $r < s$ or $h_r = 2$, $h_u = 0$ at $u \neq r$, then

$$\omega_{0,\dots,0,\underset{r}{1},\dots,\underset{s}{1},0,\dots,0}(y,t) = \sum_{m,n=1}^{p} b_{mr}(y,t)b_{ns}(y,t)\nu_{mn}(t) , \qquad (5)$$

$$\omega_{0,\dots,0,\underset{r}{2},0,\dots,0}(y,t) = \sum_{m,n=1}^{p} b_{mr}(y,t)b_{nr}(y,t)\nu_{mn}(t) \qquad (6)$$

represent the elements $\sigma_{rs}(y,t)$ $(r,s = 1,\dots,q)$ of the matrix $\sigma(y,t)$ $= b(y,t)\nu(t)b(y,t)^T$, since the second order semiinvariants of the random vector $W(t)$ at given t are the elements of its covariance matrix

$$k(t) = k(t_0) + \int_{t_0}^{t} \nu(\tau)d\tau . \qquad (7)$$

Using formula (1.2.17) for the derivative of the product of two functions and the obtained formula for $\chi(b(y,t)^T \lambda;t)$ we find

$$\left\{ \frac{\partial^{|r|}}{\partial(i\lambda_1)^{r_1}\dots(i\lambda_p)^{r_p}} \left[i\lambda^T a(y,t) + \chi(b(y,t)^T \lambda;t) \right] e^{i\lambda^T y} \right\}_{\lambda=0}$$

$$= \sum_{s=1}^{p} r_s a_s(y,t)y_1^{r_1}\dots y_s^{r_s-1}\dots y_p^{r_p}$$

$$+\sum_{k=2}^{|r|}\sum_{|h|=k}\sum_{h_1=0}^{r_1}\cdots\sum_{h_p=0}^{r_p}C_{r_1}^{h_1}\ldots C_{r_p}^{h_p}\omega_{h_1,\ldots,h_p}(y,t)y_1^{r_1-h_1}\ldots y_p^{r_p-h_p}\,,$$

$$\tag{8}$$

$$\left\{\frac{\partial^{|r|}}{\partial(i\lambda_1)^{r_1}\ldots(i\lambda_p)^{r_p}}\left[i\lambda^T a(y,t)+\chi(b(y,t)^T\lambda;t)\right]e^{i\lambda^T(y-m)}\right\}_{\lambda=0}$$

$$=\sum_{s=1}^{p}r_s a_s(y,t)(y_1-m_1)^{r_1}\ldots(y_s-m_s)^{r_s-1}\ldots(y_p-m_p)^{r_p}$$

$$+\sum_{k=2}^{|r|}\sum_{|h|=k}\sum_{h_1=0}^{r_1}\cdots\sum_{h_p=0}^{r_p}C_{r_1}^{h_1}\ldots C_{r_p}^{h_p}$$

$$\times\omega_{h_1,\ldots,h_p}(y,t)(y_1-m_1)^{r_1-h_1}\ldots(y_p-m_p)^{r_p-h_p}\,.\tag{9}$$

In the case of the normal white noise $V(t)$ $\kappa_{s_1,\ldots,s_q}(t)=0$ at $|s|>2$. Consequently, $\omega_{h_1,\ldots,h_p}(y,t)=0$ at $|h|>2$. Taking into account that $\omega_{h_1,\ldots,h_p}(y,t)$ represent the elements of the matrix $\sigma(y,t)=b(y,t)\nu(t)b(y,t)^T$ at $|h|=2$ formulae (8) and (9) take in this case the form

$$\left\{\frac{\partial^{|r|}}{\partial(i\lambda_1)^{r_1}\ldots(i\lambda_p)^{r_p}}\left[i\lambda^T a(y,t)+\chi(b(y,t)^T\lambda;t)\right]e^{i\lambda^T y}\right\}_{\lambda=0}$$

$$=\sum_{s=1}^{p}r_s a_s(y,t)y_1^{r_1}\ldots y_s^{r_s-1}\ldots y_p^{r_p}$$

$$+\frac{1}{2}\sum_{s=1}^{p}r_s(r_s-1)\sigma_{ss}(y,t)y_1^{r_1}\ldots y_s^{r_s-2}\ldots y_p^{r_p}$$

$$+\sum_{q=2}^{p}\sum_{s=1}^{q-1}r_s r_q\sigma_{sq}(y,t)y_1^{r_1}\ldots y_s^{r_s-1}\ldots y_q^{r_q-1}\ldots y_p^{r_p}\,,\tag{10}$$

$$\left\{\frac{\partial^{|r|}}{\partial(i\lambda_1)^{r_1}\ldots(i\lambda_p)^{r_p}}\left[i\lambda^T a(y,t)+\chi(b(y,t)^T\lambda;t)\right]e^{i\lambda^T(y-m)}\right\}_{\lambda=0}$$

$$=\sum_{s=1}^{p}r_s a_s(y,t)(y_1-m_1)^{r_1}\ldots(y_s-m_s)^{r_s-1}\ldots(y_p-m_p)^{r_p}$$

$$+\frac{1}{2}\sum_{s=1}^{p}r_s(r_s-1)\sigma_{ss}(y,t)(y_1-m_1)^{r_1}\ldots(y_s-m_s)^{r_s-2}\ldots(y_p-m_p)^{r_p}$$

$$+ \sum_{q=2}^{p} \sum_{s=1}^{q-1} r_s r_q \sigma_{sq}(y,t)(y_1 - m_1)^{r_1} \ldots (y_s - m_s)^{r_s - 1}$$

$$\ldots (y_q - m_q)^{r_q - 1} \ldots (y_p - m_p)^{r_p} . \tag{11}$$

4.2. Semiinvariants Methods

In order to make easier the calculation of the derivatives with respect to $i\lambda_1$, $\ldots, i\lambda_p$ in the integrand we use the formula of differentiation of the product of two functions (1.2.17). As a result we obtain

$$\frac{\partial^{|r|}}{\partial(i\lambda_1)^{r_1} \ldots (i\lambda_p)^{r_p}} \left[i\lambda^T a(y,t) + \chi(b(y,t)^T \lambda; t) \right]$$

$$\times e^{i\lambda^T (y-m)} e^{i\lambda^T m - \ln g_1(\lambda; t)}$$

$$= \sum_{h_1=0}^{r_1} \ldots \sum_{h_p=0}^{r_p} C_{r_1}^{h_1} \ldots C_{r_p}^{h_p} \frac{\partial^{|h|}}{\partial(i\lambda_1)^{h_1} \ldots (i\lambda_p)^{h_p}}$$

$$\times \left[i\lambda^T a(y,t) + \chi(b(y,t)^T \lambda; t) \right]$$

$$\times e^{i\lambda^T (y-m)} \frac{\partial^{|r|-|h|}}{\partial(i\lambda_1)^{r_1-h_1} \ldots (i\lambda_p)^{r_p-h_p}} e^{i\lambda^T - \ln g_1(\lambda; t)} . \tag{12}$$

But (Subsection 3.8.3)

$$e^{i\lambda^T m} - \ln g_1(\lambda; t) = \exp \left\{ -\sum_{k=2}^{\infty} \sum_{|\nu|=k} \frac{(i\lambda_1)^{\nu_1} \ldots (i\lambda_p)^{\nu_p}}{\nu_1! \ldots \nu_p!} \kappa_\nu \right\}$$

$$= 1 + \sum_{s=1}^{\infty} \frac{(-1)^s}{s!} \left[\sum_{k=2}^{\infty} \sum_{|\nu|=k} \frac{(i\lambda_1)^{\nu_1} \ldots (i\lambda_p)^{\nu_p}}{\nu_1! \ldots \nu_p!} \kappa_\nu \right]^s$$

$$= 1 + \sum_{k=2}^{\infty} \sum_{|\nu|=k} (i\lambda_1)^{\nu_1} \ldots (i\lambda_p)^{\nu_p}$$

$$\times \sum_{s=1}^{[k/2]} \frac{(-1)^s}{s!} \sum_{q_1 + \cdots + q_s = \nu} \frac{\kappa_{q_1} \ldots \kappa_{q_s}}{q_{11}! \ldots q_{1p}! \ldots q_{sp}!} , \tag{13}$$

where $\nu = [\nu_1 \ldots \nu_p]$, $q_1 = [q_{11} \ldots q_{1p}]$, $\ldots$, $q_s = [q_{s1} \ldots q_{sp}]$ are the multi-indexes. Differentiating this formula h_1 times with respect to $i\lambda_1$, $\ldots$, h_p times with respect to $i\lambda_p$ putting this $\lambda_1 = \cdots = \lambda_p = 0$, and introducing the multi-indexs $h = [h_1 \ldots h_p]$ we obtain at $|h| \geq 2$

$$\left\{ \frac{\partial^{|h|}}{\partial(i\lambda_1)^{h_1} \ldots (i\lambda_p)^{h_p}} e^{i\lambda^T - \ln g_1(\lambda;t)} \right\}_{\lambda=0}$$

$$= h_1! \ldots h_p! \sum_{s=1}^{[|h|/2]} \frac{(-1)^s}{s!} \sum_{q_1+\cdots+q_s=\nu} \frac{\kappa_{q_1} \ldots \kappa_{q_s}}{q_{11}! \ldots q_{1p}! \ldots q_{s1}! \ldots q_{sp}!}. \tag{14}$$

All the first derivatives corresponding to $|h| = 1$ are equal to zero. Substituting expression (14) into (12) we shall have

$$\left\{ \frac{\partial^{|r|}}{\partial(i\lambda_1)^{r_1} \ldots (i\lambda_p)^{r_p}} \left[i\lambda^T a(y,t) + \chi(b(y,t)^T \lambda;t) \right] e^{i\lambda^T y - \ln g_1(\lambda;t)} \right\}_{\lambda=0}$$

$$= \left\{ \frac{\partial^{|r|}}{\partial(i\lambda_1)^{r_1} \ldots (i\lambda_p)^{r_p}} \left[i\lambda^T a(y,t) + \chi(b(y,t)^T \lambda;t) \right] e^{i\lambda^T (y-m)} \right\}_{\lambda=0}$$

$$+ \sum_{h_1=0}^{r_1} \cdots \sum_{h_p=0}^{r_p} \sum_{|h|=2}^{|r|-1} \frac{r_1! \ldots r_p!}{(r_1-h_1)! \ldots (r_p-h_p)!} \sum_{s=1}^{[|h|/2]} \frac{(-1)^s}{s!}$$

$$\times \sum_{q_1+\cdots+q_s=h} \frac{\kappa_{q_1} \ldots \kappa_{q_s}}{q_{11}! \ldots q_{1p}! \ldots q_{s1}! \ldots q_{sp}!} \left\{ \frac{\partial^{|r|-|h|}}{\partial(i\lambda_1)^{r_1-h_1} \ldots (i\lambda_p)^{r_p-h_p}} \right.$$

$$\times \left. \left[i\lambda^T a(y,t) + \chi(b(y,t)^T \lambda;t) \right] e^{i\lambda^T (y-m)} \right\}_{\lambda=0}. \tag{15}$$

In the special case of the normally distributed white noise V in (5.4.6) the derivatives with respect to $i\lambda_1$, $\ldots$, $i\lambda_p$ in the integrand are expressed by (11).

4.3. Orthogonal Expansions Methods

Let us consider formula (8) presented in the following form:

$$\left\{ \frac{\partial^{|r|}}{\partial(i\lambda_1)^{r_1} \ldots \partial(i\lambda_p)^{r_p}} \left[\lambda^T a(y,t) + \chi(b(y,t)^T \lambda;t) \right] e^{i\lambda^T y} \right\}_{\lambda=0}$$

$$= \sum_{s=1}^{p} a_s(y,t) r_s y_1^{r_1} \ldots y_s^{r_s-1} \ldots y_p^{r_p}$$

$$+\sum_{k=2}^{|r|}\sum_{|h|=k}\sum_{h_1=0}^{r_1}\cdots\sum_{h_p=0}^{r_p}\frac{r_1!\ldots r_p!}{h_1!\ldots h_p!(r_1-h_1)!\ldots(r_p-h_p)!}$$

$$\times\omega_{h_1,\ldots,h_p}(y,t)y_1^{r_1-h_1}\ldots y_p^{r_p-h_p}.\tag{16}$$

Noticing that

$$\frac{r_1!\ldots r_p!}{(r_1-h_1)!\ldots(r_p-h_p)!}y_1^{r_p-h_p}\ldots y_p^{r_p-h_p}=\frac{\partial^{|h|}}{\partial y_1^{h_1}\ldots\partial y_p^{h_p}}y_1^{r_1}\ldots y_p^{r_p},$$

$$r_s y_1^{r_1}\ldots y_s^{r_s-1}\ldots y_p^{r_p}=\frac{\partial}{\partial y_s}y_1^{r_1}\ldots y_p^{r_p},$$

we get

$$\left\{q_\kappa(\partial/i\partial\lambda)\left[i\lambda^T a(y,t)+\chi(b(y,t)^T\lambda;t)\right]e^{i\lambda^T y}\right\}_{\lambda=0}$$

$$=\sum_{s=1}^{p}a_s(y,t)\frac{\partial q_\kappa(y)}{\partial y_s}+\sum_{k=2}^{|\kappa|}\sum_{|h|=k}\sum_{h_1,\ldots,h_p=0}\frac{\omega_{h_1,\ldots,h_p}(y,t)}{h_1!\ldots h_p!}\frac{\partial^{|h|}q_\kappa(y)}{\partial y_1^{h_1}\ldots\partial y_p^{h_p}}.\tag{17}$$

In the special case of the normal white noise V using (10) instead of (8) we get

$$\left\{q_\kappa(\partial/i\partial\lambda)\left[i\lambda^T a(y,t)+\chi(b(y,t)^T\lambda;t)\right]e^{i\lambda^T y}\right\}_{\lambda=0}$$

$$=-\sum_{s=1}^{p}a_s(y,t)\frac{\partial q_\kappa(y)}{\partial y_s}+\frac{1}{2}\sum_{s,u=1}^{p}\sigma_{su}(y,t)\frac{\partial^2 q_\kappa(y)}{\partial y_s\partial y_u}.\tag{18}$$

4.4. Quasimoments Methods

In particular, using Hermite polynomial expansion (17) takes the form

$$\left\{q_\kappa(\partial/i\partial\lambda)\left[i\lambda^T a(y,t)+\chi(b(y,t)^T\lambda;t)\right]e^{i\lambda^T y}\right\}_{\lambda=0}$$

$$=\left\{G_\kappa(\partial/i\partial\lambda-m)\left[i\lambda^T a(y,t)+\chi(b(y,t)^T\lambda;t)\right]e^{i\lambda^T y}\right\}_{\lambda=0}$$

$$=\left\{G_\kappa(\partial/i\partial\lambda)\left[i\lambda^T a(y,t)+\chi(b(y,t)^T\lambda;t)\right]e^{i\lambda^T(y-m)}\right\}_{\lambda=0}$$

$$= -\sum_{s=1}^{p} a_s(y,t)\frac{\partial}{\partial y_s}G_\kappa(y-m)$$

$$+\sum_{l=2}^{|\kappa|}\sum_{|h|=l}\sum_{h_1,\ldots,h_p=0}\frac{\omega_{h_1,\ldots,h_p}(y,t)}{h_1!\ldots h_p!}\frac{\partial^k}{\partial y_1^{h_1}\ldots\partial y_p^{h_p}}G_\kappa(y-m). \qquad (19)$$

By formula (9) of Appendix 2 for the derivatives of the polynomials G_κ we get finally

$$\left\{q_\kappa(\partial/i\partial\lambda)\left[i\lambda^T a(y,t)+\chi(b(y,t)^T\lambda;t)\right]e^{i\lambda^T y}\right\}_{\lambda=0}$$

$$=\sum_{s=1}^{p}\kappa_s a_s(y,t)G_{\kappa-e_s}(y-m)$$

$$+\sum_{l=2}^{|\kappa|}\sum_{|h|=l}\sum_{h_1,\ldots,h_p=0}^{k}C_{\kappa_1}^{h_1}\ldots C_{\kappa_p}^{h_p}\omega_{h_1,\ldots,h_p}(y,t)$$

$$\times C_{\kappa-h_1 e_1-\cdots-h_p e_p}(y-m)\,, \qquad (20)$$

where e_s is as before the multi-index whose components are all equal to zero except the sth which is equal to unity. Formula (18) takes in this case the form

$$\left\{q_\kappa(\partial/i\partial\lambda)\left[i\lambda^T a(y,t)+\chi(b(y,t)^T\lambda;t)\right]e^{i\lambda^T y}\right\}_{\lambda=0}$$

$$=\sum_{s=1}^{p}\kappa_s a_s(y,t)G_{\kappa-e_s}(y-m)+\frac{1}{2}\sum_{s=1}^{p}\kappa_s(\kappa_s-1)\sigma_{ss}(y,t)$$

$$\times G_{\kappa-2e_s}(y-m)+\sum_{u=2}^{p}\sum_{s=1}^{u-1}\kappa_s\kappa_u\sigma_{su}(y,t)G_{\kappa-e_s-e_u}(y-m)\,. \qquad (21)$$

4.5. Ellipsoidal Approximation Method

Using the results of Appendix 7 in (Pugachev and Sinitsyn 2000) we have the following formulae for typical EAM integrals

$$I_1^{EAM} = E_{w_1} Y_1^{r_1} \ldots Y_p^{r_p} p_{p,\nu}(U) = \int\limits_{-\infty}^{\infty} y_1^{r_1} \ldots y_p^{r_p} p_{p,\nu}(u) w_1(u) dy$$

$$= \begin{cases} 2^\nu \nu! R_{r,\nu} & \text{at } \nu \leq [\,|r|/2\,] \\ 0 & \text{at } \nu > [\,|r|/2\,], \end{cases} \tag{22}$$

where

$$R_{r,l} = \sum_{|h|=2l} C_{r_1}^{h_1} \ldots C_{r_p}^{h_p} \alpha_{r_1-h_1,\,\ldots,\,r_p-h_p}^{w_1} \mu_{h_1,\,\ldots,\,h_p}^{w_1}, \tag{23}$$

$$R_{r,l} = 0 \text{ at } l < 0 \text{ or at } l > |r|,$$

$$I_2^{EAM} = E_{w_1} p_{p,\nu}(u) q_{p,\kappa}^{(\rho)} Y_1^{r_1} \ldots Y_p^{r_p} = \int\limits_{-\infty}^{\infty} y_1^{r_1} \ldots y_p^{r_p} p_{p,\nu}(u) q_{p,\kappa}^{(\rho)} w_1(u) dy$$

$$= \sum_{l=\beta_1}^{\beta_2} \omega_l(\nu, \kappa, \rho) R_{r,l}, \tag{24}$$

where

$$\omega_l(\nu, \kappa, \rho) = \frac{l!(p-2)!!}{(p+2\kappa-2)!!} \sum_{m=\beta_3}^{\beta_4} \frac{C_\nu^m C_m^{\kappa-\rho-l-\nu+2m}}{(l-m)!}$$

$$\times 2^{\kappa+3m-\nu-l-2\rho} \frac{(p+2l+2\nu-2m)!!}{(p+2l)!!} \tag{25}$$

$$\beta_1 = |\nu - \kappa + \rho|, \quad \beta_2 = \min([\,|r|/2\,], \nu + \kappa - \rho),$$

$$\beta_3 = [\,(l+\nu-\kappa+\rho+1)/2\,], \quad \beta_4 = \min(\nu, l, l+\nu+\rho-\kappa). \tag{26}$$

5. Basic Characteristics of Some Continuous Typical Linear Systems

Operators, Weighting and Transfer Functions for Some Continuous Typical Linear Systems

Table A.5.1.

System	Weighting Function $g(t,\tau)$, $t > \tau$	Transfer Function $\Phi(s)$
1. $y(t) = Kx(t)$, $K > 0$	$K\delta(t-\tau)$	K
2. $y(t) = \int_{t_0}^{t} x(\xi)d\xi$, $\dfrac{dy}{dt} = x$	$\mathbf{1}(t-\tau)$	$1/s$
3. $y = \dfrac{dx}{dt}$	$\delta'(t-\tau)$	s
4. $a\dfrac{dy}{dt} \pm y = bx$, $a > 0$	$\dfrac{b}{a}e^{\mp\frac{(t-\tau)}{a}}$	$\dfrac{b}{as \pm 1}$
5. $a^2\dfrac{d^2y}{dt^2} + y = bx$, $a > 0$	$\dfrac{b}{a}\sin\dfrac{(t-\tau)}{a}$	$\dfrac{b}{a^2s^2+1}$
6. $a^2\dfrac{d^2y}{dt^2} + 2\varepsilon a\dfrac{dy}{dt} + y = bx$, $a > 0, 1 > \varepsilon > 0$	$\dfrac{b}{a\sqrt{1-\varepsilon^2}}e^{-\frac{\varepsilon}{a}(t-\tau)}$ $\times\sin\dfrac{\sqrt{1-\varepsilon^2}}{a}(t-\tau)$	$\dfrac{b}{a^2s^2+2\varepsilon as+1}$
7. $a^2\dfrac{d^2y}{dt^2} + 2\varepsilon a\dfrac{dy}{dt} + y = bx$, $\varepsilon > 1$	$\dfrac{b}{a\sqrt{\varepsilon^2-1}}\left[e^{-\frac{2\varepsilon_1}{a}(t-\tau)} - e^{-\frac{2\varepsilon_2}{a}(t-\tau)}\right]$, $2\varepsilon_{1,2} = 1 \pm \sqrt{\varepsilon^2-1}$	$\dfrac{b}{a^2s^2+2\varepsilon as+1}$
8. $y = b\dfrac{dx}{dt} + x$	$b\delta'(t-\tau) + \delta(t-\tau)$	$bs+1$
9. $y = b^2\dfrac{d^2x}{dt^2}$ $+2\varepsilon b\dfrac{dx}{dt} + x, b, \varepsilon > 0$	$b^2\delta''(t-\tau)$ $+2\varepsilon b\delta'(t-\tau) + \delta(t-\tau)$	b^2s^2 $+2\varepsilon bs+1$
10. $a\dfrac{dy}{dt} + y = b\dfrac{dx}{dt} + x$	$\dfrac{b}{a}\delta(t-\tau)$ $+\left(1-\dfrac{b}{a}\right)e^{-\frac{(t-\tau)}{a}}$	$\dfrac{1+bs}{1+as}$

*Transfer Functions and Frequency
Characteristics for Some Continuous Linear Systems*
Table A.5.2.

Transfer Function $\Phi(s)$	Gain-Frequency Characteristic $A(\omega)$	Phase-Frequency Characteristic $\psi(\omega)$
1. K	K	0
2. $1/s$	$1/\omega$	$-\pi/2$
3. s	ω	$\pi/2$
4. $\dfrac{b}{as+1}$	$\dfrac{b}{\sqrt{a^2\omega^2+1}}$	$-\operatorname{arctg}(a\omega)$
5. $\dfrac{b}{as-1}$	$\dfrac{b}{\sqrt{a^2\omega^2+1}}$	$\operatorname{arctg}(a\omega)-\pi$
6. $\dfrac{b}{a^2s^2+1}$	$\dfrac{b}{\left|1-a^2\omega^2\right|}$	0 at $\omega<1/a$ π at $\omega>1/a$
7. $\dfrac{b}{a^2s^2+2\varepsilon as+1}$	$\dfrac{b}{\sqrt{(1-a^2\omega^2)^2+4a^2\varepsilon^2\omega^2}}$	$-\operatorname{arctg}\dfrac{2a\varepsilon\omega}{1-a^2\omega^2}$
8. $bs+1$	$\sqrt{b^2\omega^2+1}$	$\operatorname{arctg}(b\omega)$
9. $b^2s^2+2\varepsilon bs+1$	$\sqrt{(1-b^2\omega^2)^2+4\varepsilon^2b^2\omega^2}$	$\operatorname{arctg}\dfrac{2b\varepsilon\omega}{1-b^2\omega^2}$

Some Typicl Models of Stochastic Discrete Systems
Table A.5.3.

Model	Equations
1. Nonlinear Autoregression (AR): • l order • first order	$\begin{aligned} Y_{k+l} &= \varphi_k(Y_k,\,\ldots,\,Y_{k+l-1})+\\ &\quad+\psi_k(Y_k,\,\ldots,\,Y_{k+l-1})V_k\\ Y_{k+1} &= \varphi_k(Y_k)+\psi_k(Y_k)V_k \end{aligned}$

Table A.5.3 (continued)

Model	Equations
2. Linear AR:	
$\bullet$ l order	$$Y_{k+l} = \sum_{r=0}^{l-1} a_{rk} Y_{k+r} + \psi_k V_k$$
$\bullet$ first order	$$Y_{k+1} = c_k Y_k + q_k V_k$$
$\bullet$ l order with Moving Average (ARMA)	$$Y_{k+l} = \sum_{r=0}^{l-1} a_{rk} Y_{k+r} +$$ $$+ A_k + \sum_{s=0}^{l} b_{sk} V_{k+s}$$
3. Linear Integrated AR (IAR)	
$\bullet$ l order	$$\Delta_\alpha Y_{k+l} = \sum_{r=0}^{l-1} a_{rk} \Delta_\alpha Y_{k+r} + \psi_k V_k$$
$\bullet$ l order with Moving Average (ARIMA)	$$\Delta_\alpha Y_{k+l} = \sum_{r=0}^{l-1} a_{rk} \Delta_\alpha Y_{k+r} +$$ $$+ \sum_{s=0}^{l} b_{sk} V_k$$
4. Nonlinear ARIMA of l order	$$Y_{k+l} = \sum_{r=0}^{l-1} a_{rk} \varphi_{k+r}(Y_{k+r})$$ $$+ \sum_{s=0}^{l} b_{sk} V_{k+s}$$
5. Stationary linear ARIMA of l order	$$A(\nabla) Y_k = c + B(\nabla) V_k$$ $$A(\nabla) = I\nabla^p - \sum_{r=0}^{l-1} a_r \nabla^r,$$ $$B(\nabla) = \sum_{s=-N}^{l} b_s \nabla^s,$$ $$\nabla u_k = u_{k+1}$$
6. Stochastic Nonlinear AR	$$Y_{k+1} = \Phi_k(Y_k) + \Psi_k(Y_k) V_k$$

Operators, Weighting (Green) and Transfer Functions for Some Typical Linear One-Dimensional Systems with the Distributed Parameters

Table A.5.4.

System	Weighting (Green) Function $g(\xi,\xi_1,t,\tau)$, $t > \tau$	Transfer Function $\Phi(\xi,\xi_1,s)$
1. $y(t) = x(t-\Delta)$ $\forall t > \tau, y(t) = 0, \forall t < \tau,$ $\Delta > 0$	$\delta(t-\tau-\Delta)$	$\exp\{-\Delta s\}$
2. $x(t) = q(0,t),$ $y(t) = q(l,t),$ $\dfrac{\partial^2 q(\xi,t)}{\partial \xi^2} - \dfrac{1}{\gamma^2}\dfrac{\partial q(\xi,t)}{\partial t} = 0,$ $q(\infty,t) = 0, \Delta_1 = l^2/\gamma^2$	$\dfrac{1}{2}\sqrt{\dfrac{\Delta_1}{\pi(t-\tau)^3}}\, e^{-\frac{\Delta_1}{4(t-\tau)}}$	$\exp\{-\sqrt{\Delta_1 s}\}$
3. $x(t) = -\lambda\dfrac{\partial q(0,t)}{\partial \xi},$ $y(t) = q(0,t),$ $\dfrac{\partial^2 q(\xi,t)}{\partial \xi^2} - \dfrac{1}{\gamma^2}\dfrac{\partial q(\xi,t)}{\partial t} = 0,$ $q(\infty,t) = 0, \Delta_2 = \lambda^2/\gamma^2$	$\sqrt{\dfrac{\Delta_2}{\pi(t-\tau)}}$	$\dfrac{1}{\sqrt{\Delta_2 s}}$
4. $x(t) = -\lambda\dfrac{\partial q(0,t)}{\partial \xi}$ $+\alpha q(0,t),\ y(t) = q(0,t),$ $\dfrac{\partial^2 q(\xi,t)}{\partial \xi^2} - \dfrac{1}{\gamma^2}\dfrac{\partial q(\xi,t)}{\partial t} = 0,$ $q(\infty,t) = 0, \beta_1 = 1/\alpha,$ $\alpha_1 = \lambda^2/\alpha^2\gamma^2$	$\dfrac{\beta_1}{\alpha_1}\left[\sqrt{\dfrac{\alpha_1}{\pi(t-\tau)^3}} -e^{-\frac{(t-\tau)}{\alpha_1}} \times \mathrm{erf}\sqrt{\dfrac{(t-\tau)}{\alpha_1}}\,\right]$	$\dfrac{\beta_1}{\sqrt{\alpha_1 s}+1}$

Table A.5.4 (continued)

System	Weighting (Green) Function $g(\xi, \xi_1, t, \tau)$, $t > \tau$	Transfer Function $\Phi(\xi, \xi_1, s)$
5. $x(t) = q(0,t),$ $\dot{y}(t) = q(l,t),$ $\dfrac{\partial^2 q(\xi,t)}{\partial \xi^2} - \dfrac{1}{\gamma^2}\dfrac{\partial q(\xi,t)}{\partial t} = 0,$ $q(\infty,t) = 0, \Delta_1 = l^2/\gamma^2$	$\operatorname{erfc}\sqrt{\dfrac{\Delta_1}{4(t-\tau)}}$	$\dfrac{\exp\left\{-\sqrt{\Delta_1 s}\right\}}{s}$
6. $x(t) = -\lambda\dfrac{\partial q(0,t)}{\partial \xi},$ $\dot{y}(t) = q(0,t),$ $\dfrac{\partial^2 q(\xi,t)}{\partial \xi^2} - \dfrac{1}{\gamma^2}\dfrac{\partial q(\xi,t)}{\partial t} = 0,$ $q(\infty,t) = 0, \Delta_2 = \lambda^2/\gamma^2$	$2\sqrt{\dfrac{(t-\tau)}{\pi\Delta_2}}$	$\dfrac{1}{s\sqrt{\Delta_2 s}}$
7. $x(t) = -\lambda\dfrac{\partial q(0,t)}{\partial \xi}$ $+\alpha q(0,t), \dot{y}(t) = q(0,t),$ $\dfrac{\partial^2 q(\xi,t)}{\partial \xi^2} - \dfrac{1}{\gamma^2}\dfrac{\partial q(\xi,t)}{\partial t} = 0,$ $q(\infty,t) = 0, \beta_1 = 1/\alpha,$ $\alpha_1 = \lambda^2/\alpha^2\gamma^2$	$1 - e^{(t-\tau)/\alpha_1}$ $\times\operatorname{erfc}\sqrt{\dfrac{(t-\tau)}{\alpha_1}}$	$\dfrac{\beta_1}{s\left(\sqrt{\alpha_1 s}+1\right)}$
8. $\dfrac{\partial y}{\partial t} - a^2\dfrac{\partial^2 y}{\partial \xi^2} = x,$ $y(\xi,0) = y_0(\xi),$ $\displaystyle\int_{-\infty}^{\infty} \mid y(\xi,t)\mid d\xi < \infty,$ $t \geq 0, -\infty \leq \xi \leq \infty, a \neq 0$	$\dfrac{1}{2a\sqrt{\pi(t-\tau)}}$ $\times\exp\left[-\dfrac{(\xi-\xi_1)^2}{4a^2(t-\tau)}\right]$	$\dfrac{1}{2a^2 s_1}e^{-s_1\mid\xi-\xi_1\mid},$ $s_1 = \sqrt{s}/a$

System	Weighting (Green) Function $g(\xi, \xi_1, t, \tau)$, $t > \tau$	Transfer Function $\Phi(\xi, \xi_1, s)$
9. $\dfrac{\partial y}{\partial t} - a^2 \dfrac{\partial^2 y}{\partial \xi^2} = x,$ $y(\xi, 0) = y_0(\xi),$ $\dfrac{\partial y}{\partial \xi}(0, t) = y^0(t),$ $t \geq 0, 0 \leq \xi \leq \infty, a \neq 0$	$\dfrac{1}{2a\sqrt{\pi(t-\tau)}}$ $\times \left[e^{-\frac{(\xi-\xi_1)^2}{4a^2(t-\tau)}} \right.$ $\left. + e^{-\frac{(\xi+\xi_1)^2}{4a^2(t-\tau)}} \right]$	$\dfrac{1}{2a^2 s_1} \left[e^{-s_1(\xi-\xi_1)} \right.$ $\left. + e^{-s_1(\xi+\xi_1)} \right],$ $s_1 = \sqrt{s}/a$
10. $\dfrac{\partial y}{\partial t} - a^2 \dfrac{\partial^2 y}{\partial \xi^2} = x,$ $y(\xi, 0) = y_0(\xi),$ $y(0, t) = y^0(t),$ $y(\infty, t) = 0,$ $t \geq 0, 0 \leq \xi \leq \infty, a \neq 0$	$\dfrac{1}{2a\sqrt{\pi(t-\tau)}}$ $\times \left[e^{-\frac{(\xi-\xi_1)^2}{4a^2(t-\tau)}} \right.$ $\left. - e^{-\frac{(\xi+\xi_1)^2}{4a^2(t-\tau)}} \right]$	$\dfrac{1}{2a^2 s_1} \left[e^{-s_1|\xi-\xi_1|} \right.$ $\left. - e^{-s_1|\xi+\xi_1|} \right],$ $s_1 = \sqrt{s}/a$
11. $\dfrac{\partial^2 y}{\partial t^2} - a^2 \dfrac{\partial^2 y}{\partial \xi^2} = x,$ $y(\xi, 0) = y_0(\xi),$ $\dfrac{\partial y(\xi, 0)}{\partial t} = y_1(\xi),$ $t \geq 0, -\infty < \xi < \infty, a \neq 0$	$\dfrac{1}{2a} \mathbf{1}[a(t-\tau) - |\xi-\xi_1|]$	$\dfrac{1}{2as} e^{-\frac{s}{a}|\xi-\xi_1|}$
12. $\dfrac{\partial^2 y}{\partial t^2} - a^2 \dfrac{\partial^2 y}{\partial \xi^2} = x,$ $y(\xi, 0) = y_0(\xi),$ $\dfrac{\partial y(\xi, 0)}{\partial t} = y_1(\xi),$ $y(0, t) = y^0(t),$ $t \geq 0, \xi \geq 0, a \neq 0$	$\dfrac{1}{2a} \mathbf{1}\{[\xi-\xi_1+a(t-\tau)]$ $- \mathbf{1}[\xi-\xi_1-a(t-\tau)]$ $- \mathbf{1}[\xi+\xi_1+a(t-\tau)]$ $+ \mathbf{1}[\xi+\xi_1-a(t-\tau)]\}$	$\dfrac{1}{2as} \left[e^{-\frac{s}{a}|\xi-\xi_1|} \right.$ $\left. - e^{-\frac{s}{a}|\xi+\xi_1|} \right]$

6. Structural Transformation Rules

R u l e 1: two knots may always change places (Fig.A.6.1).

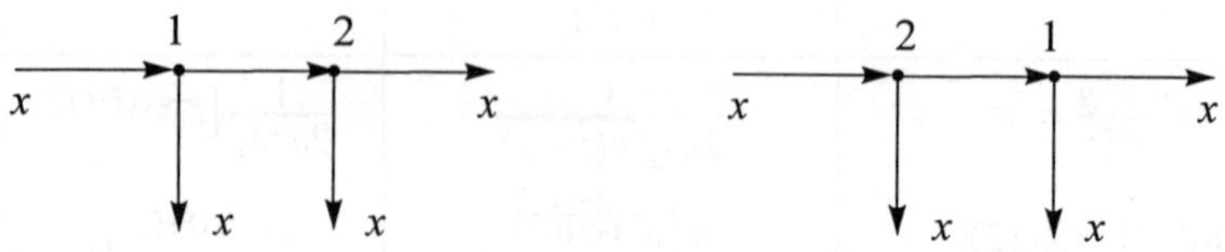

Fig. A.6.1

R u l e 2: two summators may always change places (Fig.A.6.2).

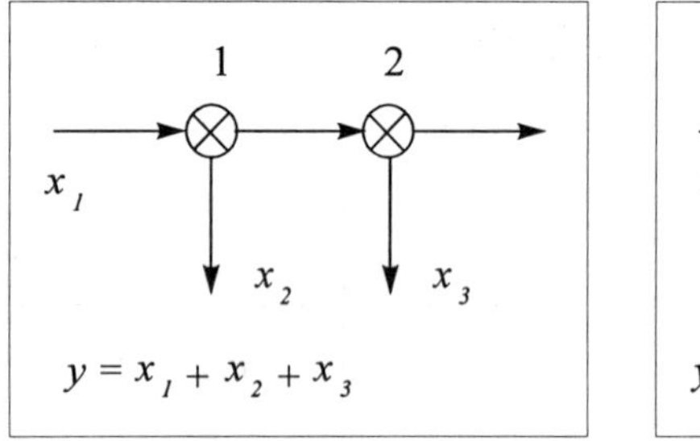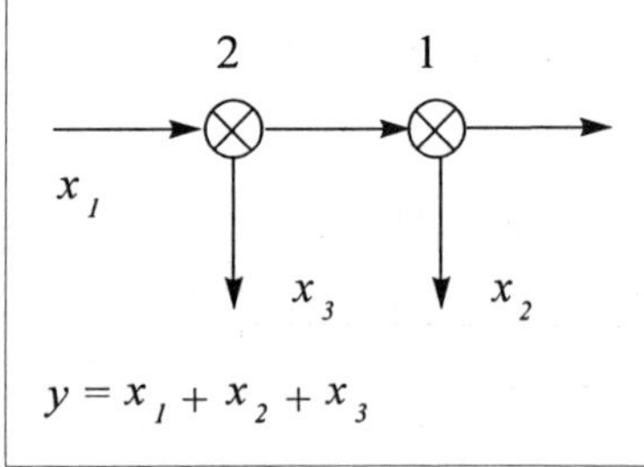

Fig. A.6.2

R u l e 3: while transferring the knot through the summator along the signal direction or the summator through the knot opposite signal direction the line of the connection between the lateral branches directed along the signal in the direct chain and which contains the amplifier with the gain equal to -1 should be added (Fig.A.6.3).

R u l e 4: while transferring the summator throuh the knot along the signal direction (or the knot through the summator opposite signal direction) a rigid positive connection between the lateral braches oppositely goes in the direct chain should be added (Fig.A.6.4).

R u l e 5: while transferring the knot through the linear system along the signal direction it is necessary to include into the branch the inverse linear system (Fig.A.6.5).

R u l e 6: while transferring the knot through the linear system in the opposite signal direction it is necessary to include into the branch the same linear system (Fig.A.6.6).

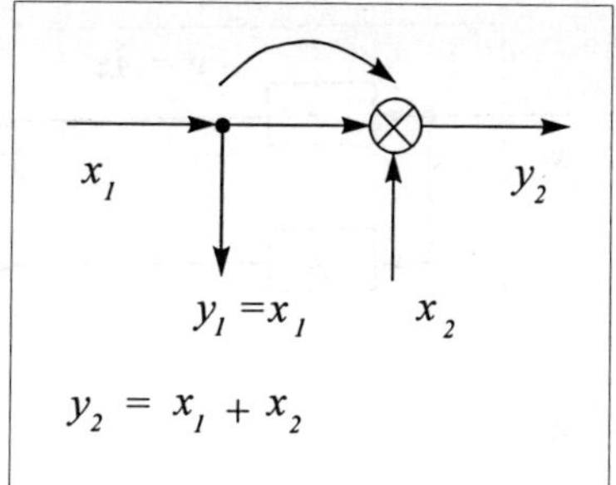
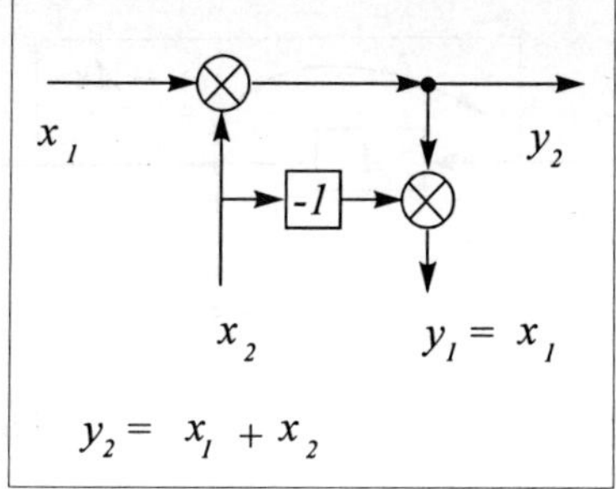

Fig. A.6.3

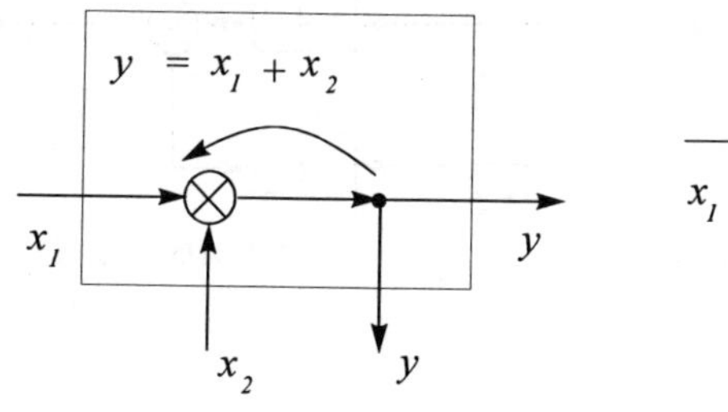
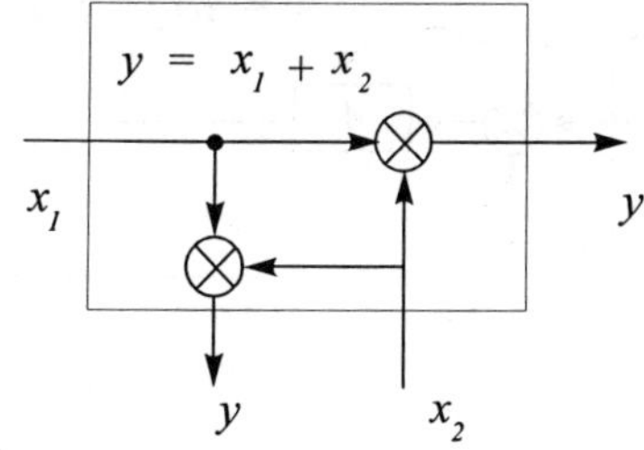

Fig. A.6.4

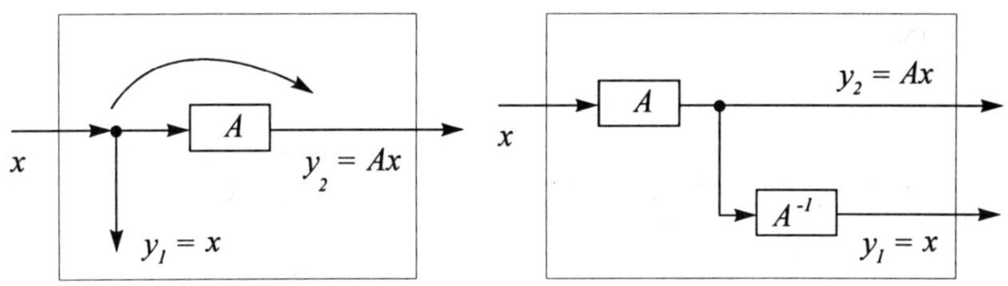

Fig. A.6.5

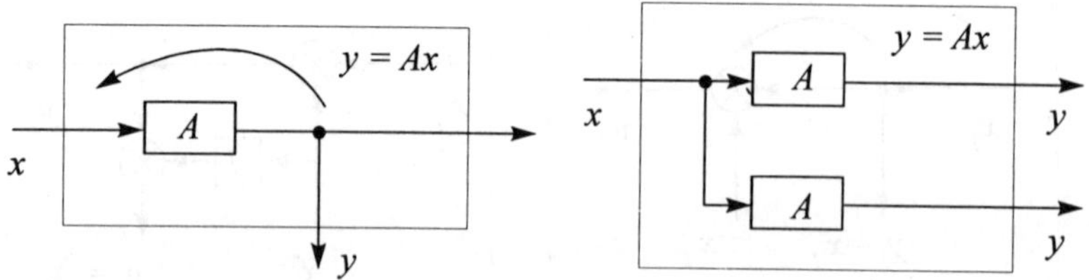

Fig. A.6.6

R u l e 7: while transferring the summator through the linear system along signal direction it is necessary to include into the line of the second input of the summator the same linear system (Fig.A.6.7).

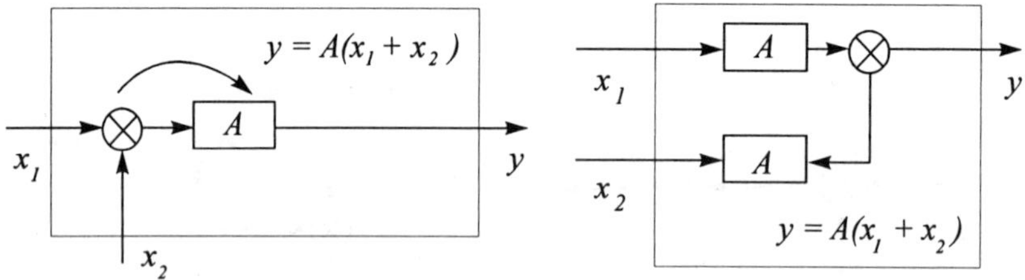

Fig. A.6.7

R u l e 8: while transferring the summator through the linear system in the opposite direction of the signal it is necessary to include into the line of the second input of the summator the inverse system (Fig.A.6.8).

R u l e 9: the branches of the parallel connection may change places (Fig.A.6.9).

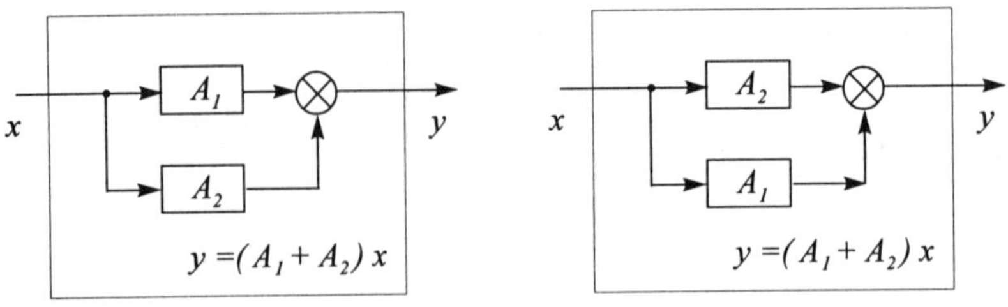

Fig. A.6.9

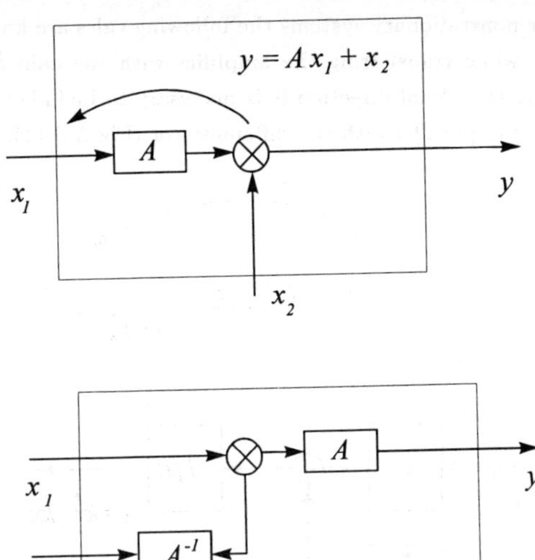

Fig. A.6.8

R u l e 10: in the system with the feedback we may change the places of
the system which is in the direct chain and the system included into the chain of
the feedback substituting here both systems by the correspondent inverse systems
(Fig.A.6.10).

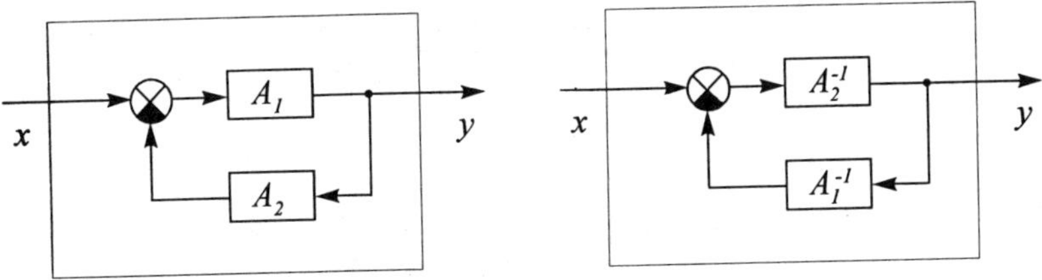

Fig. A.6.10

R u l e 11: the sequentially connected stationary linear systems also including
the inertialess stationary and nonstationary linear amplifiers may change places.

For the arbitrary nonstationary linear systems rules 1–10 remain. For the special cases of the linear nonstationary systems the following rules are known.

R u l e 12: while transferring the amplifier with the gain $k(t)$ through the differentiator along the signal direction it is necessary to include the amplifier with the gain $\dot{k}(t)/k(t)$ in parallel with the differentiator (Fig.A.6.11).

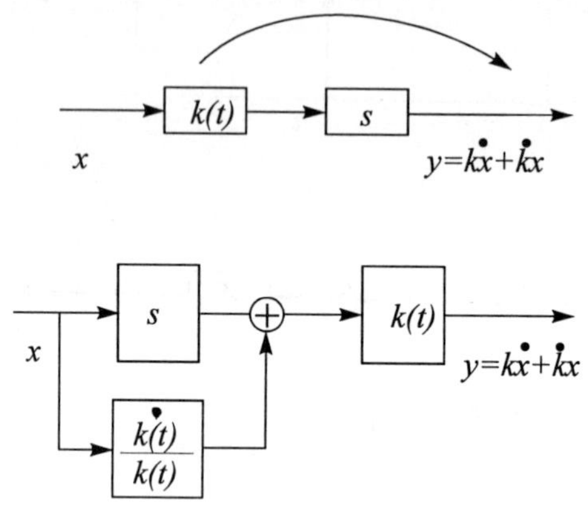

Fig. A.6.11

R u l e 13: while transferring the amplifier with the $k(t)$ through the differentiator in the opposite direction of the signal it is necessary to include the amplifier with the gain $-\dot{k}(t)/k(t)$ in parallel with the differentiator (Fig.A.6.12).

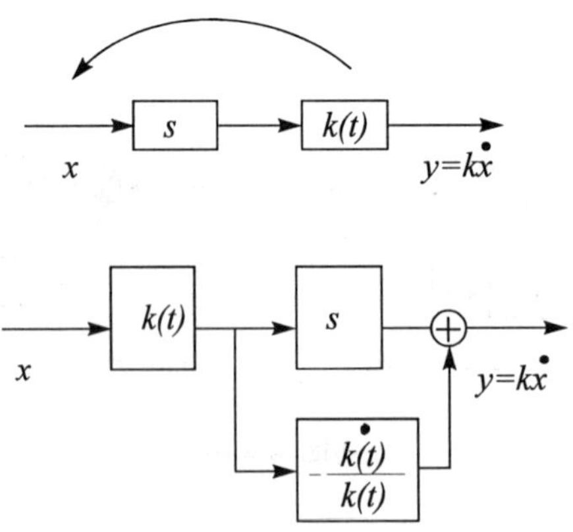

Fig. A.6.12

R u l e 14: while transfering the amplifier with the gain $k(t)$ through the integrator along the signal direction it is necessary to close the integrator by negative feedback which contains the amplifier with the gain $\dot{k}(t)/k(t)$ (Fig.A.6.13).

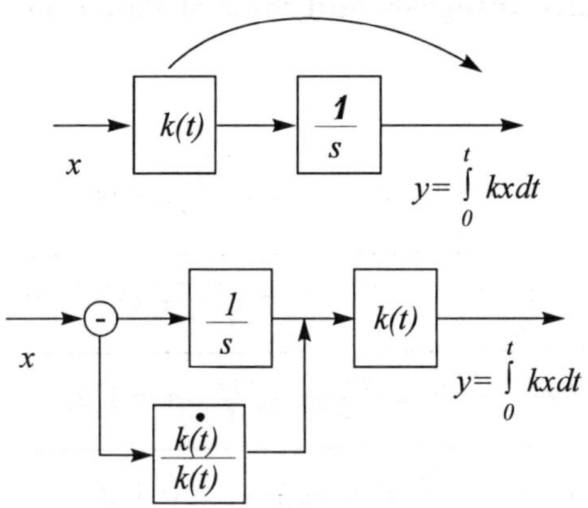

Fig. A.6.13

R u l e 15: while transferring the amplifier with the gain $k(t)$ through the integrator in the opposite direction of the signal it is necessary to close the integrator by the negative feedback which contains the amplifier with the gain $-\dot{k}(t)/k(t)$ (Fig.A.6.14).

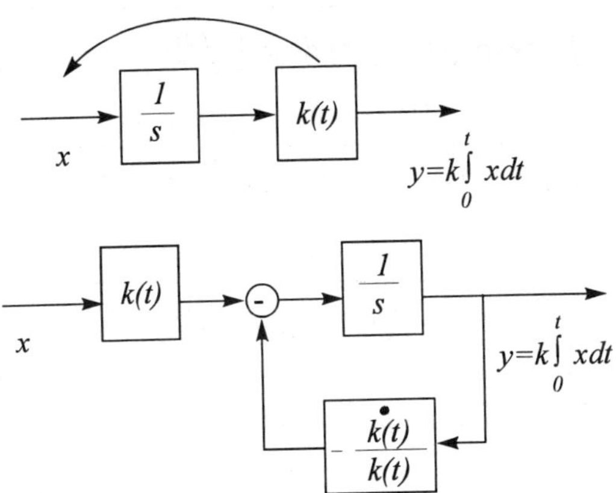

Fig. A.6.14

R u l e 16: the connection lines on the structural connections may be arbitrarily deformed without change of the order of the branches and systems which are placed on them.

7. Some Definite Integrals and Special Functions

Definite Integrals
Table A.7.1

1.	$$\int_{-\infty}^{\infty} e^{\eta t - ct^2/2}\, dt = \sqrt{\tfrac{2\pi}{c}}\, e^{\eta^2/2c}$$
2.	$$\int_{-\infty}^{\infty} e^{-\eta^T t - t^T Ct/2}\, dt = \sqrt{\tfrac{(2\pi)^n}{|C|}}\, e^{\eta^T C^{-1}\eta/2}$$
3.	$$\int_{0}^{\infty} t^{2n} e^{-ct^2/2}\, dt = \tfrac{(2n-1)!!}{2c^n}\sqrt{\tfrac{2\pi}{c}}, \quad n = 1, 2, \ldots, c > 0$$
4.	$$\int_{0}^{\infty} t^{2n+1} e^{-ct^2/2}\, dt = \tfrac{n!}{2^n c^{n+1}}, \quad n = 1, 2, \ldots, c > 0$$
5.	$$\int_{0}^{\infty} \exp\left(\eta t - \tfrac{c}{2}t^2\right) dt = \sqrt{\tfrac{\pi}{2c}}\, e^{\eta^2/2c}\, \mathrm{erfc}\left(-\tfrac{\eta}{\sqrt{2c}}\right), \quad (\mathrm{Re}\, c > 0)$$
6.	$$\int_{-\infty}^{\infty} t \exp\left(\eta t - \tfrac{c}{2}t^2\right) dt = \tfrac{\eta}{c}\sqrt{\tfrac{2\pi}{c}}\, e^{\eta^2/2c}, \quad (\mathrm{Re}\, c > 0)$$
7.	$$\int_{0}^{\infty} t \exp\left(\eta t - \tfrac{c}{2}t^2\right) dt = \tfrac{1}{c} + \tfrac{2\eta}{c}\sqrt{\tfrac{2\pi}{c}}\, e^{\eta^2/2c}\, \mathrm{erfc}\left(-\tfrac{\eta}{\sqrt{2c}}\right)$$ $$(\mathrm{Re}\, c > 0,\ |\arg \eta| < \pi)$$
8.	$$\int_{-\infty}^{\infty} t^2 \exp\left(\eta t - \tfrac{c}{2}t^2\right) dt = \tfrac{1}{c}\sqrt{\tfrac{2\pi}{c}}\left(1 + \tfrac{\eta^2}{c}\right) e^{\eta^2/2c}$$ $$(\mathrm{Re}\, c > 0,\ |\arg \eta| < \pi)$$
9.	$$\int_{0}^{\infty} t^2 \exp\left(\eta t - \tfrac{c}{2}t^2\right) dt = \tfrac{\eta}{c^2} + \sqrt{\tfrac{2\pi}{c^5}}\, \tfrac{c+\eta^2}{4}\, e^{\eta^2/2c}\, \mathrm{erfc}\left(-\tfrac{\eta}{\sqrt{2c}}\right)$$ $$(\mathrm{Re}\, c > 0,\ |\arg q| < \pi)$$
10.	$$\int_{-\infty}^{\infty} e^{-q^2 t^2} \sin[p(t+\lambda)]\, dt = \tfrac{\sqrt{\pi}}{q}\, e^{-p^2/4q^2} \sin p\lambda$$
11.	$$\int_{-\infty}^{\infty} e^{-q^2 t^2} \cos[p(t+\lambda)]\, dt = \tfrac{\sqrt{\pi}}{q}\, e^{-p^2/4q^2} \cos p\lambda$$
12.	$$\int_{-\infty}^{\infty} e^{i\mu t + t^2/2}\, \tfrac{d^n}{dt^n} e^{-t^2/2} = \sqrt{2\pi}\,(i)^n\, e^{\mu^2/2}\, \tfrac{d^n}{d\mu^n} e^{-\mu^2}$$

Special Functions and Polynomials

Table A.7.2

1.	$\operatorname{erf} z = \frac{2}{\sqrt{\pi}} \int\limits_0^z e^{-t^2} dt, \;\; \operatorname{erf} z = -\operatorname{erf}(-z),$ $\operatorname{erf}(0) = 0, \;\; \operatorname{erf}(\infty) = 1$
2.	$\operatorname{erfc} z = \frac{2}{\pi} \int\limits_z^\infty e^{-t^2} dt = 1 - \operatorname{erf} z$ $\operatorname{erfc}(-z) = 2 - \operatorname{erfc} z, \;\; \operatorname{erfc}(0) = 1, \operatorname{erfc}(\infty) = 0$
3.	$\Phi(z) = \frac{1}{\sqrt{2\pi}} \int\limits_0^z e^{-t^2/2} dt = \frac{1}{2} - \operatorname{erf}(z/\sqrt{2})$ $\Phi(0) = 0, \;\; \Phi(\infty) = \frac{1}{2}, \;\; \Phi(-z) = -\Phi(z)$
4.	$\Gamma(z) = \int\limits_{-}^{\infty} e^{-t} t^{z-1} dt \quad (\operatorname{Re} z > 0)$ $\Gamma(z+1) = z\Gamma(z), \;\; \Gamma(z)\Gamma(-z) = -\frac{\pi}{z \sin \pi z}, \;\; \Gamma(z)\Gamma(1-z) = \frac{\pi}{\sin \pi z}$ $\Gamma(nz) = \sqrt{\frac{n^{2nz-1}}{(2\pi)^{n-1}}} \Gamma(z)\Gamma\left(z + \frac{1}{n}\right) \cdots \Gamma\left(z + \frac{n-1}{n}\right)$
5.	$J_m(z) = \left(\frac{z}{2}\right)^m \sum\limits_{k=0}^{\infty} \frac{(-1)^k}{k!\Gamma(m+k+1)} \left(\frac{z}{2}\right)^{2k}$ $(|\arg z| < \pi, \quad m = 0, \pm 1 \pm 2, \ldots)$ $J_m(z) = \frac{1}{\pi} \int\limits_0^\pi \cos(mt - z \sin t) dt = \frac{(-i)^m}{\pi} \int\limits_0^\pi e^{iz \cos t} \cos mt \, dt$ $(m = 0, 1, 2, \ldots)$
6.	$N_m(z) = \frac{1}{\sin m\pi} [J_m(z) \cos m\pi - J_{-m}(z)]$ $(m \neq 0, \pm 1, \pm 2, \ldots)$
7.	$J_{m+1/2}(z) = \sqrt{\frac{2}{\pi}} \, z^{m+1/2} \left(-\frac{1}{z}\frac{d}{dz}\right)^m \frac{\sin z}{z} \;\; (m = 1, 2, \ldots)$
8.	$T_n(z) = \cos(n \arccos z) = \frac{n}{2} \sum\limits_{m=0}^{[n/2]} (-1)^m \frac{(n-m-1)!}{m!(n-2m)!} (2z)^{n-2m}$ $= \frac{(-2)^n n!}{(2n)!} \sqrt{1 - z^2} \frac{d^n}{dz^n} (1 - z^2)^{n-1/2}$ $T_{n+1}(z) = 2z T_n(z) - T_{n-1}(z) \;\; (n = 1, 2, \ldots)$
9.	$U_n(z) = \sin(n \arccos z) = \frac{\sqrt{1-z^2}}{n} \frac{d}{dz} T_n(z), \;\; (n = 1, 2, \ldots)$

10.	$$L_n(z) = (n!)^2 \sum_{m=0}^{n} (-1)^m \frac{z^m}{(m!)^2(n-m)!} = e^z \frac{d^n}{dz^n}(z^n e^{-z}),$$ $$(n = 1, 2, \ldots)$$ $$L_{n+1}(z) = (2n+1-z)L_n(z) - n^2 L_{n-1}(z)$$ $$\frac{dL_{n+1}(z)}{dz} = (n+1)\left[\frac{dL_n(z)}{dz} - L_n(z)\right]$$
11.	$$H_n(z) = n! \sum_{m=0}^{[n/2]} (-1)^m \frac{(2z)^{n-2m}}{m!(n-2m)!} = (-1)^n e^{z^2} \frac{d^n}{dz^n}(e^{-z^2})$$ $$H_{n+1}(z) = 2z H_n(z) - 2n H_{n-1}(z)$$ $$\frac{dH_n(z)}{dz} = 2n H_{n-1}(z), \ (n = 1, 2, \ldots)$$
12.	$$P_n(z) = 2^{-n} \sum_{m=0}^{[n/2]} (-1)^m \frac{(2n-2m)!}{m!(n-m)!(n-2m)!} z^{n-2m}$$ $$= \frac{1}{2^n n!} \frac{d^n}{dz^n}(z^2-1)^n, \ (n = 1, 2, \ldots)$$ $$P_{n+1}(z) = \frac{2n+1}{n+1} z P_n(z) - \frac{n}{n+1} P_{n-1}(z) = z P_n(z) + \frac{z^2-1}{n+1}\frac{dP_n}{dz}$$

BIBLIOGRAPHICAL NOTES

Chapter 1. Dynamical Systems and Their Characteristics

Sections 1.1–1.4 are mainly based on (Pugachev 1965, 1974a, Pugachev and Sinitsyn 1987, 1990, 2000). For the detailed study of deterministic and stochastic stability see (Hasminski 1980, Kushner 1967, Voronov 1979, 1985). Frequency response methods are given in (Butkovskiy 1982, 1983, James et al. 1947, MacFralane 1979, Silyak 1969, 1978, Vidyasagar 1978, Voronov 1965, 1966, 1970, 1979, 1985).

Section 1.5. Sampled-data (discrete or impulse) systems are considered in detail in (Äström 1970, Kashyap and Rao 1976, Pugachev 1974a).

Section 1.6. For deterministic structural theory see (Barkin 1982, Desoer and Vidyasagar 1975, Pugachev 1965, 1974a, Voronov 1965, 1966, 1970, 1979, 1985, Zadeh and Desoer 1963). Notions of structural theory of stochastic systems are given on the basis of (Pugachev and Sinitsyn 2000).

Examples, Problems and Appendices 5–7 are taken from (Butkovskiy 1982, 1983, Korn and Korn 1968, Pugachev and Sinitsyn 1987, 1990, 2000, Voronov 1965, 1966, 1970, 1979, 1985).

Chapter 2. Random Variables, Processes and Functions

Sections 2.1 and 2.2. The detailed treatment of probability spaces and the theory of probability distributions are given in (Cramér and Leadbetter 1967, Doob 1953, Gikhman and Skorokhod 1979, Kolmogorov 1933, Loève 1977, Shiryaev 1995). For the limit theorems address (Jacod and Shiryaev 1987).

Section 2.3. Theory of conditional probabilities is studied in (Doob 1953, Dynkin 1965, Loéve 1977, Wong 1971).

Sections 2.4 and 2.5. For the detailed treatment of probabilities in finite and infinite products of spaces address (Doob 1953, Gikhman and Skorokhod 1979, Kolmogorov 1933, Loève 1977, Shiryaev 1995).

Section 2.6. The Kolmogorov theorems concerning regular conditional distribution are considered more extensively in (Doob 1953, Gikhman and Skorokhod 1979, Kolmogorov 1933, Loève 1977).

Section 2.7. The detailed treatment of the probability measures of random function is given in (Billingsley 1968, Cramér and Leadbetter 1967, Doob 1953, Gikhman and Skorokhod 1979). Probabilities of events connected with random function are considered in (Cooper 1981, Cramér and Leadbetter 1967, Feller 1968, Feynman and Hibbs 1965, Freidlin and Venttell 1984, Gelenbe and Pijolle 1987, Hasminski 1980, Hida 1980, Kunita 1990, Kushner 1967, Medhi 1984, Pugachev 1965).

Sections 2.8–2.10. Separability, continuity, differentiability and other analytical properties of random functions are studied in Chapter 4 and in (Cramér and Leadbetter 1967).

Examples and Problems are due to (Bharucha–Reid 1960, Davenport and Root 1958, Karlin and Taylor 1975, 1986, Laning and Bettin 1956, Parzen 1962, Prabhu 1965, Pugachev 1965, Pugachev and Sinitsyn 1987, 1990, 2000).

Chapter 3. Characteristic Functions, Functionals and Canonical Expansions

Sections 3.1–3.5. Properties of expectation, moments, characteristic functions and functionals are studied more extensively in (Cramér and Leadbetter 1967, Doob 1953, Gikhman and Skorokhod 1979, Loève 1977, Shiryaev 1995). The moments theory of random variables with values in topological linear spaces is originally stated.

In Sections 3.6 and 3.7 only elements of the theory of distributions in infinite-dimensional spaces are presented. For more detailed treatment refer to (Balakrishnan 1971, Billingsley 1968, Curtain and Pritchard 1978, Gikhman and Skorokhod 1979, Kuo 1965, Pugachev and Sinitsyn 2000, Vakhania et al. 1987).

Section 3.8 is written on the basis of (Cramér 1946, Gajdarov 1966, Pugachev 1981, Pugachev and Sinitsyn 1987, 2000).

The general theory of the canonical expansions especially in the topological linear spaces is originally stated in Section 3.9. The practical problems of the canonical expansions derivation are given on the basis of (Pugachev 1965, Pugachev and Sinitsyn 2000).

Examples and Problems are taken from (Adamian 1983, Dragan 1980, James et al. 1947, Laning and Bettin 1956, Loève 1977, Pugachev 1965, Pugachev and Sinitsyn 1987, 1990, 2000).

Chapter 4. Stochastic Integrals, Spectral and Integral Canonical Representations

Sections 4.1–4.7 contains the systematic treatment of convergence, continuity, differentiability and integrability based only on the measure theory. Special attention is paid to the general theory of the stochastic Itô, Stratonovich etc. integrals and the generalized random functions. For the stochastic calculus based on martingales address to (Elliott 1982, Gikhman and Skorokhod 1979, Liptser and Shiryaev 1989, McKean 1969, McShane 1974, Rao 1979). For stochastic calculus in manifolds see (Emery 1989).

The general theory of the integral canonical representations especially in the topological linear spaces is originally stated in Section 4.8. The integral canonical representations derivation are given on the basis of (Karhunen 1946, 1947, Pugachev 1965, Pugachev and Sinitsyn 2000).

Examples, Problems and Appendix 1 are taken from (Kloeden and Platen 1999, Pugachev 1965, Pugachev and Sinitsyn 1987, 1990, 1999, 2000, Soise 1994).

Chapter 5. General Theory of Stochastic Systems and Its Applications

Sections 5.1–5.4. The theory of stochastic differential equations goes back to the works (Bernstein 1934, 1938, Feller 1936, Gikhman 1947, 1950a,b, Itô 1944, 1951a,b, Kolmogorov 1933, Pugachev 1944). For standard mathematical treatises on stochastic differential equations see (Gikhman and Skorokhod 1972, Ikeda and Watanabe 1981, 1989). Applied mathematical texts in stochastic differential equations are the following (Arnold 1974, 1998, Friedman 1975, 1976, Gard 1988, Horsthemke and Lefever 1984, Pugachev and Sinitsyn 1985, 1987, 1990, 2000, Sobczyk 1991, van Kampen 1981, Wong 1971). Differential equations containing random functions are stadied in (Arnold 1998, Bharucha–Reid 1979, Bunke 1972, McShane 1974, Protter 1990, Soong 1973, van Kampen 1981). Some existence and uniqueness proofs are given in (Gikhman and Skorokhod 1972, Ikeda and Watanabe 1989, Pugachev and Sinitsyn 2000). For nonlipschitzian conditions see (Balakrishnan 1977, Veretennikov 1980). Special attention is paid to the scalar and the vector Itô formulae of the differentiation and their new generalizations for the arbitrary processes with the independent increments. Basic rules for the change of variables and transformations of equations to stochastic Itô, Stratonovich etc. equations are given.

Section 5.5 contains the elements of numerical integration of the Itô stochastic differential equations. For systematic study of modern numerical analysis of stochastic differential equations refer to (Kloeden and Platen 1992, 1994, 1995, 1997, 1999). See also (Artemiev and Averina 1997, Bratley et al. 1987, Cyganowski 1996, Cyganowski et al. 1998, Ogorodnikov and Prigarin 1996).

Sections 5.6 and 5.7 are devoted to the systematic derivation of the equations for the one- and the multi-dimensional characteristic functions and densities in the stochastic systems described by the differential and/or difference equations. In contrast to standard treatises on the arbitrary processes with the independent increments are considered.The Lipschitz sufficient conditions of the existence of the multi-dimensional distributions are proved. Special attention is paid to the practical methods of derivation of corresponding equations. The results due to (Pugachev and Sinitsyn 1987, 1990, 2000). See also (Arnold 1974, 1998, Feller 1936, 1938, Friedman 1975, 1976, Gard 1988, Gikhman and Skorokhod 1972, Ikeda and Watanabe 1981, 1989, Risken 1989, Soize 1994, Stratonovich 1963, 1968). For the invariant measure methods for the Fokker–Plank–Kolmogorov equation address Problems 5.16–5.21 and to (Kloeden and Platen 1999, Moschyk and Sinitsyn 1991, 1992a,b). Direct numerical methods for solving the Fokker–Plank–Kolmogorov, the Feller–Kolmogorov and the Pugachev equations are used only for small dimensional stochastic systems.

Section 5.8 is based on (Pugachev and Sinitsyn 1990, 2000). For structural dynamical systems and the systems with random structure see (Arnold 1974, 1998, Bolotin 1969, Clough 1975, Crandall 1958, 1963, Crandall and Mark 1973, Elishakoff 1983, Ibrahim 1985, Kazakov 1977, Kazakov and Artemiev 1980, Lin 1967, Soise 1994).

Section 5.9. See (Baras et al 1983, Benson 1980, Chandasekhar 1954, Elworthy 1982, Feynman and Hibbs 1965, Gardiner 1983, Kunita 1990, Mohammed 1984, Prato and Sobczyk 1992, Rosovski 1990, Skorokhod 1984).

Section 5.10. Is due to (Pugachev and Sinitsyn 1987, 1990). An original simple derivation of the Zakai equations for the nonnormalized conditional density and characteristic function is given. See also (Jarwinski 1970, Kallianpur 1980, Liptser and Shiryaev 1977, 1978).

Examples and Problems are taken from (Kloeden and Platen 1999, Pugachev and Sinitsyn 1987, 1990, 2000, Silyijanova 1975, 1982, Sinitsyn 1974, 1986, 1992, 1993, 1996a,b,c, 1997a,b Moschuk and Sinitsyn 1990, 1991, 1992a,b, 1993).

Chapter 6. Methods of Linear Stochastic Systems Theory and Their Applications

Sections 6.1 and 6.2 are written on the basis of (Pugachev 1965, Pugachev and Sinitsyn 1987, 1990, 2000). Methods for the discrete stochastic systems are given as a specific case of continuous ones. For details concerning stochastic linear discrete system see (Åström 1970, Kashyap and Rao 1976). Tables of integrals (6.1.27) are given in (James et al. 1947, Newton et al. 1957, Pugachev 1965). Direct numerical analysis based on spectral and correlation characteristics is given in all standard mathematical libraries. See also (Forsythe et al. 1977, Kahaner et al. 1989). For details concerning the stochastic infinite-dimensional linear systems refer (Bolotin 1969, Brockett 1970, Clough 1975, Crandall 1950, 1963, 1973, D'Azzo and Houpis 1975, Desoer and Vidyasagar 1975, Elishakoff 1983, Hasminski 1980, Hida 1980, McFralane 1979, Mohammed 1984, Prato and Sabczyk 1992, Pugachev and Sinitsyn 1992, 2000, Rosovski 1990, Schultz and Melsa 1967, Skorokhod 1984, Wong 1971, Zadeh and Desoer 1963).

Sections 6.3 and 6.4 present the most complete systematic treatment of the finite and the infinite-dimensional linear stochastic systems theory based on the canonical expansions and integral canonical representations. The treatment is due to (Pugachev and Sinisyn 2000). Some additional material is in (Adomian 1983, Dostupov 1970, Dragan 1980, Marchenko 1973, Marchenko and Sherbak 1975). For calculation of the Radom–Nikodym derivatives by the canonical expansions see (Pugachev and Sinistyn 2000) and by functional methods see (Arato 1982).

Section 6.5 includes the applications of the described methods to the offline spectral and correlation analysis and the shaping filter design. Specialized software for analytical modeling for linear systems is described in (Pugachev et al. 1995).

Section 6.6 contains the simple presentation of the Kalman–Bucy and the Liptser–Shiryaev online filters. For more extensive study refer to (Balakrishnan 1987, Brammer 1989, Chen 1994, Chui and Chen 1987, Grewal and Andrews 1993, Mendel 1987, Ruymgaart and Soong 1988).

Examples and Problems are taken from (Aoki 1967, Äström 1970, Bolotin 1969, Clough and Penzien 1975, Crandall 1958, 1963, 1973, Horsthemke and Lefever 1984, Kallianpur 1980, Law and Kelton 1991, Mikchailov 1986, Ogata 1959, Pugachev 1965, 1974a,b, Pugachev and Sinitsyn 1987, 1990, 2000, Sobczyk 1991, Soong 1973, Stratonovich 1963, 1968).

Chapter 7. Methods of Nonlinear Stochastic Systems Theory and Their Applications

The main results of Section 7.1 are due to (Pugachev and Sinitsyn 1987, 1990, 2000). Special attention is paic to various forms of infinite set of equations for moments in case of Eqs. (5.4.6). Systems with the parametric noises are studied in detail.

Section 7.2 contains the systematic presentation of the following methods: normal approximation, statistical linearization (by Booton and Kazakov), equivalent linearization and normalization. It is based on (Booton 1954, Kazakov 1956, 1965, 1975, 1977, 1983, Pugachev 1944, 1961, 1965, 1974a, Pugachev and Sinitsyn 1987, 1990, 2000, Pugachev and Silyianova 1975, 1978, 1982, Sinitsyn 1974). See also (Bolotin 1969, Roberts and Spanos 1990, Sawaragi et al. 1962, Sinitsyn 1986, 1992, 1996a, 1997a, Sinitsyn and Karpenko 1996, Sinitsyn et al. 1998). Software "Sts-Analysis" for PC is described in (Pugachev et al. 1991, 1995, Pugachev and Sinitsyn 1990, 2000). For NAM based on the canonical expansions see (Sinitsyn 1971, 1974). NAM for Eqs. (5.4.7) and for the infinite-dimensional stochastic systems is due to (Pugachev and Sinitsyn 1992, 1993, 1996a, 2000). Methods of Section 7.2 are the basis of different principles of the equaivalence of stochastic disturbances in the composite dynamical systems. See, for example, (Sinitsyn 1996b,c).

Section 7.3 contains some practical arguments for the parametrization of the multi-dimensional distributions in the finite- and the infinte-dimensional systems. The problem of practical reducing the number of equations for distribution parameters is considered.

Section 7.4. is devoted to the method of moments (MM), semiinvariants methods (SM) and moments-semiinvariants methods (MSM) for the variaous nonlinear stochastic systems. The results concerning MM and their modifications in the normal stochastic systems are due to (Pugachev 1944, 1965, 1974a). For MM in case of Eqs. (5.4.6), (5.4.7) see (Pugachev and Sinitsyn 1985, 1987, 1990, 2000). SM equations for the normal stochastic systems are given in (Dashevski 1967, 1968, 1976, Dashevski and Liptser 1967a,b,). For SM equations in case of Eqs. (5.4.6) and (5.4.7) see (Pugachev and Sinitsyn 1985, 1987, 1990). Some specific SM cases are due to (Fox 1974, Kubo 1962, Kuznetzov et al. 1953, 1954a,b, Leonov 1964, Malachov 1978, Stratonovich 1961, 1985). SM and MSM modifications for the normal stochastic systems are considered in (Kashkarova and Shin 1986, Konashenkova and Shin 1990). SM and MSM software realizations for polynomial Eqs. (5.4.6) are considered in (Dashevski 1976, Pugachev and Sinitsyn 1993).

Section 7.5 is dedicated to the orthogonal expansions methods (OEM) and quasimoments methods (QM). OEM for the normal stochastic systems and one-dimensional densities is due to (Ahlbehrendt and Kempe 1984, Pugachev 1944, Pugachev and Sinitsyn 1985, 1987, 1990, Semenov 1977). OEM generalization for Eqs. (5.4.6) and (5.4.7) is given (Pugachev and Sinitsyn 1985, 1990, 2000, Sinitsyn 1992, 1996a, 1997a). QM for normal stochastic systems Eq (5.4.6) is given in (Boguslavski 1969, Kuznetzov et al. 1960). Equations for QM in case of Eqs. (5.4.6) and (5.4.7) is in (Pugachev and Sinitsyn 1985, 1987, 1990, 2000). Method of the Fourier series for normal stochastic systems Eq. (5.4.6) is given in (Moschuk 1994).

Section 7.6 is written on the basis of (Pugachev et al. 1995, Pugachev and Sinitsyn 1990, 2000, V.Sinitsyn 1991, 1992).

Section 7.7 is due to (Pugachev 1965, Pugachev and Sinitsyn 2000). Special attention is paid to the canonical representations of nonlinear transformation reducible to linear one. For the applied methods based on the canonical exponsions method see (Adamian 1983, Dostupov 1974, Sinitsyn 1974).

Section 7.8 contains the applications of Chapter 7 methods. Subsection 7.8.1 is due to (Pugachev 1965, Pugachev and Sinitsyn 2000), Subsection 7.8.2 – due to (Sinitsyn 1966b,c). Subsection 7.8.3 – due to (Sinitsyn 1986), Subsection 7.8.4 – due to (Pugachev 1974a), Subsection 7.8.5 – due to (Artemiev et al. 1996), Subsection 7.8.6 – due to (Pugachev at al. 1991). For the stochastic functional series see (Dostupov 1970, Pugachev 1965, Pupkov 1965, Pupkov et al. 1976).

Section 7.9 is devoted to application of Chapter 7 methods to the theory and software for the discrete conditionally optimal nonlinear (Pugachev's) filters. These filters turn out to be more simple than the extended Kalman filters. Nowdays Pugachev's theory of conditionally optimal nonlinear estimation is well developed and has wide applications. The results are due to (Pugachev 1979, Pugachev and Sinitsyn 1985, 1990, Sinitsyn 1996b, Sinitsyn et al. 1998).

REFERENCES

Adamian, G. (1983) *Stochastic Systems*. Academic Press, New York.

Ahlbehrendt, N. and V. Kempe (1984) *Analyse Stochastischer Systeme*. Akademie-Verlag, Berlin.

Andreeva, E.V., T.D.Konashenkova, E.J.Maisheva, O.S.Ogneva, M.V.Petrova and V.I.Shin (1992). Modified quasimoment and moment-semiinvariant methods of analysis of multi-dimensional stochastic systems and their software (in Russian). *Informatics and Computer Systems*. Nauka, Moscow, 2, 160–171.

Andronov, A.A., A.A.Witt and L.S.Pontryagin (1933) On statistical research of dynamic systems (in Russian). *Journ. Exper. and Theor. Phys.*, **3**, 3, 165–180.

Aoki, M. (1967) *Optimization of Stochastic Systems* . Academic Press, New York.

Arato, M. (1982) *Linear Stochastic Systems with Constant Coefficients. A Statistical Approach*. Springer-Verlag, Berlin.

Arnold, L. (1974) *Stochastic Differential Equetions*. Wiley, New York.

Arnold, L. (1998) *Random Dynamical Systems*. Springer, Berlin.

Artemiev, S.S. and A.T.Averina (1997) *Numerical Analysis of Systems of Ordinary and Stochastic Differential Equations*. VSP, Utrecht.

Artemiev, S.S., Michailichenko I.G. and I.N. Sinitsyn (1996) *Statistical Simulation of Financial Transactions* (in Russian). Computer Center of Syberian Division, Russian Academy of Sciences, Novosibirsk.

Äström, K.J. (1970) *Introduction to Stochastic Control Theory*. Academic Press, New York.

Balakrishnan, A.V. (1971) *Introduction to Optimization Theory in a Hilbert Space*. Springer–Verlag, Berlin.

Balakrishnan, A.V. (1977) On a class of stochastic differential equation which do not satisfy Lipschitz conditions. In *New Trends in System Analysis*. Volume 2 of *Lecture Notes in Control and Information*. Sc., 27–35, Springer.

Balakrishnan, A.V. (1987) *Kalman Filtering Theory*. Optimization Software. New York.

Baras, Y.S., G.L.Blankenship and W.E.Hopkins (1983) Existence Uniqueness and asymptotic behaviour of solutions to a class of Zakai equations with unbounded coefficients. *IEEE*, 28, 203–214.

Barkin, A.I. (1982) *Estimates of Quality of Nonlinear Control Systems* (in Russian). Nauka, Moscow.

Bateman, H. and F.Erdélyi (1953) *Higher Transcendental Functions*. McGraw-Hill, New York.

Belopolskaya, T. and Ju.L.Daleskii (1990) *Stochastic Equations on Manifolds*. Kluwer.

Benson, S.W. (1980) *The Foundations of Chemical Kinetics*. McGraw-Hill, New York.

Berstein, S.N. (1934) Principes de la théorie des equations différentielles stochastiques. *Travaux de l'Institut Physicow Mathématique de Stekloff*, 5, 95–124.

Berstein S.N. (1938) Equations différentielles stochastiques. *Actual. Scient. et Industr.*, 738, Conf. Intern. Sci. Math. Univ. Geneve. German, Paris, 5–31.

Bharucha–Reid, A.T. (1960) *Elements of the Theory of Markov Processes and Their Applications*. McGraw–Hill, New York.

Bharucha–Reid, A.T. (1979) *Approximate Solution of Random Equations*. North Holland, Amsterdam.

Billingsley, P. (1968) *Convergence of Probability Measures*. John Wiley, New York.

Boguslavskii, I.A. (1969) Statistical analysis of a multi-dimensional dynamic system using Hermite polynomials of many variables (in Russian). *Avtom. and Telemekh.* 7, 36-51.

Bolotin, V.V. (1969) *Statistical Methods in Structural Mechanics*. Holden Day, Inc., San Francisco.

Booton, R.C. (1954) Nonlinear control systems with random inputs. *Trans. IRE*. V.CT-1, 9–18.

Brammer, K. and G.Siffling (1989) *Kalman–Bucy Filters*. Artech House, Norwood, Mass.

Bratley, P., B.L.Fox and L.Schrage (1987) *A Guide to Simulation* (2nd ed.). Springer, Berlin.

Brockett, R.W. (1970) *Finite Dimensional Linear Systems*. Wiley, New-York.

Bunke, H. (1972) *Gewöhnliche Differentialgleichungen mit zufälligen Parametern*. Akademie-Verlag, Berlin.

Butkovskiy, A.G. (1982) *Green Functions and Transfer Functions Handbook*. Ellis Hordwood, New York.

Butkovskiy, A.G. (1983) *Structural Theory of Distributed Systems*. Ellis Hordwood: John Wiley, Chichester.

Chandrasekhar, S. (1954) Stochastic problems in physics and astronomy. *Rev. Modern Phys.*, **15**, 2–59.

Chen, G (ed.) (1994) *Approximate Kalman Filtering*. World Scientific, Singapore.

Chui, C.K. and G.Chen (1987) *Kalman Filtering with Real-Time Applications*. Springer–Verlag, New York.

Clough, R.W. and J.Penzien (1975) *Dynamics of Structures*. McGraw–Hill, New York.

Cooper, R.B. (1981) *Introduction to Queueing Theory*. North Holland, New York.

Cramér, H. (1946) *Mathematical Methods of Statistics*. Princeton University Press.

Cramér, H. and M.Leadletter (1967) *Stationary and Related Stochastic Process. Sample Function Properties and Their Applications*. John Wiley, New York.

Crandall, S.H. (1958) *Random Vibrations*. Vol. I, Technical Press, Cambridge, Massachusetts.

Crandall, S.H. (1963) *Random Vibration*. Vol. II, MIT Press, Cambridge.

Crandall, S.H. and D.W.Mark (1973) *Random Vibration in Mechanical Systems*. Academic Press, New York.

Curtain, R.F. and A.J.Pritchard (1978) *Infinite Dimensional Linear Systems Theory*. Springer–Verlag, New York.

Cyganowski, S.O. (1996) *Solving Stochastic Differential Equations with Maple*. MAPLE Tech. 3, 38.

Cyganowski, S.O., P.E. Kloede and T.Pohl (1998) *Maple for Stochastic Differential Equations*. WIAS Berlin, Preprint Nr. 453, Available: Postrscript 467 KB, http://www/wias-berlin.de/WIAS_publ_preprints_nr453.AB.

Daletskii, Ju.L. and M.G.Krein (1970) *Stability of the Solutions of Differential Equations in Banach Space* (in Russian). Nauka, Moscow.

Dashevskii, M.L. (1967) Approximate analysis of the accuracy of nonstationary, nonlinear systems using the method of semiinvariants (in Russian) *Avtom. and Telemekh.* 11, 62–81.

Dashevskii, M.L. (1968) Equations for the semiinvariants of a nonlinear dynamic system (in Russian). *Avtom. and Telemekh.*, 10, 63–71.

Dashevskii, M.L. (1976) Technical realizations of the moment-semiinvariant method for analysis of random processes (in Russian) *Avtom. and Telemekh.*, 10, 23–26.

Dashevskii, M.L. and R.S.Liptser (1967a) An approximate analyisis of nonlinear nonstationary dynamic systems (in Russian). *Avtom. and*

Telemekh., 8, 32–43.

Dashevskii, M.L. and R.S.Liptser (1967b) An approximate analysis of accuracy of nonstationary nonlinear systems by semiinvariants methods (in Russian). *Avtom. and Telemekh.*, 11, 62–81.

Davenport, W.B. and W.L.Root (1958) *An Introduction to the Theory of Random Signals and Noise.* McGraw–Hill, New York.

D'Azzo, J.J. and C.H.Houpis (1975) *Linear Control Systems Analysis and Design : Conventional and Modern.* McGraw–Hill, New York.

Desoer, C.A. and M.Vidyasagar (1975) *Feedback systems : Input-output Properties.* Academic Press, New York.

Doob, J.L. (1953) *Stochastic Processes.* John Wiley, Chapman & Hall, London.

Dostupov, B.G. (1970) *Statistical Methods in Design of Nonlinear Automatic Control Systems* (in Russian). Mashinostroyenie, Moscow.

Dragan, J.P. (1980) *The Structure and Models for Representation of Stochastic Signals* (in Russian). Naukova Dumka, Kiev.

Duncan, D.B. (1953) Response of linear time dependent systems to random intpus. *Journ. Appl. Phys.* **24**, 609–611.

Edgeworth, F.I. (1905) The low of error. *Trans. Cambridge Philos.* Soc., **20**, 36–66, 113–141.

Einstein, A. (1956) *Investigations on the Theory of the Brownian Motion.* Dover. New York.

Elishakoff, I. (1983) *Probabilitic Methods in the Theory of Structures.* John Wiley and Sons, New York.

Elliott, R.J. (1982) *Stochastic Calculus and Applications.* Springer, Berlin.

Emery, M. (1989) *Stochastic Calculus in Manifolds.* Springer, Berlin.

Feller, W. (1936) Zur theorie der stochastischen Prozesse (Existenz – und Eindentig Keits – sätze. *Math. Ann.*, 113, 113–160.

Feller, W. (1968) *An Introduction to Probability Theory and Its Applications* (3nd. ed.). V.1, Wiley, New York.

Feller, W. (1971) *An Introduction to Probability Theory and Its Applications* (2nd ed.). V.2., Wiley, New York.

Feynman, R.P., W.Feller and A.R.Hibbs (1965) *Quantum Mechanics and Path-Integrals.* McGraw–Hill, New York.

Forsythe, G.E., M.A.Malcolm and C.B.Moler (1977) *Computer Methods for Mathematical Computations.* Prentice–Hall, Englewood–Cliffs, New York.

Fox, R.F. (1974) Applications of cumulant techniques to multiplicative stochastic process. *J.Math. Phys.*, **15**, 9, 1479–1483.

Freidlin, M.I. and A.Ventzell (1984) *Random Perturbations of Dynamical Systems*. Springer–Verlag, Berlin.

Friedman, A. (1975, 1976) *Stochastic Differential Equations and Applications*. Academic Press, New York.

Fuller, A.T. (1969) Analysis of nonlinear stochastic systems by means of the Fokker–Plank equation. *Int. J. Control*, 9, 603-655.

Gard, T.C. (1988) *Introduction to Stochastic Differential Equations*. Marcel Dekker, New York.

Gardiner, C.W. (1983) *Handbook of Stochastic Methods. For Physics Chemistry and the Natural Sciences*. Springer, Berlin.

Geidarov, T.G. (1966) Expansions of n-dimensional distribution laws in edgeworth series (in Russian). *Izv. AN Azerb. SSR. Series Phys–Techn. and Math. Sciences.* **4**, 237-246.

Gelenbe, E and G.Pujolle (1987) *Introduction to Queueing Networks*. John Wiley, New York.

Gibbs, J.W. (1928) *The Collected Words of Willard Gibbs: in two volumes*. Longmans, Green and Co, New York.

Gikhman, I.I. (1947) About one scheme of random processes formation (in Russian). *Dokl, Acad. Nauk SSSR*, **58**, 6, 961–964.

Gikhman, I.I. (1950a) About some differential equations with random functions (in Russian). *Ukr. Math. J.*, **2**, 4, 45–69.

Gikhman, I.I. (1950b) About theory of differential equations of random processes (in Russian). *Ukr. Math. J.*, **2**, 4, 36–63.

Gikhman, I.I. and A.V.Skorokhod (1979) *The Theory of Stochastic Processes*. V. I–III. Springer, Berlin.

Grewall, M.S. and A.P.Andrews (1993) *Kalman Filtering. Theory and Practice*. Prentice Hall, A Simon & Schuster Company. Engleword Cliffs, New Jersey, 07632.

Hasminsi, R.Z. (1980) *Stochastic Stability of Differential Equations*. Sijthoff and Noordhoff, Alpen naan den Rijn.

Hida, T. (1980) *Brownian Motion*. Springer, Berlin.

Horsthemke, W. and R.Lefever (1984) *Noise Induced Transitions*. Springer, Berlin.

Ibrahim, R.A. (1985) *Parametric Random Vibration*. John Wiley and Sons, New York.

Ikeda, N. and S. Watanabe (1989) *Stochastic Differential and Diffusion Processes* (2nd ed). North–Holland.

Itô, K. (1944) Stochastic integral. *Proc. Imp. Acad. Tokio.* **20**, 519–524.

Itô, **K**. (1951a) On a formula concerning stochastic differentials. *Nagoya Math. J.* **3**, 55–65.

Itô, **K**. (1951b) On stochastic differential equations. *Mem. Amer. Math. Soc.* **4**, 1–51.

Itô, **K**. (1984) *Introduction to Probability Theory*. Cambridge University Press.

Jacod, J. and A.N.Shiryaev (1987) *Limit Theorems for Stochastic Processes*. Springer, Berlin.

James, H.M., N.B.Nichols and R.S.Phillips (1947) *Theory of Servomechanisms*. McGraw–Hill, New York.

Jazwinski, A. (1970) *Stochastic Processes and Filtering Theory*. Academic Press, London.

Kahaner, D., C.Moler and S.Nash (1989) *Numerical Methods and Software*. Prenfice–Hall, Englewood Cliffs, New York.

Kallianpur, G. (1980) *Stochastic Filtering Theory*. Springer, Berlin.

Karhunen, K. (1946) Zur Spectraltheorie Stochastisher Prozesse. *Ann. Acad. Sci. Fennicae.* A-34, 7–79.

Karhunen, K. (1947) Uber lineare Methoden in der Wahrshein lich keitsrechnung. *Ann. Acad. Sci. Feunicae.* A.1. **37**, 3–79.

Karlin, S. and H.M.Taylor (1975) *A First Course in Stochastic Processes* (2nd ed). Academic Press, New York.

Karlin, S. and H.M.Taylor (1981) *A Second Course in Stochastic Processes*. Academic Press, New York.

Kashkarova, A.G. and V.I.Shin (1986) Modified semiinvariant methods of analysis of stochastic systems (in Russian). *Avtom. and Tekcmekh.*, **2**, 69-79.

Kashyap, P.L. and F.R.Rao (1976) *Dynamic Stochastic Models from Empirical Data*. Academic Press, New York.

Kazakov, I.E. (1956) Aproximate probability analysis of the exactness of performance of essentially nonlinear systems (in Russian). *Avtom. and Telemekh.*, **17**, 5, 385–409.

Kazakov, I.E. (1965) Statistical analysis of systems with multi-dimensional nonlinearities (in Russian). *Avtom. and Telemekh.*, **26**, 3, 463–469.

Kazakov, I.E. (1975) *Statistical Theory of Control Systems in Space of States* (in Russian). Nauka, Moscow.

Kazakov, I.E. (1977) *Statistical Dynamics of Systems with Variable Structure* (in Russian). Nauka, Moscow.

Kazakov, I.E. (1980) *The Optimization of Dynamic Systems of Random Structure* (in Russian). Nauka, Moscow.

Kazakov, I.E. and V.M.Artemjev (1980) *The Optimization of Dynamic Systems of Random Structure* (in Russian). Nauka, Moscow.

Kazakov, I.E. and S.V.Malchikov (1983) *Analysis of Stochastic Systems in States Space* (in Russian). Nauka, Moscow.

Kerstan, J., R.Matthes and J.Meche (1978) *Infinitely Divisible Point Processes*. Academic–Verlag, Berlin.

Kloeden, P.E. and E.Platen (1992, 1995) *Numerical Solution of Stochastic Differential Equations*. Springer, Berlin.

Kloeden, P.E., E.Platen and H.Schurz (1994, 1997) *Numerical Solution of SDE's Through Computer Experiments*. Universitext, Springer.

Kolmogorov, A.N. (1931) Über die analitische Methoden in der Wahrscheinlichkeitsrechnung. *Math. Ann.* **104**, 415–458.

Kolmogorov, A.N. (1933) *Grundbegriffe der Wahrscheinlichkeitsrechnung*. Springer, Berlin.

Konashenkova, T.D. and V.I.Shin (1990) An approximate method of determining the moments of phase coordinates of multi-dimensional stochastic systems (in Russian). *Avtom. and Telemekh.*, 1, 43–52.

Korn, G.A. and T.M.Korn (1968) *Mathematical Handbook*. McGraw–Hill, New York.

Kubo, R. (1962) Generalized cumulant expansion method. *Journ Phys. Soc. of Japan*, **17**, 7, 1100–1120.

Kunita, H. (1990) *Stochastic Flows and Stochastic Differential Equations*. Cambridge University Press, Cambridge.

Kuo, H.H. (1965) *Gaussian Measures in B- and H-spaces*. Springer–Verlag, Berlin.

Kushner, H.Y. (1967) *Stochastic Stability and Control*. Academic Press, New York.

Kushner, H.Y. (1971) *Introduction to Stochastic Control*. Holt, New York.

Kushner, H.Y. (1977) *Probabillity Methods for Applications in Stochastic Control and for Elliptical Equations*. Academic Press, New York.

Kushner, H.Y. and P.G.Dupuis (1992) *Numerical Methods for Stochastic Control Problems in Continuous Time*. Springer.

Kuznetsov, P.I., R.L.Stratonovich and V.I.Tichonov (1953) The passages of random functions through nonlinear systems (in Russian). *Avtom. and Telemekh.*, **14**, 4, 375–391.

Kuznetsov, P.I., R.L.Stratonovich and V.I.Tichonov (1954a) Quasimoment functions in theory of random processes (in Russian). *Dokl. Acad. Nauk SSSR.*, **94**, 4, 1615–1618.

Kuznetsov, P.I., R.L.Stratonovich and V.I.Tichonov (1954b) Correlation functions in theory of Brown motion. Generalizations of Fokker–Planck equations (in Russian). *Jour. Exper. and Tech. Phys.*, **26**, 2, 189–207.

Kuznetsov, P.I., R.L.Stratonovich and V.I.Tichonov (1960) Quasimoment functions in theory of random processes (in Russian). *Probability Theory and Its Applications*, **5**, 1, 84–102.

Laning, J. and R.H.Battin (1956) *Random Processes in Automatic Control*. McGraw–Hill, New York.

Law, A.M. and W.D.Kelton (1991) *Simulation Modeling and Analysis* (2nd ed). McGraw–Hill, New York.

Leonov, Ju.P. (1964) *Some Applications of Higher Simiinvariants to Theory of Stationary Random Processes* (in Russian). Nauka, Moscow.

Lin, Y.K. (1967) *Probability Theory of Structural Dynamics*. McGraw–Hill, New York.

Liptser, R.S. and A.N.Shiryaev (1977, 1978) *Statistics Random Processes. I General Theory, II. Applications*. Springer–Verlag, New York.

Liptser, R.S. and A.N.Shiryaev (1989) *Theory of Martingales*. Kluwer Acad. Publ., Dordrecht.

Loève M. (1977) *Probability Theorie (4th ed.)*. Springer, Berlin.

MacFralane, A.G. (ed) (1979) *Frequency–response Methods in Control Systems*. IEEE Pressbook, New York.

Malachov, A.N. (1978) *Cumulant Analysis of Random Nongaussian Processes and Their Transformations* (in Russian). Sov. Radio, Moscow.

Malchikov, S.V. (1970) Approximate methode of determining the phase coordinate distributions of nonlinear automatic systems (in Russian). *Avtom. and Telemekh.*, 5, 43–51.

Malchikov, S.V. (1973a) An approximate methods of statistical analysis of dynamic systems containing a multiplicative nonlinear element (in Russian). *Avtom. and Telemekh.*, 10, 33–38.

Malchikov, S.V. (1973b) Determination of output variables of multi-dimensional nonlinear system (in Russian). *Avtom. and Telemekh.*, 11, 16–21.

Marchenko, B.G. (1973) *The Methods of Stochastic Integral Representations and Its Applications in Radiotechnique* (in Russian). Naukova Dumka, Kiev.

Marchenko, B.G. and L.N.Cherbak (1975) *Linear Random Processes and Their Applications* (in Russian). Naukova Dumka, Kiev.

McKean, H.P. (1969) *Stochastic Calculus*. Academic Press, New York.

McShane, E.J. (1974) *Stochastic Calculus and Stochastic Models*. Academic Press, New York.

Medhi, J. (1984) *Stochastic Models in Queueing Theory*. Academic Press, London.

Mendel, I.M. (1987) *Kalman Filtering and Other Digital Estimation Techniques*. IEEE Individual and Learning Package, IEEE Press, Piscataway, New York.

Milstein, G.N. (1995) *Numerical Integration of Stochastic Differential Equations*. Mathematics and its Applicatons, Kluwer.

Michailov, F.A. (1986) *The Theory and Methods of Researching Nonstationary Linear Systems* (in Russian). Nauka, Moscow.

Mohammed, S.A. (1984) *Stochastic Functional Differential Equations*. Pitman, New York.

Moshuk, N.K. and I.N.Sinitsyn (1990) About stochastic nonholonomic systems (in Russian). *Appl. Mech. and Math.*, **54**, 2, 213–223.

Moshuk, N.K. and I.N.Sinitsyn (1991) On stationary distributions in nonlinear stochastic differential systems. *Quart. J. Mech. and Math.*, 4, 571–579.

Moshuk, N.K. and I.N.Sinitsyn (1992a) The distributions with invariant measure in mechanical stochastic systems (in Russian). *Dokl. Akad. Nauk. Russia*, **322**, 4, 662–667.

Moshuk, N.K. and I.N.Sinitsyn (1992b) An approximate method of analysis of stochastic mechanical systems (in Russian). *Appl. Mech. and Math.*, **56**, 6, 907-917.

Moshuk, N.K. and I.N.Sinitsyn (1993) About fluctuations in random media of a solid with fixed point (in Russian). *Mechanics of Solids*, 1, 39–44.

Moshuk, N.K. (1994) An approximate of determining finite-dimensional distributions of state vector in stochastic differential systems (in Russian). *Avtom. and Telemekh.*, 1, 72–87.

Newton, G.C., L.A.Gould and J.E.Kaiser (1957) *Analytical Design of Linear Feedback Controls*. Wiley, London.

Ogata, K. (1967) *State Space Analysis of Control Systems*. Prentice–Hall, Englewood Cliffs, New Gersey.

Ogorodnikov, V.A. and S.M.Prigarin (1996) *Numerical Modelling of Random Processes and Fields. Algorithms and Applications*. VSP, Utrecht.

Parzen, E. (1962) *Stochastic Processes*. Holden–Day, San Francisco.

Prabhu, N.U. (1965) *Stochastic Processes–Basic Theory and its Applications*. MacMillan, New York.

Prato, G. and J.Zabczyk (1992) *Stochastic Equations in Infinite Dimensions*. Cambridge University Press, Cambridge.

Protter, P. (1990) *Stochastic Integration and Differential Equations*. Springer, Berlin.

Pugachev, V.S. (1944) Random functions determined by ordinary differential equations (in Russian). *Trudy Academii N.Ye.Zhukovskogo*, 118, 3–36.

Pugachev, V.S. (1961) Application of Markov process theory to automatic systems accuracy (in Russian). *Izvestiya Academii Nauk SSSR, Energetica i Avtomatica*, 3, 46–57.

Pugachev, V.S. (1965) *Theory of Random Fucntions and Its Application to Control Problems*. Pergamon Press, London.

Pugachev, V.S. (1967) On distribution of crossings number of a random process (in Russian). *Trudy I Union–Wide Sumposium of Statistical Problems in Technical Cybernetics, Sborn. Nonlinear and Optimal Systes*. Nauka, Moscow.

Pugachev, V.S. (1971) Stochastic systems and their connections (in Russian). *Dokl. Acad. Nauk SSSR*, **197**, 6, 1288–1290.

Pugachev, V.S. (1973) Normal stochastic systems (in Russian). *Dokl. Acad. Nauk SSSR*, **208**, 3, 566–569.

Pugachev, V.S. (ed.) (1974a) *The Foundations of Automatic Control* (in Russian). Nauka, Moscow.

Pugachev, V.S. (1974b) An approximate research of stochastic systems by normalization method (in Russian). *Avtom. and Telemekh.*, 5, 74–81.

Pugachev, V.S. (1980) Finite-dimensional distributions of processes defined by stochastic differential equations, and the extrapolotion of such processes (in Russian). *Dokl. Acad. Nauk SSSR.*, **251**, 1, 40–43.

Pugachev, V.S. (1984) *Probability Theory and Mathematical Statistics for Engineers*. Pergamon Press, Oxford.

Pugachev, V.S. (1994a) About some measures in functions space with values in Hilbert space (in Russian). *Dokl. Acad. Nauk Russia.*, **335**, 2, 153–156.

Pugachev, V.S. (1994b) Stochastic differential equations in Hilbert spaces (in Russian). *Dokl. Acad. Nauk Russia.*, **336**, 6, 741-744.

Pugachev, V.S. (1995a) Stochastic differential equations in Hilbert spaces (in Russian). *Differential Equations.*, 3, 456–464.

Pugachev, V.S. (1995b) Stochastic Differential Equations in Banach Spaces with a Basis (in Russian). *Dokl. Acad. Nauk Russia.*, **342**, 5, 592–595.

Pugachev V.S. and I.N.Sinitsyn (1985, 1990) *Stochastic Differential Systems. Analysis and Filtering* (in Russian). Nauka, Moscow. Translation of 1st edition: John Wiley & Sons, Chichester.

Pugachev, V.S. and I.N.Sinitsyn (1986) The directions of the development of mathematical software for stochastic systems research (in Russian). *Informatics: Problems, Prospects*. Nauka, Moscow, 30–38.

Pugachev, V.S. and I.N.Sinitsyn (1992) Approximate methods of statistical analysis of distributed stochastic systems (in Russian). *Informatics and Computer Systems*. Nauka, Moscow, 2, 146-154.

Pugachev, V.S. and I.N.Sinitsyn (1996a) Stochastic differential equations in Hilbert spaces for some random functions of vector argument (in Russian). *Dokl. Acad. Nauk of Russia*, **346**, 3, 1–4.

Pugachev, V.S. and I.N.Sinitsyn (1996b) Stochastic differential equations with random variable structure in Banach spaces (in Russian). *Vestnik of Moscow University. Series 1, Mathematics, Mechanics*, 6, 86–89.

Pugachev, V.S. and I.N.Sinitsyn (1999) *Lectures on Functional Analysis and Applications*. World Scientific, Singapore.

Pugachev, V.S. and I.N.Sinitsyn (2000) *Theory of Stochastics Systems* (in Russian). Logos, Moscow.

Pugachev, V.S., I.N.Sinitsyn, A.A.Cherednichenko, V.I.Shin and V.I.Sinitsyn (1995) Mathematical software for analysis of multi-dimensional nonlinear stochastic systems (in Russian). *Avtom. and Telemekh.*, 1, 87–97.

Pupkov, K.A. (1965) *Statistical Estimate of Nonlinear Systems of Automatic Control* (in Russian). Mashinostroyenie, Moscow.

Pupkov, K.A., V.I.Kapalin and A.S.Yuschenko (1976) *Functional Series in Nonlinear Systems Theory* (in Russian). Nauka, Moscow.

Rao, M.M. (1979) *Stochastic Processes and Integration*. Sijthoff and Noordhoff, Alpen haan den Rijn.

Risken, H. (1989) *The Fokker–Planck Equation* (2nd edition). Springer–Verlag, Berlin.

Roberts, J.B. and P.D.Spanos (1990) *Random Vibration and Statistical Linearization*. Yohn Wiley and Sons, New York.

Rozovskii, B.L. (1990) *Stochastic Evolution Equations. Linear Theory and Applications to Nonlinear Filtering*. Kluwer.

Ruymgaart, P.A. and T.T.Soong (1988) *Mathematics of Kalman–Bucy Filtering*. Springer–Verlag, Berlin.

Sawaragi, Y., N.Sugui and Y.Sunahara (1962) *Statistical Studies on Nonlinear Control Systems*. Nippon Printing and Publishing Company, Osaka, Japan.

Schultz, D.G. and J.L.Melsa (1967) *State Functions and Linear Control Systems*. McGraw–Hill, New York.

Semenov, V.V. (1977) The equations of a generalized characteristic function of a state vector of automatic system (in Russian). *Analytical Methods of Regulators Synthesis. Saratov Polytechn. Institute*, Saratov, 2, 3–36.

Siljak, D.D. (1969) *Nonlinear Systems*. Wiley, New York.

Siljak, D.D. (1978) *Large-Scale Dynamic Systems*. North Holland, New York.

Sinlyianova, I.D. (1975) Approximate research of one-channel queueing system by normalization method (in Russian). *Avtom. and Telemekh.*, 11, 23–27.

Sinlyianova, I.D. (1982) The finite-dimensional distributions of the outputs of one class of nonlinear systes. *Probl. Control and Inform. Theory*, **11**, 6, 407–418.

Sinitsyn, V.I. (1989) A new approximate method of finding one-dimensional distribution of vector process determined by stochastic differential equation (in Russian). *Dokl. Acad. Nauk SSSR*, **309**, 3, 541–544.

Sinitsyn, V.I. (1991) Finding multi-dimensional distributions of a process determined by stochastic differential equation by ellipsoidal approximation method (in Russian). *Dokl. Acad. Nauk SSSR*, **320**, 2, 280–283.

Sinitsyn, V.I. (1992) Ellipsoidal approximation method in problems of analysis and filtering of processes in stochastic systems (in Russian). *Informatics and Computer Systems*. Nauka, Moscow, 2, 154–160.

Sinitsyn, I.N. (1974) Methods of statistical linearization (Survey) (in Russian). *Avtom. and Telemekh.*, 5, 74–94.

Sinitsyn, I.N. (1986) Stochastic hereditary control systems. *Probl. Control and Inform. Theory*. **15**, 4, 287–298.

Sinitsyn, I.N. (1992) *Lectures on PC-based Nonlinear Stochastic Mechanical Systems Research*. Ustav Thermomenchaniky ĈAV, Praha.

Sinitsyn, I.N. (1993) Finite-dimensional distributions with invariant measure in stochastic mechanical systems (in Russian). *Dokl. Acad. Nauk SSSR*, **238**, 3, 308–310.

Sinitsyn, I.N. (1996a) Stationary fluctuations of a rigid body with fixed point in nongaussian random media (in Russian). *Mechanics of Solids*, 6, 14–22.

Sinitsyn, I.N. (1996b) From the experience of teaching statistical bases of informatics in technical universities (in Russian). *Informatics and Computer Systems*. Nauka, Moscow, 8, 68–73.

Sinitsyn, I.N. (1996c) Parallel methods in nonlinear statistical dynamics (in Russian). *Proc. EUROMECH-96. 2nd European Nonlinear Oscillations Conference*. 1, 407–410.

Sinitsyn, I.N. (1997a) Finite-dimensional distributions with invariant measure in nonlinear stochastic diffential systems (in Russian). *Sbornik "Algorithms of Control and Identification"*, Dialogue of Moscow University, Moscow, 129–140.

Sinitsyn, I.N. (1997b) Parallel Simulation Technologies for Stochastic Systems. *Lecture Notes in Computer Science*, 1277, Springer, Berlin, 383–388.

Sinitsyn, I.N., E.R.Korepanov and V.I.Shin (1998) Methods, algorithms and software tools for CAE of stochastic control systems. *Trans. of EUROSIM'98*, Helsinki, 344-348.

Sinitsyn, I.N. and A.P.Karpenko (1996) Combined parallel statistical and analytical modeling methods, algorithms and software for dynamical stochastic systems. *Proc. of EUROSIM'96 Intern. Cont. "HPCN challenges in telecomp. and telecom: Parallel simulation of complex systems and large–scale applications"*. Elsevier Sciences Publishers B.V., 187–194.

Shiryaev, A.N. (1995) *Probability (2nd ed)*. Springer–Verlag, New York.

Skorokhod, A.V. (1964) *Random Processes with Independent Increments* (in Russian). Nauka, Moscow.

Skorokhod, A.V. (1982) *Studies in the Theory of Stochastic Processes*. Dover, New York.

Skorokhod, A.V. (1983) *Stochastic Equations for Complex Systems* (in Russian). Nauka, Moscow.

Sobczyk, K. (1991) *Stochastic Differential Equations: With Applications to Physics and Engineering*. Volume 40 of *Math. Appl. (East European Ser.)*, Kluwer.

Soise, C. (1994) *The Fokker–Plank Equation for Stochastic Dynamical Systems and Its Explicit Steady Solutions*. World Scientific, Singapore.

Soong, T.T. (1973) *Random Differential Equations in Science and Engineering*. Volume 103 of *Math. Sci. Engrg*. Academic Press, New York.

Stratonovich, R.L. (1963) *Topics in the Theory of Random Noise*. Gordon and Breach, New York.

Stratonovich, R.L. (1966) A New Representation for Stochastic Integrals and Equatios. *SIAM Journal of Control*, 4(2), 362–371.

Stratonovich, R.L. (1968) *Conditional Markov Processes and Their Applications to Theory of Optimal Control*. Amer. Elsevier, New York.

Tichonov, V.I. (1970) *Crossings of Random Processes* (in Russian). Nauka, Moscow.

Vakhania, N.N., V.I. Tarieladze and S.A.Chobanyan (1987) *Probability Distributions on Banach Spaces*. Reidel.

van Kampen, N.G. (1981) *Stochastic Processes in Physics and Chemistry*, Volume **888** of *Lecture Notes in Math*. North Holland.

Veretennikov, A.J. (1980) On the strong solutions of stochastic differential equations in Russian). *Theory of Probability and Its Applications*, **24**, 354–366.

Vidyasagar, M. (1978) *Nonlinear systems analysis*. Prentice Hall, Englewood Cliffs, New Jersey.

Voronov, A.A. (1965, 1966, 1970) *Fundamentals of Automatic Control Theory*. Parts 1–3 (in Russian). Energy, Moscow.

Voronov, A.A. (1979) *Stability, Controlability, Observability* (in Russian). Nauka, Moscow.

Voronov, A.A. (1985) *Introduction into Dynamics of Complex Control Systems* (in Russian). Nauka, Moscow.

Wong, E.W. (1971) *Stochastic Processes in Information and Dynamical Systems*. McGraw–Hill, New York.

Zakai, M. (1969) On the optimal filtering of diffusion processes. *Z. Wahrsch. Verw. Geviete*, 11, 230-243.

Zadeh, L.A. and C.A.Desoer (1963) *Linear System Theory*. McGraw–Hill, New York.

NOTATIONS

$A,\,B,\,C$	events, sets, matrices, operators
$\mathcal{A},\,\mathcal{B},\,\mathcal{C},\ldots$	fields of events, sets, algebras, σ-algebra
$X,\,Y,\,Z,\ldots$	random variables
$x,\,y,\,z,\ldots$	realizations of random variables $X,\,Y,\,Z,\ldots$
$f(x)$	density of a random variable X
$F(x)$	distribution function of a random variable X
$f(x,y)$	joint density of random variables $X,\,Y$
$F(x,y)$	joint distribution function of random variables $X,\,Y$
$f(x\mid y)$	conditional density of a random variable X at a given value y of a random variable Y
$F(x\mid y)$	conditional distribution function of a random variable X at a given value y of random variable Y
$m_x = EX$	expectation of a random variable X
$X^0 = X - m_x$	a centered random variable X
$D_x = DX$	variance of a random variable X
k_{xy}	covariance of random variables $X,\,Y$ of a random variable X
K_{xy}	cross-covariance (operator) of random variables $X,\,Y$
Γ_{xy}	initial mixed second order moment (operator) of random variables $X,\,Y$
$\alpha_r = \alpha_{r_1,\ldots,r_n}$	r^{th} order initial moment (operator) of a random variable
$\mu_r = \mu_{r_1,\ldots,r_n}$	r^{th} order central moment (operator) of a random variable
$\kappa_r = \kappa_{r_1,\ldots,r_n}$	r^{th} order semiinvariant of a random variable
$P(A)$	probability of event A
$(\Omega,\mathcal{S},P)$	probability space
$\mu_x(A) = P(X \in A)$	probability measure
$g(\lambda)$	characteristic function (functional) of a random variable
$m_{Y\mid x} = E[Y \mid x]$	conditional expectation of a random variable Y at a given value x of a random variable X, the expectation of Y depending on a nonrandom variable x (regression)
$P(A\mid B)$	conditional probability

$P_{\mathcal{B}}(A \mid \omega)$	conditional probability relative to σ-algebra $\mathcal{B}$
$\mu_y(B \mid x)$	conditional probability measure (CPM) of a random variable Y relative to a random variable X
$X(t) = x(t, \omega)$	random function
$\mu_{t_1, \ldots, t_n}$	joint probability measure of values $X_{t_1}, \ldots, X_{t_n}$ at all n, $t_1, \ldots, t_n$
$K_x(t_1, t_2)$	covariance function operator of a random function $X(t)$
$K_{xy}(t_1, t_2)$	cross-covarince function of $X(t)$, $Y(t)$
$\Gamma_x(t_1, t_2)$	initial second order moment (operator) of a random function $X(t)$
$\Gamma_{xy}(t_1, t_2)$	the mixed second order moment (operator) of random functions $X(t)$, $Y(t)$
$s_x(\lambda)$	spectral and density function of a stationary random function $X(t)$
$k_x(\tau)$	covariace function of a stationary random function $X(t)$
$\alpha_r^x(t_1, \ldots, t_n)$	r^{th} order initial moment (operator) of a random function $X(t)$
$\mu_r^x(t_1, \ldots, t_n)$	r^{th} order central moment (operator) of a random function $X(t)$
$c_\nu(t_1, \ldots, t_n)$	coefficients of the consistent orthogonal expansions (quasimoments) for n-dimensional densities f_n of a random function $X(t)$
$\alpha_r^x(t_1^{(1)}, \ldots, t_{r_1}^{(1)}, \ldots, t_1^{(n)}, \ldots, t_{r_n}^{(n)})$	mixed initial moments (operators) of random functions $X_1(t) \ldots X_n(t)$
$\mu_r^x(t_1^{(1)}, \ldots, t_{r_1}^{(1)}, \ldots, t_1^{(n)}, \ldots, t_{r_n}^{(n)})$	mixed central moments (operators) of random functions $X_1(t), \ldots, X_n(t)$
$g_n^x(\lambda_1, \ldots, \lambda_n; t_1, \ldots, t_n)$	n-dimensional charactristic function of a random function $X(t)$
$f_n^x(x_1, \ldots, x_n; t_1, \ldots, t_n)$	n-dimensional density of a random function $X(t)$